Nuclear Magnetic Resonance

Volume 44

Specialist Periodical Reports Online

Application for Free Access to Electronic Chapters

Customers purchasing a print volume are now entitled to free site-wide access to the electronic version of that title.
(For the definition of a site, please consult **www.rsc.org/subagree**)

To apply for free access, please contact **ejournals@rsc.org** with proof of purchase (a scanned copy is acceptable) containing details of:

- **Title of book**
- **Volume number**
- **ISBN**
- **Address of your organisation**

By providing these details you will be agreeing to the terms of the Electronic Information Licence Agreement at
www.rsc.org/subagree

ROYAL SOCIETY OF **CHEMISTRY**

A Specialist Periodical Report

Nuclear Magnetic Resonance

Volume 44

Editor
K. Kamieńska-Trela, *Polish Academy of Sciences, Warszawa, Poland*

Authors
A. E. Aliev, *University College London, UK*
Joanna Bandorowicz-Pikula, *Nencki Institute of Experimental Biology, Warsaw, Poland*
Szymon Buda, *Jagiellonian University, Krakow, Poland*
Gerardino D'Errico, *Department of Chemistry, University of Naples, Italy*
Angel C. De Dios, *Georgetown University, Washington, USA*
Patrick Groves, *University of Gdansk, Poland*
Cynthia J. Jameson, *University of Illinois, Chicago, USA*
Jarosław Jaźwiński, *Institute of Organic Chemistry, Poland*
Tsunenori Kameda, *National Institute of Agrobiological Sciences, Japan*
Krystyna Kamieńska-Trela, *Institute of Organic Chemistry, Poland*
Jozef Kowalewski, *Stockholm University, Sweden*
Shigeki Kuroki, *Tokyo Institute of Technology, Japan*
Hiromichi Kurosu, *Nara Women's University, Kitauoyahigash, Japan*
Aatto Laaksonen, *Stockholm University, Sweden*
R. V. Law, *Imperial College London, UK*
Daniel Łowicki, *Jagiellonian University, Krakow, Poland*
Sandeep Kumar Mishra, *Indian Institute of Science, India*
Jacek Mlynarski, *Jagiellonian University, Krakow, Poland*
Francesca Mocci, *University of Cagliari, Italy*
Luigi Paduano, *Department of Chemistry, University of Naples, Italy*
Slawomir Pikula, *Nencki Institute of Experimental Biology, Warsaw, Poland*
M. J. W. Prior, *Nottingham University, UK*
W. Schilf, *Institute of Organic Chemistry, Poland*
Peter J. Simpson, *Imperial College London, UK*
N. Suryaprakash, *Indian Institute of Science, India*
Takeshi Yamanobe, *Nara Women's University, Kitauoyahigash, Japan*
Hidekazu Yasunaga, *Kyoto Institute of Technology, Japan*

ROYAL SOCIETY OF CHEMISTRY

THE QUEEN'S AWARDS FOR ENTERPRISE: INTERNATIONAL TRADE 2013

Print ISBN: 978-1-78262-052-5
PDF eISBN: 978-1-78262-275-8
ISSN: 0305-9804
DOI: 10.1039/9781782622758

A catalogue record for this book is available from the British Library

Published by The Royal Society of Chemistry,
Thomas Graham House, Science Park, Milton Road,
Cambridge CB4 0WF, UK

Registered Charity Number 207890

For further information see our web site at www.rsc.org

Printed and bound in the United States of America

Dedicated to the memory of Krystyna Kamieńska-Trela who sadly passed away in February 2015 during the compilation of this work.

Preface

DOI: 10.1039/9781782622758-FP007

It is my great pleasure to introduce Volume 44 of the SPR on NMR. As its predecessors the current volume aims to provide comprehensive coverage of the NMR literature. In the present case, the relevant literature appears between June 2013 and May 2014.

This is the last volume edited by me and therefore I would like to summarize briefly the history of the Specialist Periodical Reports on NMR. First published in 1972, this series has provided an essential digest of the NMR literature for more than four decades. In his foreword to the first volume Prof. Robin Harris declared that the series would be a discussion on the phenomena of NMR and that articles will be critical surveys of the literature. His line was continued with a great success by Prof. Abraham who was Senior Reporter from 1975 to 1978 and by Prof. Graham Webb who started editing the SPR on NMR in 1978. In 2008, after 30 years, Prof. Webb retired and I was invited to take over the editorship of the series. Vol. 45 and the subsequent volumes will be edited by Dr V. Ramesh, who will also collect a new team of co-workers.

It is a pleasure for me to thank all the reporters for their valuable and interesting accounts, as well as for their timely submission. Some of them, like Profs Jameson, De Dios, Wójcik, Kuroki, Kurosu, Law, Prior, Ramanathan, Simpson and Schilf, have been writing their excellent contributions for many years and I believe that the readers as well as myself deeply appreciate their longstanding efforts. I would also like to acknowledge the skilful help of my co-editor Dr Jacek Wójcik in editing Volumes 41–43.

My sincere thanks are also due to the production team at Thomas Graham House for their support in the realisation of this volume.

K. Kamieńska-Trela

CONTENTS

Cover

Symbols and abbreviations

These lists contain the symbols and abbreviations most frequently used in this volume, but they are not expected to be exhaustive. Some specialized notation is only defined in the relevant chapter. An attempt has been made to standardize usage throughout the volume as far as is feasible, but it must be borne in mind that the original research literature certainly is not standardized in this way, and some difficulties may arise from this fact. Trivial use of subscripts *etc.* is not always mentioned in the symbols listed below. Some of the other symbols used in the text, *e.g.* for physical constants such as h or π, or for the thermodynamic quantities such as **H** or **S**, are not included in the list since they are considered to follow completely accepted usage.

Symbols

a_N	hyperline (electron–nucleus) interaction constant
A	(i) hyperfine (electron–nucleus) interaction constant (ii) parameter relating to electric field effects on nuclear shielding
B	(i) magnetic induction field (magnetic flux density) (ii) parameter relating to electric field effects on nuclear shielding
B_0	static magnetic field of NMR or ESR spectrometer
B_1, B_2	r.f. magnetic fields associated with ν_1, ν_2
C_x	spin-rotation coupling constant of nucleus X (used sometimes in tensor form): $C^2 = 1/3(C_{\|\|}^2 + 2C_{\perp}^2)$
$C_{\|\|}$, $C_{\perp}$	components of **C** parallel and perpendicular to a molecular symmetry axis
D	(i) self-diffusion coefficient (ii) zero-field splitting constant
D	rotational diffusion tensor
$D_{\|\|}$, $D_{\perp}$	components of D parallel and perpendicular to a molecular symmetry axis
D_{int}	internal diffusion coefficient
D_0	overall isotropic diffusion coefficient
E	electric field
E_n	eigenvalue of $\mathscr{H}$ (or a contribution to $\mathscr{H}$)
g	nuclear or electronic g-factor
G	magnetic field gradient
H_{ij}	element of matrix representation of $\mathscr{H}$
$\mathscr{H}$	Hamiltonian operator-subscripts indicate its nature
I_i	nuclear spin operator for nucleus i
I_{ix}, I_{iy}, I_{iz}	components of I_i
I	(i) ionization potential (ii) moment of inertia

^nJ	nuclear spin–spin coupling constant through n bonds (in Hz). Further information may be given by subscripts or in brackets. Brackets are used for indicating the species of nuclei coupled, *e.g.* $J\,(^{13}C,\,^{1}H)$ or additionally, the coupling path, *e.g.* J(POCF)
J_r	reduced splitting observed in a double resonance experiment
J	rotational quantum number
^nK	reduced nuclear spin–spin coupling constant (see the notes concerning ^nJ)
m_i	eigenvalue of I_{iz} (magnetic component quantum number)
M_0	equilibrium macroscopic magnetization of a spin system in the presence of B_0
M_x, M_y, M_z	components of macroscopic magnetization
M_n	the number of average mol. wt.
P_A	valence p orbital of atom A
P_i	fractional population (or rotamers *etc.*)
P_{uv}	element of bond-order, charge-density matrix
q	electric field gradient
Q	(i) nuclear quadrupole moment (ii) quality factor for an r.f. coil
s_A	valence *s*-orbital of atom A
$S_A^2(0)$	electron density in S_A at nuclear A
S	(i) singlet state (ii) electron (or, occasionally, nuclear spin) *cf.* I (iii) ordering parameter for oriented systems (iv) overlap integral between molecular orbitals
t	elapsed time
T	(i) temperature (ii) triplet state
T_c	coalescence temperature for an NMR spectrum
T_g	the glass transition temperature (of a polymer)
T_1^X	spin–lattice relaxation time of the X nuclei (further subscripts refer to the relaxation mechanism)
T_2^X	spin–spin relaxation time of the X nucleus (further subscripts refer to the relaxation mechanism)
T_1'	inhomogeneity contribution to dephasing time for M_x or M_y
T_1^*	total dephasing time for M_x or M_y; $(T_2^*)^{-1} = T_2^{-1} + (T_2')^{-1}$
T_3	decay time following 90_0–τ–90_{90} pulse sequences
T_{1p}^X,T_{2p}^X	spin–lattice and spin–spin relaxation time of the X nuclei in the frame of reference rotating with B_1
T_{1D}	dipolar spin–lattice relaxation time
X_i	mole fraction of compound
Z_A	atomic number of atom A

α	(i) nuclear spin wavefunction (eigenfunction of I_z) for a spin −1/2 nucleus (ii) polarizability
β	nuclear spin wavefunction (eigenfunction of I_z) for a spin $-\frac{1}{2}$ nucleus
γ_X	magnetogyric ratio of nucleus X
δ_X	chemical shift of a nucleus of element X (positive when the sample resonates to high frequency of the reference). Usually in p.p.m.
δ_{ij}	Kronecker delta (=1 if $i=j$, and =0 otherwise)
$\delta(r_{KA})$	Dirac delta operator
Δ	(i) time between field gradient pulses (ii) spectral width
ΔJ	anisotropy in $J(\Delta J = J_{\|\|}J_{\perp}$, for axial symmetry)
Δn	population difference between nuclear states
$\Delta\delta$	change of difference in δ
$\Delta_{v\frac{1}{2}}$	full width (in Hz) of a resonance line at half-height
$\Delta\sigma$	(i) anisotropy in σ ($\Delta\sigma = \sigma_{\|\|}-\sigma_{\perp}$, for axial symmetry) (ii) differences in σ for two different situations
$\Delta\chi$	(i) susceptibility anisotropy ($\Delta_{\chi} = \chi_{\|\|}-\chi_{\perp}$, for axial symmetry) (ii) differences in electronegativities
ε_r	relative permittivity
ε_0	permittivity of a vacuum
η	(i) nuclear Overhauser effect (ii) asymmetry factor (*e.g.* in e^2qQ/h) (iii) refractive index (iv) viscosity
μ	magnetic dipole moment
μ_0	permeability of a vacuum
μ_B	Bohr magneton
μ_N	nuclear magneton
ν_i	Larmor precession frequency of nucleus i (in Hz)
ν_0	(i) spectrometer operating frequency (ii) Larmor precession frequency (general, or of bare nucleus)
ν_1	frequency of 'observing' r.f. magnetic field
ν_2	frequency of 'irradiating' r.f. magnetic field
σ_i	shielding parameter of nucleus i (used sometimes in tensor form). Usually in p.p.m. Subscripts may alternatively indicate contributions to σ.
$\sigma_{\|\|}$, $\sigma_{\perp}$	components of σ parallel and perpendicular to a molecular symmetry axis
σ^d	diagrammatic contribution to σ
σ^p	paramagnetic contribution to σ
τ	(i) pre-exchange lifetime of molecular species (ii) time between r.f. pulses (general symbol)
τ_c	correlation time

τ_{coll}	mean time between molecular collisions in the liquid state
τ_j	angular momentum correlation time
τ_p	pulse duration
τ_t	translational magnetic relaxation correlation time
χ	(i) magnetic susceptibility (ii) electronegativity (iii) nuclear quadrupole coupling constant $(= e_2qQ/h)$
ω	carrier frequency in rad s^{-1}
ω_i, ω_0, ω_1, ω_2	as for ν_i, ν_0, ν_1, ν_2 but in rad s^{-1}
ω_m	modulation angular frequency (in rad s^{-1})
ω_r	sample rotation (rad^{-1})

Abbreviations

(a) Physical properties

a.f.	audiofrequency
a.u.	atomic unit
a.m.	amplitude modulation
b.c.c.	body-centred cubic
c.m.c.	critical micelle concentration
e.d.	electron diffraction
e.f.g.	electric field gradient
f.c.c.	face-centred cubic
f.m.	frequency modulation
h.c.p.	hexagonal close-packed
h.f.	hyperfine
i.d.	inside diameter
i.f.	intermediate frequency
l.c.	liquid crystalline
mol.wt.	molecular weight
o.d.	outside diameter
p.p.m.	parts per million
r.f.	radiofrequency
r.m.s.	root mean square
s.h.f.	super-high frequency
u.h.f.	ultra-high frequency
ADC	analogue-to-digital converter
AEE	average excitation energy approximation
AQ	acquire
ARP	adiabatic rapid passage
BIRD	bilinear rotation decoupling
CCPPA	coupled cluster polarization propagator approximation
CH-COSY	carbon–hydrogen correlation spectroscopy
CHESS	chemical shift selection

CHF	coupled Hartree–Fock molecular orbital calculations
CIDEP	chemically induced dynamic electron polarization
CIDNP	chemically induced dynamic nuclear polarization
COSY	correlation spectroscopy
CP	cross polarization
CPMG	Carr–Purcell pulse sequence. Meiboom–Gill modification
CSA	chemical shielding anisotropy
CSI	chemical shift imaging
CW	continuous wave
DAC	digital-to-analogue converter
DD	dipole–dipole (interaction or relaxation mechanism)
DEPT	distortionless enhancement by polarization transfer
DLB	differential line broadening
DNP	dynamic nuclear polarization
DQ	double quantum
DQF	double quantum filter
ECOSY	exclusive correlation spectroscopy
EHT	extended Hückel molecular orbital theory
ENDOR	electron–nucleus double resonance
EOM	equations of motion
ESR	electron spin resonance
EXSY	exchange spectroscopy
FC	Fermi contact
FID	free induction decay
FLASH	fast low angle shot
FPT	finite perturbation theory
FT	Fourier transform
GIAO	gauge included atomic orbitals
HMQ	heteronuclear multiquantum
HOHAHA	homonuclear Hartman–Hahn
HRPA	higher random phased approximation
IDESS	improved depth selective single surface coil spectroscopy
IGLO	individual gauge for different localized orbitals
INADE-QUATE	incredible natural abundance double quantum transfer experiment
INDO	intermediate neglect of differential overlap
INDO/S	intermediate neglect of differential overlap calculations for spectroscopy
INDOR	internuclear double resonance
INEPT	insensitive nuclei enhanced by polarization transfer
IR	infrared
ISIS	image selected *in vivo* spectroscopy
LIS	lanthanide induced shift
LORG	local origin
LSR	lanthanide shift reagent
MASS	magic angle sample spinning
MBPT	many body perturbation theory

MEM	maximum entropy method
MINDO	modified INDO
MQ	multiple quantum
MQC	multiple quantum coherence
MQF	multiple quantum filter
NMR	nuclear magnetic resonance
NOE	nuclear Overhauser enhancement
NOESY	nuclear Overhauser enhancement spectroscopy
NQCC	nuclear quadrupole coupling constant
NQR	nuclear quadrupole resonance
PFG	pulsed field gradient
PRE	proton relaxation enhancement
QF	quadrupole moment/field gradient
QPD	quadrature phase detection
REX	relativistically extended Hückel molecular orbital theory
ROESY	rotating frame Overhauser enhancement spectroscopy
RPA	random phase approximation
SCPT	self consistent perturbation theory
SD	spin dipolar
SECSY	spin echo correlation spectroscopy
SEFT	spin echo Fourier transform
SLITDRESS	slice interleaved depth resolved surface coil spectroscopy
SOPPA	second order polarization propagator approach
SPI	selective population inversion
SPT	selective population transfer
SR	spin rotation (interaction or relaxation mechanism)
TART	tip angle reduced T1 imaging
TOCSY	total correlation spectroscopy
UV	ultraviolet
WAHUHA	Waugh, Huber and Häberlen (cycle of pulses)
ZQ	zero quantum
ZQC	zero quantum coherence

(b) Chemical species[a]

acac	acetylacetonato
ACTH	adrenocorticotropic hormone (corticotropin)
ADP	adenosine diphosphate
AMP	adenosine monophosphate
ATP	adenosine triphosphate
BSA	bovine serum albumin
CMP	cytidine monophosphate
cp	cyclopentadienyl
DAP	dodecylammonium propionate

[a] Lower case initials are used when the species is a ligand.

DME	1,2-dimethoxyethane
DMF	dimethylformamide
DML	dimyristoyl-lecithin
DMS	dimethylsiloxane
DMSO	dimethyl sulfoxide
DNA	deoxyribonucleic acid
DPG	2,3-diphosphoglycerate
DPI	dipalmitoyl-lecithin
dpm	dipivaloylmethanato
DPPH	diphenylpicrylhydrazyl
DSS	2,2-dimethyl-2-silapentane-5-sulfonate (usually as the sodium salt)
DTBN	di-t-butyl nitroxide
EBBA	*N*-(*p*-ethoxybenzylidene)-*p*-butylaniline
EDTA	ethylenediaminetetra-acetic acid
EVA	ethylene-vinyl acetate
fod	1,1,1,2,2,3,3-heptafluoro-7,7-dimethyloctane-4,6-dionato
HAB	4,4′-bis(heptyl)azoxybenzene
HMPA	hexamethylphosphoramide
HOAB	*p*-n-heptyloxyazoxybenzene
IHP	inositolhexaphosphate
KDP	potassium dihydrogen phosphate
MBBA	*N*-(*p*-methoxybenzylidene)-*p*-butylaniline
NADH(P)	nicotinamide adenine dinucleotide (phosphate)
NMF	*N*-methylformamide
PAA	*p*-azoxyanisole
PBA	pyrene butyric acid
PBLG	poly(L-benzyl μ-glutamate)
PC	phosphatidyl choline (lecithin)
PCB	polychlorinated biphenyl
PDMS	polydimethylsiloxane
PMA	poly(methacrylic acid)
PMMA	poly(methyl methacrylate)
POM	poly(oxymethylene)
PS	phosphatidylserine
PTFE	polytetrafluoroethylene
PVC	poly(vinyl chloride)
PVF	poly(vinyl fluoride)
PVP	poly(vinyl pyrrolidone)
RNA	ribonucleic acid (tRNA, transfer RNA)
SDS	sodium dodecyl sulfate
TAB	trimethylammonium bromide
TCNQ	tetracyanoquinodimethane
TFA	trifluoroacetic acid
THF	tetrahydrofuran
TMS	tetramethylsilane
UTP	uridine triphosphate

Amino-acid residues

Ala	alanine	Leu	leucine
Arg	arginine	Lys	lysine
Asn	asparagine	Met	methionine
Asp	aspartic acid	Phe	phenylalanine
Cys	cysteine	Pro	proline
Gln	glutamine	Ser	serine
Glu	glutamic acid	Thr	threonin
Gly	glycine	Trp	tryptophan
His	histidine	Tyr	tyrosine
Hyp	hydroxyproline	Val	valine
Ile	isoleucine		

Books and reviews

W. Schilf
DOI: 10.1039/9781782622758-00001

The section lists all books and reviews, with NMR and EPR as the principal theme, that were known to the compiler for the period under review. In most cases the source of references is the Chemical Abstracts listings for the period from June 2013 to May 2014. References to technical bulletins, company or institute house journals, special reports, *etc.*, are normally omitted. The compilation is sectionalized for the reader's convenience. The foreign language reviews are grouped together according to language and the English title is provided.

Books

1 H. Günther, *NMR Spectroscopy: Basic Principles, Concepts, and Applications in Chemistry*, Wiley-VCH Verlag GmbH & Co. KGaA, Weinheim, Germany, 3rd edn, 2013.

Regular reviews series

2 S. Abbrent and S. Greenbaum, Recent Progress in NMR Spectroscopy of Polymer Electrolytes for Lithium Batteries, *Curr. Opin. Colloid Interface Sci.*, 2013, **18**, 228.
3 D. P. Acharya and P. G. Hartley, Progress in Microemulsion Characterization, *Curr. Opin. Colloid Interface Sci.*, 2012, **17**, 274.
4 P. D. Adams, D. Baker, A. T. Brunger, R. Das, F. DiMaio, R. J. Read, D. C. Richardson, J. S. Richardson and T. C. Terwilliger, Advances, Interactions, and Future Developments in the CNS, Phenix, and Rosetta Structural Biology Software Systems, *Annu. Rev. Biophys.*, 2013, **42**, 265.
5 A. Alia, F. Buda, H. J. M. de Groot and J. Matysik, Solid-State NMR of Nanomachines Involved in Photosynthetic Energy Conversion, *Annu. Rev. Biophys.*, 2013, **42**, 675.
6 B. Alonso and C. Marichal, Solid-State NMR Studies of Micelle-Templated Mesoporous Solids, *Chem. Soc. Rev.*, 2013, **42**, 3808.
7 J. M. Anna, C. R. Baiz, M. R. Ross, R. McCanne and K. J. Kubarych, Ultrafast Equilibrium and Non-Equilibrium Chemical Reaction Dynamics Probed with Multidimensional Infrared Spectroscopy, *Int. Rev. Phys. Chem.*, 2012, **31**, 367.
8 M. R. Antoniewicz, Tandem Mass Spectrometry for Measuring Stable-Isotope Labeling, *Curr. Opin. Biotechnol.*, 2013, **24**, 48.
9 T. Asakura, Y. Suzuki, Y. Nakazawa, K. Yazawa, G. P. Holland and J. L. Yarger, Silk Structure Studied with Nuclear Magnetic Resonance, *Prog. Nucl. Magn. Reson. Spectrosc.*, 2013, **69**, 23.
10 S. Asami and B. Reif, Proton-Detected Solid-State NMR Spectroscopy at Aliphatic Sites: Application to Crystalline Systems, *Acc. Chem. Res.*, 2013, **46**, 2089.

Institute of Organic Chemistry, Polish Academy of Sciences, Poland.
E-mail: schilf@icho.edu.pl

11 S. E. Ashbrook and D. M. Dawson, Exploiting Periodic First-Principles Calculations in NMR Spectroscopy of Disordered Solids, *Acc. Chem. Res.*, 2013, **46**, 1964.
12 L. S. Azevedo, F. P. Moraes, M. M. Xavier, E. O. Pantoja, B. Villavicencio, J. A. Finck, A. M. Proenca, K. B. Rocha and W. Filgueira de Azevedo Jr., Recent Progress of Molecular Docking Simulations Applied to Development of Drugs, *Curr. Bioinf.*, 2012, 7, 352.
13 B. A. Berkowitz, D. Bissig, O. Dutczak, S. Corbett, R. North and R. Roberts, MRI Biomarkers for Evaluation of Treatment Efficacy in Preclinical Diabetic Retinopathy, *Expert Opin. Med. Diagn.*, 2013, 7, 393.
14 D. Bernin and D. Topgaard, NMR Diffusion and Relaxation Correlation Methods. New Insights in Heterogeneous Materials, *Curr. Opin. Colloid Interface Sci.*, 2013, **18**, 166.
15 I. Bertini, C. Luchinat, G. Parigi and E. Ravera, SedNMR: On the Edge between Solution and Solid-State NMR, *Acc. Chem. Res.*, 2013, **46**, 2059.
16 M. Bjerring, S. Jain, B. Paaske, J. M. Vinthe and N. C. Nielsen, Designing Dipolar Recoupling and Decoupling Experiments for Biological Solid-State NMR Using Interleaved Continuous Wave and RF Pulse Irradiation, *Acc. Chem. Res.*, 2013, **46**, 2098.
17 F. Blanc, M. Leskes and C. P. Grey, In Situ Solid-State NMR Spectroscopy of Electrochemical Cells: Batteries, Supercapacitors, and Fuel Cells, *Acc. Chem. Res.*, 2013, **46**, 1952.
18 C. Bonhomme, C. Gervais and D. Laurencin, Recent NMR Developments Applied to Organic-Inorganic Materials, *Prog. Nucl. Magn. Reson. Spectrosc.*, 2014, 77, 1.
19 L. Borisjuk, H. Rolletschek and T. Neuberger, Nuclear Magnetic Resonance Imaging of Lipid in Living Plants, *Prog. Lipid Res.*, 2013, **52**, 465.
20 R. C. Breton and W. F. Reynolds, Using NMR to Identify and Characterize Natural Products, *Nat. Prod. Rep.*, 2013, **30**, 501.
21 J. Burt, W. Levason and G. Reid, Coordination Chemistry of the Main Group Elements with Phosphine, Arsine and Stibine Ligands, *Coord. Chem. Rev.*, 2014, **260**, 65.
22 S. J. Butler and D. Parker, Anion Binding in Water at Lanthanide Centres: From Structure and Selectivity to Signalling and Sensing, *Chem. Soc. Rev.*, 2013, **42**, 1652.
23 D. A. Case, Chemical Shifts in Biomolecules, *Curr. Opin. Struct. Biol.*, 2013, **23**, 172.
24 D. Castillo-Gonzalez, G. Perez-Machado, F. Pallardo, T.-M. Garrigues-Pelufo and M.-A. Cabrera-Perez, Computational Tools in the Discovery of New G-Quadruplex Ligands with Potential Anticancer Activity, *Curr. Top. Med. Chem.*, 2012, **12**, 2843.
25 J. Cerkovnik and B. Plesnicar, Recent Advances in the Chemistry of Hydrogen Trioxide (HOOOH), *Chem. Rev.*, 2013, **113**, 7930.
26 S. P. Chauhan, N. R. Sheth, I. S. Rathod, B. N. Suhagia and R. B. Maradia, Analysis of Betalains from Fruits of Opuntia Species, *Phytochem. Rev.*, 2013, **12**, 35.
27 H. Chen, S. Viel, F. Ziarelli and L. Peng, ^{19}F NMR: A Valuable Tool for Studying Biological Events, *Chem. Soc. Rev.*, 2013, **42**, 7971.
28 C.-Y. Cheng and S. Han, Dynamic Nuclear Polarization Methods in Solids and Solutions to Explore Membrane Proteins and Membrane Systems, *Annu. Rev. Phys. Chem.*, 2013, **64**, 507.
29 J.-H. Cheng, Q. Dai, D.-W. Sun, X.-A. Zeng, D. Liu and H.-B. Pu, Applications of Non-Destructive Spectroscopic Techniques for Fish Quality and Safety Evaluation and Inspection, *Trends Food Sci. Technol.*, 2013, **34**, 18.

30 M. Cifelli, V. Domenici and C. A. Veracini, Recent Advancements in Understanding Thermotropic Liquid Crystal Structure and Dynamics by Means of NMR Spectroscopy, *Curr. Opin. Colloid Interface Sci.*, 2013, **18**, 190.

31 G. Comellas and C. M. Rienstra, Protein Structure Determination by Magic-Angle Spinning Solid-State NMR, and Insights into the Formation, Structure, and Stability of Amyloid Fibrils, *Annu. Rev. Biophys.*, 2013, **42**, 515.

32 G. Communie, M. R. Jensen, R. W. Ruigrok and M. Blackledge, Intrinsically Disordered Proteins Implicated in Paramyxoviral Replication Machinery, *Curr. Opin. Virol.*, 2014, **5**, 72.

33 M. Concistrè, O. G. Johannessen, E. Carignani, M. Geppi and M. H. Levitt, Magic-Angle Spinning NMR if Cold Samples, *Acc. Chem. Res.*, 2013, **46**, 1914.

34 N. Cox, D. A. Pantazis, F. Neese and W. Lubitz, Biological Water Oxidation, *Acc. Chem. Res.*, 2013, **46**, 1588.

35 F. Dalitz, M. Cudaj, M. Maiwald and G. Guthausen, Process and Reaction Monitoring by Low-Field NMR Spectroscopy, *Prog. Nucl. Magn. Reson. Spectrosc.*, 2012, **60**, 52.

36 G. M. Daubner, A. Clery and F. H.-T. Allain, RRM-RNA Recognition: NMR or Crystallography. . .and New Findings, *Curr. Opin. Struct. Biol.*, 2013, **23**, 100.

37 G. De Paepe, Dipolar Recoupling in Magic Angle Spinning Solid-State Nuclear Magnetic Resonance, *Annu. Rev. Phys. Chem.*, 2012, **63**, 661.

38 H. Desvaux, Non-Linear Liquid-State NMR, *Prog. Nucl. Magn. Reson. Spectrosc.*, 2013, **70**, 50.

39 T. Didenko, J. J. Liu, R. Horst, R. C. Stevens and K. Wuthrich, Fluorine-19 NMR of Integral Membrane Proteins Illustrated with Studies of GPCRs, *Curr. Opin. Struct. Biol.*, 2013, **23**, 740.

40 Y. Ding, Y. Yao and F. M. Marassi, Membrane Protein Structure Determination in Membrana, *Acc. Chem. Res.*, 2013, **46**, 2182.

41 U. H. N. Durr, R. Soong and A. Ramamoorthy, When Detergent Meets Bilayers: Birth and Coming of Age of Lipid Bicelles, *Prog. Nucl. Magn. Reson. Spectrosc.*, 2013, **69**, 1.

42 F. J. Fernandez and M. C. Vega, Technologies to Keep an Eye on: Alternative Hosts for Protein Production in Structural Biology, *Curr. Opin. Struct. Biol.*, 2013, **23**, 365.

43 M. Foston, Advances in Solid-State NMR of Cellulose, *Curr. Opin. Biotechnol.*, 2014, **27**, 176.

44 M. Fragai, C. Luchinat, G. Parigi and E. Ravera, Conformational Freedom of Metalloproteins Revealed by Paramagnetism-Assisted NMR, *Coord. Chem. Rev.*, 2013, **257**, 2652.

45 J. M. Franck, A. Pavlova, J. A. Scott and S. Han, Quantitative CW Overhauser Effect Dynamic Nuclear Polarization for the Analysis of Local Water Dynamics, *Prog. Nucl. Magn. Reson. Spectrosc.*, 2013, **74**, 33.

46 P. A. Frey and A. D. Hegeman, Chemical and Stereochemical Actions of UDP-Galactose 4-Epimerase, *Acc. Chem. Res.*, 2013, **46**, 1417.

47 D. P. Frueh, A. C. Goodrich, S. H. Mishra and S. R. Nichols, NMR Methods for Structural Studies of Large Monomeric and Multimeric Proteins, *Curr. Opin. Struct. Biol.*, 2013, **23**, 734.

48 M. Halabalaki, K. Vougogiannopoulou, A. L. Skaltsounis and E. Mikros, Recent Advances and New Strategies in the NMR-Based Identification of Natural Products, *Curr. Opin. Biotechol.*, 2014, **25**, 1.

49 S. Gayen, Q. Li and C. Kang, Solution NMR Study of the Transmembrane Domain of Single-Span Membrane Proteins: Opportunities and Strategies, *Curr. Protein Pept. Sci.*, 2012, **13**, 585.

50 G. Geudtner, P. Calaminici, J. Carmona-Espindola, J. M. del Campo, V. D. Dominguez-Soria, R. F. Moreno, G. U. Gamboa, A. Goursot, A. M. Koster, J. U. Reveles, T. Mineva, J. M. Vasquez-Perez, A. Vela, B. Zuninga-Gutierrez and D. R. Salahub, deMon2k, *Wiley Interdiscip. Rev.: Comput. Mol. Sci.*, 2012, **2**, 548.
51 A. Goldbourt, Biomolecular Magic-Angle Spinning Solid-State NMR: Recent Methods and Applications, *Curr. Opin. Biotechnol.*, 2013, **24**, 705.
52 H. Gong, R. Peng and Z. Liu, Carbon Nanotubes for Biomedical Imaging: The Recent Advances, *Adv. Drug Delivery Rev.*, 2013, **65**, 1951.
53 T. Gopinath, K. R. Mote and G. Veglia, Sensitivity and Resolution Enhancement of Oriented Solid-State NMR: Application to Membrane Proteins, *Prog. Nucl. Magn. Reson. Spectrosc.*, 2013, **75**, 50.
54 J.-F. Halet and C. Lapinte, Charge Delocalization vs Localization in Carbon-Rich Iron Mixed-Valence Complexes: A Subtle Interplay between the Carbon Spacer and the (dppe)Cp*Fe Organometallic Electrophore, *Coord. Chem. Rev.*, 2013, **257**, 1584.
55 M. R. Hansen, R. Graf and H. W. Spiess, Solid-State NMR in Macromolecular Systems: Insights on How Molecular Entities Move, *Acc. Chem. Res.*, 2013, **46**, 1996.
56 M. C. Heffern, L. M. Matosziuk and T. J. Meade, Lanthanide Probes for Bioresponsive Imaging, *Chem. Rev.*, 2014, **114**, 4496.
57 J. Hell, W. Kneifel, T. Rosenau and S. Bohmdorfer, Analytical Techniques for the Elucidation of Wheat Bran Constituents and Their Structural Features with Emphasis on Dietary Fiber – A Review, *Trends Food Sci. Technol.*, 2014, **35**, 102.
58 H. Hiroaki, Recent Applications of Isotopic Labeling for Protein NMR in Drug Discovery, *Expert Opin. Drug Discovery*, 2013, **8**, 523.
59 J. C. Hoch, M. W. Maciejewski, M. Mobli, A. D. Schuyler and A. S. Stern, Nonuniform Sampling and Maximum Entropy Reconstruction in Multidimensional NMR, *Acc. Chem. Res.*, 2014, **47**, 708.
60 M. Hong and K. Schmidt-Rohr, Magic-Angle-Spinning NMR Techniques for Measuring Long-Range Distances in Biological Macromolecules, *Acc. Chem. Res.*, 2013, **46**, 2154.
61 T. Hussain and Q. T. Nguyen, Molecular Imaging for Cancer Diagnosis and Surgery, *Adv. Drug Delivery Rev.*, 2014, **66**, 90.
62 S. Jaldoppagari, S. Balakrishnan, A. H. Hegde, N. L. Taradal and P. S. Narayan, Interactions of Polyphenols with Plasma Proteins: Insights from Analytical Techniques, *Curr. Drug Metab.*, 2013, **14**, 456.
63 M. Jaszunski, A. Antusek, P. Gorbacz, K. Jackowski, W. Makulski and M. Wilczek, The Determination of Accurate Nuclear Magnetic Dipole Moments and Direct Measurement of NMR Shielding Constants, *Prog. Nucl. Magn. Reson. Spectrosc.*, 2012, **67**, 49.
64 B. Jayaram and P. Dhingra, Towards Creating Complete Proteomic Structural Databases of Whole Organisms, *Curr. Bioinf.*, 2012, **7**, 424.
65 M. R. Jensen, R. W. H. Ruigrok and M. Blackledge, Describing Intrinsically Disordered Proteins at Atomic Resolution by NMR, *Curr. Opin. Struct. Biol.*, 2013, **23**, 426.
66 G. Jeschke, Conformational Dynamics and Distribution of Nitroxide Spin Labels, *Prog. Nucl. Magn. Reson. Spectrosc.*, 2013, **72**, 42.
67 M. A. Johnson and D. R. Bundle, Designing a New Antifungal Glycoconjugate Vaccine, *Chem. Soc. Rev.*, 2013, **42**, 4327.
68 Q. Kaas and D. J. Craik, NMR of Plant Proteins, *Prog. Nucl. Magn. Reson. Spectrosc.*, 2013, **71**, 1.

69 J. Kaerger and R. Valiullin, Mass Transfer in Mesoporous Materials: The Benefit of Microscopic Diffusion Measurement, *Chem. Soc. Rev.*, 2013, **42**, 4172.
70 I. A. Kaltashov, C. E. Bobst, S. N. Nguyen and S. Wang, Emerging Mass Spectrometry-Based Approaches to Probe Protein-Receptor Interactions: Focus on Overcoming Physiological Barriers, *Adv. Drug Delivery Rev.*, 2013, **65**, 1020.
71 G. Kedarnath and V. K. Jain, Pyridyl and Pyrimidyl Chalcogen (Se and Te) Compounds: A Family of Multi Utility Molecules, *Coord. Chem. Rev.*, 2013, **257**, 1409.
72 O. Y. Kim, J. H. Lee and G. Sweeney, Metabolomic Profiling as a Useful Tool for Diagnosis and Treatment of Chronic Disease: Focus on Obesity, Diabetes and Cardiovascular Diseases, *Expert Rev. Cardiovasc. Ther.*, 2013, **11**, 61.
73 E. Kish-Trier and C. P. Hill, Structural Biology of the Proteasome, *Annu. Rev. Biophys.*, 2013, **42**, 29.
74 T. Kjellman and V. Alfredsson, The Use of *in situ* and *ex situ* Techniques for the Study of the Formation Mechanism of Mesoporous Silica Formed with Non-Ionic Triblock Copolymers, *Chem. Soc. Rev.*, 2013, **42**, 3777.
75 M. J. Knight, I. C. Felli, R. Piarattelli, L. Emsley and G. Pintacuda, Magic Angle Spinning NMR of Paramagnetic Proteins, *Acc. Chem. Res.*, 2013, **46**, 2108.
76 M. Kornreich, R. Avinery and R. Beck, Modern X-Ray Scattering Studies of Complex Biological Systems, *Curr. Opin. Biotechnol.*, 2013, **24**, 716.
77 D. Kruk, A. Herrmann and E. A. Roessler, Field-Cycling NMR Relaxometry of Viscous Liquids and Polymers, *Prog. Nucl. Magn. Reson. Spectrosc.*, 2012, **63**, 33.
78 A. Kryshelnitsky, D. Reichert and K. Saalwaechter, Solid-State NMR Approaches to Internal Dynamics of Proteins: From Picoseconds to Microseconds and Seconds, *Acc. Chem. Res.*, 2013, **46**, 2028.
79 X. Kubiak, J. Dairou, J.-M. Dupret and F. Rodrigues-Lima, Crystal Structure of Arylamine N-Acetyltransferases Insights into the Mechanisms of Action and Substrate Selectivity, *Expert Opin. Drug Metab. Toxicol.*, 2013, **9**, 349.
80 N. Kurochkina and U. Guha, SH3 Domains: Modules of Protein-Protein Interactions, *Biophys. Rev.*, 2013, **5**, 29.
81 K. Kuwata, Logical Design of Medical Chaperone for Prion Diseases, *Curr. Top. Med. Chem.*, 2013, **13**, 2432.
82 D. Lalli and P. Turano, Solution and Solid State NMR Approaches to Draw Iron Pathways in the Ferritin Nanocage, *Acc. Chem. Res.*, 2013, **46**, 2676.
83 T. Lam, P. Pouliot, P. K. Avti, F. Lesage and A. K. Kakkar, Superparamagnetic Iron Oxide Based Nanoprobes for Imaging and Theranostics, *Adv. Colloid Interface Sci.*, 2013, **199–200**, 95.
84 B. P. Lankadurai, E. G. Nagato and M. J. Simpson, Environmental Metabolomics: An Emerging Approach to Study Organism Responses to Environmental Stressors, *Environ. Rev.*, 2013, **21**, 180.
85 D. Laurencin and M. E. Smith, Development of ^{43}Ca Solid State NMR Spectroscopy as a Probe of Local Structure in Inorganic and Molecular Materials, *Prog. Nucl. Magn. Reson. Spectrosc.*, 2013, **68**, 1.
86 J. R. Lewandowski, Advances in Solid-State Relaxation Methodology for Probing Site-Specific Protein Dynamics, *Acc. Chem. Res.*, 2013, **46**, 2018.
87 Z. Li, Y. Yang, J. Zhan, L. Dai and Y. Zhou, Energy Functions in de Novo Protein Design: Current Challenges and Future Prospects, *Annu. Rev. Biophys.*, 2013, **42**, 315.

88 A. Loquet, B. Habenstein and A. Lange, Structural Investigations of Molecular Machines by Solid-State NMR, *Acc. Chem. Res.*, 2013, **46**, 2070.
89 X. Lu, T. Akasaka and S. Nagase, Carbide Cluster Metallofullerenes: Structure, Properties, and Possible Origin, *Acc. Chem. Res.*, 2013, **46**, 1627.
90 N. W. Lutz and P. J. Cozzone, Principles of Multiparametric Optimization for Phospholipidomics by ^{31}P NMR Spectroscopy, *Biophys. Rev.*, 2013, **5**, 295.
91 M. I. Malik and H. Pasch, Novel Developments in the Multidimensional Characterization of Segmented Copolymers, *Prog. Polym. Sci.*, 2014, **39**, 87.
92 R. Mallol, M. A. Rodriguez, J. Brezmes, L. Masana and X. Correig, Human Serum/Plasma Lipoprotein Analysis by NMR: Application to the Study of Diabetic Dyslipidemia, *Prog. Nucl. Magn. Reson. Spectrosc.*, 2013, **70**, 1.
93 M. D. Mantle, NMR and MRI Studies of Drug Delivery Systems, *Curr. Opin. Colloid Interface Sci.*, 2013, **18**, 214.
94 I. Maslennikov and S. Choe, Advances in NMR Structures of Integral Membrane Proteins, *Curr. Opin. Struct. Biol.*, 2013, **23**, 555.
95 D. Massiot, R. J. Messinger, S. Cadars, M. Deschamps, V. Montouillout, N. Pellerin, E. Veron, M. Allix, P. Florian and F. Fayon, Topological, Geometric, and Chemical Order in Materials: Insights from Solid—State NMR, *Acc. Chem. Res.*, 2013, **46**, 1975.
96 A. F. Mironov, Lanthanide Porphyrin Complexes, *Russ. Chem. Rev.*, 2013, **82**, 333.
97 J. Mitchell, T. C. Chandrasekera and L. F. Gladden, Numerical Estimation of Relaxation and Diffusion Distributions in Two Dimensions, *Prog. Nucl. Magn. Reson. Spectrosc.*, 2012, **62**, 34.
98 J. Mitchell, T. C. Chandrasekera, D. J. Holland, L. F. Gladden and E. J. Fordham, Magnetic Resonance Imaging in Laboratory Petrophysical Core Analysis, *Phys. Rep.*, 2013, **526**, 165.
99 J. Mitchell, L. F. Gladden, T. C. Chandrasekera and E. J. Fordham, Low-Field Permanent Magnets for Industrial Process and Quality Control, *Prog. Nucl. Magn. Reson. Spectrosc.*, 2014, **76**, 1.
100 K. Moebius, W. Lubitz and A. Savitsky, High-Field EPR on Membrane Proteins – Crossing the Gap to NMR, *Prog. Nucl. Magn. Reson. Spectrosc.*, 2013, **75**, 1.
101 S. Mori, K. Oishi, A. V. Faria and M. I. Miller, Atlas-Based Neuroinformatics via MRI: Harnessing Information from Past Clinical Cases and Quantitative Image Analysis for Patient Care, *Annu. Rev. Biomed. Eng.*, 2013, **15**, 71.
102 L. J. Mueller and M. F. Dunn, NMR Crystallography of Enzyme Active Sites: Probing Chemically Detailed, Three-Dimensional Structure in Tryptophan Synthase, *Acc. Chem. Res.*, 2013, **46**, 2008.
103 D. T. Murray, N. Das and T. A. Cross, Solid State NMR Strategy for Characterizing Native Membrane Protein Structures, *Acc. Chem. Res.*, 2013, **46**, 2172.
104 Q. Z. Ni, E. Daviso, T. V. Can, E. Markhasin, S. K. Jawla, T. M. Swager, R. J. Temkin, J. Herzfeld and R. G. Griffin, High Frequency Dynamic Nuclear Polarization, *Acc. Chem. Res.*, 2013, **46**, 1933.
105 C. Nymberg, T. Jia, B. Ruggeri and G. Schumann, Analytical Strategies for Large Imaging Genetic Datasets: Experiences from the IMAGEN Study, *Ann. N. Y. Acad. Sci.*, 2013, **1282**(Adduction Reviews), 92.
106 S. J. Opella, Structure Determination of Membrane Proteins in Their Native Phospholipid Bilayer Environment by Rotationally Aligned Solid-State NMR Spectroscopy, *Acc. Chem. Res.*, 2013, **46**, 2145.

107 S. J. Opella, Structure Determination of Membrane Proteins by Nuclear Magnetic Resonance Spectroscopy, *Annu. Rev. Anal. Chem.*, 2013, **6**, 305.

108 C. M. Paquete and R. O. Louro, Unveiling the Details of Electron Transfer in Multicenter Redox Proteins, *Acc. Chem. Res.*, 2014, **47**, 56.

109 T. Parella and J. F. Espinosa, Long-Range Proton-Carbon Coupling Constants: NMR Methods and Applications, *Prog. Nucl. Magn. Reson. Spectrosc.*, 2013, **73**, 17.

110 S. Parthasarathy, Y. Nishiyama and Y. Ishii, Sensitivity and Resolution Enhanced Solid-State NMR for Paramagnetic Systems and Biomolecules under Very Fast Magic Angle Spinning, *Acc. Chem. Res.*, 2013, **46**, 2127.

111 M. E. Patarroyo, A. Bermudez and A. Moreno-Vranich, Towards the Development of a Fully Protective Plasmodium Falciparum Antimalarial Vaccine, *Expert Rev. Vaccines*, 2012, **11**, 1057.

112 R. Priori, R. Scrivo, J. Brandt, M. Valerio, L. Casadei, G. Valesini and C. Manetti, Metabolomics in Rheumatic Diseases: The Potential of an Emerging Methodology for Improved Patient Diagnosis, Prognosis, and Treatment Efficacy, *Autoimmun. Rev.*, 2013, **12**, 1022.

113 P. Pyykko, Relativistic Effects in Chemistry: More Common then You Though, *Annu. Rev. Phys. Chem.*, 2012, **63**, 45.

114 R. N. Rai, R. S. B. Reddi and U. S. Rai, Developments and Future Directions of Phase Diagram, Physicochemical and Optical Studies of Binary Organic Complexes, *Prog. Cryst. Growth Charact. Mater.*, 2013, **59**, 73.

115 A. Ramanathan, A. Savol, V. Burger, C. S. Chennubhotla and P. K. Agarwal, Protein Conformational Populations and Functionally Relevant Substates, *Acc. Chem. Res.*, 2014, **47**, 149.

116 R. Ramautar, R. Berger, J. van der Greef and T. Hankemeier, Human Metabolomics: Strategies to Understand Biology, *Curr. Opin. Chem. Biol.*, 2013, **17**, 841.

117 P. Ramm Sander, P. Hau, S. Koch, K. Schuetze, U. Bogdahn, H. R. Kalbitzer and L. Aigner, Stem Cell Metabolic and Spectroscopic Profiling, *Trends Biotechnol.*, 2013, **31**, 204.

118 P. Reiss and U. Koert, Ion-Channels: Goals for Function-Oriented Synthesis, *Acc. Chem. Res.*, 2013, **46**, 2773.

119 A. Roehrich and G. Drobny, Solid-State NMR Studies of Biomineralization Peptides and Proteins, *Acc. Chem. Res.*, 2013, **46**, 2136.

120 A. Rosato, R. Tejero and G. T. Montelione, Quality Assessment of Protein NMR Structures, *Curr. Opin. Struct. Biol.*, 2013, **23**, 715.

121 A. J. Rossini, A. Zagdoun, M. Lelli, A. Lesage, C. Coperet and L. Emsley, Dynamic Nuclear Polarization Surface Enhanced NMR Spectroscopy, *Acc. Chem. Res.*, 2013, **46**, 1942.

122 E. A. Rossler, S. Stapf and N. Fatkullin, Recent NMR Investigations on Molecular Dynamics of Polymer Melts in Bulk and in Confinement, *Curr. Opin. Colloid Interface Sci.*, 2013, **18**, 173.

123 Y. Y. Rusakov and L. B. Krivdin, Modern Quantum Chemical Methods for Calculating Spin-Spin Coupling Constants: Theoretical Basis and Structural Application in Chemistry, *Russ. Chem. Rev.*, 2013, **82**, 99.

124 A. Schlessinger, N. Khuri, K. M. Giacomini and A. Sali, Molecular Modeling and Ligand Docking for Solute Carrier (SLC) Transporters, *Curr. Top. Med. Chem.*, 2013, **13**, 843.

125 M. Schonhoff, NMR Studies of Sorption and Adsorption Phenomena in Colloidal Systems, *Curr. Opin. Colloid Interface Sci.*, 2013, **18**, 201.

126 R. W. Schurko, Ultra-Wideline Solid-State NMR Spectroscopy, *Acc. Chem. Res.*, 2013, **46**, 1985.
127 S. E. Sebastian, N. Harrison and G. G. Lonzarich, Towards Resolution of the Fermi Surface in Underdoped High-T_c Superconductors, *Rep. Prog. Phys.*, 2012, **75**, 102501.
128 C. Seger, S. Sturm and H. Stuppner, Mass Spectrometry and NMR Spectroscopy: Modern High-End Detectors for High Resolution Separation Techniques – State of the Art in Natural Product HPLC-MS, HPLC-NMR, and CE-MS Hyphenations, *Nat. Prod. Rep.*, 2013, **30**, 970.
129 I. Sengupta, P. S. Nadaud and C. P. Jaroniec, Protein Structure Determination with Paramagnetic Solid-State NMR Spectroscopy, *Acc. Chem. Res.*, 2013, **46**, 2117.
130 S. Sengupta and F. Wuerthner, Chlorophyll J-Aggregates: From Bioinspired Dye Stacks to Nanotubes, Liquid Crystals, and Biosupramolecular Electronics, *Acc. Chem. Res.*, 2013, **46**, 2498.
131 Y. E. Shapiro, NMR Spectroscopy on Domain Dynamics in Biomacromolecules, *Prog. Biophys. Mol. Biol.*, 2013, **112**, 58.
132 C. Shen and E. J. New, Promising Strategies for Gd-Based Responsive Magnetic Resonance Imaging Contrast Agents, *Curr. Opin. Chem. Biol.*, 2013, **17**, 158.
133 A. D. Sherry and Y. Wu, The Importance of Water Exchange Rates in the Design of Responsive Agents for MRI, *Curr. Opin. Chem. Biol.*, 2013, **17**, 167.
134 C. Simmler, J. G. Napolitano, J. B. McAlpine, S.-N. Chen and G. F. Pauli, Universal Quantitative NMR Analysis of Complex Natural Samples, *Curr. Opin. Biotechnol.*, 2014, **25**, 51.
135 P. Simon and Y. Gogotsi, Capacitive Energy Storage in Nanostructured Carbon-Electrolyte Systems, *Acc. Chem. Res.*, 2013, **46**, 1094.
136 M. Storr, H. J. Vogel and R. Schicho, Metabolomics: Is It Useful for Inflammatory Bowel Diseases?, *Curr. Opin. Gastroenterol.*, 2013, **29**, 378.
137 Y. Su, G. Liu, B. Xie, D. Fu and D. Wang, Crystallization Features of Normal Alkanes in Confined Geometry, *Acc. Chem. Res.*, 2014, **47**, 192.
138 M. Tang, G. Comellas and C. M. Rienstra, Advanced Solid-State NMR Approaches for Structure Determination of Membrane Proteins and Amyloid Fibrils, *Acc. Chem. Res.*, 2013, **46**, 2080.
139 M. Tariq, S. Ali and N. Khalid, Activity of Homogeneous and Heterogeneous Catalysts, Spectroscopic and Chromatographic Characterization of Biodiesel: A Review, *Renewable Sustainable Energy Rev.*, 2012, **16**, 6303.
140 E. L. Thomas, J. A. Fitzpatrick, S. J. Malik, S. D. Taylor-Robinson and J. D. Bell, Whole Body Fat: Content and Distribution, *Prog. Nucl. Magn. Reson. Spectrosc.*, 2013, **73**, 56.
141 F. V. Toukach and V. P. Ananikov, Recent Advances in Computational Predictions of NMR Parameters for the Structure Elucidation of Carbohydrates: Methods and Limitations, *Chem. Soc. Rev.*, 2013, **42**, 8376.
142 V. Tugarinov, Indirect Use of Deuterium in Solution NMR Studies of Protein Structure and Hydrogen Bonding, *Prog. Nucl. Magn. Reson. Spectrosc.*, 2014, **77**, 49.
143 R. Tycko, NMR at Low and Ultralow Temperatures, *Acc. Chem. Res.*, 2013, **46**, 1923.
144 R. Tycko and R. B. Wickner, Molecular Structures of Amyloid and Prion Fibrils: Consensus versus Controversy, *Acc. Chem. Res.*, 2013, **46**, 1487.

145 S. J. Ullrich and C. Glaubitz, Perspectives in Enzymology of Membrane Proteins by Solid-State NMR, *Acc. Chem. Res.*, 2013, **46**, 2164.
146 A. V. Vasilyev, Electrophilic Activation of Acetylene Compounds in Bronsted Superacids. Reactions of Vinyl Type Cations, *Russ. Chem. Rev.*, 2013, **82**, 187.
147 V. I. Volkov and A. A. Marinin, NMR Methods for Studying Ion and Molecular Transport in Polymer Electrolytes, *Russ. Chem. Rev.*, 2013, **82**, 248.
148 A. J. Wand, The Dark Energy of Proteins Comes to Light: Conformational Entropy and Its Role in Protein Function Revealed by NMR Relaxation, *Curr. Opin. Struct. Biol.*, 2013, **23**, 75.
149 C. A. Waudby, H. Launay, L. D. Cabrita and J. Christodoulou, Protein Folding on the Ribosome Studied Using NMR Spectroscopy, *Prog. Nucl. Magn. Reson. Spectrosc.*, 2013, **74**, 57.
150 M. Weingarth and M. Baldus, Solid-State NMR-Based Approaches for Supramolecular Structure Elucidation, *Acc. Chem. Res.*, 2013, **46**, 2037.
151 H. Weingartner, NMR Studies of Ionic Liquids: Structure and Dynamics, *Curr. Opin. Colloid Interface Sci.*, 2013, **18**, 183.
152 K. Wilczewska, A. Kot-Wasik and J. Namiesnik, LC-MS and LC-NMR as Complementary Techniques for the Determination of Pharmaceuticals in Dosage Formulations, *Crit. Rev. Anal. Chem.*, 2013, **43**, 148.
153 A. Winter, A. P. Higeruelo, M. Marsh, A. Sigurdardottir, W. R. Pitt and T. L. Blundell, Biophysical and Computational Fragment-Based Approaches to Targeting Protein-Protein Interactions: Applications in Structure-Guided Drug Discovery, *Q. Rev. Biophys.*, 2012, **45**, 383.
154 W. Xu, K. Qiao and Y. Tang, Structural Analysis of Protein-Protein Interactions in Type I Polyketide Synthases, *Crit. Rev. Biochem. Mol. Biol.*, 2013, **48**, 98.
155 S. Xue, J. Qiao, F. Pu, M. Cameron and J. J. Yang, Design of a Novel Class of Protein-Based Magnetic Resonance Imaging Contrast Agents for the Molecular Imaging of Cancer Biomarkers, *Wiley Interdiscip. Rev.: Nanomed. Nanobiotechnol.*, 2013, **5**, 163.
156 K. Yamasaki, T. Kigawa, M. Seki, K. Shinozaki and S. Yokoyama, DNA-Binding Domains of Plant-Specific Transcription Factors: Structure, Function, and Evolution, *Trends Plant Sci.*, 2013, **18**, 267.
157 S. Yan, C. L. Suiter, G. Hou, H. Zhang and T. Polenova, Probing Structure and Dynamics of Protein Assemblies by Magic Angle Spinning NMR Spectroscopy, *Acc. Chem. Res.*, 2013, **46**, 2047.
158 G. Yu, X. Yan, C. Han and F. Huang, Characterization of Supramolecular Gels, *Chem. Soc. Rev.*, 2013, **42**, 6697.
159 J.-X. Yu, R. R. Hallac, S. Chiguru and R. P. Mason, New Frontiers and Developing Applications in ^{19}F NMR, *Prog. Nucl. Magn. Reson. Spectrosc.*, 2013, **70**, 25.
160 N. D. Yuliana, M. Jahangir, R. Verpoorte and Y. H. Choi, Metabolomics for the Rapid Dereplication of Bioactive Compounds from Natural Sources, *Phytochem. Rev.*, 2013, **12**, 293.
161 G.-P. Zhou and R.-B. Huang, The pH-Triggered Conversion of the PrP(c) to PrP(sc.), *Curr. Top. Med. Chem.*, 2013, **13**, 1152.
162 Z. Zhou and Z.-R. Lu, Gadolinium-Based Contrast Agents for Magnetic Resonance Cancer Imaging, *Wiley Interdiscip. Rev.: Nanomed. Nanobiotechnol.*, 2013, **5**, 1.
163 J. J. Ziarek and B. F. Volkman, NMR in the Analysis of Functional Chemokine Interactions and Drug Discovery, *Drug Discovery Today: Technol.*, 2012, **9**, e293.

Edited books and symposia

164 A. Alaaeddine, C. Negrell and B. Ameduri, Recent Advances on New Fluorinated Copolymers Based on Carbonate and Oligo(Ethylene Oxide) by Radical Copolymerization, in *Advances in Fluorine-Containing Polymers*, ed. S. Iacono, D. Smith, S. Kettwich and D. Boday, ACS Symposium Series, American Chemical Society, 2012, vol. 1106, p. 141.

165 R. M. Alonso-Salces, M. V. Holland, C. Guillou and K. Heberger, Quality Assessment of Olive Oil by ^{1}H-NMR Fingerprinting, in *Olive Oil: Constituents, Quality, Health Properties and Bioconversions*, ed. B. Dimitrios, InTech, Rijeka, Croatia, 2012, p. 185.

166 T. C. Alves, I. Jarak and R. A. Carvalho, NMR Methodologies for Studying Mitochondrial Bioenergetics, in *Mitochondrial Bioenergetics*, ed. C. M. Palmeira and A. J. Moreno, Methods in Molecular Biology, Springer, New York, NY, United States, 2012, vol. 810, p. 281.

167 S. Anderson, D. Luffer-Atlas and M. P. Knadler, Predicting Human Biotransformation Pathways, in *Encyclopedia of Drug Metabolism and Interactions*, ed. A. V. Lyubimov, A. D. Rodrigues and M. A. Sinz, John Wiley & Sons, Inc., Hoboken, N.J., 2012, vol. 6, p. 177.

168 I. M. Armitage, T. Drakenberg and B. Reilly, Use of (113)Cd NMR to Probe the Native Metal Binding Sites in Metalloproteins: An Overview, in *Cadmium: From Toxicity to Essentiality*, ed. A. Sigel, H. Sigel and K. O. Roland, Metal Ions in Life Sciences, Springer, 2013, vol. 11, p. 117.

169 A. Arora, Solution NMR Spectroscopy for the Determination of Structures of Membrane Proteins in a Lipid Environment, in *Lipid-Protein Interactions: Methods and Protocols*, ed. J. H. Kleinschmidt, Methods in Molecular Biology, Springer, New York, NY, United States, 2013, vol. 974, p. 389.

170 J. D. Ballin and G. M. Wilson, Role and Applications of Electrostatic Effects on Nucleic Acid Conformational Transitions and Binding Processes, in *Application of Thermodynamics to Biological and Materials Science*, ed. T. Mizutani, InTech, Rijeka, Croatia, 2011, p. 129.

171 E. Barbet-Massin and G. Pintacuda, Biomolecular Solid-State NMR/Basics, in *NMR of Biomolecules: Towards Mechanistic Systems Biology*, ed. I. Bertini, K. S. McGreevy and G. Parigi, Wiley-Blackwell, Weinheim, Germany, 2012, p. 347.

172 G. Bevilacqua, V. Biancalana, Y. Dancheva and L. Moi, Optical Atomic Magnetometry for Ultra-Low-Field NMR Detection, in *Annual Reports on NMR Spectroscopy*, ed. G. A. Webb, Elsevier Ltd., 2013, vol. 78, p. 103.

173 B. Bluemich, Stochastic Excitation, in *Encyclopedia of NMR*, ed. R. K. Harris and R. E. Wasylishen, John Wiley & Sons, Ltd., Chichester, UK, 2012, vol. 8, p. 4758.

174 R. Boelens and R. Kaptein, Homonuclear Three-Dimensional NMR of Biomolecules, in *Encyclopedia of NMR*, ed. R. K. Harris and R. E. Wasylishen, John Wiley & Sons, Ltd., Chichester, UK, 2012, vol. 4, p. 1910.

175 B. B. Bonev, High-Resolution Solid-State NMR of Lipid Membranes, in *Advances in Planar Lipid Bilayers and Liposomes*, ed. A. Iglic and J. Genova, A Tribute to Marin D. Mitov, Academic Press, 2013, vol. 17, p. 299.

176 E. Bordignon, Site-Directed Spin Labeling of Membrane Proteins, in *EPR Spectroscopy: Applications in Chemistry and Biology*, ed. M. Drescher and G. Jeschke, Topics in Current Chemistry, Springer GmbH, vol. 321, p. 121.

177 G. Brunklaus, H.-W. Spiess and H. Ecker, Solid State NMR: A Versatile Tool in Solid State Chemistry and Materials Science, in *Methods in Physical*

Chemistry, ed. R. Schaefer and P. C. Schmidt, Wiley-VCH Verlag GmbH & Co. KGaA, Weinheim, Germany, 2012, vol. 1, p. 87.
178 D. L. Bryce and R. E. Wasylishen, Quadrupolar Nuclei in Solids: Influence of Different Interactions on Spectra, in *Encyclopedia of NMR*, ed. R. K. Harris and R. E. Wasylishen, John Wiley & Sons, Ltd., Chichester, UK, 2012, vol. 7, p. 3871.
179 P. Caboni, N. G. Ntalli, C. E. Bueno and L. E. Alche, Isolation and Chemical Characterization of Components with Biological Activity Extracted from Azadirachta Indica and Melia Azedarach, in *Emerging Trends in Dietary Components for Preventing and Combating Disease*, ed. B. Patil, G. K. Jayaprakasha, K. N. C. Murthy and N. Seeram, ACS Symposium Series, American Chemical Society, 2012, vol. 1093, p. 51.
180 W. R. Carper, Molten Salts, in *Encyclopedia of NMR*, ed. R. K. Harris and R. E. Wasylishen, John Wiley & Sons, Ltd., Chichester, UK, 2012, vol. 5, p. 2739.
181 C. M. Castro, H. Lee and R. Weissleder, Cancer Cell Detection and Molecular Profiling Using Diagnostic Magnetic Resonance, in *Biosensors and Molecular Technologies for Cancer Diagnostics*, ed. K. E. Herold and A. Rasooly, CRC Press, Boca Raton, Fla., 2012, p. 731.
182 K. Chen and N. Tjandra, The Use of Residual Dipolar Coupling in Studying Proteins by NMR, in *NMR of Proteins and Small Biomolecules*, ed. G. Zhu, Topics in Current Chemistry, Springer GmbH, 2012, vol. 326, p. 47.
183 A. Ciulli, Biophysical Screening for the Discovery of Small-Molecule Ligands, in *Protein-Ligand Interactions: Methods and Applications*, ed. M. A. Williams and T. Daviter, Methods in Molecular Biology, Springer, New York, NY, United States, 2013, vol. 1008, p. 357.
184 Y. Cohen, L. Avram, T. Evan-Salem, S. Slovak, N. Shemesh and L. Frish, Diffusion NMR in Supramolecular Chemistry and Complex Systems, in *Analytical Methods in Supramolecular Chemistry*, ed. C. A. Schalley, Wiley-VCH Verlag GmbH & Co. KGaA, Weinheim, Germany, 2nd edn, 2012, vol. 1, p. 197.
185 R. Consonni and M. Veronesi, NMR Screening Methods in Fragment-Based Drug Discovery, in *Structure – Activity Relationship Studies in Drug Development by NMR Spectroscopy*, ed. Atta-ur-Rahman and M. I. Choudhary, Bentham Science Publishers Ltd., 2011, vol. 1 p. 67.
186 J. Courtieu, P. Lesot, A. Meddour, D. Merlet and C. Aroulanda, Chiral Discrimination Using Chiral Ordering Agents, in *Encyclopedia of NMR*, ed. R. K. Harris and R. E. Wasylishen, John Wiley & Sons, Ltd., Chichester, UK, 2012, vol. 1, p. 578.
187 N. C. Da Costa and S. Eri, Chemical Characterization, in *Handbook of Meat, Poultry and Seafood Quality*, ed. L. M. L. Nollet, Wiley-Blackwell, Ames, Iowa, 2nd edn, 2012, p. 76.
188 P. Dais and A. Spyros, Nuclear Magnetic Resonance, in *Chemical Analysis of Food: Techniques ad Applications*, ed. Y. Pico, Elsevier Inc., Waltham, Mass., 2012, p. 91.
189 B. Davis, Screening Protein-Small Molecule Interactions by NMR, in *Protein-Ligand Interactions: Methods and Applications*, ed. M. A. Williams and T. Daviter, Methods in Molecular Biology, Springer, New York, NY, United States, 2013, vol. 1008, p. 389.
190 A. Di Domizio and F. Peri, Sugar-Based Inhibitors of Ras Activation: Biological Activity and Identification of Ras-Inhibitor Binding Interface, in *Inhibitors of the Ras Superfamily G-Proteins, Part A*, ed. F. Tamonai, Enzymes, Elsevier, San Diego, Calif., vol. 33, p. 95.

191 K. B. Dillon, Nuclear Quadrupole Resonance Spectroscopy, in *Spectroscopic Properties of Inorganic and Organometallic Compounds*, ed. J. Yarwood, R. Douthwaite and S. Duckett, Royal Society of Chemistry, 2012, vol. 43, p. 216.
192 K. B. Dillon, Nuclear Quadrupole Resonance Spectroscopy, in *Spectroscopic Properties of Inorganic and Organometallic Compounds*, ed. J. Yarwood, R. Douthwaite and S. Duckett, Royal Society of Chemistry, 2013, vol. 44, p. 28.
193 M. L. Di Vona, E. Sgreccia and S. Tosto, Diffusion in Solid Proton Conductors: Theoretical Aspects and Nuclear Magnetic Resonance Analysis, in *Solid State Proton Conductors: Properties and Applications in Fuel Cells*, ed. P. Knauth and M. L. Di Vona, John Wiley & Sons Ltd., Chichester, UK, 2012, p. 25.
194 M. Ernst and B. H. Meier, Adiabatic Polarization-Transfer Methods in MAS Spectroscopy, in *Encyclopedia of NMR*, ed. R. K. Harris and R. E. Wasylishen, John Wiley & Sons, Ltd., Chichester, UK, 2012, vol. 1, p. 44.
195 T. W.-M. Fan, Considerations of Sample Preparation for Metabolomics Investigation, in *Methods in Pharmacology and Toxicology*, ed. T. W.-M. Fan, R. M. Higashi and A. N. Lane, Handbook of Metabolomics, Springer, New York, N.Y., 2012, vol. 17, p. 7.
196 C. Fanali, P. Dugo, L. Mondello, G. D'Orazio and S. Fanali, Recent Development in High-Performance Liquid Chromatography, in *Food Analysis by HPLC*, ed. L. M. L. Nollet and F. Toldra, CRC Press, Boca Raton, Fla., 3rd edn, 2013, p. 1.
197 X. Feng, K. Yang, J. Hu, T. Xu and Y. Cheng, Nuclear Magnetic Resonance Techniques in the Analysis of Pamam Dendrimer-Based Drug Delivery Systems, in *Dendrimer-Based Drug Delivery Systems: From Theory to Practice*, ed. Y. Cheng, John Wiley & Sons, Inc., Hoboken, N.J., 2012, p. 439.
198 C. Fernandez and M. Pruski, Probing Quadrupolar Nuclei by Solid-State NMR Spectroscopy: Recent Advances, in *Solid State NMR*, ed. J. C. C. Chan, Topics in Current Chemistry, Springer GmbH, 2012, vol. 306, p. 119.
199 R. Freeman and E. Kupce, Concepts on Projection-Reconstruction, in *Novel Sampling Approaches in Higher Dimensional NMR*, ed. M. Billeter and V. Orekhov, Topics in Current Chemistry, Springer GmbH, 2012, vol. 316, p. 1.
200 E. Freisinger and M. Vasak, Cadmium in Metallothioneins, in *Cadmium: From Toxicity to Essentiality*, ed. A. Sigel, H. Sigel and K. O. Roland, Metal Ions in Life Sciences, Springer, 2013, vol. 11, p. 339.
201 C. P. Frizzo, A. Z. Tier, I. M. Gindri, L. Buriol, M. A. Villetti, N. Zanatta and M. A. P. Martins, Nanostructure Evaluation of Ionic Liquid Aggregates by Spectroscopy, in *Nanoscale Spectroscopy with Applications*, ed. S. M. Musa, CRC Press, 2013, p. 215.
202 D. Gajan, G. Lapadula, M. Schwarzwalder and C. Coperet, Solid-State NMR: A Key Tool for the Understanding at a Molecular Level of Well-Defined Heterogeneous Catalysts and Surface Chemistry on Top of Oxide Materials, in *Spectroscopic Properties of Inorganic and Organometallic Compounds*, ed. J. Yarwood, R. Douthwaite and S. Duckett, Royal Society of Chemistry, 2012, vol. 43, p. 57.
203 L. Gales and A. M. Damas, Three-Dimensional Structure of Xanthones, in *Chemistry and Pharmacology of Naturally Occurring Bioactive Compounds*, ed. G. Brahmachari, CRC Press, Boca Raton, Fla., 2013, p. 349.
204 B. Gierczyk, Application of Transition Metals NMR in Supramolecular Chemistry, in *From Molecules to Functional Architecture Supramolecular Interactions*, ed. V. I. Rybachenko, East Publisher House, Donetsk, Ukraine, 2012, p. 269.

205 E. P. Gilbert, A. Lopez-Rubio and M. J. Gidley, Characterisation Techniques in Food Materials Science, in *Food Material Science and Engineering*, ed. B. Bhandari and Y. H. Roos, Wiley-Blackwell, Chichester, UK, 2012, p. 52.
206 P. Giraudeau and S. Akoka, Fast and Ultrafast Quantitative 2D NMR: Vital Tools for Efficient Metabolomic Approaches, in *Metabolomics Coming of Age with Its Technological Diversity*, ed. D. Rolin, Advances in Botanical Research, Elsevier Ltd., 2013, vol. 67, p. 99.
207 S. Gloeggler, B. Bluemich and S. Appelt, NMR Spectroscopy for Chemical Analysis at Low Magnetic Fields, in *Modern NMR Methodology*, ed. H. Heise, Topics in Current Chemistry, Springer, 2013, vol. 335, p. 1.
208 G. D. Goldhaber-Pasillas, Y. H. Choi and R. Verpoorte, New Methods of Analysis and Investigation of Terpenoid Indole Alkaloids, in *New Light on Alkaloid Biosynthesis and Future Prospects*, ed. N. Giglioli-Guivarc'h, Advances in Botanical Research, Elsevier Ltd., 2013, vol. 68, p. 233.
209 S. Gorog, The Role of Analytical Chemistry in the Safety of Drug Therapy, in *Analytical Techniques for Clinical Chemistry: Methods and Applications*, ed. S. Caroli and G. Zaray, John Wiley & Sons, Inc., Hoboken, N.J., 2012, p. 179.
210 J. M. Griffin and S. E. Ashbrook, Solid-State NMR of High-Pressure Silicates in the Earth's Mantle, in *Annual Reports on NMR Spectroscopy*, ed. G. A. Webb, Elsevier Ltd., 2013, vol. 79, p. 241.
211 A. Gruenberg, H. Breitzke and G. Buntkowsky, Solid State NMR of Immobilized Catalysts and Nanocatalysts, in *Spectroscopic Properties of Inorganic and Organometallic Compounds*, ed. J. Yarwood, R. Douthwaite and S. Duckett, Royal Society of Chemistry, 2012, vol. 43, p. 289.
212 U. L. Guenther, Dynamic Nuclear Hyperpolarization in Liquids, in *Modern NMR Methodology*, ed. H. Heise, Topics in Current Chemistry, Springer, 2013, vol. 335, p. 23.
213 P. Guerry and T. Herrmann, Comprehensive Automation for NMR Structure Determination of Proteins, in *Protein NMR Techniques*, ed. A. Shekhtman and D. S. Burz, Methods in Molecular Biology, Springer, New York, NY, United States, 2012, vol. 831, p. 429.
214 X. Guo and E. Lankmayr, Hyphenated Techniques in Gas Chromatography, in *Advanced Gas Chromatography: Progress in Agricultural, Biomedical and Industrial Applications*, ed. M. A. Mohd, InTech, Rijeka, Croatia, 2012, p. 3.
215 R. Haensel, S. Foldynova-Trantirkova, V. Doetsch and L. Trantirek, Investigation of Quadruplex Structure under Physiological Conditions Using in-Cell NMR, in *Quadruplex Nucleic Acids*, ed. J. B. Chaires and D. Graves, Topics in Current Chemistry, Springer, 2013, vol. 330, p. 47.
216 A. K. Haghi and G. E. Zaikov, A Promising New Class of Polymer Material with Particular Application in Nanotechnology, in *Nanopolymers and Modern Materials: Preparation, Properties, and Applications*, ed. O. V. Stoyanov, A. K. Haghi and G. E. Zaikov, CRC Press, 2014, p. 223.
217 M. Hayward, M. D. Bacolod, Q. P. Han, M. Cajina and Z. Zou, Techniques to Facilitate the Performance of Mass Spectrometry: Sample Preparation, Liquid Chromatography, and Non-Mass-Spectrometric Detection, in *Mass Spectrometry in Drug Metabolism and Disposition: Basic Principles and Applications*, ed. M. S. Lee and M. Zhu, John Wiley & Sons, Inc., Hoboken, N. J., 2011, p. 353.
218 S. Henning, D. Edelhoff, S. Leick, H. Rehage and D. Suter, Permeability and Stability of Microcapsules, in *Magnetic Resonance in Food Science*, ed. J. van Duynhoven, P. S. Belton, G. A. Webb and H. van As, Special Publication – Royal Society of Chemistry, Royal Society of Chemistry, 2013, vol. 343, p. 122.

219 T. Herrmann and B. Elena-Herrmann, Applications of Projection NMR Techniques, in *Annual Reports on NMR Spectroscopy*, ed. G. A. Webb, Elsevier Ltd., 2013, vol. 78, p. 55.
220 H. D. W. Hill and G. A. Gray, Spectrometers: A General Overview, in *Encyclopedia of NMR*, ed. R. K. Harris and R. E. Wasylishen, John Wiley & Sons, Ltd., Chichester, UK, 2012, vol. 8, p. 4652.
221 P. E. Hoggan and A. Bouferguene, Why Specific ETOs Are Advantageous for NMR and Molecular Interactions, in *Advances in the Theory of Quantum Systems in Chemistry and Physics*, ed. P. E. Hoggan, E. J. Brändas, J. Maruani, P. Piecuch and G. Delgado-Barrio, Progress in Theoretical Chemistry and Physics, Springer, 2012, vol. 22, p. 83.
222 K. Holderna-Natkaniec, R. Jakubas and I. Natkaniec, Internal Dynamics of the Ferroelectric $(C_3N_2H_5)_5$-Bi_2Cl_{11} Studied by ^1H NMR and IINS Methods, in *Ferroelectrics*, ed. M. Lallart, In-Tech, Rijeka, Croatia, 2011, p. 41.
223 G. Holdgate, S. Geschwindner, A. Breeze, G. Davies, N. Colclough, D. Temesi and L. Ward, Biophysical Methods in Drug Discovery from Small Molecule to Pharmaceuticals, in *Protein-Ligand Interactions: Methods and Applications*, ed. M. A. Williams and T. Daviter, Methods in Molecular Biology, Springer, New York, NY, United States, 2013, vol. 1008, p. 327.
224 R. V. Hosur, D. Kumar, S. Chakraborty, M. K. Rout, J. G. Reddy and P. Mishra, Proteomics, Protein Folding and Self-Association: New Insights from Recent Advances in Protein NMR Methodologies, in *Biomolecular Forms and Functions*, ed. M. Bansal and N. Srinivasan, World Scientific Publishing Co. Pte. Ltd., Singapore, Singapore, 2013, p. 184.
225 S. G. Hyberts, H. Arthanari and G. Wagner, Applications of Non-Uniform Sampling and Processing, in *Novel Sampling Approaches in Higher Dimensional NMR*, ed. M. Billeter and V. Orekhov, Topics in Current Chemistry, Springer GmbH, 2012, vol. 316, p. 125.
226 K. Ishida and Y. Nakai, Review of NMR Studies on Iron-Based Superconductors, in *Iron-based Superconductors: Materials, Properties and Mechanisms*, ed. N.-L. Wang, H. Hosono and P. Dai, Pan Stanford Publishing, 2012, p. 275.
227 D. M. Jacobs, E. van Velzen and V. Mihaleva, Evaluation of Approaches for Quantitative Targeted Profiling of Complex Compositions Using 1D ^1H NMR Spectroscopy, in *Magnetic Resonance in Food Science*, ed. J. van Duynhoven, P. S. Belton, G. A. Webb and H. van As, Special Publication – Royal Society of Chemistry, Royal Society of Chemistry, 2013, vol. 343, p. 3.
228 M. C. Jarvis, Cellulose, Crystallinity: Perspectives from Spectroscopy and Diffraction, in *Stability of Complex Carbohydrate Structures: Biofuels, Foods, Vaccines and Shipwrecks*, ed. S. E. Harding, Special Publication – Royal Society of Chemistry, Royal Society of Chemistry, 2013, vol. 341, p. 125.
229 M. R. Jensen, P. Bernado, R. W. H. Ruigrok and M. Blackledge, Structural Disorder within Sendai Virus Nucleoprotein and Phosphoprotein, in *Flexible Viruses: Structural Disorder in Viral Proteins*, ed. V. N. Uversky and S. Longhi, John Wiley & Sons, Inc., Hoboken, N.J., 2012, p. 95.
230 C. Jones, Stability and Degradation Pathways of Polysaccharide and Glycoconjugate Vaccines, in *Stability of Complex Carbohydrate Structures: Biofuels, Foods, Vaccines and Shipwrecks*, ed. S. E. Harding, Special Publication – Royal Society of Chemistry, Royal Society of Chemistry, 2013, vol. 341, p. 56.
231 C. E. Kaiser, V. Gokhale, D. Yang and L. H. Hurley, Gaining Insights into the Small Molecule Targeting of the G-Quadruplex in the c-MYC Promoter Using NMR and an Allele-Specific Transcriptional Assay, in *Quadruplex*

Nucleic Acids, ed. J. B. Chaires and D. Graves, Topics in Current Chemistry, Springer, 2013, vol. 330, p. 1.
232 D. Kalnin, Methods of Studying Cocoa Butter and Bloom, in *Cocoa Butter and Related Compounds*, ed. N. Garti and N. R. Widlak, AOCS Monograph Series on Oilseeds, AOCS Press, 2012, vol. 6, p. 307.
233 P. B. Karadakov and G. A. Webb, Shielding: Overview of Theoretical Methods, in *Encyclopedia of NMR*, ed. R. K. Harris and R. E. Wasylishen, John Wiley & Sons, Ltd., Chichester, UK, 2012, vol. 8, p. 4457.
234 K. Kato, Structural Glycomic Approaches to Molecular Recognition Events on Cell Surfaces, in *Biochemical Roles of Eukaryotic Cell Surface Macromolecules*, ed. P. R. Sudhakaran and A. Surolia, Advances in Experimental Medicine and Biology, Springer, 2012, vol. 749, p. 15.
235 K. Kazimierczuk, M. Misiak, J. Stanek, A. Zawadzka-Kazimierczuk and W. Kozminski, Generalized Fourier Transform for Non-Uniform Sampled Data, in *Novel Sampling Approaches in Higher Dimensional NMR*, ed. M. Billeter and V. Orekhov, Topics in Current Chemistry, Springer GmbH, 2012, vol. 316, p. 79.
236 B. Kirchner, P. J. di Dio and J. Hutter, Real-World Predictions from Ab Initio Molecular Dynamics Simulations, in *Multiscale Molecular Methods in Applied Chemistry*, ed. B. Kirchner and J. Vrabec, Topics in Current Chemistry, Springer GmbH, 2012, vol. 307, p. 109.
237 H. Koller and M. Weiss, Solid State NMR of Porous Materials: Zeolites and Related Materials, in *Solid State NMR*, ed. J. C. C. Chan, Topics in Current Chemistry, Springer GmbH, 2012, vol. 306, p. 189.
238 P. Kosobucki and B. Buszewski, Isotachophoresis, in *Electromigration Techniques: Theory and Practice*, ed. B. Buszewski, E. Dziubakiewicz and M. Szumski, Springer Series in Chemical Physics, Springer GmbH, 2013, vol. 105, p. 93.
239 E. Kupce, NMR with Multiple Receivers, in *Modern NMR Methodology*, ed. H. Heise, Topics in Current Chemistry, Springer, 2013, vol. 335, p. 71.
240 M. F. L'Annunziata, Flow-Cell Analysis, in *Handbook of Radioactivity Analysis*, ed. M. F. L'Annunziata, Elsevier Ltd., Oxford, UK, 3rd edn, 2012, p. 1117.
241 P. Leinweber, J. Kruse, C. Baum, M. Arcand, J. D. Knight, R. Farrell, K.-U. Eckhardt, K. Kiersch and G. Jandl, Advances in Understanding Organic Nitrogen Chemistry in Soils Using State-of-the-Art Analytical Techniques, in *Advances in Agronomy*, ed. D. L. Sparks, Elsevier, 2013, vol. 119, p. 83.
242 D. D. Le Pevelen, Physical and Spectrometric Analysis: An Overview of Chiral Physical Analysis, in *Comprehensive Chirality*, ed. E. M. Carreira and H. Yamamoto, Elsevier B.V., Amsterdam, Neth., 2012, vol. 8, p. 625.
243 J. R. Lewandowski and L. Emsley, Protein Dynamics in the Solid State, in *NMR of Biomolecules: Towards Mechanistic Systems Biology*, ed. I. Bertini, K. S. McGreevy and G. Parigi, Wiley-Blackwell, Weinheim, Germany, 2012, p. 367.
244 H. Li and H. Sun, NMR Studies of Metalloproteins, in *NMR of Proteins and Small Biomolecules*, ed. G. Zhu, Topics in Current Chemistry, Springer GmbH, 2012, vol. 326, p. 69.
245 S. Li and F. Deng, Recent Advances of Solid-State NMR Studies on Zeolites in *Annual Reports on NMR Spectroscopy*, ed. G. A. Webb, Elsevier Ltd., 2013, vol. 78, p. 1.
246 Y. Lin, Y. Huang, S. Cai and Z. Chen, Intermolecular Zero Quantum Coherence in NMR Spectroscopy in *Annual Reports on NMR Spectroscopy*, ed. G. A. Webb, Elsevier Ltd., 2013, vol. 78, p. 209.

247 L. Liu, G. Jiang, F. Ning, Y. Yu, L. Zhang and Y. Tu, Well Logging in Gas Hydrate-Bearing Sediment: A Review, in *Natural Resources and Sustainable Development II*, ed. J. Wu, J. Yang, N. Nakagoshi, X. Lu and H. Xu, Advanced Materials Research, Trans Tech Publications Ltd., Durnten-Zurich, Switzerland, 2012, vol. 524–527, p. 1660.
248 R. P. Mailavaram and M. Devarakonda, NMR – A Gate Way to Drug Discovery, in *Structure – Activity Relationship Studies in Drug Development by NMR Spectroscopy*, ed. Atta-ur-Rahman and M. I. Choudhary, Bentham Science Publishers Ltd., 2011, vol. 1, p. 1.
249 F. Manna, F. Rossi and R. Fioravanti, Analytical Techniques and Quality Control of Pharmaceuticals, in *Analytical Techniques for Clinical Chemistry: Methods and Applications*, ed. S. Caroli and G. Zaray, John Wiley & Sons, Inc., Hoboken, N.J., 2012, p. 245.
250 J. P. Mathews, R. Winans, R. P. Rodgers and A. Sharma, Advances in Coal Analyses and Simulations 1989–2012, in *Proceedings – Annual International Pittsburgh Coal Conference 2012, 29th*, Pittsburgh Coal Conference, University of Pittsburgh, 2012, vol. 2, p. 1484.
251 M. Monajjemi and V. S. Lee, Quantum Calculation in Prediction the Properties of Single-Walled Carbon Nanotubes, in *Electronic Properties of Carbon Nanotubes*, ed. J. M. Marulanda, InTech, Rijeka, Croatia, 2011, p. 575.
252 I. L. Moudrakovski, Recent Advances in Solid-State NMR of Alkaline Earth Elements, in *Annual Reports on NMR Spectroscopy*, ed. G. A. Webb, Elsevier Ltd., 2013, vol. 79, p. 129.
253 H. Mueller, M. Etzkorn and H. Heise, Solid-State NMR Spectroscopy of Proteins, in *Modern NMR Methodology*, ed. H. Heise, Topics in Current Chemistry, Springer, 2013, vol. 335, p. 121.
254 N. C. Nielsen, C. Kehlet, S. J. Glaser and N. Khaneja, Optimal Control Methods in NMR Spectroscopy, in *Encyclopedia of NMR*, ed. R. K. Harris and R. E. Wasylishen, John Wiley & Sons, Ltd., Chichester, UK, 2012, vol. 6, p. 3161.
255 N. C. Nielsen, L. A. Strassoe and A. B. Nielsen, Dipolar Recoupling, in *Solid State NMR*, ed. J. C. C. Chan, Topics in Current Chemistry, Springer GmbH, 2012, vol. 306, p. 1.
256 S. R. Obach, D. K. Dalvie and G. S. Walker, Identification of Drug Metabolites, in *Encyclopedia of Drug Metabolism and Interactions*, ed. A. V. Lyubimov, A. D. Rodrigues and M. A. Sinz, John Wiley & Sons, Inc., Hoboken, N.J., 2012, vol. 3, p. 121.
257 L.-O. Oehman and U. Edlund, Aluminum-27 NMR in Solutions, in *Encyclopedia of NMR*, ed. R. K. Harris and R. E. Wasylishen, John Wiley & Sons, Ltd., Chichester, UK, 2012, vol. 1, p. 90.
258 M. Oromi-Farrus, J. P. Minguell and R. Canela, Characterization and Quantification of Phosphonates by NMR, in *Comunicaciones Presentadas a las Jornadas del Comite Espanol de la Detergencia*, Comite Espanol de la Detergencia, Tensioactivos y Afines, 2013, vol. 43, p. 42.
259 G. Ortega, M. Pons and O. Millet, Protein Functional Dynamics in Multiple Timescales as Studied by NMR Spectroscopy, in *Dynamics of Proteins and Nucleic Acids*, ed. T. Karabencheva-Christova, Advances in Protein Chemistry and Structural Biology, Elsevier, 2013, vol. 92, p. 219.
260 T. Parella, Multidimensional Methods for the Measurement of Long-Range Proton-Carbon Coupling Constants at Natural Abundance, in *Encyclopedia of NMR*, ed. R. K. Harris and R. E. Wasylishen, John Wiley & Sons, Ltd., Chichester, UK, 2012, vol. 5, p. 2806.
261 M. V. Petoukhov and D. I. Svergun, Small Angle X-Ray Scattering/Small Angle Neutron Scattering as Methods Complementary to NMR, in *NMR of*

Biomolecules: Towards Mechanistic Systems Biology, ed. I. Bertini, K. S. McGreevy and G. Parigi, Wiley-Blackwell, Weinheim, Germany, 2012, p. 563.

262 G. Pintacuda and G. Kervern, Paramagnetic Solid-State Magic-Angle Spinning NMR Spectroscopy, in *Modern NMR Methodology*, ed. H. Heise, Topics in Current Chemistry, Springer, 2013, vol. 335, p. 157.

263 A. Piserchio, Expression and Purification of Src-Family Kinases for Solution NMR Studies, in *Protein NMR Techniques*, ed. A. Shekhtman and D. S. Burz, Methods in Molecular Biology, Springer, New York, NY, United States, 2012, vol. 831, p. 111.

264 A. Piserchio, K. N. Dalby and R. Ghose, Assignment of Backbone Resonances in a Eukaryotic Protein Kinase – ERK2 as a Representative Example, in *Protein NMR Techniques*, ed. A. Shekhtman and D. S. Burz, Methods in Molecular Biology, Springer, New York, NY, United States, 2012, vol. 831, p. 359.

265 R. Pohmann, Physical Basics of NMR, in *In vivo NMR Imaging: Methods and Protocols*, ed. L. Schröder and C. Faber, Methods in Molecular Biology, Springer, New York, NY, United States, 2011, vol. 771, p. 3.

266 A. Polimeno, V. Barone and J. H. Freed, Stochastic Methods for Magnetic Resonance Spectroscopies, in *Computational Strategies for Spectroscopy: From Small Molecules to Nano Systems*, ed. V. Barone, John Wiley & Sons, Inc., Hoboken, N.J., 2012, p. 549.

267 V. H. Pomin, Unravelling Glycobiology by NMR Spectroscopy, in *Glycosylation*, ed. S. Petrescu, InTech, Rijeka, Croatioa, 2012, p. 63.

268 P. S. Pregosin, NMR Diffusion Methods in Inorganic and Organometallic Chemistry, in *Spectroscopic Properties of Inorganic and Organometallic Compounds*, ed. J. Yarwood, R. Douthwaite and S. Duckett, Royal Society of Chemistry, 2012, vol. 42, p. 248.

269 T. Qureshi and N. K. Goto, Contemporary Methods in Structure Determination of Membrane Proteins by Solution NMR, in *NMR of Proteins and Small Biomolecules*, ed. G. Zhu, Topics in Current Chemistry, Springer GmbH, 2012, vol. 326, p. 123.

270 A. W. Ravna and I. Sylte, Homology Modeling of Transporter Proteins (Carriers and Ion Channels), in *Homology Modeling: Methods and Protocols*, ed. A. J. W. Orry and R. Abagyan, Methods in Molecular Biology, Springer, New York, NY, United States, 2012, vol. 857, p. 281.

271 M. D. Reily, S. Huang, J. Caceres-Cortes, K. Johnson and Y.-Z. Shu, Nuclear Magnetic Resonance Spectroscopy for the Study of Drug Metabolism, in *Encyclopedia of Drug Metabolism and Interactions*, ed. A. V. Lyubimov, A. D. Rodrigues and M. A. Sinz, John Wiley & Sons, Inc., Hoboken, N.J., 2012, vol. 5, p. 331.

272 T. Richard, H. Temsamani, E. Cantos-Villar and J.-P. Monti, Application of LC-MS and LC-NMR Techniques for Secondary Metabolite Identification, in *Metabolomics Coming of Age with Its Technological Diversity*, ed. D. Rolin, Advances in Botanical Research, Elsevier Ltd., 2013, vol. 67, p. 67.

273 A. Roehrich, J. Ash, A. Zane, D. L. Masica, J. J. Gray, G. Goobes and G. Drobny, Solid-State NMR Studies of Biomineralization Peptides and Proteins, in *Proteins at Interfaces III State of the Art*, ed. T. Horbett, J. L. Brash and W. Norde, ACS Symposium Series, American Chemical Society, 2012, vol. 1120, p. 77.

274 D. Rolin, C. Deborde, M. Maucourt, C. Cabasson, F. Fauvelle, D. Jacob, C. Canlet and A. Moing, High-Resolution 1H-NMR Spectroscopy and beyond to Explore Plant Metabolome, in *Metabolomics Coming of Age with Its*

Technological Diversity, ed. D. Rolin, Advances in Botanical Research, Elsevier Ltd., 2013, vol. 67, p. 1.
275 A.-L. Rollet and C. Bessada, NMR Studies of Molten Salt and Room Temperature Ionic Liquids, in *Annual Reports on NMR Spectroscopy*, ed. G. A. Webb, Elsevier Ltd., 2013, vol. 78, p. 149.
276 M. G. Sajilata and R. S. Singhal, Quality Indices for Spice Essential Oils, in *Handbook of Herbs and Spices, vol. 1*, ed. K. V. Peter, Woodhead Publishing Series in Food Science, Technology and Nutrition, Woodhead Publishing Ltd., 2012, vol. 227, p. 42.
277 K. Sato, S. Nakazawa, S. Nishida, R. D. Rahimi, T. Yoshino, Y. Morita, K. Toyota, D. Shiomi, M. Kitagawa and T. Takui, Novel Applications of ESR/EPR Quantum Computing/Quantum Information Processing, in *EPR of Free Radicals in Solids II: Trends in Methods and Applications*, ed. A. Lund and M. Shiotani, Progress in Theoretical Chemistry and Physics, Springer, 2013, vol. 25, p. 163.
278 F. Savorani, G. Tomasi and S. B. Engelsen, Alignment of 1D NMR Data Using the Icoshift Tool: A Tutorial, in *Magnetic Resonance in Food Science*, ed. J. van Duynhoven, P. S. Belton, G. A. Webb and H. van As, Special Publication – Royal Society of Chemistry, Royal Society of Chemistry, 2013, vol. 343, p. 14.
279 U. Schieborr, S. Sreeramulu and H. Schwalbe, NMR Structure Determination of Protein-Ligand Complexes, in *NMR of Biomolecules: Towards Mechanistic Systems Biology*, ed. I. Bertini, K. S. McGreevy and G. Parigi, Wiley-Blackwell, Weinheim, Germany, 2012, p. 549.
280 V. Schurig and D. Kreidler, Gas-Chromatographic Enantioseparation of Unfunctionalized Chiral Hydrocarbons: An Overview, in *Chiral Separation: Methods and Protocols*, ed. K. E. Scriba, Methods in Molecular Biology, Springer, New York, NY, United States, 2013, vol. 970, p. 45.
281 A. Schwenk, Steady-State Techniques for Low Sensitivity and Slowly Relaxing Nuclei, in *Encyclopedia of NMR*, ed. R. K. Harris and R. E. Wasylishen, John Wiley & Sons, Ltd., Chichester, UK, 2012, vol. 8, p. 4719.
282 O. Segman, Z. Wiesman and L. Yarmolinsky, Methods and Technologies Related to Shea Butter Chemophysical Properties and to the Delivery of Bioactives in Chocolate and Related Products, in *Cocoa Butter and Related Compounds*, ed. N. Garti and N. R. Widlak, AOCS Monograph Series on Oilseeds, AOCS Press, 2012, vol. 6, p. 417.
283 D. Selent and D. Heller, In-Situ Techniques for Homogeneous Catalysis, in *Catalysis from Principles to Applications*, ed. M. Beller, A. Renken and R. van Santen, Wiley-VCH Verlag GmbH & Co. KGaA, Weinheim, Germany, 2012, p. 465.
284 J. N. Shoolery, Quantitative Measurements, in *Encyclopedia of NMR*, ed. R. K. Harris and R. E. Wasylishen, John Wiley & Sons, Ltd., Chichester, UK, 2012, vol. 7, p. 3901.
285 L. O. Sillerud and R. S. Larson, Advances in Nuclear Magnetic Resonance for Drug Discovery, in *Bioinformatics and Drug Discovery*, ed. R. S. Larson, Methods in Molecular Biology, Springer, New York, NY, United States, 2012, vol. 910, p. 195.
286 V. R. Solomon and H. Lee, Utilization of NMR-Based Techniques in Anticancer Drug Development, in *Structure – Activity Relationship Studies in Drug Development by NMR Spectroscopy*, ed. Atta-ur-Rahman and M. I. Choudhary, Bentham Science Publishers Ltd., 2011, vol. 1, p. 102.
287 Y. Song, Laplace Inversion for NMR Data Analysis, in *Magnetic Resonance in Food Science*, ed. J. van Duynhoven, P. S. Belton, G. A. Webb and H. van As,

Special Publication – Royal Society of Chemistry, Royal Society of Chemistry, 2013, vol. 343, p. 25.
288 O. Soubias and K. Gawrisch, Rhodopsin-Lipid Interactions Studied by NMR, in *G Protein Coupled Receptors Modeling, Activation, Interactions and Virtual Screening*, ed. P. M. Conn, Methods in Enzymology, Elsevier, 2013, vol. 522, p. 209.
289 A. J. Storaska and R. R. Neubig, NMR Methods for Detection of Small Molecule Binding to RGS4, in *G Protein Coupled ReceptorsModeling, Activation, Interactions and Virtual Screening*, ed. P. M. Conn, Methods in Enzymology, Elsevier, 2013, vol. 522, p. 133.
290 S. Sun, Solid-State NMR Spectroscopy of Protein Complexes, in *Protein NMR Techniques*, ed. A. Shekhtman and D. S. Burz, Methods in Molecular Biology, Springer, New York, NY, United States, 2012, vol. 831, p. 303.
291 D. E. K. Sutherland and M. J. Stillman, Mammalian Metallothioneins, in *Brain Diseases and Metalloproteins*, ed. D. R. Brown, CRC Press, 2012, p. 81.
292 A. Suzuki and T. Oku, Electronic Structure and Magnetic Properties of N@C_{60} – SWCNT, in *Electronic Properties of Carbon Nanotubes*, ed. J. M. Marulanda, InTech, Rijeka, Croatia, 2011, p. 423.
293 K. H. Sze, Q. Wu, H. S. Tse and G. Zhu, Dynamic Nuclear Polarization: New Methodology and Applications, in *NMR of Proteins and Small Biomolecules*, ed. G. Zhu, Topics in Current Chemistry, Springer GmbH, 2012, vol. 326, p. 215.
294 M. Szumski, Equipment [for Capillary Electrophoresis], in *Electromigration Techniques: Theory and Practice*, ed. B. Buszewski, E. Dziubakiewicz and M. Szumski, Springer Series in Chemical Physics, Springer GmbH, 2013, vol. 105, p. 27.
295 L. Tenori, Metabolomics, in *NMR of Biomolecules: Towards Mechanistic Systems Biology*, ed. I. Bertini, K. S. McGreevy and G. Parigi, Wiley-Blackwell, Weinheim, Germany, 2012, p. 467.
296 Y. Y. Tong, E. Oldfield and A. Wieckowski, Platinum Nanoparticle Electrode Surfaces: Nuclear Magnetic Resonance Spectroscopy, in *Encyclopedia of Surface and Colloid Science*, ed. P. Somasundaran, Taylor & Francis, Boca Raton, Fla., 2nd edn, 2012, p. 4698.
297 P. Turano, Heme Acquisition by Hemophores: A Lesson from NMR, in *NMR and EPR Techniques*, ed. K. M. Kadish, K. M. Smith and R. Guilard, Handbook of Porphyrin Science: With Applications to Chemistry, Physics, Materials Science, Engineering, Biology and Medicine, World Scientific Publishing Co. Pte. Ltd., Singapore, Singapore, 2010, vol. 6, p. 339.
298 R. Tycko and S. E. Barrett, Optically Pumped NMR of Semiconductors and Two-Dimensional Electron Systems, in *Encyclopedia of NMR*, ed. R. K. Harris and R. E. Wasylishen, John Wiley & Sons, Ltd., Chichester, UK, 2012, vol. 6, p. 3153.
299 M. Vijayan, The Legacy of G.N. Ramachandran and the Development of Structural Biology in India, in *Biomolecular Forms and Functions*, ed. M. Bansal and N. Srinivasan, World Scientific Publishing Co. Pte. Ltd., Singapore, Singapore, 2013, p. 1.
300 O. Vinogradova and J. Qin, NMR as a Unique Tool in Assessment an Complex Determination of Weak Protein-Protein Interactions, in *NMR of Proteins and Small Biomolecules*, ed. G. Zhu, Topics in Current Chemistry, Springer GmbH, 2012, vol. 326, p. 35.
301 K. Wahala, S. Deb and T. Hase, The Structure of Isoflavones by 1D and 2D Homonuclear and Heteronuclear NMR Spectroscopy, in *Isoflavones: Chemistry,*

Analysis, Function and Effects, ed. V. R. Preedy, Food and Nutritional Components in Focus, Royal Society of Chemistry, 2013, vol. 5, p. 94.
302 F. A. Walker, NMR and EPR Spectroscopy of Paramagnetic Metalloporphyrins and Heme Proteins, in *NMR and EPR Techniques*, ed. K. M. Kadish, K. M. Smith and R. Guilard, Handbook of Porphyrin Science: With Applications to Chemistry, Physics, Materials Science, Engineering, Biology and Medicine, World Scientific Publishing Co. Pte. Ltd., Singapore, Singapore, 2010, vol. 6, p. 1.
303 J. A. Wand, V. R. Moorman and K. W. Harpole, A Surprising Role for Conformational Entropy in Protein Function, in *Dynamics in Enzyme Catalysis*, ed. J. Klinman and S. Hammes-Schiffer, Topics in Current Chemistry, Springer Verlag, 2013, vol. 337, p. 69.
304 B. Wen, Metabonomics in Understanding Drug Metabolism and Toxicology, in *Encyclopedia of Drug Metabolism and Interactions*, ed. A. V. Lyubimov, A. D. Rodrigues and M. A. Sinz, John Wiley & Sons, Inc., Hoboken, N.J., 2012, vol. 4, p. 281.
305 M. N. Wilson, A Century of Superconducting Technology, in *Advances in Cryogenic Engineering: Transactions of the International Cryogenic Materials Conference – ICMC, vol. 58*, ed. U. Balachandran, K. Amm, A. Ghosh, K. T. Hartwig, P. Lee, M. Osofsky, S. Pamidi, C. Park, W. Stautner, M. Sumption, R. Walsh and T. Zhang, AIP Conference Proceedings, American Institute of Physics, 2012, vol. 1435, p. 11.
306 S. Xiao, C. V. Finkielstein and D. G. S. Capelluto, The Enigmatic Role of Sulfatides: New Insights into Cellular Functions and Mechanisms of Protein Recognition, in *Lipid-Mediated Protein Signaling*, ed. D. G. S. Capelluto, Advances in Experimental Medicine and Biology, Springer, 2013, vol. 991, p. 27.
307 Y. Xu and S. Matthews, TROSY NMR Spectroscopy of Large Soluble Proteins, in *Modern NMR Methodology*, ed. H. Heise, Topics in Current Chemistry, Springer, 2013, vol. 335, p. 97.
308 M. C. Yappert, Compositional Analysis of Phospholipids by Mass Spectrometry and Phosphorus-31 Nuclear Magnetic Resonance Spectroscopy, in *Methods in Pharmacology and Toxicology*, ed. T. W.-M. Fan, R. M. Higashi and A. N. Lane, Handbook of Metabolomics, Springer, New York, N.Y., 2012, vol. 17, p. 393.
309 A. I. Zenchuk, Profiles of Multiple Quantum NMR Coherences in a Gas of Spin-1/2 Molecules in Nanopores, in *Annual Reports on NMR Spectroscopy*, ed. G. A. Webb, Elsevier Ltd., 2013, vol. 79, p. 1.
310 X. Zhao, Protein Structure Determination by Solid-State NMR, in *NMR of Proteins and Small Biomolecules*, ed. G. Zhu, Topics in Current Chemistry, Springer GmbH, 2012, vol. 326, p. 187.
311 E. Zomer, A. A. Klyosov and D. Platt, Development of a Galactomannan Polysaccharide as a Vehicle to Improve the Efficacy of Chemotherapeutics, in *Glycobiology and Drug Design*, ed. A. A. Klyosov, ACS Symposium Series, American Chemical Society, 2012, vol. 1102, p. 69.
312 E. R. P. Zuiderweg, E. B. Bertelsen, A. Rousaki, M. P. Mayer, J. E. Gestwicki and A. Ahmad, Allostery in the Hsp70 Chaperone Proteins, in *Molecular Chaperones*, ed. S. Jackson, Topics in Current Chemistry, Springer GmbH, 2013, vol. 328, p. 99.
313 E. Zurek and J. Autschbach, Density Functional Calculations of NMR Chemical Shifts in Carbon Nanotubes, in *Computational Nanoscience*, ed. E. Bichoutskaia, RSC Theoretical and Computational Chemistry Series, Royal Society of Chemistry, 2011, vol. 4, p. 279.

Reviews in periodicals

314 C. C. Abbott, A. Jaramillo, C. E. Wilcox and D. A. Hamilton, Antipsychotic Drug Effects in Schizophrenia: A Review of Longitudinal fMRI Investigations and Neural Interpretations, *Curr. Med. Chem.*, 2013, **20**, 428.

315 A. Akki, A. Gupta and R. G. Weiss, Magnetic Resonance Imaging and Spectroscopy of the Murine Cardiovascular System, *Am. J. Physiol.*, 2013, **304**, H633.

316 C. Albanese, O. C. Rodriguez, J. VanMeter, S. T. Fricke, B. R. Rood, Y. C. Lee, S. S. Wang, S. Madhavan, Y. Gusev, E. F. Petricoin *et al.*, Preclinical Magnetic Resonance Imaging and Systems Biology in Cancer Research, *Am. J. Pathol.*, 2013, **182**, 312.

317 M. Alfeld and J. A. C. Broekaert, Mobile Depth Profiling and Sub-Surface Imaging Techniques for Historical Paintings – A Review, *Spectrochim. Acta, Part B*, 2013, **88**, 211.

318 C. Al-Jassar, H. Bikker, M. Overduin and M. Chidgey, Mechanistic Basis of Desmosome-Targeted Diseases, *J. Mol. Biol.*, 2013, **425**, 4006.

319 M. Alzweiri, D. G. Watson and J. A. Parkinson, Metabonomics as a Clinical Tool of Analysis: LC-NS Approaches, *J. Liq. Chromatogr. Relat. Technol.*, 2012, **36**, 94.

320 H. Amiri, K. Saeidi, P. Borhani, A. Manafirad, M. Ghavani and V. Zerbi, Alzheimer's Disease: Pathophysiology and Applications of Magnetic Nanoparticles as MRI Theranostic Agents, *ACS Chem. Neurosci.*, 2013, **4**, 1417.

321 H. Antelmann and C. J. Hamilton, Bacterial Mechanisms of Reversible Protein S-Thiolation: Structural and Mechanistic Insights into Mycoredoxins, *Mol. Microbiol.*, 2012, **86**, 759.

322 A. Arda, M. A. Berbis, P. Blasco, A. Canales, F. J. Canada, M. C. Fernandez-Alonso, F. Marcelo and J. Jimenez-Barbero, Recent Advances on the Application of NMR Methods to Study the Conformation and Recognition Properties of Carbohydrates, *Carbohydr. Chem.*, 2012, **38**, 192.

323 T. Asakura, Y. Suzuki, Y. Nakazawa, G. P. Holland and J. L. Yarger, Elucidating Silk Structure Using Solid-State NMR, *Soft Matter*, 2013, **9**, 11440.

324 G. Audran, P. Bremond, J.-M. Franconi, S. R. A. Marque, P. Massot, P. Mellet, E. Parzy and E. Thiaudiere, Alkoxyamines: A New Family of Pro-Drugs Against Cancer. Concept for Theranostics, *Org. Biomol. Chem.*, 2014, **12**, 719.

325 D. R. Baer, M. H. Engelhard, G. E. Johnson, J. Laskin, J. Lai, K. Mueller, P. Manusamy, S. Thevuthasan, H. Wang, N. Washton, A. Elder, B. L. Baisch, A. Karakoti, S. V. N. T. Kuchibhatla and D. Moon, Surface Characterization of Nanomaterials and Nanoparticles: Important Needs and Challenging Opportunities, *J. Vac. Sci. Technol., A*, 2013, **31**, 050820.

326 Y. Bang, Resonant Impurity Scattering in the Unconventional Superconductors, *EPJ Web Conf.*, 2012, **23**, 00010.

327 M. Baranac-Stojanovic, New Insight into the Anisotropic Effects in Solution-State NMR Spectroscopy, *RSC Adv.*, 2014, **4**, 308.

328 J. J. Barchi Jr., Mucin-Type Glycopeptide Structure in Solution: Past, Present, and Future, *Biopolymers*, 2013, **99**, 713.

329 G. Baverel, B. Knouzy, C. Gauthier, M. El Hage, B. Ferrier, G. Martin and A. Duplany, Use of Precision-Cut Renal Cortical Slices in Nephrotoxicity Studies, *Xenobiotica*, 2013, **43**, 54.

330 J. K. P. Begley, T. W. Redpath, P. J. Bolan and F. J. Gilbert, *In vivo* Proton Magnetic Resonance Spectroscopy of Breast Cancer: A Review of the Literature, *Breast Cancer Res.*, 2012, **14**, 207.
331 K. M. Bennett, J. F. Bertram, S. C. Beeman and N. Gretz, The Emerging Role of MRI in Quantitative Renal Glomerular Morphology, *Am. J. Physiol.*, 2013, **304**(Pt. 2), F1252.
332 D. L. Beveridge, T. E. Cheatham and M. Mezei, The ABC of Molecular Dynamics Simulations on B-DNA, Circa 2012, *J. Biosci.*, 2012, **37**, 379.
333 C. A. Bewley and S. Shahzad-ul-Hussan, Characterizing Carbohydrate-Protein Interactions by Nuclear Magnetic Resonance Spectroscopy, *Biopolymers*, 2013, **99**, 796.
334 A. Bhaumik, C. Luchinat, G. Parigi, E. Ravera and M. Rinaldelli, NMR Crystallography on Paramagnetic Systems. Solved and Open Issues, *CrystEngComm*, 2013, **15**, 8639.
335 K. Bingol and R. Bruschweiler, Multidimensional Approaches to NMR-Based Metabolomics, *Anal. Chem.*, 2014, **86**, 47.
336 B. Blasiak, F. C. J. M. van Veggel and B. Tomanek, Applications of Nanoparticles for MRI Cancer Diagnosis and Therapy, *J. Nanomater.*, 2013, 148578.
337 T. G. Bloemberg, J. Gerretzen, A. Lunshof, R. Wehrens and L. M. C. Buydens, Warping Methods for Spectroscopic and Chromatographic Signal Alignment: A Tutorial, *Anal. Chim. Acta*, 2013, **781**, 14.
338 O. Y. Bo and J. J. Chou, The Minimalist Architectures of Viroporins and Their Therapeutic Implications, *Biochim. Biophys. Acta, Biomembr.*, 2014, **1838**, 1058.
339 B. Bonnard and O. Cots, Geometrical Numerical Methods and Results in the Contrast Imaging Problem in Nuclear Magnetic Resonance, *Math. Models Methods Appl. Sci.*, 2014, **24**, 187.
340 T. Braeuniger and M. Jansen, Solid-state NMR Spectroscopy of Quadrupolar Nuclei in Inorganic Chemistry, *Z. Anorg. Allg. Chem.*, 2013, **639**, 859.
341 M. Bruno, S. Bancheva, S. Rosselli and A. Maggio, Sesquiterpenoids in Subtribe Centaureinae (Cass.) Dumort (Tribe Cardueae, Asteraceae): Distribution, ^{13}C NMR Spectral Data and Biological Properties, *Phytochemistry (Elsevier)*, 2013, **95**, 19.
342 A. Buchwald, M. Vanooteghem, E. Gruyaert, H. Hilbig and N. De Belie, Purdocement: Application of Alkali-Activated Slag Cement in Belgium in the 1950s, *Mater. Struct.*, 2015, **48**, 501.
343 B. P. Burke and S. J. Archibald, Macrocyclic Coordination Chemistry, *Annu. Rep. Prog. Chem., Sect. A: Inorg. Chem.*, 2013, **109**, 232.
344 O. Cala, F. Guilliere and I. Krimm, NMR-Based Analysis of Protein-Ligand Interactions, *Anal. Bioanal. Chem.*, 2014, **406**, 943.
345 X. Cao, J. Yang and J. Mao, Characterization of Kerogen Using Solid-State Nuclear Magnetic Resonance Spectroscopy: A Review, *Int. J. Coal Geol.*, 2013, **108**, 83.
346 C. M. Castro, A. A. Ghazani, J. Chung, H. Shao, D. Issadore, T.-J. Yoon, R. Weissleder and H. Lee, Miniaturized Nuclear Magnetic Resonance Platform for Detection and Profiling of Circulating Tumor Cells, *Lab Chip*, 2014, **14**, 14.
347 L. Cegelski, REDOR NMR for Drug Discovery, *Bioorg. Med. Chem. Lett.*, 2013, **23**, 5767.
348 R. Chadha and S. Bhandari, Drug-Excipient Compatibility Screening-Role of Thermoanalytical and Spectroscopic Techniques, *J. Pharm. Biomed. Anal.*, 2014, **87**, 82.

349 A. Chalkias, V. Fanos, A. Noto, M. Castren, A. Gulati, H. Svavarsdottir, N. Iacovidou and T. Xanthos, ^{1}H NMR-Metabolomics: Can They be a Useful Tool in Our Understanding of Cardiac Arrest?, *Resuscitation*, 2014, **85**, 595.
350 N. Chalouhi, P. Jabbour, V. Magnotta and D. Hasan, The Emerging Role of Ferumoxytol-Enhanced MRI in the Management of Cerebrovascular Lesions, *Molecules*, 2013, **18**, 9670.
351 Y.-G. Chang, R. Tseng, N.-W. Kuo and A. Li Wang, Nuclear Magnetic Resonance Spectroscopy of the Circadian Clock of Cyanobacteria, *Integr. Comp. Biol.*, 2013, **53**, 93.
352 T. Charpentier, NMR of Amorphous Materials, *EPJ Web Conf.*, 2012, **30**, 04004.
353 T. Charpentier, M. C. Menziani and A. Pedone, Computational Simulation of Solid State NMR Spectra: A New Era in Structure Determination of Oxide Glasses, *RSC Adv.*, 2013, **3**, 10550.
354 S. Chaudhary and A. Pal, Conductometric and Spectroscopic Study of Interaction of Cationic Surfactants with 3-Methyl-1-pentylimidazolium Hexafluorophosphate, *J. Mol. Liq.*, 2014, **190**, 10.
355 E. A. Chekhovich, M. N. Makhonin, A. I. Tartakovskii, A. Yacoby, H. Bluhm, K. C. Nowack and L. M. K. Vandersypen, Nuclear Spin Effects in Semiconductor Quantum Dots, *Nat. Mater.*, 2013, **12**, 494.
356 J.-H. Chen and M.-Y. Su, Clinical Application of Magnetic Resonance Imaging in Management of Breast Cancer Patients Receiving Neoadjuvant Chemotherapy, *BioMed Res. Int.*, 2013, 348167.
357 Q. Chen, C. Zhang, J. Zhao and Q. Ouyang, Recent Advances in Emerging Imaging Techniques for non-Destructive Detection of Food Quality and Safety, *TrAC, Trends Anal. Chem.*, 2013, **52**, 261.
358 M. R. Chierotti and R. Gobetto, NMR Crystallography. The Use of Dipolar Interactions in Polymorph and Co-Crystal Investigation, *CrystEngComm*, 2013, **15**, 8599.
359 K. H. Chow, A. I. Mansour, I. Fan, G. D. Morris, Z. Salman, T. Dunlop, W. A. MacFarlane, R. F. Kiefl, H. Saadaoui, O. Mosendz, B. Kardasz, B. Heinrich, J. Jung, C. D. P. Levy, M. R. Pearson, T. J. Parolin, D. Wang, M. D. Hossain, Q. Song and M. Smadella, Observing Cross-Relaxation with Depth-Resolved ^{8}Li β-NMR, *Phys. Procedia*, 2012, **30**, 16.
360 K.-C. Chu and C.-Y. Wu, Carbohydrate-Based Synthetic Vaccines: Does the Synthesis of Longer Chains of Carbohydrate Make This a Step over Closer?, *Future Med. Chem.*, 2012, **4**, 1767.
361 M. Cifelli and C. A. Veracini, Translational Diffusion in Thermotropic Smectic Phases, *Mol. Cryst. Liq. Cryst.*, 2013, **576**, 127.
362 G. Ciric-Marjanovic, I. Pasti, N. Gavrilov, A. Janosevic and S. Mentus, Carbonised Polyaniline and Polypyrrole: Towards Advanced Nitrogen-Containing Carbon Materials, *Chem. Pap.*, 2013, **67**, 781.
363 G. M. Clore, Interplay between Conformational Selection and Induced Fit in Multidomain Protein-Ligand Binding Probed by Paramagnetic Relaxation Enhancement, *Biophys. Chem.*, 2014, **186**, 3.
364 G. Contarini and M. Povolo, Phospholipids in Milk Fat: Composition, Biological and Technological Significance, and Analytical Strategies, *Int. J. Mol. Sci.*, 2013, **14**, 2808.
365 A. J. L. Cooper, Possible Treatment of End-Stage Hyperammonemic Encephalophaty by Inhibition of Glutamine Synthetase, *Metab. Brain Dis.*, 2013, **28**, 119.

366 C. L. Cooper, T. Cosgrove, J. S. van Duijneveldt, M. Murray and S. W. Prescott, The Use of Solvent Relaxation NMR to Study Colloidal Suspensions, *Soft Matter*, 2013, **9**, 7211.
367 Y. Cordeiro, D. Foguel and J. L. Silva, Pressure-Temperature Folding Landscape in Proteins Involved in Neurodegenerative Diseases and Cancer, *Biophys. Chem.*, 2013, **183**, 9.
368 J. G. Correia, K. Johnson and U. Wahl, Nuclear Radioactive Techniques Applied to Materials Research, *Radiochim. Acta*, 2012, **100**, 127.
369 U. Costantino, F. Costantino, F. Elisei, L. Latterini and M. Nocchetti, Coupling Physical Chemical Techniques with Hydrotalcite-Like Compounds to Exploit Their Structural Features and New Multifunctional Hybrids with Luminescent Properties, *Phys. Chem. Chem. Phys.*, 2013, **15**, 13254.
370 R. C. Craddock, S. Jbabdi, C.-G. Yan, J. T. Vogelstein, F. X. Castellanos, A. Di Martino, C. Kelly, K. Heberlein, S. Colcombe and M. P. Milham, Imaging Human Connectomes at the Macroscale, *Nat. Methods*, 2013, **10**, 524.
371 T. A. Cross, V. Ekanayake, J. Paulino and A. Wright, Solid State NMR: The Essential Technology for Helical Membrane Protein Structural Characterization, *J. Magn. Reson.*, 2014, **239**, 100.
372 S. Dalapati, S. Jana and N. Guchhait, Anion Recognition by Simple Chromogenic and Chromo-Fluorogenic Salicylidene Schiff Base or Reduced-Schiff Base Receptors, *Spectrochim. Acta, Part A*, 2014, **129**, 499.
373 F. Dall'Antonia, T. Pavkov-Keller, K. Zangger and W. Keller, Structure of Allergens and Structure Based Epitope Predictions, *Methods*, 2014, **66**, 3.
374 E. Danieli and B. Bluemich, Single-Sided Magnetic Resonance Profiling in Biological and Materials Science, *J. Magn. Reson.*, 2013, **229**, 142.
375 G.-L. Davies, I. Kramberger and J. J. Davis, Environmentally Responsive MRI Contrast Agents, *Chem. Commun.*, 2013, **49**, 9704.
376 N. R. de Almeida, A. Beatriz, A. C. Micheletti and E. J. de Arruda, Ozonized Vegetable Oils and Therapeutic Properties: A Review, *Orbital: Electron. J. Chem.*, 2012, **4**, 313.
377 G. de Jong and R. Bischoff, Analysis of Biopharmaceuticals, *TrAC, Trends Anal. Chem.*, 2013, **48**, 40.
378 Z. Demirel and A. Sukatar, Cyanobacterial Toxin, *Internet J. Toxicol.*, 2012, **8**. Avail. URL: http://www.ispub.com/journal/the-internet-journal-of-toxicology/volume-8-issue-2/cyanobacterial-toxin.html.
379 J. Deng, F. Zhao, J. Li, H. Liu and H. Jiang, Discovery and Early Evaluation of New Chemical Entities, *J. Chin. Pharm. Sci.*, 2012, **21**, 369.
380 G. de Paepe, Hyperpolarization in Magnetic Resonance Special Focus on High Field MAS Dynamic Nuclear Polarization, *EPJ Web Conf.*, 2012, **30**, 05001.
381 E. G. Derouane, J. C. Vedrine, R. R. Pinto, P. M. Borges, L. Costa, M. A. N. D. A. Lemos, F. Lemos and F. R. Ribeiro, The Acidity od Zeolites: Concepts, Measurements and Relation to Catalysis: A Review on Experimental and Theoretical Methods for the Study of Zeolite Acidity, *Catal. Rev.: Sci. Eng.*, 2013, **55**, 454.
382 I. Diaz-Moreno, J. M. Garcia-Heredia, K. Gonzalez-Arzola, A. Diaz-Quintana and M. A. De la Rosa, Recent Methodological Advances in the Analysis of Protein Tyrosine Nitration, *ChemPhysChem*, 2013, **14**, 3095.
383 X. Ding, X. Zhao and A. Watts, G-Protein-Coupled Receptor Structure, Ligand Binding and Activation as Studied by Solid-State NMR Spectroscopy, *Biochem. J.*, 2013, **450**, 443.

384 S. Divya, D. Thyagarajan and G. Sujatha, Magnetic Resonance Imaging Technology for Process Control and Quality Maintenance in Food Quality Operation, *Int. J. Eng. Technol.*, 2013, **4**, 441. Avail. URL: http://www.enggjournals.com/ijet/docs/IJET12-04-06-037.pdf.
385 J. C. do Nascimento, V. F. De Paula, J. M. David and J. P. David, Occurrence, Biological Activities and ^{13}C NMR Data of Amides from Piper (Piperaceae), *Quim. Nova*, 2012, **35**, 2288.
386 S. J. Dorazio, A. O. Olatunde, P. B. Tsitovich and J. R. Morrow, Comparison of Divalent Transition Metal Ion ParaCEST MRI Contrast Agents, *JBIC, J. Biol. Inorg. Chem.*, 2014, **19**, 191.
387 G. W. Duncan, M. J. Firbank, J. T. O'Brien and D. J. Burn, Magnetic Resonance Imaging: a Biomarker for Cognitive Impairment in Parkinson's Disease?, *Mov. Disord.*, 2013, **28**, 425.
388 J. Duyn, MR Susceptibility Imaging, *J. Magn. Reson.*, 2013, **229**, 198.
389 M. Dvoyashkin, A. Wang, A. Katihar, J. Zang, G. I. Yucelen, S. Nair, D. S. Sholl, C. R. Bowers and S. Vasenkov, Signatures of Normal and Anomalous Diffusion in Nanotube Systems by NMR, *Microporous Mesoporous Mater.*, 2013, **178**, 119.
390 B. H. Ebdrup, H. Noerbak, S. Borgwardt and B. Glenthoej, Volumetric Changes in the Basal Ganglia after Antipsychotic Monotherapy: A Systematic Review, *Curr. Med. Chem.*, 2013, **20**, 438.
391 M. Edgar, Physical Methods and Techniques NMR Spectroscopy, *Annu. Rep. Prog. Chem., Sect. B: Org. Chem.*, 2012, **108**, 292.
392 M. Edmundson, N. T. K. Thanh and B. Song, Nanoparticles Based Stem Cell Tracking in Regenerative Medicine, *Theranostics*, 2013, **3**, 573.
393 A. E. Eggers, A Serotonin Hypothesis of Schizophrenia, *Med. Hypotheses*, 2013, **80**, 791.
394 J. Ehling, T. Lammers and F. Kiessling, Non-Invasive Imaging for Studying Anti-Angiogenic Therapy Effects, *Thromb. Haemostasis*, 2013, **109**, 375.
395 J. J. Eicher, J. L. Snoep and J. M. Rohwer, Determining Enzyme Kinetics for Systems Biology with Nuclear Magnetic Resonance Spectroscopy, *Metabolites*, 2012, **2**, 818.
396 T. Einfalt, O. Planinsek and K. Hrovat, Methods of Amorphization and Investigation of the Amorphous State, *Bull. Chem. Soc. Jpn.*, 2013, **86**, 305.
397 J. Elguero, I. Alkorta, R. M. Claramunt, P. Cabildo, P. Cornago, M. A. Farran, M. A. Garcia, C. Lopez, M. Perez-Torrabla, D. S. Maria and D. Sanz, Structure of NH-Benzazoles (1H-Benzimidazoles, 1H- and 2H-Indazoles, 1H- and 2H-Benzotriazoles, *Chem. Heterocycl. Compd.*, 2013, **49**, 177.
398 J. J. Ellinger, R. A. Chylla, E. L. Ulrich and J. L. Markley, Databases and Software for NMR-Based Metabolomics, *Curr. Metabolomics*, 2013, **1**, 28.
399 L. Emsley, Advances in Magnetic Resonance: From Stems Cells to Catalytic Surfaces, *J. Am. Chem. Soc.*, 2013, **135**, 8089.
400 A.-H. M. Emwas, R. M. Salek, J. L. Griffin and J. Merzaban, NMR-Based Metabolomics in Human Disease Diagnosis: Applications, Limitations, and Recommendations, *Metabolomics*, 2013, **9**, 1048.
401 M. Espy, A. Matlashov and P. Volegov, SQUID-Detected Ultra-Low Field MRI, *J. Magn. Reson.*, 2013, **229**, 127.
402 S. Essig and D. Menche, Stereochemistry and Total Synthesis of Complex Myxobacterial Macrolides, *Pure Appl. Chem.*, 2013, **85**, 1103.
403 G. Ferey, M. Haouas, T. Loiseau and F. Taulelle, Nanoporous Solids: How Do They Form? An in Situ Approach, *Chem. Mater.*, 2014, **26**, 299.

404 T. Fergoug and Y. Bouhadda, Determination of Hassi Messaoud Asphaltene Aromatic Structure from ^{1}H & ^{13}C NMR Analysis, *Fuel*, 2014, **115**, 521.
405 C. Fernandes, M. E. Tiritan and M. Pinto, Small Molecules as Chromatographic Tools for HPLC Enantiomeric Resolution: Pirkle-Type Chiral Stationary Phases Evolution, *Chromatographia*, 2013, **76**, 871.
406 R. Fernando, J. Downs, D. Maples and A. Ranjan, MRI-Guided Monitoring of Thermal Dose and Targeted Drug Delivery for Cancer Therapy, *Pharm. Res.*, 2013, **30**, 2709.
407 R. Flukiger, Overview of Superconductivity and Challenges in Applications, *CERN Doc. Server: Prepr.*, 2013, 1.
408 C. Fotakis, K. Kokkotou, P. Zoumpoulakis and M. Zervou, NMR Metabolite Fingerprinting in Grape Derived Products: An Overview, *Food Res. Int.*, 2013, **54**, 1184.
409 C. Foti, K. Alsante, G. Cheng, T. Zelesky and M. Zell, Tools and Workflow for Structure Elucidation of Drug Degradation Products, *TrAC, Trends Anal. Chem.*, 2013, **49**, 89.
410 H. A. Gad, S. H. El-Ahmady, M. I. Abou-Shoer and M. M. Al-Azizi, Application of Chemometrics in Authentication of Herbal Medicines: A Review, *Phytochem. Anal.*, 2013, **24**, 1.
411 D. Gagne and N. Doucet, Structural and Functional Importance of Local and Global Conformational Fluctuations in the RNase A Superfamily, *FEBS J.*, 2013, **280**, 5596.
412 X. Gao, Recent Advances in Computational Methods for Nuclear Magnetic Resonance Data Processing, *Genomics, Proteomics Bioinf.*, 2013, **11**, 29.
413 L. Garcia-Alvarez, J. H. Busto, J. M. Peregrina, R. M. A. Fernandez, A. Avenoza and J. A. Oteo, Nuclear Magnetic Resonance Applied to Antimicrobial Drug Susceptibility, *Future Microbiol.*, 2013, **8**, 537.
414 D. Garnier, N. Jabado and J. Rak, Extracellular Vesicles as Prospective Carriers of Oncogenic Protein Signatures in Adult and Paediatric Brain Tumors, *Proteomics*, 2013, **13**, 1595.
415 D. Garriga, C. Ferrer-Orta, J. Querol-Audi, B. Oliva and N. Verdaguer, Role of Motif B Loop in Allosteric Regulation of RNA-Dependent RNA Polymerization Activity, *J. Mol. Biol.*, 2013, **425**, 2279.
416 A. Gautam and F. C. J. M. van Veggel, Synthesis of Nanoparticles, Their Biocompatibility, and Toxicity Behavior for Biomedical Applications, *J. Mater. Chem. B*, 2013, **1**, 5186.
417 A. Gautier, Structure Determination of α-Helical Membrane Proteins by Solution-State NMR: Emphasis on Retinal Proteins, *Biochim. Biophys. Acta, Bioenerg.*, 2014, **1837**, 578.
418 P. C. Geiszler, D. P. Auer and C. A. Daykin, The Journey from Metabolic Profiling to Biomarkers: The Potential of NMR Spectroscopy Based Metabolomics in Neurodegenerative Disease Research, *Curr. Metabolomics*, 2013, **1**, 160.
419 S. Geninatti-Crich, A. Deagostino, A. Toppino, D. Alberti, P. Venturello and S. Aime, Boronated Compounds for Imaging Guided BNCT Applications, *Anti-Cancer Agents Med. Chem.*, 2012, **12**, 543.
420 V. Gervais, S. Campagne, J. Durand, I. Muller and A. Milon, NMR Studies of a New Family of DNA Binding Proteins: The THAP Proteins, *J. Biomol. NMR*, 2013, **56**, 3.
421 A. S. Ghasemi, M. Rezaei and M. Molla, NMR Study of Gases Adsorption on Single-Walled Carbon Nanotubes. A Review, *Int. J. ChemTech Res.*, 2013, **5**, 1585.

422 G. Giachin, I. Biljan, G. Ilc, J. Plavec and G. Legname, Probing Early Misfolding Events in Prion Protein Mutants by NMR Spectroscopy, *Molecules*, 2013, **18**, 9451.
423 F. Giannini, L. E. H. Paul and J. Furrer, Insights into the Mechanism of Action and Cellular Targets of Ruthenium Complexes from NMR Spectroscopy, *Chimia*, 2012, **66**, 775.
424 H. G. Gika and I. D. Wilson, Global Metabolic Profiling for the Study of Alcohol-Related Disorders, *Bioanalysis*, 2014, **6**, 59.
425 S. Gloggler, J. Colell and S. Appelt, Para-Hydrogen Perspectives in Hyperpolarized NMR, *J. Magn. Reson.*, 2013, **235**, 130.
426 K. Grant, M. L. Lindenberg, H. Shebel, Y. Pang, H. K. Agarwal, M. Bernardo, K. A. Kurdziel, B. Turkbey and P. L. Choyke, Functional and Molecular Imaging of Localized and Recurrent Prostate Cancer, *Eur. J. Nucl. Med. Mol. Imaging*, 2013, **40**(Suppl. 1), 48.
427 E. L. Green, L. L. Lumata, J. S. Brooks, P. Kuhns, A. Reyes, S. E. Brown and M. Almeida, ^{1}H and ^{195}Pt NMR Study of the Parallel Two-Chain Compound $Per_2[Pt(mnt)_2]$, *Crystals*, 2012, **2**, 1116.
428 P. Guerry, L. Mollica and M. Blackledge, Mapping Protein Conformational Energy Landscapes Using NMR and Molecular Simulation, *ChemPhysChem*, 2013, **14**, 3046.
429 V. M. Gun'ko, I. N. Savina and S. V. Mikhalovsky, Cryogels: Morphological, Structural and Adsorption Characterization, *Adv. Colloid Interface Sci.*, 2013, **187–188**, 1.
430 J. Guo, W. Yang and C. Wang, Magnetic Colloidal Supraparticles: Design, Fabrication and Biomedical Applications, *Adv. Mater.*, 2013, **25**, 5196.
431 T. Gutmann, I. del Rosal, B. Chaudret, R. Poteau, H.-H. Limbach and G. Buntkowsky, From Molecular Complexes to Complex Metallic Nanostructures – ^{2}H Solid-State NMR Studies of Ruthenium-Containing Hydrogenation Catalysts, *ChemPhysChem*, 2013, **14**, 3026.
432 C. Guzzi, J. C. Munoz-Garcia, P. M. Enriquez-Navas, J. Rojo, J. Angulo and P. M. Nieto, NMR Studies on Carbohydrate Interactions with DC-SIGN towards a Quantitative STD Analysis, *Pure Appl. Chem.*, 2013, **85**, 1771.
433 J. A. Hamilton, NMR Reveals Molecular Interactions and Dynamics of Fatty Acid Binding to Albumin, *Biochim. Biophys. Acta, Gen. Subj.*, 2013, **1830**, 5418.
434 H. Hampel, S. Lista, S. J. Teipel, F. Garaci, R. Nistico, K. Blennow, H. Zetterberg, L. Bertram, C. Duyckaerts and H. Bakardjian, Perspective on Future Role of Biological Markers in Clinical Therapy Trials of Alzheimer's Disease: A Long-Range Point of View beyond 2020, *Biochem. Pharmacol.*, 2014, **88**, 426.
435 K. G. Hartman, L. E. McKnight, M. A. Liriano and D. J. Weber, The Evolution of S100B Inhibitors for the Treatment of Malignant Melanoma, *Future Med. Chem.*, 2013, **5**, 97.
436 L. C. Heather, X. Wang, J. A. West and J. L. Griffin, A Practical Guide to Metabolomic Profiling as a Discovery Tool for Human Heart Disease, *J. Mol. Cell. Cardiol.*, 2013, **55**, 2.
437 S. Hirose, M. Misawa and A. Kawamoto, Magnetic and Electric Properties of Organic Conductors Probed by ^{13}C-NMR Using Selective-Site Substituted Molecules, *Crystals*, 2012, **2**, 1034.
438 H. G. Hocking, K. Zangger and T. Madl, Studying the Structure and Dynamics of Biomolecules by Using Soluble Paramagnetic Probes, *ChemPhysChem*, 2013, **14**, 3082.

439 I. T. Holder, M. Drescher and J. S. Hartig, Structural Characterization of Quadruplex DNA with in-Cell EPR Approaches, *Bioorg. Med. Chem.*, 2013, **21**, 6156.
440 S. T. Islam and J. S. Lam, Topological Mapping Methods for is α-Helical Bacterial Membrane Proteins–an Update and a Guide, *MicrobiologyOpen*, 2013, **2**, 350.
441 H. Jiang, X. Zeng, N. He, Y. Deng, G. Lu and K. Li, Preparation and Biomedical Applications of Gold-Coated Magnetic Nanocomposites, *J. Nanosci. Nanotechnol.*, 2013, **13**, 1617.
442 A. Kalemis, B. M. A. Delattre and S. Heinzer, Sequential Whole-Body PET/MR Scanner: Concept, Clinical Use, and Optimisation after Two Years in the Clinic. The Manufacturer's Perspective, *Magn. Reson. Mater. Phys., Biol. Med.*, 2013, **26**, 5.
443 H.-W. Kao, S.-W. Chiang, H.-W. Chung, F. Y. Tsai and C.-Y. Chen, Advanced MR Imaging of Gliomas: An Update, *BioMed Res. Int.*, 2013, 970586.
444 J. Karger, In-Depth Study of Surface Resistances in Nanoporous Materials by Microscopic Diffusion Measurement, *Microporous Mesoporous Mater.*, 2014, **189**, 126.
445 S. V. Kathuria, A. Chan, R. Graceffa, N. R. Paul, M. C. Robert, T. C. Irving, B. Perot and O. Bilsel, Advances in Turbulent Mixing Techniques to Study Microsecond Protein Folding Reactions, *Biopolymers*, 2013, **99**, 888.
446 K. Kazimierczuk, J. Stanek, A. Zawadzka-Kazimierczuk and W. Kozminski, High-Dimensional NMR Spectra for Structural Studies of Biomolecules, *ChemPhysChem*, 2013, **14**, 3015.
447 B. Keil and L. L. Wald, Massively Parallel MRI Detector Arrays, *J. Magn. Reson.*, 2013, **229**, 75.
448 M. M. Kerssens, C. Sprung, G. T. Whiting and B. M. Weckhuysen, Selective Staining of Zeolite Acidity: Recent Progress and Future Perspectives on Fluorescence Microscopy, *Microporous Mesoporous Mater.*, 2014, **189**, 136.
449 S. H. Kim, C. M. Lee and K. Kafle, Characterization of Crystalline Cellulose in Biomass: Basic principles, Applications, and Limitations of XRD, NMR, IR, Raman, and SFG, *Korean J. Chem. Eng.*, 2013, **30**, 2127.
450 R. Kimmich, NMR Diffusometry: Molecular Dynamics in Complex Systems Probed over Many Decades of Time, *EPJ Web Conf.*, 2012, **30**, 05003.
451 D. J. Kojetin and T. P. Burris, Small Molecule Modulation of Nuclear Receptor Conformational Dynamics: Implications for Function and Drug Discovery, *Mol. Pharmacol.*, 2013, **83**, 1.
452 H. Komatsu, M. Ota, R. L. Smith and H. Inomata, Review of CO_2-CH_4 Clathrate Hydrate Replacement Reaction Laboratory Studies – Properties and Kinetics, *J. Taiwan Inst. Chem. Eng.*, 2013, **44**, 517.
453 L. Konermann, S. Vahidi and M. A. Sowole, Mass Spectrometry Methods for Studying Structure and Dynamics of Biological Macromolecules, *Anal. Chem.*, 2014, **86**, 213.
454 I. V. Koptyug, Spin Hyperpolarization in NMR to Address Enzymatic Processes in vivo, *Mendeleev Commun.*, 2013, **23**, 299.
455 J. Kragelj, V. Ozenne, M. Blackledge and M. R. Jensen, Conformational Propensities of Intrinsically Disordered Proteins from NMR Chemical Shifts, *ChemPhysChem*, 2013, **14**, 3034.
456 G. Kramer-Marek, J. Gore and M. Korc, Molecular Imaging in Pancreatic Cancer – A Roadmap for Therapeutic Decisions, *Cancer Lett.*, 2013, **341**, 132.
457 T. Krasia-Christoforou and T. K. Georgiou, Polymeric Theranostics: Using Polymer-Based Systems for Simultaneous Imaging and Therapy, *J. Mater. Chem. B*, 2013, **1**, 3002.

458 D. M. Lakshmana and K. G. Lalitha, Molecular Docking Studies of Thiazole Schiff's Bases as HIV1 – Protease Inhibitors, *J. Curr. Chem. Pharm. Sci.*, 2011, **1**, 52. Avail. URL: http://www.sadgurupublications.com/ContentPaper/2011/7_116_1%281%292011.pdf.
459 S. Langereis, T. Geelen, H. Gruell, G. J. Strijkers and K. Nicolay, Paramagnetic Liposomes for Molecular MRI and MRI-Guided Drug Delivery, *NMR Biomed.*, 2013, **26**, 728.
460 M. P. Ledbetter and D. Budker, Zero-Field Nuclear Magnetic Resonance, *Phys. Today*, 2013, **66**, 44.
461 T. Lehmann and E. Topchiy, Contributions of NMR to the Understanding of the Coordination Chemistry and DNA Interactions of Metallo-Bleomycins, *Molecules*, 2013, **18**, 9253.
462 S. Lei and R. Powers, NMR Metabolomics Analysis of Parkinson's Disease, *Curr. Metabolomics*, 2013, **1**, 191.
463 D. Letourneau, A. Lefebvre, P. Lavigne and J.-G. LeHoux, STARD5 Specific Ligand Binding: Comparison with STARD1 and STARD4 Subfamilies, *Mol. Cell. Endocrinol.*, 2013, **371**, 20.
464 V. Leuzzi, M. Mastrangelo, R. Battini and G. Cioni, Inborn Errors of Creatine Metabolism and Epilepsy, *Epilepsia*, 2013, **54**, 217.
465 M. H. Levitt, Spectroscopy of Light-Molecule Endofullerenes, *Philos. Trans. R. Soc., A*, 2013, **371**, 20120429.
466 H. Li and H. Sun, In-Cell NMR: An Emerging Approach for Monitoring Metal-Related Events in Living Cells, *Metallomics*, 2014, **6**, 69.
467 D. M. L. Lilburn, G. E. Pavlovskaya and T. Meersmann, Perspectives of Hyperpolarized Noble Gas MRI beyond ^{3}He, *J. Magn. Reson.*, 2013, **229**, 173.
468 B. Lizarbe, A. Benitez, G. A. Pelaez Brioso, M. Sanchez-Montanes, P. Lopez-Larrubia, P. Ballesteros and S. Cerdan, Hypothalamic Metabolic Compartmentation During Appetite Regulation as Revealed by Magnetic Resonance Imaging and Spectroscopy Methods, *Front. Neuroenerg.*, 2013, **5**, 6.
469 G. J. Lee, P. Attri, E. H. Choi, Y.-W. Kwon, I. Krasnikov and A. Seteikin, Optical and Structural Properties of Nanobiomaterials, *J. Nanosci. Nanotechnol.*, 2014, **14**, 221.
470 T. Lehmann and E. Topchiy, Contributions of NMR to the Understanding of the Coordination Chemistry and DNA Interactions of Metallobleomycins, *Molecules*, 2013, **18**, 9253.
471 D. Letourneau, A. Lefebvre, P. Lavigne and J.-G. LeHoux, STARD5 Specific Ligand Binding: Comparison with STARD1 and STARD4 Subfamilies, *Mol. Cell. Endocrinol.*, 2013, **371**, 20.
472 L. Li, W. Jiang, K. Luo, H. Song, F. Lan, Y. Wu and Z. Gu, Superparamagnetic Iron Oxide Nanoparticles as MRI Contrast Agents for Non-Invasive Stem Cell Labeling and Tracking, *Theranostics*, 2013, **3**, 595.
473 G. Lippens, Molecular Interactions by NMR Spectroscopy, *EPJ Web Conf.*, 2012, **30**, 03001.
474 D. Liu, K. Lu, C. Poon and W. Lin, Metal-Organic Frameworks as Sensory Materials and Imaging Agents, *Inorg. Chem.*, 2014, **53**, 1916.
475 C. T. Lohans and J. C. Vederas, Structural Characterization of Thioether-Bridged Bacteriocins, *J. Antibiot.*, 2014, **67**, 23.
476 M. Long and W.-Q. Yu, Review of Nuclear Magnetic Resonance Studies on Iron-Based Superconductors, *arXiv.org, e-Print Arch., Condens. Matter*, 29 Jul 2013, 1. Avail. URL: http://arxiv.org/ftp/arxiv/papers/1307/1307.7528.pdf.
477 G. Lopez, A. Alaaeddine and B. Ameduri, Radical Telomerization of Fluorinated Alkenes with Dialkyl Hydrogenophosphonates, *Polym. Chem.*, 2013, **4**, 3636.

478 Y. Lu, X. Wang, X. Chen and G. Zhao, Computational Methods for DNA-Binding Protein and Binding Residue Prediction, *Protein Pept. Lett.*, 2013, **20**, 346.
479 Y. Lvovsky, E. W. Stautner and T. Zhang, Novel Technologies and Configurations of Superconducting Magnets for MRI, *Supercond. Sci. Technol.*, 2013, **26**, 093001.
480 L. Ma and W.-Q. Yu, Review of Nuclear Magnetic Resonance Studies on Iron-Based Superconductors, *Chin. Phys. B*, 2013, **22**, 087414.
481 P. M. MacDonald, Q. Saleem, A. Lai and H. H. Morales, NMR Methods for Measuring Lateral Diffusion in Membranes, *Chem. Phys. Lipids*, 2013, **166**, 31.
482 S. Macholl, D. Tietze and G. Buntkowsky, NMR Crystallography of Amides, Peptides and Protein-Ligand Complexes, *CrystEngComm*, 2013, **15**, 8627.
483 P. K. Madhu, Heteronuclear Spin Decoupling in Solid-State Nuclear Magnetic Resonance: Overview and Outlook, *Israel J. Chem.*, 2014, **54**, 25.
484 L. Maeler, Solution NMR Studies of Peptide-Lipid Interaction in Model Membranes, *Mol. Membr. Biol.*, 2012, **29**, 155.
485 A. D. Maher, V. Solo and C. D. Rae, Magnetic Resonance-Based Metabolomics for Understanding Neurological Disorders: Current Status and Statistical Considerations, *Curr. Metabolomics*, 2013, **1**, 2.
486 G. Manegold-Brauer, A. K. Bellin, S. Tercanli, O. Lapaire and V. Heinzelmann-Schwarz, The Special Role of Ultrasound for Screening, Staging and Surveillance of Malignant Ovarian Tumors: Distinction from other Methods of Diagnostic Imaging, *Arch. Gynecol. Obstet.*, 2014, **289**, 491.
487 M. F. Marcone, S. Wang, W. Albabish, S. Nie, D. Somnarain and A. Hill, Diverse Food-Based Applications of Nuclear Magnetic Resonance (NMR) Technology, *Food Res. Int.*, 2013, **51**, 729.
488 D. Marion, An Introduction to Biological NMR Spectroscopy, *Mol. Cell. Proteomics*, 2013, **12**, 3006.
489 C. B. Marshall, D. Meiri, M. J. Smith, M. T. Mazhab-Jafari, G. M. C. Gasmi-Seabrook, R. Rottapel, V. Stambolic and M. Ikura, Probing the GTPase Cycle with Real-Time NMR: GAP and GEF Activities in Cell Extracts, *Methods*, 2012, **57**, 473.
490 D. Massiot, Techniques of Solid-State NMR, *EPJ Web Conf.*, 2012, **30**, 02002.
491 Y. Matsumoto and A. Jasanoff, Metalloprotein-Based MRI Probes, *FEBS Lett.*, 2013, **587**, 1021.
492 P. L. McCormack, Natalizumab: A Review of Its Use in the Management of Relapsing-Remitting Multiple Sclerosis, *Drugs*, 2013, **73**, 1463.
493 A. McNally, Magnetic Sensors Nanoparticles Detect Infection, *Nat. Nanotechnol.*, 2013, **8**, 315.
494 D. Mehlhorn, R. Valiullin, J. Kaerger, K. Cho and R. Ryoo, Exploring Mass Transfer in Mesoporous Zeolites by NMR Diffusometry, *Materials*, 2012, **5**, 699.
495 R. Meier, D. Kruk and E. A. Roessler, Intermolecular Spin Relaxation and Translation Diffusion in Liquids and Polymer Melts: Insight from Field-Cycling ^{1}H NMR Relaxometry, *ChemPhysChem*, 2013, **14**, 3071.
496 T. Michel, E. Destandau and C. Elfakir, New Advances in Countercurrent Chromatography and Centrifugal Partition Chromatography: Focus on Coupling Strategy, *Anal. Bioanal. Chem.*, 2014, **406**, 957.
497 R. Micura, C. Kreutz, K. Breuker and S. A. Woodson, A Personal Perspective on Chemistry-Driven RNA Research, *Biopolymers*, 2013, **99**, 1114.
498 J. L. Miller, Diamond Defects Enable Nanoscale Nuclear Magnetic Resonance, *Phys. Today*, 2013, **66**, 12.

499 L. J. Miller and M. Dong, The Orthosteric Agonist-Binding Pocket in the Prototypic Class B G-Protein-Coupled Secretin Receptor, *Biochem. Soc. Trans.*, 2013, **41**, 154.
500 R. L. Mills, J. Latoski, G. Zhao, K. Akhtar, Z. Chang, J. He, X. Hu, G. Wu, G. Chu and Y. Lu, Identification of New Hydrogen States, *Phys. Essays*, 2011, **24**, 95.
501 M. Mohammadi and K. V. Sharp, Experimental Techniques for Bubble Dynamics Analysis in Microchannels: A Review, *J. Fluids Eng.*, 2013, **135**, 011202/1.
502 Yu. B. Monakhova, T. Kuballa and D. W. Lachenmeier, Chemometric Methods in NMR Spectroscopic Analysis of Food Products, *J. Anal. Chem.*, 2013, **68**, 755.
503 A. Morin, P. Sliz and A. Wittinghofer, Structural Biology Computing: Lessons for the Biomedical Research Sciences, *Biopolymers*, 2013, **99**, 809.
504 J. H. Morrissey, E. Tajkhorshid, S. G. Sligar and C. M. Rienstra, Tissue Factor/Factor VIIa Complex: Role of the Membrane Surface, *Thromb. Res.*, 2012, **129**(Suppl. 2), S8.
505 C. Mousoulis, X. Xu, D. A. Reiter and C. P. Neu, Single Cell Spectroscopy: Noninvasive Measures of Small-Scale Structure and Function, *Methods*, 2013, **64**, 119.
506 D. D. Mowrey, M. N. Kinde, Y. Xu and P. Tang, Atomistic Insights into Human Cys-Loop Receptors by Solution NMR, *Biochim. Biophys. Acta, Biomembr.*, 2015, **1848**, 307.
507 R. Mueller, R. Kanungo, A. Menjoge, M. Kiyono-Shimobe, W. J. Koros, S. A. Bradley, D. B. Galloway, J. J. Low, S. Prabhakar and S. Vasenkov, Sorbate Transport in Carbon Molecular Sieve Membranes and FAU/EMT Intergrowth by Diffusion NMR, *Materials*, 2012, **5**, 302.
508 R. Nakabayashi and K. Saito, Metabolomics for Unknown Plant Metabolites, *Anal. Bioanal. Chem.*, 2013, **405**, 5005.
509 Y. Nakai, T. Iye, S. Kitagawa, K. Ishida, S. Kasahara, T. Shibauchi, Y. Matsuda, H. Ikeda and T. Terashima, Normal-State Spin Dynamics in the Iron-Pnictide Superconductors $BaFe_2(As_{1-x}P_x)_2$ and $Ba(Fe_{1-x}Co_x)_2As_2$ Probed with NMR Measurements, *Phys. Rev. B: Condens. Matter Mater. Phys.*, 2013, **87**, 174507.
510 M. Nakamura, A. Ikezaki and M. Takahashi, Metal-Porphyrin Orbital Interactions in Paramagnetic Iron Complexes Having Planar and Deformed Porphyrin Ring, *J. Chin. Chem. Soc.*, 2013, **60**, 9.
511 B. Natalini, R. Sardella, A. Gioiello, F. Ianni, A. Di Michele and M. Marinozzi, Determination of Bile Salt Critical Micellization Concentration on the Road to Drug Discovery, *J. Pharm. Biomed. Anal.*, 2014, **87**, 62.
512 T. R. Nayak, L. K. Krasteva and W. Cai, Multimodality Imaging of RNA Interference, *Curr. Med. Chem.*, 2013, **20**, 3664.
513 J. L. Neira, NMR as a Tool to Identify and Characterize Protein Folding Intermediates, *Arch. Biochem. Biophys.*, 2013, **531**, 90.
514 B. A. D. Neto, J. R. Diniz and A. L. de Oliveira, Ionic Liquid Effect Probed by Nuclear Magnetic Resonance: NMR Approaches to Catalytic Reactions Performed in Ionic Liquids, *Curr. Org. Chem.*, 2013, **17**, 273.
515 C. A. Ng, A. M. Torres, G. Pages, P. W. Kuchel and J. I. Vandenberg, Insights into hERG K^+ Channel Structure and Function from NMR Studies, *Eur. Biophys. J.*, 2013, **42**, 71.
516 E. N. Nikolova, H. Zhou, F. L. Gottardo, H. S. Alvey, I. J. Kimsey and H. M. Al-Hashimi, A Historical Account of Hoogsteen Base-Pairs in Duplex DNA, *Biopolymers*, 2013, **99**, 955.

517 M. Nilges, Molecular Modelling Applied to NMR Structure Determination, *EPJ Web Conf.*, 2012, **30**, 01003.
518 S. Nishimura, A. Takagaki and K. Ebitani, Characterization, Synthesis and Catalysis of Hydrotalcite-Related Materials for Highly Efficient Materials Transformations, *Green Chem.*, 2013, **15**, 2026.
519 A. Obenaus, Neuroimaging Biomarkers for Epilepsy: Advances and Relevance to Glial Cells, *Neurochem. Int.*, 2013, **63**, 712.
520 L. A. O'Dell, The WURST Kind of Pulses in Solid-State NMR, *Solid State Nucl. Magn. Reson.*, 2013, **55–56**, 28.
521 A. C. O'Farrell, S. D. Shnyder, G. Marston, P. L. Coletta and J. H. Gill, Non-Invasive Molecular Imaging for Preclinical Cancer Therapeutic Development, *Br. J. Pharmacol.*, 2013, **169**, 719.
522 J. Oliveira, N. Mateus and V. de Freitas, Previous and Recent Advances in Pyranoanthocyanins Equilibria in Aqueous Solution, *Dyes Pigm.*, 2014, **100**, 190.
523 T. Paget, N. Haroune, S. Bagchi and E. Jarroll, Metabolomics and Protozoan Parasites, *Acta Parasitol.*, 2013, **58**, 127.
524 M. S. A. Palmnas and H. J. Vogel, The Future of NMR Metabolomics in Cancer Therapy: Towards Personalizing Treatment and Developing Targeted Drugs?, *Metabolites*, 2013, **3**, 373.
525 A. Papon, T. C. L. Guy, K. Saalwaechter, J. Oberdisse, S. Merabia, D. Long, P. Sotta, H. H. Frielinghaus, A. Radulescu, B. Deme, L. Nolrez, H. Montes and F. Lequeux, Studying Model Samples to Understand Mechanical Properties of Filled Elastomers, *KGK, Kautsch. Gummi Kunstst.*, 2013, **66**, 52.
526 H. Pasch, A. C. Makan, H. Chirowodza, N. Ngaza and W. Hiller, Analysis of Complex Polymers by Multidetector Field-Flow Fractionation, *Anal. Bioanal. Chem.*, 2014, **406**, 1585.
527 D. N. Patel, L. Li, C.-L. Kee, X. Ge, M.-Y. Low and H.-L. Koh, Screening of Synthetic PDE-5 Inhibitors and Their Analogues as Adulterants: Analytical Techniques and Challenges, *J. Pharm. Biomed. Anal.*, 2014, **87**, 176.
528 M. A. Patel and H. Kadiker, Interpretation of Structure in Organic Chemistry by Miscellaneous Technique, *Int. J. Pharm. Res. Bio-Sci.*, 2012, **1**, 145. Avail. URL: http://www.ijprbs.com/issuedocs/2012/06/7-34-9-IJPRBS-879.pdf.
529 S. Patra, D. Maity, R. Gunupuru, P. Agnihotri and P. Paul, Calixarenes: Versatile Molecules as Molecular Sensors for Recognition Study, *J. Chem. Sci.*, 2012, **124**, 1287.
530 H. D. Paz-Simon, A. Chemtob, F. Crest, C. Croutxe-Barghorn, L. Michelin, L. Vidal, S. Rigolet and B. Lebeau, Thick Mesostructured Films via Light Induced Self-Assembly, *RSC Adv.*, 2012, **2**, 11944.
531 E. A. Perigo and R. V. Martin, Magnetic Field and Gradient Standards Using Permanent Magnets: Design Considerations, Construction and Validation by Nuclear Magnetic Resonance, *IEEE Trans. Magn.*, 2013, **49**(Pt. 1), 4717.
532 F. A. Perras, J. Viger-Gravel, K. M. N. Burgess and D. L. Bryce, Signal Enhancement in Solid-State NMR of Quadrupolar Nuclei, *Solid State Nucl. Magn. Reson.*, 2013, **51–52**, 1.
533 J. Perry, P. Jefferson and J. A. Tainer, Developing Advanced X-Ray Scattering Methods Combined with Crystallography and Computation, *Methods*, 2013, **59**, 363.
534 M. V. Petoukhov and D. I. Svergun, Applications of Small-Angle X-Ray Scattering to Biomacromolecular Solutions, *Int. J. Biochem. Cell Biol.*, 2013, **45**, 429.

535 M. M. Petrinovic and B. Kuennecke, Neuroimaging Endophenotypes in Animal Models of Autism Spectrum Disorders: Lost or Found in Translation?, *Psychopharmacology*, 2014, **231**, 1167.

536 K. Petzold-Welcke, K. Schwikal, S. Daus and T. Heinze, Xylan Derivatives and Their Application Potential – Mini-Review of Own Results, *Carbohydr. Polym.*, 2014, **100**, 80.

537 V. C. Pierre, M. J. Allen and P. Caravan, Contrast Agents for MRI: 30+ Years and where Are We Going?, *JBIC, J. Biol. Inorg. Chem.*, 2014, **19**, 127.

538 D. A. T. Pires, M. P. Bemquerer and C. J. do Nascimento, Some Mechanistic Aspects on Fmoc Solid Phase Peptide Synthesis, *Int. J. Pept. Res. Ther.*, 2014, **20**, 53.

539 V. H. Pomin, NMR Chemical Shifts in Structural Biology of Glycosaminoglycans, *Anal. Chem.*, 2014, **86**, 65.

540 Y. Qi, G. Feng, Z. Huang and W. Yan, The Application of Superparamagnetic Iron Oxide-Labeled Mesenchymal Stem Cells in Cell-Based Therapy, *Mol. Biol. Rep.*, 2013, **40**, 2733.

541 C. Qian, G. Zabow and A. Koretsky, Engineering Novel Detectors and Sensors for MRI, *J. Magn. Reson.*, 2013, **229**, 67.

542 J. Qiao, S. Xue, F. Pu, N. White, J. Jiang, Z.-R. Liu and J. J. Yang, Molecular Imaging of EGFR/HER2 Cancer Biomarkers by Protein MRI Contrast Agents, *JBIC, J. Biol. Inorg. Chem.*, 2014, **19**, 259.

543 B. Rameev, G. Mozzhukhin and B. Aktas, Magnetic Resonance Detection of Explosives and Illicit Materials, *Appl. Magn. Reson.*, 2012, **43**, 463.

544 R. A. Ramli, W. A. Laftah and S. Hashim, Core-Shell Polymers: A Review, *RSC Adv.*, 2013, **3**, 15543.

545 W. Rauf and J. M. Brown, Reactive Intermediates in Catalytic Alkenylation; Pathways for Mizoroki-Heck, Oxidative Heck and Fujiwara-Moritani Reactions, *Chem. Commun.*, 2013, **49**, 8430.

546 S. Reckel and S. Hiller, Perspectives of Solution NMR Spectroscopy for Structural and Functional Studies of Integral Membrane Proteins, *Mol. Phys.*, 2013, **111**, 843.

547 C. Reddy, P. Vishnu, V. Kumar and J. Sivaraman, Linkers in the Structural Biology of Protein-Protein Interactions, *Protein Sci.*, 2013, **22**, 153.

548 S. Redhu and A. Jindal, Molecular Modelling: A New Scaffold for Drug Design, *Int. J. Pharm. Pharm. Sci.*, 2013, **5**(Suppl. 1), 5. Avail. URL: http://www.ijppsjournal.com/Vol5Suppl1/6209.pdf.

549 C. Reichhardt and L. Cegelski, Solid-State NMR for Bacterial Biofilms, *Mol. Phys.*, 2014, **112**, 887.

550 E. Rennella and B. Brutscher, Fast Real-Time NMR Methods for Characterizing Short-Lived Molecular States, *ChemPhysChem*, 2013, **14**, 3059.

551 S. Rezzi, S. Collino, L. Goulet and F.-P. Martin, Metabonomic Approaches to Nutrient Metabolism and Future Molecular Nutrition, *TrAC, Trends Anal. Chem.*, 2013, **52**, 112.

552 C. Rischpler, S. G. Nekolla, I. Dregely and M. Schwaiger, Hybrid PET/MR Imaging of the Heart: Potential, Initial Experiences, and Future Prospects, *J. Nucl. Med.*, 2013, **54**, 402.

553 S. L. Robinette, J. C. Lindon and J. K. Nicholson, Statistical Spectroscopic Tools for Biomarker Discovery and Systems Medicine, *Anal. Chem.*, 2013, **85**, 5297.

554 C. H. Roeder, S. Dieleman, F. M. van der Veen and D. Linden, Systematic Review of the Influence of Antipsychotics on the Blood Oxygenation Level-Dependent Signal of Functional Magnetic Resonance Imaging, *Curr. Med. Chem.*, 2013, **20**, 448.

555 A. Rogberg-Munoz, D. M. Posik, M. V. Ripoli, A. H. Falomir Lockhart, P. Peral-Garcia and G. Giovambattista, Recent Patents for Detecting the Species of Origin in Animal Feedstuff, and Raw and Processed Meat Products, *Recent Pat. Food, Nutr. Agric.*, 2013, **5**, 3.
556 R. T. M. Rosales, Potential Clinical Applications of Bimodal PET-MRI or SPECT-MRI Agents, *J. Labelled Compd. Radiopharm.*, 2014, **57**, 298.
557 R. S. Rosenson and J. A. Underberg, Systematic Review: Evaluating the Effect of Lipid-Lowering Therapy on Lipoprotein and Lipid Values, *Cardiovasc. Drugs Ther.*, 2013, **27**, 465.
558 M. Rowinska-Zyrek, M. Skilandat and R. K. O. Sigel, Hexaamminecobalt(III) – Probing Metal Ion Binding Sites in Nucleic Acids by NMR Spectroscopy, *Z. Anorg. Allg. Chem.*, 2013, **639**, 1313.
559 K. Saalwaechter, Robust NMR Approaches for the Determination of Homonuclear Dipole-Dipole Coupling Constants in Studies of Solid Materials and Biomolecules, *ChemPhysChem*, 2013, **14**, 3000.
560 K. Saalwaechter, W. Chasse and J.-U. Sommer, Structure and Swelling of Polymer Networks: Insights from NMR, *Soft Matter*, 2013, **9**, 6587.
561 J. Sagi, G-Quadruplexes Incorporating Modified Constituents: A Review, *J. Biomol. Struct. Dyn.*, 2014, **32**, 477.
562 T. Saio and C. G. Kalodimos, NMR Disentangles a Dynamic Disaggregase Machinery, *Nat. Struct. Mol. Biol.*, 2013, **20**, 409.
563 M. F. Santarelli, V. Positano, L. Menichetti, L. Landini and L. Landini, Cardiovascular Molecular Imaging: New Methodological Strategies, *Curr. Pharm. Des.*, 2013, **19**, 2439.
564 E. U. Saritas, P. W. Goodwill, L. R. Croft, J. J. Konkle, K. Lu, B. Zheng and S. M. Conolly, Magnetic Particle Imaging (MPI) for NMR and MRI Researchers, *J. Magn. Reson.*, 2013, **229**, 116.
565 H. Schmidbaur, H. G. Raubenheimer and L. Dobrzanska, The Gold-Hydrogen Bond, Au-H, and the Hydrogen Bond to Gold, Au···H-X, *Chem. Soc. Rev.*, 2014, **43**, 345.
566 M. Schnabelrauch, D. Scharnweber and J. Schiller, Sulfated Glycosaminoglycans as Promising Artificial Extracellular Matrix Components to Improve the Regeneration of Tissues, *Curr. Med. Chem.*, 2013, **20**, 2501.
567 L. Schroder, Xenon for NMR Biosensing–Inert but Alert, *Phys. Medica*, 2013, **29**, 3.
568 R. Schuhmacher, R. Krska, W. Weckwerth and R. Goodacre, Metabolomics and Metabolite Profiling, *Anal. Bioanal. Chem.*, 2013, **405**, 5003.
569 D. Schwarz, I. Orf, J. Kopka and M. Hagemann, Recent Applications of Metabolomics toward Cyanobacteria, *Metabolites*, 2013, **3**, 72.
570 A. Schwerk, F. D. S. Alves, P. J. W. Pouwels and T. van Amelsvoort, Metabolic Alterations Associated with Schizophrenia: A Critical Evaluation of Proton Magnetic Resonance Spectroscopy Studies, *J. Neurochem.*, 2014, **128**, 1.
571 J. Seitz, J. G. Vineberg, E. S. Zuniga and I. Ojima, Fluorine-Containing Taxoid Anticancer Agents and Their Tumor-Targeted Drug Delivery, *J. Fluorine Chem.*, 2013, **152**, 157.
572 A. Sekhar and L. E. Kay, NMR Paves the Way for Atomic Level Descriptions of Sparsely Populated, Transiently Formed Biomolecular Conformers, *Proc. Natl. Acad. Sci. U. S. A.*, 2013, **110**, 12867.
573 J. Seliger, V. Zagar and T. Asaji, NQR Investigation and Characterization of Cocrystals and Crystal Polymorphs, *Hyperfine Interact.*, 2013, **222**, 1.
574 S. R. Shaikh, J. J. Kinnun, X. Leng, J. A. Williams and S. R. Wassall, How Polyunsaturated Fatty Acids Modify Molecular Organization in

Membranes: Insight from NMR Studies of Model Systems, *Biochim. Biophys. Acta, Biomembr.*, 2015, **1848**, 211.
575 B. A. Shainyan and E. Kleinpeter, Silacyclohexanes and Silaheterocyclohexanes – Why Are They so Different from Other Heterocyclohexanes?, *Tetrahedron*, 2013, **69**, 5927.
576 R. A. Shank, N. A. Foroud, P. Hazendonk, F. Eudes and B. A. Blackwell, Current and Future Experimental Strategies for Structural Analysis of Trichothecene Mycotoxins – A Prospectus, *Toxins*, 2011, **3**, 1518.
577 P. Shanmugasundaram, V. Viswanath, A. S. K. Sankar and V. Ravichandiran, Metabolomics: A Cancer Diagnostic Tool, *J. Pharm. Res.*, 2012, **5**, 5210.
578 X. Shen, C. Xu and L. Ye, Molecularly Imprinted Polymers for Clean Water: Analysis and Purification, *Ind. Eng. Chem. Res.*, 2013, **52**, 13890.
579 H. Shi, Z. Liang, T. Sema, A. Naami, P. Usubharatana, R. Idem, C. Saiwan and P. Tontiwachwuthikul, Part 5a: Solvent Chemistry: NMR Analysis and Studies for Amine-CO_2-H_2O Systems with Vapor-Liquid Equilibrium Modeling for CO_2 Capture Processes, *Carbon Manage.*, 2012, **3**, 185.
580 L. Shi and L. Gan, Open-Cage Fullerenes as Tailor-Made Container for a Single Water Molecule, *J. Phys. Org. Chem.*, 2013, **26**, 766.
581 H. Shokrollahi, Contrast Agents for MRI, *Mater. Sci. Eng., C*, 2013, **33**, 4485.
582 E. Sievanen and E. Kolehmainen, High-Resolution Solid-State NMR Spectroscopy of Steroids and Their Derivatives, *Appl. Spectrosc. Rev.*, 2013, **48**, 50.
583 G. Siligardi, R. Hussain, S. G. Patching and M. K. Phillips-Jones, Ligand- and Drug-Binding Studies of Membrane Proteins Revealed through Circular Dichroism Spectroscopy, *Biochim. Biophys. Acta, Biomembr.*, 2014, **1838**, 34.
584 D. Sinnaeve, The Stejskal-Tanner Equation Generalized for any Gradient Shape – An Overview of Most Pulse Sequences Measuring Free Diffusion, *Concepts Magn. Reson., Part A*, 2012, **40A**, 39.
585 B. N. Siriwardena-Mahanama and M. J. Allen, Strategies for Optimizing Water-Exchange Rates of Lanthanide-Based Contrast Agents for Magnetic Resonance Imaging, *Molecules*, 2013, **18**, 9352.
586 J. Sirker, The Luttinger Liquid and Integrable Models, *arXiv.org, e-Print Arch., Condens. Matter*, 11 Jul 2012, 1-23, arXiv:1207.2582v1 [cond-mat.str-el]. Avail. URL: http://arxiv.org/PS_cache/arxiv/pdf/1207/1207.2582v1.pdf.
587 L. J. Sitole, A. A. Williams and D. Meyer, Metabonomic Analysis of HIV – Infected Biofluids, *Mol. BioSyst.*, 2013, **9**, 18.
588 L. Skrbek, Quantum Turbulence, *J. Phys.: Conf. Ser.*, 2011, **318**, 012004.
589 D. Sleiman, V. Goldschmidt, P. Barraud, R. Marquet, J.-C. Paillart and C. Tisne, Initiation of HIV-1 Reverse Transcription and Functional Role of Nucleocapsid-Mediated tRNA/Viral Genome Interactions, *Virus Res.*, 2012, **169**, 324.
590 L. Sobczyk, B. Czarnik-Matusewicz, M. Rospenk and M. Obrzud, Proton Transfer Equilibria and Critical Behavior of H-Bonding, *J. At., Mol., Opt. Phys.*, 2012, 217932.
591 L. Sobczyk, M. Obrzud and A. Filarowski, H/D Isotope Effects in Hydrogen Bonded Systems, *Molecules*, 2013, **18**, 4467.
592 C. Song, T. Rohmer, M. Tiersch, J. Zaanen, J. Hughes and J. Matysik, Solid-State NMR Spectroscopy to Probe Photoactivation in Canonical Phytochromes, *Photochem. Photobiol.*, 2013, **89**, 259.
593 Y.-Q. Song, Magnetic Resonance of Porous Media (MRPM). A Perspective, *J. Magn. Reson.*, 2013, **229**, 12.
594 H. W. Spiess, Magic Angle Spinning NMR of Macromolecular and Supramolecular Systems, *Israel J. Chem.*, 2014, **54**, 16.

595 C. Studholme and F. Rousseau, Quantifying and Modelling Tissue Maturation in the Living Human Fetal Brain, *Int. J. Dev. Neurosci.*, 2014, **32**, 3.
596 H. Suit, Proton: The Particle, *Int. J. Radiat. Oncol., Biol., Phys.*, 2013, **87**, 555.
597 N. Sun, Y. Liu, L. Qin, H. Lee, R. Weissleder and D. Ham, Small NMR Biomolecular Sensors, *Solid-State Electron.*, 2013, **84**, 13.
598 S. Sun, Chemical Synthesis of Monodisperse Magnetic Nanoparticles for Sensitive Cancer Detection, *J. Inorg. Organomet. Polym. Mater.*, 2014, **24**, 33.
599 U. K. Sundekilde, L. B. Larsen and H. C. Bertram, NMR-Based Milk Metabolomics, *Metabolites*, 2013, **3**, 204.
600 F. Sussman, M. C. Villaverde, J. L. Dominguez and U. H. Danielson, On the Active Site Protonation State in Aspartic Proteases: Implications for Drug Design, *Curr. Pharm. Des.*, 2013, **19**, 4257.
601 A. Sutrisno and Y. Huang, Solid-State NMR: A Powerful Tool for Characterization of Metal-Organic Frameworks, *Solid State Nucl. Magn. Reson.*, 2013, **49–50**, 1.
602 V. Swarnkar, V. Yadav, S. Kumar, P. Khandelwal and N. K. Gill, An Overview on Isolation, Characterization and Separation of Impurities of API by Using HPLC, LC-MS and LC-NMR Technique, *Inventi Impact: Pharm Anal. Qual. Assur.*, 2013, **1**, 17.
603 A. Szulc, B. Galinska-Skok, N. Waszkiewicz, D. Bibulowicz, B. Konarzewska and E. Tarasow, Proton Magnetic Resonance Spectroscopy Changes after Antipsychotic Treatment, *Curr. Med. Chem.*, 2013, **20**, 414.
604 K. Tanaka and Y. Chujo, Design of Functionalized Nanoparticles for the Applications in Nanobiotechnology, *Adv. Powder Technol.*, 2014, **25**, 101.
605 F. Taulelle, B. Bouchevreau and C. Martineau, NMR Crystallography Driven Structure Determination. Nanoporous Materials, *CrystEngComm*, 2013, **15**, 8613.
606 P. Thanasekaran, C.-Y. Huang and K.-L. Lu, Synthesis, Structure, and Dynamic Behavior of Discrete Metallacyclic Rotors, *Chem. Lett.*, 2013, **42**, 776.
607 F. L. Theiss, G. A. Ayoko and R. L. Frost, Removal of Boron Species by Layered Double Hydroxides: A Review, *J. Colloid. Interface Sci.*, 2013, **402**, 114.
608 H. This, Solutions are Solutions, and Gels are Almost Solutions, *Pure Appl. Chem.*, 2013, **85**, 257.
609 H. Tidow and P. Nissen, Structural Diversity of Calmodulin Binding to Its Target Site, *FEBS J.*, 2013, **280**, 5551.
610 A. P. Tikunov, J. H. Winnike, K. Tech, R. E. Jeffries, C. T. A. Semelka, J. Martin, R. McClelland, L. M. Graves and J. M. MacDonald, Fluxomics by NMR Spectroscopy from Cells to Organisms Focusing on Liver, *Curr. Metabolomics*, 2013, **1**, 128.
611 V. Tisato, A. Norcio, R. Voltan, C. Coleghini, D. Zella and P. Secchiero, MDM2 Non-Genotoxic Inhibitors as Innovative Therapeutic Approaches for the Treatment of Pediatric Malignancies, *Curr. Med. Chem.*, 2013, **20**, 2226.
612 Y. Tokunaga, H. Sakai, S. Kambe, H. Chudo, M. Osaka, S. Miwa, T. Nishi, M. Nakada, A. Itoh and Y. Homma, Solid-State NMR Study on Actinide Dioxides, *MRS Proceedings, Online Proc. Libr.*, 2012, **1444**, DOI: 10.1557/opl.2012.1001.
613 D. Tomasi and N. D. Volkow, Brain Activation and Neurochemistry, *Proc. Natl. Acad. Sci. U. S. A.*, 2013, **110**, 10888.
614 M. P. Toraskar, N. S. Singasane, J. B. Pichake and V. Kadam, Fragment Based Drug Discovery – A Tool for Drug Discovery, *Int. J. Drug Des. Discovery*, 2013, **4**, 1083.
615 E. Torrontegui, S. Ibanez, S. Martinez-Garaot, M. Modugno, A. del Campo, D. Guery-Odelin, A. Ruschhaupt, X. Chen and J. G. Muga, Shortcuts to Adiabaticity, *Adv. At., Mol., Opt. Phys.*, 2013, **62**, 117.

616 A. B. Uryupin and A. S. Peregudov, Application of NMR Techniques to the Determination of the Composition of Tobacco, Coffee, and Tea Products, *J. Anal. Chem.*, 2013, **68**, 1021.
617 A. Valori, P. J. McDonald and K. L. Scrivener, The Morphology of C-S-H: Lessons from ^{1}H Nuclear Magnetic Resonance Relaxometry, *Cem. Concr. Res.*, 2013, **49**, 65.
618 H. Van As and J. van Duynhoven, MRI of Plants and Foods, *J. Magn. Reson.*, 2013, **119**, 25.
619 J. C. Vedrine, G. J. Hutchings and C. J. Kiely, Molybdenum Oxide Model Catalysts and Vanadium Phosphates as Actual Catalysts for Understanding Heterogeneous Catalytic Partial Oxidation Reactions: A Contribution by Jean-Claude Volta, *Catal. Today*, 2013, **217**, 57.
620 G. Veglia, Protein Dynamics Catch Them if You Can, *Nat. Chem. Biol.*, 2013, **9**, 410.
621 G. Veglia and A. Cembran, Role of Conformational Entropy in the Activity and Regulation of the Catalytic Subunit of Protein Kinase A, *FEBS J.*, 2013, **280**, 5608.
622 R. Ventura-Clapier and H. A. Coller, A Step toward System Metabolomics, *Physiol. Genomics*, 2012, **44**, 383.
623 N. Verdaguer and C. Ferrer-Orta, Conformational Changes in Motif D of RdRPs as Fidelity Determinant, *Structure*, 2012, **20**, 1448.
624 E. Vinogradov, A. D. Sherry and R. E. Lenkinski, CEST: From Basic Principles to Applications, Challenges and Opportunities, *J. Magn. Reson.*, 2013, **229**, 155.
625 G. Voicu, E. Andronescu, A. M. Grumezescu, K.-S. Huang, A. Ficai, C.-H. Yang, C. Bleotu and M. C. Chifiriuc, Antitumor Activity of Magnetite Nanoparticles: Influence of Hydrocarbonated Chain of Saturated Aliphatic Monocarboxylic Acids, *Curr. Org. Chem.*, 2013, **17**, 831.
626 D. E. Volk, C. W. Tuthill, M.-A. Elizondo-Riojas and D. G. Gorenstein, NMR Structural Studies of Thymosin α1 and β-Thymosins, *Ann. N. Y. Acad. Sci.*, 2012, **1270**, 73.
627 N. Vora and P. Beniwal, Basic Principles of ^{13}C-NMR, *Int. J. Pharm. Technol.*, 2012, **4**, 1987. Avail. URL: http://www.ijptonline.com/wp-content/uploads/2009/10/1987-1999.pdf.
628 T. N. Vu and K. Laukens, Getting Your Peaks in Line: A Review of Alignment Methods for NMR Spectral Data, *Metabolites*, 2013, **3**, 259.
629 G. W. Vuister, R. H. Fogh, P. M. S. Hendrickx, J. F. Doreleijers and A. Gutmanas, An Overview of Tools for the Validation of Protein NMR Structures, *J. Biomol. NMR*, 2014, **58**, 259.
630 I. E. Wachs, Catalysis Science of Supported Vanadium Oxide Catalysts, *Dalton Trans.*, 2013, **42**, 11762.
631 J. L. Wagstaff, S. L. Taylor and M. J. Howard, Recent Developments and Applications of Saturation Transfer Difference Nuclear Magnetic Resonance (STD NMR) Spectroscopy, *Mol. BioSyst.*, 2013, **9**, 571.
632 G. Wang, Database-Guided Discovery of Potent Peptides to Combat HIV-1 or Superbugs, *Pharmaceuticals*, 2013, **6**, 728.
633 G. Wang, B. Mishra, R. F. Epand and R. M. Epand, High Quality 3D Structures Shine Light on Antibacterial, Anti-Biofilm and Antiviral Activities of Human Cathelicidin LL-37 and Its Fragments, *Biochim. Biophys. Acta, Biomembr.*, 2014, **1838**, 2160.
634 G. Wang, Z.-T. Zhang, B. Jiang, X. Zhang, C. Li and M. Liu, Recent Advances in Protein NMR Spectroscopy and Their Implications in Protein Therapeutics Research, *Anal. Bioanal. Chem.*, 2014, **406**, 2279.

635 H.-B. Wang, L.-P. Liu and X.-Y. Wang, ^{13}C-NMR Data of Daphnane Diterpenoids, *Magn. Reson. Chem.*, 2013, **51**, 580.
636 P. Wang and A. Moore, Application of in vivo Magnetic Resonance Imaging for Islet Transplantation, *Micro Nanosyst.*, 2013, **5**, 224.
637 Y. Wang, C. Xu and H. Ow, Commercial Nanoparticles for Stem Cell Labeling and Tracking, *Theranostics*, 2013, **3**, 544.
638 B. Warth, M. Sulyok and R. Krska, LC-MS/MS-Based Multibiomarker Approaches for the Assessment of Human Exposure to Mycotoxins, *Anal. Bioanal. Chem.*, 2013, **405**, 5687.
639 W. E. Washburn and J. E. Birdwell, Updated Methodology for Nuclear Magnetic Resonance Characterization of Shales, *J. Magn. Reson.*, 2013, **233**, 17.
640 E. D. Watt and C. M. Rienstra, Recent Advances in Solid-State Nuclear Magnetic Resonance Techniques to Quantify Biomolecular Dynamics, *Anal. Chem.*, 2014, **86**, 58.
641 A. G. Webb, Radiofrequency Microcoils for Magnetic Resonance Imaging and Spectroscopy, *J. Magn. Reson.*, 2013, **229**, 55.
642 A. Wegner, T. Cordes, A. Michelucci and K. Hiller, The Application of Stable Isotope Assisted Metabolomics in Biomedicine, *Curr. Biotechnol.*, 2012, **1**, 88.
643 F. W. Wehrli, Magnetic Resonance of Calcified Tissues, *J. Magn. Reson.*, 2013, **229**, 35[online computer file].
644 Z. Wei and M. Zhang, Decorating Proteins with "Sweets" Is a Flexible Matter, *Structure*, 2013, **21**, 1.
645 T. Weidner and D. G. Castner, SFG Analysis of Surface Bound Proteins: A Route towards Structure Determination, *Phys. Chem. Chem. Phys.*, 2013, **15**, 12516.
646 G. Widmalm, A Perspective on the Primary and Three-Dimensional Structures of Carbohydrates, *Carbohydr. Res.*, 2013, **378**, 123.
647 M. D. Williams, R. Reeves, L. S. Resar and H. H. Hill, Jr., Metabolomics of Colorectal Cancer: Past and Current Analytical Platforms, *Anal. Bioanal. Chem.*, 2013, **405**, 5013.
648 D. S. Wishart, Characterization of Biopharmaceuticals by NMR Spectroscopy, *TrAC, Trends Anal. Chem.*, 2013, **48**, 96.
649 C. Witte and L. Schroder, NMR of Hyperpolarised Probes, *NMR Biomed.*, 2013, **26**, 788.
650 J.-L. Wolfender, S. Rudaz, Y. H. Choi and H. K. Kim, Plant Metabolomics: From Holistic Data to Relevant Biomarkers, *Curr. Med. Chem.*, 2013, **20**, 1056.
651 G. Xie, S. Zhang, X. Zheng and W. Jia, Metabolomics Approaches for Characterizing Metabolic Interactions between Host and Its Commensal Microbes, *Electrophoresis*, 2013, **34**, 2787.
652 R. Xing, G. Liu, J. Zhu, Y. Hou and X. Chen, Functional Magnetic Nanoparticles for Non-Viral Gene Delivery and MR Imaging, *Pharm. Res.*, 2014, **31**, 1377.
653 H.-j. Yang, Z.-l. Zhang, L.-y. Ren, W.-l. Tang and W. Zheng, Isolation and Structure Analysis of an Active Metabolite from Eupenicillium sp. E-UN41, *Tianran Chanwu Yanjiu Yu Kaifa*, 2012, **24**, 1172.
654 M. Yang, T. Soga and P. J. Pollard, Oncometabolites: Linking Altered Metabolism with Cancer, *J. Clin. Invest.*, 2013, **123**, 3652.
655 J. Yates, Prediction of NMR Parameters in the Solid-State, *EPJ Web Conf.*, 2012, **30**, 04002.
656 P. L. Yeagle, Non-Covalent Binding of Membrane Lipids to Membrane Proteins, *Biochim. Biophys. Acta, Biomembr.*, 2014, **1838**, 1548.
657 J. Yu, J. R. Horsley and A. D. Abell, The Influence of Secondary Structure on Electron Transfer in Peptides, *Aust. J. Chem.*, 2013, **66**, 848.

658 J. Yu, D.-Y. Huang, M. Z. Yousaf, Y.-L. Hou and S. Gao, Magnetic Nanoparticle-Based Cancer Therapy, *Chin. Phys. B*, 2013, **22**, 027506.
659 J. Yu, A. Tahmasebi, Y. Han, F. Yin and X. Li, A Review on Water in Low Rank Coals: The Existence, Interaction with Coal Structure and Effects on Coal Utilization, *Fuel Process. Technol.*, 2013, **106**, 9.
660 Z. Yu and H. Mao, Non-B DNA Structures Show Diverse Conformations and Complex Transition Kinetics Comparable to RNA or Proteins-A Perspective from Mechanical Unfolding and Refolding Experiments, *Chem. Rec.*, 2013, **13**, 102.
661 A. Zhang, H. Sun, H. Xu, S. Qiu and X. Wang, Cell Metabolomics, *OMICS*, 2013, **17**, 495.
662 A.-h. Zhang, H. Sun, S. Qiu and X.-j. Wang, NMR-Based Metabolomics Coupled with Pattern Recognition Methods in Biomarker Discovery and Disease Diagnosis, *Magn. Reson. Chem.*, 2013, **51**, 549.
663 H. Zhang, D. Fan, J. Deng, X. Ma, C. Zhu and W. Xue, Review Separation and Purification of Proteins with Chromatography Applications, *J. Pure Appl. Microbiol.*, 2013, **7**, 7.
664 J. Zhang, C. P. Jones and A. R. Ferre-D'Amare, Global Analysis of Riboswitches by Small-Angle X-Ray Scattering and Calorimetry, *Biochim. Biophys. Acta, Gene Regul. Mech.*, 2014, **1839**, 1020.
665 A. Zheng, S. Huang, Q. Wang, H. Zhang, F. Deng and S. Liu, Progress in Development and Application of Solid-State NMR for Solid Acid Catalysis, *Cuihua Xuebao*, 2013, **34**, 436.
666 H.-X. Zhou and T. A. Cross, Modeling the Membrane Environment Has Implications for Membrane Protein Structure and Function: Influenza A M2 Protein, *Protein Sci.*, 2013, **22**, 381.
667 X.-l. Zhu and J. Tian, Application of Magnetic Nanoparticles in Stem Cells, *Zhongguo Jiaoxing Waike Zazhi*, 2012, **20**, 1013.
668 Y. Zhu and Q. Fang, Analytical Detection Techniques for Droplet Microfluidics – A Review, *Anal. Chim. Acta*, 2013, **787**, 24.
669 D. Zuccaccia, L. Belpassi, A. Macchioni and F. Tarantelli, Ligand Effects on Bonding and Ion Pairing in Cationic Gold(I) Catalysts Bearing Unsaturated Hydrocarbons, *Eur. J. Inorg. Chem.*, 2013, **1013**, 4121.
670 E. R. P. Zuiderweg and A. Rousaki, Gradient-Enhanced TROSY Described with Cartesian Product Operators, *Concepts Magn. Reson., Part A*, 2011, **38A**, 280.

Reviews and books in foreign languages

Chinese

671 J.-l. Ai, M. Tian and Q-x. Yang, Advance on the C_{21} Steroidal Glycosides from Cynanchum Otophyllum, *Guizhou Shifan Daxue Xuebao, Ziran Kexueban*, 2012, **30**, 106.
672 X. Cao, B. Xiang and C.-y. Zhang, Review on Relationship between the Molecular and Crystal Structure of Explosives and Their Sensitivities, *Hanneng Cailiao*, 2012, **20**, 643.
673 J. Chen, W.-j. Yuan and P.-w. Hao, Influence of Microcosmic Index on Thermal Stability Performance of Asphalt, *Chang'an Daxue Xuebao, Ziran Kexueban*, 2012, **32**, 12.
674 L. Chen, C. Li, J. Yang, X. Yang, J. Chu, H. Pu and Z. Luo, Investigation on Chemical Composition and Structure of Petroleum Resins, *Guangzhou Huagong*, 2012, **40**, 51.
675 P. Chen and X.-l. Chu, Development of Rapid Analysis Technologies in Crude and Heavy Oil, *Fenxi Ceshi Xuebao*, 2012, **31**, 1191.

676 G. Cheng, R. Song, J. Miao and J. Qian, Characterization of Interfacial Structure of Polymer Nanocomposites, *Cailiao Daobao*, 2012, **26**, 130.
677 P. Cheng, S. Jiao, K. Xu and C. Wang, Progress in Enantioselective Recognition Based on Chiral Crown Ether, *Youji Huaxue*, 2013, **33**, 280.
678 X.-p. Dong, Y.-r. Pang, L.-y. Yang, F. Wang and Y.-x. Chang, Latest Development of Non-Destructive Testing During Hydrogen Absorption and Desorption of Hydrogen Storage Alloys, *Xiyou Jinshu Yu Yingzhi Hejin*, 2013, **41**, 49.
679 C. Feng, X. Wang and D. Li, Application Progress of Solid ^{29}Si, ^{27}Al NMR in the Research of Cement-Based Materials, *Hejishu*, 2014, **37**, 010502.
680 H.-w. Gu, Y.-p. Qi, N. Xu, J.-h. Ding, Y.-b. An and H.-w. Chen, Nuclear Magnetic Resonance Spectroscopy and Mass Spectrometry-Based Metabolomics for Cancer Diagnosis, *Fenxi Huaxue*, 2012, **40**, 1933.
681 H.-j. Guo, H.-x. Zhao and H. Bai, Spectroscopic Characteristics of Steroidal Saponins and Homoisoflavonoids from Polygonatum Odoratum, *Zhongyiyao Xuebao*, 2012, **40**, 41.
682 Q.-y. Guo, J. Xin, X. Zhang, B. Yu, Z.-s. Zhao, H.-L. Li, H. Li, X.-m. Wang and W. Liao, Progress in Study of Molecular Mechanism in Water Diffusion Weighted MRI, *Zhongguo Linchuang Yixue Yingxiang Zazhi*, 2013, **24**, 496.
683 F.-y. He, K.-w. Deng, J.-l. Zeng, R.-w. Dai, Z.-s. Xia, W.-l. Liu and J.-l. Shi, Current Development of Rapid High-Throughput Determination Technology for Total Components of Traditional Chinese Medicines and Formula and Synthetic Immunity Chip Method, *Zhongguo Zhongyao Zazhi*, 2012, **37**, 3164.
684 K. He, G.-f. Li, Z.-d. Wu, M. Chen, J.-m. Xie and Y.-s. Yan, Recent Progress of Analytical Methods for Glyphosate, *Lihua Jianyan Huaxue Fence*, 2011, **47**, 991.
685 L. Hu and W. Guan, Recent Development in Targeted MRI Negative Contrast Based on Ultra-Small Superparamagnetic Iron Oxide Particles, *Zhongguo Zhongliu Linchuang*, 2013, **40**, 423.
686 L.-l. Huang, W.-s. Cheng, Y.-d. Chen and F. Hou, Research Progress in Ginseng Fingerprint, *Zhongcaoyao*, 2013, **44**, 241.
687 Q.-l. Huang, Y. Tang, W. Liu and G.-m. Lu, Research of NMR of Nucleus Amygdalae and Cerebral Function of Patients with Post-Traumatic Stress Disorder, *Zhonghua Jingshenke Zazhi*, 2012, **45**, 185.
688 T. Huang, W. Zhang, C. Quan and H.-m. Li, Review on Quantitative Nuclear Magnetic Resonance, *Huaxue Shiji*, 2012, **34**, 327.
689 Y.-L. Huo, J.-s. Hu and H.-x. Li, Review on the Methods of Fat Extraction and Fatty Acid Analysis, *Hebei Lianhe Daxue Xuebao, Ziran Kexueban*, 2013, **35**, 104.
690 L.-t. Jia, S.-j. Yang and S.-y. Bai, Spectral Characteristics of Main Active Compounds from Trillium Tschonoskii Maxim, *Yaoxue Yanjiu*, 2013, **32**, 351.
691 D.-d. Lei, Y.-f. Guan, G.-d. Wang, N. Dong and G.-h. Zhao, Application of Solid ^{13}C Nuclear Magnetic Resonance (NMR) in Food Science, *Shipin Gongye Keji*, 2013, **34**, 392.
692 B.-y. Li and Z.-q. Zhang, Progress in Detecting Mild Cognitive Impairment by Neuroimaging, *Yixue Zongshu*, 2014, **20**, 504.
693 G.-h. Li, F.-s. Zhang and H.-g. Li, Research Progress on Methods for Determination of Phosphorus Components in Animal Manure, *Turang Tongbao*, 2013, **44**, 760.
694 S. Li and X.-y. Feng, A Mini Review on NMR Approaches for Coal Analysis, *Bopuxue Zazhi*, 2013, **30**, 148.
695 C.-l. Liu, Q.-g. Meng and Y.-g. Ye, Applications of Solid State NMR in the Studies of Gas Hydrate, *Bopuxue Zazhi*, 2012, **29**, 465.

696 D. Liu and X. Cai, Decoding Mechanisms of Crystallization Technology Underlying Genetic Information Transduction, *Shengming De Huaxue*, 2013, **33**, 67.
697 G. Liu, C. Tao, W. Bai, W. Zhao and J. Wang, Advances on Determination of Ethyl Carbamate in Fermented Food, *Zhongguo Niangzao*, 2012, **31**, 1.
698 H.-W. Liu, K.-F. Yang, T. D. Mishima, M. B. Santos, K. Nagase and Y. Hirayama, Resistively Detected Nuclear Magnetic Resonance: Recent Development, *Wuli Xuebao*, 2012, **61**, 147302.
699 X. Liu, N. Sheng, F. Li, H. Huo and L. Zhang, LC-MS-NMR Technique and Its Application in Drug Metabolism and Structure Identification, *Zhongguo Yiyuan Yaoxue Zazhi*, 2012, **32**, 972.
700 Y. Liu, S.-g. Wu, Z. Yin, G.-h. Wu and L.-f. Cao, Application of Fingerprint Technique in Tea Research, *Chaye Kexue*, 2013, **33**, 13.
701 Y.-y. Liu and T.-y. Fan, Progress in Research of Nanomedicine-Based Gadolinium Contrast Agents for Magnetic Resonance Imaging, *Zhongguo Xinyao Zazhi*, 2013, **22**, 787.
702 Z.-q. Liu, M.-g. Wang, M. Xue and J.-y. Li, Application of Cyclodextrin in Separation and Analysis of Chiral Enantiomers, *Fenxi Ceshi Jishu Yu Yiqi*, 2012, **18**, 14.
703 Z.-t. Liu and S.-h. Yuan, Analysis on Application Trends of PET/MRI in Nuclear Cardiology, *He Dianzixue Yu Tance Jishu*, 2013, **33**, 263.
704 Z.-j. Luo, Y. Dai, H.-y. Pu and C.-y. Li, Separation and Purification of Asphaltenes and Research Techniques of Its Molecular Structure, *Hebei Huagong*, 2012, **35**, 47.
705 X.-w. Meng, D. Wang and B.-z. Shen, Progress of Researching on Alzheimer's Disease by MRI, *Xiandai Shengwuyixue Jinzhan*, 2013, **13**, 970.
706 H.-y. Pu, C.-j. Diao, Z.-j. Luo, L.-p. Chen and C.-y. Li, Separation and Analysis Methods of Petroleum Components, *Liaoning Huagong*, 2012, **41**, 1322.
707 J. Ren, J. Li, F. Shi, X. Wang, J.-h. He, Y.-c. Xu, N.-x. Zhang, B. Xiong and J.-k. Shen, Progress in the Fragment-Based Drug Discovery, *Yaoxue Xuebao*, 2013, **48**, 14.
708 P.-c. Sun, S.-y. Zhao, Y.-y. Wang, X. He and R.-c. Zhang, Solid-State NMR Studies on Mutilscale Structure and Dynamics of Macromolecules, *Gaofenzi Tongbao*, 2013, **1**, 72.
709 Q. Sun, X.-m. Guan, M.-f. Gou and J.-p. Zhu, Study Progress of C-S-H by Solid-State NMR, *Guisuanyan Tongbao*, 2013, **32**, 440.
710 Z. Sun, Y. Yu, M. Pang, P. Yang, W. Yu and H. Cao, Applications and Outlook of ^{1}H Low Field NMR Probing into Cement-Based Materials, *Cailiao Daobao*, 2011, **25**, 110.
711 H. Tian and Y.-m. Zhang, Structural Determination of Polysaccharides and Its Application, *Zhongguo Shipin Tianjiaji*, 2012, **2**, 177.
712 L. Wang, Q.-l. Wang, H.-s. Wang and H. Wang, Overview of the Effect of Mechanical Behavior on YBCO Performance, *Diwen Wuli Xuebao*, 2013, **35**, 224.
713 S.-n. Wang, S.-c. Yu, X.-d. Xu, S.-j. She and S.-h. Fan, Triterpenoid Saponins and Flavonoids from Clinopodium Linn. : Chemical Component Analyses and NMR Spectral Features, *Bopuxue Zazhi*, 2013, **30**, 447.
714 W.-h. Wang, T.-l. Tian, L. Ding and H.-y. Li, Synthesis of Novel Complexes of Lanthanide with Tripodal Schiff Base and Their Interactions with DNA, *Xitu*, 2012, **33**, 90.

715 X. Wang, Y.-d. Sun, Q. Zhang and J. Chen, Analytical Methods for Structure and Composition of Coke Deposited on Catalysts, *Fenxi Ceshi Jishu Yu Yiqi*, 2013, **19**, 6.
716 X.-l. Wang, H.-n. Liu and X.-l. Chen, Fibrovascularization of Different Types of Hydroxyapatite Orbital Implants, *Zhongguo Zuzhi Gongcheng Yanjiu*, 2013, **17**, 2265.
717 Y. Wang and X. Jiang, Research Development of Study Methods of Interaction between Drug Molecules and HSA, *Guangdong Huagong*, 2012, **39**, 76.
718 Z.-j. Wang and L.-b. Wang, Progress in Determining Stereo-Configuration of Chiral Compounds by Using Computational Chemistry, *Zhongguo Yaowu Huaxue Zazhi*, 2013, **23**, 235.
719 Z. Xu and G.-y. Tang, Progress on Evaluation of Marrow Fat Content for Bone Quality in Osteoporosis, *Tongji Daxue Xuebao, Yixueban*, 2012, **33**, 121.
720 T.-t. Yang, X. Duan, S.-z. Jin and Q.-f. Jiang, Application of LC-NMR in Pharmaceutical Analysis, *Xiandai Yaowu Yu Linchuang*, 2012, **27**, 635.
721 Z.-d. Zhai and X.-m. Wu, Modern Isotope Analysis Technologies for the Authentication of Natural Vinegars, *Zhongguo Tiaoweipin*, 2012, **37**, 6.
722 J. Zhang, Q. Luo, Q.-n. Guo, S.-h. Chen and X. Zhou, Magnetic Resonance Molecular Imaging with Hyperpolarized ^{129}Xe, *Bopuxue Zazhi*, 2013, **30**, 279.
723 J. Zhang, Y. Ren, X. Yong and Q. Shi, Progress of ^{27}Al NMR Studies on Hydroxyl Polyaluminum Solution, *Zhongguo Wuji Fenxi Huaxue*, 2013, **3**, 17.
724 L. Zhang, Y. Shi, S.-x. Bian and Q.-y. Guo, Application Progress in BOLD-MRI of Liver Fibrosis, *Zhongguo Linchuang Yixue Yingxiang Zazhi*, 2013, **24**, 120.
725 L.-l. Zhang, Y.-m. Wang, M. Xu, D.-m. Wu and J.-h. Chen, Research Progress in Analytical Methods of Vegetable Tannins, *Linchan Huaxue Yu Gongye*, 2012, **32**, 107.
726 Y.-b. Zhang, Z.-z. Chen, H.-d. Yu and J.-c. Zhang, Recent Progress in the Synthesis of 10-HAD, *Guangzhou Huagong*, 2013, **41**, 20.
727 Z.-f. Zhang and J. Yang, Solid-State NMR Studies on Amyloid Fibrils: Recent Progresses, *Bopuxue Zazhi*, 2013, **30**, 157.
728 Y. Zhao and L. Ding, Applications of Nuclear Magnetic Resonance in Plant Metabolics Research, *Xiandai Yiqi Yu Yiliao*, 2013, **19**, 21.
729 A. Zheng, S. Huang, Q. Wang, H. Zhang, F. Deng and S. Liu, Progress in Development and Application of Solid-State NMR for Solid Acid Catalysis, *Cuihua Xuebao*, 2013, **34**, 436.
730 Q. Zhou, X. Yu, G. Tu, Q. Wang, W. Hu and H. Li, Advances in Metabonomics and Its Applications, *Shengwu Jishu Tongbao*, 2013, **1**, 49.

Czech

731 K. Benesova, S. Belakova, R. Mikulikova and Z. Svoboda, Survey of Analytical Methods for the Phytic Acid Determination, *Kvasny Prum.*, 2013, **59**, 127.
732 A. Gregorova, H. Cizkova, J. Mazac and M. Voldrich, Authenticity Assessment of Spirit Vinegar. Part I. Production and Analysis Methods, *Kvasny Prum.*, 2012, **58**, 309.

French

733 S. Caldarelli, NMR, a Tool for the Molecular Characterization of Food Products, *Actual. Chim.*, 2012, **364–365**, 69.
734 F. Herman, I. Salliot-Maire and S. Sable, From Molecule to Patients: The NMR at the Heart of Pharmaceutical Research, *Actual. Chim.*, 2012, **364–365**, 42.

735 J.-Y. Le Gall, Methods of Analyzing the Genome and Its Expression: Medical Applications, *Bulletin de l'Academie Nationale de Medecine*, 2012, **196**, 151.

German

736 P. Adelhelm, M. Armbruster and F. Kraus, Solid State Chemistry 2012, *Nachr. Chem.*, 2013, **61**, 252.
737 P. Bayer and D. Grum, Contrast Agents for Magnetic Resonance Tomography. How Does the Contrast Work?, *GIT Labor-Fachz.*, 2013, **57**, 502.
738 R. Geyer, NMR Spectroscopy in Plant Breeding: Technology, Possibilities and Applications, *VDLUFA – Schriftenreihe*, 2012, **68**, (Nachhaltigkeitsindikatoren fuer die Landwirtschaft), 794.
739 S. Henkel and S. Beutel, Determination of pH Value in Biotechnology, *Chem. Ing. Tech.*, 2013, **85**, 872.
740 T. Madl, Integrative Structural Biology. Integration of NMR and SAXS/SANS in Structural Biology, *BIOspectrum*, 2013, **19**, 386.
741 D. Schleuter, S. Ueberlein and E. Brunner, Biominerals and Biomaterials: Investigated through Solid-State NMR Spectroscopy, *GIT Labor-Fachz.*, 2013, **57**, 297.
742 J. Skibsted, T. F. Sevelsted, S. L. Poulsen and T. T. Tran, Studies on Guest-Ion Incorporation in Portland Cement – Part 1, *ZKG Int.*, 2013, **66**, 66.
743 B. Stadler and G. Henkelmann, Fats and Oils in the Area of Bioenergy, *VDLUFA – Schriftenreihe*, 2012, **68**, (Nachhaltigkeitsindikatoren fuer die Landwirtschaft), 867.

Hungarian

744 M. Hollosi, K. Knapp, Z. Majer, G. Szilvagyi and E. Vass, Application of VCD Spectroscopy in the Structure Elucidation of Peptides, Peptidomimetics and Chiral Transition Metal Complexes, *Magy. Kem. Foly., Kem. Kozl.*, 2012, **118**, 145.
745 R. Szalay, M. Mortl, G. Pongor, Z. Kolos, J. Rohonczy and D. Knausz, Latest Achievements in the Chemistry of Silylated Acid Amide Derivatives with Respect to Current Research at the Eotvos Lorand University, Budapest, *Magy. Kem. Foly., Kem. Kozl.*, 2012, **118**, 168.
746 C. Szantay, Jr., Z. Szakacs, Z. Beni and Z. Dubrovay, The Role and Applicability of NMR Spectroscopy in Modern Pharmaceutical Industry, *Magy. Kem. Foly., Kem. Kozl.*, 2013, **119**, 16.

Italian

747 G. M. Nicosia, Analyses Using Nuclear Magnetic Resonance (NMR). Overview of the Technology, Prospects and Application in Refineries, *Rivista dei Combustibili e dell'Industria Chimica*, 2012, **66**, 11.

Japanese

748 K. Akasaka, Dynamism of Proteins Explored by High-Pressure NMR, *Koatsuryoku no Kagaku to Gijutsu*, 2013, **23**, 3.
749 I. Aoki, Molecular Imaging by MRI. From Functional Contrast Media to Theranostics, *Farumashia*, 2013, **49**, 671.
750 M. Arai, Quantitative Analysis of Protein-Ligand Interactions by NMR, *Seibutsu Butsuri*, 2013, **53**, 305.
751 A. Asada and M. Kitamura, Molecular Chain Structures and Magnetic Resonance of Natural Rubbers, *Kagaku to Kyoiku*, 2013, **61**, 476.

752 M. Fujioka, Review on Application of Nuclear Magnetic Resonance for the Field on Cement and Concrete, *Konkurito Kogaku*, 2012, **50**, 372.
753 Y. Goda, QNMR and Its Application to Reagents in the Japanese Pharmacopoeia, *Pharm Tech Jpn.*, 2012, **28**, 2795.
754 H. Hiroaki, NMR Methods for Observing Protein-Ligand Interaction, *Seikagaku*, 2013, **85**, 679.
755 M. Ishijima, M. Kubota, L. Ning, L. Liu, H. Kaneko, I. Futami, R. Sadatsuki, S. Haneda, A. Yusup, Y. Shimura *et al.*, The Frontiers of the Pathophysiology, Diagnosis, and Treatment of Knee Osteoarthritis, *Juntendo Iji Zasshi*, 2013, **59**, 138.
756 J. Kikuchi, Usefulness of Nuclear Magnetic Resonance in Pharmaceutical Developments and Research, *Yakuzaigaku*, 2013, **73**, 24.
757 R. Kitahara, S. Kitazawa and M. Urushitani, Pressure-Induced Misfolding of RNA Recognition Motif 1 of TDP-43, *Seibutsu Butsuri*, 2013, **53**, 321.
758 M. Kitamura, T. Hayano and K. Yoshizaki, Re-Design and Evaluation of the NMR Probes for Living Organs and Tissues, *Shikoku Igaku Zasshi*, 2013, **69**, 67.
759 Y. Kobashikawa, M. Yokogawa and F. Inagaki, Advanced Sample Production Method for Protein NMR Spectroscopy, *Seibutsu Butsuri*, 2013, **53**, 268.
760 Y. Kobayashi, T. Inokuma and Y. Takemoto, Development of Innovative Hydrogen-Bond-Donor Catalysts Based on Heterocyclic Scaffolds and Their Applications to Asymmetric Reactions, *Yuki Gosei Kagaku Kyokaishi*, 2013, **71**, 491.
761 H. Kotegawa, Superconducting Seen in NMR "NMR for Various Superconductors", *Bussei Kenkyu*, 2012, **97**, 898.
762 H. Kotegawa and T. C. Kobayashi, Indenter Type of Pressure Cell and Its Applications, *Koatsuryoku no Kagaku to Gijutsu*, 2012, **22**, 214.
763 K. Moribe, K. Higashi and K. Yamamoto, Physicochemical Characterization of Nano Formulation, *Bunseki*, 2012, **3**, 135.
764 T. Murakami and N. Fukutsu, Structure Elucidation of Impurities in Pharmaceuticals by HPLC Hyphenated Techniques, *Chromatography*, 2012, **33**, 179.
765 S. Nagamoto, Y. Yamamura and K. Saito, Cyclic Oligomers via Hydrogen Bonds in Organic Liquids, *Netsu Sokutei*, 2013, **40**, 107.
766 Y. Nakamura, K. Yamamura and H. Fujii, Application of Pulse NMR for Adhesives, *Setchaku no Gijutsu*, 2012, **32**, 9.
767 K.-i. Nakashima, M. Oyama and M. Iinuma, Chemical Constituents of Melicope Species (Rutaceae), *Gifu Yakka Daigaku Kiyo*, 2013, **62**, 38.
768 M. Osawa, S. Imai, K. Takeuchi and I. Shimada, Solution NMR Study on Functional Mechanism of Membrane Proteins, *Seibutsu Butsuri*, 2013, **53**, 236.
769 R. Ota, R. Miyoshi and K. Mitsuhashi, Examination of Reaction Progress for the Light Curing Resin, *Toso Kogaku*, 2012, **47**, 331.
770 H. Saji, Molecular Imaging in Drug Development Study, *Farumashia*, 2013, **49**, 639.
771 S. Sakurai, Red-Time Monitoring of Chemical Reactions by the NMR Based on Microtechnology, *Kagaku to Kyoiku*, 2012, **60**, 204.
772 Y. Suzuki, NMR Study of Bombyx Mori Silk Fibroin for Understanding the Fibrillation Mechanism, *Sen'i Gakkaishi*, 2012, **68**, P382.
773 S.-i. Takada, Characterization of the Stability of Highly Concentrated Particle Dispersions, *Funtai Kogaku Kaishi*, 2012, **49**, 448.
774 H. Takahashi, Investigation of Active Sites in Ligand Molecules by NMR: Epitope Mapping by DIRECTION Method, *Kagaku*, 2013, **68**, 39.
775 H. Takamura, I. Kadota and D. Uemura, Synthetic Study of the Polyol Natural Product Symbiodinolide toward the Structural Elucidation, *Yuki Gosei Kagaku Kyokaishi*, 2014, **72**, 2.

776 S.-i. Tate, Dynamical Aspects of Protein Structures Revealed by Newly Established NMR Approaches, *Seikagaku*, 2013, **85**, 638.
777 T. Ueda, Dynamic Behavior of Guest Molecules Confined by Nanospace as Studied by NMR, *Hyomen*, 2011, **49**, 245.
778 T. Yamaoka, In vivo Tracking and Viability Measurement of Transplanted Stem Cells Using Novel Polymeric MRI Contrast Agent, *Drug Delivery System*, 2013, **28**, 24.
779 K. Yoshida, Development of High-Pressure High-Temperature NMR Probe and Dynamics, Structure and Reactions in Supercritical Water, *Koatsuryoku no Kagaku to Gijutsu*, 2012, **22**, 153.
780 H. Yoshimizu, Application of NMR Spectroscopy to the Characterization of Gas Separation Membranes, *Sen'i Gakkaishi*, 2012, **68**, P323.

Korean

781 J. W. Min and W. B. Im, Phosphor for LED: Structural Analysis-Technology and Research Trend, *Seramisutu*, 2013, **16**, 67.

Polish

782 E. Blonska-Sikora, J. Oszczudlowski, Z. Witkiewicz and D. Widel, Glutathione: Methods of Sample Preparation for Chromatography and Capillary Electrophoresis, *Chemik*, 2012, **66**, 929.
783 A. Jakubowska, Ion Biding at Interfaces and Specific Ion Effects, *Wiad. Chem.*, 2012, **66**, 193.
784 D. Krasowska, 2,2′-Dihydroxy-1,1′-Binaphtyl (BINOL) and Its Derivatives. Selected Synthesis Methods and Applications, *Wiad. Chem.*, 2013, **67**, 55.
785 P. Mlynarz, W. Barg, S. Deja and R. Jankowska, Applications of Metabolomic in COPD Diagnosing, *Pol. Merkuriusz Lek.*, 2012, **33**, 207.
786 B. Pacholczyk, A. Fabianska, R. Kusinska, P. Potemski, R. Kordek and S. Jankowski, Analysis of Cancer Tissues by Means of Spectroscopic Methods, *Wspolczesna Onkol.*, 2012, **16**, 290.

Russian

787 S. V. Dushina, V. A. Sharnin and V. V. Aleksandriiskii, New Possibilities of Solvation-Thermodynamic Approach to Describe the Role of Solvent in Complexation Reactions, *Izv. Vyssh. Uchebn. Zaved., Khim. Khim. Tekhnol.*, 2012, **55**, 3.
788 V. M. Gruznov, Methods of Direct Detection of Explosives, *Gorenie Plazmokhim.*, 2012, **10**, 69.
789 I. G. Meerovich, M. V. Gulyaev, G. A. Meerovich, M. S. Belov, V. M. Derkacheva, O. V. Dolotova, V. B. Loshchenov, A. Y. Baryshnikov and Y. A. Pirogov, Investigation of Contrast Agents for MRI Based on Phthalocyanine Derivatives, *Ross. Khim. Zh.*, 2013, **57**, 110.

Slovak

790 P. Patlevic, F. Dorko, P. Svorc, Jr., J. Vaskova and L. Vasko, Counterfeit Drugs – How to Reveal Them?, *Chem. Listy*, 2013, **107**, 37.

Theoretical and physical aspects of nuclear shielding

Cynthia J. Jameson*[a] and Angel C. De Dios[b]
DOI: 10.1039/9781782622758-00046

This chapter is a review of the literature published between June 2013 and May 2014 on physical and theoretical aspects of nuclear magnetic shielding. Theoretical aspects include (a) general theory, for example, newly developed approaches in relativistic theory of nuclear shielding, the relation between the spin-rotation tensor and shielding in relativistic theory, NMR shielding for electronically degenerate states and benchmark studies (b) *ab initio* and DFT calculations, both relativistic and non-relativistic. Physical aspects include (a) anisotropy of the shielding tensor, (b) shielding surfaces and rovibrational averaging, (c) isotope shifts, (d) intermolecular effects on nuclear shielding, and (e) absolute shielding scales.

1 Theoretical aspects of nuclear shielding

1.1 General theory

Aucar shows that by using the path integral formalism that was developed within quantum field theory, (as opposed to the wavefunction-based formalism that is used in quantum chemistry) polarization propagators have the same formal definition within both the relativistic and non-relativistic domains, providing a unified view and new insights on the relationship between spin and time-reversal operators.[1] In this article he explains why the non-relativistic limit is obtained by scaling the velocity of light to infinity, and how within the relativistic regime the negative energy states contribute to electron correlation, and how to consider both quantum electrodynamics (QED) and electron correlation effects on the same ground. Although QED effects are not yet implemented in calculations of NMR quantities, it had been suggested by Pyykkö and Zhao that the QED effects on NMR parameters in molecules containing heavy atoms could be of the same size as solvent effects.[2] With the generalization of polarization propagators to the relativistic framework[3] diamagnetic and paramagnetic contributions which arise from completely different electronic mechanisms within the non-relativistic regime have become unified. They are produced by one and the same mechanism which is still not completely understood in our non-relativistic and perturbative way of thinking. In a proper relativistic framework the spin is no longer a good quantum number, so spin–orbit effects cannot be used to explain heavy atom effects in a molecule within the full relativistic formalism. Despite this, we likely will continue to

[a]*Department of Chemistry, University of Illinois at Chicago, 845 W. Taylor St., Chicago, Illinois 60607-7061, USA. E-mail: cjjames@uic.edu*
[b]*Department of Chemistry, Georgetown University, 37th and O Streets, NW, Washington, D. C. 20057-2222, USA. E-mail: dediosa@georgetown.edu*

think about and explain NMR parameters in terms of spin–orbit contributions, diamagnetic and paramagnetic terms, and other familiar crutches.

In 1950, Ramsey developed the general non-relativistic theory of the nuclear magnetic shielding tensor, and in the same paper showed that for linear molecules, the paramagnetic term (for the gauge origin taken at the center of mass of the molecule) is directly related to the experimentally measurable spin-rotational magnetic interaction constant of the molecule.[4] In 1964 Flygare derived equations relating the spin-rotation constants and nuclear magnetic shielding for any type of molecule and showed how the identity could be useful in obtaining shielding values from the spin-rotation constants that arise from high-resolution microwave spectroscopic measurements in the gas phase.[5,6] Experimental values of ^{19}F spin rotation tensors for many molecules have been reported, thus providing multiple anchor points for the ^{19}F absolute shielding scale. Agreement was excellent (within experimental error bars of the spin-rotation tensors) between the absolute shieldings individually derived from the spin-rotation tensors and those absolute shielding values that arise from using one ^{19}F nuclear site (^{19}F in HF molecule, for example) as a reference to convert the simultaneous measurements of ^{19}F chemical shifts in a large number of molecules in the zero-density limit relative to ^{19}F in SiF_4 (for example).[7–9] Absolute shielding scales for other nuclei have likewise been established by combining the spin-rotation-derived paramagnetic shielding contribution with theoretically calculated diamagnetic shielding, and refinements are reported annually in Section 2.5 of this chapter. For molecules containing only light nuclei, where non-relativistic theory is expected to be a reasonable description, these absolute shieldings, σ_0(300 K), provide stringent tests for validating theoretical methods of calculating nuclear magnetic shielding and shielding surfaces for rovibrational corrections and thermal averaging.[10] For shielding of heavy nuclei or for shielding of even light nuclei in molecules containing heavy atoms, using the experimental (relativistic) spin-rotation and relativistic diamagnetic shielding (in those formulations where diamagnetic shielding can be explicitly separated out) does not fix the problem because the identity relation between the paramagnetic part of the shielding and the spin-rotation derived in non-relativistic theory no longer holds in relativistic theory. What is needed, and recently has become available, is a general relativistic treatment for both shielding and spin-rotation tensors which will provide a direct mapping between the two tensors. Aucar *et al.* established a theoretical expression for the spin-rotation tensor in the case of relativistic electrons while treating nuclear motion non-relativistically.[11] The authors applied this theory to HX molecules (X = H, F, Cl, Br, I),[12] established a comparison of the relativistic effects on the nuclear magnetic shielding and the spin rotation tensors in these molecules, and examined the validity of Flygare's relation between these two properties for these molecular systems. Flygare's identity relation derived from non-relativistic theory is anticipated to fail for HX with heavy X; the authors find that for H in this series of molecules, the relation is an approximation for molecules containing atoms of the 4th row or heavier, and for X = Br and I the Flygare relation does not hold at all because operators that describe relativistic effects on the magnetic interaction with B_0 in the shielding tensor have no counterpart in the spin-rotation tensor and the spin contribution to the spin–orbit effect is different in the two properties.[12] In addition, Aucar *et al.* also considered small corrections (Breit interaction effects) in their nuclear spin rotation theory and applied these to the HX series,[13] and Malkin *et al.* applied the Aucar theory to Sn spin-rotation and nuclear magnetic shielding in SnH_4, $Sn(Me)_4$ and $SnCl_4$.[14] and to spin-rotation and nuclear magnetic shielding in the HCl molecule.[15]

With the same objectives, Xiao and Liu[16] developed a relativistic molecular Hamiltonian that describes electrons relativistically and nuclei quasi-relativistically. After transforming into the body-fixed frame of reference, the body-fixed relativistic Hamiltonian is used to formulate, among other molecular electronic property tensors, a relativistic theory of the spin-rotation tensor for semi-rigid non-linear molecules, which includes Aucar's formulation[11] as a special case. Using the formulation by Xiao and Liu, a formal relationship between the two tensors can be written if the shielding tensor is formulated through the external field-dependent unitary transformation (EFUT) ansatz. This treatment is also applied to linear molecules.[17] For linear molecules, there is no molecular rotation about the line of centers and no paramagnetic term in the component of shielding along this direction in non-relativistic theory, so the paramagnetic component of the shielding along this direction has to be calculated by four-component relativistic theory. A relativistic mapping between nuclear magnetic shielding and spin-rotation tensors is proposed as follows:[16] The previous non-relativistic identity relation between paramagnetic shielding tensor vu component at equilibrium molecular geometry for the Kth nucleus and the spin-rotation uv tensor component, where I^0_{vv} is the principal inertia tensor, g_K is the g factor for the Kth nucleus, and μ_n is the nuclear magneton, is as follows:

$$\sigma^{K,p}_{vu,\mathrm{eq}}(\text{"expt"}) = -\frac{I^0_{vv}}{2g_K\mu_\mathrm{n}} M^{K,p}_{uv,\mathrm{eq}}(\mathrm{expt}) \tag{1}$$

The relativistic relation between these tensors is designated by Xiao and Liu as follows:

$$\sigma^{K,p}_{vu,\mathrm{eq}} = -\frac{I^0_{vv}}{2g_K\mu_\mathrm{n}} M^{K,p}_{uv,\mathrm{eq}} + \Delta^{K,p}_{vu,\mathrm{eq}} \tag{2}$$

where the correction, *i.e.*, the difference between the electronic paramagnetic contributions to the nuclear shielding and the spin-rotation tensors, which they designate as the 'relativistic effect in the paramagnetic mapping', $\Delta^{K,p}_{vu,\mathrm{eq}}$, can be directly evaluated in the EFUT approach and formally written in sum over states form.[16] Therefore, Xiao, Zhang and Liu enumerate three relativistic corrections to the absolute shielding:[18] (a) the relativistic effect in the paramagnetic mapping as described above, (b) the relativistic effect in the nuclear shielding diamagnetism, and (c) the relativistic effect in the spin-rotation diamagnetism.

In our earlier attempts to include relativistic corrections in the absolute shielding (for example, for Se and Te in ref. 19), by using the experimental value of the spin-rotation tensor, we included (c) the relativistic effect in the spin-rotation diamagnetism, and by using a relativistic value for the shielding of the free (Se or Te) atom,[20] we included (b) the relativistic effect in the nuclear shielding diamagnetism approximately, but we failed to include (a) the relativistic effect in the paramagnetic mapping. Unfortunately, it has been recently shown by Xiao and Liu that the latter correction predominates over the other two.[18] Indeed, recent 4-component calculations of the spin-rotation tensor and the absolute isotropic shielding for Se and Te in SeF_6 and TeF_6 by Ruud *et al.*[21] show that this is also the case for these systems.

Xiao, Zhang and Liu also find (using four-component DFT with either LDA or PP86 functionals) that the relativistic effect in the paramagnetic mapping leads to a correction of 4.5 ppm for the absolute shielding of ^{19}F in HF.[18] If this is correct, the ^{19}F absolute shieldings would all be shifted by this much. This may not be the final word, however, because Xiao *et al.* have used DFT in their calculations. Teale *et al.* have previously demonstrated the inaccuracy of DFT in a systematic non-relativistic study, in which DFT calculations (using LDA and 13 other functionals) lead to mean absolute errors in ^{19}F spin-rotation constants which are 10–30 times as large as CCSD(T) in comparison with experiments and similarly deficient compared to CCSD(T) in shielding calculations.[10]

New two-component relativistic formalisms have been proposed. Yoshizawa and Hada developed a new formalism for Foldy–Wouthuysen transformation including a vector potential which can be used to introduce restricted magnetic balance (RMB) in the Douglas–Kroll–Hess (DKH) method.[22] They also developed a way of using GIAOs in the DKH method by separating the GIAO function into two functions. First applications of this new DKH2 formulation to shielding in noble gas atoms Ne to Xe, HX molecules (X = F to I), and H_2X molecules (X = O to Te) appear to provide results that are comparable to 4-component calculations by Manninen and Vaara.[23] An efficient method of treating scalar relativistic effects on shielding based on the spin-free exact-two-component theory (X2C) in its one-electron variant has been proposed by Cheng, Gauss and Stanton.[24] Increased efficiency in comparison to 4-component methods arises from a focus on spin-free contributions and from elimination of the small component. They start from 4-component theory, by separating the spin-free and spin-dependent terms for the matrix representation of the Dirac equation in terms of a RMB GIAO basis. The resulting spin-free 4-component matrix elements required for calculating nuclear magnetic shielding are then used to construct the Hamiltonian for the corresponding spin-free exact-two-component theory in its one-electron variant and its perturbed counterpart using analytic derivative theory. They suggest that for calculations of (a) heavy atom shielding in cases where both spin-free relativistic and spin–orbit effects are generally important, or for calculations of (b) light atom shieldings in cases where the bonding between the light and heavy elements is dominated by p-type orbitals of the light element, the spin-dependent terms are small and a pure spin-free relativistic treatment turns out to be useful; scalar-relativistic effects can be obtained rigorously at a highly correlated level, CCSD(T) for example, while the computationally expensive spin–orbit contributions may be treated in a perturbative manner or *via* additivity schemes. With this new formalism, they carry out coupled-cluster calculations for ^{129}Xe in xenon fluorides and ^{17}O shielding in $[MO_4]^{2-}$ complexes (M = Cr, Mo and W), but this new method is not applicable for systems like ^{1}H shielding in HX (X = F, Cl, Br, I) where spin–orbit contributions dominate. The experimental ^{17}O chemical shifts also include deshielding medium effects, of course, which are not included in the theoretical calculations. When converted to absolute shielding, all experimental values are deshielded compared to CCSD(T) relativistic values. In these systems, electron correlation effects are very large, especially for $[CrO_4]^{2-}$ and are greater than relativistic effects. For ^{129}Xe in XeF_2, XeF_4 and octahedral XeF_6, The calculated isotropic chemical shifts relative to free Xe atom compare favorably with the gas phase data for these molecules. Unfortunately the authors did not report their calculated $\delta_{\parallel}$ and $\delta_{\perp}$. It would have been interesting to see to what extent $\delta_{\parallel}$ differed from zero and to what extent $(\delta_{\parallel} - \delta_{\perp})$ differed from $(3/2)\delta_{iso}$, in the theoretical results for this case. The individual components $\delta_{\parallel}$ and $\delta_{\perp}$ had been measured for ^{129}Xe in XeF_2 by Wasylishen *et al.* in the solid state relative to free Xe atom,[25] although there are intermolecular effects in the experiment that could be of the order of 300 ppm (deshielded).

Autschbach reports on a previously neglected term from the response of the exchange–correlation (XC) potential which has recently been included in the relativistic NMR module of the ADF package.[26] The XC response markedly improves calculated proton chemical shifts for hydrogen halides. Mercury chemical shifts for mercury dihalides are also noticeably altered. The term vanishes in the absence of spin–orbit coupling. The new results of fully relativistic calculations are compared with ZORA. While absolute shielding values for Hg are not accurately predicted with ZORA, the ZORA chemical shifts agree well with those from fully relativistic calculations. Autschbach provides a review of relativistic calculations of magnetic resonance parameters, with selected applications of relativistic-DFT.[27]

For the theoretical calculations of NMR chemical shifts for paramagnetic systems, one needs to carry out calculations of the nuclear magnetic shielding tensor, the hyperfine coupling tensor, and sometimes also the electronic g tensor. In the absence of zero-field splitting (ZFS) and when the $2S + 1$ degenerate ground state is well separated from excited energy levels, and the electronic g tensor is isotropic and equal to the pure electron-spin value $g_e = 2.0023\ldots$, then the NMR chemical shift in paramagnetic systems can be obtained as a function of temperature with the high-temperature limit expression. In that special case, the isotropic chemical shift only depends on the isotropic part of the hyperfine coupling tensor (Fermi contact mechanism only) and the nuclear magnetic shielding. In the general case of an anisotropic g tensor and a non-vanishing zero field splitting, the theoretical prediction of the temperature-dependent NMR chemical shift for paramagnetic systems requires accurate calculations of the nuclear magnetic shielding tensor, the *g* tensor, and the anisotropic hyperfine tensor (including the traceless spin-dipolar part). Recently Soncini *et al.* have formulated the theoretical treatment of the NMR chemical shift in an electronic state of arbitrary degeneracy,[28–30] for an open-shell molecule in a pure spin state, subject to a zero-field splitting. Four-component matrix Dirac–Kohn–Sham calculations of these quantities have been carried out for ^{1}H in a system with no zero-field splitting by Malkin *et al.*[31] In the 4-component relativistic domain, the *g* tensor and the hyperfine coupling tensor are first order properties. They found that differences between the present 4-component and earlier ZORA calculations on the same system are not large except in 2 out of 9 sites, but both calculations have large deviations from experiment. An interesting and important point was made by Malkin *et al.*: The extraction of the spin density at the position of a particular nucleus is problematic since in 2- or 4-component relativistic theory, the Fermi contact contribution to the hyperfine coupling tensor is not defined solely by the spin density. This means that the development of more accurate theoretical methods for paramagnetic NMR has to be accompanied by development of new interpretation tools and more advanced models for interpretation of the experimental data.

Chirality is an important aspect of chemistry and life. Unlike electronic, infrared and Raman spectroscopy, NMR is said to be blind to chirality, that is, the parameters determining an NMR spectrum (nuclear shielding, spin–spin coupling, nuclear quadrupole coupling, are identical for a molecule and its mirror image. Distinguishing enantiomers by NMR requires placing the molecule in a chiral environment, *e.g.*, a chiral solvent, a chiral cage, a helical bundle, *etc.* Various suggestions have been made about how, with the introduction of an external electric field into the NMR experiment, chirally sensitive polarizability (the nuclear magnetic shielding polarizability) could discriminate the shielding of one enantiomer from another in an isotropic medium.[32,33] The typical values of these shielding polarizabilities (first and second derivatives of the shielding tensor with respect to an external electric field) have been found to be disappointingly small for the observation of this effect.[34–36] Buckingham has recently proposed an additional mechanism.[37] Chiral molecules possessing a permanent electric dipole moment exhibit a temperature-dependent orientational contribution to the electric polarization in addition to that arising from shielding polarizability. This arises from the torque exerted on the molecule, through the antisymmetric part of the nuclear magnetic shielding tensor of nucleus N, by the magnetic field $\boldsymbol{B}_0$ and the nuclear moment, $\boldsymbol{m}$(N) the orthogonal precessing magnetic moment of this nucleus following a $\pi/2$ pulse. Buckingham has derived the equations for this electric polarization in terms of the antisymmetric parts of the nuclear shielding tensor and estimates that this temperature-dependent contribution to the electric polarization can be 1000 times larger than that due to the shielding polarizability; it is dependent on the polar molecule being able to follow the precessing nuclear magnetization. It may be large enough an effect to make it possible to record a positive or negative signal for every line in the NMR spectrum; enantiomers will give equal but opposite signals, just as in electronic circular dichroism spectroscopy. In principle, the effect is quite general since all chiral molecules except those that belong to D_n symmetry group will have a non-vanishing electric dipole moment and all chiral molecules have a non-vanishing antisymmetric component. The electric polarization varies as $\boldsymbol{B}_0^2$ and is proportional to the molecule's permanent electric dipole moment.

Parity-violation (PV) shifts in properties of chiral molecules are expected to be very small and, so far, not observed experimentally. Robert and Barra have suggested that experimental detection of NMR frequency differences between left- and right-handed enantiomers is a feasible route towards a first experimental observation of parity violation in chiral molecules, a tiny (mHz) but fundamental nuclear-spin dependent effect.[38–41] A recent contribution towards this goal is a ZORA-DFT calculation of the isotropic ^{183}W shielding in a set of likely candidates: chiral W compounds such as NWBrClF and NWBrFI.[42] An important finding is that the PV NMR frequency splittings are very sensitive to electronegativity of the ligands (a strongly heterogeneous electronic environment leads to a strong chiral field) showing an increase by three orders of magnitude from NWBrClI to NWHFI.

1.2 *Ab initio* and DFT calculations

Relativistic effects on shielding are expected to be significant when the observed nucleus itself is that of a heavy atom. In this reporting period, particular attention has been cast on relativistic calculations of the shielding of ^{205}Pb in PbTe,[43] ^{205}Pb in plumbacyclopentadienylidenes,[44] of ^{199}Hg in $Hg(SCN)_2$, $Hg(CN)_2$, $Hg(SeCN)_2$, $Hg(CO_2CH_3)_2$, HgF_2, $HgCl_2$, $HgBr_2$, HgI_2, Hg_2Cl_2, $K[Hg(SeCN)_3]$, $Hg_2(NO_3)_2 \cdot 2H_2O$,[45] of ^{183}W in WF_6,[21] of ^{129}Xe in Xe atom,[22] ^{129}Xe in XeF_2, XeF_4, XeF_6,[22,24] of ^{125}Te in PbTe,[43] in H_2Te,[22] and in TeF_6,[21] of ^{119}Sn in SnH_4, $Sn(Me)_4$,[14] of ^{95}Mo in MoF_6,[21] of $^{79/81}$Br in organic bromo compounds,[46] of ^{77}Se in SeF_6,[21] and ^{77}Se in phosphine selenides,[47] of ^{51}V in divanadium-substituted polyoxometalates,[48] of ^{33}S in SF_6,[21] and ^{33}S in H_2S[22] and of the halogen nuclei in HX (X = Cl, Br, I).[12,15,22]

Relativistic effects on shielding arising from heavy atom neighbors are empirically classified into (1) HALA heavy atom effect on magnetic shielding of the neighboring light atom (heavy atom–light atom effect) and (2) HAHA (heavy atom–heavy atom) effect. Examples of compounds where HALA effects are significant are halogen-containing compounds where shieldings of ^{1}H or ^{13}C having neighboring Cl, Br, or I atoms are found to be atypical, or coordination compounds of transition metals with organic ligands where ^{13}C, ^{15}N or ^{17}O shieldings in the complex are significantly different from the free ligand. HAHA effects are inherent to a much broader series of compounds containing two or more heavy atoms, and the contribution of spin–orbit interaction for each heavy atom should be taken into account in addition to the scalar relativistic effects. In this reporting period, relativistic calculations have been carried out that provide examples of the HALA effect for ^{1}H in in HCl,[15] ^{1}H in HCl, HBr, HI,[12,24,26] ^{1}H in H_2S, H_2Se, H_2Te,[12] for ^{11}B in monoiodo derivatives of closo-carboranes[49] for ^{13}C in 6 organic compounds with bromine and iodine substituents,[50] ^{13}C in 3,6-dihalogenated-*N*-alkyl carbazoles,[51] ^{13}C in carbene complexes of Pd and Zr,[52] for ^{15}N in an irridium complex,[53] for ^{17}O in $[MO_4]^{2-}$ (M = Cr, Mo, W),[24] and in polyoxometalates such as $[M_6O_{19}]^{n-}$, $[MW_5O_{19}]^{n-}$, $[M_2W_4O_{19}]^{n-}$, $[(MeO)MW_5O_{18}]^{n-}$, $[(\mu\text{-O})(TiW_5O_{18})_2]^{6-}$, $[\{(\mu\text{-HO})ZrW_5O_{18}\}_2]^{6-}$,[54] for ^{13}C and ^{31}P in organometallic molecular squares $[Au\{Co(P_2C_2tBu_2)_2\}]_4$.[55] Relativistic calculations have been carried out that provide examples of the HAHA effect for ^{119}Sn and ^{35}Cl in $SnCl_4$[14] and for ^{31}P and ^{77}Se in phosphine selenides.[47] The calculations mentioned above in the present section were carried out including spin–orbit and scalar relativistic effects at the DFT zeroth-order regular approximation level of theory (SO-ZORA), with the following exceptions: 4-component calculations were used in ref. 12, 14, 21 and 50, the spin-free exact 2-component approach of Cheng, Gauss and Stanton was used in ref. 24, and the two-component Douglas–Kroll–Hess method was used in ref. 22 and 48.

Non-relativistic calculations of nuclear shielding in single molecules in this reporting period are primarily DFT calculations using the popular B3LYP functional. In some cases a variety of other functionals were tried, in addition. In a study of ^{1}H ^{13}C and ^{19}F shielding in 1,1-difluoro and *cis*- and *trans*-1,2-difluoroethylenes, for which gas phase data are available, DFT calculations were carried out to optimize the molecular geometry (using BLYP and B3LYP functionals), to obtain the harmonic vibrational frequencies (B3LYP and BLYP) and the anharmonic frequencies (B3LYP), and to obtain the shieldings of all nuclei at the equilibrium geometry (using the BHandH and the VSXC functionals).[56] The results were compared to isotropic shieldings at the HF, MP2, CCSD and CCSD(T) levels. The DFT-VSXC predicted carbon isotropic shieldings in difluoroethylenes are better than DFT-BH and H but the reverse situation was observed for fluorine shieldings. When rovibrational corrections from the literature, which had markedly improved the agreement of CCSD(T) calculations with experiment, were applied to the DFT results, agreement with experiment became much worse.

DFT calculations (8 different functionals were tried) for the calculations of the $^{47/49}Ti$ chemical shifts in half-metalocene complexes of Ti(IV): $[CpTiCl_3]$, $[Cp^*TiCl_3]$ $[Cp_2TiCl_2]$, $[Cp^*{}_2TiCl_2]$ and 10 derivatives (Cp = cyclopentadiene, Cp* = pentamethyl cyclopentadiene) were overestimated in magnitude to varying degrees in all of the calculations.[57] Nevertheless, the experimentally observed trend of Ti deshielding when replacing Cp by Cp*, or when substituting Cp/Cp* by Cl, is reproduced correctly in all cases. ZORA calculations were used to show that relativistic effects are minor. The chemical shifts change in magnitude by less than 9 ppm for Cp/Cp* complexes upon inclusion of scalar relativistic effects. In another study, DFT-B3LYP calculations of ^{77}Se shielding, using Slater-type basis sets of quadruple zeta with four polarization functions, led to an excellent linear correlation between the experimentally observed ^{77}Se chemical shifts in aromatic selenoic, selenothioic, and diselenoic acid salts and the calculated shielding differences for 9 Se sites and the reference Me_2Se over a range of 1700 ppm.[58] Neglected medium effects and relativistic effects may have been small (or cancelling in taking shielding differences for the comparison). The ^{103}Rh shielding in 7 complexes were calculated using DFT-B3LYP; Rh was described by means of a relativistically-adjusted effective core potential for the inner electrons and its associated double-ζ basis set for the outer ones, complemented with a set of f-polarization functions. Geometry was optimized for the isolated molecule using DFT with PBE0 and M06 functionals. Rh–P bond distances emerge as the decisive parameters governing the ^{103}Rh chemical shifts.[59]

Intramolecular donor acceptor complexes have been well characterized by the shieldings of donor and acceptor nuclei in Si←N.[60] The ^{29}Si and ^{15}N shieldings in silatranes were calculated by *ab initio* (CCSD, MP2) and DFT (B3PW91, PBE0, M06-2X) methods. The shieldings serve as a measure of strength of the dative Si←N bond; the ^{15}N chemical shifts of the donor nitrogen show a linear relation to the Si–N distance. Similarly, the experimental P–Sn distances are conspicuously short and shorter than the sum of van der Waals radii, suggesting Sn←P dative bonds[61] in four related sterically restricted peri-substituted acenaphthenes [Acenap(SnR_3)(P^iPr_2)] containing mixed tin–phosphorus moieties in the proximal 5,6-positions. Geometry optimization used DFT-B3LYP with a polarizable continuum model (PCM) for the solution; the optimized Sn–Cl bond lengths are in good agreement with those observed experimentally in the solid. Natural bond orbital (NBO) analysis suggests weaker Sn–Cl bonding as a result of the interaction of the P lone pair with the Sn center to form the Sn←P dative bond. The ^{31}P and ^{119}Sn solid-state NMR spectra were obtained for the compounds but no shielding calculations were carried out in these fairly large systems.

The ^{17}O and ^{33}S nuclear shielding of the SO_3 isolated molecule as well as possible dimer and trimer structures were calculated using DFT with B3LYP, B971, PBE0, CAM-B3LYP, and HCTH functionals, to discover the nature of the molecular species that may be present in liquid sulfur trioxide.[62] The experimental ^{17}O spectrum for the liquid was more complicated than expected and temperature-dependent. Such a temperature dependence suggests that the liquid SO_3 can be described as a mixture of 'free' SO_3 molecules and some unknown mixture of $(SO_3)_n$ aggregates, where n can be 2, 3 or even higher. The authors made a systematic computational study of $(SO_3)_n$, $n = 1$, 2, 3 species. It was concluded that the observed ^{17}O NMR signals in the liquid originate mostly from the SO_3 monomer and a cyclic $(SO_3)_3$ trimer, in which the out-of-the-ring oxygen atoms contribute to an averaged signal observed as a single peak. The presence of significant $(SO_3)_2$ dimers in the liquid was considered unlikely. The ^{15}N shielding in 5 azines: pyridine, pyridazine, pyrimidine, pyrazine and 1,3,5-triazine have been calculated using DFT.[63] The best results were obtained with the use of the Keal–Tozer KT2 functional which provided fastest convergence to the complete basis set limit. Two ways of accounting for solvation effects on shielding using a polarizable continuum model (PCM) model or a supermolecule calculation with 1 to

$4H_2O$ were compared. Smallest absolute errors in comparison with experiment could be found using supermolecule calculations with $4H_2O$ molecules in the solvation shell.

When the spin state of a paramagnetic system is not unequivocally known from some independent measurement, can computational methods provide a ranking of the energy of all spin states in the correct order and then predict the NMR parameters accordingly? It has been suggested by Bagno *et al.* that DFT-B3LYP calculations are capable of ranking the energy of the spin states of Fe(II,III) complexes with good accuracy at a moderate computational cost, and that this method also yields calculated paramagnetic NMR spectra for the protons for each spin state, which are sufficiently different to allow for a reliable identification.[64]

2 Physical aspects of nuclear shielding

2.1 Anisotropy of the shielding tensor

Theoretical and experimental methods of studying NMR shielding tensors continue to be applied to systems that seem unyielding to existing approaches. For instance, precise measurements of proton shielding tensors in the solid state are often daunting because of the abundance of proton sites and the resulting lack of resolution. Recently, hydrogen shielding tensors for hydrides in transition metal complexes have been measured.[65] By replacing with deuterium the protons of the other ligands in these complexes, the often buried resonance from the hydride can now easily be recorded. Tensor spans of about 80 to 110 ppm have been observed for the highly shielded (about 50 ppm more shielded than the protons in tetramethylsilane) hydride in complexes containing either Ir or Rh. Relativistic computations suggest that spin–orbit contributions are significant in these systems, and are in fact partly the reason why hydrides in these systems are substantially more shielded than other lighter metal hydrides. Another way by which shielding tensors at specific hydrogen sites can be measured is by making use of a multi-dimension experiment that utilizes symmetry-based pulse sequences. This allows for resonances from protons attached to either ^{13}C or ^{15}N to be simultaneously observed and resolved. This technique has been demonstrated to be feasible in extracting the proton shielding tensors in the amino acid histidine.[66] Lastly, with fast magic angle spinning, a two-dimensional solid state experiment that makes use of a recoupling sequence to measure shielding anisotropy in the indirect dimension is now available. This new method has been applied to the study of proton shielding tensors of hydrogen bonded –OH sites in tyrosine-HCl and citric acid.[67] It has also been reported that the C_{70} fullerene cage causes an anisotropy of about 10 ppm on the ^1H shielding tensor of an H_2 molecule encapsulated inside the cage.[68]

Another area of challenge in the study of NMR shielding tensors concerns quadrupolar nuclei. ^{35}Cl solid state NMR experiments are especially demanding because of the substantial quadrupolar interactions present when the ^{35}Cl nucleus resides in a site far from being spherically symmetric. Unfortunately, it is the lack of symmetry that can result in an observable chemical shift anisotropy. Nevertheless, with high field magnets and piecewise acquisition of wideline ^{35}Cl solid state NMR spectra, several ^{35}Cl NMR nuclear shielding tensors have been recently obtained.[69] The observed spans in systems, in which Cl acts as a ligand bridging either Ni, Rh or Pd atoms, range from 450 to 750 ppm. ^{35}Cl, ^{14}N and ^{195}Pt solid state NMR experiments have likewise been applied to the study of both pink and green forms of Magnus salt, $[(Pt(NH_3)_4][PtCl_4]$.[70] While the ^{195}Pt shielding tensors in the pink complex are similar to those of the starting materials, K_2PtCl_4 and $[Pt(NH_3)_4]Cl_2$, the corresponding tensors for the green complex are significantly different. The Pt sites in the green complex happen to be more deshielded and have larger tensor spans. By examining the crystal structures and performing *ab initio* shielding calculations, this difference has now been attributed to the presence of Pt–Pt interactions in the green complex. The green complex structure is characterized by an alternate stacking of square planar Pt complexes while the pink complex structure is now believed to be lacking such a columnar arrangement. ^{93}Nb, ^{35}Cl and ^{81}Br shielding tensors have been determined for the following niobium halide cluster compounds: $[Nb_6Br_{12}(H_2O)_6][HgBr_4]$ $12H_2O$ and $[Nb_6Cl_{12}(OH)_2(H_2O)_4]$ $4H_2O$.[71] The ^{93}Nb shielding tensors in both compounds are nearly axially symmetric with tensor spans ranging from 600 to 1200 ppm. In vanadium(V) complexes, ^{51}V isotropic shielding and shielding tensor anisotropy are both found to be quite sensitive to the presence of ancillary ligands indicating that these ligands change appreciably the electronic environment around the transition metal.[72] A combination of solid state ^{115}In NMR measurements and theoretical studies has been performed on indium(I) compounds.[73] Both the principal components and the orientation of the ^{115}In shielding tensors in these systems are demonstrated to be sensitive to molecular structure and symmetry. In this narrow range of compounds, agreement between experimental and nonrelativistic calculated isotropic shielding values is reached. On the other hand, the calculated tensor spans differ substantially from experiment. For instance, in the case of InI, the calculated span is 334 ppm, which is three times bigger than the experimental value of 110 ppm. ^{89}Y solid state NMR experiments have been used to characterize a series of the silicides YT_2Si_2 (where T = Co, Ni, Cu, Ru, Rh, or Pd).[74] Within this series, a strong correlation is seen between the observed Knight shifts and the valence electron count. Changes in line widths measured as widths at half-height seem to be unsystematic and are most probably influenced by structural or electronic disorder within the materials. However, an anomalous behavior in YRu_2Si_2 is especially noted. In this compound, ^{89}Y not only has a large Knight shift, but also a significantly larger Knight anisotropy. Related to these various efforts, Schurko has written a conspectus summarizing the recent progress in ultra-wideline solid state NMR spectroscopy which now affords measurements of NMR resonances from nuclei with very large quadrupolar splittings.[75]

^{17}O NMR shielding tensors have been reported for ^{17}O-labeled salicylic acid and acetylsalicylic acid.[76] In order to reproduce the experimental

^{17}O tensors, it is necessary to take into account the concerted double proton transfer occurring within the carboxylic acid dimer. ^{43}Ca shielding tensors have been obtained for a variety of calcium-containing compounds.[77] By performing shielding calculations on these systems, it is demonstrated that the ^{43}Ca shielding tensor can serve as a sensitive probe for subtle changes in crystalline structure. The sensitivity of both isotropic value and tensor span to the Ca–O distance in $Ca(OH)_2$ and $CaCrO_4$ is especially mentioned. ^{33}S is expected to have only small quadrupolar splittings in sulfates because of the inherent high symmetry of the ion. The anisotropy of the ^{33}S shielding may therefore contribute significantly to the shape of stationary ^{33}S spectra of sulfates.[78] A range of about 20–60 ppm has been observed for the tensor span of ^{33}S shielding in these systems.

The ^{13}C and ^{15}N chemical shift tensors of the carbonyl carbon and the amide nitrogen, respectively, have long been shown to be influenced by hydrogen bonding. Altheimer and Mehta have looked at a series of seven dipeptides consisting of alanine, valine and isoleucine residues.[79] The results match earlier predictions and data that indicate that for the carbonyl chemical shift tensor, the intermediate component is most sensitive to hydrogen bond distance. However, no clear correlation between hydrogen bond distance and principal components is observed for the carboxyl chemical shift tensor. Albeit the carboxyl chemical shift tensor is very sensitive to hydrogen bonding, in systems where the hydrogen bonding environment involves more than one partner, it is difficult to separate the various contributions. For the amide N tensor, both σ_{11} and σ_{33} are shown to be sensitive to hydrogen bond distance.

^{13}C chemical tensors have been recently measured for a series of anthracene derivatives.[80] This survey reveals that these compounds have significant lattice distortions as suggested by the observation that the number of isotropic peaks seen in the high resolution spectra is greater than the number of chemically inequivalent nuclei in the molecule. This lattice distortion can result in a change of about 8 ppm in one of the ^{13}C shielding tensor components. The span of the ^{207}Pb shielding tensor in compounds formed by lead-based pigments and fatty acids has been suggested as a way of distinguishing lead azelate from either lead stearate or palmitate.[81] The tensor span in the azelate is less than 1000 ppm while for both stearate and palmitate, the span is observed to be greater than 2000 ppm. ^{207}Pb shielding tensors have likewise been demonstrated to be quite useful in discriminating coordination environments around Pb, as illustrated in the cases of ethylenediamine and -diphosphine adducts of $(2,6\text{-}(CH_3)_2C_6H_3S)_2Pb$.[82] Lastly, orientation distributions for an ethylene molecule in the oblate large cages of the Structure I clathrate hydrate have been obtained *via* molecular dynamics simulations at various temperatures. Based on these distributions, lineshapes have been calculated predicting a temperature-dependent lineshape for static ^{13}C NMR spectra in this sample.[83]

The ^{31}P shielding anisotropy parameters (full tensor values) have been reported for Cu(I), Ag(I), and Au(I) complexes of $[Co(P_2C_2R_2)_2]^-$ anions (diphosphacyclobutadiene sandwich anions),[84] and also for different ^{31}P

environments in Au(I) and Ag(I) complexes of diphosphacyclobutadiene cobaltate sandwich anions, namely compounds [Au $(PMe_3)_4$][Au{Co-$(P_2C_2Ag_2)_2$}$_2$] and [K([18]crown-6) $(thf)_2$][Au{Co-$(P_2C_2Ag_2)_2$}$_2$]. The anisotropies of the latter range from 2890 to 400 ppm, and a highly symmetric or highly mobile [Au $(PMe_3)_4]^+$ cation is observed at −35.4 ppm relative to phosphoric acid standard.[85] The chemical shift anisotropy for ^{75}As in $[AsPh_4]$Br, has been reported, span = 80(10) ppm but the shielding anisotropy is not a significant contributor to the lineshape which is dominated by the nuclear quadrupole coupling. Also, the ^{121}Sb in $KSb(OH)_6$ anisotropy has been obtained, span = 150(10) ppm.[86]

2.2 Shielding surfaces and rovibrational averaging

The incorporation of dynamics into shielding computations is desired since such approach captures more closely experiments done at ambient temperature. This task, however, demands so much more in terms of computational effort. Knowledge of how shielding can change with motion is summarized on a shielding surface. Prior knowledge of this surface allows for extracting the effects of motion without additional shielding computations. Unfortunately, as the system grows in size so does the number of dimensions of the shielding surface rendering the construction of a shielding surface prohibitive. Thus, for larger systems, taking snapshots of a system and calculating the shielding at each of these snapshots provides a cheaper route to motionally-averaged chemical shifts. Dracinsky and Hodgkinson have recently incorporated path integral molecular dynamics (PIMD) into shielding computations.[87] By using values averaged over about 40 snapshots, with each snapshot taken at a tenth of a picosecond, a closer agreement with experiment is obtained for the ^{13}C chemical shifts in glycine, alanine, methyl-xylopyranoside, pentaerythritol, and thymine. A closer examination of measured shielding as well as electric field gradient tensors in sodium metaphosphate glass, combined with GIPAW computations, suggests that a Gaussian isotropic model could be used to describe the lineshapes seen in both ^{17}O and ^{31}P solid state NMR spectra.[88] By measuring solid state ^{27}Al and ^{29}Si silicon NMR spectra and performing GIPAW calculations using glass structural models on silicate and aluminosilicate glasses of various compositions, the following trends have become evident.[89] Both ^{27}Al and ^{29}Si become deshielded by about 5 ppm for each Al added in the second sphere while both ^{27}Al and ^{29}Si become shielded with a reduction in the average Si/Al–O–T bond angle.

By altering the unit cell volume in a GIPAW calculation, insights on how shielding is affected by changes in the crystal structure can be drawn. To determine if crystals of Tyr-(D)Ala-Phe-Gly change structure upon dehydration, ^{1}H shieldings have been computed using a smaller unit cell, about 10% less in volume compared to the X-ray determined structure of the hydrated sample. The computation with the smaller unit cell yields values that do not correlate as well with experiment as the calculated values obtained using the X-ray structure of the hydrated form. The same holds for a larger unit cell, prompting the conclusion that the peptide does not change its structure appreciably upon dehydration.[90] The structures in the solid state of enkephalins, Tyr-Gly-Gly-Phe-Leu and Tyr-Gly-Gly-Phe-Met, have likewise been recently refined by energy optimizing the positions of all atoms while preserving the geometry of the unit cell.[91] Computed ^{13}C shielding tensor components from the optimized structure correlate better with experimental values than those calculated from original crystal data. It should be noted that in this particular case, the improvement is dramatically more evident if individual tensor components are considered and not just the isotropic values. A similar treatment has been applied to refine the crystal structures of two sodium diphosphates ($Na_4P_2O_7$ and $Na_3HP_2O_7 \cdot H_2O$).[92] In these systems, both chemical shift and electric field gradient tensors of ^{23}Na prove to be very useful, in addition to the ^{31}P shielding tensors.

The sensitivity of ^{19}F chemical shifts of fluoride to the neighboring ion is exploited in the characterization of fluoridation of peralkaline and peraluminous glasses.[93] In peralkaline glasses, a fluoride ion can find itself in either one of the following six different environments: F–Ca, Si–F–Na, Al–F–Ca, Al–F–Al, Al–F–Na, and F–Na, while in peraluminous glasses only the following have been observed: Si–F–Na, Al–F–Ca, Al–F–Al, and Al–F–Na, that is, fluorides exclusively surrounded by alkali or alkali-earth metal ions are not present. GIPAW shielding calculations have been implemented to explain the ^{19}F solid state NMR spectra of NbF_5 and TaF_5,[94] and of GaF_3, InF_3, TlF and several AlF_3 polymorphs.[95] ^{23}Na resonances in the solid state appear to be sensitive to the crystalline form and degree of hydration as exemplified in the drug sodium valproate.[96] Solid-state ^{23}Na MAS measurements have been performed for the first time, yielding chemical shifts at −29.1 ppm in $NaUO_3$, 15.1 ppm inNa_4UO_5, and −14.1 and −19 ppm for the Na1 (8-fold coordinated) and Na2 (7-fold coordinated) sites in $Na_2U_2O_7$, respectively.[97] DFT-GIPAW calculations of the NMR parameters on the nonmagnetic compounds Na_4UO_5 and $Na_2U_2O_7$ using PBE functional have permitted the differentiation between the two Na sites of the $Na_2U_2O_7$ structure. Core electrons were described by norm-conserving pseudopotentials generated from both nonrelativistic and relativistic all-electron atomic calculations by Troullier and Martins. Theoretical and experimental studies of newly synthesized ferrocene-stabilized silicon cations establish that the ^{29}Si shielding in these compounds becomes deshielded with increasing Fe···Si distance.[98] This correlation, however, manifests only after first grouping the silicon cations into subsets defined by the type, alkyl *versus* aryl, of substituents. Through acquiring solid state ^{31}P and ^{43}Ca solid state NMR spectra, and performing GIPAW shielding calculations on a series of calcium phosphonates, several structural trends have been obtained.[99] The ^{43}Ca isotropic shielding emerges to be sensitive to the Ca–O distances as well as the nature of the oxygenated ligands. On the other hand, the anisotropy of the ^{31}P shielding tensor, specifically its asymmetry, depends on the protonation state. Solid state ^{13}C NMR spectra have been obtained for a series of cucurbituril macrocycles CB[n], $n = 5$, 6, 7 and 8, where n refers to the number of glycoluril units linked together by bridging methylenes to form the ring.[100] In this series, ^{13}C

nuclei in both CH and CH_2 moieties become deshielded with increasing cucurbituril size.

The effects of lanthanum on ^{11}B, ^{17}O and ^{29}Si NMR chemical shifts in borosilicate glass have been investigated.[101] With increasing amounts of lanthanum, ^{29}Si resonances move toward greater shielding and become wider. The greater shielding is attributed to a decrease in the Si–O–Si bond angle while the broader resonance is due to an increase in disorder upon addition of lanthanum. The ^{17}O resonances, a more direct probe for the lanthanum sites, suggest a random distribution of lanthanum ions throughout the lattice, ruling out a clustering of the rare-earth metal ions. Lastly, the ^{11}B signals report the transformation of tetrahedral boron sites to trigonal ones as rare-earth metal ions are introduced. Lanthanide doping, in this case, addition of Er^{3+} on $NaYF_4$ nanoparticles, has been characterized by solid state ^{19}F, ^{23}Na and ^{89}Y NMR experiments and DFT shielding calculations.[102] Similar to the case of borosilicate glass, the dopant ions are also found to be randomly distributed. Resonances are broadened due to a larger distribution of chemical shifts, but the lattice constants are only slightly changed. Solid state ^{27}Al NMR experiments and GIPAW calculations have been employed to study in greater detail the structure of $LaSrAl_3O_7$.[103] In this particular instance, the observed quadrupolar coupling reveals a difference between the average structure suggested by both diffraction data and the local structure around the Al sites.

The crystal structure of one polymorph, form 4, of the drug 4-[4-(2-adamantylcarbamoyl)-5-*tert*-butyl-pyrazol-1-yl]benzoic acid has been resolved by performing plane-wave shielding calculations on structures suggested by a crystal structure prediction software.[104] In this scheme, ^{1}H NMR chemical shifts are measured in the solid state and the structure (among low-energy candidates) that produces calculated shifts closest to experiment is presumed to depict the crystal structure of the drug. The NMR-derived structure agrees well with powder X-ray diffraction results. The all atom rmsd is 0.284 Å between the two structures. Taking this approach a step further, starting from structural models for a silicate layer and aided by GIPAW calculations, a general method of determining structures of molecularly ordered but noncrystalline silicate frameworks has been proposed.[105] Applied to a surfactant-directed layered silicate, the derived structures also reproduce satisfactorily $^{2}J(^{29}Si–O–^{29}Si)$ scalar couplings. Of course, NMR chemical shifts can also be used to validate structures obtained from powder diffraction data. The comparison of experimental and calculated ^{1}H and ^{13}C chemical shifts, as illustrated in the case of a 1:1 cocrystal of indomethacin and nicotinamide, certainly serves as an excellent and independent way of validating powder crystal structures.[106] Ripmeester and Wasylishen have recently edited a special issue of *CrystEngComm* devoted to NMR crystallography,[107] from which some papers have been separately cited in this report.

Solid state ^{17}O NMR experiments combined with GIPAW calculations have been executed to characterize brownmillerite, $Ba_2In_2O_5$.[108] *Via* the results of the theoretical calculations, the two observed ^{17}O resonances have now been assigned: One resonance (179 ppm referenced to water) is due to the combination of the O ions connecting the InO_4 tetrahedra and the O ions in equatorial sites in octahedral InO_6 coordination while another signal is attributed to the axial O ions linking the four- and six-fold coordinated In^{3+} ions.

The van der Waals-including density functional vdW-DF[109–111] has been formulated particularly for describing intermolecular effects. This functional has been used to calculate shielding for CO_2, H_2, H_2O molecules in the pores of a metal organic framework to better describe the intermolecular interactions between the adsorbed molecules and the atoms of the solid framework.[112] Another approach is the use of semi-empirical dispersion correction schemes in which the total energy of the system is modified by adding a contribution from an empirical dispersion correction (pairwise vdW interactions between each of the atoms in the system as a function of separation with empirical parameters) to the value obtained using Kohn Sham DFT to give a dispersion-corrected (DFT-D) energy.[113,114] These schemes have been introduced into periodic calculations performed using the GGA PBE functional in the CASTEP DFT code using GIPAW (gauge-including projector-augmented wave approach) and compared against the case of no dispersion correction to predict the structure of aluminophosphates.[115] It was found that DFT-D geometry optimizations provide structures in better agreement with experimental diffraction measurements, although the ^{31}P and ^{27}Al nuclear shieldings and the Al quadrupole coupings obtained were not very different for structures optimized with and without the semi-empirical dispersion corrections.

Calculations using model clusters that simulate the local solid-state environment are an alternative to the periodic-boundary approach using GIPAW. More accurate chemical-shift and quadrupolar parameters result when the cluster size is sufficient to reflect the nature of local structure on the electronic state. Relativistic calculations of NMR shielding tensors for a series of solid mercury compounds, using the cluster approach have been reported by Alkan and Dybowski.[45] The authors systematically evaluated the effect of molecular-cluster size on NMR parameters of a suite of ^{199}Hg-containing materials. They examined the effect of application of the frozen core approximation (FCA) in calculating NMR chemical shielding using molecular clusters, and examined the effect of neglect of various parts of the relativistic Hamiltonian in NMR calculations by treating the molecular clusters at different levels of theory. Compounds studied were $Hg(SCN)_2$, $Hg(CN)_2$, $Hg(SeCN)_2$, $Hg(CO_2CH_3)_2$, HgF_2, $HgCl_2$, $HgBr_2$, HgI_2, Hg_2Cl_2, $K[Hg(SeCN)_3]$, and $Hg_2(NO_3)_2 \cdot 2H_2O$. The authors also investigated the use of frozen core approximation for atoms outside the first coordination shell. All calculations were done using ZORA with scalar relativistic and spin–orbit effects. Non-relativistic results are quite poor, as are ZORA calculations with only scalar relativistic corrections. Their findings verify for relativistic systems what had been found in non-relativistic treatments of extended systems using cluster models: Clusters have to be constructed to include the nearest atomic neighbors of the nucleus in question, and the next nearest neighbors, (*i.e.*, need not only the first coordination shell but also at least the second coordination shell), must reflect the correct symmetry around the nucleus in question, have to include complete molecular units where possible, have to compensate for charges that were brought in when including atoms with net charge in the first and second coordination shells. Saturation of any dangling bonds to decrease charge on the cluster is important in using clusters to model the local structure in solids. Dangling bonds may be saturated with the cheapest unit (H atoms usually, sometimes CH_3). It is important to test against experimental values of the full tensor, not just the isotropic value.

A cluster approach has also been applied to prediction of ^{11}B and ^{15}N spectra of heterofullerenes formed from systematic substitution of BN units for CC units in C_{70},[116] and ^{13}C spectra of graphene, hydrogenated graphene (graphane) and fluorinated graphene (fluorographene).[117] In these cases, the cluster is taken to be a fragment; the fragment may be either flat or curved and relaxed structures separated from the system of interest. DFT-B3LYP calculations are carried out for a set of finite, increasing size fragments. In the heterofullerenes, the shielding anisotropies of the heteroatoms are found to be dependent on the local structure of the substitution site and also on the curvature.[116] Flakes made up of concentric hexagon-shaped (zig–zag boundary) and crenellated (armchair) fragments were investigated in the graphene and functionalized graphenes.[117] Regardless of the boundary structure, the interior carbon signals should eventually converge to similar shieldings as the size of the flake is increased. The results reveal that as the system size increases, the innermost carbon sites have signals that converge towards smaller chemical shift values for the graphene fragments, whereas for graphane and fluorographene, the overall trends are opposite, that is toward larger chemical shifts. The observed trends imply that the coupling of the electronic ground state to the accessible excited states becomes progressively smaller for the inner atoms of the graphene fragments upon increasing the system size, and that the symmetry-allowed, low-lying excitations occur at the perimeter of graphene flakes. In contrast, the strong coupling to the excited states persists for the inner carbon sites of graphane and fluorographene. The conclusions are the same when other functionals are used in the calculations.

Several reviews on both experimental and theoretical methods for shielding in the solid state have been published during this reporting period. The usefulness of plane-wave shielding calculations in characterizing oxide glasses is highlighted in a review by Charpentier and coworkers.[118] The use of periodic shielding calculations as a tool to analyze and explain solid state NMR spectra of disordered inorganic materials has been summarized by Ashbrook and Dawson.[119] Solid state NMR studies of microporous crystalline materials have also been recently reassessed by Ashbrook and coworkers.[120] Lastly, a more extensive overview of the capabilities of solid-state DFT simulations has been made available by Hasnip and coworkers.[121]

2.3 Isotope shifts

Modeling isotope shifts for systems involving heavy atoms is particularly difficult. The presence of heavy elements requires not only an extensive basis set, but also electron correlation as well as relativistic effects. These requirements apply not only to the construction of shielding surfaces but also to the potential surfaces. Not only should shielding derivatives be accurate, but also the average bond lengths and angles. What leads to a quantitative agreement with experiment for isotope-induced effects in these systems is a high-level relativistic shielding computation combined with a high quality potential energy surface, as illustrated in the case of secondary isotope effects on ^{77}Se and ^{13}C chemical shifts in CSe_2.[122] For both nuclei, electron correlation seems to be a factor much more important than relativity, suggesting that cheaper approximations for relativistic corrections may suffice.

Theoretical studies of $^{35/37}$Cl and $^{16/18}$O induced isotope shifts on ^{103}Rh and ^{195}Pt have been reported for a series of aquachloro and chlorohydroxido complexes of rhodium and platinum.[123] At the ZORA-SO/PW91/QZ4P/TZ2P level of theory, a change of 1 ppm is achieved by simply changing either the M–Cl or M–O bond lengths by a femtometer. Although trends are reproduced qualitatively, a quantitative agreement with experiment has yet to be achieved. Reasons provided for the failure to reproduce accurately the experimentally observed isotope shifts include solvent effects, which apparently are becoming more pronounced with higher absolute charges on the complexes. Incorporating solvent effects by adding two molecules of water has a significant impact on the computed values, but no improvement in accurately reproducing experimentally observed values is attained, indicating that additional solvent molecules may still be necessary. Nevertheless, the computations demonstrate that for these systems, the changes in chemical shifts are also largely dominated by changes in average bond lengths upon isotopic substitution.

Deuterium-induced isotope effects on ^{13}C NMR chemical shifts have been obtained from path integral molecular dynamics simulations (PIMD).[87] Instead of calculating the shielding at various snapshots during a simulation, shielding derivatives with respect to bond lengths and angles are first calculated. These are then combined with the average values of the average geometric parameters extracted from the simulation. It has been demonstrated that unlike a classical molecular dynamics simulation, PIMD is able to reproduce the experimental values for $^{1/2}$H-induced isotope effects on ^{13}C chemical shifts in glycine, alanine, methyl-xylopyranoside, pentaerythritol, and thymine. An increase in vibrational anharmonicity can lead to larger rovibrational corrections to the average geometry of a molecule. The measured deuterium-induced isotope effects on the ^{13}C shielding of the C2 carbon in 1,1,1,3,3,3-hexafluoro-2-propanol changes from 341 ppb to 364 ppb when the compound is complexed *via* a hydrogen bond with triethylamine.[124] These changes are accompanied by a decrease in the observed C–H stretching frequency. For this alcohol, calculations that take into account only the bond length contributions of isotope effects reproduce qualitatively the experimental values. Lastly, changes in ^{13}C chemical shifts upon deuterium replacement of an N–H proton have been measured for anions of 1,8-bis(4-toluenesulphonamido)naphthalenes.[125]

2.4 Intermolecular effects on nuclear shielding

Although plane-wave calculations deal with the system as a periodic infinite lattice, paying attention to some features in the crystal structure can lead to the identification of specific factors that affect shielding in the condensed state. For example, examining closely the NMR spectra of two crystalline forms of the organic blue-emitting compound, tris(8-hydroxyquinoline) aluminum(III), and comparing plane-wave calculation results to those obtained from an isolated molecule, shows that the difference in the observed ^{13}C NMR spectra between the two forms is not due to intramolecular conformations, but is due to a variation in the intermolecular packing of the molecules.[126] ^{1}H and ^{13}C chemical shifts of alanine tri- and tetra-peptides have been revisited.[127] In these peptides, a better agreement with experiment is reached with GIPAW calculations that make use of X-ray structures in which proton positions, specifically those participating in hydrogen bonds, have been previously optimized. Shielding tensors of alkaline earth metals ^{43}Ca and ^{87}Sr appear to be sensitive to the presence of N in its coordination sphere in metal carboxylates that contain an amino group, as exemplified by *p*-aminosalicylate.[128] GIPAW calculations have been performed to illustrate the dependence of ^{43}Ca shielding on the Ca···N distance; it becomes shielded at the rate of 36 ppm $Å^{-1}$. ^{13}C chemical shifts of carboxylates are shown to be sensitive to binding to Mg^{2+}.[129] In magnesium benzoate or salicylate, binding to Mg^{2+}, in which the carboxylate serves as a bridge between two Mg^{2+} ions, can cause as much as 8 ppm deshielding on the carboxyl carbon. On the other hand, ^{25}Mg NMR chemical shifts report on the type of framework present. ^{25}Mg nuclei participating in either one- or two-dimensional frameworks are about 8 ppm more shielded than when ^{25}Mg nuclei are found simply as a part of an isolated molecular ion in the crystal. Finally, a review on theoretical and experimental solid state NMR studies of alkaline earth metals is now available.[130]

^{13}C solid state NMR has proven to be quite informative in characterizing the inclusion of small organic compounds such as methanol in clathrate hydrates.[131] The sensitivity of ^{13}C chemical shifts to the environment enables the assignment of peaks according to the type of cages in the clathrate hydrate structure an organic compound is occupying. In the case of methane, for example, several resonances can be observed within −4 to −8 ppm (relative to tetramethylsilane), which arise from the different cages Structure I (sI) hydrates provide. This sensitivity is even highlighted in the case of propane hydrates formed in the presence of methanol. In this particular case, the ^{13}C resonances from propane molecules that occupy the same $5^{12}6^{4}$ cages begin to split since some of these larger cages are now sharing a wall with a smaller cage that has incorporated methanol. The appearance of doublet structures, on the other hand, in the low temperature solid state magic angle spinning ^{13}C NMR spectra of CH_3F in Structure II (sII) and hexagonal structure (sH) is taken as a reflection of the fact that these cages are nonspherical.[132] At temperatures higher than 250 K, these doublets collapse into one peak, indicating migration of CH_3F between cages. Siuda and Sadlej have calculated chemical shifts for carbon dioxide sI and sII clathrate hydrates.[133] In this work, clathrates have been modeled by clusters of water molecules representing various cages. These clusters are derived from known neutronographic data, but geometry optimizations are performed for the hydrogen positions as well as for the encapsulated carbon dioxide molecule. Calculated ^{13}C shielding constants for CO_2 occupying 5^{12}, $5^{12}6^2$, and $5^{12}6^4$ cages are −0.9, 2.0 and 0.5 ppm (relative to an isolated CO_2 molecule), respectively. The corresponding changes for ^{17}O are −18.9, −6.2 and −2.4 ppm.

Barbitals exist in various polymorphs that can differ in the type of hydrogen bonding network. It is therefore not surprising that either solid state ^{13}C or ^{15}N NMR spectra of this compound can easily distinguish between these various forms.[134] Weak intramolecular C–H···N and C–H···S hydrogen bonding involving vinyl protons have been shown to be responsible for the relatively large difference in proton chemical shifts and one-bond C–H coupling constants observed for the terminal vinyl protons in vinyl pyrroles that have amino and sulfanyl groups on positions 2 and 5.[135] One of the observed ^{19}F resonances in yttrium fluoride nanoparticles with the general formula $(H_3O)Y_3F_{10} \cdot xH_2O$ is reported to have a shielding tensor span that is twice as big as the span observed in bulk YF_3.[136] Therefore, it has been suggested that the observed larger span is due to H–O–H···F hydrogen bonding in the nanoparticle. By examining the ^{31}P nucleus in dry propylphosphonic acid functionalized SBA-15 mesoporous silica, it has been made clear that phosphonic acid groups are hydrogen bonded to adjacent phosphonic acid groups and not to surface SiOH groups.[137] The observed decrease in shielding of ^{31}P sites in octacalcium phosphate (OCP) upon inclusion of citrate has been taken as evidence of a disruption of the extended hydrogen bonding network among the phosphates.[138] This has been supported by a favorable comparison between observed and calculated ^{13}C chemical shifts of citrate obtained from a structural model of OCP-citrate. Interestingly, the ^{17}O NMR spectrum of OCP-citrate resembles better the spectrum from bone mineral than hydroxyapatite does. A theoretical and experimental survey of organic acid–base complexes shows the expected dramatic difference in ^{15}N chemical shifts between protonated and hydrogen-bonded N sites.[139] By using ^{17}O-enriched H_2O in hydrating layered double oxides, an economical ^{17}O labeling of layered double hydroxides (LDHs) can be attained.[140] With substantial enhancement in signal, both Mg_3OH and Mg_2AlOH environments can now be easily observed and quantified, providing insights regarding local structure inside layered double oxides.

Calculations augmented by the conductor-like screening solvation model (COSMO) seem to improve agreement between computed and experimental ^{1}H and ^{13}C chemical shifts in 2,5-dicarboxy-1-methylpyridinium inner salt.[141] Moreover, the model that shows better agreement with experiment involves an intermolecular asymmetric hydrogen bond, with the proton closer to the carboxyl group at position 5. Similar improvements are seen with COSMO on the calculations of ^{1}H and ^{13}C chemical shifts in 2,3-diethoxycarbonyl-1-methypyridinium iodide, where DMSO and $CHCl_3$ are the solvents.[142] COSMO has also been applied to reproducing observed ^{1}H and ^{13}C chemical shifts in an aqueous solution of dimethylphenyl betaine hydrate.[143]

To pinpoint the intermolecular interactions responsible for observed changes in the shielding tensor components, the use of clusters allows for comparison between calculated shieldings for an isolated molecule and those in the presence of neighbors. The clusters can be of increasing size or these could be formed by accounting for one neighboring molecule at a time. For instance, in closely examining the ^{13}C shielding tensors in theobromine, it has become apparent that two conventional hydrogen bonds influence the chemical shift tensors of the carbonyl carbon C2 in theobromine, while for the other carbonyl carbon, C6, one hydrogen bond plus stacking interactions have substantial effects, and for C8, a weak hydrogen bond, C–H···O, affects its tensor.[144] ^{1}H shieldings of guest molecules in cucurbituril macrocycles have been shown to be useful in deciphering the type of interactions between the guest and the macrocycle. Hydrogen bonding causes deshielding while C–H···π interactions lead to upfield proton shifts.[145]

Quantitatively evaluating the contributions of dynamics, solvent, electron correlation and relativity on ^{129}Xe shielding necessitates a high level of theory for each contribution. Standara and coworkers have recently attempted to achieve this in interpreting the observed ^{129}Xe chemical shift of a benzene-dissolved sample of Xe encapsulated in C_{60}.[146] Unfortunately, the basis sets that Standara and coworkers used for both Xe and C_{60} are not large enough, thereby casting serious doubts on all of their findings. The C_{60} atoms are provided only with a split-valence basis set, and the basis sets used for Xe in the relativistic computations are likewise small and are different from those employed in non-relativistic calculations. ^{3}He shieldings for the He_2 dimer have been calculated at the CCSD level of theory.[147] Results are not significantly different from those obtained at the HF level of theory. Moreover, the calculated ^{3}He shieldings using the VSXC functional for He encapsulated inside various fullerenes are no closer to experiment than the HF calculated values.

Relativistic effects have been demonstrated to be noteworthy in explaining the observed ^{13}C NMR chemical shifts of the carbon covalently bound to an iodine in a series of cocrystals of three different diiodobenzene molecules and various onium halide (Cl^- or Br^-) salts.[148] Greater deshielding of the ^{13}C nucleus is found to correlate with an increasing C–I bond length in these systems. Halogen bonding causes an increase in the C–I bond length and thus, it is now clear why more positive chemical shifts for the C site bound to I are usually observed upon formation of cocrystals that involve halogen bonding.[149] Through molecular dynamics simulations and comparisons between calculated and experimental proton shifts in water, structural differences between low- and high-salt aqueous solutions have been inferred.[150] Using an aqueous solution of LiI, the simulations suggest that at high concentration (9 M), water molecules begin to bridge Li^+ ions. In electronically complex materials such as paramagnetic Li-ion battery positive electrodes, it is extremely difficult to interpret NMR spectra. An approach that decomposes the observed Fermi-contact shifts into a simple sum of contributions from each metal-NMR observed nucleus pathway has been demonstrated to be well suited for explaining the observed $^{6/7}$Li and ^{31}P NMR chemical shifts in these systems.[151]

Theoretical shielding studies at the BLYP/TZP//BLYP/TZP level of encapsulated fullerenes predict that upon encapsulation, for example of C_{60} in C_{180}, the ^{13}C sites of C_{60} become shielded by about 20 ppm while the ^{13}C sites of the encapsulating C_{180} become deshielded by about 9 ppm.[152] For the cocrystal of indomethacin and nicotinamide, a combination of GIPAW calculations, nuclear-independent chemical shifts (NICS), and computations of shielding for isolated molecules, can separate the effects of hydrogen bonding from those of ring currents on ^{1}H shieldings.[153] The effects from hydrogen bonding and ring currents in this system are apparently of the same order of magnitude. NICS calculations have also been performed on coronene, circumcoronene and dicircumcoronene to model observed NMR chemical shifts of adsorbed species inside porous carbon.[154] Calculated NICS depend on the position of the probe with respect to the aromatic ring. Gauging the aromaticity *via* this method can be problematic when dealing with multi-ring systems. Gershoni-Poranne and Stanger have suggested an in-plane scan with a constant height of 1.7 Å above the ring as a way of identifying local and global ring currents in conjugated multi-ring systems.[155] With this methodology, for example, the calculated NICS values above the center bond of a naphthalene ring suggest a single naphthalene ring current and not a sum of two benzene molecules. On the experimental side, fast magic angle spinning (100 KHz) has been successfully applied in two-dimensional indirectly detected heteronuclear experiments that probe ^{13}C and ^{15}N sites of the host–guest 5,10,15-tris(pentafluorophenyl)corrole: toluene system.[156] Host–guest interactions are demonstrated to cause a shielding effect on the ^{13}C and ^{1}H nuclei of the methyl group in toluene. This shielding is attributed to ring-current effects arising from the corrole macrocycle. Lastly, calculated NICS maps for hexabenzocoronene have been used to interpret solid state ^{1}H spectra of H_2O trapped in mesoporous carbon materials.[157]

Several reviews related to intermolecular effects on chemical shifts have been published during this period. Conventional views on the anisotropic effects of different functional groups and ring systems on NMR chemical shifts have been revisited in a recent discussion by Baranac-Stojanović.[158] The sensitivity of ^{7}Li (for batteries) to its environment and oxidation state, as well as ^{11}B (for capacitors), and ^{13}C and ^{195}Pt (for fuel cells) have been highlighted in a review on *in situ* solid-state NMR spectroscopy of electrochemical cells.[159] A review on the use of NMR chemical shifts in biologically relevant molecules as probes of structure and dynamics has been recently written by Case.[160] Advances in experimental solid state NMR techniques probing paramagnetic metalloproteins have been summarized in an account by Knight and coworkers.[161] Spiess likewise comments on recent advances in multidimensional solid state NMR techniques that now allow for characterizing non-covalent interactions in supramolecular systems.[162]

2.5 Absolute shielding scales

With the ability of incorporating both solvent effects and dynamics, as well as adding relativistic corrections, and performing calculations at a high level of theory, it is now possible to derive accurate absolute shielding constants solely from *ab initio* calculations. The following table summarizes the results obtained when such an approach is applied to alkaline earth metal ions dissolved in water (Table 1).[163]

The above values are obtained using non-relativistic values calculated at the CCSD level of theory combined with relativistic corrections derived from the Dirac–Hartree–Fock method, plus dynamic and solvent effects estimated from cluster calculations involving water molecules that have been generated from Car–Parrinello molecular dynamics simulations.

The hyperfine structures in the rotational spectra of $H_2{}^{33}S$ and $^{33}SO_2$ have been recently resolved, paving the way for extraction of spin rotation constants.[164] Using the new constants, an absolute shielding value of ^{33}S for an isolated molecule H_2S at its equilibrium geometry has been obtained, 737 ppm. With rovibrational corrections at 300 K, the shielding is 716 ppm. These numbers, however, do not contain relativistic corrections which are predicted to be substantial (about 20 ppm) for ^{33}S in H_2S. Given the theoretical approach that has recently been established by Aucar *et al.*[11] and Xiao *et al.*[16] for making the total relativistic corrections to the relation between the spin-rotation tensor and the nuclear shielding tensor, see eqn (2), including the relativistic effect in the paramagnetic mapping between the spin-rotation and the absolute shielding tensors, we have to consider the 737 and 716 ppm values for σ_e and σ_0(300 K) for ^{33}S in H_2S molecule to be approximate.

Non-relativistic calculations using the coupled-cluster method CCSD and CCSD(T), CCSD with a non-iterative perturbative triples correction, were carried out for shielding tensors in H_2Se and H_2Te molecules.[165] The results of DFT calculations with BLYP, B3LYP, PBE and KT2 functionals were compared to these. As had been found in the earlier comprehensive study of Teale *et al.*,[10] even the best-performing KT2 functional still gives isotropic shielding values for ^{77}Se and ^{125}Te that are less than the benchmark CCSD(T) result and the anisotropies are too large. Four-component relativistic DFT calculations were carried out using the KT2 functional and the corresponding DFT/KT2 results were subtracted from this to find a relativistic correction. The authors then added these relativistic corrections to the CCSD(T) non-relativistic values to get absolute shieldings at the equilibrium molecular geometry.

Table 1 Absolute shielding values for alkaline earth metal ions in aqueous solution.[163]

Nucleus	Absolute shielding (ppm)
^{9}Be	112.18
^{25}Mg	602.48
^{43}Ca	1259.39
^{87}Sr	3536.01
$^{135/137}Ba$	6854.27

Vibrational corrections calculated at the CCSD level were then applied to arrive at $\sigma_0(300\ K) = 2447.02$ ppm for ^{77}Se and 33.15 ppm for ^{1}H in H_2Se as their final values. Using the same procedure for H_2Te, they report $\sigma_0(300\ K) = 4809.17$ ppm for ^{125}Te and 39.05 ppm for ^{1}H as their final absolute shielding values for H_2Te.

In an *ab initio* and relativistic DFT study of spin-rotation and NMR shielding constants in XF_6 molecules, X = S, Se, Te, Mo, and W, spin-rotation and absolute shielding constants of XF6 molecules (X = S, Se, Te, Mo, W) based on *ab initio* coupled cluster and four-component relativistic density-functional theory (DFT) calculations were carried out by Ruud *et al.*[21] To obtain the best theoretical estimate for the spin-rotation and nuclear magnetic shielding tensors, they assumed that the different contributions to the computed properties are additive. Hence, for comparison with available experimental data, they use the sum of the best nonrelativistic result (CCSD value obtained with the largest basis set) and the DFT relativistic correction. The results show that the relativistic contributions to the spin-rotation and shielding constants are large both for the heavy elements as well as for the ^{19}F nuclei. In most cases, incorporating the computed relativistic corrections significantly improves the agreement between the calculated results and the well-established experimental values for the ^{19}F isotropic spin-rotation constants and their anisotropic components and also for the isotropic spin-rotation constants ^{77}Se and ^{125}Te. This suggests that the calculations are reliable also for ^{33}S, ^{95}Mo and ^{183}W, for which accurate and reliable experimental data are not available. For ^{33}S, ^{77}Se, ^{125}Te, ^{95}Mo, and ^{183}W in the hexafluorides, the relativistic corrections to the absolute isotropic shielding are found to be 42.6, 265.4, 852.0, 593.6, and 4014.6 ppm, respectively. For the heavy nuclei, the breakdown of the non-relativistic identity relationship between the spin-rotation constant and the paramagnetic contribution to the shielding tensor discussed in Section 1.1 causes a significant error in the total absolute shielding constants in the heavy atom as well as for the ^{19}F nuclei in these molecules.

Abbreviations

ADF	Amsterdam density functional
CCSD	Coupled-cluster singles and doubles
CCSD(T)	CCSD model augmented by perturbative corrections for triple excitations
COSMO	Conductor-like screening solvation model
DFT	Density functional theory
DKH	Douglas–Kroll–Hess theory
EFUT	External field dependent unitary transformation
GIAO	Gauge-including atomic orbitals
GIPAW	Gauge-including projector augmented waves
HAHA	Heavy atom heavy atom relativistic effect
HALA	Heavy atom light atom relativistic effect
HF	Hartree–Fock
LDA	Local density approximation

MAS	Magic angle spinning
MP2	Second order Møller–Plesset
NBO	Natural bond orbital
NICS	Nucleus-independent chemical shift
NMR	Nuclear magnetic resonance
PBE	Perdew–Burke–Ernzerhof functional
PCM	Polarizable continuum solvent model
PIMD	Path integral molecular dynamics
PV	Parity violation
RMB	Restricted magnetic balance
SO	Spin–orbit
QED	Quantum electrodynamics
vdW	van der Waals
vdW-DF	van der Waals-including density functional
X2C	Exact two-component relativistic theory
XC	Exchange correlation
ZFS	Zero-field splitting
ZORA	Zeroth-order regular approximation

References

1 G. A. Aucar, *Phys. Chem. Chem. Phys.*, 2014, **16**, 4420–4438.
2 P. Pyykkö and L. B. Zhao, *J. Phys. B*, 2003, **36**, 1469–1478.
3 G. A. Aucar and J. Oddershede, *Int. J. Quantum Chem.*, 1993, **47**, 425–435.
4 N. F. Ramsey, *Phys. Rev.*, 1950, **78**, 699–703.
5 W. H. Flygare, *J. Chem. Phys.*, 1964, **41**, 793–800.
6 W. H. Flygare, *Chem. Rev.*, 1974, **74**, 653–687.
7 D. K. Hindermann and C. D. Cornwell, *J. Chem. Phys.*, 1968, **48**, 4148–4154.
8 C. J. Jameson, A. K. Jameson and P. M. Burrell, *J. Chem. Phys.*, 1980, **73**, 6013–6020.
9 C. J. Jameson, A. K. Jameson and J. Honarbakhsh, *J. Chem. Phys.*, 1984, **81**, 5266–5267.
10 A. M. Teale, O. B. Lutnas, T. Helgaker, D. J. Tozer and J. Gauss, *J. Chem. Phys.*, 2013, **138**, 024111.
11 I. A. Aucar, S. S. Gómez, M. C. Ruiz de Azúa and C. G. Giribet, *J. Chem. Phys.*, 2012, **136**, 204119.
12 I. A. Aucar, S. S. Gómez, J. I. Melo, C. C. Giribet and M. C. Ruiz de Azúa, *J. Chem. Phys.*, 2013, **138**, 134107.
13 I. A. Aucar, S. S. Gómez, C. G. Giribet and M. C. Ruiz de Azúa, *J. Chem. Phys.*, 2013, **139**, 094112.
14 E. Malkin, S. Komorovsky, M. Repisky, T. B. Demissie and K. Ruud, *J. Phys. Chem. Lett.*, 2013, **4**, 459–463.
15 M. Jaszunski, M. Repisky, T. B. Demissie, S. Komorovsky, E. Malkin, K. Ruud, P. Garbacz, K. Jackowski and W. Makulski, *J. Chem. Phys.*, 2013, **139**, 234302.
16 Y. Xiao and W. Liu, *J. Chem. Phys.*, 2013, **138**, 134104.
17 Y. Xiao and W. Liu, *J. Chem. Phys.*, 2013, **139**, 034113.
18 Y. Xiao, Y. Zhang and W. Liu, *J. Chem. Theory Comput.*, 2014, **10**, 600–608.
19 C. J. Jameson and A. K. Jameson, *Chem. Phys. Lett.*, 1987, **135**, 254–259.
20 D. Kolb, W. R. Johnson and P. Shorer, *Phys. Rev. A: At., Mol., Opt. Phys.*, 1982, **26**, 19–31.

21 K. Ruud, T. B. Demissie and M. Jaszunski, *J. Chem. Phys.*, 2014, **140**, 194308.
22 T. Yoshizawa and M. Hada, *Chem. Phys. Lett.*, 2013, **580**, 145–151.
23 P. Manninen and J. Vaara, *J. Chem. Phys.*, 2006, **124**, 137101.
24 L. Cheng, J. Gauss and J. F. Stanton, *J. Chem. Phys.*, 2013, **139**, 054105.
25 M. A. M. Forgeron, R. E. Wasylishen and G. H. Penner, *J. Phys. Chem. A*, 2004, **108**, 4751–4758.
26 J. Autschbach, *Mol. Phys.*, 2013, **111**, 2544–2554.
27 J. Autschbach, *Philos. Trans. R. Soc. A*, 2014, **372**, 20120489.
28 W. Van Den Heuvel and A. Soncini, *Phys. Rev. Lett.*, 2012, **109**, 073001.
29 A. Soncini and W. Van Den Heuvel, *J. Chem. Phys.*, 2013, **138**, 021103.
30 W. Van Den Heuvel and A Soncini, *J. Chem. Phys.*, 2013, **138**, 054113.
31 S. Komorovsky, M. Repisky, K. Ruud, O. L. Malkina and V. G. Malkin, *J. Phys. Chem. A*, 2013, **117**, 14209–14219.
32 A. D. Buckingham, *Chem. Phys. Lett.*, 2004, **398**, 1–5.
33 A. D. Buckingham and P. Fischer, *Chem. Phys.*, 2006, **324**, 111–116.
34 P. Lazzeretti, A. Soncini and R. Zanasi, *Theor. Chem. Acc.*, 2008, **119**, 99–106.
35 G. Monaco and R. Zanasi, *Chirality*, 2011, **23**, 752–755.
36 S. Pelloni, F. Faglioni and P. Lazzeretti, *Rend. Lincei Sci. Fis. Nat.*, 2013, **24**, 283–289.
37 A. D. Buckingham, *J. Chem. Phys.*, 2014, **140**, 011103.
38 A. L. Barra, J. B. Robert and L. Wiesenfeld, *Phys. Lett. A*, 1986, **115**, 443–447.
39 A. L. Barra, J. B. Robert and L. Wiesenfeld, *Europhys. Lett.*, 1988, **5**, 217–222.
40 J. B. Robert and A. L. Barra, *Chirality*, 2001, **13**, 699–702.
41 R. Berger, in *Relativistic Electronic Structure Theory, Part: 2, Applications*, ed. P. Schwerdtfeger, Elsevier, Netherlands, 2004, pp. 188–288.
42 S. Nahrwold, R. Berger and P. Schwerdtfeger, *J. Chem. Phys.*, 2014, **140**, 024305.
43 R. E. Taylor, F. Alkan, D. Koumoulis, M. P. Lake, D. King, C. Dybowski and L. Bouchard, *J. Phys. Chem. C*, 2013, **117**, 8959–8967.
44 T. Kawamura, M. Abe, M. Saito and M. Hada, *J. Comput. Chem.*, 2014, **35**, 847–853.
45 F. Alkan and C. Dybowski, *Phys. Chem. Chem. Phys.*, 2014, **16**, 14298–308.
46 A. Gryff-Keller, S. Molchanov and A. Wodynski, *J. Phys. Chem. A*, 2014, **118**, 128–133.
47 K. A. Chernyshev, L. B. Krivdin, S. V. Fedorov, S. N. Arbuzova and N. I. Ivanova, *Russ. J. Org. Chem.*, 2013, **49**, 1420–1427.
48 K. Uehara, T. Miyachi, T. Nakajima and N. Mizuno, *Inorg. Chem.*, 2014, **53**, 3907–3918.
49 D. Hnyk, J. Holub, A. Ruzicka, Z. Padelkova and M. Buehl, *Struct. Chem.*, 2013, **24**, 927–932.
50 T. B. Demissie, M. Repisky, S. Komorovsky, J. Isaksson, J. S. Svendsen, H. Dodziuk and K. Ruud, *J. Phys. Org. Chem.*, 2013, **26**, 679–687.
51 K. Radula-Janik, T. Kupka, K. Ejsmont, Z. Daszkiewicz and S. P. A. Sauer, *Magn. Reson. Chem.*, 2013, **51**, 630–635.
52 V. H. Gessner, F. Meier, D. Uhrich and M. Kaupp, *Chem. – Eur. J.*, 2013, **19**, 16729–16739.
53 J. Vicha, M. Straka, M. L. Munzarova and R. Marek, *J. Chem. Theory Comput.*, 2014, **10**, 1489–1499.
54 M. Pascual-Borras, X. Lopez, A. Rodriguez-Fortea, R. J. Errington and J. M. Poblet, *Chem. Sci.*, 2014, **5**, 2031–2042.
55 J. Malberg, M. Bodensteiner, D. Paul, T. Wiegand, H. Eckert and R. Wolf, *Angew. Chem., Int. Ed.*, 2014, **53**, 2771–2775.
56 F. Nozirov, T. Kupka and M. Stachów, *J. Chem. Phys.*, 2014, **140**, 144303.

57 M. Srebro and J. Autschbach, *Chem. – Eur. J.*, 2013, **19**, 12018–12033.
58 T. Murai, D. Nishi, S. Hayashi and W. Nakanishi, *Bull. Chem. Soc. Jpn.*, 2014, **87**, 677–692.
59 M. A. Ortuno, L. Castro and M. Bühl, *Organometallics*, 2013, **32**, 6437–6444.
60 E. F. Belogolova and V. F. Sidorkin, *J. Phys. Chem. A*, 2013, **117**, 5365–5376.
61 K. S. A. Arachchige, P. S. Camacho, M. J. Ray, B. A. Chalmers, F. R. Knight, S. E. Ashbrook, M. Bühl, P. Kilian, A. M. Z. Slawin and J. D. Woollins, *Organometallics*, 2014, **33**, 2424–2433.
62 W. Makulski, A. Tulewicz and A. Les, *Magn. Reson. Chem.*, 2014, **52**, 106–110.
63 V. A. Semenov, D. O. Samul'tsev and L. B. Krivdin, *Russ. J. Org. Chem.*, 2014, **50**, 381–388.
64 A. Borgogno, F. Rastrelli and A. Bagno, *Dalton Trans.*, 2014, **43**, 9486–9496.
65 P. Garbacz, V. V. Terskikh, M. J. Ferguson, G. M. Bernard, M. Kedziorek and R. E. Wasylishen, *J. Phys. Chem. A*, 2014, **118**, 1203–1212.
66 G. Hou, R. Gupta, T. Polenova and A. J. Vega, *Isr. J. Chem.*, 2014, **54**, 171–183.
67 H. K. Miah, D. A. Bennett, D. Iuga and J. J. Titman, *J. Magn. Reson.*, 2013, **235**, 1–5.
68 S. Mamone, M. Concistre, I. Heinmaa, M. Carravetta, I. Kuprov, G. Wall, M. Denning, X. Lei, J. Y. Chen, Y. Li, Y. Murata, N. J. Turro and M. H. Levitt, *ChemPhysChem*, 2013, **14**, 3121–3130.
69 K. E. Johnston, C. A. O'Keefe, R. M. Gauvin, J. Trebosc, L. Delevoye, J. Amoureux, N. Popoff, M. Taoufik, K. Oudatchin and R. W. Schurko, *Chem. – Eur. J.*, 2013, **19**, 12396–12414.
70 B. E. G. Lucier, K. E. Johnston, W. Xu, J. C. Hanson, S. D. Senanayake, S. Yao, M. W. Bourassa, M. Srebro, J. Autschbach and R. W. Schurko, *J. Am. Chem. Soc.*, 2014, **136**, 1333–1351.
71 B. Peric, R. Gautier, C. J. Pickard, M. Bosiocic, M. S. Grbic and M. Pozek, *Solid State Nucl. Magn. Reson.*, 2014, **59–60**, 20–30.
72 O. Goncharova-Zapata, P. B. Chatterjee, G. Hou, L. L. Quinn, M. Li, J. Yehl, D. C. Crans and T. Polenova, *CrystEngComm*, 2013, **15**, 8776–8783.
73 H. Hamaed, K. E. Johnston, B. F. T. Cooper, V. V. Terskikh, E. Ye, C. L. B. Macdonald, D. C. Arnold and R. W. Schurko, *Chem. Sci.*, 2014, **5**, 982–995.
74 C. Hoeting, H. Eckert, S. F. Matar, U. C. Rodewald and R. Poettgen, *Z. Naturforsch., B: J. Chem. Sci.*, 2014, **69**, 305–312.
75 R. W. Schurko, *Acc. Chem. Res.*, 2013, **46**, 1985–1995.
76 X. Kong, M. Shan, V. Terskikh, I. Hung, Z. Gan and G. Wu, *J. Phys. Chem. B*, 2013, **117**, 9643–9654.
77 C. M. Widdifield, I. Moudrakovski and D. L. Bryce, *Phys. Chem. Chem. Phys.*, 2014, **16**, 13340–13359.
78 P. J. Pallister, I. L. Moudrakovski, G. D. Enright and J. A. Ripmeester, *Crystengcomm*, 2013, **15**, 8808–8822.
79 B. D. Altheimer and M. A. Mehta, *J. Phys. Chem. A*, 2014, **118**, 2618–2628.
80 C. L. Hoop and R. J. Iuliucci, *Solid State Nucl. Magn. Reson.*, 2013, **53**, 1–12.
81 J. Catalano, Y. Yao, A. Murphy, N. Zumbulyadis, S. A. Centeno and C. Dybowski, *Appl. Spectrosc.*, 2014, **68**, 280–286.
82 A. J. Rossini, A. W. Macgregor, A. S. Smith, G. Schatte, R. W. Schurko and G. G. Briand, *Dalton Trans.*, 2013, **42**, 9533–9546.
83 H. Mohammadi-Manesh, S. Alavi, T. K. Woo and B. Najafi, *J. Iran. Chem. Soc.*, 2013, **10**, 659–667.
84 J. Malberg, T. Wiegand, H. Eckert, M. Bodensteiner and R. Wolf, *Eur. J. Inorg. Chem.*, 2014, **2014**, 1638–1651.

85 J. Malberg, T. Wiegand, H. Eckert, M. Bodensteiner and R. Wolf, *Chem. – Eur. J.*, 2013, **19**, 2356–2369.
86 A. Faucher, V. V. Terskikh and R. E. Wasylishen, *Solid State Nucl. Magn. Reson.*, 2014, **61–62**, 54–61.
87 M. Dracinsky and P. Hodgkinson, *Chem. – Eur. J.*, 2014, **20**, 2201–2207.
88 F. Vasconcelos, S. Cristol, J. Paul, L. Delevoye, F. Mauri, T. Charpentier and G. Le Caer, *J.Phys.: Condens. Matter*, 2013, **25**, 255402.
89 E. Gambuzzi, A. Pedone, M. C. Menziani, F. Angeli, D. Caurant and T. Charpentier, *Geochim. Cosmochim. Acta*, 2014, **125**, 170–185.
90 T. Pawlak, P. Paluch, K. Trzeciak-Karlikowska, A. Jeziorna and M. J. Potrzebowski, *CrystEngComm*, 2013, **15**, 8680–8692.
91 T. Pawlak and M. J. Potrzebowski, *J. Phys. Chem. B*, 2014, **118**, 3298–3309.
92 F. A. Perras, I. Korobkov and D. L. Bryce, *CrystEngComm*, 2013, **15**, 8727–8738.
93 A. Baasner, B. C. Schmidt, R. Dupree and S. L. Webb, *Geochim. Cosmochim. Acta*, 2014, **132**, 151–169.
94 M. Biswal, M. Body, C. Legein, A. Sadoc and F. Boucher, *J. Solid State Chem.*, 2013, **207**, 208–217.
95 A. Sadoc, M. Biswal, M. Body, C. Legein, F. Boucher, D. Massiot and F. Fayon, *Solid State Nucl. Magn. Reson.*, 2014, **59–60**, 1–7.
96 N. M. Dicaire, F. A. Perras and D. L. Bryce, *Can. J. Chem.*, 2014, **92**, 9–15.
97 A. L. Smith, P. E. Raison, L. Martel, T. Charpentier, I. Farnan, D. Prieur, C. Hennig, A. C. Scheinost, R. J. M. Konings and A. K. Cheetham, *Inorg. Chem.*, 2014, **53**, 375–382.
98 K. Muether, P. Hrobarik, V. Hrobarikova, M. Kaupp and M. Oestreich, *Chem. – Eur. J.*, 2013, **19**, 16579–16594.
99 S. Sene, B. Bouchevreau, C. Martineau, C. Gervais, C. Bonhomme, P. Gaveau, F. Mauri, S. Begu, P. H. Mutin, M. E. Smith and D. Laurencin, *CrystEngComm*, 2013, **15**, 8763–8775.
100 D. Bardelang, A. Brinkmann, C. I. Ratcliffe, J. A. Ripmeester, V. V. Terskikh and K. A. Udachin, *CrystEngComm*, 2014, **16**, 3788–3795.
101 F. Angeli, T. Charpentier, E. Molieres, A. Soleilhavoup, P. Jollivet and S. Gin, *J. Non Cryst. Solids*, 2013, **376**, 189–198.
102 A. A. Arnold, V. Terskikh, Q. Y. Li, R. Naccache, I. Marcotte and J. A. Capobianco, *J. Phys. Chem. C*, 2013, **117**, 25733–25741.
103 C. Ferrara, C. Tealdi, A. Pedone, M. C. Menziani, A. J. Rossini, G. Pintacuda and P. Mustarelli, *J. Phys. Chem. C*, 2013, **117**, 23451–23458.
104 M. Baias, J. Dumez, P. H. Svensson, S. Schantz, G. M. Day and L. Emsley, *J. Am. Chem. Soc.*, 2013, **135**, 17501–17507.
105 D. H. Brouwer, S. Cadars, J. Eckert, Z. Liu, O. Terasaki and B. F. Chmelka, *J. Am. Chem. Soc.*, 2013, **135**, 5641–5655.
106 D. V. Dudenko, P. A. Williams, C. E. Hughes, O. N. Antzutkin, S. P. Velaga, S. P. Brown and K. D. M. Harris, *J. Phys. Chem. C*, 2013, **117**, 12258–12265.
107 J. A. Ripmeester and R. E. Wasylishen, *CrystEngComm*, 2013, **15**, 8598–8598.
108 R. Dervisoglu, D. S. Middlemiss, F. Blanc, L. A. Holmes, Y. Lee, D. Morgan and C. P. Grey, *Phys. Chem. Chem. Phys.*, 2014, **16**, 2597–2606.
109 M. Dion, H. Rydberg, E. Schröder, D. C. Langreth and B. I. Lundqvist, *Phys. Rev. Lett.*, 2004, **92**, 246401.
110 T. Thonhauser, V. R. Cooper, S. Li, A. Puzder, P. Hyldgaard and D. C. Langreth, *Phys. Rev. B: Condens. Matter Mater. Phys.*, 2007, **76**, 125112.
111 D. C. Langreth, B. I. Lundqvist, S. D. Chakarova-Käck, V. R. Cooper, M. Dion, P. Hyldgaard, A. Kelkkanen, J. Kleis, L. Kong, S. Li, P. G. Moses, E. Murray, A. Puzder, H. Rydberg, E. Schröder and T. Thonhauser, *J. Phys. Condens. Matter*, 2009, **21**, 084203.

112 M. G. Lopez, P. Canepa and T. Thonhauser, *J. Chem. Phys.*, 2013, **138**, 154704.
113 A. Tkatchenko and M. Scheffler, *Phys. Rev. Lett.*, 2009, **102**, 073005.
114 S. Grimme, *J. Comput. Chem.*, 2006, **27**, 1787.
115 S. Sneddon, D. M. Dawson, C. J. Pickard and S. E. Ashbrook, *Phys. Chem. Chem. Phys.*, 2014, **16**, 2660–2673.
116 M. Anafcheh and R. Ghafouri, *Monatsh. Chem.*, 2014, **145**, 411–419.
117 N. Ozcan, J. Vahakangas, P. Lantto and J. Vaara, *ChemPhysChem*, 2014, **15**, 1799–1808.
118 T. Charpentier, M. C. Menziani and A. Pedone, *RSC Adv.*, 2013, **3**, 10550–10578.
119 S. E. Ashbrook and D. M. Dawson, *Acc. Chem. Res.*, 2013, **46**, 1964–1974.
120 S. E. Ashbrook, D. M. Dawson and V. R. Seymour, *Phys. Chem. Chem. Phys.*, 2014, **16**, 8223–8242.
121 P. J. Hasnip, K. Refson, M. I. J. Probert, J. R. Yates, S. J. Clark and C. J. Pickard, *Philos. Trans. R. Soc., A*, 2014, **372**, 20130270.
122 P. Lantto, S. Kangasvieri and J. Vaara, *Phys. Chem. Chem. Phys.*, 2013, **15**, 17468–17478.
123 J. C. Davis, M. Bühl and K. R. Koch, *J. Phys. Chem. A*, 2013, **117**, 8054–8064.
124 U. Pal, S. Sen and N. C. Maiti, *J. Phys. Chem. A*, 2014, **118**, 1024–1030.
125 M. Pietrzak, E. Grech, J. Nowicka-Scheibe and P. E. Hansen, *Magn. Reson. Chem.*, 2013, **51**, 683–688.
126 F. Suzuki, T. Fukushima, M. Fukuchi and H. Kaji, *J. Phys. Chem. C*, 2013, **117**, 18809–18817.
127 T. Asakura, K. Yazawa, K. Horiguchi, F. Suzuki, Y. Nishiyama, K. Nishimura and H. Kaji, *Biopolymers*, 2014, **101**, 13–20.
128 K. M. N. Burgess, Y. Xu, M. C. Leclerc and D. L. Bryce, *Inorg. Chem.*, 2014, **53**, 552–561.
129 K. M. N. Burgess, Y. Xu, M. C. Leclerc and D. L. Bryce, *J. Phys. Chem. A*, 2013, **117**, 6561–6570.
130 I. L. Moudrakovski, *Annu. Rep. NMR Spectrosc.*, 2013, **79**, 129–240.
131 K. Shin, K. A. Udachin, I. L. Moudrakovski, D. M. Leek, S. Alavi, C. I. Ratcliffe and J. A. Ripmeester, *Proc. Natl. Acad. Sci. U. S. A.*, 2013, **110**, 8437–8442.
132 A. T. Trueba, M. C. Kroon, C. J. Peters, I. L. Moudrakovski, C. I. Ratcliffe, S. Alavi and J. A. Ripmeester, *J. Chem. Phys.*, 2014, **140**, 214703.
133 P. Siuda and J. Sadlej, *Chem. Phys.*, 2014, **433**, 31–41.
134 N. Zencirci, U. J. Griesser, T. Gelbrich, D. C. Apperley and R. K. Harris, *Mol. Pharm.*, 2014, **11**, 338–350.
135 A. V. Afonin, D. V. Pavlov, A. I. Albanov, O. A. Tarasova and N. A. Nedolya, *Magn. Reson. Chem.*, 2013, **51**, 414–423.
136 B. E. G. Lucier, K. E. Johnston, D. C. Arnold, J. Lemyre, A. Beaupre, M. Blanchette, A. M. Ritcey and R. W. Schurko, *J. Phys. Chem. C*, 2014, **118**, 1213–1228.
137 N. Bibent, T. Charpentier, S. Devautour-Vinot, A. Mehdi, P. Gaveau, F. Henn and G. Silly, *Eur. J. Inorg. Chem.*, 2013, **2013**, 2350–2361.
138 E. Davies, K. H. Mueller, W. C. Wong, C. J. Pickard, D. G. Reid, J. N. Skepper and M. J. Duer, *Proc. Natl. Acad. Sci. U. S. A.*, 2014, **111**, E1354–E1363.
139 J. S. Stevens, S. J. Byard, C. C. Seaton, G. Sadiq, R. J. Davey and S. L. M. Schroeder, *Phys. Chem. Chem. Phys.*, 2014, **16**, 1150–1160.
140 L. Zhao, Z. Qi, F. Blanc, G. Yu, M. Wang, N. Xue, X. Ke, X. Guo, W. Ding, C. P. Grey and L. Peng, *Adv. Funct. Mater.*, 2014, **24**, 1696–1702.

141 P. Barczynski, A. Komasa, A. Katrusiak, Z. Dega-Szafran, L. Nowaczyk, M. Ratajczak-Sitarz and M. Szafran, *Spectrochim. Acta, Part A*, 2014, **121**, 586–595.
142 P. Barczynski, M. Ratajczak-Sitarz, L. Nowaczyk, A. Katrusiak, Z. Dega-Szafran, A. Komasa and M. Szafran, *Spectrochim. Acta, Part A*, 2013, **115**, 208–218.
143 M. Szafran, K. Ostrowska, A. Katrusiak and Z. Dega-Szafran, *Spectrochim. Acta, Part A*, 2014, **128**, 844–851.
144 K. Bouzkova, M. Babinsky, L. Novosadova and R. Marek, *J. Chem. Theory Comput.*, 2013, **9**, 2629–2638.
145 I. A. Raja, V. V. Gobre, R. V. Pinjari and S. P. Gejji, *J. Mol. Model.*, 2014, **20**, 2138.
146 S. Standara, P. Kulhanek, R. Marek and M. Straka, *J. Comput. Chem.*, 2013, **34**, 1890–1898.
147 T. Kupka, M. Stachow, L. Stobinski and J. Kaminsky, *Magn. Reson. Chem.*, 2013, **51**, 463–468.
148 J. Viger-Gravel, S. Leclerc, I. Korobkov and D. L. Bryce, *J. Am. Chem. Soc.*, 2014, **136**, 6929–6942.
149 C. M. Widdifield, G. Cavallo, G. A. Facey, T. Pilati, J. Lin, P. Metrangolo, G. Resnati and D. L. Bryce, *Chem. – Eur. J.*, 2013, **19**, 11949–11962.
150 C. Allolio, N. Salas-Illanes, Y. S. Desmukh, M. R. Hansen and D. Sebastiani, *J. Phys. Chem. B*, 2013, **117**, 9939–9946.
151 D. S. Middlemiss, A. J. Ilott, R. J. Clement, F. C. Strobridge and C. P. Grey, *Chem. Mater.*, 2013, **25**, 1723–1734.
152 G. Casella, A. Bagno and G. Saielli, *Phys. Chem. Chem. Phys.*, 2013, **15**, 18030–18038.
153 D. V. Dudenko, J. R. Yates, K. D. M. Harris and S. P. Brown, *CrystEngComm*, 2013, **15**, 8797–8807.
154 A. C. Forse, J. M. Griffin, V. Presser, Y. Gogotsi and C. P. Grey, *J. Phys. Chem. C*, 2014, **118**, 7508–7514.
155 R. Gershoni-Poranne and A. Stanger, *Chem. – Eur. J.*, 2014, **20**, 5673–5688.
156 T. Kobayashi, K. Mao, P. Paluch, A. Nowak-Krol, J. Sniechowska, Y. Nishiyama, D. T. Gryko, M. J. Potrzebowski and M. Pruski, *Angew. Chem., Int. Ed.*, 2013, **52**, 14108–14111.
157 Y. Xu, T. Watermann, H. Limbach, T. Gutmann, D. Sebastiani and G. Buntkowsky, *Phys. Chem. Chem. Phys.*, 2014, **16**, 9327–9336.
158 M. Baranac-Stojanović, *RSC Adv.*, 2014, **4**, 308–321.
159 F. Blanc, M. Leskes and C. P. Grey, *Acc. Chem. Res.*, 2013, **46**, 1952–1963.
160 D. A. Case, *Curr. Opin. Struct. Biol.*, 2013, **23**, 172–176.
161 M. J. Knight, I. C. Felli, R. Pierattelli, L. Emsley and G. Pintacuda, *Acc. Chem. Res.*, 2013, **46**, 2108–2116.
162 H. W. Spiess, *Isr. J. Chem.*, 2014, **54**, 16–24.
163 A. Antusek, P. Rodziewicz, D. Kedziera, A. Kaczmarek-Kedziera and M. Jaszunski, *Chem. Phys. Lett.*, 2013, **588**, 57–62.
164 T. Helgaker, J. Gauss, G. Cazzoli and C. Puzzarini, *J. Chem. Phys.*, 2013, **139**, 244308.
165 M. Jaszunski and M. Olejniczak, *Mol. Phys.*, 2013, **111**, 1355–1363.

Applications of nuclear shielding

Shigeki Kuroki,*[a] Tsunenori Kameda*[b] and Hidekazu Yasunaga*[c]

DOI: 10.1039/9781782622758-00076

This chapter reports a selection of recent publications on the application of nuclear shielding, related NMR chemical shifts and NMR lineshapes, which have been published from 1 June 2013 to 31 May 2014. The reports include not only experimental NMR results but also calculated ones. The number of the papers in this field is rapidly increasing, covering practically all magnetically active nuclei. Therefore, a selection of related publications was made and generally it is restricted to journal original articles and reviews published in English. The survey of the papers starts from those devoted to the group 1 elements (^{1}H, ^{2}H, ^{3}H, ^{6,7}Li, ^{23}Na, ^{39}K, ^{85,87}Rb, ^{133}Cs) and continues along the periodic table up to the group 18 elements (^{21}Ne, ^{39}Ar, ^{129,131}Xe).

1 Introduction

The report introduces studies on the application of nuclear shielding, covering the chemical shifts and lineshapes of NMR spectra, which were made in the world from 1 June 2013 to 31 May 2014. The shieldings of particular nuclear species are described in the following section according to their position in the periodic table. Although there is a great number of articles on the research made by NMR spectroscopies and related theoretical calculations during the period, we have space limitation in this chapter. Therefore, the coverage of this chapter is restricted to widely available and common journals, which are written and published in English, as a general rule.

The NMR experiments reported in the chapter for most elements were made in the course of physical, chemical or biological studies during the period. The simple chemical assignments, chemical structure determinations and related studies for synthesised and natural products are excluded from the text as possible. The review and preprint articles are given when they provide valuable information.

2 Group 1 (^{1}H, ^{2}H, ^{3}H, ^{6,7}Li, ^{23}Na, ^{39}K, ^{85,87}Rb, ^{133}Cs)

2.1 Hydrogen (^{1}H) ($I = 1/2$)

The shielding effect of fluoroaromatic compounds on nuclei was studied by using nucleus-independent chemical shift (NICS) calculations at the

[a]*Tokyo Institute of Technology, Department of Organic and Polymeric Materials, 2-12-1 Ookayama, Meguro-ku, Tokyo 152-8552, Japan. E-mail: skuroki@polymer.titech.ac.jp*

[b]*National Institute of Agrobiological Sciences, 1-2 Ohwashi, Tsukuba 305-8634, Japan. E-mail: kamedat@affrc.go.jp*

[c]*Kyoto Institute of Technology, Department of Biobased Materials Science, Gosyokaido-tyo, Matugasaki, Sakyo-ku, Kyoto 606-8585, Japan. E-mail: yasunaga@kit.ac.jp*

GIAO/B3LYP/6-311++G(2d,p) level.[1] It was found and demonstrated by ^{1}H NMR measurements on BINAP bearing C_6F_5 groups that the chemical shift of a proton located above the fluoroaromatic ring is shifted upfield owing to a relatively weak shielding effect of the ring. The poly[2,7-(9,9-bis(octyl)-fluorene)-*alt*-benzothiadiazole] (F8BT), poly[2,7-(9,9-bis(2′-ethylhexyl)-fluorene)-*alt*-thiophene-*S*,*S*-dioxide] (PFTSO2) and poly[2,7-(9,9-bis(2′-ethylhexyl)-fluorene)-*alt*-1,4-phenylene] (PFP) containing, benzothiadiazole (BT), thiophene-*S*,*S*-dioxide and phenylene groups, respectively, were investigated by ^{1}H and ^{13}C NMR spectroscopies.[2] The obtained NMR spectra and corresponding shielding tensors of the ^{1}H and ^{13}C nuclei were interpreted by density functional theory (DFT) with complete geometry optimisation using B3LYP functional. The DFT analysis provides the relationship between the structure and NMR, optical, electronic properties. This conformation of 3-(1,3-diphenylpropan-2-yl)-4-methyl-6-phenylisoxazolo[3,4-*d*]pyridazin-7(6*H*)-one in $CDCl_3$ solution was evidenced by distinct phenyl-shielding effects on the ^{1}H NMR signals.[3] The isoxazolo[3,4-*d*]pyridazin ring system is virtually planar, whereas the *N*-bonded Ph group is inclined to the former by an ring–ring angle of 55.05(3)°. The 1D and 2D Very Fast Magic-Angle Spinning (VF MAS) NMR experiments with sample rotation up to 55 kHz were applied to study both the dihydrate form and the anhydrous form of Tyr-(D)Ala-Phe-Gly.[4] It was shown by NMR measurements employing both homo-nuclear (^{1}H–^{1}H BABA and ^{13}C–^{13}C SHANGHAI) and hetero-nuclear 2D correlations (^{1}H–^{13}C and ^{1}H–^{15}N) that removing water from the crystal lattice of this tetrapeptide in thermal processes does not destroy its subtle pseudo-cyclic structure. The geometry of the peptides and the ^{13}C σ_{ii} principal elements of the NMR shielding tensor parameters and ^{1}H isotropic NMR shifts were calculated using gauge including projector augmented wave (GIPAW) method. The ^{1}H NMR spectrum of diquat (6,7-dihydrodipyrido[1,2-*b*:1′,2′-*e*] pyrazine-5,8-diium or DQ^{2+}) encapsulated by cucurbit[*n*]uril ($n = 7$, 8) hosts was obtained using the density functional theory, and theoretical calculations showed that both CB[7] or CB[8] host possesses strong affinity toward DQ^{2+}.[5] It was revealed that H_α protons on bi-pyridinium rings of DQ^{2+}@CB[8] complex are de-shielded owing to C=O···H interactions and aromatic H_β and H_δ protons of DQ^{2+} within the CB[8] cavity exhibit significant shielding. The effects of stress-shielding on both viscoelastic properties and microstructure of collagen fibers in the Achilles tendon were studied by proton double-quantum filtered (^{1}H-DQF) NMR spectroscopy.[6] The ^{1}H-DQF NMR results demonstrated significantly reduced residual dipolar coupling in the Achilles tendons and evaluated the alterations of the collagen fiber orientation that contributed to the function of tendinous tissue. Pyrazolo[4,3-*c*]pyrazoles and three regioisomers of their N-Me derivatives were studied by ^{1}H, ^{13}C and ^{15}N NMR spectroscopies and the NMR spectroscopic data for all isomers were interpreted by DFT calculations of nuclear shielding constants and indirect spin–spin coupling constants[7] The N-Me isomers were used as model compounds to investigate a potential N1–H/N4–H, N2–H/N4–H and N1–H/N5–H tautomerism of 3,6-substituted pyrazolo[4,3-*c*]pyrazoles by using low-temperature NMR spectroscopy, and the stabilities of individual tautomeric forms were calculated by DFT methods. Oligostyrene–perylenediimide–oligostyrene coil–rod–coil molecule, which does not show self-assembly of the perylene segment in solvents that are selective for oligostyrene, was investigated by ^{1}H NMR.[8] The chemical shift of Ph-CH in ^{1}H NMR spectrum shows a small shift from 7.15 to 7.08 ppm due to the ring current of perylene shielding the chemical environment of Ph-CH in solution. Characteristic NMR spectra of proton transfer in protonated water clusters, $H^+(H_2O)_n$ complexes (n = 2–5), were studied using *ab initio* calculations at the RIMP2/TZVP level and BOMD simulations.[9] The results of RIMP2/TZVP calculations showed that the proton moving on the oscillatory shuttling path is characterised by the ^{1}H NMR shielding constant ($\sigma^{corr}{}_{H+}$) varying in a narrow range, whereas $\sigma^{corr}{}_{H+}$ changes exponentially with R_{O-H} due to the proton moving on the structural diffusion path. ^{1}H NMR spectra of *N*-substituted epoxy-fused cyclopentane-3,5-diol derivatives were computationally investigated with density functional calculations in order to unravel the shielding effect of the epoxy ring on the ^{1}H NMR chemical shifts.[10] Both the ^{1}H NMR chemical shifts and spin–spin coupling constants were calculated by the WP04/cc-pVTZ//B3LYP/6-31+G(d) methodology, and reproducing the best experimental results on epoxy-fused cyclopentane derivatives was found. A solid-state NMR study of H_2 molecules confined inside the cavity of C_{70} fullerene cages over a wide range of temperatures (300 K to 4 K) was made.[11] It was found that the magnitude of the ^{1}H NMR chemical shift anisotropy is consistent with DFT estimations of the chemical shielding tensor field within the cage and the ground state of endohedral *ortho*-H_2 in C_{70} is doubly degenerate and polarised transverse to the principal axis of the cage. The conformation of 4′-substituted neutral/protonated furfurylidenanilines and *trans*-styrylfurans was calculated by DFT methods and ^{1}H and ^{13}C substituent induced shifts (SIS) were calculated comparing with experimental values.[12] The calculated shielding constants were shown to be sensitive to the substituent effect through a linear fit with substituent's Hammett constants The structures of *N*-mercapto-4-formylcarbostyril produced by Pseudomonas fluorescens strain G308 and 2-(2-hydroxyphenyl)thiazole-4-carbaldehyde isolated from *Pseudomonas protegens* Pf-5 were revised by using computational predictions of the ^{1}H and ^{13}C NMR magnetic shielding tensors.[13] Molecules were proposed to originate from the non-ribosomal peptide synthetase enzymes involved in the siderophores enantio-pyochelin (or pyochelin) biosynthetic pathways. The 1-[(pyridin-2-yl amino)Me] pyrrolidine-2,5-dione (SFAP) was studied by ^{1}H and ^{13}C NMR spectroscopies determining the structure property relation *via* chemical shifts and magnetic shielding effects.[14] The theoretical calculations were also performed using HF/6-31+G/6-311++G(d,p), B3LYP/6-31+G/6-311++G(d,p) model. The 2,2′-[(1*E*,2*E*)-hydrazine-1,2-diylidenedi(1*E*)eth-1-yl-1-ylidene]diphenol and its dimeric, binuclear Ni(II), Pd(II) and Pt(II) metal complexes were studied by ^{1}H and ^{13}C NMR spectroscopies and ^{1}H and ^{13}C shielding tensors for crystal structure were calculated with GIAO/DFT/B3LYP/6-311++G(d,p) methods in $CDCl_3$.[15] The vibrational band assignments were performed for the samples at B3LYP/6-311++G(d,p) theory level combined with scaled quantum mechanics force field (SQMFF) methodology and their antibacterial activities were tested. The basic physico-chemical properties of the synthetic mesoporous materials synthesised from polymers and silica gels were characterised by ^{1}H and ^{13}C solid-state MAS NMR measurements and quantum chemical calculations of the magnetic shielding effect of the surface on the solvent molecules were performed.[16] The theoretical calculation of ^{1}H NICS maps obtained shows that the electronic ring currents localised in aromatic groups on the surface may result in the proton chemical shifts towards high-field values. The structure of 2,5-dicarboxy-1-methylpyridinium inner salt (1) was studied by B3LYP/6-311G(d,p) calculations and NMR spectroscopy.[17] The hydrogen bonds in the molecules optimised by the B3LYP/6-311G(d,p) approach in trimer (2) and dimer (3) are slightly longer than in the crystal. The experimental ^{1}H and ^{13}C NMR chemical shifts (δ_{exp}) of the inner salt in D_2O and the calculated magnetic isotropic shielding constants (σ_{calc}) for the optimised monomer (4a) solvated in water show a linear correlations as $\delta_{exp} = a + b\sigma_{calc}$. The structures of dimethylphenyl betaine crystallizes as monohydrate (1), in which water molecules mediate in hydrogen bonds between the carboxylate groups, monomer (2) and dimer (4) hydrates in vacuum, and the structure of monomer (3) in an aqueous environment were optimised by the B3LYP/6-311++G(d,p) approach.[18] It was found that the correlations between the experimental ^{1}H and ^{13}C NMR chemical shifts (δ_{exp}) of 1 in D_2O and the magnetic isotropic shielding constants (σ_{calc}) calculated by the GIAO/B3LYP/6-311G++(d,p) approach, using the screening solvation model (COSMO), $\delta_{exp} = a + b\sigma_{calc}$, for optimised molecule 3 in water solution are linear. Asymmetric phase-transfer catalysis with homo- and hetero-chiral quaternary ammonium salts were studied through the theoretical calculations of ^{1}H NMR spectra of the two isomers for different model systems.[19] The results indicated that the upfield shift of signals for the benzylic protons in the heterochiral form could be explained by an increase in the shielding effect of the aromatic parts of the system around these protons. Diketopiperazines were studied by ^{1}H NMR spectroscopy and they were found to adopt a folded conformation according to the upfield chemical shift of the *cis*-proton due to a shielding effect.[20] Cobalt(II) and zinc(II) tetranitrooctaethylporphyrins (tn-OEP) complexes, Co^{II}(tn-OEP) and Zn^{II}(tn-OEP), linear 1D polymeric chains of the six-coordinated Co(II) complexes and five-coordinate porphyrin dimers $[Zn^{II}(tn\text{-}OEP)]_2 \cdot L$ were studied by ^{1}H NMR spectroscopy.[21] While the large downfield shift and the peak broadening of the axial ligand resonances were discerned in the ^{1}H NMR of $[Co^{II}(tn\text{-}OEP)] \cdot L$, which is due to the contiguity to the paramagnetic Co(II) center, the upfield shift of the ^{1}H NMR resonances of the axial ligand L was observed for $[Zn^{II}(tn\text{-}OEP)]_2 \cdot L$ due to the shielding effect of porphyrin ring current. ^{1}H chemical shifts of inclusion complexes between CB[*n*] ($n = 6, 7$) or their inverted iCB[*n*] diastereomer hosts and 1,6-hexyldiammonium (HDA) or *p*-xylyldiammonium (XYL) cationic guests were obtained from the d. functional calculations.[22] It was revealed that guest protons within the host cavity not participating in hydrogen bonding interactions, exhibit shielded signals as compared with those of isolated XYL or HDA. ^{1}H and ^{13}C NMR studies of 4-methoxy-4′-nitrobiphenyl (4M4′NBPL) were made by utilising HF and DFT/B3LYP with 6-31G(d,p) as basis set, and a good correlation between the experimental ^{1}H and ^{13}C NMR chemical shifts in chloroform solution and calculated GIAO shielding tensors was found.[23] The linear polarisability and first order hyperpolarisability of 4M4′NBPL indicate that the compound is a candidate of nonlinear optical materials. Dimerisation of 2-naphthalenecarbonitrile (2-NpCN) mediated by cucurbit[8]uril (CB[8]) combining monomers in anti-head-to-head (A), anti-head-to-tail (B) and *syn*-head-to-tail (C) fashion was investigated employing the DFT.[24] The calculated ^{1}H NMR spectra revealed that the 2-NpCN dimer exhibits large shielding for aromatic protons and the protons attached to cubane moiety lead to down-field signals. The structure of (−)-Encecanescin (1) obtained from Eupatorium aschembornianum was studied by ^{1}H and ^{13}C NMR spectroscopies using 2D NMR techniques (gHSQC, gHMBC and NOESY).[25] A Monte Carlo random search using molecular mechanics followed by the geometry optimisation of each minimum energy structure using DFT calculations at the B3LYP/6-31G* level and a Boltzmann analysis of the total energies were used to give molecular models describing the conformational behaviour of 1. The correlations between the experimental ^{1}H and ^{13}C NMR chemical shifts (δ_{exp}) and the relation among the experimental data and GIAO/B3LYP/6-311++G(d,p)-calculated magnetic isotropic shielding tensors (σ_{calc}) for conformers 2 and 3 were reported as $\delta_{exp} = a + b\sigma_{calc}$. ^{1}H-NMR chemical shifts of all-*syn*-1,2,4,5-tetrafluorocyclohexane, all-*syn*-1,2,3,4-tetrafluorocyclohexane and their complexes with benzene were calculated at the BHandH/6-311+G(2d,p) level and the observed shielding of certain resonances on going from CD_2Cl_2 to toluene-d_8 solution was qualitatively reproduced in these model calculations using B3LYP/def2-TZVP optimised geometries.[26] The ^{1}H chemical shifts of 48 amides in DMSO solvent were assigned and it was reported that the solvent shifts $\Delta\delta$ (DMSO–$CDCl_3$) are 1–2 ppm for the NH protons, whereas they are smaller and neg. (−0.1 to −0.2 ppm) for close range protons.[27] The effects of substituents F, Cl, Me on the NH proton shifts were also reproduced. It was found that the electric field coefficient for the protons in DMSO is 90% of that in $CDCl_3$, and the observed deshielding is due to the electric field effect.

2.2 Deuterium (^{2}H) ($I = 1$)

The acetylacetonato Ni-(II) complexes: [Ni-$(acac)_2$] and *trans*-[Ni-$(acac)_2(X)_2$]nH/D_2O, X = H_2O, D_2O, NH_3, MeOH, $PMePh_2$, PMe_2Ph or $[dppe]_{1/2}$, $n = 0$ or 1, dppe = 1,2-bis-(diphenylphosphino)-ethane and *cis*-[Ni-$(F_6\text{-acac})_2(D_2O)_2$], F_6-acac = hexafluoroacetylonato, were studied by ^{2}H and ^{13}C solid-state MAS NMR spectroscopies.[28] The ^{2}H MAS NMR was used to probe the local H bonding network in [Ni-$(acac)_2(D_2O)_2$]-D_2O and *cis*-[Ni-$(F_6\text{-acac})_2(D_2O)_2$]. The size of the anisotropy (d), which is the sum of the chemical shift anisotropy (CSA) and the paramagnetic electron-nuclei dipolar coupling, is much more descriptive than the isotropic shift and the CSA was found to constitute up to 1/3 of the total anisotropy.

Inter-cage dynamics in structure I, II and H fluoromethane hydrates was studied by NMR and molecular dynamics simulations.[29] ^{2}H NMR powder lineshapes of a CD_3F/THF sII and CD_3F/TBME sH hydrate, display the evidence that the populations of CH_4 and CH_3F in the D and D′ cages were in a state of rapid exchange. The di-μ-oxo dimanganese(IV) complex with salen, where salen is (*R*,*R*)-*N*,*N*′-bis(3,5-di-*tert*-butylsalicylidene)-1,2-cyclohexanediamine, was studied by ^{1}H, ^{2}H NMR, CD spectroscopies and DFT calculation.[30] It was found that the dimerisation reaction is highly diastereoselective in the presence of the *t*-Bu group at the 3/3′ position as a determinant steric factor and a less-covalent Ti–O bonding causes a distortion of the di-μ-oxo dititanium(IV) core structure, generating a completely different framework for interligand interaction. The zwitterions $\{ON(CH^+)NO\}M^-(CH_2SiMe_3)_2$ (M = Ga, 2; In, 3) obtained from hydropyrimidine salan (salan = *N*,*N*′-dimethyl-*N*,*N*′-bis[(2-hydroxyphenyl)methylene]-1,2-diaminoethane) proteo-ligands with a rigid backbone $\{ON(CH_2)NO\}H_2$ by the reaction with $M(CH_2SiMe_3)_3$ (M = Ga, In) were studied by ^{2}H NMR spectroscopy and DFT calculations.[31] It was shown that the formation of the Ga/In zwitterion involves rapid release of $SiMe_4$ followed by evolution of H_2 and the systems $\{ON(CH_2)NO\}H_2 + M(CH_2SiMe_3)_3$ (M = Al, Ga, In) initially lead to the formation of the neutral monophenolate dihydrocarbyl species through a single protonolysis.

2.3 Tritium (^{3}H) ($I = 1/2$)

A ^{3}H NMR spectrum of [3′-^{3}H]-4-(2′-chloro-6′-hydroxyphenyl)-2-thioxo-3,4-dihydro-1*H*-indeno[1,2-*d*]pyrimidin-5(2*H*)-one on 13.5 MBq at 75 μM was recorded.[32] The specific activity of the compound is 0.9 TBq $mmol^{-1}$. [3α-^{3}H]-17α-hydroxypregnenolone was synthesised and [3β-^{3}H]-isomer in RP-HPLC purified product was identified by ^{3}H NMR.[33]

2.4 Lithium (^{6,7}Li) ($I = 1, 3/2$)

Monoclinic lithium titanate was studied by ^{6,7}Li NMR spectroscopies and three lines found in the ^{7}Li NMR spectra differ in quadrupole splitting frequencies ν_Q and are assigned to three crystallographic sites of lithium: Li_1 ($\nu_Q \sim$ 27 kHz); Li_2 ($\nu_Q \sim$ 59 kHz); Li_3 ($\nu_Q \sim$ 6 kHz) according to *ab initio* electric field gradient (EFG) calculations.[34] It was assumed based on the analysis of ^{6}Li NMR spectra that there is mixed occupancy of lithium and titanium sites in the corresponding layers of the crystal structure of low-temperature lithium titanate. Li_2[L1], Li_2[L2] and pyridine adduct of Li_2[L1], $(py)_2Li_2$[L1], where [L1] = 2-tetrahydrofuranyl-*N*,*N*-bis(2-methylene-4-methyl-6-*tert*-butylphenolate) and [L2] = 2-tetrahydrofuranyl-*N*,*N*-bis(2-methylene-4,6-*tert*-butylphenolate) were studied by multi-nuclear solution NMR and solid-state ^{6}Li and ^{7}Li NMR spectroscopies.[35] The behaviour of Li^+ ions on electrochemical Li extension/insertion process was investigated by ^{7}Li NMR measurements for $LiCoO_2$ cathode in a plastic cell battery.[36] It was found that the peak position of the ^{7}Li NMR shift for Li_xCoO_2 moves as a function of Li content. ^{7}Li NMR spectroscopy was used to investigate the stoichiometry and stability of a Li^+ complex with N^1,N^2-bis(pyridin-2-ylmethylene)ethane-1,2-diamine (L^1), N^1,N^3-bis(pyridin-2-ylmethylene)propane-1,3-diamine (L^2) and N^1,N^4-bis(pyridin-2-ylmethylene) butane-1,4-diamine (L^3), and the complexation of Mn^{2+}, Cd^{2+} and Zn^{2+} ions with L^1, L^2 and L^3 in acetonitrile (AN) and nitromethane (NM) solutions.[37] The formation constants of the resulting complexes were evaluated from computer fitting of the mole ratio data to an equation that relates the observed chemical shifts to the formation constant and the stability of the complexes were found to vary in the order $Zn^{2+} > Cd^{2+} > Mn^{2+} > Li^+$. The complex formation of Li^+ with 15-crown-5 (15C5, L) and 18-crown-6 (18C6, L) in a hydrophobic ionic liquid (RTIL) 1-butyl-3-methylimidazolium bis[(trifluoromethyl)sulfonyl]imide ([BMIM][N(Tf)$_2$]) were studied with ^{7}Li NMR technique at 27–57 °C.[38] The $\log K$(Li + L) and ΔH^o(Li + L) values at 25 °C were determined as 1.66 (0.03), ~0 kJ mol^{-1} for 15C5 and 2.0 (0.1), −17.0 (0.5) kJ mol^{-1} for 18C6, respectively. The ^{6,7}Li NMR spectra and ^{7}Li spin–lattice relaxation rate were measured on Li_2ZrO_3 and the ^{7}Li NMR lines are attributed to corresponding structural positions of lithium Li_1 and Li_2 by comparing the EFG components with those obtained in the first-principles calculations of the charge density.[39] The complex-formation reaction between Li^+ and the tridentate N-donor ligand terpyridine in the ionic liquids [emim][NTf$_2$] and [emim][ClO$_4$] was studied ^{7}Li NMR experiments and the NMR data implicate the formation of $[Li(terpy)_2]^+$.[40] DFT calculations showed that partial coordination of terpyridine involving the coordination of a solvent anion is excluded. Lithium storage in TiO_2(B) was investigated by ^{6}Li MAS NMR.[41] The Band A at 0 ppm *vs.* LiCl in the ^{6}Li NMR spectra and ΔS measurements indicates that ionic lithium is stored at surface domain of Li_xTiO_2(B) up to $x = 0.18$, and the Band B at 2.5 ppm significantly increases with lithium insertion in the region for $0.25 < x < 0.5$, which indicates that lithium is mainly inserted into the site lying between bridging oxygens in Li_xTiO_2(B). Band D is observed at more positive chemical shift than Band B in the region for $x > 0.75$ for Li_x TiO_2(B). Lithium ephedrates and norcarane-derived lithium amino alkoxides were investigated by ^{6}Li NMR spectroscopy and it was revealed that the lithium amino alkoxides are tetrameric.[42] The low-temperature ^{6}Li NMR spectra and DTF computations show stereoisomerically pure homoaggregates displaying resonances consistent with an S_4-symmetric cubic core rather than the alternative D_{2d} core. The ^{7}Li and ^{31}P NMR studies in $LiCoPO^{tetra}{}_4$ microcrystals and olivine $LiCoPO_4$, which is transformed from $LiCoPO^{tetra}{}_4$ by an annealing process, were made.[43] The paramagnetic shift caused by the Co spin of Li_xCoO_2 ($x > 1$) was appreciable by ^{6,7}Li solid-state MAS NMR, and there are five major minor signals at ~ −16, −6, 3, 185 and 1100 ppm with the intensity ratio of 2 : 4 : 4 : 2 : 0.5–1.0 to the major signal at 0 ppm.[44] The analysis of temperature-dependent ^{7}Li shifts and also ^{6}Li T_1 experiments confirm that the observed shifts for these minor peaks are ascribed to the paramagnetic interaction and the ^{7}Li–^{7}Li two-dimensional correlation experiments show that the three minor Li atoms, whose shifts are 3, −6 and −16 ppm, exist in the same Li layer. Structural changes in amorphous Li_xSi were studied as a function of Li flux and total charge using ^{7}Li solid-state NMR spectroscopy and it was found that the peak shifts in solid-state ^{7}Li NMR spectra vary as a function of the Li flux during the initial lithiation and stay constant for the samples of the same Li flux regardless of the total charge applied to the electrodes.[45] $LiCo_{1-2x}Ni_xMn_xO_2$ ($0<x\leq0.5$) was studied by ^{7}Li MAS NMR technologies and the structural-chemical shift regularities were established by systematic variation of the ratio of diamagnetic Co^{3+} to paramagnetic Ni/Mn ions with variable valences.[46] ^{7}Li NMR probes the local structure of Li with respect to transition metal ions located in two adjacent layers and it was demonstrated that for $LiNi_{0.5}Mn_{0.5}O_2$ has only paramagnetic Ni^{2+} and Mn^{2+} ions. $[(C_{20}H_{10}{}^{4-})(LiK_5)^{6+}(C_{20}H_{10}{}^{4-})]^{2-}$ and $[(C_{20}H_{10}{}^{4-})(Li_3K_3)^{6+}(C_{20}H_{10}{}^{4-})]^{2-}$ was studied by ^{7}Li NMR and the engagement of the hub-site of $C_{20}H_{10}{}^{4-}$ in Li binding is accompanied by unprecedented shifts up to −24 ppm in ^{7}Li NMR spectra.[47] ^{6,7}Li MAS NMR studies on a Li–S battery and its sulfur components were made in order to understand the mechanism of polysulfide formation and the interactions of sulfur and polysulfides with a host matrix and electrolyte.[48] The electrochemical activation process of the so-called alkali-treated soft C (ASC) was examined by *ex situ* ^{7}Li NMR measurements and it was proved that the insertion of Li^{+} into the pore structure of ASC is associated with the activation process in the electrolyte dissolving Li salt.[49] Organic–inorganic hybrid polymer electrolytes based on poly(propylene glycol)-*block*-poly(ethylene glycol)-*block*-poly-(propylene glycol)bis(2-aminopropyl ether), polyacrylonitrile (PAN), 3-(glycidyloxypropyl)trimethoxysilane (GLYMO) and 3-(aminopropyl)trimethoxysilane (APTMS) complexed with $LiClO_4$ were systematically investigated by a variety of techniques including multinuclear (^{7}Li, ^{13}C, ^{29}Si) solid-state NMR, differential scanning calorimetry (DSC), Fourier transform IR spectroscopy (FTIR), AC impedance, linear sweep voltammetry (LSV) and charge–discharge measurement.[50] The ^{7}Li NMR measurements revealed that there are the strong correlation of the lithium cation and the polymer matrix, and lithium has two local environments. Site occupation and structural change in Li substituted $AgNbO_3$ were probed by ^{7}Li NMR and characteristic changes in the NMR spectra depending on Li content were observed.[51]

2.5 Sodium (^{23}Na) ($I = 3/2$)

The ^{23}Na, ^{19}F and ^{89}Y solid-state NMR measurements and DFT calculations were made to characterise the structure of cubic $NaYF_4$ nanoparticles with and without Er^{3+} doping.[52] The ^{23}Na MAS NMR spectra obtained at various magnetic fields and 3QMAS spectra at ultrahigh field are characteristic of a solid solution in which cation sites are statistically occupied by Na^{+} or Y^{3+} ions, and ^{23}Na NMR spectra are broadened as a result of isotropic chemical shift distribution, whereas the quadrupolar products appear to be small, which is in good agreement with DFT calculations using CASTEP. Phase separation of alkali halide (MX) ($M = Li^{+}$, Na^{+} and K^{+} and $X = Cl^{-}$ and Br^{-})-acetonitrile (AN)–water mixtures by cooling was investigated by ^{23}Na and ^{7}Li NMR measurements, small-angle neutron scattering (SANS) and IR.[53] The chemical shifts and spin–lattice relaxation rates of ^{23}Na and ^{7}Li NMR for the mixtures suggested the different solvation structure of Li^{+} and Na^{+}. ^{23}Na NMR studies of the Na–K eutectic alloy embedded into porous glass with 7 nm pores were made and it was shown that melting of Na_2K confined nanoparticles is a continuous process with smooth changes in the Knight shift of a narrow resonance line and nuclear spin relaxation between those in the crystalline and liquid states.[54] The metalation, demetalation, transmetalation and/or site migration processes in the mixtures of $[LuNa(R_B)(CH_3COO)]\cdot 2^{i}PrOH$, $[LuNa(L_B)(Cl)]$, $[LaNa(R_B)(Cl)]$ and $[LaNa(L_B)(Cl)]$ with different lanthanide(III) salts were studied by ^{23}Na NMR spectroscopy.[55] Here, the [1 + 1] Schiff base macrocyclic ligands H_2L_A and H_3L_B, synthesised by condensation of 3,3′-(3-oxapentane-1,5-diyldioxy)bis(2-hydroxybenzaldehyde) (H_2L') with 1,5-diamino-3-azamethylpentane (A′) and *N*,*N*-bis(2-aminoethyl)-2-hydroxybenzylamine · 3HCl (HB′ · 3HCl), respectively, and their related polyamines H_2R_A and H_3R_B, exhibit enhanced stability toward hydrolysis and greater flexibility. The $NaNTf_2$ (or NaTFSI) in *N*-propyl-*N*-methylpyrrolidinium bis(fluorosulfonyl)imide (or C_3mpyrFSI) ionic liquid were investigated by ^{23}Na NMR and the chemical shifts and spectral linewidths indicate a complex coordination of the Na^{+} ion.[56] ^{23}Na MAS, double-rotation (DOR) and multiple-quantum MAS (MQMAS) NMR spectra provide useful information on the number of molecules in the asymmetric unit, the local coordination geometry of the Na cations, and the presence of amorphous phases. Then, ^{23}Na and ^{13}C solid-state NMR study of sodium valproate was made.[57]

2.6 Potassium (^{39}K) ($I = 3/2$)

The structures of K-rich feldspar, $(K > Na)AlSi_3O_8$ were studied by high-resolution MAS multinuclear magnetic resonance spectroscopies, leading to ^{29}Si, ^{27}Al and ^{23}Na spectra at 9.4 T and ^{27}Al, ^{39}K and ^{23}Na spectra at 19.6 T.[58] Long-range disordered structures show non-random disorder at the medium-range scale, triclinic-like distortions with four sets of T sites for framework atoms, two sets of M sites for alkali atoms and Al–O–(K, Na) multi-site correlations. An NMR crystallographic approach incorporating multinuclear solid-state NMR, X-ray structure determinations and DFT were used to characterise the H bonding arrangements in benzoic acid and the corresponding Group I alkali metal hydrogen dibenzoates systems.[59] ^{7}Li, ^{39}K, ^{87}Rb and ^{133}Cs NMR spectroscopies and DFT calculations provided trends and a corroboration of the H bond geometry.

2.7 Rubidium ($^{85,87}Rb$) ($I = 5/2, 3/2$)

The structures of mixed $(NH_4)_{2-x}Rb_xZnCl_4$ ($x = 0$, 1 and 2) single crystals were studied by ^{1}H and ^{87}Rb NMR spectroscopies.[60] The NMR results confirmed that there are two crystallographically inequivalent NH_4 sites in $(NH_4)_2ZnCl_4$ and two crystallographically inequivalent Rb sites in Rb_2ZnCl_4. RbRE-$[T\text{-}(CN_2)_4]$ (RE = La, Pr, Nd, Gd; T = Si, Ge) was investigated by ^{73}Ge, ^{87}Rb and ^{139}La solid-state NMR measurements and DFT calculations[61] Luminescence properties of Eu^{3+}, Ce^{3+} and Tb^{3+} doped samples were also reported. The H bonding arrangements in benzoic acid and the corresponding Group I alkali metal hydrogen dibenzoates systems were studied by using ^{7}Li, ^{39}K, ^{87}Rb and ^{133}Cs NMR spectroscopies and DFT calculations as described at §2.6.[59]

2.8 Cesium (^{133}Cs) ($I = 7/2$)

The binding properties of cryptophane-222, cryptophane-223 and cryptophane-233 to cesium and thallium cations were studied by ^{133}Cs and ^{205}Tl NMR spectroscopies.[62] The study revealed that the cavity size of the cryptophane is not the main parameter for efficient binding, whereas the number of phenolate moieties surrounding the cryptophane backbone seems to be pivotal for the complexation. Mixed-ion metaphosphate glasses $A_{1-x}B_xPO_3$ ($0 \leq x \leq 1$) based on the cation combinations Cs–Li, Rb–Li and Cs–Ag were analysed by ^{31}P, ^{87}Rb, ^{109}Ag and ^{133}Cs NMR spectroscopies.[63] The results from ^{7}Li–^{133}Cs (SEDOR) experiments indicated that a random mixture of Cs and Li in Cs–Li metaphosphate glasses and it was found that the substitution by the bigger Cs^{+} ion causes a strong reduction in Ag^{+} mobility. The H bonding arrangements in benzoic acid and the corresponding Group I alkali metal hydrogen dibenzoates systems were studied by using ^{7}Li, ^{39}K, ^{87}Rb and ^{133}Cs NMR spectroscopies and DFT calculations as described at §2.6.[59] A UMo powellite glass–ceramic was studied by ^{133}Cs solid state NMR spectroscopy and SEM, and it was shown that the Cs to inhabit the glass phase.[64] The relationship between the intrinsic crystallographic $C_{60}{}^{3-}$ orientational disorder of Cs_3C_{60} and the molecular Jahn–Teller (JT) effect dynamics in the Mott insulating state was studied by ^{133}Cs and ^{13}C MAS NMR spectroscopies.[65] It was revealed that the JT-effect dynamics are fast on the NMR time-scale of 10^{-5} s despite the presence of the frozen-in $C_{60}{}^{3-}$ merohedral disorder and the number of inequivalent ^{13}C and ^{133}Cs sites is increased by cooling to sub-liquid-nitrogen temperatures.

3 Group 2 (^{9}Be, ^{25}Mg, ^{43}Ca, ^{87}Sr, ^{135,137}Ba)

3.1 A review on ^{25}Mg, ^{43}Ca, ^{87}Sr and ^{135,137}Ba

Recent specifics and the experimental challenges of the solid-state NMR spectroscopies of ^{25}Mg, ^{43}Ca, ^{87}Sr and ^{135,137}Ba, which are characterised by low intrinsic sensitivity due to their small magnetic moments and low natural abundance, were discussed in a context of contemporary applications to inorganic, organometallic and biological materials as a review.[66]

3.2 Beryllium (^{9}Be) ($I = 3/2$)

The structures of 1-tris(pyrazolyl)borate beryllium halides TpBeX (X = Cl, Br, I, F), the pyrazole adduct of TpBeF and of 1-tris(pyrazolyl)borate beryllium hydride, deuteride and azide TpBeX (X = H, ^{2}H, N_3; Tp = 1-trispyrazolylborate) were studied by ^{9}Be NMR spectroscopy and the formation of heteroleptic coordination complexes of beryllium was discussed.[67] The complexation behaviours of Be^{2+} with cyclo-μ-imido triphosphate anions, $cP_3O_{9-n}(NH)_n^{3-}$ ($n=1$, 2) were investigated by ^{9}Be and ^{31}P NMR techniques to clarify the coordination structures of the complexes.[68] It was shown that $cP_3O_{9-n}(NH)_n$ ($n=1$, 2) ligands form ML, ML_2 and M_2L complexes with Be^{2+} ions. The stability constant of the beryllium complexes was also estimated.

3.3 Magnesium (^{25}Mg) ($I=5/2$)

Mg^{2+} ions in magnesium coordination environments in benzoate and salicylate complexes were investigated by ^{25}Mg solid-state NMR and GIPAW DFT computations.[69] The ultrahigh-field solid-state NMR experiments revealed a relatively large range of values for the ^{25}Mg quadrupolar coupling constant, $C_q(^{25}Mg)$. $MgCl_2$ was demonstrated as a non-nucleophilic and cheap Mg^{2+} source in combination with Al Lewis acids ($AlCl_3$, $AlPh_3$ and $AlEtCl_2$) to formulate a series of Mg^{2+} electrolytes and the electrolyte systems ($MgCl_2$–$AlCl_3$, $MgCl_2$–$AlPh_3$ and $MgCl_2$–$AlEtCl_2$) were studied by multinuclear NMR (^{1}H, ^{27}Al{^{1}H} and ^{25}Mg{^{1}H}) spectroscopies and electrochemical analysis.[70] The molecular structures of the three electrolytes sharing a common Mg^{2+}-dimer mono-cation, $[(\mu\text{-Cl})_3Mg_2(THF)_6]^+$, along with $AlCl_4^-$, $AlPh_3Cl^-$ or $AlEtCl_3^-$, respectively were characterised.

3.4 Calcium (^{43}Ca) ($I=7/2$)

Multinuclear solid-state NMR including ^{43}Ca MAS NMR and GIPAW DFT calculations were applied to the study of the crystal structure of $CaC_2O_4 \cdot H_2O$ and the sensitivity of ^{43}Ca NMR as a structural probe was demonstrated.[71] $Ca(C_6H_5–PO_3H)_2$, $Ca(C_6H_5–PO_3) \cdot 2H_2O$, $Ca(C_4H_9–PO_3H)_2$ and $Ca(C_4H_9–PO_3) \cdot H_2O$ were studied by ^{1}H, ^{13}C, ^{31}P and ^{43}Ca solid-state NMR spectroscopies and GIPAW calculation approach.[72] The NMR parameters calculated from the structural models were found to be in good agreement with the experimental values and this indicates the high accuracy of the DFT-optimised structures. The correlations between the ^{43}Ca isotropic chemical shift/^{31}P asymmetry parameter and the structural features around the phosphonate were reported. The ^{2}H and variable-temperature ^{43}Ca MAS NMR study of Ca-exchanged hectorite and the molecular behaviour at the smectite–H_2O interface was made and it was indicated that Ca^{2+} occurs as proximity-restricted outer-sphere hydration complexes between -120 and $+25$ °C in a two-layer hydrate and at $T \leq -50$ °C in a 2 : 1 water/solid paste.[73] Commonly-available calcium compounds such as calcium tartrate tetrahydrate, calcium formate, $CaCl_2 \cdot 2H_2O$, $Ca(OH)_2$, $CaCO_3$ and *etc.* were analysed ^{43}Ca solid-state NMR and GIPAW DFT computations, and the ^{43}Ca EFG and chemical shift tensors were also discussed.[74] A series of calcium and strontium aryl carboxylate complexes was investigated by ^{43}Ca and ^{87}Sr solid-state NMR experiments in a magnetic field of 21.1 T.[75] The ^{43}Ca isotopic enrichment enabled the characterisation of the ^{43}Ca EFG and chemical shift tensors of the two calcium sites in calcium *p*-aminosalicylate, and the NMR results and DFT calculations showed that ^{43}Ca isotropic chemical shift is sensitive to the Ca–N distance.

3.5 Strontium (^{87}Sr) ($I = 9/2$)

A series of calcium and strontium aryl carboxylate complexes was investigated by ^{43}Ca and ^{87}Sr solid-state NMR experiments in a magnetic field of 21.1 T as described at §3.4.[75] It was found that the ^{87}Sr EFG tensor is sensitive to the presence or absence of nitrogen in the first coordination sphere and the $|C_Q(^{87}Sr)|$ values are sensitive to the coordination number about Sr^{2+}.

3.6 Barium (^{135}Ba) ($I = 3/2$)

$Ba_3Co_2O_6(CO_3)_{0.7}$ was studied by ^{135}Ba NMR.[76] It was reported that marginally itinerant electrons possess large entropy and low-lying excitations with a Wilson ratio $R_W = 116$.

4 Group 3 (^{45}Sc, ^{89}Y, ^{139}La, ^{141}Pr, ^{171}Yb, ^{153}Eu)

4.1 Scandium (^{45}Sc) ($I = 7/2$)

Complexes such as $[ScCl_2([18]aneO_4S_2)][FeCl_4]$, $[ScI_2([18]aneO_4S_2)]I$, $[YCl_2(18\text{-crown-}6)]_3[Y_2Cl_9]$, $[YCl_2([18]aneO_4S_2)][FeCl_4]$, $[LaI_3(15\text{-crown-}5)]$, $[LaI_2(18\text{-crown-}6)(MeCN)]I$, $[LuI(18\text{-crown-}6)(MeCN)_2]I_2$, $[Lu(15\text{-crown-}5)(MeCN)_2(OH_2)]I_3$, $[LaI_3([18]aneO_4S_2)]$, $[LaI([18]aneO_4S_2)(OH_2)]I_2$, $[LaI_3([18]aneO_4Se_2)]$ and $[LuI_2([18]aneO_4Se_2)]I$ were studied by ^{1}H, ^{45}Sc and ^{77}Se NMR spectroscopies, microanalysis, IR and X-ray.[77] $ScCl_3(L_3)$ [L_3 = 1,4,7-Me$_3$-1,4,7-triazacyclononane (Me$_3$tacn), 1,4,7-$(C_{10}H_{21})_3$-1,4,7-triazacyclononane (decyl$_3$-tacn), $HN(CH_2CH_2SC_{10}H_{21})_2$ (decyl-SNS), $HN(CH_2CH_2S$ *t*-Bu$)_2$ (*t*-Bu-SNS)] were characterised by ^{1}H, ^{13}C{^{1}H} and ^{45}Sc NMR spectroscopies, IR, microanalysis and solid-state and solution XAFS spectroscopies.[78] The ^{45}Sc NMR and Sc K-edge XAFS spectroscopic studies of the reaction of $[ScCl_3(Me_3\text{-tacn})]$ with $AlMe_3$ suggested that alkylation at the Sc(III) center does not occur, instead retaining the Cl_3N_3 coordination environment and most likely forming Sc–Cl–$AlMe_3$ bridging interactions. It was similarly shown that alkylation at the Sc(III) center does not occur for $[ScCl_3(\text{decyl-SNS})]$ with $AlMe_3$, too. Diethylenetriamine-*N*,*N*,*N*′,*N*″,*N*″-pentaacetic acid (DTPA) and 1,4,7,10-tetraazacyclododecane-1,4,7,10-tetraacetic acid (DOTA) scandium(III) complexes were investigated by ^{45}Sc NMR spectroscopy and the stability constants of scandium(III) complexes ($\log K_{ScL}$ 27.43 and 30.79 for DTPA and DOTA, respectively) were determined from potentiometric and ^{45}Sc NMR spectroscopic data.[79] The correlation between the local structure and the EFG at the Sc site in ScO_6 in Sc_2O_3, $Ba_3Sc_4O_9$ and $BaSc_2O_4$ was studied by ^{45}Sc NMR spectroscopy and DFT calculations comparing experimental and calculated results.[80] The C_q results obtained from the DFT calculation are in good agreement with those from the NMR measurements.

4.2 Yttrium (^{89}Y) ($I = 1/2$)

$NaYF_4$ was studied by ^{23}Na, ^{19}F and ^{89}Y solid-state NMR measurements and DFT calculations as described at §2.5.[52] The ^{89}Y NMR spectra consist of a single broad pattern resulting from the chemical shift distribution being correlated to the coordination environment of the Y^{3+} cations. $[Cp^*_2Y(\mu\text{-}OOCAr^{Me})(\mu\text{-}H)AlR_2]$ ($Cp^* = C_5Me_5$, $Ar^{Me} = C_6H_2Me_3$-2,4,6, $R = CH_2SiMe_3$, Me) was characterised by ^{89}Y NMR spectroscopy.[81] The ^{89}Y DEPT45 NMR chemical shifts are $\delta = -88.1$ ppm ($R = CH_2SiMe_3$) and $\delta = -86.3$ ppm (R = Me). The structural transformation of $Ba_3YIr_2O_9$ under an applied pressure of 8 GPa at 1273 K was reported and the ^{89}Y NMR shift has no T dependence in the range of 4–120 K and its spin-lattice relaxation rate varies linearly with T in the range of 8–45 K.[82] The cation disorder in $Y_2Ti_{2-x}Sn_xO_7$ pyrochlore ceramics was studied by ^{89}Y and ^{119}Sn NMR spectroscopies and calculations of NMR parameters in solids.[83] YT_2Si_2 (T = Co, Ni, Cu, Ru, Rh, Pd) was analysed by using ^{89}Y solid-state NMR spectra.[84] The results showed that large Knight shifts arises from unpaired conduction electron spin density near the Fermi edge and the Knight shift decreases with increasing valence electron count, reflecting the sensitivity of this parameter to electronic properties. Y_2RuB_6, Y_2ReB_6, Y_2RhSi_3, YRh_2Si and YPdSi were studied ^{89}Y solid-state NMR spectroscopy and the NMR signal shifts were correlated with the density of states of the s electrons at the Fermi level.[85] A non-ambiguous assignment of the NMR signals to the various sites in the crystal structures was obtained. ^{89}Y solid-state NMR spectroscopy was utilised to investigate TiNiSi type tetrelides YTX (space group Pnma) with T = Ni, Ir and X = Si, Ge, Sn, Pb.[86] The ^{89}Y Knight shifts and a monotonic dependence on the tetrelide Pauling electronegativity were observed. Proton to ^{89}Y cross polarisation (CP) MAS spectra of Y^{3+} in a frozen aqueous solution were acquired using the AMUPol biradical as a polarising agent by indirect dynamic nuclear polarisation (DNP) methods, and the ^{89}Y and ^{1}H NMR signals from hydrated Y-doped zirconate ceramics were also detected.[87]

4.3 Lanthanum (^{139}La) ($I = 7/2$)

The structure of $La_{12}GdEuB_6Ge_2O_{34}$ was studied by ^{139}La NMR and IR spectroscopies and the local symmetry of La in fine crystalline $La_{12}G$ $dEuB_6Ge_2O_{34}$ were assessed at room temperature.[88] The ^{139}La NMR spectra of the $LaBO_3$, $La(BO_2)_3$ and $La_{12}GdEu(BO_3)_6(GeO_4)_2O_8$ borates show a signal at (1080 ± 40) ppm, which is absent in the spectrum of La_2GeO_5 and the shape of the NMR spectra of $La_{12}GdEu(BO_3)_6(GeO_4)_2O_8$ and $LaBO_3$ is characterised by the 2nd-order quadrupole splitting. The LaCoGe with the tetragonal CeFeSi-type structure were investigated by ^{59}Co and ^{139}La NMR and NQR measurements.[89] It was concluded that spin-density-wave (SDW)-type ordering occurs, where magnetic correlations are of the checkerboard type in the *ab*-plane and have a long periodicity along the *c*-axis. Tetracyanamidometallates RbRE-[T-$(CN_2)_4$] (RE = La, Pr, Nd, Gd; T = Si, Ge) were studied by ^{73}Ge, ^{87}Rb and ^{139}La solid state NMR measurements and DFT calculations to verify the homoleptic $[Ge\text{-}(CN_2)_4]^{4-}$ ion.[90] The Mn spin correlations for $LaMnO_3$ crystal were studied near the O′–O phase transition at $T_{JT} = 750$ K up to 950 K by ^{17}O and ^{139}La NMR spectroscopies.[91] It was found that the measured local hyperfine fields originate from the electron density transferred from the e_g and t_{2g} orbitals to the 2s(O) and 6s(La) orbits, respectively. The crystal structure of $La_{10}W_2O_{21}$ was investigated by ^{139}La NMR spectroscopy, convergent beam electron diffraction (CBED), TEM and other techniques.[92] The La/W mixed site was confirmed by ^{139}La NMR.

4.4 Praseodymium (^{141}Pr) ($I = 5/2$)

The influence of microwave irradiation on the recovery of nanocrystalline PrF_3 powders was analysed by ^{141}Pr NMR at 1.5 K.[93]

4.5 Ytterbium (^{171}Yb) ($I = 1/2$)

The ^{171}Yb and ^{195}Pt NMR studies on $YbPtGe_2$ were reported and a well-resolved NMR line with small shift was assigned to divalent ^{171}Yb.[94] The model with di- and trivalent subsets of Yb species located on the Yb2 and Yb1 site of the $YbPtGe_2$ lattice was proposed.

4.6 Europium (^{153}Eu) ($I = 5/2$)

$EuGa_4$ was studied by ^{153}Eu and ^{69,71}Ga NMR spectroscopies and ^{153}Eu zero-field NMR signal with well-resolved nuclear quadrupole splitting was observed.[95]

5 Group 4 (^{47,49}Ti, ^{91}Zr, ^{177}Hf)

5.1 Titanium (^{47,49}Ti) ($I = 5/2, 7/2$)

The MS_2 (M = Zr, Ti, W, Mo and Ta) was studied by ^{33}S, ^{47,49}Ti, ^{91}Zr and ^{95}Mo solid-state NMR spectroscopies and quantum chemical calculations.[96] The NMR parameters obtained were shown to be sensitive to local geometric and electronic environments. Metal NMR shielding and EFG tensors were examined by quantum-chemical calculations for Ti(IV) complexes and the Ti shielding constants and EFG tensors for a variety of olefin (co)polymerisation catalysts were analysed.[97] A specific dependence of the Ti EFG tensors on the exact arrangement and type of surrounding bonds was also demonstrated.

5.2 Zirconium (^{91}Zr) ($I = 5/2$)

The MS_2 (M = Zr, Ti, W, Mo and Ta) was studied by ^{33}S, ^{47,49}Ti, ^{91}Zr and ^{95}Mo solid-state NMR spectroscopies and quantum chemical calculations. as described at §5.1.[96]

5.3 Hafnium (^{177}Hf) ($I = 7/2$)

The NMR measurements of the magnetic dipole moments of two high-K isomers in ^{177}Hf and in ^{180}Hf by the method of online nuclear orientation and the results on the angular distributions of γ transitions in the decay of the ^{177}Hf isotope were reported.[98]

6 Group 5 (^{51}V, ^{93}Nb)

6.1 Vanadium (^{51}V) ($I = 7/2$)

The $[(V^{5+}O)_2(bdeah)]$ or $[(V^{4.5+}O)_2(bdeas)S_2]$ (S = DMSO or MeOH) obtained by the reaction of $NaVO_3$ or $[VO(acac)_2]$ with 2,5-bis((bis(2-hydroxyethyl)amino)methyl)hydroquinone, H_6bdeah, were studied by variable temperature UV-visible and ^{51}V NMR spectroscopies as well as EPR.[99] The thermodynamic parameters of this intramolecular electron transfer were determined using the results obtained by UV-visible ($\Delta H = -36 \pm 2$ kJ mol^{-1} and $\Delta S = -129 \pm 5$ J mol^{-1} K^{-1}) and ^{51}V NMR spectroscopy ($\Delta H = -37 \pm 2$ kJ mol^{-1} and $\Delta S = -109 \pm 5$ J mol^{-1} K^{-1}). The structural and electronic properties as well as the catalytic activity toward sulfoxidation of $VO_2L^NS^2$-$HNEt_3$ (L^NS^2 = 2,2′-pyridine-2,6-diyl-bis-1,1′-diphenylethanethiol) and $VOL^{N2}S^2$ ($L^{N2}S^2$ = 2,2′-2,2′-bipyridine-6,6′-diyl-bis-1,1′-diphenylethanethiol) were investigated by DFT calculations, ^{51}V NMR and EPR spectroscopies.[100] The structure and its stability of the *cis*-oxo peroxo V^V intermediate species were partly explained. [V^{IV}O{Hdfmp(inh)$_2$}(H_2O)], [V^{IV}O{Hdfmp(nah)$_2$}(H_2O)], [V^{IV}O{Hdfmp(bhz)$_2$}(H_2O)], [V^VO(OMe)(MeOH){Hdfmp(inh)$_2$}], [V^V O(OMe)(MeOH){Hdfmp(nah)$_2$}], [V^VO(OMe)(MeOH){Hdfmp(bhz)$_2$}], [V^VO_2{H_2dfmp(inh)$_2$}]$_n \cdot 5H_2O$, K[V^VO_2{Hdfmp(nah)$_2$}] and K[V^V O_2{Hdfmp(bhz)$_2$}] prepared as the Schiff bases H_3dfmp(L)$_2$ obtained by the condensation of 2,6-diformyl-4-methylphenol and hydrazones [L = isonicotinoylhydrazide (inh), nicotinoylhydrazide (nah) and benzoylhydrazide (bhz)] were studied IR, UV-visible, EPR, ^1H, ^{13}C and ^{51}V NMR spectroscopies and DFT calculation.[101] The V^{IV} hyperfine coupling constants of V^{IV} compounds and ^{51}V NMR chemical shifts of several V^V-species were measured and the intermediates involved in the catalytic process were established by UV-visible, ^{51}V NMR and DFT studies. $VCl(PMe_3)_2(N^tBu)_2$, $VCl(PEt_3)_2(N^tBu)_2$, $VCl(PMe_2Ph)_2(N^tBu)_2$, $VCl(Py)_2(N^t Bu)_2$, $VBr(PMe_3)_2(N^tBu)_2$, $VI(PMe_3)_2(N^tBu)_2$, $VPh(PMe_3)_2(N^tBu)_2$, $[V(PMe_3)_3(N^tBu)_2][Al(PFTB)_4]$, $[V(PEt_3)_2(N^tBu)_2][Al(PFTB)_4]$ and $[V(DMAP)(PEt_3)_2(N^tBu)_2][Al(PFTB)_4]$ (DMAP = dimethylaminopyridine, PFTB = perfluoro-*tert*-butoxide) were studied by X-ray diffraction, ^{13}C and ^{51}V NMR spectroscopies and V $L_{3,2}$-edge X-ray absorption near-edge structure (XANES) spectroscopy.[102] The results obtained by ^{51}V NMR shift, ^{13}C NMR $\Delta\delta_\alpha$b and L_3-edge energy to the nature of the supporting phosphine ligand showed that the electronic structure is dominated by π bonding to the imides, and localisation of electron density at the nitrogen atoms of the imides is dictated by the cone angle and donating ability of the axial neutral supporting ligands. $K_3[V_2O_2(O_2)_4(IO_3)] \cdot H_2O$ was investigated by ^{51}V NMR spectroscopy and the signals corresponding to $[VO(O_2)_2(H_2O)]^-$, $[V_2O_2(OH)(O_2)_4]^{3-}$ and $H_2VO_4^-$ species were obtained.[103] [V^VO(hap-dahp)] obtained through the reaction of [V^{IV}O(acac)$_2$] with H_3hap-dahp (Schiff base derived from 2-hydroxyacetophenone and 1,3-diamino-2-hydroxypropane) in methanol under aerobic conditions and the encapsulated oxidovanadium(V) complex [V^VO(hap-dahp)]-Y in the cavity of zeolite Y were studied by elemental analysis, IR, electronic, ^1H and ^{51}V NMR spectroscopies, scanning electron micrographs and X-ray diffraction patterns.[104] $(TMA)_9[V_3Nb_{12}O_{42}] \cdot 18H_2O$ was studied by X-ray crystallography, ESI-MS and ^{51}V liquid and solid-state NMR spectroscopies and the ^{51}V NMR spectra indicate two major peaks corresponding to one VO_4 and two VO_5 sites.[105] The synergic effect of combined addition of V_2O_5 and P_2O_5 on structural and sealing properties of glasses with nominal composition (mol%) $27SiO_2$–$23SrO$–$32BaO$–$4Al_2O_3$–$10B_2O_3$–$(4\text{-}x)P_2O_5$-xV_2O_5

was reported as studied by ^{29}Si, ^{31}P, ^{51}V, ^{27}Al and ^{11}B MAS NMR spectroscopies.[106] The structures vanadium(v) cation in mixed acid based electrolyte solution existing as di-nuclear $[V_2O_3Cl_2 \cdot 6H_2O]^{2+}$ at higher vanadium concentrations (≥1.75 M) were analysed by DFT based computational modeling and ^{51}V and ^{35}Cl NMR spectroscopies.[107] The biosorption potential of different types of carbohydrate biomass was investigated by ESR, ^{51}V NMR spectroscopy, CD and UV-visible absorption measurements.[108] Three related oxidovanadium(IV) complexes, $[V^{IV}O(SO_4)(H_2O)_2(NN)]$, with the phenanthroline derivatives (NN) [1,2,5]thiadiazolo[3,4-*f*][1,10]phenanthroline (tdzp), 1,10-phenanthroline-5,6-dione (phendione) and 5,6-epoxy-5,6-dihydro-1,10-phenanthroline (epoxyphen) were characterised by elemental analysis, electrospray ionisation mass spectrometry (ESI-MS), conductometric measurements, FTIR, UV/visible, electronic paramagnetic resonance (EPR) spectroscopy and ^{51}V-NMR spectroscopy.[109] A series of vanadium(v) complexes with *o*-dioxolene (catecholato) ligands and an ancillary ligand, (*N*-(salicylideneaminato)ethylenediamine) (hensal) were investigated using ^{51}V solid-state MAS NMR spectroscopy.[110] The solid-state ^{51}V NMR parameters of V(v) complexes with a related potentially tetradentate ancillary ligand (*N*-salicylidene-*N'*-(2-hydroxyethyl)ethylenediamine) ($h_2$shed) were previously shown to be associated with the size of the HOMO–LUMO gap in the complex and it was shown that the modification of the ancillary ligand does not affect the observed trend between complexes ranging from catechols with electron-rich to electron-poor substituents. Antidiabetic drugs metformium and decavanadate were investigated by NMR and IR spectroscopies.[111] It was found that there are only small differences between the metformium and Na^+ decavanadate materials but the decavanadate-metformin material affect differently the water pool and water organisation near the interface of reverse micelles formed. A V-containing solution prepared from methanolysis of VCl_4 was added to the TiO_2 suspension and the resulting colloidal counterpart of the TiO_2-supported V_2O_5 system was characterised by X-ray diffraction, TEM, FTIR, Raman and ^{51}V solid-state MAS NMR spectroscopies.[112] ^{51}V MAS NMR showed that V is partially inserted in the anatase lattice, while the surface is covered with a denser V–O–V network. The $[VO(L^{5\text{-}Br})(OMe)]$ (1) or $[VO(L^{3\text{-}OMe})(OMe)]$ (2) prepared from sugar-modified Schiff-base ligands derived from benzyl 2-deoxy-2-salicylideneamino-α-D-glucopyranoside ($H_2L^{5\text{-}Br}$ and $H_2L^{3\text{-}OMe}$) and $[VO(L^{5\text{-}Br})(OMe)(MeOH)]$ (1-MeOH) or $[VO(L^{3\text{-}OMe})(OMe)(MeOH)]$ (2-MeOH) was studied by ^{1}H and ^{51}V NMR spectroscopies and DFT calculations, and the difference between the ^{51}V NMR chemical shifts of the two diastereomers for 1 or 2 is 30 or 28 ppm, respectively.[113] The chemical speciation of 2-amino-*N*-hydroxypropanamide (β-alaninohydroxamic acid, HL) and V(v) in aqueous solution was studied using the results obtained by the calculations of the thermodynamic properties and the ^{51}V NMR chemical shifts of the species formed at equilibrium and the experimental ^{51}V NMR measurements.[114] The ^{51}V NMR chemical shifts were calculated by DFT approach accounting for relativistic corrections and solvent effects and all the tautomers of the 1 : 1 and 1 : 2 VO_2^+/β-ala complexes with different degrees of protonation were calculated. The NMR chemical shift parameters obtained by DFT calculations are in good agreement with the experimental data. $[V^VO(L^1)(OMe)]$ and $[V^VO(L^2)(OMe)]$, with ONNO donor ethylenediamine-bis(phenolate) ligands (H_2L^1 (3) and H_2L^2 (4)) were studied by ^{51}V NMR spectroscopy.[115] The thermodynamic model of the vanadate system under neutral to basic conditions and the thermodynamic equilibrium constants determined by ^{51}V NMR spectroscopy and potentiometric titrations were reported.[116] $Na_{17}[(Co(OH_2)Co_2V W_9O_{34})_2(VW_6O_{26})]\cdot 31H_2O$ was investigated by X-ray crystallography, elemental analysis, FTIR, UV/Vis and ^{51}V NMR spectroscopies and by cyclic voltammetry.[117] The stability studies and ^{51}V NMR and X-ray crystallographic analyses showed that the polyoxoanion transforms into the tetracobalt-containing sandwich-type polyoxoanion. The effects of vanadyl, vanadate and decavanadate in the actin structure and function were studied partly by ^{51}V-NMR spectroscopy.[118] An efficient method of synthesis of the vanadium nitride at low temperature was evaluated, a mechanism for the crystallisation process was proposed and the structural transformations of guanidinium *m*-vanadate (GmV) with increasing temperatures were studied by periodic DFT calculations, ^{51}V MAS NMR, XRD, FTIR and elemental analytical methods.[119] The results showed that the transformation is energetically favorable and the formation of a VO_xN_{1-x} solid solution is feasible. The dioxidovanadium(V) complexes with hydrazone Schiff-base ligands derived from salicylaldehyde and ω-hydroxy functionalised carbonic acid hydrazides with three different chain lengths were investigated by ^{51}V solid-state MAS NMR spectroscopy.[120] It was found that the variation of the chemical shift anisotropy parameters dependent on differences in the supramolecular structure of the compounds, such as hydrogen bonding or crystal packing. The vanadate-3-hydroxy-1,2-dimethyl-pyridinone (Hdhp) system was studied by pH-potentiometry and ^{51}V NMR spectroscopy and simple mono (in two different protonation states) and bis complexes (in three different protonation states) a tris complex, three dinuclear and a trinuclear complex were found based on the ^{51}V NMR spectra measured at different pH, various metal ion concentrations and metal-to-ligand ratios.[121] The kinetics of CO_2 absorption into 30% wt/wt K_2CO_3 solutions containing 0.1–0.5 M $K_4V_2O_7$ was studied at temperatures of 40, 60 and 75 °C, and the V(V) speciation diagrams were developed as a function of CO_2 loading by using ^{51}V NMR spectroscopy.[122] The three bis-peroxo-oxovanadate species containing histidine peptides were studied by IR spectroscopy, ^1H, ^{13}C and ^{51}V NMR spectroscopies and the geometry optimisations and IR spectra in gas phase were carried out with the DFT (B3LYP functional) method and the spectra of the peptides under study were obtained with the gauche-including AO (GIAO) method as implemented in the GAUSSIAN 09 package.[123] The electronic implications and structure–activity relationships in the case of histidine peptide-containing bis-peroxo-oxovanadate structures were discussed. The (imido)-vanadium-(V) dichloride complexes containing 1,3-imidazolin-2-iminato or 1,3-imidazolidin-2-iminato ligands V-(NR′)-Cl_2 (L) [R′ = 2,6-$Me_2C_6H_3$, L = 1,3-$R_2(CHN)_2C{=}N$ (1a–c,e) or 1,3-$R_2(CH_2N)_2C{=}N$ (2a–d),

$R = t$-Bu, 2,6-$Me_2C_6H_3$, 2,6-$^iPr_2C_6H_3$, C_6H_5, 2,6-$(Ph_2CH)_2$-4-MeC_6H_2; L = 1,3-(2,6-$^iPr_2C_6H_3)_2(CHN)_2C{=}N$, R′ = 1-adamantyl (Ad, 3c), C_6H_5 (4c); L = 1,3-(2,6-$^iPr_2C_6H_3)_2(CH_2N)_2C{=}N$, R′ = Ad (5c)] were studied and a good correlation between the catalytic activity and the ^{51}V NMR chemical shift was found for the (arylimido)-vanadium precatalysts.[124] ^1H and $^{13}C\{^1H\}$ NMR spectra for $[VO\{OPr^i\}_{3-n}\ \{L\}_n]$ {where, $n = 1$–3 and LH = $C_9H_{16}C{=}NOH$ (1–3) and $(CH_3)_2C{=}NOH$ (4–6)} suggested that the oximato moieties are monodentate in solution which was further confirmed by the ^{51}V NMR signals, appeared in the region expected for tetra-coordinated oxo-vanadium atoms.[125] The stoichiometry and thermodynamic stability of Cu(II), V(IV/V), Fe(II)/(III) and Ga(III) complexes of salicylaldehyde semicarbazone (SSC, HL) and its 5-bromo derivative (Br-SSC, HL) were determined by pH-potentiometry, UV-Vis spectrophotometry, EPR, ^1H and ^{51}V NMR spectroscopies.[126] The formation of mono-ligand complexes such as ML|, MLH_{-1}|, MLH_{-2}| was found with Cu(II), V(IV/V) and bis-ligand species of Fe(II)/(III) and Ga(III) such as ML_2|, ML_2H_{-1}| and ML_2H_{-2}| were detected. $[V^VO_2$(acpy-bhz)], $[V^VO_2$(acpy-inh)] and $[V^V O_2$(acpy-nah], and the corresponding polymer-supported (PS) dioxidovanadium(V) complexes, PS-i.m.$[V^VO_2$(acpy-bhz)], PS-i.m.$[V^VO_2$(acpy-inh)] and PS-i.m.$[V^VO_2$(acpy-nah)], and their catalytic processes were studied by UV-Vis and ^{51}V NMR spectroscopies.[127] The coordination modes and the solution structure of 5-amino-1,10-phenanthroline (5-NH_2-phen), and the coordination reaction between peroxovanadium(V) complex $[OV(O_2)_2(D_2O)]^-$/$[OV(O_2)_2(HOD)]^-$ and 5-NH_2-phen were studied by ^1H, ^{13}C and ^{51}V NMR with variable temperature COSY and heteronuclear single quantum coherence (HSQC).[128] The solution structures of the peroxo-vanadate species were proposed based on experimental NMR information and confirmed by theoretical calculations, and the results of DFT calculations indicated that solvation plays an important role in the coordination interactions. The substitution effects of pyrazolylpyridine (pzpy) on the coordination reaction equilibrium and the interactions between a series of pzpy-like ligands and biperoxidovanadate $[OV(O_2)_2(D_2O)]^-$/ $[OV(O_2)_2(HOD)]^-$, (bpV) were explored using a combination of ^1H, ^{13}C and ^{51}V HSQC and variable temperature NMR spectroscopies.[129] $(K_5, H_3O)[SiV_3W_9O_{40}H]\cdot xH_2O$ of the vanadium tri-substituted α-Keggin dodecatungstosilicate were analysed by vibrational, EPR and ^{51}V NMR spectroscopies.[130] The single crystal X-ray diffraction data and ^{51}V solid-state NMR spectra confirmed the occurrence of a single vanadium site in a cubic structure due to rotational disorder of the Keggin ion, and EPR experiments confirmed strong interaction of V^{4+} with two V^{5+} ions, in accordance with insertion of a V_3 subunit into the lacunary Keggin ion. Vanadium(V) complexes with chiral tridentate Schiff base ligands and six-coordinate complexes with additional monoanionic bidentate benzohydroxamate co-ligand were investigated by IR, CD, UV-visible, 1D and 2D (COSY, gHSQC and NOESY) ^1H and ^{51}V NMR spectroscopies.[131] $[V^{IV}O$(sal-iah)$(H_2O)]$, $K[V^V O_2$(sal-iah)$]\cdot H_2O$, PS-$[V^{IV}O$(sal-iah)$(H_2O)]$ and PS-$[V^VO(OMe)$(sal-iah)] were studied by IR, electronic, ESR, EPR, ^1H and ^{51}V NMR spectroscopies and FE-SEM.[132] ^1H and ^{51}V NMR spectra of $K[V^VO_2$(sal-iah)$]\cdot H_2O$ are compatible with the existence of expected dioxide and oxidomethoxido species.

6.2 Niobium (^{93}Nb) ($I = 9/2$)

^{93}Nb solid-state NMR spectroscopy was employed to investigate the atomic and electronic structures in Ni–Nb based metallic glass and the ^{93}Nb NMR isotropic metallic shift of $Ni_{60}Nb_{35}Sn_5$ is ~100 ppm lower than that of $Ni_{60}Nb_{35}Zr_5$.[133] Hexanuclear niobium halide cluster compounds with a $[Nb_6X_{12}]^{2+}$ (X = Cl, Br) diamagnetic cluster core were studied by a combination of experimental solid-state NMR/NQR techniques and PAW/GIPAW calculations.[134] The analysis of geometrical relations between computed EFG and chemical shift tensors with respect to structural features of cluster units was carried out. The compounds possess large positive chemical shifts and the experimental determined δ_{iso}(^{93}Nb) values are in the range from 2400–3000 ppm.

7 Group 6 (^{53}Cr, ^{95}Mo, ^{183}W)

7.1 Chromium (^{53}Cr) ($I = 3/2$)

^{53}Cr and ^{63,65}Cu NMR spectra for $CuCr_2O_4$ were measured at various magnetic fields and temperatures and the microscopic evidence of orbital ordering in $CuCr_2O_4$ was obtained from a dipolar hyperfine field, NQR and magnetic anisotropy analysis of the linewidth broadening of the ^{53}Cr and ^{63,65}Cu NMR spectra.[135]

7.2 Molybdenum (^{95}Mo) ($I = 5/2$)

$[MoO_2(O_2)(CN)_2]$-PAN [PAN = poly(acrylonitrile)] (PANMo) was studied by IR, UV-visible and ^{13}C and ^{95}Mo NMR spectroscopies, CHN and EDX analysis, TGA–DTG and SEM.[136] The catalysts $[MoO_2(O_2)(L)_2]^{2-}$MR [L = L-valine (MRVMo) or DL-alanine (MRAMo) and MR = Merrifield resin] were studied by ^{13}C NMR and ^{95}Mo NMR spectroscopies.[137] The electron localisation of polyoxomolybdates in ε-Keggin structure of $[Me_3NH]_6[H_2Mo_{12}O_{28}(OH)_{12}\{MoO_3\}_4]\cdot 2H_2O$, $[PMo_{12}O_{36}(OH)_4\{La(H_2O)_{2.75}Cl_{1.25}\}_4]\cdot 27H_2O$ and $[H_2Mo_{12}O_{30}(OH)_{10}\{Ni(H_2O)_3\}_4]\cdot 14H_2O$ was studied by ^{95}Mo solid-state NMR spectroscopy and DFT calculation. and the NMR parameters of the chemical shift and quadrupole interactions for ^{95}Mo were estimated.[138]

7.3 Tungsten (^{183}W) ($I = 1/2$)

The hydrolysis of a series of unactivated dipeptides in the presence of $(Me_4N)_2[W_5O_{18}Zr(H_2O)_3]$ (ZrW_5) was studied by kinetic experiments and ^{1}H, ^{13}C and ^{183}W NMR spectroscopies, and the binding of His-Ser to ZrW_5 was examined by UV/Vis and NMR spectroscopies.[139] It was indicated that pD His-Ser chelates the Zr^{IV} through its imidazole nitrogen, amine nitrogen and amide carbonyl oxygen. The bifunctional activation of CO_2 and 1,2-phenylenediamine (1) caused at tungsten-oxo moiety in $TBA_2[WO_4]$ (2, TBA = tetrabutylammonium) and adducts 1–2 and 2-$(CO)_n$ ($n = 1$ and 2) formed by the reactions of 2 with 1 and CO_2, respectively were confirmed by ^{1}H, ^{13}C and ^{183}W NMR spectroscopies.[140] The α-$[AsPW_{18}O_{62}]^{6-}$ and α-$[H_4XW_{18}O_{62}]^{7-}$ (X = P^V, As^V) were studied by ^{183}W and ^{31}P NMR spectroscopies and 2-dimensional ^{183}W INADEQUATE experiments.[141] The effects of P/As substitutions, generating unsymmetrical structures, on the NMR $\delta_{W\text{-}183}$, $\delta_{P\text{-}31}$ $^2J_{W\text{-}O\text{-}W}$ and $^2J_{W\text{-}O\text{-}P}$ parameters were discussed in relation to their bond length and bond angle alteration. The DFT calculation within the two-component quasi-relativistic zeroth-order regular approximation (ZORA) was used to predict parity violation shifts in ^{183}W NMR shielding tensors of chiral, tetrahedrally bonded W complexes NWXYZ (X, Y, Z = H, F, Cl, Br or I).[142]

8 Group 7 (^{55}Mn, ^{99}Tc)

8.1 Manganese (^{55}Mn) ($I = 5/2$)

The magnetic structure and metal-insulator transition in antiferromagnetic (AFM) $BaMn_2As_2$ and $Ba_{1-x}K_xMn_2As_2$ single crystals were investigated by ^{55}Mn and ^{75}As NMR measurements and a ^{55}Mn zero-field NMR (ZFNMR) spectrum was observed.[143] It was confirmed that the G-type AFM structure from the field dependence of the ^{55}Mn spectra and ^{75}As NMR spectra below T_N. ^{55}Mn and ^{145}Nd NMR measurements on $La_{1-x}Nd_xMn_2Si_2$ ($x = 0.35, 0.4$) were made and the ^{55}Mn NMR frequencies of both compounds are 160–163 MHz at 1.4 K and are about 166 MHz at 77 K without external magnetic field.[144] The temperature dependence of resonance frequency was measured for $La_{0.65}Nd_{0.35}Mn_2Si_2$ in the temperature from 1.4 K up to 88 K. The influence of substitution of Fe ions for manganese on the structure, phase transitions, magnetoresistance, ^{55}Mn NMR and ^{57}Fe Mossbauer spectra for $La_{0.6}Sr_{0.3}Mn_{1.1-x}Fe_xO_3$ ($x = 0$–0.15) was studied and the ^{55}Mn NMR spectra confirmed the high frequency electron double exchange between Mn^{3+} and Mn^{4+} ions and irregularity of their surrounding by other ions and defects.[145] The structure, lattice imperfection and properties of $La_{0.6-x}Nd_xSr_{0.3}Mn_{1.1}O_{3-\delta}$ ($x = 0$–0.4) were investigated ^{55}Mn NMR spectroscopy and it was revealed that there are a high-frequency electronic double exchange (Mn^{3+}–$O^{2-}$$Mn^{4+}$)$\leftrightarrow$($Mn^{4+}$–$O^{2-}$$Mn^{3+}$) and an inhomogeneity of the magnetic and charge states of manganese due to the heterogeneous environment of the Mn ions by other ions and defects.[146] A decrease in the ferromagnetic phase content with an increase in the Nd concentration was determined from the $4\pi N\chi_{ac}(T)$ and the ^{55}Mn NMR results. The influence of excess manganese on structure defects and functional properties of $(Nd_{0.7}Sr_{0.3})_{1-x}Mn_{1+x}O_3$ ($x = 0$–0.20) ceramics annealed at 1200 and 1500 °C was studied by X-ray powder diffraction, resistance, magnetic susceptibility, magnetisation, ^{55}Mn NMR spectroscopy and magnetoresistance measurement methods.[147] The high-frequency electron $Mn^{3+} \leftrightarrow Mn^{4+}$ exchange and the uneven distribution of ions and vacancies were indicated by broad asymmetric ^{55}Mn NMR spectra.

8.2 Technetium (^{99}Tc) ($I = 9/2$)

The ^{99}Tc chemical shifts based on the (^{99m}Tc)$(CO)_3$(NNO) complex conjugated to the antitumor agent 2-(4′-aminophenyl)benzothiazole were reported and the thermal and solvent effects were studied computationally by quantum-chemical methods, using the DFT calculations at the level BPW91/aug-cc-pVTZ for the Tc and BPW91/IGLO-II for the other atoms.[148] The ^{99}Tc NMR chemical shifts of the complex in gas phase and solution using different solvation models were also calculated.

9 Group 8 (^{57}Fe)

9.1 Iron (^{57}Fe) ($I = 1/2$)

Local methods based on hyperfine interactions determined by ^{57}Fe NMR and the Mossbauer spectroscopy, bentonite/iron oxide system prepared by isothermal calcination of powder composed of bentonite clay and precursor containing ferric acetate was analysed.[149] ^{57}Fe NMR experiment was applied to investigate the magnetoelectric hexaferrite $Ba_{0.5}Sr_{1.5}Zn_2Fe_{12}O_{22}$ (BSZFO) and $Ba_{0.5}Sr_{1.5}Zn_2$ $(Fe_{0.92}Al_{0.08})_{12}O_{22}$ (BSZFAO) were investigated. The spin-canting angles of all NMR peaks were measured to assign each peak to corresponding Fe^{3+} sites.[150] The ^{57}Fe NMR spectrum of $FeCl_3$ in various temperature and magnetic fields fits well with the theoretical prediction for an antiferromagnet with a magnetic anisotropy field of less than 70 mT in the rib-plane.[151] The magnetoelectric properties of hexaferrite $Ba_{0.5}Sr_{1.5}Zn_2Fe_{12}O_{22}$ are significantly improved by Al substitution and thermal annealing. From the measuring the enhancement factor of ^{57}Fe NMR, the direct microscopic evidence that the magnetic moments of the L and S blocks are rotated by a magnetic field in such a way as to increase the net magnetic moment of a magnetic unit, even after the field is removed, was found.[152]

10 Group 9 (^{59}Co, ^{103}Rh)

10.1 Cobalt (^{59}Co) ($I = 7/2$)

^{59}Co NMR Knight Shift, which is directly related to the microscopic spin susceptibility, was measured in order to determine the superconducting paring state in the ferromagnetic superconductor UCoGe, which shows to be almost constant, but does not decrease below a superconducting transition.[153] Cobalt-containing Fischer–Tropsch synthesis catalysts supported on different alumina modifications was studied by an internal-field ^{59}Co NMR.[154] The low-temperature magnetic properties of LaCoGe with the tetragonal CeFeSi-type structure were investigated by ^{59}Co NMR and NQR measurements.[155] ^{59}Co NMR measurements under pressure in UCoAl with a quantum critical endpoint of the metamagnetic transition from the paramagnetic phase to the ferromagnetic (FM) phase.[156] Partially gapped Fermi surfaces in single-crystal $La_3Co_4Sn_{13}$ was investigated by means of the specific heat and ^{59}Co NMR spectroscopy.[157] Multiple well-resolved NMR peaks of ^{23}Na and ^{59}Co are observed in order to investigate the Na atomic ordering and the local Co magnetism in a metallic antiferromagnet made of $Na_{0.825}CoO_2$, which suggests a superlattice structure forming in this Na-vacant cobaltate.[158]

10.2 Rhodium (^{103}Rh) ($I = 1/2$)

^{103}Rh NMR chemical shifts were computed at the GIAO-B3LYP level of DFT for a number of $[Rh(COD)(PnP)]^+$ complexes [COD = 1,5-cyclooctadiene, P = chelating bis(phosphine) including bis(dimethylphosphino) ethane (dmpe), bis(diphenylphosphino)ethane (dmpe), MeDUPHOS, DIOP, BINAP and others].[159] ^{103}Rh NMR was used to characterise the complex $[(Ph_3P)_2Rh(H)_2(mu\text{-}PTC)W(CO)_4(P(4\text{-}MeC_6H_4)_3)]$, which shows pi–pi interactions between the bridging pyridyl and a phenyl group from a phosphine on each metal.[160] The novel rhodium (Rh) catalysts [{nbd-$(CH_2)_4$-X}RhR] (1, X = PPh2, R = Cl; 2, X = NPh2, R = Cl; 3, X = PPh2, R = triphenylvinyl; nbd = 2,5-norbornadiene) were synthesised, and their catalytic activities were examined for the polymerisation of phenylacetylene (PA) and its derivatives. ^{103}Rh NMR spectroscopy together with DFT calculations (B3LYP/6-31G*-LANL2DZ) were carried out to characterise the complex.[161] Zero-point vibrationally averaged (r_g^0) structures were computed at the PBE0/SDD/6-31G* level for the $[(Rh^{35}Cl_n{}^{37}Cl_{(5-n)})(H_2O)]^{2-}$ ($n = 0$–5), *cis*$[(Rh^{35}Cl_n{}^{37}Cl_{(4-n)})(H_2O)_2]^-$ ($n = 0$–4) and fac$[(Rh^{35}Cl_n{}^{37}Cl_{(3-n)})(H_2O)_3]$($n =$ 0–3) isotopologues and isotopomers. Magnetic shielding constants, computed at the ZORA-SO/PW91/QZ4P/TZ2P level, were used to evaluate the corresponding $^{35/37}$Cl isotope shifts on the ^{103}Rh NMR spectra.[162]

11 Group 10 (^{195}Pt)

11.1 Platinum (^{195}Pt) ($I = 1/2$)

^{195}Pt NMR spectra of D_2O solutions containing equimolar amounts of $[Pt(bipy)_2]^{2+}$ and the heterocyclic N donor ligands pyridine, 3-methylpyridine and 4-methylpyridine gave evidence for the formation of at least one species, while quinoline appears to give only a single product under the same conditions.[163] Series of aqua-hydroxo-nitrato complexes $[Pt(L)_x(NO_3)_{6-x}]$ (L = H_2O or OH^-; $x = 0, \ldots, 6$) were found to exist in nitric acid (6–15.8 M) solutions of $H_2[Pt(OH)_6]$. The pair additivity model of chemical shifts and statistical theory were used to assign signals in NMR spectra to particular $[Pt(L)_x(NO_3)_{6-x}]$ species.[164] Behaviour of one and two Cisplatin molecules confined in capped and opened single-walled carbon nanotubes (CNTs) was studied by means of ^{195}Pt chemical shift tensors calculations using GIAO method. NMR calculations reveal that Platinum chemical shift is sensitive to CNT's diameter and is linearly correlated to confinement energy. Pt-195 chemical shift measurement may be a direct method to access to the diameter of the encapsulating CNT's and to control the amount of drug molecule transported by this CNT.[165] The GIAO-PBE0/SARC-ZORA(Pt)boolean OR 6-31 + G(d)(E) (E = main group element) computational protocol without including relativistic and spin–orbit effects was offered for the accurate prediction of the ^{195}Pt NMR chemical shifts of a series of *cis*-(amine)$_2$PtX$_2$ (X = Cl, Br, I) anticancer agents (in total 42 complexes) and *cis*-diacetylbis(amine) platinum(II) complexes (in total 12) in solutions employing the Polarizable Continuum Model (PCM) solvation model, thus contributing to the difficult task of computation of ^{195}Pt NMR.[166]

12 Group 11 (^{63,65}Cu, ^{107,109}Ag)

12.1 Copper (^{63,65}Cu) ($I = 3/2, 3/2$)

The transition-metal complexes of the tripodal tristibine $N(CH_2\text{-}2\text{-}C_6H_4SbMe_2)_3$ were characterised by elemental analysis, ESI+ mass spectrometry, IR spectroscopy and ^{63}Cu as well as ^{1}H, $^{13}C\{^{1}H\}$ and ^{55}Mn NMR spectroscopy. NMR spectroscopic analysis suggested that the ligand also adopts this rigid conformation in solution.[167] Detailed ^{63}Cu and ^{29}Si NMR above the critical magnetic field, $H_{c1} = 23.4$ T, in order to understand the nature of the two-dimensional Bose–Einstein condensed (BEC) phase in $BaCuSi_2O_6$.[168] The spin Hamiltonian parameters (g factors and hyperfine structure constants) and the Knight shifts for Cu^{2+} in the spin-Pierels compound $CuGeO_3$ of rhombic and tetragonal phases are theoretically investigated from the high order perturbation formulas of these parameters for a 3d(9) ion under rhombically and tetragonally elongated octahedra in a uniform way. The approximately linear relationships between Knight shifts and g factors are found.[169] The results of magnetic susceptibility and ^{63}Cu, ^{65}Cu and ^{51}V NMR measurements on a high-quality powder sample of vesignieite $BaCu_3V_2O_8(OH)_2$, a candidate for the spin-1/2 kagome antiferromagnet were reported.[170] The temperature dependent lattice parameters and electronic properties of the solid solution $Cu_2ZnGeSe_{4-x}S_x$ were reported in combination with ^{63}Cu NMR, and an attempt will be made to relate the nature of the electronic phase transformation to a superionic phase transformation and a changing covalent character of the lattice upon anion substitution in this class of materials.[171]

12.2 Silver ($^{107,109}Ag$) ($I = 1/2$, 1/2)

^{109}Ag NMR spectra of hydride-centered heptanuclear silver clusters, $[Ag^{-7}(H)\{E_2P(OR)_2\}_6]$ (R = Pr-i, E = Se: R = Et; E = S) and $[Ag^{-8}(H)\{E_2P(OR)_2\}_6](PF_6)$ (R = Pr-i, E = Se: R = Et; E = S) were observed.[172] ^{109}Ag NMR measurements were performed on two powder samples of Ag nanoparticles with average sizes of 20 nm and 80 nm. The measurements carried out in an external field of 9.4 T and in the temperature range 10 K < T < 280 K. The ^{109}Ag NMR spectra for both samples are close to Lorentzian shapes and turn out to be mixtures of homogeneous and inhomogeneous lines.[173] ^{109}Ag and ^{1}H NMR spectroscopies of novel cluster anions $[Ag(mu\text{-}H)_4\{Re_2(mu\text{-}H)(CO)_8\}_2]^-$ and $[Ag(mu\text{-}H)_4\{Re_4(mu\text{-}H)_3(CO)_{16}\}_2]^-$ showed that each Ag atom is surrounded by four hydrido ligands, bridging Re–Ag interactions.[174]

13 Group 12 ($^{111,113}Cd$, ^{199}Hg)

13.1 Cadmium ($^{111,113}Cd$) ($I = 1/2$, 1/2)

Two calcium binding proteins: calbindin D-9k and calmodulin were focused. For calbindin D-9k and its mutants, ^{113}Cd NMR has been useful both to follow actual changes in the metal binding sites and the cooperativity in the metal binding. Ligand binding to calmodulin was studied extensively with ^{113}Cd NMR showing that the metal binding sites are not directly involved in the ligand binding. The ^{113}Cd chemical shifts are, however, exquisitely sensitive to minute changes in the metal ion environment.[175] The solid state ^{113}Cd NMR spectrum indicated that novel metal–organic framework $Na_2(Me_2NH_2)_4[Cd_7L_{10}]\text{-}9H_2O\text{-}17Me_2NH\text{-}2DMF$ has three 6-, 7- and 8-coordinated Cd^{2+} ions with chemical shifts at delta

−25.3, −41.2 and −83 ppm, respectively.[176] The structure of Cd^{2+} complexes with Schiff bases derived from L-arginine (L-Arg) and 2-hydroxy carbonyl compounds were determined by heteronuclear NMR measurement (^{13}C, ^{15}N and ^{113}Cd) methods in DMSO solution and in the solid state. The large set of spectral parameters: chemical shifts, homo- and heteronuclear coupling constants, were used in structure study.[177]

13.2 Mercury (^{199}Hg) ($I = 1/2$)

A dissymmetric thiosemicarbazone ligand containing a hydrazonequinoline arm, $H_2AMeTsQ$ and its chloride salt, $[H_3AMeTsQ]Cl$, were synthesised. ^{199}Hg NMR spectroscopy together with X-ray diffraction indicated that the complexes form monomeric and dimeric structures and even a coordination polymer.[178] ^{119}Hg NMR was applied to characterise the complexes of $[M\{SeC_4H(R\text{-}4,6)_2N_2\}_2(tmeda)_n]$ ($M/R/n = Hg/H/1$ or Hg/Me/0; tmeda = *N*,*N*,*N*′,*N*′-tetramethylethylenediamine).[179]

14 Group 13 (^{11}B, ^{27}Al, ^{71}Ga, ^{115}In, $^{203,205}Tl$)

14.1 Boron (^{11}B) ($I = 3/2$)

NBS 3-methylindole reacted with various imidazoles to give the (indol-2-yl)imidazolium salts, which were converted in aqueous solution into the 2-(imidazolium-3-yl)-3-methylindolates by base. These conjugated ylides, which represent a subclass of mesomeric betaines-are the exclusively detectable form in the ^{11}B, ^{15}N and ^{7}Li NMR spectra taken in DMSO-d(6) and DFT calculations.[180] The effect of combined addition of P_2O_5 and V_2O_5 on structural and sealing properties of glasses with nominal composition (mol%) $27SiO_2–23SrO–32BaO–4Al_2O_3–10B_2O_3–(4\text{-}x)P_2O_5\text{-}xV_2O_5$ is reported in the present study. ^{11}B, ^{29}Si, ^{31}P, ^{51}V and $^{27}A1$ MAS-NMR were used to characterise the local environment in the glasses and glass ceramics.[181] The effects of nanoconfinement on the dehydrogenation rate and reaction pathways of the eutectic $LiBH_4–Mg(BH_4)_2$ were comprehensively investigated by means of thermal analysis, mass spectroscopy and solid state ^{11}B MAS NMR.[182] A DFT study was performed to investigate electronic and magnetic properties of armchair BNNTs with Stone–Wales defects based on ^{11}B and ^{15}N NMR parameters and NICS indices. The smallest ^{11}B chemical shielding appears around 68.6–69.3 ("Parallel" orientation of the defect site) or 71.6–72.1 ppm ("Diagonal" orientation of the defect site) arising from the boron surrounded by three different rings. Moreover, CS tensors are shown to be quite sensitive to the curvature at the corresponding site.[183] ^{11}B and ^{13}C NMR spectra of so-far experimentally unknown carbon-rich cationic closo-carboranes $C_3Bn_3H_n^{+}$ ($n = 5, 6, 7, 10, 12$) were calculated at the GIAO-MP2 level and subsequently analysed to reveal the nature of bonding in these potentially weakly coordinating cations. All previous rules derived for understanding ^{11}B NMR spectra of borane derivatives was applied to realistically account for the corresponding shieldings.[184] 6-Substituted nido-$B_{10}H_{13}$ decaboranes were characterised by ^{11}B NMR.[185] ^{11}B NMR chemical shifts of gas-phase structure of *arachno*-4,6-$(CH_2)_2B_7H_9$ and *arachno*-4,6-$S_2B_7H_9$ were achieved by calculating the appropriate NMR chemical shifts using GIAO-MP2 methods and the IGLO-II basis set for various geometries, both experimental and calculated.[186] The B3LYP-computed ^{11}B and ^{1}H chemical shifts for $[(Cp^{*}RuCO)_3(mu_3\text{-}H)BH]$ ($Cp^{*} = eta_5\text{-}C_5Me_5$)precisely follow the experimentally measured values.[187] All possible derivatives of fluorinated phenylboronic acids were studied by ^{11}B, ^{1}H, ^{13}C, ^{19}F and ^{17}O NMR and theoretical calculations.[188] Considerations about stereochemistry of $[(1\text{-}HOOC\text{-}1,2\text{-}C_2B_9H_{10})(1', 2'\text{-}C_2B_9H_{11})3,3'\text{-}Co]$ ion and $[(HOOC)_2\text{-}(1,2\text{-}C_2B_9H_{10})_2\text{-}3,3'\text{-}Co]$ ion are supported by geometry optimisations and calculations of ^{11}B NMR shifts at the GIAO-DFT level.[189] From ^{11}B NMR experiments, unprecedented experimental information regarding the B–B bond was obtained.[190] The structure and thermal properties of the cured phenylboronic acid modified aryl boron-containing phenolic resin were investigated by ^{11}B NMR, FTIR, XPS and TGA.[191] The kinetic modeling of self-hydrolysis of 10–20 wt% $NaBH_4$ at 25–80 °C was investigated from ^{11}B NMR measurements and pH data.[192]

14.2 Aluminum (^{27}Al) ($I = 5/2$)

^{27}Al MAS NMR was applied to two fields (8.45 T and 14.1 T) to follow the optimisation of a sol–gel process to produce yttrium aluminium garnet, $Y_3Al_5O_{12}$ (YAG), at moderate temperatures, which indicated to be a highly sensitive tool to determine the presence of the impurity phase, yttrium aluminium perovskite, $YAlO_3$.[193] The structures of aluminophosphate materials could be determined from $^{27}Al/^{31}P$ solid-state NMR experiments since this approach does not require the isolated spin pairs which have been important for determining structures of silicate materials by ^{29}Si solid-state NMR. This algorithm was tested on a series of 27 zeolite framework types found in the International Zeolite Association's zeolite structure database.[194] Various first-row transition metal cations (Cr^{3+}, Mn^{2+}, Fe^{3+}, Co^{2+}, Ni^{2+}, Cu^{2+} and Zn^{2+}) were introduced to zeolite beta using ion exchange procedures. Both aluminum and transition metal sites were studied by ^{27}Al NMR, UV-Vis spectroscopy and XPS.[195] Based on deconvolution of the ^{27}Al NMR spectra and quantitative ^{13}C NMR spectra, the structure of Al detoxicification by simple carboxylic acid, the interaction between aluminium ion and three dicarboxylic acids (oxalic acid, malonic acid and succinic acid) under an acidic condition was elucidated.[196] Solid-state ^{27}Al NMR spectroscopy and periodic DFT calculations were used to highlight the differences between local and average order in the $LaSrAl_3O_7$ crystal. The ^{27}Al isotropic chemical shifts and quadrupolar coupling constants were computed by employing the GIPAW and PAW formalisms, respectively.[197] ^{27}Al NMR spectroscopy was used to characterise species present in the alkaline aqueous (tetraoctylammonium hydroxide) and alcoholic aluminate solutions.[198] ^{27}Al NMR was used to probe the nature of the phases that formed during the conversion of methane at high temperature on the Mo/HZSM-5 catalyst.[199] ^{27}Al and ^{7}Li MAS NMR and high-resolution X-ray diffraction to determine the disposition of Al cations during the formation of low temperature cubic lithium lanthanum zirconium oxide garnet ($Li_7La_3Zr_2O_{12}$).[200] Sorption of the trivalent metals Eu^{3+} and Y^{3+} on a natural kaolinite mineral was investigated with solid-state ^{27}Al and ^{1}H NMR. Eu^{3+} and Y^{3+} ion additions to the kaolinite samples had no influence on recorded ^{27}Al spectra, which appeared a center peak at 6.15 ppm attributable to octahedrally coordinated aluminum in the gibbsite-like sheets of kaolinite. However, the ^{27}Al measurements were imperative to perform as potential mineral dehydroxylation was monitored from the ^{27}Al spectra before and after drying treatment and normalisation of ^{1}H spectra was based on the acquired ^{27}Al data.[201] $^{27}Al/^{119}Sn$ MAS NMR measurement was applied to the nanocrystalline $Al_{0.4}Mg_{0.2}Sn_{0.4}O_{1.6}$ and $Al_{0.25}Mg_{0.38}Sn_{0.38}O_{1.5}$, which indicates that the calcination results in the formation of $[AlO_4]$ and $[AlO_5]$ units, in addition to the initial $[AlO_6]$ environment, and in local disorder around the Sn atoms. The electrochemical performance was studied by galvanostatic cycling against Li metal.[202] Mesoporous MCM-22 zeolite was prepared by treating MCM-22 with sodium hydroxide solution through an organic amine-assisted reversible structural change. The acid sites related to framework aluminum cations were almost intact after mesopore creation, which is evidenced by pyridine or ammonia adsorption–desorption and ^{27}Al NMR investigations.[203] High-field ^{27}Al NMR and first principle calculations indicated that triisobutylaluminum reacts with silica yielding three different Al sites, in which a quadruply grafted dimeric surface species and two incorporated $Al(O)_x$ species ($x=4$ or 5).[204] Solid alkoxides, which were yielded from aluminum reacted with excess alcohol in the presence of mercuric chloride catalyst, were obtained in high yields and analysed by ^{27}Al, ^{1}H and ^{13}C NMR, X-ray diffraction, ICP-MS and elemental analysis.[205] The zeolites, such as LTA and MFI type silica materials, were generated from collected silica waste, which are analysed by ^{27}Al and ^{29}Si NMR, FT-IR, XRD and SEM.[206] ^{27}Al NMR was used for the characterisation of complex of heteronucleic-mu-oxoisopropoxide $[SnO_2AlB(O\text{-}Pri)_4]$.[207] ^{27}Al NMR was applied to the hollow mesoporous aluminosilicate spheres with middle strong acidity, together with TEM, SEM, XRD, N-2 adsorption–desorption and NH3-TRD. The results showed that the hollowing process was associated with a progressive mass redistribution and changes of pores structures.[208] The hierarchical zeolite Y, which is hydrothermally synthesised using an organosilane surfactant, was characterised by ^{27}Al MAS NMR and N-2 adsorption–desorption isotherms.[209] Both ^{27}Al and ^{29}Si NMR studies were applied to the sub 40 nm nanocrystalline FAU type zeolites with germanium incorporated into the framework.[210] The high-silica mordenite zeolite by hydrothermal transformation of the Na-RUB-18 layered silicate, which was conducted by the addition of $Al[OCH(CH_3)_2]_3$ or $Na_2Al_2O_4$ into the gel after 7 days of hydrothermal treatment of a non-aluminum gel, was analysed by ^{27}Al and ^{29}Si NMR.[211] The structural modifications of amorphous alumina thin films with deposition temperature (T_d) was analysed by very high-field (20 T) solid state ^{27}Al NMR spectroscopy.[212] ^{27}Al MAS NMR spectra of the cement–limestone blends suggested that dissolved aluminate species in the cement–limestone blends readily react with carbonate ions from the limestone filler, forming calcium monocarboaluminate hydrate.[213] ^{27}Al MAS NMR spectra showed Al exists in both tetrahedral and five-fold coordination, with the latter directly coordinated to hydroxyl O atoms as well as framework O atoms.[214] The ^{27}Al and ^{13}C NMR chemical shifts of the supermolecular-polarizable continuum model (SM-PCM) models were calculated using the GIAO method at the HF/6-311+G(d,p) level, which are in good agreement with the experimental data.[215] ^{27}Al MAS NMR spectra of calcined mesoporous silicoaluminophosphate observed three major signals around 0, 34 and 60 ppm corresponding to aluminium species in different coordination environments.[216] A new gamma-Al-13 Keggin, in which the Al-13 Keggin structures are selectively transformed through $\varepsilon \rightarrow \delta \rightarrow \gamma$ isomerisation process, was structurally analysed by ^{27}Al NMR and single-crystal XRD.[217] By means of ^{27}Al NMR studies, magnetic excitations and the hybridisation effect between Pr 4f and conduction electrons in a pair of cubic compounds $PrTr_2Al_{20}$ (Tr = Ti, V) were investigated.[218] The characterisation of the host–guest interactions includes a comprehensive solid-state NMR spectroscopic study based on a full assignment of ^{1}H and ^{13}C high-resolution spectra using CRAMPS decoupling schemes to enhance ^{1}H resolution combined with advanced 2D HETCOR (^{1}H–^{13}C, ^{1}H–^{27}Al and ^{1}H–^{14}N) spectra at high magnetic fields. In spite of a pronounced dynamical disorder of the guests, a preferred binding of the acetone *via* a NH center dot center dot center dot OC hydrogen bond for the NH_2 and the NHCHO anchor groups could bu identifed by analysing trends in the C-13 isotropic chemical shifts.[219] Al-30 is a new kind of polymeric aluminum species with 30 aluminum atoms. Al-13 aggregates and monomeric Al under high temperature circumstance on Al-30 formation was studied by ^{27}Al NMR.[220] ^{27}Al NMR spectroscopy was used to verify the effect of Ferron on the sensitivity of total aluminum (Al-T) determination.[221] Structure of the biosilica was characterised by solid-state ^{27}Al NMR spectroscopy. This result indicated that an amorphous aluminosilicate phase is formed where the aluminum exists as four- and sixfold-coordinated species.[222] The AlH_3 adducts of TMEDA ($Me_2NCH_2CH_2NMe_2$), DIOX ($O(CH_2CH_2)_2O$), TEA (Et_3N), BDMA ($PhNMe_2$) and TMPDA ($Me_2NCH_2CH_2CH_2NMe_2$) were characterised by ^{27}Al NMR spectroscopy and by DFT calculations.[223] Ordered mesoporous aluminum silicates with strong acid sites were characterised by ^{27}Al NMR.[224] Al MAS NMR was applied to the characterisation of TAPO materials, particularly centered in the AFI-structured TAPO-5.[225] The alkylation of ethylbenzene with ethanol over HZSM-5 zeolite catalysts was carried out using a riser simulator reactor at different reaction temperatures and contact times. The relationship between alkylation activity and extra-framework aluminium species was investigated by ^{27}Al MAS NMR.[226] The incorporation mechanisms of fluorine into peralkaline was investigated by ^{27}Al MAS NMR, which indicated the spectra of the peralkaline glasses show only four-coordinated Al, while in the peraluminous glasses similar to 5% of the Al was found to be five-coordinated. No changing of the amount of five-coordinate Al with increasing fluorine was found, but the environment of the five-coordinate Al becomes more symmetric with increasing fluorine.[227] The number of H^+ sites on MFI and CHA zeolites measured by direct chemical titration was less than the number of framework Al (Al-f) atoms determined from

^{27}Al MAS NMR spectra (H^+: Al-f = 0.7–1.0).[228] Chemical structure of alumina sample constituted by η and γ-Al_2O_3 crystallites was confirmed by ^{27}Al MAS NMR and ATR-FTIR.[229] In order to know about the distribution of aluminium atoms in amorphous aluminosilicates, ^{27}Al NMR experiment for Al/SiO_2 materials prepared by two successive grafting steps was carried out, which indicates the presence of five-coordinated aluminium species.[230] ^{27}Al NMR confirmed the formation of ester group and the cross-link with Al^{3+} in the modified starch (OS-starch-Al) prepared from normal cornstarch, octenylsuccinic anhydride and Al-$2(SO_4)_3$.[231] ^{27}Al MAS NMR was used for the characterisation of highly ordered mesoporous Al-SBA-15 materials with high Al/Si ratios (0.2, 0.5 and 1.0).[232] The ^{27}Al MAS NMR band of stoichiometric dicalcium silicate, $Ca_2Si_{0.972}Al_{0.028}O_{3.986}\sqrt{0.014}$ located close to similar to −70 ppm, which is ascribed to tetrahedral environments, in agreement with the proposed aliovalent Si/Al atomic substitution mechanism.[233] ^{27}Al MAS-NMR of single-site CeAlPO-5 with Al/Ce ratios 25, CeAlPO-5(25), showed two unusual peaks at −20.78 and −71.35 ppm due to delocalisation of cerium unpaired electron.[234] ^{27}Al MAS NMR was used to characterise the influences of alumina binder on the performance of Ni/H-ZSM-5 for hydrodeoxygenation of cyclohexanone.[235] Aluminium chemical environments in aluminosilicate glasses with compositions $60SiO_2$–$20Al_2O_3$–20CaO, $78SiO_2$–$11Al_2O_3$–$11Na_2O$ and $60SiO_2$–$10Al_2O_3$–$10Na_2O$–20CaO were investigated by ^{27}Al MAS NMR and MQMAS NMR.[236] ^{27}Al NMR characterisation was applied to M^{2+}-doped alunainate spinels (M = Co or Ni) prepared by a polymeric route leading to pure phases for synthesis temperatures equal to 800 or 1200 °C.[237] In order to acid properties of ZSM-5 type zeolite from trimethylphosphine oxide (TMPO) as a probe molecule, the amounts of the loaded TMPO were quantified by the signal intensities of ^1H MAS NMR spectra, which are compared with the numbers of Brønsted acid sites estimated by the Si/Al ratio of the framework derived from ^{29}Si MAS NMR spectra and the content of fourfold-coordinated Al from ^{27}Al MAS NMR spectra.[238] Well-crystallised synthetic saponites, which have a higher Al(IV)/Al(VI) ratio than the poorly crystallised samples, were investigated by ^{27}Al MAS NMR spectra.[239] Solid-state ^{27}Al MAS NMR experiment was applied to a new intermediate, $Li_{4-x}Al_x(NH)_{2-2x}N_{2x}$, is observed during the decomposition of $LiAlH_4$–$LiNH_2$ (1 : 1.5), (1 : 2) and (1 : 2.5), formed from Li_2NH and Al prior to the formation of Li_3AlN_2. $Li_{4-}xAl_x(NH)_{2-2x}N_{2x}$ and chemical shift of δ(Al) = 125 ppm was observed.[240] In order to explore the fluorine substitution process and the influence of F^- on the chemical behaviour of Keggin polynuclear Al species, the structures and reactivity of fluoridated Keggin-aluminum tridecamers (K-Al-13) in aqueous solution were studied using DFT. The consistency between the computational and experimental ^{19}F and ^{27}Al NMR chemical shifts were validated the suitable computational method for the present clusters.[241] ^{27}Al NMR spectra of rare-earth aluminum intermetallic compound $CeAl_2$ show a typical powder pattern for a nuclear spin I of 5/2 with the second-order nuclear quadrupole interaction at high temperature and an additional large dipolar broadening between the 4f electron spins of cerium and the ^{27}Al

nuclear spins at low temperature. The ^{27}Al NMR Knight shift follows the same temperature dependence as the magnetic susceptibility, suggesting that the ^{27}Al NMR Knight shift originates from the transferred hyperfine field of the Ce 4f electron spins with the hyperfine coupling constant of $A = +5.7$ kOe/μB.[242] A detailed analysis of the silica-supported chloro alkyl aluminum co-catalysts structure by solid-state ^{27}Al NMR at high-field (17.6 T and 20.0 T) and ultrafast spinning rates allows to detect six sites, characterised by a distribution of quadrupolar interaction principal values C_Q and isotropic chemical shifts.[243] ^{27}Al NMR spectra of dinuclear aluminum complexes bearing multidentate aliphatic aminoethanol-based ligands demonstrated that they exist as both five-coordinate Al(III) species in benzene-d(6) solution.[244] Biaxial Q-shearing of ^{27}Al 3QMAS NMR spectra in the analysis of structural defects of aluminium units in aluminosilicates was investigated, which demonstrated that the combination of various shearing transformations of the recorded ^{27}Al 3QMAS NMR spectra enables an understanding of the broadening processes of the correlation signals of disordered framework aluminosilicates, for which a wide distribution of ^{27}Al MAS NMR chemical shifts and quadrupolar parameters (*i.e.*, second-order quadrupolar splitting and quadrupole-induced chemical shifts) can be expected.[245] The average conditional stability constant ($\log K$) corresponding to the interaction between the aluminum ion (Al^{3+}) and Inogashira fulvic acid (IFA) at pH 3 was calculated to be 2.002.04 (M − 1) (bidentatemonodentate) from the results of calibration curve method for ^{27}Al NMR spectra.[246] Structure of Na_2O–Al_2O_3–SiO_2 glasses and melts, containing 75 mol% SiO_2 and different Al/(Al + Na) ratios was investigated by using ^{27}Al NMR, as well as Raman spectroscopy, ^{23}Na and ^{29}Si 1D MAS NMR spectroscopy. Results confirm the role change of Na^+ cations from network modifier to charge compensator in the presence of Al^{3+} ions.[247] There are three independent Al atoms in the R3 structure model in single crystal of pure aluminoborate PKU-1 ($Al_3B_6O_{11}(OH)_5$-$n$$H_2O$), and Al_3 locates in a quite distorted octahedral environment, which was evidenced by ^{27}Al NMR results.[248] The Ce/Al–SiO_2 catalysts were fabricated under various aerosol-spraying temperatures, and the temperature effects on the structural were investigated by ^{27}Al MAS NMR analysis.[249] A new design of a non-magnetic high-pressure anvil celAl for NMR experiments at Giga-Pascal pressures was presented, which uses a micro-coil inside the pressurised region for high-sensitivity NMR. ^{27}Al NMR test measurements showed a resolution of better than 4.5 ppm, and an almost maximum possible signal-to-noise ratio.[250] ^{27}Al MAS NMR was used for the characterisation of palladium based catalysts prepared by aqueous ion exchange with dealuminated NaY-zeolite support.[251] Paramagnetic ^{27}Al NMR spectroscopic studies were undertaken to investigate the influence of the lanthanoid amide complexes of the type $[Ln(NMe_2)_3(LiCl)_3]$ with an excess amount of $AlMe_3$ on the solvent chemical shifts.[252] At low magnetic field (7 T), the individual spinning sideband lineshapes of the outer satellite transitions in the ^{27}Al NMR spectrum of kaolinite are sensitive enough to differentiate information coming from the two aluminum sites in kaolinite. Thus, the isotropic chemical shift, the quadrupolar constant and asymmetry parameter of each site can be obtained by accurately fitting the full ^{27}Al MAS spectrum acquired at low magnetic field.[253] ^{27}Al, ^{29}Si and ^{11}B MAS NMR spectra confirmed the values of Weibull characteristic strength and of average flexural strengths for all the glass–ceramics are higher than those reported for G-18 commercial glass (51 MPa), with Weibull modulus varying in the range 11.6–34.4 towards good mechanical reliability.[254] The ^{27}Al MAS NMR spectra reveal four-, five- and six-coordinated aluminum in medium-range order of $0.5Al_2O_3$-$x SiO_2$ glasses ($1 \leq x \leq 6$). For $x \geq 3$, the Al species detected are essentially independent of sample composition indicating a constant structural environment of Al. In contrast, for $x = 1$ and 2, an increase in the ^{27}Al isotropic chemical shifts suggests an increasing number of Al.[255] Broad ^{27}Al NMR resonances of fast-ion-conducting cubic-garnet $Li_{7-3x}Al_x^{3+}La_3Zr_2O_{12}$ resulted reflecting the distribution of different isotropic chemical shifts and quadrupole coupling constants. From an energetic point of view, there is evidence that Al^{3+} could also occupy the 48g site with its almost regular octahedral coordination sphere.[256] In order to intrinsic ductile *versus* brittle properties of alloys connected with bonding characters, ^{27}AI isotropic metallic shifts of Al–Li solid solutions were measured by ^{27}AI NMR spectroscopy. Previously observed anomalous elastic properties upon Li alloying were found to be closely related to ^{27}AI metallic shifts which were associated with s electron density of states at the Fermi level on Al sites.[257] Lack of sharp structure in the first-order quadrupole pattern and the absence of second-order quadrupole structure in the ^{27}Al NMR data of AlH_4 anions in $KAlH_4$ suggest asymmetry ($\eta \neq 0$) and/or variations in the anion electric field gradients from structural disorder.[258] The surface hydroxyl groups of alumina dehydroxylated at 500 °C were studied by a combination of one- and two-dimensional homo- and heteronuclear ^1H and ^{27}Al NMR spectroscopy at high magnetic field.[259] ^{27}Al NMR peak of complex formed by AI(III) and phosphocreatine [Al(PCr)(H_2O)] in aqueous solution appeared at 48.92 ppm indicated a tetrahedral molecule.[260] A variety of phosphated zeolite H-ZSM-5 samples were investigated by using a combination of ^{27}Al–^{31}P CP and 3Q MAS NMR, as well as FTIR spectroscopy, single pulse 27Al, 29Si, 31P, ^1H–^{31}P CP MAS NMR, STXM and N-2 physisorption.[261] By deploying single pulse ^{27}Al ^{31}P MAS NMR, ^{27}Al–^{31}P HETCOR, ^{27}Al MQ MAS NMR spectroscopy, TPD of pyridine monitored by FT-IR spectroscopy and STXM, the interplay and influence of acidity, thermal treatment and phosphorus on the structure and acidity of phosphorus–zeolite H-ZSM-5 were established.[262] Although the structural characteristics zeolite β synthesised by hydrothermally ageing zeolite β nanoparticles, as determined *via* N-2-sorption, DRIFT, XRD and SEM, do not show clear discrepancies, an extensive study of the aluminium coordination (^{27}Al MAS NMR) and the acidic characteristics (*in situ* FT-IR with NH_3 and CO as probe) reveal a difference between the two nanoparticles based zeolites.[263] ^{27}Al MAS NMR was used for the characterisations of two kinds of mesoporous ZSM-5 synthesised using a hydrothermal methodology by utilising different soft templates, namely, dimethyl octadecyl [3-(trimethoxysilyl) propyl] ammonium chloride ([$(CH_3O)_3SiC_3H_6N(CH_3)_2C_{18}H_{37}$]Cl, TPOAC) and hexadecyl trimethyl ammonium bromide ($C_{16}H_{33}(CH_3)_3NBr$, CTAB).[264] ^{27}Al NMR spectra of organotemplate-free, seed-directed and rapid synthesis of Al-rich zeolite MTT (denoted as ZJM-6) with improved catalytic performance in isomerisation of *m*-xylene to *p*-xylene showed that the zeolitic product has good crystallinity, uniform crystals, high surface area, tetrahedral Al^{3+} species and abundant acidic sites.[265]

14.3 Gallium (^{71}Ga) ($I = 3/2$)

Multinuclear (^{1}H, ^{13}C and ^{71}Ga) magnetic resonance spectroscopy (1D and 2D), DFT calculations and luminescence techniques were used to study 8-hydroxyquinoline-5-sulfonate (8-HQS) and its complexes with Ga(III) in aqueous solutions. This study combined the high sensitivity of luminescence techniques and the selectivity of multinuclear NMR spectroscopy with the structural details accessible through DFT calculations, and aims to obtain a complete understanding of the complexation between the Ga^{3+} ion and 8-HQS, and how this influences the luminescence behaviour.[266] Buffering of Lewis acidic chlorometallate ionic liquids is a useful tool to modify their properties for electrochemical and catalytic applications. Lewis acidic chlorogallate(III) ionic liquids containing the 1-octyl-3-methylimidazolium cation, buffered with sodium chloride, were studied using ^{71}Ga NMR spectroscopy and cyclic voltammetry.[267] In order to insight into the relation between coordination changes and refractive and dielectric response, highlighting the role of non-bridging-oxygens and interphase, ^{71}Ga NMR investigation of Ga-containing germanosilicate glass was carried out.[268] A new design of a non-magnetic high-pressure anvil celAl for NMR experiments at Giga-Pascal pressures was presented, which uses a micro-coil inside the pressurised region for high-sensitivity NMR. ^{71}Ga NMR test measurements showed a resolution of better than 4.5 ppm, and an almost maximum possible signal-to-noise ratio.[269] ^{71}Ga solid state NMR of chalcogenide and chalco-halide glasses containing gallium showed an unusual line shape. This line shape was ascribed to the preferential local symmetry existing around gallium atoms in these glasses which are characterised by the covalent character of their chemical bonds. Then, the local environment of gallium in Ga_2S_3–GeS_2–CsCl glasses was studied by ^{71}Ga NMR as a function of the ionic cesium chloride CsCl concentration.[270] The substitution of gallium cation into β-tricalcium phosphate (β-TCP; Ca-3$(PO_4)_2$) was investigated, and the Ca sites involved successfully determined, using ^{71}Ga (natural abundance) NMR and 2D ^{71}Ga{^{31}P} rotary-resonance-recoupled heteronuclear multiple-quantum correlation.[271] ^{71}Ga NMR was applied to study of complexation of the parent NO_2A-*N*-(alpha-amino)propionic acid chelator and of a low molecular weight (model) amide conjugate with Ga^{3+}.[272] ^{71}Ga NMR spectroscopic studies of fluorination of complexes [$GaCl_3$(L)] (L = Me-3-tacn, BzMe$_2$-tacn) led to sharp resonances with ^{19}F–^{71}Ga coupling evident.[273]

14.4 Indium (^{115}In) ($I = 9/2$)

In the antiferromagnetic state at 4.5 K, ^{115}In NMR spectra of antiferromagnet URhIn5 with the Neel temperature $T_N = 98$ K in the zero external field indicated a commensurate antiferromagnetic structure.[274] ^{115}In NMR spectroscopic studies of fluorination of complexes [$InCl_3$(L)] (L = Me-3-tacn, $BzMe_2$-tacn) led to sharp resonances with ^{19}F–^{115}In coupling evident.[275] The electronic structure calculations of ^{115}In and ^{1}H solid-state NMR spectroscopy of $Ca_{54}In_{13}B_{4-x}H_{23+x}$ $(2.4<x<4)$ were carried out.[276] The indium electric field gradient (EFG) and chemical shift (CS) tensor parameters for low oxidation-state indium compounds were determined from the ^{115}In solid-state NMR spectra.[277]

14.5 Thallium (^{203,205}Tl) ($I = 1/2, 1/2$)

^{205}Tl NMR was used for the studies of binding properties of water-soluble cryptophane toward thallium cation, in basic solution. The results that water-soluble cryptophane-222 (2), cryptophane-223 (3) and cryptophane-233 (4), bearing zero, one and two propylenedioxy linkers, respectively, efficiently bind thallium cation under similar experimental conditions were shown.[278]

15 Group 14(^{13}C, ^{29}Si, ^{117,119}Sn, ^{207}Pb)

15.1 Carbon (^{13}C) ($I = 1/2$)

^{1}H and ^{13}C NMR chemical shifts of iron porphyrin complexes are determined mainly by the spin densities at the peripheral carbon and nitrogen atoms caused by the interaction between paramagnetic iron 3d and porphyrin MOs.[279] The author overviews the role of NMR chemical shifts in structural biology of glycosaminoglycans (GAGs), or derived oligosaccharides, across the last recent years of the current glycomics age, during which NMR spectroscopy has been giving a unique contribution toward structure and function determination of carbohydrates.[280] A description of the utility of deuteration in protein NMR is provided with an emphasis on quantitative evaluation of the effects of deuteration on a number of NMR parameters of proteins: (1) chemical shifts, (2) scalar coupling constants, (3) relaxation properties (R_1 and R_2 rates) of nuclei directly attached to one or more deuterons as well as protons of Me groups in a highly deuterated environment, (4) scalar relaxation of ^{15}N and ^{13}C nuclei in ^{15}N–D and ^{13}C–D spin systems as a measure of hydrogen bonding strength and (5) NOE-based applications of deuteration in NMR studies of protein structure.[281] This review summarises the longstanding challenges in the development of molecules having a direct "link" between N-heterocyclic carbene carbon and heavier group 14 elements such as germanium, tin and lead. Structural parameters such as bond length between carbene carbon and the metal, geometry of the metal, oxidation state of the metal, the changes in the chemical shift value in multi nuclear (^{13}C and ^{119}Sn) NMR and the reaction chemical of carbene–germanium/tin adducts are discussed.[282]

15.2 Silicon (^{29}Si) ($I = 1/2$)

The surface structure and adjacent interior of commercial available silicon nanopowder (np-Si) was studied using multinuclear, solid-state

NMR spectroscopy. From a combination of ^{1}H, ^{29}Si and ^{2}H MAS NMR results and quantum mechanical ^{29}Si chemical shift calculations, silicon atoms on the surface of "as-received" np-Si were found to exist in a variety of chemical structures, with apparent populations in the order (a) $(Si–O–)_3Si–H$ > (b) $(Si–O–)_3SiOH$ > (c) $(HO–)_nSi(Si)_m(–OSi)_{4-m-n}$ ≈ (d) $(Si–O–)_2Si(H)OH$ > (e) $(Si–O–)_2Si(–OH)_2$ > (f) $(Si–O–)_4Si$, where Si stands for a surface silicon atom and Si represents another silicon atom that is attached to Si by either a Si–Si bond or a Si–O–Si linkage.[283] Five selected compounds of Li-NASICON, $Li_3Zr_{2-y}Si_{2-4y}P_{1+4y}O_{12}$ $(0<y<0.5)$, were synthesised by sol–gel chemistry to obtain pure polycrystalline powder and then analysed by different physicochemical characterisations such as coupled DTA–TGA (thermogravimetric analysis), XRPD (X-ray powder diffraction), CIS (complex impedance spectroscopy) and ^{7}Li, ^{31}P and ^{29}Si MAS NMR.[284] This article reports on the influence of strontium for calcium substitution in diopside-Ba disilicate glass–ceramics on stabilisation thermal parameters and improvement adhesion to interconnect material of SOFCs. ^{29}Si MAS NMR and ^{27}Al MAS NMR spectra of the glasses revealed irrelevant chemical shifts for silicon and aluminum atoms upon changing the SrO content.[285] The effects of intramolecular N→Si coordination and electronic and conformational factors on the chemical shift of ^{29}Si nucleus in silacyclobutane (siletane) derivatives were studied by quantum-chemical methods.[286] DFT calculations are performed to investigate the electronic features of the structures of fluorinated polysilanes Si_nF_n ($n=4$, 6, 8, 10, 12, 20, 24, 28, 30, 32, 36, 50 and 60). ^{29}Si CS parameters detect equivalent electronic environment for silicon atoms within Si_nH_n polysilanes with $n \leq 20$ while ^{29}Si NMR pattern indicates a few separated peaks for Si_nH_n polysilanes with $n \geq 20$.[287] The gas-phase measurements of ^{19}F and ^{29}Si chemical shifts and scalar spin–spin couplings in SiF_4 and $(SiF_3)_2O$ were reported. The density dependence of Si and F nuclei chemical shifts is strictly linear for both compounds.[288] The silica gels, derived from water glass solution with pH adjusted at 3.0 and 9.9, were revisited to investigate their constitution, although water glass has been studied for last tens of decades on gelation. Solid-state NMR spectroscopy was applied to the nuclei ^{1}H and ^{29}Si, by the use of MAS, ^{29}Si CP/MAS, and modern techniques such as 2D HETCOR (two dimensional heteronuclear correlation) and variable-contact time CP techniques.[289] The ^{29}Si chemical shift anisotropies are investigated for calcium silicate hydrates.[290] The synthesis and structures of four silaoxazolinium salts and their application to the catalytic Mukaiyama aldol reaction are described. The structures of the silaoxazolinium salts were determined by X-ray crystal analysis as well as ^{29}Si NMR spectra.[291] The multinuclear silicon complexes $[\{H_2ClSi(\mu\text{-}pz^*)_2\}_2SiH_2]$ (1) and $[Cl_2Si(\mu\text{-}pz^{RR})_2SiCl_2]$ (2, R = Me; 3, R = Ph) were formed by the reaction of dichlorosilane (for 1) or hexachlorodisilane with 1-trimethylsilyl-3,5-dimethylpyrazole (for 1 and 2) and the 3,5-di-Ph analog (for 3). The structural and electronic features of the compounds synthesised in this work were analysed with single-crystal X-ray diffraction and ^{29}Si CP/MAS NMR spectroscopy combined with quantum chemical computations to investigate their ^{29}Si chemical shift anisotropy principal tensor components.[292] The large scale Targejia hot spring type Cs deposit in Tibet is unique and still active today with the distribution of hot springs being controlled by tectonic movements of the Tibetan Plateau. The ore bodies of the Targejia Cs deposit are mainly composed of geyserite, a Cs-containing opal. In this study, a combination of X-ray diffraction, IR spectroscopy and ^{29}Si, ^{1}H, ^{133}Cs MAS NMR were used to study the location and mobilisation of Cs in geyserites.[293] The electronic structures and molecular properties of pristine and N-doped zigzag (6,0) and armchair (4,4) SiC nanotubes (SiCNTs) were studied by DFT. Their results reveal that the influence of different N-doping on the properties of the electronic structure of the (6,0) and (4,4) SiCNTs could be detected by the values of ^{13}C and ^{29}Si chemical shifts.[294] Dynamic equilibrium related to change of the coordination number of the Si atom in bis[*N*-(dimethylamino)imidato-*N′*,*O*]silacycloalkanes in solution were studied by theoretical calculations and experimental measurement of the ^{29}Si NMR chemical shifts.[295] The synthesis and structural study of 23 new bis-, tris-, tetra-(dithiazinan-2-yl)silanes and stannanes is reported. The compounds were obtained by reaction of 5-alkyl-[1,3,5]-dithiazinanes (alkyl = Me, ^{i}Pr, ^{t}Bu) with ^{t}BuLi followed by the addition of R'_nSiCl_{4-n} or R'_nSnCl_{4-n} (n = 0–2, R′ = Me and Ph) in THF. Structures were determined by ^{1}H, ^{13}C, ^{29}Si or ^{119}Sn NMR, mass spectrometry and X-ray diffraction analyses.[296] Solid-state NMR is a well-established tool for the structural characterisation and dynamics study of various multifunctional materials. This review intends to cover the recent process for versatile solid-state NMR studies of topological structures, crystallisation processes, host–guest interactions, acidities and catalytic reaction mechanisms of various zeolites.[297] The purpose of this systematic experimental and theoretical study is to deeply understand the unique bonding situation in ferrocene-stabilised silylium ions as a function of the substituents at the Si atom and to learn about the structure parameters that determine the ^{29}Si NMR chemical shift and electrophilicity of these strong Lewis acids.[298] Careful NMR measurements on a very lightly-doped reference Si sample provide a convenient highly precise and accurate secondary chemical shift reference standard for ^{29}Si MAS NMR applicable over a wide temperature range.[299] The ^{29}Si NMR chemical shifts of ferrocene-stabilised silicon cations span a wide range depending on the substituents at the silicon atom. The Lewis acidity of silicon cations and their performance as catalysts cannot be correlated with ^{29}Si NMR chemical shifts as well as resonances of adducts with Lewis base probes, not even for a subset of silicon Lewis acids.[300] Silicon and aluminum chemical environments in silicate and aluminosilicate glasses with $60SiO_2 \cdot 20Na_2O \cdot 20CaO$ (CSN), $60SiO_2 \cdot 20Al_2O_3 \cdot 20CaO$ (CAS), $78SiO_2 \cdot 11Al_2O_3 \cdot 11Na_2O$ (NAS) and $60SiO_2 \cdot 10Al_2O_3 \cdot 10Na_2O \cdot 20CaO$ (CASN) were investigated by ^{27}Al and ^{29}Si solid state MAS and multiple quantum MAS (MQMAS) NMR experiments.[301] Aiming at the identification of an efficient computational protocol for the accurate NMR assessment of organosilanes in low-polarity organic solvents, ^{29}Si NMR chemical shifts of a selected set of such species relevant in organic synthesis were calculated relative to tetramethylsilane using selected density functional and perturbation theory methods.[302] ^{29}Si NMR spectra were recorded for a series of uranium complexes containing silicon, and the data were combined with results in the literature to determine if any trends exist between chemical shift and structure, ligand type, or oxidation state.[303] Ferrocenylsilanes with various functions at silicon (chlorine, vinyl, ethynyl) were prepared and studied by multinuclear magnetic resonance methods (^{1}H, ^{13}C, ^{29}Si NMR spectroscopy).[304]

15.3 Tin (^{117,119}Sn) ($I = 1/2, 1/2$)

A new class of bis-di-organotin (IV) compounds were synthesised in good yields (~80%) by a 1-step five molecules reaction in which intervene two molecules equivalent of leucine, one of 5,5′-methylene-bis-salicylaldehyde and either two of di-*n*-butyltin (IV) or diphenyltin (IV) oxides, that independently provided two bis-di-organotin derivatives. The structures of these two compounds were established by ^{1}H, ^{13}C and ^{119}Sn NMR, as well as by IR and elemental analysis.[305] The reaction of $SnCl_4$ with β-chlorovinyl aldehydes in anhydrated CH_2Cl_2 gave octahedral complexes $SnCl_4 \cdot 2L$ (L = aldehyde). The adducts were characterised in solution using multinuclear (^{1}H, ^{13}C and ^{119}Sn) NMR and IR spectroscopy.[306] The complexation chemistry of the stannylene $Sn\{N[SiMe_3]_2\}_2$, 1st reported by Lappert in the 1970s, was studied by ^{119}Sn NMR chemical shift measurements.[307] The synthesis of octahedral complexes $[SnCl_4L_2]$ (L = $R_2NP(O)(OCH_2CF_3)$(O-*p*-tolyl)): $R_2N = Me_2N$ (1), Et_2N (2), $CH_2(CH_2CH_2)_2N$ (3) and $O(CH_2CH_2)_2N$ (4), or L = $R_2NP(O)(OCH_2CF_3)$(O-*p*-$PhNO_2$): $R_2N = Me_2N$ (5), Et_2N (6) and $O(CH_2CH_2)_2N$ (7) is described. The new adducts were characterised by multinuclear (^{31}P, ^{19}F, ^{119}Sn) NMR, IR spectroscopy and elemental analyses.[308] Five new diorganotin(IV) compounds were prepared by reacting diorganotin dichlorides R_2SnCl_2 (R = Me, *n*-Bu and Ph) with sodium salts of the tridentate NO_2 ligands (*E*)-3-hydroxy-2-((2-hydroxybenzylidene)amino)propanoic acid (L^1H_2Na) and (*E*)-3-hydroxy-2-((1-(2-hydroxyphenyl)ethylidene)amino)propanoic acid (L^2H_2Na). The molecular structures of the resulting diorganotin(IV) compounds were established by elemental analysis and a combination of IR and NMR (^{1}H, ^{13}C, ^{119}Sn) spectroscopy.[309] A series of 2-R'_3E (E = Si or Sn; R′ = Me or Ph) derivatives of 5-R-[1,3,5]-dithiazinanes (R = Me, ^{i}Pr, ^{t}Bu) are reported, as well as some of their N-BH_3 adducts. Structures were determined by ^{29}Si, ^{119}Sn, ^{11}B, ^{13}C and ^{1}H NMR and X ray diffraction analyses.[310] The synthesis and structural study of 23 new bis-, tris-, tetra-(dithiazinan-2-yl)silanes and stannanes is reported. The compounds were obtained by reaction of 5-alkyl-[1,3,5]-dithiazinanes (alkyl = Me, ^{i}Pr, ^{t}Bu) with ^{t}BuLi followed by addition of R'_nSiCl_{4-n} or R'_nSnCl_{4-n} ($n = 0$–2, R′ = Me and Ph) in THF. Structures were determined by ^{1}H, ^{13}C, ^{29}Si or ^{119}Sn NMR, mass spectrometry and X-ray diffraction analyses.[311] Dilithiostannoles, inverse-sandwich stannacyclopentadiene compounds $[(THF)_2Li_2(\mu\text{-}\eta^5{:}\eta^5\text{-}SnC_5\text{-}2,5\text{-}R^1_2\text{-}3,4\text{-}R^2_2)]$ (3a–d; R^1 = $SiMe_3$, Si^tBuMe_2, Ph, Et; R^2 = Ph, Et), which are aromatic tin-containing ring compounds, were proposed to have stannylene character, as judged from their NMR analysis. The authors report on the synthesis of silyl-substituted dilithiostannoles, which were characterised by ^{7}Li and ^{119}Sn NMR spectroscopy and X-ray diffraction analysis.[312]

15.4 Lead (^{207}Pb) ($I = 1/2$)

The authors present an alternative approach to separating contributions to the NMR shift originating from the Knight shift and chemical shielding by a combination of experimental solid-state NMR results and *ab initio* calculations. Here, the authors analysed ^{207}Pb and ^{125}Te NMR spin–lattice relaxation rates and NMR shifts for bulk and nanoscale PbTe.[313] Coordination complexes of $(2,6\text{-}Me_2C_6H_3S)_2Pb$ (1) with flexible bidentate ligands were prepared to explore new bonding environments for Pb(II) thiolates. Static ^{207}Pb solid-state NMR spectra of $[(2,6\text{-}Me_2C_6H_3S)_2Pb]_2$(tmeda), $[(2,6\text{-}Me_2C_6H_3S)_2Pb]_3$(dmpe) and $[(2,6\text{-}Me_2C_6H_3S)_2Pb]_2$(dppe) [tmeda = *N*,*N*,*N*′,*N*′-tetramethylethylenediamine; dmpe = 1,2-bis(dimethylphosphino)ethane; dppe = 1,2-bis(diphenylphosphino)ethane] and $[Ph_4As][(PhS)_3Pb]$ were acquired with cross polarisation (CP)-CPMG and frequency swept pulse (WURST)-CPMG pulse sequences. The ^{207}Pb SSNMR spectra reveal that the lead chemical shift anisotropies (CSA) vary greatly between the different Pb sites, and are generally large in magnitude.[314] The authors investigated temperature distribution in a solid state NMR sample rotor under MAS experiments by analysing the ^{207}Pb chemical shift of solid lead nitrate ($Pb(NO_3)_2$).[315] The authors carried out a series of zeroth-order regular approximation (ZORA)-DFT and ZORA-time-dependent (TD)-DFT calculations for molecular geometries, NMR chemical shifts, nucleus-independent chemical shifts (NICS) and electronic transition energies of plumbacyclopentadienylidenes stabilised by several Lewis bases, $(Ph)_2({}^{t}BuMe_2Si)_2C_4PbL_1L_2$ (L_1, L_2 = THF, Pyridine, N-heterocyclic carbene) and their model molecules.[316] Soap formation in traditional oil paintings occurs when heavy-metal-containing pigments, such as lead white, $2PbCO_3 \cdot Pb(OH)_2$, and lead tin yellow type I, Pb_2SnO_4, react with fatty acids in the binding medium. The factors that trigger soap formation and the mechanism(s) of the process are not yet well understood. To elucidate these issues, chemical and structural information is necessary, which can be obtained using solid-state ^{207}Pb and ^{13}C NMR.[317]

16 Group 15 ($^{14,15}N$, ^{31}P)

16.1 Nitrogen ($^{14,15}N$) ($I = 1$, 1/2)

The identification of intermolecular hydrogen bonding interactions in a nicotinamide palmitic acid cocrystal and a 50% wt/wt acetaminophen-polyvinylpyrrolidone solid dispersion are reported using advanced solid-state MAS NMR methods. The application of a novel ^{14}N–^{1}H HMQC experiment, where coherence transfer is achieved *via* through-space couplings, is shown to identify specific hydrogen bonding motifs.[318] Under sample spinning, the ^{14}N overtone signal is split into five (0, ± 1, ± 2) overtone sidebands separated by the spinning frequency. For a powder sample spinning at the magic angle, the $+2\omega_r$ sideband is dominant while the others show significantly lower signal intensities. The resultant

MAS powder patterns show characteristic quadrupolar lineshapes from which the ^{14}N quadrupolar parameters and isotropic chemical shift can be determined.[319] The Raman (1400–100 cm^{-1}) and IR (4000–400 cm^{-1}) of solid hexachlorocyclotriphosphazene, $P_3N_3Cl_6$ (HCCTP) were recorded. The ^{14}N and ^{31}P chemical shifts were predicted with B3LYP/6-311+G(2d,p) calculations using the GIAO technique with solvent effect modeled using the PCM method.[320] Electronic structures and magnetic properties on nitride mixed metal cluster within fullerene cage of $Sc_xY_{3-x}N$ at $C_{80}(CF_3)_n$ at $n=0$, 2 and 8 were studied for controlling spin gate in the NMR quantum computer. Addition of trifluoromethyl groups on the fullerene carbon cage encapsulated with nitride mixed metal cluster provided influence of optical properties at excited state and magnetic parameters of chemical shift in ^{45}Sc, ^{89}Y, ^{14}N, ^{19}F and ^{13}C-NMR.[321] The DFT-based method with periodic boundary conditions and addition of a pair-wised empirical correction for the London dispersion energy (DFT-D) was used to study the NMR quadrupolar interaction (coupling constant C_q and asymmetry parameter η_q) of ^{14}N nuclei in a homologous series of tetra-*n*-alkylammonium halides $(C_xH_{2x+1})_4N^+X^-$ ($x=1$–4), (X = Br, I). These ^{14}N quadrupolar properties are particularly challenging for the DFT-D computations because of their very high sensitivity to tiny geometrical changes, being negligible for other spectral property calculations as, for example, NMR ^{14}N chemical shift.[322] Dynamic nuclear polarisation (DNP) was used to obtain MAS $^{14}N^{OT}$ (nitrogen-14 overtone) solid-state NMR spectra from several model amino acids, with both direct and indirect observation of the $^{14}N^{OT}$ signal.[323] The ^{13}C and ^{15}N CP/MAS NMR measurements were accompanied by the proper theoretical description of the solid-phase environment, as provided by the DFT in the pseudopotential plane-wave scheme, and employed in refining the atomic coordinates of the crystal structures of thiamine chloride hydrochloride and of its monohydrate.[324] The authors report the dramatic impact of the local environment and in particular that of the interplay between solvent and substituents on the influence of solvent on the N coordination shift $\Delta^{15}N_{coord}$.[325] The five Schiff bases obtained by condensation of dehydroacetic acid, *p*-phenylenediamine and derivatives of salicylaldehyde were studied by ^{13}C and ^{15}N CP/MAS NMR methods to find the structure of intramolecular hydrogen bridges.[326] Au(III) and Pd(II) chloride complexes with 2-benzylpyridine (2bzpy) – (2bzpyH)[$AuCl_4$], [Au(2bzpy)Cl_3] and *trans*-[$Pd(2bzpy)_2Cl_2$], as well as Au(III) chloride organometallics with monoanionic form of 2bzpy, deprotonated in the benzyl side group at the *ortho*-carbon C(2′) (2bzpy*) – [Au(2bzpy*)Cl_2], were studied by ^{1}H, ^{13}C and ^{15}N NMR[327] The work presented herein describes an emerging approach for the analysis of unfractionated and low molecular weight heparins, as well as porcine and human-derived HS. This approach uses the untapped potential of ^{15}N NMR to characterise these preparations through detection of the NH resonances of *N*-sulfo-glucosamine residues.[328] ^{1}H, ^{13}C and ^{15}N NMR studies of five nitrobenzene-1,2-diamines in solution and solid state were achieved and the experimental chemical shifts and coupling constants agree with the theoretical values obtained at the B3LYP/6-311++G(dp) computational level using the geometries fully optimised with the hybrid HF/DFT B3LYP method and the 6-31G(d) basis set.[329] There is considerable interest in determining amide-^{15}N chemical shift anisotropy (CSA) tensors from biomolecules and understanding their variation for structural and dynamics studies using solution and solid-state NMR spectroscopy and also by quantum chemical calculations.[330] ^{15}N-labeled polyaniline (PANI) is heat treated at temperatures from 200 to 1000 °C in a nitrogen atmosphere. The carbonisation of PANI is determined using solid state ^{13}C and ^{15}N NMR.[331] ^{1}H, ^{13}C and ^{15}N NMR chemical shifts were measured for 2-aminopyridine *N*-oxide, its eleven derivatives, and 3-Cl and 3-Br substituted 4-nitropyridine *N*-oxides.[332] The adsorption behaviour of the amino acid glycine in mesoporous silica was investigated using a combination of quantum chemical calculations and NMR spectroscopic experiments.[333] Mirtazapine (±)-1,2,3,4,10,14*b*-hexahydro-2-methylpyrazino(2,1-*a*)pyrido(2,3-*c*)(2)benzazepine is a compound with antidepressant therapeutic effects. It is the 6-aza derivative of the tetracyclic antidepressant mianserin (±)-2-methyl-1,2,3,4,10,14*b*-hexahydrodibenzo[*c*,*f*]pyrazino[1,2-*a*]azepine. The optimised geometry, energies, nonlinear optical properties, vibrational frequencies, ^{13}C, ^{1}H and ^{15}N NMR chemical shift values of mirtazapine were determined using the DFT/B3LYP method.[334] A DFT study was performed to investigate electronic and magnetic properties of armchair (4, 4), (5, 5) and (6, 6) BNNTs with Stone–Wales defects based on ^{11}B and ^{15}N NMR parameters and NICS indexes.[335] 2D [^{1}H, ^{15}N] HSQC NMR spectroscopy was used to monitor the reaction of fully ^{15}N-labeled [{*trans*-PtCl$(NH_3)_2$}$_2$(μ-*trans*-Pt$(NH_3)_2${$NH_2(CH_2)_6NH_2$}$_2$)]$^{4+}$ (Triplatin, BBR3464 or 1,0,1/t,t,t (^{15}N-1)) with the self-complementary 10-mer DNA duplex 5′-{d(ACGTATACGT)$_2$} (duplex I) at pH 5.4 and 298 K.[336] While chemical shift prediction has played an important role in aspects of protein NMR that include identification of secondary structure, generation of torsion angle constraints for structure determination and assignment of resonances in spectra of intrinsically disordered proteins, interest has arisen more recently in using it in alternate assignment strategies for crosspeaks in ^{1}H–^{15}N HSQC spectra of sparsely labeled proteins.[337] Solid state ^{13}C and ^{15}N chemical shifts of uracil and imidazole were calculated using a 2-layer ONIOM approach at 32 levels of theory.[338] Papaverine can be easily oxidised to papaverinol, papaveraldine and 2,3,9,10-tetramethoxy-12-oxo-12*H*-indolo[2,1-*a*]isoquinolinium chloride. On addition of alkali solution the latter compound forms 2-(2-carboxy-4,5-dimethoxyphenyl)-6,7-dimethoxyisoquinolinium inner salt. Together with these structures the interesting 13-(3,4-dimethoxyphenyl)-2,3,8,9-tetramethoxy-6*a*-12*a*-diazadibenzo[*a*,*g*]fluorenylium chloride is discussed, which is formed in the Gadamer–Schulemann reaction of papaverine as a side product. This letter reports the ^{15}N NMR spectra of the above mentioned compounds.[339] The ^{1}H, ^{13}C and ^{15}N NMR chemical shifts of ten 4-nitropyridine *N*-oxide derivatives are assigned.[340] The disulfide bond generation system in *E. coli* is led by a periplasmic protein, DsbA, and an integral membrane protein, DsbB. Here the authors present a solid-state NMR study of a 41 kDa membrane protein complex DsbA/DsbB precipitated in the presence of native lipids to investigate conformational changes and dynamics that occur upon transient complex formation within the electron transfer pathway.[341] Effect of solvation on the accuracy of DFT quantum-chemical calculations of ^{15}N NMR chemical shifts of pyrrole, *N*-methylpyrrole and pyridine was studied.[342] The existence of ether-bridged condensation products in urea-formaldehyde (UF) resins is still disputed in the literature as these products have never been isolated or fully characterised. Spectroscopic chemical shifts of the proposed ether-bridged condensation product in ^{15}N, ^{13}C, ^{1}H NMR spectroscopy were assigned. Furthermore, individual peak assignments are provided for the methylolurea hemiformal moiety.[343] Pyridinylazolato (*N–N′*) ruthenium(II) complexes [(*N–N′*)RuCl(PMe_3)$_3$] were obtained in high yields by treating the corresponding functionalised azolylpyridines with [$RuCl_2(PMe_3)_4$] in the presence of a base. ^{15}N NMR spectroscopy was used to elucidate the electronic influence of the substituents attached to the azolyl ring.[344] NMR chemical shift is a molecular property that can be computed from 1st principles. By using the authors' combined fragmentation method (CFM), one is able to accurately compute this property for small proteins. Without nonbonded interactions, the root mean square errors (RMSEs) compared to the full calculations for ^{1}H, ^{13}C, ^{5}N, ^{17}O and ^{33}S were 0.340, 0.649, 3.052, 6.928 and 0.122 ppm, respectively, while with the inclusion of nonbonded interactions the RMSEs for ^{1}H, ^{13}C, ^{15}N, ^{17}O and ^{33}S were 0.038, 0.253, 0.681, 3.480 and 0.052 ppm, respectively.[345] The characteristic ^{15}N and $^{1}H_N$ NMR chemical shifts and $^{1}J(^{15}N–^{1}H)$ coupling constants of various symmetric and unsymmetrical substituted 1,4-dihydropyridine derivatives are described.[346] Protonation sites of a chelating mono-protic ionic liquid (PIL) comprising *N*-hexylethylenediaminium depending on the counter anion were studied by ^{15}N and ^{13}C NMR spectroscopy and DFT calculation.[347] Histidine usually exists in three different forms (including biprotonated species, neutral τ and π tautomers) at physiological pH in biological systems. The different protonation and tautomerisation states of histidine can be characteristically determined by ^{13}C and ^{15}N chemical shifts of imidazole ring.[348] The selectively ^{15}N labeled azido-1,2,4-triazine 2^*A and azidopyrimidine 4^*A were synthesised by treating hydrazinoazines with ^{15}N-labeled nitrous acid. The synthesised compounds were studied by ^{1}H, ^{13}C and ^{15}N NMR spectroscopy in DMSO, TFA and DMSO/TFA solutions, where the azide-tetrazole equilibrium could give two tetrazoles (T, T′) and one azide (A) isomer for each compound.[349] Recently, new families of compounds based on amantadine- and aryl-substituted isoxazole were discovered to inhibit the S31N channel activity and reduce replication of S31N-harboring viruses. The authors use solid-state NMR spectroscopy to investigate the effects of one of these isoxazole compounds, WJ352, on the conformation of the S31N TM segment and the dynamics of the proton-selective residue, His37.[350] Silatranes $XSi(OCH_2CH_2)_3N$ exhibit a good linear relationship between their experimental and calculated (IGLO and GIAO) values of the NMR chemical shifts of ^{15}N, δ_N and the lengths of dative bonds Si←N, d_{SiN}, determined in the gas phase (ED, CCSD), solutions (COSMO PBE0, B3PW91) and crystals (X-ray).[351] ^{15}N NMR of pyridine was used to study Lewis and Brønsted sites at the acid surface of magnesium hydroxide fluoride at two different pyridine loadings with ratios of excess pyridine compared to pyridine molecules bound to Lewis and Bronsted sites of about 1 : 1 (highly loaded) and 0.15 : 1 (low loaded), respectively.[352] ^{15}N NMR relaxation studies, analyses of NMR data to include chemical shifts, residual dipolar couplings (RDC), NOEs and H^{N}–H^{α} coupling constants and molecular dynamics (MD) simulations were used to characterise the behaviour of lysozyme from bacteriophage lambda (λ lysozyme) in solution.[353] The impact of pressure on the backbone ^{15}N, ^{1}H and ^{13}C chemical shifts in N-terminally acetylated α-synuclein was evaluated over a pressure range 1–2500 bar.[354] An approach towards accurate NMR measurements of deuterium isotope effects on the chemical shifts of all backbone nuclei in proteins (^{15}N, $^{13}C_{\alpha}$, ^{13}CO, $^{1}H_{\alpha}$) and $^{13}C_{\beta}$ nuclei arising from ^{1}H-to-D substitutions at amide nitrogen positions is described.[355] The authors introduce a Python-based program that utilises the large database of ^{13}C and ^{15}N chemical shifts in the Biological Magnetic Resonance Bank to rapidly predict the amino acid type and secondary structure from correlated chemical shifts.[356] The coupled effect of ionic strength (50–400 mM) and pH (2–8) on ionisation and conformation equilibrium of lysozyme was studied using NMR spectroscopy.[357] The interactions between histidine (I) and metal species play essential roles in a wide range of important biological processes including enzymic catalysis and signal transduction. Here, solid-state NMR techniques were employed to det. the interaction between I and Zn(II) in the pH range of 3.5–14.[358] An analysis is presented of secondary deuterium isotope effects on ^{15}N and ^{13}C chemical shifts of the methylamine Schiff base of 4,6-dimethoxysalicylaldehyde.[359] 2,4,6-Triazido-*s*-triazine, 2,4,6-triazidopyrimidine and six different 2,4,6-triazidopyridines were studied by ^{15}N NMR spectroscopy.[360] 2-Aminopyridinyl iron(III) complex was grafted on chlorosilane modified mesoporous MCM-41 to give MCM-Py-Fe(III). The immobilisation was confirmed by FT-IR, ^{13}C, ^{29}Si, ^{15}N CP/MAS NMR, nitrogen adsorption–desorption study and elemental analysis.[361] Sequence specific resonance assignment of proteins forms the basis for variety of structural and functional proteomics studies by NMR. In this context, an efficient standalone method for rapid assignment of backbone (^{1}H, ^{15}N, $^{13}C^{\alpha}$ and $^{13}C'$) resonances of proteins was presented.[362] Fused five-membered nitrogen heterocycles comprise a very important group of compounds frequently utilised in pharmaceutical applications. In this study, the authors report the first systematic synthesis of substituted pyrazolo[4,3-*c*]pyrazoles and three regioisomers of their N-Me derivatives. All compounds were fully characterised by NMR spectroscopy in solution and selected compounds also were studied by X-ray diffraction in the solid state.[363] The molecular origin of the selectivity of N-donor ligands, such as alkylated bis(triazinyl)pyridines (BTPs), for actinide complexation in the presence of lanthanides is still largely unclear. NMR investigations of an $Am(PrBTP)_3^{3+}$ complex with a ^{15}N labeled ligand showed that it exhibits large differences in ^{15}N chemical shift for coordinating N-atoms in comparison to both lanthanide(III) complexes and the free ligand.[364]

The reaction of 3,5-dimethyl-4-formylpyrazoles, bearing various substituents at N-1 atom, with propane-1,3-diselenol and 2-hydroxypropane-1,3-diselenol in the presence of TMSCl proceeds without heating to chemoselectively give hitherto unknown 2-(pyrazol-4-yl)-1,3-diselenane hydrochlorides in high yields. The latter are easily transformed to the corresponding free bases – 2-(pyrazol-4-yl)-1,3-diselenanes. The ^{15}N chemical shifts of the pyrazole ring in 2-(pyrazol-4-yl)-1,3-diselenanes obtained by 2D HMBC-gp (^{15}N–^{1}H) technique are indicative of the N-2 atom protonation in hydrochlorides.[365] The three possible isomeric forms of disubstituted 1,2,3-triazole (1,4- or 1,5- or 2,4-disubstituted derivatives) were characterised and distinguished by routine ^{1}H/^{15}N gHMBC experiments at ^{15}N natural abundance. The calculated (GIAO-B3LYP/6-311++G**) ^{15}N NMR chemical shifts showed good agreement with the experimental values, further supporting their unambiguous structural characterisation.[366] DFT/Becke–Lee–Yang–Parr (B3LYP) and gauge-including AO (GIAO) calculations were performed on a number of 1,2,4-triazole derivatives, and the optimised structural parameters were employed to ascertain the nature of their predominant tautomers. ^{13}C and ^{15}N NMR chemical shifts of 3-substituted 1,2,4-triazole-5-thiones and their propargylated derivatives were calculated *via* GIAO/DFT approach at the B3LYP level of theory with geometry optimisation using a 6-311++G** basis set.[367] Two polymorphic forms of Et 3-{3-[((2R)-3-{[2-(2,3-dihydro-1*H*-inden-2-yl)-1,1-dimethylethyl]-amino}-2-hydroxypropyl)-oxy]-4,5-difluorophenyl} propanoate hydrochloride, an investigational pharmaceutical compound, are characterised using spectroscopic and diffractometric techniques. Detailed solid-state NMR studies observing the ^{1}H, ^{13}C, ^{15}N, ^{19}F and ^{35}Cl nuclei are performed to characterise the subtle structural differences between the two forms.[368] The global fold of human cannabinoid type 2 (CB_2) receptor in the agonist-bound active state in lipid bilayers was investigated by solid-state ^{13}C and ^{15}N MAS NMR, in combination with chemical shift prediction from a structural model of the receptor obtained by microsecond-long molecular dynamics (MD) simulations.[369] Water-oxidation catalysts (WOCs) can potentially be improved by installing pendant electron-donor groups that may also be proton donors or acceptors. The authors have modified one of the most well-studied WOCs with alkoxy or hydroxy substituents on the bidentate bipyridine ligand (*N*,*N*), thereby forming [(terpy)Ru^{II}(*N*,*N*)X] (X = Cl, H_2O; terpy = 2,2′; 6′,2″-terpyridine). A combination of NMR spectroscopy (particularly ^{15}N chemical shift data), UV/visible spectroscopy, X-ray diffraction and oxygen evolution data point to interesting and beneficial effects of an oxygenated group proximal to X.[370] In this paper, the solution structure of native "liquid silk" was determined with solution NMR, especially for tandem repeated sequences with (GAGXGA)$_n$ (X = S, Y, V) and GAASGA motifs in the B. mori silk fibroin. The assignment of the ^{13}C, ^{15}N and ^{1}H solution NMR spectra for the repetitive sequence motifs was achieved, and the chemical shifts were obtained.[371] The properties of nitrogen centers acting either as hydrogen-bond or Bronsted acceptors in solid molecule acid–base complexes were probed by N 1s XPS as well as ^{15}N solid-state

NMR spectroscopy and are interpreted with refer to local crystallographic structure information provided by X-ray diffraction (XRD).[372] The thermo- and solvo-driven chiroptical switching process observed in specific polycarbodiimides occurs in a concerted fashion with large deviations in specific optical rotation (OR) and CD Cotton effect as a consequence of varying populations of two distinct polymer conformations. These two conformations are clearly visible in the ^{15}N NMR and IR spectra of the ^{15}N-labeled poly-(^{15}N-(1-naphthyl)-*N*′-octadecylcarbodiimide) (Poly-3) and poly-(^{15}N-(1-naphthyl)-^{15}N′-octadecylcarbodiimide) (Poly-5).[373] Two-dimensional ^{15}N chemical shift/^{1}H chemical shift and three-dimensional ^{1}H–^{15}N dipolar coupling/^{15}N chemical shift/^{1}H chemical shift MAS solid-state NMR correlation spectra of the filamentous bacteriophage Pf1 major coat protein show single-site resolution in noncrystalline, intact-phage preparations.[374] Two novel reduced dimensionality (RD) tailored HN(C)N experiments are proposed to facilitate the backbone resonance assignment of proteins both in terms of its accuracy and speed. These experiments exploit the linear combination of backbone ^{15}N and ^{13}C′/^{13}C$^{\alpha}$ chemical shifts simultaneously to achieve higher peak dispersion and randomness along their respective F_1 dimensions.[375] ^{1}H, ^{13}C, ^{15}N and ^{17}O NMR chemical shifts are used for the characterisation of the intramolecular interactions in several nitramines of the Me_2N–G–NO_2 type.[376] While antimicrobial peptides (AMPs) have been widely investigated as potential therapeutics, high-resolution structures obtained under biological relevant conditions are lacking. Here, the high-resolution NMR structures of the homologous 22-residue-long AMPs, piscidin 1 (I) and piscidin 3 (II) were determined in fluid-phase 3 : 1 phosphatidylcholine/phosphatidylglycerol (PC/PG) and 1 : 1 phosphatidylethanolamine/phosphatidylglycerol (PE/PG) bilayers to identify molecular features important for membrane destabilisation in bacterial cell membrane mimics. Structural refinement of ^{1}H–^{15}N dipolar couplings and ^{15}N chemical shifts measured by oriented sample solid-state NMR provided structural and orientational information of high precision and accuracy about these interfacially bound α-helical peptides.[377] Tris(pentafluorophenyl)corrole and its ^{15}N-enriched isotopomer were studied in [D_8]toluene solution by 1D and 2D variable-temperature NMR techniques to establish the mechanisms of tautomerisation of the NH protons inside the interior of the corrole macrocycle.[378] In this study, the authors examined PriB′s interaction with ssDNA (oligo-dT35, -dT15 and -dT7) using heteronuclear NMR analysis. Interestingly, ^{1}H or ^{15}N chemical shift changes of the PriB main-chain showed two distinct modes using oligo-dT35.[379] The 231-residue capsid (CA) protein of human immunodeficiency virus type 1 (HIV-1) spontaneously self-assembles into tubes with a hexagonal lattice that is believed to mimic the surface lattice of conical capsid cores within intact virions. The authors report the results of solid-state NMR measurements on HIV-1 CA tubes that provide new information regarding changes in molecular structure that accompany CA self-assembly, local dynamics within CA tubes, and possible mechanisms for the generation of lattice curvature.[380] A review with 351 references. It collects the ^{15}N and ^{31}P NMR data for *ca.* 390 and *ca.* 340, respectively, complexes of transition metal ions such as Ni(0), Pd(0), Pt(0), Ni(II), Pd(II), Pt(II), Pd(IV), Pt(IV), Co(−1), Rh(−1), Rh(I), Ir(I), Co(III), Rh(III), Ir(III), Fe(−2), Ru(−2), Fe(II), Ru(II), Os(II), Cu(I), Ag(I), Au(I), Au(III), Zn(II), Cd(II) and Hg(II) with nitrogen-containing heterocycles (azines, azoles and azoloazines like purines or 1,2,4-triazolo-[1,5*a*]-pyrimidines) and their phosphorus analogs (phosphinines and phospholes).[381] Complexation of rhodium(II) tetraacetate and rhodium(II) tetrakistrifluoroacetate with a set of heteroaromatic *N*-oxides containing additional functional groups was investigated by means of DFT calculations, and ^{1}H, ^{13}C and ^{15}N NMR spectroscopy in $CDCl_3$ solutions.[382] An NMR-based relative binding affinity measurement method has been developed in which differences in the binding affinities of different hosts toward a particular guest ($\Delta\log K_{ass}$ values) are measured in the same solution. It was found for symmetric receptor molecules from the same compound family that there is a correlation between the acetate binding affinity of a receptor and the ^{15}N chemical shift of the nitrogen atoms of its binding centers.[383] Based on the present data, the authors recommend GIAO-DFT-KT3/pcS-3//pc-2 as one of the most effective locally dense basis set schemes for the calculation of ^{15}N NMR chemical shifts.[384] A reaction of di-Et 2,4,6-trimethyl-1,4-dihydropyridine-3,5-dicarboxylate with 1, 2 and more equilibrant of *N*-bromosuccinimide (NBS) in methanol was studied by NMR spectroscopy at a temperature interval ranging from 25 to 40 °C.[385] The authors present an alternative model for the interpretation of the stochastic processes underlying these fluctuations of resonance frequencies, using bovine pancreatic trypsin inhibitor (BPTI) as a model protein. Time correlation functions of ^{15}N amide chemical shifts computed from molecular dynamics simulations (MD) of BPTI were analysed in terms of a transiently fractional diffusion process.[386] Pyridylmethylamines or pma are versatile platforms for different catalytic transformations. Five pma-ligands and their respective Pd complexes were studied by liquid state NMR. By comparing ^{1}H, ^{13}C and ^{15}N chemical shifts for each pma/pma-Pd couple, a general trend for the metallacycle atoms concerns variations of the electronic distribution at the pendant arm, especially at the nitrogen atom of the ligand.[387] The authors report the ^{13}C (carbonyl and Cα) and ^{15}N (amine and amide) solid-state NMR isotropic chemical shifts in a series of seven such isostructural porous dipeptides as well as shift tensor data for the carbonyl and amide sites.[388] The crystal structure of a powder pharmaceutical cocrystal, theophylline–nicotinamide (1:1) crystal complex, is determined for the first time by using a combination of X-ray powder diffraction (XRPD), 1D solid state NMR, as well as DFT calculations.[389] The authors report an improved synthetic pathway, which enables the synthesis of safranine analogs containing electron-donating Ph substituents. NMR analysis of the safranine series demonstrates that the ^{15}N chemical shift at the N(5) position correlates with the 2-electron reduction midpoint potential.[390] Although almost 200 years-old, several unknown aspects remain to be explored of colchicine, the unique available drug for acute flares of gout. In this article, the authors report DFT studies of geometry, energy and NMR; ^{1}H, ^{13}C and ^{15}N NMR chemical shifts and some spin–spin coupling constants, including the complete analysis of the saturated part (ring B); the assignment of both enantiomers by NMR using a chiral solvating agent; solid-state NMR experiments of the different forms of natural and racemic colchicine, and IR and vibrational CD (VCD) studies of these same forms.[391] The experimental and theoretical properties of ligands consisting of pyridine and phenanthroline derivatives were studied. The results show a very high correlation between the experimental and theoretical spectroscopic properties of the ligands such as the IR, NMR chemical shift and UV.[392] Effects of solvation on the accuracy of the calculation of ^{15}N chemical shifts in the azine series were analysed at the DFT/GIAO level of theory.[393] The ^{15}N NMR chemical shifts of possible nitrogen-containing moieties at edges and defects of graphene are studied by using the first-principles method. Computational results show that pyridine-like and graphite-like N can be rather easily identified using the ^{15}N NMR technique, in agreement with experiment.[394] The icosahedral bacteriophage T7 is a 50 MDa double-stranded DNA (dsDNA) virus that infects *Escherichia coli*. The authors apply the solid-state MAS NMR technique to study a uniformly ^{13}C and ^{15}N labeled wild-type T7 phage.[395] *Ortho*-substituted and *para*-substituted aminophenyl benzothiazoles were synthesised and characterised using NMR spectroscopy. A detailed analysis of the ^{1}H, ^{13}C and ^{15}N NMR chemical shifts and DFT calculated structures of the compounds are described.[396] The authors review and demonstrate how CD- and solid-state NMR spectroscopic approaches can be used to follow the reconstitution of peptides into membranes and to describe some of their fundamental characteristics. ^{15}N chemical shift solid-state NMR spectroscopy indicates that the hydrophobic domain of p24 as well as a designed sequence of 19 hydrophobic amino acid residues adopt transmembrane alignments in phosphatidylcholine membranes.[397] Homonuclear correlation NMR experiments are commonly used in the high-resolution structural studies of proteins. While $^{13}C/^{13}C$ chemical shift correlation experiments utilising dipolar recoupling techniques are fully utilised under MAS, correlation of the chemical shifts of ^{15}N nuclei in proteins has been a challenge.[398] A procedure for the simultaneous acquisition of {HNCOCANH & HCCCONH} chemical shift correlation spectra employing sequential ^{1}H data acquisition for moderately sized proteins is presented. The suitability of the approach for obtaining sequential resonance assignments, including complete ^{15}N, $^{1}H^{N}$, ^{13}CO, $^{13}C^{\alpha}$, $^{13}C^{\beta}$ and $^{1}H^{\alpha}$ chemical shift information, is demonstrated experimentally for a ^{13}C and ^{15}N labeled sample of the C-terminal winged helix (WH) domain of the minichromosome maintenance (MCM) complex of Sulfolobus solfataricus.[399]

16.2 Phosphorus (^{31}P) ($I = 1/2$)

The nature of the short-range structural disorder and its dependence on the fictive temperature are investigated in a phosphate glass of compound $60ZnO \cdot 40P_2O_5$ using ^{31}P two-dimensional phase adjusted spinning sidebands NMR (2D PASS NMR) spectroscopy. The 2D PASS experiment is shown to be a robust method for determining the correlations between the ^{31}P isotropic shift and chemical shift tensor parameters and their distributions in this glass that allows for a remarkably detailed quantification of structural disorder.[400] The mixed glass former (MGF) effect (MGFE) is defined as a nonlinear and nonadditive change in the ionic conductivity with changing glass former compound at constant modifier compound. In this study, sodium borophosphate $0.35Na_2O + 0.65[xB_2O_3 + (1-x)P_2O_5]$, $0 \leq x \leq 1$, glasses which were shown to exhibit a positive MGFE were prepared and examined using Raman and ^{11}B and ^{31}P MAS NMR spectroscopies.[401] Five selected compounds of Li-NASICON, $Li_3Zr_{2-y}Si_{2-4y}P_{1+4y}O_{12}$ $(0<y<0.5)$, were synthesised by sol–gel chemistry to obtain pure polycrystalline powder and then analysed by different physicochemical characterisations such as coupled DTA–TGA (thermogravimetric analysis), XRPD (X-ray powder diffraction), CIS (complex impedance spectroscopy) and MAS-NMR.[402] This work addresses the question of the ability of the molecular dynamics/density functional theory (MD/DFT) approach to reproduce sequence trend in ^{31}P chemical shifts in the backbone of nucleic acids.[403] ^{31}P NMR measurements were employed to monitor the stoichiometry and stability of complexes between $HgCl_2$, $HgBr_2$ and HgI_2 with 4-methylbenzoylmethylenetriphenylphosphorane (L) in acetonitrile (AN), dimethylsulfoxide (DMSO), DMF and THF.[404] Two novel compounds: polynuclear catena-poly[bis(μ_3-*O,O′*-di-cyclo-hexyldithiophosphato-*S,S,S′*)digold(I)] (AuAu) (1) and crystalline bis(*O,O′*-di-cyclo-hexylthiophosphoryl)disulfide, $(cyclo\text{-}C_6H_{11}O)_2P(S)SS(S)P(O\text{-}cyclo\text{-}C_6H_{11})_2$ (2) were prepared using heterogeneous reaction between freshly precipitated binuclear Cd(II) dithiophosphate (Dtph), $[Cd_2\{S_2P(O\text{-}cyclo\text{-}C_6H_{11})_2\}_4]$ and $H[AuCl_4]$ in 2 M HCl. The isolated 1 and 2 (the fixation mode of Au from the solution and the oxidised form of Dtph groups, respectively) were studied by single-crystal X-ray diffraction, ^{13}C and ^{31}P CP/MAS NMR spectroscopy and simultaneous thermal analysis.[405] The ^{31}P NMR chemical shift of carbene–phenylphosphinidene adducts gives a good indication of the relative π-accepting properties of carbenes.[406] The analysis of free fatty acid (FFA) mixtures is a very important but, even nowadays, challenging task. It was convincingly shown that FFA may be analysed by ^{31}P NMR subsequent to derivatisation with 2-chloro-4,4,5,5-tetramethyldioxaphospholane (CTDP).[407] The *O,O′*-diethyl dithiophosphate complex of tetraphenylantimony(V) $[Sb(C_6H_5)_4\{S_2P(OC_2H_5)_2\}]$ (I) and its benzene-solvated form $I.1/2C_6H_6$ (II) were synthesised and studied by high-resolution solid-state ^{13}C and ^{31}P NMR.[408] A series of mitochondria targeted α-aminophosphonates combining a diethoxyphosphoryl group and an alkyl chain-connected triphenylphosphonium bromide tail were designed and synthesised, and their pH-sensitive ^{31}P NMR properties and biological activities *in vitro* and *in vivo* were evaluated.[409] Adhesive silks spun by aquatic caddisfly (order Trichoptera) larvae are used to build both intricate protective shelters and food harvesting nets underwater. In this study, the authors use ^{13}C and ^{31}P solid-state NMR and wide angle X-ray diffraction (WAXD) as tools to elucidate molecular protein structure of caddisfly larval silk from the species Hesperophylax consimilis.[410] Acceptor numbers (AN) of several halogenated and perfluorohalogenated solvents were determined using ^{31}P NMR of triethylphosphine oxide (Et_3PO) as a probe.[411] The ligand 1,3-bis(diphenylphosphino)-1*H*-indole, (L1) with palladium promotes Suzuki–Miyaura coupling reaction of chloroarenes and benzyl chlorides with arylboronic acids. ^{31}P chemical shift values of the substituted indolylphosphines appear to show a correlation with observed trend in catalytic efficiency.[412] Lipid membrane dynamics in cell sized vesicles (CSVs) in water is characterised by the solution-state NMR. CSV was successfully prepared by natural swelling method using osmotic effect. Broad ^{1}H NMR signal and large ^{31}P NMR chemical shift anisotropy show the slow tumbling of the lipid with the rotational correlation time 0.2–2.8 s in CSV of 10–20 μm-diams.[413] ^{11}B and ^{31}P MAS NMR spectroscopy are used to quantify short and medium range order in a series of borate-rich K, Rb and Cs borophosphate glasses, with a $B_2O_3 : P_2O_5$ ratio of 5.[414] The conjugated frustrated phosphane/borane Lewis pairs formed by 1,1-carboboration of substituted diphenylphosphino acetylene, undergo a synergistic 1,1-addition reaction to Bu isocyanide with formation of new B–C and P–C bonds to the former isonitrile carbon atom. The different modes of isocyanide binding to the FLPs in the solid state were characterised using X-ray crystal structure analyses and comprehensive ^{11}B and ^{31}P solid-state MAS NMR experiments.[415] The stepwise protonation constants of two linear triphosphate ligand anions, triphosphate and di-imidotriphosphate were investigated by potentiometric titration and the intrinsic chemical shifts of the stepwise protonated species of these anions were determined from the pH-dependence of the ^{31}P NMR chemical shifts.[416] The main objective of the current study was to investigate penetration of cell penetrating peptides (CPPs: TAT, R_8, R_{11} and YKA) through skin intercellular lipids using ^{31}P solid-state MAS NMR.[417] Detailed kinetic studies on the hydrolysis of glycylserine (Gly-Ser) and glycylglycine (Gly-Gly) in the presence of the dimeric zirconium(IV)-substituted Keggin type polyoxometalate $(Et_2NH_2)_8[\{\alpha\text{-}PW_{11}O_{39}Zr(\mu\text{-}OH)(H_2O)\}_2]\cdot 7H_2O$ (1) were performed by a combination of ^{1}H, ^{13}C and ^{31}P NMR spectroscopy.[418] Barium monofluorophosphate, $BaPO_3F$, was prepared in a polycrystalline form by fast precipitation, whereas single crystals were obtained using a long-lasting gel growth method. The samples were characterised by powder- and single-crystal X-ray diffraction, which revealed orthorhombic symmetry for the polycrystalline material and monoclinic symmetry for the single crystals. Solid-state ^{19}F and ^{31}P MAS NMR spectra of polycrystalline $BaPO_3F$ acquired at magnetic fields of 7.05 T and 14.09 T resolve resonances from two distinct ^{19}F sites and ^{31}P sites, in accordance with the local symmetry of the OD description.[419] ^{31}P NMR was employed to study the effects of proton irradiation on $KTiOPO_4$ (KTP) in view of the previously studied paramagnetic impurity doping effects.[420] Hexachlorocyclotriphosphazene $N_3P_3Cl_6$ and *gem*-disubstituted cyclotriphosphazene derivatives $N_3P_3Cl_4X_2$ (X = Ph, PhS, PhNH) were reacted with 4-isopropylaniline to give geminal tetra and hexa substituted compounds. The compounds were separated by column chromatography on silica gel and analysed by elemental analysis, mass spectrometry and ^{31}P and ^{1}H NMR spectroscopies, and also crystal structures were determined by X-ray crystallography.[421] The purpose of this research was to synthesise BATCCP hybrid monomers and to evaluate the effect of solvent purity on the BATCCP production. BATCCP monomers were prepared by an interfacial procedure in a H_2O/toluene system as a function of time with the assistance of a phase transfer catalyst, tetraoctylammonium bromide. ^{1}H and ^{31}P NMR confirmed the production of BATCCP monomer by the appearance of chemical shifts at 7.18 and 5.35 ppm in the ^{1}H NMR and 23.4 and 13.9 ppm in the ^{31}P NMR, respectively.[422] ^{31}P and ^{13}C NMR studies of the toxic chemical warfare agent VX [*O*-Et *S*-(2-*N*,*N′*-(diisopropylamino)-ethyl) methylphosphonothioate] have been used to understand the reactivity in a wide variety of matrixes.[423] Raman and ^{31}P NMR spectral parameters were investigated in a series of synthetic B-type carbonated apatites (CAps).[424] The structures of four Ca–phosphonate phases are reported here: $Ca(C_6H_5{-}PO_3H)_2$ (1), $Ca(C_6H_5{-}PO_3)\cdot 2H_2O$ (2), $Ca(C_4H_9{-}PO_3H)_2$ (3) and $Ca(C_4H_9{-}PO_3)\cdot H_2O$ (4). Structural models were obtained *ab initio* by using a combined synchrotron powder diffraction, solid-state NMR and gauge including projector augmented wave (GIPAW) calculation approach.[425] The combination of solid-state MAS NMR spectroscopy and first-principles calculations is used to elucidate the structure of an as-prepared microporous AlPO (STA-2), in which the template (bis-diazabicyclooctane-butane) is charge balanced by hydroxyl groups coordinated to framework aluminum species.[426] The structure of roller-quenched fast ion conductive glasses (FICs) $(Ag_3PO_4)_x(AgI)_{1-x}$ $(0.15 \leq x \leq 0.50)$ was studied by ^{109}Ag and ^{31}P solid-state NMR spectroscopies.[427] Conformational and relativistic effects on the ^{31}P and ^{77}Se chemical shifts of phosphine selenides were analysed in terms of the ZORA-GIAO-B1PW91/TZP approach.[428] Intermolecular and intramolecular hydrogen bond (H-bond) and proton transport in acid–base complexes and amphoteric molecules consisting of phosphonic acid groups and nitrogenous heterocyclic rings are investigated by DFT calculations and ^{1}H NMR and ^{31}P NMR spectroscopy.[429] ^{31}P NMR spectroscopy was applied to the evaluation of water-tolerant Lewis acid catalysis of various metal triflates and chlorides in water using trimethylphosphine oxide (TMPO) as a probe molecule.[430] The effect of noncovalent interactions such as C–H···O and O–H···O–H bonds and coordination to Pb^{2+} on the ^{31}P NMR chemical shift tensor in the titled compounds was studied experimental and simulated theoretical using the density functional theory gauge-invariant AO (DFT-GIAO) approach.[431] The complexation behaviours of Be^{2+} with cyclo-μ-imido triphosphate anions, $cP_3O_{9-n}(NH)n^{3-}$ $(n=1, 2)$, were investigated by both ^{9}Be and ^{31}P NMR techniques at $-2.3°$ in order to clarify the coordination structures of the complexes.[432] 1,4-Benzodioxan-6-amino substituted cyclophosphazenes were synthesised to provide insight into the reaction mechanism for nucleophilic substitution. All compounds were fully characterised by elemental and mass analyses, ^{1}H and ^{31}P NMR spectra and crystallography.[433] Acid properties of ZSM-5 type zeolite were studied by solid-state NMR using trimethylphosphine oxide (TMPO) as a probe molecule. The ^{31}P chemical shifts at 90–60 ppm are attributed to TMPO adsorbed on Bronsted acid sites, whereas the signals between 60 and 40 ppm are ascribed to TMPO adsorbed on other sites than Bronsted acid sites.[434] The effects of doped low-valence cations on the properties of the SnP_2O_7 proton conductor at ambient temperature were studied from changes in solid-state NMR spectra and nuclear magnetic relaxation times.[435] A review with 351 refs. It collects the ^{15}N and ^{31}P NMR data for *ca.* 390 and *ca.* 340, respectively, complexes of transition metal ions such as Ni(0), Pd(0), Pt(0), Ni(II), Pd(II), Pt(II), Pd(IV), Pt(IV), Co(−1), Rh(−1), Rh(I), Ir(I), Co(III), Rh(III), Ir(III), Fe(−2), Ru(−2), Fe(II), Ru(II), Os(II), Cu(I), Ag(I), Au(I), Au(III), Zn(II), Cd(II) and Hg(II) with nitrogen-containing heterocycles (azines, azoles and azoloazines like purines or 1,2,4-triazolo-[1,5*a*]-pyrimidines), and their phosphorus analogs (phosphinines and phospholes).[436] *N*-Diphenylphosphino-4-methylpiperidine sulfide was synthesised and characterised by elemental analysis, ^{1}H NMR, ^{31}P NMR, IR and X-ray single-crystal determination.[437] The structural aspects of the glass-to-crystal transition in the technological important ion conducting glass ceramic system $Li_{1+x}Al_xGe_{2-x}(PO_4)_3$ ($0 \leq x \leq 0.75$) were examined by complementary multinuclear solid state nuclear magnetic single and double-resonance experiments.[438] *N*-Diphenylphosphino-4-methylpiperidine selenide was prepared and characterised by elemental analysis, ^{1}H NMR, ^{31}P-{^{1}H} NMR, IR and X-ray single-crystal determination.[439] The kinetics of the reaction $[Rh(H_2O)_6]^{3+} + H_3PO_4 \leftrightarrow [Rh(H_2O)_5H_2PO_4]^{2+} + H_3O^+$ was studied by ^{31}P NMR; $E_a = 142 \pm 12$ kJ mol^{-1}, $\log A = 17 \pm 2$. An empirical dependence of the ^{31}P NMR chemical shift on the equivalent pH value was found.[440] ^{31}P NMR spectroscopic measurement with TMPO was applied to evaluate the Lewis acid catalysis of various metal triflates in water.[441]

17 Group 16 (^{17}O, ^{77}Se, ^{125}Te)

17.1 Oxygen (^{17}O) ($I = 5/2$)

The authors examine several computational strategies for the prediction of the ^{17}O NMR shielding constants for a selection of organic acids and peracids in aqueous solution.[442] Herein the authors report a detailed ^{1}H and ^{17}O relaxometric investigation of Mn(II) complexes with cyclen-based ligands such as 2-(1,4,7,10-tetraazacyclododecan-1-yl)acetic acid (DO1A), 2,2′-(1,4,7,10-tetraazacyclododecane-1,4-diyl)diacetic acid (1,4-DO2A), 2,2′-(1,4,7,10-tetraazacyclododecane-1,7-diyl)diacetic acid (1,7-DO2A) and 2,2′,2″-(1,4,7,10-tetraazacyclododecane-1,4,7-triyl)triacetic acid (DO3A).[443] The group of 155 substituted phenylboronic acids and their derivatives: esters, boroxines and benzoxaboroles, were investigated by ^{17}O NMR spectroscopy. The influence of substituents of phenylboronic acids on the ^{17}O chemical shift was evaluated.[444] To understand how Li^+ interacts with individual carbonate molecules in nonaqueous electrolytes, the authors conducted natural abundance ^{17}O NMR measurements on electrolyte solutions of 1 M $LiPF_6$ in a series of binary solvent mixtures of ethylene carbonate (EC) and di-Me carbonate (DMC).[445] NMR chemical shift is a molecular property that can be computed from 1st principles. Without nonbonded interactions, the root mean square errors (RMSEs) compared to the full calculations for ^{1}H, ^{13}C, ^{15}N, ^{17}O and ^{33}S were 0.340, 0.649, 3.052, 6.928 and 0.122 ppm, respectively, while with the inclusion of nonbonded interactions the RMSEs for ^{1}H, ^{13}C, ^{15}N, ^{17}O and ^{33}S were 0.038, 0.253, 0.681, 3.480 and 0.052 ppm, respectively.[446] The authors report solid-state NMR characterisation of the ^{17}O quadrupole coupling (QC) and chemical shift (CS) tensors in five site-specifically ^{17}O-labeled samples of salicylic acid and *o*-acetylsalicylic acid (Aspirin).[447] The grafting of an oxo chloro trisalkyl tungsten derivative on silica dehydroxylated at 700 °C was studied by several techniques that showed reaction *via* W–Cl cleavage, to afford a well-defined precatalyst for alkene metathesis. ^{17}O labeling of the oxo moiety of a series of related molecular and supported tungsten oxo derivatives was achieved, and the corresponding ^{17}O MAS NMR spectra were recorded.[448] The nuclear isotropic shielding constants $\sigma(^{13}C)$ and $\sigma(^{17}O)$ of isomers of retinoic acid and retinal in gas-phase and in chloroform, acetonitrile, methanol and water solutions were investigated *via* Monte Carlo simulation and quantum mechanics calculations using the GIAO-B3LYP/6-311++G(2d,2p) approach.[449] The authors report the synthesis of the ligand H_2MeNO2A (1,4-bis(carboxymethyl)-7-methyl-1,4,7-triazacyclononane) and a detailed experimental and computational study of the hyperfine coupling constants (HFCCs) on the inner-sphere H_2O molecules of [Mn(MeNO2A)] and related Mn^{2+} complexes relevant as potential contrast agents in magnetic resonance imaging (MRI). Nuclear magnetic relaxation dispersion (NMRD) profiles, ^{17}O NMR chemical shifts and transverse relaxation rates of aqueous solutions of [Mn(MeNO2A)] were recorded to determine the parameters governing the relaxivity in this complex and the ^{17}O and ^{1}H HFCCs.[450] ^{1}H, ^{13}C, ^{15}N and ^{17}O NMR chemical shifts are used for the characterisation of the intramolecular interactions in several nitramines of the Me_2N–G–NO_2 type.[451] Structural characterisation of Brownmillerite $Ba_2In_2O_5$ was achieved by an approach combining experimental ^{17}O solid-state NMR spectroscopy, DFT energetics and GIPAW NMR calculations.[452] Nuclear Magnetic Relaxation Dispersion (NMRD) profiles and ^{17}O NMR chemical shifts and transverse relaxation rates of aqueous solutions of $[Mn(H_2O)_6]^{2+}$ were recorded to determine the parameters governing the relaxivity in this complex and the ^{17}O and ^{1}H hyperfine coupling constants. (HFCCs).[453] The authors report a theoretical analysis on ^{17}O NMR chemical shifts for a family of prototypical polyoxometalate anions.[454]

17.2 Selenium (^{77}Se) ($I = 1/2$)

The coordination environments of Se atoms in binary Ge_xSe_{100-x} glasses with $5 \leq x \leq 30$ were studied using a novel, two-dimensional ^{77}Se NMR spectroscopic technique.[455] This review intends to cover the synthesis of both mono- and di-chalcogenides of 2-, 3- and 4-pyridyl and 2-pyrimidyl groups, their spectroscopic and structural studies. The ^{77}Se and ^{125}Te NMR chemical shifts of some pyridyl and pyrimidyl compounds are discussed.[456] The authors describe the synthesis and biomimetic activity of a series of *N,N*-disubstituted thiones and selones that contain an imidazole pharmacophore. The ^{77}Se NMR chemical shifts for the selones show large upfield shift, thus confirming the zwitterionic structure in solution.[457] A new method for the assessment of the π-acceptor strength of N-heterocyclic carbenes is presented. The ^{77}Se chemical shifts of the easily available selenium carbene adducts 1 · Se-7 · Se correlate with the π-acceptor character of the respective carbenes. The δ-(^{77}Se) values cover a range of almost 800 ppm, with increasing π-acidity leading to a downfield shift of the signal.[458] A series of new enantiomeric *N*-substituted benzisoselenazol-3(2*H*)-ones were prepared from 2-(chloroseleno)benzoyl chloride and 9-amino-deoxyquinine (both 9-native and 9-epi) as well as the other chiral primary amines. The ^{77}Se NMR parameters for the obtained benzisoselenazolones were measured and theoretical calculated at the DFT level of theory using B97-2 hybrid functional and cc-pVTZ basis set.[459] Conformational and relativistic effects on the ^{31}P and ^{77}Se chemical shifts of phosphine selenides were analysed in terms of the ZORA-GIAO-B1PW91/TZP approach.[460] The series of eight 5-(TeY)-6-(SePh)acenaphthenes (Y = Fp (2), Tol (3), An-p (4), An-o (5), Tp (6), Mes (7), Tip (8), Nap (9)) were prepared and structurally characterised by X-ray crystallography, solution and ^{77}Se and ^{125}Te solid-state NMR spectroscopy and DFT/B3LYP calculations.[461] Combination of the dithiol *N,N′*-bis(2-mercaptoethyl)isophthalamide, abbreviated as $BDTH_2$ and as 1, with excess H_2SeO_3 in aqueous acidic (pH ≈ 1) conditions resulted in preparation of BDT(S–Se–S) (6), with a ^{77}Se NMR chemical shift of $\delta = 675$ ppm, and oxidised BDT.[462]

17.3 Tellurium (^{125}Te) ($I = 1/2$)

A review intends to cover the synthesis of both mono- and di-chalcogenides of 2-, 3- and 4-pyridyl and 2-pyrimidyl groups, their spectroscopic and structural studies. The ^{77}Se and ^{125}Te NMR chemical shifts of some pyridyl and pyrimidyl compounds are discussed.[463] The authors present an alternative approach to separating contributions to the NMR shift originating from the Knight shift and chemical shielding by a combination of experimental solid-state NMR results and *ab initio* calculations. The authors analysed ^{207}Pb and ^{125}Te NMR spin–lattice relaxation rates and NMR shifts for bulk and nanoscale PbTe.[464] The list of classical tellurides with organic cations was expanded by the synthesis of single-crystalline salts $[Li_2(en)_n](Te_x)$ (en = ethylenediamine) with $n = 4$, $x = 1$ (1), $n = 4$, $x = 2$ (2) and $n = 4.5$, $x = 3$ (3), $[K(18c6)]_2(Te_x)$ (18c6 = 1,4,7,10,13,16-hexaoxacyclooctadecane) with $x = 2$ (4), 4 (5), $[K(18c6)]_2[K(en)]_2(Te_2)_2$ (6) and $[K([2.2.2]crypt)]_2(Te_2)$ (7) ([2.2.2]crypt = 4,7,13,16,21,24-hexaoxa-1,10-diazabicyclo[8.8.8]-hexacosane). ^{125}Te NMR chemical shifts of the presented series of en solvated lithium tellurides are presented.[465] Complex tellurides, such as doped PbTe, GeTe and their alloys, are among the best thermoelectric materials. Knowledge of the change in ^{125}Te NMR chemical shift due to bonding to dopant or "solute" atoms is useful for determination of phase component, peak assignment and analysis of local bonding.[466] The series of eight 5-(TeY)-6-(SePh)acenaphthenes (Y = Fp (2),

Tol (3), An-p (4), An-o (5), Tp (6), Mes (7), Tip (8), Nap (9)) were prepared and structurally characterised by X-ray crystallography, solution and ^{77}Se and ^{125}Te solid-state NMR spectroscopy and DFT/B3LYP calculations.[467]

18 Group 17 (^{19}F, $^{35,37}Cl$, ^{81}Br)

18.1 Fluorine (^{19}F) ($I = 1/2$)

α-$LaZr_2F_{11}$ was synthesised by solid state reaction. ^{19}F 1D MAS NMR spectra of α-$LaZr_2F_{11}$ are in agreement with the proposed structural model. Assignment of the ^{19}F resonances to the corresponding crystallographic sites was performed on the basis of both their relative intensities and their correlation patterns in a ^{19}F 2D dipolar-based double-quantum recoupling MAS NMR spectrum.[468] The comprehensive investigation reported herein provides compelling evidence that anion–π interactions are the main driving force in the formation of self-assembled Fe(II)-templated metallacycles with bptz [3,6-bis(2-pyridyl)-1,2,4,5-tetrazine] in high yields. It was demonstrated by X-ray crystallography, ^1H NMR, solution and solid-state MAS ^{19}F NMR spectroscopies, CV and MS studies that $[X]^- = [BF_4]^-$, $[ClO_4]^-$ and $[Y]^- = [SbF_6]^-$, $[AsF_6]^-$, $[PF_6]^-$ template molecular squares $[Fe_4(bptz)_4(CH_3CN)_8][X]_8$ and pentagons $[Fe_5(bptz)_5(CH_3CN)_{10}][Y]_{10}$, respectively.[469] In this contribution the gas-phase measurements of ^{19}F and ^{29}Si chemical shifts and scalar spin–spin couplings in SiF_4 and $(SiF_3)_2O$ were reported.[470] The authors evaluate the utility a novel amino acid, L-*O*-(perfluoro-*t*-butyl)-homoserine (pFtBSer), that can easily be synthesised and incorporated into peptides and provides greatly enhanced sensitivity over currently used ^{19}F biomolecular NMR probes.[471] The structural and vibrational properties of a potential anticancer agent, the 5-difluoromethyl-1,3,4-thiadiazole-2-amino derivative and its tautomeric forms were studied by using the available experimental IR and ^1H, ^{13}C and ^{19}F NMR spectra and theoretical calculations based DFT.[472] The authors combine solid-state NMR spectroscopy and DFT calculations to study the structure and reveal the F patterning in graphite fluorides C_2F_x intercalated with MeCN.[473] Ligating simple amino alcohol or amino ester monomers containing enantiotopic fluorine substituents to the C-terminus of a helical peptide places the fluorine atoms in diastereotopic environments, and gives two distinct and easily identifiable signals in the ^{19}F NMR spectrum.[474] The authors prepared by solid-phase methods, chromatographic purified and characterised three analogs of the ten-amino acid-residue, membrane-active, lipopeptaibiotic trichogin GA IV, each containing a single (4-fluorophenyl)alanine in position 3, 7, or 10, where it replaces the hydrophobic residue Leu^3, Leu^7 or Ile^{10}, respectively. The fluorine probe based on the observation that the ^{19}F-NMR technique has been extensively utilised to analyse peptide-membrane interactions in biological systems.[475] The 3M Company product Novec® 71IPA DL, a mixture of methoxyperfluorobutane, methoxyperfluoroisobutane and 4.5 wt% iso-Pr alcohol, was found to be very stable at ambient temperature, producing fluoride at the rate of $\sim$1 ppm per year. This paper identifies the ^1H and ^{19}F NMR chemical shifts, multiplicities and coupling constants of reactants and the major products that result from aging the mixture in sealed Pyrex NMR tubes for periods up to 1.8 years at temperatures from 26 °C to 102 °C.[476] Reaction of perfluoroquinoline 1 and perfluoroisoquinoline 2 with benzylamine gave mono- and di-aminated quinoline and isoquinoline systems, respectively, depending upon the reaction conditions by selective S_NAr processes. Comparison with observed ^{19}F and ^{13}C NMR shifts give excellent correlations, indicating that ^{19}F and ^{13}C GIAO-NMR computations are powerful tools in structurally identifying polyfunctional, polycyclic perfluoroheteroaromatic compounds and aiding NMR resonance assignment.[477] Variation in the position of CF_3 groups in several aromatic Group-14 compounds was studied by ^{19}F-NMR spectroscopy.[478] Fluorine in polytetrafluoroethylene (PTFE) changes its states and transfers to titania by comilling and annealing of a titania–PTFE mixture. XPS, ^{19}F MAS NMR, FT-IR, Raman spectra, TEM and EDX analyses consistently indicated the oxidative decomposition of PTFE, inducing partial fluorination of titania.[479] Barium monofluorophosphate, $BaPO_3F$, was prepared in a polycrystalline form by fast precipitation, whereas single crystals were obtained using a long-lasting gel growth method. Solid-state ^{19}F and ^{31}P MAS NMR spectra of polycrystalline $BaPO_3F$ acquired at magnetic fields of 7.05 T and 14.09 T resolve resonances from two distinct ^{19}F sites and ^{31}P sites, in accordance with the local symmetry of the OD description.[480] Optically active 1-fluoroindan-1-carboxylic acid (FICA) was designed and prepared as its' Me ester for determining the absolute configuration of chiral molecules by both ^{1}H and ^{19}F NMR spectroscopies.[481] The allosteric regulation of substrate channeling in tryptophan synthase involves ligand-mediated allosteric signaling that switches the α- and β-subunits between open (low activity) and closed (high activity) conformations. This switching prevents the escape of the common intermediate, indole, and synchronises the α- and β-catalytic cycles. ^{19}F NMR studies of bound α-site substrate analogs, *N*-(4′-trifluoromethoxybenzoyl)-2-aminoethyl phosphate (F6) and *N*-(4′-trifluoromethoxybenzenesulfonyl)-2-aminoethyl phosphate (F9), were found to be sensitive NMR probes of β-subunit conformation.[482] DNA local structures with bulged bases and mismatched base pairs as well as ordinary full-matched base pairs by using ^{19}F NMR spectroscopy with ^{19}F-labeled oligodeoxynucleotides (ODNs) were monitored.[483] Chemical shifts in ^{19}F and ^{13}C NMR spectra of substituted pentafluorobenzenes are calculated by Hartree–Fock and DFT methods.[484] A novel method for the formation of highly stable BF_3-induced *J*-aggregates by interaction between *meso*-tetrakis(4-sulfonatophenyl)porphyrin ($TPPS_4$) and $BF_3O(C_2H_5)_2$ is reported. The aggregates were characterised by NMR (^{1}H, ^{11}B and ^{19}F), optical absorption, cyclic voltammetry, FT-IR and fluorescence spectroscopic techniques.[485] The ^{19}F isotropic chemical shifts (δ_{iso}) of two isomorphic compounds, NbF_5 and TaF_5, which involve six nonequivalent F sites, were experimental determined from the reconstruction of 1-dimensional ^{19}F MAS NMR spectra.[486] It was shown by the example of TiF_4 complexes with α-picolinic acid (C_5H_5NCOOH, picH) and its esters (picEt and $picSiMe_3$) in CH_3CN that the ^{19}F NMR chemical shifts can be used for establishing the basic properties of nonequivalent donor sites of chelating ligands.[487] The authors demonstrated the use of ^{23}Na, ^{19}F and ^{89}Y solid-state NMR together with DFT calculations to characterise the structure of cubic $NaYF_4$ nanoparticles with and without Er^{3+} doping.[488] The structural and electronic properties of linear and branched perfuorobutanesulfonate (PFBS) in its anionic, acidic, and potassium or sodium salt forms were studied in a polarisable continuum model(PCM) of methanol solvent with the B3LYP functional and the 6-31G(d,p) basis set. The ^{19}F chemical shifts and ^{19}F–^{19}F *J*-coupling constants were determined in a PCM of methanol solvent with GIAO B3LYP/6-31++G(d,p).[489] On the basis of the varying amplitude and patterns of the ^{19}F NMR chemical shift of C_6F_5X (X = F, Cl, Br, I) in the presence of chloride anions, bonding models of $C_6F_5X \cdot Cl^-$ complexes were tentatively established, and the relevant binding constants were obtained.[490] The structure and spectroscopic properties of selected models of B-type carbonate defects in apatite locally compensated by fluoride or hydroxyl ions were investigated using first-principles quantum mechanical calculations. Theoretical IR absorption spectra and ^{13}C and ^{19}F NMR chemical shifts are determined.[491] The relationship between the experimental ^{19}F isotropic chemical shift and the ^{19}F isotropic shielding calculated using the gauge including projector augmented-wave (GIPAW) method with PBE functional is investigated in the case of GaF_3, InF_3, TlF and several AlF_3 polymorphs.[492] The method of continuous variation in conjunction with ^{1}H and ^{19}F NMR spectroscopies was used to characterise lithium and sodium enolates solvated by *N,N,N′,N′*-tetramethylethyldiamine (TMEDA) and THF.[493] The first scale of solvent hydrogen-bond basicity applicable to amphiprotic solvents is established by means of a new method that compares the ^{19}F NMR chemical shifts of 4-fluorophenol and 4-fluoroanisole in hydrogen-bond-acceptor solvents.[494] Three fluorinated Mo–Cu–thiolate isomers, $[Ph_4P]_2[S_2MoS_2Cu(n\text{-}SPh_F)]$, [$n$-$SPh_F$ = 2-fluorothiophenol (1a), 3-fluorothiophenol (1b) and 4-fluorothiophenol (1c)] were synthesised and spectroscopically characterised. The ^{19}F-NMR signal of the F atom in the benzene has different chemical shift for each isomer, which is highly influenced by the local environment that can be manipulated by different solvents and solutes.[495] The Au(III) complex $Au(OAc^F)_2(tpy)$ (1, $OAc^F = OCOCF_3$; tpy = 2-*p*-tolylpyridine) undergoes reversible dissociation of the OAc^F ligand trans to C, as seen by ^{19}F NMR.[496]

18.2 Chlorine ($^{35,37}Cl$) ($I = 3/2, 1/2$)

A novel crystalline form of the boron-containing antibacterial drug (*S*)-3-(aminomethyl)-7-(3-hydroxypropoxy)benzo[*c*][1,2]oxaborol-1(3*H*)-ol hydrochloride is studied by solid-state NMR (SSNMR) and single-crystal X-ray diffraction techniques. ^{11}B and ^{35}Cl chemical shift and quadrupolar parameters are obtained using the analysis of MAS spectra and are found to be accurately reproduced using DFT calculations.[497] Two polymorphic forms of Et 3-{3-[((2*R*)-3-{[2-(2,3-dihydro-1*H*-inden-2-yl)-1,1-dimethylethyl]-amino}-2-hydroxypropyl)-oxy]-4,5-difluorophenyl} propanoate hydrochloride, an investigational pharmaceutical compound, are characterised using spectroscopic and diffractometric techniques. Detailed solid-state NMR studies observing the ^{1}H, ^{13}C, ^{15}N, ^{19}F and ^{35}Cl nuclei are performed to characterise the subtle structural differences between the two forms.[498] A series of HCl salts of active pharmaceutical ingredients (APIs) were characterised *via* ^{35}Cl solid-state NMR spectroscopy and first-principles plane-wave DFT calculations of ^{35}Cl NMR interaction tensors.[499]

18.3 Bromine (^{81}Br) ($I = 3/2$)

The authors report a multifaceted experimental and computational study of halogen bonds in the solid state. A series of cocrystals of three different diiodobenzene molecules and various onium halide (Cl^- or Br^-) salts, designed to exhibit moderately strong halogen bonds ($C–I\cdots X^-$) in the absence of competing hydrogen bonds, was prepared and characterised by single-crystal X-ray diffraction. Interestingly, a wide range of geometries about the halide anion are observed $^{35/37}$Cl and $^{79/81}$Br solid-state NMR spectroscopy is applied to characterise the nuclear quadrupolar coupling constants (C_Q) and asymmetry parameters (η_Q) for the halogen-bonded anions at the center of bonding environments ranging from approximately linear to distorted square planar to octahedral.[500]

19 Group 18 (^{21}Ne, ^{39}Ar, ^{129,131}Xe)

19.1 Neon (^{21}Ne) ($I = 3/2$)

Theoretical modeling of noble gas interaction with furan as a simple heterocyclic ring was performed. The predicted ^{21}Ne and ^{39}Ar NMR chemical shifts for the Ne–furan and Ar–furan complexes calculated with pcS-3 and aug-pcS-3 basis sets were sensitive to the presence of the aromatic furan ring. Their results indicate a higher sensitivity of the neon and argon NMR probes than the previously reported ^{3}He NMR spectroscopic parameters in studies of small heterocyclic rings containing the oxygen atom.[501]

19.2 Argon (^{39}Ar) ($I = 7/2$)

Theoretical modeling of noble gas interaction with furan as a simple heterocyclic ring was performed. The predicted ^{21}Ne and ^{39}Ar NMR chemical shifts for the Ne–furan and Ar–furan complexes calculated with pcS-3 and aug-pcS-3 basis sets were sensitive to the presence of the aromatic furan ring. Their results indicate a higher sensitivity of the neon and argon NMR probes than the previously reported ^{3}He NMR spectroscopic parameters in studies of small heterocyclic rings containing the oxygen atom.[501]

19.3 Xenon (^{129,131}Xe) ($I = 1/2, 3/2$)

The influence of the layer charge on the microstructure was studied for a series of three hybrid pillared interlayered clays based on the organic dication Me_2DABCO^{2+} and charge reduced synthetic fluorohectorites. To get a detailed picture of the local arrangements within the interlayer space, multinuclear solid-state NMR spectroscopy was performed in conjunction with high-resolution ^{129}Xe MAS NMR, temperature-dependent wide-line 1D and 2D ^{129}Xe NMR and Ar/Ar(l) and Xe/Xe(l) physisorption measurements.[502] H-bonding-like interactions between AH acids and Ne, Ar, Kr and Xe are examined by analysis of predicted A–H···Rg geometry, energy, A–H stretching vibration (CCSD(T) level) as well as assumed proton affinity, pK_a, polarisability and chemical hardness of the complex components. The A–H···Xe complexes can be divided into stronger and weaker ones using the ^{129}Xe chemical shift as the criterion.[503] The chemical shift of xenon (at natural abundance) dissolved in a variety of 1-butyl-3-methylimidazolium-based ionic liquids (ILs) was measured with ^{129}Xe NMR spectroscopy.[504] The isotropic ^{129}Xe NMR chemical shift (CS) in Xe at C_{60} dissolved in liquid benzene was calculated by piecewise approximation to faithfully simulate the experimental conditions and to evaluate the role of different physical factors influencing the ^{129}Xe NMR CS.[505] The liquid organisation of linear, branched and cyclic alkanes was studied using ^{129}Xe as a NMR probe. ^{129}Xe chemical shifts were experimental determined for xenon dissolved in a total of 21 alkanes.[506] Structural changes in a porous hybrid inorganic–organic ZIF-8 compound were explored using hyperpolarised ^{129}Xe NMR of adsorbed xenon at various temperatures.[507] The mesoscopic structure of 30 distinct ionic liquids was probed by ^{129}Xe NMR spectroscopy.[508] *In situ* high pressure ^{129}Xe NMR spectroscopy in combination with volumetric adsorption measurements were used for the textural characterisation of different carbon materials with well-defined porosity including microporous carbide-derived carbons, ordered mesoporous carbide-derived carbon and ordered mesoporous CMK-3.[509]

References

1 T. Korenaga, F. Kobayashi, K. Nomura, T. Sakai and K. Shimada, *J. Fluorine Chem.*, 2013, **156**, 1–4.

2 L. L. G. Justino, M. Luisa Ramos, P. E. Abreu, A. Charas, J. Morgado, U. Scherf, B. F. Minaev, H. Agren and H. D. Burrows, *J. Phys. Chem. C*, 2013, **117**, 17969–17982.

3 C. F. Campana, J. Mirzaei, C. Koerner, C. Gates and N. R. Natale, *Acta Crystallogr., Sect. E: Struct. Rep. Online*, 2013, **69**, o1680–o1681.

4 T. Pawlak, P. Paluch, K. Trzeciak-Karlikowska, A. Jeziorna and M. J. Potrzebowski, *CrystEngComm*, 2013, **15**, 8680–8692.

5 S. R. Peerannawar and S. P. Gejji, *J. Mol. Model.*, 2013, **19**, 5113–5127.

6 K. Ikoma, M. Kido, M. Nagae, T. Ikeda, T. Shirai, K. Ueshima, Y. Arai, R. Oda, H. Fujiwara and T. Kubo, *J. Orthop. Res.*, 2013, **31**, 1708–1712.

7 S. S. Kadam, L. Maier, I. Kostakis, N. Pouli, J. Tousek, M. Necas, P. Marakos and R. Marek, *Eur. J. Org. Chem.*, 2013, **2013**, 6811–6822.

8 M. R. Islam and P. R. Sundararajan, *Eur. Polym. J.*, 2013, **49**, 2042–2051.

9 C. Lao-ngam, M. Phonyiem, S. Chaiwongwattana, Y. Kawazoe and K. Sagarik, *Chem. Phys.*, 2013, **420**, 50–61.

10 C. Ozen, A. Yildirim, N. S. Tuzun, P. Haslak, E. Dogan, E. Acar and V. Aviyente, *Mol. Phys.*, 2013, **111**, 3147–3155.

11 S. Mamone1, M. Concistrè1, I. Heinmaa, M. Carravetta1, I. Kuprov1, G. Wall, M. Denning, X. Lei, J. Y.-C. Chen, Y. Li, Y. Murata, N. J. Turro and M. H. Levitt, *ChemPhysChem*, 2013, **14**, 3121–3130.

12 D. Santos-Carballal, R. Suardiaz, R. Crespo-Otero, L. Gonzalez and C. S. Perez, *J. Mol. Model.*, 2013, **19**, 4591–4601.

13 L. Ye, P. Cornelis, K. Guillemyn, S. Ballet, C. Christophersen and O. Hammerich, *Nat. Prod. Commun.*, 2014, **9**, 789–794.
14 M. Boobalan, S. Ramalingam, M. Amaladasan, D. Tamilvendan, G. Venkatesa Prabhu and M. Bououdina, *J. Mol. Struct.*, 2014, **1072**, 153–172.
15 S. Alyar, N. Ozbek, S. O. Yildirim, I. S. Butcher and J. Ray, *Spectrochim. Acta, Part A*, 2014, **130**, 198–207.
16 Y. Xu, T. Watermann, H.-H. Limbach, T. Gutmann, D. Sebastiani and G. Buntkowsky, *Phys. Chem. Chem. Phys.*, 2014, **16**, 9327–9336.
17 P. Barczynski, A. Komasa, A. Katrusiak, Z. Dega-Szafran, L. Nowaczyk, M. Ratajczak-Sitarz and M. Szafran, *Spectrochim. Acta, Part A*, 2014, **121**, 586–595.
18 M. Szafran, K. Ostrowska, A. Katrusiak and Z. Dega-Szafran, *Spectrochim. Acta, Part A*, 2014, **128**, 844–851.
19 G. P. Petrova, H.-B. Li, K. Maruoka and K. Morokuma, *J. Phys. Chem. B*, 2014, **118**, 5154–5167.
20 M. Nakao, Y. Toriuchi, S. Fukayama and S. Sano, *Chem. Lett.*, 2014, **43**, 340–342.
21 S. Dey, S. A. Ikbal and S. P. Rath, *New J. Chem.*, 2014, **38**, 1458–1470.
22 I. A. Raja, V. V. Gobre, R. V. Pinjari and S. P. Gejji, *J. Mol. Model.*, 2014, **20**, 1–17.
23 K. Govindarasu and E. Kavitha, *Spectrochim. Acta, Part A*, 2014, **122**, 130–141.
24 S. R. Peerannawar, S. S. Rao and S. P. Gejji, *J. Mol. Model.*, 2014, **20**, 1–9.
25 B. Reyes-Trejo, D. Guerra-Ramírez, H. Zuleta-Prada, R. Santillán, M. E. Sánchez-Mendoza, J. Arrieta and L. Reyes, *Molecules*, 2014, **19**(4), 4695–4707.
26 R. A. Cormanich, A. Durie, R. Bjornsson, R. Rittner, D. O'Hagan and M. Bühl, *Helv. Chim. Acta*, 2014, **97**, 797–807.
27 R. J. Abraham, L. Griffiths and M. Perez, *Magn. Reson. Chem.*, 2014, **52**, 395–408.
28 A. Lennartson, L. U. Christensen, C. J. McKenzie and U. G. Nielsen, *Inorg. Chem.*, 2014, **53**, 399–408.
29 A. T. Trueba, M. C. Kroon, C. J. Peters, I. L. Moudrakovski, C. I. Ratcliffe, S. Alavi and J. A. Ripmeester, *J. Chem. Phys.*, 2014, **140**, 214703.
30 T. Kurahashi, M. Hada and H. Fujii, *Inorg. Chem.*, 2014, **53**, 1070–1079.
31 N. Maudoux, J. Fang, T. Roisnel, V. Dorcet, L. Maron, J.-F. Carpentier and Y. Sarazin, *Chem. - Eur. J.*, 2014, **20**, 7706–7717.
32 J. Malmquist, A. Bernlind and S. Lindberg, *J. Labelled Compd. Radiopharm.*, 2013, **56**, 536–537.
33 Y. Tian, Y. Hong, S. J. Bonacorsi, A. Balog and S. Gong, *J. Labelled Compd. Radiopharm.*, 2014, **57**, 1–11.
34 Y. V. Baklanova, I. Y. Arapova, I. R. Shein, L. G. Maksimova, K. N. Mikhalev and T. A. Denisova, *J. Struct. Chem.*, 2013, **54**, 111–118.
35 R. K. Dean, A. M. Reckling, H. Chen, L. N. Dawe, C. M. Schneider and C. M. Kozak, *Dalton Trans.*, 2013, **42**, 3504–3520.
36 K. Shimoda, M. Murakami, D. Takamatsu, H. Arai, Y. Uchimoto and Z. Ogumi, *Electrochim. Acta*, 2013, **108**, 343–349.
37 H. Keypour, M. H. Zebarjadian, M. Rezaeivala, M. Shamsipur and S. J. Sabounchei, *J. Iran. Chem. Soc.*, 2013, **10**, 1137–1143.
38 K. I. Popov, A. G. Vendilo, V. Chistov, H. Ronkkomaki, M. Lajunen, V. Privalov and J. Dikareva, *Polyhedron*, 2013, **63**, 50–54.

39 Y. V. Baklanova, I. Y. Arapova, A. L. Buzlukov, A. P. Gerashenko, S. V. Verkhovskii, K. N. Mikhalev, T. A. Denisova, I. R. Shein and L. G. Maksimova, *J. Solid State Chem.*, 2013, **208**, 43–49.
40 K. Pokorny, M. Schmeisser, F. Hampel, A. Zahl, R. Puchta and R. van Eldik, *Inorg. Chem.*, 2013, **52**, 13167–13178.
41 K. Hoshina, Y. Harada, H. Inagaki and N. Takami, *J. Electrochem. Soc.*, 2014, **161**, A348–A354.
42 A. M. Bruneau, L. Liou and D. B. Collum, *J. Am. Chem. Soc.*, 2014, **136**, 2885–2891.
43 S.-H. Baek, R. Klingeler, C. Neef, C. Koo, B. Buechner and H.-J. Grafe, *Phys. Rev. B: Condens. Matter Mater. Phys.*, 2014, **89**, 134424.
44 M. Murakami, Y. Noda, Y. Koyama, K. Takegoshi, H. Arai, Y. Uchimoto and Z. Ogumi, *J. Phys. Chem. C*, 2014, **118**, 15375–15385.
45 M. Song, S. P. V. Nadimpalli, V. A. Sethuraman, M. J. Chon, P. R. Guduru and L.-Q. Wang, *J. Electrochem. Soc.*, 2014, **161**, A915–A919.
46 R. Stoyanova, S. Ivanova, E. Zhecheva, A. Samoson, S. Simova, P. Tzvetkova and A.-L. Barra, *Phys. Chem. Chem. Phys.*, 2014, **16**, 2499–2507.
47 A. S. Filatov, A. V. Zabula, S. N. Spisak, A. Yu. Rogachev and M. A. Petrukhina, *Angew. Chem., Int. Ed.*, 2014, **53**, 140–145.
48 M. U. M. Patel, I. Arcon, G. Aquilanti, L. Stievano, G. Mali and R. Dominko, *ChemPhysChem*, 2014, **15**, 894–904.
49 M. Morita, R. Arizono, N. Yoshimoto and M. Egashira, *J. Appl. Electrochem.*, 2014, **44**, 447–453.
50 Y.-C. Pan, D. Saikia, J. Fang, L.-D. Tsai, G. T. K. Fey and H.-M. Kao, *RSC Adv.*, 2014, **4**, 13293–13303.
51 K. Nakamura, Y. Michihiro, C. Moriyoshi, Y. Kuroiwa and S. Wada, *Solid State Ionics*, 2014, **262**, 202–205.
52 A. A. Arnold, V. Terskikh, Q.-Y. Li, R. Naccache, I. Marcotte and J. A. Capobianco, *J. Phys. Chem. C*, 2013, **117**, 25733–25741.
53 H. Haramaki, T. Shimomura, T. Umecky and T. Takamuku, *J. Phys. Chem. B*, 2013, **117**, 2438–2448.
54 E. V. Charnaya, M. K. Lee, C. Tien, L. J. Chang, Z. J. Wu, Y. A. Kumzerov and A. S. Bugaev, *Phys. Rev. B: Condens. Matter Mater. Phys.*, 2013, **87**, 155401.
55 V. Peruzzo, A. Lanza, F. Nestola, P. A. Vigato and S. Tamburini, *Inorg. Chim. Acta*, 2014, **416**, 226–234.
56 H. Yoon, H. Zhu, A. Hervault, M. Armand, D. R. MacFarlane and M. Forsyth, *Phys. Chem. Chem. Phys.*, 2014, **16**, 12350–12355.
57 N. M. Dicaire, F. A. Perras and D. L. Bryce, *Can. J. Chem.*, 2014, **92**, 9–15.
58 L. Sanchez-Munoz, J. Sanz, I. Sobrados and Z. Gan, *Am. Mineral.*, 2013, **98**, 2115–2131.
59 G. J. Rees, S. P. Day, A. Lari, A. P. Howes, D. Iuga, M. B. Pitak, S. J. Coles, T. L. Threlfall, M. E. Light, M. E. Smith, D. Quigley, J. D. Wallis and J. V. Hanna, *CrystEngComm*, 2013, **15**, 8823–8839.
60 A. R. Lim and K.-Y. Lim, *J. Solid State Chem.*, 2013, **200**, 227–231.
61 M. Kalmutzki, D. Enseling, J. E. C. Wren, S. Kroeker, V. V. Terskikh, T. Juestel and H.-J. Meyer, *Inorg. Chem.*, 2013, **52**, 12372–12382.
62 T. Brotin, S. Goncalves, P. Berthault, D. Cavagnat and T. Buffeteau, *J. Phys. Chem. B*, 2013, **117**, 12593–12601.
63 J. Schneider, J. Tsuchida and H. Eckert, *Phys. Chem. Chem. Phys.*, 2013, **15**, 14328–14339.
64 E. R. Vance, J. Davis, K. Olufson, D. J. Gregg, M. G. Blackford, G. R. Griffiths, I. Farnan, J. Sullivan, D. Sprouster, C. Campbell and J. Hughes, *J. Nucl. Mater.*, 2014, **448**, 325–329.

65 A. Potocnik, A. Y. Ganin, Y. Takabayashi, M. T. McDonald, I. Heinmaa, P. Jeglic, R. Stern, M. J. Rosseinsky, K. Prassides and D. Arcon, *Chem. Sci.*, 2014, **5**, 3008–3017.

66 I. L. Moudrakovski, *Annu. Rep. NMR Spectrosc.*, 2013, **79**, 129–240.

67 D. Naglav, D. Bläser, C. Wölper and S. Schulz, *Inorg. Chem.*, 2014, **53**, 1241–1249.

68 H. Maki, M. Tsujito, H. Nariai and M. Mizuhata, *Magn. Reson. Chem.*, 2014, **52**, 69–81.

69 K. M. N. Burgess, Y. Xu, M. C. Leclerc and D. L. Bryce, *J. Phys. Chem. A*, 2013, **117**, 6561–6570.

70 T. Liu, Y. Shao, G. Li, M. Gu, J. Hu, S. Xu, Z. Nie, X. Chen, C. Wang and J. Liu, *J. Mater. Chem. A*, 2014, **2**, 3430–3438.

71 H. Colas, L. Bonhomme-Coury, C. C. Diogo, F. Tielens, F. Babonneau, C. Gervais, D. Bazin, D. Laurencin, M. E. Smith, J. V. Hanna, M. Daudon and C. Bonhomme, *CrystEngComm*, 2013, **15**, 8840–8847.

72 S. Sene, B. Bouchevreau, C. Martineau, C. Gervais, C. Bonhomme, P. Gaveau, F. Mauri, S. Bégu, P. H. Mutin, M. E. Smith and D. Laurencin, *CrystEngComm*, 2013, **15**, 8763–8775.

73 G. M. Bowers, J. W. Singer, D. L. Bish and R. J. Kirkpatrick, *Am. Mineral.*, 2014, **99**, 318–331.

74 C. M. Widdifield, I. Moudrakovski and D. L. Bryce, *Phys. Chem. Chem. Phys.*, 2014, **16**, 13340–13359.

75 K. M. N. Burgess, Y. Xu, M. C. Leclerc and D. L. Bryce, *Inorg. Chem.*, 2014, **53**, 552–561.

76 H. Igarashi, Y. Shimizu, Y. Kobayashi and M. Itoh, *Phys. Rev. B: Condens. Matter Mater. Phys.*, 2014, **89**, 054431.

77 M. J. D. Champion, P. Farina, W. Levason and G. Reid, *Dalton Trans.*, 2013, **42**, 13179–13189.

78 S. A. Bartlett, G. Cibin, A. J. Dent, J. Evans, M. J. Hanton, G. Reid, R. P. Tooze and M. Tromp, *Dalton Trans.*, 2013, **42**, 2213–2223.

79 M. Pniok, V. Kubicek, J. Havlickova, J. Kotek, A. Sabatie-Gogova, J. Plutnar, S. Huclier-Markai and P. Hermann, *Chem. – Eur. J.*, 2014, **20**, 7944–7955.

80 I. Oikawa and H. Takamura, *Dalton Trans.*, 2014, **43**, 9714–9721.

81 C. Schädle, A. Fischbach, E. Herdtweck, K. W. Törnroos and R. Anwander, *Chem. – Eur. J.*, 2013, **19**, 16334–16341.

82 T. Dey, A. V. Mahajan, R. Kumar, B. Koteswararao, F. C. Chou, A. A. Omrani and H. M. Ronnow, *Phys. Rev. B: Condens. Matter Mater. Phys.*, 2013, **88**, 134425.

83 S. E. Ashbrook and D. M. Dawson, *Acc. Chem. Res.*, 2013, **46**, 1964–1974.

84 C. Höting, H. Eckert, S. F. Matar, U. C. Rodewald and R. Pöttgen, *Z. Naturforsch., B: J. Chem. Sci.*, 2014, **69**(3), 305–312.

85 C. Höting, H. Eckert, F. Haarmann and R. Pöttgen, *Z. Anorg. Allg. Chem.*, 2014, **640**, 1303–1308.

86 C. Höting, H. Eckert, F. Haarmann, F. Winter and R. Pöttgen, *Dalton Trans.*, 2014, **43**, 7860–7867.

87 F. Blanc, L. Sperrin, D. Lee, R. Dervisoglu, Y. Yamazaki, S. M. Haile, G. De Paepe and C. P. Grey, *J. Phys. Chem. Lett.*, 2014, **5**, 2431–2436.

88 V. A. Krut'ko, V. P. Tarasov, G. A. Bandurkin and M. G. Komova, *Russ. J. Inorg. Chem.*, 2013, **58**, 1090–1096.

89 K. Karube, T. Hattori, K. Ishida, N. Tamura, K. Deguchi and N. K. Sato, *J. Phys. Soc. Jpn.*, 2013, **82**, 083712.

90 M. Kalmutzki, D. Enseling, J. E. C. Wren, S. Kroeker, V. V. Terskikh, T. Jüstel and H.-J. Meyer, *Inorg. Chem.*, 2013, **52**, 12372–12382.

91 A. Trokiner, S. Verkhovskii, A. Gerashenko, Z. Volkova, O. Anikeenok, K. Mikhalev, M. Eremin and L. Pinsard-Gaudart, *Phys. Rev. B: Condens. Matter Mater. Phys.*, 2013, **87**, 125142.
92 M.-H. Chambrier, A. Le Bail, F. Giovannelli, A. Redjaimia, P. Florian, D. Massiot, E. Suard and F. Goutenoire, *Inorg. Chem.*, 2014, **53**, 147–159.
93 E. M. Alakshin, R. R. Gazizulin, A. V. Klochkov, S. L. Korableva, T. R. Safin, K. R. Safiullin and M. S. Tagirov, *Opt. Spectrosc.*, 2014, **116**, 721–723.
94 R. Sarkar, R. Gumeniuk, A. Leithe-Jasper, W. Schnelle, Y. Grin, C. Geibel and M. Baenitz, *Phys. Rev. B: Condens. Matter Mater. Phys.*, 2013, **88**, 201101.
95 M. Yogi, S. Nakamura, N. Higa, H. Niki, Y. Hirose, Y. Onuki and H. Harima, *J. Phys. Soc. Jpn.*, 2013, **82**, 103701.
96 S. Andre and Y.-N. Huang, *Bopuxue Zazhi*, 2013, **30**, 461–487.
97 M. Srebro and J. Autschbach, *Chem. - Eur. J.*, 2013, **19**, 12018–12033.
98 S. Muto, N. J. Stone, C. R. Bingham, J. R. Stone, P. M. Walker, G. Audi, C. Gaulard, U. Köster, J. Nikolov, K. Nishimura, T. Ohtsubo, Z. Podolyak, L. Risegari, G. S. Simpson, M. Veskovic and W. B. Walters, *Phys. Rev. C: Nucl. Phys.*, 2014, **89**, 044309.
99 C. Drouza, M. Vlasiou and A. D. Keramidas, *Dalton Trans.*, 2013, **42**, 11831–11840.
100 N. Hall, M. Orio, A. Jorge-Robin, B. Gennaro, C. Marchi-Delapierre and C. Duboc, *Inorg. Chem.*, 2013, **52**, 13424–13431.
101 M. R. Maurya, C. Haldar, A. Kumar, M. L. Kuznetsov, F. Avecilla and J. C. Pessoa, *Dalton Trans.*, 2013, **42**, 11941–11962.
102 H. S. La Pierre, S. G. Minasian, M. Abubekerov, S. A. Kozimor, D. K. Shuh, T. Tyliszczak, J. Arnold, R. G. Bergman and F. D. Toste, *Inorg. Chem.*, 2013, **52**, 11650–11660.
103 M. Simunekova, P. Schwendt, J. Chrappova, L. Smrcok, R. Cerny and W. Beek, *Cent. Eur. J. Chem.*, 2013, **11**, 1352–1359.
104 M. R. Maurya, M. Bisht, N. Chaudhary, F. Avecilla, U. Kumar and H.-F. Hsu, *Polyhedron*, 2013, **54**, 180–188.
105 J.-H. Son, C. A. Ohlin, E. C. Larson, P. Yu and W. H. Casey, *Eur. J. Inorg. Chem.*, 2013, **2013**, 1748–1753.
106 K. Sharma, G. P. Kothiyal, L. Montagne, F. O. Mear and B. Revel, *Int. J. Hydrogen Energy*, 2013, **38**, 15542–15552.
107 M. Vijayakumar, W. Wang, Z. Nie, V. Sprenkle and J.-Z. Hu, *J. Power Sources*, 2013, **241**, 173–177.
108 J. Garcia, J. C. Gonzalez, M. I. Frascaroli, S. Garcia, P. Blanes, I. Correia, J. C. Pessoa and L. F. Sala, *Can. J. Chem.*, 2013, **91**, 186–195.
109 J. Benítez, I. Correia, L. Becco, M. Fernández, B. Garat, H. Gallardo, G. Conte, M. L. Kuznetsov, A. Neves, V. Moreno, J. C. Pessoa and D. Gambino, *Z. Anorg. Allg. Chem.*, 2013, **639**, 1417–1425.
110 O. Goncharova-Zapata, P. B. Chatterjee, G. Hou, L. L. Quinn, M. Li, J. Yehl, D. C. Crans and T. Polenova, *CrystEngComm*, 2013, **15**, 8776–8783.
111 A. Chatkon, P. B. Chatterjee, M. A. Sedgwick, K. J. Haller and D. C. Crans, *Eur. J. Inorg. Chem.*, 2013, **2013**, 1859–1868.
112 M. Epifani, R. Diaz, C. Force, E. Comini, T. Andreu, R. R. Zamani, J. Arbiol, P. Siciliano, G. Faglia and J. R. Morante, *J. Phys. Chem. C*, 2013, **117**, 20697–20705.
113 G. Mohammadnezhad, M. Böhme, D. Geibig, A. Burkhardt, H. Görls and W. Plass, *Dalton Trans.*, 2013, **42**, 11812–11823.
114 H. A. Duarte, N. Vankova, I. P. Ferreira, E. B. Paniago and T. Heine, *J. Phys. Chem. B*, 2013, **117**, 11670–11680.

115 M. Debnath, A. Dutta, S. Biswas, K. K. Das, H. M. Lee, J. Vicha, R. Marek, J. Marek and M. Ali, *Polyhedron*, 2013, **63**, 189–198.
116 N. McCann, M. Wagner and H. Hasse, *Dalton Trans.*, 2013, **42**(7), 2622–2628.
117 H. Lv, J. Song, Y. V. Geletii, W. Guo, J. Bacsa and C. L. Hill, *Eur. J. Inorg. Chem.*, 2013, **2013**, 1720–1725.
118 S. Ramos, J. J. G. Moura and M. Aureliano, *Spectroscopy of Biological Molecules*, Advances in Biomedical Spectroscopy, 2013, vol. 7, pp. 261–265.
119 E. F. de Souza, C. A. Chagas, T. C. Ramalho, V. Teixeira da Silva, D. L. M. Aguiar, R. S. Gil and R. B. de Alencastro, *J. Phys. Chem. C*, 2013, **117**, 25659–25668.
120 A. Buchholz, S. Nica, R. Debel, A. Fenn, H. Breitzke, G. Buntkowsky and W. Plass, *Inorg. Chim. Acta*, 2014, **420**, 166–176.
121 T. Jakusch, E. A. Enyedy, K. Kozma, Z. Paar, A. Benyei and T. Kiss, *Inorg. Chim. Acta*, 2014, **420**, 92–102.
122 N. J. Nicholas, G. da Silva, S. Kentish and G. W. Stevens, *Ind. Eng. Chem. Res.*, 2014, **53**, 3029–3039.
123 F. J. Melendez, A. Degollado, M. E. Castro, N. A. Caballero, J. A. Guevara-Garcia and T. Scior, *Inorg. Chim. Acta*, 2014, **420**, 149–158.
124 K. Nomura, B. K. Bahuleyan, S. Zhang, P. M. Veeresha Sharma, S. Katao, A. Igarashi, A. Inagaki and M. Tamm, *Inorg. Chem.*, 2014, **53**, 607–623.
125 A. Chaudhary, N. Sharma, M. Nagar, S. M. Mobin, P. Mathur and R. Bohra, *J. Sol-Gel Sci. Technol.*, 2014, **70**, 464–472.
126 E. A. Enyedy, G. M. Bognar, N. V. Nagy, T. Jakusch, T. Kiss and D. Gambino, *Polyhedron*, 2014, **67**, 242–252.
127 M. R. Maurya, N. Chaudhary, A. Kumar, F. Avecilla and J. C. Pessoa, *Inorg. Chim. Acta*, 2014, **420**, 24–38.
128 X.-Y. Yu, L. Deng, H. Tao, B. Jiang and X. Li, *J. Coord. Chem.*, 2014, **67**, 315–322.
129 X.-Y. Yu, L. Deng, B. Zheng, B.-R. Zeng, P. Yi and X. Xu, *Dalton Trans.*, 2014, **43**, 1524–1533.
130 R. d. P. F. Bonfim, L. C. de Moura, J.-G. Eon, O. Mentre, H. Vezin and S. Caldarelli, *J. Solid State Chem.*, 2014, **213**, 9–16.
131 G. Romanowski, J. Kira and M. Wera, *Polyhedron*, 2014, **67**, 529–539.
132 M. R. Maurya and N. Kumar, *J. Mol. Catal. A: Chem.*, 2014, **383–384**, 172–181.
133 C. C. Yuan, Y.-F. Yang and X. K. Xi, *J. Appl. Phys.*, 2013, **114**, 213511.
134 B. Peric, R. Gautier, C. J. Pickard, M. Bosiocic, M. S. Grbic and M. Pozek, *Solid State Nucl. Magn. Reson.*, 2014, **59–60**, 20–30.
135 E. Jo, B. Kang, C. Kim, S. Kwon and S. Lee, *Phys. Rev. B: Condens. Matter Mater. Phys.*, 2013, **88**, 094417.
136 J. J. Boruah, S. P. Das, R. Borah, S. R. Gogoi and N. S. Islam, *Polyhedron*, 2013, **52**, 246–254.
137 J. J. Boruah, S. P. Das, S. R. Ankireddy, S. R. Gogoi and N. S. Islam, *Green Chem.*, 2013, **15**, 2944–2959.
138 T. Iijima, T. Yamase, M. Tansho, T. Shimizu and K. Nishimura, *J. Phys. Chem. A*, 2014, **118**, 2431–2441.
139 H. G. T. Ly, G. Absillis, S. R. Bajpe, J. A. Martens and T. N. Parac-Vogt, *Eur. J. Inorg. Chem.*, 2013, **2013**, 4601–4611.
140 K. Kamata, T. Kimura, H. Sunaba and N. Mizuno, *Catal. Today*, 2014, **226**, 160–166.
141 M. Haouas, I.-M. Mbomekalle, N. Vila, P. de Oliveira and F. Taulelle, *Inorg. Chem.*, 2014, **53**, 5568–5574.
142 S. Nahrwold, R. Berger and P. Schwerdtfeger, *J. Chem. Phys.*, 2014, **140**, 024305.

143 S. Yeninas, A. Pandey, V. Ogloblichev, K. Mikhalev, D. C. Johnston and Y. Furukawa, *Phys. Rev. B: Condens. Matter Mater. Phys.*, 2013, **88**, 241111.
144 K. Shimizu, M. Matsumoto, M. Yamaguchi, B. Emre, I. Dincer and Y. Elerman, *J. Alloys Compd.*, 2014, **584**, 640–645.
145 A. V. Pashchenko, V. P. Pashchenko, Y. F. Revenko, V. K. Prokopenko, A. A. Shemyakov, G. G. Levchenko, N. E. Pismenova, V. V. Kitaev, Y. M. Gufan, A. G. Sil'cheva and V. P. Dyakonov, *J. Magn. Magn. Mater.*, 2014, **369**, 122–126.
146 A. V. Pashchenko, V. P. Pashchenko, V. K. Prokopenko, Y. F. Revenko, N. G. Kisel, V. I. Kamenev, A. G. Sil'cheva, N. A. Ledenev, V. V. Burkhovetskii and G. G. Levchenko, *Phys. Solid State*, 2014, **5**, 955–966.
147 A. V. Pashchenko, V. P. Pashchenko, V. K. Prokopenko, Y. F. Revenko, Y. S. Prylipko, N. A. Ledenev, G. G. Levchenko, V. P. Dyakonov and H. Szymczak, *Acta Mater.*, 2014, **70**, 218–227.
148 D. T. Mancini, E. F. Souza, M. S. Caetano and T. C. Ramalho, *Magn. Reson. Chem.*, 2014, **52**, 129–137.
149 P. Kristan, V. Chlan, H. Stepankova, R. Reznicek, K. Kouril, J. Stepanek, K. Polakova, V. Prochazka, J. Cuda and I. Medrik, *J. Nanomater.*, 2013, **2013**, 179794.
150 S. Kwon, D. Y. Yoon, S. Lee, Y. S. Chai, S. H. Chun and K. H. Kim, *Phys. Rev. B: Condens. Matter Mater. Phys.*, 2013, **88**, 064404.
151 B. Kang, C. Kim, E. Jo, S. Kwon and S. Lee, *J. Magn. Magn. Mater.*, 2014, **360**, 1–5.
152 S. Kwon, B. Kang, C. Kim, E. Jo, S. Lee, Y. S. Chai, S. H. Chun and K. H. Kim, *J. Phys.: Condens. Matter*, 2014, **26**, 146004.
153 T. Hattori, K. Karube, Y. Ihara, K. Ishida, K. Deguchi, N. K. Sato and T. Yamamura, *Phys. Rev. B: Condens. Matter Mater. Phys.*, 2013, **88**, 085127.
154 A. S. Andreev, O. B. Lapina, J. B. D. de Lacaillerie and A. A. Khassin, *J. Struct. Chem.*, 2013, **54**, S102–S110.
155 K. Karube, T. Hattori, K. Ishida, N. Tamura, K. Deguchi and N. K. Sato, *J. Phys. Soc. Jpn.*, 2013, **82**, 083712.
156 H. Kotegawa, H. Nohara, H. Tou, T. Matsuda, E. Yamamoto, Y. Haga, Z. Fisk, Y. Onuki, D. Aoki and J. Flouquet, *J. Korean Phys. Soc.*, 2013, **63**, 341–344.
157 H. F. Liu, C. N. Kuo, C. S. Lue, K. Z. Syu and Y. K. Kuo, *Phys. Rev. B: Condens. Matter Mater. Phys.*, 2013, **88**, 115113.
158 B. L. Young, P. Y. Chu, J. Y. Juang, G. J. Shu and F. C. Chou, *Phys. Rev. B: Condens. Matter Mater. Phys.*, 2013, **88**, 064418.
159 M. A. Ortuno, L. Castro and M. Buhl, *Organometallics*, 2013, **32**, 6437–6444.
160 R. M. Mampa, M. A. Fernandes and L. Carlton, *Transition Met. Chem.*, 2013, **38**, 219–224.
161 N. Onishi, M. Shiotsuki, T. Masuda, N. Sano and F. Sanda, *Organometallics*, 2013, **32**, 846–853.
162 J. C. Davis, M. Buhl and K. R. Koch, *J. Phys. Chem. A*, 2013, **117**, 8054–8064.
163 G. Cavigliasso, R. Stranger, W. K. C. Lo, J. D. Crowley and A. G. Blackman, *Polyhedron*, 2013, **64**, 238–246.
164 D. Vasilchenko, S. Tkachev, I. Baidina and S. Korenev, *Inorg. Chem.*, 2013, **52**, 10532–10541.
165 Z. Hosni, R. Bessrour and B. Tangour, *J. Comput. Theor. Nanosci.*, 2014, **11**, 318–323.
166 A. C. Tsipis and I. N. Karapetsas, *Dalton Trans.*, 2014, **43**, 5409–5426.
167 S. L. Benjamin, W. Levason and G. Reid, *Organometallics*, 2013, **32**, 2760–2767.

168 S. Kramer, N. Laflorencie, R. Stern, M. Horvatic, C. Berthier, H. Nakamura, T. Kimura and F. Mila, *Phys. Rev. B: Condens. Matter Mater. Phys.*, 2013, **87**, 180405.
169 M. Q. Kuang, S. Y. Wu, X. F. Hu and B. T. Song, *Phys. B*, 2013, **417**, 13–16.
170 M. Yoshida, Y. Okamoto, M. Takigawa and Z. Hiroi, *J. Phys. Soc. Jpn.*, 2013, **82**, 013702.
171 W. G. Zeier, C. P. Heinrich, T. Day, C. Panithipongwut, G. Kieslich, G. Brunklaus, G. J. Snyder and W. Tremel, *J. Mater. Chem. A*, 2014, **2**, 1790–1794.
172 C. W. Liu, Y. R. Lin, C. S. Fang, C. Latouche, S. Kahlal and J. Y. Saillard, *Inorg. Chem.*, 2013, **52**, 2070–2077.
173 K. Son and Z. Jang, *J. Korean Phys. Soc.*, 2013, **62**, 292–296.
174 D. Donghi, D. Maggioni, G. D'Alfonso and T. Beringhelli, *J. Organomet. Chem.*, 2014, **751**, 462–470.
175 I. M. Armitage, T. Drakenberg and B. Reilly, *Cadmium: From Toxicity to Essentiality*, 2013, **11**, 117–144.
176 R. Luo, H. Xu, H. X. Gu, X. Wang, Y. Xu, X. Shen, W. W. Bao and D. R. Zhu, *CrystEngComm*, 2014, **16**, 784–796.
177 B. Kolodziej, E. Grech, W. Schilf, B. Kamieriski, A. Pazio and K. Woiniak, *J. Mol. Struct.*, 2014, **1063**, 145–152.
178 D. G. Calatayud, E. Lopez-Torres and M. A. Mendiola, *Eur. J. Inorg. Chem.*, 2013, 80–90.
179 R. K. Sharma, A. Wadawale, G. Kedarnath, B. Vishwanadh and V. K. Jain, *Inorg. Chim. Acta*, 2014, **411**, 90–96.
180 N. Pidlypnyi, J. C. Namyslo, M. H. H. Drafz, M. Nieger and A. Schmidtt, *J. Org. Chem.*, 2013, **78**, 1070–1079.
181 K. Sharma, G. P. Kothiyal, L. Montagne, F. O. Mear and B. Revel, *Int. J. Hydrogen Energy*, 2013, **38**, 15542–15552.
182 Z. Zhao-Karger, R. Witter, E. G. Bardaji, D. Wang, D. Cossement and M. Fichtner, *J. Mater. Chem. A*, 2013, **1**, 3379–3386.
183 R. Ghafouri and M. Anafcheh, *Superlattices Microstruct.*, 2013, **55**, 33–44.
184 D. Hnyk and E. G. Jayasree, *J. Comput. Chem.*, 2013, **34**, 656–661.
185 D. Naoufal, Z. Laila, O. Yazbeck, H. Hamad, G. Ibrahim, R. Aoun, A. Safa, M. E. L. Jamal and A. Kanj, *Main Group Chem.*, 2013, **12**, 39–48.
186 D. A. Wann, P. D. Lane, H. E. Robertson, J. Holub and D. Hnyk, *Inorg. Chem.*, 2013, **52**, 4502–4508.
187 K. Yuvaraj, D. K. Roy, K. Geetharani, B. Mondal, V. P. Anju, P. Shankhari, V. Ramkumar and S. Ghosh, *Organometallics*, 2013, **32**, 2705–2712.
188 B. Gierczyk, M. Kazmierczak, L. Popenda, A. Sporzynski, G. Schroeder and S. Jurga, *Magn. Reson. Chem.*, 2014, **52**, 202–213.
189 J. Nekvinda, V. Sicha, D. Hnyk and B. Gruner, *Dalton Trans.*, 2014, **43**, 5106–5120.
190 F. A. Perras and D. L. Bryce, *Chem. Sci.*, 2014, **5**, 2428–2437.
191 S. J. Wang, X. L. Jing, Y. Wang and J. J. Si, *Polym. Degrad. Stab.*, 2014, **99**, 1–11.
192 L. Yu, P. Pellechia and M. A. Matthews, *Int. J. Hydrogen Energy*, 2014, **39**, 442–448.
193 R. S. Azis, D. Holland, M. E. Smith, A. Howes, M. Hashim, A. Zakaria, J. Hassan, N. M. Saiden and M. K. Ikhwan, *J. Aust. Ceram. Soc.*, 2013, **49**, 74–80.
194 D. H. Brouwer and K. P. Langendoen, *CrystEngComm*, 2013, **15**, 8748–8762.
195 D. Esquivel, A. J. Cruz-Cabeza, C. Jimenez-Sanchidrian and F. J. Romero-Salguero, *Microporous Mesoporous Mater.*, 2013, **179**, 30–39.

196 M. Etou, T. Kurisaki, Y. Okaue, H. Wakita and T. Yokoyama, *Anal. Sci.*, 2013, **29**, 843–848.
197 C. Ferrara, C. Tealdi, A. Pedone, M. C. Menziani, A. J. Rossini, G. Pintacuda and P. Mustarelli, *J. Phys. Chem. C*, 2013, **117**, 23451–23458.
198 N. Goudarzi, *J. Mol. Struct.*, 2013, **1046**, 57–63.
199 V. T. T. Ha, A. Sarioglan, A. Erdem-Senatalar and Y. Ben Taarit, *J. Mol. Catal. A: Chem.*, 2013, **378**, 279–284.
200 A. A. Hubaud, D. J. Schroeder, B. Key, B. J. Ingram, F. Dogan and J. T. Vaughey, *J. Mater. Chem. A*, 2013, **1**, 8813–8818.
201 N. Huittinen, P. Sarv and J. Lehto, *Appl. Clay Sci.*, 2013, **80–81**, 182–188.
202 I. Issac, R. Heinzmann, M. Kaus, Z. Zhao-Karger, H. Gesswein, T. Bergfeldt, V. S. K. Chakravadhanula, C. Kubel, H. Hahn and S. Indris, *J. Mater. Chem. A*, 2013, **1**, 13842–13852.
203 Y. J. Ji, H. Xu, D. R. Wang, L. Xu, P. Ji, H. H. Wu and P. Wu, *ACS Catal.*, 2013, **3**, 1892–1901.
204 A. Kermagoret, R. N. Kerber, M. P. Conley, E. Callens, P. Florian, D. Massiot, C. Coperet, F. Delbecq, X. Rozanska and P. Sautet, *Dalton Trans.*, 2013, **42**, 12681–12687.
205 M. M. Khosravi, M. B. Andrus, S. R. Burt and B. F. Woodfield, *Polyhedron*, 2013, **62**, 18–25.
206 R. Kumar, *Bull. Chem. Soc. Ethiop.*, 2013, **27**, 377–386.
207 Y. K. Kim, K. P. Rajesh and J. S. Yu, *J. Hazard. Mater.*, 2013, **260**, 350–357.
208 J. H. Li, D. Zhang and X. Li, *Chin. J. Inorg. Chem.*, 2013, **29**, 2049–2056.
209 B. Y. Liu, F. Chen, L. M. Zheng, J. H. Ge, H. X. Xi and Y. Qian, *RSC Adv.*, 2013, **3**, 15075–15084.
210 P. S. Mueller and S. C. Larsen, *Microporous Mesoporous Mater.*, 2013, **180**, 229–234.
211 F. S. O. Ramos, E. C. O. Munsignatti and H. O. Pastore, *Microporous Mesoporous Mater.*, 2013, **177**, 143–150.
212 V. Sarou-Kanian, A. N. Gleizes, P. Florian, D. Samelor, D. Massiot and C. Vahlas, *J. Phys. Chem. C*, 2013, **117**, 21965–21971.
213 T. F. Sevelsted, D. Herfort and J. Skibsted, *Cem. Concr. Res.*, 2013, **52**, 100–111.
214 V. R. Seymour, E. C. V. Eschenroeder, M. Castro, P. A. Wright and S. E. Ashbrook, *CrystEngComm*, 2013, **15**, 8668–8679.
215 A. K. Singh, R. Yadav and A. Sakthivel, *Microporous Mesoporous Mater.*, 2013, **181**, 166–174.
216 W. J. Shi, X. Y. Jin, S. N. Dong and S. P. Bi, *Geochim. Cosmochim. Acta*, 2013, **121**, 41–53.
217 S. E. Smart, J. Vaughn, I. Pappas and L. Pan, *Chem. Commun.*, 2013, **49**, 11352–11354.
218 Y. Tokunaga, H. Sakai, S. Kambe, A. Sakai, S. Nakatsuji and H. Harima, *Phys. Rev. B: Condens. Matter Mater. Phys.*, 2013, **88**, 085124.
219 J. Wack, R. Siegel, T. Ahnfeldt, N. Stock, L. Mafra and J. Senker, *J. Phys. Chem. C*, 2013, **117**, 19991–20001.
220 C. Q. Ye, Z. Bi and D. S. Wang, *Colloids Surf., A*, 2013, **436**, 782–786.
221 J. Zhang, F. Chang, Y. J. Ren and Q. Y. Shi, *Appl. Spectrosc.*, 2013, **67**, 985–992.
222 S. Machill, L. Kohler, S. Ueberlein, R. Hedrich, M. Kunaschk, S. Paasch, R. Schulze and E. Brunner, *BioMetals*, 2013, **26**, 141–150.
223 T. D. Humphries, K. T. Munroe, A. Decken and G. S. McGrady, *Dalton Trans.*, 2013, **42**, 6953–6964.
224 S. Li, H. Zhou, J. Fan and L. P. Xiao, *Chin. J. Inorg. Chem.*, 2013, **29**, 896–902.

225 A. Alfayate, M. Sanchez-Sanchez and J. Perez-Pariente, *Microporous Mesoporous Mater.*, 2014, **190**, 334–345.
226 S. Al-Khattaf, C. D'Agostino, M. N. Akhtar, N. Al-Yassir, N. Y. Tan and L. F. Gladden, *Catal. Sci. Technol.*, 2014, **4**, 1017–1027.
227 A. Baasner, B. C. Schmidt, R. Dupree and S. L. Webb, *Geochim. Cosmochim. Acta*, 2014, **132**, 151–169.
228 S. A. Bates, W. N. Delgass, F. H. Ribeiro, J. T. Miller and R. Gounder, *J. Catal.*, 2014, **312**, 26–36.
229 M. Benitez-Guerrero, L. A. Perez-Maqueda, P. E. Sanchez-Jimenez and J. Pascual-Cosp, *Microporous Mesoporous Mater.*, 2014, **185**, 167–178.
230 M. Caillot, A. Chaumonnot, M. Digne, C. Poleunis, D. P. Debecker and J. A. van Bokhoven, *Microporous Mesoporous Mater.*, 2014, **185**, 179–189.
231 F. D. Chang, X. W. He, X. Fu, Q. Huang and Y. F. Qiu, *Food Chem.*, 2014, **152**, 177–183.
232 S. F. Chen, J. L. Li, Y. H. Zhang, Y. X. Zhao, K. Y. Liew and J. P. Hong, *Catal. Sci. Technol.*, 2014, **4**, 1005–1011.
233 A. Cuesta, M. A. G. Aranda, J. Sanz, A. G. de la Torre and E. R. Losilla, *Dalton Trans.*, 2014, **43**, 2176–2182.
234 S. Devika, B. Sundaravel, M. Palanichamy and V. Murugesan, *J. Nanosci. Nanotechnol.*, 2014, **14**, 3187–3192.
235 X. B. Du, X. J. Kong and L. G. Chen, *Catal. Commun.*, 2014, **45**, 109–113.
236 E. Gambuzzi, A. Pedone, M. C. Menziani, F. Angeli, D. Caurant and T. Charpentier, *Geochim. Cosmochim. Acta*, 2014, **125**, 170–185.
237 M. Gaudon, L. C. Robertson, E. Lataste, M. Duttine, M. Menetrier and A. Demourgues, *Ceram. Int.*, 2014, **40**, 5201–5207.
238 S. Hayashi, K. Jimura and N. Kojima, *Microporous Mesoporous Mater.*, 2014, **186**, 101–105.
239 H. P. He, T. Li, Q. Tao, T. H. Chen, D. Zhang, J. X. Zhu, P. Yuan and R. L. Zhu, *Am. Mineral.*, 2014, **99**, 109–116.
240 L. H. Jepsen, D. B. Ravnsbaek, C. Grundlach, F. Besenbacher, J. Skibsted and T. R. Jensen, *Dalton Trans.*, 2014, **43**, 3095–3103.
241 X. Y. Jin, H. Wu, X. Y. Jiang and H. Zhang, *Phys. Chem. Chem. Phys.*, 2014, **16**, 10566–10572.
242 K. Kang and M. Lee, *Curr. Appl. Phys.*, 2014, **14**, 383–388.
243 A. Kermagoret, R. N. Kerber, M. P. Conley, E. Callens, P. Florian, D. Massiot, F. Delbecq, X. Rozanska, C. Coperet and P. Sautet, *J. Catal.*, 2014, **313**, 46–54.
244 S. H. Kim, D. Ahn, M. J. Go, M. H. Park, M. Kim, J. Lee and Y. Kim, *Organometallics*, 2014, **33**, 2770–2775.
245 L. Kobera, J. Brus, P. Klein, J. Dedecek and M. Urbanova, *Solid State Nucl. Magn. Reson.*, 2014, **57–58**, 29–38.
246 T. Kurisaki, M. Etou, Y. Okaue, H. Wakita and T. Yokoyama, *Polyhedron*, 2014, **72**, 135–139.
247 C. Le Losq, D. R. Neuville, P. Florian, G. S. Henderson and D. Massiot, *Geochim. Cosmochim. Acta*, 2014, **126**, 495–517.
248 Q. Q. Li, C. Lin, Z. Y. Zhou, J. Ju, G. B. Li, J. H. Lin, W. L. Gao, R. H. Cong and T. Yang, *Inorg. Chem.*, 2014, **53**, 5600–5608.
249 L. Y. Lin and H. L. Bai, *Appl. Catal., B*, 2014, **148**, 366–376.
250 T. Meier, T. Herzig and J. Haase, *Rev. Sci. Instrum.*, 2014, **85**, 4870798.
251 H. Najar, M. S. Zina and A. Ghorbel, *React. Kinet., Mech. Catal.*, 2014, **111**, 775–790.
252 A. Nieland, A. Mix, B. Neumann, H. G. Stammler and N. W. Mitzel, *Eur. J. Inorg. Chem.*, 2014, **2014**, 51–57.

253 M. Paris, *Am. Mineral.*, 2014, **99**, 393–400.
254 A. A. Reddy, A. Goel, D. U. Tulyaganov, M. Sardo, L. Mafra, M. J. Pascual, V. V. Kharton, E. V. Tsipis, V. A. Kolotygin and J. M. F. Ferreira, *J. Mater. Chem. A*, 2014, **2**, 1834–1846.
255 J. J. Ren, L. Zhang and H. Eckert, *J. Phys. Chem. C*, 2014, **118**, 4906–4917.
256 D. Rettenwander, P. Blaha, R. Laskowski, K. Schwarz, P. Bottke, M. Wilkening, C. A. Geiger and G. Amthauer, *Chem. Mater.*, 2014, **26**, 2617–2623.
257 C. L. Shi and X. K. Xi, *Intermetallics*, 2014, **51**, 64–67.
258 E. G. Sorte, S. B. Emery, E. H. Majzoub, T. Ellis-Caleo, Z. L. Ma, B. A. Hammann, S. E. Hayes, R. C. Bowman and M. S. Conradi, *J. Phys. Chem. C*, 2014, **118**, 5725–5732.
259 M. Taoufik, K. C. Szeto, N. Merle, I. Del Rosal, L. Maron, J. Trebosc, G. Tricot, R. M. Gauvin and L. Delevoye, *Chem. – Eur. J.*, 2014, **20**, 4038–4046.
260 T. Tenorio, D. C. N. Lopes, A. M. Silva, J. M. Ramos and C. D. Buarque, *Spectrochim. Acta, Part A*, 2014, **118**, 260–270.
261 H. E. van der Bij, L. R. Aramburo, B. Arstad, J. J. Dynes, J. Wang and B. M. Weckhuysen, *ChemPhysChem*, 2014, **15**, 283–292.
262 C. J. Van Oers, K. Gora-Marek, K. Sadowska, M. Mertens, V. Meynen, J. Datka and P. Cool, *Chem. Eng. J.*, 2014, **237**, 372–379.
263 H. E. van der Bij and B. M. Weckhuysen, *Phys. Chem. Chem. Phys.*, 2014, **16**, 9892–9903.
264 Q. Y. Wang, S. T. Xu, J. R. Chen, Y. X. Wei, J. Z. Li, D. Fan, Z. X. Yu, Y. Qi, Y. L. He, S. L. Xu, C. Y. Yuan, Y. Zhou, J. B. Wang, M. Z. Zhang, B. L. Su and Z. M. Liu, *RSC Adv.*, 2014, **4**, 21479–21491.
265 Q. M. Wu, X. Wang, X. J. Meng, C. G. Yang, Y. Liu, Y. Y. Jin, Q. Yang and F. S. Xiao, *Microporous Mesoporous Mater.*, 2014, **186**, 106–112.
266 M. L. Ramos, A. R. E. de Sousa, L. L. G. Justino, S. M. Fonseca, C. F. G. C. Geraldes and H. D. Burrows, *Dalton Trans.*, 2013, **42**, 3682–3694.
267 K. R. Seddon, G. Srinivasan, M. Swadzba-Kwasny and A. R. Wilson, *Phys. Chem. Chem. Phys.*, 2013, **15**, 4518–4526.
268 A. Paleari, V. N. Sigaev, N. V. Golubev, E. S. Ignat'eva, S. Bracco, A. Comotti, A. Azarbod and R. Lorenzi, *Acta Mater.*, 2014, **70**, 19–29.
269 T. Meier, T. Herzig and J. Haase, *Rev. Sci. Instrum.*, 2014, **85**, 4870798.
270 Y. Ledemi, B. Bureau, G. Le Caer, L. Calvez, C. Roiland, G. Tricot, P. Florian, V. Nazabal and D. Massiot, *J. Non-Cryst. Solids*, 2014, **383**, 216–221.
271 A. T. Grigg, M. Mee, P. M. Mallinson, S. K. Fong, Z. H. Gan, R. Dupree and D. Holland, *J. Solid State Chem.*, 2014, **212**, 227–236.
272 M. F. Ferreira, G. Pereira, J. P. Andre, M. I. M. Prata, P. M. T. Ferreira, J. A. Martins and C. F. G. C. Geraldes, *Dalton Trans.*, 2014, **43**, 8037–8047.
273 R. Bhalla, C. Darby, W. Levason, S. K. Luthra, G. McRobbie, G. Reid, G. Sanderson and W. J. Zhang, *Chem. Sci.*, 2014, **5**, 381–391.
274 H. Sakai, S. Kambe, Y. Tokunaga, Y. Matsumoto, N. Tateiwa, Y. Haga and Z. Fisk, *Phys. Rev. B: Condens. Matter Mater. Phys.*, 2013, **88**, 045123.
275 R. Bhalla, C. Darby, W. Levason, S. K. Luthra, G. McRobbie, G. Reid, G. Sanderson and W. J. Zhang, *Chem. Sci.*, 2014, **5**, 381–391.
276 T. V. Blankenship, B. H. Chen and S. E. Latturner, *Chem. Mater.*, 2014, **26**, 3202–3208.
277 H. Hamaed, K. E. Johnston, B. F. T. Cooper, V. V. Terskikh, E. Ye, C. L. B. Macdonald, D. C. Arnold and R. W. Schurko, *Chem. Sci.*, 2014, **5**, 982–995.
278 T. Brotin, S. Goncalves, P. Berthault, D. Cavagnat and T. Buffeteau, *J. Phys. Chem. B*, 2013, **117**, 12593–12601.

279 M. Nakamura, A. Ikezaki and M. Takahashi, *J. Chin. Chem. Soc.*, 2013, **60**, 9–21.
280 V. H. Pomin, *Anal. Chem.*, 2014, **86**, 65–94.
281 V. Tugarinov, *Prog. Nucl. Magn. Reson. Spectrosc.*, 2014, **77**, 49–68.
282 G. Prabusankar, A. Sathyanarayana, P. Suresh, C. Naga Babu, K. Srinivas and B. P. R. Metla, *Coord. Chem. Rev.*, 2014, **269**, 96–133.
283 R. A. Faulkner, J. A. DiVerdi, Y. Yang, T. Kobayashi and G. E. Maciel, *Materials*, 2013, **6**, 18–46.
284 W. Belam, *J. Alloys Compd.*, 2013, **551**, 267–273.
285 A. A. Reddy, D. U. Tulyaganov, M. J. Pascual, V. V. Kharton, E. V. Tsipis, V. A. Kolotygin and J. M. F. Ferreira, *Int. J. Hydrogen Energy*, 2013, **38**, 3073–3086.
286 K. A. Chernyshev, B. A. Gostevskii, A. I. Albanov and L. B. Krivdin, *Russ. J. Org. Chem.*, 2013, **49**, 34–41.
287 M. Anafcheh and R. Ghafouri, *Phys. E*, 2013, **48**, 13–20.
288 W. Makulski, *J. Mol. Struct.*, 2013, **1036**, 168–173.
289 J. Li, S. Hayakawa, Y. Shirosaki and A. Osaka, *J. Sol-Gel Sci. Technol.*, 2013, **65**, 135–142.
290 P. Rejmak, J. S. Dolado, M. J. Stott and A. Ayuela, *J. Phys. Chem. C*, 2013, **117**, 8374–8380.
291 A. C. S. Reddy, Z. Chen, T. Hatanaka, T. Minami and Y. Hatanaka, *Organometallics*, 2013, **32**, 3575–3582.
292 F. Bitto, K. Kraushaar, U. Boehme, E. Brendler, J. Wagler and E. Kroke, *Eur. J. Inorg. Chem.*, 2013, **2013**, 2954–2962.
293 B. Zhou, E. Ren, B. L. Sherriff and Y. Yao, *Am. Mineral.*, 2013, **98**, 907–913.
294 A. Mahdaviani, M. D. Esrafili and A. Esrafili, *Superlattices Microstruct.*, 2013, **60**, 179–191.
295 K. A. Chernyshev, B. A. Gostevskii and L. B. Krivdin, *Russ. J. Org. Chem.*, 2013, **49**, 832–837.
296 P. Montes-Tolentino, R. Colourado-Peralta, L. A. Martinez-Chavando, E. Mijangos, A. M. Duarte-Hernandez, G. V. Suarez-Moreno, R. Contreras and A. Flores-Parra, *J. Organomet. Chem.*, 2014, **751**, 591–603.
297 S.H. Li and F. Deng, *Annu. Rep. NMR Spectrosc.*, 2013, **78**, 1–54.
298 K. Muether, P. Hrobarik, V. Hrobarikova, M. Kaupp and M. Oestreich, *Chem. – Eur. J.*, 2013, **19**, 16579–16594.
299 E. M. Meintjes, W. W. Warren and J. P. Yesinowski, *Solid State Nucl. Magn. Reson.*, 2013, **55–56**, 91–94.
300 A. R. Noedling, K. Muether, V. H. G. Rohde, G. Hilt and M. Oestreich, *Organometallics*, 2014, **33**, 302–308.
301 E. Gambuzzi, A. Pedone, M. C. Menziani, F. Angeli, D. Caurant and T. Charpentier, *Geochim. Cosmochim. Acta*, 2014, **125**, 170–185.
302 C. Zhang, P. Patschinski, D. S. Stephenson, R. Panisch, J. H. Wender, M. C. Holthausen and H. Zipse, *Phys. Chem. Chem. Phys.*, 2014, **16**, 16642–16650.
303 C. J. Windorff and W. J. Evans, *Organometallics*, 2014, **33**, 3786–3791.
304 B. Wrackmeyer and E. V. Klimkina, *Z. Naturforsch., B: J. Chem. Sci.*, 2014, **69**, 704–714.
305 R. Hernandez-Altamirano, V. Y. Mena-Cervantes, T. E. Chavez-Miyauchi, D. A. Nieto-Alvarez, M. A. Dominguez-Aguilar, L. S. Zamudio-Rivera, V. Barba, F. J. Fernandez-Perrino, S. Perez-Miranda and H. I. Beltran, *Polyhedron*, 2013, **52**, 301–307.
306 T. Barhoumi-Slimi, M. A. K. Sanhoury, M. T. Ben Dhia and M. R. Khaddar, *Phosphorus, Sulfur Silicon Relat. Elem.*, 2013, **188**, 1220–1227.

307 L. Broeckaert, J. Turek, R. Olejnik, A. Ruzicka, M. Biesemans, P. Geerlings, R. Willem and F. De Proft, *Organometallics*, 2013, **32**, 2121–2134.
308 T. Barhoumi-Slimi, M. A. K. Sanhoury, M. T. Ben Dhia and M. R. Khaddar, *Phosphorus, Sulfur Silicon Relat. Elem.*, 2013, **188**, 1827–1834.
309 T. S. Basu Baul, P. Kehie, O. B. Chanu, A. Duthie and H. Hopfl, *J. Organomet. Chem.*, 2013, **733**, 36–43.
310 R. Colourado-Peralta, C. Guadarrama-Perez, L. A. Martinez-Chavando, J. C. Galvez-Ruiz, A. M. Duarte-Hernandez, G. V. Suarez-Moreno, A. Vasquez-Badillo, S. A. Sanchez-Ruiz, R. Contreras and A. Flores-Parra, *J. Organomet. Chem.*, 2014, **751**, 579–590.
311 P. Montes-Tolentino, R. Colourado-Peralta, L. A. Martinez-Chavando, E. Mijangos, A. M. Duarte-Hernandez, G. V. Suarez-Moreno, R. Contreras and A. Flores-Parra, *J. Organomet. Chem.*, 2014, **751**, 591–603.
312 T. Kuwabara, J.-D. Guo, S. Nagase, M. Minoura, R. H. Herber and M. Saito, *Organometallics*, 2014, **33**, 2910–2913.
313 R. E. Taylor, F. Alkan, D. Koumoulis, M. P. Lake, D. King, C. Dybowski and L.-S. Bouchard, *J. Phys. Chem. C*, 2013, **117**, 8959–8967.
314 A. J. Rossini, A. W. MacGregor, A. S. Smith, G. Schatte, R. W. Schurko and G. G. Briand, *Dalton Trans.*, 2013, **42**, 9533–9546.
315 M. Kitamura and A. Asano, *Anal. Sci.*, 2013, **29**, 1089–1093.
316 T. Kawamura, M. Abe, M. Saito and M. Hada, *J. Comput. Chem.*, 2014, **35**, 847–853.
317 J. Catalano, Y. Yao, A. Murphy, N. Zumbulyadis, S. A. Centeno and C. Dybowski, *Appl. Spectrosc.*, 2014, **68**, 280–286.
318 A. S. Tatton, T. N. Pham, F. G. Vogt, D. Duga, A. J. Edwards and S. P. Brown, *Mol. Pharmaceutics*, 2013, **10**, 999–1007.
319 L. A. O'Dell and A. Brinkmann, *J. Chem. Phys.*, 2013, **138**, 064201.
320 W. M. Zoghaib, J. Husband, U. A. Soliman, I. A. Shaaban and T. A. Mohamed, *Spectrochim. Acta, Part A*, 2013, **105**, 446–455.
321 A. Suzuki and T. Oku, *Phys. B*, 2013, **428**, 18–26.
322 E. Dib, B. Alonso and T. Mineva, *J. Phys. Chem. A*, 2014, **118**, 3525–3533.
323 A. J. Rossini, L. Emsley and L. A. O'Dell, *Phys. Chem. Chem. Phys.*, 2014, **16**, 12890–12899.
324 J. Czernek, T. Pawlak, M. J. Potrzebowski and J. Brus, *Chem. Phys. Lett.*, 2013, **555**, 135–140.
325 R. Kleinmaier, S. Arenz, A. Karim, A.-C. C. Carlsson and M. Erdelyi, *Magn. Reson. Chem.*, 2013, **51**, 46–53.
326 W. Schilf, B. Kamienski and K. Uzarevic, *J. Mol. Struct.*, 2013, **1031**, 211–215.
327 D. Niedzielska, T. Pawlak, M. Bozejewicz, A. Wojtczak, L. Pazderski and E. Szlyk, *J. Mol. Struct.*, 2013, **1032**, 195–202.
328 D. J. Langeslay, C. N. Beecher, A. Naggi, M. Guerrini, G. Torri and C. K. Larive, *Anal. Chem.*, 2013, **85**, 1247–1255.
329 M. Bella, V. Milata, C. Lopez, R. M. Claramunt, I. Alkorta and J. Elguero, *Spectrosc. Lett.*, 2013, **46**, 91–99.
330 M. K. Pandey and A. Ramamoorthy, *J. Phys. Chem. B*, 2013, **117**, 859–867.
331 S. Kuroki, Y. Hosaka and C. Yamauchi, *Carbon*, 2013, **55**, 160–167.
332 A. Puszko, K. Laihia, E. Kolehmainen and Z. Talik, *Struct. Chem.*, 2013, **24**, 333–337.
333 N. Folliet, C. Gervais, D. Costa, G. Laurent, F. Babonneau, L. Stievano, J.-F. Lambert and F. Tielens, *J. Phys. Chem. C*, 2013, **117**, 4104–4114.
334 S. G. Sagdinc and A. E. Sahinturk, *Spectrochim. Acta, Part A*, 2013, **104**, 222–234.
335 R. Ghafouri and M. Anafcheh, *Superlattices Microstruct.*, 2013, **55**, 33–44.

336 R. A. Ruhayel, S. J. Berners-Price and N. P. Farrell, *Dalton Trans.*, 2013, **42**, 3181–3187.
337 J. H. Prestegard, S. C. Sahu, W. K. Nkari, L. C. Morris, D. Live and C. Gruta, *J. Biomol. NMR*, 2013, **55**, 201–209.
338 H. Shaghaghi, H. P. Ebrahimi, N. Bahrami Panah and M. Tafazzoli, *Solid State Nucl. Magn. Reson.*, 2013, **51–52**, 31–36.
339 A. Czyrski, U. Girreser and T. Hermann, *J. Mol. Struct.*, 2013, **1036**, 111–114.
340 A. Puszko, K. Laihia, E. Kolehmainen and Z. Talik, *J. Struct. Chem.*, 2013, **54**, 259–263.
341 L. J. Sperling, M. Tang, D. A. Berthold, A. E. Nesbitt, R. B. Gennis and C. M. Rienstra, *J. Phys. Chem. B*, 2013, **117**, 6052–6060.
342 K. A. Chernyshev, V. A. Semenov and L. B. Krivdin, *Russ. J. Org. Chem.*, 2013, **49**, 379–383.
343 E. J. Kibrik, O. Steinhof, G. Scherr, W. R. Thiel and H. Hasse, *J. Polym. Res.*, 2013, **20**, 1–10.
344 K. Muller, Y. Sun, A. Heimermann, F. Menges, G. Niedner-Schatteburg, C. van Wuellen and W. R. Thiel, *Chem. – Eur. J.*, 2013, **19**, 7825–7834.
345 H.-J. Tan and R. P. A. Bettens, *Phys. Chem. Chem. Phys.*, 2013, **15**, 7541–7547.
346 I. Goba and E. Liepinsh, *Magn. Reson. Chem.*, 2013, **51**, 391–396.
347 M. Watanabe, S. Takemura, S. Kawakami, E. Syouno, H. Kurosu, M. Harada and M. Iida, *J. Mol. Liq.*, 2013, **183**, 50–58.
348 S. Li, L. Zhou, Y. Su, B. Han and F. Deng, *Solid State Nucl. Magn. Reson.*, 2013, **54**, 13–17.
349 T. S. Shestakova, Z. O. Shenkarev, S. L. Deev, O. N. Chupakhin, I. A. Khalymbadzha, V. L. Rusinov and A. S. Arseniev, *J. Org. Chem.*, 2013, **78**, 6975–6982.
350 J. K. Williams, D. Tietze, J. Wang, Y. Wu, W. F. DeGrado and M. Hong, *J. Am. Chem. Soc.*, 2013, **135**, 9885–9897.
351 E. F. Belogolova and V. F. Sidorkin, *J. Phys. Chem. A*, 2013, **117**, 5365–5376.
352 F. Hemmann, I. Agirrezabal-Telleria, E. Kemnitz and C. Jaeger, *J. Phys. Chem. C*, 2013, **117**, 14710–14716.
353 L. J. Smith, A. M. Bowen, A. Di Paolo, A. Matagne and C. Redfield, *ChemBioChem*, 2013, **14**, 1780–1788.
354 J. Roche, J. Ying, A. S. Maltsev and A. Bax, *ChemBioChem*, 2013, **14**, 1754–1761.
355 D. Zhang and V. Tugarinov, *J. Biomol. NMR*, 2013, **56**, 169–182.
356 K. J. Fritzsching, Y. Yang, K. Schmidt-Rohr and M. Hong, *J. Biomol. NMR*, 2013, **56**, 155–167.
357 P. Kukic, F. O'Meara, C. Hewage and J. E. Nielsen, *Chem. Phys. Lett.*, 2013, **579**, 114–121.
358 L. Zhou, S. Li, Y. Su, X. Yi, A. Zheng and F. Deng, *J. Phys. Chem. B*, 2013, **117**, 8954–8965.
359 A. Filarowski and P. E. Hansen, *Z. Phys. Chem.*, 2013, **227**, 917–927.
360 S. V. Chapyshev, E. N. Ushakov and A. V. Chernyak, *Magn. Reson. Chem.*, 2013, **51**, 562–568.
361 F. Adam, C.-W. Kueh and E.-P. Ng, *J. Porous Mater.*, 2013, **20**, 1337–1343.
362 D. Kumar, *J. Struct. Funct. Genomics*, 2013, **14**, 109–118.
363 S. S. Kadam, L. Maier, I. Kostakis, N. Pouli, J. Tousek, M. Necas, P. Marakos and R. Marek, *Eur. J. Org. Chem.*, 2013, **2013**, 6811–6822.
364 C. Adam, P. Kaden, B. B. Beele, U. Muellich, S. Trumm, A. Geist, P. J. Panak and M. A. Denecke, *Dalton Trans.*, 2013, **42**, 14068–14074.
365 L. K. Papernaya, A. A. Shatrova, E. P. Levanova, A. I. Albanov, L. V. Klyba, E. V. Rudyakova and G. G. Levkovskaya, *Heteroat. Chem.*, 2013, **24**, 466–475.

366 M. Corredor, J. Bujons, A. Messeguer and I. Alfonso, *Org. Biomol. Chem.*, 2013, **11**, 7318–7325.
367 U. D. Phalgune, K. Vanka and P. R. Rajamohanan, *Magn. Reson. Chem.*, 2013, **51**, 767–774.
368 F. G. Vogt, G. R. Williams, M. N. Johnson and R. C. B. Copley, *Cryst. Growth Des.*, 2013, **13**, 5353–5367.
369 T. Kimura, K. Vukoti, D. L. Lynch, D. P. Hurst, A. Grossfield, M. C. Pitman, P. H. Reggio, A. A. Yeliseev and K. Gawrisch, *Proteins: Struct., Funct., Bioinf.*, 2014, **82**, 452–465.
370 D. C. Marelius, S. Bhagan, D. J. Charboneau, K. M. Schroeder, J. M. Kamdar, A. R. McGettigan, B. J. Freeman, C. E. Moore, A. L. Rheingold, A. L. Cooksy *et al.*, *Eur. J. Inorg. Chem.*, 2014, **2014**, 676–689.
371 Y. Suzuki, T. Yamazaki, A. Aoki, H. Shindo and T. Asakura, *Biomacromolecules*, 2014, **15**, 104–112.
372 J. S. Stevens, S. J. Byard, C. C. Seaton, G. Sadiq, R. J. Davey and S. L. M. Schroeder, *Phys. Chem. Chem. Phys.*, 2014, **16**, 1150–1160.
373 J. F. Reuther and B. M. Novak, *J. Am. Chem. Soc.*, 2013, **135**, 19292–19303.
374 S. H. Park, C. Yang, S. J. Opella and L. J. Mueller, *J. Magn. Reson.*, 2013, **237**, 164–168.
375 D. Kumar, *J. Magn. Reson.*, 2013, **237**, 85–91.
376 R. Gawinecki, E. Kolehmainen, R. Dobosz, H. L. Khouzani and S. Chandrasekaran, *J. Iran. Chem. Soc.*, 2014, **11**, 17–25.
377 B. S. J. Perrin, Y. Tian, R. Fu, C. V. Grant, E. Y. Chekmenev, W. E. Wieczorek, A. E. Dao, R. M. Hayden, C. M. Burzynski, R. M. Venable *et al.*, *J. Am. Chem. Soc.*, 2014, **136**, 3491–3504.
378 S. Szymanski, P. Paluch, D. T. Gryko, A. Nowak-Krol, W. Bocian, J. Sitkowski, B. Koszarna, J. Sniechowska, M. J. Potrzebowski and L. Kozerski, *Chem. – Eur. J.*, 2014, **20**, 1720–1730.
379 S. Fujiyama, Y. Abe, T. Takenawa, T. Aramaki, S. Shioi, T. Katayama and T. Ueda, *Biochim. Biophys. Acta, Proteins Proteomics*, 2014, **1844**, 299–307.
380 M. J. Bayro, B. Chen, W.-M. Yau and R. Tycko, *J. Mol. Biol.*, 2014, **426**, 1109–1127.
381 L. Pazderski, *Annu. Rep. NMR Spectrosc.*, 2013, **80**, 33–179.
382 R. Glaszczka and J. Jazwinski, *J. Mol. Struct.*, 2014, **1061**, 150–159.
383 S. A. Kadam, K. Haav, L. Toom, T. Haljasorg and I. Leito, *J. Org. Chem.*, 2014, **79**, 2501–2513.
384 D. O. Samultsev, V. A. Semenov and L. B. Krivdin, *Magn. Reson. Chem.*, 2014, **52**, 222–230.
385 M. Petrova, R. Muhamadejev, B. Cekavicus, B. Vigante, A. Plotniece, A. Sobolev, G. Duburs and E. Liepinsh, *Heteroat. Chem.*, 2014, **25**, 114–126.
386 P. Calligari and D. Abergel, *J. Phys. Chem. B*, 2014, **118**, 3823–3831.
387 A. Requet, O. Colin, F. Bourdreux, S. M. Salim, S. Marque, C. Thomassigny, C. Greck, J. Farjon and D. Prim, *Magn. Reson. Chem.*, 2014, **52**, 273–278.
388 B. D. Altheimer and M. A. Mehta, *J. Phys. Chem. A*, 2014, **118**, 2618–2628.
389 P. Li, Y. Chu, L. Wang, R. M. Wenslow, K. Yu, H. Zhang and Z. Deng, *CrystEngComm*, 2014, **16**, 3141–3147.
390 G. Raju, S. Singh, A. C. Mutter, B. H. Everson, J. F. Cerda and R. L. Koder, *Tetrahedron Lett.*, 2014, **55**, 2487–2491.
391 A. Virgili, M. M. Quesada-Moreno, J. R. Aviles-Moreno, J. J. Lopez-Gonzalez, M. Angeles Garcia, R. M. Claramunt, M. R. Torres, M. L. Jimeno, F. Reviriego, I. Alkorta *et al.*, *Helv. Chim. Acta*, 2014, **97**, 471–490.
392 A. A. Adeniyi and P. A. Ajibade, *Spectrochim. Acta, Part A*, 2014, **128**, 540–551.

393 V. A. Semenov, D. O. Samul'tsev and L. B. Krivdin, *Russ. J. Org. Chem.*, 2014, **50**, 381–388.
394 X. Wang, Z. Hou, T. Ikeda and K. Terakura, *J. Phys. Chem. C*, 2014, **118**, 13929–13935.
395 G. Abramov and A. Goldbourt, *J. Biomol. NMR*, 2014, **59**, 219–230.
396 G. K. Pierens, T. K. Venkatachalam and D. Reutens, *Magn. Reson. Chem.*, 2014, **52**, 453–459.
397 P. Kemayo Koumkoua, C. Aisenbrey, E. Salnikov, O. Rifi and B. Bechinger, *J. Pept. Sci.*, 2014, **20**, 526–536.
398 Y. Nishiyama, M. Malon, Y. Ishii and A. Ramamoorthy, *J. Magn. Reson.*, 2014, **244**, 1–5.
399 C. Wiedemann, P. Bellstedt, C. Herbst, M. Goerlach and R. Ramachandran, *J. Biomol. NMR*, 2014, **59**, 211–217.
400 D. C. Kaseman, T. Endo and S. Sen, *J. Non-Cryst. Solids*, 2013, **359**, 33–39.
401 R. Christensen, G. Olson and S. W. Martin, *J. Phys. Chem. B*, 2013, **117**, 2169–2179.
402 W. Belam, *J. Alloys Compd.*, 2013, **551**, 267–273.
403 J. Precechtelova, M. L. Munzarova, J. Vaara, J. Novotny, M. Dracinsky and V. Sklenar, *J. Chem. Theory Comput.*, 2013, **9**, 1641–1656.
404 S. J. Sabounchei, M. Pourshahbaz, M. H. Zebarjadian, H. Keypour and M. Bordbar, *J. Mol. Struct.*, 2013, **1034**, 189–192.
405 A. V. Ivanov, E. V. Korneeva, I. A. Lutsenko, A. V. Gerasimenko, O. N. Antzutkin, A.-C. Larsson and V. I. Sergienko, *J. Mol. Struct.*, 2013, **1034**, 152–161.
406 O. Back, M. Henry-Ellinger, C. D. Martin, D. Martin and G. Bertrand, *Angew. Chem., Int. Ed.*, 2013, **52**, 2939–2943.
407 M. Eibisch, T. Riemer, B. Fuchs and J. Schiller, *J. Agric. Food Chem.*, 2013, **61**, 2696–2700.
408 M. A. Ivanov, A. V. Gerasimenko, A. V. Ivanov, P. M. Solozhenkin, V. V. Sharutin, A.-C. Larsson, O. N. Antzutkin and M. A. Pushilin, *Russ. J. Inorg. Chem.*, 2013, **58**, 197–205.
409 M. Culcasi, G. Casano, C. Lucchesi, A. Mercier, J.-L. Clement, V. Pique, L. Michelet, A. Krieger-Liszkay, M. Robin and S. Pietri, *J. Med. Chem.*, 2013, **56**, 2487–2499.
410 J. B. Addison, N. N. Ashton, W. S. Weber, R. J. Stewart, G. P. Holland and J. L. Yarger, *Biomacromolecules*, 2013, **14**, 1140–1148.
411 X. R. Zhao, Q. J. Shen and W. J. Jin, *Chem. Phys. Lett.*, 2013, **566**, 60–66.
412 D. Saha, R. Ghosh and A. Sarkar, *Tetrahedron*, 2013, **69**, 3951–3960.
413 Y. Takechi, H. Saito and E. Okamura, *Chem. Phys. Lett.*, 2013, **570**, 136–140.
414 V. K. Michaelis, P. Kachhadia and S. Kroeker, *Phys. Chem. Glasses: Eur. J. Glass Sci. Technol., Part B*, 2013, **54**, 20–26.
415 O. Ekkert, G. G. Miera, T. Wiegand, H. Eckert, B. Schirmer, J. L. Petersen, C. G. Daniliuc, R. Froehlich, S. Grimme, G. Kehr *et al.*, *Chem. Sci.*, 2013, **4**, 2657–2664.
416 H. Maki, M. Tsujito and T. Yamada, *J. Solution Chem.*, 2013, **42**, 1063–1074.
417 P. R. Desai, A. R. Cormier, P. P. Shah, R. R. Patlolla, A. K. Paravastu and M. Singh, *Eur. J. Pharm. Biopharm.*, 2014, **86**, 190–199.
418 H. G. T. Ly, G. Absillis and T. N. Parac-Vogt, *Dalton Trans.*, 2013, **42**, 10929–10938.
419 B. Stoeger, M. Weil and J. Skibsted, *Dalton Trans.*, 2013, **42**, 11672–11682.
420 S.-H. Kim and C. E. Lee, *Solid State Commun.*, 2013, **168**, 28–30.
421 A. Uslu, S. Sahin-Un, A. Kilic, S. Yilmaz, F. Yuksel and F. Hacivelioglu, *Inorg. Chim. Acta*, 2013, **405**, 140–146.

422 T. N. Thompson, S. Ramos-Hunter, J. Robertson and N. Y. Arnett, *Tetrahedron Lett.*, 2013, **54**, 5311–5313.
423 W. R. Creasy, D. J. McGarvey and C. A. S. Brevett, *J. Phys. Chem. C*, 2013, **117**, 22677–22682.
424 J.-D. P. McElderry, P. Zhu, K. H. Mroue, J. Xu, B. Pavan, M. Fang, G. Zhao, E. McNerny, D. H. Kohn, R. T. Franceschi *et al.*, *J. Solid State Chem.*, 2013, **206**, 192–198.
425 S. Sene, B. Bouchevreau, C. Martineau, C. Gervais, C. Bonhomme, P. Gaveau, F. Mauri, S. Begu, P. H. Mutin, M. E. Smith *et al.*, *CrystEngComm*, 2013, **15**, 8763–8775.
426 V. R. Seymour, E. C. V. Eschenroeder, M. Castro, P. A. Wright and S. E. Ashbrook, *CrystEngComm*, 2013, **15**, 8668–8679.
427 J. Ren and H. Eckert, *J. Phys. Chem. C*, 2013, **117**, 24746–24751.
428 K. A. Chernyshev, L. B. Krivdin, S. V. Fedorov, S. N. Arbuzova and N. I. Ivanova, *Russ. J. Org. Chem.*, 2013, **49**, 1420–1427.
429 L. Xie, H. Liu, S. Han, B. Yue and L. Yan, *J. Phys. Chem. B*, 2013, **117**, 16345–16355.
430 Y. Koito, K. Nakajima, R. Hasegawa, H. Kobayashi, M. Kitano and M. Hara, *Catal. Today*, 2014, **226**, 198–203.
431 I. G. Shenderovich, *J. Phys. Chem. C*, 2013, **117**, 26689–26702.
432 H. Maki, M. Tsujito, H. Nariai and M. Mizuhata, *Magn. Reson. Chem.*, 2014, **52**, 69–81.
433 H. Ibisoglu, S. Besli, F. Yuksel, I. Un and A. Kilic, *Inorg. Chim. Acta*, 2014, **409**, 216–226.
434 S. Hayashi, K. Jimura and N. Kojima, *Microporous Mesoporous Mater.*, 2014, **186**, 101–105.
435 M. Nishida, T. Tanaka, H. Fukaya and W. Kanematsu, *Magn. Reson. Chem.*, 2014, **52**, 163–171.
436 L. Pazderski, *Annu. Rep. NMR Spectrosc.*, 2013, **80**, 33–179.
437 H. Saracoglu, O. Sarioz and S. Oznergiz, *J. Mol. Struct.*, 2014, **1063**, 170–177.
438 C. Schroeder, J. Ren, A. C. M. Rodrigues and H. Eckert, *J. Phys. Chem. C*, 2014, **118**, 9400–9411.
439 H. Saracoglu, O. Sarioz and S. Oznergiz, *Mol. Cryst. Liq. Cryst.*, 2014, **591**, 47–63.
440 A. V. Belyaev, S. N. Vorob'eva, M. A. Fedotov and N. B. Kompan'kov, *Russ. J. Inorg. Chem.*, 2014, **59**, 503–507.
441 Y. Koito, K. Nakajima, H. Kobayashi, R. Hasegawa, M. Kitano and M. Hara, *Chem. – Eur. J.*, 2014, **20**, 8068–8075.
442 A. Baggioli, O. Crescenzi, M. J. Field, F. Castiglione and G. Raos, *Phys. Chem. Chem. Phys.*, 2013, **15**, 1130–1140.
443 G. A. Rolla, C. Platas-Iglesias, M. Botta, L. Tei and L. Helm, *Inorg. Chem.*, 2013, **52**, 3268–3279.
444 B. Gierczyk, M. Kazmierczak, G. Schroeder and A. Sporzynski, *New J. Chem.*, 2013, **37**, 1056–1072.
445 X. Bogle, R. Vazquez, S. Greenbaum, A. Cresce and K. Xu, *J. Phys. Chem. Lett.*, 2013, **4**, 1664–1668.
446 H.-J. Tan and R. P. A. Bettens, *Phys. Chem. Chem. Phys.*, 2013, **15**, 7541–7547.
447 X. Kong, M. Shan, V. Terskikh, I. Hung, Z. Gan and G. Wu, *J. Phys. Chem. B*, 2013, **117**, 9643–9654.
448 N. Merle, G. Girard, N. Popoff, A. De Mallmann, Y. Bouhoute, J. Trebosc, E. Berrier, J.-F. Paul, C. P. Nicholas, I. Del Rosal *et al.*, *Inorg. Chem.*, 2013, **52**, 10119–10130.
449 G. Colherinhas, T. L. Fonseca, M. A. Castro, K. Coutinho and S. Canuto, *J. Chem. Phys.*, 2013, **139**, 094502.

450 V. Patinec, G. A. Rolla, M. Botta, R. Tripier, D. Esteban-Gomez and C. Platas-Iglesias, *Inorg. Chem.*, 2013, **52**, 11173–11184.
451 R. Gawinecki, E. Kolehmainen, R. Dobosz, H. L. Khouzani and S. Chandrasekaran, *J. Iran. Chem. Soc.*, 2014, **11**, 17–25.
452 R. Dervisoglu, D. S. Middlemiss, F. Blanc, L. A. Holmes, Y.-L. Lee, D. Morgan and C. P. Grey, *Phys. Chem. Chem. Phys.*, 2014, **16**, 2597–2606.
453 D. Esteban-Gomez, C. Cassino, M. Botta and C. Platas-Iglesias, *RSC Adv.*, 2014, **4**, 7094–7103.
454 M. Pascual-Borras, X. Lopez, A. Rodriguez-Fortea, R. J. Errington and J. M. Poblet, *Chem. Sci.*, 2014, **5**, 2031–2042.
455 D. C. Kaseman, I. Hung, Z. Gan and S. Sen, *J. Phys. Chem. B*, 2013, **117**, 949–954.
456 G. Kedarnath and V. K. Jain, *Coord. Chem. Rev.*, 2013, **257**, 1409–1435.
457 G. Roy, P. N. Jayaram and G. Mugesh, *Chem. – Asian J.*, 2013, **8**, 1910–1921.
458 A. Liske, K. Verlinden, H. Buhl, K. Schaper and C. Ganter, *Organometallics*, 2013, **32**, 5269–5272.
459 M. Zielinska-Blajet, P. J. Boratynski, J. Palus and J. Skarzewski, *Tetrahedron*, 2013, **69**, 10223–10229.
460 K. A. Chernyshev, L. B. Krivdin, S. V. Fedorov, S. N. Arbuzova and N. I. Ivanova, *Russ. J. Org. Chem.*, 2013, **49**, 1420–1427.
461 M. W. Stanford, F. R. Knight, K. S. A. Arachchige, P. S. Camacho, S. E. Ashbrook, M. Buehl, A. M. Z. Slawin and J. D. Woollins, *Dalton Trans.*, 2014, **43**, 6548–6560.
462 D. Burriss, W. Zou, D. Cremer, J. Walrod and D. Atwood, *Inorg. Chem.*, 2014, **53**, 4010–4021.
463 G. Kedarnath and V. K. Jain, *Coord. Chem. Rev.*, 2013, **257**, 1409–1435.
464 R. E. Taylor, F. Alkan, D. Koumoulis, M. P. Lake, D. King, C. Dybowski and L.-S. Bouchard, *J. Phys. Chem. C*, 2013, **117**, 8959–8967.
465 G. Thiele, N. Lichtenberger, R. Tonner and S. Dehnen, *Z. Anorg. Allg. Chem.*, 2013, **639**, 2809–2815.
466 B. Njegic, E. M. Levin and K. Schmidt-Rohr, *Solid State Nucl. Magn. Reson.*, 2013, **55–56**, 79–83.
467 M. W. Stanford, F. R. Knight, K. S. A. Arachchige, P. S. Camacho, S. E. Ashbrook, M. Buehl, A. M. Z. Slawin and J. D. Woollins, *Dalton Trans.*, 2014, **43**, 6548–6560.
468 C. Martineau, C. Legein, M. Body, O. Peron, B. Boulard and F. Fayon, *J. Solid State Chem.*, 2013, **199**, 326–333.
469 H. T. Chifotides, I. D. Giles and K. R. Dunbar, *J. Am. Chem. Soc.*, 2013, **135**, 3039–3055.
470 W. Makulski, *J. Mol. Struct.*, 2013, **1036**, 168–173.
471 B. C. Buer, B. J. Levin and E. N. G. Marsh, *J. Pept. Sci.*, 2013, **19**, 308–314.
472 E. Romano, M. F. Ladetto and S. A. Brandan, *Comput. Theor. Chem.*, 2013, **1011**, 57–64.
473 A. Vyalikh, L. G. Bulusheva, G. N. Chekhova, D. V. Pinakov, A. V. Okotrub and U. Scheler, *J. Phys. Chem. C*, 2013, **117**, 7940–7948.
474 S. J. Pike, M. De Poli, W. Zawodny, J. Raftery, S. J. Webb and J. Clayden, *Org. Biomol. Chem.*, 2013, **11**, 3168–3176.
475 C. Peggion, B. Biondi, C. Battistella, M. De Zotti, S. Oancea, F. Formaggio and C. Toniolo, *Chem. Biodiversity*, 2013, **10**, 904–919.
476 H. C. Knachel, V. Benin, W. E. Moddeman, J. C. Birkbeck, T. A. Kestner and T. L. Young, *Magn. Reson. Chem.*, 2013, **51**, 407–413.
477 M. A. Fox, G. Pattison, G. Sandford and A. S. Batsanov, *J. Fluorine Chem.*, 2013, **155**, 62–71.

478 B. Y. Xue and K. B. Dillon, *Helv. Chim. Acta*, 2013, **96**, 1078–1084.
479 M. Senna, A. Duevel, V. Sepelak, J. Shi, K. L. Da Silva, V. Laporte, S. Lebedkin, C. Kuebel, D. Wang, D. Schuenemann *et al.*, *J. Phys. Chem. C*, 2013, **117**, 15272–15278.
480 B. Stoeger, M. Weil and J. Skibsted, *Dalton Trans.*, 2013, **42**, 11672–11682.
481 T. Takahashi, H. Kameda, T. Kamei, J. Koyanagi, F. Pichierri, K. Omata, M. Ishizaki and H. Nakamura, *Tetrahedron: Asymmetry*, 2013, **24**, 1001–1009.
482 D. Niks, E. Hilario, A. Dierkers, H. Ngo, D. Borchardt, T. J. Neubauer, L. Fan, L. J. Mueller and M. F. Dunn, *Biochemistry*, 2013, **52**, 6396–6411.
483 K. Tanabe, T. Tsuda, T. Ito and S. Nishimoto, *Chem. – Eur. J.*, 2013, **19**, 15133–15140.
484 D. S. Fadeev, I. P. Chuikov and V. I. Mamatyuk, *J. Struct. Chem.*, 2013, **54**, 180–185.
485 N. Venkatramaiah, B. Ramakrishna, R. Venkatesan, F. A. Almeida Paz and J. P. C. Tome, *New J. Chem.*, 2013, **37**, 3745–3754.
486 M. Biswal, M. Body, C. Legein, A. Sadoc and F. Boucher, *J. Solid State Chem.*, 2013, **207**, 208–217.
487 E. G. Il'in and A. V. Tyuremnov, *Russ. J. Inorg. Chem.*, 2013, **58**, 1330–1334.
488 A. A. Arnold, V. Terskikh, Q. Y. Li, R. Naccache, I. Marcotte and J. A. Capobianco, *J. Phys. Chem. C*, 2013, **117**, 25733–25741.
489 X. Q. Ran and J. D. Goddard, *Can. J. Chem.*, 2013, **91**, 1272–1280.
490 X. Q. Yan, X. R. Zhao, H. Wang and W. J. Jin, *J. Phys. Chem. B*, 2014, **118**, 1080–1087.
491 H. Yi, E. Balan, C. Gervais, L. Segalen, M. Blanchard and M. Lazzeri, *Phys. Chem. Miner.*, 2014, **41**, 347–359.
492 A. Sadoc, M. Biswal, M. Body, C. Legein, F. Boucher, D. Massiot and F. Fayon, *Solid State Nucl. Magn. Reson.*, 2014, **59–60**, 1–7.
493 L. L. Tomasevich and D. B. Collum, *J. Am. Chem. Soc.*, 2014, **136**, 9710–9718.
494 C. Laurence, J. Legros, P. Nicolet, D. Vuluga, A. Chantzis and D. Jacquemin, *J. Phys. Chem. B*, 2014, **118**, 7594–7608.
495 B. K. Maiti, T. Aviles, I. Moura, S. R. Pauleta and J. J. G. Moura, *Inorg. Chem. Commun.*, 2014, **45**, 97–100.
496 E. Langseth, A. Nova, E. A. Traaseth, F. Rise, S. Oeien, R. H. Heyn and M. Tilset, *J. Am. Chem. Soc.*, 2014, **136**, 10104–10115.
497 F. G. Vogt, G. R. Williams and R. C. B. Copley, *J. Pharm. Sci.*, 2013, **102**, 3705–3716.
498 F. G. Vogt, G. R. Williams, M. N. Johnson and R. C. B. Copley, *Cryst. Growth Des.*, 2013, **13**, 5353–5367.
499 M. Hildebrand, H. Hamaed, A. M. Namespetra, J. M. Donohue, R. Fu, I. Hung, Z. Gan and R. W. Schurko, *CrystEngComm*, 2014, **16**, 7334–7356.
500 J. Viger-Gravel, S. Leclerc, I. Korobkov and D. L. Bryce, *J. Am. Chem. Soc.*, 2014, **136**, 6929–6942.
501 M. Nieradka and T. Kupka, *Chem. Heterocycl. Compd.*, 2014, **50**, 429–437.
502 C. D. Keenan, M. M. Herling, R. Siegel, N. Petzold, C. R. Bowers, E. A. Roessler, J. Breu and J. Senker, *Langmuir*, 2013, **29**, 643–652.
503 D. Gaszowski and M. Ilczyszyn, *Chem. Phys. Lett.*, 2013, **556**, 59–64.
504 F. Castiglione, R. Simonutti, M. Mauri and A. Mele, *J. Phys. Chem. Lett.*, 2013, **4**, 1608–1612.
505 S. Standara, P. Kulhanek, R. Marek and M. Straka, *J. Comput. Chem.*, 2013, **34**, 1890–1898.
506 P. Morgado, R. Bonifacio, L. F. G. Martins and E. J. M. Filipe, *J. Phys. Chem. B*, 2013, **117**, 9014–9024.

507 M.-A. Springuel-Huet, A. Nossov, F. Guenneau and A. Gedeon, *Chem. Commun.*, 2013, **49**, 7403–7405.
508 P. Morgado, K. Shimizu, J. M. S. S. Esperanca, P. M. Reis, L. P. N. Rebelo, J. N. C. Lopes and E. J. M. Filipe, *J. Phys. Chem. Lett.*, 2013, **4**, 2758–2762.
509 M. Oschatz, H. C. Hoffmann, J. Pallmann, J. Schaber, L. Borchardt, W. Nickel, I. Senkovska, S. Rico-Frances, J. Silvestre-Albero, S. Kaskel *et al.*, *Chem. Mater.*, 2014, **26**, 3280–3288.

Theoretical aspects of indirect spin–spin couplings

Jarosław Jaźwiński
DOI: 10.1039/9781782622758-00150

The most important works concerning the calculations of indirect spin–spin coupling constants and their application to structural analysis have been reviewed. The survey covers the period from June 2013 to May 2014. Materials have been classified into a few chapters concerning in turn compounds exhibiting conformational or rotational flexibility, indirect spin–spin couplings *via* pnicogen and hydrogen bonds, through-space interactions, new methods, reviews and benchmark calculations, and finally various topics difficult to classify. The survey was ended with an index of couplings and a list of 82 references. The survey includes calculation methods, basis sets, and software used for the calculations. Most of reported works concern organic molecules.

1 Introduction

Among various nuclear magnetic resonance parameters, spin–spin indirect coupling constants *J* belong to the most important. Spin–spin coupling patterns in ^{1}H and high resolution ^{13}C NMR spectroscopy enable one to establish the connections between atoms in organic molecules; the analysis of these patterns is fundamental in organic chemistry. The relationship between coupling constants and geometric parameters of the molecule provides information on compound conformation. The coupling phenomenon is the basis of multidimensional NMR techniques (COSY, HSQC, HMBC) giving insight into structures of macromolecules, including the structures of proteins. An inverse technique, allowing the acquisition of NMR spectra of low-abundant nuclei and nuclei having low resonance frequencies, utilises the phenomenon of spin–spin coupling. Finally, indirect spin–spin coupling is a measurable parameter providing insight into the electronic structure of molecules. The comparison of computed and experimental indirect spin–spin coupling constants enables one to verify the correctness of model structures in question.

Indirect spin–spin coupling constant, abbreviated as spin–spin coupling, coupling or simply *J* in the present survey, can be computed either at the non-relativistic or relativistic level. According to Ramsey's non-relativistic theory[1] indirect spin–spin coupling in a molecule in solution consists of four contributions: Fermi contact (FC), diamagnetic spin–orbit (DSO), paramagnetic spin–orbit (PSO), and spin–dipole (SD). Most of the common program packages compute all four components of spin–spin coupling constants routinely, providing *J* values as their

Institute of Organic Chemistry, ul. Kasprzaka 44/52, 01-224 Warszawa, Poland. E-mail: jaroslaw.jazwinski@icho.edu.pl

sum. It is worth knowing that the calculation cost is different for each of these terms. The division of couplings into four contributions is not valid at the relativistic level, although corresponding equivalents have been defined. In a solid phase and a liquid-crystalline solution, spin–spin couplings exhibit anisotropic properties. General information on the calculation of indirect spin–spin couplings, theoretical backgrounds, experimental examples and explanation of acronyms can be found in a few monographs. Helgaker, Jaszuński, and Pecul[2] summarized indirect spin–spin calculation methods; Autschbach and Zheng[3] discussed calculations of NMR parameters at the relativistic level; Vaara *et al.*[4] discussed the anisotropic character of indirect spin–spin couplings; Tomasi, Mennuci, and Cammi[5] summarised solvation models useful in the calculations.

The present review is a continuation of our previous works,[6–9] and covers the period from June 2013 to May 2014 (the last search of databases was performed at the end of June 2014) collecting the most important works on the calculations of indirect spin–spin coupling constants. Most of the works concern organic chemistry; a few are related to inorganic molecules, calculations in the solid state, theoretical methods, and benchmark calculations. The works have been roughly divided into a few chapters using their subject as a criterion. The first part gathered the works concerning the compounds exhibiting conformational or rotational flexibility, the second one the works on indirect spin–spin couplings *via* pnicogen and hydrogen bonds, and through-space interactions. The following chapters group the papers reporting new methods, reviews and benchmark calculations. The last two parts include the works on the calculation at relativistic level and various works difficult to classify. Finally, an index of couplings and a list of 82 references are enclosed.

Most calculations of coupling constants have been performed using commonly available program packages, Gaussian 03[10] and Gaussian 09[11] (the most popular ones), Dalton,[12] ACES II,[13] GAMESS,[14] CFOUR,[15] Amsterdam Density Functional (ADF),[16] and DIRAC.[17] The last two programs facilitate the calculations at the relativistic theory levels. Some calculations were performed using a combination of two programs, one for the geometry optimisation and the other for spin–spin coupling calculations. Apart from computing NMR parameters, numerous works include the analysis of electron densities in molecules by Quantum Theory of Atom in Molecule (QTAIM, AIM),[18–21] Electron Localization Function (ELF),[22] and Natural Bond Orbital (NBO)[23] approaches. For these purposes, the AIMAll,[24] NBO-5[25] within GAMESS code,[26] and TOPMOD[27] programs were applied. Indirect spin–spin coupling constants between X and Y through *n* bonds are denoted as $^{n}J(X,Y)$, the couplings *via* hydrogen, lithium and halogen bonds as $^{nh}J(X,Y)$, $^{nli}J(X,Y)$, and $^{nx}J(X,Y)$, respectively. The couplings across pnicogen bond and trough-space couplings are denoted ^{np}J and ^{TS}J. In most cases, atom mass numbers are omitted in the text. If not marked otherwise, the following isotopes are assumed (in alphabetical order): ^{13}C, ^{35}Cl, ^{19}F, ^{1}H, ^{199}Hg, ^{7}Li, ^{15}N, ^{17}O, ^{31}P, ^{77}Se, ^{29}Si, ^{129}Xe, and ^{125}Te.

2 Application of calculated indirect spin–spin couplings to compounds exhibiting conformational or rotational flexibility

He *et al.*[28] have performed density functional theory (DFT) calculations of vicinal J coupling constants involving the backbone torsional angle for the protein GB3 using the recently developed automatic fragmentation quantum mechanics/molecular mechanics (AF-QM/MM) approach.[29] The calculated values based on the NMR structure appeared to be more accurate than those based on high-resolution X-ray structure. The authors explained this effect by different approaches to structure determination: NMR structure was refined using a large number of residual dipolar couplings, whereas the hydrogen atoms were added to the X-ray structure in idealized positions. By comparing the three Karplus equations, the authors demonstrated that hydrogen bonding, substituent and electrostatic effects could have a significant impact on vicinal J couplings even though they depend mostly on the intervening dihedral angles. The root-mean-square deviations of the calculated $^3J(H^N,H^\alpha)$, $^3J(H^N,C^\beta)$, $^3J(H^N,C')$ values based on the NMR structure were 0.52, 0.25, and 0.35 Hz, respectively. The calculations of J couplings have been performed at the PW91PW91/IGLO-III theory level.

Hamann *et al.*[30] have calculated $^2J(C,H)$, $^3J(H,H)$ and $^3J(C,H)$ couplings for two compounds isolated from *Lindera melissifolia* plant, melissifolianes A and B (Fig. 1), in order to assign their relative configurations. All calculations (geometry optimisation and couplings) have been performed at the MPW1PW91/6-31G(d,p) theory level.

Widmalm *et al.*[31] have studied conformational sampling for a set of ten α- or β-(1→6)-linked oligosaccharides using explicit solvent Hamiltonian replica exchange (HREX) simulations and NMR spectroscopy techniques. In this work the authors have evaluated $^3J(H,H)$, $^2J(C,H)$ and $^2J(C,C)$ using simulated conformations of oligosaccharides and a set of Karplus-like equations.

Rittner *et al.*[32] have performed the conformational study of alanine and valine methyl esters (Fig. 2a), which do not show zwitterionic structures

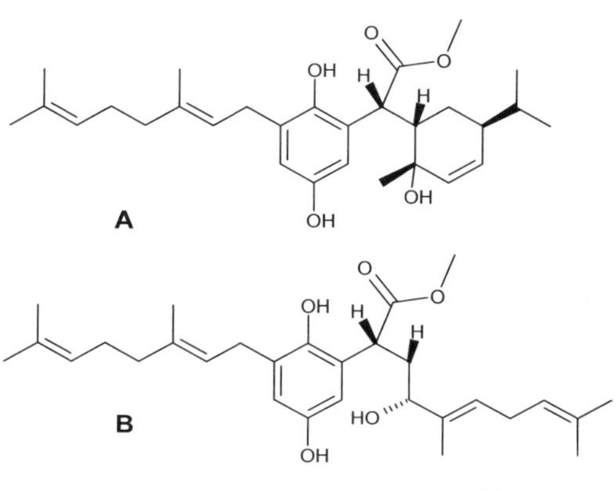

Fig. 1 Melissifolianes A and B.[30]

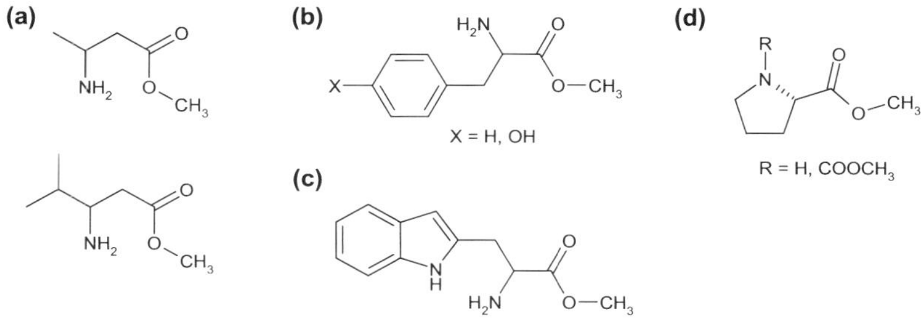

Fig. 2 Methyl esters of alanine and valine (a), tyrosine (b), tryptophane (c), proline and its acetylated derivative (d) studied by Rittner and co-workers.[32–35]

in solution, by ^{1}H NMR and theoretical calculations. The authors computed a fully relaxed potential energy surface at the B3LYP/cc-pVDZ theoretical level, by scanning dihedral angles in the molecules in question from 0° to 360° in 10° steps. Conformers exhibiting energy minima were re-optimised at the B3LYP/aug-cc-pVDZ theoretical level. For these conformers ^3J(H,H) coupling constants have been calculated within the IEF-PCM solvation model, using the B3LYP functional with the aug-cc-pVDZ basis set placed on N and O atoms and the EPR-III basis set for H and C atoms. Computed couplings were averaged taking into account their populations. The ^3J(H,H) spin–spin coupling constants and theoretical calculations were found to be in agreement.

Continuing these investigations, Rittner *et al.* have studied conformational equilibria and intramolecular interactions in phenylalanine and tyrosine methyl esters,[33] tryptophan methyl ester,[34] and in L-proline methyl ester and its *N*-acetylated derivative[35] (Fig. 2b–d). The authors varied dihedral angles in molecules exploring potential energy surfaces and searching for stable conformers of compounds. For selected structures, they have computed ^3J(H,H) coupling constants using B3LYP functional and either the EPR-III basis set for all atoms or a combination of the EPR-III basis set for C and H, and the aug-cc-pVDZ basis set for N and O atoms.

Freitas and co-workers[36] have studied conformational behaviour of 2,2-difluoroethylamine $CF_2H–CH_2NH_2$ and its hydrochloride $CF_2H–CH_2N^+H_3Cl^-$ by NMR spectroscopy and theoretical methods. The authors analysed various conformers taking into account the possibility of hydrogen bonds, and computed ^3J(H,H) and ^3J(H,F) couplings. The calculations were performed at the BHandH/EPR-III// MP2/6-311++g(d,p) theory level.

In the next work, Freitas and co-workers[37] explored the conformation of 3-fluoropiperidinium cation and its analogues (Fig. 3a). To analyse how ^3J(F,N) and ^{1h}J(F,H(N)) couplings would behave if subjected to strong H-bond or F/N$^+$ attraction, the authors computed these couplings as the dependence on shortening the N–C–C and C–C–F angles in (2-fluoroethyl)amine and (2-fluoroethyl)ammonium cation from the equilibrium structures. The calculations were performed at the BHandH/EPR-III theory level.

Continuing the study on fluoroammonium compounds, Freitas and co-workers[38] have investigated the ^1J(C,F) coupling constants in α- and

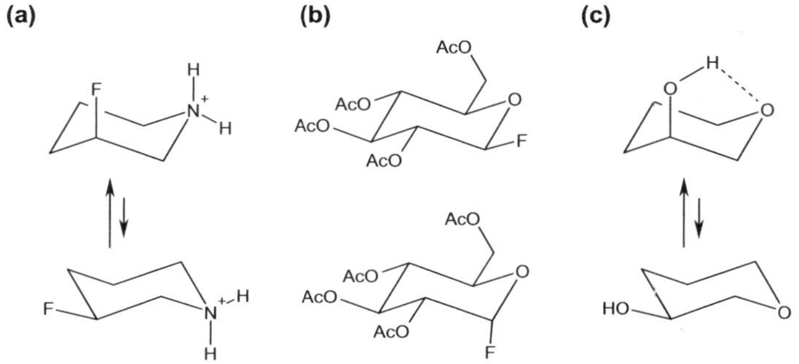

Fig. 3 3-Fluoropiperidinium cation (a), glucopyranosyl fluoride tetraacetate (b) and 3-hydroxytetrahydropyran (c) studied by Freitas and co-workers[37,38] and Tormena and co-workers.[39]

β-D-glucopyranosyl fluoride tetraacetate (Fig. 3b), both in nonpolar and polar solution. The authors analysed the phenomenon called "reverse fluorine Perlin-like effect", corresponding to larger $^1J(C,F_{ax})$ than $^1J(C,F_{eq})$. As part of this work, the calculations of $^1J(C,F)$ were carried out at the ωB97X-D/6-311+G(d,p) level of theory.

Tormena and co-workers[39] have explored the conformational preferences of 3-hydroxytetrahydropyran (Fig. 3c) using infrared and nuclear magnetic resonance spectroscopic data in solvents of different polarities and theoretical methods. The authors concluded that the axial conformation was more stable in the isolated phase and a nonpolar solvent, while the equatorial conformation was more stable than the axial in polar media. The occurrence of intramolecular hydrogen-bonded O–H···O in the axial conformer was detected from infrared spectra in a non-polar solvent at different concentrations. In this work, the authors computed $^3J(H,H)$ applying the SOPPA(CCSD)/EPR-III method.

Krivdin and co-workers[40] applied a double perturbation theory (DPT) at the second order level of approximation formalism to examine the dihedral angle dependence of the Fermi-contact (FC) contribution to nuclear spin–spin coupling constants. The authors have derived an analytical expression relating the FC term of $^3J(H,H)$ across the aliphatic single carbon–carbon bond to the dihedral angle describing inner rotation around the C–C bond in the ten-electron ten-orbital moiety H–C–C–H. In particular, the authors have shown that extrema of $^3J(H,H)$ are observed at $\varphi = \pi n$ ($n = 0, \pm 1, \pm 2, \ldots\ldots$), which provides a theoretical background for a semiempirical Karplus equation.

Wrackmeyer and co-workers[41] have performed the calculations of $^2J(P,C)$, $^2J(P,Si)$ and $^1J(P,N)$ in various amino- and silylaminophosphines including a phosphenium cation and an amino(imono)phosphine applying the B3LYP/6-311+G(d,p) theory level for both geometry optimisation and NMR parameter computing. Both magnitude and sign of the two-bond coupling constants $^2J(P–N–C)$ and $^2J(P–N–Si)$, known to be sensitive towards the respective conformation, have been well reproduced by the calculations, whereas $^1J(P,N)$, values have been slightly overestimated.

3 Indirect spin–spin couplings *via* pnicogen and hydrogen bonds, through-space interactions and van der Waals complexes

3.1 Compounds forming pnicogen bonds, their dimers and complexes

Several papers on pnicogen bonds (called also "pnictogen bonds" in one work[42]) have been published.[42–50] Typically, the authors have optimised the geometry of the compounds in question at the MP2/aug′-cc-pVTZ theory level and analysed electron densities applying AIM (atom-in-molecule), NBO (natural bond orbitals) and ELF (electron localization function) methodologies. Most of the spin–spin coupling constants have been evaluated using the equation-of-motion coupled cluster singles and doubles (EOM CCSD) method in the CI (configuration interaction)-like approximation with all electrons correlated. The qzp basis set was placed on C, N, O, and F, and the qz2p basis set on P, S and Cl atoms. In the case of hydrogen, the qz2p basis set was applied to H atoms involved in hydrogen bonds and the Dunning cc-pVDZ basis set to the remaining H atoms. The hybrid basis set was placed on B atoms. Calculations were performed with Gaussian 09 program suite (geometry optimisation) and ACES II (spin–spin coupling computing).

Alkorta *et al.*[42] have investigated the electronic structure of 3,4-dimethyl-1-cyanophosphole and its dimer (Fig. 4a) applying NMR spectroscopy and theoretical methodologies. In this work, the authors computed spin–spin coupling constants ${}^{n}J(\mathrm{C,P})$ ($n = 1$–3), ${}^{m}J(\mathrm{H,P})$ ($m = 2$, 4), ${}^{4}J(\mathrm{H,H})$, ${}^{1}J(\mathrm{C,N})$, ${}^{2}J(\mathrm{N,P})$, and couplings related to pnicogen bonds, ${}^{1\mathrm{p}}J(\mathrm{P,P})$ and ${}^{2\mathrm{p}}J(\mathrm{C,P})$. The most interesting couplings, *via* pnicogen bonds, varied from *ca.* 100 to 160 Hz for ${}^{1\mathrm{p}}J(\mathrm{P,P})$ and from *ca.* 10 to 18 Hz for ${}^{2\mathrm{p}}J(\mathrm{C,P})$. The calculations were performed at the B3LYP/aug′-cc-pVTZ theory level, using either X-ray geometry or the geometry optimised by the MP2/aug′-cc-pVTZ method.

Similar investigations[43] have been carried out on H_2XP:PCX complexes (X = F, Cl, OH, NC, CN, CCH, CH_3, and H, Fig. 4b), in search of complexes stabilized by P–P pnicogen bonds. Three conformations of complexes were considered. Depending on the conformation and X substituent, computed ${}^{1\mathrm{p}}J(\mathrm{P,P})$ coupling constants varied from *ca.* 7 to 210 Hz. The authors have found a correlation between ${}^{1\mathrm{p}}J(\mathrm{P,P})$ and P–P distances.

In the next work Alkorta and co-workers[44] explored equilibrium structures with P···Cl pnicogen bonds or halogen bonds on the potential

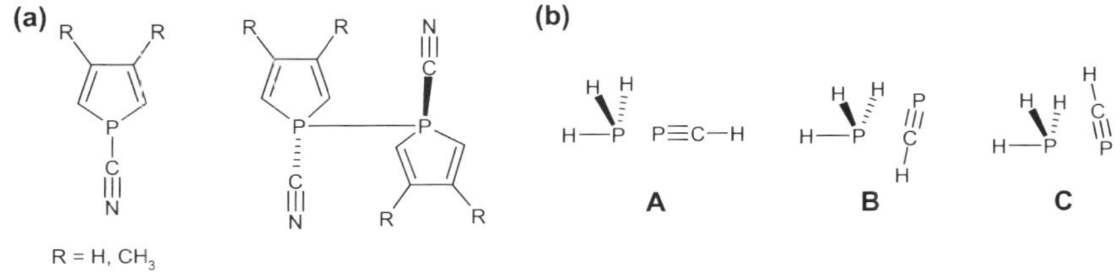

Fig. 4 3,4-Dimethyl-1-cyanophosphole and its dimer having a pnicogen bond (a), and three arrangements of an H_2XP:PCX complex (X = H) (b).

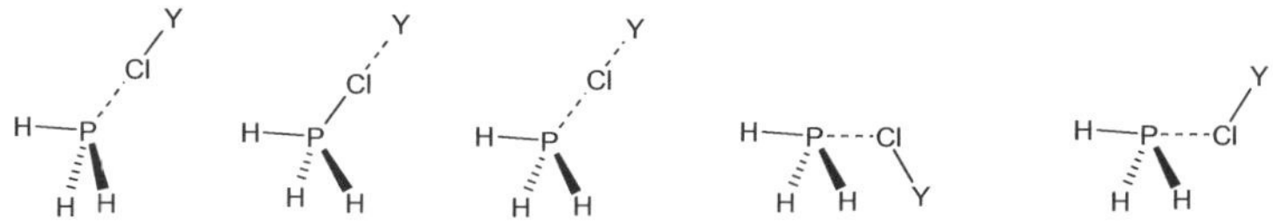

Fig. 5 Pnicogen-bonded and halogen-bonded complexes $H_2FP{:}ClY$.

energy surfaces of $H_2FP{:}ClY$ complexes ($Y = F$, NC, Cl, CN, CCH, CH_3, and H, Fig. 5). Three different types of halogen-bonded complexes with traditional, chlorine-shared, and ion-pair bonds, and two different pnicogen-bonded complexes have been identified. The authors concluded that spin–spin coupling constants $^{1x}J(P,Cl)$ for complexes with traditional halogen bonds increased with decreasing P–Cl distance, reached a maximum value for complexes with chlorine-shared halogen bonds, and then decreased and changed sign when the bond was an ion-pair bond. The $^{1p}J(P,Cl)$ coupling constants across pnicogen bonds tended to increase with decreasing P–Cl distance.

Similar studies have been carried out by Alkorta and co-workers[45] to investigate the pnicogen-bonded complexes $H_2YP{:}X^-$, for $X,Y = Cl$, NC, F, CCH, and CH_3. In the course of their work, the $^{1}J(P,A)$ couplings and $^{1p}J(P,A)$ couplings across pnicogen bonds have been computed ($A = C$, Cl, F, N). The authors concluded that spin–spin coupling constants $^{1p}J(P,A)$ differentiated between shorter ion–molecule pnicogen bonds with partial covalent character and longer P···A ion–molecule pnicogen bonds. Similarly, coupling constants $^{1}J(P,A)$ differentiated between longer covalent P–A bonds with partial ion-molecule character and shorter P–A covalent bonds.

The next work of Alkorta and co-workers[46] concerned the halogen-bonded complexes $H_2XP{:}ClF$ and $H_2XP{:}Cl_2$ ($X = F$, Cl, OH, NC, CN, CCH, CH_3, H). These complexes contained the halogen bonds instead of pnicogen bond, but they were very close to the above-mentioned compounds. In the course of their work, the authors have computed one-bond spin–spin couplings $^{1}J(F,Cl)$ and $^{1}J(Cl,Cl)$, and couplings across halogen, $^{1x}J(P,Cl)$, $^{2x}J(P,F)$, and $^{2x}J(P,Cl)$. It was found that computed spin–spin coupling constants for complexes with chlorine-shared halogen bonds did not exhibit the usual dependencies on distance: $^{2x}J(P,F)$ and $^{2x}J(P,Cl)$ for complexes with chlorine-shared halogen bonds did not correlate with P–F and P–Cl distances, respectively. In contrast, $^{1x}J(P,Cl)$ values for $H_2XP{:}ClF$ correlated best with the Cl–F distance. $^{1}J(F,Cl)$ and $^{1}J(Cl,Cl)$ for complexes with chlorine-shared halogen bonds correlated linearly with the distance between P and the proximal Cl atom.

Two papers concerned complexes of compounds containing sp^2 hybridized phosphorus atom. Alkorta and co-workers[47] have investigated by theoretical methods $(H_2C{=}PX)_2$ complexes ($X = F$, Cl, OH, CN, NC, CCH, H, CH_3, and BH_2, Fig. 6a). The authors considered various arrangements of $H_2C{=}PX$ monomers in dimers and have calculated the corresponding $^{1p}J(P,P)$ coupling constants across pnicogen P–P bonds. Depending on conformations and X substituent, the computed $^{1p}J(P,P)$ coupling constants varied from 31.2 to 330.1 Hz. These coupling constants increase quadratically as the P–P distance decreases.

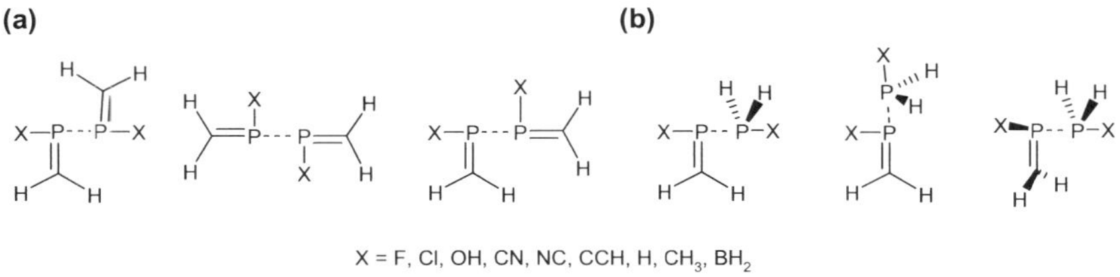

Fig. 6 Various arrangements of $(H_2C{=}PX)_2$ (a) and $(H_2C{=}PX):PXH_2$ (b) complexes having pnicogen bonds.

The same authors[48] have investigated $(H_2C{=}PX):PXH_2$ complexes (X = F, Cl, OH, CN, NC, CCH, H, CH_3, and BH_2), applying similar methodology. The various conformations of the species present in the complex shown in Fig. 6b have been taken into consideration (Fig. 6b). Depending on the conformer, the coupling constant across the pnicogen bond ^{1p}J(P–P) varied from *ca.* 47 to 637 Hz for A, from 175 to 365 Hz for B and from 40 to 126 Hz for C. Computed spin–spin coupling constants correlated with P–P distances. Complexes $(PH_2X)_2$ tended to have larger spin–spin coupling constants and shorter P–P distances than $(H_2C{=}PX):PXH_2$, which in turn exhibited larger coupling constants and shorter P–P distances than $(H_2C{=}PX)_2$. Complexes having similar P–P distances showed similar values of ^{1p}J(P,P).

In the next paper Alkorta and co-workers[49] reported a study on complexes formed between PO_2X (X = F and Cl) as the Lewis acids and a series of nitrogen bases ZN (NH_3, $H_2C{=}NH$, NH_2F, NP, NCH, NCF, NF_3, and N_2). As part of this work, the coupling constants across pnicogen bonds, ^{1p}J(P,N) have been computed. Depending on the compounds, the computed ^{1p}J(P,N) couplings varied from 12.2 to 65.3 Hz. It was found that they were dominated by the Fermi-contact term, which is an excellent approximation to total *J*. The ^{1p}J(P,N) values were small at long distances, increased as the distance decreased, but then decreased at short P–N distances. At the shortest distances, ^{1p}J(P,N) approach to the ^{1}J(P,N) value for the molecules PO_2NC and PO_2NH_2.

Alkorta and co-workers[50] have investigated the pnicogen bonded complexes formed between the acids $H_3P{=}X$ (X = O, S, NH, CH_2) and the bases NH_3, NCH, N_2, PH_3, and PCH. The work included the calculations of ^{1p}J(P,P), ^{1p}J(P,N), and ^{1}J(P,A) where (A = O, N, S, C). The authors concluded that ^{1p}J(P,N) and ^{1p}J(P,P) values correlate best with the corresponding P–N and P–P distances, whereas ^{1}J(P–A) values do not correlate with P–A distances. Decreasing ^{1}J(P,A) values correlated linearly with increasing complex binding energies.

3.2 Investigations of hydrogen bonds and related structures

Sadlej and co-workers[51] have calculated nuclear shielding and intramolecular and intermolecular spin–spin coupling constants ^{1}J(O,C), ^{2}J(O,O), ^{1}J(O,H), ^{2}J(H,H), ^{2h}J(O,O), and ^{1h}J(H,H) for cages forming sI and sII clathrate hydrates of carbon dioxide. Three cages consisting of 20, 24 and 28 water molecules, respectively, have been considered.

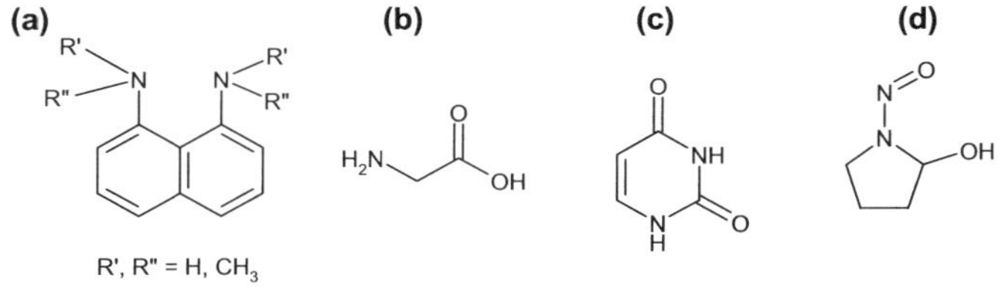

Fig. 7 Molecules of *N*-methylated 1,8-diaminonaphthalenes (a), glycine (b), uracil (c), and α-hydroxy-*N*-nitrosamine (d) subjected to theoretical investigations.

Their structures have been optimised at the B3LYP/aug-cc-PVDZ theory level; NMR parameters were computed using the same functional and the HuzzIII-su3 basis set.

Malkina *et al.*[52] have explored indirect nuclear coupling through a hydrogen bond in the free-base and protonated forms of a complete series of *N*-methylated 1,8-diaminonaphthalenes (Fig. 7a). The authors have analysed the relationships between these couplings and structural features of compounds in question. Experimental ^{2h}J(N,N) varied from 0 to 10.6 Hz, whereas PP86/III-IGLO computed values ranged from 0.49 to 10.55 Hz. The correlation between the sum of C1 and C8 chemical shifts and ^{2h}J(N,N) has been examined. The calculations of coupling constants were performed with the modified deMon software;[53] the geometry of compounds was optimised at the B3LYP/6-31G(d,p) theory level.

Chaudhuri *et al.*[54] have investigated the influence of hydrogen bond formation on the nuclear magnetic resonance parameters for the binary (1 : 1) and ternary (1 : 2) glycine–HCN complexes (Fig. 7b) in the gas phase using high-level density functional theory with the B3LYP/6-31++G(2d,2p)//B3LYP/6-31++G(d,p) model of quantum chemistry. The calculation method was selected as a good compromise between efficiency and accuracy. In this work, one-bond couplings ^{1}J(N,H), ^{1}J(N,C), ^{1}J(O,H), ^{1}J(C,O), ^{1}J(C,H), and couplings across hydrogen bonds ^{1h}J(N,H), ^{1h}J(O,H), ^{2h}J(O,N), ^{2h}J(O,C), ^{2h}J(N,C), and ^{2h}J(N,N) were computed.

Roohi and co-workers[55] have computed one- and two bond spin–spin coupling constants, ^{1}J, ^{1h}J, and ^{2h}J across X–H···O (X = C, N) hydrogen bonds and shielding constants of bridging hydrogens for complexes formed from interaction between the α-hydroxy-*N*-nitrosamine and four preferential binding sites of the uracil (Fig.7c and d). The calculations were carried out at the B3LYP/6-311++G(2d,2p)//MP2/6-311++G(2d,2p) theory level; the ^{1}J(X,H) (X = C, N, O), ^{1}J(Y,C) (Y = C, N, O), ^{1}J(N,O), ^{1h}J(O,H), ^{2h}J(N,O), and ^{2h}J(O,O) couplings have been computed. The authors noted very good correlations between NMR spin–spin coupling constant as well as isotropic shielding constant and the binding energy, H-bond distance, red-shift of vibration frequency, charge transfer energy, and electron density at H-bond critical point.

Alkorta and co-workers[56] have performed *ab initio* EOM-CCSD/(qzp, qz2p) calculations on complexes with intermolecular hydrogen bonds involving ^{15}N and ^{17}O, and molecules with and without intramolecular hydrogen bonds involving these nuclei. The calculations included two-, three-, and four-bond coupling constants. The authors analysed inter- and intramolecular $^{2h}J(X-Y)$ couplings across X–H–Y hydrogen bonds looking for the influence of resonance-assisted hydrogen bonds (RAHB) on the magnitude of couplings. All Ramsey terms of spin–spin couplings have been computed and discussed.

Zakarianejad *et al.*[57] have studied molecular interactions between uracil and nitrous acid HNO_2 (Fig. 7c) using B3LYP, B3PW91, and MP2 methods with different basis sets. The optimized geometries, harmonic vibrational frequencies, charge transfer, topological properties of electron density, nucleus-independent chemical shift (NICS), and one- and two-bonds spin–spin coupling constants have been calculated. The authors considered twenty complexes in total; for eight complexes they computed $^{1}J(O,H)$, $^{1}J(N,H)$, $^{1}J(O,C)$, $^{1h}J(O,H)$, $^{1h}J(N,H)$, $^{2h}J(O,O)$, $^{2h}J(N,N)$, and $^{2h}J(N,O)$. Spin–spin coupling constants including four Ramsey terms were computed at the B3LYP/6-311++G(2p,2d) theory level.

Masoodi *et al.*[58] have explored by theoretical methods the influence of cation–π and anion–π interactions on some NMR data of *s*-triazine-HF hydrogen bonded (Fig. 8a). The $^{1}J(H,F)$, $^{1h}J(H,F)$, and $^{2h}J(N,F)$ coupling constants including four Ramsey terms were computed using the PBE0 and PBE1KCIS functionals combined with the 6-311++G(d,p) basis set. The authors concluded that $^{1h}J(N,H)$ of anionic complexes was smaller than that of the cationic ones. Anion–π interactions amplified the magnitude of $^{2h}J(N,F)$; the opposite was true for the cation–π interaction. The $^{1}J(H,F)$ value was diminished due to the hydrogen bond formation. In contrast to anionic complexes, $^{1}J(H,F)$ decreased by increasing the distance between the cation and the centre of *s*-triazine.

Xu[59] has investigated by theoretical methods the multi-hydrogen bonded systems (Fig. 8b) with the solvated electrons. Computed couplings adopted the following values: $^{1}J(H_t,H_c)$ from −0.7 to +22.4 Hz, $^{1}J(H_c,H_c)$ from *ca.* 160 to 195 Hz, and $^{2h}J(H_t,H_t)$ from −9.2 to +1 Hz (H_c and H_t denote central and terminal hydrogen atom, respectively). The computing of spin–spin couplings was carried out at the B3LYP/6-311++G** theory level.

Esrafili *et al.*[60] have studied the interplay between conventional and unconventional lithium bonds interactions in NCLi···NCLi···XC≡CX and CNLi···CNLi···XC≡CX (X = H, F, Cl, Br, OH, CH_3, and OCH_3) complexes by *ab initio* calculations. Cooperative effects were observed when Li···N(C) and Li···π bonds existed in the same complex. The authors

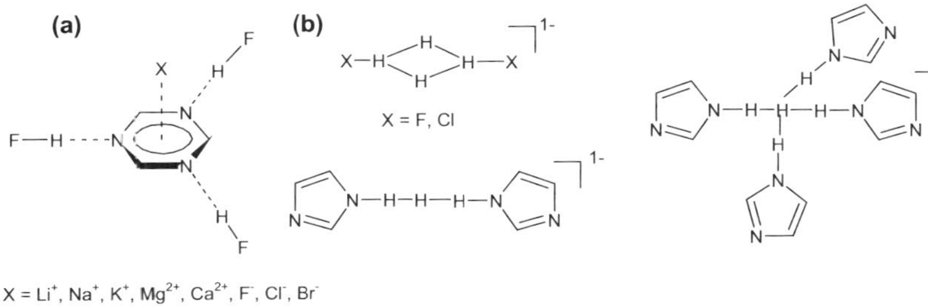

Fig. 8 Cation–π and anion–π complexes of hydrogen bonded *s*-triazine (a) and multi-hydrogen bonded system studied by theoretical methods (b).

analysed these effects in terms of geometric, energetic and electron charge density properties of the complexes, and found an excellent linear correlation between the cooperative energies and the calculated spin–spin couplings across the lithium bonds, $^{1li}J(Li,C)$ and $^{1li}J(Li,N)$. The calculations were performed at the MP2/6-31++G** and CCSD(T)/6-31++G** theory levels, using the GAMESS package.

3.3 Through-space indirect spin–spin couplings

Woollins *et al.*[61] have investigated through space interactions in eight 5-(aryltelluro)-6-(phenylselenyl)acenaphthenes (Fig. 9a) using spin–spin coupling as a probe. The authors carried out solution- and solid state NMR studies, X-ray analysis and computed J(Te, Se) at the relativistic ZORA-SO/BP86/TZZP//B3LYP/SDD/6-31G(d) theory level obtaining $J(^{125}Te,^{77}Se)$ from −460 to −585 Hz. Because experimental $J(^{125}Te,^{77}Se)$ ranged from −688 to −748 Hz, the calculations provided significantly underestimated J values.

Tormena and co-workers[62] have studied through-space interactions between F and H nuclei in 2-fluorobenzaldehyde (Fig. 9b) by NMR spectroscopy and theoretical calculations. The authors discussed unusual transmission pathways for J(F,H) in detail. The calculations performed at the SOPPA(CCSD)/EPR-III//MP2/aug-cc-pVTZ theory level provided J(H,F) of 0.55 Hz (*cis*) and −2.58 Hz (*trans*); the value of 0.48 Hz was measured in dmso-d_6 solution.

Eckert *et al.*[63] have investigated no-bond $^{TS}J(P,P)$ couplings arising from the transmission of nuclear spin polarization through interaction of proximal nonbonded electron pairs in the solid state for a series of closely related substituted P,P-3-ferrocenophanes and model systems (Fig. 9c). The experimental isotropic coupling constants J^{iso} ranged between 12 and 250 Hz. The investigations included the DFT calculation of $^{TS}J(P,P)$ and J-anisotropy ΔJ. Their dependences on intramolecular internuclear distances and relative orbital orientations were examined by DFT-calculations on suitable models. The dependence of $^{TS}J(P,P)$ on the P···P distance was found to be exponential, in agreement with the literature. The computing of $^{TS}J(P,P)$ was carried out at the B3LYP/TZVP theory level.

Tormena and co-workers[64] have studied through-space transmission of J(F,N) couplings in a set of fluoroximes. The SOPPA(CCSD) calculations with the EPR-III basis set placed for F and N atoms and aug-cc-pVTZ for

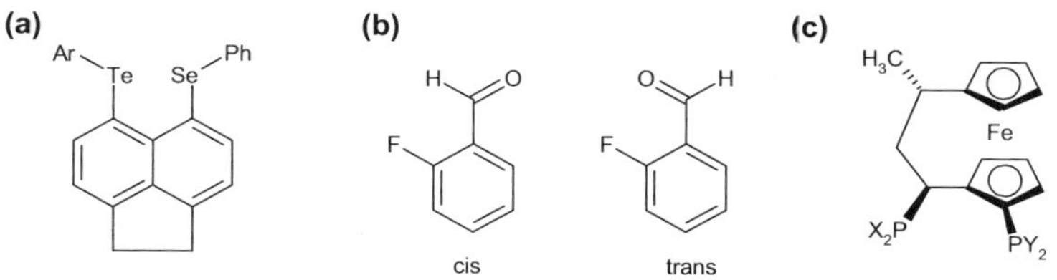

Fig. 9 The molecules of 5-(aryltelluro)-6-(phenylselenyl)acenaphthenes (a), 2-fluorobenzaldehydes (b), and P,P-3-ferrocenophanes (c) exhibiting through-space indirect spin–spin coupling.

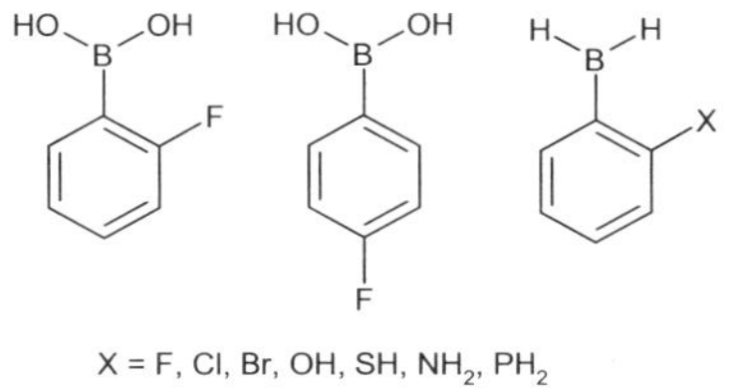

Fig. 10 Monosubstituted phenylboronic acids and phenylboranes.

the remaining atoms revealed that the FC term is the most important contribution to the through-space transmission of J(F,N) coupling constants. The authors proposed a new rationalization for this coupling, based on the overlap matrix (S_{ij}) between fluorine and nitrogen lone pairs obtained from NBO analyses. An expression was proposed to take into account the influence of the electronic density (D_{ij}) between coupled nuclei, as well as the s% character at the site of the coupling nuclei of bonds and non-bonding electron pairs involved in D_{ij}. Finally, the authors proposed the term "through overlapping orbital coupling" instead of the "through-space coupling" for this effect.

Freitas and co-workers[65] have performed conformational analysis on intramolecular interactions in monosubstituted phenylboranes and phenylboronic acids (Fig. 10) applying NBO and QTAIM methodology. Among other things, the authors considered the possibility of through-space and through-hydrogen bond spin–spin couplings. Finally, the calculations of the ^{1h}J(F,H(O)) and through-space ^{1TS}J(F,B) couplings in 2-fluorophenylboronic acid have been performed at the BHandH/EPR-III theory level. Computed ^{1h}J(F,H) varied from −18.9 to −21.9 Hz, whereas the magnitude of measured values ranged from 3.0 to 6.0 Hz.

4 New methods, benchmark calculations and reviews

Bryce and co-workers[66] have outlined the theory describing homonuclear indirect nuclear spin–spin coupling (J) interactions between pairs of quadrupolar nuclei. The theory has been supported by numerical calculations. The expected first-order multiplets for pairs of magnetically equivalent (A2), chemically equivalent (AA), and non-equivalent (AX) quadrupolar nuclei were given.

Autschbach and co-workers[67] have presented a method for a subsystem-based calculation of indirect nuclear spin–spin coupling tensors. This approach was based on the frozen-density embedding scheme within density-functional theory and was an extension of a previously reported subsystem-based approach for the calculation of nuclear magnetic resonance shielding tensors. The method was particularly useful for the inclusion of environmental effects in the calculation of nuclear spin–spin coupling constants. According to this method, the computationally expensive response calculation had to be performed only for the subsystem of interest. As an example, the authors have demonstrated the results for methylmercury halides which exhibited an exceptionally large shift of the ^{1}J(Hg,C) upon coordination of dimethylsulfoxide solvent molecules.

Kupka and co-workers[68] have reported a theoretical prediction of nuclear magnetic shieldings and indirect spin–spin coupling constants in 1,1-, *cis*- and *trans*-1,2-difluoroethylenes, using various theory levels. The calculations included ^nJ(H,H) ($n=2, 3$), ^mJ(C,H), ^mJ(C,F) ($m=1, 2$), ^1J(C,C), ^2J(F,F), and ^3J(H,F) coupling constants. Coupling constants have been computed using BH and H density functional combined with the pcJ-n and aug-pcJ-n basis sets, and CBS (complete basis set limit) estimations using two-parameter fit. For the completeness, the calculations at the CCSD/pcJ-n theory levels were carried out. The authors discussed in detail all Ramsey terms of coupling constants, as well as the performance of various theory levels. In all calculations, the B3LYP/6-311++G(3df,2pd) geometry of compounds was applied.

San Fabián *et al.*[69] have determined optimized shifting and/or scaling factors for calculated ^1J(C,H) spin–spin coupling constants for 35 combinations of representative functionals (PBE, B3LYP, B3P86, B97-2 and M06-L) and basis sets (TZVP, HIIIsu3, EPR-III, aug-cc-pVTZ-J, ccJ-pVDZ, ccJ-pVTZ, ccJ-pVQZ, pcJ-2 and pcJ-3) using 68 organic molecular systems with 88 ^1J(C,H) couplings. Model compounds contained different types of hybridized carbon atoms. The authors examined density functional theory assessment for the determination of ^1J(C,H) coupling constants comparing the computed and experimental values, and concluded that the use of shifting constants for obtaining the calculated coupling improves the results substantially.

Ananikov and co-workers[70] have summarised recent advances in computational predictions of NMR parameters for the structure elucidation of carbohydrates focusing on methods and their limitations. One part of this work concerned the computation of NMR coupling constants. In this chapter the authors summarised the calculation methods and software used for the prediction of *J*(H,H), *J*(C,H), and *J*(C,C) couplings in a set of mono-, oligo- and polysaccharides, and discussed the results.

5 The calculation including relativistic effects

Autschbach[71] has outlined some basic concepts of relativistic quantum chemistry and recent developments of relativistic methods for the calculation of the molecular properties, including important for NMR spectroscopy, nuclear magnetic resonance shielding, indirect nuclear spin–spin coupling and electric field gradients (nuclear quadrupole coupling). The author analysed the performance of density functional theory (DFT) and its applications for heavy-element systems. Finally, the author has reviewed selected applications of DFT in relativistic calculation of magnetic resonance parameters.

Krivdin and co-workers[72] have performed four-component relativistic calculations of ^1J(Se,C) spin–spin coupling constants in a series of 13 selenium heterocycles and their parent open-chain selenides. The authors concluded that relativistic effects play an essential role in the selenium–carbon coupling mechanism and could result in a contribution of as much as 15–25% of the total values of the ^1J(Se,C). In the overall contribution of the relativistic effects to the total values of ^{1}J(Se,C), the scalar relativistic corrections (negative in sign) by far dominate over the spin–orbit ones (positive in sign), the latter being of less than 5%, as compared to the former (*ca.* 20%). Non-relativistic calculations of coupling constants have been performed at the SOPPA(CC2) level using MP2/3-311G(d,p) geometries; relativistic calculations were carried out within the four-component Dirac's method at the DFT level using the DIRAC code.

Pecul and co-workers[73] have performed the calculations of ^{1}J(C,C) and ^{1}J(C,H) spin–spin coupling constants using DFT methods for a set of aliphatic hydrocarbon derivatives substituted with I, At, Cd, and Hg in order to evaluate the substituent and relativistic effects. The authors attempted to estimate heavy-atom-on-light-atom effects (HALA) on spin–spin coupling constants and to explore the factors influencing the HALA effect on these properties, taking into an account carbon hybridization and the nature of the heavy atom. Several methods were applied, Dirac–Kohn–Sham method (density functional theory with four-component Dirac–Coulomb Hamiltonian), DFT with two- and one-component Zeroth Order Regular Approximation (ZORA) Hamiltonians, and scalar non-relativistic effective core potentials with the non-relativistic Hamiltonian.

Krivdin and co-workers[74] have performed benchmark calculations using a number of second order polarization propagator based wave function methods, SOPPA, SOPPA(CC2) and SOPPA(CCSD), and density functional theory (DFT) based methods, B3LYP, PBE0, KT2, and KT3. One-bond ^{1}J(Si,H) spin–spin couplings in a series of halosilanes SiH_nX_{4-n} (X = F, Cl, Br, I) have been calculated, both at the non-relativistic and full four-parameter Dirac's relativistic levels taking into account vibrational corrections. The authors concluded that at the non-relativistic level, the wave function methods showed much better results as compared with those of DFT, whereas at the DFT level the PBE0 showed the best performance. Relativistic effects and vibrational corrections noticeably improve wave function methods results but worsen DFT results. The calculations of spin–spin couplings were performed applying the aug-cc-pVTZ-J basis set for Si and H, and the 6-311G** basis set for all halogens. Relativistic calculations were performed within the four-component Dirac's method at the DFT-PBE0 level.

Krivdin *et al.*[75] have performed a high-level correlated calculation of spin–spin coupling constants involving tellurium taking into account relativistic effects, vibrational corrections and solvent effects for medium sized organotellurium molecules, tellurophene and divinyl telluride. The ^{2}J(Te,H) and ^{3}J(Te,H) coupling constants were computed using SOPPA, SOPPA(CC2), SOPPA(CCSD), DFT B3LYP, DFT PBE0 and relativistic scalar ZORA methods. All non-relativistic calculations were performed using the aug-cc-pVTZ-J basis set for H atoms involved in coupling, the cc-pVDZ for all uncoupled atoms and av3z-J basis set for Te, developed by the authors. Relativistic corrections to coupling constants were calculated using the ZORA-DFT-B3LYP/TZ2P level. All the calculated values agreed satisfactorily with the experimental data.

Jokisaari and co-workers[76] have explored coupling constants in the van der Waals-bonded ^{129}Xe–^{129}Xe dimer. Because the spin–spin coupling constant J(Xe,Xe) cannot be determined experimentally due to the magnetic equivalence of the two nuclei, the authors computed the anisotropy of the coupling tensor ΔJ, which can be evaluated from experimental data. The calculations of ΔJ(Xe,Xe) as a function of the internuclear distance were performed both at the nonrelativistic level (HF, DFT B3LYP, with the Dalton program) and using fully relativistic four-component methodology implemented in the Dirac programme. In all the calculations, the primitive 27s25p21d1f basis set developed[77] for the hyperfine properties of van der Waals-bonded Xe was applied. The assumption of equilibrium interatomic distance of 4.363 Å allowed to calculate J(Xe,Xe). Depending on the method of calculation, this value ranged from *ca.* 28 to 64 Hz.

6 Various works

Mitoraj and co-workers[78] have explored the physical nature of interactions in Zn(II) complexes with 2,2′-bipyridyl. The work included the calculations of ^{1}J(C,H) using several theory levels (BP86/TZ2P, X3LYP/TZ2P, and PBE0/TZ2P) in order to characterize the CH close contact.

Uhlig and co-workers[79] have synthesized and investigated by ^{1}H, ^{13}C and ^{119}Sn NMR spectroscopy, X-ray diffractions and theoretical methods a series of novel tin chlorides R_nSnCl_{4-n} and respective hydrides R_nSnH_{4-n} displaying a range of substituted phenyl residues, as well as naphthyl moieties (R = Ph, *o*-tolyl, 2,4-xylyl, 2,6-xylyl, *p*-biphenyl, 1-naphthyl, 2-naphthyl). In this work, some $^{1}J(^{13}$C,^{119}Sn) couplings were computed using the MPW1PW91 functional and the Iglo-II basis set.

Ajibade and co-workers[80] have used computational method to gain insight to the correlation between the chemistry of the hydrolysis and the anticancer activities of eight Ru(II)-based complexes. In this work, one bond spin–spin coupling constants between Ru and C, Cl, P and O were calculated at the B3LYP/3-21G theory level.

Aviyente *et al.*[81] have explored by computational methods epoxy-fused cyclopentane derivatives. In this work, the authors computed both ^{1}H NMR chemical shifts and spin–spin coupling constants at the WP04/cc-pVTZ//B3LYP/6-31+G(d) theory level assuming molecules in solution. This method was found to reproduce the best experimental results on epoxy-fused cyclopentane derivatives.

Karaboga *et al.*[82] have obtained and characterised *p*-benzophenoneoxycarbonylphenyl acrylate (Fig. 11) by means of experimental and theoretical methods. In this work the authors computed all ^{1}J(C,H) and

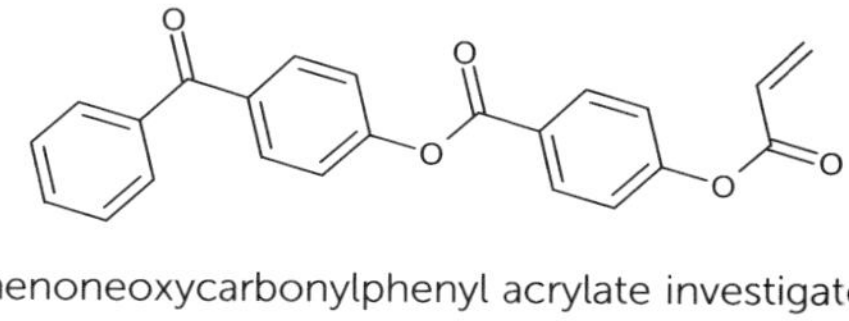

Fig. 11 *p*-Benzophenoneoxycarbonylphenyl acrylate investigated by Karaboga *et al.*

^1J(C,C) coupling constants in this molecule applying the B3LYP/6-311++G(d,p) theory level.

7 Index

7.1 Spin–spin couplings across one or more bonds

C: C,C,[31,55,68,70,73,82] C,F,[38,68] C,H,[30,31,54,55,69,68,70,73,78,82] C,N,[42,54,55] C,O,[51,54,55,57] C,P,[41,42,45,50] C,Ru,[80] C,Sn;[79] **Cl:** Cl,Cl,[46] Cl,F,[46] Cl,P,[45] Cl,P,[44] Cl,Ru,[80] C,Se;[72] **F:** F,F,[68] F,H,[36,58,68] F,N,[37] F,P;[45] **H:** H,H,[28,30–36,39,40,42,59,68,70,81] H,N,[28,54,55] H,O,[54,55,57] H,P,[42] H,Si,[74] H,Te;[75] **N:** N,P;[41,42,45,49,50] **O:** O,O,[51] O,P,[50] O,Ru;[80] **P:** P,Ru,[80] P,S,[50] P,Si.[41]

7.2 Spin–spin couplings across hydrogen bonds and similar bonds including couplings across H, Li and halogen atoms

Across hydrogen bonds (^{nh}J):
C: C,N,[54] C,O;[54] **F:** F,H,[37,58,65] F,N;[58] **H:** H,H;[59,51] H,N,[54,57] H,O;[54,55,57] **N:** N,N,[52,54,56,57] N,O;[54–57] **O:** O,O;[51,55–57] **Xe:** Xe,Xe.[76]
Across lithium bonds (^{nli}J):
C: C,Li;[60] **Li:** Li,N.[60]
Across pnicogen bonds (^{np}J):
C: C,P;[42,45] **Cl:** Cl,P;[45] **F:** F,P;[45] **N:** N,P;[45,49,50] **P:** P,P.[42,43,48,47,50]
Across halogen bonds (^{nX}J):
Cl: Cl,P;[44,46] **F:** F,P.[46]
Through space (^{TS}J):
B:[65]; **F:** F,H,[62] F,N;[64] **P:** P,P;[63] **Se:** Se,Te.[61]

References

1 N. F. Ramsey, *Phys. Rev.*, 1953, **91**, 303.
2 T. Helgaker, M. Jaszuński and M. Pecul, *Prog. Nucl. Magn. Reson. Spectrosc.*, 2008, **53**, 249.
3 J. Autschbach and S. Zheng, *Ann. Rep. NMR Spectr.*, 2009, **67**, 1.
4 J. Vaara, J. Jokisaari, R. E. Wasylishen and D. L. Bryce, *Prog. Nucl. Magn. Reson. Spectrosc.*, 2002, **41**, 233.
5 J. Tomasi, B. Mennuci and R. Cammi, *Chem. Rev.*, 2005, **105**, 1999.
6 J. Jaźwiński, Theoretical Aspects of Spin-Spin Coupling Constants, in *Nuclear Magnetic Resonance*, ed. K. Kamieńska-Trela, Specialist Periodical Reports, The Royal Society of Chemistry, Cambridge, 2011, vol. 40, p. 134.
7 J. Jaźwiński, Theoretical Aspects of Spin-Spin Coupling Constants, in *Nuclear Magnetic Resonance*, ed. K. Kamieńska-Trela and J. Wójcik, Specialist Periodical Reports, The Royal Society of Chemistry, Cambridge, 2012, vol. 41, p. 119.
8 J. Jaźwiński, Theoretical Aspects of Spin-Spin Coupling Constants, in *Nuclear Magnetic Resonance*, ed. K. Kamieńska-Trela and J. Wójcik, Specialist Periodical Reports, The Royal Society of Chemistry, Cambridge, 2013, vol. 42, p. 152.
9 J. Jaźwiński, Theoretical Aspects of Spin-Spin Coupling Constants, in *Nuclear Magnetic Resonance*, ed. K. Kamieńska-Trela and J. Wójcik, Specialist Periodical Reports, The Royal Society of Chemistry, Cambridge, 2014, vol. 43, p. 159.
10 M. J. Frisch, G. W. Trucks, H. B. Schlegel, G. E. Scuseria, M. A. Robb, J. R. Cheeseman, J. A. Montgomery, T. Vreven Jr., K. N. Kudin, J. C. Burant,

J. M. Millam, S. S. Iyengar, J. Tomasi, V. Barone, B. Mennucci, M. Cossi, G. Scalmani, N. Rega, G. A. Petersson, H. Nakatsuji, M. Hada, M. Ehara, K. Toyota, R. Fukuda, J. Hasegawa, M. Ishida, T. Nakajima, Y. Honda, O. Kitao, H. Nakai, M. Klene, X. Li, J. E. Knox, H. P. Hratchian, J. B. Cross, C. Adamo, J. Jaramillo, R. Gomperts, R. E. Stratmann, O. Yazyev, A. J. Austin, R. Cammi, C. Pomelli, J. W. Ochterski, P. Y. Ayala, K. Morokuma, G. A. Voth, P. Salvador, J. J. Dannenberg, V. G. Zakrzewski, S. Dapprich, A. D. Daniels, M. C. Strain, O. Farkas, D. K. Malick, A. D. Rabuck, K. Raghavachari, J. B. Foresman, J. V. Ortiz, Q. Cui, A. G. Baboul, S. Clifford, J. Cioslowski, B. B. Stefanov, G. Liu, A. Liashenko, P. Piskorz, I. Komaromi, R. L. Martin, D. J. Fox, T. Keith, M. A. Al-Laham, C. Y. Peng, A. Nanayakkara, M. Challacombe, P. M. W. Gill, B. Johnson, W. Chen, M. W. Wong, C. Gonzalez and J. A. Pople, *Gaussian 03, Revision B.5*, Gaussian, Inc., Pittsburgh PA, 2003.

11 M. J. Frisch, G. W. Trucks, H. B. Schlegel, G. E. Scuseria, M. A. Robb, J. R. Cheeseman, G. Scalmani, V. Barone, B. Mennucci, G. A. Petersson, H. Nakatsuji, M. Caricato, X. Li, H. P. Hratchian, A. F. Izmaylov, J. Bloino, G. Zheng, J. L. Sonnenberg, M. Hada, M. Ehara, K. Toyota, R. Fukuda, J. Hasegawa, M. Ishida, T. Nakajima, Y. Honda, O. Kitao, H. Nakai, T. Vreven, J. A. Montgomery, Jr., J. E. Peralta, F. Ogliaro, M. Bearpark, J. J. Heyd, E. Brothers, K. N. Kudin, V. N. Staroverov, R. Kobayashi, J. Normand, K. Raghavachari, A. Rendell, J. C. Burant, S. S. Iyengar, J. Tomasi, M. Cossi, N. Rega, J. M. Millam, M. Klene, J. E. Knox, J. B. Cross, V. Bakken, C. Adamo, J. Jaramillo, R. Gomperts, R. E. Stratmann, O. Yazyev, A. J. Austin, R. Cammi, C. Pomelli, J. W. Ochterski, R. L. Martin, K. Morokuma, V. G. Zakrzewski, G. A. Voth, P. Salvador, J. J. Dannenberg, S. Dapprich, A. D. Daniels, Ö. Farkas, J. B. Foresman, J. V. Ortiz, J. Cioslowski and D. J. Fox, *Gaussian 09, Revision A.1*, Gaussian Inc., Wallingford CT, 2009.

12 C. Angeli, K. L. Bak, V. Bakken, O. Christansen, R. Cimiraglia, S. Coriani, D. Dahle, E. K. Dalskov, T. Enevoldsen, B. Fernandez, C. Hättig, K. Hald, A. Halkier, H. Heiberg, T. Hekgaker, H. Hettema, H. J. A. Jensen, D. Jonsson, P. Jørgensen, S. Kirpekar, W. Klopper, R. Kobayashi, H. Koch, A. Ligabue, O. B. Lutnaes, K. V. Mikkelsen, P. Norman, J. Olsen, M. J. Packer, T. B. Pedersen, Z. Rinkevicius, E. Rudberg, T. A. Ruden, K. Ruud, P. Salek, A. Sanchez de Meras, T. Saue, S. P. A. Sauer, B. Schimmelpfennig, K. O. Sylvester-Hvid, P. R. Taylor, O. Vahtras, D. J. Wilson and H. Agren, *Dalton: An Electronic Structure Program*, Relase 2.0, 2005.

13 J. F. Stanton, J. Gauss, J. D. Watts, M. Nooijen, N. Oliphant, S. A. Perera, P. G. Szalaj, W. J. Lauderdale, S. R. Gwaltney, S. Beck, A. Balkova, D. E. Bernhold, K.-K. Baeck, P. Tozyczko, H. Sekino, C. Huber and R. J. Barlett; *ACES II, a program product of the Quantum Theory Project; University of Florida, Gainesville*, FL. Integral package include VMOL (J. Almlof and P. L. Taylor), VPROPS (P. R. Taylor) ABACUS (T. Helgaker, H. J. A. Jensen, P. Jorgensen, J. Olsen and P. R. Taylor). Brillourin-Wigner perturbation theory was implemented by J. Pittner.

14 M. W. Schmidt, K. K. Baldridge, J. A. Boatz, S. T. Elbert, M. S. Gordon, J. H. Jensen, S. Koseki, N. Matsunaga, K. A. Nguyen, S. J. Su, T. L. Windus, M. Dupuis and J. A. Montgomery, *J. Comput. Chem.*, 1993, **14**, 1347.

15 J. F. Stanton, J. Gauss, M. E. Harding, P. G. Szalay with contributions from A. A. Auer, R. J. Bartlett, U. Benedikt, C. Berger, D. E. Bernholdt, Y. J. Bomble, L. Cheng, O. Christiansen, M. Heckert, O. Heun, C. Huber, T.-C. Jagau, D. Jonsson, J. Jusélius, K. Klein, W. J. Lauderdale, D. A. Matthews, T. Metzroth, L. A. Mück, D. P. O'Neill, D. R. Price, E. Prochnow, C. Puzzarini,

K. Ruud, F. Schiffmann, W. Schwalbach, S. Stopkowicz, A. Tajti, J. Vázquez, F. Wang and J. D. Watts and the integral packages MOLECULE (J. Almlöf and P. R. Taylor), PROPS (P. R. Taylor), ABACUS (T. Helgaker, H. J. Aa. Jensen, P. Jørgensen, J. Olsen), and ECP routines by A. V. Mitin, C. van Wüllen, *CFOUR, a quantum chemical program package. For the current version, see* http://www.cfour.de.
16 E. J. Baerends, J. Autschbach, D. Bashford, A. Bérces, F. M. Bickelhaupt, C. Bo, P. M. Boerrigter, L. Cavallo, D. P. Chong, L. Deng, R. M. Dickson, D. E. Ellis, M. van Faassen, L. Fan, T. H. Fischer, C. Fonseca Guerra, A. Ghysels, A. Giammona, S. J. A. van Gisbergen, A. W. Götz, J. A. Groeneveld, O. V. Gritsenko, M. Grüning, F. E. Harris, P. van den Hoek, C. R. Jacob, H. Jacobsen, L. Jensen, G. van Kessel, F. Kootstra, M. V. Krykunov, E. van Lenthe, D. A. McCormack, A. Michalak, M. Mitoraj, J. Neugebauer, V. P. Nicu, L. Noodleman, V. P. Osinga, S. Patchkovskii, P. H. T. Philipsen, D. Post, C. C. Pye, W. Ravenek, J. I. Rodríguez, P. Ros, P. R. T. Schipper, G. Schreckenbach, M. Seth, J. G. Snijders, M. Solà, M. Swart, D. Swerhone, G. te Velde, P. Vernooijs, L. Versluis, L. Visscher, O. Visser, F. Wang, T. A. Wesolowski, E. M. van Wezenbeek, G. Wiesenekker, S. K. Wolff, T. K. Woo, A. L. Yakovlev and T. Ziegler, *ADF2009.01, SCM*, Theoretical Chemistry, Vrije Universiteit, Amsterdam, The Netherlands, http://www.scm.com.
17 R. Bast, H. J. Aa. Jensen, T. Saue and L. Visscher, with contributions from V. Bakken, K. G. Dyall, S. Dubillard, U. Ekström, E. Eliav, T. Enevoldsen, T. Fleig, O. Fossgaard, A. S. P. Gomes, T. Helgaker, J. K. Lærdahl, J. Henriksson, M. Iliaš, Ch. R. Jacob, S. Knecht, C. V. Larsen, H. S. Nataraj, P. Norman, G. Olejniczak, J. Olsen, J. K. Pedersen, M. Pernpointner, K. Ruud, P. Sałek, B. Schimmelpfennig, J. Sikkema, A. J. Thorvaldsen, J. Thyssen, J. van Stralen, S. Villaume, O. Visser, T. Winther and S. Yamamoto, DIRAC, a relativistic ab initio electronic structure program, Release DIRAC11 (2011).
18 R. F. W. Bader, *Chem. Rev.*, 1991, **91**, 893.
19 R. F. W. Bader, *Atoms in Molecules, A Quantum Theory*, Oxford University Press, Oxford, UK, 1990.
20 P. L. A. Popelier, *Atoms In Molecules, An Introduction*, Prentice Hall, Harlow, England, 2000.
21 C. F. Matta and R. J. Boyd, *The Quantum Theory of Atoms in Molecules: From Solid State to DNA and Drug Design*, Wiley-VCH, Weinheim, Germany, 2007.
22 B. Silvi and A. Savin, *Nature*, 1994, **371**, 683.
23 A. E. Reed, L. A. Curtiss and F. Weinhold, *Chem. Rev.*, 1988, **88**, 899.
24 T. A. Keith, AIMAll (version 11.08.23), TK Gristmill Software, Overland Park KS, USA, 2011.
25 E. D. Glendening, J. K. Badenhoop, A. E. Reed, J. E. Carpenter, J. A. Bohmann, C. M. Morales and F. Weinhold, NBO 5.G; University of Wisconsin: Madison, WI, 2004.
26 M. W. Schmidt, K. K. Baldridge, J. A. Boatz, S. T. Elbert, M. S. Gordon, J. H. Jensen, S. Koseki, N. Matsunaga, K. A. Nguyen, S. J. Su, T. L. Windus, M. Dupuis and J. A. Montgomery, Gamess, version 11; Iowa State University: Ames, IA, 2008.
27 S. Noury, X. Krokidis, F. Fuster and B. Silvi, TopMod Package, 1997.
28 B. Wang, X. He and K. M. Merz, *J. Chem. Theory Comput.*, 2013, **9**, 4653.
29 X. He, B. Wang and K. M. Merz, *J. Phys. Chem. B*, 2009, **113**, 10380.
30 J. Oh, J. J. Bowling, Y. Zou, A. G. Chittiboyina, R. J. Doerksen, D. Ferreira, T. D. Leininger and M. T. Hamann, *Biochim. Biophys. Acta*, 2013, **1830**, 4229.
31 D. S. Patel, R. Pendrill, S. S. Mallajosyula, G. Widmalm and A. D. MacKerell, Jr., *J. Phys. Chem. B*, 2014, **118**, 2851.

32 R. A. Cormanich, L. C. Ducati, C. F. Tormena and R. Rittner, *J. Phys. Org. Chem.*, 2013, **26**, 849.
33 R. A. Cormanich, L. C. Ducati, C. F. Tormena and R. Rittner, *Spectrochim. Acta, Part A*, 2014, **123**, 482.
34 C. J. Duarte, R. A. Cormanich, L. C. Ducati and R. Rittner, *J. Mol. Struct.*, 2013, **1050**, 174.
35 C. B. Braga, L. C. Ducati, C. F. Tormena and R. Rittner, *J. Phys. Chem. A*, 2014, **118**, 1748.
36 J. M. Silla, C. J. Duarte, R. A. Cormanich, R. Rittner and M. P. Freitas, *Beilstein J. Org. Chem.*, 2014, **10**, 877.
37 J. M. Silla, W. G. D. P. Silva, R. A. Cormanich, R. Rittner, C. F. Tormena and M. P. Freitas, *J. Phys. Chem. A*, 2014, **118**, 503.
38 J. M. Silla, M. P. Freitas, R. A. Cormanich and R. Rittner, *J. Org. Chem.*, 2014, **79**, 6385.
39 D. C. Solha, T. M. Barbosa, R. V. Viesser, R. Rittner and C. F. Tormena, *J. Phys. Chem. A*, 2014, **118**, 2794.
40 I. L. Rusakova and L. B. Krivdin, *Phys. Chem. Chem. Phys.*, 2013, **15**, 18195.
41 B. Wrackmeyer, C. Köhler and J. Weidinger, *Z. Anorg. Allg. Chem.*, 2014, **640**, 195.
42 D. Sanz, R. M. Claramunt, F. Mathey, I. Alkorta, G. Sánchez-Sanz and J. Elguero, *C. R. Chimie*, 2013, **16**, 937.
43 J. E. Del Bene, I. Alkorta and J. Elguero, *Theor. Chem. Acta*, 2014, **133**, 1464.
44 J. E. Del Bene, I. Alkorta and J. Elguero, *J. Phys. Chem. A*, 2014, **118**, 2360.
45 J. E. Del Bene, I. Alkorta and J. Elguero, *J. Phys. Chem. A*, 2014, **118**, 3386.
46 I. Alkorta, J. Elguero and J. E. Del Bene, *J. Phys. Chem. A*, 2014, **118**, 4222.
47 J. E. Del Bene, I. Alkorta and J. Elguero, *J. Phys. Chem. A*, 2013, **117**, 6893.
48 J. E. Del Bene, I. Alkorta and J. Elguero, *J. Phys. Chem. A*, 2013, **117**, 11592.
49 I. Alkorta, J. Elguero and J. E. Del Bene, *J. Phys. Chem. A*, 2013, **117**, 10497.
50 I. Alkorta, G. Sánchez-San and J. Elguero, *J. Phys. Chem. A*, 2014, **118**, 1527.
51 P. Siuda and J. Sadlej, *Chem. Phys.*, 2014, **433**, 31.
52 A. Křístkova, J. R. Asher, V. G. Malkin and O. L. Malkina, *J. Phys. Chem. A*, 2013, **117**, 9235.
53 D. R. Salahub, R. Foumier, P. Mlynarski, I. Papai, A. St-Amant and J. Ushio, in *Density Functional Methods in Chemistry*, ed. J. K. Labanowski and J. W. Andzelm, Springer, Berlin, 1991, p. 77. Modified version: V. G. Malkin, O. L. Malkina and A. Křístková, *deMon-NMR* program, version 2011.
54 A. M. da Silva, A. Ghosh and P. Chaudhuri, *J. Phys. Chem. A*, 2013, **117**, 10274.
55 H. Roohi and R. Nokhostin, *Struct. Chem.*, 2014, **25**, 483.
56 J. Elguero, I. Alkorta and J. E. Del Bene, *Mol. Phys.*, 2014, **112**, 107.
57 B. Makiabadi, M. Zakarianejad, S. Bagheri, H. R. Masoodi and R. S. Aghaie, *Int. J. Quantum Chem.*, 2013, **113**, 2361.
58 H. R. Masoodi, S. Bagheri, M. Mohammadi, M. Zakarianezhad and B. Makiabadi, *Chem. Phys. Lett.*, 2013, **588**, 36.
59 J. Xu, *Bull. Korean Chem. Soc.*, 2013, **34**, 3265.
60 M. D. Esrafili, P. Fatehi and M. Solimannejad, *J. Mol. Graphics Model.*, 2014, **49**, 129.
61 M. W. Stanford, F. R. Knight, K. S. Athukorala Arachchige, P. Sanz Camacho, S. E. Ashbrook, M. Bühl, A. M. Z. Slawin and J. D. Woollins, *Dalton Trans.*, 2014, **43**, 6548.
62 D. C. Favaro, R. H. Contreras and C. F. Tormena, *J. Phys. Chem. A*, 2013, **117**, 7939.

63 T. Wiegand, H. Eckert, J. Ren, G. Brunklaus, R. Fröhlich, C. G. Daniliuc, G. Lübbe, K. Bussmann, G. Kehr, G. Erker and S. Grimme, *J. Phys. Chem. A*, 2014, **118**, 2316.
64 D. C. Favaro, R. H. Contreras and C. F. Tormena, *Magn. Reson. Chem.*, 2013, **51**, 334.
65 J. M. Silla, R. A. Cormanich, R. Rittner and M. P. Freitas, *Beilstein J. Org. Chem.*, 2013, **9**, 1127.
66 F. A. Perras and D. L. Bryce, *J. Magn. Reson.*, 2014, **242**, 23.
67 A. W. Götz, J. Autschbach and L. Visscher, *J. Chem. Phys.*, 2014, **140**, 104107.
68 F. Nozirov, T. Kupka and M. Stachów, *J. Chem. Phys.*, 2014, **140**, 144303.
69 J. San Fabián, J. M. García de la Vega, R. Suardíaz, M. Fernández-Oliva, C. Pérez, R. Crespo-Oterod and R. H. Contrerase, *Magn. Reson. Chem.*, 2013, **51**, 775.
70 F. V. Toukach and V. P. Ananikov, *Chem. Soc. Rev.*, 2013, **42**, 8376.
71 J. Autschbach, *Philos. Trans. R. Soc., A*, 2014, **372**, 20120489.
72 Y. Yu. Rusakov, I. L. Rusakova and L. B. Krivdin, *Magn. Reson. Chem.*, 2014, **52**, 214.
73 A. Wodyński and M. Pecul, *J. Chem. Phys.*, 2014, **140**, 024319.
74 Y.Y. Rusakov and L. B. Krivdin, *Magn. Reson. Chem.*, 2013, **51**, 557.
75 Y. Yu. Rusakov, L. B. Krivdin, F. F. Østerstrøm, S. P. A. Sauer, V. A. Potapov and S. V. Amosova, *Phys.Chem. Chem. Phys.*, 2013, **15**, 13101.
76 J. Jokisaari and J. Vaara, *Phys. Chem. Chem. Phys.*, 2013, **15**, 11427.
77 J. Vaara, M. Hanni and J. Jokisaari, *J. Chem. Phys.*, 2013, **138**, 104313.
78 I. Cukrowski, J. H. de Lange and M. Mitoraj, *J. Phys. Chem. A*, 2014, **118**, 623.
79 C. Zeppek, J. Pichler, A. Torvisco, M. Flock and F. Uhlig, *J. Organomet. Chem.*, 2013, **740**, 41.
80 A. A. Adeniyi and P. A. Ajibade, *Spectrochim. Acta, Part A*, 2013, **115**, 426.
81 C. Özen, A. Yildirım, N. Ş. Tüzün, P. Haslak, E. Doğan, E. Acar and V. Aviyente, *Mol. Phys.*, 2013, **111**, 3147.
82 F. Karaboga, U. Soykan, M. Dogruer, B. Ozturk, G. Yildirim, S. Cetin and C. Terzioglu, *Spectrochim. Acta, Part A*, 2013, **113**, 80.

Applications of spin–spin couplings

Krystyna Kamieńska-Trela
DOI: 10.1039/9781782622758-00170

In this chapter, which is devoted to applications of spin–spin couplings, the literature published between June 2013 and May 2014 has been reviewed. The data included has been arranged according to (i) the increasing atomic number of the nuclei involved, and (ii) the number of the bonds separating them. The spectrum of the studied compounds in which scalar spin–spin couplings have been applied is quite impressive. It covers small organic compounds, organometallic complexes and minerals as well as very large and complex biological systems. In a number of the reviewed papers measurements of the couplings between quadrupolar nuclei have been reported. One section of this chapter covers the papers devoted to residual dipolar couplings whose role in the constitutional, configurational and conformational analysis of not only large biological systems but also small organic molecules is constantly increasing. In many papers experimental data has been combined with theoretical calculations, which allowed the authors to gain a deeper insight into the electron distribution within a given chemical bond and significantly facilitated the conformational analysis of the studied compounds.

1 Introduction

The material in this chapter covers the period from 1 June 2013 to 31 May 2014. It has been arranged as was done previously,[1] *i.e.* according to (i) the increasing atomic number of the nuclei involved, and (ii) the number of the bonds separating them. We follow the IUPAC[2] recommendations with one notable exception, namely, the nucleus with the smaller mass is given first. For the sake of simplicity the following symbols are used throughout the paper: H for ^{1}H, D – ^{2}H, T – ^{3}H, Li – ^{6}Li, Be – ^{9}Be, B – ^{11}B, C – ^{13}C, N – ^{15}N, O – ^{17}O, F – ^{19}F, Al – ^{27}Al, Si – ^{29}Si, P – ^{31}P, S – ^{33}S, V – ^{51}V, Mn – ^{55}Mn, Fe – ^{57}Fe, Co – ^{59}Co, Cu – ^{65}Cu, As – ^{75}As, Se – ^{77}Se, Br – ^{79}Br, Y – ^{89}Y, Nb – ^{93}Nb, Mo – ^{95}Mo, Ru – ^{99}Ru, Tc – ^{99}Tc, Rh – ^{103}Rh, Ag – ^{109}Ag, Cd – ^{113}Cd, In – $^{113/115}$In, Sn – ^{119}Sn, Sb – ^{121}Sb, Te – ^{125}Te, I – ^{127}I, Cs – ^{133}Cs, W – ^{183}W, Os – ^{187}Os, Pt – ^{195}Pt, Hg – ^{199}Hg, Tl – ^{205}Tl, Pb – ^{207}Pb. All the other isotopes are described explicitly.

In the didactic review Facelli and Ferraro[3] have provided an overall panorama of the potential of using high resolution NMR for structure elucidation. The direct and indirect methods have been discussed that correlate NMR parameters describing the high resolution NMR spectra of liquids and solid systems with their structural parameters. In the part devoted to indirect methods this article has discussed the intrinsic dependence of *J* couplings and shieldings on the atomic coordinates of the atoms in a molecule and its surroundings including chemical bonding, conformation, and stereo configurations.

An extensive review on the recognition and understanding of through-space NMR *J* couplings in small organic, organometallic, and coordination compounds has been published by Hierso.[4]

Institute of Organic Chemistry, Polish Academy of Sciences, Kasprzaka 44/52, Warsaw 01-224, Poland. E-mail: kkt@icho.edu.pl

In a review written by Genet and co-workers[5] on electron-deficient phosphines the authors have briefly discussed the relationship between the σ-donor ability of a phosphine group and the magnitude of $^1J_{PSe}$ in the phosphineselenides. As was established by Allen and Taylor[6] some time ago, an increase in this coupling indicates an increase in the σ character of the phosphorus lone-pair orbital (*i.e.*, a less basic phosphine).

A short account on topological, geometric, and chemical order in materials studied by solid-state NMR has been published by Massiot and co-workers.[7] The authors emphasize that by application of the indirect spin–spin couplings one can characterize a variety of materials, ranging from crystalline compounds to amorphous or glassy materials. In particular couplings across two bonds, $^2J_{Si–O–Si}$ and $^2J_{P–O–P}$, are useful in this respect.

Recent advances and new strategies in the NMR-based identification of natural products have been briefly reviewed by Halabalaki *et al.*[8] It is needless to say that the homo- and heteronuclear, and residual dipolar couplings couplings play a crucial role in these studies.

Tugarinov[9] has reviewed the utility of deuteration in protein NMR. This includes the effects of deuteration on scalar couplings in proteins such as the differential couplings in glycine methylene groups serving as predictors of backbone geometry in the left- and right-handed helical regions of the Ramachandran map. However, the coupling data of this type are very scarce.

Recent advances in computational predictions of NMR parameters for the structure elucidation of carbohydrates have been reviewed by Toukach and Ananikov.[10] The review covers, among others, the prediction of H–H, H–P, H–C and C–C couplings essential for structural studies.

Widmalm[11] in a short review based mostly on the results of his laboratory has presented a perspective on structures of carbohydrates. He has underlined the central position of the solution state NMR spectroscopy in these studies starting with analysis of the primary structure of glycans (components and sequence), followed by conformational and dynamics analysis to the study of interaction with proteins. He has also anticipated that among other NMR parameters DFT calculated vicinal proton–proton, proton–carbon and carbon–carbon scalar couplings will aid these investigations.

A brief review devoted to new NMR methods for characterizing the dynamic conformations and interactions of oligosaccharides including indirect spin–spin couplings and residual dipolar couplings has been published by Kato and co-workers.[12]

Shapiro[13] has reviewed in detail studies of domain dynamics in biomacromolecules with NMR spectroscopy, paying special attention to residual dipolar couplings. Since RDCs are limited to a report on the ps–ms motions, different implementations of these couplings have been discussed in combination with other NMR data such as PCSs or spin relaxation, as well as in combination with the data obtained from other techniques such as SAXS/WAXS, and SANS. Different approaches for averaging the data calculated for the molecular ensembles and for the quantitative description of biomacromolecule motion have been also outlined.

Residual dipolar couplings that reflect the mutual orientation of individual chemical bonds or larger structural fragments are very important parameters carrying information about the transient states populated by intrinsically disordered proteins. In two reviews, by Sklenář and coworkers[14] and by Konrat,[15] the development and recently proposed NMR-based strategies including RDCs to characterize transient states have been discussed.

In a comprehensive review Gautier[16] has highlighted the potential of NMR techniques in contributing to the understanding of membrane proteins. ^{1}D and ^{2}D couplings have been included among other NMR parameters.

Paramagnetically induced residual dipolar couplings, the paramagnetic relaxation enhancements and pseudocontact shifts are the most important paramagnetic effects that originate from paramagnetic self-alignment. They provide long-range structural restraints for orienting and positioning individual proteins within a protein–protein complex. For this purpose a large variety of paramagnetic metal binding tags has been introduced. Hass and Ubbink[17] have reviewed the recent developments in structural characterization of protein–protein complexes using anisotropic paramagnetic NMR restraints obtained with such tags.

Vuister *et al.*[18] have overviewed nine tools recently used for the validation of protein NMR structures. Two of them, CING (V1.0) and VIVALDI (V1), use RDCs in restraint analysis.

In a short review Pomin[19] has presented the results of structural studies of glycosaminoglycans families including chondroitin sulfate, dermatan, heparin, heparan sulfate, and hyaluronic acid. He has pointed out the pivotal contribution of NMR parameters, particularly scalar and residual dipolar couplings, to the progress in elucidation of the structural, dynamical, conformational and intermolecular binding aspects of the carbohydrates.

A concise review on NMR of glycans has been published by Battistel *et al.*[20] The authors have discussed the structural information obtained from traditional NMR techniques, such as chemical shifts, NOEs/ROEs, and couplings, as well as that extracted from residual dipolar couplings (RDCs). They have also surveyed advantages of using ^{13}C isotope labelling that allows gathering one-bond $^{13}C–^{13}C$ couplings with a recently improved constant-time COSY technique, in addition to the commonly measured $^{1}H–^{13}C$ RDCs.

A review of novel NMR methods devoted to measurements of heteronuclear long-range proton–carbon couplings, $^{n}J_{HC}$ $(n>1)$ in small molecules has been written by Parella and Espinosa.[21] The experiments have been arranged according to terms of NMR pulse scheme and cross-peak nature. The authors have thoroughly discussed problems connected with the sign determination and measurement of very small coupling values involving protonated and non-protonated carbons.

A review on applications of Fourier Transform to processing of nonuniformly (sparsely) data sets devoted to the reconstruction of high-resolution multidimensional NMR spectra has been written by Kazimierczuk *et al.*[22] The authors indicate that FT provides the possibility to acquire NMR spectra of ultra-high dimensionality and/or resolution which allow easy resonance assignment and precise determination of spectral parameters, *e.g.*, couplings. One section of this review has been devoted to determination of spin–spin couplings in proteins.

2 New methods

A number of broadband homonuclear decoupled CLIP/CLAP-RESET experiments for the simplified measurement of heteronuclear one-bond couplings with highest resolution to low natural abundance nuclei like ^{13}C and ^{15}N has been designed by Reinsperger and Luy.[23] The experiments are based on an isotope-selective BIRD filter element and use a recently reported improved version of Zangger–Sterk data chunking. The concatenated FIDs result in multiplets in which most homonuclear splittings are removed while the heteronuclear one-bond couplings are retained. Couplings can be extracted in an IPAP fashion without scaling of subspectra by the use of optimized coherence transfer elements like the COB-INEPT. The authors indicate that the proposed pulse sequences have high potential to perform a reliable (semi-)automated measurement of one-bond couplings in isotropic and partially aligned samples.

Broadband proton-decoupled CLIP/CLAP-HSQC experiments for the accurate determination of one-bond heteronuclear couplings and, by extension, for the reliable measurement of small residual dipolar couplings have been also reported by Kövér and co-workers.[24] The combination of an isotope-selective BIRD(d) filter module with a nonselective ^{1}H inversion pulse has been employed by the authors to refocus proton–proton coupling evolution prior to the acquisition of brief chunks of free induction decay that are subsequently assembled to reconstruct the fully-decoupled signal evolution. As a result, the cross-peaks obtained are split only by the heteronuclear one-bond coupling along the F2 dimension, allowing couplings to be extracted by measuring simple frequency differences between singlet maxima. As model compounds to test the experiments, ^{13}C-labelled [C-1]-methyl-α,β-D-glucopyranoside, tetrasodium-(1-methyl-2,3,4-tri-*O*-sulfonato-6-deoxy-6-*C*-sulfonatomethyl-α-D-glucopyranoside) and D-sucrose have been used.

A new technique, named BASHD-*J*-resolved-HMBC devoted to measurements of $^{n}J_{HC}$ and $^{n}J_{HH}$ in spin systems containing a methine group directly bonded to a methyl group, such as, *e.g.* –CH*a*–CH*b*(CH_3)–CH*c*– has been designed by Furihata and Tashiro.[25] Band selective homo decoupled pulse and *J*-scaling pulse into HMBC have been incorporated into this method. High resolution cross peaks can be observed along the F1 axis by *J*-scaling pulse, and band selective homo decoupled pulse simplified multiplet signals.

A number of papers devoted to extraction of small long-range proton-couplings from the spectra of natural products have been published by Parella and co-workers.[26–30] The effects of phase modulation due to homonuclear proton–proton couplings in HSQMBC-IPAP and

HMBC-IPAP experiments have been estimated by Saurí and Parella.[26] The authors have demonstrated that accurate values of small proton–carbon couplings, $^{n}J_{HC}$, can be extracted even for phase-distorted cross-peaks obtained from a selHSQMBC experiment applied simultaneously on two mutually *J*-coupled protons. They have also shown that HMBC-COSY experiments could be an excellent complement to HMBC for the measurement of small $^{n}J_{HC}$ values. Another set of experiments denoted as CLIP-HSQMBC designed by this group of authors[27] allows the user very easy, direct and accurate measurement of long-range proton–carbon couplings in organic molecules and natural products. The *J* value can be extracted directly from the analysis of resolved in-phase ^{1}H multiplets that show an additional splitting arising from the proton–carbon coupling. In cases of unresolved peaks, a simple fitting analysis using the internal satellite lines as a reference is performed. Addition of a spinlock period results in a CLIP-HSQMBC-TOCSY experiment that is suitable for the measurement of very small coupling values or to observe correlations from overlapped resonances. An NMR method to enhance the sensitivity and resolution in band-selective long-range heteronuclear correlation spectra has been proposed by Castañar *et al.*[28,30] In particular, it has been shown by the authors that the ^{1}H-*HO*modecoupled *B*and-*S*elective (HOBS) HSQMBC experiment represents a new way to measure heteronuclear couplings from the simplified in-phase doublets generated along the detected dimension. It has been shown by the authors by the use of cyclosporine as the model compound that peptides are excellent targets in this case because NH, Hα and other aliphatic protons resonate in characteristic regions of the ^{1}H spectrum, and there is usually no *J* interference between them. A spin-state-selective method has been proposed by Saurí *et al.*[29] which allows one the efficient measurement of both $^{1}J_{HAC}$ and $^{1}J_{HBC}$ couplings in a diastereotopic CHAHB methylene spin system. Additionally, the magnitude and the sign of the geminal $^{2}J_{HAHB}$ coupling as well as $^{1}J_{HC}$ values for CH and CH_3 multiplicities can be extracted from the same spectrum. As model compounds strychnine, 5-methylene-2-norbornene and progesterone have been used by the authors. It should also added that the designed method allows determination of residual dipolar $^{1}D_{HC}$ and $^{2}D_{HH}$ couplings in a small molecule weakly aligned in a PMMA swollen gel.

Several papers devoted to determination of magnitudes and relative signs of ^{1}H–^{19}F couplings in fluorinated organic compounds have been published by Espinosa and co-workers.[31,32] In one of them[31] a novel methodology based on 1D- and 2D TOCSY experiments has been described. The magnitude of the ^{1}H–^{19}F coupling is measured from the displacement between the relayed peaks associated with the α or β spin state of the fluorine, and its relative sign was derived from the sense of the displacement. A new NMR approach based on selective 1D HSQC experiments for the fast determination of small proton–fluorine couplings has been proposed by Espinosa and Broughton.[32] The authors have taken into account advantage of the large one-bond carbon–fluorine couplings to selectively excite the component of the carbon signal of a CF group associated with a particular spin state of the fluorine. Comparison of the $^{19}F(\alpha)$ and $^{19}F(\beta)$ proton sub-spectra has allowed them a simple and accurate measurement of the magnitude and relative sign of the proton–fluorine couplings. The signs of these small couplings could be determined unambiguously since $^{1}J_{CF}$ couplings taken as the reference are always negative. The H–F couplings measured for three model compounds are in good agreement with those DFT calculated including their signs.

The effectiveness of hetero-COSY, HETCOR, HMQC, and HSQC two-dimensional NMR pulse sequences for detection of ^{19}F–^{1}H correlations by scalar coupling has been evaluated by Marchione *et al.*[33] by the use of monofluorinated and polyfluorinated test compounds; these were 2-fluoro-4-methyl-1-(1-methylethyl)cyclohexane (fluoromenthol), and 1,1,1,2,2,3,4,5,5,5-decafluoropentane. All four sequences were amenable, to some degree, to adjustment to observe larger or smaller couplings preferentially. It is worth mentioning that J_{HF} couplings in fluoromenthol vary from about 0 Hz ($^{4}J_{HF}$) up to about 50 Hz ($^{2}J_{HF}$).

The RES-TOCSY experiment for accurate determination of very small hidden within line widths heteronuclear $^{n}J_{HF}$ ($n = 3$–6) has been reported by Suryaprakash and co-workers.[34] The coupling values and their signs have been determined from the displacement of cross sections of the 2D spectrum and from the slopes of their displacement vectors, respectively. In order to demonstrate the usefulness of the designed technique 2-fluoropyridine,1,1,1-trifluoromethylheptan-2-ol,2-fluoroacetanilide and epiflurohydrin have been applied as model compounds. The proposed pulse sequence does not require the probe containing the fluorine channel for detection and can be routinely employed for the direct extraction of magnitudes and signs of $^{n}J_{HF}$ couplings in all the fluorine containing molecules.

Merlet and co-workers[35] have proposed a new 2D gradient encoded SERF experiment (GET-SERF) for the measurement of couplings between a selected fluorine site and all the proton nuclei of the molecule. The usefulness of the method has been demonstrated for measurement of all the long and short range H–F scalar and residual dipolar couplings of diflunisal, a difluorinated anti-inflammatory drug, in isotropic and anisotropic media.

A set of modified HSQC experiments designed for the study of ^{13}C-enriched small molecules has been published by Jeannerat and co-workers.[36] It includes an improved sensitivity-enhanced HSQC experiment eliminating signal artifacts because of high-order ^{13}C magnetization terms generated at high ^{13}C enrichment. A broadband homonuclear ^{13}C decoupling sequence based on Zangger and Sterk's method simplifies the complex ^{13}C–^{13}C multiplet structure in the F1 dimension of HSQC. When recording spectra at high resolution, the combination with a multiple-site modulation of the selective pulse outperforms the constant time HSQC in terms of sensitivity and reliability. Finally, two pulse sequences reintroducing selected J_{CC} couplings with selective pulses facilitate their assignments and measurements either in the splitting of the resulting doublets or by modulation of the signal amplitude. In order to demonstrate the applicability of the designed experiments the authors used a sample of uniformly 92% ^{13}C-enriched cholesterol for which all carbon–carbon couplings have been extracted including long-range ones.

The performance of homonuclear coupling sign edited (HCSE) experiment applied to detection of signed carbon–carbon couplings has been discussed by Blechta and Schraml.[37] The authors used a set of already measured samples of nine monosubstituted benzenes and have shown that coupling sign detection is insensitive to the settings of carbon–carbon polarization transfer delays. They have also compared the results of HCSE and SLAP experiments and indicated that the two methods may complement each other in detection of signed carbon–carbon couplings.

It has been demonstrated by Thureau *et al.*[38] that homonuclear scalar *J* couplings in the solid-state can be measured by the application of the principle of frequency-selective spin-echoes. Singly or doubly frequency selective pulses have been used to generate either a *J*-modulated experiment (S) or a reference experiment (S0). The combination of these two distinct experiments provides experimental data that, in favourable cases, are insensitive to incoherent relaxation effects, and which can be used to estimate homonuclear *J*-couplings in multiple spin-systems. As a model compound uniformly ^{13}C and ^{15}N labelled sample of L-histidine has been used for which $^{n}J_{CC}$ ($n = 1$–3) and J_{NN} couplings have been precisely measured: $^{1}J_{CC}$ of 74.4 ± 0.7, $^{2}J_{CC}$ of 6.0 ± 0.1, $^{3}J_{CC}$ of 3.9 ± 0.1 and $^{2}J_{NN}$ of 0.9 ± 0.07 Hz. The obtained for the solid state coupling values are in excellent agreement with those reported in solution. Additionally, the authors have emphasized that in future work it will be most probably possible to measure also the signs of the couplings by performing the same experiment as that one used by them, but at a spinning-angle slightly different than the magic-angle.

Several F–F and F–C 2D correlation experiments, which were already known or which were modified for the use with fluorine detection by the authors have been presented by Aspers *et al.*[39] It included ^{19}F–^{13}C *J*-HMBC, ^{19}F–^{13}C XLOC, ^{19}F–^{13}C HMQC and ^{19}F–^{19}F COSY-10 experiments which have been applied for the determination of the size and sign of the $^{n}J_{CF}$– and $^{n}J_{FF}$ couplings in 2-bromotetrafluoroethyl trifluorovinylether used as a model compound for perfluorinated compounds. It has been indicated by the authors that due to the large variation in the size of the $^{n}J_{CF}$ and $^{n}J_{FF}$ couplings, scaling of the couplings proved to be an important feature.

Widmalm and co-workers[40] have modified the *J*-HMBC experiment for measuring heteronuclear long-range couplings by suppressing the interfering $^{1}J_{CC}$ couplings in site-specifically ^{13}C labelled oligosaccharides. The modified experiment increases the number of conformationally important couplings that can be obtained from a single preparation of a molecule.

Colell *et al.*[41,42] have continued their theoretical and experimental studies on the phenomena in low-field ^{1}H NMR spectroscopy by the use of parahydrogen induced polarization (PHIP). The authors have demonstrated that low-field PHIP experiments in the inverse weak coupling regime allow one to determine precisely $\delta\nu$ chemical shifts and J couplings. They have proved experimentally that observed splittings are related to $\delta\nu$ in a nonlinear way. As a model compound 1-(*tert*-butyldiphenylsilyl)-2-(ethoxy)ethene with natural abundance of ^{13}C and ^{29}Si isotopes has been chosen. PHIP-enhanced NMR in the millitesla regime allowed the authors to measure NMR parameters for this molecule in a single scan including homo- and heteronuclear couplings.

Farjon and co-workers[43] have introduced the sensitivity optimized ^{1}H^{N}-X SOFAST-HMBC technique, which is a phase sensitive echo/anti-echo HMBC with selective pulses applied to ^{1}H. The technique allows the natural abundance measurement of $^{2h}J_{NH\cdots OC}$ in peptides with values lower than 1 Hz.

Sattler and co-workers[44] have presented the BEST-selective long-range HNN-COSY experiment (BESTsellr HNN-COSY) with selective ^{15}N inversion and refocusing pulses. The method yields an overall eightfold signal enhancement, thus enabling the direct detection of H-bonds for weak and dynamic base pairs in nucleic acids and the measurement of the corresponding $^{2h}J_{NN}$.

Bax and co-workers[45] have applied a band-selective homonuclear (BASH) ^{1}H decoupling pulses during acquisition of the ^{1}H FID. This efficient procedure removes scalar and residual dipolar couplings between protons. When applied in both dimensions of a homonuclear 2D NOESY experiment it enhances spectral resolution in the HN–Hα region of spectra of peptides and proteins. Application of BASH for removal of RDCs between HN and aliphatic protons in weakly aligned proteins increases resolution in the ^{1}H–^{15}N HSQC spectrum and allows measurement of RDCs in samples that are relatively strongly aligned. The approach has been demonstrated for measurement of RDCs in protonated ^{15}N/^{13}C-enriched ubiquitin, aligned in Pf1, yielding improved fitting to the ubiquitin structure.

Tolman and co-workers[46] have developed a double resonance technique called *S*elective *I*nversion by *S*ingle *T*ransition *C*ross *P*olarization (SIST-CP). They have applied this technique for the specific inversion of amide N–H spin pairs allowing the selective activation and measurement of ^{1}H$_{N}$–^{13}C′ long-range RDCs in strongly aligned proteins.

Perras and Bryce[47] have developed J-resolved solid state NMR experiments designed for the use with half-integer quadrupolar nuclei. These experiments provide simple doublets, greatly simplifying the analysis of the NMR data. The usefulness of the method has been demonstrated using a ^{55}Mn ($S = 5/2$) J-resolved experiment on powdered dimanganese decacarbonyl making it possible to obtain $^{1}J_{MnMn} = 113.0 \pm 0.5$ Hz with sub-hertz precision.

3 One-bond couplings to hydrogen

Optimized shifting and/or scaling factors for calculating $^{1}J_{HC}$ couplings have been calculated by San Fabián *et al.*[48] for 35 combinations of representative functionals (PBE, B3LYP, B3P86, B97-2 and M06-L) and basis sets (TZVP, HIIIsu3, EPR-III, aug-cc-pVTZ-J, ccJ-pVDZ, ccJ-pVTZ, ccJ-pVQZ, pcJ-2 and pcJ-3) using 68 organic molecular systems with 88 $^{1}J_{HC}$ couplings including different types of hybridized carbon atoms. The calculated $^{1}J_{HC}$ values have been compared with the experimental ones. The statistical analysis has been carried out over five sets of data: (a) the whole data set of couplings (88 values), (b) the couplings involving sp^3 hybridized carbon atoms (30 values), (c) the couplings with sp^2 hybridized carbon atoms (27 values), (d) couplings involving sp hybridized carbon atoms (10 values), and (d) coupling values that involve aromatic carbon atoms (21 values). The authors indicate that the use of shifting constants for obtaining the calculated coupling substantially improves the results, and most models become qualitatively similar.

One-bond H–C-couplings in tetramethyl species EMe_4 (E = C, Si, Ge, Sn, Pb) measured in vapour and liquid states have been reported by Nazarski and Makulski.[49] Such gas-phase $J_{0,HC}$s, obtained as J values extrapolated to a zero-pressure limit, have been found to be only slightly different from the J_{HC}s measured for liquids. The former data show linear dependence on the density of the gaseous solvents used. The authors have also demonstrated that current DFT calculations allow one a successful prediction of the experimental NMR J-couplings and other molecular properties as well as their interpretation in terms of electronic and geometric structure. In particular, LDBS calculations of J_{HC}s, carried out with an IGLO-II basis sets for the C and H atoms and def2-QZVPPD for the E atoms afforded J values in good agreement with the experiment.

The dependence of the carbon shielding and the $^{1}J_{HC}$ coupling on the C–H bond distance and the H–C–H angle over the range of distances and angles has been calculated by Dračínský and Hodgkinson[50] by the use of the path integral molecular dynamics (PIMD) simulations approach. The averaged NMR parameters have been determined by convolution of the bond distance and valence angle distributions obtained from PIMD with the distance and angle dependence of the shift and coupling. The deuterium isotope effect calculations based on the PIMD probability distributions performed by the authors for dichloromethane have yielded the calculated data very close to the experimental values for both the carbon chemical shift and the $^{1}J_{HC}$ coupling.

The anharmonic contribution to the C–H group when the O–H group of 1,1,1,3,3,3-hexafluoro-2-propanol is engaged into an intermolecular H-bond with the amines has been studied by Pal *et al.*[51] by the use of NMR spectroscopy and quantum mechanical calculations. It has been established by the authors that involving the C2–OH group in hydrogen bond formation (O–H as hydrogen donor) results in weakening the geminal C2–H bond, which is reflected in the $^{1}J_{HC2}$ coupling, and the equilibrium distances of the C–H bond.

Variable-temperature NMR studies have been performed by Dunlop-Brière *et al.*[52] for $[Cp_2TiCH_2CHMe(SiMe_3)][B(C_6F_5)_4]$ complex yielding the relevant chemical shift data, and a set of $^{1}J_{HC}$ and $^{3}J_{HH}$ couplings. An analysis of these results provided unambiguous proof that this compound exists in the form of two equilibrating β- and γ-agostic isomers with the latter prevailing. Additionally, the performed EXSY measurements have shown that the β-H atom in this molecule undergoes facile exchange with the three hydrogen atoms of the β-methyl group (β-H/γ-H exchange) but not with the two α-H atoms.

Hansen and Kay[53] have presented NMR experiments and a simple strategy for studies of conformationally excited states of proteins based on measurement of histidine $^{13}C_{\gamma}$, $^{13}C_{\delta 2}$ chemical shifts and $^{1}J_{H\varepsilon C\varepsilon}$ couplings. The methodology has been used to obtain pK_a values and tautomer distributions for histidine residues of an invisible on-pathway folding intermediate of the colicin E7 immunity protein.

A full set of 40 $^{1}J_{HC}$ and 40 $^{1}D_{HC}$ has been measured by Hutin *et al.*[54] for a three-layer stack aggregate of a linear porphyrin tetramer. The NMR and small-angle X-ray scattering data have allowed to elucidate the solution-phase structure of this well-defined aggregate.

$^{1}J_{HC}$ coupling in the CHC(O) fragment of 2,2-dichloro-*N*-cyclohexyl-*N*-methylacetamide and 2-chloro-*N*,*N*-dicyclohexylacetamide has been measured in various solvents by Santos *et al.*[55] in their studies of conformational equilibria in these compounds. However, the observed solvent effects have been rather meagre, of 1 Hz only.

A decrease of the calculated $^{1}J_{HC}$ coupling values has been observed by Cukrowski *et al.*[56] from 177.06 Hz in ZnL to 173.87 Hz in ZnL_3 complex (L = L = bpy, 2,2′-bipyridyl) during their theoretical studies on the factors determining the stability of ZnL, ZL_2 and ZnL_3 complexes. The result has been interpreted by the authors in terms of charge depletion from σ(C–H) bonds.

$^{1}J_{HN}$ couplings of 69.3 and 18.0 Hz to the ^{15}N of ring A and ^{15}N of ring B, respectively, have been observed by Szymański *et al.*[57] in the spectrum of the ^{15}N-enriched isotopomer of tris(pentafluorophenyl)corrole measured at 201 K. According to the authors, this result is consistent with the supposition that even at such low temperature the parent corrole occurs in the form of a rapid dynamic equilibrium of two tautomers; in one of them the proton is located on atom NA, and in the other on atom NB.

It has been demonstrated by Xiang and Zweckstetter[58] that $^{1}J_{HN}$ scalar couplings and hydrogen/deuterium exchange can be a valuable source of information on a transient intramolecular hydrogen bond in the folding nucleus of the coiled-coil structure of the GCN4 leucine zipper. Their studies have shown that generally $^{1}J_{HN}$s are a sensitive tool for the detection of intramolecular hydrogen bonds in challenging biomolecules where the effective/useable concentration is low. This includes peptides at natural abundance but also uniformly labelled biomolecules that might be limited to low concentrations because of precipitation or aggregation.

The exchange rates of the NH_3^+ amino protons and the labile $NH_{\varepsilon 2}$ and $NH_{\delta 1}$ protons of the imidazole ring by indirect detection through nitrogen-15 as a function of temperature ($272\ K < T < 293\ K$) and pH ($1.3 < pH < 4.9$) of uniformly nitrogen-15- and carbon-13-labelled l-histidine · HCl · H_2O have been determined by Sehgal *et al.*[59] Exchange rates up to $8.5 \times 10^4\ s^{-1}$ could be determined (*i.e.*, lifetimes as short as 12 ms). The three chemical shifts δHi of the invisible exchanging protons Hi and the three one-bond scalar couplings $^{1}J_{HiN}$ could also be determined accurately. The obtained coupling values are: $^{1}J_{H^{3+}N} = 73.2 \pm 0.2$ Hz, $^{1}J_{H\varepsilon 2N} = 98.5 \pm 0.3$ Hz, and $^{1}J_{H\delta 1N} = 97.2 \pm 1.0$ Hz.

Two types of ammonium cations in ammonium 12-phosphotungstate hydrate, $(NH_4)_3PW_{12}O_{40} \cdot nH_2O$, have been detected by Chernyak *et al.*[60] by the use of ^{1}H and ^{15}N MAS NMR spectroscopy; one type of ammonium cations is in the bulk of crystallites, while the other type is on their surface. $^1J_{H14N}$ of 57 Hz and $^1J_{H15N}$ of 78 Hz have been measured for the first type of species and $^1J_{H14N}$ of 51 Hz and $^1J_{H15N}$ of 75 Hz for the latter one.

Deuterium isotope effects on ^{13}C chemical shifts have been investigated by Pietrzak and co-workers[61] in anions of 1,8-bis(4-toluene-sulphonamido)naphthalenes together with *N,N*-(naphthalene-1,8-diyl)-bis(2,2,2-trifluoracetamide) all with bis(1,8-dimethylamino)napthaleneH$^+$ as counterion. These compounds represent both "static" and equilibrium cases. NMR assignments of the former have been revised. Two of the studied compounds, *N,N*-(naphthalene-1,8-diyl)bis(2,2,2-trifluoracetamide) and 2,4,5,7-tetranitro-1,8-bis(4-toluenesulphonamido)naphthalene, are tautomeric as judged from the one bond H–N coupling reported by Klimkiewicz *et al.*[62]

A series of organosilicon amides, 1-[2,6-bis(di-iso-propyl)phenyl]-2,2-diphenyl-1-aza-2-silaindol, 1-[2,6-bis(di-iso-propyl)phenyl]-2-phenyl-2-hydrido-1-aza-2-silaindol, 1-*tert*-butyl-2,2-diphenyl-1-aza-2-silaindol, and 1-*tert*-butyl-2-phenyl-2-hydrido-1-aza-2-silaindol has been synthesized by Novák *et al.*[63] All these compounds have been fully characterized by the use of ^{1}H, ^{13}C and ^{29}Si NMR; for two of them *i.e.* for 1-*tert*-butyl-2-phenyl-2-hydrido-1-aza-silole and for 1-[2,6-bis(diisopropyl)phenyl]-2-phenyl-2-hydrido-1-aza-silole $^1J_{HSi}$ couplings of 209 and 212 Hz, respectively, have been measured providing clear evidence for the existence of one directly bonded proton to silicon.

Si–H and Si–C bond cleavage reactions of silane and phenylsilanes with $Mo(PMe_3)_6$ leading to silyl, hypervalent silyl, silane, and disilane complexes have been studied by Zuzek and Parkin.[64] Evidence that one of the obtained compounds, $Mo(PMe_3)_4(SiH_3)_2H_2$, is a silyl-hydride and not a silane complex has been provided by the observation of distinct quintet signals in the ^{1}H NMR spectrum at δ −4.80 and 4.02 in a 1:3 ratio, of which the former has a value of $^2J_{HP} = 26$ Hz and the latter $^3J_{HP} = 8$ Hz. In accord with the silyl-hydride assignment, the signal attributable to the SiH_3 groups exhibits coupling to silicon ($^1J_{HSi} = 157$ Hz).

The reaction between the coordinatively unsaturated phosphorus dication $[((Ph_3P)_2C)(i\text{-}Pr_2N)P]^{2+}$ and water has been carried out by Tay *et al.*[65] yielding the addition product $[((Ph_3P)C)(iPr_2N)P(H)OH]^{2+}$. Its structure has been confirmed by the observation of the $^1J_{HP}$ coupling of 642 Hz, which is the value indicating the oxidation of the central phosphorus from P^{III} to P^{V}. A similar result was obtained when methanol instead of water was used; $^1J_{HP}$ of 590 Hz and $^3J_{HP} = 14.5$ Hz (for the OMe fragment) couplings have been found for the corresponding addition product, $[((Ph_3P)C)(i\text{-}Pr_2N)P(H)OMe]^{2+}$. On the basis of these results the authors have concluded that the synthesized by them dication is the first non-metallic compound capable of facile oxidative addition of equilmolar amounts of water at room temperature and in the absence of any additives (*e.g.*, boranes).

A large $^1J_{HP}$ coupling of 357 Hz observed by Sues *et al.*[66] in the $^{31}P\{^1H\}$ and ^1H NMR spectra of the $[RuCp^*(Ph_2PCHCHPPh_2)(HPPh_2)][HCl_2]$ complex has provided evidence that an exposition of the parent complex, $RuCp^*(Ph_2PCHCHPPh_2)(PPh_2)$, to acid under moisture-free conditions has resulted in the protonation of the latter at the phosphide centre. An additional proof has been obtained by the observation of two-bond H–P coupling to the other phosphorus nuclei, $^2J_{HP} = 6$ Hz.

The cleavage of the N–H bonds of ammonia, alkylamines, and arylamines by oxidative addition to a geometrically distorted tricoordinate phosphorus compound, which displays an unusual planar T-shaped geometry at phosphorus with the tridentate O,N,O-binding motif occupying three adjacent coplanar sites, has been studied by McCarthy *et al.*[67] According to the authors, it represents a rare well-characterized example of intermolecular N–H addition to a σ^3-phosphorus compound; for the addition product of ammonia the large $^1J_{HP}$ coupling of 812 Hz has been observed. Its provides clear evidence that the hydrogen is directly attached to the phosphorous atom. The $^1J_{HP}$ couplings of similar values have been found for the addition products of alkyl and aryl amines.

One-bond coupling $^1J_{HY}$ of 24.5 Hz has been observed by Schädle *et al.*[68] in the spectra of two adduct complexes $[Cp^*_2Y(m\text{-}OOCArMe)(m\text{-}H)AlR_2]$ ($Cp^* = C_5Me_5$, $ArMe = C_6H_2Me_3$-2,4,6, $R = CH_2SiMe_3$ or Me). The observed coupling between the hydrido ligand and the yttrium metal centre is somewhat smaller than that reported for half-metallocene yttrium hydrides, such as $[\{(C_5Me_4R)YH_2\}n]$ (thf)x ($R = SiMe_3$, $n = 5$, $x = 0$, $^1J_{HY} = 35.8$ Hz; R = Me, $n = 5$, $x = 2$, $^1J_{HY} = 32.8$ Hz)[69] and $[\{(1,3\text{-}Me_2C_5H_3)_2YH\ (thf)\}_2]$ ($^1J_{HY} = 28$ Hz),[70] and significantly smaller than in other monomeric metallocene-yttriumhydride complexes, such as $[Cp^*_2YH\ (thf)]$ ($^1J_{HY} = 82$ Hz)[71] and $[(C_5Me_4SiMe_3)_2YH\ (thf)]$ ($^1J_{HY} = 74.8$ Hz).[72]

A bis(diphenylphosphino)methane-ligated trinuclear silver hydride nanocluster, $[Ag_3((Ph_2P)_2CH_2)_3(\mu^3\text{-}H)](BF_4)_2$, featuring three silver(I) ions coordinated to a μ^3-hydride, has been isolated and structurally characterized by Zavras *et al.*[73] by the use of various spectroscopic techniques including NMR spectroscopy. Simulation of the μ^3-hydride resonance with one-bond $^1J_{HAg}$ couplings of 86 and 75 Hz for the coupling of the hydride with ^{109}Ag and ^{107}Ag, respectively, and with a two-bond $^2J_{HP}$ coupling of 21 Hz for the coupling of the hydride with the six equivalent phosphorus nuclei has yielded a simulated resonance that matches the resonance at 4.75 ppm in the experimental ^1H NMR spectrum of this compound. The simulated pattern is the summation of the individual statistical contributions of the $^{109}Ag_3$, $^{109}Ag_2{}^{107}Ag$, $^{109}Ag^{107}Ag_2$, and $^{107}Ag_3$ isotopomers.

The structure of a stable Sn(II) dihydride adduct $Ph_3PCMe_2 \cdot SnH_2 \cdot W(CO)_5$ synthesized by Rivard and co-workers[74] by the use of the ylidic Wittig ligand $Ph_3P{=}CMe_2$ has been confirmed unequivocally by the presence of $^1J_{H119Sn}$ and $^1J_{H117Sn}$ couplings in its ^1H NMR spectrum, 1030 Hz and 991 Hz, respectively; $^1J_{H119Sn}$ coupling of 1030 Hz has been also observed in the ^{119}Sn NMR spectrum of the compound.

DFT calculations performed for $HIrCl_2(PMe_3)_2$ by Garbacz *et al.*[75] by the use of the ZORA relativistic method with the TZ2P basis set have predicted the values of 112 and 122 Hz for $^1J_{H191Ir}$ and $^1J_{H193Ir}$ couplings, respectively. However, no indication of spin–spin coupling to either ^{191}Ir ($I = 3/2$, $\Xi = 1.718\%$, natural abundance 37.3%) or ^{193}Ir ($I = 3/2$, $\Xi = 1.871\%$, natural abundance 62.7%) could be observed experimentally. This has been interpreted by the authors in terms of self-decoupling of the Ir nuclei due to a large nuclear quadrupolar coupling and the resultant short iridium T1.

The largest ever reported $^1J_{HPt}$ coupling of 2669 Hz has been observed by Rivada-Wheelaghan *et al.*[76] in the ^{1}H NMR spectrum of the 14-electron Pt(II) complex, [PtH(IPr)$_2$][BAr^F], where IPr = 1,3-bis(2,6-diisopropylphenyl)imidazol-2-ylidene. Low-temperature NMR experiments performed for this compound have not provided evidence of agostic interactions in CD_2Cl_2 solutions. Instead, two species have been detected in the ^{1}H NMR spectrum at −50 °C in *ca.* 3 : 2 ratio, both revealing signals in the hydride region, one at −28.50 ppm with $^1J_{HPt} = 1902$ Hz and the other at −34.19 ppm with $^1J_{HPt} = 2712$ Hz. The large coupling to Pt has been assigned to [PtH(IPr)$_2$][BAr^F], and the smaller one to its CD_2Cl_2 adduct, [PtH(IPr)$_2$][BAr^F]·CD_2Cl_2.

4 One-bond couplings not involving hydrogen

A mixed 2 : 2 aggregate of lithiated (*S*)-*N*-ethyl-3-methyl-1-(triisopropylsilyloxy)-butan-2-amine has been synthesized by Williard and co-workers[77] from (*S*)-valinol and cyclopentyllithium and its structure confirmed by X-ray and NMR spectroscopy. The aggregate adopts a ladder structure in the solid state, and the ladder-type mixed 2 : 2 aggregate is also the major species in a toluene-d$_8$ solution containing an approximately 1 : 1 molar ratio of the lithiated chiral amide to cyclopentyllithium. This has been confirmed by the observation of the quintet at 24.7 ppm corresponding to the methine carbon of c-pentylLi with $^1J_{LiC} = 10.3$ Hz, the result consistent with the carbon atom interacting with two ^{6}Li atoms. Both the multiplicity and coupling value comply with the Bauer–Winchester–Schleyer rule. The authors have also reported that identical dimeric, laddertype, mixed aggregates are the major species in toluene-d$_8$ solution for all of the different alkyllithium reagents such as *n*-butyllithium, *sec*-butyllithium and iso-propyllithium.

Bryce and co-workers have continued their experimental and theoretical studies[47,78,79] on one-bond boron–boron couplings. They have shown that the application of a double-quantum filter is necessary to observe the *J* splittings and that, under these conditions, only a simple doublet is expected.[47] Interestingly, the splitting is amplified if the spins are magnetically equivalent, which has made it possible to measure *J* couplings very precisely and unambiguously prove the symmetry of the molecule. The results of ^{11}B DQF-*J*-resolved solid-state NMR experiments performed on bis(pinacolato)diboron, bis(catecholato)diboron, and its NHC complexed analogue, yielded $^1J_{BB} = 120 \pm 2$, 136 ± 1 and 106.8 ± 0.4 Hz, respectively. The *J* coupling obtained for bis(catecholato)diboron using the DQF-*J*-resolved experiment is an order of magnitude more precise than that obtained using DOR NMR and does not require specialized hardware. The DQF method has been furthermore applied to measure the $^{1}J_{BB}$ couplings for a series of eight diboro variously substituted compounds.[78]

The obtained ^{1}J values have varied from 98 Hz in tetrakis(pyrrolidino) diborane up to 136 Hz in bis(catecholato)diboron, and they have been found to correlate well with the orbital energy of the B–B s-bonding natural bond orbital as well as the hybridisation states of the boron atoms in the bond. An increase in the p character of the bond by electron-donating ligands or *via* the formation of a sp^2–sp^3 diboron compound weakens the bond, increases the bond length, and decreases the $^{1}J_{BB}$ couplings. The theory describing homonuclear indirect nuclear spin–spin coupling (J) interactions between pairs of quadrupolar nuclei has been outlined and supported by numerical calculations in a separate paper.[79]

$^{1}J_{BP}$ couplings of about 50 Hz extracted from ^{11}B observed 2D J-resolved spectra have been reported by Liedtke *et al.*[80] for two new examples of the vicinal frustrated Lewis pair (FLP) compounds, $Ph_2P(Me_3Si)$-$C{=}C(CH_2CH_2Pmes_2)B(C_6F_5)_2$ and $Ph_2P(Me_3Si)C{=}C(CH_2CH_2P(O)mes_2)$-$B(C_6F_5)_2$; a $^{2}J_{BP}$ value of 27.7 Hz has been found for the NO cycloaddition product to $Ph_2P(Me_3Si)C{=}C(CH_2CH_2Pmes_2)B(C_6F_5)_2$.

One-bond CC couplings, $^{1}J_{CC}$, and $^{n}J_{HC}$ couplings across one, two and three bonds have been measured by Rittner and co-workers[81] for twelve 3-monosubstituted 2-methylpropenes, and the experimental values have been compared with those calculated theoretically, generally good agreement being observed between these two sets of data. No influence of substituents on $^{1}J_{CC}$'s has been observed, whereas the $^{1}J_{HcC3}$ values (where Hc is attached to the C3 carbon bearing the substituent) have exhibited good correlations with Taft's sigma (σ_I inductive parameter) and the Swain–Lupton Field Effect (F) ($R = 0.957$).

Regioselective syntheses of $[^{13}C]_4$-labelled sodium 1-carboxy-2-(2-ethylhexyloxycarbonyl)ethanesulfonate and sodium 2-carboxy-1-(2-ethylhexyloxycarbonyl)ethanesulfonate from $[^{13}C]_4$-maleic anhydride have been performed by Blakemore and co-workers,[82] and the structures of the obtained compounds have been confirmed by analysis of the INADEQUATE spectrum and $^{1}J_{CC}$ couplings, respectively.

Coupled-cluster calculations (CCSD/cc-pVDZ and CCSD/cc-pVQZ//CCSD/cc-pVDZ) of the parent cyclobutadiene and fifteen-substituted analogues have been performed by Pittman *et al.*;[83] for cyclobutadiene itself and its tetramethyl, tetraboryl and tetrafluoroderivatives one-bond carbon–carbon and proton–carbon couplings have been calculated and used for estimation of the percent p-character in ring C-to-C and C-to-substituent bonds.

Almost a full set of $^{1}J_{CC}$ couplings across single and triple bonds has been determined by Štefko *et al.*[84] for a series of donor–acceptor substituted alkynes in which the length of the π-conjugated polyyne spacer between the *N*,*N*-diisopropylanilino donor and the 1,1,4,4-tetracyanobuta-1,3-diene (TCBD) acceptor was systematically varied.

It has been demonstrated by Afonin *et al.*[85] that in ^{1}H and ^{13}C NMR spectra of *N*-substituted dimethylketimines chemical shifts of protons and carbon atoms of the methyl groups in the *cis*-position with respect to the unshared electron pair of the nitrogen are larger than those of the CH_3 groups in the *trans*-position by 0.2–0.4 and 8–11 ppm respectively. This effect is accompanied by the reduction of the corresponding spin–spin CC coupling across one bond by 10 Hz. The experimental trends in the variation of the spectral parameters are well reproduced by *ab initio* quantum-chemical calculations. The established stereochemical dependence of the chemical shifts of ^{1}H and ^{13}C can be a useful tool for the configuration assignment in various compounds with a C=N bond.

An analysis of the 2D-INADEQUATE NMR data performed by Yamada *et al.*[86] for ^{13}C enriched Sc2@C66 fullerene has shown that this compound contains a C2v(4059)-C66 cage consisting of two sets of unsaturated linear triquinanes (ULTs), in which three pentagons abut one another and two scandium ions are located within the folds of each of the ULT units. Conclusive structural elucidation has been accomplished by single crystal XRD.

$^{1}J_{C14N} = 17$ Hz and $^{1}J_{111Cd14N} = 92$ Hz couplings have been determined by Saidi *et al.*[87] for the unique N-bonded thiocyanate ligand in new polymeric hybrid organic–inorganic material catenapoly[(bis-glycinium-k^2 O:O)-di-μ-thiocyanate- k^2 N:S; k^2 S:N- cadmium (II) by the use of ^{111}Cd and ^{13}C MAS NMR spectroscopy.

Two hydrogen atom free 3,3′-bi(1,2,4-oxadiazole) derivatives, 5,5′-bis(trinitromethyl)-3,3′-bi(1,2,4-oxadiazole) and 5,5′-bis(fluorodinitromethyl)-3,3′-bi(1,2,4-oxadiazole) have been synthesized by Klapötke and co-workers[88] and characterised by multinuclear magnetic spectroscopy; for both compounds sets of spin–spin couplings have been reported. Thus, the $C(NO_2)_3$ groups of 5,5′-bis(trinitromethyl)-3,3′-bi(1,2,4-oxadiazole) have shown a septet at 117.5 ppm due to $^{1}J_{CN}$ coupling of 9.4 Hz, while in the spectrum of 5,5′-bis(fluorodinitromethyl)-3,3′-bi(1,2,4-oxadiazole) a doublet of quintets at 111.9 ppm with couplings of $^{1}J_{CF} = 299$ Hz and $^{1}J_{CN} = 9.6$ Hz has been observed.

An effect similar to the well-known reverse Perlin effect has been observed by Silla *et al.*[89] on the $^{1}J_{CF}$ couplings of α- and β-D-glucopyranosyl fluoride tetracetate, both in nonpolar and polar solution; $^{1}J_{CFax}$ of −229.6 Hz and $^{1}J_{CFeq}$ of −220.7 Hz have been found in cyclohexane solution. This has been called by the authors "reverse fluorine Perlin-like effect", and shown to be ruled by dipolar interactions rather than by hyperconjugation. The reverse fluorine Perlin-like effect does not have a general relationship with the anomeric effect, and it can be useful to determine the structure and stereochemistry of organofluorine compounds.

The computational study of the one-bond carbon–silicon couplings has been performed at the second-order polarization propagator approximation (SOPPA) level by Rusakova *et al.*[90] in a series of 60 various silanes and compared with the experimental $^{1}J_{CSi}$ values taken from the literature. A special attention has been paid by the authors to the main factors affecting the accuracy of the calculation including the level of theory, the quality of the basis set, and the contribution of solvent and relativistic effects. The reported in the paper values have been obtained at the SOPPA(CC2) level, whereas for the calculations of solvent and relativistic effects the SOPPA(CC2) DFT-PBE0 set has been used. The calculated solvent and relativistic corrections are in most cases of *ca.* 1 Hz and in a few cases only are considerably larger than that.

The chemical behaviour of a prototype boryl(phosphino)carbene has been studied by Lavigne *et al.*[91] Among others, thermal rearrangement of this compound yielded the corresponding alkene whose structure has been corroborated by the presence of $^{1}J_{CP}$ coupling of 61.2 Hz typical for phosphaalkenes. Another compound, phosphorylalkene, has been obtained from the reaction of the carbene with benzaldehyde; its structure has been confirmed by the presence of $^{1}J_{CP} = 25.6$ Hz and $^{2}J_{CP} = 2.4$ Hz couplings, corresponding to the ethylenic carbon atoms and a large coupling value $^{3}J_{HP} = 44.5$ Hz, which is in good agreement with a vinylic proton in a *trans* position to the phosphorus atom.

The ^{13}C NMR spectra of two phosphinoamides synthesized by Gholivand and Dorosti[92] exhibit obvious differences between their $^{1}J_{CP}$ couplings, 128.3 Hz in *N*-diphenylphosphino-*N'*-phenylurea *vs.* 439.2 Hz in *N*(diphenylphosphino)-2-pyrazinecarboxamide. A natural bond orbital analysis was performed by the authors to clarify the electronic behaviour of these molecules.

Extensive multinuclear NMR spectroscopic and X-ray crystallographic studies of electronic and steric effects of phosphonous acid ligands, $W(CO)_5PMe_3$, $W(CO)_5PEt_3$, $W(CO)_5P\text{-}iPr_3$, $W(CO)_5PPh_3$, $W(CO)_5P(OMe)_3$, $W(CO)_5P(OPh)_3$, $W(CO)_5P(NMe_2)_3$, $W(CO)_5PPh(NEt_2)_2$ and $W(CO)_5PF_3$, and their chlorophosphite ligand precursors in a series of tungsten pentacarbonyl complexes have been performed by Gray and co-workers.[93] This included measurements and analysis of $^{1}J_{CP}$ and $^{1}J_{PW}$ couplings; the influence of electronegativity of substituents has been observed on the latter.

Four-component relativistic calculations of $^{1}J_{CSe}$ couplings have been performed by Rusakov *et al.*[94] for a series of selenium heterocycles and their parent open-chain selenides, and the calculated data has been compared with the experimental values published in the literature. It has been found that relativistic effects play an essential role in the selenium–carbon coupling mechanism and could result in a contribution of as much as 15–25% of the total values of the one-bond selenium–carbon spin–spin couplings. In the overall contribution of the relativistic effects to the total values of $^{1}J_{CSe}$, the scalar relativistic corrections (negative in sign) by far dominate over the spin–orbit ones (positive in sign), the latter being of less than 5%, as compared to the former (*ca.* 20%). Solvent effects on the $^{1}J_{CSe}$ values calculated within the polarizable continuum model for the solvents with different dielectric constants (ε 2.2–78.4) are almost negligible, decreasing negative $^{1}J_{CSe}$ in absolute value by only about 1 Hz.

^{13}C, ^{31}P, and ^{77}Se NMR spectroscopic data has been obtained by Wrackmeyer and co-workers[95] for a large series of 4,5-[1,2-dicarba-*closo*-dodecaborano-(12)]-1,3,2-diselenaphospholanes which included 2-(1-methylethyl)-, 2-cyclohexyl-, 2-(1,1-dimethylethyl)-, 2-[3,5-dimethylphenylmethyl]-, 2-phenyl-, 2-ethoxy-, 2-diethylamino-, 2-dichlorophosphinomethyl- and 2-(2-dichlorophosphinoethyl) derivatives; for all these compounds a set of couplings including those across one C–Se and P–Se bonds has been reported.

Continuing their studies based on measurements of the nuclear spin relaxation rates running *via* the SC2 mechanism (scalar relaxation of the second kind), Gryff-Keller and co-workers[96] have presented the results obtained for three bromo compounds: $CBrCl_3$, $(CH_3)_3CBr$, and CBr_4. A careful separation of saturation-recovery curves, measured for the signals of ^{13}C nuclei at 7.05 and 11.7 T on two components, has provided the longitudinal SC2 relaxation rates of carbon signals in ^{79}Br and ^{81}Br containing isotopomers of the investigated compounds. This data has enabled the authors to experimentally determine carbon–bromine couplings and relaxation rates of quadrupole bromine nuclei, both observables being hardly accessible by direct measurements. The obtained for $(CH_3)_3CBr$, $CBrCl_3$ and CBr_4 absolute experimental $^1J_{CBr}$ values are 32, 170 and 150 Hz, respectively. The theoretical calculations have indicated that all three couplings are of negative sign.

$^1J_{C109Ag}=184$ Hz and $^1J_{C107Ag}=159$ Hz to the α carbon have been observed by Pranckevicius and Stephan[97] in the spectrum of the $Li[Ag(IDippBF_3)_2]$ complex, where IDipp-BF_3 = 1,3-(2,6-iPr$_2C_6H_3)_2$-2-$BF_3(C_3HN_2)$-4-yl).

Two papers devoted to the compounds in which Si/S and Sn/S stabilizing contacts are observed in the solid state have been published by Flores-Parra and co-workers.[98,99] One series of the compounds consists of 2-R'_3E (E = Si or Sn; R′ = Me or Ph) derivatives of 5-R-[1,3,5]-dithiazinanes (R = Me, iso-Pr, *tert*-Bu), as well as some of their N–BH_3 adducts,[98] the other contains 23 new bis-, tris-, tetra-(dithiazinan-2-yl)silanes and stannanes.[99] $^1J_{CSn}$ couplings with the methyl carbon in the compounds of both series containing the Me_3Sn fragment are considerably larger than the corresponding coupling in Me_3Sn-cyclohexane, 351.2–370.5 Hz and 299.4 Hz, respectively. According to the authors, this result corroborates the presence of the Sn···S coordination bonds also in solution.

$^1J_{CSn}$ couplings between 600 and 700 Hz have been measured by Chagas *et al.*[100] for a series of organotin(IV) derivatives of bis(pyrimidin-2ylthio)methane (*ptm*), 1,2-bis(pyrimidin-2-ylthio)ethane (*pte*) and 1,6-bis(pyrimidin-2-ylthio)hexane (*pth*). The observed coupling values have been invoked by the authors as evidence that the following compounds, *i.e.* [$Sn_2Cl_2Ph_6$(*ptm*)], [$Sn_2Cl_2Ph_6$(*pte*)], [$Sn_2Cl_2Ph_6$(*pth*)] and [$SnClPh_3$(*pth*)] are five-coordinated.

$^1J_{CPb}$ couplings ranging from 735 to 1049 Hz have been measured by Kawamura *et al.*[101,102] for substituted plumbacyclopentadienylidene, $(Ph)_2(t\text{-}BuMe_2\text{-}Si)_2C_4PbL_2$, stabilized by various Lewis bases (L = tetrahydrofuran, pyridine, N-heterocyclic carbene), and the experimental data has been compared with that calculated by the use of the ZORA-DFT methods; the corresponding values are −1106.2 and −1164.9 Hz, respectively. The authors indicate that the relativistic correction is about 60% of the total calculated *J* value.

A large set of spectral parameters: chemical shifts, as well as homo- and heteronuclear couplings has been determined by Kołodziej *et al.*[103] for five Schiff bases derived from L-arginine (L-Arg) and 2-hydroxy carbonyl compounds providing crucial information on their structure. Among others, rather rare one-bond couplings $^1J_{NCd}$ have been reported for two cadmium complexes synthesized from L-Arg, cadmium acetate hydrate and two aldehydes: salicylaldehyde (Cd(salArg)), $^1J_{NCd} = 173$ Hz, and 5-chlorosalicylaldehyde (Cd(5ClsalArg)), $^1J_{NCd} = 164$ Hz. Additionally, for both complexes $^3J_{CCd}$ couplings have been measured.

$^1J_{FP}$ couplings varying from 722 to 908 Hz have been measured by Allefeld *et al.*[104] for a series of zwitterionic, hexacoordinated phosphates $[F_5P(CH_2NMe_2\text{–}CH_2NMe_2)]$, $[(C_2F_5)F_4P(CH_2NMe_2\text{–}CH_2NMe_2)]$, $[(C_2F_5)_2F_3P(CH_2NMe_2\text{–}CH_2NMe_2)]$ and $[(C_2F_5)_3F_2P(CH_2NMe_2\text{–}CH_2NMe_2)]$ providing useful information on the geometric arrangement of substituents around the phosphorous atom.

A series of functional phosphates $[P(C_2F_5)_3F_2X]^-$ (X = OH, OPh, OEt, OAc, OCH_2CF_3 and OC_2H_4OH) has been synthesized by Bader *et al.*[105] from the phosphorane adduct $[P(C_2F_5)_3F_2(dmap)]$ (dmap = 4-(dimethylamino)pyridine) and characterized extensively by the use of NMR spectroscopy. It has been found that the $^1J_{FP}$ coupling decreases significantly going from $[P(C_2F_5)_3F_2(dmap)]$ to $[P(C_2F_5)_3F_2OH][PPh_4]$ from $J = 986$ to 846 Hz.

The ^{19}F isotropic chemical shifts (δ_{iso}) of two isomorphic compounds, NbF_5 and TaF_5, which involve six nonequivalent fluorine sites, have been experimentally determined by Biswal *et al.*[106] from the reconstruction of 1D ^{19}F MAS NMR spectra. In parallel, the corresponding ^{19}F chemical shielding tensors have been calculated using the GIPAW method for both experimental and DFT-optimized structures. Additionally, for three of the ^{19}F NMR lines of NbF_5, distorted multiplets, arising from ^1J-coupling and residual dipolar coupling between the ^{19}F and ^{93}Nb nuclei, have been simulated yielding $^1J_{FNb}$-coupling of 350, 355 and 350 Hz for the corresponding fluorine sites, respectively.

$^1J_{SiP}$ couplings of 186.4, 131.8 and 116.4 Hz have been observed by Hansen *et al.*[107] for the labile, zwitterionic "halfparent" phosphasilene L′Si=PH (L′=CH[(C=CH$_2$)CMe(N-2,6-i-Pr$_2$C$_6$H$_3$)$_2$] and its two stable 4-dimethylaminopyridine and 1,3,4,5-tetramethylimidazol-2-ylidene adducts, respęctively. Interestingly, the $^1J_{HP}$ couplings determined for these compounds are almost identical, 143.0, 144.1 and 144.1 Hz, respectively.

$^1J_{PP}$ couplings ranging from 172.8 to 225.0 Hz have been measured by Hey-Hawkins and co-workers[108] for a series of *trans*, *trans*-4,5-dicarba-*closo*-dodecaboranyl-1,2,3-triphospholanes, new phosphorus-containing, five-membered P,P,P heterocycles; for the P,N,P compound, a 1-aza-2,5-diphospholane $^2J_{HP}$ coupling of 18.5 Hz has been determined.

The preparation and comprehensive characterization of new small Group 15 element cages, the first diaza-tetraphospha-[2.1.1]bicyclohexane ($[Ter_2N_2P_2\text{-}(P_2Ph_2)]$) and tetraaza-diphospha [2.1.1]bicyclohexane ($[Ter_2N_2P_2(N_2Ph_2)]$) where Ter = 2,6-dimesitylphenyl, have been reported by Hinz *et al.*[109] Interestingly, chemical shifts and the $^1J_{PP}$ (negative) and $^2J_{PP}$ (positive) couplings measured for $[Ter_2N_2P_2\text{-}(P_2Ph_2)]$ compare well with those determined for the related $(PhP)_4N(cHex)$.

$^1J_{PP}$ coupling of 303 Hz has been reported by Kilian and co-workers[110] for phosphino-phosphonium chloride $[Acenap(PiPr_2)(PPh)]^+Cl^-$

(Acenap = acenaphtene) and only insignificantly smaller, $^1J_{PP}$ of 250.4 Hz has been observed in [Acenap(PiPr$_2$)(PPh)Mo(CO)$_4$Cl] complex.[111] This result has been used as evidence that the one-bond P–P bond has been retained upon the complex formation.

A set of the following organophosphorus(III)–tellurium heterocycles has been synthesized and characterized by Nordheider *et al.*:[112] Te$_m$(P^{III}R)$_n$ (m = 1: n = 2, R = OMes* (Mes* = supermesityl or 2,4,6-tri-*tert*-butylphenyl); n = 3, R = adamantyl (Ad); n = 4, R = ferrocene (Fc); m = n = 3: R = trityl (Trt), Mes or by the installation of a P$^V{}_2$N$_2$ anchor in RP^{III}[TeP^V-(*t*BuN)(m-N*t*Bu)]$_2$ (R = Ad, *t*Bu). The chemical shift and coupling values including $^1J_{PP}$ and $^1J_{PTe}$ couplings have been determined by iterative fitting of the NMR spectra of the studied compounds.

The $^1J_{PSe}$ couplings have been measured by Tohmé *et al.*[113] for five new selenophosphines bearing organometallic Fe(κ^2-dppe)(η^5-C$_5$Me$_5$)C≡C– substituents, [Fe(κ^2-dppe)(η^5-C$_5$Me$_5$)(C≡C-1,4-C$_6$H$_4$)]P(Se)Ph$_2$, [Fe(κ^2-dppe)(η^5-C$_5$Me$_5$)(C≡C-1,3-C$_6$H$_4$)]P(Se)Ph$_2$, [Fe(κ^2-dppe)(η^5-C$_5$Me$_5$)(C≡C-1,4-C$_6$H$_4$)]$_3$P(Se), [Fe(κ^2-dppe)(η^5-C$_5$Me$_5$)(C≡C-1,3-C$_6$H$_4$)]$_3$P(Se) and [Fe(κ^2-dppe)(η^5-C$_5$Me$_5$)(C≡C)]P(Se)Ph$_2$, with a goal to determine electronic parameters of the parent metallo-phosphines. DFT calculations performed by the authors have indicated that $^1J_{PSe}$'s can be applied for this purpose in spite of the existence of the low-lying excited states in selenophosphines.

One-bond spin–spin couplings involving selenium of seven different types, $^1J_{XSe}$, X = ^{1}H, ^{13}C, ^{15}N, ^{19}F, ^{29}Si, ^{31}P, and ^{77}Se, have been calculated by Rusakova *et al.*[114] in a series of 14 representative compounds at the SOPPA(CCSD) level taking into account relativistic corrections evaluated both at the RPA and DFT levels of theory in comparison with experiment. Relativistic corrections were found to play a major role in the calculation of $^1J_{XSe}$ reaching as much as almost 170% of the total value of $^1J_{SeSe}$ and up to 60–70% for the rest of $^1J_{XSe}$. The most 'relativistic' $^1J_{SeSe}$ demonstrates a marked Karplus-type dihedral angle dependence with respect to the mutual orientation of the selenium lone pairs providing a powerful tool for stereochemical analysis of selenoorganic compounds.

$^1J_{PSe}$ couplings of about 730 Hz have been determined by Schaarschmidt *et al.*[115] for the parent 2-diphenylphosphoroselenoyl-(2-(*E*)-phenylvinyl)ferrocene and its 4-chlorophenyl, 4-methoxyphenyl, 4-methylphenyl, 4-cyanophenyl and 4-acetylphenyl derivatives. The coupling values do not vary significantly within the series and are very close to that reported for Ph$_3$P=Se.[116] This led the authors to the conclusion that the electronic influence of substituents is limited to the ferrocene and hence no interaction with the phosphorous atom could be observed. This opens the possibility to modify the electronic characteristics of both donor functionalities (alkene and phosphine) independently from each other.

The σ-donor properties of a series of the bicyclic tertiary phosphines known as phobanes have been compared using the $^1J_{PSe}$ values for the PhobP(=Se)Bu derivatives.[117] The compounds studied were the four isomers of 9-butylphosphabicyclo[3.3.1]nonane, *s*-PhobPBu, where Bu = *n*-butyl, *sec*-butyl, isobutyl, *tert*-butyl, seven isomers of

9-butylphosphabicyclo[4.2.1]nonane (α_5-PhobPBu, where Bu = *n*-butyl, *sec*-butyl, isobutyl, *tert*-butyl and α_7-PhobPBu, where Bu = *n*-butyl, isobutyl, *tert*-butyl). The *trans* structure of the subsequently obtained platinum complexes of a general structure, [$PtCl_2(PhobPR)_2$] was assigned on the basis of the values of $^1J_{PPt}$, which lie in the range of 2200–2600 Hz, typical of *trans*-[$PtCl_2(PR_3)_2$].

The $^1J_{PRh}$ coupling of 119.6 Hz measured by Comanescu and Iluc[118] in [LRh(nbd)][BF_4] complex (L = $iPr_2PC_6H_4(CH_2)_2C_6H_4P$-$_i$$Pr_2$ and nbd = norbornadiene) has been found to be in agreement with previously reported values for wide-angle diphosphine ligands.

$^1J_{PAg}$ couplings determined by Arenas *et al.*[119] for two silver complexes, [PhB{CH_2PPh_2}$_3$]Ag(PPh_3) and [PhB{CH_2P(*p*-$CF_3C_6H_4)_2$}$_3$]Ag(PPh_3), have been of a considerable help in their structure elucidation. Thus, $^1J_{P109Ag}$ = 397.4 Hz and $^1J_{P107Ag}$ = 338.6 Hz have been measured for the PPh_3 ligand and $^1J_{P109Ag}$ = 215 Hz and $^1J_{P107Ag}$ = 192.3 Hz have been determined for the coordinated PPh_2 moiety of [PhB{CH_2PPh_2}$_3$]Ag(PPh_3). Similar coupling values have been found for [PhB{CH_2P(*p*-$CF_3C_6H_4)_2$}$_3$]Ag(PPh_3).

Two mononuclear copper(I) complexes [Cu{Co(P_2C_2 *t*$Pent_2)_2$}($PPh_3)_2$] and [Cu{Co(P_2C_2 *t*$Bu_2)_2$}($PPh_3)_2$], and three novel silver(I) complexes [Ag{Co(P_2C_2*t*$Pent_2)_2$}($PMe_3)_3$], [Ag{Co(P_2C_2*t*$Pent_2)_2$}($PMe_3)_2$], and [Ag_2{Co(P_2C_2*t*$Pent_2)_2$}($PMe_3)_5$]SbF_6 have been synthesized by Malberg *et al.*[120] and analyzed by variable-temperature ^{1}H, ^{31}P{^{1}H}, and ^{13}C{^{1}H} NMR spectroscopy. Characteristic ^{31}P chemical shifts and indirect spin–spin couplings allowed the authors a clear distinction between different bonding patterns realized in this class of compounds. Thus, for example, in the ^{31}P{^{1}H} NMR spectrum of the bis(trimethylphosphane) complex, [Ag{Co(P_2C_2*t*$Pent_2)_2$}($PMe_3)_2$], measured at 193 K in [D_8]THF $^1J_{P109Ag}$ = 377.1 and $^1J_{P107Ag}$ = 328.2 Hz couplings have been determined for the coordinated Ag($PMe_3)_2$ fragment and $^1J_{P107/109Ag}$ = 336 Hz for the phosphorus atom (P1) of the [Co-(P_2C_2*t*$Pent_2)_2$]$^-$ moiety coordinated to silver. Similar $^1J_{PAg}$ coupling values have been extracted for the Ag($PMe_3)_2$ unity from the solid state MAS NMR spectrum, $^1J_{P109Ag}$ = 360 and $^1J_{P107Ag}$ = 313 Hz. $^1J_{P63/65Cu}$ = 1085 Hz (average value) and a residual dipolar coupling $d(^{31}P,^{63/65}Cu)$ = 100 Hz determined at 9.4 T have been estimated for the [Cu{Co(P_2C_2 *t*$Bu_2)_2$}($PPh_3)_2$] complex from the analysis of its ^{31}P{^{1}H} NMR MAS spectrum.

$^1J_{P\ 115In}$ of 250 Hz has been estimated by Cummins *et al.*[121] for the indium derivative $(XL)_2InP(SnPh_3)_2$ (XL = $S(CH_2)_2NMe_2$) by the use of CP MAS spectroscopy; $^1J_{P117/119Sn}$ of 993 Hz has been observed for this compound in solution. Interestingly, $^1J_{P115In}$ = 361 Hz and $^1J_{P117/119Sn}$ = 808 Hz coupling values have been predicted by the ADF-ZORA calculations. A series of the dinitrosyl and nitrito complexes have been synthesized by Alvarez *et al.*[122] as the result of reactions between the unsaturated ditungsten complexes [$W_2Cp_2(\mu$-$PPh_2)_2(CO)_x$] (x = 1, 2) and nitric oxide, and for all these compounds $^1J_{PW}$ couplings ranging from 224 to 381 Hz have been determined; for example, $^1J_{PW}$ of 337 and 342 Hz, have been found for *cis*- and *trans*-dinitrosyl [$W_2Cp_2(\mu$-$PPh_2)_2(NO)_2$] complexes, respectively. The $^1J_{PPt}$ of 3688 Hz for *cis*-[$PtCl_2$(Hdppba)$_2$] and $^1J_{PRh}$ of −128 Hz for *trans*-[RhCl(CO)(Hdppba)$_2$] complexes, (Hdppba = 4-(diphenylphosphanyl)benzoic acid), have been measured by Golecki and Kersting.[123] The couplings of the same magnitude have been found for novel pentanuclear *cis*-[$(Zn_2L)_2$(μ-dppba)$_2$PtX$_2$]$(ClO_4)_2$ and *trans*-[$(Zn_2L)_2$(μ-dppba)$_2$RhCl(CO)] $(ClO_4)_2$ complexes (L = macrocyclic ligand and X = Cl, Br, I) synthetized from *cis*-[PtCl$_2$(Hdppba)$_2$] and *trans*-[RhCl(CO)(Hdppba)$_2$], respectively, providing evidence that the *cis* and *trans* stereochemistry of the mononuclear starting materials has been maintained in the pentanuclear products.

A series of papers devoted to synthesis and structural studies of a variety of platinum(II) complexes has been published by Nabavizadeh *et al.*[124–127] In all cases analysis of one-bond phosphorous–platinum couplings and relevant long-range couplings played a crucial role. This included two new binuclear cycloplatinated(II) complexes, of general formula [Pt$_2$Me$_2$(C^N)$_2$(*m*-dppac)] (C^N = deprotonated 2-phenylpyridine (ppy) and/or deprotonated benzo{h}quinoline (bhq); dppac = 1,10-bis(diphenylphosphino)acetylene);[124] the observed $^1J_{PPt}$ coupling values of 2049 and 2104 Hz, respectively strongly suggest that dppac is acting as a spacer ligand between the two PtMe(ppy) and/or PtMe(bhq) moieties and each P atom is *trans* arranged to coordinating C atom of the phenyl ring of the ligand. In the case of *trans* arrangement to the nitrogen atom considerably larger $^1J_{PPt}$ couplings would be observed.[125–127] In a consequence the methyl ligands are *trans* located to the coordinated N atom of ppy or bhq ligands.

$^1J_{PPt}$ coupling values have been found to be extremely useful in establishing the coordination geometry around the metal centres of the Pt complexes obtained by Anderson and Spencer[128] from the electron-poor ligand 1,3-((C$_6$F$_5$)$_2$PO)$_2$C$_6$H$_4$(POCOPH) and platinum starting material. Thus, the ^{31}P NMR data of [(POCOPH)Pt(nb)]$_2$ where nb = norbornene revealed one phosphorus environment, with a large P–Pt coupling of 4623 Hz indicative of phosphorus coordination *trans* to the norbornene. In the spectrum of *cis, trans*-[(POCOPH)PtClMe]$_2$ complex the following $^1J_{PPt}$ couplings have been found: 5494 Hz for P-*trans*-Cl, 4028 Hz for P-*trans*-P and 2066 Hz for P-*trans*-CH$_3$.

NMR data has been extensively used by Zucca *et al.*[129] for structure elucidation of a large series synthesized by them Pt complexes such as, for example:

[Pt(bipy-H)(Me)(PPh$_3$)], [Pt(bipy-H)(Me)(PMe$_3$)], [Pt(bipy-H)(Me)(PCy$_3$)], (*ax*) [Pt(bipy-H)(Me)$_2$(I)(PPh$_3$)], (*eq*) [Pt(bipy-H)(Me)$_2$(I)(PMe$_3$)], (*ax*) [Pt(bipy-H)(Me)$_2$(I)(PMe$_3$)], (*ax*) [Pt(bipy-H)(Me)$_2$(I)(PCy$_3$)] and (*ax*) [Pt(L′-H)(Me)$_2$(I)(PPh$_3$)] where L′ = 6-methyl-2,2′-bipyridine, for all the studied compounds $^1J_{PPt}$, $^3J_{HPt}$ and $^2J_{HPt}$ couplings have been measured and carefully analysed.

In particular, different geometries between (*ax*) [Pt(bipy-H)(Me)$_2$(I)(PPh$_3$)] and (*eq*) [Pt(bipy-H)(Me)$_2$(I)(PMe$_3$)] have been supported by the $^1J_{PPt}$ coupling values; $^1J_{PPt}$ of 992 Hz has been found for the former compound and 1467 Hz for the other one. The [Pt(bipy-H)(Me)(PPh$_3$)] and

[Pt(bipy-H)(Me)(PMe_3)] complexes on the other hand show rather similar $^1J_{PPt}$ values (2229 and 2112 Hz, respectively).

Two homoligated Pt(II) complexes of *N*-phenyl-*N'*-3-(2-diphenylphosphanylethylthio)phenylurea have been synthesized by McGuirk *et al.*[130] during their studies on the small molecule regulation of intermolecular hydrogen bonding interactions for the *in situ* control of catalytic activity. Two resonances, one at 8.84 ppm with $^1J_{PPt} = 3203$ Hz and the other at 44.63 ppm with $^1J_{PPt} = 3557$ Hz), correlating to the phosphorus-bound ligand and the fully chelating ligand, respectively, have been observed for the semiopen complex. In the ^{31}P{1H} NMR spectrum of the fully closed complex only a single, sharp resonance at 47.05 ppm with $^1J_{PPt} = 3110$ Hz, consistent with equivalent, chelated ligands has been found. The relatively large P–Pt coupling values in both complexes are consistent with the predicted *cis* configurations of the ligands.

$^1J_{PPb}$ coupling of 1895(100) Hz measured from the ^{31}P MAS SSNMR spectra has been reported by Rossini *et al.*[131] for the (2,6-$Me_2C_6H_3S)_2Pb]_2$(dppe) complex where dppe = bis(diphenylphosphino)ethane.

Three hydroxide bridged dinuclear metallacycles, containing either Zn(II) or Cd(II) atoms, of a general structure [M_2(μ-OH)(μ-L)$_2$](ClO_4)$_3$ have been synthesized by Reger and co-workers;[132] for the cadmium derivative [Cd_2(μ-OH)(μ-L)$_2$](ClO_4)$_3$ with L = *m*-bis[bis(3,5-dimethyl-1-pyrazolyl)-methyl]benzene $^1J_{111Cd113Cd} = 173$ Hz has been observed in its ^{113}Cd NMR spectrum. This is an important result since it provides unequivocal proof for the dinuclear structure of the studied metallacycles in solution.

5 Two-bond couplings involving hydrogen

Steric effect on energetics and dynamics of ground- and excited-state conformations in bis(9-fluorenyl)methane and its derivative 9-(9-ethylfluorenyl)-9'-fluorenylmethane has been studied by Boo *et al.*[133] The NMR data of 9-(9'-ethylfluorenyl)-9-fluorenylmethane has indicated that the central two protons in the CH_2 group are slightly different in the chemical shifts ($\delta = 2.50$ and $\delta = 2.54$) and the couplings ($^2J_{HH} = 4.74$ Hz and $^2J_{HH} = 4.71$ Hz). This has been taken by the authors as an indication that the Cs tg' conformer in this compound is not abundant.

$^2J_{HH}$ and $^1J_{HC}$ couplings for the siderofore containing two citric acid moieties connected with diaminobutane linker complexed with Fe(III) have been calculated by Dubey and Heinonen.[134] The calculations have been performed with Gaussian 09 at B3LYP/6-311G(d,p) level; the $^2J_{HH}$ couplings have been found to be mainly due to the Fermi contact term. Their values depend on the angle between the two protons and increase algebraically with the increasing angle.

A complete NMR assignment including sets of $^nJ_{HH}$ ($n = 2$, 3) couplings has been presented by Barbosa *et al.*[135] for three synthetic cycloheptenone intermediates for the guaiane sesquiterpenes and by Wang *et al.*[136] for two new monoterpene glucoside derivatives isolated from *Lomatogonium carinthiacum* (Wulf) Reichb.

The new eudesmanoid 1β-angeloyloxyeudesm-7-ene-4β,9α-diol and the known dirhamnosyl flavonoid lespidin have been isolated by Villanueva-Cañongo *et al.*[137] from the aerial parts of *Senecio polypodioides* while from the roots, the known 7β-angeloyloxy-1-methylene-8α-pyrrolizidine and sarracine *N*-oxide, as well as the new neosarracine *N*-oxide, have been obtained. The structure of eudesmanoid 1β-angeloyloxyeudesm-7-ene-4β,9α-diol and sarracine *N*-oxide has been elucidated by spectral means. Full sets of ^{1}H chemical shifts and proton–proton couplings across two, three and four bonds for the remaining two compounds as well as for sarracine and its *N*-oxide have been obtained by the use of 1D and 2D NMR experiments and by the iterative full spin analysis performed by the use of the PERCH NMR software.

Comprehensive conformational studies of the anticancer agent pironetin have been performed by Bañuelos-Hernández[138] by an extensive use of experimental and calculated $^{2,3}J_{HH}$ couplings in combination with molecular modelling. As a result, two solvent-dependent conformational families (L and M) have been revealed for the optimum conformations.

The impurity present in the drug landiolol, an ultra-short-acting β-blocker, has been established by Štujber *et al.*[139] by the extensive use of NMR and mass spectroscopy as being [(4*S*)-2,2-dimethyl-1,3-dioxolan-4-yl]methyl 3-{4-[(2*S*)-2-hydroxy-3-(3-{4-[(2*S*)-2-hydroxy-3-[(2-{[(morpholin-4-yl)carbonyl]amino}ethyl)amino]propoxy]phenyl}-*N*-(2-{[(morpholin-4-yl)-carbonyl]amino}ethyl)propanamido)propoxy]phenyl}propanoate. The authors have found that the compound exists in two rotameric forms at room temperature; for both of them ^{1}H and ^{13}C NMR data have been reported including proton–proton couplings across two and three bonds.

An analysis of the NMR data performed by Ward *et al.*[140] for the thorium complexes, {N[*o*-(NCH_2P-i-Pr_2)C_6H_4]$_3$}ThCl and {N[*o*-(NCH_2PiPr_2)C_6H_4]$_3$}Th(μ-OC)Co$(CO)_3$, has shown that the methylene protons in these compounds are diastereotopic, appearing as two doublets at 3.89 ppm ($^{2}J_{HH} = 12$ Hz) and 3.23 ppm ($^{2}J_{HH} = 12$ Hz) ppm in the ^{1}H NMR spectrum of the {N[*o*-(NCH_2P-i-Pr_2)C_6H_4]$_3$}ThCl complex, and at 3.55 ppm ($^{2}J_{HH} = 12$ Hz) and 3.05 ppm ($^{2}J_{HH} = 12$ Hz) in the spectrum of the isocarbonyl complex. This result has been interpreted by the authors in terms of phosphine coordination to the thorium centre.

The tautomeric equilibria of 2-pyridoyl-, 3-pyridoyl-, and 4-pyridoyl benzoyl β-diketones in the liquid and solid state have been determined by Hansen *et al.*[141] by the use of deuterium isotope effects on ^{1}H and ^{13}C NMR chemical shifts and spin–spin J_{HC} couplings. In particular, the two-bond and three-bond experimental and calculated $^{2,3}J_{HC}$ couplings have been applied to confirm the equilibrium positions in the solution state.

Configurational assignments of conformationally restricted two meroterpenoids, melissifolianes A and B, isolated from the drupes of endangered species *Lindera melissifolia*, have been performed by Hamann and co-workers.[142] The relative configuration of compound A has been established using the combination of NOE-based distance restraints and the comparisons of experimental and calculated $^{3}J_{HH}$ couplings. The relative configuration of a racemic mixture of melissifoliane B has been elucidated utilizing *J*-based analysis combined with theoretical calculations.

The application of *J*-based analysis to either sp^3 quaternary carbon centres or to consecutive methylene units, has been reported by Cen-Pacheco *et al.*[143] for a highly flexible natural product, nivariol, a new pentacyclic polyether isolated from a Macaronesian endemic species of *Laurencia Viridis*. The obtained results allowed the authors to support a DFT-NMR-predicted stereostructure. Further examples of application of a *J*-based configuration analysis by the use of long-range ^1H–^{13}C couplings in structural studies of natural products include the 34-membered macrodiolide luminaolide isolated by Maru *et al.*[144] from the crustose coralline algae (CCA) *Hydrolithon reinboldii*, the stereostructural studies performed by Bifulco and co-workers[145] on two new oxygenated polyketides, plakilactones G and H, isolated from the marine sponge *Plakinastrella mamillaris* collected at the Fiji Islands, and elucidation of the structure of a large series of nitrogen-containing dihydro-β-agarofuran derivatives isolated by Luo *et al.*[146] from *Tripterygium wilfordii*, whose dried roots are a widely used traditional Chinese medicine.

A method for the direct C–N coupling of imidazoles based on electrooxidative C–H functionalization of aromatic and benzylic compounds has been developed by Morofuji *et al.*[147] The stereochemistry of some compounds obtained has been determined by JBCA and NOE analyses.

The (PCP)Ir(H)(CO)($^{13}C{\equiv}CPh$) complex (PCP = κ^3-C_6H_3-2,6-($CH_2PtBu_2)_2$) has been synthesized by Hackenberg *et al.*[148] during their studies on the phenol-catalyzed addition of phenylacetylene to (PCP)Ir(CO) complex. A very large $^2J_{HC}$ coupling of 30.2 Hz observed for a hydride resonance has been invoked by the authors as a proof that the acetylide and hydride ligands are mutually *trans* arranged. The CO addition to (PCP)Ir(H)(C≡CPh) resulted, on the other hand, in the formation of the *cis* complex, since the small $^2J_{HC}$ of 5.3 Hz only has been found for this compound.

A series of substituted pyrazolo[4,3-*c*]pyrazoles and three regioisomers of their *N*-methyl derivatives have been synthesized by Marek and co-workers,[149] and all compounds were fully characterized by NMR spectroscopy in solution. In particular, long-range couplings, $^{2,3}J_{HC}$ and $^{2,3}J_{HN}$, turned out to be a useful tool in discrimination of three regioisomers of the studied compounds. It is worth noticing that the experimental coupling values are in most cases in agreement with those DFT calculated.

The NMR spectroscopic data for a series of thirty-four 3-acylpyrido[1,2-*a*]pyrimidinium salts has been reported by Girreser *et al.*,[150] including numerous long-range HC and HN couplings in the heteroaromatic ring system. The authors have indicated that besides the usually large proton–carbon couplings over three bonds, the large couplings over two bonds in the vicinity of the positively charged heteroatom have to be considered for this class of compounds as well.

An unusually low geminal coupling between the phosphorus atom and the CHF proton, $^2J_{HP}$ of 0.3 Hz in THF-d_8, has been observed by Opekar *et al.*[151] in diethyl fluoronitromethylphosphonate, a previously unknown compound, synthesized by electrophilic fluorination of diethyl nitromethylphosphonate with Selectfluor. In comparison, $^2J_{HP}$ of 13.6 Hz has been reported for fluorobisphosphonate.

A paper devoted to superbasic alkyl-substituted bisphosphazene proton sponges has been published by Sundermeyer and co-workers.[152] For the salt of one of the studied compounds, *i.e.* for 1,8-bis(tricyclopentylphosphazenyl)naphthalene · HPF_6, the simultaneous coupling of the acidic NH proton with both phosphorous atoms has been observed, $^2J_{HP} = 2.4$ Hz (δ(NH) = 13.32). This result has been invoked by the authors as an indication that a low-barrier hydrogen bond exists in the studied sponges. However, in the spectra of the remaining mono-protonated compounds described in the paper only broad NH signals have been observed. In the spectrum of the bisprotonated sponge, 1,8 bis(trispyrrolidinophosphazenyl)naphthalene · $2HN(SO_2CF_3)_2$, on the other hand, the acidic protons exhibit a doublet with coupling to only one phosphorus atom ($^2J_{HP} = 6.5$ Hz).

A diagnostic virtual triplet at −9.22 ppm with $^2J_{HP} = 84.4$ Hz has been observed by Duan *et al.*[153] in the ^{1}H NMR spectrum of [HPd-($OCOCF_3$)(*R*)-BINAP] hydride obtained from the chiral palladium catalyst [Pd-$(OCOCF_3)_2$(*R*)-BINAP]. This coupling value is roughly the mean of those of hydrogen coupling with *cis*-phosphorus ($J = 5$–16 Hz) and hydrogen coupling with *trans*-phosphorus ($J \approx 160$ Hz), which strongly suggests that the hydrogen in Pd–H couples with phosphorus with an angle between the *cis* and *trans* positions. This has been ascribed by the authors to the equilibrium between the reversible reductive elimination of TFA and some other intermediates.

The structures of a series of rhodium complexes, $[Rh(L)(CO)_2H]$, of various bisphosphine and bisphosphite ligands, which are equipped with an integral anion binding site, have been studied by Dydio *et al.*[154] Small values of *ca.* 10 Hz or less found for the phosphorus-hydride coupling at rt have indicated that all bidentate ligands in the complexes under study are coordinated predominantly in the *ee* fashion (the expected averaged values of $^2J_{HP}$ for the *ee* and *ea* are around 2 and 100 Hz, respectively).

The structures of some new stable symmetric addition products of ethylenediamine to propa-1,2-dienylphosphonates synthesized by Khusainova *et al.*[155] have been confirmed by the observation of $^2J_{HP}$ and $^4J_{HP}$ couplings of *ca.* 12 and 2 Hz, respectively; these were diethyl 2-(2-aminoethylamino)prop-1-en-1-ylphosphonate and some tetraalkyl 2,2′-[ethane-1,2-diylbisimino]bis-[prop-1-en-1-ylphosphonates].

A rare example of a mononuclear Ir hydride complex bearing an *ortho*-metalated phosphine ligand, $[Ir(H)(\kappa^5P,P,P,P,C\text{-}1)]PF_6$, has been obtained by Gloaguen *et al.*[156] by coordination of tris((2-diphenylphosphino)benzyl)-phosphine to the cationic Ir(I) precursor $[Ir(COE)_2(acetone)_2]PF_6$, where COE = cyclooctene. A complex hydride signal observed in the ^{1}H NMR spectrum of this compound at δ −5.9 ppm was resolved as a pseudo doublet of quartets with one large H–P coupling, $^2J_{HP} = 115.4$ Hz (*trans*) and a much smaller coupling, $^2J_{HP} = 10.8$ Hz (*cis*). The ^{13}C NMR spectrum revealed a doublet at 120.6 ppm, with a $^2J_{CP}$ value of 41.6 Hz, indicative of a direct Ir–carbon bond.

A series of tris(pentafluorophenyl)corrole (TPFC) tin(IV) and tin(II) complexes have been prepared by Yun *et al.*[157] and studied by a variety of techniques including ^{1}H, ^{19}F, and ^{119}Sn NMR. $^2J_{HSn}$ of 50.2 Hz, which arises from coupling with the two magnetically equivalent tin atoms observed for the resonance of the CH_2 hydrogens of (TPFC)$SnCH_2Sn$(TPFC), has indicated unambiguously a methylene-bridged dicorrole structure. $^2J_{H117Sn}$ of 89 Hz and $^2J_{H119Sn}$ of 92 Hz have been measured for the (TPFC)$SnCH_3$ complex and assigned to the axial methyl protons.

Safari *et al.*[158] have measured the value of 70 Hz for $^2J_{HSn}$ in the solution of tribenzyltin(IV)(morpholine-1-carbodithioate), thus confirming the five coordinated Sn atom in this promising antitumor agent.

The first example of a high-level correlated calculation of spin–spin couplings involving tellurium has been presented by Rusakov *et al.*[159] The HTe spin–spin couplings across two and three bonds have been calculated at the SOPPA and DFT levels for tellurophene and divinyl telluride and compared with the experimental values measured in $CDCl_3$ solution. A new full-electron basis set, av3z-J, for tellurium derived from the "relativistic" Dyall's basis set, dyall.av3z, and specifically optimized for the correlated calculations of spin–spin couplings involving tellurium has been designed by the authors. Interestingly, vibrational and solvent corrections are of small importance, whereas conformational averaging has played a crucial role in the performed calculations of the H–Te couplings.

Two $^2J_{HPt}$ couplings of *ca.* 80 Hz have been observed by Pörschke and co-workers[160] in the spectrum of bispidine (3,7-diazabicyclo[3.3.1]nonane) platinum complex, $(C_7H_{14}N_2)PtCl_2$, evidencing persistent coordination of the bispidine at Pt.

6 Two-bond couplings not involving hydrogen

Complexes of unsymmetrically substituted 2-^{13}C-imidazolium salts with NMR-active metals (Ag, Hg, Pt) have been synthesized by Emsermann *et al.*[161] and analyzed with respect to the scalar couplings between the metal centre or C2 and the other ring atoms. Among others, $^2J_{C2C4}$ and $^2J_{C2C5}$ couplings of about 3 Hz have been observed for the parent 2-^{13}C-imidazolium salts such as, for example, 5-(4-chlorophenyl)-4-(4-fluorophenyl)-1-(4-methylphenyl)-3-(2-phenylethyl)(2-^{13}C)-1*H*-imidazolium chloride where $^2J_{C2C4}$ and $^2J_{C2C5}$ of 2.2 and 3.3 Hz, respectively, have been found. However, no useful nontrivial metal-to-ligand couplings could be measured in any of the investigated 1,3,4,5-tetrasubstituted imidazol-2-ylidene ligands due to an unfavourable S/N ratio.

A series of 1-silacyclopent-2-ene derivatives has been synthesized by Khan *et al.*[162] and characterized by the use of ^{29}Si NMR chemical shifts and $^nJ_{CSi}$ couplings ($n = 1$, 2).

Very good agreement has been observed by Wrackmeyer *et al.*[163] between the experimental and calculated two-bond $^2J_{C\text{–}N\text{–}P}$ and $^2J_{Si\text{–}N\text{–}P}$ couplings obtained for a variety of various amino- and silylaminophosphines including a phosphenium cation and an amino(imino)phosphine. These are interesting results since they can be used for reliable structural assignments of preferred conformations or fast dynamic processes in the studied and similar to them compounds. In particular,

$^{2}J_{C-N-P}$s by analogy to $^{2}J_{H-N-P}$ couplings are large and positive when the phosphorous lone pair adopts the *syn*-position, and are small and of either sign in the case of the *anti* position. A series of 1,3,2-dioxaphospholanes, such as, for example, 2-chloro-4,5-[1,2-dicarba-*closo*-dodecaborano(12)]-1,3,2-dioxaphospholane, 1,2-bis{4,5-[1,2-dicarba-*closo*-dodecaborano(12)]-1,3,2-dioxaphospholan-2-yl}oxy-1,2-dicarba-*closo*-dodecaborane(12) and 1-{4,5-[1,2-dicarba-*closo*-dodecaborano(12)]-1,3,2-dioxaphospholan-2-yl}oxy-2-hydroxy-1,2-dicarba-*closo*-dodecaborane-(12), has been synthesized by Wrackmeyer and co-workers[164] by using 1,2-hydroxy-1,2-dicarba-*closo*-dodecaborane(12) and the corresponding dilithium salt [1,2-(LiO)2-1,2-$C_2B_{10}H_{10}$] and characterized by the use of $^{2}J_{CP}$ couplings which vary between 7.5 and 12.5 Hz depending on the substituent. The theoretically calculated couplings are in reasonably good agreement with those determined experimentally; the calculations indicate that they are of a negative sign.

A detailed NMR signal assignment of the ^{15}N, ^{13}C, and ^{1}H spectra of urea-formaldehyde resins has been presented by Steinhof *et al.*[165] A crucial role in this work played information on the $^{2}J_{NN}$ coupling and its dependence on the chemical environment; J_{NN} splitting could be easily observed due to the use of 98% ^{15}N-labelled urea.

A series of complexes of the type [(dippe)Ni-(η^2-$C_6F_{6\text{-}n}H_n$)] where $n = 0$–2 and dippe = 1,2-bis(disopropylphosphino)ethane has been prepared by Arévalo *et al.*[166] in their studies on homogeneous catalytic hydrodefluorination of fluoroaromatics. All these complexes as well as the obtained products have been characterized mainly by the use of NMR spectroscopy and X-ray. Thus, for example, the following NMR data including spin–spin couplings has been obtained for the [(dippe)Ni(F)-C_6F_4H)] complex. In the rt ^{31}P{^{1}H} NMR spectrum two doublets of doublets have been observed; one group of signals located at δ 70.4 ($^{2}J_{FP} = 30$ Hz, $^{2}J_{PP} = 10$ Hz) has been assigned to the phosphorus *trans* to C, the other at δ 65.6 ($^{2}J_{FP} = 34$ Hz, $^{2}J_{PP} = 10$ Hz) to the phosphorus *trans* to F. The ^{19}F NMR spectrum has shown four multiplets located at δ −115.6, −141.2, −162.6, and −164.4 for the four nonequivalent fluorine atoms at the aromatic ring and a multiplet with $^{2}J_{FP} = 34$ Hz at δ −345.4 assigned to the Ni–F resonance.

An unprecedented bicyclic 1,3,2λ3,4λ5-diazadiphosphetidine has been synthesized by Hering *et al.*[167] as a result of the reaction between 2SiMe$_3$[GaCl$_4$] and [(Me$_3$Si)$_2$NPCl][GaCl$_4$]. In the ^{31}P NMR spectrum of this compound two resonances have been detected, one for the NP(Cl)N group at 142.6 ppm, and the other for the tetracoordinated P^{V} at 23.9 ppm, with both signals being split into doublets with a $^{2}J_{PP}$ coupling of 64.5 Hz, which is in good agreement with the few examples of 1,3,2λ3,4λ5-diazaphosphetidines. Interestingly, the $^{2}J_{PP}$ couplings of similar magnitude have been found for two open chain compounds, [i-Pr$_2$NPNPPh$_3$][GaCl$_4$] and [(SiMe$_3$)$_2$NPNPPh$_3$][GaCl$_4$], 67.5 and 59 Hz, respectively.

The formation of the diamagnetic (*t*PCH=CHP)CoCl(CO) complex obtained by Barrett and Iluc[168] from (*t*PCH=CHP)CoCl (*t*PCH=CHP = 2,2′-bis(di-iso-propylphosphino)-*trans*-stilbene) has been confirmed by observation of *trans*-$^2J_{PP}$ coupling of 175 Hz between two inequivalent phosphorous atoms in the former compound. The presence of a CO ligand has been confirmed with a corresponding chemical shift at 204.93 ppm.

Weller and co-workers[169,170] have continued their studies on the electronic and steric effects in the stoichiometric dehydrocoupling of secondary and primary phosphine–boranes. Among others, three complexes of a general structure [Rh(dpp3)(H)(σ,η-PR_2BH_3)(η^1-$H_3B\cdot PR_2H$)]-[$BAr^F{}_4$], where dpp3 = bis-1,3-(diphenylphosphino)propane, $Ar^F = 3,5\text{-}(CF_3)_2C_6H_3$ and R = 3,5-bis(trifluoromethyl)phenyl; 4-trifluoromethylphenyl; 4-methoxyphenyl have been synthesized and their structure established by the use of NMR spectroscopy.[170] The spectra of all three complexes are very similar and a detailed analysis of the $^{31}P\{^1H\}$ NMR spectrum performed for the complex with the 3,5-bis(trifluoromethyl)phenyl substituent has provided useful information on the structure of all of them. First of all, it has been shown that each of the four phosphorous atoms occurs in different environments. Two of the ^{31}P peaks are broadened significantly compared to the other two, suggesting these phosphorus atoms are bound to a quadrupolar boron centre. One of these peaks has shown both a large *trans* two-bond PP coupling, $^2J_{PP}$ of 244 Hz, and a coupling to ^{103}Rh, $^1J_{PRh}$ of 75 Hz, while the other is a broad singlet. The other two signals are sharper and consequently assigned to the two ^{31}P environments of the $Ph_2P(CH_2)_3PPh_2$ ligand. One of these sharper resonances ($\delta = 29.5$, ddd, $^1J_{PRh} = 130$ Hz, $^2J_{PP} = 35$ Hz, $^2J_{PP} = 21$ Hz) has been assigned to the phosphorus atom *trans* to the weakly bound β-B-agostic interaction on the basis of the larger ^{103}Rh coupling, while the other signal ($\delta = 11.3$, ddd, $^1J_{PRh} = 103$, $^2J_{PP} = 244$, $^2J_{PP} = 35$ Hz) could be assigned to the phosphorus atom *trans* to the coordinated phosphido ligand. In the ^1H NMR spectrum one broad, relative integral 3H, signal has been observed at δ −0.78, indicative of a $Rh\cdots H_3B$ σ interaction in which the B–H bonds are undergoing rapid site exchange on the NMR spectroscopic time scale between terminal and bridging sites. A broad, relative integral ^1H, resonance at δ −6.12 has been assigned by the authors to a static β-B-agostic B–H interaction. Cooling of the solution to 0 °C has led to the resolution of this signal as doublet, $^2J_{HP} = 65$ Hz, indicating its *trans* disposition to a phosphine. The results presented above are very similar to those published by Huertos and Weller[169] for the [Rh(dpp3)(H)(σ,η-PPh_2BH_3)(η^1-$H_3B\cdot PPh_2H$)][$BAr^F{}_4$] complex.

The reaction of the first isolable benzo-1,2-diselenete, 4,5-dicyano-3,6-diethylbenzo-1,2-diselenete with tetrakis(triphenylphosphine)palladium in benzene performed by Kimura *et al.*[171] yielded a dinuclear palladium complex, bis(4,5-dicyano-3,6-diethylbenzodiselenolato)bis-(triphenylphosphine)dipalladium whose structure has been confirmed, among others, by the analysis of $^2J_{PSe}$ couplings measured for this compound. The selenium peaks at $\delta = 349.7$ ppm which have been assigned by the authors to bridged selenium have shown two couplings: $^2J_{PSe} = 120$ and 103 Hz and the peaks at $\delta = 605.0$ ppm due to the outer selenium atom have been observed as a doublet signal, $^2J_{PSe} = 37$ Hz. These couplings have been also reflected in the ^{31}P spectrum. On the basis of these results it has been suggested that the central four-membered ring of the studied compound may have an asymmetric character.

The contribution of silaalkylphosphonic acids $Me_3SiCH_2P(O)(OH)_2$ and $Me_3SiC(CH_3)_2P(O)(OH)$ as ligands has been demonstrated for the first time by Shankar *et al.*[172] by the isolation of new diorganotin(IV) phosphonates $Et_2Sn\{OP\text{-}(O)(OH)CH_2SiMe_3\}(OSO_2Me)$, $(Et_2Sn)_6\{O_3PC\text{-}(CH_3)_2SiMe_3\}_4(OSO_2Me)_4$, and $Et_2Sn(O_3PCH_2SiMe_3)$. The structures of these new compounds have been confirmed by the observation of two-bond P–O–Sn couplings; $^2J_{P\text{–}O\text{–}Sn}$ of 147.6 Hz has been found for $(Et_2Sn)_6\{O_3PC(CH_3)_2SiMe_3\}_4(OSO_2Me)_4$ and $^2J_{P\text{–}O\text{–}Sn}$ of 141 Hz for $Et_2Sn(O_3PCH_2SiMe_3)$.

A series of new compounds containing an annelated dicarba-*closo*-dodecaborane(12) unit has been obtained by Wrackmeyer and coworkers[173] by oxidation of 1,3,2-diselenaphospholanes either with sulphur or selenium. The obtained compounds, such as, for example, 2-(1-methylethyl)-2-thio-4,5-[1,2-dicarba-closododecaborano(12)]-1,3-diselena-2-phospholane, 2-(1-methylethyl)-2-seleno-4,5-[1,2-dicarba-closo-dodecaborano(12)]-1,3-diselena-2-phospholane, 2-ethoxy-2-thio-4,5-[1,2-dicarba-*closo*-dodecaborano(12)]-1,3-diselena-2-phospholane or 2-ethoxy-2-seleno-4,5-[1,2-dicarba-*closo*-dodecaborano(12)]-1,3-diselena-2-phospholane have been fully characterized by ^1H, ^{11}B, ^{13}C, ^{31}P and ^{77}Se NMR spectroscopy; this included all relevant couplings, among others $^1J_{PSe}$ and $^2J_{SeSe}$ ones, the latter being of *ca.* 34 Hz.

2-D ^{183}W INADEQUATE experiments have been systematically employed by Haouas *et al.*[174] to unequivocally verify structures, assign all resonances, and determine precisely $^2J_{WOW}$ scalar couplings in the Dawson-type arsenato–phosphato–tungstate $\alpha\text{-}[AsPW_{18}O_{62}]^{6-}$ and the seven other already known polyanions $\alpha\text{-}[H_4AsW_{18}O_{62}]^{7-}$, $\alpha\text{-}[As_2W_{18}O_{62}]^{6-}$, $\alpha\text{-}[H_4PW_{18}O_{62}]^{7-}$, $\alpha\text{-}[P_2W_{18}O_{62}]^{6-}$, $\beta\text{-}[As_2W_{18}O_{62}]^{6-}$, $\beta\text{-}[P_2W_{18}O_{62}]^{6-}$ and $\gamma\text{-}[As_2W_{18}O_{62}]^{6-}$. $^2J_{WOW}$ couplings of *ca.* 21 Hz and $^2J_{POW}$ of *ca.* 1.5 Hz have been found for the new compound, which is in agreement with the data published already in the literature. The obtained data has been discussed in terms of the effects of P/As substitutions on the NMR observables δ_W, δ_P, $^2J_{W\text{–}O\text{–}W}$, and $^2J_{W\text{–}O\text{–}P}$ in relation to the bond length and bond angle alteration. It has been indicated by the authors that the NMR results presented by them can serve as representative reference data for understanding the relationship between structure and NMR observables in polyoxotungstates.

7 Three-bond hydrogen–hydrogen couplings

Three-bond hydrogen–hydrogen couplings have remained over the years the most useful parameter in the structural studies on organic compounds including those of natural origin.

The *cis* configuration has been ascribed to one of the two [2,8]-dicarba Vc1.1 isomers by van Lierop *et al.*[175] on the basis of the value of 10.8 Hz measured for vicinal proton–proton coupling across the double bond in this peptide.

$^3J_{HNH\alpha}$ couplings have been applied by Gomez-Monterrey *et al.*[176] in an NMR analysis of four novel peptides, G protein coupled receptor kinase 2 inhibitors, performed in water and dodecylphosphocholine micelle solutions.

$^{3}J_{HNH\alpha}$ and $^{1}D_{C\alpha C\beta}$ couplings have been used by Bax and co-workers[177] to validate the high quality of the single-conformer structural model obtained for ubiquitin in solution.

Aliev and co-workers[178] have proposed a new approach for force field optimization called AMBER99SB-ILDNP for reproducing dynamics characteristics of the side chains in proteins and improving prediction of their motionally averaged structural properties. As the source of experimental data motionally averaged values of $^{3}J_{HH}$ couplings have been used together with ^{13}C NMR spin–lattice relaxation times T_1 of backbone and side-chain carbons.

The conformational studies of alanine and valine methyl esters, which do not show zwitterionic structures in solution, has been studied by Rittner and co-workers.[179] by ^{1}H NMR and theoretical calculations. The $^{3}J_{HH}$ couplings have been measured for valine methyl ester in solvents of different dielectrical constants (ε) and compared with the theoretically calculated J values for the predicted conformers. An analysis of the obtained data has led the authors to the conclusion that the interplay between steric hindrance and hyperconjugation is the responsible force for determining the conformational preferences of alanine and valine methyl esters.

Double-headed nucleoside with a hemiacetal linker between the $C_{2'}$ and the $C_{4'}$-positions, with two uracil nucleobases organized on a bicyclic skeleton has been synthesized for the first time by Nielsen and co-workers.[180] With the aid of $^{3}J_{HH}$ it has been found that the new compound is conformationally restricted and is locked in an N-type conformation.

The *p*-nitrophenyl β-L-arabinofuranoside has been synthesized by Kaeothip *et al.*[181] as the substrate for novel β-L-arabinofuranosidase using inter- and intra-molecular glycosylation methodologies. The anomeric configuration of the new compound has been confirmed by the $^{3}J_{H1H2}$ coupling (4.0 Hz) and the δ_C value of C1.

The structures of three new iridoid glycosides isolated from *Gardenia jasminoides* var. *radicans* have been elucidated by Zhou and co-workers[182] on the basis of $^{3}J_{HH}$ couplings and other NMR parameters supported with MS spectra, glycoside hydrolysis, and sugar derivatization.

The deviations of the measured values of $^{3}J_{H(C)H(O)}$ from the value of 5.5 ± 0.5 Hz, which indicates free rotation around the H–C–O–H bond, have been used by Vilén and Sandström[183] as probes of possible interactions between trehalose and lactose in solution.

It has been demonstrated by Riihinen *et al.*[184] that computer-assisted ^{1}H iterative full spin analysis (HiFSA), called by them ^{1}H-NMR fingerprinting, can replace and assist the repetitive and tedious 2D NMR identification protocol required for both known and new compounds, respectively. Using this approach the authors have fully interpreted the complex ^{1}H-NMR fingerprints of eight quercetin glycosides (QGs) isolated from the berries of V. *vitis-idaea* obtaining full sets of ^{1}H chemical shifts and proton–proton couplings. The method applied has allowed to reveal for the first time the presence of Q-3-*O*-β-glucopyranoside and

Q-3-*O*-β-glucuronopyranoside in the berries of *V. vitis-idaea*, and unambiguously confirm the structures of Q-3-*O*-[4″-(3-hydroxy-3-methyl-glutaroyl)]-α-rhamnopyranoside, Q-3-*O*-α-rhamnopyranoside, Q-3-*O*-β-galactopyranoside, Q-3-*O*-α-arabinofuranoside, Q-3-*O*-β-xylopyranoside and Q-3-*O*-α-arabinopyranoside.

Photochemistry of a series of α-diketones placed as 1α-pyruvoyl tethers on D-glucopyranose and D-glucopyranosiduronic acid skeletons has been studied by Alvarez-Dorta *et al.*[185] Highly unusual photoproducts from Norrish type II photoelimination and Norrish–Yang photocyclization initiated by the excited internal carbonyl group have been obtained and their structure has been established by the authors *via* an extensive use of vicinal proton–proton couplings. For other examples of structure elucidation of peptides, proteins, oligonucleotides and sugars with the aid of $^3J_{HH}$ couplings see Table 1.

Yoon and co-workers,[202] using $^3J_{HH}$ couplings and NOE correlations, have identified L-olivose and D-boivinose in new glycosylated derivatives of macrolide antibiotic YC-17 produced by *S. venezuelae* mutant strain YJ003/pLOLV2.

Table 1 Peptides, proteins, oligonucleotides and sugars for which the solution structure (2D or 3D) has been determined with $^3J_{HH}$'s used as restraints or comparables.

Name	*a*	*b*	Reference
Proteins and peptides			
L-proline methyl ester and its *N*-acetylated derivative	1	2	186
Two glycosylated analogues of oxytocin and vasopressin	9	7	187
2-HexaPY, two hexarepeats	12	8	188
α-CTx TxIB,a novel α-conotoxin from *Conus textile*	16	7	189
cis- and *trans*-[2,8]-, and *trans*-[3,16]-dicarba-cystino Vc1.1 α-conotoxin analogues	16	17 ÷ 23	175
Tk-AMP-X2, the wheat antimicrobial peptide	28	29	190
Aβ40, monomer–dimer equilibrium	40	28	191
Aβ42, monomer–dimer equilibrium	42	25	191
Aβ42-M35ox(S), monomer–dimer equilibrium	42	32	191
Aβ43, monomer	43	10	192
Ros87 H42A	87	60	193
Lc-LTP2, lipid transfer protein from *Lens culinaris*	93	69	194
Oligonucleotides			
Two acylated dinucleotides	2	1	195
Carbohydrates			
A naphthyl xyloside and its fluorinated analogues in a variety of solvents	1	5 ÷ 7 each	196
The L-IdoA2S residue, in eight heparin-like trisaccharides	1	4 each	197
β-L-Fucp-(1→6)-α-D-Glcp-OMe	2	2	198
Streptomycin	2	5	199
Ko-(2→6)-GlcNAc-ol (Na+) from *A. calcoaceticus* NCTC 10305 LPS	2	12	200
DgαDt-Kop-(2→6)-GlcNAc-ol	2	12	200
A novel branching pattern in the LPS expressed by *H. influenzae* strain 1232	14	12	201

[a] The number of amino acid residues or nucleotides or sugar units.
[b] The total number of vicinal proton–proton couplings measured.

Configurations of the four triterpene epoxy lactones, 3-oxo-, 29-nor-3-oxo-, 3β-acetoxy- and 3β-hydroxy-7β,8β-epoxy-5α-lanost-24-en-30,9α-olides isolated from a *Penares* sp. sponge collected from Vietnamese waters have been established by Kolesnikova *et al.*[203] by the use of vicinal proton–proton couplings, NOESY and CD data, and confirmed by X-ray crystallographic analysis of 29-nor-3-oxo-7β,8β-epoxy-5α-lanost-24-en-30,9α-olide.

Further examples of application of $^3J_{HH}$ couplings in structural studies of natural products include *trans*-dehydrocrotonin, a bioactive 19-*nor*-diterpenoid clerodane type, isolated by Soares *et al.*[204] from *Croton cajucara* Benth, and two new closely related indoloditerpenes, emindole SB beta-mannoside, substituted with a dimethylallyl containing side chain and a mannose moiety, and 27-*O*-methylasporyzin isolated by Harms *et al.*[205] from a marine-derived fungal strain of *Dichotomomyces cejpii*.

An extensive use of $^3J_{HH}$ couplings has been made by Lv *et al.*[206] in structural elucidation of nineteen new highly oxygenated norbisabolane sesquiterpenoids, phyllanthacidoid acid methyl ester, and C-T, isolated from *Phyllanthus acidus Skeels*, together with two known ones, phyllanthusols A and B, whose sugar moiety has been revised as glucosamine-*N*-acetate, rather than the previously assigned mannosamine-*N*-acetate. All of the isolated compounds except for phyllanthacidoid acid methyl ester are glycosides, whose saccharide moieties possess a pentaoxy cyclohexane (scyllo quercitol) connecting with glucosamine-*N*-acetate or glucosyl moieties, which are first examples in natural products. Some of these Phyllanthacidoids N-R with 8*R* configurations and/or 5,8-diketal skeleton, are unprecedented structures among norbisabolane sesquiterpenoids. Phyllanthacidoids S and T have been found to possess the unusual tricyclo[3.1.1.1] oxygen bridge skeleton formed by a diketal system, of which the relative configurations of the aliphatic chain have been assigned on the basis of $^{2,3}J_{HC}$ coupling values.

Vicinal proton–proton couplings have been extensively used by Mabou *et al.*[207] for elucidation of a new triterpenoid bidesmoside (leptocarposide) possessing an acyl group in their glycosidic moiety, isolated from the *n*-butanol-soluble fraction of the whole plant of *Ludwigia leptocarpa* (Nutt) Hara (Onagraceae together with the known luteolin-8-*C*-glucoside (2) and 1-*O*-β-D-glucopyranosyl-(2*S*,3*R*,8*E*)-2-[(2′*R*)-2-hydroxy-palmitoylamino]-8-octadecen-1,3-diol (3), and by Wu and co-workers[208] in their studies on salprzelactone, a seco-norabietane diterpenoid with an unprecedented carbon skeleton.

Structural analysis of several 3-methylflavones including the natural derivatives 3-methylluteolin and 3-methylapigenin by the use of ^{1}H and ^{13}C NMR spectroscopy has been performed by Rocha *et al.*;[209] for all the compounds ^{1}H and ^{13}C NMR data has been reported including vicinal and geminal H–H couplings. The authors have also corrected the erroneous ^{1}H NMR data which has been published in the literature for some derivatives.

$^3J_{HH}$ couplings have been used by Li *et al.*[210] for the identification of two biflavonoids: GB2a and GB2a glucoside from *Garcinia xanthochymus* extracts, in a new rapid screening of α-amylase inhibitors using enzyme-immobilized magnetic nanoparticles.

The structures of four new prenylated flavonoids, two cudraflavanones and two cudraflavones, isolated from the roots of *Cudrania tricuspidata* together with eight known compounds, have been established by Chen *et al.*[211] on the basis of NMR, HRESIMS and CD spectroscopies; for all four new compounds $^2J_{HH}$ and $^3J_{HH}$ couplings have been reported.

Complete NMR assignments of a new bioactive rotameric (3→8) biflavonoid isolated from the bark of *Garcinia hombroniana* have been performed by Jamila *et al.*[212] The variable NMR studies in different solvents have shown that the compound identified as (2*R*, 3*S*) volkensiflavone-7-*O*-rhamnopyranoside exists in the form of two conformers; their structures have been confirmed by analysis of proton–proton couplings across three and four bonds. In particular, the *trans* configuration at the C2 and C3 carbons of ring C of the flavanone unit has been confirmed by the presence of $^3J_{HH}$ coupling of 12.0 Hz.

A detailed study of the alkaloid content of *Narcissus triandrus*, wild daffodils of the section Ganymedes from the Iberian Peninsula, as a source of mesembrane alkaloids has been undertaken by Pigni *et al.*[213] In particular, structural elucidation of three new compounds 2-oxomesembrenone, 7,7*a*-dehydromesembrenone and 2-oxoepimesembranol, together with the identification of 5 major common mesembrane alkaloids has been performed by means of NMR spectroscopy including extensive use of the $^3J_{HH}$ couplings.

The structures of a new pyrrolizidine alkaloid, named crotavitelin, isolated from fruits of *Crotalaria vitellina*, Fabaceae (Papilionoideae) by Bezerra *et al.*[214] and six new caffeoyl derivatives possessing a rare octulosonic acid skeleton isolated from the flower heads of Roman chamomile (*Chamaemelum nobile*) by Zhao *et al.*[215] have been elucidated by the use of NMR including extensive analysis of the proton–proton couplings across two and three bonds.

Three new trace bisindole alkaloids geleganimines A and B and geleganamide have been isolated by Qu *et al.*[216] from the aerial parts of *Gelsemium elegans*. Their structures have been elucidated by NMR spectroscopy, particularly from their two- and three-bond proton–carbon couplings, and electronic circular dichroism; for all three compounds geminal and vicinal proton–proton couplings have been measured and carefully analysed. The described compounds are the first bisindole alkaloids discovered from the genus *Gelsemium*.

Structural studies of 26 analogues of natural quaternary benzo[*c*]phenanthridine alkaloids have been performed by Yang *et al.*[217] by the use of vicinal proton–proton couplings.

3β-Chloro-5α,6α-epoxycholestane and 3β-acetoxy-5α,6α-epoxycholestane have been synthesized by Shamsuzzaman *et al.*,[218] and their conformations in solution derived from the analysis of vicinal proton–proton couplings have been compared with those obtained from the X-ray analysis.

Vicinal proton–proton couplings have provided useful information on the configuration of three new cytochalasans, trichalasins E, F and H, isolated by Chen *et al.*[219] together with four known analogues, trichalasin C, aspochalasin K, trichalasin G and aspergillin PZ from one endophytic fungus *Trichoderma gamsii* inhabiting in the traditional medicinal plant Panax notoginseng (BurK.) F.H. Chen. Trichalasin E contains a unique hydroperoxyl group, which is the first report in all known analogues, whereas trichalasin H possesses the rare 6/5/6/6/5 pentacyclic skeleton with 12-oxatricyclo [6.3.1.02,7] moiety.

$^3J_{HH}$ couplings have been measured by Cheng *et al.*[220] for four new cembranoids, isolated from the soft coral *Sarcophyton* sp. together with eight known compounds including five cembranoids, two carotenoids, and a tetra-substituted quinone. The determined couplings have allowed unambiguous configuration assignment of the double bonds present in the studied compounds.

Extensive theoretical and experimental studies have been performed by Virgili *et al.*[221] on colchicine, the unique drug for acute flares of gout known already for almost 200 years. This included measurements of geminal and vicinal proton–proton couplings of the central seven-membered ring B which, in combination with DFT calculations and the Karplus equation, has allowed the authors to determine its conformation.

$^3J_{HH}$ couplings have been also of some help in stereochemical assignments of (±)-sorbiterrin, a novel sorbicillin derivative featuring a bridged [3.3.1] ring system, whose total synthesis has been described by Qi *et al.*[222]

Further examples of application of $^3J_{HH}$ couplings in the structural studies of natural products include investigation of degradation products of salbutamol performed by Garrido and co-workers,[223] of melissifoliane A and B, *Xylaria feejeensis* by Handelsman and co-workers,[224] of a variety of polyhydroxylated novel coumarin derivatives by Svinyarov *et al.*,[225] and of six glyceroglycolipids isolated from *Cucurbita moschata* by Jiang *et al.*[226]

The HR-MAS NMR technique combined with chemometrics has been applied by de Oliveira *et al.*[227] in order to evaluate the metabolic profiles of leaves and roots of T-35S 'Swingle' citrumelo plants, transformed with a heterologous Δ1-pyrroline-5-carboxylate synthetase (VaP5CSF129A) gene to produce constitutively high levels of endogenous free proline, in comparison with those of the unmodified control plants. ^{1}H and ^{13}C chemical shift assignments and $^3J_{HH}$ couplings of the compounds identified in the leaves and roots of 'Swingle' citrumelo have been reported by the authors. The compounds found were, among others, proline valine, maleic acid, α and β-glucose, sucrose and ethanol.

A series of simple amino amides has been synthesized and analyzed by Faggi *et al.*[228] with a goal to study the conformational effects introduced by the corresponding side chains. The careful analysis of the experimental $^3J_{HH}$ couplings measured in several solvents at ambient temperature has been supported by DFT calculations leading the authors to the conclusion that the aromatic derivatives are characterized by a preferred folding of the side chain toward the polar amino amide moiety. These trends have been explained by a stabilizing NH···π interaction.

Two analogues of the 2-(1-adamantyl)-3-hydroxybutyric acid, 2-(3-homoadamantyl)-3-hydroxybutyric acid, and 2-(1-homoadamantyl)-3-hydroxybutyric acid, have been prepared by Matković *et al.*[229] as mixtures of diastereoisomers, and the relative configurations of their diastereoisomers have been determined by comparison of the $^{3}J_{H2,H3}$ couplings. The differences between these couplings in isomer pairs *threo* and *erythro* are of about 6Hz for all three compounds.

The structures of two aldazines: [1,1′-(1*E*,1′*E*)-hydrazine-1,2-diylidenebis(methan-1-yl-1-ylidene)dinaphthalen-2-ol] (Lumogen) and 2,2′-(1*E*,1′*E*)-hydrazine-1,2-diylidenebis(methan-1-yl-1-ylidene)diphenol (salicylaldazine) in the solid state and in solution have been studied by Silva *et al.*;[230] for Lumogen ^{1}H NMR spectra including $^{3}J_{HH}$ couplings measured at various temperatures have been reported.

Good agreement has been observed by Lomas[231] between the experimental $^{3}J_{HH}$ couplings determined for butane-1,4-diol, pentane-1,4-diol, (*S*,*S*)-hexane-2,5-diol, 2,5-di-methylhexane-2,5-diol and cyclohexane-1,4-diols (*cis* and *trans*) in benzene and their values calculated by the use of generalized Karplus equations from the relative conformer populations and the geometries of the studied compounds. The solvent dependence of the couplings measured for butane-1,4-diol in various solvents has been attributed by the author to conformational changes occurring in this compound.

Comprehensive conformational studies by the extensive use of vicinal proton–proton couplings on a set of *gem*-disubstituted methylphenylpiperidines have been performed by Xiao *et al.*;[232] the compounds have been synthesized in the context of discovery of NK1 antagonists which in turn led to re-design and an efficient synthesis of a potent NK1 antagonist with excellent *in vivo* activity and rodent and monkey pharmacokinetic profiles.

^{1}H and ^{13}C NMR spectroscopic data including $^{3}J_{HH}$ couplings of 20 new non-symmetrical compounds containing a 4-(*N*,*N*-dimethylamino)- or 4-(pyrrolidin-1-yl)pyridinium moiety and a 3-nitro-, 3-amino-, or 3-hydroxyphenyl ring, linked by *p*-xylene, 4,4′-dimethylbiphenyl, 1,2-bis(*p*-tolyl)ethane, or 1,4-bis(*p*-tolyl)butane has been reported by Schiaffino-Ortega *et al.*[233]

The ^{1}H and ^{13}C NMR spectra of low-molecular reference monomers, poly(diphenyleneoxidephthalide) and periodic copoly(arylenephthalide) derivatives such as AB, ABB, and ABBB (where A = terphenylenephthalide and B = diphenyleneoxidephthalide) have been assigned by Kraikin *et al.*[234] An analysis of $^{3}J_{HH}$ couplings confirmed the assignments made on the basis of ^{1}H–^{1}H COSY, ^{1}H–^{13}C HSQC and HMBC NMR techniques.

The consequence of electronic modulation of the aromatic group on the conformation and reactivity of α,β-unsaturated MacMillan-type iminium salts has been reported by Gilmour and co-workers.[235] The conformer population of six such compounds has been established on the basis of the $^{3}J_{HH}$ couplings using the Diez–Altona–Donders equation.

Selected $^{3}J_{HH}$ couplings combined with theoretical calculations have been applied by Duarte *et al.*[236] in their studies on the conformational equilibrium of L-tryptophan methyl ester (Trp-OMe) in several solvents. No significant influence of solvents on the couplings has been observed, which indicates that there are no conformational population variations for the side chain of the Trp-O-Me depending on the solvent.

An analysis of experimental ${}^{3}J_{HH}$ couplings measured in solvents of different polarities combined with theoretical calculations performed for 3-hydroxytetrahydropyran in the isolated phase and those which included the solvent effect has been applied by Solha *et al.*[237] in order to establish the conformational preferences in this molecule. It has been shown by the authors that the axial conformation of this compound is more stable in the isolated phase and in a nonpolar solvent, while the equatorial conformation is more stable than the axial in polar media.

A template-based approach to form composite macrocyclic peptidomimetics by unique large ring-forming Friedel-Crafts alkylations of tryptophan and tyrosine has been reported by Lawson *et al.*[238] NMR spectroscopy including analysis of vicinal proton–proton couplings combined with molecular mechanics simulations have been used to solve the solution structure of a new macrocyclic pyrroloindoline.

The ${}^{1}H$ data including ${}^{3}J_{HH}$ and ${}^{2}J_{HH}$ couplings has been reported by Peterson *et al.*[239] for eight 1-adamantyl substituted derivatives of 1,4-,1,6- and 1,7-dihydroxynaphthalenes. The most probable torsion angles in aliphatic fragment of these compounds have been evaluated by the use of vicinal proton–proton couplings. The obtained data and peculiarities of the molecular structure of the studied compounds should be useful for prompt characterization of compounds containing adamantane, dihydroxynaphthalenes or naphthoquinone units.

The complete ${}^{1}H$ and ${}^{13}C$ assignments of eight bioactive indeno[1,2-*b*]indole-10-one derivatives have been accomplished by Al Chab *et al.*[240] by the of NMR spectroscopy; for all these compounds ${}^{3}J_{HH}$ couplings have been reported.

A versatile strategy featuring Colvin rearrangement, hydrozirconation, Sonogashira cross-coupling reaction, and Z-selective Wittig olefination, has been successfully developed by Dayaker *et al.*[241] for the construction of a conjugated *E*,*E*,*Z*-triene subunit flanked, on both sides, by two allylic hydroxyl groups. This chemical pattern is found in many lipid metabolites and analogues with appealing biological applications. The configuration of the double bonds in the studied compounds has been confirmed by an analysis of ${}^{3}J_{HH}$ couplings.

A full set of the ${}^{1}H$ and ${}^{13}C$ NMR data including vicinal proton–proton couplings for a series of 12 new nonsymmetrical bispyridinium choline kinase inhibitors has been reported by Rubio-Ruiz *et al.*[242]

A comparative study between a series of *para*-alkylaminophenyl and *ortho*-alkylaminophenyl benzothiazoles (alkyl = CH_3, C_2H_5, C_3H_7, C_4H_9, $CH(CH_3)_2$ and $CH_2CH=CHCH_3$) using NMR and DFT calculations has been performed by Pierens *et al.*;[243] for all these compounds ${}^{1}H$ and ${}^{13}C$ NMR data including a full set of ${}^{3}J_{HH}$ couplings has been obtained. A comparison of the proton chemical shift values has revealed significant differences in the observed chemical shift values for the NH protons indicating the presence of a hydrogen bond in all *ortho*-substituted compounds as compared to the *para* compounds. The splitting pattern of the methylene proton next to the amino group has also shown significant ${}^{3}J_{HH}$ coupling to the amino proton consistent with the notion of the existence of slow exchange and hydrogen bond in the *ortho*-substituted compounds. These results have been furthermore confirmed by DFT calculations which yielded a near planar low energy conformer for all the *ortho*-aminophenyl benzothiazoles and revealed a hydrogen bond from the amine proton to the nitrogen of the thiazole ring. It has been concluded by the authors that the *ortho* amino substituted aminophenyl benzothiazole may be possibly used as Alzheimer's imaging agents because the binding to amyloid fibrils prefers planar compounds.

Analysis of the experimental and calculated $^3J_{H9H10}$ couplings for diastereoisomers of the parent 6,8-dioxo-10-phenyl-7-azaspiro[4.5]decane-9-carbonitrile, its *N*-methyl derivative and a series of aryl substituted compounds performed by Kirillov *et al.*[244] has allowed to assign a configuration around C9C10 bond: $^3J_{HH}$ of about 5 and $^3J_{HH}$ of 13 Hz have been found for the *cis* and *trans* compound, respectively.

Vicinal proton–proton couplings in combination with theoretical calculations have been extensively used by Gonçalves *et al.*[245] in their studies on the conformational isomerism of 1-chloro- and 1-bromo-2-propanol. Conformers with the X–C–C–O (X = Cl and Br) fragment in the *gauche* orientation were found to be strongly prevalent both in the gas phase and solution.

DFT calculations of chemical shifts and vicinal proton–proton couplings of two series of *N*-substituted epoxy-fused cyclopentane-3,5-diols have been performed by Özen *et al.*[246] and the theoretical results compared with the experimental data obtained by the authors. The $^3J_{HH}$ couplings between 5 and 8 Hz have been found for the series where the epoxy ring is at the opposite site of the NHR group (the dihedral angle close to 160°) and of *ca.* 0.5 Hz only for the group where ω is about 90°.

Two representative ladder-type acetylene-bridged perylenediimide dimers bearing long alkyl chain solubilizing groups, bis[1-ethynyl-*N*,*N*′-bis(1-hexylheptyl)-perylene-3,4:9,10-tetracarboxylic diimide] and 1,1′-ethynyl-bis[*N*,*N*′-bis(1-hexylheptyl)-perylene-3,4:9,10-tetracarboxylic diimide] have been synthesized and thoroughly characterized by Castellano and co-workers[247] by the use of NMR spectroscopy. This was a nontrivial task because of severe broadening of the proton signals and greater than expected numbers of the observed ^{13}C resonances. However, at slightly elevated temperatures (320–330 K), some protons became progressively better resolved as a doublet, exhibiting a scalar coupling of 8.0 Hz, a value typical for 3-bond coupled aromatic protons.

A comprehensive conformational analysis by the use of vicinal proton–proton couplings measured in a variety of solvents and at different temperatures has been performed by Roberts and co-workers[248] for 1,2-disubstituted ethane system – *N*,*N*,*N*-trimethyl-(3,3-dimethylbutyl)-ammonium iodide. The authors have come to the conclusion that the compound, which contains both an organic and a highly polar substituent and therefore is soluble in a large variety of solvents, exists predominantly in the *trans* form (about 90% in contrast to the previously assumed 100%) in a wide range of protic and aprotic solvents. This behaviour makes *N*,*N*,*N*-trimethyl-(3,3-dimethylbutyl)ammonium iodide a possibly useful '*trans*-standard' in conformational analysis, much more so than 1,2-ditert-butylethane, which has a poor solubility in many solvents. Another paper published by this group of authors[249] has been devoted to estimation of the fraction of gauche conformers of *N,N*-dimethylsuccinamic acid (1) and its Li^+, Na^+, K^+, Mg^{2+}, Ca^{2+}, and $N(Bu)_4{}^+$ salts in DMSO and D_2O solution by comparing the experimental vicinal proton–proton couplings determined by ^1H NMR spectroscopy with those calculated using the Haasnoot, de Leeuw, and Altona (HLA) equation. In DMSO, the gauche preferences have been found to increase with decreasing Ahrens ionic radius of the metal counterion. The same trend has not been seen in D_2O, where the gauche fraction for all of the metallic salts was estimated to be approximately statistical or less. The authors have stressed that their results provide crucial information on the importance of metal chelation on the conformation of organic molecules in polar aprotic media, which has implications for protein folding.

The reaction of 5,10,15,20-tetraphenyl-*p*-benziporphyrin, sodium tetrachloroaurate(III) dihydrate, and potassium carbonate in dichloromethane, carried out by Szyszko *et al.*[250] yielded gold(III) 5,10,15,20-tetraphenyl-21-carbaporphyrin owing to the contraction of *p*-phenylene to cyclopentadiene. This molecule is the very first representative of a true 5,10,15,20-tetraphenyl-21-carbaporphyrin complex where four trigonal donor atoms are involved in equatorial coordination. The ^1H NMR spectrum of this Au complex resembles the basic pattern of aromatic carbaporphyrinoids with an AB spin system at $d = 8.78$, H(7,18), and 8.67 ppm, H(8,17) with a coupling, $^3J_{HH}$ of 4.9 Hz, which is typical of the pyrrole ring in porphyrinoids.

The resonance for the CH_2OH atoms in two dimolybdenum thiolate-bridged complexes, $[Mo_2Cp_2(\mu\text{-}SMe)_2(\mu\text{-}SCH_2CH_2OH)(RNC)_2]^+$ (R = *t*-Bu or Xyl) synthesized by Le Goff *et al.*[251] could be easily identified as the only one exhibiting a triplet between 4.33 and 3.71 ppm ($^3J_{HH} \approx 7.0$ Hz), which was broadened upon OH coupling to give, in some cases, a multiplet.

It has been demonstrated by Perrin *et al.*[252] that hydrogenation of pyridine to piperidine catalyzed by $[1,2,4\text{-}(Me_3C)_3C_5H_2]_2CeH$ ($[1,2,4\text{-}(Me_3C)_3C_5H_2] = Cp'$) proceeds from $[1,2,4(Me_3C)_3C_5H_2]_2Ce$(2-pirydyl), isolated from the reaction of pyridine with Cp'_2CeH. The ^1H NMR spectra including $^3J_{HH}$ couplings of variously substituted 2-pyridyl derivatives of Cp'_2Ce(2-pirydyl) have been recorded in order to facilitate the spectral assignments in this parent metallocene.

8 Three-bond couplings to hydrogen

The $^3J_{HC}$ values measured by Garson and co-workers[253] for acremine P have been diagnostic in determining the stereochemistry of this compound. The $^3J_{HH}$ couplings have been also used in this study to obtain the relative configuration of six new acremine metabolites.

Brockerman *et al.*[254] have measured three $^3J_{H\gamma1C\gamma2}$ couplings of selectively $^{13}C_{\gamma2}$ labelled threonins and two $^3J_{H\gamma C\alpha}$ couplings of selectively $^{13}C_\alpha$ labelled serins in *Bacillus circulans* xylanase. Although the Karplus equation for such $^3J_{H(O)C}$ couplings has not been parameterized, the measured coupling values suggest that five hydroxyl protons are at least moderately restrained and predominantly in *trans* conformations relative to the coupled ^{13}C nucleus.

Karplus-type dependence of vicinal proton–carbon couplings combined with *ab initio* molecular orbital (MO) calculations has been applied by Sasanuma and Touge[255] in order to derive conformational characteristics of poly(L-lactide) and poly(DL-lactide); $^{3}J_{HC}$ couplings have been measured in a variety of solvents for ^{13}C-enriched (*S*)-methyl 2-acetoxypropanoate used as the model compound. Another paper published by Sasanuma *et al.*[256] has been devoted to conformational analysis of poly(butylene terephthalate). The refined rotational isomeric state scheme combined with *ab initio* molecular orbital (MO) calculations and NMR experiments has been applied by the authors on its model compound, tetramethylene glycol dibenzoate. It included analysis of $^{3}J_{HH}$ and $^{3}J_{HC}$ couplings measured in a variety of solvents and at various temperatures; the experimental coupling values have been compared with those theoretically calculated.

It has been demonstrated by Klika[257] that an analysis of DFT calculated NMR parameters such as the chemical shifts of ^{1}H, ^{13}C and other nuclei as well as pertinent spin–spin scalar $^{3}J_{HH}$ and $^{3}J_{HC}$ couplings allows one to discriminate between 6- and 7-membered ring formation in the case of 3,4-dihydro-2*H*-3-hydroxymethyl-1,4-benzoxazine and 2,3,4,5-tetrahydro-1,5-benzoxazepine-3-ol. It has also been suggested by the author that the methodology proposed by him can also be applied to such cases in the literature where assigned structures may be in doubt.

A series of (2-hydrazono-4-oxothiazolidin-5-ylidene)acetates has been synthesized by Hassan *et al.*[258] from the reaction between 1,4-disubstituted thiosemicarbazides with dimethyl acetylenedicarboxylate. The structures of the obtained compounds have been established by extensive use of NMR spectroscopy including a careful analysis of spin–spin couplings; among others, $^{3}J_{HC}$ couplings across double bonds present in some compounds allowed to establish their configuration.

Annular tautomerism of a series of 3(5)-aminopyrazoles containing a cyano, thiocyanato, or aryl substituent in the 4-position has been studied by Emelina *et al.*[259] by ^{1}H and ^{13}C NMR in solution, cross-polarization and magic-angle spinning ^{13}C NMR in the solid state, and *ab initio* quantum chemical calculations. The structure of one of the compounds studied, 5-amino-1-benzyl-4-cyano-3-phenyl-1*H*-pyrazole, has been confirmed taking into account couplings between the *ortho* protons in the phenyl ring on C3 in the pyrazole ring, $^{3}J_{HC} = 3.6$ Hz, and between protons in the 1-CH_2 group and C5, $^{3}J_{HC} = 2.5$ Hz.

Vicinal proton–carbon and proton–proton couplings have provided useful information on the structure of the antibiotic strepturidin isolated by Pesic *et al.*[260] from the microorganism *Streptomyces albus* DSM 40763.

The enol–enol tautomerism of ethyl 2-benzoyl-5-(2-furyl)-3-hydroxy-penta-2,4-dienoate and ethyl 2-acetyl-3-hydroxy-5-phenyl-penta-2,4-dienoate in different solvents has been studied by Michalik and co-workers[261] with the aid of $^{2}J_{OHC}$, $^{3}J_{OHC}$ and $^{4}J_{H,OH}$ couplings and BP86/TZVP DFT computations. It has been established by the authors that the percentage of the keto form in the tautomeric equilibrium increases with increasing solvent polarity.

Addition of aqueous HBr to hydrophobically shielded arylalkynes HC≡C-aryl and H_3C–C≡C-aryl, carrying 2,6-di- and 2,4,6-tri-*tert*-butylphenyl as the aryl group has been studied by Knorr *et al.*[262] The strongly retarded HBr addition to H_3C–C≡C-aryl in moist chloroform produced more *E* than *Z* product, whereas a thermodynamic *E*/*Z* ratio of 10 : 87 was found in moist acetic acid. The *E*/*Z* configurational assignments of the obtained products have been deduced from the $^3J_{HC}$ couplings of C-1′ to 2-H (7.2 Hz and 3.6 Hz, respectively).

$^3J_{HF}$ couplings have provided crucial information on the conformational preferences of 3-fluoropiperidinium cation and the parent 3-fluoropiperidine studied by Freitas and co-workers.[263] The axial predominance of both compounds in polar solution is consistent with the $^3J_{H2F}$ couplings: 38.1 and 12.3 Hz (in D_2O), as well as 36.2 and 12.3 Hz (in CD_3OH) have been found for the cation; 36.0 and 13.9 Hz (in CD_3OH) have been determined for the parent compound, whereas rather small $^3J_{HF}$ values are predicted for the equatorial conformer by theoretical calculations. The effect of adding a fluorine atom in 2-fluoroethylamine hydrochloride on the conformational isomerism of the resulting 2,2-difluoroethylamine chloride has been studied by the same group of authors.[264] The analysis has been carried out by the use of $^3J_{HH}$ and $^3J_{HF}$ couplings measured in D_2O solution, in the gas phase and in implicit water through theoretical calculations. It has been estimated by the authors that the conformer containing both fluorines *gauche* to the ammonium group (*gg*) is practically the single form in the gas phase and, even in water solution, this conformer is calculated to amount to 90%. The experimental couplings for the 2,2-difluoroethylammonium cation determined in D_2O solution are $^3J_{HH} = 2.6$ Hz and $^3J_{HF} = 16.4$ Hz, and the mean calculated values for the *gg* conformer in implicit water are 1.3 Hz and 16.8 Hz [$(2.8 + 30.7)/2 = 16.8$], while the corresponding values for the *ag* conformer are 5.5 Hz and 12.4 Hz. Interestingly, no significant double *gauche* effect has been found in the parent 2,2-difluoroethylamine, since conformers possessing two fluorine atoms in the *gauche* relationship with the amino group (*gauche–gauche*, *gg*) are estimated to be similarly populated to those conformers with only one single fluorine *gauche* to the nitrogen atom (*anti-gauche*, *ag*).

A highly stereoselective synthesis of chiral optically active α,α-difluoro-β-amino sulfinic and sulfonic acids by fluoroalkylation of chiral imines using 2-$PySO_2CF_2H$ reagent has been reported by Prakash *et al.*;[265] for one of the products, $BzHNCH(Me_2)CF_2SO_2NHCH(Bz)CO_2Me$, a conformer population analysis has been performed using $^3J_{HF}$ and $^3J_{HH}$ couplings and the obtained results confronted with the conformational behaviour of its nonfluorinated analogue.

The conformational effects of mono-, di-, and trifluorination, as well as hydroxylation, on substituted azepanes have been investigated by Patel *et al.*[266] by the use of ^1H NMR spectroscopy and computational modeling in chloroform; for all studied compounds $^3J_{HH}$ couplings have been measured, and for those containing the fluorine substituent also $^3J_{HF}$

couplings have been reported. It has been concluded by the authors that a single fluorine atom, when introduced diastereoselectively on a model azepane ring, can bias its ring to one major conformation but multiple fluorinations may not lead to additive conformational control and can result in complex conformational structures.

Chaume *et al.*[267] have incorporated trifluoromethyl substituted oxazolidine-type pseudoprolines (CF_3-ΨPro) into peptide chains and measured their NMR parameters in a series of peptides. The values obtained fall in the range of 282.7–287.5 Hz for $^1J_{CF}$, 30.7–37.4 Hz for $^2J_{CF}$, and 4.6–6.05 Hz for $^3J_{HF}$.

Reactions of phosphine-derived carbenes $C_3H_2(NP\text{-}t\text{-}Bu_2)_2$ and $C_3H_2(NPN\text{-i-}Pr_2)_2$ with 9-borabicyclo[3.3.1]nonane (9-BBN) carried out by Wang and Stephan[268] have resulted in ring-expansion reactions to generate novel intramolecular frustrated Lewis pairs (FLPs). The ^{1}H NMR spectrum of the t-Bu_2 derivative has been found to be consistent with inequivalent carbene backbone C–H groups as two distinct resonances at $\delta = 5.67$ ($^3J_{HH} = 6.3$ Hz and $^3J_{HP} = 2.0$ Hz) and $\delta = 6.01$ ($^3J_{HH} = 6$ Hz and $^3J_{HP} = 1$ Hz) have been observed for this compound. The analogous spectral data has been found for the N-i-Pr_2 derivative.

$^1J_{HP}$ coupling of 545.9 Hz assigned to a one-bond coupling between P and H and $^3J_{HP}$ of 13.0 Hz assigned to a three-bond coupling between P and two *ortho* hydrogen atoms on the phenyl ring attached to a phosphorus atom have been observed by Srungavruksham and Baskar[269] in the proton-coupled ^{31}P NMR spectrum of $[Ph_3Sb(HPh\text{-}PO_2)_2]_2$. The authors have indicated that this is the first report on the observation of such a three-bond P–H coupling in a ^{31}P NMR spectrum.

Vicinal $^3J_{HAg}$ coupling of 8.57 Hz between the imine proton and the silver atom has been reported by Kay *et al.*[270] for the four-coordinate Ag complex of the ligand tris[4-(2-thienyl)-3-aza-3-butenyl]amine (TTME), $[Ag(TTME)](PF_6)\cdot H_2O$. It has been determined from the selectively decoupled INEPT spectrum of this compound.

Two new Hg(II) and Cd(II) perchlorate complexes of 2,6-bis([(2-pyridylmethyl) amino]methyl)pyridine, $[HgL(ClO_4)_2]$ and $[CdL(ClO_4)_2]$ have been isolated by Bebout and co-workers[271] as bicapped distorted square pyramidal racemates; the structures of both these compounds have been extensively studied by the use of heteronuclear couplings, $^3J_{HHg}$ and $^3J_{HCd}$, respectively.

9 Three-bond couplings not involving hydrogen

Wright and co-workers[272] have re-parametrizated Karplus curves for $^3J_{C\gamma CO}$ and $^3J_{C\gamma N}$ couplings describing χ_1 angle in proteins. The general equation was proposed in the form: $^3J = A\cos^2(\theta+\delta) + B\cos(\theta+\delta) + C$, along with separate parameters for aromatic residues Phe, Trp, and Tyr, and His, Ile, Thr and Val. The analysis is based on the eight complexes of *E. coli* dihydrofolate reductase (DHFR). The couplings are extremely sensitive measures of side chain χ_1 rotamers. The same authors[273] have also studied the 3D structure of *Homo sapiens* DHFR and found that despite a low level of sequence identity of both enzymes, their 3D structures are very similar. An analysis of χ_1 rotamer populations in human enzyme derived from measurements of $^3J_{C\gamma CO}$ and $^3J_{C\gamma N}$ indicates that many of the side chains that contact helix αF exhibit rotamer averaging which may facilitate the conformational change. Interestingly, in binary complexes of human DHFR the χ_1 of the Phe31 side chain depends upon the presence of the substrate or product in the active site. For other examples of structure elucidation of proteins and carbohydrates with the aid of heteronuclear vicinal couplings see Table 2.

A series of pyranochromene chalcones and flavanones synthesized by Pawar and Koorbanally[274] have been characterized by the authors by the use of NMR spectroscopy including $^3J_{HH}$ couplings; for the compounds containing the fluorine substituents also $^nJ_{CF}$ couplings ($n = 1$–3) have been reported.

The couplings $^1J_{CP} = 34.2$ Hz and $^3J_{CP} = 10.2$ Hz for the bridging carbon atoms adjacent to the phosphorus atoms have been observed by Kaub *et al.*[275] in the $^{13}C\{^1H\}$ NMR spectrum of the gold complex, [{(FcCSS)Au}$_2$(dppp)], where FcCSS = ferrocene dithiocarboxylate and dppp = bis(diphenylphosphino)propane; for the carbon atom in the centre of the dppp bridge a small $^2J_{CP}$ coupling of 2.5 Hz has been found.

The mixture of *E* and *Z* isomers of 3β,28-diacetoxylup-20(29)-en-29-yltriphenylphosphonium bromide has been obtained by Tsepaeva *et al.*[276] during their attempt to introduce a triphenylphosphonium fragment into betulin molecule *via* reaction of triphenylphosphine with 30-bromolup-20(29)-ene-3β,28-diyl diacetate. The structures of these two isomers could be easily assigned on the basis of $^3J_{19CP}$ couplings reported by the authors in Experimental: $^3J_{19CP}$ of 7.7 Hz obviously corresponds to the *E* isomer and $^3J_{19CP}$ of 16.4 Hz to the *Z* one. Unfortunately, the authors have not reported three-bond coupling between the phosphorous atom and the carbon of the methyl group attached to the double bond which could confirm this assignment unambiguously.

Reactions of 5-*tert*-butyl-2,2,2-trichloro-, 2,2,2-tribromo-5-*tert*-butyl-, and 2,2-dibromo-5-*tert*-butyl-2-fluoro-1,3,2λ5-benzodioxaphospholes with aryl- and alkylacetylenes have been carried out by Mironov *et al.*[277] yielding 2-halo-1,2λ5-benzoxaphosphinine 2-oxides quantitatively. The structures of the obtained compounds, which may be regarded as

Table 2 Proteins and carbohydrates for which heteronuclear vicinal couplings have been used as restraints or comparables.

Name	*a*	Reference
Peptides and proteins		
Lc-LTP2, lipid transfer protein from *Lens culinaris*	$^3J_{CC}$, $^3J_{CN}$	194
Aβ40, monomer–dimer equilibrium	$^3J_{HN}$	191
Tk-AMP-X2, the wheat antimicrobial peptide	$^3J_{HN}$	190
Carbohydrates		
β-L-Fucp-(1 → 6)-α-D-Glcp-OMe	$^3J_{HC}$	198
Fluorinated analogues of naphthyl xyloside in a variety of solvents	$^{1-5}J_{HF}$	196

a Type of vicinal heteronuclear coupling measured.

phosphorus analogues of natural heterocyclic compounds, coumarin and chromene, have been characterized by the use of NMR spectroscopy, the $^3J_{CP}$ and $^3J_{HC}$ couplings being of particular aid. The major products are 4-aryl-7-*tert*-butyl-2,6-dichloro-, 4-aryl-2-bromo-7-*tert*-butyl-, and 4-aryl-7-*tert*-butyl-2-fluoro-1,2λ5-benzoxaphosphinine 2-oxides.

NMR spectroscopy has been extensively used by Musin *et al.*[278] in order to confirm the structure of spirophosphorane, 2-diethylamino-1″,1‴-dipropyl-2λ5-trispiro[1,3,2-benzodioxaphospholane-2,2′-[1,3,2]dioxaphospholane-4′,3″ : 5′,3‴-bis(indole)]-2″,2‴-dione, obtained during their studies on the effect of the substituent attached to the phosphorus atom on the reaction of aminophosphines with 1-alkylisatins. This included analysis of $^3J_{CP}$ couplings as well the H–P couplings across four and five bonds.

The $^1J_{C(inner\ carbon)Pt}$ coupling of 983 ± 7 Hz and $^3J_{C(outer\ carbon)Pt}$ of 32.5 ± 0.7 Hz, observed by Wang *et al.*[279] in the spectra of three novel Pt(II) 2-*N* substituted *N*-confused porphyrines, (2-aza-2-benzyloxycarbonylmethyl-5,10,15,20-tetraphenyl-21-carbaporphyrinato-*N*,*N*′,*N*″)platinum(II), (2-aza-2-*p*-cyanobenzyl-5,10,15,20-tetraphenyl-21-carbaporphyrinato-*N*,*N*′,*N*″)platinum(II) and (2-aza-2-ptrifluoromethylbenzyl-5,10,15,20-tetraphenyl-21-carbaporphyrinato-*N*,*N*′,*N*″)platinum(II) have been used by the authors as the proof that the Pt atom in the studied compounds is coordinated to three nitrogen atoms and to the unsubstituted β-carbon atom of the pyrrole ring.

10 Couplings over more than three bonds and through space

$^4J_{HH}$ scalar couplings were among key NMR parameters used by Bare and Sherman[280] in proving the possibility of formation of template-assembled but cation free thymidylic acid- and deoxyguanylic acid-tetrads in polar protic solvent.

Tang and co-workers,[281] using the genetic and enzymatic data, have established the biosynthetic pathway of the antifungal drug, griseofulvin. The polyketide skeleton of this compound is produced by ascomycetes such as *Penicillium aethiopicum* in several steps. Each intermediate was characterized with UV, MS and NMR including $^4J_{HH}$ couplings.

The total asymmetric synthesis of (3*R*,4*S*,5*S*)-4-acetamido-5-amino-3-(pentan-3-yloxy)cyclohex-1-enecarboxylic acid, the 4-epi-stereoisomer of oseltamivir carboxylate, has been presented by Sartori *et al.*[282] The stereostructure of this compound as well as its conformational behaviour in the aqueous solution has been performed on the basis of 1D and 2D NMR experiments combined with molecular dynamics simulations. The large vicinal inter-proton coupling, $^3J_{5,6\beta}$ of 10.8 Hz, and NOE contacts have indicated that the cyclohexene ring of this compound adopts mainly two halfchair conformations, namely the $^5\mathbf{H}_4$ and $^4\mathbf{H}_5$ conformers with strong prevalence of the $^5\mathbf{H}_4$ conformation. However, the presence of long-range W-type couplings $^4J_{2,6\beta}$, $^5J_{3,6\beta}$ and $^5J_{3,6\alpha}$ (about 3 Hz each) indicates that either a contribution of additional boat-like conformers or considerably deformed halfchair conformations might be present, though to a lesser extent.

The Diels–Alder cycloaddition reaction with cyclopentadiene carried out by Barhoumi-Slimi *et al.*[283] with *Z*-ethyl-2-chloro-3-formylacrylate yielded two diastereomers *endo*(COOEt)-*exo*(CHO) and *exo*(COOEt)-*endo*(CHO), whereas a mixture of *Z* and *E* of ethyl 3-chloro-2-formyl-4,4,4-trifluorobut-2-enoate afforded the four diastereomers. These new cycloadducts have been characterized by NMR, IR, and mass spectrometry. Structural assignments for different isomers have been made on the basis of NMR chemical shifts and couplings using 1D, 2D, and heteroNOE NMR techniques. In particular, $^{3}J_{H1n/H6n}$ of 1 Hz has confirmed that H6n and H1n are perpendicular involving a *cis* relationship, and $^{4}J_{H6n/H7b} = 2$ Hz characteristic of a stereospecific long range W coupling observed between H6n and the bridge proton H7b, has confirmed the axial position of H6n and suggesting that the dihedral angle is close to 90°.

A series of halogenated transformation products of pharmaceuticals and related environmental contaminants have been characterized by Larive *et al.*[284] by the use of ^{1}H and ^{13}C NMR including $^{3,4}J_{HH}$ couplings; this included chloro- and bromoderivatives of salicylic acid and halogenated analogues of naproxen, gemfibrozil, diclofenac, tertiary octylphenol and bisphenol A.

Proton–proton couplings across three and four bonds have been measured by Mastoor *et al.*[285] for a large series of derivatives of 8-quinolinol and their salts.

An analysis of the complex multiplets of the olefinic protons and allylic fragment in the ^{1}H NMR spectra of two cork suberin acids performed by Santos and Graça[286] yielded the $^{3}J_{HH}$ and $^{4}J_{HH}$ coupling values typical of *Z* configuration on the double bond providing unambiguous evidence that the studied compounds are methyl (*Z*)-18-hydroxyoctadecenoate and dimethyl (*Z*)-1,18-octadecenodioate.

Long-range couplings $^{5}J_{Hortho,OMe}$ of *ca.* 0.2 Hz have been measured by Alvarez-Cisneros *et al.*[287] in a series of methoxyindoles, methoxycoumarins and methoxyflavones by the modified *J* doubling in the frequency domain method, and negative signs have been assumed for all of them. DFT calculations performed by the use of the modified aug-cc-pVTZ basis set have shown that there is good agreement between the Fermi contact term and the experimental *J* values. The through-space coupling mechanism has been invoked by the authors to interpret the obtained results.

Long range $^{5}J_{HH}$/or through-space coupling of 1.4 Hz has been observed by Wrackmeyer *et al.*[288] in the ^{1}H NMR spectrum of the 4,5-[1,2-dicarba-*closo*-dodecaborano(12)]-1,3,2-dithiaphospholane.

An analysis of long-range $^{n}J_{HN}$ ($n = 4$–6) and $^{n}J_{CN}$ ($n = 2$–4) couplings has been proposed by Khalymbadzha *et al.*[289] as a tool for efficient estimation of the azide-tetrazole equilibrium in a series of 2-azidopyrimidines.

A full set of ^{1}H and ^{13}C data including proton–proton couplings across three and four bonds has been measured by Momin *et al.*[290] for a large series of fluoro-2-styrylchromones and methoxy-2-styrylchromones; for compounds containing fluoro substituents also $^{3,4}J_{HF}$ and $^{1-4}J_{CF}$ couplings have been reported and analysed.

Systematic NMR studies on all possible fluoro derivatives of phenylboronic acid have been carried out by Gierczyk *et al.*;[291] the presented NMR data included a full set of H, C, F, B, and O NMR chemical shifts and couplings, $^{n}J_{HH}$ ($n = 3$–4), $^{n}J_{HF}$ ($n = 3$–5), $^{n}J_{CF}$ ($n = 1$–5) and $^{n}J_{FF}$ ($n = 3$–5). Interestingly, for 2,4,6-trifluoro- and 2,3,5,6-tetrafluoro-derivatives also $^{4}J_{OF}$ coupling of *ca.* 25 Hz has been observed.

Tris(pentafluorophenyl)corrole and its ^{15}N-enriched isotopomer have been studied by Szymański *et al.*[57] in toluene-D_8 solution by 1D and 2D variable-temperature NMR techniques in order to establish the mechanisms of tautomerization of the NH protons inside the interior of the corrole macrocycle. The chemical shifts of the pyrrole protons in the parent compound, the *J*-couplings $^{3}J_{HH}$ between the CH pyrrole protons, and the $^{4}J_{CHNH}$ couplings as well as $^{6}J_{CHF}$ between the CH protons and the F nuclei were evaluated in iterative fits to the CH spectra at low temperatures. The estimated values of $^{6}J_{FCH}$ and $^{4}J_{CHNH}$ have not exceeded 1.5 Hz. They have been evaluated from envelopes of unresolved patterns in the analysed spectra. Nevertheless, the obtained values of $^{6}J_{CHF}$ have favourably compared with the dynamically averaged values estimated from the relevant resolved patterns in the motionally averaged proton spectra at elevated temperatures. The observation of $^{6}J_{CHF}$ in the case of proton H17 has allowed the authors its unambiguous assignment and distinguishing from proton H18, which is not coupled to fluorine atoms. The assignments of the ^{1}H resonances have been crucial for the elucidation of the tautomerization processes reflected in VT ^{1}H NMR spectra. It is worth noticing that three transfer processes of the NH protons among the pyrrole nitrogen atoms have been identified for the studied compounds, of which one is NMR-inaccessible as it involves rapid jumps in an unsymmetrical proton sponge arrangement formed by a pyridine- and pyrrole-type nitrogen atoms.

The preferred conformation, all-*gauche*, of 3-fluoro-1,2-propanediol has been established by Freitas and co-workers[292] by an extensive use of experimental and calculated chemical shifts and $^{n}J_{HH}$ and $^{n}J_{HF}$ ($n = 2$–4) couplings in which the diastereotopic hydrogen atoms are involved. It has been indicated by the authors that the hydrogen bonding between OH protons and proton acceptors (O or F) should not be considered to be the main force of the *gauche* effect; it should be rather assigned to traditional electronic delocalizations (hyperconjugation) from σCH and σCC as electron donors to σ*CO and σ*CF as electron acceptors. This conclusion is in agreement with the fact that neither $J_{H(O),F}$ nor $J_{H(O),H(O)}$ coupling has been observed for this compound.

A coupling between the fluorine nuclei and methylene protons on C3, $^{4}J_{HF}$ of ~2.4–2.5 Hz, has been reported by Boltacheva *et al.*[293] for 1-(2-amino-6-methylpyrimidin-4-yl)-5-difluoromethyl-3-phenyl-4,5-dihydro-1*H*-pyrazol-5-ol synthesized by them during their studies on the reactions of fluoroalkyl-substituted lithium 1,3-diketonates with various hetarylhydrazines.

The ^{1}H NMR spectra of 2-fluorobenzaldehyde in nonpolar and polar solvents have been measured by Tormena and-workers.[294] The $^{TS}J_{FHf}$ coupling of 0.48 Hz (where TS stands for "through-space") has been observed in DMSO-d_6 solution only and unequivocal proof of this result has been obtained from $^1H,^1H$ COSY and $^1H,^{19}F$ long-range correlation experiments. The obtained results have been confronted with theoretically calculated spin–spin couplings leading the authors to the conclusion that $^{TS}J_{FHf}$ coupling in polar solvents is amenable to measurement only in the *trans* conformer. The mechanisms for transmission pathways to such unusual spin–spin coupling have been rationalized in terms of the molecular electronic structure.

A set of experimental J_{HH}, J_{HF}, J_{FF} and J_{CF} couplings across one up to five bonds measured in different solvents and their DFT calculated values has been reported by Pérez-Torralba *et al.*[295] for two novel tetrafluorinated 1,5-benzodiazepinones, *i.e.* 6,7,8,9-tetrafluoro-4-methyl-1,3-dihydro-2*H*-1,5-benzodiazepin-2-one and 6,7,8,9-tetrafluoro-1,4-dimethyl-1,3-dihydro-2*H*-1,5-benzodiazepin-2-one. In the spectrum of the latter the J_{HF} coupling of about 4.5 Hz (J_{HF9} calcd. = 6.6 Hz) between the protons of the *N*-methyl group and F9 has been observed. A related coupling of 4.8 Hz has been reported in the literature for 2-fluoroacetophenone. It should be noted that the nature of this coupling is not obvious, it can be either through five bonds or through space.

It has been shown by Gryff-Keller and Szczeciński[296] that properly selected DFT-based methods are able to reproduce quite well the experimental ^{13}C NMR parameters for fluoroarenes as well as for their tricarbonylchromium complexes. In the case of ^{13}C NMR chemical shifts a good choice is the DFT/B3LYP/6-311++G(2d,p) calculation method while for C–F indirect spin–spin couplings the DFT/BHandH/6-311++G(2d,p) method yields much better results. An especially interesting result obtained by the authors is the good reproduction of the experimental values of the small through-space C–F couplings and their dependence on the conformation of the complex demonstrated for (η^6-fluorobenzene)tricarbonylchromium.

Orientational order properties of two nematogens containing a fluoro- and isothiocyanate-substituted biphenyl moiety have been studied by Calucci *et al.*[297] by means of ^{13}C NMR spectroscopy which included a detailed analysis of both ^{13}C chemical shift anisotropies and $^nJ_{CF}$ couplings (n =1–4) combined with theoretical calculations. The orientational order parameters obtained by NMR have been critically compared by the authors with those obtained by dielectric spectroscopy.

$^4J_{F1F8}$ indirect spin–spin couplings in a series of variously substituted *peri*-difluoronaphthalenes have been calculated by Contreras *et al.*[298] and compared with the experimental values published by Mallory *et al.*[299] some time ago. It has been concluded by the authors that the PSO and FC contributions play the main role in the coupling transmission mechanism, whereas the SD and DSO terms can be almost neglected. The obtained agreement between the calculated and experimental J values has not been satisfactory, the calculated couplings being always overestimated by comparison with the experimental ones. Nevertheless, total calculated SSCCs have reproduced correctly the experimental trend.

The new NH-acid *N*-(2,3,4,5,6-pentafluorophenyl)-*N*-nonafluoro*tert*-butylamine, $HN(C_6F_5)(C(CF_3)_3)$, bearing two different sterically demanding and strongly electronwithdrawing perfluorinated amine substituents has been synthesized by Kögel *et al.*[300] A triplet with a chemical shift of −69.6 ppm in benzene-d_6 and a coupling of 7.1 Hz have been observed by the authors for the CF_3 groups in the ^{19}F NMR spectrum of this compound and assigned to a coupling through-space with the aromatic *ortho*-fluorine atoms.

The results of a detailed solid state NMR spectroscopic investigation of P–P no-bond spin–spin couplings in variously substituted P,P-[3]ferrocenophane-based bisphosphanes have been published by Wiegand *et al.*[301] The isotropic couplings $^{TS}J_{PP}$, measured by 1D and 2D solid-state NMR experiments, have been found to range between 12 and 100 Hz; $^{TS}J_{PP}$ of 249 Hz has been reported for 1,8-(bisdiphenylphosphino)-naphthalene. Such values cannot be assigned to the couplings across four bonds since the typical $^{4}J_{PP}$ values are of the order of 5 Hz or less. Thus, the splittings observed in the spectra of the studied compounds should be described as through-bond couplings, as previously also concluded from liquid-state NMR. This conclusion has been corroborated by the DFT calculations performed for the model compounds.

Very large through-space couplings, formally $^{4}J_{PP}$, varying from 163 up to 199 Hz have been reported by Kilian and co-workers[110] for four heteroleptic bis(phosphines), Acenap($PiPr_2$)(PPhH), Acenap($PiPr_2$)(PFcH), Bis(phosphine), Acenap($PiPr_2$)(PPhMe) and Acenap($PiPr_2$)(PFcMe), where Acenap = acenaphthene-5,6-diyl and F = ferrocenyl. These coupling values are comparable in magnitude to $^{1}J_{PP}$ couplings observed in salts, *peri*-substituted phosphino-phosphonium salts $[Acenap(PiPr_2)(PR)]^+Cl^-$ (R = Ph; R = Fc; R = NMe_2; R = iPr) (303–412 Hz).

In continuation of their studies on through-space couplings between the chalcogen atoms in *peri* substituted naphtalenes and acetnaphtenes Woollins and co-workers[302] have synthesized a series of mixed RPhTe, RPhSe acenaphthenes with a goal to investigate how substituents R at the phenyl rings affect the bonding interactions and the coupling value between formally non bonded Te and Se atoms. The compounds studied were: 5-(4-fluorophenyltelluro)-6-(phenylselenyl)-, 5-(4-methylphenyltelluro)-6-(phenylselenyl)-, 5-(4-methoxyphenyltelluro)-6-(phenylselenyl)-, 5-(2-methoxyphenyltelluro)-6-(phenylselenyl)-, 5-(2-methoxyphenyltelluro)-6-(phenylselenyl)-, 5-(4-*tert*-butylphenyltelluro)-6-(phenylselenyl)-, 5-(2,4,6-trimethylphenyltelluro)-6-(phenylselenyl)-, 5-(2,4,6-tri-iso-propylphenyltelluro)-6-(phenylselenyl)- and 5-(naphthyltelluro)-6-(phenylselenyl)acenaphthenes. $^{TS}J_{SeTe}$ couplings, which change from −688 to −748 Hz do not vary significantly from that reported for the parent 5-(phenyltelluro)-6-(phenylselenyl)-acenaphthene, −716 Hz. The $^{TS}J_{SeTe}$ coupling for 5-(2,4,6-tri-iso-propylphenyltelluro)-6-(phenylselenyl)-acetnaphtene has been measured also in the solid state agreeing well with that found in solution, −688 and −694 Hz, respectively. It is worth noticing that compared with experiment, the computed at the ZORA-SO/BP86/TZ2P level) couplings are significantly underestimated (−460 Hz to −585 Hz).

The indirect spin–spin coupling between two antimony atoms, $J_{SbSb} = 2.5 \pm 0.5$ kHz, has been determined by Orlova *et al.*[303] for the bournonite mineral ($CuPbSbS_3$) on the basis of the temperature dependent studies of its nuclear quadrupole resonance spectrum. According to the authors this interaction is possibly transferred *via* the Sb–S–Cu–S–Sb bonds.

11 Couplings through hydrogen and pnicogen bonds

The relationship between spin–spin couplings, $^{1h}J_{H\cdots Y}$ and $^{2h}J_{H\cdots Y}$, and interaction energy and electronic density at corresponding hydrogen bond critical points and H-bonds distances in uracil and nitrous acid (U–NA) [$C_4N_2O_2H_4ANO_2H$] systems have been investigated by Makiabadi *et al.*[304] by the use of B3LYP, B3PW91, and MP2 methods with different basis sets.

One- and two bond spin–spin couplings, ^1J, ^{1h}J, and ^{2h}J across X–H···O hydrogen bonds and shielding constants of bridging hydrogens have been computed by Roohi and Nokhostin[305] for complexes formed from interaction between the α-hydroxy-*N*-nitrosamine (NP) and four preferential binding sites of the uracil (U) at the B3LYP/6-311++G(2d,2p)// MP2/6-311++G(2d,2p) level of theory. All complexes are stabilized by two HU···ONP and HNP···OU hydrogen bonds. Very good correlations have been found by the authors between NMR spin–spin couplings, $^{1h}J_{H\cdots O}$ and $^{2h}J_{X\cdots Y}$ as well as isotropic shielding constant and the binding energy, H-bond distance, red-shift of vibration frequency, charge transfer energy, and electron density at H-bond critical point.

A stereocontrolled synthesis of all-*cis*-1,2,4,5-tetrafluoro-3-phenylcyclohexane has been performed by Durie *et al.*,[306] resulting in the orthogonal conformation of the aryl and cyclohexyl rings in this compound and its an *ortho*-nitro derivative. The theoretical calculations predicted the presence of $^{1h}J_{HF}$ and $^{2h}J_{CF}$ NMR couplings relayed through hydrogen bonding in both compounds. However, the experimental ^1H NMR data has shown only small J_{HF} couplings of *ca.* 1 Hz, which can be interpreted also in terms of five-bond coupling. The experimental $^{2h}J_{CF}$ couplings are considerably larger and therefore much more convincing; $^{2h}J_{CF}$ of 2.4 Hz has been observed for the parent compound and $^{2h}J_{CF}$ of 7.3 has been found for its nitro derivative.

J_{HPt} couplings of 67.0 Hz and 80.6 Hz have been determined by Baya *et al.*[307] in [Pt(C_6F_5)(bzq)L] and [Pt(C_6F_5)(bzq)L′] complexes, respectively (bzq = 7,8-benzoquinolinate; L = 8-hydroxyquinoline; L′ = 2-methyl-8-hydroxyquinoline). This result has been invoked by the authors as evidence of the existence of Pt···H–O hydrogen bonds in solution in these compounds.

A comprehensive theoretical study has been carried out by Malkina and co-workers[308] on spin–spin couplings through a hydrogen bond in the free-base and protonated forms of the complete series of [$^{15}N_2$]-*N*-methylated 1,8-diaminonaphthalenes. The theoretically predicted couplings have been compared with the experimental values obtained earlier by Lloyd-Jones *et al.*[309] The authors have demonstrated that the overlap of the coupling pathway orbitals plays a dominating role in the spin–spin coupling mechanism. They have also re-examined the correlation of the sum of the ^{13}C NMR shifts of the naphthalene ring C(1,8) carbons directly attached to the interacting nitrogens with the J_{NN} values and the degree of methylation found earlier by Lloyd-Jones *et al.*[309]

Theoretical studies of resonance assisted hydrogen bonds in open-chain and cyclic structures of malonaldehyde enol has been performed by Trujillo *et al.*[310] Among others, NMR parameters such as $^{2h}J_{OO}$ couplings and chemical shifts have been studied by the authors.

It has been shown by Elguero *et al.*[311] that intermolecular couplings $^{2h}J_{XY}$ across X–H···Y hydrogen bonds for X, Y = N, O in studied by them diazamalonaldehyde and azamalonaldehyde and their saturated counterparts are very well approximated by FC terms independent of whether the hydrogen-bonded complex is neutral or positively charged, or if the hydrogen bond is symmetric or asymmetric. Interestingly, these couplings are not sensitive to the presence of resonance-assisted hydrogen bonds (RAHBs). However, the $^{2h}J_{OO}$ and $^{2h}J_{OS}$ couplings in malonaldehyde and its sulphur analogue, respectively, are the PSO term dominated and sensitive to RAHBs. The *ab initio* equation-of-motion coupled cluster singles and doubles calculations (EOM-CCSD/(qzp, qz2p)) in the configuration interaction (CI)-like approximation studies performed by the authors indicate that coupling in these systems depends on s-electron densities in ground and exited states.

Intensive but unsuccessful efforts have been undertaken by Sanz *et al.*[312] to determine experimentally $^{1p}J_{PP}$ coupling in a dimer of 3,4-dimethyl-1-cyanophosphole by the use of the different solid state (CPMAS) experiments. The dimer presents a P···P interaction of the pnictogen type due to the presence of σ-holes. The calculated $^{1p}J_{PP}$ coupling values at the B3LYP level are +163.2 Hz and +101.8 Hz at 3.21 Å (minimum) and 3.38 Å (X-ray) P···P distances, respectively.

Alkorta, del Bene and Elguero have continued their extensive theoretical studies on pnicogen bonded compounds; this included complexes of PO_2X (X = F, Cl) with nitrogen bases,[313] cyclic trimers $(PH_2X)_3$ with X = F, Cl, OH, NC, CN, CH_3, H, and BH_2,[314] and complexes with pnicogen bonds involving sp^2 hybridized phosphorus atoms: $(H_2C{=}PX)_2$ with X = F, Cl, OH, CN, NC, CCH, H, CH_3, and BH_2,[315] complexes $H_2C{=}(X)P{:}PXH_2$, for X = F, Cl, OH, CN, NC, CCH, H, CH_3, and BH_2.[316] Further studies of this group covered the influence of substituent effects on the formation of P···Cl pnicogen bonds or halogen bonds,[317] pnicogen bonds between $X{=}PH_3$ (X = O, S, NH, CH_2) and phosphorus and nitrogen bases,[318] *ab initio* MP2/aug′-cc-pVTZ calculations[319] on complexes $H_2XP{:}PCX$, for X = F, Cl, OH, NC, CN, CCH, CH_3, and H, in search of complexes stabilized by P···P pnicogen bonds, and the theoretical investigations of the pnicogen-bonded complexes $H_2YP{:}X^-$ (X,Y = Cl, NC, F, C≡CH, and CH_3).[320] These papers are only briefly mentioned here, the more detailed review on them are presented in the preceding chapter by Jaźwiński.

12 Residual dipolar couplings

Sabo *et al.*[321] have presented ORIUM: *O*ptimized *R*DC-based *I*terative and *U*nified *M*odel-free analysis, a method for extracting structural and dynamic information from $^1D_{HN}$ and $^1D_{H\alpha C\alpha}$ measured for proteins. In the new method previously used concepts, the model free analysis (MFA) and the direct interpretation of dipolar couplings (DIDCs), are unified into one compact protocol. Calculations of the dynamic average orientation and associated amplitudes of molecular motion for inter-nuclear vectors required RDC data sets from at least five independent alignment media. Montalvao *et al.*[322] have postulated tensor-free calculations of residual dipolar couplings for the study of protein dynamics. Such an approach should be considered for conformationally heterogeneous states. However, the authors did not propose their own solution for this idea. Luchinat and co-workers[323] have modified the program REFMAC5 (from CCP4) to allow the simultaneous use of solution protein NMR data, residual dipolar couplings and paramagnetic data, with complementary X-ray crystallographic data. If both sets of data produce a consistent single protein structure, the reliability of the derived atomic model increases. The joint refinement can also reveal differences between solid-state and solution conformations of molecules.

Navarro-Vázquez and co-workers[324] have tested cromolyn–brine mixtures as a water-compatible NMR orienting medium. Using a variety of samples regarding their polarity and structural complexity, the optimum degree of alignment is achieved allowing to record D_{HC} with good accuracy. In addition, doping of brine allows to conduct anisotropic and isotropic experiments in the same sample. Other examples of peptides, proteins, oligonucleotides and sugars whose structure was solved with the aid of dipolar couplings are listed in Table 3.

Fu and Wand[338] have presented the elevated (at 2500 bar) pressure-induced structural changes in ubiquitin detected on the basis of the deviations between the observed H-N RDCs and those calculated using the ambient pressure solution structure. In addition, they have confirmed that the spontaneous alignment of Pf1 phage as the medium remains stable at the high pressure. However, the latter observation was not corroborated by Sibille *et al.*[339] who have checked the robustness of alignment media to high pressure. They have found that among media used for measuring residual dipolar couplings a C12E5/*n*-hexanol alcohol mixture remained stable to 2500 bar, whereas a reversible transition was observed for Pf1 phage and 6HB nanotubes between 300 and 900 bar, and the effective alignment was abolished for phospholipid bicelles beyond 400 bar.

$^{1}D_{HN}$ in terms of *Q*-factors have been used by Kukic *et al.*[340] to validate the ensembles of conformations generated in MD simulations with replica-averaged chemical shift restraints. The ensembles represent the conformational fluctuations of, respectively, *Drosophila melanogaster* and human calmodulin in calcium-bound state.

Backbone $^{1}D_{HN}$s have been measured by Clore and co-workers[341] for the perdeuterated wild-type HIV-1 capsid (CA_{FL}, residues 1–231), the disulfide-linked mutant (CA_{FL}^{V181C}), the monomeric mutant ($CA_{FL}^{W184/M185A}$), and the four C-terminal domain constructs, $CA_{144-231}$ and $CA_{146-231}$ and the corresponding disulfidelinked-linked mutants ($CA_{144-231}^{V181C}$ and $CA_{146-231}^{V181C}$). Experimental RDCs supplemented by NMR relaxation measurements, X-ray scattering (SAXS/WAXS) data and analytical ultracentrifugation have been used to explore the

Table 3 Proteins, peptides, oligonucleotides and carbohydrates for which the solution structure has been determined with RDCs used as restraints or comparables.

Name	*a*	*b*	*c*	Reference
Proteins and peptides				
Aβ_{10-35}	26	15	$^1D_{HC}$	325
Aβ40, monomer–dimer equilibrium	40	29	$^1D_{HN}$	192
Aβ42, monomer–dimer equilibrium	42	30	$^1D_{HN}$	192
Aβ42-M35ox(S), monomer–dimer equilibrium	42	32	$^1D_{HN}$	192
The Nβ of CPT1C, the non-inhibitory state of the regulatory domain of the enzyme carnitine palmitoyltransferase 1	50	71	$^1D_{HN}$, $^1D_{CC}$, $^1D_{CN}$	326
Hepatitis C virus p7 in 1,2-dihexanoyl-*sn-glycero*-3-phosphocholine	63	53	$^1D_{HN}$	327
The MBD3 MBD	70	53	$^1D_{HN}$	328
Intrinsically disordered ACTR$_{1018-1088}$ and its 8 mutants	71	*ca.* 15	$^1D_{HN}$	329
Octanoyl-AcpP	75+	64	$^1D_{HN}$	330
Ubiquitin in the squalamine medium	76	332	$^1D_{HC}$, $^1D_{HN}$, $^1D_{CC}$, $^1D_{CN}$	178
The A-domain (1-84) of human HMGB1	84	123	$^1D_{HN}$, $^1D_{CN}$, $^2D_{HC}$	331
Ros87 H42A	87	40	$^1D_{HN}$	194
AcpP=FabA crosslinked construct	75+ 171	53	$^1D_{HN}$	330
TatB1–101 in DPC micelles	101	75	$^1D_{HN}$	332
p15^{PAF}, the proliferating-cell-nuclear-antigen associated factor, intrinsically disordered	111	86	$^1D_{HN}$	333
K18 construct, fragment of hTau40	130	130	$^1D_{HN}$	334
α-synuclein, intrinsically disordered	140	104	$^1D_{HN}$	334
K32 construct, fragment of hTau40	198	167, 168, 139	$^1D_{HN}$, $^2D_{HC}$, $^4D_{HH}$	334
hTau40, intrinsically disordered	441	228	$^1D_{HN}$	334
The c-Abl$_{83-534}$ · imatinib complex	452	120	$^1D_{HN}$	335
The c-Abl$_{83-534}$ · GNF-5 complex	452	86	$^1D_{HN}$	335
The c-Abl$_{83-534}$ · imatinib/GNF-5 complex	452	126	$^1D_{HN}$	335
HbCO	2×141 + 2×146	298	$^1D_{HN}$	336
Oligonucleotides				
E-SS construct, the ssRNA–helix junction	60	25	$^1D_{HC}$	337
Carbohydrates				
Streptomycin	2	20	$^1D_{HC}$	200

[a] Number of residues or nucleotides or sugar units.
[b] The total number of residual dipolar couplings measured or used.
[c] Types of residual dipolar couplings measured or used.

conformational space sampled by the monomeric and dimeric species of the proteins. The same group[342] have also studied the influence of DNA binding on the large fragment of the HIV-1 Gag polyprotein, residues 133 to 432. With the aid of $^{1}D_{HN}$ it was shown that nucleic acids bind exclusively to the nucleocapsid domain and the four structural units, namely, the N- and C-terminal domains of capsid, and the N- and C-terminal Zn knuckles of nucleocapsid have the same structures as their individually isolated counterparts, and tumble semi-independently of one another in the absence of nucleic acids.

Nguyen *et al.*[330] have studied in detail interaction of *E. coli* acyl carrier protein, AcpP, with fatty acid 3-hydroxyacyl-ACP dehydratase, FabA. The RDCs were used to identify the residues at the binding interface of the covalently connected AcpP–FabA complex.

Williams, Jr. and co-workers[328] have measured $^{1}D_{HN}$ for methylcytosine-binding domain, MBD3, free and complexed with hydroxymethylated, methylated, and unmethylated DNA. The observed values show that the backbone structure of MBD3 does not change between complexes.

A comprehensive ^{1}H and ^{13}C chemical shift analysis combined with the evaluation of large residual dipolar couplings (RDCs) has been performed by Damjanovic *et al.*[343] for the single molecule magnet, tris(octabutoxyphthalocyaninato)diterbium complex. The authors emphasize that this is the first report of RDCs in SMMs published in the literature. The determined coupling values were between −78 and +99 Hz for the ^{13}C–^{1}H vectors of CH bonds and up to −109 Hz for ^{1}H–^{1}H vectors of geminal hydrogen atoms (magnetic field of 14.09 T, temperature 295 K).

Strain-induced fixation of alignment of PDMS/$CDCl_3$ gels and variable-angle assisted scaling strategies for tuning optimum range of alignment have been applied by Bharatam and co-workers[344] to record unambiguous one-bond C–H RDCs as well as the corresponding J_{HC} scalar couplings in a model multiple-stereo centre molecule, Rifamycin-S. The analysis aided by the refined RDCs thus obtained, allowed the authors to establish conformation as well as relative configuration of Rifamycin-S, from its 256 configurational possibilities.

Novel NMR methods for the measurement of heteronuclear residual dipolar couplings (RDCs) in molecules with severely overlapping NMR resonances have been designed by Uhrín and co-workers[345] which, in combination with other methods allowed them to determine 31 RDCs for α-D-cellobiose and 24 RDCs for β-D-cellobiose. The interpretation of the data in the approximation of a rigid disaccharide structure, using RDCs and interglycosidic ^{3}J couplings yielded conformation that is very close to that determined by the use of X-ray crystallography. Additionally, the RDC analysis in which rigid fragments rather than a complete molecule have been taken into consideration has shown that the glycosidic bond of cellobiose is not completely rigid and exhibits low-level flexibility. The sources of this flexibility have been discussed by the authors, who suggested that it is associated with the ψ more than the ϕ angle.

The very large splitting of 19.6 ± 2.3 Hz has been observed by Ren *et al.*[346] in ^{1}H-MRS (7 T) spectrum *in vivo* for lactate methyl and metine signals in fatigued human forearm muscle. Since the expected $^{3}J_{HH}$ value for this compound is *ca.* 7 Hz, the additional splitting has been attributed to the contribution from residual dipolar coupling. However, the information is untrustworthy since the metine signal has been observed as a doublet.

$^{1}D_{HC}$ residual dipolar couplings have been used by Norrehed *et al.*[347] for characterization of the solution conformations of *N,N′*-bis(benzhydryl)bispidine modulated by protonation. It has been found that the *chair–chair* conformation prevails for the free base and for the monoprotonated species, whereas the diprotonated bispidine exists as a mixture of *chair–boat* and *boat–boat* conformers; the latter constitutes up to 70% of the bispidine population.

Acknowledgement

The author expresses her gratitude to Dr Jacek Wójcik DSc for reviewing papers devoted to biochemical problems, preparing all the Tables and proofreading the chapter.

References

1 K. Kamieńska-Trela and J. Wójcik, *Nucl. Magn. Reson.*, 2014, **43**, 183.
2 K. J. Harris, J. Kowalewski and S. Cabral de Meneze, *Pure Appl. Chem.*, 1997, **69**, 2489.
3 J. C. Facelli and M. B. Ferraro, *Concepts Magn. Reson.*, 2013, **42 A**, 261.
4 J.-C. Hierso, *Chem. Rev.*, 2014, **114**, 4838.
5 J. P. Genet, T. Ayad and V. Ratovelomanana-Vidal, *Chem. Rev.*, 2014, **114**, 2824.
6 D. W. Allen and B. F. Taylor, *J. Chem. Soc., Dalton Trans.*, 1982, 51.
7 D. Massiot, R. J. Messinger, S. Cadars, M. Deschamps, V. Montouillout, N. Pellerin, E. Veron, M. Allix, P. Florian and F. Fayon, *Acc. Chem. Res.*, 2013, **46**, 1975.
8 M. Halabalaki, K. Vougogiannopoulou, E. Mikros and A. L. Skaltsounis, *Curr. Opin. Biotechnol.*, 2014, **25**, 1.
9 V. Tugarinov, *Prog. Nucl. Magn. Reson. Spectrosc.*, 2014, **77**, 49.
10 F. V. Toukach and V. P. Ananikov, *Chem. Soc. Rev.*, 2013, **42**, 8376.
11 G. Widmalm, *Carbohydr. Res.*, 2013, **378**, 123.
12 Y. Zhang, T. Yamaguchi and K. Kato, *Chem. Lett.*, 2013, **42**, 1455.
13 Y. E. Shapiro, *Prog. Biophys. Mol. Biol.*, 2013, **112**, 58.
14 J. Nováček, L. Žídek and V. Sklenář, *J. Magn. Reson.*, 2014, **241**, 41.
15 R. Konrat, *J. Magn. Reson.*, 2014, **241**, 74.
16 A. Gautier, *Biochim. Biophys. Acta*, 2014, **1837**, 578.
17 M. A. S. Hass and M. Ubbink, *Curr. Opin. Struct. Biol.*, 2014, **24**, 45.
18 G. W. Vuister, R. H. Fogh, P. M. S. Hendrickx, J. F. Doreleijers and A. Gutmanas, *J. Biomol. NMR*, 2014, **58**, 259.
19 V. H. Pomin, *Prog. Biophys. Mol. Biol.*, 2014, **114**, 61.
20 M. D. Battistel, H. F. Azurmendi, B. Yu and D. I. Freedberg, *Prog. Nucl. Magn. Reson. Spectrosc.*, 2014, **79**, 48.
21 T. Parella and J. F. Espinosa, *Prog. Nucl. Magn. Reson. Spectrosc.*, 2013, **73**, 17.
22 K. Kazimierczuk, M. Misiak, J. Stanek, A. Zawadzka-Kazimierczuk and W. Koźmiński, *Top. Curr. Chem.*, 2012, **316**, 79.
23 T. Reinsperger and B. Luy, *J. Magn. Reson.*, 2014, **239**, 110.

24 I. Timári, L. Kaltschnee, A. Kolmer, R. W. Adams, M. Nilsson, C. M. Thiele, G. A. Morris and K. E. Kövér, *J. Magn. Reson.*, 2014, **239**, 130.
25 K. Furihata and M. Tashiro, *Magn. Reson. Chem.*, 2014, **52**, 27.
26 J. Saurí and T. Parella, *Magn. Reson. Chem.*, 2013, **51**, 509.
27 J. Saurí, T. Parella and J. F. Espinosa, *Org. Biomol. Chem.*, 2013, **11**, 4473.
28 L. Castañar, J. Saurí, P. Nolis, A. Virgili and T. Parella, *J. Magn. Reson.*, 2014, **238**, 63.
29 J. Saurí, L. Castañar, P. Nolis, A. Virgili and T. Parella, *J. Magn. Reson.*, 2014, **242**, 33.
30 L. Castañar, P. Nolis, A. Virgili and T. Parella, *Chem. – Eur. J.*, 2013, **19**, 17283.
31 J. F. Espinosa, *J. Org. Chem.*, 2013, **78**, 12844.
32 J. F. Espinosa and H. Broughton, *Eur. J. Org. Chem.*, 2013, 6973.
33 A. A. Marchione, R. J. Dooley and B. Conklin, *Magn. Reson. Chem.*, 2014, **52**, 183.
34 Lokesh, S. R. Chaudhari and N. Suryaprakash, *Chem. Phys. Lett.*, 2014, **602**, 40.
35 M. E. Di Pietro, C. Aroulanda and D. Merlet, *J. Magn. Reson.*, 2013, **234**, 101.
36 M. Foroozandeh, P. Giraudeau and D. Jeannerat, *Magn. Reson. Chem.*, 2013, **51**, 808.
37 V. Blechta and J. Schraml, *Magn. Reson. Chem.*, 2013, **51**, 743.
38 P. Thureau, G. Mollica, F. Ziarelli and S. Viel, *J. Magn. Reson.*, 2013, **231**, 90.
39 R. L. E. G. Aspers, K. A. M. Ampt, P. Dvortsak, M. Jaeger and S. S. Wijmenga, *J. Magn. Reson.*, 2013, **231**, 79.
40 R. Pendrill, O. W. Sørensen and G. Widmalm, *Magn. Reson. Chem.*, 2014, **52**, 82.
41 J. Colell, P. Türschmann, S. Glöggler, P. Schleker, T. Theis, M. Ledbetter, D. Budker, A. Pines, B. Blümich and S. Appelt, *Phys. Rev. Lett.*, 2013, **110**, 137602.
42 P. Türschmann, J. Colell, T. Theis, B. Blümich and S. Appelt, *Phys. Chem. Chem. Phys.*, 2014, **16**, 15411.
43 A. Altmayer-Henzien, V. Declerck, D. J. Aitken, E. Lescop, D. Merlet and J. Farjon, *Org. Biomol. Chem.*, 2013, **11**, 7611.
44 A. Dallmann, B. Simon, M. M. Duszczyk, H. Kooshapur, A. Pardi, W. Bermel and M. Sattler, *Angew. Chem., Int. Ed.*, 2013, **52**, 10487.
45 J. F. Ying, J. L. Roche and A. Bax, *J. Magn. Reson.*, 2014, **241**, 97.
46 L. Arbogast, A. Majumdar and J. R. Tolman, *J. Magn. Reson.*, 2013, **235**, 26.
47 F. A. Perras and D. L. Bryce, *J. Am. Chem. Soc.*, 2013, **135**, 12596.
48 J. San Fabián, J. M. G. de la Vega, R. Suardíaz, M. Fernández-Oliva, C. Pérez, R. Crespo-Otero and R. H. Contreras, *Magn. Reson. Chem.*, 2013, **51**, 775.
49 R. B. Nazarski and W. Makulski, *Phys. Chem. Chem. Phys.*, 2014, **16**, 15699.
50 M. Dračínský and P. Hodgkinson, *Chem. – Eur. J.*, 2014, **20**, 2201.
51 U. Pal, S. Sen and N. C. Maiti, *J. Phys. Chem. A*, 2014, **118**, 1024.
52 A. F. Dunlop-Brière, M. C. Baird and P. H. M. Budzelaar, *J. Am. Chem. Soc.*, 2013, **135**, 17514.
53 A. L. Hansen and L. E. Kay, *Proc. Natl. Acad. Sci. U. S. A.*, 2014, **111**, E1705.
54 M. Hutin, J. K. Sprafke, B. Odell, H. L. Anderson and T. D. W. Claridge, *J. Am. Chem. Soc.*, 2013, **135**, 12798.
55 M. F. Santos, C. B. Braga, T. C. Rozada, E. A. Basso and B. C. Fiorin, *Spectrochim. Acta, Part A*, 2014, **129**, 148.
56 I. Cukrowski, J. H. de Lange and M. Mitoraj, *J. Phys. Chem. A*, 2014, **118**, 623.
57 S. Szymański, P. Paluch, D. T. Gryko, A. Nowak-Król, W. Bocian, J. Sitkowski, B. Koszarna, J. Śniechowska, M. J. Potrzebowski and L. Kozerski, *Chem. – Eur. J.*, 2014, **20**, 1720.

58 S. Xiang and M. Zweckstetter, *J. Magn. Reson.*, 2014, **243**, 93.
59 A. A. Sehgal, L. Duma, G. Bodenhausen and P. Pelupessy, *Chem. – Eur. J.*, 2014, **20**, 6332.
60 A. V. Chernyak, A. V. Chub, E. A. Sanginov and P. Y. Barzilovich, *Russ. Chem. Bull. Int. Ed.*, 2013, **62**, 1798.
61 M. Pietrzak, E. Grech, J. Nowicka-Scheibe and P. E. Hansen, *Magn. Reson. Chem.*, 2013, **51**, 683.
62 J. Klimkiewicz, L. Stefaniak, E. Grech, J. Nowicka-Scheibe and G. A. Webb, *J. Phys. Org. Chem.*, 1996, **9**, 781.
63 M. Novák, L. Dostál, M. Alonso, F. De Proft, A. Růžička, A. Lyčka and R. Jambor, *Chem. – Eur. J.*, 2014, **20**, 2542.
64 A. A. Zuzek and G. Parkin, *J. Am. Chem. Soc.*, 2014, **136**, 8177.
65 M. Q. Y. Tay, Y. Lu, R. Ganguly and D. Vidović, *Chem. – Eur. J.*, 2014, **20**, 6628.
66 P. E. Sues, A. J. Lough and R. H. Morris, *J. Am. Chem. Soc.*, 2014, **136**, 4746.
67 S. M. McCarthy, Y. C. Lin, D. Devarajan, J. W. Chang, H. P. Yennawar, R. M. Rioux, D. H. Ess and A. T. Radosevich, *J. Am. Chem. Soc.*, 2014, **136**, 4640.
68 C. Schädle, A. Fischbach, E. Herdtweck, K. W. Törnroos and R. Anwander, *Chem. – Eur. J.*, 2013, **19**, 16334.
69 T. Shima, M. Nishiura and Z. Hou, *Organometallics*, 2011, **30**, 2513.
70 W. J. Evans, D. K. Drummond, T. P. Hanusa and R. J. Doedens, *Organometallics*, 1987, **6**, 2279.
71 K. H. den Haan, Y. Wielstra and J. H. Teuben, *Organometallics*, 1987, **6**, 2053.
72 Y. Takenaka and Z. Hou, *Organometallics*, 2009, **28**, 5196.
73 A. Zavras, G. N. Khairallah, T. U. Connell, J. M. White, A. J. Edwards, R. J. Mulder, P. S. Donnelly and R. A. J. O'Hair, *Inorg. Chem.*, 2014, **53**, 7429.
74 A. K. Swarnakar, S. M. McDonald, K. C. Deutsch, P. Choi, M. J. Ferguson, R. McDonald and E. Rivard, *Inorg. Chem.*, 2014, **53**, 8662.
75 P. Garbacz, V. V. Terskikh, M. J. Ferguson, G. M. Bernard, M. Kędziorek and R. E. Wasylishen, *J. Phys. Chem. A*, 2014, **118**, 1203.
76 O. Rivada-Wheelaghan, M. Roselló-Merino, M. A. Ortuño, P. Vidossich, E. Gutiérrez-Puebla, A. Lledós and S. Conejero, *Inorg. Chem.*, 2014, **53**, 4257.
77 C. Su, R. Hopson and P. G. Williard, *J. Am. Chem. Soc.*, 2013, **135**, 14367.
78 F. A. Perras and D. L. Bryce, *Chem. Sci.*, 2014, **5**, 2428.
79 F. A. Perras and D. L. Bryce, *J. Magn. Reson.*, 2014, **242**, 23.
80 R. Liedtke, F. Scheidt, J. Ren, B. Schirmer, A. J. P. Cardenas, C. G. Daniliuc, H. Eckert, T. H. Warren, S. Grimme, G. Kehr and G. Erker, *J. Am. Chem. Soc.*, 2014, **136**, 9014.
81 I. T. A. Schuquel, L. C. Ducati, C. F. Tormena, M. P. de Freitas, D. G. de Kowalewski and R. Rittner, *J. Mol. Struct.*, 2014, **1068**, 170.
82 A. L. Barsamian, M. J. Perkins, J. A. Field and P. R. Blakemore, *J. Labelled Compd. Radiopharm.*, 2014, **57**, 397.
83 H. Xu, S. Saebo and C. U. Pittman Jr., *Struct. Chem.*, 2014, **25**, 635.
84 M. Štefko, M. D. Tzirakis, B. Breiten, M. O. Ebert, O. Dumele, W. B. Schweizer, J. P. Gisselbrecht, C. Boudon, M. T. Beels, I. Biaggio and F. Diederich, *Chem. – Eur. J.*, 2013, **19**, 12693.
85 A. V. Afonin, D. V. Pavlov, I. A. Ushakov, E. P. Levanova and G. G. Levkovskaya, *Russ. J. Org. Chem.*, 2013, **49**, 1117.
86 M. Yamada, H. Kurihara, M. Suzuki, J. D. Guo, M. Waelchli, M. M. Olmstead, A. L. Balch, S. Nagase, Y. Maeda, T. Hasegawa, X. Lu and T. Akasaka, *J. Am. Chem. Soc.*, 2014, **136**, 7611.

87 K. Saidi, S. Kamoun and H. F. Ayedi, *Ionics*, 2014, **20**, 517.
88 M. A. Kettner, K. Karaghiosoff, T. M. Klapötke, M. Sućeska and S. Wunder, *Chem. – Eur. J.*, 2014, **20**, 7622.
89 J. M. Silla, M. P. Freitas, R. A. Cormanich and R. Rittner, *J. Org. Chem.*, 2014, **79**, 6385.
90 I. L. Rusakova, Y. Y. Rusakov and L. B. Krivdin, *Magn. Reson. Chem.*, 2014, **52**, 413.
91 F. Lavigne, E. Maerten, G. Alcaraz, N. Saffon-Merceron and A. Baceiredo, *Chem. – Eur. J.*, 2014, **20**, 297.
92 K. Gholivand and N. Dorosti, *Monatsh. Chem.*, 2013, **144**, 1417.
93 S. D. Hastings, H. Byrd, L. N. Gray, M. J. Jablonsky, J. L. Freeman and G. M. Gray, *Eur. J. Inorg. Chem.*, 2013, 2900.
94 Y. Y. Rusakov, I. L. Rusakova and L. B. Krivdin, *Magn. Reson. Chem.*, 2014, **52**, 214.
95 B. Wrackmeyer, E. V. Klimkina and W. Milius, *Eur. J. Inorg. Chem.*, 2014, 1929.
96 D. Kubica, A. Wodyński, A. Kraska-Dziadecka and A. Gryff-Keller, *J. Phys. Chem. A*, 2014, **118**, 2995.
97 C. Pranckevicius and D. W. Stephan, *Chem. – Eur. J.*, 2014, **20**, 6597.
98 R. Colorado-Peralta, C. Guadarrama-Pérez, L. A. Martínez-Chavando, J. C. Gálvez-Ruiz, A. M. Duarte-Hernández, G. V. Suárez-Moreno, A. Vásquez-Badillo, S. A. Sánchez-Ruiz, R. Contreras and A. Flores-Parra, *J. Organomet. Chem.*, 2014, **751**, 579.
99 P. Montes-Tolentino, R. Colorado-Peralta, L. A. Martínez-Chavando, E. Mijangos, A. M. Duarte-Hernández, G. V. Suárez-Moreno, R. Contreras and A. Flores-Parra, *J. Organomet. Chem.*, 2014, **751**, 591.
100 R. C. R. Chagas, J. R. da Silveira Maia and V. P. Ferraz, *Main Group Met. Chem.*, 2012, **34**, 131.
101 M. Saito, T. Akiba, M. Kaneko, T. Kawamura, M. Abe, M. Hada and M. Minoura, *Chem. – Eur. J.*, 2013, **19**, 16946.
102 T. Kawamura, M. Abe, M. Saito and M. Hada, *J. Comput. Chem.*, 2014, **35**, 847.
103 B. Kołodziej, E. Grech, W. Schilf, B. Kamieński, A. Pazio and K. Woźniak, *J. Mol. Struct.*, 2014, **1063**, 145.
104 N. Allefeld, B. Neumann, H. G. Stammler, G. V. Röschenthaler, N. Ignat'ev and B. Hoge, *Chem. – Eur. J.*, 2014, **20**, 7736.
105 J. Bader, N. Ignat'ev and B. Hoge, *Inorg. Chem.*, 2014, **53**, 7547.
106 M. Biswal, M. Body, C. Legein, A. Sadoc and F. Boucher, *J. Solid State Chem.*, 2013, **207**, 208.
107 K. Hansen, T. Szilvási, B. Blom, E. Irran and M. Driess, *Chem. – Eur. J.*, 2014, **20**, 1947.
108 A. Kreienbrink, S. Heinicke, T. T. D. Pham, R. Frank, P. Lönnecke and E. Hey-Hawkins, *Chem. – Eur. J.*, 2014, **20**, 1434.
109 A. Hinz, A. Schulz and A. Villinger, *Chem. – Eur. J.*, 2014, **20**, 3913.
110 M. J. Ray, A. M. Z. Slawin, M. Bühl and P. Kilian, *Organometallics*, 2013, **32**, 3481.
111 M. J. Ray, M. Bühl, L. J. Taylor, K. S. A. Arachchige, A. M. Z. Slawin and P. Kilian, *Inorg. Chem.*, 2014, **53**, 8538.
112 A. Nordheider, T. Chivers, O. Schön, K. Karaghiosoff, K. S. A. Arachchige, A. M. Z. Slawin and J. D. Woollins, *Chem. – Eur. J.*, 2014, **20**, 704.
113 A. Tohmé, H. Sahnoune, T. Roisnel, V. Dorcet, J. F. Halet and F. Paul, *Organometallics*, 2014, **33**, 3385.
114 I. L. Rusakova, Y. Y. Rusakov and L. B. Krivdin, *Magn. Reson. Chem.*, 2014, **52**, 500.

115 D. Schaarschmidt, A. Hildebrandt, S. Bock and H. Lang, *J. Organomet. Chem.*, 2014, **751**, 742.
116 D. W. Allen, I. W. Nowell and B. F. Taylor, *J. Chem. Soc., Dalton Trans.*, 1985, 2505.
117 J. M. Lister, M. Carreira, M. F. Haddow, A. Hamilton, C. L. McMullin, A. G. Orpen, P. G. Pringle and T. E. Stennett, *Organometallics*, 2014, **33**, 702.
118 C. C. Comanescu and V. M. Iluc, *Inorg. Chem.*, 2014, **53**, 8517.
119 I. Arenas, M. Á. Fuentes, E. Álvarez, Y. Díaz, A. Caballero, S. Castillón and P. J. Pérez, *Inorg. Chem.*, 2014, **53**, 3991.
120 J. Malberg, T. Wiegand, H. Eckert, M. Bodensteiner and R. Wolf, *Eur. J. Inorg. Chem.*, 2014, 1638.
121 C. C. Cummins, C. Huang, T. J. Miller, M. W. Reintinger, J. M. Stauber, I. Tannou, D. Tofan, A. Toubaei, A. Velian and G. Wu, *Inorg. Chem.*, 2014, **53**, 3678.
122 M. A. Alvarez, M. E. García, D. García-Vivó, S. Melón, M. A. Ruiz and A. Toyos, *Inorg. Chem.*, 2014, **53**, 4739.
123 M. Golecki and B. Kersting, *Z. Anorg. Allg. Chem.*, 2013, **639**, 2473.
124 S. M. Nabavizadeh, H. Sepehrpour, R. Kia and A. L. Rheingold, *J. Organomet. Chem.*, 2013, **745–746**, 148.
125 S. J. Hoseini, S. M. Nabavizadeh, S. Jamali and M. Rashidi, *Eur. J. Inorg. Chem.*, 2008, 5099.
126 S. Jamali, S. M. Nabavizadeh and M. Rashidi, *Inorg. Chem.*, 2008, **47**, 5441.
127 S. Jamali, S. M. Nabavizadeh and M. Rashidi, *Inorg. Chem.*, 2005, **44**, 8594.
128 B. G. Anderson and J. L. Spencer, *Chem. – Eur. J.*, 2014, **20**, 6421.
129 A. Zucca, L. Maidich, L. Canu, G. L. Petretto, S. Stoccoro, M. A. Cinellu, G. J. Clarkson and J. P. Rourke, *Chem. – Eur. J.*, 2014, **20**, 5501.
130 C. M. McGuirk, C. L. Stern and C. A. Mirkin, *J. Am. Chem. Soc.*, 2014, **136**, 4689.
131 A. J. Rossini, A. W. Macgregor, A. S. Smith, G. Schatte, R. W. Schurko and G. G. Briand, *Dalton Trans.*, 2013, **42**, 9533.
132 D. L. Reger, A. E. Pascui, P. J. Pellechia and M. D. Smith, *Inorg. Chem.*, 2013, **52**, 11638.
133 B. H. Boo, M. Lee, K. S. Jeon and S. J. Kim, *J. Phys. Chem. A*, 2014, **118**, 2269.
134 A. Dubey and O. Heinonen, *BioMetals*, 2013, **26**, 1003.
135 L. R. Barbosa, Y. W. Vieira, V. Lacerda Jr., K. T. de Oliveira, R. B. dos Santos, S. J. Greco, A. C. Neto, E. V. R. de Castro and T. J. Brocksom, *Magn. Reson. Chem.*, 2014, **52**, 318.
136 Q. Wang, N. Dai and N. Han, *Magn. Reson. Chem.*, 2014, **52**, 511.
137 C. Villanueva-Cañongo, N. Pérez-Hernández, B. Hernández-Carlos, E. Cedillo-Portugal, P. Joseph-Nathan and E. Burgueño-Tapia, *Magn. Reson. Chem.*, 2014, **52**, 251.
138 A. E. Bañuelos-Hernández, J. A. Mendoza-Espiniza, R. Pereda-Miranda and C. M. Cerda-García-Rojas, *J. Org. Chem.*, 2014, **79**, 3752.
139 M. Štujber, S. Beldar, C. Rabong, I. Beseda, M. Breza and T. Liptaj, *Magn. Reson. Chem.*, 2014, **52**, 122.
140 A. L. Ward, W. W. Lukens, C. C. Lu and J. Arnold, *J. Am. Chem. Soc.*, 2014, **136**, 3647.
141 P. E. Hansen, E. V. Borisov and J. C. Lindon, *Spectrochim. Acta, Part A*, 2015, **136**, 107.
142 J. Oh, J. J. Bowling, Y. Zou, A. G. Chittiboyina, R. J. Doerksen, D. Ferreira, T. D. Leininger and M. T. Hamann, *Biochim. Biophys. Acta*, 2013, **1830**, 4229.
143 F. Cen-Pacheco, J. Rodríguez, M. Norte, J. J. Fernández and A. H. Daranas, *Chem. – Eur. J.*, 2013, **19**, 8525.

144 N. Maru, T. Inuzuka, K. Yamamoto, M. Kitamura, P. J. Schupp, K. Yamada and D. Uemura, *Tetrahedron Lett.*, 2013, **54**, 4385.
145 S. Di Micco, A. Zampella, M. V. D'Auria, C. Festa, S. De Marino, R. Riccio, C. P. Butts and G. Bifulco, *Beilstein J. Org. Chem.*, 2013, **9**, 2940.
146 Y. Luo, X. Pu, G. Luo, M. Zhou, Q. Ye, Y. Liu, J. Gu, H. Qi, G. Li and G. Zhang, *J. Nat. Prod.*, 2014, **77**, 1650.
147 T. Morofuji, A. Shimizu and J.-i. Yoshida, *J. Am. Chem. Soc.*, 2014, **136**, 4496.
148 J. D. Hackenberg, S. Kandu, T. J. Emge, K. Krogh-Jespersen and A. S. Goldman, *J. Am. Chem. Soc.*, 2014, **136**, 8891.
149 S. S. Kadam, L. Maier, I. Kostakis, N. Pouli, J. Toušek, M. Nečas, P. Marakos and R. Marek, *Eur. J. Org. Chem.*, 2013, 6811.
150 U. Girreser, U. Bluhm, B. Clement and D. Heber, *Magn. Reson. Chem.*, 2013, **51**, 714.
151 S. Opekar, R. Pohl, P. Beran, L. Rulíšek and P. Beier, *Chem. – Eur. J.*, 2014, **20**, 1453.
152 J. F. Kögel, X. Xie, E. Baal, D. Gesevičius, B. Oelkers, B. Kovačević and J. Sundermeyer, *Chem. – Eur. J.*, 2014, **20**, 7670.
153 Y. Duan, L. Li, M. W. Chen, C. B. Yu, H. J. Fan and Y. G. Zhou, *J. Am. Chem. Soc.*, 2014, **136**, 7688.
154 P. Dydio, R. J. Detz, B. de Bruin and J. N. H. Reek, *J. Am. Chem. Soc.*, 2014, **136**, 8418.
155 N. G. Khusainova, E. P. Pogodina and A. V. Gerasimov, *Russ. J. Org. Chem.*, 2013, **49**, 1399.
156 Y. Gloaguen, L. M. Jongens, J. N. H. Reek, M. Lutz, B. de Bruin and J. I. van der Vlugt, *Organometallics*, 2013, **32**, 4284.
157 L. Yun, H. Vazquez-Lima, H. Fang, Z. Yao, G. Geisberger, C. Dietl, A. Ghosh, P. J. Brothers and X. Fu, *Inorg. Chem.*, 2014, **53**, 7047.
158 M. Safari, M. Yousefi, H. A. Jenkins, M. B. Torbati and A. Amanzadeh, *Med. Chem. Res.*, 2013, **22**, 5730.
159 Y. Y. Rusakov, L. B. Krivdin, F. F. Østerstrøm, S. P. A. Sauer, V. A. Potapov and S. V. Amosova, *Phys. Chem. Chem. Phys.*, 2013, **15**, 13101.
160 H. Cui, R. Goddard, K. R. Pörschke, A. Hamacher and M. U. Kassack, *Inorg. Chem.*, 2014, **53**, 3371.
161 J. Emsermann, A. J. Arduengo, III and T. Opatz, *Synthesis*, 2013, **45**, 2251.
162 E. Khan, O. L. Tok and B. Wrackmeyer, *Appl. Organomet. Chem.*, 2014, **28**, 280.
163 B. Wrackmeyer, C. Köhler and J. Weidinger, *Z. Anorg. Allg. Chem.*, 2014, **640**, 195.
164 B. Wrackmeyer, E. V. Klimkina and W. Milius, *Eur. J. Inorg. Chem.*, 2014, 233.
165 O. Steinhof, É. J. Kibrik, G. Scherr and H. Hasse, *Magn. Reson. Chem.*, 2014, **52**, 138.
166 A. Arévalo, A. Tlahuext-Aca, M. Flores-Alamo and J. J. García, *J. Am. Chem. Soc.*, 2014, **136**, 4634.
167 C. Hering, M. Hertrich, A. Schulz and A. Villinger, *Inorg. Chem.*, 2014, **53**, 3880.
168 B. J. Barrett and V. M. Iluc, *Inorg. Chem.*, 2014, **53**, 7248.
169 M. A. Huertos and A. S. Weller, *Chem. Sci.*, 2013, **4**, 1881.
170 T. N. Hooper, M. A. Huertos, T. Jurca, S. D. Pike, A. S. Weller and I. Manners, *Inorg. Chem.*, 2014, **53**, 3716.
171 T. Kimura, T. Nakahodo, H. Fijihara and E. Suzuki, *Inorg. Chem.*, 2014, **53**, 4411.
172 R. Shankar, N. Singla, M. Asija, G. Kociok-Köhn and K. C. Molloy, *Inorg. Chem.*, 2014, **53**, 6195.

173 B. Wrackmeyer, E. V. Klimkina and W. Milius, *J. Organomet. Chem.*, 2013, **747**, 140.
174 M. Haouas, I. M. Mbomekallé, N. Vila, P. de Oliveira and F. Taulelle, *Inorg. Chem.*, 2014, **53**, 5568.
175 B. J. van Lierop, S. D. Robinson, S. N. Kompella, A. Belgi, J. R. McArthur, A. Hung, C. A. MacRaild, D. J. Adams, R. S. Norton and A. J. Robinson, *ACS Chem. Biol.*, 2013, **8**, 1815.
176 I. Gomez-Monterrey, A. Carotenuto, E. Cipolletta, M. Sala, E. Vernieri, A. Limatola, A. Bertamino, S. Musella, P. Grieco, B. Trimarco, E. Novellino, G. Iaccarino and P. Campiglia, *Biopolymers*, 2014, **101**, 121.
177 A. S. Maltsev, A. Grishaev, J. Roche, M. Zasloff and A. Bax, *J. Am. Chem. Soc.*, 2014, **136**, 3752.
178 A. B. Aliev, M. Kulke, H. S. Khaneja, V. Chudasama, T. D. Sheppard and R. M. Lanigan, *Proteins*, 2014, **82**, 195.
179 R. A. Cormanich, L. C. Ducati, C. F. Tormena and R. Rittner, *J. Phys. Org. Chem.*, 2013, **26**, 849.
180 S. Kumar, S. I. Steffansen, N. Albæk and P. Nielsen, *Tetrahedron*, 2014, **70**, 583.
181 S. Kaeothip, A. Ishiwata, T. Ito, S. Fushinobu, K. Fujita and Y. Ito, *Carbohydr. Res.*, 2013, **382**, 95.
182 F. M. Qin, L. J. Meng, H. L. Zou and G. X. Zhou, *Chem. Pharm. Bull.*, 2013, **61**, 1071.
183 E. M. Vilén and C. Sandström, *Molecules*, 2013, **18**, 9735.
184 K. R. Riihinen, V. V. Mihaleva, T. Gödecke, P. Soininen, R. Laatikainen, J. M. Vervoort, D. C. Lankin and G. F. Pauli, *Phytochem. Anal.*, 2013, **24**, 476.
185 D. Alvarez-Dorta, E. I. León, A. R. Kennedy, A. Martín, I. Pérez-Martín, C. Riesco-Fagundo and E. Suárez, *Chem. – Eur. J.*, 2014, **20**, 2663.
186 C. B. Braga, L. C. Ducati, C. F. Tormena and R. Rittner, *J. Phys. Chem. A*, 2014, **118**, 1748.
187 E. A. Lubecka, E. Sikorska, A. Marcinkowska and J. Ciarkowski, *J. Pept. Sci.*, 2014, **20**, 406.
188 L. Russo, L. Raiola, M. A. Campitiello, A. Magrì, R. Fattorusso, G. Malgieri, G. Pappalardo, D. La Mendola and C. Isernia, *Molecules*, 2013, **18**, 11467.
189 S. L. Luo, D. T. Zhangsun, C. I. Schroeder, X. P. Zhu, Y. Y. Hu, Y. Wu, M. M. Weltzin, S. Eberhard, Q. Kaas, D. J. Craik, J. M. McIntosh and P. Whiteaker, *FASEB J.*, 2014, **28**, 1842.
190 A. A. Berkut, D. R. Usmanova, S. Peigneur, P. B. Oparin, K. S. Mineev, T. I. Odintsova, J. Tytgat, A. S. Arseniev, E. V. Grishin and A. A. Vassilevski, *J. Biol. Chem.*, 2014, **289**, 14331.
191 D. J. Rosenman, C. R. Connors, W. Chen, C. Y. Wang and A. E. García, *J. Mol. Biol.*, 2013, **425**, 3338.
192 A. E. Conicella and N. L. Fawzi, *Biochemistry*, 2014, **53**, 3095.
193 M. Palmieri, L. Russo, G. Malgieri, S. Esposito, I. Baglivo, A. Rivellino, B. Farina, I. de Paola, L. Zaccaro, D. Milardi, C. Isernia, P. V. Pedone and R. Fattorusso, *J. Inorg. Biochem.*, 2014, **131**, 30.
194 A. K. Gizatullina, E. I. Finkina, K. S. Mineev, D. N. Melnikova, I. V. Bogdanov, I. N. Telezhinskaya, S. V. Balandin, Z. O. Shenkarev, A. S. Arseniev and T. V. Ovchinnikova, *Biochem. Biophys. Res. Commun.*, 2013, **439**, 427.
195 D. Mellal, M. Fonvielle, M. Santarem, M. Chemama, Y. Schneider, L. Iannazzo, E. Braud, M. Arthur and M. Ethève-Quelquejeu, *Org. Biomol. Chem.*, 2013, **11**, 6161.

196 J. Rönnols, S. Manner, A. Siegbahn, U. Ellervik and G. Widmalm, *Org. Biomol. Chem.*, 2013, **11**, 5465.
197 J. C. Muñoz-García, F. Corzana, J. L. de Paz, J. Angulo and P. M. Nieto, *Glycobiology*, 2013, **23**, 1220.
198 R. Pendrill, E. Säwén and G. Widmalm, *J. Phys. Chem. B*, 2013, **117**, 14709.
199 C. D. Blundell, M. J. Packer and A. Almond, *Bioorg. Med. Chem.*, 2013, **21**, 4976.
200 U. Zähringer, K. Kawahara and P. Kosma, *Carbohydr. Res.*, 2013, **378**, 63.
201 V. Vitiazeva, J. J. Li, D. W. Hood, E. R. Moxon and E. K. H. Schweda, *Carbohydr. Res.*, 2013, **378**, 114.
202 P. B. Shinde, A. R. Han, J. Cho, S. R. Lee, Y. H. Ban, Y. J. Yoo, E. J. Kim, E. Kim, M. C. Song, J. W. Park, D. G. Lee and Y. J. Yoon, *J. Biotechnol.*, 2013, **168**, 142.
203 S. A. Kolesnikova, E. G. Lyakhova, A. I. Kalinovsky, M. A. Pushilin, S. S. Afiyatullov, E. A. Yurchenko, S. A. Dyshlovoy, C. V. Minh and V. A. Stonik, *J. Nat. Prod.*, 2013, **76**, 1746.
204 B. A. Soares, C. L. Firme, M. A. M. Maciel, C. R. Kaiser, E. Schilling and A. J. Bortoluzzi, *J. Braz. Chem. Soc.*, 2014, **25**, 629.
205 H. Harms, V. Rempel, S. Kehraus, M. Kaiser, P. Hufendiek, C. E. Müller and G. M. König, *J. Nat. Prod.*, 2014, **77**, 673.
206 J. J. Lv, S. Yu, Y. F. Wang, D. Wang, H. T. Zhu, R. R. Cheng, C. R. Yang, M. Xu and Y. J. Zhang, *J. Org. Chem.*, 2014, **79**, 5432.
207 F. D. Mabou, P. L. F. Tebou, D. Ngnokam, D. Harakat and L. Voutquenne-Nazabadioko, *Magn. Reson. Chem.*, 2014, **52**, 32.
208 H. L. Jiang, X. Z. Wang, J. Xiao, X. H. Luo, X. J. Yao, Y. Y. Zhao, Y. J. Chen, P. Crews and Q. X. Wu, *Tetrahedron*, 2013, **69**, 6687.
209 D. H. A. Rocha, D. C. G. A. Pinto, R. S. G. R. Seixas and A. M. S. Silva, *Magn. Reson. Chem.*, 2014, **52**, 47.
210 Y. F. Li, Y. Chen, C. Y. Xiao, D. Chen, Y. X. Xiao and Z. A. Mei, *J. Chromatogr. B*, 2014, **960**, 166.
211 L. Chen, Y. Duan, C. Li, Y. Wang, X. Tong, Y. Dai and X. Yao, *Magn. Reson. Chem.*, 2013, **51**, 842.
212 N. Jamila, M. Khairuddean, S. N. Khan and N. Khan, *Magn. Reson. Chem.*, 2014, **52**, 345.
213 N. B. Pigni, S. Ríos-Ruiz, F. J. Luque, F. Viladomat, C. Codina and J. Bastida, *Phytochemicals*, 2013, **95**, 384.
214 D. A. C. Bezerra, J. F. Tavares, P. F. dos Santos, M. V. S. C. Branco, M. F. Agra, F. L. Subrinho, R. Braz-Filho and M. S. da Silva, *Magn. Reson. Chem.*, 2013, **51**, 497.
215 J. Zhao, S. I. Khan, M. Wang, Y. Vasquez, M. H. Yang, B. Avula, Y. H. Wang, C. Avonto, T. J. Smillie and I. A. Khan, *J. Nat. Prod.*, 2014, **77**, 509.
216 J. Qu, L. Fang, X. D. Ren, Y. Liu, S. S. Yu, L. Li, X. Q. Bao, D. Zhang, Y. Li and S. G. Ma, *J. Nat. Prod.*, 2013, **76**, 2203.
217 X. J. Yang, Y. Yao, Y. Y. Qin, Z. Hou, R. Yang, F. Miao and L. Zhou, *Chem. Pharm. Bull.*, 2013, **61**, 731.
218 Shamsuzzaman, H. Khanam, A. Mashrai, M. Ahmad, Y. N. Mabkhot, W. Frey and N. Siddiqui, *J. Cryst. Growth*, 2013, **384**, 135.
219 L. Chen, Y. T. Liu, B. Song, H. W. Zhang, G. Ding, X. Z. Liu, Y. C. Gu and Z. M. Zou, *Fitoterapia*, 2014, **96**, 115.
220 Z. B. Cheng, Q. Liao, Y. Chen, C. Q. Fan, Z. Y. Huang, X. J. Xu and S. Yin, *Magn. Reson. Chem.*, 2014, **52**, 515.
221 A. Virgili, M. M. Quesada-Moreno, J. R. Avilés-Moreno, J. J. López-González, M. Ángeles-García, R. M. Claramunt, M. R. Torres, M. L. Jimeno, F. Reviriego, I. Alkorta and J. Elguero, *Helv. Chim. Acta*, 2014, **97**, 471.

222 C. Qi, T. Qin, D. Suzuki and J. A. Porco, Jr., *J. Am. Chem. Soc.*, 2014, **136**, 3374.
223 B. C. Garrido, M. L. C. Silva, R. M. Borges, M. C. Padilha and F. R. A. Neto, *Biomed. Chromatogr.*, 2013, **27**, 1630.
224 E. G. Baraban, J. B. Morin, G. M. Phillips, A. J. Phillips, S. A. Strobel and J. Handelsman, *Tetrahedron Lett.*, 2013, **54**, 4058.
225 I. Svinyarov and M. G. Bogdanov, *Eur. J. Med. Chem.*, 2014, **78**, 198.
226 Z. G. Jiang, R. C. Shi, H. D. Zhang and Y. C. Li, *J. Liq. Chromatogr. Relat. Technol.*, 2014, **37**, 974.
227 C. S. de Oliveira, E. F. Carlos, L. G. E. Vieira, L. M. Lião and G. B. Alcantara, *Magn. Reson. Chem.*, 2014, **52**, 422.
228 E. Faggi, S. V. Luis and I. Alfonso, *RSC Adv.*, 2013, **3**, 11556.
229 M. Matković, S. Vukelić, R. Cirimotić, G. Kragol, K. Malčanov and K. Mlinarić-Majerski, *Mol. Diversity*, 2013, **17**, 817.
230 A. M. S. Silva, V. L. M. Silva, R. M. Claramunt, D. Santa María, M. B. Ferraro, F. Reviriego, I. Alkorta and J. Elguero, *Magn. Reson. Chem.*, 2013, **51**, 530.
231 J. S. Lomas, *Magn. Reson. Chem.*, 2014, **52**, 87.
232 D. Xiao, C. Wang, H.-C. Tsui, A. Palani, R. Aslanian and A. V. Buevich, *Tetrahedron Lett.*, 2013, **54**, 6199.
233 S. Schiaffino-Ortega, A. Espinosa, M. A. Gallo, L. C. López-Cara and A. Entrena, *Magn. Reson. Chem.*, 2014, **52**, 40.
234 V. A. Kraikin, A. A. Fatykhov, I. I. Sakhipova, E. A. Sedova, A. E. Egorov and S. N. Salazkin, *Magn. Reson. Chem.*, 2013, **51**, 621.
235 M. C. Holland, S. Paul, W. B. Schweizer, K. Bergander, C. Mück-Lichtenfeld, S. Lakhdar, H. Mayr and R. Gilmour, *Angew. Chem., Int. Ed.*, 2013, **52**, 7967.
236 C. J. Duarte, R. A. Cormanich, L. C. Ducati and R. Rittner, *J. Mol. Struct.*, 2013, **1050**, 174.
237 D. C. Solha, T. M. Barbosa, R. V. Viesser, R. Rittner and C. F. Tormena, *J. Phys. Chem. A*, 2014, **118**, 2794.
238 K. V. Lawson, T. E. Rose and P. G. Harran, *Tetrahedron*, 2013, **69**, 7683.
239 I. V. Peterson, N. M. Svirskaya, A. A. Kondrasenko and A. I. Rubaylo, *Magn. Reson. Chem.*, 2013, **51**, 762.
240 F. Al Chab, B. Fenet, M. Le Borgne, J. Jose, N. Pinaud, J. Guillon and L. Ettouati, *Magn. Reson. Chem.*, 2013, **51**, 837.
241 G. Dayaker, T. Durand and L. Balas, *Chem. – Eur. J.*, 2014, **20**, 2879.
242 B. Rubio-Ruiz, A. Espinosa, A. E. Guadix and A. Conejo-Garcia, *Magn. Reson. Chem.*, 2014, **52**, 470.
243 G. K. Pierens, T. K. Venkatachalam and D. Reutens, *Magn. Reson. Chem.*, 2014, **52**, 453.
244 N. F. Kirillov, E. A. Nikiforova and S. N. Shurov, *Russ. J. Org. Chem.*, 2014, **50**, 829.
245 K. M. S. Gonçalves, D. R. Garcia, T. C. Ramalho, J. D. Figueroa-Villar and M. P. Freitas, *J. Phys. Chem. A*, 2013, **117**, 10980.
246 C. Özen, A. Yildirim, N. Ş. Tüzün, P. Haslak, E. Doğan, E. Acar and V. Aviyente, *Mol. Phys.*, 2013, **111**, 3147.
247 M. Myahkostupov, V. Prusakova, D. G. Oblinsky, G. D. Scholes and F. N. Castellano, *J. Org. Chem.*, 2013, **78**, 8634.
248 A. T. Liu, M. Nag, W. R. Carroll and J. D. Roberts, *Magn. Reson. Chem.*, 2013, **51**, 701.
249 H. W. H. Lai, A. T. Liu, B. U. Emenike, W. R. Carroll and J. D. Roberts, *J. Phys. Chem. A*, 2014, **118**, 1965.
250 B. Szyszko, K. Kupietz, L. Szterenberg and L. Latos-Grażyński, *Chem. – Eur. J.*, 2014, **20**, 1376.

251 A. Le Goff, D. Vénec, C. Le Roy, F. Y. Pétillon, P. Schollhammer and J. Talarmin, *Inorg. Chem.*, 2014, **53**, 2200.
252 L. Perrin, E. L. Werkema, O. Eisenstein and R. A. Andersen, *Inorg. Chem.*, 2014, **53**, 6361.
253 Suciati, J. A. Fraser, L. K. Lambert, G. K. Pierens, P. V. Bernhardt and M. J. Garson, *J. Nat. Prod.*, 2013, **76**, 1432.
254 J. A. Brockerman, M. Okon and L. P. McIntosh, *J. Biomol. NMR*, 2014, **58**, 17.
255 Y. Sasanuma and D. Touge, *Polymer*, 2014, **55**, 1901.
256 Y. Sasanuma, Y. Wagai, N. Suzuki and D. Abe, *Polymer*, 2013, **54**, 3904.
257 K. D. Klika, *ISRN Spectrosc.*, 2012, 697156.
258 A. A. Hassan, A. A. Aly, T. I. M. Bedair, A. B. Brown and T. I. El-Emary, *J. Heterocycl. Chem.*, 2014, **51**, 44.
259 E. E. Emelina, A. A. Petrov and D. V. Filyukov, *Russ. J. Org. Chem.*, 2014, **50**, 412.
260 A. Pesic, B. Steinhaus, S. Kemper, J. Nachtigall, H. J. Kutzner, G. Höfle and R. D. Süssmuth, *J. Antibiot.*, 2014, **67**, 471.
261 A. Arrieta, H. Jiao, A. Spannenberg and D. Michalik, *Tetrahedron*, 2013, **69**, 8872.
262 R. Knorr, E. C. Rossmann, M. Knittl and P. Böhrer, *Tetrahedron*, 2014, **70**, 5332.
263 J. M. Silla, W. G. D. P. Silva, R. A. Cormanich, R. Rittner, C. F. Tormena and M. P. Freitas, *J. Phys. Chem. A*, 2014, **118**, 503.
264 J. M. Silla, C. J. Duarte, R. A. Cormanich, R. Rittner and M. P. Freitas, *Beilstein J. Org. Chem.*, 2014, **10**, 877.
265 G. K. S. Prakash, C. Ni, F. Wang, Z. Zhang, R. Haiges and G. A. Olah, *Angew. Chem., Int. Ed.*, 2013, **52**, 10835.
266 A. R. Patel, L. Hunter, M. M. Bhadbhade and F. Liu, *Eur. J. Org. Chem.*, 2014, 2584.
267 G. Chaume, J. Simon, C. Caupène, N. Lensen, E. Miclet and T. Brigaud, *J. Org. Chem.*, 2013, **78**, 10144.
268 T. Wang and D. W. Stephan, *Chem. – Eur. J.*, 2014, **20**, 3036.
269 N. K. Srungavruksham and V. Baskar, *Eur. J. Inorg. Chem.*, 2013, 4345.
270 D. L. Kay, L. J. Daumann, G. R. Hanson and L. R. Gahan, *Polyhedron*, 2013, **64**, 151.
271 B. J. Carra, S. M. Berry, R. D. Pike and D. C. Bebout, *Dalton Trans.*, 2013, **42**, 14424.
272 L. M. Tuttle, H. J. Dyson and P. E. Wright, *Biochemistry*, 2013, **52**, 3464.
273 L. M. Tuttle, H. J. Dyson and P. E. Wright, *Biochemistry*, 2014, **53**, 1134.
274 S. S. Pawar and N. A. Koorbanally, *Magn. Reson. Chem.*, 2014, **52**, 279.
275 C. Kaub, T. Augenstein, T. O. Bauer, E. Rothe, L. Esmezjan, V. Schünemann and P. W. Roesky, *Inorg. Chem.*, 2014, **53**, 4491.
276 O. V. Tsepaeva, V. F. Mironov, B. I. Khairutdinov and Y. F. Zuev, *Russ. J. Org. Chem.*, 2014, **50**, 919.
277 V. F. Mironov, A. V. Nemtarev, E. N. Varaksina, A. A. Shtyrlina, A. T. Gubaidullin, I. A. Litvinov and A. B. Dobrynin, *Russ. J. Org. Chem.*, 2014, **50**, 864.
278 L. I. Musin, A. V. Bogdanov, D. B. Krivolapov, A. B. Dobrynin, I. A. Litvinov and V. F. Mironov, *Russ. J. Org. Chem.*, 2014, **50**, 822.
279 Y. C. Wang, M. Y. Chen, H. Y. Wu, J. H. Chen, S. S. Wang and J. Y. Tung, *Polyhedron*, 2014, **67**, 73.
280 G. A. L. Bare and J. C. Sherman, *J. Org. Chem.*, 2013, **78**, 8198.
281 R. A. Cacho, Y. H. Chooi, H. Zhou and Y. Tang, *ACS Chem. Biol.*, 2013, **8**, 2322.

282 A. Sartori, L. Dell'Amico, L. Battistini, C. Curti, S. Rivara, D. Pala, P. S. Kerry, G. Pelosi, G. Casiraghi, G. Rassu and F. Zanardi, *Org. Biomol. Chem.*, 2014, **12**, 1561.
283 T. M. Barhoumi-Slimi, K. Essalah, M. A. K. Sanhoury, M. Ourévitch and M. M. El Gaied, *Struct. Chem.*, 2014, **25**, 799.
284 K. Solakyildirim, D. N. Bulloch and C. K. Larive, *Magn. Reson. Chem.*, 2014, **52**, 310.
285 S. Mastoor, S. Faizi, R. Saleem and B. S. Siddiqui, *Magn. Reson. Chem.*, 2014, **52**, 115.
286 S. Santos and J. Graça, *Phytochem. Anal.*, 2014, **25**, 192.
287 C. Alvarez-Cisneros, M. A. Muñoz, O. R. Suárez-Castillo, N. Pérez-Hernández, C. M. Cerda-García-Rojas, M. S. Morales-Ríos and P. Joseph-Nathan, *Magn. Reson. Chem.*, 2014, **52**, 491.
288 B. Wrackmeyer, E. V. Klimkina and W. Milius, *Dalton Trans.*, 2014, **43**, 5021.
289 I. A. Khalymbadzha, T. S. Shestakova, S. L. Deev, V. L. Rusinov, O. N. Chupakhin, Z. O. Shenkarev and A. S. Arseniev, *Russ. Chem. Bull. Intern. Edit.*, 2013, **62**, 521.
290 M. Momin, D. Ramjugernath and N. A. Koorbanally, *Magn. Reson. Chem.*, 2014, **52**, 521.
291 B. Gierczyk, M. Kaźmierczak, Ł. Popenda, A. Sporzyński, G. Schroeder and S. Jurga, *Magn. Reson. Chem.*, 2014, **52**, 202.
292 L. A. F. Andrade, J. M. Silla, C. J. Duarte, R. Rittner and M. P. Freitas, *Org. Biomol. Chem.*, 2013, **11**, 6766.
293 N. S. Boltacheva, L. N. Dianova, P. A. Slepukhin, V. I. Filyakova, V. A. Bakulev and V. N. Charushin, *Russ. J. Org. Chem.*, 2014, **50**, 767.
294 D. C. Favaro, R. H. Contreras and C. F. Tormena, *J. Phys. Chem. A*, 2013, **117**, 7939.
295 M. Pérez-Torralba, R. M. Claramunt, M. Á. Garcia, C. López, M. C. Torralba, M. R. Torres, I. Alkorta and J. Elguero, *Beilstein J. Org. Chem.*, 2013, **9**, 2156.
296 A. Gryff-Keller and P. Szczeciński, *RSC Adv.*, 2014, **4**, 27290.
297 L. Calucci, E. Carignani, M. Geppi, S. Macchi, B. Mennucci and S. Urban, *J. Phys. Chem. B*, 2014, **118**, 3469.
298 R. H. Contreras, T. Llorente, L. C. Ducati and C. F. Tormena, *J. Phys. Chem. A*, 2014, **118**, 5068.
299 F. B. Mallory, C. W. Mallory, K. E. Butler, M. B. Lewis, A. Q. Xia, E. D. Luzik, Jr., L. E. Fredenburgh, M. M. Ramanjulu, Q. N. Van and M. M. Francl, *J. Am. Chem. Soc.*, 2000, **122**, 4108.
300 J. F. Kögel, L. H. Finger, N. Frank and J. Sundermeyer, *Inorg. Chem.*, 2014, **53**, 3839.
301 T. Wiegand, H. Eckert, J. Ren, G. Brunklaus, R. Fröhlich, C. G. Daniliuc, G. Lübbe, K. Bossmann, G. Kehr, G. Erker and S. Grimme, *J. Phys. Chem. A*, 2014, **118**, 2316.
302 M. W. Stanford, F. R. Knight, K. S. A. Arachchige, P. S. Camacho, S. E. Ashbrook, M. Bühl, A. M. Z. Slawin and J. D. Woollins, *Dalton Trans.*, 2014, **43**, 6548.
303 Y. Orlova, R. R. Gainov, A. V. Dooglav and I. N. Pen'kov, *JETP Lett.*, 2013, **97**, 413.
304 B. Makiabadi, M. Zakarianejad, S. Bagheri, H. R. Masoodi and R. S. Aghaie, *Int. J. Quantum Chem.*, 2013, **113**, 2361.
305 H. Roohi and R. Nokhostin, *Struct. Chem.*, 2014, **25**, 483.
306 A. J. Durie, T. Fijiwara, R. Cormanich, M. Bühl, A. M. Z. Slawin and D. O'Hagan, *Chem. – Eur. J.*, 2014, **20**, 6259.
307 M. Baya, Ú. Belío and A. Martín, *Inorg. Chem.*, 2014, **53**, 189.

308 A. Křístková, J. R. Asher, V. G. Malkin and O. L. Malkina, *J. Phys. Chem. A*, 2013, **117**, 9235.
309 G. C. Lloyd-Jones, J. N. Harvey, P. Hodgson, M. Murray and R. L. Woodward, *Chem. - Eur. J.*, 2003, **9**, 4523.
310 C. Trujillo, G. Sánchez-Sanz, I. Alkorta, J. Elguero, O. Mó and M. Yáñez, *J. Mol. Struct.*, 2013, **1048**, 138.
311 J. Elguero, I. Alkorta and J. E. Del Bene, *Mol. Phys.*, 2014, **112**, 107.
312 D. Sanz, R. M. Claramunt, F. Mathey, I. Alkorta, G. Sánchez-Sanz and J. Elguero, *C. R. Chim.*, 2013, **16**, 937.
313 I. Alkorta, J. Elguero and J. E. Del Bene, *J. Phys. Chem. A*, 2013, **117**, 10497.
314 I. Alkorta, J. Elguero and J. E. Del Bene, *J. Phys. Chem. A*, 2013, **117**, 4981.
315 J. E. Del Bene, I. Alkorta and J. Elguero, *J. Phys. Chem. A*, 2013, **117**, 6893.
316 J. E. Del Bene, I. Alkorta and J. Elguero, *J. Phys. Chem. A*, 2013, **117**, 11592.
317 J. E. Del Bene, I. Alkorta and J. Elguero, *J. Phys. Chem. A*, 2014, **118**, 2360.
318 I. Alkorta, G. Sánchez-Sanz, J. Elguero and J. E. Del Bene, *J. Phys. Chem. A*, 2014, **118**, 1527.
319 J. E. Del Bene, I. Alkorta and J. Elguero, *Theor. Chem. Acc.*, 2014, **133**, 1464.
320 J. E. Del Bene, I. Alkorta and J. Elguero, *J. Phys. Chem. A*, 2014, **118**, 3386.
321 T. M. Sabo, C. A. Smith, D. Ban, A. Mazur, D. Lee and C. Griesinger, *J. Biomol. NMR*, 2014, **58**, 287.
322 R. Montalvao, C. Camilloni, A. De Simone and M. Vendruscolo, *J. Biomol. NMR*, 2014, **58**, 233.
323 M. Rinaldelli, E. Ravera, V. Calderone, G. Parigi, G. N. Murshudov and C. Luchinat, *Acta Crystallogr.*, 2014, **D70**, 958.
324 E. Troche-Pesqueira, M. M. Cid and A. Navarro-Vázquez, *Org. Biomol. Chem.*, 2014, **12**, 1957.
325 K. S. Usachev, A. V. Filippov, O. N. Antzutkin and V. V. Klochkov, *Eur. Biophys. J. Biophys. Lett.*, 2013, **42**, 803.
326 S. Samanta, A. J. Situ and T. S. Ulmer, *Biopolymers*, 2014, **101**, 398.
327 G. A. Cook, L. A. Dawson, Y. Tian and S. J. Opella, *Biochemistry*, 2013, **52**, 5295.
328 J. M. Cramer, J. N. Scarsdale, N. M. Walavalkar, W. A. Buchwald, G. D. Ginder and D. C. Williams Jr., *J. Biol. Chem.*, 2014, **289**, 1294.
329 V. Iešmantavičius, M. R. Jensen, V. Ozenne, M. Blackledge, F. M. Poulsen and M. Kjaergaard, *J. Am. Chem. Soc.*, 2013, **135**, 10155.
330 C. Nguyen, R. W. Haushalter, D. J. Lee, P. R. L. Markwick, J. Bruegger, G. Caldara-Festin, K. Finzel, D. R. Jackson, F. Ishikawa, B. O'Dowd, J. A. McCammon, S. J. Opella, S. C. Tsai and M. D. Burkart, *Nature*, 2014, **505**, 427.
331 J. Wang, N. Tochio, A. Takeuchi, J. Uewaki, N. Kobayashi and S. Tate, *Biochem. Biophys. Res. Commun.*, 2013, **441**, 701.
332 Y. Zhang, L. Wang, Y. F. Hu and C. W. Jin, *Biochim. Biophys. Acta*, 2014, **1838**, 1881.
333 A. De Biasio, A. I. de Opakua, T. N. Cordeiro, M. Villate, N. Merino, N. Sibille, M. Lelli, T. Diercks, P. Bernadó and F. J. Blanco, *Biophys. J.*, 2014, **106**, 865.
334 M. Schwalbe, V. Ozenne, S. Bibow, M. Jaremko, L. Jaremko, M. Gajda, M. R. Jensen, J. Biernat, S. Becker, E. Mandelkow, M. Zweckstetter and M. Blackledge, *Structure*, 2014, **22**, 238.
335 L. Skora, J. Mestan, D. Fabbro, W. Jahnke and S. Grzesiek, *Proc. Natl. Acad. Sci. U. S. A.*, 2013, **110**, E4437.
336 J. S. Fan, Y. Zheng, W. Y. Choy, V. Simplaceanu, N. T. Ho, C. Ho and D. W. Yang, *Biochemistry*, 2013, **52**, 5809.

337 C. D. Eichhorn and H. M. Al-Hashimi, *RNA*, 2014, **20**, 782.
338 Y. N. Fu and A. J. Wand, *J. Biomol. NMR*, 2013, **56**, 353.
339 N. Sibille, M. Dellarole, C. Royer and C. Roumestand, *J. Biomol. NMR*, 2014, **58**, 9.
340 P. Kukic, C. Camilloni, A. Cavalli and M. Vendruscolo, *J. Mol. Biol.*, 2014, **426**, 1826.
341 L. Deshmukh, C. D. Schwieters, A. Grishaev, R. Ghirlando, J. L. Baber and G. M. Clore, *J. Am. Chem. Soc.*, 2013, **135**, 16133.
342 L. Deshmukh, R. Ghirlando and G. M. Clore, *Angew. Chem., Int. Ed.*, 2014, **53**, 1025.
343 M. Damjanovic, K. Katoh, M. Yamashita and M. Enders, *J. Am. Chem. Soc.*, 2013, **135**, 14349.
344 V. M. R. Kakita, K. Rachineni and J. Bharatam, *J. Mol. Struct.*, 2013, **1053**, 122.
345 N. G. A. Bell, G. Rigg, S. Masters, J. Bella and D. Uhrín, *Phys. Chem. Chem. Phys.*, 2013, **15**, 18223.
346 J. M. Ren, A. D. Sherry and C. R. Malloy, *Magn. Reson. Med.*, 2013, **70**, 610.
347 S. Norrehed, M. Erdèlyi, M. E. Light and A. Gogoll, *Org. Biomol. Chem.*, 2013, **11**, 6292.

Nuclear spin relaxation in liquids and gases

Jozef Kowalewski
DOI: 10.1039/9781782622758-00235

The review covers the progress in the field of NMR relaxation in fluids during the period from June 2013 through May 2014. The emphasis is on comparatively simple liquids and solutions of physico-chemical and chemical interest, in analogy with the previous periods, but selected biophysics-related topics and relaxation-related work on more complex systems (macromolecular solutions, liquid crystalline systems, glassy and porous materials) are also covered. The first part of the chapter is concerned with general, physical and experimental aspects of nuclear spin relaxation, while the second part is concentrated on applications.

1 Introduction

The aim of this report is to review the progress in the field of NMR relaxation in fluids. The outline of this chapter follows closely that of the previous four volumes.[1–4] The emphasis is on comparatively simple liquids and solutions of physico-chemical and chemical interest, in analogy with the previous periods, but selected biophysics-related topics and relaxation-related work on more complex systems (macromolecular solutions, liquid crystalline systems, glassy and porous materials) are also covered. The period under review is from June 2013 through May 2014. Some earlier work, overlooked in the last year's chapter, is also included.

The concept of relaxation in physical chemistry implies development back to equilibrium after a perturbation. The equilibrium state in NMR refers to a macroscopic sample of nuclear spins in a magnetic field $\mathbf{B}_0$. Such equilibrium is characterized by the presence of a macroscopic magnetization vector $\mathbf{M}$ oriented along the direction of the magnetic field. By manipulating either the $\mathbf{B}_0$ field or the resonant radiofrequency fields, the experimenter has a possibility to create a great variety of non-equilibrium states. In the simplest cases, this amounts to changing the orientation of the magnetization vector. According to the early, phenomenological description in the form of Bloch equations,[5] the longitudinal and transverse components of the vector return after such a perturbation to their equilibrium values following simple exponential processes with time constants, denoted T_1 (longitudinal or spin–lattice relaxation time) and T_2 (transverse or spin–spin relaxation time). A more complete theory describes the NMR relaxation as a complicated network of processes, where the simple "return to equilibrium" courses of action compete with related phenomena converting various type of non-equilibrium states into each other.[6] Studies of relaxation phenomena of

Department of Materials and Environmental Chemistry, Arrhenius Laboratory, Stockholm University, SE-106 91 Stockholm, Sweden.
E-mail: jozef.kowalewski@mmk.su.se

this general type are a rich source of information on the strength of various interactions, which can be translated into structural information (*e.g.* the dipole–dipole interaction related to internuclear distances) and on the molecular dynamic processes modulating the interactions.

The outline of this chapter is as follows. Section 2 covers the general, physical and experimental aspects of nuclear spin relaxation in liquids and is further divided in eight subsections. The first two (2.1, 2.2) are fairly general and the following three (2.3–2.5) discuss more specific aspects of spin-1/2 systems. The next three subsections cover quadrupolar nuclei (2.6), paramagnetic systems (2.7) and phenomena related to hyperpolarization (2.8), respectively. Section 3 deals with applications of NMR relaxation in liquids, starting with pure liquids and continuing with solutions of low-molecular weight compounds. The last two subsections include a selection of work on solutions of biological macromolecules and other complex systems. Following volumes 42 and 43, the latter subsection is further divided in six different topic areas. Finally, the progress in the field of relaxation in gases is described in Section 4.

2 General, physical and experimental aspects of nuclear spin-relaxation

2.1 General aspects

Spin relaxation phenomena are usually described by the semiclassical theory developed by Wangsness, Bloch and Redfield[7,8] and known as the WBR theory or Redfield theory. The semiclassical nature of the theory implies that the spin system is treated quantum mechanically, while the remaining degrees of freedom (such as molecular rotations) are treated classically. Few years ago, Segnorile and Zamar[9] studied the issue of quantum decoherence (loss of system phase memory) in proton NMR of nematic liquid crystals. The spin dynamics – and the decay of the free induction decay – was found to be governed by several different processes, partly of purely quantum nature. During the period under the present review, the same group reported a related work concerned with the Jeener–Broekaert experiment on liquid crystals.[10]

As mentioned in the introduction, the (linear) Bloch equations constitute the simplest possible description of spin relaxation phenomena. Erker and Augustine[11] considered the case of non-linear Bloch equations, corrected for the effect of radiation damping and inhomogeneous broadening. Stöckmann and Dubbers[12] generalized the Bloch equations to include the evolution of polarization vectors of various ranks in arbitrary multipole fields. The paper makes extensive use of irreducible tensors and provides a concise derivation of the relaxation effects.

Measurements of spin–lattice relaxation rate as a function of the magnetic field/resonance frequency are commonly referred to as nuclear magnetic relaxation dispersion, NMRD. Measurements of this kind, when performed over a broad range, are an invaluable source of information on frequency-resolved molecular motions in complex biological or colloidal systems. Several years ago, Halle[13] proposed a theory for relaxation of quadrupolar spins in dynamically heterogeneous systems where the quadrupolar interaction of aqueous ^{2}H or ^{17}O spins was modulated by exchange between an isotropic site and a number of immobilized anisotropic sites, characterized by a residual quadrupolar coupling. The exchange was allowed between any of the anisotropic sites and the isotropic one, but not between the different anisotropic sites. The model treats the exchange process not only as means to mix spin populations but also as the direct cause of spin relaxation. In newer literature, the model is known as exchange-mediated orientational randomization (EMOR). The EMOR theory is based on solving the stochastic Liouville equation and is valid even outside of the perturbation regime, covered by the Redfield theory. In a recent work, Nilsson and Halle[14] presented a more rigorous derivation of the EMOR approach to longitudinal relaxation dispersion of spin $I = 1$ systems, removing also several restrictions and allowing for biaxial electric field gradient tensor, internal motions and multiple sites with different exchange rates. Several limiting cases and approximations were also described and assessed. Even more recently, Chang and Halle[15] developed the EMOR relaxation theory for dipole-coupled spin 1/2 pairs. Both the like and unlike spins were considered, as well as different types of exchange processes and cases of selective as well as non-selective excitation. For the experimentally important dilute regime, where populations of the immobilized anisotropic sites are very low, the authors derived simple analytical expressions generalizing the Solomon equations.[16]

Gryff-Keller and co-workers[17] pursued their interest in the scalar relaxation of the second kind (SC2). The spin–lattice relaxation rate of a spin-1/2 nucleus (I spin) scalar-coupled to a quadrupolar nucleus (S-spin) is given by[6]:

$$\frac{1}{T_{1,\mathrm{SC2}}} = \frac{(2\pi J_{IS})^2}{2} \frac{\tau_e}{1 + (\omega_I - \omega_S)^2 \tau_e^2} \tag{1}$$

where J_{IS} is the scalar coupling constant in Hz, τ_e is the spin–lattice relaxation time of the S-spin while ω_I and ω_S are the Larmor frequencies for the two species. The SC2 mechanism is usually inefficient because the expression in the denominator tends to be very large. A well-known exception from this rule is the case of ^{13}C (I spin) coupled to bromine isotopes ^{79}Br and ^{81}Br (S spin). In particular ^{79}Br has a magnetogyric ratio very close to ^{13}C, which results in a low value for the denominator even at high magnetic fields. Gryff-Keller *et al.*[17] showed that the interpretation of the SC2 relaxation in carbon–bromine systems requires the knowledge of accurate value of the bromine magnetogyric ratio as well as shielding constants of the two spin species and proposed a methodology appropriate for quantum chemical calculations of the bromine shielding.

I turn next to the relaxation-related work on long-lived spin states (LLSs), such as the nuclear singlet states, discovered some time ago by Carravetta, Levitt and Johannessen.[18–20] The relaxation times of the singlet states (T_S) in two-spin systems, or of the LLSs (T_{LSS}) in general, have attracted continuing attention. Their long lifetimes depend on the fact that the dipole–dipole (DD) interaction between the two involved spins is not operative as a relaxation mechanism. DeVience and co-workers[21] proposed a broadly applicable technique to create nuclear spin singlet states in organic molecules. The method makes use of a novel pulse sequence to produce a spin-lock induced crossing (SLIC) of the singlet and triplet energy levels. The LLSs were produced for strongly coupled proton pairs in a tripeptide in solutions and their lifetimes were measured. Theis *et al.*[22] proposed an improved version of SLIC, employing composite and adiabatic pulse excitation. Claytor and co-workers[23] used these advanced methods and demonstrated the possibility of measuring the lifetimes of the long-living $^{13}C_2$ states by detecting the double labelled species at natural abundance, where the $^{13}C_2$ pairs only appear in $(1.1\%)^2$ of molecules. The model compounds used were diphenyl-$^{13}C_2$-acetylene, diethyl-$^{13}C_2$-oxalate and 2,3-$^{13}C_2$-diacetylene. Applications of the LLSs in $^{13}C_2$ pairs was also discussed in the context of storage of hydrogen hyperpolarization[24,25] (see Section 2.8 below). Laustsen *et al.*[26] used the optimal control methods to create pulses that for near-equivalent spins accomplished transfer in and out of the singlet state (using the M2S and S2M scheme developed by Pileio and co-workers[27]) with maximum efficiency and robustness against variation of the spin system Hamiltonian. The goal of the study was related to extending the lifetime of the hyperpolarization of nuclear spins (see Section 2.8) for enhancing the sensitivity of the MRI.

Meier and co-workers[28] demonstrated that substances containing rapidly rotating methyl groups could exhibit LLSs in solution. The states become long-lived through fast internal rotation of the CH_3 groups, which imposes an approximate symmetry on the fluctuating spin interactions. The evidence of the occurrence of the LLS was provided by experiments where the LLS was populated at liquid helium temperature and transferred into solution at room temperature and at high field using the dissolution apparatus designed for dynamic nuclear polarization (DNP, see Section 2.8). After the dissolution, the cross-relaxation of the LLS generated strong antiphase NMR signals. The LLSs were also studied in intrinsically disordered proteins (IDPs) by Fernandes *et al.*[29] The lifetimes of the LLSs in solution were measured and compared with the usual spin relaxation rates for ^{15}N and ^{1}H. The LLS relaxation experiments were integrated with 2D protein NMR spectroscopy methods and it was shown that they could probe motions and interactions in biomolecules. Nagashima and co-workers[30] studied the lifetime of the LLSs for a tripeptide in stretched hydrogels. By stretching the gel, one can introduce a controlled degree of anisotropy (and thus residual dipolar coupling), which turned out to have a small effect on the LLS lifetimes if the WALTZ-decoupling was used during relaxation intervals.

Emondts and co-workers[31] studied LLS lifetimes for heteronuclear (^{13}C–^{1}H) spin pairs at zero magnetic field. In the case of ^{13}C labelled formic acid, they observed singlet lifetimes of 37 s, about three times longer than the T_1 of the dipole polarization of the triplet state. Pravdivtsev *et al.*[32] studied the LLSs, created by means of photochemically induced DNP in β-CH_2 protons of aromatic aminoacids. The LLS lifetimes were measured as a function of the magnetic field using the field cycling technique (see Section 2.3 below). The authors investigated also the effects of deuteration of the α-CH and paramagnetic impurities.

The coherence counterparts of the LLSs are called long-lived coherences (LLC). The LLSs have long lifetimes, while the LLCs correspond to narrow NMR lines, typically significantly narrower than the "usual" coherences. Singh *et al.*[33] reported a study of the behaviour of the LLCs in the presence of paramagnetic relaxation agents. They found that the LLCs relaxed faster than the single quantum coherences in the presence of paramagnetic ions, which was explained in terms of external random fields and their correlation.

2.2 Models for molecular dynamics

NMR relaxation and its field dependence are a very important source of experimental information on dynamics of molecular motions. This information is conveyed through spectral density functions, which in turn are related to time-correlation functions (TCFs), fundamental quantities in the theory of liquid state. In most cases, characterizing the molecular dynamics through NMR relaxation studies requires the identification of the relaxation mechanism (for example the dipole–dipole interaction between a pair of spins) and models for the spectral densities/correlation functions.[6] During the period covered by this review, such model development was concerned with both small and large molecules.

The spin interaction of very large importance for NMR relaxation is the dipole–dipole interaction. The DD interaction can be of intra- or intermolecular nature, depending on whether the interacting spins reside in the same or different molecules. Most NMR relaxation studies are concentrated on the intramolecular case, but some recent work has discussed the intermolecular proton relaxation and demonstrated clearly that it can be very important in some situations. We begin this section by reviewing the work of this kind. Henritzi and co-workers[34] reported molecular dynamics (MD) simulations for polymer melts, aiming at interpretation of ^{1}H and ^{2}H spin–lattice relaxation dispersion data. For the proton case, they separated the relaxation rates into intra- and intermolecular contributions. For sufficiently low molecular weights, intramolecular spectral densities from MD simulations were found to follow the Cole–Davidson and Kohlrausch–Williams–Watts models.[35] The intramolecular dipolar spectral densities for ^{1}H were found to be proportional to their counterparts relevant for the quadrupolar relaxation of ^{2}H. The intermolecular spectral densities relevant for ^{1}H showed asymptotic power law behaviour at low frequencies. Kulagina and co-workers[35] formulated a theory of the proton FID – and of the transverse relaxation – for linear flexible chain polymers. The theory predicted the dependence of T_2 on the average molecular weight and temperature. By comparison with experimental data it was shown that the topological structure of the polymer changed with the molecular weight. Another polymer paper I wish to mention here is the work by Fatkullin *et al.*,[37] who presented a theory for double-quantum (DQ) NMR in this kind of systems, including both the intra- and intermolecular DD interaction. The work used similar approach as employed by the same group in an earlier study of free induction decay (FID) in polymer melts.[38] The authors showed that also the DQ NMR kinetic curve was dependent on the intermolecular DD interaction. Faux *et al.*[39] reported related work, MD as well as Monte Carlo (MC) simulations along with analytical theories, for protons in water as a neat fluid as well as in confined environments, represented by two-dimensional and quasi-two-dimensional diffusion.

I continue by referring to some papers not directly concerned with NMR, but dealing with generalizations of the simple hydrodynamic description of rotational motion in liquids. The simplest form of TCF used in NMR relaxation theory is a decaying exponential, characterized by a time constant denoted as rotational correlation time, related to the size of the solute and the solvent viscosity.[6] In order to simulate long-time behaviour of more complicated, but still dilute, systems with hydrodynamic interactions, one can use the technique of Brownian dynamics, BD.[40] The limitations of BD simulations for concentrated macromolecular systems can be overcome, using the simulation technique introduced by Durlofsky *et al.*[41] and denoted Stokesian dynamics (SD). During the period covered by this review, Ando and co-workers[42] investigated the case of concentrated macromolecular solutions, such as occurring *e.g.* in cells, and proposed a modification of the SD description corresponding to screening the long-range hydrodynamic interactions. In another study, Dlugosz and Antosiewicz[43] used the rigid-body Brownian dynamics simulations in order to elucidate the dynamic effects of a crowded (volume-occupied) environment, such as anisotropy of diffusional motion. They used simulations for hen egg white lysozyme molecule as an example and found that the anisotropy of rotational motion of the protein increased with an increasing degree of crowding.

The Stokes–Einstein–Debye description pertains for molecules treated as rigid bodies. Molecules in general do not conform to this characteristics and NMR relaxation is an important source of information on internal motions. To obtain this information, one has to assume a motional model of some kind. The simple models for methyl group dynamics, described as random jumps or rotational diffusion, were presented long time ago by Woessner.[44] I wish to mention here the recent work by Vugmeyster and Ostrovsky[45] who studied the restricted dynamics of methyl groups in proteins as revealed by deuteron NMR, even though the paper is mainly concentrated on the solid state situation. The authors proposed a dynamic model where the restricted motion was decomposed into two independent processes: the jumps between the potential wells and intra-well fluctuations, assuming time scale separation between the two modes. The model was applied to analyse the relaxation and line-shape data for the villin headpiece subdomain protein over the temperature range 300–31 K.

The most common way of dealing with non-rigid (macro) molecular systems is the "model free" analysis proposed by Lipari and Szabo[46] and even earlier, in a slightly different formulation, by Wennerström *et al.*[47] The Lipari–Szabo model assumes uncorrelated internal and global motions. The spectral densities are expressed in terms of a global and a local correlation time and a generalized order parameter, measuring the spatial extent of the internal motions. The situation when these two types of motion are explicitly coupled can be described using the slowly relaxing local structures (SRLS) approach of Polimeno and Freed.[48] During the period under review here, three papers have been concerned with development of the SRLS approach for the $^{15}N-^{1}H$ bond as the dynamic probe. In the first of these studies, Shapiro and Meirovitch[49] introduced the time-correlation function perspective on the NMR relaxation in proteins. They solved the SRLS Smoluchowski equation using typical best-fit parameters (such as obtained for two residues, K19 and A34 in the GB3 protein) as input. The approach was found useful to illustrate the relation between the SRLS and the Lipari–Szabo approach and the validity ranges of more specific dynamic models for internal motions. In the second paper by the same authors, the concept of eigenmodes as descriptors in the context of SRLS was introduced.[50] An eigenmode is a weighted exponential with the decay constant given by an eigenvalue of the Smoluchowski operator and the weighting factor determined by the corresponding eigenvector. Under simple limiting conditions, two eigenmodes (associated with the protein and local probe motions) dominate. Also in this approach, the two residues in the GB3 protein were used as examples. Finally, in the most recent paper, the same two residues were used as models for understanding the protein dynamics in terms of yet another descriptor: the pattern of motions that dissipate the power from the spin system to the bath. Here, the authors introduced a novel concept, called *j*-fraction, representing the relative contribution of an eigenmode to a given value of the spectral density function at a specific frequency.[51] In addition to these three papers, Meirovitch presented a review of the SRLS approach.[52]

Xia and co-workers[53] proposed a generalization of the Lipari–Szabo model allowing for including two slowly exchanging macrostates, each exhibiting its own fast local motions in addition to molecular tumbling. The expressions for NMR relaxation rates simplify in two limits: when the conformational exchange is much slower than the tumbling (the common situation) or much faster than the tumbling. The authors used HIV-1 protease as an example and used stochastic simulations based on Markov state models (MSM) constructed from atomistic replica exchange molecular dynamics (REMD). This allowed generating microsecond trajectories at several different temperatures which were used as references for evaluating the time correlation functions to be compared with the exchanging macrostates model. Even though the work does not explicitly mention NMR relaxation, I wish to refer here to the paper by Gur *et al.*,[54] who chose another approach to very long simulations for proteins. They used the Anton supercomputer technology to obtain an atomistic MD trajectory – longer than a millisecond – for bovine pancreatic trypsin inhibitor (BPTI) and used the data to evaluate a simple, coarse-grained model of beads and springs known as anisotropic network model.[55] The same long trajectory was also used by Calligari and Abergel.[56] They investigated the issue of fluctuations of resonance frequency shifts and their relation with protein conformational exchange. Rather than using the simple two-site exchange model, the authors developed an approach allowing for multiple time scale dynamics contributing to the ^{15}N amide chemical shift correlation functions. The TCFs from MD were analysed in terms of transiently fractional diffusion process, which was found to agree with the MD simulations and to yield the exchange contribution to R_2 in qualitative agreement with experiments. Maciejewski and co-workers[57] reported Brownian dynamics simulations for three-domain proteins. The domains were connected in beads-on-a-string fashion with various repulsive potentials between the central and the terminal domains. The simulated TCFs were fitted to the extended model-free formalism,[58] which was able to capture the interdependence of the domain motions. A number of papers proposed improvement of force-fields for MD simulations and tested them against the Lipari–Szabo order parameters.[59–62]

An important application of the Lipari–Szabo order parameters derived from relaxation studies is to estimate the conformational entropy.[63] Genheden and co-workers [64] discussed the issue of conformational entropy and order parameters using long MD simulations for several proteins as a starting point. They found that the order parameters and conformational entropies calculated over 10–100 ns windows were typically well-behaved on a per-residue level, while the total conformational entropy evaluated as the sum over residues, was more difficult to estimate. Kasinath *et al.*[65] probed the microscopic origin of the link between conformational dynamics and conformational entropy using MD simulations for a number of proteins. They demonstrated that the motions of methyl-bearing side chains were sufficiently coupled to those of other side chains to be excellent reporters of the side chain conformational entropy. Marsh[66] proposed the use of relative accessible surface area in monomeric proteins as an indicator of conformational entropy/flexibility.

Emani and co-workers[67] reported a study of molecular motions in RNA, following up earlier work from the same group.[68,69] In the recent paper, they elucidated the collective molecular motions and the conformational landscape of RNA using a subset of possible structures to calculate solution ^{13}C relaxation rates for multiple residues in different domains. The experimental data were matched within rather small error limits.

Two papers have been concerned with the software used for analysis of NMR relaxation data in proteins. Dosset *et al.*[70] discussed various conventions associated with the Euler angles, describing coordinate frame rotations, in different commonly used programs (such as ROTDIF, TENSOR2 *etc.*). A new version of ROTDIF, equipped with a graphical user interface, was described by Berlin and co-workers.[71]

2.3 Experimental aspects: T_1 and T_2

In this section, I review the papers related to experimental aspects of measurement of recovery/decay of transverse and longitudinal magnetization aiming at evaluating T_1 or T_2 (or the corresponding rates, $R_1 = 1/T_1$ and $R_2 = 1/T_2$, respectively). Measurements of the spin–lattice relaxation time in the rotating frame ($T_{1\rho}$) are also included here. Papers dealing with cross-relaxation and other multispin phenomena will be reviewed in the next section. Work centred at studies of conformational exchange through experiments of T_2 or $T_{1\rho}$-type will be covered in Section 2.5.

I begin with a couple of papers from Lucio Frydman's group. Smith and co-workers[72] proposed a method for ultrafast measurements of T_1, designed for spins with fast longitudinal relaxation. The method measures T_1 in a single scan with a full chemical shift resolution, making use of spatially encoded relaxation and detection scheme related to echo planar spectroscopic imaging. The new method was compared with the approach of Loening *et al.*,[73] which also is a single scan technique, but designed for relatively slowly relaxing spins. In the second paper, Shemesh and co-workers[74] investigated the longitudinal relaxation enhancement, LRE, for protons both downfield and upfield from the water resonance, obtained by switching from broadband to selective excitation. By leaving the water signal unperturbed, one can make it act as a relaxation "sink" resulting in significant LRE and possibility of faster pulsing in the acquisition of both exchangeable and non-labile proton signals in tissue. In a similar vein, Gil *et al.*[75] proposed novel pulse sequences designed for intrinsically disordered proteins, based on starting with ^{13}C polarization and ending with ^{13}C detection. Also here, the authors optimized the technique by making use of ^{1}H LRE. The concept was also used in the study by Bermel *et al.*,[76] describing new multidimensional NMR experiments combining ^{13}C detection, non-uniform sampling (NUS) and the LRE. The experiments were designed for the automatic assignment of IDP spectra.

Ishima[77] described a novel method to monitor performance of ^{15}N longitudinal relaxation experiments for proteins in solution at high magnetic fields. The new tool consists of measuring the difference between the R_1 values between the two components of the ^{1}H-coupled amide ^{15}N magnetizations. Under the conditions of suppressing the interference (cross-correlation) phenomena between the DD interaction and the chemical shielding anisotropy (CSA), the difference is expected to vanish. Insufficient cancellation of the phenomenon due to instrumental imperfections can thus be visualized in experiments and simulations. Yuwen and Skrynnikov[78] reported a new implementation of the proton-decoupled CPMG experiment for measuring ^{15}N T_2 in disordered proteins, characterized by slow transverse relaxation. The new method allows one to extend the range of sampled relaxation periods to more than 500 ms, which leads to improved accuracy of the T_2 determination. Kasinath *et al.*[79] proposed a novel ^{13}C labelling strategy for specific introduction of ^{13}C in a single *ortho* position in aromatic amino acid side chains and applied it to calcium-saturated calmodulin.

For some experiments involving longer pulse sequences, the relaxation is also a source of problems (relaxation losses). Kovacs and Gossert[80] described an improvement of the frequently used HC(CCO)NH experiment[81] which is shorter and where the relaxation losses are thus minimized. Blechta and Schraml[82] discussed transverse relaxation effects in modern techniques for measurement of small ^{13}C–^{13}C couplings. For large and slowly tumbling biomolecules, the efficiency of the transverse relaxation and the associated large NMR linewidths are often a problem.

Wand *et al.*[83] proposed long time ago methods for encapsulating the proteins within the aqueous core of a reverse micelle and dissolving the micelles in low viscosity solvents such as small hydrocarbons. Dodevski and co-workers[84] reported recently an optimized reverse micelle surfactant system for measurements of this kind.

Charlier and co-workers[85] described a method for high resolution NMR relaxometry, *i.e.* measurements of field dependence of T_1, applied to ^{15}N in proteins. The key feature of the technique is the use of fast (about 50 ms) shuttling of the sample between the high field (14.1 T) and a range of lower fields (down to 0.5 Tesla) in a conventional high field superconducting magnet, using a dedicated probehead. The polarization of the sample and NMR detection is carried out at the high field with a rather small loss in sensitivity and resolution compared to the standard probe. The approach is similar to that of Clarkson *et al.*,[86] but the shuttling is faster and the range of covered low fields is broader. The method was applied to ^{15}N-labelled ubiquitin. Still within the relaxometry, I wish to mention a couple of interesting development in fast field-cycling (FFC) measurements. The concept of field-cycling refers to experiments where the preparation of the spin system, the relaxation process under study, and detection of the NMR signal occur at different B_0 fields.[6] The FFC methodology allows obtaining a whole set of field-dependent T_1 data in a short time and using a single apparatus. Kresse and co-workers[87] described a FFC instrument allowing to extend the low-field end to few microtesla, corresponding to about 100 Hz proton resonance frequency. Broche *et al.*[88] proposed a development of the FFC in the context of MRI. The FFC MRI technology is difficult and requires optimisation of the experiment as well as data processing. The authors proposed an enhanced algorithm, derived from the so-called two point method of Edelstein and co-workers,[89] that allowed estimating the T_1 using only one image per field, taking advantage of the fact that the equilibrium magnetisation is proportional to the magnetic field strength and needs to be measured only once.

Relaxation measurements in large molecules are typically carried out in the form of two- (or higher) dimensional experiments. One special way of performing three-dimensional experiments, proposed long time ago by Bodenhausen and Ernst[90] is called the accordion spectroscopy. Among other application, the accordion spectroscopy allows efficient measurements of relaxation data. Harden and Frueh[91] described a MATLAB-based software environment called SARA (Software for Accordion Relaxation Analysis). Advantages and limitations of different fitting methods and protocols were discussed. Ahlner and co-workers[92] presented a new software (PINT) for integration of peak volumes in 2D spectra, designed for, among others, evaluation of relaxation rates.

Next, I wish to mention here the work by Pinto *et al.*[93] who presented a study of dendrimers where differences in transverse relaxation rates between different layers of the polymer (faster relaxation at the core compared to the periphery) were exploited in T_2-edited 1D and 2D NMR experiments. The experiments allowed for stepwise filtering of the internal nuclei (which could be ^{1}H, ^{13}C or ^{31}P) according to their topology within the dendritic structure. The resulting filtered spectra could be conceived as corresponding to virtual hollow dendrimers.

Farooq and co-workers[94] proposed a method for fast parameter optimization using rapid CPMG (RCPMG) pulsing during the acquisition. The method involves a combination of the inversion recovery (with a selective 90° pulse) and the RCPMG detection for estimation of the slowest ^{13}C spin–lattice relaxation rate in the range of interest. This approach, used at high field but accumulating all the ^{13}C signal intensities into a single peak, was applied to optimization of relaxation delays for samples of lysozyme and riverine dissolved organic matter.

Next, I turn to the discussion of new hardware design for relaxation experiments at medium and low magnetic field. Liu *et al.*[95] described a cesium atomic magnetometer operating near room temperature and able to detect NMR signals at very low magnetic field. The device was shown to be able to measure ^{1}H T_1 for aqueous solutions of a Gd(III) contrast agent at essentially zero field. Vesanen and co-workers[96] reported T_1 and T_2 measurements for an agarose gel sample in an ultra-low field (50 μT–52 mT) magnetic resonance imaging (MRI) device and showed that they could be used for temperature measurement. Borowiak *et al.*[97] presented a battery-driven low-field NMR unit and demonstrated that it could be used for relaxation measurements in thermally polarized samples. Horch *et al.*[98] described a low-field (118 mT) apparatus for relaxation measurements under high pressure, designed for porous materials. Dalitz *et al.*[99] described design and applications of a flow probe, meant to be used in a commercial 0.47 Tesla magnet in the context of reaction monitoring in process analytics. Wu and co-workers[100] reported design and fabrication of a low-field NMR probe based on multilayer planar microcoil, suitable for T_2 measurements. Dudova and co-workers[101] described a compact low-field NMR relaxometer-spectrometer designed for biological samples, *e.g.* to be used to measure glucose concentration in blood from T_1 and the NMR spectrum from a human finger. Diaz-Diaz *et al.*[102] presented an improved low-cost embeddable miniature NMR sensor capable of measuring water loss and porosity refinement in cement-based materials, by analysing the transverse magnetization decay from CPMG experiments. Related device, designed for quality control of edible oil, was described by Lu *et al.*[103] Guo and co-workers[104] proposed an apparatus with a related purpose – degradation control for hydraulic fluid – using a sensitive spot unilateral magnet, again suitable for CPMG (and inversion-recovery prepared CPMG) measurements of T_2 distributions.

Low-field NMR experiments are usually designed to differentiate various sample properties through relaxation or diffusion experiments. However, differentiating between signals with different relaxation properties may also be of interest at high fields. With this fact in mind, Lopez *et al.*[105] discussed relaxation-selective pulses for fast relaxing systems. They designed and optimized pulses with property of suppressing magnetization possessing a selected value of T_2 or T_1. The selectivity of the pulses was demonstrated at high field in aqueous solutions with varying concentrations of paramagnetic ions and for use in solvent suppression.

Hormann *et al.*[106] described the experimental setup for measuring the FID in the Earth magnetic field (which is very homogeneous) and showed how this could be used to get information on material porosity through T_2 distribution studies. Walsh and co-workers[107] designed a small-diameter (about 5 cm) NMR logging tool for groundwater studies. Tang *et al.*[108] described the design of a coaxial NMR probe, suitable for NMR diffusion and relaxation correlation experiments. The device was tested placed in the bore of a 1 Tesla superconducting magnet. Deng *et al.*[109] described a new approach to two-dimensional relaxation measurements in flowing fluids. The issue of NMR detection of explosives and illicit materials was the topic of a book[110] where several chapters were concerned with hardware design and with relaxation-related subjects. Mitchell and co-workers[111] reviewed the technology of low-field permanent magnets, useful in the area of industrial process and quality control.

Several authors were concerned with CPMG measurements in inhomogeneous magnetic fields, often encountered in work on porous media where internal gradients can be large. Ideally, these experiments can yield a T_2 distribution, which in turn gives information of the pore size distribution. Mandal and co-workers[112] proposed an improvement of the performance of the CPMG sequence under such conditions by using broadband excitation and refocusing pulses. They introduced a new class of such pulses, called the axis-matching excitation pulses, which resulted in a significant increase of the signal-to-noise ratio. The issue of separation of the "clean" transverse relaxation and effects of diffusion in the field gradient was dealt with in a series of papers by Mitchell and co-workers.[113–115] The methodology proposed is applicable at high magnetic fields where the problems caused by gradient are particularly serious. Related work was also reported by Xie *et al.*[116] and by Muncaci and Ardelean.[117]

Xiao, Zhang and co-workers[118,119] presented a three-dimensional experimental protocol providing information on of T_1, T_2 and internal gradients (diffusion) effects. The 3D data set was subjected to a 3D inverse Laplace transform (ILT), yielding 3D distribution functions and their projections. By making the experiments at two magnetic fields (corresponding to proton resonance frequencies of 23 MHz and 2 MHz), they revealed the scaling of the gradient effects with the field strength. The algorithm for 3D ILT was further discussed in another paper from the same group.[120] Celik *et al.*[121] dealt with a similar topic and proposed a new approach to the issue of stabilizing the ILT of a multiexponential decay signal, by introducing the second dimension followed by projection technique. In yet another related study, Tan *et al.*[122] compared different ILT algorithms for 3D NMR data. Zhang and Blümich[123] proposed a combination of 2D Laplace transform-based NMR with MRI, in order to analyse pore geometry and diffusion of fluids in porous media. In this exploratory study, they analysed spatially resolved diffusion-T_2 correlation maps from glass beads and mortar. Finally, Bernin and Topgaard[124] reviewed recent developments in experimental methodology and data analysis approaches for NMR diffusion and relaxation correlation in heterogeneous materials.

NMR for geophysical applications can also take the form of surface NMR. Measurements of this type are carried out at Earth magnetic field using very large coils/loops (40–100 m diameter), positioned on the ground in the area of interest, and huge "sample" volumes. The method is used, for example, to monitor underground water reservoirs. Grunewald and Walsh[125] described the applications of the CPMG technique in this type of experiments. In another paper from the same group, the authors proposed an improved scheme for spin-echo T_2 measurements, including phase cycling.[126] In yet another related study, Walbrecker *et al.*[127] described relaxation experiments at the Earth magnetic field in a laboratory setting. They investigated the effects of prepolarization, introduced to increase the sensitivity, on the measured relaxation rates.

Finally, I wish to mention some interesting recent work on the MRI-based techniques for creating spatially-resolved T_1 or T_2 maps. Wang and co-workers[128] proposed new methods for directly imaging T_1, T_2 as well as the proton density, compared them with earlier propositions and discussed the minimum number of acquisitions required. The new techniques were validated on phantoms as well as in human brain studies. Kellman *et al.*[129] investigated adiabatic inversion pulses as the tool for improving the inversion-recovery based T_1 mapping under realistic power limitations for human MRI. The inversion efficiency of adiabatic pulses was found to be sensitive to transverse relaxation and the tangent/hyperbolic tangent pulse design turned out to be most efficient. Andronesi and co-workers[130] proposed new methods for clinical imaging of the relaxations times in the rotating frame ($T_{1\rho}$, $T_{2\rho}$), applicable to the human brain. Two papers have dealt with the "relaxation MRI" in a very different context of porous media. Marica and co-workers[131] proposed a technique designed for heterogeneous porous media, with the ability to determine the T_2* (which stands for effective T_2, shortened by field inhomogeneity effects) for each individual pixel. The images of this kind have the potential to aid estimating the local fluid content. Mitchell[132] reported methods for rapid measurements of heterogeneity in sandstones by creating two-dimensional correlations of distributions of T_1 and $\Delta\chi_{app}$ (the latter symbol denotes the magnetic susceptibility difference between solid and liquid) through the ILT processing. The new methods allowed also for determining the spatially-resolved T_1 maps.

2.4 Relaxation in multi-spin systems

In this section, I review the work on nuclear multi-spin relaxation phenomena – the work where one of the involved spins belongs to an electron will be covered in Section 2.7. Several interesting papers by the group of Wiktor Kozminski have dealt with high-dimensionality experiments making use of non-uniform sampling (NUS). Stanek and co-workers[133] described a band-selective aromatic-aliphatic C,C-edited four-dimensional experiment for measuring proton–proton nuclear Overhauser enhancement spectra (4D NOESY). The method was tested on a 20kDa protein. The same group proposed also a similar 4D ^{13}C(aromatic), ^{13}C(ribose)-edited NOESY experiment designed for

$^{13}C,^{15}N$-labelled RNAs.[134] Stanek *et al.*[135] presented in addition a novel pulse sequence for probing local backbone structure in IDPs. Making again use of high-dimensionality experiments and NUS, the authors designed a method to measure interference between the $^{15}N–^{1}H$ dipolar interaction and ^{13}C chemical shielding anisotropy (CSA) in highly overlapped spectra typical for the IDPs. The high-dimensionality NUS spectroscopy for structural studies of biomolecules was also reviewed by Kazimierczuk and co-workers.[136] Another review that has to mentioned in this section is that by Vögeli,[137] covering measurements of exact NOE (eNOE) which enable the determination of multiple-state 3D protein structures.

Kaderavek *et al.*[138] presented several spectral density mapping protocols developed for analysis of dynamics in disordered proteins. The rationale for the work is the fact that the Lipari–Szabo method, based on the statistical independence of the local and global motions, is not valid for the IDP and that the commonly used reduced density mapping approach[139,140] may have problems for this category of systems. In the new protocols, the usual ^{15}N data (T_1, T_2, NOE) are supplemented by measurements of longitudinal and transverse DD-CSA cross-correlated relaxation rates (CCRRs).

Most work on the nuclear Overhauser enhancement deals with the intramolecular effects. Gabl and co-workers[141,142] concentrated their interest on the intermolecular NOE in liquids. They developed theoretical models allowing to calculate the intermolecular NOE at full distance resolution, *i.e.* allowing to estimate the contributions to the enhancement coming from spin pairs at a certain distance. The theory was applied to a model system of ionic liquids.

2.5 Relaxation and chemical exchange

Chemical exchange can influence NMR spectra and relaxation in several ways. If the exchange process is very fast, with the lifetime comparable to the rotational correlation time, then the exchange can compete with rotational modulation of anisotropic interactions (*e.g.* the DD interaction). The exchange lifetime then acts as a correlation time, the case discussed long time ago by Wennerström.[143] On a slower time scale, the exchange can modulate isotropic interactions, such as chemical shift. These processes affect mainly the transverse relaxation and can be suppressed by the design of experiments measuring T_2 (see below). Finally, even slower chemical exchange processes can cause magnetization flow competing on a more macroscopic level with intrinsic relaxation processes, as described already in 1958 by McConnell in his modification of Bloch equation.[144]

During the period covered by this review, several papers have been devoted to two types of methods relevant for this chapter. I begin by covering the studies related to the T_2 dispersion experiments, where the relaxation rate is measured using different delays between the 180° pulses in the CPMG experiment. At slow pulsing rate, the measurements contain relaxation rate contribution from chemical shift modulation by chemical exchange. At fast pulsing rate, this contribution is suppressed.

Weininger and co-workers[145] discussed ^{13}C L-TROSY CPMG dispersion experiments as applied to slow aromatic ring flips. Using the protein BPTI as the model system, the authors studied the carbons δ and ε in the Y21 residue characterized by single peaks. The plot of R_2 rates *vs.* CPMG frequency (inverse delay between the 180° pulses) displayed an anomalous upside-down turned dispersion. The observation was explained in terms of strong coupling effects in the nearly-degenerate spin systems and it was demonstrated that such dispersion profiles could in favourable cases be given quantitative interpretation. Sekhar *et al.*[146] studied ^{15}N CPMG relaxation dispersion profiles for the four-helix bundle FF module. The module folds to the native N state *via* a transiently formed and sparsely populated intermediate I. The authors used their data to define a length scale relevant for the I–N transition providing an average measure of the structural units participating in the transition. Ban and co-workers[147] discussed the constant time (CT) CPMG dispersion experiments where kinetic and structural information could be obtained at atomic resolution. In particular the authors studied the case of fast processes, requiring high refocusing frequencies and concluded that errors could in such cases be largely reduced by performing a separate measurement of the intrinsic relaxation rate $R_{2,0}^{CT\text{-}CPMG}$, making use of large amplitude spin-lock technique. The area of CPMG dispersion studies attracted also some software development. For this purpose, Sugase *et al.*[148] proposed the software package GLOVE, that alternately minimizes the global and local parameters, incorporates a sophisticated Monte Carlo minimization technique and implements also random search methods. Moreover, Mazur and co-workers[149] presented a new software tool for the analysis of the CPMG dispersion data. As an alternative to the CPMG dispersion experiments, some authors prefer to study the conformational exchange on the micro- to millisecond range by spin-lock dispersion experiments, where the $R_{1\rho}$ relaxation rate is measured as a function of the spin-lock power. Weininger *et al.*[150] proposed an experiment of this kind designed for ^1H spins in methyl CHD_2 groups. The new experiment allowed for improving the characterization of fast exchange processes and was shown to be practically free from the effects originating in the rotating-frame Overhauser effect (ROE). The same group developed also a related experiment designed for the aromatic carbons in selectively ^{13}C labelled proteins.[151] The new experiment uses the ^{13}C longitudinal and transverse relaxation optimized (L-TROSY) approach. It was demonstrated by measuring exchange parameters for the tyrosine Y23 ring flips in BPTI at elevated temperature. Aramini and co-workers[152] studied a protein labelled with 5-fluorotryptophan and demonstrated that ^{19}F relaxation dispersion data could be used to demonstrate the presence of conformational dynamics and the oligomerization state of the protein.

The second group of investigations discussed here is related to the chemical exchange saturation transfer (CEST) and related methods, sometimes denoted as *z*-spectra. Briefly, when a weak ("invisible") signal corresponding to a low-populated state is saturated, it is possible to read out the saturation effect through intensity changes of the much stronger signal in exchange with the weak one. In a typical experiment, the intensity of a strong signal is monitored as a function of the frequency offset of the saturating rf field. Bouvignies and co-workers[153] proposed the extension of the CEST methodology to ^{13}C resonances of uniformly ^{13}C labelled proteins. The experiments were demonstrated to provide insights into side-chain packing and dynamics. Hansen and Kay[154] developed a CEST experiment designed for quantifying the ionization/tautomeric states and pK_a values of histidine side chains in conformationally excited protein states. It was demonstrated that the approach could provide important details on intermediate states in the folding process. Zhao *et al.*[155] described the ^{13}C CEST and weak spin-lock $T_{1\rho}$ relaxation dispersion experiments designed to characterize the slow chemical exchange in nucleic acids. The method was applied to base and sugar carbons in a 47-nucleotide fluoride riboswitch. Xu and co-workers[156] presented a methodology for fast scanning of exchangeable proton sites over a broad range of saturating frequency offsets involving only two signal excitations. The method, called ultrafast *z*-spectroscopy (UFZ) uses spatial encoding of indirect evolution times, in a manner similar to the work of Lucio Frydman's group some time ago.[157] Döpfert *et al.*[158] proposed an extension of the UFZ method, including, among other features, slice selection for studies of heterogeneous samples and meant to be used in the context of high-throughput screening of paramagnetic CEST (PARACEST) contrast agents.

2.6 Quadrupolar interactions

Quadrupolar nuclei, characterized by the spin quantum number $I \geq 1$, have often rather broad NMR lines in solution, which is one of the reasons for less frequent studies. Briefly, the high efficiency of the quadrupolar relaxation is related to large strength of the interaction: the quadrupole coupling constants (QCC) are typically orders of magnitudes larger than the dipolar coupling constant. During the period under review here, only few methodological papers within the field of quadrupolar relaxation have been published. Stobbe and Beaulieu[159] proposed a novel pulse sequence making use of relaxation of ^{23}Na ($I = 3/2$) in the context of sodium MRI. The new method improves the contrast between the sodium images in different tissue models as well as real tissues. Tugarinov[160] published a review on indirect use of deuterium in solution NMR studies of protein structure and hydrogen bonding. I choose to mention this work here, though it strictly speaking does not belong in this section since it deals with relaxation properties of the ^{15}N or ^{13}C spins carrying one or more deuterons and not with relaxation of the ^{2}H spins themselves.

2.7 Paramagnetic systems in solution

Paramagnetic effects on nuclear spin relaxation remain a topic attracting a lot of interest. In this section, I review the work with emphasis on theoretical/methodological aspects, while more application-oriented papers will be mentioned later on in this review.

Paramagnetic species have a very strong effect on nuclear spin relaxation, because of the huge magnetic moment associated with unpaired electron spin. The most important effects arise usually through DD interaction between the electron and nuclear spins, with transverse relaxation of ^{17}O (where the scalar part of the hyperfine interaction normally dominates) as an exception (see Section 3.2 below). Two cases can be discerned: the inner-sphere and the outer-sphere relaxation. One speaks about the inner-sphere case when the nuclear spin resides at least intermittently in the first coordination sphere of the paramagnetic transition metal (or lanthanide) ion. The measured paramagnetic relaxation enhancement, PRE, in the case of inner-sphere longitudinal relaxation is simply related to the solution composition, exchange lifetime and the in-complex relaxation rate, as shown long time ago by Luz and Meiboom.[161] In the case of outer-sphere relaxation, the mutual translational diffusion of the paramagnetic species and the molecule carrying the nuclear spin is of central importance.

Kruk and co-workers[162] reported a study of the latter case, where the paramagnetic species was the nitroxide radical 4-oxo-TEMPO-d_{16}, with the nitrogen atom being either ^{14}N ($I=1$) or ^{15}N ($I=1/2$), and the NMRD was measured for protons residing in the solvent propylene glycol at different temperatures. In addition, the authors measured also the X-band ESR spectra for the same systems at a lower radical concentration. The ESR and NMRD data were submitted to a joint analysis. The ESR lineshapes gave access to the nitrogen hyperfine tensor component and the rotational correlation time for the radical molecules while the NMRD data were interpreted using a recently presented theory.[163] The theory is concerned with the DD interaction between the solvent protons and the electron spin in the radical, modulated by relative translational diffusion, molecular rotation and electron spin relaxation. The consistency of the unified analysis was evaluated (by controlling the agreement between, on the one hand, the rotational correlation times obtained for both sets of data, and, on the other hand, the agreement of translational diffusion coefficients with the literature data) and found satisfactory.

Bhattacharya and co-workers[164] reported another theoretical study of TEMPO-based radical systems, more specifically biradicals with photochromic couplers allowing switching between the anti-ferro- and ferromagnetic arrangements. The goal was to evaluate these materials as possible MRI contrast agents and the tools included density functional theory (DFT) calculations of the exchange coupling constants and zero-field splitting (ZFS) of the triplet (ferromagnetically coupled) states.

Several papers have dealt with theoretical aspects of electron spin relaxation in transition metal and lanthanide complexes. The topic is of central importance for understanding the PRE. I choose to discuss these papers here even if they do not explicitly discuss the nuclear spin relaxation. Håkansson[165] reported an X-band ESR study of Cu(II)-porphyrin in fluid and frozen solution. The experimental data were interpreted using a simulation method based on stochastic Liouville equation (SLE). The analysis provided the porphyrin rotational correlation time (in the nanosecond range) combined with a fast local motion, as well as an accurate set of spin Hamiltonian parameters. Fries[166] studied electron spin relaxation in Gd(III)-based contrast agents. He simulated the quantum-mechanical time correlation functions of the electronic spin operators assuming the presence of static and transient ZFS and using the so-called Grenoble method for solving the SLE. The effects of electron spin relaxation on the inner- and outer-sphere relaxivity were also discussed. Fries and Belorizky[167] reported also a quantitative interpretation of the very fast electron spin relaxation in systems containing lanthanide(III) ions other than Gd(III). The relaxation mechanism was assumed to be related to the Hamiltonians for the static molecule-fixed ligand field Hamiltonian and its transient analogue, originating from vibrations-distortions. The electronic relaxation of the Ln(III) ions was defined by the decays of TCFs of the longitudinal and transverse components of the total angular momentum *J*. The calculations of the TCFs were based on the close analogy between the fluctuations of the ligand field Hamiltonian and the ZFS Hamiltonian as described in the Grenoble model in the case of Gd(III).

Two papers have described combinations of quantum chemical calculations and molecular dynamics simulations related to electronic and nuclear spin relaxation in paramagnetic systems. Lasoroski *et al.*[168] reported calculations of the hyperfine coupling tensors of the water molecules coordinated to a gadolinium complex (used as MRI contrast agent) along a trajectory calculated by *ab initio* molecular dynamics. The validity of the so-called point-dipole approximation, assuming the whole spin density to be located at the metal ion, was evaluated. They found that the hydrogen bonding of the inner sphere water protons resulted in their non-equivalence at short time scales. The influence of the very fast motions on the anisotropic (dipolar) contribution to the hyperfine couplings could be probed by introducing a small set of collective variables. Mares and co-workers[169] continued their earlier simulation work on aqueous solution of Ni(II).[170,171] In the most recent contribution,[169] they used an MD trajectory based on empirical polarizable force field and performed quantum chemical calculations of hyperfine couplings, *g*- and ZFS tensors and nuclear shielding terms at the snapshot configurations. Thus obtained fluctuations of the molecular properties were used to estimate the Curie-type paramagnetic relaxation contribution (which is a paramagnetic analogue of the CSA relaxation in the diamagnetic systems) to the NMR relaxation of aqueous protons.

I turn next to some experimental studies on small paramagnetic complexes, which I judge to be methodologically interesting. Esteban-Gomez *et al.*[172] reported proton NMRD profiles and ^{17}O chemical shifts and transverse relaxation data for aqueous solutions of $[Mn(H_2O)_6]^{2+}$. The data were analysed providing water exchange rate and hyperfine coupling constants (HFCCs) for both ^{1}H and ^{17}O. The experimental data were complemented by DFT calculations of the HFCCs for both $[Mn(H_2O)_6]^{2+}$ and its analogue with additional twelve water molecules. The ^{1}H relaxivity displays a substantial scalar contribution at low field, related to slow electron relaxation and a rather slow exchange rate for the inner-sphere water molecules. Gale and co-workers[173] described a simple method for estimating the hydration state of the Mn(II) ion in coordination complexes, as well as metalloproteins. The linewidth of bulk $H_2{}^{17}O$ was measured in the presence and the absence of Mn(II) as a function of temperature. The transverse ^{17}O relaxivity measured at high field (≥7 T) and at micromolar Mn(II) concentrations, was demonstrated to provide the number of inner-sphere water molecules in the complexes.

In the paper mentioned above (Section 2.1), Singh *et al.*[33] reported an investigation of paramagnetic relaxation of ^1H long-lived coherences. The effect of paramagnetic ions was analysed in terms of correlated external random field contributions. It was found that the LLCs relax in the presence of paramagnetic ions faster than ordinary single-quantum coherences. The results were compared with the paramagnetic enhancement data for more usual relaxation rates (R_1, R_2, $R_{1\rho}$) and with the relaxation rate of long-lived states.

Yang and co-workers[174] studied spin–spin and spin–lattice relaxation of protons in dispersions of superparamagnetic nanoparticles, called ferrofluids. In analogy with an earlier study from the same group,[175] the measurements were performed at 102 μT using a SQUID device. In the present work, the T_1 experiments were also carried out at a higher magnetic field, by using a prepolarization field. The latter measurements are not very clearly described or explained.

Next, I move to new experimental methods and protocols concerned with paramagnetic relaxation in biomolecular systems. Ciofi-Baffoni and co-workers[176] proposed a new NMR experiment for paramagnetic metalloproteins, designed to identify residues in the vicinity of the metal centre with their fast nuclear spin relaxation. The method is a modification of the HSQC (heteronuclear single quantum coherence) scheme, including an inversion-recovery filter and optimization of the INEPT segments. Gu *et al.*[177] described a novel Gd(III)-based paramagnetic relaxation reagent, to be added to protein solutions with the purpose of probing the protein surface through the outer-sphere PRE. Clore[178] reviewed the applications of the PRE measurements for studies of sparsely-populated states of paramagnetically-tagged (by a metal complex or a nitroxide) multidomain proteins, in the context of interplay between conformational selection and induced fit in the ligand binding. Jeschke[179] reviewed the applications of nitroxide spin labels attached to biomacromolecules, providing the PRE as well as DNP effects. The emphasis was on the conformational dynamics and distribution of the spin labels. Eggimann *et al.*[180] presented a protocol for modelling moderate resolution backbone structures for helical proteins, making use of PREs as structural constraints. Wunderlich and co-workers[181] proposed a new paramagnetic nitroxide tag, designed for nucleic acids. Zhang *et al.*[182] reviewed the novel NMR tools for characterizing dynamic conformations and interactions of oligosaccharides, including the PRE effects from lanthanoid tagging and spin labelling. Finally, Alberti and co-workers[183] described the use of the PRE for the measurements of cell surface biomarkers. More specifically, they proposed a method, denoted R-ELISA (relaxometric enzyme linked in cells suspension assay) making use of the enzymatic disruption of liposomes loaded with paramagnetic species.

I shall return to application-oriented papers on relaxation in paramagnetic systems in Section 3.2–3.5.

2.8 Hyperpolarization and relaxation

In this section, I review the methodological work on hyperpolarization, if it also refers to relaxation phenomena. Three types of hyperpolarization technology will be covered. We start with the dynamic nuclear polarization (DNP) and continue with hyperpolarization through parahydrogen (PHIP) and through optically pumped xenon. Koptyug[184] presented recently a review of hyperpolarization techniques applied to enzymatic processes and covering the first two approaches.

Double resonance, involving electron and nuclear spins, leads to the phenomenon of dynamic nuclear polarization and carries promises of huge sensitivity enhancement of the NMR methods, both in the liquid and the solid state. Here, I limit my interest to liquids. Briefly, the DNP in solution can be obtained by three techniques. One of the methods makes use of polarization of electron spin at low field and low temperature (solid-state mechanism), followed by sudden sample dissolution in warm solvent and transfer to the high field NMR magnet (dissolution DNP). One other option is direct double resonance at high field, making use of powerful microwave sources (gyrotrons) and based on the Overhauser-mechanism in the liquid state. Obtaining efficient polarization in liquids at high field is technically difficult and the third option is based on either working at low field or shuttling the liquid sample between a low-field polarizing magnetic field and the high detection field.

I go through the progress of DNP in the last year in that order. Donovan *et al.*[185] described experiments in which both ^{13}C and ^{1}H in a sample were sequentially hyperpolarized under cryogenic conditions. After rapid dissolution of the sample, time-dependent polarization gains, persistent on the time scale of ^{13}C T_1, were observed for protons bound to ^{13}C. The phenomenon was explained by a model based on heteronuclear cross-relaxation in the dissolved state. In another paper from the same group, Harris and co-workers[186] investigated the potential of using hyperpolarized water for biomolecular NMR studies. The initial step was the low-temperature hyperpolarization of proton spins in a 3 : 2 water/glycerol sample, containing some TEMPO radical, by irradiation at the ESR frequency. Next, the sample was dissolved and transferred rapidly to the NMR probe, where it was used to dissolve a biomolecular material. As a consequence of chemical exchange of the polarized solvent with amine and amide groups in the biomolecule, the polarization was transferred to the signals of these groups. Further, the non-equilibrium population of these protons could be transferred to the ^{15}N sites, through an Overhauser-type mechanism. Marco-Rius *et al.*[187] described the experiments based on the opposite principle, *i.e.* using the low temperature and radical irradiation to hyperpolarize ^{13}C in [1,4-$^{13}C_2$]fumarate and subsequent nuclear Overhauser transfer to the solvent (water) protons. In analogy with the polarization transfer from optically pumped ^{129}Xe gas proposed several years ago by Navon and co-workers,[188] the experiment was denoted SPINOE (spin polarization-induced NOE). The same group reported (in a paper mentioned already in Section 2.1) an investigation of hyperpolarized long-lived singlet states of ^{13}C in [1,2-$^{13}C_2$]pyruvate in human blood and in a living mouse.[25]

The cryogenically hyperpolarized water was also used by Ardenkjaer-Larsen and co-workers[189] who dissolved it in hot D_2O, injected the resulting solvent (after reduction of the radical) into a phantom or a living animal, and performed an MRI scan. In the phantom experiments, the T_1 of the dilute water protons at elevated temperatures was several tens of seconds. The experiment is meant to facilitate MR angiography. The same group proposed also a modification of the dissolution DNP hardware in order to allow larger volume samples.[190]

Chen and Hilty[191] proposed a variant of the hyperpolarized NMR spectroscopy allowing for obtaining 2D correlations, *e.g.* between ^{13}C and ^{1}H shifts. The method applies a flow NMR probe and uses Hadamard spectroscopy to obtain a HSQC spectrum, which requires a correction procedure to account for variations in concentration and T_1. Swisher *et al.*[192] worked with hyperpolarized ^{13}C in [1,2-$^{13}C_2$]pyruvate and proposed the use of stimulated echo acquisition mode (STEAM) approach. In this way, they were able to measure the conversion rates to different metabolites as well as multiple T_1 values. In a similar vein, Pagès and co-workers[193] studied transmembrane exchange of hyperpolarized ^{13}C-urea in human erythrocytes. The measurements were carried out on up to 2 minutes time scale and the data analyzed in terms of exchange and relaxation processes. Reed *et al.*[194] used also the ^{13}C hyperpolarized urea to perform ^{13}C MRI experiments on rats. They were able to obtain a spatial map of T_2. Using the urea labelled both in ^{13}C and ^{15}N, they demonstrated that the ^{13}C spin–spin relaxation was slowed down by suppressing the strong relaxation pathway from scalar interaction between ^{13}C and ^{14}N. Nonaka and co-workers[195] proposed a new chemical probe for hyperpolarized NMR measurements: [^{15}N,D_9]trimethylphenylammonium. The molecule shows a remarkably long T_1 (816 s at 14.1 T) which makes it suitable for use as hyperpolarization agent. They demonstrated also several chemical application of the material.

Let us now turn to the work on hyperpolarization obtained by continuous irradiation of ESR frequencies for radical in solution, making direct use the cross-relaxation of the electron and nuclear spins, known as Overhauser DNP (ODNP). The phenomenon was discussed in the review by Jeschke[179] on conformational dynamics and distribution of nitroxide radicals, mentioned already in the previous section. In another review in the same journal, Franck *et al.*[196] covered recent development in quantitative Overhauser DNP applied for studies of water dynamics in the vicinity of nitroxide radical spin probe attached to a targeted site in a macromolecule. These experiments are typically performed using X-band ESR (0.35 Tesla, around 10 GHz ESR frequency), corresponding to 15 MHz ^{1}H Larmor frequency. In a recent paper from the same group, Cheng and co-workers[197] applied the site-specific quantification of translational diffusion of hydration water to investigating protein structure, topology and immersion depth at lipid membrane interface. In yet another study, Franck *et al.*[198] proposed combining the ODNP with one-dimensional and two-dimensional relaxometry techniques for porous material. This allowed for selective amplification of the relaxation dynamics of the fluid in the local porous volume containing the paramagnetic probe. Other applications of the ODNP methods in complex systems will be mentioned in Section 3.5.

Valentine and co-workers[199] proposed merging the ODNP technology at 0.35 Tesla with encapsulating proteins in reverse micelles dispersed in low viscosity hydrocarbon solvents. Using these solvents rather than water greatly reduced the heating effects of microwaves. The nitroxide radicals were introduced in the reverse micelle system in three ways: attached to the protein, embedded in the reverse micelle shell and free in the aqueous core.

Luchinat and co-workers[200] discussed the question whether it might be possible to replace the nitroxides as the paramagnetic ODNP agents by metal ion complexes. The discussion was based on the "classical" theory of Overhauser DNP, presented long time ago by Hausser and Stehlik.[201] More specifically, the authors re-analyzed the NMRD profiles of slowly-relaxing paramagnetic aqua ions and complexes using the Solomon–Bloembergen–Morgan theory (see Section 2.7) in order to estimate the coupling factor of Hausser and Stehlik, responsible for maximum enhancement attainable in solution. The conclusion was that under certain conditions, the metal complexes might indeed compete with nitroxides.

The para-hydrogen (p-H_2) induced polarization (PHIP) is an alternative, less-frequently used, hyperpolarization technique. Briefly, very high observable nuclear polarization can be created by catalyzed hydrogenation reaction with para-hydrogen, if the reaction breaks the symmetry of the p-H_2. The reaction can be carried out inside the NMR magnet (the PASADENA method[202]) or outside of it (the ALTADENA method[203]). More recently, Adams and co-workers[204] discovered that the polarization transfer from p-H_2 to nuclear spins can occur without the incorporation of the para-hydrogen into substrate, in a process denoted Signal Amplification by Reversible Exchange, SABRE. During the period covered by this review, some papers have been concerned with spin relaxation aspects of PHIP. Kiryutin and co-workers[205] investigated the mechanism of the PHIP transfer from the original strongly aligned protons to other nuclear spins at low external magnetic field. The highly polarized state was created at high field (7 Tesla) and the field was varied by moving the whole NMR probe between the homogeneous spot of the superconducting magnet and variable positions in the fringe field. The results were an oscillatory time evolution of the polarization transfer, consistent with the coherent transfer process, while the cross-relaxation effects were found unimportant. Reineri and co-workers[206] performed their experiment in both the ALTADENA and PASADENA manner, using 1-^{13}C-acetylenedicarboxylic acid as the substrate and a Rh(I)-based catalyst. The comparison of the two approaches and DFT calculation for possible reaction intermediates were applied to account for the obtained polarization effects. Barskiy *et al.*[207] reported a SABRE-type study of pyridine and pyridine-d_5 with an iridium complex as catalyst. The experiment was performed at 9.4 Tesla, a much higher field than what is typically used in the PHIP method. Real time kinetic studies of the polarization build-up showed that the rates obtained for various species were in close agreement with the corresponding ^{1}H T_1 data. A tentative conclusion was that the results were consistent with SPINOE-type[188] effect.

The third basic technique for creating hyperpolarized nuclear spin magnetization makes use of spin exchange optical pumping. This method is commonly applied to hyperpolarize ^{129}Xe ($I = 1/2$) in the gas phase. The principles of the technique as applied in the MRI context were reviewed recently by Palaniappan and co-workers,[208] who put particular emphasis on the CEST approach (called in this case HYPERCEST because of combination with xenon hyperpolarization). Several papers described different technique for preparation of hyperpolarized ^{129}Xe for biomedical MRI applications.[209–211] I shall return to some results of these works in Section 4. Finally, Abrams and co-workers[212] proposed a hyperpolarization scheme for nuclear spins in fluids, based on optical pumping of paramagnetic nitrogen-vacancy (NV) centres in nanostructured diamond. The authors derived the Solomon-type[16] equations for population kinetics between different energy levels in the combined NV-nuclear spin system and the steady-state polarization of the nuclear spins. The effects of the position of the NV centre with respect to the surface and of the surface characteristics were also discussed.

3 Selected applications of nuclear spin relaxation in liquids

Here, I am going to review applications of nuclear spin relaxation in pure liquids (Section 3.1), liquid solutions (Section 3.2–3.4) and in more complicated fluid systems (Section 3.5).

3.1 Pure liquids

Neat liquids are, in a way, difficult objects for NMR relaxation studies. The simple modelling of reorientational motion as small-step rotational diffusion is based on hydrodynamics (large body immersed in continuum solvent) and becomes problematic if we deal with a liquid consisting of molecules of a single kind. Deviations from the models based on few discrete correlation times can therefore be expected.

Bauer and co-workers[213] investigated neat 4-methyl-3-heptanol using broadband dielectric spectroscopy as well as ^{1}H and ^{2}H spin–lattice relaxation measurements over a broad temperature range. The dielectric data for this compound differ from many other monohydroxyalcohols and the combination with NMR data was used to explain the differences. Bock *et al.*[214] reported also a combined study of dielectric and NMR (^{2}H, ^{31}P) spectroscopy for the neat glassformer tripropyl phosphate (TPP), as well as its mixtures with polystyrene (see Section 3.3 below). I also wish to mention here another paper from the same group,[215] even though it does not include NMR measurements. Here, the reorientational dynamics in molecular liquids – in the temperature range from boiling point to the glass transition – was studied by dynamic light scattering.

Some work on NMR relaxation studies of neat room-temperature ionic liquids (RTILs) has been reported. The field was subject to a review by Weingärtner.[216] Daniel and co-workers[217] reported a study of proton spin–lattice relaxation over a broad range of magnetic fields for an ionic liquid with a quarternary ammonium ion with three long hydrocarbon chains as a cation and chloride anion. The relaxation was found bi-exponential, with the slower component assigned to the methyl groups and the faster one to the methylene units. The relaxation rate *vs.* frequency curves showed additional structure at high fields, assigned to the cross-relaxation between the ^{1}H and ^{35}Cl spins.

Finishing this section on neat liquid systems, I wish to mention few papers on polymer melts. Rössler and co-workers presented two reviews of different aspects of the field. One of them[218] was concerned with field-cycling studies of intermolecular proton relaxation and translational diffusion in polymer melts, as well as other types of liquids. The other review[219] described the recent field cycling work, combined with the so-called frequency-temperature superposition approach, in the context of polymer melts in bulk and in confinement. The same group reported also an FFC-based investigation of the translational diffusion coefficients for a number of polymer melts, measured over a range of the molecular mass.[220] Extracting the diffusion coefficient from the universal power law at the low field and correcting for the segmental dynamics, the authors were able to present the temperature-independent iso-frictional diffusion constant revealing the mass dependence of the collective dynamics. In a related MD simulation study, the same laboratory investigated the molecular mass dependence of rank-two orientational correlation functions, providing access to different dynamic process (segmental motion, Rouse dynamics, entanglement dynamics) in poly(propylene oxide) melts.[221] The simulations were compared with experimental FFC data. Kadayakkara *et al.*[222] studied ^{19}F spin–lattice relaxation for linear and cyclic perfluoropolyethers at three (high) magnetic fields and different temperatures. The R_1 in these systems was found to be dominated by the dipole–dipole interaction and chemical shift anisotropy. The importance of these findings for rational design of ^{19}F MRI agents and methods was discussed.

The earlier paper by Litvinov and co-workers, dealing with chain entanglement in polyethylene melts, has caused some debate.[223,224]

3.2 Electrolyte solutions

The organization of this section is as follows: I begin with reviewing work on aqueous or organic solvent solutions of compounds containing elements from the main groups of the periodic table, followed by diamagnetic solutions in RTILs (here, it is the solvents that consist of ions) and by diamagnetic transition metal compounds. Some work on metal complexes is included here, even though they are stable and electrically neutral. Selected work on paramagnetic ions and their complexes is reviewed last.

Masuda and co-workers[225] studied the proton location and transfer in the N–H···N hydrogen bond in protonated ^{15}N-labelled 1,8-bis(dimethylamino) naphthalene ($DMANH^{+}$) in the presence of different counterions. They measured ^{13}C and ^{15}N spin–lattice relaxation rates in acetonitrile or aqueous solution. From the ^{13}C data, they determined the reorientational dynamics of the molecular skeleton. The ^{15}N data were then interpreted in terms of the effective NH distances, assuming the limit of the proton transfer time much shorter than the rotational correlation time of the NH axis. The variation of the NH distance was assigned to intermolecular interactions with counterions and/or solvent. The interpretation was corroborated by Car–Parrinello molecular dynamics simulations. Nazarski[226] reported ^{13}C T_1 measurements in ammonium cations of a tetraaza-macrocyclic system with an amino-pendant arm in aqueous solution at high and low pH. The data were used for structural analysis of the protonated species. Ciftja *et al.*[227] reported a systematic study of ^{13}C T_1 for several amines and aminoacids of possible interest in the context of CO_2 capture, the corresponding carbamates and carbonate/bicarbonate ions. The purpose of the investigation was to specify a suitable protocol for speciation and quantitative analysis of the CO_2 adsorbent systems. Related work on carbon relaxation in relation to CO_2 capture studies was described by Yamada and co-workers.[228] Richardson and co-workers[229] reported a study of spin–lattice relaxation for ^1H, ^{19}F and ^7Li in solutions of $LiBF_4$ in propylene carbonate at high or intermediate magnetic field, at different temperatures and salt concentrations. They assumed that the main relaxation mechanism for all three nuclear species was dipolar (which the reviewer finds somewhat remarkable for ^7Li with its $I = 3/2$) and discussed the rotational and translational motion as the source of spin relaxation. I also wish to mention here the work by Stirnemann and co-workers[230] on the mechanism of acceleration and retardation of water dynamics in dilute and concentrated ionic solutions. The work consists of MD simulations as well as analytical modelling and can be of interest in the context of NMR relaxation studies.

I turn next to the systems where the room-temperature ionic liquids are used as solvent. Murgia and co-workers[231] studied the phase diagram and the mesoscopic structures in mixtures of water and 1-butyl-3-methylimidazolium (C4mim) trifluoroborate. Analysis of diffusion coefficients and ^{11}B relaxation times revealed that ion-pair dissociation was a progressive process starting at water mole fraction of about 0.2 and ending at high water content. Umecky *et al.*[232] studied the effects of dissolved water on Li^+ solvation in the ionic liquid with the 1-ethyl-3-methylimidazolium (C2mim) cations and bis(trifluoromethane sulfonyl)amide (TFSA) anions. They measured proton and ^7Li chemical shifts, self-diffusion coefficients for various species and ^7Li longitudinal relaxation times at two magnetic fields and various water contents. This allowed to evaluate the one-jump distances of Li^+ in the RTIL solutions. Dias and co-workers[233] investigated also the C2mim – TFSA system in mixtures with benzene and fluorinated benzene derivatives, reporting among others intermolecular proton NOEs. In addition, the authors performed also MD simulations for the same systems. Lin *et al.*[234] studied related Li^+-doped imidazolium-based RTIL systems, making use (among other techniques) of heteronuclear ^1H–^7Li NOE experiments. Simons and co-workers[235] studied the influence of Zn^{2+} and water on the transport properties in a different type of RTIL, pyrrolidinium dicyanamide. They reported variable-temperature ^{13}C T_1 data as a function of sample composition and were able to draw conclusions on both the mobility of various species and the chemical environment of the anion.

Next, I turn to relaxation studies of diamagnetic transition metal systems. Zhang and Dorn[236] reported a general discussion of NMR relaxation in systems of metal (such as ^{139}La, ^{45}Sc, ^{89}Y) ions or clusters, encapsulated in fullerene cages. Cmoch and co-workers[237] investigated adducts of nitrogen-containing ligands with dimeric rhodium(II) complexes. They measured, among other properties, the ligand ^{1}H T_1 data in chloroform solution and found that they decreased significantly upon binding to the rhodium complexes. In addition, the NOE between the protons of the rhodium complexes and the ligands were also reported. Wang and co-workers[238] described a study of platinum(II) substituted porphyrin. They measured proton spectra at two magnetic fields and noticed that the intensities of satellites due to couplings to ^{195}Pt were reduced at the higher field (14.1 Tesla) compared to the lower (9.4 Tesla), which was explained in terms of more efficient platinum spin relaxation due to the CSA mechanism. Bera and co-workers[239] studied dynamics of H-atom exchange in stable *cis*-dihydrogen/hydride complexes of ruthenium(II), through measurements of ^{1}H T_1. The η^2-H_2 and hydride proton gave rise to only one signal over the investigated temperature range, with very fast relaxation at the T_1 minimum. The data allowed estimation of the H–H distance in the η^2-H_2 ligand.

I now move on to paramagnetic metal complexes. Upadhyay *et al.*[240] studied ruthenium(III) compounds containing clusters of six ruthenium atoms. The Ru(III) ion has an odd number of electrons which leads to paramagnetism. In the cluster, the spins of the six ions are subject to antiferromagnetic exchange interactions, which results in rather narrow NMR lines, the possibility to observe ^{1}H–^{1}H NOEs and measurably long (on the order of a millisecond) ^{1}H T_1 values. The paper by Daniel *et al.*,[217] mentioned above in the context of neat RTILs, discussed also properties of the material doped with 1% (v/v) of the analogue with a paramagnetic counterion, $FeCl_4^-$. This gives rise to a magnetic ionic liquid with unusual properties, *e.g.* viscosity varying with the magnetic field strength. The authors reported NMRD data for that system, measured over a broad range of magnetic fields. Chierotti and co-workers[241] investigated non-covalent interactions occurring between a transition metal fluoride and a transition metal hydride, making use of ^{1}H and ^{19}F relaxation measurements as well as NOESY and HOESY experiments.

Three papers have dealt with manganese(II) systems. Gale and co-workers[173] proposed a simple method to estimate the inner-sphere hydration state of the Mn(II) ion in coordination complexes and metalloproteins. The method makes use of the ^{17}O linewidth measurements for bulk water in the presence and in the absence of Mn(II), which allows the determination of transverse ^{17}O relaxivity. Doing this as a function of temperature and finding the maximum yields a quantity which is directly proportional to the number of inner-sphere water ligands. Molnar *et al.*[242]

synthesized picolinate-containing macrocyclic Mn(II) complexes and characterized them using proton NMRD and ^{17}O NMR measurements. The analysis of the data provided a very high water exchange rate of the inner-sphere water and an unusually high ^{17}O hyperfine coupling constant. The experimental work was supplemented by DFT calculations. A similar approach was also applied to other Mn(II) complexes in another paper from the same group.[243]

Numerous papers have been concerned with complexes of lanthanides. The work explicitly devoted to optimization of the MRI contrast agents is not included here, unless judged to be of more general chemical interest. The metal-based MRI contrast agents were a topic of a special issue of the Journal of Biological Inorganic Chemistry (vol.19, issue 2, February 2014), introduced by a paper by Pierre and co-workers.[244] The lanthanide contrast agents have essentially two modes of action: the Gd(III) complexes give rise to the PRE, while other lanthanides operate through the PARACEST phenomenon, introduced in Section 2.5 above. The advantages of the latter approach were reviewed by Soesbe *et al.*[245]

Hung and co-workers[246] studied systems consisting of Gd(III) attached to carbon nanomaterials, *e.g.* graphene. A library of a number of such "gadographenes" was prepared and the materials were characterized through ^{1}H NMRD measurements. The NMRD profiles were interpreted using the so-called modified Florence model,[247,248] incorporating the Lipari–Szabo approach. The observed variations in relaxivity could be explained by changes of the hydration number, water residence time, molecular tumbling rate and local motions. Jaccard *et al.*[249] reported an investigation of dynamic aggregation of a mid-size Gd(III) complex, leading to a complex with three metal ions. A combined analysis of the concentration-dependent ^{1}H NMRD profiles and ^{17}O NMR data was presented, making use of different theoretical approaches. Delli Castelli and co-workers[250] described a related (^{1}H NMRD and ^{17}O NMR) study of the novel macrocyclic Gd(III) complex, denoted GdHPDO3A, combined with high resolution ^{1}H NMR spectroscopy of all other lanthanide(III) ions complexed by the same macrocycle. The data were interpreted in terms of different conformations of the macrocyclic ring. Baranyai *et al.*[251] studied related macrocyclic, DO3A-based, complexes of Gd(III) and Eu(III), meant to function as pH-responsive contrast MRI agents. Also in this work, the relaxation experiments (^{1}H NMRD and ^{17}O NMR) were combined with other physico-chemical techniques: luminescence lifetime determination and high-resolution NMR for the Eu(III) systems.

Neil and co-workers[252] studied lanthanide(III) complexes of a nonadentate macrocyclic ligands based on triazacyclononane. They reported structures of the complexes, obtained by X-ray crystallography, as well as the variable field, high resolution ^{1}H T_1 measurements for the macrocycle protons. Knowledge of the distances to the metal centres allowed the determination of the effective magnetic moments, rotational correlation times and electron spin relaxation rates for six lanthanide ions. Babailov *et al.*[253] reported a study of lanthanide complexes with dithiophosphinate and 1,10-phenantroline ligands. The structure of the complexes was investigated using a combination of the lanthanide-induced shifts and the PREs. Fusaro and Luhmer[254] reported a study of Pr(III) complexes with ^{17}O-enriched macrocycle (DOTA). They employed variable-temperature ^{17}O NMR of the ligand oxygens at high magnetic field and at different pH values. The study provided the first quantitative data characterizing the interchange of the oxygen atoms in a DOTA-lanthanide complex. Finally, Soesbe and co-workers[255] studied Dy(III) complexes with DOTA derivatives with a variable number glycinate side-arms. They combined transverse proton relaxivity measurement at high field and ^{17}O NMR studies, with the goal of fine tuning the water exchange rate for enhanced T_2 contrast in MRI.

3.3 Non-electrolyte solutions

The structure of this section is as follows: we start by discussing binary mixtures of solvents, followed by solutions of small organic molecules. Papers using NOE measurements belong in principle here, but I have decided to omit articles where NOE data are used in a routine way for spectral assignment and similar purposes. Later on in the section, I go through works on nuclear spin relaxation in host–guest systems, oligosaccharides and peptide solutions.

I begin here by mentioning work on binary mixtures of water and organic solvents. Sauer and co-workers[256] investigated dynamics of glass-forming mixtures of water and propylene glycol in very broad temperature range using, among other NMR techniques, measurements of ^{2}H spin–lattice relaxation rates. Combining NMR data with results from broadband dielectric spectroscopy and triplet solvation dynamics allowed characterization of molecular motions over several decades of relevant correlation times. Uyama and co-workers[257] reported a study of mixtures of water with glycerol and with 1,3-butanediol. They measured, among other properties, the proton T_2 for aqueous protons and for the CH_2 groups in the alcohols as a function of alcohol concentration at 253 K. Depending on the alcohol content, the samples either remained single-phase liquid or separated into ice and a liquid phase. For the latter case, the data were interpreted in terms of interaction between the alcohol and ice. Rivlin *et al.*[258] studied aqueous solution of formaldehyde at different temperatures and pH, using ^{1}H relaxation measurements (T_1, T_2, $T_{1\rho}$), as well as CEST. They concluded that the T_2 and $T_{1\rho}$ values of the water signal were reduced compared to the pure water case by chemical exchange with the hydroxyl protons of hydrated formaldehyde (methylene glycol). Though not strictly within the limit of this review, I wish also to mention two recent theoretical papers concerned with dynamics of water reorientation near hydrophobic[259] and amphiphilic[260] solutes.

Non-aqueous binary glass formers were studied by Micko *et al.*[261] They worked on mixtures of toluene with polychlorinated biphenyl or picoline (with different isotopic compositions) at low temperatures and measured, among other properties, the ^{2}H spin–lattice relaxation rates. The analysis of the data allowed conclusions on glassy dynamics. Bauer *et al.*[213] reported a related study on mixtures of alcohol isomers, 4-methyl-3-heptanol and 2-ethyl-1-hexanol, making use of ^{1}H and ^{2}H relaxation experiments.

Three papers have dealt with NMR relaxation of small organic molecules dissolved in organic solvents. Kathmann *et al.*[262] reported ^{14}N and ^{2}H spin–lattice relaxation measurements for cyclic amines (pyridine, 2,6-lutidine, 2,2,6,6-tertamethyl piperidine) in toluene and dichloromethane solutions. The rotational correlation times were derived from NMR data and compared with MD simulations. Alemany[263] discussed highly resolved ^{13}C spectra for long-chain *n*-alkanes with 12–22 carbons, dissolved in benzene or toluene. Besides the chemical shifts, also the T_1 data were obtained and related to segmental motions. Ekinci and co-workers[264] studied ^{1}H T_1 data for the chiral *N*-benzil-2-isobutyl aza-15-crown-5-ether derivative as a function of temperature. The rotational correlation times and their activation energy were derived.

Some authors have been concerned with host–guest chemistry and related problems. Takacs and co-workers[265] investigated chloromethane complexes of cryptophane-A and its analogue with butoxy groups. They used several NMR technique, among other ^{13}C T_1 and CPMG relaxation dispersion as well as proton NOESY/ROESY, and were able to obtain detailed information on structural and dynamic properties of the systems. D'Acquarica *et al.*[266] reported ^{1}H ROESY and selective proton relaxation rate measurements for chiral resorcine[4]arene derivatives complexes with 2′-deoxycytidine and pyrimidine nucleoside analogues. Kuz'mina and co-workers[267] studied complex formation of crown ethers with amino acids. They measured ^{1}H T_1 data for the α-protons for glycine, leucine and norleucine in solution in the presence of crown ethers and drew conclusions concerning the host–guest interactions and hydrogen bonding. Gierczyk and Pankiewicz[268] reported studies of supramolecular complexes between xenon and podand polyoxoethylene ligands, using among others heteronuclear NOE between ^{129}Xe and selectively irradiated proton signal. Alvim and co-workers[269] studied the interaction of a fluorescent chalcone derivative with bovine serum albumin, using measurements of ^{1}H and ^{13}C relaxation data for the dye. The NMR experiments were complemented by DFT and molecular docking calculations.

Two papers have dealt with hydrogen molecule encapsulated in a fullerene cavity, interacting with a nitroxide. Garbuio *et al.*[270] investigated two molecules in which the endofullerene and a nitroxide radical were connected by a folded 3_{10}-helical peptide. The proton T_1 data and the *para–ortho* conversion rate for the incarcerated H_2 were reported and interpreted in terms of the Solomon–Bloembergen theory combined with DFT calculations of molecular structures. Rastrelli and co-workers[271] studied related systems making use of DFT calculations to obtain geometry and hyperfine coupling constants. Quantum versus classical descriptions of roto-translational dynamics of the encaged H_2 were compared and the Solomon–Bloembergen theory was used to estimate the PRE due to interaction with the unpaired electron.

Some authors have reported NMR relaxation data for carbohydrate solutions. Pendrill *et al.*[272] described a study of conformation and dynamics in a $1\rightarrow6$ linked disaccharide in water solution by means of ^{13}C relaxation experiments and MD simulations. Moderate differences in the

T_1 data were explained by axially symmetric tumbling in combination with slightly different librational mobilities. Soltesova and co-workers[273] studied the dynamics of the exocyclic hydroxymethyl groups in a polysaccharide using ^{13}C relaxation (T_1, T_2, NOE) at multiple magnetic fields. The data were analyzed using the standard and extended Lipari–Szabo models, as well as a conformational jump model. Wu *et al.*[274] described a study of structure and dynamics in a lipopolysaccharide, reporting ^{1}H NOESY data and MD-simulations. Shinar and co-workers[275] reported CEST data for hydroxyl groups in a naturally occurring polysaccharide in aqueous solution.

Peptides fall on the borderline between this section and the next one. I wish to mention one example of the peptide work here. Björnerås and co-workers[276] studied membrane interaction properties of two single residue variants of the 17-amino acid neuropeptide dynorphin A. Along with circular dichroism spectra and other NMR data, the authors reported PRE results caused by two paramagnetic probes (a nitroxide and the Mn(II) ion) and were able to demonstrate large differences between the two variants interacting with phospholipid bicelles.

3.4 Biological macromolecules in solution

NMR relaxation work on solutions of biological macromolecules is very common and this section is not meant to provide a comprehensive coverage of the field, but rather to give a selection of illustrative examples. This selection is not based on biochemical significance but rather on physico-chemical interest as reflected in the personal taste of the reviewer. The methodological work already discussed in Chapter 2 is not mentioned again in this section.

A number of reviews, covering dynamics of biomolecules in general and proteins in particular, has appeared during the period covered by this review. Konrat[277] and Ringkjoeping Jensen *et al.*[278] reviewed the work on IDPs. Shapiro[279] described the progress in the field of domain dynamics in biomacromolecules. Wang *et al.*[280] described the recent advances in protein NMR with the emphasis on therapeutic research. Vögeli and co-workers[137,281] covered the field of so-called exact NOEs (eNOEs). Veglia and Cembran[282] presented a review of the progress in the field of dynamics and conformational entropy in protein kinases. Guerry *et al.*[283] presented the field of mapping protein conformational landscapes, combining the NMR data (relaxation, residual dipolar couplings) and molecular simulations. Al-Hashimi[284] reviewed NMR studies of nucleic acid dynamics.

Several authors reported investigations of protein dynamics on the pico- and nano-second time scales. A standard approach to the protein backbone dynamics in this range is to measure ^{15}N T_1, T_2 and NOE and to interpret the data using the Lipari–Szabo model-free approach, while ^{13}C or ^{2}H relaxation measurements provide information on the side-chain motions. The Lipari–Szabo order parameters can also be derived from MD simulations. Some authors compared the experimentally derived S^2 with the MD-derived counterparts. Smith and co-workers[285] reported a study of this kind for the backbone of lysozyme from a bacteriophage and found the MD data to confirm the Lipari–Szabo analysis. Zerbetto *et al.*[286] presented a similar experimental work for another protein (Rho-GTPase Binding Domain, RBD) with a propensity to dimerize. The results were analysed with the Lipari–Szabo model, as well as with the recently developed integrated SRLS/MD approach. Chen and Tjandra[287] studied a two-domain retroviral capsid protein using ^{15}N relaxation rates and the T_1/T_2 ratio as a function of protein concentration. The data were interpreted using an extended version of the Lipari–Szabo approach (allowing for anisotropic overall motion) and yielded information on the interdomain motion. Smith and co-workers[288] reported the conventional ^{15}N relaxation data along with the DD-CSA CCRRs for the complex between a ubiquitin ligase and a molecular chaperone. The partners were found to move largely independently. Similar combination of relaxation techniques was also applied by Haba *et al.*[289] in a study of transient structure and dynamics in the disordered C-terminal domain of a protein denoted WIP. Gümral and co-workers[290] studied ^{15}N relaxation in the human protein, β_2-microglobulin, and one of its mutants. The data were analysed using reduced spectral density mapping and supplemented by hydrodynamic calculations and MD-modelling. Rasheed *et al.*[291] reported similar measurements for a dimethylarginine dimethylaminohydrolase (DDAH) enzyme, combined with MD simulations and time-resolved fluorescence measurements. Information was obtained on the ligand-dependent dynamics at the active site of the enzyme. Meli and co-workers[292] studied dynamic aspects of drug–protein recognition, using ^{15}N relaxation experiments and MD simulations.

Roche and co-workers[293] studied the effect of high pressure on an intrinsically disordered protein, α-synuclein. ^{15}N $R_{1\rho}$ relaxation rates showed a modest and rather uniform increase at the pressure of 2500 bar, compared to 1 bar. This was explained in terms of the small increase of the solvent viscosity and the backbone dynamics unaffected by pressure.

Wilson and co-workers[294] investigated a two-domain protein denoted Rin1 using both the ^{15}N relaxation for the backbone and ^{2}H relaxation rates (R_1, $R_{1\rho}$) for CH_2D methyl groups. The nitrogen data were analysed using the reduced spectral density approach and the deuteron relaxation was interpreted in terms of rapid rotation of the methyl groups superimposed on the slower dynamics of the methyl symmetry axis. Horowitz *et al.*[295] studied another isotopic variety of methyl groups, $^{13}CHD_2$ and reported ^{13}C relaxation measurements in *S*-adenosylmethionine-dependent methyltransferases. The dynamic information was discussed in relation to the carbon–oxygen hydrogen bonding. Stafford and co-workers[296] described ^{13}C relaxation measurements (R_1 and the transverse CCRR) of dynamics of side chain amide and carboxyl groups in a ribonuclease. The data were analysed using the Lipari–Szabo model and thus determined residue-specific S^2 values were compared with MD simulations. Robustelli *et al.*[297] presented an extensive discussion of the usefulness of MD simulations as a compliment to experimental relaxation work.

NMR relaxation measurements on biomolecules can also be used to study dynamic processes on the slower time scale ranging from micro- to milliseconds, using the transverse relaxation dispersion methods (see Section 2.5). Palmer[298] reviewed the field of relaxation dispersion methods. Sekhar and Kay[299] covered the applications of measurements of this kind to the description of sparsely populated, transiently formed biomolecular conformers. I choose to begin exemplifying the work of this kind by quoting a number of papers combining measurements of the pico- to nanosecond and micro- to millisecond dynamics. Saio and co-workers[300] reported solution structure, energetics and dynamics of trigger factor chaperone molecules in complex with alkaline phosphatase captured in an unfolded state. The proton NOEs were used for structure determination, while ^{15}N relaxation data provided information on the rapid scale dynamics. Moreover, the slower motions were characterized by CPMG dispersion measurements. Muthu *et al.*[301] applied similar methodology to a diamagnetic complex between the protein nitrophorin 2 and nitric oxide. Measurements were reported at different pH and the complex was found to be rather rigid. Tzeng and Kalodimos[302] reported ^{15}N relaxation measurements and CPMG dispersions for both ^{15}N and methyl ^{13}C resonances. They demonstrated allosteric inhibition of transient conformational states in catabolite activation protein. Favretto and co-workers[303] used a related approach to study human fatty acid binding protein interacting with some ligands. Besides the ^{15}N relaxation experiments, the authors reported also isotope-edited intermolecular ^{1}H NOE measurements. Boehr *et al.*[304] reported a study of pico- to nano- and micro- to millisecond dynamics of dihydrofolate reductase and found that a distal mutation had a significant effect on the backbone rigidity as well as on sampling of higher energy conformations. Huan and co-workers[305] studied structure and dynamics of a ligand binding to Eph receptors. The ^{15}N relaxation measurements aiming at understanding short time scale dynamics were complemented by MD simulations. The systems were found to lack the intermediate micro- to millisecond dynamics, but the hydrogen/deuterium exchange experiments demonstrated dynamics on the second – hour time range. Audin and co-workers[306] studied a large exosome complex and chose a different strategy. The fast dynamics of the methyl groups was investigated through intra-methyl ^{1}H–^{1}H dipolar cross-correlated spin relaxation rates in highly deuterated, methyl protonated protein (as proposed some time ago by Tugarinov *et al.*[307]) while the slower motions were followed using multiple-quantum relaxation dispersion experiments of Korzhnev and co-workers.[308]

I wish to quote some more examples of applications of transverse relaxation dispersion data. Li and co-workers[309] (with erratum[310]) used the ^{1}H R_2 dispersion experiments to directly observe and characterize intermediate states in the dimerization of the protein E-cadherin. Sanchez-Medina *et al.*[311] used relaxation dispersion spectroscopy (for ^{1}H, ^{15}N and ^{13}C) to probe the energy landscape and to characterize intermediates for the fast-folding protein denoted gpW. They identified a sparsely populated state with the lifetime of 250 μs, interconverting with the native state. In another paper from the same group, Shi and Kay[312] studied the allosteric pathway regulating the activity of the HslV protease, explored through methyl-TROSY ^{13}C relaxation dispersion experiments. Libich and co-workers[313] combined several relaxation-based NMR experiments to probe the transient, not directly visible, state of substrate (peptide Aβ40) binding to a prototypical chaperonin (GroEL). The NMR techniques used were line-broadening (^{15}N ΔR_2), ^{15}N CPMG dispersion and the ^{15}N dark-state exchange saturation transfer (DEST), an experiment related to CEST.

Some authors have worked with ^{19}F labelled proteins in solution. Kitevski-LeBlanc and co-workers[314] prepared a sample of calmodulin enriched with 3-fluorophenylalanine. They reported ^{19}F CPMG dispersion experiments in order to characterize a transition to a thermal folding intermediate. Ye *et al.*[315] worked with two other fluorine-labelled amino-acids and used ^{19}F relaxation experiments to measure cytoplasmic viscosity and to characterize nonspecific protein–protein interaction in living cells. Yue and co-workers[316] reported ^{19}F relaxation data for a fluorocarbon covalently conjugated to albumin *via* flexible linkers.

Next, I turn to the paramagnetic relaxation effects in protein systems. I begin with the works where these are combined with the "diamagnetic" type effects of the type discussed above. Castaneda and co-workers[317] investigated K11 linked diubiquitin systems. They reported ^{15}N relaxation data for both the distal and proximal chains and obtained their relative orientations. In addition, they attached a paramagnetic nitroxide spin label to the distal ubiquitin and derived interdomain distance and orientation from the PRE in the proximal chain. Anthis and Clore[318] reported a related study of a two-domain protein calmodulin. They measured ^{15}N T_1 and T_2 for several linker mutants and estimated effective correlation times for each domain. Making use of a paramagnetic tag at a particular position in one of the domains they performed interdomain PRE measurements and were able to draw conclusions concerning the effect of the linker length on the interdomain association. In another related study, Papouskova *et al.*[319] applied similar methodology to a partially disordered subunit of RNA polymerase from a model organism. Almeida and co-workers[320] presented an investigation of interactions of the enzyme superoxide reductase with its redox partners. Again, they reported the ^{15}N relaxation data for the systems (with the high-spin iron ion substituted by the diamagnetic Zn(II)) and combined them with the PRE measurements. The difference compared to the studies mentioned above was that the paramagnetic tag was a caged lanthanide compound. Zhuang *et al.*[321] studied the outer membrane protein G using the PRE-based approach to visualize the conformational dynamics of one of the loops and to calculate the conformational ensembles. In addition, they also reported the ^{15}N CPMG relaxation dispersion data.

The use of PRE measurements was also the topic of a couple of reviews. Fragai and co-workers[322] reviewed paramagnetism-assisted NMR studies of conformational freedom of metalloproteins while Luna *et al.*[323] described how PREs could be useful for examining weak protein–protein interactions in the context of codon recognition. Van Ingen and Bonvin[324] discussed information-driven modelling of large macromolecular assemblies using NMR data in general and PREs in particular. I wish to mention also some additional original studies making use of the

PRE effects. Schilder and co-workers[325] investigated encounter complexes between cytochrome c peroxidase and cytochrome c carrying a paramagnetic spin label. Burmann *et al.*[326] reported a related study of conformation and dynamics of a periplasmic chaperone and its complexes with outer-membrane proteins.

Another field where the PRE phenomenon proved very useful are studies of IDPs. Iesmantavicius and co-workers[327] used this methodology to study the propensity to form transient helices while, in another paper from the same group, Ozenne *et al.*[328] studied the unfolded states of globular proteins. Bozoky *et al.*[329] investigated multisite interaction of various IDP segments.

Next, I go back to the diamagnetic protein systems and review work dealing with protein-model membrane interactions. Mäler[330] reviewed solution NMR studies of cell-penetrating peptides in model membrane systems. Roche and co-workers[331] showed by solutions NMR spectroscopy (including among others ^{1}H NOE measurements and ^{15}N relaxation experiments) that a six-helix bundle formed of envelope glycoprotein binds to zwitterionic detergents and phospholipid vesicles. Kim *et al.*[332] reported ^{19}F NMR spectra and relaxation data for the β_2 adrenergic receptor containing a trifluoromethyl tag, reconstituted in detergent micelles. The results were interpreted in terms of several states in equilibrium. Pandey and co-workers[333] presented a study of changes in ^{15}N CSA tensors as a consequence of interactions between cytochrome-P450 and cytochrome b_5 in a membrane environment. The tensors were determined using the ^{15}N CSA – NH dipolar cross-correlated transverse relaxation rates. Sommer *et al.*[334] characterized the membrane immersion properties of the oxidized and reduced forms of the FATC domain of a kinase, making use the PRE effects due to paramagnetic tagging of the micelle components or to the Mn(II) ions in the solvent. The NMR results were compared with CD spectroscopy and MD simulations.

Some papers have dealt with relaxation studies of biomolecules other than proteins. Kang and co-workers[335] reported solution NMR structure and conformational dynamics of a 59-nucleotide riboswitch. ^{13}C relaxation measurements, along with residual dipolar coupling data, revealed interhelical flexibility. Bonneau and Legault[336] investigated the RNA interactions with metal cations by mapping the binding sites through Mn(II)-induced PRE effects. Widmalm[337] reviewed solution NMR work, including cross-relaxation studies, applied for primary and three-dimensional structure of biologically important polysaccharides.

Finishing this section, I want to mention some work on water in biomolecular systems in solution. Wierzuchowska and Blicharska[338] studied proton relaxation and CPMG relaxation dispersion for water protons in solutions of some proteins. The dependences on the protein concentration were also obtained. Persson and Halle[339] analysed the millisecond all-atom MD trajectory (produced by Shaw *et al.*[340]) for the protein bovine pancreatic trypsin inhibitor and compared the results with the experimental NMRD data as interpreted using the exchange-mediated orientational randomization model. I shall return to the issue of water dynamics in biological gels in subsection 3.5.4.

3.5 Other complex systems

In this section, I collect the relaxation work on complex systems other than biological molecules in solution. The concept of complex systems includes here multicomponent homogeneous mixtures, surfactant/colloidal systems, solutions of synthetic polymers, gels, liquids in porous media (and related heterogeneous systems), and systems containing nanoparticles.

3.5.1 Complex mixtures. I begin with citing some papers related to NMR of crude oil. Korb and co-workers[341] studied bulk crude oil (with and without asphaltene, the fraction insoluble in *n*-alkanes but soluble in toluene) by a variety of NMR techniques, including field-cycling relaxometry. The data were interpreted in terms of field-dependent distributions of ^{1}H spin–lattice relaxation rates, which showed a connection between the hydrocarbon dynamics and the chain length. Similar work was also published by the same group in another journal.[342] Asphaltene itself was subject to an investigation by Majumdar *et al.*,[343] who investigated, among other properties, the proton relaxation behaviour of asphaltene solutions in chloroform. Stapf and co-workers[344] studied interaction between asphaltene and other crude oil components (maltene) making use of ^{19}F relaxation in fluorine-containing tracer molecules, such as *e.g.* perfluorobenzene or perfluorooctane. Morgan *et al.*[345] proposed a new way to classify oil through ^{1}H T_2 measurements at low field, combined with determination of the relative hydrogen index. Fridjonsson and co-workers[346] developed methods for determining the water droplet size for water-in-crude oild emulsions, based on NMR at low field and making use of relaxation properties.

Other complex mixtures have also attracted attention. Del Coco *et al.*[347] applied one-dimensional NOESY and transverse-relaxation filter CPMG spectra at high magnetic field to investigate the urine metabolome of sick patients and healthy controls. Constantino and co-workers[348] used low-field NMR (CPMG measurements, among others) to analyze the oil content and oil quality in oilseeds, aiming at the selection of suitable starting material for biodiesel production. Ueda *et al.*[349] studied the effects of various forms of hydroxypropyl methylcellulose acetate succinate on drug crystallization and dissolution, exemplified by among others carbamazepine. The mobility of the drug in solution was characterized by ^{1}H relaxation time measurements at high field. Hsieh and co-workers[350] used NMR relaxometry measurements to characterize the biological cellulose degradation. In particular, the emphasis was on the cellulase inhibition by high concentration of monosaccharides.

3.5.2 Surfactant systems. Surfactant-containing systems have attracted some attention, not least in the form of reviews by Rosenholm,[351] by Monduzzi *et al.*[352] and by Schönhoff.[353]

I begin this section by referring to the work on surfactants in solution and micellar and related systems. Popova and co-workers[354] reported relaxation times measurements for ^{23}Na for sodium octanoate in aqueous solution and aqueous dispersions of silica. Concentration and temperature dependencies of the spin–lattice and spin–spin relaxation times were investigated. In another paper from the same group,[355] the authors studied the behaviour of sodium lauroyl sarcosinate in solution and mixed binary systems using measurements of ^{1}H relaxation data under high resolution conditions. Cardoso and Sabadini[356] investigated the interactions between the head groups of *n*-alkyl glycosides and water molecules in the range from low concentration to above the critical micelle concentration, using T_2 measurements for aqueous protons. Keymeulen *et al.*[357] described measurements of the PRE at high magnetic field for systems containing uranyl-salophen receptor incorporated in micelles, formed by cetyltrimethylammonium chloride or bromide. The data were interpreted in terms of the localization and orientation of the receptor inside the micelles. Bernewitz and co-workers[358] reported also the PRE measurements (in this case at the magnetic field of 0.47 Tesla) for double emulsions, using a polar Gd(III) complex as the relaxation agent.

Three papers by Chinese groups have described work involving gemini surfactants. Wang and Wang[359] studied interactions between anionic sulfonate gemini surfactant and a triblock copolymer, using proton relaxation data and 2D NOESY measurements. Wu *et al.*[360] reported investigations of interactions between a novel gemini surfactant and bovine serum albumin (BSA) using proton relaxation time measurements. Similar work was also described in another paper by the same group.[361]

Next, I move on to work on bicelles, vesicles and related systems. Matsumori[362] studied interaction between model drugs and bicelles, making use of NOE measurements and paramagnetic relaxation methods. Martins and co-workers[363] and de Sousa *et al.*[364] investigated the influence of drugs on the properties of bilayers, in the form of liposomes, using ^{1}H spin–lattice relaxation rates, among other techniques. Song and co-workers[365] reported effects of specific ions on the translational diffusion of water in the vicinity of large hydrophilic lipid vesicle surfaces, using DNP measurements governed by the Overhauser mechanism. Bahamonde-Padilla and co-workers have presented two papers on interactions involving lyotropic liquid crystal aggregates in aqueous solutions. In the earlier one,[366] they studied effects of Li^{+} ions on the deuteron quadrupolar couplings and ^{2}H T_1 in the liquid crystalline system consisting of deuterated sodium dodecyl sulphate, decanol, salts and water. In the second study,[367] they used the same liquid crystal and studied its interactions with small quinones. In both papers, the NMR data were combined with MD simulations.

3.5.3 Synthetic polymers in solution. Some authors have worked with solutions of synthetic polymers. Bock, Kahlau and co-workers have reported a series of investigations of asymmetric binary glasses formed by tripropyl phosphate (TPP) and deuterated polystyrene (PS) over the full concentration range, using a combination of NMR and dielectric spectroscopy. The temperature interval was large, extending from the range where the mixtures behaved as liquids down to the solid-like glassy states. In the first paper, the emphasis was on the cooperative nature of the second relaxation (β-relaxation) peak in the dielectric spectra.[368] The persistence of the β-process was confirmed by ^{31}P relaxation experiments on TPP in the glassy range. The corresponding ^{2}H data on the polystyrene component were consistent with the lack of the β-process in the neat polystyrene, while the signature of that process was visible for the PS component in the mixtures. In the next paper, the authors performed in addition depolarized light scattering and differential scanning calorimetry on the same systems.[369] In the third paper, the NMR methodology was further extended to include 2D ^{31}P exchange spectra at low temperatures.[214]

Van Lokeren and co-workers[370] reported a study of polymer mixtures in solution using pulsed field-gradient spin echo NMR. They proposed a quantification approach in which the fractions of different components were obtained explicitly including relaxation effects in the Stejskal–Tanner equation.

Some authors have worked with block copolymers in solution. Yokoyama and Yusa[371] studied water-soluble complexes formed by poly(ethylene glycol)-containing triblock copolymers and poly(methacrylic acid), using proton spin–spin relaxation measurements along with light scattering and electron microscopy. Weiss and co-workers[372] used ^{1}H relaxation to investigate self-assembly (into micelles) of a double thermo-responsive diblock copolymer in dilute aqueous solution. The diblock copolymer micelles were also studied, again using ^{1}H relaxation measurements, by Nakai *et al.*[373]

A couple of papers have been concerned with fluorinated polymers. Twum and co-workers[374] made use of ^{19}F relaxation measurements (under high resolution conditions) along with other NMR techniques, to study end groups and branching structures of copolymers of vinylidene fluoride and hexafluoropropylene. In the paper mentioned above (Section 3.3), Kadayakkara *et al.*[222] reported an investigation of ^{19}F relaxation mechanisms in linear and cyclic perfluoropolyethers, based on high resolution measurements at three different magnetic field between 7 and 14 Tesla and several temperatures. The R_1 relaxation was found to be dominated by the dipole–dipole and CSA mechanism.

A couple of papers have dealt with dendrimers. Wang and co-workers[375] reported a study of host–guest behaviour of poly(amidoamine) dendrimers with various functionalities interacting with paramagnetic TEMPO derivatives. They measured the PRE effects on the polymer protons. Neelov *et al.*[376] investigated second- and fourth-generation lysine dendrimers by MD simulations, combined with ^{1}H relaxation rates measurements.

Finishing this subsection, I wish to mention some studies of polymeric, water-soluble and biocompatible systems. D'Amelio and co-workers[377] investigated molecular properties of a polysaccharide chitosan derivative modified with lactitol moieties. They evaluated the structure and dynamics of the side chains making use of NOE and proton relaxation measurements, combined with MD simulations. Watanabe *et al.*[378] studied thermo-responsive behaviour of an amphiphilic poly(asparagine) derivative in water, using several NMR techniques. The deuteron NMR signal from added D_2O was subject to 2D T_1–T_2 and T_2–T_2 correlation experiments using the ILT methodology. The results demonstrated the presence of several water reservoirs at the elevated temperature corresponding to gel formation.

3.5.4 Gels and related system. When polymers dispersed in a liquid are cross-linked, gels are formed. I review here the works dealing with relaxation/dynamics of the liquid component in gels (the matrix components are usually studied using the solid state techniques). One paper has dealt with an inorganic gel (calcium–silica hydrate, C–S–H), important in the context of cement-based materials.[379] The authors used MD simulations and ^{35}Cl NMR studies to characterize the transport properties.

Numerous papers have dealt with water contained in gels based on synthetic polymers. Sierra-Martin *et al.*[380] reviewed the field of temperature-responsive microgels (networks of microscopic filaments) of poly(*N*-isopropylacylamide), studied by means of solvent relaxation NMR, among other techniques. Bai and co-workers[381] reported water proton T_1 in sodium polyacrylate solutions and gels. In dilute solutions, the relaxation rate showed a linear dependence on the polymer concentration while the T_1 was less influenced by the polymer in the gel state. Zhu *et al.*[382] studied the water–polymer interactions in cross-linked polyurethane coatings using aqueous ^{1}H T_2 distribution data. Related work was also reported for the swelling-deswelling process of a polyampholyte hydrogel.[383] Compan and co-workers[384] reported a study of oxygen permeability in acrylic-based hydrogels, making use of the paramagnetic relaxation enhancement for aqueous protons due to the dissolved oxygen gas. Paramagnetic relaxation effects in coacervate hydrogels were also investigated by Ortony and co-workers.[385] They used the Overhauser DNP caused by block-specific nitroxide spin labels, along with ESR spectra and other techniques. Liu *et al.*[386] described a low-field investigation of the distribution of aqueous proton relaxation times in soy hull pectic polysaccharide gel and the influence of added D-glucono-1,5-lactone and sucrose.

A number of papers have been concerned with water in protein-based gels. Kaieda and co-workers[387] reported a study of the mobility of core water in *bacillus subtilis* spores. Two scenarios have been proposed for the spore's core region: the gel scenario and the glass scenario. The authors reported ^{2}H longitudinal relaxation dispersion data in the presence and in the absence of Mn(II) ions and found that the data clearly supported the gel scenario. In another paper from the same group, Kaieda *et al.*[388] studied the cross-linking process leading to protein gels. The work is centred on small-angle X-ray scattering, but comparisons are presented with the earlier NMRD data. He and co-workers[389] compared the gel-forming ability and gel properties of some proteins, including the effects of high pressure. Among a variety of physical methods, they reported the multiexponential fitting of the aqueous ^{1}H CPMG data. Oztop *et al.*[390] studied effects of divalent ions, Mn(II) and Ca(II), on heat-set whey protein gels. The water uptake and ion release by the gels was monitored through relaxometric experiments. Lopez-Cebral and co-workers[391] proposed a new hydrogel (incorporating the protein albumin) as a drug delivery system, making use of saturation transfer difference spectra to study the binding between albumin and model drugs. Luchinat *et al.*[392] studied water and protein dynamics in systems similar to gels, but formed through protein sedimentation under centrifugation in NMR rotors for solid-state NMR experiments. They studied the hydration of bovine serum albumin through NMRD experiments, showing that the relaxation properties in these systems were very different from proteins in concentrated aqueous solutions and similar to those of slightly re-hydrated lyophilized proteins.

Finishing this subsection, I wish to review a couple of papers on organogels. Kowalczuk and co-workers[393] described a sugar derivative able to form gels with ethylene glycol or 1,3-propanediol as the mobile organic liquid component. In order to characterize the materials, they reported among others the proton NMRD measurements. Salami *et al.*[394] studied the probe mobility in native phosphocaseinate suspensions and in the corresponding concentrated gels. The probe they proposed was a dendrimer and the dynamics was measured through translational diffusion and CPMG measurements.

3.5.5 Porous media and related systems.

Next, we turn to the work on liquids confined in solid porous materials or otherwise staying close to solid surfaces. I use here a non-unique and somewhat arbitrary distinction between "chemical" and "geological" samples and start with the former.

We begin with the work where the studied fluid was different than water. Some papers have dealt with gases contained in the porous materials. Yager and co-workers[395] reported relaxation time measurements for ^{3}He adsorbed in the pores of mesoporous molecular sieve MCM-41 at low temperatures (down to 1.7 K) and at a range of frequencies. They observed the $T_1 \propto \omega^{1/2}$ behaviour, characteristic for one-dimensional diffusion. Gul-E-Noor *et al.*[396] presented an investigation of ^{13}C spin-lattice relaxation in ^{13}CO and $^{13}CO_2$ adsorbed in metal–organic frameworks. The measurements were carried out over a wide temperature range and at two (high) magnetic fields. The data were used to discuss the relaxation mechanisms in the systems. Carbon-13 relaxation was also applied to study molecular motions of CO_2 adsorbed on an intrinsically microporous polymer, functionalized by tetrazole, by Moore *et al.*[397] Ji and co-workers[398] studied interactions and diffusion of methane and HD in microporous zeolite structures. The results for methane were in agreement with Monte Carlo simulations while the data for hydrogen deuteride proved consistent with neutron scattering.

Two studies have been concerned with catalytic reaction on porous catalysts. Mitchell and co-workers[399] used FFC relaxometry to probe the slow dynamics of liquid molecules imbibed in the Pd/Al_2O_3 catalyst. The dispersion curves showed a different behaviour for water on the one hand and organic liquids (2-butanone, 2-propanol) on the other hand. D'Agostino *et al.*[400] reported a study of aerobic oxidation of 1,3-propanediols over gold supported on titania. They measured ^{1}H T_1 and T_2

(along with diffusion coefficients) for the diol at high field using methanol and methanol/water mixtures as solvent. The results indicated that the NMR diffusometry and relaxometry have significant potential as tools for catalyst characterization.

In a somewhat related study, Chen and co-workers[401] reported an investigation of transverse relaxation rate distribution for metalorganic frameworks imbibed with dimethylsulfoxide. The measurements, using a portable NMR system (NMR Mouse), were shown to be useful as high throughput screening technique. Perepukhov *et al.*[402] described specific features of proton relaxation of hydrocarbon (decane) and water in the pore space of silicates characterized by the presence of paramagnetic species on the surface. Han and co-workers[403] studied a more complicated liquid – a room-temperature ionic liquid – confined in mesoporous silica. They measured ^{1}H T_1 and T_2 at high field and drew conclusions regarding rotational dynamics of the liquid. The review by Rössler *et al.*[219] mentioned in Section 3.1 above, described also some work on polymer melts in confinement.

Several related papers have been concerned with water in porous media. I begin here with the case of inorganic porous media. Sattig and Vogel[404] reported deuteron spin–lattice relaxation study of D_2O confined in the nanopores of the MCM-41 silica matrix at low temperatures (150–235 K). The measurements indicated the occurrence of two dynamic crossovers, related to the formation of crystalline or glassy water species. Chemmi *et al.*[405] presented an investigation of the impact of wettability on the moisture transport at mesoscale in porous media. They made use of NMRD measurements to probe the proton dynamics on control pore glasses with and without methyl groups on the surface. Daiko[406] studied temperature dependence of aqueous ^{1}H T_1 (along with impedance) in nanoporous inorganic glasses and thin films. Porion and co-workers[407] studied water dynamics within dense clay sediments making use of ^{2}H multiple quantum relaxation measurements. Keating[408] presented a study of relations between pore size, pore surface-to-area ratio and proton relaxation rates for water on glass beads with specially prepared surfaces. Totland and co-workers[409] investigated thin films of water between glass plates and the effects of addition of 1-butanol. Fantazzini and co-workers[410] applied the measurements of the aqueous proton T_2 distributions to investigate the cleaned skeletons obtained from living corals. Franck *et al.*[198] studied the behaviour of water, doped with a paramagnetic TEMPO derivative, in a narrow glass capillary, using the measurements of Overhauser DNP. This allowed to selectively amplify only the relaxation dynamics of a specific pool of water molecules. Several authors reported low-field NMR relaxation investigations of relevance for cement research.[411–416]

Next, I turn to work on water confined in porous media of carbon-based materials. Bardenhagen and co-workers[417] studied fluid distribution and pore wettability for water and dimethylsufoxide in carbon xerogels. They used ^{1}H relaxation measurements and related the results to the liquid exchange between micro- and mesopores. Wakai *et al.*[418] reported a study of the hydration process of a nafion membrane by ^{1}H

NMR, including both the chemical shifts and spin–lattice relaxation measurements. The experimental data were analysed using a newly developed method. Polyelectrolyte-fluoropolymer blend membranes were studied using proton relaxometric techniques by Hou and co-workers[419] while Park *et al.*[420] used related methodology to study water in multiblock copolymer membranes. Victor and co-workers[421] reported a relaxation study of translational dynamics of water at the phospholipid interface. The NMRD profile for residual water protons (in D_2O) was found to be a logarithmic function of the proton Larmor frequency at high field and independent of the frequency at low field, which could be interpreted using relaxation theory for two-dimensional diffusion. A number of authors have reported relaxometric work on wood and related systems.[422–426]

Next, I present some examples of NMR relaxation measurements in the context of geological problems. Some of the work mentioned in subsection 3.5.1 on complex mixtures, concerned with crude oil, covered also measurements on oil in reservoir rocks and oil-asphaltene systems.[341,344] Other work on crude oil in porous media was reported by Benamsili *et al.*[427] and by Yang and co-workers,[428] while related work on porous coal was presented by Zou *et al.*[429] Finishing this subsection, I wish to mention other geology-related NMR relaxation work. Daigle *et al.*[430] applied NMR relaxometry on samples from the Integrated Ocean Drilling Expedition 333. Works on other materials from well-drilling were reported by Mitchell and co-workers,[431] Hübner[432] and Eslami *et al.*[433] Dlugosch and co-workers[434] used NMR relaxation data in the context of hydraulic conductivity for coarse-grained unconsolidated material. NMR-based analysis of soil-water characteristics was also presented by Tian *et al.*,[435] Dlubac and co-workers[436] and Keating and Falzone.[437] Hydrological problems can also be studied by surface NMR, where proton signals from subterranean water reservoirs are detected in the earth magnetic field using very large detection coils. As recent examples of relaxation work of this kind I wish to mention the papers by Dlugosch *et al.*[438] and by Müller-Petke and co-workers.[439]

3.5.6 Nanoparticle systems. The work on nano-object, interesting in the present context, can be roughly divided into studies of magnetic nanoparticles of possible applicability as MRI contrast agent and other investigations. I begin with reviewing the latter category. Cooper and co-workers[440] investigated competition between two polymers for adsorption on the silica nanoparticle surface, using solvent (water) proton relaxation properties along with small angle neutron scattering. Kim and Zukoski[441] studied also interactions between silica nanoparticles and a polymer (poly(ethylene glycol)) of variable molecular weight in concentrated solution in D_2O. More specifically, they related the shape of the proton FID, originating from chain protons and obtained in different types of NMR experiments, to the segmental mobility. Kedir *et al.*[442] used 2D NOESY experiments (along with NMR diffusion measurements) to characterize the spatial proximity between backbone protons and amide protons in nanoparticles obtained by cross-linking high-molecular weight partially hydrolysed poly(acrylamide) with aluminum(III). Lau and co-workers[443] reported a liquid-state NMR study of interactions between silver nanoparticles and natural organic matter in the presence of capping agents (citrate or poly(vinylpyrrolidone)), using proton relaxation measurements for methylene and methine signals. Jespersen *et al.*[444] studied organic–inorganic hybrids, denoted nanoscale ionic materials and prepared from functionalized nanoparticles neutralized by oligomeric polymer counterions. They used ^{13}C relaxation data (along with diffusion measurements) to probe the canopy dynamics. The work by Korb *et al.*,[341] mentioned above in subsections 3.5.1 and 3.5.5, addressed also the issue of proton relaxation in crude oil in the presence of asphaltene nanoaggregates. Uccello-Barretta and co-workers[445] measured selective relaxation rates for ^{1}H NMR signals from a model drug, interacting with polymers dispersed in solution and aggregated in nanoparticles. Kato *et al.*[446] studied another type of nanoscale objects: single-walled carbon nanotubes dispersed in water and interacting with surfactants. The transverse relaxation rates of the surfactant protons were strongly enhanced by the presence of the nanotubes.

I turn now to work concerned with ^{19}F relaxation in nanoparticles built of fluorine-containing polymers, aimed at the use as ^{19}F MRI contrast agents. Wang and co-workers[447] studied fluorine-containing core cross-linked star polymer nanoparticles, reporting among others ^{19}F relaxation times as a function of solution pH. De Vries *et al.*[448] reported another ^{19}F relaxometric investigation of nanoparticles, in the form of emulsions of perfluoro-15-crown-5 in aqueous solution in the presence of gadolinium(III) complexes as relaxation agents. ^{19}F relaxation rates were measured at five different (high) magnetic fields and were combined with proton NMRD experiments.

In the final paragraphs of this section, I review some physicochemically interesting work on water proton relaxation in liquid systems containing magnetic nanoparticles, objects of dimensions below around hundred nm. A number of papers have dealt with Fe_3O_4 (magnetite) nanoparticles in aqueous environment. Wang and co-workers[449] reported proton spin–lattice relaxation rates in such a system at very low magnetic field, using a SQUID device. SQUID was also used by Kenouche *et al.*[450] in their work on monodisperse Fe_3O_4 nanoparticles of different sizes. Other groups studied magnetite nanoparticles decorated with polymers such as polypyrrole[451] or block copolymers.[452]

Other types of magnetic nanomaterials have also been investigated. Menelaou and co-workers[453] described proton relaxation work on magnetite with one of the iron atoms replaced by nickel, $NiFe_2O_4$, coated with oleyamine. Issa and co-workers[454] studied temperature-dependent NMR relaxation in magnetite-based nanoparticle systems where some iron atoms were substituted by manganese, zink and a small amount of gadolinium. Xu *et al.*[455] developed an analytical relationship between NMR relaxation rates and interparticle spacing in polymer nanocomposites filled with clay nanoparticles carrying paramagnetic impurities. Scorciapino and co-workers[456] investigated diamagnetic Bi_2S_3

nanoparticles, capped with oleic acid and oleyamine in the presence of a surfactant. The systems were studied by NMR at high resolution, after adding the $MnCl_2$ paramagnetic probe in order to estimate the insertion depth of the capping agents. Finally, Calucci *et al.*[457] measured relaxation enhancement of water protons by gadolinium(III)-doped boron nitride nanotubes.

4 Nuclear spin relaxation in gases

Some of the papers concerned with hyperpolarization of ^{129}Xe (or ^{83}Kr), mentioned in Section 2.8, contained data on spin–lattice relaxation rates for the gaseous species.[208–211] Likewise, some of the work on gases adsorbed in porous media, mentioned in subsection 3.5.5, should be referred to also here.[395–398] In addition to these works, only one paper has dealt with relaxation in gases: Horch *et al.*[98] compared (in the paper mentioned in Section 2.3) the ^{1}H relaxation data for methane in gaseous state and adsorbed on various adsorbents.

Acknowledgements

This work was supported by the Swedish Research Council (Grant No 613-2011-3311).

References

1 J. Kowalewski, *Nuclear Spin Relaxation in Liquids and Gases*, Royal Society for Chemistry, 2011, vol. 40, pp. 205–253.
2 J. Kowalewski, *Nuclear Spin Relaxation in Liquids and Gases*, Royal Society for Chemistry Publishing, Cambridge, 2012, vol. 41, pp. 196–243.
3 J. Kowalewski, *Nuclear Spin Relaxation in Liquids and Gases*, Royal Society for Chemistry Publishing, Cambridge, 2013, vol. 42, pp. 230–275.
4 J. Kowalewski, *Nuclear Spin Relaxation in Liquids and Gases*, Royal Society for Chemistry Publishing, Cambridge, 2014, vol. 43, pp. 230–285.
5 F. Bloch, *Phys. Rev.*, 1946, **70**, 460–474.
6 J. Kowalewski and L. Mäler, *Nuclear Spin Relaxation in Liquids*, Taylor and Francis, New York, 2006.
7 R. K. Wangsness and F. Bloch, *Phys. Rev.*, 1953, **89**, 728–739.
8 A. G. Redfield, *Adv. Magn. Reson.*, 1965, **1**, 1–32.
9 H. H. Segnorile and R. C. Zamar, *J. Chem. Phys.*, 2011, **135**, 244509.
10 C. J. Bonin, C. E. Gonzalez, H. H. Segnorile and R. C. Zamar, *J. Chem. Phys.*, 2013, **139**, 144907.
11 J. C. Erker and M. P. Augustine, *J. Magn. Reson.*, 2014, **238**, 106–114.
12 H. J. Stöckmann and D. Dubbers, *New J. Phys.*, 2014, **16**, 053050.
13 B. Halle, *Prog. NMR Spectrosc.*, 1996, **28**, 137–159.
14 T. Nilsson and B. Halle, *J. Chem. Phys.*, 2012, **137**, 054503.
15 Z. W. Chang and B. Halle, *J. Chem. Phys.*, 2013, **139**, 144203.
16 I. Solomon, *Phys. Rev.*, 1955, **99**, 559–565.
17 A. Gryff-Keller, S. Molchanov and A. Wodynski, *J. Phys. Chem. A*, 2014, **118**, 128–133.
18 M. Carravetta and M. H. Levitt, *J. Am. Chem. Soc.*, 2004, **126**, 6228–6229.
19 M. Carravetta, O. G. Johannessen and M. H. Levitt, *Phys. Rev. Lett.*, 2004, **92**, 153003.

20 M. Carravetta and M. H. Levitt, *J. Chem. Phys.*, 2005, **122**, 214505.
21 S. J. DeVience, R. L. Walsworth and M. S. Rosen, *Phys. Rev. Lett.*, 2013, **111**, 173002.
22 T. Theis, Y. Feng, T. Wu and W. S. Warren, *J. Chem. Phys.*, 2014, **140**, 014201.
23 K. Claytor, T. Theis, Y. S. Feng and W. Warren, *J. Magn. Reson.*, 2014, **239**, 81–86.
24 Y. S. Feng, T. Theis, X. F. Liang, Q. Wang, P. Zhou and W. S. Warren, *J. Am. Chem. Soc.*, 2013, **135**, 9632–9635.
25 I. Marco-Rius, M. C. D. Tayler, M. I. Kettunen, T. J. Larkin, K. N. Timm, E. M. Serrao, T. B. Rodrigues, G. Pileio, J. H. Ardenkjaer-Larsen, M. H. Levitt and K. M. Brindle, *NMR Biomed.*, 2013, **26**, 1696–1704.
26 C. Laustsen, S. Bowen, M. S. Vinding, N. C. Nielsen and J. H. Ardenkjaer-Larsen, *Magn. Reson. Med.*, 2014, **71**, 921–926.
27 G. Pileio, J. T. Hill-Cousins, S. Mitchell, I. Kuprov, L. J. Brown, R. C. D. Brown and M. H. Levitt, *J. Am. Chem. Soc.*, 2012, **134**, 17494–17497.
28 B. Meier, J. N. Dumez, G. Stevanato, J. T. Hill-Cousins, S. S. Roy, P. Håkansson, S. Mamone, R. C. D. Brown, G. Pileio and M. H. Levitt, *J. Am. Chem. Soc.*, 2013, **135**, 18746–18749.
29 L. Fernandes, C. Guerniou, I. Marin-Montesinos, M. Pons, F. Kateb and P. R. Vasos, *Magn. Reson. Chem.*, 2013, **51**, 729–733.
30 K. Nagashima, D. K. Rao, G. Pages, S. S. Velan and P. W. Kuchel, *J. Biomol. NMR*, 2014, **59**, 31–41.
31 M. Emondts, M. P. Ledbetter, S. Pustelny, T. Theis, B. Patton, J. W. Blanchard, M. C. Butler, D. Budker and A. Pines, *Phys. Rev. Lett.*, 2014, **112**, 077601.
32 A. N. Pravdivtsev, A. V. Yurkovskaya, H. Zimmermann, H. M. Vieth and K. L. Ivanov, *Phys. Chem. Chem. Phys.*, 2014, **16**, 7584–7594.
33 M. Singh, C. Srinivas, M. Deb and N. D. Kurur, *ChemPhysChem*, 2013, **14**, 3977–3981.
34 P. Henritzi, A. Bormuth and M. Vogel, *Solid State Nucl. Magn. Reson.*, 2013, **54**, 32–40.
35 C. J. F. Böttcher and P. Bordewijk, *Theory of Electron Polarization*, Elsevier, Amsterdam, 1973, vol. 2.
36 T. P. Kulagina, V. A. Varakina and A. N. Kuzina, *Russ. J. Phys. Chem. B*, 2014, **8**, 391–395.
37 N. Fatkullin, C. Mattea and S. Stapf, *J. Chem. Phys.*, 2013, **139**, 194905.
38 N. Fatkullin, A. Gubaidullin, C. Mattea and S. Stapf, *J. Chem. Phys.*, 2012, **137**, 224907.
39 D. A. Faux, P. J. McDonald, N. C. Howlett, J. S. Bhatt and S. V. Churakov, *Phys. Rev. E: Stat., Nonlinear, Soft Matter Phys.*, 2013, **87**, 062309.
40 D. L. Ermak and J. A. McCammon, *J. Chem. Phys.*, 1978, **69**, 1352–1360.
41 L. Durlofsky, J. F. Brady and G. Bossis, *J. Fluid Mech.*, 1987, **180**, 21–49.
42 T. Ando, E. Chow and J. Skolnick, *J. Chem. Phys.*, 2013, **139**, 121922.
43 M. Dlugosz and J. M. Antosiewicz, *J. Chem. Theory Comput.*, 2014, **10**, 481–491.
44 D. E. Woessner, *J. Chem. Phys.*, 1962, **36**, 1–4.
45 L. Vugmeyster and D. Ostrovsky, *J. Chem. Phys.*, 2014, **140**, 075101.
46 G. Lipari and A. Szabo, *J. Am. Chem. Soc.*, 1982, **104**, 4546–4559.
47 H. Wennerström, B. Lindman, O. Söderman, T. Drakenberg and J. B. Rosenholm, *J. Am. Chem. Soc.*, 1979, **101**, 6860–6864.
48 A. Polimeno and J. H. Freed, *J. Phys. Chem.*, 1995, **99**, 10995–11006.
49 Y. E. Shapiro and E. Meirovitch, *J. Chem. Phys.*, 2013, **139**, 084107.
50 Y. E. Shapiro and E. Meirovitch, *J. Chem. Phys.*, 2013, **139**, 225104.

51 Y. E. Shapiro and E. Meirovitch, *J. Chem. Phys.*, 2014, **140**, 155102.
52 E. Meirovitch, *Isr. J. Chem.*, 2014, **54**, 47–59.
53 J. C. Xia, N. J. Deng and R. M. Levy, *J. Phys. Chem. B*, 2013, **117**, 6625–6634.
54 M. Gur, E. Zomot and I. Bahar, *J. Chem. Phys.*, 2013, **139**, 121912.
55 A. R. Atilgan, S. R. Durrell, R. L. Jernigan, M. C. Demirel, O. Keskin and I. Bahar, *Biophys. J.*, 2001, **80**, 505–515.
56 P. Calligari and D. Abergel, *J. Phys. Chem. B*, 2014, **118**, 3823–3831.
57 M. Maciejewski, P. N. Barlow and N. Tjandra, *J. Comput. Chem.*, 2014, **35**, 518–525.
58 G. M. Clore, A. Szabo, A. Bax, L. E. Kay, P. C. Driscoll and A. M. Gronenborn, *J. Am. Chem. Soc.*, 1990, **112**, 4989–4991.
59 L. P. Lee, D. J. Cole, C. K. Scylaris, W. L. Jorgensen and M. C. Payne, *J. Chem. Theory Comput.*, 2013, **9**, 2981–2991.
60 J. Huang and A. D. MacKerell, *J. Comput. Chem.*, 2013, **34**, 2135–2145.
61 X. D. Xiao, T. Zhu, C. G. Ji and J. Z. H. Zhang, *J. Phys. Chem. B*, 2013, **117**, 14885–14893.
62 P. E. M. Lopes, J. Huang, J. Shim, Y. Luo, H. Li, B. Roux and A. D. MacKerell, *J. Chem. Theory Comput.*, 2013, **9**, 5430–5449.
63 M. Akke, R. Brüschweiler and A. G. Palmer, *J. Am. Chem. Soc.*, 1993, **115**, 9832–9833.
64 S. Genheden, M. Akke and U. Ryde, *J. Chem. Theory Comput.*, 2014, **10**, 432–438.
65 V. Kasinath, K. A. Sharp and A. J. Wand, *J. Am. Chem. Soc.*, 2013, **135**, 15092–15100.
66 J. A. Marsh, *J. Mol. Biol.*, 2013, **425**, 3250–3263.
67 P. S. Emani, M. F. Bardaro, W. Huang, S. Aragon, G. Varani and G. P. Drobny, *J. Phys. Chem. B*, 2014, **118**, 1726–1742.
68 P. S. Emani, G. L. Olsen, G. Varani and G. P. Drobny, *J. Phys. Chem. A*, 2011, **115**, 12055–12069.
69 P. S. Emani, G. L. Olsen, G. Varani and G. P. Drobny, *J. Phys. Chem. A*, 2012, **116**, 7253–7260.
70 P. Dosset, P. Barthe, M. Cohen-Gonsaud, C. Roumestand and H. Demene, *J. Biomol. NMR*, 2013, **57**, 305–311.
71 K. Berlin, A. Longhini, T. K. Dayie and D. Fushman, *J. Biomol. NMR*, 2013, **57**, 333–352.
72 P. E. S. Smith, K. J. Donovan, O. Szekely, M. Baias and L. Frydman, *ChemPhysChem*, 2013, **14**, 3138–3145.
73 N. M. Loening, M. J. Thrippleton, J. Keeler and R. G. Griffin, *J. Magn. Reson.*, 2003, **164**, 321–328.
74 N. Shemesh, J. N. Dumez and L. Frydman, *Chem. – Eur. J.*, 2013, **19**, 13002–13008.
75 S. Gil, T. Hosek, Z. Solyom, R. Kümmerle, B. Brutscher, R. Pierattelli and I. C. Felli, *Angew. Chem., Int. Ed.*, 2013, **52**, 11808–11812.
76 W. Bermel, I. C. Felli, L. Gonnelli, W. Kozminski, A. Piai, R. Pierattelli and A. Zawadzka-Kazimierczuk, *J. Biomol. NMR*, 2013, **57**, 353–361.
77 R. Ishima, *J. Biomol. NMR*, 2014, **58**, 113–122.
78 T. Yuwen and N. R. Skrynnikov, *J. Magn. Reson.*, 2014, **241**, 155–169.
79 V. Kasinath, K. G. Valentine and A. J. Wand, *J. Am. Chem. Soc.*, 2013, **135**, 9560–9563.
80 H. Kovacs and A. Gossert, *J. Biomol. NMR*, 2014, **58**, 101–112.
81 S. Grzesiek, J. Anglister and A. Bax, *J. Magn Reson., Ser. B*, 1993, **101**, 114–119.
82 V. Blechta and J. Schraml, *Magn. Reson. Chem.*, 2013, **51**, 743–749.

83 A. J. Wand, M. R. Ehrhardt and P. F. Flynn, *Proc. Natl. Acad. Sci. U. S. A.*, 1998, **95**, 15299–15302.
84 I. Dodevski, N. V. Nucci, K. G. Valentine, G. K. Sidhu, E. S. O'Brien, A. Pardi and A. J. Wand, *J. Am. Chem. Soc.*, 2014, **136**, 3465–3474.
85 C. Charlier, S. N. Khan, T. Marquardsen, P. Pelupessy, V. Reiss, D. Sakellariou, G. Bodenhausen, F. Engelke and F. Ferrage, *J. Am. Chem. Soc.*, 2013, **135**, 18665–18672.
86 M. W. Clarkson, M. Lei, E. Z. Eisenmesser, W. Labeikovsky, A. Redfield and D. Kern, *J. Biomol. NMR*, 2009, **45**, 217–225.
87 B. Kresse, A. F. Privalov, A. Herrmann, M. Hofmann, E. A. Rössler and F. Fujara, *Solid State Nucl. Magn. Reson.*, 2014, **59-60**, 45–47.
88 L. M. Broche, P. J. Ross, K. J. Pine and D. J. Lurie, *J. Magn. Reson.*, 2014, **238**, 44–51.
89 W. A. Edelstein, J. M. S. Hutchison, G. Johnson and T. Redpath, *Phys. Med. Biol.*, 1980, **25**, 751–756.
90 G. Bodenhausen and R. R. Ernst, *J. Magn. Reson.*, 1981, **45**, 367–373.
91 B. J. Harden and D. P. Frueh, *J. Biomol. NMR*, 2014, **58**, 83–99.
92 A. Ahlner, M. Carlsson, B. H. Jonsson and P. Lundström, *J. Biomol. NMR*, 2013, **56**, 191–202.
93 L. F. Pinto, R. Riguera and E. Fernandez-Megia, *J. Am. Chem. Soc.*, 2013, **135**, 11513–11516.
94 H. Farooq, D. Courtier-Murias, R. Soong, H. Masoom, W. Maas, M. Fey, R. Kumar, M. Monette, H. Stronks, M. J. Simpson and A. J. Simpson, *Magn. Reson. Chem.*, 2013, **51**, 129–135.
95 G. B. Liu, X. F. Li, X. P. Sun, J. W. Feng, C. H. Ye and X. Zhou, *J. Magn. Reson.*, 2013, **237**, 158–163.
96 P. T. Vesanen, K. C. J. Zevenhoven, J. O. Nieminen, J. Dabek, L. T. Parkkonen and R. J. Ilmoniemi, *J. Magn. Reson.*, 2013, **235**, 50–57.
97 R. Borowiak, N. Schwaderlapp, F. Huethe, T. Lickert, E. Fischer, S. Bär, J. Hennig, D. von Elverfeldt and J. B. Hövener, *Magn. Reson. Mater. Phys., Biol. Med.*, 2013, **26**, 491–499.
98 C. Horch, S. Schlayer and F. Stallmach, *J. Magn. Reson.*, 2014, **240**, 24–33.
99 F. Dalitz, L. Kreckel, M. Maiwald and G. Guthausen, *Appl. Magn. Reson.*, 2014, **45**, 411–425.
100 W. P. Wu, H. Yi, D. Chen, R. S. Lu, T. Yuan, J. D. Chen and Z. H. Ni, *Microsyst. Technol.*, 2014, **20**, 419–425.
101 D. S. Dudova, D. E. Protasov and E. A. Protasov, *Appl. Magn. Reson.*, 2013, **44**, 781–790.
102 F. Diaz-Diaz, P. F. D. Cano-Barrita, B. J. Balcom, S. E. Solis-Najera and A. O. Rodriguez, *Sensors*, 2013, **13**, 15985–15999.
103 R. S. Lu, X. L. Zhou, W. P. Wu, Y. Y. Zhang and Z. H. Ni, *Appl. Magn. Reson.*, 2014, **45**, 461–469.
104 P. Guo, W. He and J. C. Garcia-Naranjo, *Sensors*, 2014, **14**, 6797–6805.
105 C. J. Lopez, W. Lu and J. D. Walls, *J. Magn. Reson.*, 2014, **242**, 95–106.
106 V. Hormann, M. Veevaete, R. Goedecke and H. W. Fischer, *Appl. Magn. Reson.*, 2013, **44**, 803–815.
107 D. Walsh, P. Turner, E. Grunewald, H. Zhang, J. J. Butler, E. Reboulet, S. Knobbe, T. Christy, J. W. Lane, C. D. Johnson, T. Munday and A. Fitzpatrick, *Groundwater*, 2013, **51**, 914–926.
108 Y. Q. Tang, M. D. Hürlimann, S. Mandal, J. Paulsen and Y. Q. Song, *J. Appl. Phys.*, 2014, **115**, 073903.
109 F. Deng, L. Z. Xiao, G. Z. Liao, F. R. Zong and W. L. Chen, *Appl. Magn. Reson.*, 2014, **45**, 179–192.

110 *Magnetic Resonance Detection of Explosives and Illicit Materials (NATO Science for Peace and Security Series B: Physics and Biophysics)*, ed. T. Apih, B. Ramaev, G. Mozzhukhin and J. Barras, Springer, Dordrecht, 2013.
111 J. Mitchell, L. F. Gladden, T. C. Chandrasekera and E. J. Fordham, *Prog. NMR Spectrosc.*, 2014, **76**, 1–60.
112 S. Mandal, V. D. M. Koroleva, T. W. Borneman, Y. Q. Song and M. D. Hürlimann, *J. Magn. Reson.*, 2013, **237**, 1–10.
113 J. Mitchell, T. C. Chandrasekera and L. F. Gladden, *Microporous Mesoporous Mater.*, 2013, **178**, 20–22.
114 J. Mitchell, T. C. Chandrasekera and L. F. Gladden, *J. Chem. Phys.*, 2013, **139**, 074205.
115 L. M. Schwartz, D. L. Johnson, J. Mitchell, T. C. Chandrasekera and E. J. Fordham, *Phys. Rev. E: Stat., Nonlinear, Soft Matter Phys.*, 2013, **88**, 032813.
116 R. H. Xie, L. Z. Xiao, K. J. Dunn and S. Q. Fu, *J. Geophys. Eng.*, 2013, **10**, 054008.
117 S. Muncaci and I. Ardelean, *Appl. Magn. Reson.*, 2013, **44**, 837–848.
118 L. Z. Xiao, H. B. Liu, F. Deng, Z. F. Zhang, T. L. An, F. R. Zong, V. Anferov and S. Anferova, *Microporous Mesoporous Mater.*, 2013, **178**, 90–93.
119 Z. F. Zhang, L. Z. Xiao, H. B. Liu, F. Deng, X. Li, T. L. An, V. Anferov and S. Anferova, *Appl. Magn. Reson.*, 2013, **44**, 849–857.
120 Z. F. Zhang, L. Z. Xiao, G. Z. Liao, H. B. Liu, W. Xu, Y. Wu and S. J. Jiang, *Appl. Magn. Reson.*, 2013, **44**, 1335–1343.
121 H. Celik, M. Bouhrara, D. A. Reiter, K. W. Fishbein and R. G. Spencer, *J. Magn. Reson.*, 2013, **236**, 134–139.
122 M. J. Tan, P. Wang and K. Y. Mao, *J. Appl. Geophys.*, 2014, **103**, 12–30.
123 Y. Zhang and B. Blümich, *J. Magn. Reson.*, 2014, **242**, 41–48.
124 D. Bernin and D. Topgaard, *Curr. Opin. Colloid Interface Sci.*, 2013, **18**, 166–172.
125 E. Grunewald and D. Walsh, *Geophys. Res. Lett.*, 2013, **40**, 6346–6350.
126 E. Grunewald, R. Knight and D. Walsh, *Geophysics*, 2014, **79**, EN15–EN23.
127 J. O. Walbrecker, R. Knight and E. Grunewald, *Geophysics*, 2014, **79**, EN39–EN48.
128 G. Wang, A. M. El-Sharkawy and P. A. Bottomley, *J. Magn. Reson.*, 2014, **242**, 243–255.
129 P. Kellman, D. A. Herzka and M. S. Hansen, *Magn. Reson. Med.*, 2014, **71**, 1428–1434.
130 O. C. Andronesi, H. Bhat, M. Reuter, S. Mukherjee, P. Caravan and B. R. Rosen, *NeuroImage*, 2014, **89**, 92–109.
131 F. Marica, F. G. Goora and B. J. Balcom, *J. Magn. Reson.*, 2014, **240**, 61–66.
132 J. Mitchell, *J. Magn. Reson.*, 2014, **240**, 52–60.
133 J. Stanek, M. Nowakowski, S. Saxena, K. Ruszczynska-Bartnik, A. Ejchart and W. Kozminski, *J. Biomol. NMR*, 2013, **56**, 217–226.
134 J. Stanek, P. Podbevsek, W. Kozminski, J. Plavec and M. Cevec, *J. Biomol. NMR*, 2013, **57**, 1–9.
135 J. Stanek, S. Saxena, L. Geist, R. Konrat and W. Kozminski, *Angew. Chem., Int. Ed.*, 2013, **52**, 4604–4606.
136 K. Kazimierczuk, J. Stanek, A. Zawadzka-Kazimierczuk and W. Kozminski, *ChemPhysChem*, 2013, **14**, 3015–3025.
137 B. Vögeli, *Prog. NMR Spectrosc.*, 2014, **78**, 1–46.
138 P. Kaderavek, V. Zapletal, A. Rabatinova, L. Krasny, V. Sklenar and L. Zidek, *J. Biomol. NMR*, 2014, **58**, 193–207.
139 R. Ishima and K. Nagayama, *J. Magn. Reson., Ser. B*, 1995, **108**, 73–76.

140 N. A. Farrow, O. Zhang, A. Szabo, D. A. Torchia and L. E. Kay, *J. Biomol. NMR*, 1995, **6**, 153–162.

141 S. Gabl, O. Steinhauser and H. Weingärtner, *Angew. Chem., Int. Ed.*, 2013, **52**, 9242–9246.

142 S. Gabl, C. Schröder, D. Braun, H. Weingärtner and O. Steinhauser, *J. Chem. Phys.*, 2014, **140**, 184503.

143 H. Wennerström, *Mol. Phys.*, 1972, **24**, 69–80.

144 H. M. McConnell, *J. Chem. Phys.*, 1958, **28**, 430–431.

145 U. Weininger, M. Respondek, C. Low and M. Akke, *J. Phys. Chem. B*, 2013, **117**, 9241–9247.

146 A. Sekhar, P. Vallurupalli and L. E. Kay, *Proc. Natl. Acad. Sci. U. S. A.*, 2013, **110**, 11391–11396.

147 D. Ban, A. Mazur, M. G. Carneiro, T. M. Sabo, K. Giller, L. M. I. Koharudin, S. Becker, A. M. Gronenborn, C. Griesinger and D. Lee, *J. Biomol. NMR*, 2013, **57**, 73–82.

148 K. Sugase, T. Konuma, J. C. Lansing and P. E. Wright, *J. Biomol. NMR*, 2013, **56**, 275–283.

149 A. Mazur, B. Hammesfahr, C. Griesinger, D. Lee and M. Kollmar, *Bioinformatics*, 2013, **29**, 1819–1820.

150 U. Weininger, A. T. Blissing, J. Hennig, A. Ahlner, Z. H. Liu, H. J. Vogel, M. Akke and P. Lundström, *J. Biomol. NMR*, 2013, **57**, 47–55.

151 U. Weininger, U. Brath, K. Modig, K. Teilum and M. Akke, *J. Biomol. NMR*, 2014, **59**, 23–29.

152 J. M. Aramini, K. Hamilton, L. C. Ma, G. V. T. Swapna, P. G. Leonard, J. E. Ladbury, R. M. Krug and G. T. Montelione, *Structure*, 2014, **22**, 515–525.

153 G. Bouvignies, P. Vallurupalli and L. E. Kay, *J. Mol. Biol.*, 2014, **426**, 763–774.

154 A. L. Hansen and L. E. Kay, *Proc. Natl. Acad. Sci. U. S. A.*, 2014, **111**, E1705–E1712.

155 B. Zhao, A. L. Hansen and Q. Zhang, *J. Am. Chem. Soc.*, 2014, **136**, 20–23.

156 X. Xu, J. S. Lee and A. Jerschow, *Angew. Chem.,Int. Ed.*, 2013, **52**, 8281–8284.

157 L. Frydman, T. Scherf and A. Lupulescu, *Proc. Natl. Acad. Sci. U. S. A.*, 2002, **99**, 15858–15862.

158 J. Döpfert, C. Witte and L. Schröder, *J. Magn. Reson.*, 2013, **237**, 34–39.

159 R. W. Stobbe and C. Beaulieu, *Magn. Reson. Mater. Phys., Biol. Med.*, 2014, **27**, 21–33.

160 V. Tugarinov, *Prog. NMR Spectrosc.*, 2014, **77**, 49–68.

161 Z. Luz and S. Meiboom, *J. Chem. Phys.*, 1964, **40**, 2686–2692.

162 D. Kruk, S. K. Hoffmann, J. Goslar, S. Lijewski, A. Kubica-Misztal, A. Korpala, I. Oglodek, J. Kowalewski, E. A. Rössler and J. Moscicki, *J. Chem. Phys.*, 2013, **139**, 244502.

163 D. Kruk, A. Korpala, A. Kubica, J. Kowalewski, E. A. Rössler and J. Moscicki, *J. Chem. Phys.*, 2013, **138**, 124506.

164 D. Bhattacharya, S. Shil, T. Goswami, A. Misra, A. Panda and D. J. Klein, *Comput. Theor. Chem.*, 2013, **1024**, 15–23.

165 P. Håkansson, T. Nguyen, P. B. Nair, R. Edge and E. Stulz, *Phys. Chem. Chem. Phys.*, 2013, **15**, 10930–10941.

166 P. H. Fries, *Eur. J. Inorg. Chem.*, 2012, 2156–2166.

167 P. H. Fries and E. Belorizky, *J. Chem. Phys.*, 2012, **136**, 074513.

168 A. Lasoroski, R. Vuilleumier and R. Pollet, *J. Chem. Phys.*, 2013, **139**, 104115.

169 J. Mares, M. Hanni, P. Lantto, J. Lounila and J. Vaara, *Phys. Chem. Chem. Phys.*, 2014, **16**, 6916–6924.

170 J. Mares, H. Liimatainen, T. O. Pennanen and J. Vaara, *J. Chem. Theory Comput.*, 2011, **7**, 3248–3260.

171 J. Mares, H. Liimatainen, K. Laasonen and J. Vaara, *J. Chem. Theory Comput.*, 2011, **7**, 2937–2946.
172 D. Esteban-Gomez, C. Cassino, M. Botta and C. Platas-Iglesias, *RSC Adv.*, 2014, **4**, 7094–7103.
173 E. M. Gale, J. Zhu and P. Caravan, *J. Am. Chem. Soc.*, 2013, **135**, 18600–18608.
174 H. C. Yang, S. H. Liao, C. W. Liu, H. H. Chen, H. E. Horng, K. L. Chen, M. J. Chen, C. I. Liu and L. M. Wang, *IEEE Trans. Appl. Supercond.*, 2013, **23**, 1602804.
175 H. C. Yang, C. W. Liu, S. H. Liao, H. H. Chen, M. J. Chen, K. L. Chen, H. E. Horng, S. Y. Yang and L. M. Wang, *Appl. Phys. Lett.*, 2012, **100**, 202405.
176 S. Ciofi-Baffoni, A. Gallo, R. Muzzioli and M. Piccioli, *J. Biomol. NMR*, 2014, **58**, 123–128.
177 X. H. Gu, Z. Gong, D. C. Guo, W. P. Zhang and C. Tang, *J. Biomol. NMR*, 2014, **58**, 149–154.
178 G. M. Clore, *Biophys. Chem.*, 2014, **186**, 3–12.
179 G. Jeschke, *Prog. NMR Spectrosc.*, 2013, **72**, 42–60.
180 B. L. Eggimann, V. V. Vostrikov, G. Veglia and J. I. Siepmann, *Theor. Chem. Acc.*, 2013, **132**, 1388.
181 C. H. Wunderlich, R. G. Huber, R. Spitzer, K. R. Liedl, K. Kloiber and C. Kreutz, *ACS Chem. Biol.*, 2013, **8**, 2697–2706.
182 Y. Zhang, T. Yamaguchi and K. Kato, *Chem. Lett.*, 2013, **42**, 1455–1462.
183 D. Alberti, M. van't Erve, R. Stefania, M. R. Ruggiero, M. Tapparo, S. G. Crich and S. Aime, *Angew. Chem., Int. Ed.*, 2014, **53**, 3488–3491.
184 I. V. Koptyug, *Mendeleev Commun.*, 2013, **23**, 299–312.
185 K. J. Donovan, A. Lupulescu and L. Frydman, *ChemPhysChem*, 2014, **15**, 436–443.
186 T. Harris, O. Szekely and L. Frydman, *J. Phys. Chem. B*, 2014, **118**, 3281–3290.
187 I. Marco-Rius, S. E. Bohndiek, M. I. Kettunen, T. J. Larkin, M. Basharat, C. Seeley and K. M. Brindle, *Contrast Media Mol. Imaging*, 2014, **9**, 182–186.
188 G. Navon, Y. Q. Song, T. Room, S. Appelt, R. E. Taylor and A. Pines, *Science*, 1996, **271**, 1848–1851.
189 J. H. Ardenkjaer-Larsen, C. Laustsen, S. Bowen and R. Rizi, *Magn. Reson. Med.*, 2014, **71**, 50–56.
190 S. Bowen and J. H. Ardenkjaer-Larsen, *J. Magn. Reson.*, 2014, **240**, 90–94.
191 H. Y. Chen and C. Hilty, *Anal. Chem.*, 2013, **85**, 7385–7390.
192 C. L. Swisher, P. E. Z. Larson, K. Kruttwig, A. B. Kerr, S. Hu, R. A. Bok, A. Goga, J. M. Pauly, S. J. Nelson, J. Kurhanewicz and D. B. Vigneron, *Magn. Reson. Med.*, 2014, **71**, 1–11.
193 G. Pagès, M. Puckeridge, L. F. Guo, Y. L. Tan, C. Jacob, M. Garland and P. W. Kuchel, *Biophys. J.*, 2013, **105**, 1956–1966.
194 G. D. Reed, C. von Morze, R. Bok, B. L. Koelsch, M. Van Criekinge, K. J. Smith, H. Shang, P. E. Z. Larson, J. Kurhanewicz and D. B. Vigneron, *IEEE Trans. Med. Imaging*, 2014, **33**, 362–371.
195 H. Nonaka, R. Hata, T. Doura, T. Nishihara, K. Kumagai, M. Akakabe, M. Tsuda, K. Ichikawa and S. Sando, *Nat. Commun.*, 2013, **4**, 2411.
196 J. M. Franck, A. Pavlova, J. A. Scott and S. Han, *Prog. NMR. Spectrosc.*, 2013, **74**, 33–56.
197 C. Y. Cheng, J. Varkey, M. R. Ambroso, R. Langen and S. I. Han, *Proc. Natl. Acad. Sci. U. S. A.*, 2013, **110**, 16838–16843.
198 J. M. Franck, R. Kausik and S. I. Han, *Microporous Mesoporous Mater.*, 2013, **178**, 113–118.
199 K. G. Valentine, G. Mathies, S. Bedard, N. V. Nucci, I. Dodevski, M. A. Stetz, T. V. Can, R. G. Griffin and A. J. Wand, *J. Am. Chem. Soc.*, 2014, **136**, 2800–2807.

200 C. Luchinat, G. Parigi and E. Ravera, *J. Biomol. NMR*, 2014, **58**, 239–249.
201 K. H. Hausser and D. Stehlik, *Adv. Magn. Reson.*, 1968, **3**, 79–139.
202 C. R. Bowers and D. P. Weitekamp, *J. Am. Chem. Soc.*, 1987, **109**, 5541–5542.
203 M. G. Pravica and D. P. Weitekamp, *Chem. Phys. Lett.*, 1988, **145**, 255–258.
204 R. W. Adams, J. A. Aguilar, K. D. Atkinson, M. J. Cowley, P. I. P. Elliott, S. B. Duckett, G. G. R. Green, I. G. Khazal, J. Lopez-Serrano and D. C. Williamson, *Science*, 2009, **323**, 1708–1711.
205 A. S. Kiryutin, A. V. Yurkovskaya, R. Kaptein, H. M. Vieth and K. L. Ivanov, *J. Phys. Chem. Lett.*, 2013, **4**, 2514–2519.
206 F. Reineri, S. Aime, R. Gobetto and C. Nervi, *J. Chem. Phys.*, 2014, **140**, 094307.
207 D. A. Barskiy, K. V. Kovtunov, I. V. Koptyug, P. He, K. A. Groome, Q. A. Best, F. Shi, B. M. Goodson, R. V. Shchepin, A. M. Coffey, K. W. Waddell and E. Y. Chekmenev, *J. Am. Chem. Soc.*, 2014, **136**, 3322–3325.
208 K. K. Palaniappan, M. B. Francis, A. Pines and D. E. Wemmer, *Isr. J. Chem.*, 2014, **54**, 104–112.
209 T. Hughes-Riley, J. S. Six, D. M. L. Lilburn, K. F. Stupic, A. C. Dorkes, D. E. Shaw, G. E. Pavlovskaya and T. Meersmann, *J. Magn. Reson.*, 2013, **237**, 23–33.
210 P. Nikolaou, A. M. Coffey, L. L. Walkup, B. M. Gust, N. Whiting, H. Newton, S. Barcus, I. Muradyan, M. Dabaghyan, G. D. Moroz, M. S. Rosen, S. Patz, M. J. Barlow, E. Y. Chekmenev and B. M. Goodson, *Proc. Natl. Acad. Sci. U. S. A.*, 2013, **110**, 14150–14155.
211 P. Nikolaou, A. M. Coffey, L. L. Walkup, B. M. Gust, C. D. LaPierre, E. Koehnemann, M. J. Barlow, M. S. Rosen, B. M. Goodson and E. Y. Chekmenev, *J. Am. Chem. Soc.*, 2014, **136**, 1636–1642.
212 D. Abrams, M. E. Trusheim, D. R. Englund, M. D. Shattuck and C. A. Meriles, *Nano Lett.*, 2014, **14**, 2471–2478.
213 S. Bauer, H. Wittkamp, S. Schildmann, M. Frey, W. Hiller, T. Hecksher, N. B. Olsen, C. Gainaru and R. Böhmer, *J. Chem. Phys.*, 2013, **139**, 134503.
214 D. Bock, R. Kahlau, B. Pötzschner, T. Körber, E. Wagner and E. A. Rössler, *J. Chem. Phys.*, 2014, **140**, 094505.
215 B. Schmidtke, N. Petzold, R. Kahlau and E. A. Rössler, *J. Chem. Phys.*, 2013, **139**, 084504.
216 H. Weingärtner, *Curr. Opin. Colloid Interface Sci.*, 2013, **18**, 183–189.
217 C. I. Daniel, F. V. Chavez, G. Feio, C. A. M. Portugal, J. G. Crespo and P. J. Sebastiao, *J. Phys. Chem. B*, 2013, **117**, 11877–11884.
218 R. Meier, D. Kruk and E. A. Rössler, *ChemPhysChem*, 2013, **14**, 3071–3081.
219 E. A. Rössler, S. Stapf and N. Fatkullin, *Curr. Opin. Colloid Interface Sci.*, 2013, **18**, 173–182.
220 R. Meier, A. Herrmann, M. Hofmann, B. Schmidtke, B. Kresse, A. F. Privalov, D. Kruk, F. Fujara and E. A. Rössler, *Macromolecules*, 2013, **46**, 5538–5548.
221 A. Bormuth, M. Hofmann, P. Henritzi, M. Vogel and E. A. Rössler, *Macromolecules*, 2013, **46**, 7805–7811.
222 D. K. Kadayakkara, K. Damodaran, T. K. Hitchens, J. W. M. Bulte and E. T. Ahrens, *J. Magn. Reson.*, 2014, **242**, 18–22.
223 K. Saalwächter, *Macromolecules*, 2013, **46**, 5090–5093.
224 V. M. Litvinov, M. E. Ries, T. W. Baughman, A. Henke and P. P. Matloka, *Macromolecules*, 2013, **46**, 5094–5096.
225 Y. Masuda, Y. Mori and K. Sakurai, *J. Phys. Chem. A*, 2013, **117**, 10576–10587.
226 R. B. Nazarski, *J. Inclusion Phenom. Macrocyclic Chem.*, 2014, **78**, 299–310.

227 A. F. Ciftja, A. Hartono and H. F. Svendsen, *Int. J. Greenhouse Gas Control*, 2013, **16**, 224–232.
228 H. Yamada, Y. Matsuzaki and K. Goto, *Ind. Eng. Chem. Res.*, 2014, **53**, 1617–1623.
229 P. M. Richardson, A. M. Voice and I. M. Ward, *J. Chem. Phys.*, 2013, **139**, 214501.
230 G. Stirnemann, E. Wernersson, P. Jungwirth and D. Laage, *J. Am. Chem. Soc.*, 2013, **135**, 11824–11831.
231 S. Murgia, M. Monduzzi, F. Lopez and G. Palazzo, *J. Solution Chem.*, 2013, **42**, 1111–1122.
232 T. Umecky, T. Takamuku, T. Matsumoto, E. Kawai, M. Takagi and T. Funazukuri, *J. Phys. Chem. B*, 2013, **117**, 16219–16226.
233 N. Dias, K. Shimizu, P. Morgado, E. J. M. Filipe, J. N. C. Lopes and F. V. Chavez, *J. Phys. Chem. B*, 2014, **118**, 5772–5780.
234 Y. J. Lin, L. Hao, Y. C. Lin, C. W. Kuo, P. R. Chen, T. Y. Wu and I. W. Sun, *Int. J. Electrochem. Sci.*, 2013, **8**, 8097–8114.
235 T. J. Simons, P. M. Bayley, Z. Zhang, P. C. Howlett, D. R. MacFarlane, L. A. Madsen and M. Forsyth, *J. Phys. Chem. B*, 2014, **118**, 4895–4905.
236 J. Y. Zhang and H. C. Dorn, *Fullerenes, Nanotubes, Carbon Nanostruct.*, 2014, **22**, 35–46.
237 P. Cmoch, R. Glaszczka, J. Jazwinski, B. Kamienski and E. Senkara, *Magn. Reson. Chem.*, 2014, **52**, 61–68.
238 Y. C. Wang, M. Y. Chen, H. Y. Wu, J. H. Chen, S. S. Wang and J. Y. Tung, *Polyhedron*, 2014, **67**, 73–80.
239 B. Bera, Y. P. Patil, M. Nethaji and B. R. Jagirdar, *Dalton Trans.*, 2014, **43**, 4726–4733.
240 A. Upadhyay, J. Rajpurohit, M. K. Singh, R. Dubey, A. K. Srivastava, A. Kumar, G. Rajaraman and M. Shanmugam, *Chem. – Eur. J.*, 2014, **20**, 6061–6070.
241 M. R. Chierotti, A. Rossin, R. Gobetto and M. Peruzzini, *Inorg. Chem.*, 2013, **52**, 12616–12623.
242 E. Molnar, N. Camus, V. Patinec, G. A. Rolla, M. Botta, G. Tircso, F. K. Kalman, T. Fodor, R. Tripier and C. Platas-Iglesias, *Inorg. Chem.*, 2014, **53**, 5136–5149.
243 V. Patinec, G. A. Rolla, M. Botta, R. Tripier, D. Esteban-Gomez and C. Platas-Iglesias, *Inorg. Chem.*, 2013, **52**, 11173–11184.
244 V. C. Pierre, M. J. Allen and P. Caravan, *JBIC, J. Biol. Inorg. Chem.*, 2014, **19**, 127–131.
245 T. C. Soesbe, Y. K. Wu and A. Sherry, *NMR Biomed.*, 2013, **26**, 829–838.
246 A. H. Hung, M. C. Duch, G. Parigi, M. W. Rotz, L. M. Manus, D. J. Mastarone, K. T. Dam, C. C. Gits, K. W. MacRenaris, C. Luchinat, M. C. Hersam and T. J. Meade, *J. Phys. Chem. C*, 2013, **117**, 16263–16273.
247 I. Bertini, J. Kowalewski, C. Luchinat, T. Nilsson and G. Parigi, *J. Chem. Phys.*, 1999, **111**, 5795–5807.
248 D. Kruk, T. Nilsson and J. Kowalewski, *Phys. Chem. Chem. Phys.*, 2001, **3**, 4907–4917.
249 H. Jaccard, P. Mieville, C. Cannizzo, C. R. Mayer and L. Helm, *JBIC, J. Biol. Inorg. Chem.*, 2014, **19**, 145–159.
250 D. Delli Castelli, M. C. Caligara, M. Botta, E. Terreno and S. Aime, *Inorg. Chem.*, 2013, **52**, 7130–7138.
251 Z. Baranyai, G. A. Rolla, R. Negri, A. Forgacs, G. B. Giovenzana and L. Tei, *Chem. – Eur. J.*, 2014, **20**, 2933–2944.

252 E. R. Neil, A. M. Funk, D. S. Yufit and D. Parker, *Dalton Trans.*, 2014, **43**, 5490–5504.
253 S. P. Babailov, E. N. Zapolotsky and E. S. Fomin, *Polyhedron*, 2013, **65**, 332–336.
254 L. Fusaro and M. Luhmer, *Dalton Trans.*, 2014, **43**, 967–972.
255 T. C. Soesbe, S. J. Ratnakar, M. Milne, S. R. Zhang, Q. N. Do, Z. Kovacs and A. D. Sherry, *Magn. Reson. Med.*, 2014, **71**, 1179–1185.
256 D. Sauer, B. Schuster, M. Rosenstihl, S. Schneider, V. Talluto, T. Walther, T. Blochowicz, B. Stühn and M. Vogel, *J. Chem. Phys.*, 2014, **140**, 114503.
257 M. Uyama, M. Harada, T. Tsukahara and T. Okada, *J. Phys. Chem. C*, 2013, **117**, 24873–24882.
258 M. Rivlin, U. Eliav and G. Navon, *J. Magn. Reson.*, 2014, **242**, 107–112.
259 N. Galamba, *J. Phys. Chem. B*, 2014, **118**, 4169–4176.
260 S. Banerjee and B. Bagchi, *J. Chem. Phys.*, 2013, **139**, 164301.
261 B. Micko, C. Tschirwitz and E. A. Rössler, *J. Chem. Phys.*, 2013, **138**, 154501.
262 S. M. Kathmann, H. Cho, T. M. Chang, G. K. Schenter, K. Parab and T. Autrey, *J. Phys. Chem. B*, 2014, **118**, 4883–4888.
263 L. B. Alemany, *Magn. Reson. Chem.*, 2013, **51**, 605–613.
264 A. Ekinci, M. Z. Koylu and M. Askin, *Asian J. Chem.*, 2014, **26**, 2400–2402.
265 Z. Takacs, E. Steiner, J. Kowalewski and T. Brotin, *J. Phys. Chem. B*, 2014, **118**, 2134–2146.
266 I. D'Acquarica, A. Calcaterra, F. Sacco, F. Balzano, F. Aiello, A. Tafi, N. Pesci, G. Uccello-Barretta and B. Botta, *Chirality*, 2013, **25**, 840–851.
267 N. E. Kuz'mina, V. A. Yashkir, S. V. Moiseev, O. V. Fedorov, E. V. Rakhmanov, N. S. Baleeva, A. V. Tarakanova and A. V. Anisimov, *Russ. J. Org. Chem.*, 2013, **49**, 1386–1396.
268 B. Gierczyk and R. Pankiewicz, *Cent. Eur. J. Chem.*, 2014, **12**, 624–634.
269 H. G. O. Alvim, E. L. Fagg, A. L. de Oliveira, H. C. B. de Oliveira, S. M. Freitas, M. A. E. Xavier, T. A. Soares, A. F. Gomes, F. C. Gozzo, W. A. Silva and B. A. D. Neto, *Org. Biomol. Chem.*, 2013, **11**, 4764–4777.
270 L. Garbuio, Y. J. Li, S. Antonello, J. A. Gascon, R. G. Lawler, X. G. Lei, Y. Murata, N. J. Turro and F. Maran, *Photochem. Photobiol.*, 2014, **90**, 439–447.
271 F. Rastrelli, D. Frezzato, R. G. Lawler, Y. J. Li, N. J. Turro and A. Bagno, *Philos. Trans. R. Soc., A.*, 2013, **371**, 20110634.
272 R. Pendrill, E. Säwén and G. Widmalm, *J. Phys. Chem. B*, 2013, **117**, 14709–14722.
273 M. Soltesova, J. Kowalewski and G. Widmalm, *J. Biomol. NMR*, 2013, **57**, 37–45.
274 E. L. Wu, O. Engström, S. Jo, D. Stuhlsatz, M. S. Yeom, J. B. Klauda, G. Widmalm and W. Im, *Biophys. J.*, 2013, **105**, 1444–1455.
275 H. Shinar, M. D. Battistel, M. Mandler, F. Lichaa, D. I. Freedberg and G. Navon, *Carbohydr. Res.*, 2014, **389**, 165–173.
276 J. Björnerås, A. Gräslund and L. Mäler, *Biochemistry*, 2013, **52**, 4157–4167.
277 R. Konrat, *J. Magn. Reson.*, 2014, **241**, 74–85.
278 M. Ringkjoeping Jensen, R. W. H. Ruigrok and M. Blackledge, *Curr. Opin. Struct. Biol.*, 2013, **23**, 426–435.
279 Y. E. Shapiro, *Prog. Biophys. Mol. Biol.*, 2013, **112**, 58–117.
280 G. F. Wang, Z. T. Zhang, B. Jiang, X. Zhang, C. G. Li and M. L. Liu, *Anal. Bioanal. Chem.*, 2014, **406**, 2279–2288.
281 B. Vögeli, J. Orts, D. Strotz, C. Chi, M. Minges, M. A. Walti, P. Güntert and R. Riek, *J. Magn. Reson.*, 2014, **241**, 53–59.
282 G. Veglia and A. Cembran, *FEBS J.*, 2013, **280**, 5608–5615.

283 P. Guerry, L. Mollica and M. Blackledge, *ChemPhysChem*, 2013, **14**, 3046–3058.
284 H. M. Al-Hashimi, *J. Magn. Reson.*, 2013, **237**, 191–204.
285 L. J. Smith, A. M. Bowen, A. Di Paolo, A. Matagne and C. Redfield, *ChemBioChem*, 2013, **14**, 1780–1788.
286 M. Zerbetto, R. Anderson, S. Bouguet-Bonnet, M. Rech, L. Q. Zhang, E. Meirovitch, A. Polimeno and M. Buck, *J. Phys. Chem. B*, 2013, **117**, 174–184.
287 K. Chen and N. Tjandra, *Biochemistry*, 2013, **52**, 5365–5371.
288 M. C. Smith, K. M. Scaglione, V. A. Assimon, S. Patury, A. D. Thompson, C. A. Dickey, D. R. Southworth, H. L. Paulson, J. E. Gestwicki and E. R. P. Zuiderweg, *Biochemistry*, 2013, **52**, 5354–5364.
289 N. Y. Haba, R. Gross, J. Novacek, H. Shaked, L. Zidek, M. Barda-Saad and J. H. Chill, *Biophys. J.*, 2013, **105**, 481–493.
290 D. Gümral, F. Fogolari, A. Corazza, P. Viglino, S. Giorgetti, M. Stoppini, V. Bellotti and G. Esposito, *Magn. Reson. Chem.*, 2013, **51**, 795–807.
291 M. Rasheed, C. Richter, L. T. Chisty, J. Kirkpatrick, M. Blackledge, M. R. Webb and P. C. Driscoll, *Biochemistry*, 2014, **53**, 1092–1104.
292 M. Meli, K. Pagano, L. Ragona and G. Colombo, *PLoS One*, 2014, **9**, e97153.
293 J. Roche, J. F. Ying, A. S. Maltsev and A. Bax, *ChemBioChem*, 2013, **14**, 1754–1761.
294 K. A. Wilson, J. J. Bouchard and J. W. Peng, *Biochemistry*, 2013, **52**, 6968–6981.
295 S. Horowitz, L. M. A. Dirk, J. D. Yesselman, J. S. Nimtz, U. Adhikari, R. A. Mehl, S. Scheiner, R. L. Houtz, H. M. Al-Hashimi and R. C. Trievel, *J. Am. Chem. Soc.*, 2013, **135**, 15536–15548.
296 K. A. Stafford, F. Ferrage, J. H. Cho and A. G. Palmer, *J. Am. Chem. Soc.*, 2013, **135**, 18024–18027.
297 P. Robustelli, N. Trbovic, R. A. Friesner and A. G. Palmer, *J. Chem. Theory Comput.*, 2013, **9**, 5190–5200.
298 A. G. Palmer, *J. Magn. Reson.*, 2014, **241**, 3–17.
299 A. Sekhar and L. E. Kay, *Proc. Natl. Acad. Sci. U. S. A.*, 2013, **110**, 12867–12874.
300 T. Saio, X. Guan, P. Rossi, A. Economou and C. G. Kalodimos, *Science*, 2014, **344**, 1250494.
301 D. Muthu, R. E. Berry, H. J. Zhang and F. A. Walker, *Biochemistry*, 2013, **52**, 7910–7925.
302 S. R. Tzeng and C. G. Kalodimos, *Nat. Chem. Biol.*, 2013, **9**, 462–465.
303 F. Favretto, M. Assfalg, M. Gallo, D. O. Cicero, M. D'Onofrio and H. Molinari, *ChemBioChem*, 2013, **14**, 1807–1819.
304 D. D. Boehr, J. R. Schnell, D. McElheny, S. H. Bae, B. M. Duggan, S. J. Benkovic, H. J. Dyson and P. E. Wright, *Biochemistry*, 2013, **52**, 4605–4619.
305 X. L. Huan, J. H. Shi, L. Z. Lim, S. Mitra, W. L. Zhu, H. N. Qin, E. B. Pasquale and J. X. Song, *PLoS One*, 2013, **8**, e74040.
306 M. J. C. Audin, G. Dorn, S. A. Fromm, K. Reiss, S. Schutz, M. K. Vorlander and R. Sprangers, *Angew. Chem., Int. Ed.*, 2013, **52**, 8312–8316.
307 V. Tugarinov, R. Sprangers and L. E. Kay, *J. Am. Chem. Soc.*, 2007, **129**, 1743–1750.
308 D. M. Korzhnev, K. Kloiber, V. Kanelis, V. Tugarinov and L. E. Kay, *J. Am. Chem. Soc.*, 2004, **126**, 3964–3973.
309 Y. Li, N. L. Altorelli, F. Bahna, B. Honig, L. Shapiro and A. G. Palmer, *Proc. Natl. Acad. Sci. U. S. A.*, 2013, **110**, 16462–16467.

310 Y. Li, N. L. Altorelli, F. Bahna, B. Honig, L. Shapiro and A. G. Palmer, *Proc. Natl. Acad. Sci. U. S. A.*, 2013, **110**, 19651.
311 C. Sanchez-Medina, A. Sekhar, P. Vallurupalli, M. Cerminara, V. Munoz and L. E. Kay, *J. Am. Chem. Soc.*, 2014, **136**, 7444–7451.
312 L. C. Shi and L. E. Kay, *Proc. Natl. Acad. Sci. U. S. A.*, 2014, **111**, 2140–2145.
313 D. S. Libich, N. L. Fawzi, J. F. Ying and G. M. Clore, *Proc. Natl. Acad. Sci. U. S. A.*, 2013, **110**, 11361–11366.
314 J. L. Kitevski-LeBlanc, J. Hoang, W. Thach, S. T. Larda and R. S. Prosser, *Biochemistry*, 2013, **52**, 5780–5789.
315 Y. S. Ye, X. L. Liu, Z. T. Zhang, Q. Wu, B. Jiang, L. Jiang, X. Zhang, M. L. Liu, G. J. Pielak and C. G. Li, *Chem. – Eur. J.*, 2013, **19**, 12705–12710.
316 X. Y. Yue, Y. Feng and Y. B. Yu, *J. Fluorine Chem.*, 2013, **152**, 173–181.
317 C. A. Castaneda, T. R. Kashyap, M. A. Nakasone, S. Krueger and D. Fushman, *Structure*, 2013, **21**, 1168–1181.
318 N. J. Anthis and G. M. Clore, *J. Am. Chem. Soc.*, 2013, **135**, 9648–9651.
319 V. Papouskova, P. Kaderavek, O. Otrusinova, A. Rabatinova, H. Sanderova, J. Novacek, L. Krasny, V. Sklenar and L. Zidek, *ChemBioChem*, 2013, **14**, 1772–1779.
320 R. M. Almeida, P. Turano, I. Moura, J. J. G. Moura and S. R. Pauleta, *ChemBioChem*, 2013, **14**, 1858–1866.
321 T. D. Zhuang, C. Chisholm, M. Chen and L. K. Tamm, *J. Am. Chem. Soc.*, 2013, **135**, 15101–15113.
322 M. Fragai, C. Luchinat, G. Parigi and E. Ravera, *Coord. Chem. Rev.*, 2013, **257**, 2652–2667.
323 R. E. Luna, S. R. Akabayov, J. J. Ziarek and G. Wagner, *FEBS J.*, 2014, **281**, 1965–1973.
324 H. van Ingen and A. M. J. J. Bonvin, *J. Magn. Reson.*, 2014, **241**, 103–114.
325 J. Schilder, F. Löhr, H. Schwalbe and M. Ubbink, *FEBS Lett.*, 2014, **588**, 1873–1878.
326 B. M. Burmann, C. W. Wang and S. Hiller, *Nat. Struct. Mol. Biol.*, 2013, **20**, 1265–1272.
327 V. Iesmantavicius, M. R. Jensen, V. Ozenne, M. Blackledge, F. M. Poulsen and M. Kjaergaard, *J. Am. Chem. Soc.*, 2013, **135**, 10155–10163.
328 V. Ozenne, J. K. Noel, P. O. Heidarsson, S. Brander, F. M. Poulsen, M. R. Jensen, B. B. Kragelund, M. Blackledge and J. Danielsson, *J. Mol. Biol.*, 2014, **426**, 722–734.
329 Z. Bozoky, M. Krzeminski, R. Muhandiram, J. R. Birtley, A. Al-Zahrani, P. J. Thomas, R. A. Frizzell, R. C. Ford and J. D. Forman-Kay, *Proc. Natl. Acad. Sci. U. S. A.*, 2013, **110**, E4427–E4436.
330 L. Mäler, *Adv. Drug Delivery Rev.*, 2013, **65**, 1002–1011.
331 J. Roche, J. M. Louis, A. Grishaev, J. F. Ying and A. Bax, *Proc. Natl. Acad. Sci. U. S. A.*, 2014, **111**, 3425–3430.
332 T. H. Kim, K. Y. Chung, A. Manglik, A. L. Hansen, R. O. Dror, T. J. Mildorf, D. E. Shaw, B. K. Kobilka and R. S. Prosser, *J. Am. Chem. Soc.*, 2013, **135**, 9465–9474.
333 M. K. Pandey, S. Vivekanandan, S. Ahuja, R. Huang, S. C. Im, L. Waskell and A. Ramamoorthy, *J. Phys. Chem. B*, 2013, **117**, 13851–13860.
334 L. A. M. Sommer, J. J. Janke, W. F. D. Bennett, J. Burck, A. S. Ulrich, D. P. Tieleman and S. A. Dames, *J. Phys. Chem. B*, 2014, **118**, 4817–4831.
335 M. J. Kang, C. D. Eichhorn and J. Feigon, *Proc. Natl. Acad. Sci. U. S. A.*, 2014, **111**, E663–E671.
336 E. Bonneau and P. Legault, *Biochemistry*, 2014, **53**, 579–590.
337 G. Widmalm, *Carbohydr. Res.*, 2013, **378**, 123–132.

338 D. Wierzuchowska and B. Blicharska, *Acta Phys. Pol., A*, 2014, **125**, 907–910.
339 F. Persson and B. Halle, *J. Am. Chem. Soc.*, 2013, **135**, 8735–8748.
340 D. E. Shaw, P. Maragakis, K. Lindorff-Larsen, S. Piana, R. O. Dror, M. P. Eastwood, J. A. Bank, J. M. Jumper, J. K. Salmon, Y. Shan and W. Wriggers, *Science*, 2010, **330**, 341–346.
341 J. P. Korb, A. Louis-Joseph and L. Benamsili, *J. Phys. Chem. B*, 2013, **117**, 7002–7014.
342 A. Louis-Joseph, L. Benamsili and J. P. Korb, *Microporous Mesoporous Mater.*, 2013, **178**, 23–26.
343 R. D. Majumdar, M. Gerken, R. Mikula and P. Hazendonk, *Energy Fuels*, 2013, **27**, 6528–6537.
344 S. Stapf, A. Ordikhani-Seyedlar, N. Ryan, C. Mattea, R. Kausik, D. E. Freed, Y. Q. Song and M. D. Hürlimann, *Energy Fuels*, 2014, **28**, 2395–2401.
345 V. G. Morgan, L. L. Barbosa, V. Lacerda and E. V. R. de Castro, *Ind. Eng. Chem. Res.*, 2014, **53**, 8881–8889.
346 E. O. Fridjonsson, B. F. Graham, M. Akhfash, E. F. May and M. L. Johns, *Energy Fuels*, 2014, **28**, 1756–1764.
347 L. Del Coco, M. Assfalg, M. D'Onofrio, F. Sallustio, F. Pesce, F. P. Fanizzi and F. P. Schena, *Metabolomics*, 2013, **9**, 740–751.
348 A. F. Constantino, V. Lacerda, R. B. dos Santos, S. J. Greco, R. C. Silva, A. C. Neto, L. L. Barbosa, E. V. R. de Castro and J. C. C. Freitas, *Quim. Nova*, 2014, **37**, 10–17.
349 K. Ueda, K. Higashi, K. Yamamoto and K. Moribe, *Int. J. Pharm.*, 2014, **464**, 205–213.
350 C. W. C. Hsieh, D. Cannella, H. Jorgensen, C. Felby and L. G. Thygesen, *J. Agric. Food Chem.*, 2014, **62**, 3800–3805.
351 J. B. Rosenholm, *Adv. Colloid Interface Sci.*, 2014, **205**, 9–47.
352 M. Monduzzi, S. Lampis, S. Murgia and A. Salis, *Adv. Colloid Interface Sci.*, 2014, **205**, 48–67.
353 M. Schönhoff, *Curr. Opin. Colloid Interface Sci.*, 2013, **18**, 201–213.
354 M. V. Popova, Y. S. Chernyshev and D. Michel, *Appl. Magn. Reson.*, 2014, **45**, 365–374.
355 M. V. Popova and D. Michel, *Appl. Magn. Reson.*, 2014, **45**, 353–364.
356 M. V. C. Cardoso and E. Sabadini, *Langmuir*, 2013, **29**, 15778–15786.
357 F. Keymeulen, P. De Bernardin, A. Dalla Cort and K. Bartik, *J. Phys. Chem. B*, 2013, **117**, 11654–11659.
358 R. Bernewitz, F. Dalitz, K. Köhler, H. P. Schuchmann and G. Guthausen, *Microporous Mesoporous Mater.*, 2013, **178**, 69–73.
359 R. J. Wang and Y. L. Wang, *Acta Chim. Sin.*, 2014, **72**, 41–50.
360 G. Wu, X. H. Jiang, L. M. Zhou, L. J. Yang, Y. Wang, G. Q. Xia, Z. J. Chen and M. Duan, *J. Mol. Struct.*, 2013, **1045**, 47–54.
361 Y. Wang, X. H. Jiang, L. M. Zhou, L. J. Yang, G. Q. Xia, Z. J. Chen and M. Duan, *Colloids Surf., A*, 2013, **436**, 1159–1169.
362 N. Matsumori, *J. Synth. Org. Chem. Jpn.*, 2014, **72**, 596–603.
363 V. H. G. Martins, M. R. Rodrigues, L. D. Mascarenhas, C. R. L. de Azambuja, J. L. Londono and V. R. de Lima, *J. Mol. Struct.*, 2014, **1059**, 225–231.
364 R. S. de Sousa, A. O. D. Nogueira, V. G. Marques, R. M. Clementin and V. R. de Lima, *Bioorganic Chem.*, 2013, **51**, 8–15.
365 J. Song, J. Franck, P. Pincus, M. W. Kim and S. Han, *J. Am. Chem. Soc.*, 2014, **136**, 2642–2649.
366 V. E. Bahamonde-Padilla, J. Espinoza, B. E. Weiss-Lopez, J. J. L. Cascales, R. Montecinos and R. Araya-Maturana, *J. Chem. Phys.*, 2013, **139**, 014703.

367 V. E. Bahamonde-Padilla, J. J. Lopez-Cascales, R. Araya-Maturana, M. Martinez-Cifuentes and B. E. W. Lopez, *ChemPhysChem*, 2014, **15**, 1422–1431.
368 D. Bock, R. Kahlau, B. Micko, B. Pötzschner, G. J. Schneider and E. A. Rössler, *J. Chem. Phys.*, 2013, **139**, 064508.
369 R. Kahlau, D. Bock, B. Schmidtke and E. A. Rössler, *J. Chem. Phys.*, 2014, **140**, 044509.
370 L. Van Lokeren, H. Ben Sassi, G. Van Assche and F. Ribot, *J. Magn. Reson.*, 2013, **231**, 46–53.
371 Y. Yokoyama and S. Yusa, *Polym. J.*, 2013, **45**, 985–992.
372 J. Weiss, H. Wienk, R. Boelens and A. Laschewsky, *Macromol. Chem. Phys.*, 2014, **215**, 915–919.
373 K. Nakai, M. Nishiuchi, M. Inoue, K. Ishihara, Y. Sanada, K. Sakurai and S. Yusa, *Langmuir*, 2013, **29**, 9651–9661.
374 E. B. Twum, E. F. Mccord, D. F. Lyons, P. A. Fox and P. L. Rinaldi, *Eur. Polym. J.*, 2014, **51**, 136–150.
375 F. Wang, N. M. Shao and Y. Y. Cheng, *PLoS One*, 2013, **8**, e64722.
376 I. M. Neelov, D. A. Markelov, S. G. Falkovich, M. Y. Ilyash, B. M. Okrugin and A. A. Darinskii, *Polym. Sci., Ser. C*, 2013, **55**, 154–161.
377 N. D'Amelio, C. Esteban, A. Coslovi, L. Feruglio, F. Uggeri, M. Villegas, J. Benegas, S. Paoletti and I. Donati, *J. Phys. Chem. B*, 2013, **117**, 13578–13587.
378 E. Watanabe, G. S. Boutis, H. Sato, S. Sekine and T. Asakura, *Polymer*, 2014, **55**, 278–286.
379 D. S. Hou and Z. J. Li, *J. Mater. Civ. Eng.*, 2014, **26**, 930–940.
380 B. Sierra-Martin, J. R. Retama, M. Laurenti, A. F. Barbero and E. L. Cabarcos, *Adv. Colloid Interface Sci.*, 2014, **205**, 113–123.
381 R. L. Bai, P. J. Basser, R. M. Briber and F. Horkay, *J. Appl. Polym. Sci.*, 2014, **131**, 40001.
382 H. J. Zhu, H. P. Huinink, O. C. G. Adan and K. Kopinga, *Macromolecules*, 2013, **46**, 6124–6131.
383 C. G. Lu, K. Xu, W. B. Li, P. C. Li, Y. Tan and P. X. Wang, *Chem. Res. Chin. Univ.*, 2013, **29**, 1203–1207.
384 V. Compan, S. Molla, S. Vallejos, F. Garcia, J. M. Garcia, J. Guzman and L. Garrido, *Macromol. Chem. Phys.*, 2014, **215**, 624–637.
385 J. H. Ortony, S. H. Choi, J. M. Spruell, J. N. Hunt, N. A. Lynd, D. V. Krogstad, V. S. Urban, C. J. Hawker, E. J. Kramer and S. Han, *Chem. Sci.*, 2014, **5**, 58–67.
386 H. Liu, Q. H. Li, D. S. Zhu, J. Li, J. S. Liu, P. Geng and Y. T. He, *Carbohydr. Polym.*, 2013, **98**, 542–545.
387 S. Kaieda, B. Setlow, P. Setlow and B. Halle, *Biophys. J.*, 2013, **105**, 2016–2023.
388 S. Kaieda, T. S. Plivelic and B. Halle, *Phys. Chem. Chem. Phys.*, 2014, **16**, 4002–4011.
389 J. S. He, T. H. Mu, X. S. Guo, S. M. Zhu, N. Azuma and C. Kanno, *Food Hydrocolloids*, 2013, **33**, 415–424.
390 M. H. Oztop, K. L. McCarthy, M. J. McCarthy and M. Rosenberg, *LWT--Food Sci. Technol.*, 2014, **56**, 93–100.
391 R. Lopez-Cebral, M. Martin-Pastor, P. Paolicelli, M. A. Casadei, B. Seijo and A. Sanchez, *Colloids Surf., B*, 2014, **115**, 391–399.
392 C. Luchinat, G. Parigi and E. Ravera, *ChemPhysChem*, 2013, **14**, 3156–3161.
393 J. Kowalczuk, M. Bielejewski, A. Lapinski, R. Luboradzki and J. Tritt-Goc, *J. Phys. Chem. B*, 2014, **118**, 4005–4015.

394 S. Salami, C. Rondeau-Mouro, J. van Duynhoven and F. Mariette, *J. Agric. Food Chem.*, 2013, **61**, 5870–5879.
395 B. Yager, J. Nyéki, A. Casey, B. P. Cowan, C. P. Lusher and J. Saunders, *Phys. Rev. Lett.*, 2013, **111**, 215303.
396 F. Gul-E-Noor, D. Michel, H. Krautscheid, J. Haase and M. Bertmer, *J. Chem. Phys.*, 2013, **139**, 034202.
397 J. K. Moore, M. D. Guiver, N. Y. Du, S. E. Hayes and M. S. Conradi, *J. Phys. Chem. C*, 2013, **117**, 22995–22999.
398 Y. Ji, N. S. Sullivan, Y. B. Tang and J. A. Hamida, *Materials*, 2013, **6**, 2464–2482.
399 J. Mitchell, L. M. Broche, T. C. Chandrasekera, D. J. Lurie and L. F. Gladden, *J. Phys. Chem. C*, 2013, **117**, 17699–17706.
400 C. D'Agostino, T. Kotionova, J. Mitchell, P. J. Miedziak, D. W. Knight, S. H. Taylor, G. J. Hutchings, L. F. Gladden and M. D. Mantle, *Chem. – Eur. J.*, 2013, **19**, 11725–11732.
401 J. J. Chen, X. Q. Kong, K. Sumida, M. A. Manumpil, J. R. Long and J. A. Reimer, *Angew. Chem., Int. Ed.*, 2013, **52**, 12043–12046.
402 A. M. Perepukhov, O. V. Kishenkov, S. V. Gudenko, A. V. Maksimychev, D. A. Aleksandrov, L. I. Men'shikov and S. I. Tkachenko, *Russ. J. Phys. Chem. B*, 2014, **8**, 284–292.
403 K. S. Han, X. Q. Wang, S. Dai and E. W. Hagaman, *J. Phys. Chem. C*, 2013, **117**, 15754–15762.
404 M. Sattig and M. Vogel, *J. Phys. Chem. Lett.*, 2014, **5**, 174–178.
405 H. Chemmi, D. Petit, J. P. Korb, R. Denoyel and P. Levitz, *Microporous Mesoporous Mater.*, 2013, **178**, 104–107.
406 Y. Daiko, *J. Sol-Gel Sci. Technol.*, 2014, **70**, 172–179.
407 P. Porion, A. M. Faugere and A. Delville, *J. Phys. Chem. C*, 2013, **117**, 26119–26134.
408 K. Keating, *Near Surf. Geophys.*, 2014, **12**, 243–254.
409 C. Totland, R. T. Lewis and W. Nerdal, *Langmuir*, 2013, **29**, 11055–11061.
410 P. Fantazzini, S. Mengoli, S. Evangelisti, L. Pasquini, M. Mariani, L. Brizi, S. Goffredo, E. Caroselli, F. Prada, G. Falini, O. Levy and Z. Dubinsky, *Environ. Sci. Technol.*, 2013, **47**, 12679–12686.
411 A. C. A. Muller, K. L. Scrivener, A. M. Gajewicz and P. J. McDonald, *Microporous Mesoporous Mater.*, 2013, **178**, 99–103.
412 A. Pop, C. Badea and I. Ardelean, *Appl. Magn. Reson.*, 2013, **44**, 1223–1234.
413 Y. Yu, Z. P. Sun, M. Pang and P. Q. Yang, *J. Wuhan Univ. Technol., Mater. Sci. Ed.*, 2013, **28**, 963–967.
414 A. Valori, P. J. McDonald and K. L. Scrivener, *Cem. Concr. Res.*, 2013, **49**, 65–81.
415 M. van Landeghem, J. B. D. de Lacaillerie, B. Blümich, J. P. Korb and B. Bresson, *Cem. Concr. Res.*, 2013, **48**, 86–96.
416 V. Bortolotti, R. J. S. Brown, P. Fantazzini and M. Mariani, *Microporous Mesoporous Mater.*, 2013, **178**, 108–112.
417 I. Bardenhagen, W. Dreher, D. Fenske, A. Wittstock and M. Baumer, *Carbon*, 2014, **68**, 542–552.
418 C. Wakai, T. Shimoaka and T. Hasegawa, *Anal. Chem.*, 2013, **85**, 7581–7587.
419 J. B. Hou, J. Li, D. Mountz, M. Hull and L. A. Madsen, *J. Membr. Sci.*, 2013, **448**, 292–299.
420 D. Y. Park, P. A. Kohl and H. W. Beckham, *J. Phys. Chem. C*, 2013, **117**, 15468–15477.
421 K. G. Victor, J. P. Korb and R. G. Bryant, *J. Phys. Chem. B*, 2013, **117**, 12475–12478.

422 C. Terenzi, S. V. Dvinskikh and I. Furo, *J. Phys. Chem. B*, 2013, **117**, 8620–8632.

423 P. M. Kekkonen, A. Ylisassi and V. V. Telkki, *J. Phys. Chem. C*, 2014, **118**, 2146–2153.

424 R. Ibbett, F. Wortmann, K. Varga and K. C. Schuster, *Cellulose*, 2014, **21**, 139–152.

425 M. H. Zhang, X. M. Wang and R. Gazo, *Wood Fiber Sci.*, 2013, **45**, 423–428.

426 X. Z. Meng, M. Foston, J. Leisen, J. DeMartini, C. E. Wyman and A. J. Ragauskas, *Bioresour. Technol.*, 2013, **144**, 467–476.

427 L. Benamsili, J. P. Korb, G. Hamon, A. Louis-Joseph, B. Bouyssiere, H. G. Zhou and R. G. Bryant, *Energy Fuels*, 2014, **28**, 1629–1640.

428 P. Yang, H. K. Guo and D. Y. Yang, *Energy Fuels*, 2013, **27**, 5750–5756.

429 M. J. Zou, C. T. Wei, M. Zhang, J. Shen, Y. H. Chen and Y. Qi, *Energy Fuels*, 2013, **27**, 3699–3708.

430 H. Daigle, B. Thomas, H. Rowe and M. Nieto, *J. Geophys. Res.: Solid Earth*, 2014, **119**, 2631–2650.

431 J. Mitchell, J. Staniland, R. Chassagne, K. Mogensen, S. Frank and E. J. Fordham, *J. Pet. Sci. Eng.*, 2013, **108**, 14–21.

432 W. Hübner, *J. Appl. Geophys.*, 2014, **104**, 97–105.

433 M. Eslami, A. Kadkhodaie-Ilkhchi, Y. Sharghi and N. Golsanami, *J. Pet. Sci. Eng.*, 2013, **111**, 50–58.

434 R. Dlugosch, T. Günther, M. Müller-Petke and U. Yaramanci, *Geophysics*, 2013, **78**, EN55–EN64.

435 H. H. Tian, C. F. Wei, H. Z. Wei, R. T. Yan and P. Chen, *Appl. Magn. Reson.*, 2014, **45**, 49–61.

436 K. Dlubac, R. Knight and K. Keating, *Near Surf. Geophys.*, 2014, **12**, 219–230.

437 K. Keating and S. Falzone, *Geophysics*, 2013, **78**, D461–D472.

438 R. Dlugosch, T. Günther, M. Müller-Petke and U. Yaramanci, *Near Surf. Geophys.*, 2014, **12**, 231–241.

439 M. Müller-Petke, J. O. Walbrecker and R. Knight, *Geophysics*, 2013, **78**, EN83–EN94.

440 C. L. Cooper, T. Cosgrove, J. S. van Duijneveldt, M. Murray and S. W. Prescott, *Langmuir*, 2013, **29**, 12670–12678.

441 S. Y. Kim and C. F. Zukoski, *Macromolecules*, 2013, **46**, 6634–6643.

442 A. S. Kedir, J. G. Seland, A. Skauge and T. Skauge, *Energy Fuels*, 2014, **28**, 2948–2958.

443 B. L. T. Lau, W. C. Hockaday, K. Ikuma, O. Furman and A. W. Decho, *Colloids Surf., A*, 2013, **435**, 22–27.

444 M. L. Jespersen, P. A. Mirau, E. D. Von Meerwall, H. Koerner, R. A. Vaia, N. J. Fernandes and E. P. Giannelis, *Macromolecules*, 2013, **46**, 9669–9675.

445 G. Uccello-Barretta, F. Balzano, F. Aiello, A. Senatore, A. Fabiano and Y. Zambito, *Int. J. Pharm.*, 2014, **461**, 489–494.

446 H. Kato, A. Nakamura and M. Horie, *RSC Adv.*, 2014, **4**, 2129–2136.

447 K. W. Wang, H. Peng, K. J. Thurecht, S. Puttick and A. K. Whittaker, *Polym. Chem.*, 2014, **5**, 1760–1771.

448 A. de Vries, R. Moonen, M. Yildirim, S. Langereis, R. Lamerichs, J. A. Pikkemaat, S. Baroni, E. Terreno, K. Nicolay, G. J. Strijkers and H. Grull, *Contrast Media Mol. Imaging*, 2014, **9**, 83–91.

449 N. Wang, Y. R. Jin, F. Y. Jiang, J. F. Zeng, M. Y. Gao, Y. L. Wu, H. Deng, G. L. Zheng, J. Li, Y. F. Chen, Y. Tian, Y. F. Ren, G. H. Chen and D. N. Zheng, *IEEE Trans. Appl. Supercond.*, 2013, **23**, 1602504.

450 S. Kenouche, J. Larionova, N. Bezzi, Y. Guari, N. Bertin, M. Zanca, L. Lartigue, M. Cieslak, C. Godin, G. Morrot and C. Goze-Bac, *Powder Technol.*, 2014, **255**, 60–65.
451 K. G. B. Alves, C. A. S. Andrade, S. L. Campello, R. E. de Souza and C. P. de Melo, *J. Appl. Polym. Sci.*, 2013, **128**, 3170–3176.
452 N. Rocha, J. Mendes, L. Duraes, H. Maleki, A. Portugal, C. F. G. C. Geraldes, A. Serra and J. Coelho, *J. Mater. Chem. B*, 2014, **2**, 1565–1575.
453 M. Menelaou, K. Georgoula, K. Simeonidis and C. Dendrinou-Samara, *Dalton Trans.*, 2014, **43**, 3626–3636.
454 B. Issa, I. M. Obaidat, R. H. Hejasee, S. Qadri and Y. Haik, *J. Magn. Reson. Imaging*, 2014, **39**, 648–655.
455 B. Xu, J. Leisen and H. W. Beckham, *Nanoscale*, 2014, **6**, 1318–1322.
456 M. A. Scorciapino, R. Sanna, A. Ardu, F. Orru, M. Casu, A. Musinu and C. Cannas, *J. Colloid Interface Sci.*, 2013, **407**, 67–75.
457 L. Calucci, G. Ciofani, V. Mattoli, B. Mazzolai, A. Boni and C. Forte, *J. Phys. Chem. C*, 2014, **118**, 6473–6479.

Solid state NMR spectroscopy

A. E. Aliev*[a] and R. V. Law*[b]

DOI: 10.1039/9781782622758-00294

Recent advances and applications achieved in the area of solid-state NMR based on the literature published between July 2013 and June 2014 have been collected in this contribution, which considers review articles on both general and specific aspects of solid-state NMR spectroscopy and its applications, as well as experimental and computational developments and novel applications that are of general methodological interest. Various examples of solid-state NMR applications are collected in the final section, which is divided into 11 subsections depending on the type of the material studied: (1) organic solids; (2) pharmaceutical and biomedical applications; (3) amino acids and peptides; (4) proteins; (5) membrane proteins and lipids; (6) polymers; (7) carbonaceous materials and soils; (8) organometallic and coordination compounds; (9) glasses and amorphous solids; (10) surface science and catalysis, and (11) inorganic and other related solids.

1 Introduction

This contribution outlines important recent advances and applications achieved in the area of solid-state NMR based on the literature published between July 2013 and June 2014.

The format used in this report is similar to that used in the previous years. Section 2 of the report includes review articles on both general and specific aspects of solid-state NMR spectroscopy and its applications. Experimental and computational developments and novel applications that are of methodological interest are arranged in Section 3.

Various examples of solid-state NMR applications are collected in the final Section 4. This section is divided into subsections depending on the type of the material studied: (4.1) organic solids; (4.2) pharmaceutical and biomedical applications; (4.3) amino acids and peptides; (4.4) proteins; (4.5) membrane proteins and lipids; (4.6) polymers; (4.7) carbonaceous materials and soils; (4.8) organometallic and coordination compounds; (4.9) glasses and amorphous solids; (4.10) surface science and catalysis, and (4.11) inorganic and other related solids.

2 Reviews

Although solid-state NMR provides detailed information about molecular structure and dynamics, the small spin population differences between pairs of spin states that give rise to NMR transitions make it an inherently insensitive spectroscopic technique in terms of signal acquisition.[1] Scientists have continuously addressed this issue *via* improvements in NMR hardware and probes, increases in the strength of the magnetic field, and the development of innovative pulse sequences and acquisition

[a]*University College London, 20 Gordon Street, London WC1H 0AJ, UK. E-mail: a.e.aliev@ucl.ac.uk*

[b]*Imperial College London, South Kensington Campus, London SW7 2AZ, UK. E-mail: r.law@imperial.ac.uk*

methodologies. As a result, researchers can now study NMR active nuclides previously thought to be unobservable for routine examination *via* solid-state NMR. Several factors can make it extremely challenging to detect signal or acquire spectra using solid-state NMR: (i) low gyromagnetic ratios (*i.e.* low Larmor frequencies), (ii) low natural abundances or dilution of the nuclide of interest (*e.g.* metal nuclides in proteins or in organometallic catalysts supported on silica), (iii) inconvenient relaxation characteristics (*e.g.* very long longitudinal or very short transverse relaxation times), and/or (iv) extremely broad powder patterns arising from large anisotropic NMR interactions. Efficient acquisition of broad NMR powder patterns for a variety of spin 1/2 and quadrupolar (spin > 1/2) nuclides has been investigated in detail.[1] Traditionally, researchers have used the term "wideline" NMR to refer to experiments yielding broad ^{1}H and ^{2}H solid-state NMR spectra ranging from tens of kHz to ~250 kHz in breadth. With modern FT NMR hardware, uniform excitation in these spectral ranges is relatively easy, allowing for the acquisition of high quality spectra. However, spectra that range in breadth from ~250 kHz to tens of MHz cannot be uniformly excited with conventional, high-power rectangular pulses. Rather, researchers must apply special methodologies to acquire such spectra. The term ultra-wideline NMR (UWNMR) spectroscopy was introduced to describe this set of methodologies. Schurko describes recent developments in pulse sequences and strategies for the efficient acquisition of UWNMR spectra.[1] After an introduction to anisotropically broadened NMR patterns, the author presents a brief history of methods used to acquire UWNMR spectra. New acquisition methodologies are then discussed, including the acquisition of CPMG echo trains and the application of pulses capable of broadband excitation and refocusing. Finally, several applications of UWNMR methods that use these broadband pulses have been presented.[1]

Saalwaechter's review addresses the NMR spectroscopic study of molecular structure and dynamics by way of homonuclear dipole–dipole couplings using their orientation and direct distance dependence.[2] The study of homonuclear couplings as opposed to heteronuclear couplings poses specific challenges, as two like spins cannot be independently manipulated easily, which means that simple shift-refocusing concepts by using hard pulses cannot be used to cope with potentially large chemical-shift dispersions at the high fields used nowadays. In addition, the noncommutativity of the different pair Hamiltonians in a multispin system leads to complications associated with the isolation of specific pair couplings while minimizing the influence of the other spins. Recent advances in determining homonuclear dipole–dipole coupling constants are reviewed by Saalwaechter, with the emphasis on the use of double-quantum spectroscopy approaches and their similarity to the popular heteronuclear rotational-echo double-resonance experiment. Particular emphasis is put on corrections for the influence of transverse relaxation effects on the measured data, and the handling of distribution effects as well as potential dynamic heterogeneities in complex substances.[2]

WURST pulses (wideband, uniform rate, smooth truncation) were first introduced by Kupce and Freeman as a means of achieving broadband adiabatic inversion of magnetisation for solution-state ^{13}C decoupling at high magnetic field strengths. In more recent years these pulses have found use in an increasingly diverse range of applications in solid-state NMR.[3] The article by O'Dell reviews a number of recent developments that take advantage of WURST pulses, including broadband excitation, refocusing and cross polarisation for the acquisition of ultra-wideline powder patterns, signal enhancement for half-integer and integer spin quadrupolar nuclei, spectral editing, direct and indirectly observed ^{14}N overtone MAS, and symmetry-based homonuclear recoupling. Simple mathematical descriptions of WURST pulses and some brief theory behind their operation in the adiabatic and non-adiabatic regimes are provided, and various practical considerations for their use are also discussed.[3]

Porous media have become a major focus of materials science in recent years, due to their versatile properties, such as high surface area, low specific weight, high surface functionality, and the ability to customize their surface properties.[4] Applications of porous media range from catalysis to separation media to gas storage. All of the mentioned applications involve the introduction of guest molecules into the pores. For efficient application of the materials, it is essential to know the behaviour of these introduced molecules in the confined state. Solid state NMR gives a unique insight into the dynamics, the guest–host interactions, and the binding sites of porous materials and is probably the most powerful characterization method for probing a huge variety of real-life systems. Recent results in research of microporous zeolites and periodically mesoporous silica (PMS) materials using NMR are highlighted in the review article by Werner *et al.*[4]

The link of structure and dynamics of biomolecules and their complexes to their function and to many cellular processes has driven the quest for their detailed characterization by a variety of biophysical techniques.[5] Solid-state MAS NMR spectroscopy provides rich information on the structural properties of such systems and in particular contributes invaluable information on non-soluble, large molecular-weight and non-crystalline biomolecules. The review article by Goldbourt summarizes the recent progress that has been made in the characterization of macromolecular assemblies, viruses, membrane proteins, amyloid fibrils, protein aggregates and more by solid-state MAS NMR.[5]

Solid-state NMR spectroscopy has matured to the point that it is possible to determine the structure of proteins in immobilized states, such as within microcrystals or embedded in membranes.[6] Currently, researchers continue to develop and apply NMR techniques that can deliver site-resolved dynamic information toward the goal of understanding protein function at the atomic scale. As a widely used approach, researchers have mostly measured longitudinal relaxation times, which, like in solution-state NMR, are sensitive to picosecond and nanosecond motions, and motionally averaged dipolar couplings, which provide an integral amplitude of all motions with a correlation time of up to a few microseconds. While overall Brownian tumbling in solution mostly precludes access to slower internal dynamics, dedicated solid-state NMR approaches are now emerging as powerful new options. Krushelnitsky, Reichert and Saalwachter give an overview of the classes of solid-state NMR experiments that have expanded the accessible range correlation times from microseconds to many milliseconds.[6] The measurement of relaxation times in the rotating frame, $T_{1\rho}$, now allows researchers to access the microsecond range. Using the results of the theoretical work, researchers can now quantitatively analyze this data to distinguish relaxation due to chemical shift anisotropy from that due to dipole–dipole couplings. Off-resonance irradiation allows researchers to extend the frequency range of such experiments. Krushelnitsky, Reichert and Saalwachter have built multidimensional analogues of T_2-type or line shape experiments using variants of the dipolar-chemical shift correlation (DIPSHIFT) experiment that are particularly suited to extract intermediate time scale motions in the millisecond range. The NH dipolar coupling offers a useful probe of local dynamics, especially with proton-depleted samples that suppress the adverse effect of strong proton dipolar couplings. The authors demonstrate how these techniques have provided a concise picture of the internal dynamics in a popular model system, the SH3 domain of α-spectrin. T_1-based methods have shown that large-amplitude bond orientation fluctuations in the picosecond range and slower 10 ns low-amplitude motions coexist in these structures.[6]

Solid-state MAS NMR spectroscopy is emerging as an important technique for the determination of three-dimensional structures of biological molecules and for the characterization of their dynamics.[7] While there is an established suite of MAS solid-state NMR experiments for protein structure determination in small- and medium-sized proteins, these methods face many challenges in large systems. Ladizhansky reviews recent progress in MAS NMR spectroscopy, specifically focusing on the emerging developments aimed at improving the sensitivity and resolution of solid-state NMR that are likely to determine its future applications. These developments include sample preparation and isotopic labelling strategies, fast MAS, proton detection, and paramagnetic NMR spectroscopy.[7]

Oriented solid-state NMR (O-ssNMR) spectroscopy is a major technique for the high-resolution analysis of the structure and topology of transmembrane proteins in native-like environments.[8] Unlike MAS techniques, O-ssNMR spectroscopy requires membrane protein preparations that are uniformly oriented (mechanically or magnetically) so that anisotropic NMR parameters, such as dipolar and chemical shift interactions, can be measured to determine structure and orientation of membrane proteins in lipid bilayers. Traditional sample preparations involving mechanically aligned lipids often result in short relaxation times which broaden the ^{15}N resonances and encumber the manipulation of nuclear spin coherences. The introduction of lipid bicelles as membrane mimicking systems has changed this scenario, and the more favourable relaxation properties of membrane protein ^{15}N and ^{13}C resonances make it possible to develop new, more elaborate pulse sequences for higher spectral resolution and sensitivity. Gopinath, Mote and Veglia describe recent progress in the optimization of O-ssNMR pulse sequences.[8] They present the theory behind these experiments, demonstrate their application to small and medium size proteins, and describe the technical details for setting up these new experiments on the new generation of NMR spectrometers.[8]

Many cellular events involve attachment of proteins to the surfaces of rigid or semi-rigid solid materials, such as the inorganic materials in the extracellular matrix of hard tissue, and the macromolecular scaffolds made of actin and tubulin filaments in the cytoskeleton.[9] Understanding these processes on a fundamental level will have far-reaching repercussions for the design of biomaterials, biomedical research, and biomineralization. Goobes has reviewed numerous studies on structural changes experienced by proteins as they adhere to surfaces. It was found that there are only a few examples in which detailed views of protein conformation and alignment on surfaces were measured. As pointed by Goobes, modern multidimensional solid-state NMR spectroscopy is capable of unveiling molecular details of these processes and shed light on many fundamental questions related to recognition of surfaces by biomolecules. Targeting these questions is currently at the focal point of many research fields and can lead to insights and breakthroughs in biotechnology and in biomimetic material design.[9]

In the review by Bonhomme, Gervais and Laurencin, the latest developments in solid state NMR are presented in the field of organic–inorganic (O/I) materials (or hybrid materials).[10] Such materials involve mineral and organic (including polymeric and biological) components, and can exhibit complex O/I interfaces. Hybrids are currently a major topic of research in nanoscience, and solid-state NMR is a pertinent spectroscopic tool of investigation. Its versatility allows the detailed description of the structure and texture of such complex materials. The article is divided into two main parts. In the first part, recent NMR methodological/instrumental developments are presented in connection with hybrid materials. In the second part, an exhaustive overview of the major classes of O/I materials and their NMR characterization is presented.[10]

NMR spectroscopy is a well-established analytical technique in biofuel research. Over the past few decades, lignocellulosic biomass and its conversion to supplement or displace non-renewable feedstocks has attracted increasing interest.[11] The application of solid-state NMR spectroscopy has long been seen as an important tool in the study of cellulose and lignocellulose structure, biosynthesis, and deconstruction, especially considering the limited number of effective solvent systems and the significance of plant cell wall three-dimensional microstructure and component interaction to conversion yield and rate profiles. The article by Foston reviews common and recent applications of solid-state NMR spectroscopy methods that provide insight into the structural and dynamic processes of cellulose that control bulk properties and biofuel conversion.[11]

Unlike the long-range order of ideal crystalline structures, local order is an intrinsic characteristic of real materials and often serves as the key to the tuning of their properties and their final applications.[12] Although researchers can easily assess local ordering using two-dimensional imaging techniques with resolution that approaches the atomic level, the diagnosis, description, and qualification of local older in three dimensions is much more challenging. Solid-state NMR and its panel of continually developing instruments and methods enable the local, atom-selective characterization of structures and assemblies ranging from the atomic to the nanometer length scales. By making use of the indirect *J*-coupling that distinguishes chemical bonds, researchers can use solid-state NMR to characterize a variety of materials, ranging from crystalline compounds to amorphous or glassy materials. Massiot *et al.* give examples of materials featuring either chemical disorder in a topological order or topological disorder with local chemical order. For glasses, they show that geometric and chemical contributions can be separated to the local order by identifying structural motifs with a viewpoint that extends from the atomic sole up to the nanoscale. As identified by solid-state NMR, the local structure of amorphous materials or glasses consists of well-identified structural entities up to at least the nanometer scale. Instead of speaking of disorder, they propose a new description for these structures as a continuous assembly of locally defined structures, an idea that draws on the concept of locally favoured structures (IFS) introduced by Tanaka and co-workers. This idea provides a comprehensive picture of amorphous structures based on fluctuations of chemical composition and structure over different length scales. It is expected that these local or molecular insights will allow researchers to consider key questions related to nucleation and crystallization, as well as chemically (spinodal decomposition) or density-driven (polyamorphism) phase separation, which could lead to future applications in a variety of materials.[12]

Microporous materials, having pores and channels on the same size scale as small to medium molecules, have found many important applications in current technologies, including catalysis, gas separation and drug storage and delivery.[13] Many of their properties and functions are related to their detailed local structure, such as the type and distribution of active sites within the pores, and the specific structures of these active sites. Solid-state NMR spectroscopy has a strong track record of providing the requisite detailed atomic-level insight into the structures of microporous materials, in addition to being able to probe dynamic processes occurring on timescales spanning many orders of magnitude (*i.e.* from ps to s). Ashbrook and Dawson provide a brief review of some of the basic experimental approaches used in solid-state NMR spectroscopy of microporous materials, and then discuss some more recent advances in this field, particularly those applied to the study of crystalline materials such as zeolites and metal–organic frameworks. These advances include improved software for aiding spectral interpretation, the development of the NMR crystallography approach to structure determination, new routes for the synthesis of isotopically-labelled materials, methods for the characterisation of host–guest interactions, and methodologies suitable for observing NMR spectra of paramagnetic microporous materials. The authors also discuss possible future directions, which are expected to have strong impact on the field over the coming years.[13]

In recent years, solid-state NMR spectroscopy has evolved into an important characterization tool for the study of solid catalysts and chemical processes on their surface.[14] This interest is mainly triggered by the need of environmentally benign organic transformations ("green chemistry"), which has resulted in a large number of new catalytically active hybrid materials, which are organized on the meso- and nanoscale. Typical examples of these catalysts are supported homogeneous transition metal catalysts or transition metal nanoparticles (MNPs). Solid-state NMR spectroscopy is able to characterize both the structures of these materials and the chemical processes on the catalytic surface. Gutmann *et al.* present recent trends both on the characterization of immobilized homogeneous transition metal catalysts and on the characterization of surface species on transition metal surfaces.[14]

Solid-state NMR is also used for the structural characterization and dynamics study of various multifunctional materials.[15] The review article by Li and Deng covers recent solid-state NMR studies of topology structures, crystallization processes, host–guest interactions, acidities and catalytic reaction mechanisms of various zeolites. It is shown that new insights into the relationship between zeolite structures and properties at the atomic level could benefit greatly from the new developments in solid-state NMR methodology and DFT calculations. The spatial connections, distance information and anisotropic chemical shifts of silicon sites determined by ^{29}Si MAS NMR can serve as structural constraints to resolve the topology structure of zeolites. ^{129}Xe NMR can be employed to investigate the cage and channel structure and communication in zeolites. The host–guest interactions between zeolite frameworks and confined molecules have been characterized by 2D heteronuclear correlation experiments. The acidity property of zeolites (including acid type, strength, concentration as well as distribution) can be well understood through the combination of solid-state NMR measurements and DFT calculations. In particular, information about spatial proximities among various acid sites in zeolites is available from two-dimensional double quantum MAS NMR experiments. The crystallization and catalytic reaction mechanisms can be elucidated by monitoring the evolution of crystallization and reaction processes, respectively, through solid-state NMR spectroscopy.[15]

3 Experimental and computational developments

3.1 Proton NMR experiments

When applied to biomolecules, solid-state NMR suffers from low sensitivity and resolution. The major obstacle to applying proton detection in the solid state is the proton dipolar network, and deuteration can help avoid this problem. In the past, researchers primarily focused on the investigation of exchangeable protons in these systems. Asami and Reif present NMR spectroscopic strategies that allow researchers to observe aliphatic non-exchangeable proton resonances in proteins with high sensitivity and resolution. They highlight spectroscopic approaches for obtaining resonance assignments, a prerequisite for any study of structure and dynamics of a protein by NMR spectroscopy. Because of the dilution of the proton spin system in the solid state, solution-state NMR (HCCH)-^{1}H–^{1}H type strategies cannot easily be transferred to these experiments. Instead, (^{1}H)(CCH)-^{1}H, (CCH)-^{1}H, (HCC)-^{1}H or (^{2}H)(CCH)-^{1}H type experiments were pursued. In protonated samples, distance restraints were obtained for structure calculations from samples grown in bacteria in media containing [1,3]-^{13}C-glycerol, [2]-^{13}C-glycerol, or selectively enriched glucose to dilute the ^{13}C spin system. The isotopic labelling scheme allowed to measure the long-range contacts among aliphatic protons, which can then serve as restraints for the three-dimensional structure calculation of a protein. Due to the high gyromagnetic ratio of protons, longer range contacts are more easily accessible for these nuclei than for carbon nuclei in homologous experiments.[16]

A novel method to realize rapid repetition of ^{1}H NMR experiments at ultra-fast MAS frequencies is demonstrated by Ye *et al.* The ultra-fast MAS at 110 kHz slows the ^{1}H–^{1}H spin diffusion, leading to variations of ^{1}H T_1 relaxation times from atom to atom within a molecule. The different relaxation behaviour is averaged by applying ^{1}H–^{1}H recoupling during relaxation delay even at ultra-fast MAS, reducing the optimal relaxation delay to maximize the signal to noise ratio. The way to determine optimal relaxation delay for arbitrary relaxation curve is shown. The reduction of optimal relaxation delay by rf driven recoupling (RFDR) was demonstrated on powder samples of glycine and ethenzamide with one and multi-dimensional NMR measurements.[17]

The hydrogen bond network of three polymorphs (1α, 1β, and 1γ) and one solvate form ($1\cdots H_2O$) arising from the hydration-dehydration process of the Ru(II) complex [(*p*-cymene)Ru(κN-INA)Cl-2] (where INA is isonicotinic acid), has been ascertained by means of 1D and 2D double quantum ^{1}H CRAMPS and ^{13}C CP MAS solid-state NMR experiments. The resolution improvement provided by homonuclear decoupling pulse sequences, with respect to fast MAS experiments, has been highlighted. The solid-state structure of 1γ has been fully characterized by combining powder X-ray diffraction, solid-state NMR, and periodic plane-wave first-principles calculations. None of the forms show the expected supramolecular cyclic dimerisation of the carboxylic functions of INA, because of the presence of Cl atoms as strong hydrogen bond acceptors. The hydration–dehydration process of the complex has been discussed in terms of structure and hydrogen bond rearrangements.[18]

Proton chemical shifts are a rich probe of structure and hydrogen-bonding environments in organic and biological molecules. Until recently, measurements of ^{1}H chemical-shift tensors have been restricted to either solid systems with sparse proton sites or were based on the indirect determination of anisotropic tensor components from cross-relaxation and liquid-crystal experiments. A MAS approach has been introduced that permits site-resolved determination of chemical shift anisotropy tensors of protons forming chemical bonds with labelled spin 1/2 nuclei in fully protonated solids with multiple sites, including organic molecules and proteins. This approach, originally introduced for the measurements of chemical shift tensors of amide protons, is based on three *RN*-symmetry-based experiments, from which the principal components of the ^{1}H chemical shift tensor can be reliably extracted by simultaneous triple fit of the data. The new approach was expanded to a much more challenging system involving aliphatic and aromatic protons. A review of prior work on experimental NMR spectroscopy and computational quantum chemical approaches for measurements of ^{1}H chemical shift tensors and relating these to the electronic structures is also presented. The authors describe their experimental results on U–^{13}C,^{15}N-labelled histidine and demonstrate that ^{1}H chemical shift tensors can be reliably determined for the (HN)-^{1}H–^{15}N and (HC)-^{1}H–^{13}C spin pairs in cationic and neutral forms of histidine. Finally, they demonstrate that the experimental ^{1}H(C) and ^{1}H(N) chemical-shift tensors are in agreement with DFT calculations; therefore, establishing the usefulness of the method for the characterization of structures and hydrogen-bonding environments in organic and biological solids.[19]

The paramagnetic complex bis(oxazolinylphenyl)amine-Fe(III)Cl_2 is investigated by means of solid-state proton NMR at 18.8 T (800 MHz) using MAS at 65 kHz. It has been shown that spin echoes that are excited and refocused by combs of rotor-synchronized pulses in the manner of 'Delays Alternating with Nutation for Tailored Excitation' (DANTE) allow one to characterize different chemical environments that severely overlap in conventional MAS spectra. Such sequences combine two apparently contradictory features: an overall bandwidth exceeding several MHz, and very selective irradiation of a few kHz within inhomogeneously broadened sidebands. It is shown that the experimental hyperfine interactions correlate well with DFT calculations.[20]

Sinnige *et al.* have shown that selective labelling of proteins with protonated amino acids embedded in a perdeuterated matrix, dubbed 'proton clouds', provides general access to long-range contacts between non-exchangeable side chain protons in proton-detected solid-state NMR, which is important to study protein tertiary structure. Proton-cloud labelling significantly improves spectral resolution by simultaneously reducing proton line width and spectral crowding despite a high local proton density in clouds. The approach is amenable to almost all canonical amino acids. The new method is demonstrated on ubiquitin and the β-barrel membrane protein BamA.[21]

3.2 Decoupling experiments

Sensitivity and resolution are the two fundamental obstacles to extending solid-state NMR to large protein systems. A novel long-observation-window band-selective homonuclear decoupling (LOW BASHD) scheme has been introduced that increases resolution up to a factor of 3 and sensitivity up to 1.8 by decoupling backbone α-carbon (C_α) and carbonyl (C') nuclei in U–^{13}C-labelled proteins during direct ^{13}C acquisition. This approach introduces short (<200 μs) pulse breaks into much longer (~8 ms) sampling windows to efficiently refocus the *J*-coupling interaction during detection while avoiding the deleterious effects on sensitivity inherent in rapid stroboscopic band-selective homonuclear decoupling techniques. A significant advantage of LOW-BASHD detection is that it can be directly incorporated into existing correlation methods, as illustrated for 2D CACO, NCO, and NCA correlation spectroscopy applied to the β1 immunoglobulin binding domain of protein G and 3D CBCACO correlation spectroscopy applied to the α-subunit of tryptophan synthase.[22]

Halse, Schlagnitweit and Emsley have investigated the optimal implementation of windowed LG4 decoupling for the direct acquisition of high-resolution ^{1}H solid-state NMR spectra in the moderate MAS regime ($\omega_r < 2\pi \times 35$ kHz). They have determined how the optimal value of the azimuth of the LG rotation axis is affected by the choice of windowing parameters. It was found that for both the windowless and windowed implementation of LG4, the optimal value of α is that for which the chemical-shift scaling factor is approximately 0.43. The authors then provide an analytical expression for estimating the chemical shift scaling factor as a function of both α and the windowed decoupling parameters, which allows the optimal value of α to be easily determined for any implementation of LG4. They also introduce a simple and general method for improving the performance of any phase-modulated homonuclear dipolar decoupling sequence by compensating for pulse imperfection errors that accumulate in windowed spectra due to the repeated switching of the RF irradiation. This method involves the introduction of short (<0.5 μs) compensation pulses at the start and end of each decoupling pulse, the duration and phase of which are chosen to cancel the effects of the amplitude and phase transients that arise as a result of the RF switching. The authors demonstrate how this method can be implemented for homonuclear dipolar decoupling sequences such as wPMLG and wLG4 to reduce the chemical-shift scaling factor distortions generated by phase transients such that high-quality, high-resolution ^{1}H NMR spectra can be acquired at arbitrary RF offsets.[23]

Equbal *et al.* have presented a simple refocused modification, rTPPM, of the Two-Pulse Phase-Modulation (TPPM) heteronuclear decoupling method, which improves decoupling and makes the sequence much more robust with respect to essential experimental parameters. The modified sequence is compared with the established TPPM sequence and a variety of other decoupling sequences at low to moderate MAS frequencies. Simulations are shown to compare TPPM and rTPPM with respect to various experimental parameters. The observations from simulations are corroborated with experimental findings at two spinning frequencies on U–^{13}C-glycine and U–^{13}C-L-histidine · HCl · H_2O.[24] Rapid developments in solid-state NMR methodology have boosted this technique into a highly versatile tool for structural biology. The invention of increasingly advanced rf pulse sequences that take advantage of better hardware and sample preparation have played an important part in these advances. In the development of these new pulse sequences, researchers have taken advantage of analytical tools, such as average Hamiltonian theory or lately numerical methods based on optimal control theory. Bjerring *et al.* focus on the interplay between these strategies in the systematic development of simple pulse sequences that combines continuous wave (CW) irradiation with short pulses to obtain improved rf pulse, recoupling, sampling, and decoupling performance. Their initial work on this problem focused on the challenges associated with the increasing use of fully or partly deuterated proteins to obtain high-resolution, liquid-state-like solid-state NMR spectra. The authors exploit the overwhelming presence of ^{2}H in such samples as a source of polarization and to gain structural information. The ^{2}H nuclei possess dominant quadrupolar couplings which complicate even the simplest operations, such as rf pulses and polarization transfer to surrounding nuclei. Using optimal control and easy analytical adaptations, they demonstrate that a series of rotor synchronized short pulses may form the basis for essentially ideal rf pulse performance. Using similar approaches, they design ^{2}H to ^{13}C polarization transfer experiments that increase the efficiency by one order of magnitude over standard cross polarization experiments. The new experiment significantly improves ^{1}H–^{15}N and ^{15}N–^{13}C transfers, which are key elements in the vast majority of biological solid-state NMR experiments. In addition, the authors demonstrate how interleaved sampling of spectra exploiting polarization from ^{1}H and ^{2}H nuclei can substantially enhance the sensitivity of such experiments.[25]

3.3 Cross-polarisation

Double cross polarization (DCP) has been widely used for heteronuclear polarization transfer between ^{13}C and ^{15}N in solid-state MAS NMR. However, DCP is so sensitive to experimental settings that small variations or deviations in RF fields would deteriorate its efficiency. Asymmetric simultaneous phase-inversion cross polarization (referred as aSPICP) has been presented for selective polarization transfer between low-γ ^{13}C and ^{15}N spins. The authors have demonstrated through simulations and experiments using biological solids that the asymmetric duration in the simultaneous phase-inversion cross polarization scheme leads to efficient polarization transfer between ^{13}C and ^{15}N even with large chemical shift anisotropies in the presence of B_1 field variations or mismatch of the Hartmann–Hahn conditions. This is expected to be useful for long experiments such as membrane protein studies at high fields.[26]

A simple new method is presented that yields quantitative solid-state MAS ^{13}C NMR spectra of organic materials with good signal-to-noise ratios. It achieves long (>10 ms) CP from ^{1}H without significant magnetization losses due to relaxation and with a moderate duty cycle of the rf irradiation, by multiple 1 ms CP periods alternating with ^{1}H spin–lattice relaxation periods that repolarize the protons. The new method incorporates previous techniques that yield less distorted CP MAS spectra, such as a linear variation ("ramp") of the rf field strength, and it overcomes their main limitation, which is $T_{1\rho}$ relaxation of the spin-locked ^{1}H magnetization. The ramp of the rf field strength and the asymptotic limit of cross polarization makes the spectral intensity quite insensitive to the exact field strengths used. The new multiCP pulse sequence is a "drop-in" replacement for previous CP methods and produces no additional data-processing burden. Compared to the only reliable quantitative ^{13}C NMR method for unlabelled solids previously available, namely direct-polarization NMR, the measuring time is reduced by more than a factor of 50, enabling higher-throughput quantitative NMR studies. The new multiCP technique is validated with 14 kHz MAS on amino-acid derivatives, plant matter, a highly aromatic humic acid, and carbon materials made by low-temperature pyrolysis.[27]

Nitrogen is an element of utmost importance in chemistry, biology and materials science. Of its two NMR-active isotopes, ^{14}N and ^{15}N, solid-state NMR experiments are rarely conducted upon the former, due to its low gyromagnetic ratio and broad powder patterns arising from first-order quadrupolar interactions. Harris *et al.* have proposed a methodology for the rapid acquisition of high quality ^{14}N solid-state NMR spectra that is easy to implement, and can be used for a variety of nitrogen-containing systems. They demonstrate that it is possible to dramatically enhance ^{14}N NMR signals in spectra of stationary, polycrystalline samples (*i.e.* amino acids and active pharmaceutical ingredients) by means of broadband cross polarization (CP) from abundant nuclei (*e.g.* ^{1}H). The BRoadband Adiabatic INversion Cross-Polarization (BRAIN-CP) pulse sequence is combined with other elements for efficient acquisition of ultra-wideline solid-state NMR spectra, including Wideband Uniform-Rate Smooth-Truncation (WURST) pulses for broadband refocusing, Carr–Purcell–Meiboom–Gill (CPMG) echo trains for T_2-driven S/N enhancement, and frequency-stepped acquisitions. The feasibility of utilizing the BRAIN-CP/WURST-CPMG sequence is tested for ^{14}N, with special consideration given to (i) spin-locking integer spin nuclei and maintaining adiabatic polarization transfer, and (ii) the effects of broadband polarization transfer on the overlapping satellite transition patterns. The BRAIN-CP experiments are shown to provide increases in signal-to-noise ranging from four to ten times and reductions of experimental times from one to two orders of magnitude compared to analogous experiments where ^{14}N nuclei are directly excited. Furthermore, patterns acquired with this method are generally more uniform than those acquired with direct excitation methods. The proposed method and its potential for probing a variety of chemically distinct nitrogen environments are also discussed.[28]

Polarization transfer between spin-1/2 nuclei and quadrupolar spin-1 nuclei such as ^{14}N in solid-state NMR is severely challenged by the typical presence of large quadrupole coupling interactions. This has effectively prevented the use of the abundant ^{14}N spin as a probe to structural information and its use as an element in multi-dimensional solid-state NMR correlation experiments for assignment and structural characterization. In turn, this has been a contributing factor to the extensive use of isotope labelling in biological solid-state NMR, where ^{14}N is replaced with ^{15}N. The alternative strategy of using the abundant ^{14}N spins calls for methods enabling efficient polarization transfer between ^{14}N and its binding partners. It has been shown that the recently introduced (RESPIRATION) CP transfer method can be optimized to achieve efficient $^{1}\mathrm{H} \leftrightarrow {}^{14}\mathrm{N}$ polarization transfer under MAS conditions. The method is demonstrated numerically and experimentally on powder samples of NH_4NO_3 and L-alanine.[29]

Establishing high-resolution structures of biological macromolecules in heterogeneous environments by MAS solid-state NMR is an important challenge where development of advanced experimental procedures is in high demand. Promising new methods take advantage of samples with extensive ^{2}H, ^{13}C, and ^{15}N isotope labelling, effectively diluting ^{1}H spins. In many cases, a sufficient amount of ^{1}H at exchangeable sites cannot be re-established during the purification procedure, hence it is necessary to exploit also the potential of ^{2}H as a starting point in pulse sequences, capitalizing on its short T_1 as compared to ^{13}C, and to detect carbon or proton spins as appropriate. Jain *et al.* have presented a new method that enables the required high-efficiency ^{2}H, ^{13}C, and ^{15}N polarization transfer to be accomplished under the limited ^{2}H rf power conditions using current ^{1}H, ^{2}H, ^{13}C and ^{15}N quadruple-resonance MAS NMR instrumentation.[30]

3.4 Multidimensional techniques

Two-dimensional indirectly detected through-space and through-bond ^{1}H–^{15}N solid-state NMR experiments utilizing fast MAS and homonuclear multipulse ^{1}H decoupling have been evaluated. It has been shown that remarkable efficiency of polarization transfer can be achieved at MAS rate of 40 kHz by both CP and INEPT, which makes these methods applicable for routine characterizations of natural abundance solids. The first measurement of 2D ^{1}H–^{15}N HETCOR spectrum of natural abundance surface species is also reported.[31]

Covariance spectroscopy (COV), a statistical method that provides increased sensitivity, can be applied to two-dimensional high-resolution solid-state NMR experiments, such as homonuclear spin-exchange spectroscopy. Lin and Opella describe applications of covariance (COV) spectroscopy to reconstruct spectra whose resonances have signal-to-noise ratios <10 and relatively broad linewidths. These include the use of sampling schemes based on the States method. By combining COV with other processing methods for non-uniform sampling (NUS), many different three-dimensional experiments can be performed with substantial increases in overall sensitivity. As an example, they show a three-dimensional homonuclear spin-exchange/separated-local-field (SLF) spectrum that enables the assignment of resonances and the measurement of structural restraints from a single experiment performed in a limited amount of time.[32]

Hou *et al.* have recently described a family of experiments for $R2_n^{\nu}$ Driven Spin Diffusion (RDSD) spectroscopy suitable for homonuclear correlation experiments under fast MAS conditions. In these RDSD experiments, since the broadened second-order rotational resonance conditions are dominated by the rf field strength and the phase shifts, as well as the size of reintroduced dipolar couplings, the different $R2_n^{\nu}$ sequences display unique polarization transfer behaviours and different recoupling frequency bandwidths. In the subsequent report, Hou *et al.* present a series of modified $R2_n^{\nu}$ sequences, dubbed *COmbined $R2_n^{\nu}$-Driven* (CORD), that yield broadband homonuclear dipolar recoupling and give rise to uniform distribution of cross peak intensities across the entire correlation spectrum. They report NMR experiments and numerical simulations demonstrating that these CORD spin diffusion sequences are suitable for broadband recoupling at a wide range of magnetic fields and MAS frequencies, including fast-MAS conditions ($\nu_r = 40$ kHz and above). Since these CORD sequences are largely insensitive to dipolar truncation, they are well suited for the determination of long-range distance constraints, which are indispensable for the structural characterization of a broad range of systems. Using U–^{13}C-alanine and U–^{13}C,^{15}N-histidine, Hou *et al.* show that under fast MAS conditions, the CORD sequences display polarization transfer efficiencies within broadband frequency regions that are generally higher than those offered by other existing spin diffusion pulse schemes. A 89-residue U–^{13}C,^{15}N-dynein light chain (LC8) protein has also been used to demonstrate that the CORD sequences exhibit uniformly high cross peak intensities across the entire chemical shift range.[33]

Shi *et al.* have recently presented band-selective homonuclear cross-polarization (BSH-CP) as an efficient method for CO-CA transfer in deuterated as well as protonated solid proteins. In the latest report they show how the BSH-CP CO-CA transfer block can be incorporated in a set of 3D solid-state NMR pulse schemes tailored for resonance assignment of proteins at high static magnetic fields and moderate MAS rates. Due to the achieved excellent transfer efficiency of 33% for BSH-CP, a complete set of 3D spectra needed for unambiguous resonance assignment could be rapidly recorded within 1 week for the model protein ubiquitin. Thus it is expected that BSH-CP could replace the typically used CO-CA transfer schemes in well-established 3D solid-state NMR approaches for resonance assignment of solid biomolecules.[34]

3.5 Dynamic nuclear polarization

Thurber and Tycko have reported solid-state ^{13}C and ^{1}H NMR experiments with MAS on frozen solutions containing nitroxide-based paramagnetic dopants that indicate significant perturbations of nuclear spin polarizations without microwave irradiation. At temperatures near 25 K, ^{1}H and cross-polarized ^{13}C NMR signals from ^{15}N, ^{13}C-labelled L-alanine in trinitroxide-doped glycerol/water are reduced by factors as large as six compared to signals from samples without nitroxide doping. Without MAS or at temperatures near 100 K, differences between signals with and without nitroxide doping are much smaller. They have attributed most of the reduction of NMR signals under MAS near 25 K to nuclear spin depolarization through the cross-effect dynamic nuclear polarization mechanism, in which three-spin flips drive nuclear polarizations toward equilibrium with spin polarization differences between electron pairs. When T_{1e} is sufficiently long relative to the MAS rotation period, the distribution of electron spin polarization across the nitroxide electron paramagnetic resonance lineshape can be very different from the corresponding distribution in a static sample at thermal equilibrium, leading to the observed effects. They describe three-spin and 3000-spin calculations that qualitatively reproduce the experimental observations.[35]

Thanks to instrumental and theoretical development, notably the access to high-power and high-frequency microwave sources, high-field dynamic nuclear polarization on solid-state NMR currently appears as a promising solution to enhance nuclear magnetization in many different types of systems. In MAS DNP experiments, systems of interest are usually dissolved or suspended in glass-forming matrices doped with polarizing agents and measured at low temperature (down to $\sim$100 K). Takahashi *et al.* discuss the influence of sample conditions (radical concentration, sample temperature, *etc.*) on DNP enhancements and various nuclear relaxation times which affect the absolute sensitivity of DNP spectra, especially in multidimensional experiments. Furthermore, DNP-enhanced solid-state NMR experiments performed at 9.4 T are complemented by high-field CW EPR measurements performed at the same magnetic field. Microwave absorption by the DNP glassy matrix is observed even below the glass transition temperature caused by softening of the glass. Shortening of electron relaxation times due to glass softening and its impact in terms of DNP sensitivity is discussed.[36]

Smith *et al.* present electron paramagnetic resonance experiments for which solid effect dynamic nuclear polarization transitions were observed indirectly *via* polarization loss on the electron. This use of indirect observation allows characterization of the DNP process close to the electron. Frequency profiles of the electron-detected solid effect obtained using trityl radical showed intense saturation of the electron at the usual solid effect condition, which involves a single electron and nucleus. However, higher order solid effect transitions involving two, three, or four nuclei were also observed with surprising intensity, although these transitions did not lead to bulk nuclear polarization-suggesting that higher order transitions are important primarily in the transfer of polarization to nuclei nearby the electron. Similar results were obtained for the SA-BDPA radical where strong electron-nuclear couplings produced splittings in the spectrum of the indirectly observed solid effect conditions. Observation of high order solid effect transitions supports recent studies of the solid effect, and suggests that a multi-spin solid effect mechanism may play a major role in polarization transfer *via* DNP.[37]

Dynamic nuclear polarization is shown to greatly improve the solid-state NMR analysis of synthetic polymers by allowing structural assignment of intrinsically diluted NMR signals, which are typically not detected in conventional solid-state NMR. Specifically, solid-state NMR and DNP solid-state NMR were comparatively used to study functional polymers for which precise structural elucidation of chain ends is essential to control their reactivity and to eventually obtain advanced polymeric materials of complex architecture. Results show that the polymer chain-end signals, while hardly observable in conventional solid-state NMR, could be clearly identified in the DNP solid-state NMR spectrum owing to the increase in sensitivity afforded by the DNP setup (a factor $\sim$10 was achieved here), hence providing access to detailed structural characterization within realistic experimental times. This sizable gain in sensitivity opens new avenues for the characterization of "smart" functional polymeric materials and new analytical perspectives in polymer science.[38]

Escherichia coli EmrE, a homodimeric multidrug antiporter, has been suggested to offer a convenient paradigm for secondary transporters due to its small size. It contains four transmembrane helices and forms a functional dimer. Ong *et al.* have probed the specific binding of substrates TPP+ and MTP+ to EmrE reconstituted into 1,2-dimyristoyl-*sn-glycero*-3-phosphocholine liposomes by ^{31}P MAS NMR. Their NMR data show that both substrates occupy the same binding pocket but also indicate some degree of heterogeneity of the bound ligand population, reflecting the promiscuous nature of ligand binding by multidrug efflux pumps. Direct interaction between ^{13}C-labelled TPP+ and key residues within the EmrE dimer has been probed by through-space ^{13}C–^{13}C correlation spectroscopy. This was made possible by the use of solid-state NMR enhanced by dynamic nuclear polarization through which a 19-fold signal enhancement was achieved. The data provide clear evidence for the long assumed direct interaction between substrates such as TPP+ and the essential residue E14 in transmembrane helix 1. Ong *et al.* also demonstrate the power of DNP-enhanced solid-state NMR at low temperatures for the study for secondary transporters, which are highly challenging for conventional NMR detection.[39]

3.6 Distance measurements

Distances between a spin-1/2 and a spin > 1/2 can be efficiently measured by a variety of MAS solid state NMR methods such as Rotational Echo Adiabatic Passage Double Resonance (REAPDOR), Low-Amplitude REDOR (LA-REDOR) and Rotational-Echo Saturation-Pulse Double-Resonance (R/S-RESPDOR). Nimerovsky *et al.* show that the incorporation of a phase modulation into a long quadrupolar recoupling pulse, lasting 10 rotor periods that are sandwiched between rotor-synchronized pairs of dipolar recoupling π pulses, extends significantly the range of the values of the quadrupole moments that can be accessed by the experiment. They show by a combination of simulations and experiments that the new method, phase-modulated LA-REDOR, is very weakly dependent on the actual value of the rf field, and is highly robust with respect to off-resonance irradiation. The experimental results can be fitted by numerical simulations or using a universal formula corresponding to an equal-transition-probability model. Phase-modulated LA-REDOR ^{13}C–^{11}B and ^{15}N–^{51}V dipolar recoupling experiments confirm the accuracy and applicability of this new method.[40]

The characterization of metal sites in many types of materials is essential for understanding many aspects of chemistry and biochemistry, such as catalysis, enzymatic activity, and coordination. Of the many approaches to study such sites, Goldbourt considers the utilization of dipolar recoupling in the context of MAS solid state NMR for qualitative and, in particular, quantitative distance measurements between metals and other quadrupolar nuclei to their surrounding environment. The development of techniques, their applicability towards experimental realizations, key applications to materials and biological systems, and current challenges and limitations are also considered.[41]

3.7 Theoretical developments

Mananga presents the possibility of applying the Floquet–Magnus expansion to the most useful interactions known in solid-state NMR using the magic-echo scheme. The results of the effective Hamiltonians of these theories and average Hamiltonian theory are presented.[42]

Floquet–Magnus expansion is also used to study the effect of chemical shift anisotropy in solid-state NMR of rotating solids. The chemical shift interaction is irradiated with two types of rf pulse sequences: BABA and C7. The criteria for the chemical shift anisotropy to be averaged out in each rotor period are obtained.[43]

Edwards *et al.* have demonstrated that Fokker–Planck equations in which spatial coordinates are treated on the same conceptual level as spin coordinates yield a convenient formalism for treating MAS NMR experiments. In particular, time dependence disappears from the background Hamiltonian (sample spinning is treated as an interaction), spherical quadrature grids are avoided completely (coordinate distributions are a part of the formalism) and relaxation theory with any linear diffusion operator is easily adopted from the Stochastic Liouville Equation theory. The proposed formalism contains Floquet theory as a special case. The elimination of the spherical averaging grid comes at the cost of increased matrix dimensions, but Edwards *et al.* show that this can be mitigated by the use of state space restriction and tensor train techniques. It is also demonstrated that low correlation order basis sets apparently give accurate answers in powder-averaged MAS simulations.[44]

3.8 Computational techniques

Much of the information contained within solid-state NMR spectra remains unexploited because of the challenges in obtaining high-resolution spectra and the difficulty in assigning those spectra. Recent advances that enable researchers to accurately and efficiently determine NMR parameters in periodic systems have revolutionized the application of DFT calculations in solid-state NMR spectroscopy. These advances are particularly useful for experimentalists. The use of first-principles calculations aids in both the interpretation and assignment of the complex spectral line shapes observed for solids. Furthermore, calculations provide a method for evaluating potential structural models against experimental data for materials with poorly characterized structures. Determining the structure of well-ordered, periodic crystalline solids can be straightforward using methods that exploit Bragg diffraction. However, the deviations from periodicity, such as compositional, positional, or temporal disorder, often produce the physical properties (such as ferroelectricity or ionic conductivity) that may be of commercial interest. With its sensitivity to the atomic-scale environment, NMR provides a potentially useful tool for studying disordered materials, and the combination of experiment with first-principles calculations offers a particularly attractive approach. Ashbrook and Dawson discuss some of the issues associated with the practical implementation of first-principles calculations of NMR parameters in solids. They then use two key examples to illustrate the structural insights that researchers can obtain when applying such calculations to disordered inorganic materials. First, they describe an investigation of cation disorder in $Y_2Ti_{2-x}Sn_xO_7$ Pyrochlore ceramics using ^{89}Y and ^{119}Sn NMR. Researchers have proposed that these materials could serve as host phases for the encapsulation of lanthanide- and actinide-bearing radioactive waste. In a second example, Ashbrook and Dawson discuss how ^{17}O NMR can be used to probe the dynamic disorder of H in hydroxyl-humite minerals ($n$$Mg_2SiO_4 \cdot Mg(OH)_2$), and how ^{19}F NMR can be used to understand F substitution in these systems. The combination of first-principles calculations and multinuclear NMR spectroscopy facilitates the investigation of local structure, disorder, and dynamics in solids. It is expected that applications will undoubtedly become more widespread with further advances in computational and experimental methods.[45]

Lin and Opella describe the incorporation of non-uniform sampling (NUS) compressed sensing into oriented sample solid-state NMR for stationary aligned samples and MAS solid-state NMR for unoriented 'powder' samples. Both simulated and experimental results indicate that 25–33% of a full linearly sampled data set is required to reconstruct two- and three-dimensional solid-state NMR spectra with high fidelity. A modest increase in signal-to-noise ratio accompanies the reconstruction.[46]

A combination of solid-state NMR techniques and theoretical calculations has been used by Kobayashi *et al.* to confirm that unsubstituted metal-free corroles form supramolecular systems with toluene through π–π interactions in the solid state. Ultrafast MAS enabled the use of ^{1}H–^{1}H homonuclear correlation spectroscopy to explore those intermolecular interactions for which heteronuclear correlation methods were difficult to apply.[47]

The interactions between histidine and metal species play essential roles in a wide range of important biological processes including enzymes catalysis and signal transduction. Zhou *et al.* have employed solid-state NMR techniques to determine the interaction between histidine and Zn(II) from pH 3.5 to 14. 2D homo- and heteronuclear correlation NMR experiments were utilized to extract the ^{1}H, ^{13}C, and ^{15}N chemical shifts in various histidine–Zn(II) binding complexes. Several histidine–Zn(II) binding models were proposed on the basis of experimental results as well as DFT theoretical calculations.[48]

The ruthenium-containing sensitizing dye N719 grafted on TiO_2 nanoparticles was investigated by solid-state NMR. The carbon resonances were assigned by means of ^{13}C cross-polarized dipolar dephasing experiments. It has been shown that DFT calculations of the carbon magnetic shielding tensors accurately describe the changes in chemical shifts observed upon grafting onto a titania surface *via* one or two carboxylic functions in the plane defined by the two isothiocyanate groups.[49]

Diboron compounds are a part of a relatively unexplored class of compounds. Their main use is for β-boration reactions where a boron centre is rendered nucleophilic with the use of a metal catalyst or a Lewis base to form a sp^2–sp^3 diboron compound. The reactivity of these reagents is largely dictated by the nature of the B–B bond (strength and polarity); however, no experimental methods have been used to directly probe both of these quantities. Perras and Bryce demonstrate that unprecedented experimental information regarding the B–B bond may be obtained using ^{11}B solid-state NMR spectroscopy. For example, the ^{11}B quadrupolar coupling constants can be understood on the basis of the polarization of the B–B bond. ^{11}B DQF *J*-resolved NMR spectroscopy was applied to easily and accurately measure $J(^{11}\text{B}, ^{11}\text{B})$ coupling constants with high precision. These are shown to be well correlated with the orbital energy of the B–B σ-bonding natural bond orbital as well as the hybridisation states of the boron atoms in the bond. An increase in the π character of the bond by electron-donating ligands or *via* the formation of a sp^2–sp^3 diboron compound weakens the bond, increases the bond length, and decreases the $J(^{11}\text{B}, ^{11}\text{B})$ coupling constants. These experiments provide a detailed experimental characterization of the B–B bond and may be useful in understanding the reactivity of diboron compounds and in designing new systems. The potential applicability of ^{11}B DQF *J*-resolved NMR spectroscopy towards analyzing complex mixtures of diboron compounds and towards measuring ^{11}B *J* coupling across multiple intervening bonds is also investigated.[50]

No-bond ^{31}P–^{31}P indirect dipolar couplings, which arise from the transmission of nuclear spin polarization through interaction of proximal nonbonded electron pairs have been investigated in the solid state for a series of closely related substituted P,P-[3]ferrocenophanes and model systems. Through variation and combination of ligands (phenyl, cyclohexyl, isopropyl) at the two phosphorus sites, the (PP)-P-· · · distances in these compounds can be varied from 3.49 to 4.06 Å. Thus, the distance dependence of the indirect no-bond coupling constant J_{nb}, can be studied in a series of closely related compounds. One- and two-dimensional solid-state NMR experiments serve to establish the character of these couplings and to measure the isotropic coupling constants, which were found to range between 12 and 250 Hz. To develop an understanding of the magnitude of J_{nb}, in terms of molecular structure, their dependences on intramolecular internuclear distances and relative orbital orientations is discussed by DFT-calculations on suitable models. In agreement with the literature the dependence of J_{nb} on the P–P distance is found to be exponential. However, the steepness of this curve is highly dependent on the internuclear equilibrium distance. For a quantitative description, the off-diagonal elements of the expectation value of the Kohn–Sham–Fock operator in the LMO basis for the LMOs of the two phosphorus lone-pairs is proposed. This parameter correlates linearly with the calculated J_{nb} values and possesses the same distance-dependence. In addition, the simulations indicate a distinct dependence of J_{nb} on the dihedral angle defined by the two C–P bonds providing ligation to the molecular backbone. For disordered materials or those featuring multiple sites, conformers, and/or polymorphism, a new double-quantum NMR method termed DQDRENAR can be used to conveniently measure internuclear ^{31}P–^{31}P distances. If conducted on compounds with known P–P distances, such measurements can also serve to estimate the magnitude of the anisotropy ΔJ of these no-bond indirect spin–spin couplings. The DFT results suggest that in the present series of compounds the magnitude of ΔJ is strongly correlated with that of the isotropic component, as both parameters have analogous distance dependences. Based on the above insights, the structural aspects of conformational disorder and polymorphism observed in some of the P,P-[3]ferrocenophanes are discussed.[51]

3.9 Other experimental developments

Molecular dynamics characterisations in solids can be carried out selectively using dipolar-dephasing experiments. Aliev and Courtier-Murias have shown that the introduction of a sum of Lorentzian and Gaussian functions greatly improve fittings of the "intensity *versus* time" data for protonated carbons in dipolar-dephasing experiments. The Lorentzian term accounts for remote intra- and intermolecular ^{1}H–^{13}C dipole–dipole interactions, which vary from one molecule to another or for different carbons within the same molecule. Thus, by separating contributions from weak remote interactions, more accurate Gaussian decay constants, T_{dd}, can be extracted for directly bonded ^{1}H–^{13}C dipole–dipole interactions. Reorientations of the ^{1}H–^{13}C bonds lead to the increase of T_{dd}, and by measuring dipolar-dephasing constants, insight can be gained into dynamics in solids. They have demonstrated advantages of the method using comparative dynamics studies in α and γ polymorphs of glycine, cyclic amino acids L-proline, DL-proline and *trans*-4-hydroxy-L-proline, the Ala residue in different dipeptides, as well as adamantane and hexamethylenetetramine. It was possible to distinguish subtle differences in dynamics of different carbon sites within a molecule in polymorphs and in L- and DL-forms. The presence of overall molecular motions is shown to lead to particularly large differences in dipolar-dephasing experiments. The differences in dynamics were attributed to differences in noncovalent interactions. In the case of hexamethylenetetramine, for example, the presence of C–H$\cdots$N interactions leads to nearly rigid molecules. Overall, the method allows one to gain insight into the role of noncovalent interactions in solids and their influence on the molecular dynamics.[52]

Lu *et al.* have analyzed the direct excitation of wide 1D spectra of nuclei with spin $I = 1/2$ or 1 in rotating solids submitted to pulse trains in the manner of Delays Alternating with Nutations for Tailored Excitation (DANTE), either with one short rotor-synchronized pulse of duration τ_p in each of rotor period or with N interleaved equally spaced pulses τ_p in each rotor period.[53]

Chemically informative *J* couplings between pairs of quadrupolar nuclei in dimetallic and dimetalloid coordination motifs are measured using *J*-resolved solid-state NMR experiments. It is shown that the application of a double-quantum filter is necessary to observe the *J* splittings and that, under these conditions, only a simple doublet is expected. The splitting is amplified if the spins are magnetically equivalent, making it possible to measure highly precise *J* couplings and unambiguously probe the symmetry of the molecule. This is demonstrated experimentally by chemically breaking the symmetry about a pair of boron spins by reaction with an N-heterocyclic carbene to form a β-borylation reagent. The results show that the *J* coupling is a sensitive probe of bonding in diboron compounds and that the *J* values quantify the weakening of the B–B bond which occurs when forming a sp^2–sp^3 diboron compound, which is relevant to their reactivity. Due to the prevalence of quadrupolar nuclei among transition metals, this work also provides a new approach to probe metal–metal bonding.[54]

A five-channel (^{1}H, ^{19}F, ^{31}P, ^{27}Al, ^{13}C) 2.5 mm MAS NMR probe is used in combination with three separate receivers for the parallel acquisitions of 1D and 2D NMR spectra in model fluorinated aluminophosphate and porous Al-based metal–organic framework. Possible combinations to record simultaneously spectra using this setup are presented, including (i) parallel acquisitions of quantitative 1D NMR spectra of solids containing nuclei with contrasted T_1 relaxation rates and (ii) parallel acquisitions of 2D heteronuclear NMR spectra. In solids containing numerous different NMR-accessible nuclei, the number of NMR experiments that have to be acquired to get accurate structural information is shown to be high.[55]

Mobile and cost-effective NMR spectroscopy exploiting low-field permanent magnets is a field of tremendous development with obvious applications for arrayed large scale analysis, field work, and industrial screening. So far such demonstrations have concentrated on relaxation measurements and lately high-resolution liquid-state NMR applications. With high-resolution ssNMR spectroscopy being increasingly important in a broad variety of applications, low-field MAS solid-state multinuclear NMR is introduced based on a commercial ACT 0.45 T 62 mm bore Halbach magnet along with a homebuilt FPGA digital NMR console, amplifiers, and a modified standard 45 mm wide MAS probe for 7 mm rotors. To illustrate the performance of the instrument and address cases where the low magnetic field may offer complementarity to high-field NMR experiments, applications for ^{23}Na MAS NMR have been demonstrated with enhanced second-order quadrupolar coupling effects and ^{31}P MAS NMR where reduced influence from chemical shift anisotropy at low field may facilitate determination of heteronuclear dipole–dipole couplings.[56]

Recent research in fast MAS methods has drastically improved the resolution and sensitivity of NMR spectroscopy of biomolecules and materials in solids. Recent and ongoing developments have been summarised in this area by presenting ^{13}C and ^{1}H ssNMR studies on paramagnetic systems and biomolecules under fast MAS. It has been shown how very fast MAS at the spinning speed of at least 20 kHz allows to overcome major difficulties in ^{1}H and ^{13}C high-resolution solid-state NMR of paramagnetic systems. As a result, both sensitivity and resolution by up to a few orders of magnitude can be enhanced. Using fast recycling ($\sim$ms per scan) with short ^{1}H T_1 values, ^{1}H solid-state NMR microanalysis of paramagnetic systems on the microgram scale can be performed with greatly improved sensitivity over that observed for diamagnetic systems. Solid-state ^{13}C NMR measurements for biomolecules under faster MAS at a higher field were also examined. The prospects for studying biomolecules using ^{13}C solid-state NMR under ultrafast MAS at the spinning speed of ∼100 kHz are discussed.[57]

Intranucleotide resonance of the 26mer box C/D RNA in complex with the L7Ae protein were assigned by ssNMR spectroscopy. It is expected that this investigation will promote studies of RNA in large protein–RNA complexes by solid-state NMR spectroscopy.[58]

Advanced multinuclear solid state NMR experiments were developed to probe the structure of two organometallic aluminum derivatives, which are relevant to olefin polymerization processes. For the first time, NMR observation of ^{27}Al–^{13}C covalent bonds in solids is performed with the natural abundance material. Triple-resonance (^{1}H–^{13}C–^{27}Al) and quadruple-resonance (^{1}H–^{7}Li–^{13}C–^{27}Al) heteronuclear correlation two-dimensional NMR experiments are also introduced to probe ^{27}Al–^{13}C and ^{13}C–^{7}Li proximities.[59]

4 Applications

4.1 Organic solids

Annular tautomerism of 3(5)-aminopyrazoles containing a cyano, thiocyanato, or aryl substituent in the 4-position has been studied by ^{1}H and ^{13}C NMR in solution, CP MAS ^{13}C NMR in the solid state, and DFT calculations.[60]

Three polymorphic phases of xaliproden are evidenced through thermal analysis: Form II has an orthorhombic symmetry, while Forms I and III have monoclinic unit cells. Structural characterizations of xaliproden are carried out by means of single-crystal and synchrotron powder X-ray diffraction (XRD), as well as high-resolution solid-state ^{13}C CP MAS NMR spectroscopy. The results show that polymorphism between xaliproden Forms I and III mostly deals with the degree of freedom of the CF_3 group, higher in Form III than in Form I and clarifies the status of Form III that can therefore be assigned to a well-defined polymorph of xaliproden.[61]

Eight single crystal X-ray structures, solid-state NMR spectroscopic, and theoretical studies utilizing QTAIM methodology were used to characterize the 2-acyl (alkyl in acyl = methyl, ethyl, *t*-butyl, and 1-adamantyl) amino-6-R-pyridine/4-R′-benzoic acid (R,R′ = H or Me) cocrystals. As expected among alkyl groups 1-adamantyl due to its bulkiness has the most significant effect on the relative positions of molecules in cocrystals. In addition, the subtle electronic and steric effects by the methyl substituents were observed. The ^{15}N solid-state CP MAS NMR chemical shifts prove that molecules in cocrystal are held together by hydrogen bonding. The biggest variation in the ^{15}N chemical shift of acylamino nitrogen can be related with the size of the alkyl group in acyl moiety.[62]

A combination of solid-state ^{13}C NMR tensor data and DFT computational methods is utilized to predict the conformation in disordered methyl α-L-rhamnofuranoside. This previously uncharacterized solid is found to be crystalline and consists of at least six distinct conformations that exchange on the kHz time scale. A total of 66 model structures were evaluated, and six were identified as being consistent with experimental ^{13}C NMR data. A concerted rearrangement of OH hydrogens is proposed to account for the observed dynamic disorder.[63]

An asymmetric tapered Janus bisamide supramolecule consisting of 1,4-bis[3,4,5-tris(alkan-1-yloxy)benzamido]benzene bisamide has been investigated using various solid-state NMR techniques at the atomic level.[64]

Spectroscopic strategies have been presented that tailor solid-state NMR spectroscopic methods to the challenge of supramolecular structure investigation.[65]

4.2 Pharmaceutical and biomedical applications

The anhydrous crystalline forms of Naproxen [(*S*)-(+)-2-(6-methoxy-2-naphthyl)propionic acid], (NAPRO-A) and its sodium salt (NAPRO-S), widely used anti-inflammatory drugs, have been investigated by means of 1D and 2D MAS NMR and DFT based calculations. From calculations, 1D ^{13}C CP MAS and ^{1}H CRAMPS and 2D ^{1}H–^{13}C MAS-J-HMQC, refocused INEPT, FSLG-HETCOR, and ^{1}H–^{1}H DQ-CRAMPS solid-state NMR experiments, ^{1}H and ^{13}C resonances have been fully assigned for NAPRO-A and -S. In the case of NAPRO-S, all of the nuclei belonging to the two inequivalent molecules of the asymmetric cell gave rise to distinct signals, which could be completely assigned. Interesting intermolecular ring current effects on ^{1}H chemical shifts have been experimentally observed for the two samples, even if with significant differences between the two cases. The measured and calculated proton chemical shift values showed a very good agreement for both NAPRO-A and -S, allowing to correlate the different ring current effects with the crystal structures. The comparison between the proton chemical shifts calculated in the crystal structures and *in vacuo* allowed to confirm the mainly intermolecular character of the ring current effects and to quantify them.[66]

A novel crystalline form of the boron-containing antibacterial drug (*S*)-3-(aminomethyl)-7-(3-hydroxypropoxy)benzo[*c*][1,2]oxaborol-1(3*H*)-ol hydrochloride has been studied by solid-state NMR and single-crystal X-ray diffraction techniques. After determination of the crystal structure by X-ray diffraction, solid-state NMR spectroscopy of this form is performed to obtain structural information using experimental approaches based on dipolar correlation, chemical shift analysis, and quadrupolar interaction analysis. ^{1}H solid-state NMR experiments at 16.4 T using MAS and homonuclear dipolar decoupling, 2D solid-state NMR experiments based on ^{1}H–^{13}C and ^{1}H–^{11}B dipolar heteronuclear correlation, and DFT calculations are combined to obtain a nearly complete assignment of the ^{1}H spectrum of this crystalline phase. ^{11}B and ^{35}Cl chemical shift and quadrupolar parameters are obtained using the analysis of MAS spectra and are found to be accurately reproduced using DFT calculations. NMR chemical shielding and electric field gradient parameters obtained using these methods are related to hydrogen-bonding trends in the crystal structure. The results illustrate the increasing capability of solid-state NMR techniques involving ^{1}H, ^{11}B, and ^{35}Cl solid-state NMR in the analysis of the crystal structure of a pharmaceutical compound containing covalently bonded boron.[67]

The stability of amorphous simvastatin was evaluated by probing molecular dynamics using two nondestructive techniques: solid-state NMR and dielectric relaxation spectroscopy. Glassy simvastatin was obtained by the melt quench technique. ^{13}C CP MAS NMR spectra and ^{1}H MAS NMR spectra were obtained from 293 K up to 333 K ($T_g \sim 302$ K). The ^{13}C spin lattice relaxation times in the rotating frame, $T_{1\rho}$, were measured as a function of temperature, and the correlation time and activation energy data obtained for local motions in different frequency scales revealed strong dynamic heterogeneity, which appears to be essential for the stability of the amorphous form of simvastatin. In addition, the ^{1}H MAS measurements presented evidence for mobility of the hydrogen atoms in hydroxyl groups which was assigned to noncooperative secondary relaxations. Overall, the results confirm that the simvastatin in the glass form reveals a potential use in the solid phase formulation on the pharmaceutical industry.[68]

Novel applications of solid-state ^{17}O NMR have been demonstrated in the analysis of a polymorph of diflunisal, a cocrystal of diflunisal with pyrazinamide, and amorphous dispersions of diflunisal in two polymers. The observation of ^{17}O nucleus is shown to be a highly specific and useful alternative. A novel application of ^{1}H–^{17}O CP HETCOR experiments is also demonstrated for the cocrystal and two dispersions. This experiment offers specific information about proton environments in proximity to the labelled oxygen sites.[69]

Kong *et al.* have reported solid-state NMR characterization of the ^{17}O quadrupole coupling (QC) and chemical shift (CS) tensors in five site-specifically ^{17}O-labelled samples of salicylic acid and *o*-acetylsalicylic acid (aspirin). High-quality ^{17}O NMR spectra were obtained for these important pharmaceutical compounds under both static and MAS conditions at two magnetic fields, 14.0 and 21.1 T. A total of 14 ^{17}O QC and CS tensors were experimentally determined for the seven oxygen sites in salicylic acid and aspirin. Although both salicylic acid and aspirin form hydrogen bonded cyclic dimers in the solid state, it was found that the potential curves for the concerted double proton transfer in these two compounds are significantly different. In particular, while the double-well potential curve in aspirin is nearly symmetrical, it is highly asymmetrical in salicylic acid. This difference results in quite different temperature dependencies in ^{17}O MAS spectra of the two compounds.[70]

Sodium valproate is a pharmaceutical with applications in the treatment of epilepsy, bipolar disorder, and other ailments. It can exist in many hydrated and acid-stabilized forms in the solid state. Solid-state ^{13}C and ^{23}Na NMR study of several forms of sodium valproate have been presented. ^{23}Na MAS, DOR, and MQMAS NMR spectra are shown to provide useful information on the number of molecules in the asymmetric unit, the local coordination geometry of the sodium cations, and the presence of amorphous phases. Two previously identified forms are shown to be highly similar, or identical, according to the ^{23}Na NMR data. The utility of carrying out both DOR and MQMAS NMR experiments to identify all crystallographically unique sites is demonstrated. ^{13}C CP MAS NMR spectra also provide complementary information on the number of molecules in the asymmetric unit and the crystallinity of the sample.[71]

A combined X-ray diffraction, solid-state NMR, differential scanning calorimetry, and modelling approach has been applied to study the solid-solid transition of ibuprofen sodium salt between the hydrated racemic compound (RS–H) and the anhydrous conglomerate (RS–A). For comparison, the dihydrate → anhydrous transformation of the sodium salt of the pure *S*-enantiomer of ibuprofen was also investigated by means of solid-state NMR and DSC. All the solid state studies indicate that the RS–H → RS–A transition is fully reversible under different experimental conditions. The static and dynamic disorder affecting the isobutyl fragment in RS–H, already observed by solid-state NMR, has been further investigated by single crystal XRD and computational techniques. A model for the dihydrate → anhydrous solid–solid transformations is proposed.[72]

Antiviral drug compounds based on amantadine- and aryl-substituted isoxazole have been investigated using solid-state NMR spectroscopy. Measurements of ^{15}N and ^{13}C chemical shifts, ^{13}C–^{2}H distances and ^{1}H–^{1}H NOEs have been reported.[73]

Peptides penetrating through skin intercellular lipids have been studied using solid-state ^{31}P MAS NMR. *In vitro* skin permeation studies were performed on rat skin, and sections were collected and analyzed for ^{31}P NMR signal. The results of the solid-state NMR study were in agreement with confocal microscopy studies.[74]

New crystalline forms of hydrated and anhydrous *N*-acylhydrazones have been reported. X-ray, solid-state NMR studies and theoretical calculations were conducted.[75]

Solid-state NMR ^{1}H T_1 and $T_{1\rho}$ relaxation times were used to evaluate the miscibility of amorphous solid dispersions of nifedipine (NIF) and polyvinylpyrrolidone (PVP) prepared by three different methods: melt quenching in the typical lab setting, spray drying and melt quenching in the NMR rotor while spinning. Variable-temperature solid-state NMR ^{1}H $T_{1\rho}$ relaxation measurements revealed a change in relaxation time at ~20 °C below T_g, suggesting increased molecular mobility above that temperature.[76]

Oriented solid-state NMR spectroscopy and biochemical cross-linking experiments were used to show that the ligand-free membrane protein transporter EmrE forms anti-parallel dimers with different monomer tilt angles relative to the lipid bilayer. In addition, subtle conformational changes were detected upon drug binding that emphasize the need for an atomic-resolution structure.[77]

Multinuclear and multidimensional solid-state NMR techniques were applied to perform detailed characterization of amorphous dispersions of ezetimibe adsorbed on mesoporous silica. Solid-state ^{1}H, ^{13}C, ^{19}F, and ^{29}Si NMR analysis including relaxation time measurements has been presented. Interactions between the drug and silica were investigated using 1D and 2D solid-state NMR methods based on dipolar correlation using CP and spin diffusion. ^{19}F solid-state NMR was used to assess drug physical state and study mobility. ^{19}F–^{29}Si CP methods were used to directly detect adsorbed ezetimibe. ^{1}H–^{13}C, ^{1}H–^{19}F, and ^{1}H–^{29}Si, and heteronuclear correlation and ^1H homonuclear correlation experiments were used to investigate interactions between the drug and silica.[78]

4.3 Amino acids and peptides

Histidine usually exists in three different forms (including biprotonated species, neutral τ and π tautomers) at physiological pH in biological systems. The different protonation and tautomerization states of histidine can be characteristically determined by ^{13}C and ^{15}N chemical shifts of imidazole ring. Solid-state NMR techniques have been developed for spectral editing of ^{13}C and ^{15}N sites in histidine imidazole ring, which provides a benchmark to distinguish the existing forms of histidine. The selections of $^{13}C_{\gamma}$, $^{13}C_{\delta 2}$, $^{15}N_{\delta 1}$, and $^{15}N_{\varepsilon 2}$ sites were successfully achieved based on one-bond homo- and hetero-nuclear dipole–dipole interactions. Moreover, it was demonstrated that ^1H, ^{13}C, and ^{15}N chemical shifts were roughly linearly correlated with the corresponding atomic charge in histidine imidazole ring by theoretical calculations. Accordingly, the ^1H, ^{13}C and ^{15}N chemical shifts variation in different protonation and tautomerization states could be ascribed to the atomic charge change due to proton transfer in biological process.[79]

Recent work that is providing insight into the structure and crystal recognition mechanisms of a salivary protein model system, as well as the structure and interactions of a peptide that catalyzes the formation of biosilica composites have been highlighted by Roehrich and Drobny. To develop a better understanding of the structure and interactions of proteins in biomaterials, they have used solid-state NMR techniques to determine the molecular structure and dynamics of proteins and peptides adsorbed onto inorganic crystal surfaces and embedded within biomineral composites. This work adds to the understanding of the structure and crystal recognition mechanisms of an acidic human salivary phosphoprotein, statherin.[80]

The chloroplast twin arginine translocation (cpTat) system transports highly folded precursor proteins into the thylakoid lumen using the protonmotive force as its only energy source. Hcfl 06 is one of the core components of the cpTat system. To better understand the membrane active capabilities of Hcf106, solid-state NMR spectroscopy was used to investigate properties of the amphipathic cc-helix. Solid-state ^{31}P NMR and ^2H NMR spectroscopic experiments revealed that the peptide perturbs the headgroup and the acyl chain regions of phospholipids as indicated by changes in spectral lineshape, chemical shift anisotropy, line width, and ^2H order parameters.[81]

^2H solid-state NMR was employed to critically examine domain formation in bacteria-mimetic membranes due to cationic peptide binding. The results provide insight into the extent of domain formation in bacterial membranes and the possible peptide structural requirements for this phenomenon.[82]

Using solid-state NMR, a mitochondria-penetrating peptide interacting with the model inner mitochondrial membrane was observed in order to gain insight into the mechanism of translocation. The paramagnetic relaxation effect due to Mn^{2+} ions on ^{13}C MAS NMR was used to measure the insertion depth of the peptide and its distribution in each monolayer of the membrane.[83]

A designer peptide MAX8 known to undergo self-assembly following changes in temperature, pH, and ionic strength, has demonstrated usefulness for tissue engineering and drug delivery. It is hypothesized that the self-assembled MAX8 nanofibre structure consists of closed β-hairpins aligned into antiparallel beta-sheets. Evidence from solid-state NMR spectroscopy has been reported that supports the presence of the hypothesized β-hairpin conformation within the nanofibre structure. NMR spectra also indicate the existence of a minor conformation, which is discussed in terms of previously hypothesized nanofibre physical cross-linking and possible nanofibre polymorphism.[84]

Three amphibian skin neuropeptides were studied using a combination of ^{31}P and ^{2}H solid state NMR studies in eukaryotic phospholipid models.[85]

4.4 Proteins

One of the main applications of solid-state NMR is to study the structure and dynamics of biopolymers under physiological conditions where the polypeptides undergo global motions as they do in biological membranes. The effects of NMR rf irradiations on nuclear spins are strongly influenced by these motions. For example, it has been shown that the MSHOT-Pi4 pulse sequence yields spectra with resonance line widths about half of those observed using the conventional pulse sequence when applied to membrane proteins undergoing rapid uniaxial rotational diffusion in phospholipid bilayers. In contrast, the line widths were not changed in microcrystalline samples where the molecules did not undergo global motions. Lu and Opella have demonstrated experimentally and described analytically how some Hamiltonian terms are susceptible to sample motions, and it is their removal through the critical $\pi/2$ *Z*-rotational symmetry that confers the "motion adapted" property to the MSHOT-Pi4 pulse sequence. The new approach is applicable to both stationary and MAS solid-state NMR experiments.[86]

Many biomolecular reactions take place within microseconds to milliseconds, and this timescale is therefore of central functional importance. It has been shown that R_1 relaxation dispersion experiments in solid-state MAS NMR spectroscopy make it possible to investigate the thermodynamics and kinetics of such exchange process, and gain insight into structural features of short-lived states.[87]

^{1}H-detected triple-resonance H/N/C experiments have been presented that incorporate CO-CA and CA-CB out-and-back scalar-transfer blocks optimized for robust resonance assignment in biosolids under ultra-fast MAS. The first experiment, (H)(CO)CA(CO)NH, yields ^{1}H-detected inter-residue correlations, in which the chemical shifts of the CA spins are recorded in the first indirect dimension while during the scalar-transfer delays the coherences are present only on the longer-lived CO spins. The second experiment, (H)(CA)CB(CA)NH, correlates the side-chain CB chemical shifts with the NH of the same residue. These high sensitivity experiments are demonstrated on both fully-protonated and 100%-H–N back-protonated perdeuterated microcrystalline samples of Acinetobacter phage 205 (AP205) capsids at 60 kHz MAS.[88]

Solid-state NMR is an emerging structure determination technique for crystalline and non-crystalline protein assemblies, *e.g.* amyloids. Resonance assignment constitutes the first and often very time-consuming step to a structure. A generally applicable algorithm ssFLYA has been presented for automatic assignment of protein solid-state NMR spectra. Application to microcrystals of ubiquitin and the Ure2 prion C-terminal domain, as well as amyloids of HET-s(218-289) and α-synuclein yielded 88–97% correctness for the backbone and side-chain assignments that are classified as self-consistent by the algorithm, and 77–90% correctness if also assignments classified as tentative by the algorithm are included.[89]

Two sequences have been presented which combine (^{1}H, ^{15}N) and (^{15}N, ^{13}C) selective cross-polarization steps with an efficient variant of the *J*-based homonuclear transfer scheme, in which a spin-state-selective ((*SE*)-*E*-3) block is incorporated to improve both resolution and sensitivity in the direct ^{13}C dimension. These two sequences are proposed as a part of a suite of four N–C correlation experiments allowing for the assignment of protein backbone resonances in the solid state. These experiments are illustrated under ultra-fast MAS conditions on two samples of microcrystalline dimeric human superoxide dismutase, in its diamagnetic (empty, Zn-II) and paramagnetic (Cu-II, Zn-II) states.[90]

Two-dimensional ^{15}N chemical shift/^{1}H chemical shift and three-dimensional ^{1}H–^{15}N dipolar coupling/^{15}N chemical shift/^{1}H chemical shift MAS solid-state NMR correlation spectra have been used to study the filamentous bacteriophage Pf1 major coat protein. The high sensitivity and resolution was shown to arise from ^{1}H detection was shown at 600 MHz under 50 kHz MAS using ~0.5 mg of perdeuterated and uniformly ^{15}N-labelled protein in which the exchangeable amide sites are partially or completely back-exchanged. Notably, the heteronuclear ^{1}H–^{15}N dipolar coupling frequency dimension is shown to select among ^{15}N resonances, which will be useful in structural studies of larger proteins where the resonances exhibit a high degree of overlap in multidimensional chemical shift correlation spectra.[91]

The process of resonance assignment represents a time-consuming and potentially error-prone bottleneck in structural studies of proteins by solid-state NMR. Procedures developed through the last decades for solution-state NMR are not directly applicable for solid-state NMR due to the inherently lower data quality caused by lower sensitivity and broader lines, leading to overlap between peaks. A robust automatic method has been presented, which can accurately assign protein resonances using peak lists from a small set of simple 2D and 3D solid-state NMR experiments, applicable in cases with low sensitivity. The method is demonstrated on three uniformly ^{13}C, ^{15}N labelled biomolecules with different challenges on the assignments. It is demonstrated, by using a procedure for generating artificial spectra with increasing line widths, that the new method, GAMES_ASSIGN, can handle a significant amount of overlapping peaks in the assignment. The impact of including different solid-state NMR experiments is also evaluated.[92]

Solvent ordering at an interface can be studied by multiple-quantum NMR. Quantitative studies of (H_2O)–^2H ordering require clean double-quantum (2Q) filtration and an analysis of 2Q buildup curves that accounts for relaxation and, if randomly oriented samples are used, the distribution of residual couplings. A pulse sequence with absorption mode detection is extended for separating coherences by order and measuring relaxation times such as the 2Q filtered T_2. Coherence separation is used to verify 2Q filtration and the 2Q filtered T_2 is required to extract the coupling from the 2Q buildup curve when it is unresolved. It was found that the coupling extracted from the buildup curve in (H_2O)–^2H hydrated collagen was equivalent to the resolved coupling measured in the usual 1D experiment and the 2Q to 1Q signal ratio was in accord with theory. Application to buildup curves from (H_2O)–^2H hydrated elastin revealed a large increase in the 2Q signal upon mechanical stretch that is due to an increase in the ordered water fraction while changes in the residual coupling and T_2 are small.[93]

The use of a proton-based slow MAS solid-state NMR experiment has been reported that exploits the unaveraged ^1H–^1H dipolar couplings from a membrane-bound protein. The authors have shown that the difference in the buildup rates of cross-peak intensities against the mixing time obtained from 2D ^1H–^1H rf-driven recoupling (RFDR) and NOESY experiments on a 16.7 kDa micelle-associated full-length rabbit cytochrome-b_5 (cytb5) – can provide insights into protein dynamics and could be useful to measure ^1H–^1H dipolar couplings. The experimental buildup curves compare well with theoretical simulations and are used to extract relaxation parameters.[94]

^1H-detected MAS NMR experiments facilitate structural biology of solid proteins, which requires using deuterated proteins. However, often amide protons cannot be back-exchanged sufficiently, because of a possible lack of solvent exposure. For such systems, using ^2H excitation instead of ^1H excitation can be beneficial because of the larger abundance and shorter longitudinal relaxation time, T_1, of deuterium. A new structure determination approach, "quadruple-resonance NMR spectroscopy", is presented which relies on an efficient ^2H-excitation and ^2H–^{13}C CP step, combined with ^1H detection. It has been shown that by using ^2H-excited experiments better sensitivity is possible on an SH3 sample recrystallized from 30% H_2O.[95]

Solid-state NMR studies of collagen samples of various origins confirm that the amplitude of collagen backbone and sidechain motions increases significantly on increasing the water content. This conclusion is supported by the changes observed in three different NMR observables: (i) the linewidth dependence on the ^1H decoupling frequency; (ii) ^{13}C CSA changes for the peptide carbonyl groups, and (iii) dephasing rates of ^1H–^{13}C dipolar couplings. A nearly threefold increase in motional amplitudes of the backbone librations about C–C or N–C bonds was found on increasing the added water content up to 47 wt% D_2O. On the basis of the frequencies of NMR observables involved, the timescale of the protein motions dependent on the added water content is estimated to be of the order of microseconds. This estimate agrees with that from wideline T_2 ^1H NMR measurements. Wideline ^1H NMR measurements revealed that the timescale of the microsecond motions in proteins reduces significantly on increasing the added water content, *i.e.* ~15-fold increase in protein motional frequencies is observed on increasing the added water content to 45 wt% D_2O. The observed changes in collagen dynamics is attributed to the increase in water translational diffusion on increasing the amount of added water, which leads to more frequent bound water/free water exchange on the protein surface, accompanied by the breakage and formation of new hydrogen bonds with polar functionalities of protein.[96]

A comprehensive analysis of the dynamics of the SH3 domain of chicken α-spectrin has been presented, based upon ^{15}N T_1 and on- and off-resonance $T_{1\rho}$ relaxation times obtained on deuterated samples with a partial back-exchange of labile protons under a variety of the experimental conditions, taking explicitly into account the dipolar order parameters calculated from ^{15}N–^1H dipole–dipole couplings. It is demonstrated that such a multi-frequency approach enables access to motional correlation times spanning about 6 orders of magnitude. The validity of different motional models based upon orientation autocorrelation functions has been assessed with a different number of motional components. It was found that for many residues a "two components" model is not sufficient for a good description of the data and more complicated fitting models must be considered. Surprisingly, the obtained data revealed no clearly evident correlations between secondary structure of the protein and motional parameters. No unambiguous correlations were noted between motions in different time scales along the protein backbone emphasizing the importance of the inter-residue interactions and the cooperative nature of protein dynamics.[97]

A detailed theoretical investigation has been undertaken on how dipolar couplings and relaxation data can provide information about amplitudes and time scales of local motion. The authors have shown that the measurement of dipolar couplings is crucial for obtaining accurate motional parameters, while systematic errors are found when only relaxation data are used. Based on this realization, the ability of REDOR experiment to provide accurate data has been studied. It was found that with accurate rf calibration, and explicit consideration of rf field inhomogeneities, one can obtain highly accurate absolute order parameters. It was also shown that nanosecond motion can be detected primarily in loop regions.[98]

Lopez del Amo *et al.* have shown that high magnetic fields allow to overcome the broadening of resonance lines often experienced at liquid nitrogen temperatures. For a fibril sample of the Alzheimer's disease β-amyloid peptide, the authors found similar line widths at low temperature and at room temperature. The presented results are shown to open new perspectives for structural investigations in the solid-state.[99]

Copper–zinc superoxide dismutase 1 (SOD1) is present in the protein aggregates deposited in motor neurons of amyotrophic lateral sclerosis (ALS) patients. Banci *et al.* have shown that structural information can be obtained at a site-specific level from solid-state NMR in protein aggregates. The residues that are sequentially assignable are found to be located at the putative nucleation site for fibrillar species formation, as detected by other experimental techniques.[100]

The study by Jenkins *et al.* provides a detailed secondary structural characterization of major ampullate dragline silk from Latrodectus hesperus (black widow) spiders. The combination of 2D ^{13}C–^{13}C through-space and through-bond solid-state NMR experiments are shown to provide chemical shifts that are used to determine detailed information about the amino acid motif secondary structure in black widow spider dragline silk. Individual amino acids are incorporated into different repetitive motifs that make up the majority of this protein-based biopolymer. From the solid-state NMR measurements, distinct secondary conformations have been assigned to each repetitive amino acid motif and, hence, to the amino acids that make up the motifs.[101]

Solid-state NMR studies of the structure of the 20-residue N-terminal fusion peptide (FP) in a membrane-associated construct (FP-Hairpin) have been reported. The data show that there are populations of FP-Hairpin with either a helical or 13 sheet FP conformation.[102]

Using solid-state 2D MAS NMR, Liao *et al.* have investigated the secondary structure and dynamics of full-length influenza A M2 protein forms.[103]

Solid-state MAS NMR structural studies of nitroxide and EDTA-Cu^{2+} labelled variants of a model 56 amino acid globular protein, B1 immunoglobulin-binding domain of protein G (GB1), in the microcrystalline solid phase have been reported. The authors used a set of six EDTA-Cu^{2+} tagged GB1 mutants to rapidly determine the global protein fold in a *de novo* fashion. These studies required quantitative measurements of only approximately four or five backbone amide ^{15}N longitudinal paramagnetic relaxation enhancements per residue, in the complete absence of the usual internuclear distance restraints. The presented paramagnetic solid-state NMR methodology is of general use and can be directly applied to larger proteins and protein complexes for which a significant fraction of the signals can be assigned in standard 2D and 3D MAS NMR chemical shift correlation spectra.[104]

The factors affecting spectral quality in solid-state NMR samples have been discussed by Fragai *et al.* examining cases encountered in the literature and presenting new experimental data. A discussion on resolution and sensitivity in sedimented solutes is framed in this context.[105]

A sample preparation technique known as self-crowding has been considered by Bertini *et al.* To achieve self-crowding, researchers can use a centrifugal field to create a concentration gradient or use regular ultracentrifugation to produce highly concentrated, gel-like solutions. Thus, sedimented solute NMR (SedNMR) provides a simple method to prepare biological samples for solid-stet NMR experiments with minimal perturbation.[106]

Ivanir and Goldbourt have applied solid-state MAS NMR to study CBM3b-Cbh9A (CBM3b), a cellulose binding module protein belonging to family 3b. It is a 146-residue protein having a unique nine-stranded β-sandwich fold, in which 35% of the structure is in a β-sheet conformation and the remainder of the protein is composed of loops and unstructured regions. Close to complete chemical shift assignment of the protein was obtained by combining 2D and 3D experiments using a fully labelled sample and a glycerol-labelled sample. Conformational analysis shows good correlation between the NMR-predicted secondary structure and the reported X-ray crystal structure, in particular in the structured regions.[107]

The conformation of the complement 4 binding protein (C4bp), which plays a crucial role in the inhibition of the complement cascade, has been studied by Luckgei *et al.* The solid-state NMR resonance assignment of the human C4bp C-terminal oligomerization domain has been presented.[108]

The global fold of human cannabinoid type 2 (CB_2) receptor in the agonist-bound active state in lipid bilayers has been investigated by solid-state ^{13}C- and ^{15}N MAS NMR, in combination with chemical-shift prediction from a structural model of the receptor obtained by microsecond-long MD simulations. Uniformly ^{13}C- and ^{15}N-labelled CB_2 receptor was expressed in milligram quantities. ^{13}C MAS NMR spectra were recorded without sensitivity enhancement for direct comparison of C_α, C_β, and CO bands of superimposed resonances with predictions from protein structures generated by MD. The experimental NMR spectra matched the calculated spectra reasonably well indicating agreement of the global fold of the protein between experiment and simulations. The ^{13}C chemical shift distribution of C_α resonances was shown to be very sensitive to both the primary amino acid sequence and the secondary structure of CB_2.[109]

4.5 Membrane proteins

Microsomal monoxygenase enzymes of the cytochrome-P450 family are found in all biological kingdoms, and play a central role in the breakdown of metabolic as well as xenobiotic, toxic and 70% of the drugs in clinical use. Yamamoto *et al.* showed that functional and ssNMR data can reveal interactions between the proteins in fluid lamellar phase bilayers.[110]

Inherent low sensitivity of NMR spectroscopy has been a major disadvantage, especially to study biomolecules like membrane proteins. Ultralow temperatures were used to enhance the NMR sensitivity for solid-state NMR spectroscopy. However, this has led to excessively long T_1 times. This was successfully overcome by using paramagnetic quenching agents.[111]

Despite the fact that bicelles have been shown to very successful in understanding protein-membrane function, the loss of their magnetic-alignment at low temperatures makes them unsuitable to study heat-sensitive membrane proteins like cytochrome-P450 and protein–protein complexes. Yamamoto *et al.* showed that temperature-resistant bicelles that can magnetically-align over a wide range of temperatures were shown with the structural studies of full-length microsomal cytochrome-P450 and cytochrome-b5 by solid-state NMR spectroscopy.[112]

Bicelles composed with many different types of lipids and detergents have been utilised for structural determination of membrane proteins, the NMR requirement of magnetic alignment of bicelles, however, limits the temperature range in which they can be used and subsequently their composition.[113] Because of this restriction, low-temperature experiments desirable for heat-sensitive membrane proteins have not been conducted because bicelles could not be aligned. Here, the magnetic alignment of bicelles was characterised with various compositions for a broad range of temperatures using ^{31}P static NMR spectroscopy in search of temperature-resistant bicelles. Our systematic investigation identified a temperature range of magnetic alignment for bicelles composed of phospholipids (short and long chain), cationic lipids and cholesterol. These systems successfully allowed the alignment of bicelles at low temperatures.

Ultrafast MAS experiments sensitivity can be enhanced through the use of low power sequences combined with paramagnetically enhanced relaxation times to reduce recycle delays, as well as proton detected experiments. In this work the sensitivity of ^{13}C and ^{1}H detected experiments applied to large membrane proteins reconstituted in lipids and packed in small MAS NMR rotors.[114] This is an attractive prospect for samples of limited quantity, as this allows for a reduction in the amount of protein that needs to be produced without the necessity for increased experimental time.

MAS NMR spectroscopy can be used to determine structures of membrane proteins when reconstituted in synthetic lipids, which are a mimic for the natural membrane. A number of experimentally determined interatomic distances and local torsional restraints were used to solve the structure of an oligomeric membrane protein of common seven-helical fold, Anabaena sensory rhodopsin.[115]

The nitrophorins comprise a group of, NO transporting proteins found in the saliva of the blood sucking insect Rhodnius prolixus. ssNMR was employed as an alternative method to gain structural insights on the precipitated protein by Varghese *et al.*[116] Using ^{13}C,^{15}N isotopically labelled proteins were used to show unique structural features that agreed well with other experimental data.

A considerable limitation of NMR spectroscopy is its inherent low sensitivity. Most of the measuring time is used by the spin system to return to its Boltzmann equilibrium after excitation, which is determined by ^{1}H T_1 in CP NMR experiments.[117] Paramagnetic relaxation agents such as Cu^{2+}-EDTA and Gd^{3+}-complexes accelerate this process of relaxation considerably resulting in enhanced sensitivity. Their effect on ^{1}H T_1 has been assessed on the membrane protein proteorhodopsin.

ssNMR spectroscopy has become an important technique for studying the biophysics and structure biology of proteins. Tang *et al.* present the technological advances in ssNMR approaches towards protein structure determination for large membrane proteins and amyloid fibrils.[118] In particular, the ssNMR measurements of an integral membrane protein DsbB, an *Escherichia coli* plasma membrane protein were shown. DsbB supports the formation of a charge-transfer complex between DsbB and ubiquinone during the disulfide bond transfer pathways. The high-resolution structure of the DsbA DsbB complex demonstrates that the joint use of X-ray and ssNMR restraints for membrane proteins with low-resolution crystal structure is generally applicable.

The outer membrane protein BamA is the key player in β-barrel assembly in Gram-negative bacteria. Despite the availability of high-resolution crystal structures, the dynamic behavior of the transmembrane domain and the large periplasmic extension consisting of five POlypeptide-TRansport-Associated domains remains unclear. Sinnige *et al.* used ssNMR to investigate the dynamics of aspects of the protein, showing different timescales associated with different regions of the protein.[119]

Schmidt *et al.* have demonstrated the molecular mobility of the human neuropeptideY receptor type2 reconstituted into dimyristoylphosphatidylcholine (DMPC) membranes was investigated by means of solid-state NMR spectroscopy.[120] A variety of static and MAS NMR experiment revealed the different dynamics of the protein in the membrane.

Michalek *et al.* have determined the high-resolution structure of huntingtin 1–17 using a combination of solution and solid state NMR spectroscopy both in micelles and in oriented phosphatidylcholine bilayers.[121] ssNMR was used to identify specific parts of the structure in the orientated bilayer with angular restraints was obtained using labelled with ^{15}N and ^{2}H at selected sites.

MAS NMR is used here by Miao *et al.* to examine the full length M2 protein embedded in synthetic lipid bilayers. ^{13}C–^{13}C spin diffusion spectral difference has been used to aid the structural determination of the membrane protein. The integral membrane M2 protein of Influenza A virus assembles as a tetrameric bundle to form a proton-conducting channel that is activated by low pH and is essential for the viral lifecycle. Based on known amino acid resonance assignments from amino acid specific labelled samples of truncated M2 sequences or from 3D experiments of uniformly labelled samples, some inter-residue resonances of the full length M2 protein were identified in the difference spectra of uniformly ^{13}C labelled protein that are consistent with the high resolution structure of the M2 protein.[122]

Paramagnetic relaxation enhancement (PRE) is used approach for determining long-range distance constraints in biomolecular solution NMR spectroscopy. Maltsev *et al.* showed that ^{31}P PRE solid-state NMR spectroscopy can be utilized to determine the immersion depth of spin-labeled membrane peptides and proteins. Changes in the ^{31}P NMR PRE times coupled with modelling studies can be used to describe the spin-label position/amino acid within the lipid bilayer and the corresponding helical tilt and the alpha-helical nicotinic acetylcholine receptor nAChR M2 delta transmembrane is used to show this. This PRE ssNMR technique can be used as a molecular ruler to measure membrane immersion depth.[123]

Histidine usually exists in three different forms, Li *et al.* used to show the different protonation and tautomerization states of histidine can be characteristically determined by ^{13}C and ^{15}N chemical shifts of imidazole ring. ssNMR was used to show this and was correlated with modelling studies.[124]

Leftin *et al.* misfolding and aggregation of the intrinsically disordered protein α-synuclein (α-S) are characteristic markers of late-stage

Parkinson's disease. The interaction of α-S on the membrane lipid bilayer is poorly understood. ssNMR, 1D and 2D including, ^{13}C SLF NMR and DROSS were used to probe the interaction between the bilayer and the protein. They utilised a canonical raft type formulation to determine the interaction between the protein.[125]

Ferritins are intracellular proteins that can store thousands of iron(III) ions as a solid mineral. These structures autoassemble from four-helix bundle subunits to form a hollow sphere and are a prototypical example of protein nanocages. ssNMR was used in addition to other solution state techniques to determine its structure.[126]

Kwon *et al.* used ^{2}H NMR to investigate domain formation in membranes (phosphatidylethanolamine (POPE) and anionic 1-palmitoyl-2-oleoyl-*sn-glycero*-3-phosphatidylglycerol (POPG)) with to cationic peptides. The antimicrobial peptides were (AMP(3)) of the beta-hairpin family of protegrin-1 (PG-1), and two cell-penetrating peptides (CPPs), HIV TAT and penetratin. The ^{2}H NMR showed the extent of the interaction between the bilayers and the peptides.[127]

ssNMR showed the conformation that was formed in the lipid bilayers with peptides. Here Huang *et al.* investigated NADPH-cytochrome P450 oxidoreductase (CYPOR) which is an essential redox partner of the cytochrome P450 (cyt P450) superfamily. They showed by ssNMR that CYPOR is a tilted α-helix in the lipid bilayer.[128]

Hansen *et al.* showed from molecular dynamics experiments ^{31}P NMR solid state spectra are simulated. This was used to investigate the interaction between of lipid bilayers containing the antimicrobial peptides *e.g.* alamethicin. Close agreement was found for a range of peptides between simulation and experimental results. A comment was made on the diffusion rate of the phospholipids and the effect it has the ^{31}P NMR spectra.[129]

ssNMR spectroscopy was used to examine subvirus particles generated by Hepatitis B surface antigen (HBsAg) produced from yeast share immunological determinants with mature viruses as shown by Grelard *et al.* These are used in the development of vaccines. They are composed of both proteins and 40 lipids with various degrees of saturation. This study shed insight on the structure of these subvirus particles and the role that water had in them.[130]

ssNMR was used by Fritzsching *et al.* to investigate the canonical rafts forming ternary system both DOPC/eSM/Chol and DOPC/DPPC/Chol. ^{13}C CP MAS was used in conjunction with DSC studies to investigate the differences between DPPC and egg sphingomyelin (eSM). There was reported to be a stronger interaction between cholesterol (Chol) and eSM than DPPC.[131]

Cukkemane *et al.* used ssNMR to look at the cyclic nucleotide-activated K^{+} channel (MlCNG) in lipid bilayers. Comparison of ssNMR data obtained on isotope-labelled cell membranes suggests a tight association of negatively charged lipids to the channel. Polymorphism that extends beyond the ligand binding site, and the corresponding protein segments have been associated with mutant channel types in eukaryotic systems was shown.[132] These findings illustrate the potential of ssNMR for structural investigations on large membrane-embedded proteins, even in the presence of local disorder.

Cross *et al.* have shown the utility of the application of ssNMR spectroscopy to helical membrane proteins. These are very difficult and challenging because of issues of expression and purification of these proteins is very time intensive. Here the role of the membrane environment and the influence of membrane mimics are discussed. Recent progress in expression, purification, reconstitution, sample preparation for ssNMR spectroscopy of both oriented samples and MAS samples has been shown for helical membrane protein structures.[133]

Lipopolysaccharide (LPS) is a major component of the external leaflet of bacterial outer membranes with many other functions. Ciesielski *et al.* use solid state ^{13}C and ^{31}P MAS NMR to investigate the interactions of LPS from three bacterial species with mixed lipid membranes, raft models. The dynamics and structure were investigated by ssNMR and the role of lipid rafts on these dynamics was also examined.[134]

Conformational dynamics are often inferred and quantified from the motional averaging of the NMR parameters. Here De Simone *et al.* have employed oriented solid-state NMR data, such as dipolar couplings and chemical shift anisotropy measured in lipid bicelles, to refine the conformational ensemble of these proteins in lipid membranes. They specifically investigated calcium ATPase SERCA.[135]

The low sensitivity of both the static and MAS techniques of ssNMR spectroscopy has thus far limited the routine application of multidimensional experiments to determine the structure of membrane proteins in lipid bilayers. Mote *et al.* demonstrate the advantage of using a recently developed class of experiments, polarization optimized experiments, for substantial time-savings for 2D and 3D experiments. They demonstrate this methodology using sarcolipin, a single pass membrane protein.[136]

In a series of experiments Murray *et al.* investigated the structure of small helical membrane proteins in different detergents. As their structures are highly sensitive to their environment, solid state NMR is an ideal structural technique that can be used to characterise these membrane proteins in native-like lipid bilayers and proteoliposomes.[137]

Murray *et al.* utilised ssNMR to compare the effect of solution and the presence of membranes upon the conformations of proteins as the structure of membrane proteins is known to be sensitive to the membrane mimetic environment. They compared oriented sample solid-state NMR spectral data of diacylglycerol kinase previously published with predictions of such data from recent structures of this protein.[138]

4.6 Polymers

Li *et al.* used ssNMR to examine the order–disorder phenomenon and spatial heterogeneity of chain packing, partitions of stereodefects, and molecular dynamics of alpha form of isotactic polypropylene (iPP) samples, which are synthesized by Zieglar–Natta catalysts. 1D and 2D ^{13}C MAS NMR revealed the defects in the α-form of iPP, in particular the order–disorder phenomenon in the chain-packing structures.

High-resolution ^{13}C NMR under high-power TPPM decoupling at field strengths of 110 kHz allows observation of the order–disorder phenomenon in the chain-packing structures of α-form.[139]

Hansen *et al.* used ssNMR to describe the function of synthetic and natural macromolecular systems. These critically depend on the packing and dynamics of the individual components of a given system. ssNMR provides structural information with atomic resolution, but it can also provide a way to characterise the amplitude and time scales of motions over broad ranges of length and time. These movements include molecular dynamics, rotational and translational motions of the building blocks, and also the motion of the functional species themselves, such as protons or ions.[140]

Polyacrylonitrile (PAN) and PAN/carbon nanotubes (CNTs) precursor fibers coated by spin finishes were characterized using the ^{1}H ssNMR technique, these include Hahn echo sequences and by the T_1 analysis. Ghoshal *et al.* found that the presence of CNT affects the T_1 relaxation time of the polymer in the composite fiber, however, that of relatively mobile components remains unaffected.[141]

Gradient copolymers can exhibit physical properties that are different than their block polymer analogues. Clough *et al.* describe experimental approaches based on fast and slow MAS NMR, which reveal the amount of rigid and soft phases in styrene-butadiene gradient copolymers with component specific resolution. The differences in structural mobility were shown by the different MAS techniques.[142]

Solid polymer electrolytes for lithium batteries applications are commonly prepared by dissolving a lithium salt in poly(ethylene oxide) (PEO)-based materials. Chiappone *et al.* investigated these systems by a ^{7}Li and ^{19}F NMR study yielding local dynamics and mass transport by temperature-dependent T_1 and PFG-NMR diffusion measurements.[143]

Polymeric materials showing shape-memory behaviour are attracting large interest especially in the field of biomaterials. Borsacchi *et al.* show in this work a multinuclear and multi-technique ssNMR investigation on recently developed hybrid materials formed by sol–gel cross-linked alkoxysilane terminated poly(ε-caprolactone). ^{29}Si, ^{13}C, and ^{1}H MAS NMR present in the hybrid material were characterised by ssNMR.[144]

Polymer/day nanocomposites show remarkably improved properties with respect to their microscale counterparts and pristine polymers and these were examined by Borsacchi *et al.* Due to the substantially apolar character of most of the organic polymers, natural occurring hydrophilic clays are modified into organophilic clays with consequent increase of the polymer/clay compatibility. The samples were also investigated by means of ^{29}Si, ^{13}C, and ^{1}H ssNMR, obtaining information on the structural properties of the modified clays. Moreover, by exploiting the effect of bentonite paramagnetic (Fe^{3+}) ions on ^{1}H T_1, useful information about the extent of the polymer-clay dispersion and their interfacial interactions could be obtained.[145]

The structure of the lamellar phase of aqueous pentaethylene glycol mono-*n*-dodecyl ether ($C_{12}E_5$) surfactant at various temperatures and molar fractions is studied by using united atom molecular dynamics simulations and NMR measurements by Ferreira *et al.*[146]

Gao *et al.* investigated the segmental dynamics of rigid, intermediate, and mobile molecular components in carboxyl terminated polybutadiene (CTPB)/organo-clay (C_{18}-clay) systems was characterized by fully refocused ^{1}H NMR FID. In addition, ^{1}H DQ NMR experiments allowed semiquantitative monitoring of changes in segmental dynamics near the interface.[147]

Hong *et al.* examined chain-folding (CF) structure of semicrystalline polymers, by, ^{13}C–^{13}C DQNMR was used to determine for the first time the detailed CF structure of ^{13}C CH_3-labeled isotactic poly(1-butene) (iPB1) in solution-grown crystals blended with non-labelled poly(1-butene) across a wide range of crystallization temperatures.[148]

Hu *et al.* examined the dynamical heterogeneity in a series of 4,4′-dicyclohexylmethane diisocyanate-diethyltoluenediamine-poly(tetramethylene oxide) based poly(urethane urea) (PUU) elastomers by ssNMR methods. This included ^{1}H wideline, ^{13}C T_1 and time-domain-WISE experiment, which showed the different mobilities within the polymer.[149]

Kang *et al.* investigated the chain packing, crystal thickness, molecular dynamics, and melting temperature of α-form isotactic polypropylene (iPP) drawn uniaxially at high temperatures were investigated using ssNMR and DSC. Two types of iPP samples were studied by ^{13}C NMR and centerband-only detection of exchange (CODEX) NMR. These showed the structural change in the crystalline polymers.[150]

Le *et al.* show the utility of the overall sensitivity enhancements provided by dynamic nuclear polarization (DNP) for the CPNMR characterization of polymer samples doped with biradicals and prepared either by film casting, or by glass forming using 1,1,2,2-tetrachloroethane as the solvent.[151]

Solid-state ^{13}C NMR was used to compare differences in mobility of the cell wall polysaccharides of apples after 20 weeks of storage. Ng *et al.* used two Pulse Phase Modulation (TPPM) decoupling was combined with CP and SPE experiments.[152]

Nieuwendaal *et al.* used ^{13}C CP MAS NMR to examine the semi-crystallinity, high molar mass poly(3-hexylthiophene) (P3HT). The relative degree of crystallinity was estimated for two films with different drying conditions *via* X-ray and differential scanning calorimetry. They were unable to correlate order determined by ^{13}C CP MAS with crystallinity, indicating that local order can occur in non-crystalline regions.[153]

The molecular dynamics within the crystallites of poly(ε-caprolactone), PCL, crystallized from the melt were studied by means of high-field ^{13}C and ^{1}H NMR spectroscopy, including ^{13}C DIPSHIFT and CODEX MAS experiments. Schaeler *et al.* determined the different molecular regimes within the polymer.[154]

The chemical modification of the pendant hydroxyl functional groups on cyclodextrins (CDs) significantly suppresses the hydrogen-bonding interactions between the cyclodextrin molecules and leads to the unique viscoelastic properties of hydroxypropylated polyrotaxane (HyPR). Tang *et al.* investigated the molecular dynamics of PR and HyPR using various solid-state NMR techniques. 2D ^{13}C and ^{1}H wide-line separation (WISE) NMR spectra and centerband-only detection of exchange (CODEX) NMR was used to characterize the slow dynamics of both CD and PEG molecules with frequencies directly comparable to those used in viscoelastic measurements.[155]

The structure of MQ copolymers of the general chemical formula $[(CH_3)_3SiO_{0.5}]_m[SiO_2]_n$ was characterized by means of ssMAS NMR spectroscopy. The MQ copolymers are highly branched polycyclic compounds (densely cross-linked nanosized networks). ^{29}Si NMR spectra were obtained by SPE and ^{29}Si CP MAS together with T_1 studies.[156]

The ultimate draw ratio in semicrystalline polymers, and the resultant mechanical properties of the polymer, strongly depends on entanglement density in the amorphous region of the polymer. Yao *et al.* used 2D ^{13}C exchange NMR methods have been employed to follow the spatial proximity between the methylene segments of the noncrystalline regions (mobile- and rigid-amorphous phases) with the crystalline regions (crystalline core and crystal surface) during deformation of the two polymers.[157]

4.7 Carbonaceous materials and soils

Zhou *et al.* studied the chemical composition of soil organic matter (SOM) along a 2000 year paddy soil chronosequence in eastern China by use of advanced ssNMR spectroscopy as well as FTIR, aiming to identify changes in the chemical composition of SOM over a millennium timescale.[158]

Identifying which components of organic matter are preserved over millennia in paleosols is important for understanding the stabilization mechanisms of soil organic matter (SUM) and in evaluating the portion of the terrestrial carbon stock that might be attributed to paleosols. Using advanced ssNMR techniques, the chemical composition of clay-associated SUM in a chronosequence of four paleosols were examined by Zhou *et al.*[159]

Gabrienko *et al.* have used the ^{13}C-label scrambling from the position 3 to the position 1 in the ^{13}C-labeled propene adsorbed on zinc oxide has been performed with ^{1}H and ^{13}C ssNMR *in situ*. This allowed the characterization of this mechanism.[160]

Anderson *et al.* have examined the hydrothermal stability of carbon based acid catalysts synthesized by sulfonating carbohydrates pyrolyzed at moderate temperatures has been reported previously and examined this by many techniques including ssNMR.[161]

Chemical changes in the structure of kerogen during simulated thermal maturation were investigated by 1D and 2D ssNMR ^{13}C NMR spectroscopy. The spin counting technique was used to evaluate the share of "observable" carbon atoms with increasing hydrous pyrolysis temperature.[162]

Eldridge *et al.* used ssNMR ^{13}C NMR and elemental composition data in a molecular mixing model, to estimate the molecular components of the organic matter in 16 recycled organic (RO) wastes representative of the major materials generated in the Sydney basin area.[163]

McBeath *et al.* used ssNMR ^{13}C NMR to evaluate the carbon chemistry of twenty-six biochars produced from eleven different feedstocks at production temperatures. ^{13}C NMR spectra were acquired using both CP and SPE techniques. Overall, the corresponding CP and SPE spectra were similar, although aromaticity was slightly higher and observability much higher when DP was used. The relative size and purity of the aromatic ring structures were also gauged using the ring current technique.[164]

Rabbi *et al.* used ssNMR to understand the changes in soil organic matter (SOM) composition during aggregate formation is crucial to explain the stabilization of SOM in aggregates. ^{13}C CP MAS NMR was used to determine the SOM composition in different fractions.[165]

4.8 Organometallic and coordination compounds

Lennartson *et al.* examined nine structurally related paramagnetic acetylacetonato nickel(II) complexes by solid state ^{13}C and ^{2}H NMR spectroscopy. The ^{2}H MAS NMR was used to probe the local hydrogen bonding in the structure. The size of the CSA was also examined and related to the paramagnetic centre.[166]

A series of transition-metal organometallic complexes with commonly occurring metalchlorine bonding motifs were characterized using ^{35}Cl ssNMR spectroscopy, ^{35}Cl NQR spectroscopy, and first-principles density functional theory (DFT) calculations of NMR interaction tensors. Static ^{35}Cl ultra-wideline NMR spectra were acquired in a piecewise manner at standard (9.4T) and high (21.1T) magnetic field strengths using the WURST-QCPMG pulse sequence. The EFG and CSA tensor parameters were readily extracted from analytical simulations of the spectra.[167]

The first ^{1}H, ^{13}C, ^{31}P and ^{11}B MAS NMR studies of electron-deficient carborane-containing ruthenium and osmium complexes are reported by Barry *et al.*[168]

Burgess *et al.* reported ^{13}C and ^{25}Mg ssNMR characterization of a series of metal organic frameworks (MOFs). These were benzoate and salicylate with magnesium complexes, based around the featuring Mg^{2+} ions in organic coordination environments. The C_Q for ^{25}Mg species were determined and compared with computational studies.[169]

A series of calcium and strontium complexes featuring aryl carboxylate ligands, binding to Ca^{2+} and Sr^{2+}, has been prepared and characterized by alkaline-earth (^{43}Ca and ^{87}Sr) solid-state NMR experiments in a magnetic field of 21.1 T were studied by Burgess *et al.* Determination of the C_Q of the Sr and Ca nuclei were undertaken and compared to computational calculations.[170]

Ruthenocene, and other ruthenium complexes were shown by Cluff *et al.*, some of which were labelled with ^{2}H, were adsorbed on silica surfaces by grinding the polycrystalline materials with silica. The progress was monitored by ^{1}H, ^{13}C, and ^{2}H solid-state NMR spectroscopy. The transition from the crystal lattice to the surface species that are highly mobile is proven by strongly reduced chemical shift anisotropies and diminished dipolar interactions. The ^{2}H MAS spectra of surface-adsorbed ferrocene-d_2 prove that the motion of the metallocenes on the surfaces is fast and nearly isotropic, as in solution.[171]

In this work, He *et al.* successfully prepared several prototypical ^{17}O enriched metal–organic frameworks (MOFs). Depending on the target MOF, different isotopic enrichment methods were used to effectively incorporate ^{17}O from ^{17}O-enriched H_2O. Using these ^{17}O-enriched MOFs, were able to acquire ^{17}O ssNMR spectra at a magnetic field of 21.1 T.[172]

The molecular stacking of an artificial light-harvesting antenna self-assembled from aminofunctionalized zinc-chlorins was determined by ssNMR by Pandit *et al.* in combination with computational modelling. These included ^{1}H–^{13}C HETCOR and other experiments.[173]

Smart *et al.* prepared three new N-heterocyclic-silazane compounds, and examined by ssNMR spectroscopy including ^{1}H wPMLG, ^{29}Si CP MAS, and 2D ^{1}H–^{29}Si dipolar HETCOR experiments and other techniques.[174]

4.9 Glasses and amorphous solids

Martel *et al.* investigated phase separation in glasses which occurs in silica-rich glasses however, some aspects as the first steps of the nucleation of domains with different compositions remain to be clarified as they are difficult to characterize with most analytical techniques. T_1 studies and 2D ^{29}Si–^{29}Si correlation experiments were used to characterize the Si–O–Si network after various heat treatments and are used to obtain the compositions of the two demixing phases.[175]

In this study by Aschenbrenner *et al.* they examined the glassy structure of a sugar matrix that encapsulated freeze-dried bacteria by ^{1}H T_2 measurements. The aim of this work was to evaluate whether, in case of freeze-dried bacteria, the protective effect of a surrounding glassy matrix can be ascribed to its inherent restricted molecular mobility. They examined the mobility as a function of the temperature.[176]

Silicon and aluminium chemical environments in silicate and aluminosilicate glasses were examined by Gambuzzi *et al.* They looked at the relationships between the local structural environments and the ^{29}Si and ^{27}Al NMR parameters are highlighted and showed that the isotropic chemical shift of both ^{29}Si and ^{27}Al increases of about +5 ppm for each Al added in the second sphere and both the ^{29}Si and ^{27}Al isotropic chemical shifts linearly decrease with the reduction of the average Si/Al–O–T bond angle.[177]

^{1}H, ^{23}Na and ^{29}Si ssNMR was used by Lapina *et al.* to investigate the formation of fiberglass zirconium-silicate supports and catalysts. H–D exchange was used to investigate the surface sites.[178]

Mathew *et al.* studied silicate-based bioactive glasses (BGs). These are known to depend critically on both the P content of the glass and its silicate network connectivity. However, while the bioactivity generally displays a nonmonotonic dependence on phosphate content itself, recent work suggest that it is merely the net orthophosphate content that directly links to the bioactivity. These were examined by both ^{31}P and ^{29}Si MAS NMR together with computational simulation.[179]

^{29}Si–^{1}H CP MAS were employed to probe directly the interface of a silica-filled polysiloxane elastomer subjected to γ-radiation. T_1 times in the rotating frame were used to investigate the dynamics of the degraded siloxane.[180]

4.10 Micro- and mesoporous solids

Luzgin *et al.* used ^{13}C MAS to examine the mechanism of dimethyl ether (DME) carbonylation with carbon monoxide has been studied on solid metal-containing acidic cesium salt of 12-tungstophosphoric acid, $M/Cs_2HPW_{12}O_{40}$ (M/HPA, M = Ag, Pt, and Rh). The kinetics of the reaction has been monitored with ^{1}H MAS NMR *in situ*.[181]

The joint conversion of methane and propane on Mo-containing zeolite H-BEA has been studied with ssNMR and other techniques. The involvement of ^{13}C-methane in the reaction of propane aromatization to form benzene and toluene as the products has been demonstrated by Luzgin *et al.*[182]

4.11 Surface science and catalysis

The synthesis and full characterization of platinum nanoparticles prepared are reported. In these studies, (^{13}CO) was used as a probe molecule to investigate the surface of the particles, using IR and solid-state NMR spectroscopies with magic angle spinning (MAS-NMR). Spectroscopic data suggest a modification of the electronic state of the nanoparticles between 1.2 and 2.0 nm which can be related to the presence of Knight shift.[183]

The ^{17}O labelling of the oxo moiety of a series of related molecular and silica-supported tungsten oxo metathesis catalysts was achieved by Merle *et al.*, and the corresponding ^{17}O MAS NMR spectra were recorded. Combined experimental and theoretical NMR studies yielded information on the local structure of the surface species.[184]

A microautoclave MAS NMR rotor was developed enabling *in situ* monitoring of solid–liquid–gas reactions at high temperatures and pressures. It is used in a kinetic and mechanistic study of the reactions of cyclohexanol on zeolite. The ^{13}C NMR spectra show that dehydration of ^{13}C-cyclohexanol that occurs with a proposed mechanism.[185]

4.12 Inorganic and other related solids

The exceptional upconverting properties of lanthanide-doped nanoparticles make them attractive systems with applications ranging from photovoltaics to biological labelling, imaging, and therapeutics. However, the structural understanding of the particles is poor, here, Arnold *et al.* used ^{23}Na, ^{19}F, and ^{89}Y ssNMR, including 3QMAS, together with DFT calculations to characterise the structure of cubic $NaYF_4$ nanoparticles with and without Er^{3+} doping.[186]

Nanocrystalline CaF_2 powder specimens were prepared and examined by ssNMR and powder XRD by Abdellatief *et al.* The specimen homogeneity and a detailed picture of the lattice defects can be assessed by the simultaneous analysis of the X-ray powder diffraction pattern and of the solid-state ^{19}F MAS NMR T_1 relaxometry data. The relationship between linewidths, structure and defects were discussed.[187]

A comprehensive molecular description of the inorganic–bioorganic interfaces and internal structure of the aragonitic shells of Perna canaliculus is derived by employing ssNMR spectroscopy and X-ray diffraction by Ben Shir *et al.* The nature of the $CaCO_3$ was investigated and compared with X-ray diffraction. The bioorganic–inorganic interfaces in a mollusk shell and demonstrates the unique capability of NMR.[188] Molecular-level structure–property relationships in biogenic calcium carbonates was examined by ssNMR by Ben Shir *et al.*[189]

Blanc *et al.* showed in this review the utility of multinuclear NMR spectroscopy to examine the different forms of electrochemical cells, in the form of batteries or supercapacitors and fuel cells. Here a wide variety of different nuclei are examined including ^{7}Li, ^{11}B, ^{13}C and ^{195}Pt amongst many others. Chemical transformations of the batteries or fuel cells are readily observed using NMR spectroscopy as they are placed directly into the NMR coil in the spectrometers. Oxidation and reduction processes can also be observed in real time in this manner.[190]

In this study Bracco *et al.* used exploiting ^{1}H, ^{13}C and ^{31}P MAS NMR to examine the effect of different types of ivory. Both the inorganic and organic components were examined with differences shown for the calcium phosphate part between the different ivories types. The signal assigned to the bulk PO_4^{3-} groups of the hydroxyapatite phase and of minor side peaks ascribed to unprotonated surface sites PO_x (PO, PO_2^- and PO_3^{2-}) and to protonated sites PO_xH on the surface of the nano-sized crystals of the hydroxyapatite.[191]

^{29}Si MAS NMR was used by Cappelletto *et al.* to examine calcium silicate hydrate gel (C–S–H) present in Portland cement and is the complex phase mostly responsible for the binding properties and the mechanical resistance of the material. The effects of different comb-shaped superplasticizers on the silicate structure are investigate. The analysis of ^{29}Si MAS NMR spectra shows that the additives increase the degree of polymerization and hence the average length of the silicate chains in C–S–H.[192]

The electrochemical lithiation of undoped, P-doped and B-doped nano-silicon particles has been studied during the first cycle by *ex situ* ^{6}Li and ^{7}Li MAS-NMR spectroscopy by Cattaneo *et al.* Samples were charged within pouch cells up to capacities followed by NMR analysis. Different crystalline phases occurred after higher capacitance was induced. Other effects including boron doping on the silicon nano-particles was also examined.[193]

Structural characterization of brownmillerite, $Ba_2In_2O_5$, were ssNMR was achieved by an approach combining experimental ssNMR spectroscopy, and computational calculations was carried out by Dervisoglu *et al.* A different number of resonances were observed from that previously reported.[194]

The potentiality of ^{31}P ssNMR to determine in a non-destructive way the different states of phosphates in cheeses was examined by Gobet *et al.* Sixteen semi-hard cheeses of various compositions were studied, and three fractions of phosphates (P) were distinguished according to their mobility: (1) mobile soluble P (2) mobile insoluble P and (3) immobile insoluble P. This method was shown to be of sufficient accuracy to evidence the effect of ripening on the different P fractions.[195]

The ceramic lithium ion conductor β-Li_3PS_4 has a disordered and nanoporous structure that leads to an enhancement in ionic conductivity compared to the crystalline g-phase was studied by Gobet *et al.* Multinuclear (^{1}H, ^{6}Li, ^{7}Li, ^{31}P) ssNMR spectroscopy is used to characterise thermal treatment.[196]

^{115}In ssNMR spectroscopy was applied to characterise a variety of low oxidation-state indium(I) compounds by Hamaed *et al.* ^{115}In static wideline and MAS NMR spectra of several In(I) complexes were acquired with moderate and ultra-high field NMR spectrometers (9.4 and 21.1 T, respectively). ^{115}In MAS NMR spectra were obtained with moderate and ultra-fast (>60 kHz) spinning speeds at 21.1 T of low oxidation-state indium complexes.[197]

The intermetallic compounds Y_2RuB_6, Y_2ReB_6, Y_2RhSi_3, YRh_2Si, and YPdSi were synthesised by Hoeting *et al.* These were examined by X-ray diffraction and ^{89}Y ssNMR spectra. The NMR signal shifts were correlated with the density of states of the s electrons at the Fermi level, resulting in a non-ambiguous assignment of the NMR signals to the various sites in the crystal structures.[198]

^{89}Y ssNMR spectra show large Knight shifts arising from unpaired conduction electron spin density near the Fermi edge. The silicides YT_2Si_2 (T = Co, Ni, Cu, Ru, Rh, Pd) were synthesized and studied by Hoeting *et al.*[199]

Discharged lithium-type battery cathodes are investigated by ^{6}Li MAS NMR spectroscopy has been used identify lithium products that are formed in the cathodes and differences between products formed with the different catalysts and solvents. This study by Huff *et al.* showed that there are significant differences in the products formed in the cathodes with the two different solvents.[200]

Ion mobility and transport properties of the fluorite-type solid solutions $50PbF_2.30BiF_3.20RbF$ and $50PbF_2.25BiF_3.25RbF$, $50PbF_2.30BiF_3.20RbF$ were studied by ^{19}F ssNMR spectroscopy and conductivity measurements. An analysis of ^{19}F NMR spectra made it possible to assess the temperature effect on the nature of ion motions in the fluoride sublattice, to identify types of ionic mobility, and to determine the corresponding temperature ranges.[201]

In a study by Kavun *et al.* ionic mobility and electrical transport in the $50PbF_2.30BiF_3.20NaF$ fluorite solid solution have been studied by ^{19}F and ^{23}Na NMR and impedance spectroscopy. They were able to determine the motion and mobility of the ions by this method.[202]

A multi-nuclear solid-state NMR approach by Leskes *et al.*, including ^{17}O MAS NMR, was employed to investigate the lithium–air battery, to monitor the evolution of the electrochemical products formed during cycling, and to gain insight into processes affecting capacity fading. Other processes that occurred during the cycling were monitored by ^{13}C MAS. The utility of ssNMR spectroscopy in directly detecting product formation and decomposition within the battery is demonstrated, a necessary step in the assessment of new electrolytes, catalysts, and cathode materials for the development of a viable lithium–oxygen battery.[203]

In this study, novel ^{31}P MAS NMR spectroscopy was used to investigate the mechanism of phosphate sorption on aluminium hydroxides under different environmental conditions. In addition the aluminium was analysed *in situ* by a double-resonance ^{31}P–^{27}Al REAPDOR. All of these measurements reveal that NMR spectroscopy is a useful analytical tool for studying phosphorus chemistry at environmental interfaces.[204]

ssNMR spectroscopy was used in a combination with other techniques *e.g.* Raman, X-ray diffraction to investigate the role of carbonate in apatites by McElderry *et al.* There was a correlation with ^{31}P MAS NMR linewidth and the crystal structure of the carbonated apatites as a function of carbonate content. It was proposed that two domains formed as a function of carbonate content. This is a useful tool for obtaining structural information on bone organic and mineral components and synthetic model minerals at the atomic-level.[205]

For the interrogation of precursor nucleation for vapour deposition processes like atomic layer deposition and chemical vapour deposition, a modified method for quantitative analysis of surface coverage was undertaken *via* NMR. Substrates were analysed by ^{1}H, ^{29}Si and ^{13}C ssNMR to determine the degree of coverage and interaction with the surface.[206]

Two hexanuclear niobium halide cluster compounds with a $[Nb_6X_{12}]^{2+}$ (X = CI, Br) diamagnetic cluster core, have been studied by a combination of experimental solid-state NMR/NQR techniques and PAW/GIPAW calculations. For niobium sites the NMR parameters were determined by using variable 8, field static broadband NMR measurements and additional NQR measurements. These were studied by ^{95}Mo in addition to $^{79}Br/^{81}Br$ NQR and ^{35}Cl solid-state NMR measurements.[207]

The structure of roller-quenched fast ion conductive glasses (FICs) $(Ag_3PO_4)_x(AgI)_{(1-x)}$ $(0.15 \leqq x \leqq 0.50)$ is investigated by ^{109}Ag and ^{31}P solid-state NMR spectroscopies including double-quantum-based dipolar recoupling effects nuclear alignment reduction (DQ-DRENAR) techniques.[208]

Glasses along the composition line $0.5Al_2O_3–xSiO_2$ $(1 \leq x \leq 6)$ were prepared *via* a novel sol–gel route using tetraethylorthosilicate and aluminum lactate as precursors. The structural evolution from solution to gel to glass is monitored by standard ^{27}Al and ^{29}Si MAS NMR spectroscopies, including SQMAS, DQMAS, REAPDOR and CPMG studies, both the short range and medium were examined.[209] Sol–gel preparation of mesoporous Al_2O_3–SiO_2 glasses: structural evolution monitored by ssNMR studied by Ren *et al.*[210]

Lattice disorder in hydrozincite – natural carbonate, natural hydrozincite samples and synthetic analogues were investigated by ssNMR and other techniques. Sanna *et al.* used ^{13}C MAS and CP MAS NMR spectra to show distinct signals, these studies were carried out in addition to $T_{1\rho}$ and CSA analysis. Differences between the natural and synthetic analogues were discussed.[211]

^{1}H and ^{27}Al NMR is used to reveal the motions of AlH_4 anions in $KAlH_4$. Sorte *et al.* used line-narrowing from rotations and from translational diffusion is observed in the NMR of both nuclei. The studies include ^{1}H lineshape analysis and relaxation using both T_1 and $T_{1\rho}$ times. Lack of sharp structure in the first-order quadrupolar pattern and the absence of second-order quadrupole structure in the ^{27}Al NMR data indicated asymmetry variations in the anion electric field gradients from structural disorder.[212]

Quantitative ^{17}O MAS NMR investigation by Sun *et al.* of the $^{17}O/^{16}O$ isotope exchange on pure-anatase TiO_2/A and mixed-phase TiO_2/A + R with an anatase to rutile ratio of 1 : 1 has been performed. ^{17}O MAS NMR studies of the surface exchange was evidenced.[213]

Taylor *et al.* used ^{127}I and ^{207}Pb ssNMR to study of lead iodide. This is a layered structure that experiences polytypism. The polytype contains high rates of iodine vacancies but retains its stoichiometry. This was studied by ^{127}I and ^{207}Pb T_1 relaxation methods of the polytype below 400 K.[214]

The role of plastic-crystalline succinonitrile (SN) within the electrolyte system comprising polyethylene oxide (PEO), SN and $LiBF_4$ was investigated. Voigt *et al.* used ssNMR spectroscopy and powder X-ray diffraction to determine the function of the SN molecule in the electrolyte.[215]

Vyalikh *et al.* used ssNMR to investigate a model system, fluorapatite–gelatin, of natural dental and bone tissues. ssNMR including ^{31}P, ^{19}F, and ^{1}H MAS NMR and heteronuclear correlations, found the nanocomposite to consist of crystalline apatite-like regions to be in close contact with a more dissolved (amorphous) layer containing first motifs of the apatite crystal structure as well as the organic component.[216]

Broadline static ^{51}V NMR, and other techniques are used to study the thermal behaviour of a manganese-nickel pyrovanadate complex $Mn_{2-2x}Ni_{2x}V_2O_7$. The lineshapes reveal the structure of the solid solution.[217]

Kim *et al.* used ^{17}O MAS NMR was used to provide unique insight into rare and usual oxide *e.g.* Ta_2O_5 which has been made by enriching the oxide during ion spluttering in and enriched ^{17}O atmosphere.[218]

Palke *et al.* used ^{31}P MAS NMR spectroscopy to examine flux-grown solid solutions of $La_{1-x}Ce_xPO_4$ (x between 0.027 and 0.32) having the monoclinic monazite structure, and of $Y_{1-x}M_xPO_4$ ($M = Vn^+$, Ce^{3+}, Nd^{3+}, x between 0.001 and 0.014) having the tetragonal zircon structure. Paramagnetically shifted NMR resonances are observed in all samples due to the presence of paramagnetic Vn^+, Ce^{3+}, and Nd^{3+} in the diamagnetic $LaPO_4$ or YPO_4. The increased resolution given by the paramagnetic interactions also leads to the observation of splitting of specific resonances in the ^{31}P NMR spectra that may be caused by local, small-scale distortions from the substitution of ions having dissimilar ionic radii.[219]

Thompson *et al.* examined ^{27}Al and ^{17}O MAS NMR data on several calcium aluminosilicate glasses along two silica isopleths crossing the metaluminous (SiO_2–$CaAl_2O_4$) join, prepared with varying quench rates and thus with fictive temperatures spanning approximately 100 K. They investigated the configurational changes that occur with temperature, and how these vary with composition and also the role of Al(v) species.[220]

References

1 R. W. Schurko, *Acc. Chem. Res.*, 2013, **46**, 1985–1995.

2 K. Saalwaechter, *ChemPhysChem*, 2013, **14**, 3000–3014.

3 L. A. O'Dell, *Solid State Nucl. Magn. Reson.*, 2013, **55–56**, 28–41.

4 M. Werner, N. Rothermel, H. Breitzke, T. Gutmann and G. Buntkowsky, *Isr. J. Chem.*, 2014, **54**, 60–73.

5 A. Goldbourt, *Curr. Opin. Biotechnol.*, 2013, **24**, 705–715.
6 A. Krushelnitsky, D. Reichert and K. Saalwachter, *Acc. Chem. Res.*, 2013, **46**, 2028–2036.
7 V. Ladizhansky, *Isr. J. Chem.*, 2014, **54**, 86–103.
8 T. Gopinath, K. R. Mote and G. Veglia, *Prog. Nucl. Magn. Reson. Spectrosc.*, 2013, **75**, 50–68.
9 G. Goobes, *Isr. J. Chem.*, 2014, **54**, 113–124.
10 C. Bonhomme, C. Gervais and D. Laurencin, *Prog. Nucl. Magn. Reson. Spectrosc.*, 2014, **77**, 1–48.
11 M. Foston, *Curr. Opin. Biotechnol.*, 2014, **27**, 176–184.
12 D. Massiot, R. J. Messinger, S. Cadars, M. Deschamps, V. Montouillout, N. Pellerin, E. Veron, M. Allix, P. Florian and F. Fayon, *Acc. Chem. Res.*, 2013, **46**, 1975–1984.
13 S. E. Ashbrook, D. M. Dawson and V. R. Seymour, *Phys. Chem. Chem. Phys.*, 2014, **16**, 8223–8242.
14 T. Gutmann, A. Gruenberg, N. Rothermel, M. Werner, M. Srour, S. Abdulhussain, S. Tan, Y. Xu, H. Breitzke and G. Buntkowsky, *Solid State Nucl. Magn. Reson.*, 2013, **55–56**, 1–11.
15 S. Li and F. Deng, *Annu. Rep. NMR Spectrosc.*, 2013, **78**, 1–54.
16 S. Asami and B. Reif, *Acc. Chem. Res.*, 2013, **46**, 2089–2097.
17 Y. Q. Ye, M. Malon, C. Martineau, F. Taulelle and Y. Nishiyama, *J. Magn. Reson.*, 2014, **239**, 75–80.
18 M. R. Chierotti, R. Gobetto, C. Nervi, A. Bacchi, P. Pelagatti, V. Colombo and A. Sironi, *Inorg. Chem.*, 2014, **53**, 139–146.
19 G. Hou, R. Gupta, T. Polenova and A. J. Vega, *Isr. J. Chem.*, 2014, **54**, 171–183.
20 D. Carnevale, A. Linde, G. Bauer and G. Bodenhausen, *Chem. Phys. Lett.*, 2013, **580**, 172–178.
21 T. Sinnige, M. Daniels, M. Baldus and M. Weingarth, *J. Am. Chem. Soc.*, 2014, **136**, 4452–4455.
22 J. O. Struppe, C. Yang, Y. Wang, R. V. Hernandez, L. M. Shamansky and L. J. Mueller, *J. Magn. Reson.*, 2013, **236**, 89–94.
23 M. E. Halse, J. Schlagnitweit and L. Emsley, *Isr. J. Chem.*, 2014, **54**, 136–146.
24 A. Equbal, S. Paul, V. S. Mithu, J. M. Vinther, N. C. Nielsen and P. Madhu, *J. Magn. Reson.*, 2014, **244**, 68–73.
25 M. Bjerring, S. Jain, B. Paaske, J. M. Vinther and N. C. Nielsen, *Acc. Chem. Res.*, 2013, **46**, 2098–2107.
26 Z. Zhang, R. Fu, J. Li and J. Yang, *J. Magn. Reson.*, 2014, **242**, 214–219.
27 R. L. Johnson and K. Schmidt-Rohr, *J. Magn. Reson.*, 2014, **239**, 44–49.
28 K. J. Harris, S. L. Veinberg, C. R. Mireault, A. Lupulescu, L. Frydman and R. W. Schurko, *Chem. – Eur. J.*, 2013, **19**, 16469–16475.
29 K. Basse, S. K. Jain, O. Bakharev and N. C. Nielsen, *J. Magn. Reson.*, 2014, **244**, 85–89.
30 S. K. Jain, A. B. Nielsen, M. Hiller, L. Handel, M. Ernst, H. Oschkinat, U. Akbey and N. C. Nielsen, *Phys. Chem. Chem. Phys.*, 2014, **16**, 2827–2830.
31 S. M. Althaus, K. Mao, J. A. Stringer, T. Kobayashi and M. Pruski, *Solid State Nucl. Magn. Reson.*, 2014, **57–58**, 17–21.
32 E. C. Lin and S. J. Opella, *J. Magn. Reson.*, 2014, **239**, 57–60.
33 G. Hou, S. Yan, J. Trebosc, J. P. Amoureux and T. Polenova, *J. Magn. Reson.*, 2013, **232**, 18–30.
34 C. Shi, H. K. Fasshuber, V. Chevelkov, S. Xiang, B. Habenstein, S. K. Vasa, S. Becker and A. Lange, *J. Biomol. NMR*, 2014, **59**, 15–22.
35 K. R. Thurber and R. Tycko, *J. Chem. Phys.*, 2014, **140**, 184201.

36 H. Takahashi, C. Fernandez-De-Alba, D. Lee, V. Maurel, S. Gambarelli, M. Bardet, S. Hediger, A. L. Barra and G. De Paepe, *J. Magn. Reson.*, 2014, **239**, 91–99.
37 A. A. Smith, B. Corzilius, O. Haze, T. M. Swager and R. G. Griffin, *J. Chem. Phys.*, 2013, **139**, 214201.
38 O. Ouari, P. Trang, F. Ziarelli, G. Casano, F. Aussenac, P. Thureau, D. Gigmes, P. Tordo and S. Viel, *ACS Macro Lett.*, 2013, **2**, 715–719.
39 Y. S. Ong, A. Lakatos, J. Becker-Baldus, K. M. Pos and C. Glaubitz, *J. Am. Chem. Soc.*, 2013, **135**, 15754–15762.
40 E. Nimerovsky, R. Gupta, J. Yehl, M. Li, T. Polenova and A. Goldbourt, *J. Magn. Reson.*, 2014, **244**, 107–113.
41 A. Goldbourt, *Isr. J. Chem.*, 2014, **54**, 125–135.
42 E. S. Mananga, *Solid State Nucl. Magn. Reson.*, 2013, **55–56**, 54–62.
43 E. S. Mananga, *Solid State Nucl. Magn. Reson.*, 2013, **55–56**, 63–72.
44 L. J. Edwards, D. Savostyanov, V. A. Nevzorov, M. Concistre, G. Pileio and I. Kuprov, *J. Magn. Reson.*, 2013, **235**, 121–129.
45 S. E. Ashbrook and D. M. Dawson, *Acc. Chem. Res.*, 2013, **46**, 1964–1974.
46 E. C. Lin and S. J. Opella, *J. Magn. Reson.*, 2013, **237**, 40–48.
47 T. Kobayashi, K. Mao, P. Paluch, A. Nowak-Krol, J. Sniechowska, Y. Nishiyama, D. T. Gryko, M. J. Potrzebowski and M. Pruski, *Angew. Chem., Int. Ed.*, 2013, **52**, 14108–14111.
48 L. Zhou, S. Li, Y. Su, X. Yi, A. Zheng and F. Deng, *J. Phys. Chem. B*, 2013, **117**, 8954–8965.
49 N. Salvi, J. Frey, D. Carnevale, M. Graetzel and G. Bodenhausen, *Dalton Trans.*, 2014, **43**, 6389–6395.
50 F. A. Perras and D. L. Bryce, *Chem. Sci.*, 2014, **5**, 2428–2437.
51 T. Wiegand, H. Eckert, J. Ren, G. Brunklaus, R. Froehlich, C. G. Daniliuc, G. Luebbe, K. Bussmann, G. Kehr, G. Erker and S. Grimme, *J. Phys. Chem. A*, 2014, **118**, 2316–2331.
52 A. E. Aliev and D. Courtier-Murias, *J. Phys. Chem. A*, 2013, **117**, 7855–7862.
53 X. Lu, J. Trebosc, O. Lafon, D. Carnevale, S. Ulzega, G. Bodenhausen and J. P. Amoureux, *J. Magn. Reson.*, 2013, **236**, 105–116.
54 F. A. Perras and D. L. Bryce, *J. Am. Chem. Soc.*, 2013, **135**, 12596–12599.
55 C. Martineau, F. Decker, F. Engelke and F. Taulelle, *Solid State Nucl. Magn. Reson.*, 2013, **55–56**, 48–53.
56 M. K. Sorensen, O. Bakharev, O. Jensen, H. J. Jakobsen, J. Skibsted and N. C. Nielsen, *J. Magn. Reson.*, 2014, **238**, 20–25.
57 S. Parthasarathy, Y. Nishiyama and Y. Ishii, *Acc. Chem. Res.*, 2013, **46**, 2127–2135.
58 A. Marchanka, B. Simon and T. Carlomagno, *Angew. Chem., Int. Ed.*, 2013, **52**, 9996–10001.
59 F. Pourpoint, Y. Morin, R. M. Gauvin, J. Trebosc, F. Capet, O. Lafon and J. P. Amoureux, *J. Phys. Chem. C*, 2013, **117**, 18091–18099.
60 E. Emelina, A. Petrov and D. V. Filyukov, *Russ. J. Org. Chem.*, 2014, **50**, 412–421.
61 R. K. Chandrappa, P. Ochsenbein, C. Martineau, M. Bonin, G. Althoff, F. Engelke, H. Malandrin, B. Castro, M. El Hajji and F. Taulelle, *Cryst. Growth Des.*, 2013, **13**, 4678–4687.
62 B. Osmialowski, E. Kolehmainen, K. Ejsmont, S. Ikonen, A. Valkonen, K. Rissanen and Nonappa, *J. Mol. Struct.*, 2013, **1054**, 157–163.
63 J. K. Harper, D. Tishler, D. Richardson, J. Lokvam, R. Pendrill and G. Widmalm, *J. Phys. Chem. A*, 2013, **117**, 5534–5541.
64 W. Chen, H. J. Sun and T. Miyoshi, *J. Phys. Chem. B*, 2013, **117**, 13698–13709.

65 M. Weingarth and M. Baldus, *Acc. Chem. Res.*, 2013, **46**, 2037–2046.
66 E. Carignani, S. Borsacchi, J. P. Bradley, S. P. Brown and M. Geppi, *J. Phys. Chem. C*, 2013, **117**, 17731–17740.
67 F. G. Vogt, G. R. Williams and R. C. Copley, *J. Pharm. Sci.*, 2013, **102**, 3705–3716.
68 T. G. Nunes, M. Viciosa, N. T. Correia, F. Danede, R. G. Nunes and H. P. Diogo, *Mol. Pharm.*, 2014, **11**, 727–737.
69 F. G. Vogt, H. Yin, R. G. Forcino and L. Wu, *Mol. Pharm.*, 2013, **10**, 3433–3446.
70 X. Kong, M. Shan, V. Terskikh, I. Hung, Z. Gan and G. Wu, *J. Phys. Chem. B*, 2013, **117**, 9643–9654.
71 N. M. Dicaire, F. A. Perras and D. L. Bryce, *Can. J. Chem.*, 2014, **92**, 9–15.
72 P. Rossi, E. Macedi, P. Paoli, L. Bernazzani, E. Carignani, S. Borsacchi and M. Geppi, *Cryst. Growth Des.*, 2014, **14**, 2441–2452.
73 J. K. Williams, D. Tietze, J. Wang, Y. Wu, W. F. DeGrado and M. Hong, *J. Am. Chem. Soc.*, 2013, **135**, 9885–9897.
74 P. R. Desai, A. R. Cormier, P. P. Shah, R. R. Patlolla, A. K. Paravastu and M. Singh, *Eur. J. Pharm. Biopharm.*, 2014, **86**, 190–199.
75 L. Mazur, K. N. Jarzembska, R. Kaminski, K. Wozniak, E. Pindelska and M. Zielinska-Pisklak, *Cryst. Growth Des.*, 2014, **14**, 2263–2281.
76 X. Yuan, D. Sperger and E. J. Munson, *Mol. Pharm.*, 2014, **11**, 329–337.
77 A. Gayen, J. R. Banigan and N. J. Traaseth, *Angew. Chem., Int. Ed.*, 2013, **52**, 10321–10324.
78 F. G. Vogt, K. Roberts-Skilton and S. A. Kennedy-Gabb, *Pharm. Res.*, 2013, **30**, 2315–2331.
79 S. Li, L. Zhou, Y. Su, B. Han and F. Deng, *Solid State Nucl. Magn. Reson.*, 2013, **54**, 13–17.
80 A. Roehrich and G. Drobny, *Acc. Chem. Res.*, 2013, **46**, 2136.
81 L. Zhang, L. Liu, S. Maltsev, G. A. Lorigan and C. Dabney-Smith, *Biochim. Biophys. Acta, Biomembr.*, 2014, **1838**, 413–418.
82 B. Kwon, A. J. Waring and M. Hong, *Biophys. J.*, 2013, **105**, 2333–2342.
83 L. E. Marbella, H. S. Cho and M. M. Spence, *Biochim. Biophys. Acta, Biomembr*, 2013, **1828**, 1674–1682.
84 S. R. Leonard, A. R. Cormier, X. Pang, M. I. Zimmerman, H. X. Zhou and A. K. Paravastu, *Biophys. J.*, 2013, **105**, 222–230.
85 P. J. Sherman, F. Separovic and J. H. Bowie, *Peptides*, 2014, **55**, 98–102.
86 G. J. Lu and S. J. Opella, *J. Chem. Phys.*, 2013, **139**, 084203.
87 P. Ma, J. D. Haller, J. Zajakala, P. Macek, A. C. Sivertsen, D. Willbold, J. Boisbouvier and P. Schanda, *Angew. Chem., Int. Ed.*, 2014, **53**, 4312–4317.
88 E. Barbet-Massin, A. J. Pell, K. Jaudzems, W. Franks, J. S. Retel, S. Kotelovica, I. Akopjana, K. Tars, L. Emsley, H. Oschkinat, A. Lesage and G. Pintacuda, *J. Biomol. NMR*, 2013, **56**, 379–386.
89 E. Schmidt, J. Gath, B. Habenstein, F. Ravotti, K. Szekely, M. Huber, L. Buchner, A. Boeckmann, B. H. Meier and P. Guentert, *J. Biomol. NMR*, 2013, **56**, 243–254.
90 E. Barbet-Massin, A. J. Pell, M. J. Knight, A. L. Webber, I. C. Felli, R. Pierattelli, L. Emsley, A. Lesage and G. Pintacuda, *ChemPhysChem*, 2013, **14**, 3131–3137.
91 S. H. Park, C. Yang, S. J. Opella and L. J. Mueller, *J. Magn. Reson.*, 2013, **237**, 164–168.
92 J. T. Nielsen, N. Kulminskaya, M. Bjerring and N. C. Nielsen, *J. Biomol. NMR*, 2014, **59**, 119–134.
93 T. V. Krivokhizhina and R. Wittebort, *J. Magn. Reson.*, 2014, **243**, 33–39.

94 M. K. Pandey, S. Viyekanandan, K. Yamamoto, S. Im, L. Waskell and A. Ramamoorthy, *J. Magn. Reson.*, 2014, **242**, 169–179.
95 U. Akbey, A. J. Nieuwkoop, S. Wegner, A. Voreck, B. Kunert, P. Bandara, F. Engelke, N. C. Nielsen and H. Oschkinat, *Angew. Chem., Int. Ed.*, 2014, **53**, 2438–2442.
96 A. E. Aliev and D. Courtier-Murias, *Biopolymers*, 2014, **101**, 246–256.
97 T. Zinkevich, V. Chevelkov, B. Reif, K. Saalwaechter and A. Krushelnitsky, *J. Biomol. NMR*, 2013, **57**, 219–235.
98 J. D. Haller and P. Schanda, *J. Biomol. NMR*, 2013, **57**, 263–280.
99 J.-M. Lopez del Amo, D. Schneider, A. Loquet, A. Lange and B. Reif, *J. Biomol. NMR*, 2013, **56**, 359–363.
100 L. Banci, O. Blazevits, F. Cantini, J. Danielsson, L. Lang, C. Luchinat, J. Mao, M. Oliveberg and E. Ravera, *J. Biol. Inorg. Chem.*, 2014, **19**, 659–666.
101 J. E. Jenkins, S. Sampath, E. Butler, J. Kim, R. W. Henning, G. P. Holland and J. L. Yarger, *Biomacromolecules*, 2013, **14**, 3472–3483.
102 K. Sackett, M. J. Nethercott, Z. Zheng and D. P. Weliky, *J. Mol. Biol.*, 2014, **426**, 1077–1094.
103 S. Y. Liao, K. J. Fritzsching and M. Hong, *Protein Sci.*, 2013, **22**, 1623–1638.
104 I. Sengupta, P. S. Nadaud and C. P. Jaroniec, *Acc. Chem. Res.*, 2013, **46**, 2117–2126.
105 M. Fragai, C. Luchinat, G. Parigi and E. Ravera, *J. Biomol. NMR*, 2013, **57**, 155–166.
106 I. Bertini, C. Luchinat, G. Parigi and E. Ravera, *Acc. Chem. Res.*, 2013, **46**, 2059–2069.
107 H. Ivanir and A. Goldbourt, *J. Biomol. NMR*, 2014, **59**, 185–197.
108 N. Luckgei, B. Habenstein, F. Ravotti, S. Megy, F. Penin, J. B. Marchand, F. Hill, A. Boeckmann and B. H. Meier, *Biomol. NMR Assignments*, 2014, **8**, 1–6.
109 T. Kimura, K. Vukoti, D. L. Lynch, D. P. Hurst, A. Grossfield, M. C. Pitman, P. H. Reggio, A. A. Yeliseev and K. Gawrisch, *Proteins: Struct., Funct., Bioinf.*, 2014, **82**, 452–465.
110 K. Yamamoto, U. H. Duerr, J. Xu, S. C. Im, L. Waskell and A. Ramamoorthy, *Sci. Rep.*, 2013, **3**, 2538.
111 K. Yamamoto, M. A. Caporini, S. Im, L. Waskell and A. Ramamoorthy, *J. Magn. Res.*, 2013, **237**, 175–181.
112 K. Yamamoto, M. Gildenberg, S. Ahuja, S. C. Im, P. Pearcy, L. Waskell and A. Ramamoorthy, *Sci. Rep.*, 2013, **3**, 2556.
113 K. Yamamoto, P. Pearcy and A. Ramamoorthy, *Langmuir*, 2014, **30**, 1622–1629.
114 M. E. Ward, S. Wang, S. Krishnamurthy, H. Hutchins, M. Fey, L. S. Brown and V. Ladizhansky, *J. Biomol. NMR*, 2014, **58**, 37–47.
115 S. Wang, R. A. Munro, L. Shi, I. Kawamura, T. Okitsu, A. Wada, S. Y. Kim, K. H. Jung, L. S. Brown and V. Ladizhansky, *Nat. Methods*, 2013, **10**, 1007–1012.
116 S. Varghese, F. Yang, V. Pacheco, K. Wrede, A. Medvedev, H. Ogata, M. Knipp and H. Heise, *Biochemistry*, 2013, **52**, 7031–7040.
117 S. J. Ullrich, S. Hoelper and C. Glaubitz, *J. Biomol. NMR*, 2014, **58**, 27–35.
118 M. Tang, G. Comellas and C. M. Rienstra, *Acc. Chem. Res.*, 2013, **46**, 2080–2088.
119 T. Sinnige, M. Weingarth, M. Renault, L. Baker, J. Tommassen and M. Baldus, *J. Mol. Biol.*, 2014, **426**, 2009–2021.
120 P. Schmidt, L. Thomas, P. Mueller, H. A. Scheidt and D. Huster, *Chem. – Eur. J.*, 2014, **20**, 4986–4992.
121 M. Michalek, E. S. Salnikov and B. Bechinger, *Biophys. J.*, 2013, **105**, 699–710.

122 Y. Miao, T. A. Cross and R. Fu, *J. Biomol. NMR*, 2013, **56**, 265–273.
123 S. Maltsev, S. M. Hudson, I. D. Sahu, L. Liu and G. A. Lorigan, *J.Phys. Chem. B*, 2014, **118**, 4370–4377.
124 S. Li, L. Zhou, Y. Su, B. Han and F. Deng, *Solid State Nucl. Magn. Reson.*, 2013, **54**, 13–17.
125 A. Leftin, C. Job, K. Beyer and M. F. Brown, *J. Mol. Biol.*, 2013, **425**, 2973–2987.
126 D. Lalli and P. Turano, *Acc. Chem. Res.*, 2013, **46**, 2676–2685.
127 B. Kwon, A. J. Waring and M. Hong, *Biophys. J.*, 2013, **105**, 2333–2342.
128 R. Huang, K. Yamamoto, M. Zhang, N. Popovych, I. Hung, S. C. Im, Z. Gan, L. Waskell and A. Ramamoorthy, *Biophys. J.*, 2014, **106**, 2126–2133.
129 S. K. Hansen, M. Vestergaard, L. Thogersen, B. Schiott, N. C. Nielsen and T. Vosegaard, *J. Phys. Chem. B*, 2014, **118**, 5119–5129.
130 A. Grelard, P. Guichard, P. Bonnafous, S. Marco, O. Lambert, C. Manin, F. Ronzon and E. J. Dufourc, *FASEB J.*, 2013, **27**, 4316–4326.
131 K. J. Fritzsching, J. Kim and G. P. Holland, *Biochim. Biophysim. Acta, Biomembr.*, 2013, **1828**, 1889–1898.
132 A. Cukkemane and M. Baldus, *ChemBioChem*, 2013, **14**, 1789–1798.
133 T. A. Cross, V. Ekanayake, J. Paulino and A. Wright, *J. Magn. Reson.*, 2014, **239**, 100–109.
134 F. Ciesielski, D. C. Griffin, M. Rittig, I. Moriyon and B. B. Bonev, *Biochim. Biophys. Acta, Biomemb.*, 2013, **1828**, 1731–1742.
135 A. De Simone, K. R. Mote and G. Veglia, *Biophys. J.*, 2014, **106**, 2566–2576.
136 K. R. Mote, T. Gopinath and G. Veglia, *J. Biomol. NMR*, 2013, **57**, 91–102.
137 D. T. Murray, J. Griffin and T. A. Cross, *Biochemistry*, 2014, **53**, 2454–2463.
138 D. T. Murray, C. Li, P. Gao, H. Qin and T. A. Cross, *Biophys. J.*, 2014, **106**, 1559–1569.
139 Z. Li, T. Miyoshi, M. K. Sen, T. Koga, A. Otsubo and A. Kamimura, *Macromolecules*, 2013, **46**, 6507–6519.
140 M. R. Hansen, R. Graf and H. W. Spiess, *Acc. Chem. Res.*, 2013, **46**, 1996–2007.
141 S. Ghoshal, Y. Liu, P. Gulgunje, K. Gupta, H. G. Chae, J. Leisen and S. Kumar, *J. Appl. Polym. Sci.*, 2014, **131**, 40734.
142 A. Clough, J. L. Sigle, A. Tapash, L. Gill, N. V. Patil, J. Zhou and J. L. White, *Macromolecules*, 2014, **47**, 2625–2631.
143 A. Chiappone, S. Jeremias, R. Bongiovanni and M. Schoenhoff, *J. Polym. Sci., Part B: Polym. Phys.*, 2013, **51**, 1571–1580.
144 S. Borsacchi, K. Paderni, M. Messori, M. Toselli, F. Pilati and M. Geppi, *Macromolecules*, 2014, **47**, 3544–3552.
145 S. Borsacchi, U. Sudhakaran, M. Geppi, L. Ricci, V. Liuzzo and G. Ruggeri, *Langmuir*, 2013, **29**, 9164–9172.
146 T. M. Ferreira, D. Topgaard and O. Ollila, *Langmuir*, 2014, **30**, 461–469.
147 Y. Gao, R. Zhang, W. Lv, Q. Liu, X. Wang, P. Sun, H. Winter and G. Xue, *J. Phys. Chem. C*, 2014, **118**, 5606–5614.
148 Y. l. Hong and T. Miyoshi, *ACS Macro Lett.*, 2014, **3**, 556–559.
149 W. Hu and A. J. Hsieh, *Polymer*, 2013, **54**, 6218–6225.
150 J. Kang and T. Miyoshi, *Macromolecules*, 2014, **47**, 2993–3004.
151 D. Le, G. Casano, T. N. Phan, F. Ziarelli, O. Ouari, F. Aussenac, P. Thureau, G. Mollica, D. Gigmes, P. Tordo and S. Viel, *Macromolecules*, 2014, **47**, 3909–3916.
152 J. K. Ng, Z. D. Zujovic, B. G. Smith, J. W. Johnston, R. Schroeder and L. D. Melton, *Carbohydr. Res.*, 2014, **386**, 1–6.
153 R. C. Nieuwendaal, C. R. Snyder and D. M. DeLongchamp, *ACS Macro Lett.*, 2014, **3**, 130–135.

154 K. Schaeler, A. Achilles, R. Baerenwald, C. Hackel and K. Saalwaechter, *Macromolecules*, 2013, **46**, 7818–7825.
155 C. Tang, A. Inomata, Y. Sakai, H. Yokoyama, T. Miyoshi and K. Ito, *Macromolecules*, 2013, **46**, 6898–6907.
156 S. G. Vasil'ev, V. I. Volkov, E. A. Tatarinova and A. M. Muzafarov, *Appl. Magn. Reson.*, 2013, **44**, 1015–1025.
157 Y. Yao, S. Jiang and S. Rastogi, *Macromolecules*, 2014, **47**, 1371–1382.
158 Z. Zhou, X. Cao, K. Schmidt-Rohr, D. C. Olk, S. Zhuang, J. Zhou, Z. Cao and J. Mao, *Biol. Fertil. Soils*, 2014, **50**, 571–581.
159 Z. Zhou, N. Chen, X. Cao, T. Chua, J. Mao, R. D. Mandel, E. Bettis, Arthur, III and M. L. Thompson, *Geoderma*, 2014, **223**, 54–61.
160 A. A. Gabrienko and A. G. Stepanov, *Chem. Phys. Lett.*, 2014, **607**, 21–24.
161 J. M. Anderson, R. L. Johnson, K. Schmidt-Rohr and B. H. Shanks, *Carbon*, 2014, **74**, 333–345.
162 N. Burdelnaya, D. Bushnev, M. Mokeev and A. Dobrodumov, *Fuel*, 2014, **118**, 308–315.
163 S. Eldridge, C. Chen, Z. Xu, P. Nelson, S. Boyd, I. Meszaros and K. Chan, *Waste Manage.*, 2013, **33**, 2157–2169.
164 A. V. McBeath, R. J. Smernik, E. S. Krull and J. Lehmann, *Biomass Bioenergy*, 2014, **60**, 121–129.
165 S. Rabbi, R. Linser, J. M. Hook, B. R. Wilson, P. V. Lockwood, H. Daniel and I. M. Young, *Commun. Soil Sci. Plant Anal.*, 2014, **45**, 1523–1537.
166 A. Lennartson, L. U. Christensen, C. J. McKenzie and U. G. Nielsen, *Inorg. Chem.*, 2014, **53**, 399–408.
167 K. E. Johnston, C. A. O'Keefe, R. M. Gauvin, J. Trebosc, L. Delevoye, J. P. Amoureux, N. Popoff, M. Taoufik, K. Oudatchin and R. W. Schurko, *Chem. – Eur. J.*, 2013, **19**, 12396–12414.
168 N. P. Barry, T. F. Kemp, P. J. Sadler and J. V. Hanna, *Dalton Trans.*, 2014, **43**, 4945–4949.
169 K. M. Burgess, Y. Xu, M. C. Leclerc and D. L. Bryce, *J. Phys. Chem. A*, 2013, **117**, 6561–6570.
170 K. M. Burgess, Y. Xu, M. C. Leclerc and D. L. Bryce, *Inorg. Chem.*, 2014, **53**, 552–561.
171 K. J. Cluff, N. Bhuvanesh and J. Bluemel, *Organometallics*, 2014, **33**, 2671–2680.
172 P. He, J. Xu, V. V. Terskikh, A. Sutrisno, H. Y. Nie and Y. Huang, *J. Phys. Chem. C*, 2013, **117**, 16953–16960.
173 A. Pandit, K. Ocakoglu, F. Buda, T. van Marle, A. R. Holzwarth and H. J. de Groot, *J. Phys. Chem. B*, 2013, **117**, 11292–11298.
174 K. A. Smart, M. Grellier, Y. Coppel, L. Vendier, S. A. Mason, S. C. Capelli, A. Abinati, V. Montiel-Palma, M. A. Munoz-Hernandez and S. Sabo-Etienne, *Inorg. Chem.*, 2014, **53**, 1156–1165.
175 L. Martel, D. Massiot and M. Deschamps, *J. Non-Cryst. Solids*, 2014, **390**, 37–44.
176 M. Aschenbrenner, E. Grammueller, U. Kulozik and P. Foerst, *Food Bioprocess Technol.*, 2014, **7**, 1012–1024.
177 E. Gambuzzi, A. Pedone, M. C. Menziani, F. Angeli, D. Caurant and T. Charpentier, *Geochim. Cosmochim. Acta*, 2014, **125**, 170–185.
178 O. Lapina, D. Khabibulin, E. Papulovskiy, A. Shubin and Y. Gulyaeva, *J. Struct. Chem.*, 2013, **54**, S152–S167.
179 R. Mathew, B. Stevensson, A. Tilocca and M. Eden, *J. Phys. Chem. B*, 2014, **118**, 833–844.
180 B. P. Mayer, S. C. Chinn, R. S. Maxwell and J. A. Reimer, *Polym. Degrad. Stab.*, 2013, **98**, 1362–1368.

181 M. V. Luzgin, M. S. Kazantsev, G. G. Volkova and A. G. Stepanov, *J. Catal.*, 2013, **308**, 250–257.
182 M. V. Luzgin, A. V. Toktarev, V. N. Parmon and A. G. Stepanov, *J. Phys. Chem. C*, 2013, **117**, 22867–22873.
183 S. Kinayyigit, P. Lara, P. Lecante, K. Philippot and B. Chaudret, *Nanoscale*, 2014, **6**, 539–546.
184 N. Merle, G. Girard, N. Popoff, A. De Mallmann, Y. Bouhoute, J. Trebosc, E. Berrier, J. F. Paul, C. P. Nicholas, I. del Rosal, L. Maron, R. M. Gauvin, L. Delevoye and M. Taoufik, *Inorg. Chem.*, 2013, **52**, 10119–10130.
185 A. Vjunov, M. Y. Hu, J. Feng, D. M. Camaioni, D. Mei, J. Z. Hu, C. Zhao and J. A. Lercher, *Angew.Chem., Int. Ed.*, 2014, **53**, 479–482.
186 A. A. Arnold, V. Terskikh, Q. Y. Li, R. Naccache, I. Marcotte and J. A. Capobianco, *J. Phys. Chem. C*, 2013, **117**, 25733–25741.
187 M. Abdellatief, M. Abele, M. Leoni and P. Scardi, *J. Appl. Crystal.*, 2013, **46**, 1049–1057.
188 I. Ben Shir, S. Kababya, I. Katz, B. Pokroy and A. Schmidt, *Chem. Mater.*, 2013, **25**, 4595–4602.
189 I. Ben Shir, S. Kababya and A. Schmidt, *Isr. J.Chem.*, 2014, **54**, 74–85.
190 F. Blanc, M. Leskes and C. P. Grey, *Acc. Chem. Res.*, 2013, **46**, 1952–1963.
191 S. Bracco, A. Brajkovic, A. Comotti and V. Rolandi, *Period. Mineral.*, 2013, **82**, 239–250.
192 E. Cappelletto, S. Borsacchi, M. Geppi, F. Ridi, E. Fratini and P. Baglioni, *J. Phys. Chem. C*, 2013, **117**, 22947–22953.
193 A. S. Cattaneo, S. Dupke, A. Schmitz, J. P. Badillo, M. Winter, H. Wiggers and H. Eckert, *Solid State Ionics*, 2013, **249**, 41–48.
194 R. Dervisoglu, D. S. Middlemiss, F. Blanc, L. A. Holmes, Y. L. Lee, D. Morgan and C. P. Grey, *Phys. Chem. Chem. Phys.*, 2014, **16**, 2597–2606.
195 M. Gobet, S. Buchin, C. Rondeau-Mouro, B. Mietton, E. Guichard, C. Moreau and J. L. Le Quere, *Food Anal. Methods*, 2013, **6**, 1544–1550.
196 M. Gobet, S. Greenbaum, G. Sahu and C. Liang, *Chem. Mater.*, 2014, **26**, 3558–3564.
197 H. Hamaed, K. E. Johnston, B. F. Cooper, V. V. Terskikh, E. Ye, C. L. Macdonald, D. C. Arnold and R. W. Schurko, *Chem. Sci.*, 2014, **5**, 982–995.
198 C. Hoeting, H. Eckert, F. Haarmann and R. Poettgen, *Z. Anorg. Allg. Chem.*, 2014, **640**, 1303–1308.
199 C. Hoeting, H. Eckert, S. F. Matar, U. C. Rodewald and R. Poettgen, *Z. Naturforsch., B: J. Chem. Sci.*, 2014, **69**, 305–312.
200 L. A. Huff, J. L. Rapp, L. Zhu and A. A. Gewirth, *J. Power Sources*, 2013, **235**, 87–94.
201 V. Kavun, N. Uvarov, V. Goncharuk, E. Merkulov, A. Ulikhin, I. Telin and V. I. Kharchenko, *Solid State Ionics*, 2014, **257**, 17–22.
202 V. Kavun, N. Uvarov, I. Telin, R. Yaroshenko, A. Ulikhin, A. Podgorbunskii and V. Goncharuk, *Inorg. Mater.*, 2013, **49**, 1157–1161.
203 M. Leskes, A. J. Moore, G. R. Goward and C. P. Grey, *J. Phys. Chem. C*, 2013, **117**, 26929–26939.
204 W. Li, X. Feng, Y. Yan, D. L. Sparks and B. L. Phillips, *Environ. Sci. Technol.*, 2013, **47**, 8308–8315.
205 J. D. McElderry, P. Zhu, K. H. Mroue, J. Xu, B. Pavan, M. Fang, G. Zhao, E. McNerny, D. H. Kohn, R. T. Franceschi, M. M. Holl, M. M. Tecklenburg, A. Ramamoorthy and M. D. Morris, *J. Solid State Chem.*, 2013, **206**, 192–198.
206 P. J. Pallister, S. C. Buttera and S. T. Barry, *J. Phys. Chem. C*, 2014, **118**, 1618–1627.

207 B. Peric, R. Gautier, C. J. Pickard, M. Bosiocic, M. S. Grbic and M. Pozek, *Solid State Nucl. Magn. Reson.*, 2014, **59–60**, 20–30.
208 J. Ren and H. Eckert, *J. Phys. Chem. C*, 2013, **117**, 24746–24751.
209 J. Ren, L. Zhang and H. Eckert, *J. Phys. Chem. C*, 2014, **118**, 4906–4917.
210 J. Ren, L. Zhang and H. Eckert, *J. Sol-Gel Sci. Technol.*, 2014, **70**, 482–490.
211 R. Sanna, G. De Giudici, A. M. Scorciapino, C. Floris and M. Casu, *Am. Mineral.*, 2013, **98**, 1219–1226.
212 E. G. Sorte, S. B. Emery, E. Majzoub, T. Ellis-Caleo, Z. L. Ma, B. A. Hammann, S. E. Hayes, R. C. Bowman and M. S. Conradi, *J. Phys. Chem. C*, 2014, **118**, 5725–5732.
213 X. Sun, M. Dyballa, J. Yan, L. Li, N. Guan and M. Hunger, *Chem. Phys. Lett.*, 2014, **594**, 34–40.
214 R. Taylor, P. A. Beckmann, S. Bai and C. Dybowski, *J. Phys. Chem. C*, 2014, **118**, 9143–9153.
215 N. Voigt and L. van Wuellen, *Solid State Ionics*, 2014, **260**, 65–75.
216 A. Vyalikh, P. Simon, E. Rosseeva, J. Buder, R. Kniep and U. Scheler, *J. Phys. Chem. B*, 2014, **118**, 724–730.
217 N. Zhuravlev, M. V. Rotermel, T. I. Krasnenko and R. Samigullina, *J. Struct. Chem.*, 2013, **54**, 126–129.
218 N. Kim and J. F. Stebbins, *J. Non-Cryst. Solids*, 2013, **378**, 158–162.
219 A. C. Palke, J. F. Stebbins and L. A. Boatner, *Inorg. Chem.*, 2013, **52**, 12605–12615.
220 L. M. Thompson and J. F. Stebbins, *Am. Mineral.*, 2013, **98**, 1980–1987.

NMR of proteins and nucleic acids

Peter J. Simpson[a,b]
DOI: 10.139/9781782622758-00348

The dazzling wealth of data that Nuclear Magnetic Resonance (NMR) spectroscopy can provide on biological systems leads to its application in many thousands of studies published annually. This review highlights a number of key papers which used the technique in the field of Structural Biology, appearing in peer-reviewed journals over the 12 month period up until May 2014. The first part of the chapter deals with papers which present methodological advances, in particular in sample production and new or improved algorithms for data analysis. Subsequently, a number of studies are summarised, selected for their particular biological interest or due to the challenging nature of the system under investigation. As far as possible within the constraints of space, the methods used are described. The review is thus intended to provide the reader with a snapshot of the "state of the art" of NMR in Structural Biology.

1 Introduction

This chapter reports a number of publications published recently where NMR has been key to the study of biological processes, *i.e.* in the field of Structural Biology. Those familiar with the field will be aware that, because of the wealth of information that NMR can provide, it is used routinely in an ever-increasing number of studies, with the number of peer-reviewed publications annually running into the thousands. Hence, this review can only report a smattering of studies and is aimed to provide the reader only with a snapshot of the field. As always, the chapter is split roughly into two, with Section 2 (New Methodology) highlighting reports in technique development and Sections 3–5 reporting a selection of NMR studies of biological systems. The sub-sections in New Methodology are always governed by trends in publication and interestingly, the first of these (2.1) on new sample production methods this year is relatively small, comprising short sub-sections on (2.1.1) Isotope Labelling and (2.1.2) Incorporation of Paramagnetic Centres. The second Section (2.2) Software, Web Tools and Protocols is significantly larger and of the five sub-sections covering data analysis, assignment, structure determination/validation and data acquisition, sub-section 2.2.3, "Novel Structure Calculation and/or Integrative Approaches", is by far the largest. This reflects much of the recent development in the field revolving around better utilisation of readily-accessible data, predominantly chemical shifts. Whilst chemical shift (CS) assignment is a pre-requisite of most studies, for assigning distance restraints, probing dynamics or ligand binding (see an excellent review),[1] as the NMR frequency contains information both on the local environment and motion of a nucleus, in

[a]*Department of Life Sciences, Imperial College London, Exhibition Road, London SW7 2AZ, UK*
[b]*Bruker UK Ltd, Banner Lane, Coventry CV4 9GH, UK.*
E-mail: peter.simpson@bruker.co.uk

theory it should be all that is required in many cases. Thus, many of the developments are in CS-only *de novo* structure determination, with much of Section 2.2.3 dedicated to ROSETTA-type approaches. The extraction of motional information has also recently become possible with the "random coil index" (RCI), which this year has been extended to report on sidechain dynamics.[2] Also in the New Methodology section is a short report on Approaches for Larger Biomolecules (2.3).

The second half of the chapter gives what can only be some examples of High Resolution Studies of Biomolecules (3; split crudely into protein structures, nucleic acids and protein–nucleic interactions), NMR Characterisation of Dynamics (4) and an ever-increasing section on "NMR Studies of Large Biomolecules" (5). Whilst much of the research highlighted in these sections involves some measurement of dynamics, Section 4 is included to spotlight a few key publications which illustrate the importance of motion in biological function. Several of these use the recently-developed Chemical Exchange Saturation Transfer (CEST) approach, which promises to be a powerful tool for studying conformational exchange on the ms–μs timescale along with relaxation dispersion. Finally, it remains to be said that due to space restraints, the chapter is limited primarily to solution-state studies, except in a few exceptional cases. An example of the latter (discussed in Section 2.3) uses magic angle spinning (MAS) of protein solutions to reduce the linewidth of very large systems, permitting assignment of a protein within a 1.1 MDa complex *via* 3D triple resonance experiments.

2 New methodology

2.1 Samples

2.1.1 Isotope labelling. Deuteration to varying degrees is a well-established approach for studying proteins and nucleic acids, taking advantage of the spectral simplification resulting from reducing ^{1}H density, or changes in dipolar and scalar coupling effects. These effects are discussed in a comprehensive review this year from Tugarinov.[3] For isotope labelling of Val and Leu methyls in an otherwise perdeuterated background, minimal medium can be supplemented with labelled 2-ketoisovalerate. Due to the narrow spectral range in which these groups resonate it may be advantageous to label only Val or Leu individually, especially in larger systems or those with naturally-low dispersion, such as helical membrane proteins. Lichtenecker *et al.* demonstrate that if 2-ketoisocaproate is added as a supplement, an intermediate in the conversion of 2-ketoisovalerate to L-leucine, Leu can be selectively labelled or "unlabelled".[4] The authors demonstrate this with production of ^{13}C-Val, ^{15}N-immunoglobulin binding protein G (GB1) in minimal medium using 100 mg L^{-1} $^{13}CH_3$-2-ketoisovalerate and 100 mg L^{-1} (unlabelled) 2-ketoisocaproate. As pointed out by Mas and colleagues, for very large systems significant overlap can still occur and the neighbouring prochiral methyl provides a proximal source of dipolar interaction.[5] Their solution is to use 2-acetolactate as the isotope source, permitting stereospecific labelling of Leu and Val, with suppression of incorporation into Leu by the addition of perdeuterated Leucine to the minimal medium. The spectral simplification of labelling the 468 kDa archaeal aminopeptidase TET2 with solely $^{13}CH_3$-pro-*S*-Val in an otherwise perdeuterated background is demonstrated, with concomitant improvement in methyl linewidths. An alternative is proposed by Miyanoiri and co-workers, who used the appropriately stereo-labelled Leu and Val as additives to minimal medium.[6] Whilst Leu incorporation is quite efficient, with no metabolic scrambling at 20 mg L^{-1}, Val required >100 mg L^{-1} to achieve >80% labelling. Production of the 82 kDa malate synthase G enzyme labelled in separate samples at pro-*R*- or pro-*S*- positions of Leu or Val is shown. Whilst it is now possible to specifically label any methyl group in any methyl-bearing amino acid in *E. coli*, prokaryotic expression is often not suitable for producing functional eukaryotic proteins. Miyazawa-Onami and co-workers investigated I, L, V labelling in the yeast *Kluyveromyces lactis*.[7] Adding precursors to the growth medium yielded highly perdeuterated maltose-binding protein (MBP) with good incorporation of ^{1}H and ^{13}C at the Ile δ^1 methyl. However, Val and Ile labelling from α-ketoisovalerate was poor, presumably due to the lack of suitable metabolite transporters in mitochondria, where amino acid biosynthesis takes place. The less cost-effective solution of supplementing the medium with 100 mg L^{-1} of labelled Val was successful. In a separate publication to their methyl-labelling paper (above), Lichtenecker *et al.* also describe labelling of Phe and Tyr residues by the addition of ^{13}C-α-ketoacid precursors to media used for the growth of *E. coli*.[8] These aromatics are notoriously difficult to assign, but are often important in defining the structure of a protein's hydrophobic core. They also benefit from a ^{1}H–^{13}C TROSY effect and can provide a useful site for measuring dynamics, if complications with the large ^{13}C–^{13}C *J*-coupling can be circumnavigated. Supplementing the growth medium with ~100 mg L^{-1} of phenyl- or 4-hydroxyphenylpyruvate led to high incorporation in GB1, with no observable metabolic scrambling. The precursors can be synthesised with labels incorporated at various positions, demonstrated in separate samples with $^{13}C^{\alpha}$ labelling and ^{12}C, ^{2}H phenylalanine which was ^{13}C, ^{1}H-labelled at the ring ε position, providing isolated ^{13}C–^{1}H spin pairs for relaxation studies. A different approach from Kasinath *et al.* for the latter application uses ^{13}C-4-erythrose with deuterated pyruvate, which results in ^{1}H–^{13}C labelling of Phe and Tyr solely at the δ^2 position and Trp at ε^3.[9] Utility is demonstrated with labelling of Flavodoxin and with ^{13}C relaxation measurements in the aromatics of Calmodulin.

2.1.2 Incorporation of paramagnetic centres. As will be evident from the second half of this chapter, the use of paramagnetic probes is providing insight into biological systems at a level of detail not previously imaginable. As a result, the volume of research published using restraints derived from such species has multiplied, as attested by a number of informative reviews recently. Their application in the structure determination of protein–protein complexes is discussed in a paper from one of the main proponents of their use.[10] Their utility in delineating so-called "invisible" states that proteins and nucleic acids can transiently access is the subject of two publications from Clore, with examples taken from his own laboratory's work.[11,12] One of the most straightforward applications is as a probe of solvent accessibility and surface burial due to paramagnetic relaxation enhancement (PRE) of nuclei by addition of a paramagnetic compound to the solvent. A very good overview of so-called "solvent PREs" (sPREs), both theory and practice, has been written by Hocking and colleagues.[13] A common paramagnetic co-solvent is Gd(III)-DTPA-BMA, commercially available as a contrast agent. However, the octadentate chelator leaves space for water co-ordination, and the increased relaxation experienced by this ligand can be transferred to protein nuclei by exchange, making the measurement of surface burial less reliable. For this reason, Gu *et al.* have synthesised a decadentate Gd(III) species which precludes water from the inner co-ordination sphere.[14] This proved to be a more inert paramagnetic co-solvent, as illustrated by comparison of measured- *versus* back-calculated PREs in three proteins. In most cases, a paramagnetic centre is introduced into the system either through engineering a Lanthanide (Ln)-binding site or by covalently attaching a chelating group or stable free radical site-specifically. Structural restraints derived from these, such as PREs, or where the magnetic susceptibility is anisotropic, induced pseudo-contact shifts (PCSs) and residual dipolar couplings (RDCs), can then be averaged by any relative motion between the spin label and system under study (discussed in detail in a review[15] and the effect on structural quality investigated in a paper[16]). Huang *et al.* have developed a small lanthanide chelating group which can be attached to a protein *via* a disulfide bond.[17] The poly-aromatic 4MTDA group is relatively rigid and gave large PCSs and H^N–N^H RDCs up to 13 Hz, depending on the Ln used. The penta-coordinate moiety benefits from having spare chelating sites which can be occupied by, *e.g.* a carboxylate sidechain on the protein positioned at an $i+4$ position along a helix, which further restricts its flexibility. Liu *et al.* took the approach of engineering a metal binding site for their study of the dynamics of a bacterial periplasmic binding protein.[18] A di-histidine motif was introduced within an a helix, which enabled coordination of Cu(II) bound to nitriloacetic acid (NTA) to act as a "cap" to prevent a second protein binding simultaneously. The rigidity of this simple motif enabled accurate PREs to be measured, as verified by introducing the spin label at two different sites in the protein.

2.2 Software, web tools and protocols

2.2.1 Software packages for data analysis. As noted in the introduction to a special journal edition on computational methods,[19] progress in biomolecular NMR spectroscopy has been greatly facilitated by improvements in software, algorithms and more recently web tools and databases. As will be evident from the following sections, this continues to be the case and applications span all aspects, from webservers allowing sharing and analysis of the pulse sequences used to collect data[20] to specific programs which assist in untangling the structural and dynamics information encoded in the RDC. An example of the latter is the REDCAT software which permits structural analysis of RDC data. The package has been updated by Schmidt *et al.* who describe, amongst other improvements, a new interface to the molecular visualisation software VMD and a GUI which provides an easier link to XPLOR-NIH.[21] The influence of dynamics on the RDC spans the previously "invisible" timescale from τ_c to a few tens of μs and so is of much interest, but is technically challenging to extract. Sabo and colleagues have published a more efficient procedure for obtaining the generalised RDC order parameter, S^2_{RDC} from multiple alignment measurements, termed Optimised RDC-based Iterative and Unified Model-free analysis (ORIUM).[22] The protocol uses concepts from both previously-proposed iterative schemes based around the so-called "direct interpretation of dipolar couplings" (DIDC) and "model free analysis" (MFA) approaches. In addition, a new method for estimating a lower bound on protein motion using only one set of RDCs is shown. The authors demonstrate the application to the previously well-characterised ubiquitin and GB3 proteins. Many functional processes are thought to occur on a slower (ms–μs) timescale and where these can be studied by relaxation dispersion techniques, additional information on the chemical shifts and populations of states accessed by these motions can be obtained. Central to this is accurate fitting of the relaxation dispersion curves. Sugase and colleagues present a software package "Global and local optimisation of variable expressions" (GLOVE) for this purpose.[23] Whilst there are several programs available for relaxation dispersion analysis, these rely on non-linear least-squares fitting which can become stuck in local minima of the function being searched. GLOVE gets around this by optimising global and local parameters alternately and also incorporating a Monte Carlo-based minimisation. Its application is demonstrated with the global fitting of dispersion curves obtained from a protein undergoing exchange with a 3%-populated conformation on the ms timescale. For analysis of relaxation data, Berlin *et al.* have updated the ROTDIF program.[24] ROTDIF3 can handle ^{15}N and/or ^{13}C data at multiple field strengths, is now controlled through a Java-based GUI and interfaces with ELMDOCK to allow domain orientation and docking of two-domain systems based on relaxation data. Its application is demonstrated on proteins, DNA and RNA elements. Previously, it has been shown that the amplitude of fast backbone motions can be obtained using the chemical shift deviation from random coil, using the "random coil index", RCI. Berjanskii and Wishart have been able to extend this idea to measure total sidechain motions by incorporating backbone and sidechain CS deviations.[2] Empirically optimising the shift contributions using the amplitudes of sidechain motions observed in MD trajectories gave a "sidechain RCI", for which a Python program has been made available. Per-residue comparisons showed that the RCI correlates well with sidechain RMSD in NMR ensembles, solvent accessibility and crystallographic thermal B-factors. Interestingly, the correlation with order parameters measured in sidechain methyls was poor, presumably due to the large influence of rotameric jumps which are independent of total sidechain motions.

2.2.2 Automated assignment and structure calculation. Chemical shift assignments are often required to enable NMR studies of proteins (or homologues) for which crystal structures already exist. With improvements in CS prediction, this can aid automated assignment, as noted by MacRaild and Norton, who demonstrate use of their Resonance Assignment by chemical Shift Prediction (RASP) algorithm.[25] By optimising a scoring function which combines both agreement with predicted shift and matching of only sequential C^{α} and C' shifts, 88% of residues could be assigned with >99.7% accuracy in a test set of 154 proteins. Importantly, benchmarking showed that incomplete data sets reduced assignment completeness with little reduction in accuracy. With inclusion of C^{β} shifts and manual intervention, further improvements were attained, as demonstrated by RASP's ~85% assignment of a 34 kDa protein using spectra complicated by conformational heterogeneity. Where data are missing, it is hard to be confident of assignments without considering all other possibilities, which may be computationally unfeasible in larger proteins, which themselves are inherently more likely to be incomplete. This is addressed specifically by Zuiderweg *et al.* with EZ-ASSIGN, which reduces the number of possibilities considered by initially searching for unique decapeptide assignments.[26] In automatic mode, EZ-ASSIGN compares well with existing methods, such as PINE, as tested on malate synthase G (MSG; 723 residues) and two ~40 kDa domains. However, the approach is designed for incomplete data sets analysed in a semi-automatic fashion and indeed out-performs existing methods for these cases. Aeschbacher and co-workers have analysed ^{1}H and ^{13}C chemical shifts in double-stranded RNA (dsRNA) triplets and derived statistics which they use to aid assignment of the non-labile protons.[27] Their approach uses only homonuclear TOCSY and NOESY spectra and a natural abundance ^{1}H–^{13}C HSQC in unlabelled dsRNA. For manual assignment, the bivariate normal distributions of CSs are provided and directly displayed in the assignment software SPARKY, speeding up the process for RNAs up to ~40 nt. The statistics also aid automated assignment, which the authors implement in an algorithm based on the FLYA approach previously published for automated protein assignment (*vide infra*), demonstrated here on sequences of 20–40 nt. Note that another group has demonstrated the use of a "beam search" approach utilising a neural network for assignment of RNAs from NOE connectivity pathways.[28] The FLYA approach is based on comparison of spectral data with a predicted peak network which could arise from through-space or through-bond correlations. Fully automated protein structure determination from raw spectra was originally demonstrated using FLYA for assignment and CYANA for structure calculation using NOESY and triple-resonance correlation spectra. Schmidt and Güntert have now investigated its performance using only 3D $^{15}N/^{13}C$-separated NOESY spectra.[29] On a test set of proteins up to 160 residues, FLYA achieved 64–96% correct backbone and sidechain assignments from raw NOESY peak lists, or consistently >91% with manual curation of peaks. Structures within 0.8–2.0 Å backbone RMSD were obtained in all cases from the curated peak lists with CYANA automated structure calculation, although without refinement of the peak lists, structures converged in only 50% of cases. Hence automated structure determination can be achieved for small proteins using only NOE data, although the versatility of the FLYA approach means that further improvements can be made with addition of limited through-bond spectra, as pointed out by the authors. The general applicability of the FLYA approach is further illustrated by its use in automated assignment of proteins in solid-state spectra.[30]

Fu *et al.* describe their All atom Molecular Simulation Toolkit (ALMOST) which provides a framework for determining and analysing protein and nucleic acid structures using a wide variety of data; NOE, RDC, pseudo-contact shifts (PCS), CS, chemical shift anisotropy (CSA) and small-angle X-ray scattering (SAXS).[31] Implemented with C++, at the surface level a number of protocols are available, such as structure calculation with automated NOE assignment using CANDID and CYANA protocols, CS-only structure determination using their CHESHIRE pipeline and direct structural refinement under CSs using CamShift.

2.2.3 Novel structure calculation and/or integrative approaches. In the basic form of the CS-ROSETTA algorithm, chemical shifts are used to bias ROSETTA fragment generation and model selection, with moderate success reported for the CS-only based structure determination of small proteins ($\lesssim$100 residues). Other data have been used to guide structure assembly, such as NOEs or RDCs in the so-called "RASREC-ROSETTA" implementation and the first studies using these or other hybrid methods are now being published (see examples.)[32,33] Van der Schot and co-workers have made refinements to the basic protocol, a significant difference being that CS data are used in all three stages.[34] The initial fragment selection utilises a hybrid method which performs better on regions for which no data are available, described in more detail in a separate publication.[35] Next, a version of the "RASREC" algorithm is implemented, which biases sampling towards previously better-performing conformations, but incorporating a new chemical shift scoring function. The final stage includes a reliability filter; of the 39 proteins tested, 70% which passed this filter were within 2 Å RMSD of the reference structure. It is clear that ROSETTA constitutes a viable alternative to full *de novo* structure determination of small-medium systems, with further improvements to the protocol anticipated, *e.g.* iterating it with MD simulations,[36] and with much better convergence achieved by adding even very sparse amounts of experimental data, as noted in an assessment of the technique.[37] Furthermore, the approach is not limited to proteins. Building on previous *de novo* modelling of RNA and recent improvements in back-prediction of ^{1}H (and also ^{13}C) chemical shifts,[38] Sripakdeevong and colleagues tested a CS-ROSETTA approach on a set of 23 non-canonical RNA motifs.[39] Structures were generated by pooling models from two methods, "SWA" and "FARFAR", then ranked by a ROSETTA-type energy term which included back-calculated ^{1}H shifts. Impressively, models within 0.6–2.0 Å RMSD of published structures were obtained in 18/23 cases, with 4 of the 5 outlying results being attributable potentially to dynamics in solution. Van der Werf and co-workers have also published a CS-only based method for determining RNA helical structures.[40] Different from ROSETTA approaches, "Chemical shift *de novo* structure derivations protocol employing singular value decomposition" (CHEOPS) uses ^{1}H shifts to guide re-calculation of structures built in 3DNA over multiple iterations, with back-prediction of CSs using NUCHEMICS. Structures are then subjected to an rMD step to regularise the ribose phosphate backbone. The method is applied successfully to three RNA stem-loops of known structure and in future will be expanded to DNA and potentially single-stranded nucleotides. Given the inherent difficulty in DNA/RNA structure determination, due to the low proton density and non-globular structure, these methods promise to be significant advances in the field.

ROSETTA-like methods generally become unfeasible for medium-large biomolecules as it is not possible to sample adequately the full conformational range available to a system assembled from short peptide templates. Based on their observation that all novel protein folds reported in the last decade are made up of different configurations of existing "super secondary structure motifs" (Smotifs; defined as two elements of secondary structure connected by a loop), Menon, Vallat *et al.* propose a CS-based *de novo* structure determination protocol based on assembly of such fragments.[41] SmotifCS uses TALOS+ to identify Smotifs in the protein sequence which are compared with those back-calculated by SPARTA from an Smotif fragment library. Fragments are then combined and ranked. Testing on 102 known structures, high-quality models were generated for ~50% of targets and in most cases the correct topology was at least identified. The larger fragments used by SmotifCS mean that the calculation time doesn't increase exponentially with size, hence it should be applicable to large systems, although a down-side appears to be worse performance in proteins with long loop regions. Once source of error appeared to result from the initial definition of secondary structure using TALOS+. Kumari and Dorai investigated the use of 3D correlation maps of chemical shifts for this purpose, with improved identification of loop regions achieved relative to existing methods.[42] The backbone dihedral angle prediction software TALOS+ has itself been refined, with the new release "TALOS-N" having greater reliance on the trained neural network filtering of TALOS output.[43] With improvements such as use of pentapeptide fragments, an increased CS database derived from SPARTA+ predictions and a different treatment of the Ramachandran map regions, a modest increase in the percentage of residues for which confident predictions are made (~90%) but with a much tighter acceptance criterion than TALOS+. Also, a prediction of sidechain χ^{1} rotamers is now included and a secondary structure prediction is made based on another neural network from CSs or sequence alone.

Weiner and colleagues report the incorporation of NMR restraints into the *de novo* structure determination algorithm BCL::Fold, which, akin to other methods, starts with identification of secondary structural elements which are then assembled using a Monte Carlo search of a knowledge-based energy function.[44] Utilising CS data for fragment selection and sparse RDC and NOE data during refinement, the method compared favourably with ROSETTA for secondary structural elements, although the authors comment that the latter is better for loop building and sidechain placement. BCL::Fold does, however, perform encouragingly with larger systems, five of the test set being helical membrane proteins greater than 400 residues. A similar, but even cruder, approach was investigated by Eggimann *et al.*, who modelled helical proteins using sparse PRE/EPR-based distances and RDCs.[45] Helical fragments identified with dipolar wave analysis were assembled using a simple residue-based forcefield, tested with simulated data. Similar to previous studies, RDC data alone were poor unless multiple, orthogonal sets were used, with the best results only obtained combining these with PRE data.

In addition to *de novo* structure determination, another application of ROSETTA is in refinement of NMR structures. Mao *et al.* subjected 40 NMR structures from the PDB to a ROSETTA-based refinement incorporating NMR restraints.[46] The resultant models were generally assessed by the PSVS suite to be of better structural quality, fitted the NMR data at least as well and were also consistently closer to the X-ray crystallographic structure. Consequently, the refined structures could be used to solve the X-ray structures by molecular replacement (MR) in all cases where data had been collected under similar conditions. Improved success rates in using NMR for MR have been obtained by using the ensemble to guide poorly-defined regions of the structure, which are then removed or remodelled. For this purpose, Bibby *et al.* investigated the application of a CCP4 module actually developed for *de novo* model creation, termed AMBER.[47] Similar levels of success were achieved to a previously-published algorithm (FindCore), again with modest improvements using ROSETTA remodelling, but with scope for future improvements with better handling of sidechain variability, as suggested by the authors.

Integrative approaches for structure determination are increasingly used to overcome shortfalls in NMR data, *e.g.* in defining accurately the overall shape of non-globular/multi-domain biomolecules, or where data are lacking. For example, Wu *et al.* demonstrate the use of Double Electron–Electron Resonance (DEER) measurements in the structure determination of a $\sim$30 kDa two-domain *Mycobacterium tuberculosis* protein required for survival in the phagosome.[48] The NOE, dihedral angle and H-bond data did not define the relative domain orientation, with an overall backbone RMSD of >10 Å. Addition of 428 PREs improved this to $\sim$4.2 Å, and with inclusion of only three DEER-derived distances, which are more precise than PREs, this dropped to 3.0 Å. Schieborr and colleagues address the problem of determining a structure when assignments are incomplete, for example due to line broadening from conformational exchange. Their solution uses homology modelling to guide NOE assignment in a protocol termed "Model-assisted software for

NMR structure determination" (MOTOR).[49] With NOESY peak lists, primary sequence and partial assignments, a BLAST search is used to find homologues from which initial NOE assignments are made. These are fed to a structure calculation program such as CYANA, with the output used to refine assignments in subsequent calculations. Tested on a known structure, accurate models were recreated with only 70% of assignments. Rinaldelli and colleagues have implemented terms for RDCs and PCSs into the crystallographic refinement program REFMAC5, permitting joint refinement of a structure with X-ray and NMR data.[50] NMR data are introduced after an initial stage which uses only the crystallographic data, and for three proteins tested this improved the fit of RDCs and PCSs to the structure without degrading the R_{free}. Calmodulin is known to have differences in the solution- and crystal-derived structures and indeed the R_{free} and geometric parameters were degraded by introduction of RDCs, with an elevated Q_{RDC} factor. The approach is proposed as a method for improving structures, *e.g.* in regions which are poorly defined crystallographically, or to identify where there a genuine differences in the solution and crystal state, *e.g.* due to packing forces or dynamics. A striking example of a potential application is the small C-terminal domain from the endoplasmic reticulum protein ERp29; the NMR and X-ray structures differ substantially, with two helices protruding from the core in the latter, but which are packed in to form a compact domain according to the NMR data. Yagi *et al.* addressed this by calculating a structure using a version of their PCS-ROSETTA algorithm updated to permit multiple sets of PCSs, "GPS-ROSETTA".[51] Using PCSs from samples tagged with lanthanides at four positions, the solution structure is actually very close to the X-ray model (2.4 Å C^{α} RMSD) and distinct from the NOE-based structure (6.0 Å.) The largest difference is in the position of the protruding helices, which the authors identify are incorrectly packed in the NMR structure due to mis-assignment of a Tyr hydroxyl. The accuracy of the model obtained here using only NH PCSs suggests this as a route for obtaining structures for any protein for which amides can be assigned. Finally, in a slightly different vein, Kannan *et al.* show that backbone and methyl chemical shifts can be used as restraints in replica-averaged MD simulations.[52] Akin to their previous use of chemical shifts for the direct refinement of protein structures, an energy term corresponding to $\Delta\delta$(observed − calculated) was incorporated into the molecular mechanics forcefield by back-calculation using CH3Shift. Initial benchmarking on four proteins showed that inclusion of backbone and/or methyl shifts improved ensemble-averaged back-calculation of measured NOE, *J*-coupling and CS data (obtained from 4DSPOT). The method was used to direct MD simulations on the human DNA-binding HU dimer, which imply that the clamping motions of the β-arms are inherently present and sampled in the free state.

2.2.4 Structure validation. As is evident from the preceding sections, the degree of automation in structure elucidation and the number of hybrid approaches are increasing. Hence, methods of validating such models are as important as ever and two papers which review these are timely. Rosato *et al.* provide a short review of the metrics available for assessing protein structures.[53] The standard "knowledge-based" measures (*e.g.* dihedral angle distributions, core atom packing) and "model *versus* data" measures (*e.g.* NOE violations) which many web-based tools will derive are discussed. Vuister *et al.* provide quite an in-depth overview of these (and standalone) tools, including all the standard web servers such as PSVS and iCing and more recently-reported ones such as Vivaldi and ResProx.[54] A useful description of each is provided and their performance in analysing NMR structures assessed with two models taken from the PDB. Ideally, NMR models would be cross-validated using data that are not used directly for refining the structure. This is not done commonly because of the restricted number of restraints available under normal circumstances and the current lack of an accepted practice community-wide. One laboratory has proposed previously the use of $^{13}C^{\alpha}$ chemical shifts, which are influenced primarily by local conformation and can be calculated with density functional theory. Martin and colleagues have now extended their *Che*Shift-2 protein validation web server to include consideration of $^{13}C^{\beta}$, which they show gives a more reliable indicator of sidechain χ^1 and χ^2 dihedrals.[55] Their program generates a red-orange-green (ROG) output on the PDB structure which highlights residues whose local conformation may be flawed. Furthermore, a list of suggested changes in χ^1 and χ^2 are outputted and they show for a test set of five proteins that when these corrections are made, the NOE data fit just as well. Where corrections still left discrepancies between theory and measured shifts, the authors suggest this implies an error in the backbone dihedral(s). An obvious difficulty arises when trying to compare the quality of structures obtained by different methods or, as is increasingly common, "hybrid" or "integrative" methods. Bagaria and co-workers suggest an "equivalent resolution" (e-resolution) score which, akin to ResProx, is derived from a number of existing knowledge-based metrics.[56] The e-resolution is the sum of combinations of such metrics (*e.g.* Whatcheck *Z* score, Molprobity clash score) which are weighted according to a multiple linear regression analysis of existing X-ray structures. After testing that the e-resolution correlates reasonably well with actual crystallographic resolution, the authors illustrate the applicability of the score by assessing the historical improvement in quality of PDB-deposited structures and comparing the e-resolution obtained with five different structural techniques. Tejero *et al.* have published a description of their program PDBStat.[57] Originally developed as a converter of co-ordinates and restraints between the many different structure generation software available, it has evolved into a powerful tool for handling and analysing NMR ensembles, including IUPAC re-naming, co-ordinate precision and analysis of NOE, dihedral and RDC restraints. An application in generating input for restrained CS-ROSETTA calculations from existing CNS-format files is demonstrated.

2.2.5 Non-uniform/sparse data sampling. Sparse, non-uniform sampling (NUS) of the indirect dimensions of multi-dimensional spectra can be used to dramatically increase resolution and/or reduce total acquisition time, or potentially be used to increase signal:noise by biasing the sampling towards earlier parts of the FID where the signal is stronger. Whilst the potential gains are high, data quality is influenced by the choice of sampling scheme and method of data reconstruction, as the discrete Fourier transform may no longer be appropriate for such acquisition. An excellent review from Hyberts *et al.* describes many of these considerations, with illustrations taken from their own developments in this area.[58] Hoch and colleagues have also published an overview, with focus on data reconstruction with Maximum Entropy methods.[59] Time savings become almost exponentially greater as the number of dimensions are increased, meaning >4 dimensional spectra become perfectly feasible. Nováček *et al.* make excellent use of this in their protocol for assigning large, intrinsically-disordered proteins (IDPs). Their approach uses two 5D experiments, a CACONCACO which provides all the correlations required for backbone assignment and an HC(CC_TOCSY)CACON which links all aliphatic sidechain nuclei. Taking advantage of the favourable relaxation properties of the IDP, high-resolution spectra were acquired using ~0.002 and ~0.001% sampling in 2- and 3 days spectrometer time, respectively. Coupled with direct $^{13}C'$ acquisition, their approach permitted complete backbone and 98% aliphatic sidechain assignment of the 49 kDa microtubule-associated protein 2c, whose sequence is ~10% Proline. Similar time savings were realised in 5D amide-detected experiments for assignment of another IDP in a separate publication.[60] The NUS is implemented using a Poisson disc sampling scheme and processed with a sparse multi-dimensional Fourier transform (SMFT) algorithm for automated assignment using their TSAR program. Application of this in the assignment of the IDP α-synuclein and also optimised 5D sequences are reported.[61] *N.B.* A review from Felli and Pieratelli discusses ^{13}C-detected methods for assignment of IDPs.[62]

Whilst the above applications utilise NUS to achieve digital resolutions in the indirect dimensions of high-dimensionality experiments which are not accessible realistically with "full" Nyquist sampling, Tikole and coworkers have investigated the effect of digital resolution on the quality of structures calculated with NOEs recorded in traditional $^{15}N/^{13}C$-resolved 3D spectra.[63] Spectra were simulated from deposited chemical shifts for ~400 proteins of 5–33 kDa, varying the acquisition time of the indirect ^1H dimension with NUS to keep the total experiment time constant. A number of metrics were evaluated, including peak overlap, signal:noise (using T_2-weighted acquisition – below) and similarity of the resultant CYANA model to the deposited structure. Predictably, the structural quality increased with increasing digital resolution and the authors conclude that this should be set at, or slightly above, the intrinsic linewidth of the protein, which may involve sampling 2–4× further than what would normally be feasible with traditional schemes. Utilising NUS to improve the resolution of NOESY spectra is the goal of two publications from the same laboratory. For proteins, the authors suggest a diagonal-free ^{13}C, ^{13}C-edited aliphatic–aromatic 4D NOESY experiments for which they use ~1.5% sparse acquisition to achieve high resolution in spectra of a 20 kDa homodimer.[64] A similar approach is proposed to aid sequential assignment in uniformly-labelled RNA.[65] An alternative application of NUS is in realising gains in sensitivity by biasing sampling to earlier parts of the FID where signal:noise is high, but increasing the number of scans to keep total experiment time the same. Palmer and colleagues investigate this practically in 2D ^{1}H–^{15}N and ^{1}H–^{13}C spectra of proteins and a natural product.[66] Real sensitivity gains (20–70%) were obtained in natural abundance HSQC spectra of ubiquitin, depending on the sampling bias, but the authors reiterate that this is only significant with FID acquisition times (aq) $> 1.26 \times T_2$, corresponding to high resolution in protein spectra. Thus, in standard 3D triple resonance, where aq times are short and non-decaying constant-time frequency labelling is often employed in one- or more indirect dimensions, there is little gain. Sampling schemes are also discussed and it is shown that the line broadening effects of strong T_2-weighted biasing can be ameliorated using a pattern based on a quarter sine wave.

2.3 Approaches for larger biomolecules

Whilst methyl TROSY approaches permit monitoring of biomolecules up to at least hundreds of kDa, assignment remains to be challenging. In solution, conventional NMR experiments which rely on correlation of frequencies are likely to be limited to ~100 kDa, even in favourable cases. For example, Tugarinov and co-workers report an "out-and-back" methyl detected experiment for simultaneous correlation of I, L and V methyls with the backbone C′ frequency, demonstrated on a ~70 kDa symmetric homodimer at 37 °C[67] (also, see Sinha *et al.* who look at the sensitivity of such experiments.)[68] Frueh has written an excellent review on the practical aspects of assigning challenging proteins, including NOE assignment, which is recommended for anyone contemplating such systems.[69] The same author and co-workers have also published another brief overview.[70] The caveat to the "~100 kDa" statement above may be "in solution". Mainz and co-workers have shown recently that by subjecting solution preparations of large biomolecules to MAS, the sedimentation and/or packing against the rotor walls under high centrifugal force restricts motion enough to enable solid state-like line narrowing by averaging anisotropic interactions.[71] Using ^{15}N, ^{13}C, α-subunit from the 20S *T. acidophilum* proteasome, which was also 20% protonated at amides and ^{2}H-labelled to remove ^{1}H–^{1}H dipolar coupling, ^{1}H-detected 2D and 3D triple-resonance spectra could be acquired within a complex of ~1.1 MDa. With Cu^{2+} added as a relaxation agent, reasonable quality spectra could be obtained, permitting ~50% backbone assignment despite only ~60% of amide resonances being observable due to line broadening through chemical exchange. As shown in various complexes from 360–1100 kDa, sensitivity and resolution improves with increasing molecular weight, apparently due to the more restricted motion. No additional preparation of the sample other than filling the rotor with solution is required.

There are many technical challenges which need to be overcome to study integral membrane proteins by NMR. Whilst β-barrel membrane proteins are generally more stable, less heterogeneous and have better dispersed signals than their all-helical counterparts, one difficulty can be the large extracellular loops which are commonly dynamic and exhibit different linewidths to the β-strand core. Fox and Columbus thus use an approach which effectively treats the two areas as different species.[72] For assignment of the β-strand regions, trypsin cleavage was used to remove sections of the extracellular loops, reducing the overall tumbling time and removing much of the overlap from $^{1}H-^{15}N$ spectra of the *Neisserial* eight-stranded barrel OmpA. Loop regions were assigned through a combination of reducing temperature, broadening out the structured strand regions, or assigning synthetic peptides which, due to the random coil nature of extended loops, exhibit virtually identical resonances. Recent developments in the NMR study of IMPs are reviewed by Maslennikov and Choe[73] and also by Oxenoid and Chou, with a focus on membrane channels.[74] Gautier provides a description of methods for the structure determination of α-helical membrane proteins by NMR, using his experience of retinal proteins.[75]

As pointed out by the authors of a review on the subject,[76] very large systems and/or those exhibiting low chemical shift dispersion may still be amenable to NMR analyses using sparse labelling schemes. For example, Stehle *et al.* were able to study the kinetics of light activation in $^{1}H-^{15}N$ SOFAST-HMQC spectra of ^{15}N-Trp labelled samples of the G protein-coupled receptor (GPCR) rhodopsin.[77] The use of ^{19}F NMR as a sparse label is undergoing a resurgence in biomolecular NMR (see a review on applications in membrane proteins),[78] the obvious benefits being its high signal sensitivity, the strong influence of local environment on its chemical shift and, through its fairly common use in synthetic Chemistry, readily-available hardware, including cryoprobes. Chadegani and co-workers have incorporated 5-fluorotryptophan into the 83 kDa anthrax protective antigen (PA), one of the components of the anthrax toxin from *Bacillus anthracis*. Assignments for the seven Trps were made through combined use of mutagenesis and spin-labelling, permitting the state of the protein to be monitored readily in 1d ^{19}F spectra. The crystal structure at 1.7 Å confirmed that the structure was unperturbed by the labelling. Whilst methods are available for the biosynthetic incorporation of ^{19}F in amino acid analogues, post-translational labelling through chemical conjugation provides an alternative, *e.g.* for systems not amenable to heterologous expression methods which are adaptable for isotope labelling. Mammalian membrane proteins are notoriously difficult in this regard, hence Horst and colleagues labelled Cys mutants of the β_2-adrenergic receptor (β_2AR) with 2,2,2-trifluoroethanethiol.[79] This permitted monitoring of the temperature dependence of the inactive and active states in the presence of partial- (norepinephrine) and full (formoterol) agonists using lineshape analysis and exchange spectroscopy, from which thermodynamic parameters could be derived. An essentially identical approach was used to study the same system using a trifluoromethyl derivative in a separate study from Kim and co-workers.[80]

Interestingly, under their micelle conditions, β_2AR exhibits three conformational states in slow exchange, which are perturbed to differing extents by addition of different ligands or a G-protein mimic "nanobody 80". These and a number of recent NMR studies highlight the degree of conformational plasticity in cell surface receptors, reviewed this year.[81] An alternative chemical modification is reductive methylation of lysines, which can be used to di-methylate the sidechain with ^{1}H–^{13}C. The degenerate methyl resonances provide a highly sensitive probe for measurement in ^{1}H–^{13}C HMQC spectra. For example, although not applied to particularly large systems, see applications in monitoring ligand binding to interleukin-8[82] and calmodulin in spectra recorded at sub-μM concentrations.[83]

A promising but as-yet not widely-applied method for improving the spectral properties of biomolecules involves their encapsulation in reverse micelles which, when suspended in a low-viscosity alkane, permit faster tumbling than when the molecule is suspended in standard aqueous buffer (reviewed recently.)[84] The laboratory pioneering this approach has developed a more widely-applicable surfactant system, demonstrated on proteins from ~8–80 kDa with pIs from 4.5–11, including a lipid-anchored protein, and a 27 kDa tRNA.[85] Suspending in pressurised ethane, molecules over ~20 kDa benefit from a reduced correlation time (of ~30% for MBP) which, for example, should facilitate conventional non-deuterated methods to be applied to proteins >40 kDa. The authors also point out that the decreased signal:noise which results from the reduced overall solute concentration is further compensated by the lower conductivity of the alkane, which benefits cryoprobes especially. Another benefit of the micelle/alkane medium is that it ameliorates the strong absorption of microwave frequencies by water, which gives rise to sample heating and a "skin depth" effect, where microwaves fail to penetrate through to all regions of the sample. This, therefore, facilitates solution-state dynamic nuclear polarisation (DNP), where the polarisation of free electrons by microwaves is transferred to enhance that of solute molecules, as demonstrated in another publication from the same laboratory.[86] DNP enhancements of bulk water of approaching two orders of magnitude were obtained using nitroxide radicals contained either in the buffer or micelle shell. The increased water residency at the protein surface aids transmission of some of this enhancement, as demonstrated on encapsulated flavodoxin.

3 High resolution structural studies of biomolecules

3.1 Protein structures and interactions

The RING-finger protein gp78 is an E3 ligase which catalyses ubiquitin (Ubq) transfer from its cognate E2 enzyme, Ube2g2, to form K48-linked ubiquitin chains on mammalian substrates targeted for proteasomal degradation. Das *et al.* have investigated the structural details behind the observation that the primary RING-Ube2g2 interaction is enhanced allosterically by a secondary binding site on gp78, termed G2BR, 180 residues away.[87] Comparison of their RING:Ube2g2:G2BR complex structure obtained by CSP and NOE-driven HADDOCK docking with structures of the free components reveals that binding of G2BR to Ube2g2 causes allosteric rigidification of a loop containing a Glu residue key to RING domain binding, increasing affinity ∼50-fold. Also, RING binding causes further allosteric effects which weaken the G2BR–Ube2g2 interaction, aiding release of the E2 to enable the next round of catalysis. The diversity of processes regulated by ubiquitination appears to arise from its ability to form poly-Ubq chains with different linkages and, presumably, conformations. The solution structure of Lys11-linked Ubq has been analysed in solution by Castañeda and colleagues, using segmental isotope labelling to overcome degeneracy in individual subunits of di-ubiquitin.[88] Relative domain orientation was established with RDCs and PREs and corroborated using ^{15}N relaxation-derived rotational diffusion tensors and small-angle neutron scattering (SANS). Interestingly, Lys11-linked di-Ubq adopts conformations distinct from Lys48- or Lys63-linked chains and the NMR/SANS-derived solution data were also inconsistent with both previously-determined crystal structures. For the malaria parasite *Plasmodium falciparum*, one route towards invasion of erythrocytes involves interaction between the reticulyte-binding homolgue protein 4 (PfRh4) and the complement receptor type I (CR1). Park and colleagues used a combined NMR and SAXS approach to determine the structure of the N-terminal three Complement Control Protein (CCP) domains of CR1, which they have shown previously to contain the site of PfRh4 interaction.[89] The three CCPs form an extended structure with small inter-domain interfaces which presumably aids protrusion of CCP1 clear of the glycocalyx, which the authors show by site-directed mutagenesis (SDM) to be the primary binding site. Import of peroxisomal matrix proteins into peroxisomes is mediated predominantly by binding of their receptor protein Pex5 to the membrane-associated Pex14, normally resulting from interaction of di-aromatic penta-peptide motifs WXXX(F/Y) in Pex5 with the N-terminal domain of Pex14. Neuhaus *et al.* identified a novel motif in Pex5 comprising LVAEF which cannot be substituted efficiently by canonical sequences, even those with higher affinity for Pex14.[90] Using mixed isotope labelling and conventional isotope filtering/editing experiments a structure of the complex reveals that the novel motif binds with the same α-helical conformation as the di-aromatic sequence, with the Leu and Val sidechains making hydrophobic contacts in place of the standard Trp. PREs were also used to confirm the peptide orientation on Pex14. Intriguingly, surface plasmon resonance (SPR) experiments revealed a 33× faster off-rate for the non-canonical motif, which the authors speculate is linked to its essential role in Pex14 binding; the motif could provide the first contact point before rapid dissociation permits hand-off to higher-affinity WXXX(F/Y) motifs for further processing of the receptor complex at the peroxisomal membrane. In addition to its function as an ATP-dependent molecular chaperone, the 90 kDa Heat-shock Protein (Hsp)90 has recently been implemented in the assembly of several ubiquitous eukaryotic cellular machineries, such as snoRNPs involved in RNA maturation. Back and co-workers have investigated a key interaction between Hsp90 and a co-chaperone in this process, the yeast R2TP complex, by determining the structure of the component protein Tah1 both in the free state and bound to a peptide containing the Hsp90 ^{704}EMEEVD^{709} motif.[91] Using similar methodology to the previous study, the structure reveals the expected tetratricopeptide (TPR) domain flanked by an extended C-terminal unstructured region, which the authors show is required for interaction with another R2TP protein, Pih1. An unusual 3_{10}-helical conformation of the Hsp90 motif is observed which differs from that implied by a low-resolution structure determined previously, mediated by an S–CH_3–π interaction between M^{705} and a Tyr sidechain on Tah1. The mammalian αβT-cell receptor (TCR) is a transmembrane signalling complex which recognises antigens presented by major histocompatibility complex (MHC) *via* its TCRαβ proteins, transmitting signals to the cell interior through associated subunits such as the CD3δγ heterodimer. Berry and co-workers have determined the solution structure of an avian CD3εδ/γ heterodimer, comprising the ectodomains of CD3ε bound to CD3δ/γ, a single protein which exhibits homology to both δ and γ mammalian chains.[92] Joining the heterodimer *via* a covalent linker permitted conventional structure determination, which revealed a 'skewed' topology strikingly different from mammalian complexes, apparently mediated by an unusual conformation in the base of the CD3ε G-strand, central to the dimer interface.

CREB-binding protein (CBP) mediates interactions between transcription factors such as the mixed-lineage leukaemia (MLL) protein and CREB through two distinct binding sites found within its KIX domain. As part of ongoing studies into the mechanism underlying co-operativity observed in ligand binding, Brüschweiler *et al.* have determined the structure of KIX complexed with the MLL activation domain and in the ternary complex with MLL and the phosphorylated kinase inducible domain of CREB (pKID).[93] NOE- and RDC-based structures indicate that the fold and backbone of KIX is essentially identical in the binary and ternary complexes. Co-operativity between the two ligand-binding sites appears to arise through subtle allosteric effects mediated by changes in packing and dynamics of the intervening hydrophobic core, supported by chemical-shift based analysis of Val and Ile rotameric states. The interaction of another transcription factor, the NK-κB protein RelA, with CBP has been characterised by Mukherjee *et al.*[94] Their structure of the mouse CBP TAZ1 domain bound to RelA's transcriptional activation sub-domain TA2 shows that an ~50 residue region of the latter wraps completely around TAZ1, bound in an extended conformation over multiple hydrophobic pockets. The authors suggest this is energetically possible due to the disordered nature of free TA2, as indicated by 2D spectra of the protein. The onset of spinocerebellar ataxia type-I is influenced by self-association of the transcriptional co-regulator ataxin-1, which is mediated at least in part by a domain termed AXH. Using NMR and other biophysical methods, de Chiara and colleagues show that AXH's natural propensity for dimer- and higher multimerisation is inhibited in solution by binding of a peptide comprising a region from the transcriptional repressor protein CIC.[95] A structure of the complex determined using a conventional mixed labelling strategy revealed that the peptide competes for the hydrophobic area normally shielded by the N-terminus of the other AXH protein in the dimeric interface, albeit with a different conformation. Key interactions are identified which the authors suggest could form the basis for therapeutic peptidomimetics. A putative mechanism by which the protein kinase CK2 may influence transcription by phosphorylation of the transcriptional co-repressor Silencing Mediator for Retinoid and Thyroid receptor (SMRT) has been investigated by Mikami and co-workers.[96] Surface Plasmon Resonance (SPR) and NMR titration experiments were used to demonstrate that phosphorylation of a Serine within a peptide comprising the C-terminal eight residues of SMRT increased the binding affinity for the co-repressor protein SHARP approximately 100-fold. A structure of the ^{15}N, ^{13}C-labelled SHARP SPOC domain in complex with partially- or unlabelled peptide allowed site-directed mutants to be designed to confirm the importance of phosphorylation *in vivo*.

3.2 Nucleic acid structures and their interactions

G-quadruplexes are four-stranded structures formed by G-rich DNA or RNA oligonucleotides, which in recent years have been the focus of many structural studies. This is due to their occurrence in, and apparent influence on, genomic regions associated with disease, such as telomeres and proto-oncogenes. Although having in common a core of multiple stacked G-tetrads associated through Hoogsteen H-bonds, G-quadruplexes can be formed from a variety of topologies, influenced by sequence, concentration and presence of cations. Due to this non-canonical nature and the observation that their formation in the aforementioned regions can promote anticancer activity, molecules which stabilise G-quadruplexes have been suggested as potential new therapeutics. Chung *et al.* report the first structure of the well-characterised G-quadruplex-stabilising bisquinolinium compound Phen-DC3 bound to a 24 nucleotide (nt) region of the human *c-myc* transcription factor promoter, a target of cancer therapies.[97] Using conventional NOE-based methods, with selective ^{2}H- and ^{13}C-labelling to aid assignment of the nucleotide and ligand, respectively, the structure confirms that the aromatic moieties of Phen-DC3 stack over the 5′ end, parallel to the G-tetrads. Their model suggests several places at which the ligand could be derivatised to improve affinity and further reveals that binding disrupts an A3 · A12 base pair, despite stabilising the quadruplex with a $\Delta T_m \approx +12$ °C. The structural complexity of even relatively short G-rich motifs is exemplified nicely in a study from the same laboratory; Adrian *et al.* investigated the solution structure of a G-rich quadruplex found in an 18 nt sequence from the CEB1 mini-satellite, located in the promoter region of the human sialidase-encoding sub-telomeric NEU4 gene.[98] With assignment aided by site-specific deuteration and low-enrichment ^{15}N labelling, the authors found a parallel "snapback" scaffold which associates into a dimer *via* π–π stacking of 5′-end tetrads. The topology is likely further complicated by the presence of stacking rotamers due to the symmetry at the dimer interface and ms-timescale exchange observed in the NMR spectra involving a swapping motion of bases within the first G-tract. As part of their ongoing characterisation of such topologies, the same laboratory has also studied how B-form DNA can transition into G-quadruplexes.[99] Their structures of five quadruplexes incorporating DNA duplex hairpins suggest that the latter can be used to influence the topology of the quadruplex and conversely the quadruplex can be used as a core to position "duplex arms" around the G-junction as a route towards designed DNA structures for nanotechnology. Knowledge of the determinants of G-quadruplex formation has been employed in the design of aptamers as therapeutics, the first of which ever approved by the U.S. Food and Drug Administration (FDA) being the drug Macugen, which acts against age-related macular degeneration by neutralising vascular endothelial growth factor, VEGF. Marušič *et al.* have investigated the use of locked nucleic acids (LNAs) to stabilise a 25 nt DNA aptamer for VEGF.[100] Whilst the aptamer comprising natural nucleic acids was observed to adopt multiple structures, substitution at three positions with LNAs stabilised a single fold, shown to be a monomeric parallel G-quadruplex with increased thermal stability.

The transcription factor AML1/RUNX1 binds to the Runt-binding double-stranded DNA element (RDE) of target genes *via* a small, N-terminal region termed the Runt domain. Nomura *et al.* have obtained RNA aptamers which bind AML1 Runt using systematic evolution of ligands by exponential enrichment (SELEX) with higher affinity than the RDE. An NOE- and RDC-based structure of a 22 nt aptamer was determined to understand the elements essential for Runt binding.[101] Interestingly, the aptamer forms an RNA hairpin with a base triple and AH^+-C mis-match which adopts an unusual conformation mimicking the RDE DNA structural element recognised by Runt. Substrate recognition by the *Neurospora* Varkud satellite (VS) ribozyme involves recognition of nucleotides *via* a kissing complex between stem-loops (SLs), termed SLI in the target and SLV in *VS*. Bouchard and Legault have obtained the structure of the complex using similar methods to the aforementioned study.[102] Both stem-loops adopt tight U-turn structures which present a row of three nucleotides for intermolecular Watson–Crick base pairing. The authors use their results to propose a kinetic model for ribozyme activation after the substrate recognition step.

The laboratory that has reported most of the recent NMR-based DNA G-quadruplexes (*vide supra*) have turned their expertise to G-rich RNA regions with their study of the first 18 nt from the 5′ terminal region of human telomerase RNA.[103] NMR was used to imply formation of a three-layer parallel-stranded G-quadruplex which is in equilibrium with a stacked dimeric form, which might be biologically relevant for at least some reported telomerase functions. Telomerase is the ribonucleoprotein responsible for extending the telomere which protects the 3′ end of eukaryotic chromosomes and is upregulated in most cancer cells. Cash and colleagues have determined the structure of another region of telomerase RNA, the template-adjacent pseudoknot from the *Kluyveromyces lactis* enzyme.[104] Using NOE-, dihedral angle- and RDC data the pseudoknot is shown to contain an extended pyrimidine triple helix and incorporates three separate bulged nts. Interestingly, comparison with their previously-determined structure of the pseudoknot from human telomerase revealed that despite differing sequences, base triples and junctions, the conformation of the triple helices are very similar, which the authors suggest is indicative of the importance of the motif in telomerase function. In addition to this region, the other catalytically-essential domain is the stem terminus element, which in vertebrates is termed conserved regions 4 and 5 (CR4/5). Using unlabelled, doubly-labelled and selectively-deuterated RNA, Kim and co-workers have determined the structure of the entire ($\sim$50 nt) CR4/5 domain from the vertebrate *Oryzias latipes* using NOE- and RDC restraints.[105] The element comprises a three-way junction core from which the P5 and functional P6 and P6.1 loops project in an overall Y-shape. Order tensor analysis of the RDCs reveals that the core provides a highly dynamic scaffold on the sub-ms timescale, with inter-helical amplitudes of motion estimated to be $\sim$50° and 60° for P5 and P6.1 relative to P6. Furthermore, whilst Mg^{2+} changes the average overall conformation of the element significantly, as judged by RDC data, the dynamic nature is maintained and is presumably essential to permit conformational rearrangements required to interact with the telomerase reverse transcriptase. Dynamics also appear to be important in the mode of action of a class II prequeuosine ($preQ_1$) riboswitch, which translationally regulates $preQ_1$ biosynthesis, as reported by the same laboratory.[106] Kang and colleagues used nt-selective-, selective deuteration- and uniform ^{13}C, ^{15}N-labelling to determine the structure and dynamics of a 59 nt *Streptococcus pneumoniae* riboswitch bound to $preQ_1$ in the presence of Ca^{2+}. The structure comprises an H-type pseudoknot which sequesters the Shine–Dalgarno sequence, featuring an extended P4 hairpin which is dynamic on the ms-μs timescale, as revealed by ^{13}C T_1, $T_{1\rho}$ and RDC measurements. The authors present evidence that the extended helix of P4 introduces conformational disorder to hinder pseudoknot formation in the ligand-free state, whilst upon preQ1 binding a helical twist of P4 causes blocking of ligand exit and creates contacts which stabilise the bound-state triple helix.

3.3 Protein–nucleic acid interactions

The RNA-recognition motif (RRM) is the most widely encountered accessory domain for binding single-stranded (ss)RNA, comprising a βαββαβ fold which ordinarily interacts with oligonucleotides *via* conserved aromatic-containing sequences on the β-sheet surface. Cléry and co-workers have determined the structure of a "pseudo-RRM", which lacks the conserved aromatic platform, from the splicing regulator and oncoprotein SRSF1 in complex with a short purine-rich ssRNA.[107] Unexpectedly, their conventionally-determined structure reveals that RNA is bound on the opposite face of the protein, using an α-helical motif which is conserved in pseudo-RRMs. Remarkably, this is also a site of interaction for proteins such as the kinase SRPK1, where two aromatic residues occupy the same sites as the first two bases from the RNA central GGA motif. To obtain a well-behaved construct for NMR studies, Cléry *et al.* fused the pseudo-RRM to the commonly-used GB1 tag and used

SELEX to obtain an optimal ssRNA ligand. A review of methods for optimising protein–ssRNA complexes for NMR and X-ray structure determination has been published by Foot *et al.*, illustrated with examples from their own research.[108] Another review from Carlomagno focuses on methods for larger protein–RNA complexes, including integration of data from other techniques such as SAXS/SANS.[109] Protein–nucleic acid complexes are often technically challenging because of their slow molecular tumbling time and difficulties in resonance assignment of the DNA/RNA, meaning a common approach is to model structures using data-driven methods such as HADDOCK. Such an approach was invoked by Paquet *et al.* to generate a model for the bending of DNA by the *Euryarchaeal* protein methanogenic chromosomal protein 1 (MC1), which is structurally unrelated to other known DNA-binding proteins.[110] NMR was used to map the interaction surface on both the 15 nt double-stranded (ds)DNA and protein, with PREs measured on MC1 from DNA modified with a Mn^{2+}-chelator. Initial models generated using a coarse-grained docking approach were then submitted to HADDOCK semi-flexible docking driven by the NMR and mutational data. In the model, MC1 bends the DNA about itself by $>100°$, roughly in line with electron microscopy observations. The model explains well mutagenesis data such as the importance of Arg25, which appears to neutralise the negative charges of two phosphate groups brought into close proximity by the strand distortion. A similar approach was used by Gong and co-workers to model the interaction of dsDNA with the double Tudor domain from the DNA/histone-tail binding retinoblastoma-binding protein 1 (RBBP1), a tumour and leukaemia suppressor.[111] The conventionally-determined solution structure revealed an interdigitated double-Tudor structure which lacked the aromatic cage associated with methylated histone recognition. Instead, NMR titrations and isothermal titration calorimetry (ITC) were used to confirm a relatively non-specific DNA-binding surface. From a HADDOCK-derived model, the RBBP1 binds dsDNA primarily at one lobe of the Tudor domain *via* predominantly electrostatic interactions which span both the major and minor groove. See a further example of the use of HADDOCK to model Zn^{2+}-finger proteins binding to dsRNA.[112]

4 NMR characterisation of dynamics

A number of studies this year have utilised the recently-developed chemical exchange saturation transfer (CEST) approach to detect and characterise minor, high-energy states of biomolecules which are in exchange with the native ground-state conformation. For CEST, a weak radio frequency (RF) field is scanned across a range of frequencies and if this saturates an (unobservable) minor-state resonance, exchange to the major form can cause a drop in intensity of this (observable) signal, depending on the exchange rate and population (see a highly recommended review which describes methods for measurement of chemical exchange, including CEST).[113] Lim *et al.* used ^{15}N CEST to detect the presence of two lowly-populated states in an acyl carrier protein (ACP), present at ~4% and ~1% in solution.[114] The 4% state corresponds to an unfolded species, confirmed by comparing ^{15}N shifts with the protein after urea denaturation, whilst the 1% state fits to an off-pathway folded conformation. Residues exhibiting this third state are located around the "recognition helix" known site of interaction and the authors postulate that a predisposition for structural rearrangements facilitates interaction of different catalytic enzymes by one ACP. Zhao *et al.* use the technique to characterise nucleic acid dynamics in the apo form of a 47 nt fluoride riboswitch.[115] Bound to fluoride, the riboswitch forms a compact pseudoknot, but in the free state the structure is dynamic and essentially unfolded. 2D ^{13}C-CEST experiments on a G-labelled sample reveal that the nucleotide is in exchange with a ~10%-populated species, whose ^{13}C shifts match the F-bound form. Interestingly, another paper used NMR to reveal the presence of a similar equilibrium in an adenine-sensing translational riboswitch from the bacterium *Vibrio vulnificus*.[116] For this riboswitch, the equilibrium between an adenine-binding and non adenine-binding form is strongly temperature dependent. Furthermore, the authors show that the ligand affinity is also temperature dependent and is anti-correlated with equilibrium shift towards the adenine-sensing state. Hence, the overall translation-activating efficiency is maintained over a range of temperatures and thus the temperature-dependent pre-existing conformational equilibrium is postulated to be a mechanism of adaptation to the conditions of its marine (~10 °C) and human host (37 °C) environments (discussed in a perspective.)[117] Whilst relaxation dispersion (RD) permits some characterisation of excited states, mainly through measurement of their backbone NMR frequencies, it is technically challenging to obtain sidechain data, *e.g.* because of the ^{13}C–^{13}C scalar coupling, whose modulation of R_2 (effective) with respect to change in CPMG frequency due to Hartmann–Hahn magnetisation transfer would swamp the dispersion curve. Bouvignies *et al.* show that sidechain shifts can be readily derived from CEST experiments and demonstrate it on two systems they have previously characterised with RD.[118] Using only conventional uniform ^{13}C labelling, ~90% of aliphatic shifts could be measured in a ~6%-populated state of a Fyn SH3 domain mutant and a ~2.5% "invisible" state of an α-helical FF domain, exchanging with the native conformations with a k_{ex} ~130 and ~550 s^{-1}, respectively. The ability to readily obtain sidechain chemical shifts in invisible states is particularly informative in light of the recently-developed CS-based measure of sidechain dynamics[2] (*vide supra*). In a further study, CEST experiments were used to derive the pK_a values and distribution of tautomeric forms in Histidine sidechains of an on-pathway intermediate of the *E. coli* colicin immunity protein Im7.[119] After first demonstrating that the pK_a and tautomeric state can be derived from the $^{1}J_{C\varepsilon-H\varepsilon}$ coupling and difference in $^{13}C^{\gamma}$ and $^{13}C^{\delta2}$ shifts, respectively, CEST was used to obtain these for the intermediate, populated at ~1.5%. pK_as of ~6.9 were obtained for the intermediate which, together with tautomeric distribution, implies the sidechains are solvent exposed, in contrast to the highly structured native state where the Histidines have highly shifted values, below 5. Libich and co-workers used a variation of the technique, which they term "dark-state exchange saturation transfer" (DEST), ^{15}N relaxation dispersion, line-broadening and exchange-induced CS changes to probe the transiently-bound state of the intrinsically-disordered amyloid β peptide Aβ40 interacting with the chaperone GroEL.[120] As all four types of data are influenced by the transient interaction, they were fitted simultaneously to a simple two-state model, varying the rate constant for association and fraction of free Aβ40, to derive exchange kinetics and residue-specific changes in ^{15}N R_2 and CS. The peptide binds with a lifetime of ~1 ms, with two hydrophobic regions which are predicted consensus GroEL-binding sequences exhibiting elevated R_2 when bound. Chemical shifts in the bound state suggest the peptide is essentially unstructured when associated with GroEL, in-line with previous hydrogen-exchange data, but differing from crystallographic structures and NMR trNOE studies.

The gene regulatory catabolite activator protein (CAP) binds DNA at its DBD domain only when activated allosterically by cAMP binding to its cyclic nucleic acid binding domain (CBD). Tzeng and Kalodimos used ^{15}N and ^{13}C methyl relaxation dispersion to study a mutant which is structurally identical to wild-type, but with micromolar affinity for DNA.[121] RD analysis revealed that the mutant transiently accesses a state whose chemical shifts correlate with the activated (DNA-binding) form, populated at ~7%, with DNA presumably binding this and causing a population shift. Furthermore, cyclic GMP abrogates DNA binding and this correlates with loss of relaxation dispersion. Hence, allosteric inhibition results from destabilising the low-populated "invisible" state, which the authors point out is something which could not be explained by conventional structural studies. The DNA binding affinity of the transcription factor Ets-related gene (ERG) is also regulated allosterically, by auto-inhibitory elements which flank the binding site.[122] Crystal structures in uninhibited, inhibited and DNA-bound states reveal the inhibitory elements bind remotely from the nucleotide-binding helix with minor conformational differences. However, ^{15}N dynamics and RD reveal substantial ms-μs timescale fluctuations which are quenched in the auto-inhibited state. Hence, the authors propose that the inhibitory elements act allosterically to modulate both dynamics and subtle conformational changes which temper the DNA affinity about 5-fold. There are a growing number of examples where allosteric effects are propagated by dynamics as opposed to conformational changes, reviewed recently.[123] As pointed out by McDonald and co-authors, characterising allosteric changes in dynamics requires measurement of data on a range of timescales and with different probes.[124] They use both ^{15}N relaxation dispersion and traditional dynamics measurements in ^{15}N and ^{2}H nuclei in sidechain methyls to characterise activation of flagellar motor binding, regulated by phosphorylation of the *E. coli* chemotactic sensor CheY. Upon phosphorylation, as reported by RD, there is a decrease in motion on the ms–μs timescale in CheY which is related to residues which sense and propagate the allosteric response. Whilst there is little change in backbone ns–ps dynamics, a general increase in fast-timescale motions in a number of sidechain groups is observed, perhaps to compensate entropically for the general rigidification around the phosphorylation site. Hence the modulation of dynamics is different at different timescales and not necessarily apparent from just (*e.g.*) backbone measurements. Epithelial cadherin (E-cadherin) mediates Ca^{2+}-dependent homophilic cell–cell adhesion by dimerisation through 3D domain swapping. Dimerisation has been proposed to proceed through an intermediate "X-dimer", with an interface distinct from the final conformation, but this has only been inferred from mutants. Li *et al.* used ^{15}N ZZ-exchange and rapid dilution measurements monitored by ^{1}H–^{15}N TROSY and Trp fluorescence to characterise the binding kinetics of E-cadherin and a mutant designed to disrupt the X-dimer interface.[125] Whilst the binding affinities of both species are essentially identical, the on rate of the mutant was slower by nearly 10 000-fold. Furthermore, ^{1}H amide-detected RD revealed exchange with a minor species centred around residues of the wild-type at the X-dimer interface, and not present in the mutant. Together, the authors argue that this provides strong evidence for association proceeding through the X-dimer interface mechanism.

5 NMR studies of large biomolecules

5.1 Large, soluble molecules ($\lesssim$100 kDa)

Over the past 20 years or so, detailed studies of what were traditionally thought of as "large" molecules for NMR have increased from a handful of pioneering reports published annually to the commonplace. With amide TROSY-based methodology, assignments of protein backbone atoms, with varying degrees of completeness in sidechain assignment, are being reported routinely for monomeric proteins up to 40 kDa,[126–130] or 60 kDa and higher for multimeric species and complexes (*e.g.* see[131,132] and examples below). Using the methyl-TROSY approach, a useful level of assignment can often be achieved for much larger systems (*vide infra*). One of the earliest systems for which detailed structural data were reported was the 43 kDa *E. coli* periplasmic maltose-binding protein (MBP). Investigating the effect of molecular crowding on MBP, Miklos *et al.* noticed that signal broadening in ^{1}H–^{15}N TROSY spectra caused by the macromolecular polysaccharide Ficoll was relieved upon addition of MBP's natural ligand, maltose.[133] Fluorescence-monitored binding experiments revealed that Ficoll is weakly bound by MBP and, coupled with the NMR data, suggest it is competitive for the maltose binding site. The authors hypothesise that weak polysaccharide association might be a mechanism by which apo-MBP is localised within the bacterial peptidoglycan matrix; here it is ready to receive maltose from maltoporin, which in turn relieves MBP-peptidoglycan association, to allow transport of maltose across the periplasm to its transporter to the cytoplasm. Cytochrome P450cam catalyses the hydroxylation of camphor using electrons transferred from the iron–sulfur protein putidaredoxin (Pdx). Hiruma *et al.* have determined the structure of the 58 kDa complex by both NMR and X-ray crystallography.[134] Ln tags were introduced at three places in the complex using the CLaNP-7 moiety, permitting measurement of PREs, RDCs and PCSs. Within the precision of the ensemble, the solution structure was essentially identical to the crystal structure, including most sidechains, despite only restraints to backbone atoms being used and sidechains being allowed to move in the docking (*N.B.* such methods are the subject of an excellent review.)[135] Their complex differs from a model determined previously using a lower number of RDC and PRE restraints, which the authors attribute to mobility in the nitroxide spin label which increased the error in distance measurement in the previous study. The same laboratory also characterised the final and encounter complexes formed as part of electron transfer between yeast iso-1-cytochrome *c* (Cc) and cytochrome *c* peroxidase (CcP).[136] A previous report had measured extensive PREs on the smaller Cc component by 10 different nitroxide spin-labelled CcP constructs. By assigning and monitoring the 34 kDa CcP partner, with one MTSL-labelled Cc, additional interactions arising from the encounter complex were observed, further highlighting the need to obtain as complete a set of independent data as possible. Translesion synthesis (TLS) enzymes are DNA polymerases which can replicate past DNA lesions and are recruited by proteins such as the sliding clamp proliferating cell nuclear antigen (PCNA) in yeast. Noting that the TLS enzyme Rev1 lacks the canonical PCNA-interacting protein box (PIP-box) motif, Pustovalova and co-workers investigated the putative binding of PCNA by the Rev1 BRCA1 C-terminus (BRCT) domain.[137] After determining the solution structure of the 11 kDa BRCT and assigning the 87 kDa trimeric PCNA using conventional amide TROSY-based methods, the interaction could be mapped by CSPs in ^{1}H–^{15}N TROSY spectra of both species. Interestingly, further titrations of PCNA with a PIP-box peptide showed it is bound at the same binding site and is directly competitive to BRCT, despite no sequence similarity, hinting at a mechanism for regulation, as suggested by the authors. The Abelson tyrosine kinase (c-Abl) autoinhibits its activity *via* burial of a covalently-attached myristoyl group, stabilising a closed state which sequesters the sites of protein/substrate interaction and phosphorylation. An oncogenic fusion protein relieves this inhibition, ultimately inducing two different types of leukaemia and is thus the target for two classes of therapeutics, both in use (ATP binding site inhibitors, *e.g.* imatinib) and in development (myristoyl pocket binding, *e.g.* GNF-5). Skora and colleagues have studied the dynamics and obtained structural models for a 50 kDa construct of the kinase in four states by combining RDC and SAXS data.[138] The apo form and GNF-5 inhibited form exhibit the expected "closed" state, whilst the ATP-binding site inhibitor imatinib induces an open state which is highly dynamic. Interestingly, addition of GNF-5 to this restores the closed state, resulting in a "doubly-inhibited" form which may explain how their combination is able to overcome a known drug-resistance mechanism.

5.2 Membrane proteins

Peng and co-workers were able to obtain ~90% backbone assignments for the voltage-sensor domain (VSD) from the human KCNQ1 K^{+} ion channel, a protein implicated in various diseases due to its functions in regulating heartbeat and potassium homeostasis in the inner ear.[139] After screening a variety of detergents, adequate-quality spectra were obtained in lyso-palmitoylphosphatidylglycerol (LPPG) micelles at 50 °C, permitting assignment using standard amide TROSY-based methodology of triple-labelled protein, aided by selective amino-acid labelled samples. Secondary structure and relaxation data confirm the tetra-span helical topology common to VSDs found in other channels, although some clear differences warrant the full structure determination that will presumably ensue. The transport of metabolites and ions across the outer membrane of mitochondria is mediated by a single protein, the voltage-dependent anion channel (VDAC). To resolve conflicting reports on the specificity, and ultimately selectivity, of ATP binding and passage, Villinger and co-workers monitored nucleotide titrations of the 19-stranded human VDAC1 in lauryldimethylamine oxide (LDAO) micelles with amide TROSY spectra.[140] Chemical Shift Perturbations (CSPs) and line broadening indicated a large binding region for ATP, GTP and UTP which overlaps the previously-reported NADH binding site, with preference for charged ATP, providing support for passage selectivity influenced by binding specificity. The above study from Peng *et al.* is in line with numerous reports on the importance of screening the membrane mimetic to obtain stable and, importantly, functional protein. For the C99 protein, a product of β-secretase cleavage of amyloid precursor protein and the substrate for further proteolytic cleavage into amyloidogenic β peptides by γ-secretase, Song *et al.* found little chemical shift difference between spectra in a variety of bicelles.[141] However, subtle differences were observed in the pattern of amide broadening by the water-soluble paramagnetic probe Gd(III)-DTPA, implying different insertion depths of the trans-membrane regions. This presumably alters the position of the γ-secretase cleavage site and may in fact explain, at least in part, the known dependency of the product β-42 to β-40 ratio on membrane thickness. *N.B.* a number of new bicelle compositions are reported in this study. The pH-dependent gating of the *E. coli* outer membrane protein G (OmpG) has been investigated by Zhuang and colleagues.[142] Crystal structures of the 33 kDa β-barrel at pH 5.6 and 7.5 had previously hinted at a mechanism where the extended extracellular loop 6 partially blocked the lumen at low pH and was folded away at higher pH. Using nine nitroxide spin labels, Zhuang *et al.* measured PREs and calculated ensembles of structures assuming the observed PREs were averages of multiple dynamic states. At pH 6.3, the restraints best fit three ensemble states, two of which corresponded approximately to the open and closed forms observed in the two crystal structures. At pH 7.0, the data are better represented by four ensembles and thus a much more dynamic picture for OmpA pH gating is revealed than a simple "open/closed" model. Cross-linking was used to broadly verify the modelled ensemble states. OuYang *et al.* have determined the structure of the Hepatitis C Virus (HCV) membrane protein p7, which is known to assemble into a ~42 kDa hexamer capable of selective cation conduction.[143] In dodecylphosphocholine (DPC) micelles the viroporin is comprised of solely three helices which associate only intermolecularly with other monomers at the $i + 1$, $i + 2$ and $i + 3$ positions, forming a novel "funnel" shape. Three features appear to influence selectivity: (1) a ring of Asn sidechains which are postulated to dehydrate ions, permitting them to pass through (2) a narrow hydrophobic neck of Ile residues, and (3) basic residues at the open end of the funnel which could act to bind anions, preventing ingress. ^{2}H, ^{15}N- and A, L, V-methyl-labelled samples were further used to determine intermolecular NOEs to the bound drug amantidine, providing an initial model for drug binding. Cook *et al.* provide a different picture of p7 in their study in 1,2-dihexanoyl-*sn*-*glycero*-3-phosphocholine (DHPC) micelles.[33] Their structure, calculated using a hybrid protocol combining CS-ROSETTA with RDC refinement in Xplor-NIH, reveals a monomeric p7 where the majority of helical residues associate into a two-span motif. This structure and orientation within the micelle is verified with line-broadening measurements from chelated Mn^{2+} and further highlights the remarkable complexity that can be present even in such a small membrane protein (63 residues). The α7 nicotinic acetylcholine receptor (nAChR) is a member of the family of pentameric ligand-gated ion channels, which mediate fast synaptic transmission, and is localised to areas of the brain involved in learning, memory and cognition. Using standard methods, Bondarenko *et al.* have determined the structure of a functional mutant in LDAO micelles and investigated binding of the anaesthetics halothane and ketamine.[144] Interestingly, an interaction with both anaesthetics was detected at overlapping binding sites, despite the former not being an inhibitor of α7 function. Thus binding within the channel *per se* is not enough for inhibition without further modulation of conformation or dynamics, evidence for the latter being changes in linewidth observed in the presence of ketamine. The *E. coli* EmrE transporter provides multi-drug resistance by facilitating the export of cationic polyaromatic compounds across the inner membrane, which it achieves by flipping between an "open in" and "open out" state. Morrison and Henzler-Wildman monitored the binding and dynamics of ^{2}H, ^{15}N-EmrE in 1,2-dilauroyl-*sn-glycero*-3-phosphocholine (DLPC)/DHPC- or DMPC/DHPC bicelles with a range of polyaromatics.[145] In ^{1}H–^{15}N TROSY spectra, ligand-saturated EmrE exists in equilibrium between the two states, the timescale for which they measured using ZZ-exchange or lineshape analysis. Remarkably, the exchange rate varied by nearly three orders of magnitude over a similarly-charged ligand series, and this was correlated with K_d. This suggests a mechanism for inhibition of transport, as a substrate with a high enough affinity would effectively restrict the functional gating of the transporter.

The release of neurotransmitters in eukaryotic cells requires fusion of synaptic vesicles with the pre-synaptic plasma membrane, a process mediated by the zippering of "SNARE" motifs from three proteins. Liang and colleagues have determined the structure of one of these, syntaxin-1A (syx), in its pre-fusion state, requiring NOE-, dihedral-, RDC- and PRE restraints to define the helical angles and spatial arrangement in DPC micelles.[146] The zippering helix is broken in two about the "0-layer" residue Q226, a known determinant of register in the coiled coil. This differs from the conformation found in both the SNARE fusion complex and free syx, and the authors suggest this as a pre-fusion state permitting proof-reading of the 0-layer to ensure correct SNARE assembly. Neurotransmitter release is triggered by Ca^{2+} influx into a pre-synaptic terminal, as sensed by synaptotagmin-1 (syt), which binds the cations in its two C_2 domains, A and B. Two mechanisms for instigating membrane fusion have been proposed, either through self-association of syt or direct bridging by single syt molecules, leading to small inter-membrane distances. Seven *et al.* used PREs measured in I, L, V methyl-labelled samples in nanodiscs to show that the relative domain orientation is consistent with their direct-binding model.[147] Furthermore, using NMR and cryo-EM they show that correctly-purified syt constructs are monomeric and the observed self-association may arise through nucleic acid contamination. Xu and colleagues applied NMR to test another aspect of this process, how inhibition of exocytosis by complexin-I is overcome by syt binding to the SNARE complex.[148] A current model proposes that syt binding displaces the inhibitor, so the authors monitored complexin-I using both amide- and methyl-TROSY spectra. SNARE binding is accompanied by CSPs in selected residues of complexin. Addition of syt causes line broadening but not further shift changes, implying that both proteins can be bound to SNARE simultaneously and relief of inhibition must therefore arise through a different mechanism, such as a more subtle change in conformation. Finally, in a different vein, Burmann *et al.* have characterised the residual structure and dynamics of two *E. coli* outer-membrane proteins, OmpX and the OmpA transmembrane domain, within the substrate chamber of the trimeric Skp chaperone, which conveys such proteins through the aqueous periplasm after synthesis.[149] Using chemical shifts, ^{15}N dynamics and PREs measured in the ^{2}H, ^{15}N-labelled OMPs, they are shown to be highly dynamic and devoid of secondary structure both in high denaturant concentration and when contained within Skp. However, the Skp-bound form is significantly more compact, occupying a sphere of ~20 Å. By measuring PREs on isotope-labelled Skp from MTSL-labelled OmpX and vice versa, the sphere of dynamically-interconverting OmpX is positioned centrally within the Skp trimer and fits with little re-arrangement of structure, as confirmed by only minor shift changes. Thus the OMPs are held in a different state to the molten-globule forms seen for substrates of ATP-dependent chaperones such as GroEL and Hsp90, in a dynamic ensemble which the authors suggest permits selection by the Bam complex for membrane insertion and folding by an energy-independent mechanism.

5.3 Very large systems

Lapinaite and colleagues have published a particularly impressive study of the box C/D ribonucleoprotein (RNP) from the archaeon *Pyrococcus furiosus*, a complex of ~390 kDa responsible for methylating ribosomal RNA.[150] By I, L, V-methyl labelling the three protein components of the complex, methyl TROSY spectra were used to monitor chemical shifts and PREs. Models for the entire complex could thus be calculated both with and without substrate RNA present using previously-determined structures of protein and RNA sub-units. Domain boundaries were allowed to be flexible during docking using NMR and SANS data, with further selection based on SAXS. Assignment in the bound state was aided by mutations and consideration of NOE patterns in ^{13}C NOESY spectra. The structures reveal a large conformational change upon substrate RNA binding which, in combination with monitoring of the enzymatic reaction by NMR, suggests that methylation has to occur at the two template RNA guide sites sequentially and in a specific order, as discussed in a separate perspective.[151] Combined NMR and SAXS data were also used to obtain a model for the binding of the ~45 kDa intrinsically-disordered neuronal protein Tau binding to the ~170 kDa protein Hsp90 dimer by Karagöz *et al.*[152] Chemical shifts were monitored using standard ^{15}N HSQC spectra of Tau and methyl TROSY spectra of Ile-labelled Hsp90. Hsp90 binds Tau over an extended region containing the microtubule-binding repeats, including the motifs responsible for promoting aggregation in Alzheimer's disease. To accomplish this, Hsp90 uses a >100 Å-long binding surface spanning the N-terminal and middle domains. The authors propose that the extended, multiple low-affinity contacts allow recognition of scattered hydrophobic residues present in late folding intermediates, on which Hsp90 acts preferentially, and likewise selected IDPs such as Tau which have unusual hydrophobic elements. A similar system was tackled using a Herculean battery of NMR-based experiments by Saio *et al.*: how the bacterial trigger factor (TF) interacts with emerging nascent polypeptide chains to prevent misfolding and aggregation[153] (see also perspective).[154] Various domain dissection and isotope-labelling strategies were used to assign and monitor the ~100 kDa dimeric TF binding to unfolded ~48 kDa alkaline phosphatase (PhoA). NMR titrations revealed seven distinct regions of unfolded PhoA that are bound by TF at four sites which span the substrate-binding and peptidyl prolyl isomerase domains. Intermolecular NOEs obtained from an A, I, L, M, T, V-methyl- and F, Y, W-protonated TF and other uniformly ^{15}N, ^{13}C-constructs enabled calculation of a structural model for the full ~200 kDa complex which contains the three monomeric TFs shown to be required to bind full-length unfolded PhoA. Titrations with other unfolded proteins and an aggregation-prone mutant of maltose-binding protein (MBP) indicate that the binding sites in TF are conserved and that it is able to stabilise proteins by engaging the unfolded state to prevent aggregation. A short report from Nishida and colleagues mapped the interaction sites on the IDP α-synuclein (αSN) of the 800 kDa bacterial chaperone GroEL.[155] Conventional ^{1}H–^{15}N spectra of αSN showed selective broadening at two N-terminal regions which interact with polyphenolic compounds that inhibit formation of toxic aggregates. A similar mapping of bound denatured bovine Rho protein by ^{13}C spectra of $^{13}C^{\alpha}$ amino acid selectively-labelled samples was less conclusive due to the extreme line broadening observed. Smith *et al.* used NMR to study another chaperone, heat shock cognate 70 (Hsc70) which associates with co-chaperones to facilitate refolding or, in the case of the E3 ubiquitin ligase CHIP, degradation of substrate proteins.[156] Binding of the 70 kDa CHIP dimer to Hsc70 could be mapped to the flexible C-terminal EEVD motif, known to be important in Hsp90 for recruiting co-chaperones. Surprisingly, 1d envelope ^{15}N amide relaxation measurements suggested that the correlation time of Hsc70 in the 210 kDa complex with CHIP is similar to the ~70 kDa free state, suggesting the two proteins tumble independently. With additional data from binding affinity measurements, the authors conclude that the C-terminal motif is sufficient for CHIP recruitment and an extended flexible region permits the E3 ligase to traverse a large area of the complex, facilitating ubiquitination of bound substrates.

Two publications from the Kay group utilise their methyl-TROSY technology to elucidate the mechanism of proteolysis by the *Thermoplasma acidophilum* 20S proteasome. In the first, the recently-developed $^{13}CH_3$–^{2}H-threonine labelling scheme was utilised to confirm the first step in the postulated mechanism of catalysis, which involves the N-terminal Thr amino group of the β subunit acting as a base to accept a proton from its own hydroxyl group, known to be the catalytic nucleophile.[157] 10/15 Threonines could be assigned from NOESY data and mutagenesis, permitting the pKa to be measured in Thr1 by pH titration of the 670 kDa multimer and thus confirming it has the requisite ionisation state for the proposed mechanism. A separate paper investigates the process by which the "in–out" exchange of the N-terminal residues in the α7 subunit occurs, a transition which gates entry of substrates to the catalytic site within the lumen of the proteasome.[158] The exchange rate was quantified as a function of viscogen concentration using 2D magnetisation-exchange experiments detected in the 180 kDa $^{13}CH_3$-Met labelled α7 ring, which permits estimation of the contribution of solvent *versus* intramolecular motion to the overall exchange process. After confirming that the viscogens do not interact with the proteasome (using $^{13}CH_3$-Ile labelled protein) or affect the populations of the "in–out" states, measurement of exchange rate in different viscogens was used to obtain the average size of the protein unit undergoing conformational exchange to complete the gating motion. This was found to be very small, <3.5 Å, indicating the process involves many small, local rearrangements as opposed to one or two concerted motions which would have a large effective radius. The same laboratory has also studied a bacterial proteasome by methyl-labelling the 230 kDa HslV protease, which functions analogously to the archaeal β_7 double ring.[159] 95% of I, L, M, T and V methyls could be assigned, permitting relaxation dispersion measurements and monitoring of CSPs. Akin to the *T. acidophilum* proteasome, a pH titration was used to confirm the neutral protonation state of the catalytic Threonine at the optimum pH for reaction. Significantly, a network of residues which allosterically regulate substrate proteolysis depending on binding of the partner complex HslU were mapped based on mutagenesis, chemical shift and dynamics data, entirely analogous to what was previously observed in the 20S proteasome. Audin *et al.* used I, L, V-methyl TROSY experiments to monitor dynamics and binding of the hexameric 173 kDa archaeal exosome core complex.[160] Assignments from a "divide and conquer" approach were transferred to the complex, aided by mutagenesis, permitting ~70% of methyls to be monitored. Interestingly, a number of residues near the cap interface were observed to be in slow exchange between two conformationally-similar states, as judged by NOESY spectra and thermodynamic measurements. The authors present evidence that the cap proteins select the major (∼75%-populated) form and in the bound state the conformational exchange is quenched. The ∼110 kDa homodimer MDMX protein binds the transactivation domain (TAD) of p53 through its N-terminal domain (NTD), inhibiting p53's interaction with transcription factors. Bista and co-workers used amide-TROSY spectra to monitor the interaction of different constructs of MDMX binding to p53, including the ∼400 kDa complex formed by the full-length proteins.[161] Intriguingly, the isolated NTD spectrum was different from that in the full-length MDMX, and its affinity for p53 was found to be 100x stronger by ITC. The authors were able to demonstrate by NMR and mutational analysis that full-length MDMX contains a self-inhibitory element that occludes the p53 binding site on the NTD. This may provide a mechanism of regulation of MDMX's inhibition of p53, either by an accessory protein sequestering the self-inhibitory element or *via* a known alternative splicing variant in which the sequence is absent.

The basis of the cyanobacterial circadian rhythm appears quite straightforward, being centred around three proteins; KaiA, which stimulates autophosphorylation of KaiC, and KaiB, which promotes dephosphorylation by sequestering KaiA. In reality, the picture is more complicated and much of the detail is being elucidated using NMR and methods such as fluorescence anisotropy by one laboratory in particular (see a review of their NMR-based approach.)[162] In their latest publication,[163] methyl- and amide TROSY spectra of Ile-δ1-$^{13}CH_3$-labelled KaiA and ^{2}H, ^{15}N-KaiB were used to show that KaiA promotes formation of the KaiB–KaiC complex and thus encourages *de*phosphorylation of KaiC. Loss of symmetry in these spectra also implies a change of multimerisation state of KaiB may occur with KaiA binding. Further, interaction of KaiB and the read-out protein SasA is monitored with TROSY spectra of KaiC, revealing that they compete for similar sites and KaiA also promotes displacement of SasA and thus influences the clock output. Hence, there is much underlying complexity which arises ultimately through allosteric effects and potential changes in multimerisation state in this macromolecular system, much of which is being revealed by NMR.

References

1 M. P. Williamson, *Prog. Nucl. Magn. Reson. Spectrosc.*, 2013, **73**, 1–16.

2 M. V. Berjanskii and D. S. Wishart, *J. Am. Chem. Soc.*, 2013, **135**, 14536–14539.

3 V. Tugarinov, *Prog. Nucl. Magn. Reson. Spectrosc.*, 2014, **77**, 49–68.

4 R. J. Lichtenecker, K. Weinhaupl, L. Reuther, J. Schorghuber, W. Schmid and R. Konrat, *J. Biomol. NMR*, 2013, **57**, 205–209.

5 G. Mas, E. Crublet, O. Hamelin, P. Gans and J. Boisbouvier, *J. Biomol. NMR*, 2013, **57**, 251–262.

6 Y. Miyanoiri, M. Takeda, K. Okuma, A. M. Ono, T. Terauchi and M. Kainosho, *J. Biomol. NMR*, 2013, **57**, 237–249.

7 M. Miyazawa-Onami, K. Takeuchi, T. Takano, T. Sugiki, I. Shimada and H. Takahashi, *J. Biomol. NMR*, 2013, **57**, 297–304.

8 R. J. Lichtenecker, K. Weinhaupl, W. Schmid and R. Konrat, *J. Biomol. NMR*, 2013, **57**, 327–331.
9 V. Kasinath, K. G. Valentine and A. J. Wand, *J. Am. Chem. Soc.*, 2013, **135**, 9560–9563.
10 M. A. S. Hass and M. Ubbink, *Curr. Opin. Struct. Biol.*, 2014, **24**, 45–53.
11 G. M. Clore, *Biochem. Soc. Trans.*, 2013, **41**, 1343–1354.
12 G. M. Clore, *Biophys. Chem.*, 2014, **186**, 3–12.
13 H. G. Hocking, K. Zangger and T. Madl, *ChemPhysChem*, 2013, **14**, 3082–3094.
14 X. H. Gu, Z. Gong, D. C. Guo, W. P. Zhang and C. Tang, *J. Biomol. NMR*, 2014, **58**, 149–154.
15 G. Jeschke, *Prog. Nucl. Magn. Reson. Spectrosc.*, 2013, **72**, 42–60.
16 D. Shishmarev and G. Otting, *J. Biomol. NMR*, 2013, **56**, 203–216.
17 F. Huang, Y. Y. Pei, H. H. Zuo, J. L. Chen, Y. Yang and X. C. Su, *Chem. - Eur. J.*, 2013, **19**, 17141–17149.
18 Z. Liu, Z. Gong, D. C. Guo, W. P. Zhang and C. Tang, *Biochemistry*, 2014, **53**, 1403–1409.
19 J. C. Hoch and M. Billeter, *J. Biomol. NMR*, 2014, **58**, 231–232.
20 A. G. Irvine, V. Slynko, Y. Nikolaev, R. R. P. Senthamarai and K. Pervushin, *J. Magn. Reson.*, 2014, **239**, 121–129.
21 C. Schmidt, S. J. Irausquin and H. Valafar, *BMC Bioinf.*, 2013, **14**, 302.
22 T. M. Sabo, C. A. Smith, D. Ban, A. Mazur, D. Lee and C. Griesinger, *J. Biomol. NMR*, 2014, **58**, 287–301.
23 K. Sugase, T. Konuma, J. C. Lansing and P. E. Wright, *J. Biomol. NMR*, 2013, **56**, 275–283.
24 K. Berlin, A. Longhini, T. K. Dayie and D. Fushman, *J. Biomol. NMR*, 2013, **57**, 333–352.
25 C. A. MacRaild and R. S. Norton, *J. Biomol. NMR*, 2014, **58**, 155–163.
26 E. R. P. Zuiderweg, I. Bagai, P. Rossi and E. B. Bertelsen, *J. Biomol. NMR*, 2013, **57**, 179–191.
27 T. Aeschbacher, E. Schmidt, M. Blatter, C. Maris, O. Duss, F. H. T. Allain, P. Guntert and M. Schubert, *Nucleic Acids Res.*, 2013, **41**, e172.
28 M. Szachniuk, M. Malaczynski, E. Pesch, E. K. Burke and J. Blazewicz, *J. Heuristics*, 2013, **19**, 443–464.
29 E. Schmidt and P. Guntert, *J. Biomol. NMR*, 2013, **57**, 193–204.
30 E. Schmidt, J. Gath, B. Habenstein, F. Ravotti, K. Szekely, M. Huber, L. Buchner, A. Bockmann, B. H. Meier and P. Guntert, *J. Biomol. NMR*, 2013, **56**, 243–254.
31 B. Fu, A. B. Sahakyan, C. Camilloni, G. G. Tartaglia, E. Paci, A. Caflisch, M. Vendruscolo and A. Cavalli, *J. Comput. Chem.*, 2014, **35**, 1101–1105.
32 T. Rao, J. W. Lubin, G. S. Armstrong, T. M. Tucey, V. Lundblad and D. S. Wuttke, *Proc. Natl. Acad. Sci. U. S. A.*, 2014, **111**, 214–218.
33 G. A. Cook, L. A. Dawson, Y. Tian and S. J. Opella, *Biochemistry*, 2013, **52**, 5295–5303.
34 G. van der Schot, Z. Y. Zhang, R. Vernon, Y. Shen, W. F. Vranken, D. Baker, A. Bonvin and O. F. Lange, *J. Biomol. NMR*, 2013, **57**, 27–35.
35 R. Vernon, Y. Shen, D. Baker and O. F. Lange, *J. Biomol. NMR*, 2013, **57**, 117–127.
36 S. Lindert, J. Meiler and J. A. McCammon, *J. Chem. Theory Comput.*, 2013, **9**, 3843–3847.
37 D. E. Kim, F. DiMaio, R. Y. R. Wang, Y. F. Song and D. Baker, *Proteins*, 2014, **82**, 208–218.
38 A. T. Frank, S. H. Bae and A. C. Stelzer, *J. Phys. Chem. B*, 2013, **117**, 13497–13506.

39 P. Sripakdeevong, M. Cevec, A. T. Chang, M. C. Erat, M. Ziegeler, Q. Zhao, G. E. Fox, X. L. Gao, S. D. Kennedy, R. Kierzek, E. P. Nikonowicz, H. Schwalbe, R. K. O. Sigel, D. H. Turner and R. Das, *Nat. Methods*, 2014, **11**, 413–416.
40 R. M. van der Werf, M. Tessari and S. S. Wijmenga, *J. Biomol. NMR*, 2013, **56**, 95–112.
41 V. Menon, B. K. Vallat, J. M. Dybas and A. Fiser, *Structure*, 2013, **21**, 891–899.
42 A. Kumari and K. Dorai, *J. Mol. Struct.*, 2013, **1041**, 200–212.
43 Y. Shen and A. Bax, *J. Biomol. NMR*, 2013, **56**, 227–241.
44 B. E. Weiner, N. Alexander, L. R. Akin, N. Woetzel, M. Karakas and J. Meiler, *Proteins*, 2014, **82**, 587–595.
45 B. L. Eggimann, V. V. Vostrikov, G. Veglia and J. I. Siepmann, *Theor. Chem. Acc.*, 2013, **132**, 1388.
46 B. C. Mao, R. Tejero, D. Baker and G. T. Montelione, *J. Am. Chem. Soc.*, 2014, **136**, 1893–1906.
47 J. Bibby, R. M. Keegan, O. Mayans, M. D. Winn and D. J. Rigden, *Acta Crystallogr., Sect. D: Biol. Crystallogr.*, 2013, **69**, 2194–2201.
48 K. Q. Wu, C. W. Shi, J. Li, H. P. Wang, P. Shi, L. Chen, F. M. Wu, Y. Xiong and C. L. Tian, *Protein Cell*, 2013, **4**, 893–896.
49 U. Schieborr, S. Sreeramulu, B. Elshorst, M. Maurer, K. Saxena, T. Stehle, D. Kudlinzki, S. L. Gande and H. Schwalbe, *Proteins*, 2013, **81**, 2007–2022.
50 M. Rinaldelli, E. Ravera, V. Calderone, G. Parigi, G. N. Murshudov and C. Luchinat, *Acta Crystallogr., Sect. D: Biol. Crystallogr.*, 2014, **70**, 958–967.
51 H. Yagi, K. B. Pilla, A. Maleckis, B. Graham, T. Huber and G. Otting, *Structure*, 2013, **21**, 883–890.
52 A. Kannan, C. Camilloni, A. B. Sahakyan, A. Cavalli and M. Vendruscolo, *J. Am. Chem. Soc.*, 2014, **136**, 2204–2207.
53 A. Rosato, R. Tejero and G. T. Montelione, *Curr. Opin. Struct. Biol.*, 2013, **23**, 715–724.
54 G. W. Vuister, R. H. Fogh, P. M. S. Hendrickx, J. F. Doreleijers and A. Gutmanas, *J. Biomol. NMR*, 2014, **58**, 259–285.
55 O. A. Martin, Y. A. Arnautova, A. A. Icazatti, H. A. Scheraga and J. A. Vila, *Proc. Natl. Acad. Sci. U. S. A.*, 2013, **110**, 16826–16831.
56 A. Bagaria, V. Jaravine and P. Guntert, *Comput. Biol. Chem.*, 2013, **46**, 8–15.
57 R. Tejero, D. Snyder, B. C. Mao, J. M. Aramini and G. T. Montelione, *J. Biomol. NMR*, 2013, **56**, 337–351.
58 S. G. Hyberts, H. Arthanari, S. A. Robson and G. Wagner, *J. Magn. Reson.*, 2014, **241**, 60–73.
59 J. C. Hoch, M. W. Maciejewski, M. Mobli, A. D. Schuyler and A. S. Stern, *Acc. Chem. Res.*, 2014, **47**, 708–717.
60 Z. Orban-Nemeth, M. A. Henen, L. Geist, S. Zerko, S. Saxena, J. Stanek, W. Kozminski, F. Propst and R. Konrat, *Biomol. NMR Assignments*, 2014, **8**, 123–127.
61 W. Bermel, I. C. Felli, L. Gonnelli, W. Kozminski, A. Piai, R. Pierattelli and A. Zawadzka-Kazimierczuk, *J. Biomol. NMR*, 2013, **57**, 353–361.
62 I. C. Felli and R. Pierattelli, *J. Magn. Reson.*, 2014, **241**, 115–125.
63 S. Tikole, V. Jaravine, V. Y. Orekhov and P. Guntert, *PLoS One*, 2013, **8**, e68567.
64 J. Stanek, M. Nowakowski, S. Saxena, K. Ruszczynska-Bartnik, A. Ejchart and W. Kozminski, *J. Biomol. NMR*, 2013, **56**, 217–226.
65 J. Stanek, P. Podbevsek, W. Kozminski, J. Plavec and M. Cevec, *J. Biomol. NMR*, 2013, **57**, 1–9.

66 M. R. Palmer, B. R. Wenrich, P. Stahlfeld and D. Rovnyak, *J. Biomol. NMR*, 2014, **58**, 303–314.
67 V. Tugarinov, V. Venditti and G. M. Clore, *J. Biomol. NMR*, 2014, **58**, 1–8.
68 K. Sinha, L. Jen-Jacobson and G. S. Rule, *J. Biomol. NMR*, 2013, **56**, 331–335.
69 D. P. Frueh, *Prog. Nucl. Magn. Reson. Spectrosc.*, 2014, **78**, 47–75.
70 D. P. Frueh, A. C. Goodrich, S. H. Mishra and S. R. Nichols, *Curr. Opin. Struct. Biol.*, 2013, **23**, 734–739.
71 A. Mainz, T. L. Religa, R. Sprangers, R. Linser, L. E. Kay and B. Reif, *Angew. Chem., Int. Ed.*, 2013, **52**, 8746–8751.
72 D. A. Fox and L. Columbus, *Protein Sci.*, 2013, **22**, 1133–1140.
73 I. Maslennikov and S. Choe, *Curr. Opin. Struct. Biol.*, 2013, **23**, 555–562.
74 K. Oxenoid and J. J. Chou, *Curr. Opin. Struct. Biol.*, 2013, **23**, 547–554.
75 A. Gautier, *Biochim. Biophys. Acta, Bioenerg.*, 2014, **1837**, 578–588.
76 J. H. Prestegard, D. A. Agard, K. W. Moremen, L. A. Lavery, L. C. Morris and K. Pederson, *J. Magn. Reson.*, 2014, **241**, 32–40.
77 J. Stehle, R. Silvers, K. Werner, D. Chatterjee, S. Gande, F. Scholz, A. Dutta, J. Wachtveitl, J. Klein-Seetharaman and H. Schwalbe, *Angew. Chem., Int. Ed.*, 2014, **53**, 2078–2084.
78 T. Didenko, J. J. Liu, R. Horst, R. C. Stevens and K. Wuthrich, *Curr. Opin. Struct. Biol.*, 2013, **23**, 740–747.
79 R. Horst, J. J. Liu, R. C. Stevens and K. Wuthrich, *Angew. Chem., Int. Ed.*, 2013, **52**, 10762–10765.
80 T. H. Kim, K. Y. Chung, A. Manglik, A. L. Hansen, R. O. Dror, T. J. Mildorf, D. E. Shaw, B. K. Kobilka and R. S. Prosser, *J. Am. Chem. Soc.*, 2013, **135**, 9465–9474.
81 N. Nishida, M. Osawa, K. Takeuchi, S. Imai, P. Stampoulis, Y. Kofuku, T. Ueda and I. Shimada, *J. Magn. Reson.*, 2014, **241**, 86–96.
82 K. Mobius, K. Nordsieck, A. Pichert, S. A. Samsonov, L. Thomas, J. Schiller, S. Kalkhof, M. T. Pisabarro, A. G. Beck-Sickinger and D. Huster, *Glycobiology*, 2013, **23**, 1260–1269.
83 T. S. Chavan, S. Abraham and V. Gaponenko, *Molecules*, 2013, **18**, 7103–7119.
84 N. V. Nucci, K. G. Valentine and A. J. Wand, *J. Magn. Reson.*, 2014, **241**, 137–147.
85 I. Dodevski, N. V. Nucci, K. G. Valentine, G. K. Sidhu, E. S. O'Brien, A. Pardi and A. J. Wand, *J. Am. Chem. Soc.*, 2014, **136**, 3465–3474.
86 K. G. Valentine, G. Mathies, S. Bedard, N. V. Nucci, I. Dodevski, M. A. Stetz, T. V. Can, R. G. Griffin and A. J. Wand, *J. Am. Chem. Soc.*, 2014, **136**, 2800–2807.
87 R. Das, Y. H. Liang, J. Mariano, J. Li, T. Huang, A. King, S. G. Tarasov, A. M. Weissman, X. H. Ji and R. A. Byrd, *EMBO J.*, 2013, **32**, 2504–2516.
88 C. A. Castañeda, T. R. Kashyap, M. A. Nakasone, S. Krueger and D. Fushman, *Structure*, 2013, **21**, 1168–1181.
89 H. J. Park, M. Guariento, M. Maciejewski, R. Hauhart, W. H. Tham, A. F. Cowman, C. Q. Schmidt, H. D. T. Mertens, M. K. Liszewski, D. E. Hourcade, P. N. Barlow and J. P. Atkinson, *J. Biol. Chem.*, 2014, **289**, 450–463.
90 A. Neuhaus, H. Kooshapur, J. Wolf, N. H. Meyer, T. Madl, J. Saidowsky, E. Hambruch, A. Lazam, M. Jung, M. Sattler, W. Schliebs and R. Erdmann, *J. Biol. Chem.*, 2014, **289**, 437–448.
91 R. Back, C. Dominguez, B. Rothe, C. Bobo, C. Beaufils, S. Morera, P. Meyer, B. Charpentier, C. Branlant, F. H. T. Allain and X. Manival, *Structure*, 2013, **21**, 1834–1847.

92 R. Berry, S. J. Headey, M. J. Call, J. McCluskey, C. A. Tregaskes, J. Kaufman, R. Koh, M. J. Scanlon, M. E. Call and J. Rossjohn, *J. Biol. Chem.*, 2014, **289**, 8240–8251.
93 S. Bruchweiler, R. Konrat and M. Tollinger, *ACS Chem. Biol.*, 2013, **8**, 1600–1610.
94 S. P. Mukherjee, M. Behar, H. A. Birnbaum, A. Hoffmann, P. E. Wright and G. Ghosh, *PLos Biol.*, 2013, **11**, e1001647.
95 C. de Chiara, R. P. Menon, G. Kelly and A. Pastore, *PLoS One*, 2013, **8**, e76456.
96 S. Mikami, T. Kanaba, N. Takizawa, A. Kobayashi, R. Maesaki, T. Fujiwara, Y. Ito and M. Mishima, *Structure*, 2014, **22**, 35–46.
97 W. J. Chung, B. Heddi, F. Hamon, M. P. Teulade-Fichou and A. T. Phan, *Angew. Chem., Int. Ed.*, 2014, **53**, 999–1002.
98 M. Adrian, D. J. Ang, C. J. Lech, B. Heddi, A. Nicolas and A. T. Phan, *J. Am. Chem. Soc.*, 2014, **136**, 6297–6305.
99 K. W. Lim and A. T. Phan, *Angew. Chem., Int. Ed.*, 2013, **52**, 8566–8569.
100 M. Marušič, R. N. Veedu, J. Wengel and J. Plavec, *Nucleic Acids Res.*, 2013, **41**, 9524–9536.
101 Y. Nomura, Y. Tanaka, J. Fukunaga, K. Fujiwara, M. Chiba, H. Iibuchi, T. Tanaka, Y. Nakamura, G. Kawai, T. Kozu and T. Sakamoto, *J. Biochem.*, 2013, **154**, 513–519.
102 P. Bouchard and P. Legault, *Biochemistry*, 2014, **53**, 258–269.
103 H. Martadinata and A. T. Phan, *Biochemistry*, 2014, **53**, 1595–1600.
104 D. D. Cash, O. Cohen-Zontag, N. K. Kim, K. Shefer, Y. Brown, N. B. Ulyanov, Y. Tzfati and J. Feigon, *Proc. Natl. Acad. Sci. U. S. A.*, 2013, **110**, 10970–10975.
105 N. K. Kim, Q. Zhang and J. Feigon, *Nucleic Acids Res.*, 2014, **42**, 3395–3408.
106 M. J. Kang, C. D. Eichhorn and J. Feigon, *Proc. Natl. Acad. Sci. U. S. A.*, 2014, **111**, E663–E671.
107 A. Cléry, R. Sinha, O. Anczukow, A. Corrionero, A. Moursy, G. M. Daubner, J. Valcarcel, A. R. Krainer and F. H. T. Allain, *Proc. Natl. Acad. Sci. U. S. A.*, 2013, **110**, E2802–E2811.
108 L. N. Foot, M. Feracci and C. Dominguez, *Methods*, 2014, **65**, 288–301.
109 T. Carlomagno, *J. Magn. Reson.*, 2014, **241**, 126–136.
110 F. Paquet, O. Delalande, S. Goffinont, F. Culard, K. Loth, U. Asseline, B. Castaing and C. Landon, *PLoS One*, 2014, **9**, e88809.
111 W. B. Gong, J. F. Wang, S. Perrett and Y. G. Feng, *J. Biol. Chem.*, 2014, **289**, 4882–4895.
112 R. G. Burge, M. A. Martinez-Yamout, H. J. Dyson and P. E. Wright, *Biochemistry*, 2014, **53**, 1495–1510.
113 A. G. Palmer III, *J. Magn. Reson.*, 2014, **241**, 3–17.
114 J. Lim, T. S. Xiao, J. S. Fan and D. W. Yang, *Angew. Chem., Int. Ed.*, 2014, **53**, 2358–2361.
115 B. Zhao, A. L. Hansen and Q. Zhang, *J. Am. Chem. Soc.*, 2014, **136**, 20–23.
116 A. Reining, S. Nozinovic, K. Schlepckow, F. Buhr, B. Furtig and H. Schwalbe, *Nature*, 2013, **499**, 355–359.
117 R. Micura, *Nature*, 2013, **499**, 289–290.
118 G. Bouvignies, P. Vallurupalli and L. E. Kay, *J. Mol. Biol.*, 2014, **426**, 763–774.
119 A. L. Hansen and L. E. Kay, *Proc. Natl. Acad. Sci. U. S. A.*, 2014, **111**, E1705–E1712.
120 D. S. Libich, N. L. Fawzi, J. F. Ying and G. M. Clore, *Proc. Natl. Acad. Sci. U. S. A.*, 2013, **110**, 11361–11366.
121 S. R. Tzeng and C. G. Kalodimos, *Nat. Chem. Biol.*, 2013, **9**, 462–465.

122 M. C. Regan, P. S. Horanyi, E. E. Pryor, J. L. Sarver, D. S. Cafiso and J. H. Bushweller, *Proc. Natl. Acad. Sci. U. S. A.*, 2013, **110**, 13374–13379.
123 H. N. Motlagh, J. O. Wrabl, J. Li and V. J. Hilser, *Nature*, 2014, **508**, 331–339.
124 L. R. McDonald, M. J. Whitley, J. A. Boyer and A. L. Lee, *J. Mol. Biol.*, 2013, **425**, 2372–2381.
125 Y. Li, N. L. Altorelli, F. Bahna, B. Honig, L. Shapiro and A. G. Palmer, *Proc. Natl. Acad. Sci. U. S. A.*, 2013, **110**, 16462–16467.
126 Y. Horimoto, S. L. Wang and R. Y. Yada, *Biomol. NMR Assignments*, 2014, **8**, 57–61.
127 D. M. Francis, R. Page and W. Peti, *Biomol. NMR Assignments*, 2014, **8**, 185–188.
128 A. J. Guerra and D. P. Giedroc, *Biomol. NMR Assignments*, 2014, **8**, 11–14.
129 J. Shin, G. Chakraborty and H. S. Yoon, *Biomol. NMR Assignments*, 2014, **8**, 29–31.
130 X. F. Xu and W. A. Horne, *Biomol. NMR Assignments*, 2014, **8**, 217–220.
131 J. Isaksson, E. Woestenenk, C. Sahlberg and T. Agback, *Biomol. NMR Assignments*, 2014, **8**, 81–84.
132 P. J. Sapienza and A. L. Lee, *Biomol. NMR Assignments*, 2014, **8**, 195–199.
133 A. C. Miklos, M. Sumpter and H. X. Zhou, *PLoS One*, 2013, **8**, e74969.
134 Y. Hiruma, M. A. S. Hass, Y. Kikui, W. M. Liu, B. Olmez, S. P. Skinner, A. Blok, A. Kloosterman, H. Koteishi, F. Lohr, H. Schwalbe, M. Nojiri and M. Ubbink, *J. Mol. Biol.*, 2013, **425**, 4353–4365.
135 H. van Ingen and A. Bonvin, *J. Magn. Reson.*, 2014, **241**, 103–114.
136 J. Schilder, F. Lohr, H. Schwalbe and M. Ubbink, *FEBS Lett.*, 2014, **588**, 1873–1878.
137 Y. Pustovalova, M. W. Maciejewski and D. M. Korzhnev, *J. Mol. Biol.*, 2013, **425**, 3091–3105.
138 L. Skora, J. Mestan, D. Fabbro, W. Jahnke and S. Grzesiek, *Proc. Natl. Acad. Sci. U. S. A.*, 2013, **110**, E4437–E4445.
139 D. G. Peng, J. H. Kim, B. M. Kroncke, C. L. Law, Y. Xia, K. D. Droege, W. D. Van Horn, C. G. Vanoye and C. R. Sanders, *Biochemistry*, 2014, **53**, 2032–2042.
140 S. Villinger, K. Giller, M. Bayrhuber, A. Lange, C. Griesinger, S. Becker and M. Zweckstetter, *J. Biol. Chem.*, 2014, **289**, 13397–13406.
141 Y. L. Song, K. F. Mittendorf, Z. W. Lu and C. R. Sanders, *J. Am. Chem. Soc.*, 2014, **136**, 4093–4096.
142 T. D. Zhuang, C. Chisholm, M. Chen and L. K. Tamm, *J. Am. Chem. Soc.*, 2013, **135**, 15101–15113.
143 B. OuYang, S. Q. Xie, M. J. Berardi, X. H. Zhao, J. Dev, W. J. Yu, B. Sun and J. J. Chou, *Nature*, 2013, **498**, 521–525.
144 V. Bondarenko, D. D. Mowrey, T. S. Tillman, E. Seyoum, Y. Xu and P. Tang, *Biochim. Biophys. Acta, Biomembr.*, 2014, **1838**, 1389–1395.
145 E. A. Morrison and K. A. Henzler-Wildman, *J. Biol. Chem.*, 2014, **289**, 6825–6836.
146 B. Y. Liang, V. Kiessling and L. K. Tamm, *Proc. Natl. Acad. Sci. U. S. A.*, 2013, **110**, 19384–19389.
147 A. B. Seven, K. D. Brewer, L. Shi, Q. X. Jiang and J. Rizo, *Proc. Natl. Acad. Sci. U. S. A.*, 2013, **110**, E3243–E3252.
148 J. J. Xu, K. D. Brewer, R. Perez-Castillejos and J. Rizo, *J. Mol. Biol.*, 2013, **425**, 3461–3475.
149 B. M. Burmann, C. W. Wang and S. Hiller, *Nat. Struct. Mol. Biol.*, 2013, **20**, 1265–1272.
150 A. Lapinaite, B. Simon, L. Skjaerven, M. Rakwalska-Bange, F. Gabel and T. Carlomagno, *Nature*, 2013, **502**, 519–523.

151 M. Hengesbach and H. Schwalbe, *Angew. Chem., Int. Ed.*, 2014, **53**, 1742–1744.
152 G. E. Karagöz, A. M. S. Duarte, E. Akoury, H. Ippel, J. Biernat, T. M. Luengo, M. Radli, T. Didenko, B. A. Nordhues, D. B. Veprintsev, C. A. Dickey, E. Mandelkow, M. Zweckstetter, R. Boelens, T. Madl and S. G. D. Rudiger, *Cell*, 2014, **156**, 963–974.
153 T. Saio, X. Guan, P. Rossi, A. Economou and C. G. Kalodimos, *Science*, 2014, **344**, 597–611.
154 M. Gamerdinger and E. Deuerling, *Science*, 2014, **344**, 590–591.
155 N. Nishida, M. Yagi-Utsumi, F. Motojima, M. Yoshida, I. Shimada and K. Kato, *J. Biosci. Bioeng.*, 2013, **116**, 160–164.
156 M. C. Smith, K. M. Scaglione, V. A. Assimon, S. Patury, A. D. Thompson, C. A. Dickey, D. R. Southworth, H. L. Paulson, J. E. Gestwicki and E. R. P. Zuiderweg, *Biochemistry*, 2013, **52**, 5354–5364.
157 A. Velyvis and L. E. Kay, *J. Am. Chem. Soc.*, 2013, **135**, 9259–9262.
158 M. P. Latham, A. Sekhar and L. E. Kay, *Proc. Natl. Acad. Sci. U. S. A.*, 2014, **111**, 5532–5537.
159 L. C. Shi and L. E. Kay, *Proc. Natl. Acad. Sci. U. S. A.*, 2014, **111**, 2140–2145.
160 M. J. C. Audin, G. Dorn, S. A. Fromm, K. Reiss, S. Schutz, M. K. Vorlander and R. Sprangers, *Angew. Chem., Int. Ed.*, 2013, **52**, 8312–8316.
161 M. Bista, M. Petrovich and A. R. Fersht, *Proc. Natl. Acad. Sci. U. S. A.*, 2013, **110**, 17814–17819.
162 Y. G. Chang, R. Tseng, N. W. Kuo and A. LiWang, *Integr. Comp. Biol.*, 2013, **53**, 93–102.
163 R. Tseng, Y. G. Chang, I. Bravo, R. Latham, A. Chaudhary, N. W. Kuo and A. LiWang, *J. Mol. Biol.*, 2014, **426**, 389–402.

NMR of lipids

Slawomir Pikula,*[a] Joanna Bandorowicz-Pikula[a] and Patrick Groves[b]
DOI: 10.1039/9781782622758-00385

This chapter reviews articles on NMR and lipids published between June 2013 and May 2014. The number of papers devoted to use of NMR methods in order to investigate lipid and lipid derived molecules, their structures, behavior in native and artificial membranes, interactions with proteins, peptides and low molecular weight solutes, as well as biomedical applications is constantly growing over recent years (resulting in more than 1500 articles found in Pub-Med in the examined period of time) although we included here only a selection of those papers that were accessible through the internet and peer-reviewed. The reviewed material has been arranged into several sections devoted to the structure and function of lipids in artificial and biological membranes (mostly liposomes and monolayers), their roles in membrane-related processes including lipid-mediated signal transduction, the interactions of lipids with membranous and soluble proteins, peptides and low molecular weight compounds, lipid metabolomics, visualization of lipid related processes in biomedicine, and lipid-based diagnosis including some methodological approaches.

1 Introduction

This is our third review for RSC Specialist Periodical Reports in the field of NMR of lipids. We cover the same topic and areas as before in 2013 and 2014.[1,2] Previous reviews covering NMR of lipids were written from 1995 until 2009 by Dr Elizabeth F. Hounsel. In 2010 and 2011 (volumes 41 and 42), Dr Elzbieta Swierzewska and Dr Jacek Wojcik[3,4] covered a broader area, including carbohydrates. Our contribution reviews articles on NMR and lipids published between June 2013 and May 2014. The number of papers devoted to NMR investigations of lipid and lipid-derived molecules, their structures, behavior in native and artificial membranes, interactions with membranous and soluble proteins and peptides, as well as with low molecular weight compounds, and biomedical applications continues to grow (more than 1500 entries in Pub-Med in the examined period of time). We have covered a selection of accessible through the internet and peer-reviewed papers. As previously, the review has been divided into several sections devoted to the structure and function of lipids in artificial and biological membranes, their roles in membrane-related processes including membrane fusion and lipid-mediated signal transduction, interactions of lipids with membrane and soluble proteins, peptides, antibiotics and other small molecular weight compounds, lipid metabolomics, visualization of lipid related processes in biomedicine, diagnosis and therapy, and methodological approaches.

[a]*Department of Biochemistry, Nencki Institute of Experimental Biology, 3 Pasteur St., 02-093 Warsaw, Poland. E-mail: s.pikula@nencki.gov.pl*
[b]*Department of Biomedicinal Chemistry, University of Gdansk, Wita Stwosza 63, 80-308 Gdansk, Poland*

It is worth to note the substantial number of review articles published in 2013/2014 related to the NMR methodology used to investigate lipid biochemistry, biophysics, biomedicine and biotechnology. These include discussion on recent advances in using in-cell and on-cell NMR to study lipids with atomic resolution in native cells,[5] to study the interaction of membrane proteins with their lipid environment by solid-state nuclear magnetic resonance spectroscopy (ssNMR),[6–12] including non-covalent binding of membrane lipids to membrane proteins,[13] membrane protein and peptide structure determination using solution-state NMR,[14,15] and to identify mechanisms by which the lipid membrane physical state may affect protein functions in biological,[16] as well as artificial lipid membranes.[17] Furthermore, Ragona *et al.*[18] discussed the role of lipid binding properties of proteins in modulating their ligand exchange properties and penetration of amphipathic molecules though the membranes, while Borisjuk *et al.*[19] provided insights on using NMR imaging to study lipid properties in living plants.

2 The structures and cellular functions of lipids

Important discoveries in the field of structure and properties of lipids and their partner molecules, made over the last decade, involving NMR techniques, strongly suggest that lipid and lipid-derived molecules play pivotal roles in many biological processes, including dynamic shaping of cellular membranes, organelle biogenesis, cellular signalling, secretion, fertilization, cellular proliferation and cell death. This takes into account the roles of certain lipids, such as cholesterol, sphingomyelin (SM), phosphatidylserine (PS), phosphatidylinsitol (PI) and its derivatives: phosphatidylinositol (3,4) or (4,5)-bisphosphates (PtdIns(3,4)P_2 or PtdIns(4,5)P_2), and phosphatidylinositol (3,4,5)-triphosphate (PtdIns(3,4,5)P_3), and their protein partners, lipid- and membrane-binding proteins, in the formation, stabilization and sustenance of membrane heterogeneity, both in the vertical and horizontal plane of the lipid bilayer. In addition, it is worthy mentioning that upon cell stimulation, a membrane lipid degradation cascade is induced through the activation of several phospholipases, yielding various lipid metabolites such as diacylglycerols, fatty acids, lysophospholipids and phosphatidic acid (PA), playing a role in literally every biological process in norm and pathology.[20–22]

The specific roles of lipid molecules are based on their interactions with proteins and low molecular weight molecules, as reviewed in Section 3.

2.1 The structures of lipids and lipid-derived molecules

Brash *et al.*[23] isolated and characterized two geometric allene oxide isomers synthesized from natural 9*S*-hydroperoxylinoleic acid by allene oxide synthase cytochrome P450 CYP74C3. The obtained results provide insights into the mechanisms of allene oxide cyclization, and the double bond geometry in naturally occurring allene oxides. Using high field NMR and gas chromatography-mass spectrometry techniques, Liu *et al.*[24] characterized di-hydroxylated metabolites deriving from α-linolenic acid

(18:3n-3) treated by soybean 15-lipoxygenase. They compared the inhibitory effect on blood platelet and anti-inflammatory properties of 9(*R*),16(*S*)-dihydroxy-10*E*,12*E*,14*E*-octadecatrienoic acid, 9(*S*),16(*S*)-dihydroxy-10*E*,12*E*,14*E*-octadecatrienoic acid, 9(*S*),16(*S*)-dihydroxy-10*E*,12*Z*,14*E*-octadecatrienoic acid, and 9(*R*),16(*S*)-dihydroxy-10*E*,12*Z*,14*E*-octadecatrienoic acid with protectin DX (PDX) which derives from docosahexaenoic acid by double lipoxygenation. Synthesis of a *Mycobacterium tuberculosis* tetra-acylated sulfolipid analogue and characterization of their chiral acyl chains was performed using anisotropic natural abundance deuterium (NAD) 2D-NMR spectroscopy by Lemétais *et al.*[25] Novel lipoamino acids were prepared with the coupling of sapienic acid [(*Z*)-6-hexadecenoic acid] with α-amino group of amino acids and the resulting *N*-sapienoyl amino acids were tested for their cytotoxicity activities against four cancer cell lines by Gopal *et al.*[26] Inhibitors of inducible nitric oxide synthase (iNOS), sesquiterpenoids anmindenol A and B were isolated from a marine-derived bacterium *Streptomyces* sp by Lee *et al.*,[27] while lipid-A homologs from the ustilaginomycetous yeast *Pseudozyma aphidis* ZJUDM34 by Fan *et al.*[28] Regioselective conversion of 3-keto-5α- and 3-keto-5β-steroids (*trans* A/B- and *cis* A/B-ring juncture, respectively) to the corresponding enones (Δ(1)- and Δ(4)-3-ketones) by treatment with *o*-iodoxybenzoic acid was reported by Iida *et al.*[29] Using heteronuclear single quantum correlation spectroscopy and ^{1}H-NMR spectroscopy Borchman *et al.*[30] confirmed the presence of squalene in human eyelid lipid that could be secreted from sebaceous glands. Sallans *et al.*[31] revealed structural identities of four glycosylated lipids in the oral bacterium *Streptococcus mutans* UA159.

Hidalgo and Zamora[32] found that 2-alkenal-scavenging ability of *m*-diphenols is a consequence of their chemical structure, while El-Haci *et al.*[33] characterized antimicrobial activity of *Ammodaucus leucotrichus* fruit oil from Algerian Sahara against *Escherichia coli*, *Staphylococcus aureus*, as well as the filamentous fungi, *Fusarium oxysporum* and *Aspergillus flavus*. Moreover, Bouzabata *et al.*[34] described composition and chemical variability of *Eucalyptus bosistoana* essential oil from Algerian Sahara. Recke *et al.*[35] provided structural details of important biosurfactants – microbial 2-tridecyl sophorosides. Woodcock and co-workers[36] elaborated the biomimetic nitration method for nitro-conjugated linoleic acids, endogenous nitrodiene lipids, which act as inflammatory signaling mediators.

2.2 Lipid molecules in biological processes and in artificial membranes

The engulfment function of macrophages relies on complex molecular interactions involving both lipids and proteins. In particular, the clearance of apoptotic bodies is enabled by externalization on the cell target of phosphatidylserine molecules, which activate receptors on macrophages. Justice *et al.*[37] used model lipid membranes and rat alveolar macrophages to show how lipid bilayer properties probed by small-angle X-ray scattering and solid-state ^{2}H-NMR correlate with engulfment rates measured by flow cytometry. They found that engulfment of protein-free model lipid vesicles is promoted by the presence of phosphatidylserine molecules but inhibited by ceramides. Moreover, Dutagaci *et al.*[38] reported a direct contact between ceramide and lipids, such as DMPC molecules. The lipid bilayer properties and dynamics are strongly affected also by cholesterol, as shown by Fraenza *et al.*[39] and Hsieh *et al.*,[40] resulting in changes of the order fluctuation of membranes and the diffusional motion of lipid molecules. These observations were extended for other sterols (desmosterol, lanosterol, 4-cholesten-3-ol and 4-cholesten-3-one)[41,42] and for fluorescent cholesterol analogs in lipid bilayers.[43]

Molecular dynamics simulations performed by Sodt *et al.*[44] revealed substructures within the liquid-ordered phase of lipid bilayers. These substructures turned out to be composed of saturated hydrocarbon chains packed with local hexagonal order and separated by interstitial regions enriched in cholesterol and unsaturated chains. The existence of such structures is consistent with the results of NMR measurements. Graber *et al.*,[45] on the basis of PtdIns(4,5)P_2 ionization values measured using solid-state magic angle spinning ^{31}P NMR (ss MAS^{31}P NMR) in the presence of cholesterol and divalent cations, suggested that cholesterol participates in this intermolecular hydrogen bond network, resulting in stabilization of PtdIns(4,5)P_2-enriched domains due an increased spacing between the PtdIns(4,5)P_2 headgroups.

Distance measurements and conformational analysis of *sn*-2-arachidonoylglycerol (2-AG)-membrane sample by ^{2}H–^{31}P rotational-echo double-resonance NMR were performed by Tiburu and Shen[46] in order to determine the intermolecular distances that define the location, orientation, and conformation of *sn*-2-arachidonoylglycerol (2-AG) in 1-palmitoyl-2-oleoyl-*sn-glycero*-3-phosphocholine (POPC) lipid bilayers. Farver *et al.*[47] analyzed interactions with PS molecules (in the presence or absence of lung surfactant protein B) and dynamic properties of individual lipid species in native and synthetic pulmonary surfactants, such as 1,2-dipalmitoyl-*sn-glycero*-3-phosphocholine (DPPC), 1-palmitoyl-2-oleoyl-*sn-glycero*-3-phosphatidylethanolamine (POPE), POPC and 1-palmitoyl-2-oleoyl-*sn-glycero*-3-phosphatidylglycerol (POPG).

3 Interactions of lipids with proteins, peptides and low molecular weight molecules

Many interesting reviews, as well as experimental articles, were published between June 2013 and May 2014 discussing the advantages of using NMR methods to study protein structures in lipid environments, and protein–lipid and protein–peptide interactions in natural and artificial lipid membranes. For example, Alves *et al.*[48] underlined difficulties in studying membrane protein structures by either X-ray crystallography or NMR techniques, especially adenosine 5′-triphosphate P2 receptors. Moreover, Hamilton[7] convincingly described application of NMR methods to study fatty acid binding to bovine and human albumins, concluding that utilizing fatty acid as a probe allows characterization of site-specific interactions and molecular motions within binding sites. Gautier,[15] discussing retinal protein structural studies, summarized that together with X-ray crystallography and electron microscopy, NMR spectroscopy plays an important role in solving three-dimensional structures of proteins at atomic resolution. Kovacs *et al.*[49] reported improved NMR experiments with ^{13}C-isotropic mixing for assignment of aromatic and aliphatic side chains in labeled proteins. Yeagle[13] analyzed structural data based on X-ray diffraction, electron crystallography and NMR and found over 100 specific lipid non-covalent binding sites on membrane proteins. Murray *et al.*[50] and Miao *et al.*[51] emphasized an importance of elaborating detergent optimized membrane protein reconstitution methods into liposomes for solid state NMR and spectra analyses to study structure of proteins in lipid bilayer environments and the free energy measurements of transmembrane helix–helix interactions.[52]

Solid state NMR was found to be a promising tool for elucidating also membrane-related biological phenomena,[53,54] including functioning of lipid-embedded heptahelical membrane proteins.[55] Oxenoid *et al.*[56] put forward applications of solution NMR in investigating the structure and dynamics of ion channels and solute transporters as well as proteins involved in transmitting signals across the membrane. In addition, high-pressure NMR spectroscopy has emerged as a complementary approach to other techniques for investigating various structural and thermodynamic properties of macromolecules,[57] as well as static and magic angle spinning techniques of solid-state NMR to determine structural topology of membrane proteins in lipid bilayers using polarization optimized experiments (POE).[58,59]

More details about experiments aimed to study lipid structures and properties, as well as protein–lipid, peptide–lipid and small molecular weight compound–lipid interactions are given in the Section 3.1–3.3.

3.1 Interactions with membrane proteins

3.1.1 Proteins involved in vesicular traffic and signal transduction. The solution NMR structure of micelle-bound syntaxin-1A in its prefusion conformation, was reported by Liang *et al.*,[60] and it is of particular importance to understand folding of three neuronal soluble *N*-ethylmaleimide-sensitive factor attachment protein (SNAP) receptor (SNARE) proteins: synaptobrevin 2, syntaxin-1A, and SNAP-25, into neuronal *trans*-SNARE complex fusion intermediate. Ghai *et al.*[61] characterized the interactions of FERM domain-containing sorting nexin proteins (they belong to a subfamily of PX-FERM proteins that contain in their structures Phox homology (PX) and band 4.1/ezrin/radixin/moesin (FERM) domain) with endosomal lipids and cargo molecules.

Tinoco *et al.*[62] provided structural characterization of a neuroblast-specific phosphorylated region of myristoylated alanine-rich C kinase substrate (MARCKS), a protein that interacts with actin, Ca^{2+}-calmodulin, and plasma membrane lipids and is specifically phosphorylated by protein kinase C (PKC). By using high-resolution field-cycling ^{31}P-NMR spectroscopy and a spin-labeled recombinant PH domain of Akt 1 RAC-α serine/threonine-protein kinase Gradziel *et al.*[63] provided evidence that anticancer cytotoxic amphiphiles (perifosine and miltefosine) and phosphoinositides bind to two discrete sites on the PH domain this kinase. Liou *et al.*[64] analyzed sequential interactions and protein conformational changes during the binding of chemokine CXCL-8 to its receptor CXCR1 for rational drug design.

3.1.2 Proteins involved in membrane transport of ions and other solutes. Van der Cruijsen *et al.*[65] described the roles of lipid–pore loop interface in selective passage of potassium ions *via* the voltage-gated K^+ channels, while Wylie *et al.*[66] focused on the role of transmembrane allostery in inactivation of the potassium channel KcsA. The structure of human ether-a-go-go-related gene (hERG) voltage-gated K^+ channels and the role of the hERG channel selectivity filter in the long QT syndrome was studied by Gravel *et al.*[67] Lee *et al.*,[68] using NMR techniques, characterized the structure of NHE1 isoform of the cardiac Na^+/H^+ exchanger (NHE) and compared it with an *Escherichia coli* homolog, NhaA. Ullrich *et al.*[69] described the paramagnetic doping of a seven transmembrane domain (7TM) proteins (such as bacteriorhodopsin and vertebrate rhodopsin) in lipid bilayers by Gd^{3+}-complexes suitable for solid state NMR spectroscopy.

NMR techniques were also used to study bacterial membrane proteins and protein complexes, such as the twin-arginine transport (Tat) system identified in bacteria, as well as in plant chloroplasts, as a unique system that transports proteins across membranes in their fully-folded states.[70]

3.1.3 Other membrane and cytoskeletal proteins. Rui *et al.*[71] characterized the structure and dynamics of caveolin-1 in a DMPC bilayer using NMR, fluorescence emission measurements, and molecular dynamics simulations. They found that interaction of two helices H1 and H2 located in the central portion of caveolin-1 is important for stabilization of the U-shaped conformation of protein in a lipid bilayer. One of the key protein for T cell activation is the actin-remodeling, cofilin, which is dephosphorylated on serine 3 upon T cell co-stimulation and has an essential role in formation of mature immune synapses between T cells and antigen-presenting cells. Schulte *et al.*[72] observed that a reducing milieu renders cofilin insensitive to PtdIns(3,4)P_2 and may have effect on restoration of T cell activation. Beta2-glycoprotein I (β2GPI) is the major antigen for the antiphospholipid syndrome (APS)-related antibodies. Kolyada *et al.*[73] identified the binding site on Beta2-glycoprotein I for fondaparinux, the synthetic pentasaccharide commonly used for prophylaxis and treatment of thrombosis in patients with antiphospholipid syndrome. The structural characteristic of bacterial outer-membrane lipoprotein LpoB that spans the periplasm to stimulate the peptidoglycan synthase PBP1B activity was also reported.[74] Orientation and dynamics of GWALP transmembrane peptides that often flank the transmembrane alpha helices of integral membrane proteins, were investigated by Sparks *et al.*[75]

3.2 Interactions with soluble and viral proteins

3.2.1 α-Synuclein and other membrane-binding proteins. One of the extensively investigated proteins using NMR techniques in terms of its structure and interaction with lipids is soluble α-synuclein, primarily expressed in neural tissues. It is involved in neurotransmitter release in presynaptic terminals.[76] The aberrant aggregation of α-synuclein was found to be associated with Parkinson's disease, while gene encoding α-synuclein to be mutated in several families with autosomal dominant Parkinson's disease.[77] Normal cellular functions for synuclein have not been determined yet, however some data suggest a role in the regulation of membrane stability and/or turnover. Recently, Fusco *et al.*[78] characterized three regions of α-synuclein, an N-terminal helical segment having a role of membrane anchor, an unstructured C-terminal region that is weakly associated with the membrane and a central region determining the affinity of the protein for lipids. Furthermore, solid-state ^{13}C NMR revealed annealing of raft-like membranes containing cholesterol by the intrinsically disordered α-synuclein.[79]

^{1}H Overhauser dynamic nuclear polarization (ODNP)-enhanced NMR relaxometry confirmed that that the membrane-bound region of α-synuclein forms an extended α-helix parallel to the surface of negatively charged phospholipid membranes.[80] Bartels *et al.*[81] reported that *N*-α-acetylation of α-synuclein resulted in an increase of its helical folding propensity, GM1 binding specificity and resistance of the protein molecules to aggregation. The mechanisms of interaction of α-synuclein with membranes was also analyzed by Lokappa *et al.*[82] who focused on a random coil–helix transitions of the protein and by Zhang *et al.*[83] who described the effects of calcium ions on synuclein-membrane interactions. Hellstrand *et al.*[84] described membrane lipid co-aggregation with α-synuclein fibrils, the process revealing specificity to certain lipid classes.

Egea-Jiménez *et al.*,[85] using ^{2}H-NMR, confirmed that the binding of the C2 domain of protein kinase Cα to membranes is accompanied by the formation of domains enriched in activating lipids, like phosphatidylserine and PtdIns(3,4)P_2. The conformational properties of chicken engrailed 2 homeodomain (En2HD) in aqueous solution and in membrane mimetic environments using circular dichroism, tryptophan fluorescence, and NMR spectroscopy performed by Carlier *et al.*[86] revealed that engrailed-2 homeodomain is able to insert into lipid membranes.

3.2.2 Viral proteins. Grélard *et al.*[87] determined the intimate structure of pseudoviral particles of hepatitis B subvirus using solid-state NMR, light scattering, and cryo-electron microscopy. Cheng *et al.*[88] studied the bacteriofage Pf1 coat protein structure in lipid membranes, Zazrin *et al.*[89] the architecture of hepatitis C virus E1 glycoprotein transmembrane domain, and Claridge *et al.*[90] the juxtamembrane region of the influenza M2 protein reconstituted into isotropic bicelles. Integral matrix protein M2 able to form a proton-selective channel in the viral envelope was studied by Kawano *et al.*[91] and they found that the minimal proton-conducting unit for the channel is a dimer.

3.3 Interactions with peptides

Molecular dynamics simulation of neuropeptide B and neuropeptide W, antagonists of GPR7 and GPR8 G-protein-coupled receptors that are implicated in the regulation of neuroendocrine function in the DPPC membrane bilayers, confirming previous NMR structural data, was performed by Patra *et al.*[92] Fibrillation of β amyloid peptides in the presence of phospholipid bilayers and the consequent membrane disruption were characterized by Qiang *et al.*[93] by solid state NMR spectroscopy.

Further readings in the field of membrane integral and surface proteins as well as soluble lipid-binding proteins, and various peptides are given in Table 1.[94–142]

Table 1 Structure of lipid-binding and/or interacting proteins and peptides determined by NMR spectroscopic techniques.

Protein/peptide	Lipid/membrane/organism	Ref.
Membrane integral or peripheral proteins		
G-protein-coupled human neuropeptide Y receptor type 2	Dynamics in DMPC membranes	94
CD3 ζζ (T cell receptor complex)	POPC and POPC/cholesterol/sphingomyelin bilayers	95
Apelin receptor N-terminal tail	Dodecylphosphocholine micelles	96
GluA2 ligand-binding domain of AMPA receptor	Ligand binding properties	97
Human calcium-dependent calcium channel Polycystin 2 (PC2)	EF-hand calcium binding sites	98
Cytochrome b5-cytochrome P450 complex	Detergent micelles and lipid bicelles	99
Human cytochrome P450 2U1	Interaction with the substrate	100
Huntingtin 1-17 membrane anchor	Oriented phosphatidylcholine bilayers	101
Synaptotagmin-1	Neurotransmitter release, participation in SNARE complex formation	102
Liver NADPH-cytochrome P450 oxidoreductase	Enzyme structure at the membrane	103
Pulmonary surfactant protein B	Bicellar lipid mixtures	104
Hcf106, one of the core components of the chloroplast twin arginine translocation system	POPC/MGDG multilamellar vesicles	105
Outer membrane protein BamA	Gram-negative bacteria	106
Enterobacterial common antigen	Covalent linkage with lipopolysaccharide, Gram-negative bacteria	107
Cyclic nucleotide-activated K^+ channel	Bacteria	108
Mercury transporter MerF	Lipid micelles, bacteria	109
Foreign monotopic glycosyltransferase MGS from *Acholeplasma laidlawii*	Bicelles from bacterial lipids	110
Hepatitis C virus (HCV) protein p7	DHPC micelles	111
HIV-1 matrix protein anchoring	Interaction with PtdIns(4,5)P_2	112
Soluble proteins		
Liver fatty acid-binding protein (LFABP; FABP1)	*sn*-2-monoacylglycerol	113
Human liver fatty acid binding protein (hL-FABP)	Fatty acid binding sites	114
Human amyloid precursor protein	Proteolytic cleavage by γ-secretase	115, 116
Import receptor Pex5	Interaction with peroxisomal targeting signal 1 receptor	117

Table 1 (*Continued*)

Protein/peptide	Lipid/membrane/organism	Ref.
Phospholipase C δ1 EF-hand domain	Ca^{2+}-independent binding of anionic phospholipids	118
Lipocalin prostaglandin D synthase (L-PGDS)	Ligand binding properties	119
N-terminal pleckstrin homology domain of ceramide transfer protein (CERT)	PtdIns(4)P	120
Membrane anchor (FACT domain) of phosphatidylinositol 3-kinase-related kinases	Lipids and different membrane mimetics	121, 122
Myoglobin	Interaction with palmitate	123
Ubiquitin	Interaction with the Tollip C2 and CUE domains and PtdIns(3)P	124
Lipid storage droplet protein 1 (PLIN1)	*Drosophila melanogaster*	125
Major translocator protein PopB	Membrane and chaperone recognition, *Pseudomonas aeruginosa*	126
Anti-apoptotic factor Bcl-2	interaction with the autophagy-related GABAA receptor-associated protein	127
Group VI Ca^{2+}-independent phospholipase A_2 ($iPLA_2$)	Mechanism of insertion into lipid bilayer	128
Peptides		
Prion nucleation site	POPG	129
Insect adipokinetic hormones	Phospholipid micelles	130
α-Conotoxin of marine cone snail	Antagonist of acetycholine receptors	131
Multifunctional peptide BP100	Lipid bilayers (DMPC/DMPG)	132
Fusion peptide of the paramyxovirus PIV5	POPC/POPG, DOPC/DOPG, and DOPE	133
Unrelated antimicrobial amphiphilic peptides	Peptide–lipid membrane interactions	134
Antimicrobial and amyloidlike peptide dermaseptin S9	PC and PG monolayers	135
Antimicrobial peptides, aurein 1.2 and caerin 1.1	DMPC/DMPG bilayers	136
Tachyplesin I, an antimicrobial peptide derived from horseshoe crab	Interaction with lipopolysaccharide (LPS) of the outer membrane of Gram-negative bacteria	137
Mitochondria-penetrating peptide	Model inner mitochondrial membrane	138
Cell penetrating peptide transportan 10	Charateristids of multiple membrane-bound states	139
Cecropin-melittin hybrid antimicrobial peptide BP100	Large and giant unilamellar vesicles made of PG and PC	140
Lipid clustering by cationic antimicrobial and cell-penetrating peptides in model bacterial membranes	POPE and POPG	141
Amphipathic peptides	DMPC and DMPG vesicles	142

3.4 Interactions of lipids with low molecular weight molecules and solutes

Behavior of α-tocopherol in DMPC bilayers was characterized by Marquardt *et al.*[143] Phase separation in phosphatidylcholine membranes caused by the presence of a pyrimidine, analogue of fluphenazine possessing high anti-multidrug-resistance activity, was observed by Cieślik-Boczula *et al.*[144] Santeusanio *et al.*[145] studied the effects of antiobesity and antioxidant agent, hydrogenated cardanol, on the structure of model membranes and observed that this compound is able to penetrate in a pH-dependent manner lipid micelles. Strong effects on water and membrane dynamics in lipid vesicular membranes was also described for poly(ethylene glycol)s.[146] Jiménez-Rojo *et al.*[147] reported that sphingosine induces the aggregation of imine-containing peroxidized lipid membranes, the process that may contribute to the pathogenesis of diseases like atherosclerosis and multiple sclerosis. Furthermore, Vermathen *et al.*[148] described self-assembling properties of porphyrinic photosensitizers and their effects on phospholipid bilayers and micells made of DPOC and 1,2-dihexanoyl-*sn-glycero*-3-phosphocholine (DHPC), respectively.

Bicelles are model membranes generally made of long-chain DMPC and short-chain DHPC. Lipid concentration and molar ratio boundaries for the use of isotropic bicelles and are extensively used to study membrane interactions and structure determination of membrane-associated peptides.[149] By using bicelles and lipid liposomes Furlan *et al.*[150] have shown that red wine tannins make more fluid and precipitate lipid liposomes and bicelles, suggesting interactions of tannins with membrane lipids.

The properties of an amorphous solid dispersion of a drug, cyclosporine A, were investigated by ^{1}H-NMR in solution and in membrane-bound form by ^{1}H-NMR in small unilamellar vesicles and by ^{31}P–^{2}H-NMR in dispersions of DMPC by Debouzy *et al.*[151] Münch *et al.*[152] analyzed structural features of lantibiotic NAI-107 active against Gram-positive bacteria including vancomycin-resistant enterococci and methicillin-resistant *Staphylococcus aureus.* Interactions of quercetin, 3,3′,4′,5,7-pentahydroxyflavone is claimed to exert many beneficial health effects, with liposomes and vacuolar membranes of the liverwort *Conocephalum conicum* and the lipid binding properties of genistein, a phytoestrogen that belongs to isoflavones, were performed by Pawlikowska-Pawlęga *et al.*[153,154] using liposomes, human erythrocyte membranes and internal membranes of HeLa cells. Gilardi *et al.*[155] investigated properties of LT175, a novel peroxisome proliferator-activated receptor alpha/gamma (PPARα/γ) ligand with potent insulin-sensitizing effects and reduced adipogenic properties.

4 Lipids in biomedicine

4.1 Diagnostic approaches

A growing number of diseases, accompanied by defects in the metabolism and storage of lipids or lipid-derived molecules, with problematic or difficult diagnostics, calls for the development of reliable methods of assesment of changes in tissue/cell structure as well as lipid metabolism and content. NMR techniques, including *in vivo* proton magnetic resonance spectroscopy (MRS) and functional magnetic resonance imaging (fMRI), in conjuction with metabolomic approaches, are frequently used for diagnostic purposes related to the abnormal metabolism of lipids and their derivatives.[156,157] These topics were recently reviewed for example for rheumatoid arthritis and use of magnetic resonance imaging,[158] for aging and Alzheimer's disease diagnosis,[156] for diagnosis of brain infections,[159] for the detection of coronary artery disease,[160] for neonatal and pediatric laboratory medicine,[161] and for metabolomic studies in general.[162,163]

Millon *et al.*[164] monitored plaque inflammation in atherosclerotic rabbits using a combined PET/MR scanner. Proton magnetic resonance metabolomic characterization of ovarian serous carcinoma was performed by Vettukattil *et al.*[165] for diagnostic purposes. Pan *et al.*,[166] using NMR lipidomic approach, noticed increased unsaturation of lipids in cytoplasmic lipid droplets in DAOY cancer cells in response to cisplatin treatment. Akoudad *et al.*[167] examined formation of cerebral microbleeds associated with the progression of ischemic vascular lesions, while Kostara *et al.*[168] followed the progression of coronary heart disease NMR-based lipidomic analysis of blood lipoproteins.

Further achievements in the field briefly described above are summarized in Table 2.[169–189]

4.2 Pathomechanisms

Ho *et al.*[190] examined how the association of long-chain fatty acids to a partially unfolded, extracellular protein can alter the presentation to target cells and cellular effects, using as an example human α-lactalbumin made lethal to tumor cells (HAMLET) in complex with oleic acid. Dalli *et al.*[191] described the mechanism by which 13*S*,14*S*-epoxy-maresin is converted by human macrophages to maresin 1 (MaR1), and then inhibits leukotriene A4 hydrolase (LTA4H), and shifts macrophage phenotype. It is worth adding that maresins are produced by macrophages from docosahexaenoic acid and exert potent proresolving and tissue homeostatic actions.

4.3 Drug delivery, lipid nanoparticles and therapy responses

Recently, the production and use of lipoprotein biomimetic particles, as for example thermosensitive liposomes loaded with diagnostically or therapeutically active compounds in their core or lipid-polymer hybrid nanoparticles as a new generation therapeutic delivery platform, as well as liposomes usable for MRI-guided drug delivery, was reviewed.[192–194]

Doxorubicin (DOX) conjugated to a lipophilic triphenylphosphonium (TPP) was used for mitochondrial delivery of anticancer drug to overcome drug resistance in MDA-MB-435/DOX cells by Han *et al.*[195] Liu *et al.*[196] described strategies of using superparamagnetic iron-oxide nanoparticles as contrast agents for anatomical, functional and cellular MRI, drug delivery agents, and diagnostic biosensors. Pandey *et al.*[197] developed lactoferrin-appended solid lipid nanoparticles for directed delivery of antineoplasitic agent, paclitaxel, for effective treatment of bronchogenic carcinoma with minimal side effects. Gallic acid, a naturally occurring in vegetables and fruits phenolic compound, that has been shown to have potent anti-oxidative and anti-obesity activities, was used to treat non-alcoholic fatty liver disease. Chao *et al.*[198] performed ^{1}H-NMR-based

Table 2 Biomedical investigations related to changes of lipid composition and metabolism in pathological states I comparison to norm, performed with the aid of NMR metabolomic platforms, MRI and MRS.

Target investigated	Pathology and organism	Ref.
Absolute brain metabolite concentrations	Brain development in premature-born infants	169
Analysis of fasting plasma lipids	Subclinical cerebrovascular disease (human subjects)	170
Metabolomic fingerprints	Human neonatal hypoxic ischemic encephalopathy (HIE)	171
Muscle lipid fractions	Infant Duchenne muscular dystrophy	172
Metabolomic fingerprints	Acute-on-chronic liver failure in humans with alcoholic cirrhosis, requiring intensive care	173
Serum metabolites	Women with polycystic ovary syndrome	174
Metabolomic signatures of the disease	Chronic obstructive pulmonary disease in humans	175
Reactive oxygen species	Mitochondrial dysfunction in human skeletal muscle atrophy and dysfunction with aging	176
Metabolomics (diffusion-weighted MRI)	Gaucher's disease in humans, splenic nodules	177
Chylomicron and high density lipoprotein metabolism	Endotoxemia, inflammation, vascular function, and postprandial lipid metabolism in patients with type 1 diabetes	178
Selected metabolites	Marginal vitamin B-6 deficiency in healthy human subjects	179
Lipoprotein subclasses and endogenous sex hormones	Women at midlife	180
Metabolic profile using *in vivo* ^{1}H NMR	All-*trans*-retinoic acid (ATRA)-induced craniofacial development of murine embryos	181
Hepatic triglyceride levels	Obesity, dyslipidemia, fructose metabolism, mice	182
Subfractions of lipoprotein particles	Angiographic progression of coronary artery disease	183
Hydrophilic and lipid metabolites from plasma	Rat hemorrhagic shock	184
Metabolic profiling	Influenza virus infection in obese mice	185
Serum targeted metabolite analysis	Rats with renal fibrosis	186
Metabolic profiling	Dividing cells in live rodent brain	187
^{1}H NMR-based lipidomics	Sebaceous excretions in rodents	188
Metabolomics including lipid metabolites	Hypoxia-induced insulin-producing β-cell death line INS-1	189

metabolomics to investigated the beneficial effects of gallic acid administration on nutritional hepatosteatosis. Schaffran *et al.*[199] examined boron-containing lipids with three different structures for boron neutron capture therapy in mouse tumors and found some of them lethal and some with unwanted side effects (hemorrhage). The observed unwanted effects of these particles precludes their use in boron neutron capture therapy.

4.4 Methodological approaches

In this section we refer to recent reports describing the successful application of NMR-based techniques to study lipids, lipid-binding proteins and peptides, as well as pathomechanisms or therapy responses, or to validate other methods on the basis of NMR parameters.

Liang *et al.*[200] published new method for preparation of uniform biodegradable polymer nanoparticles with controllable sizes by a hand-driven membrane-extrusion emulsification approach and provided *in vitro* proof of their anti-melanoma activity. Paramagnetic liposomes for molecular MRI and MRI-guided drug delivery with a high targeting specificity were designed by Langereis *et al.*[194] Andreozzi *et al.*[201] synthesized solid lipid nanoparticles loaded with gadolinium (1,4,7,10-tetraazacyclododecane)-1,4,7,10-tetraacetate in order to produce a new category of stable T1-weighted MRI contrast agents, while Ketkar-Atre *et al.*[202] reported methodology of using a novel lactose functionalized magnetoliposomes for *in vivo* hepatocyte MR imaging. In addition, the bifunctional liposomes from poly(2-ethyl-oxazoline)-cholesteryl methyl carbonate were constructed in order to achieve long-circulating and pH-sensitive properties.[203] Song *et al.*,[204] using phospholipase D (PLD) mediated transphosphatidylation in a biphasic reaction system, synthesized phosphatidyl saccharides for preparation of anionic nanoliposomes with enhanced stability. Furthermore, magnetic iron oxide nanoparticles for MRI tracking of adipose tissue-derived progenitor cells *in vitro* have been used by Kasten *et al.*[205]

To determine the structures of integral membrane proteins using spectroscopic methods the lipodisq nanoparticles for membrane protein spectroscopic studies, characterized by homogeneous lipid bilayer environment, were design by Zhang *et al.*[206] Fully deuterated magnetically oriented system based on fatty acid direct hexagonal phases for studies of the structure of host protonated molecules (*e.g.*, peptides, proteins) was prepared by Douliez *et al.*,[207] while new NMR technique, called aliphatic chain length by isotropic mixing (ALCHIM), to determine composition of complex lipid samples by ^{1}H-NMR spectroscopy, was developed by Sachleben *et al.*[208]

New methodological approach based on Dixon and T_1-weighted MR methods to improve fat infiltration in Duchenne muscular dystrophy patients was presented by Wokke *et al.*[209] To investigate the characteristics of nuclear Overhauser enhancement (NOE) imaging signals in the brain at 7T healthy, and C6 glioma-bearing Wistar rats were scanned using chemical exchange saturation transfer-magnetic resonance imaging (CEST-MRI) and chemical exchange saturation transfer-magnetic resonance spectroscopy (CEST-MRS) sequences by Lu *et al.*[210] These *in vivo* experiments showed that the amide proton transfer signal was larger in the tumor, whereas the NOE signal was larger in the normal white matter. The use of PGN635, a novel human monoclonal antibody that specifically targets phosphatidylserine molecules, for *in vivo* molecular MRI of tumor vasculature, was described by Zhou *et al.*[211] Last but not least, development and *in silico* evaluation of large-scale metabolite identification methods using functional group detection for metabolomics was reported by Mitchell *et al.*[212]

Abbreviations

2-AG	*sn*-2-arachidonoylglycerol
DHPC	1,2-dihexanoyl-*sn-glycero*-3-phosphocholine

DMPC	1,2-dimyristoyl-*sn-glycero*-3-phosphocholine
DMPG	1,2-dimyristoyl-*sn*-3-phosphoglycerol
DOPC	1,2-dioleoyl-*sn-glycero*-3-phosphocholine
DPPC	1,2-dipalmitoyl-*sn-glycero*-3-phosphocholine
MGDG	monogalactosyldiacylglycerol
PC	phosphatidylcholine
PE	phosphatidylethanolamine
PI	phosphatidylinostol
POPE	1-palmitoyl-2-oleoyl-*sn-glycero*-3-phosphatidylethanolamine
POPC	1-palmitoyl-2-oleoyl-*sn-glycero*-3-phosphocholine
POPG	1-palmitoyl-2-oleoyl-*sn-glycero*-3-phosphatidylglycerol
POPS	1-palmitoyl-2-oleoyl-*sn-glycero*-3-phospho-L-serine
PS	phosphatidylserine
PtdIns(3,4,5)P_3	phosphatidylinositol (3,4,5)-triphosphate
PtdIns(3,4)P_2	phosphatidylinositol (4,5)-bisphosphate
PtdIns(4,5)P_2	phosphatidylinositol (4,5)-bisphosphate
PtdIns(3)P	phosphatidylinositol 3-monophosphate
PtdIns(4)P	phosphatidylinositol 4-monophosphate
SM	sphingomyelin

References

1 S. Pikula, J. Bandorowicz-Pikula and P. Groves, *Nucl. Magn. Reson.*, 2013, **42**, 362.
2 S. Pikula, J. Bandorowicz-Pikula and P. Groves, *Nucl. Magn. Reson.*, 2014, **43**, 378.
3 E. Swiezewska and J. Wojcik, *Nucl. Magn. Reson.*, 2011, **40**, 344.
4 E. Swiezewska and J. Wojcik, *Nucl. Magn. Reson.*, 2012, **41**, 320.
5 D. I. Freedberg and P. Selenko, *Annu. Rev. Biophys.*, 2014, **43**, 171.
6 T. Gopinath, K. R. Mote and G. Veglia, *Prog. Nucl. Magn. Reson. Spectrosc.*, 2013, **75**, 50.
7 J. A. Hamilton, *Biochim. Biophys. Acta*, 2013, **1830**, 5418.
8 I. Maslennikov and S. Choe, *Curr. Opin. Struct. Biol.*, 2013, **23**, 555.
9 Y. Miao and T. A. Cross, *Curr. Opin. Struct. Biol.*, 2013, **23**, 919.
10 S. J. Ullrich and C. Glaubitz, *Acc. Chem. Res.*, 2013, **46**, 2164.
11 L. A. Baker and M. Baldus, *Curr. Opin. Struct. Biol.*, 2014, **27C**, 48.
12 D. Huster, *Biochim. Biophys. Acta*, 2014, **1841**, 1146.
13 P. L. Yeagle, *Biochim. Biophys. Acta*, 2014, **1838**, 1548.
14 L. Mäler, *Adv. Drug Delivery Rev.*, 2013, **65**, 1002.
15 A. Gautier, *Biochim. Biophys. Acta*, 2014, **1837**, 578.
16 R. M. Epand, K. D'Souza, B. Berno and M. Schlame, *Biochim. Biophys. Acta*, 2015, **1848**, 220.
17 S. R. Shaikh, J. J. Kinnun, X. Leng, J. A. Williams and S. R. Wassall, *Biochim. Biophys. Acta*, 2015, **1848**, 211.
18 L. Ragona, K. Pagano, S. Tomaselli, F. Favretto, A. Ceccon, S. Zanzoni, M. D'Onofrio, M. Assfalg and H. Molinari, *Biochim. Biophys. Acta*, 2014, **1844**, 1268.
19 L. Borisjuk, H. Rolletschek and T. Neuberger, *Prog. Lipid Res.*, 2013, **52**, 465.
20 H. M. Hankins, R. D. Baldridge, P. Xu and T. R. Graham, *Traffic*, 2015, **16**, 35.

21 M. Masoodi, O. Kuda, M. Rossmeisl, P. Flachs and J. Kopecky, *Biochim. Biophys. Acta*, 2015, **1851**, 503.
22 S. Suetsugu, S. Kurisu and T. Takenawa, *Physiol. Rev.*, 2014, **94**, 1219.
23 A. R. Brash, W. E. Boeglin, D. F. Stec, M. Voehler, C. Schneider and J. K. Cha, *J. Biol. Chem.*, 2013, **288**, 20797.
24 M. Liu, P. Chen, E. Véricel, M. Lelli, L. Béguin, M. Lagarde and M. Guichardant, *J. Lipid Res.*, 2013, **54**, 2083.
25 A. Lemétais, Y. Bourdreux, P. Lesot, J. Farjon and J. M. Beau, *J. Org. Chem.*, 2013, **78**, 7648–57.
26 S. C. Gopal, S. S. Kaki, B. V. Rao, Y. Poornachandra, C. G. Kumar and R. B. Narayana Prasad, *J. Oleo Sci.*, 2014, **63**, 717.
27 J. Lee, H. Kim, T. G. Lee, I. Yang, D. H. Won, H. Choi, S. J. Nam and H. Kang, *J. Nat. Prod.*, 2014, **77**, 1528.
28 L. L. Fan, Y. C. Dong, Y. F. Fan, J. Zhang and Q.H. Chen, *Carbohydr. Res.*, 2014, **392**, 1.
29 T. Iida, K. Omura, R. Sakiyama and M. Kodomari, *Chem. Phys. Lipids*, 2014, **178**, 45.
30 D. Borchman, M. C. Yappert, S. E. Milliner, R. J. Smith and R. Bhola, *Lipids*, 2013, **48**, 1269.
31 L. Sallans, J. L. Giner, D. J. Kiemle, J. E. Custer and E. S. Kaneshiro, *Biochim. Biophys. Acta*, 2013, **1831**, 1239.
32 F. J. Hidalgo and R. Zamora, *Food Chem.*, 2014, **160**, 118.
33 I. A. El-Haci, C. Bekhechi, F. Atik-Bekkara, W. Mazari, M. Gherib, A. Bighelli, J. Casanova and F. Tomi, *Nat. Prod. Commun.*, 2014, **9**, 711.
34 A. Bouzabata, A. Bighelli, L. Abed, J. Casanova and F. Tomi, *Nat. Prod. Commun.*, 2014, **9**, 701.
35 V. K. Recke, J. B. Park, M. Gerlitzki, R. Hausmann, C. Syldatk, V. Wray, H. Tokuda, N. Suzuki and S. Lang, *Z. Naturforsch., C: J. Biosci.*, 2013, **68**, 489.
36 S. R. Woodcock, S. R. Salvatore, G. Bonacci, F. J. Schopfer and B. A. Freeman, *J. Org. Chem.*, 2014, **79**, 25.
37 M. J. Justice, D. N. Petrusca, A. L. Rogozea, J. A. Williams, K. S. Schweitzer, I. Petrache, S. R. Wassall and H. I. Petrache, *Biophys. J.*, 2014, **106**, 598.
38 B. Dutagaci, J. Becker-Baldus, J. D. Faraldo-Gómez and C. Glaubitz, *Biochim. Biophys. Acta*, 2014, **1838**, 2511.
39 C. C. Fraenza, C. J. Meledandri, E. Anoardo and D. F. Brougham, *ChemPhysChem*, 2014, **15**, 425.
40 C. J. Hsieh, Y. W. Chen and D. W. Hwang, *Phys. Chem. Chem. Phys.*, 2013, **15**, 16634.
41 L. Mouret, G. Da Costa and A. Bondon, *Magn. Reson. Chem.*, 2014, **52**, 339.
42 M. G. Benesch, D. A. Mannock, R. N. Lewis and R. N. McElhaney, *Chem. Phys. Lipids*, 2014, **177**, 71.
43 S. Milles, T. Meyer, H. A. Scheidt, R. Schwarzer, L. Thomas, M. Marek, L. Szente, R. Bittman, A. Herrmann, T. G. Pomorski, D. Huster and P. Müller, *Biochim. Biophys. Acta*, 2013, **1828**, 1822.
44 A. J. Sodt, M. L. Sandar, K. Gawrisch, R. W. Pastor and E. Lyman, *J. Am. Chem. Soc.*, 2014, **136**, 725.
45 Z. T. Graber, A. Gericke and E. E. Kooijman, *Chem. Phys. Lipids*, 2014, **182**, 62.
46 E. K. Tiburu and L. Shen, *J. Membr. Biol.*, 2014, **247**, 231.
47 S. Farver, A. N. Smith, F. D. Mills, A. G. Egri and J. R. Long, *Biochim. Biophys. Acta*, 2015, **1848**, 203.
48 L. A. Alves, J. H. da Silva, D. N. Ferreira, A. A. Fidalgo-Neto, P. C. Teixeira, C. A. de Souza, E. R. Caffarena and M. S. de Freitas, *Int. J. Mol. Sci.*, 2014, **15**, 4531.

49 H. Kovacs and A. Gossert, *J. Biomol. NMR*, 2014, **58**, 101.
50 D. T. Murray, J. Griffin and T. A. Cross, *Biochemistry*, 2014, **53**, 2454.
51 Y. Miao, T. A. Cross and R. Fu, *J. Biomol. NMR*, 2013, **56**, 265.
52 K. S. Mineev, D. M. Lesovoy, D. R. Usmanova, S. A. Goncharuk, M. A. Shulepko, E. N. Lyukmanova, M. P. Kirpichnikov, E. V. Bocharov and A. S. Arseniev, *Biochim. Biophys. Acta*, 2014, **1838**, 164.
53 K. Nomura, E. Harada, K. Sugase and K. Shimamoto, *J. Phys. Chem. B*, 2014, **118**, 2405.
54 J. R. Banigan, A. Gayen and N. J. Traaseth, *Biochim. Biophys. Acta*, 2015, **1848**, 334.
55 S. Wang, R. A. Munro, L. Shi, I. Kawamura, T. Okitsu, A. Wada, S. Y. Kim, K. H. Jung, L. S. Brown and V. Ladizhansky, *Nat. Methods*, 2013, **10**, 1007.
56 K. Oxenoid and J. J. Chou, *Curr. Opin. Struct. Biol.*, 2013, **23**, 547.
57 Y. Fu and A. J. Wand, *J. Biomol. NMR*, 2013, **56**, 353.
58 K. R. Mote, T. Gopinath and G. Veglia, *J. Biomol. NMR*, 2013, **57**, 91.
59 M. E. Ward, S. Wang, S. Krishnamurthy, H. Hutchins, M. Fey, L. S. Brown and V. Ladizhansky, *J. Biomol. NMR*, 2014, **58**, 37.
60 B. Liang, V. Kiessling and L. K. Tamm, *Proc. Natl. Acad. Sci. U. S. A.*, 2013, **110**, 19384.
61 R. Ghai, M. Mobli and B. M. Collins, *Methods Enzymol.*, 2014, **534**, 331.
62 L. W. Tinoco, J. L. Fraga, C. D. Anobom, F. R. Zolessi, G. Obal, A. Toledo, O. Pritsch and C. Arruti, *Biochim. Biophys. Acta*, 2014, **1844**, 837.
63 C. S. Gradziel, Y. Wang, B. Stec, A. G. Redfield and M. F. Roberts, *Biochemistry*, 2014, **53**, 462.
64 J. W. Liou, F. T. Chang, Y. Chung, W. Y. Chen, W. B. Fischer and H. J. Hsu, *PLoS One*, 2014, **9**, e94178.
65 E. A. van der Cruijsen, D. Nand, M. Weingarth, A. Prokofyev, S. Hornig, A. A. Cukkemane, A. M. Bonvin, S. Becker, R. E. Hulse, E. Perozo, O. Pongs and M. Baldus, *Proc. Natl. Acad. Sci. U. S. A.*, 2013, **110**, 13008.
66 B. J. Wylie, M. P. Bhate and A. E. McDermott, *Proc. Natl. Acad. Sci. U. S. A.*, 2014, **111**, 185.
67 A. E. Gravel, A. A. Arnold, E. J. Dufourc and I. Marcotte, *Biochim. Biophys. Acta*, 2013, **1828**, 1494.
68 B. L. Lee, B. D. Sykes and L. Fliegel, *J. Mol. Cell. Cardiol.*, 2013, **61**, 60.
69 S. J. Ullrich, S. Hölper and C. Glaubitz, *J. Biomol. NMR*, 2014, **58**, 27.
70 Y. Zhang, Y. Hu, H. Li and C. Jin, *PLoS One*, 2014, **9**, e103157.
71 H. Rui, K. T. Root, J. Lee, K. J. Glover and W. Im, *Biophys. J.*, 2014, **106**, 1371.
72 B. Schulte, I. John, B. Simon, C. Brockmann, S. A. Oelmeier, B. Jahraus, H. Kirchgessner, S. Riplinger, T. Carlomagno, G. H. Wabnitz and Y. Samstag, *J. Biol. Chem.*, 2013, **288**, 29430.
73 A. Kolyada, A. De Biasio and N. Beglova, *Biochim. Biophys. Acta*, 2013, **1834**, 2080.
74 A. J. Egan, N. L. Jean, A. Koumoutsi, C. M. Bougault, J. Biboy, J. Sassine, A. S. Solovyova, E Breukink, A. Typas, W. Vollmer and J. P. Simorre, *Proc. Natl. Acad. Sci. U. S. A.*, 2014, **111**, 8197.
75 K. A. Sparks, N. J. Gleason, R. Gist, R. Langston, D. V. Greathouse and R. E. Koeppe 2nd, *Biochemistry*, 2014, **53**, 3637.
76 M. H. Polymeropoulos, C. Lavedan, E. Leroy, S. E. Ide, A. Dehejia, A. Dutra, B. Pike, H. Root, J. Rubenstein, R. Boyer, E. S. Stenroos, S. Chandrasekharappa, A. Athanassiadou, T. Papapetropoulos, W. G. Johnson, A. M. Lazzarini, R. C. Duvoisin, G. Di Iorio, L. I. Golbe and R. L. Nussbaum, *Science*, 1997, **276**, 2045.
77 M. Goedert, *Nat. Rev. Neurosci.*, 2001, **2**, 492.

78 G. Fusco, A. De Simone, T. Gopinath, V. Vostrikov, M. Vendruscolo, C. M. Dobson and G. Veglia, *Nat. Commun.*, 2014, **5**, 3827.
79 A. Leftin, C. Job, K. Beyer and M. F. Brown, *J. Mol. Biol.*, 2013, **425**, 2973.
80 C. Y. Cheng, J. Varkey, M. R. Ambroso, R. Langen and S. Han, *Proc. Natl. Acad. Sci. U. S. A.*, 2013, **110**, 16838.
81 T. Bartels, N. C. Kim, E. S. Luth and D. J. Selkoe, *PLoS One*, 2014, **9**, e103727.
82 S. B. Lokappa, J. E. Suk, A. Balasubramanian, S. Samanta, A. J. Situ and T. S. Ulmer, *J. Mol. Biol.*, 2014, **426**, 2130.
83 Z. Zhang, C. Dai, J. Bai, G. Xu, M. Liu and C. Li, *Biochim. Biophys. Acta*, 2014, **1838**, 853.
84 E. Hellstrand, A. Nowacka, D. Topgaard, S. Linse and E. Sparr, *PLoS One*, 2013, **8**, e77235.
85 A. L. Egea-Jiménez, A. M. Fernández-Martínez, Á. Pérez-Lara, A. de Godos, S. Corbalán-García and J. C. Gómez-Fernández, *PLoS One*, 2014, **9**, e95973.
86 L. Carlier, S. Balayssac, F. X. Cantrelle, L. Khemtémourian, G. Chassaing, A. Joliot and O. Lequin, *Biophys. J.*, 2013, **105**, 667.
87 A. Grélard, P. Guichard, P. Bonnafous, S. Marco, O. Lambert, C. Manin, F. Ronzon and E. J. Dufourc, *FASEB J.*, 2013, **27**, 4316.
88 X. Cheng, S. Jo, F. M. Marassi and W. Im, *Biophys. J.*, 2013, **105**, 691.
89 H. Zazrin, H. Shaked and J. H. Chill, *Biochim. Biophys. Acta*, 2014, **1838**, 784.
90 J. K. Claridge, J. Aittoniemi, D. M. Cooper and J. R. Schnell, *Biochemistry*, 2013, **52**, 8420.
91 K. Kawano, Y. Yano and K. Matsuzaki, *J. Mol. Biol.*, 2014, **426**, 2679.
92 M. C. Patra, J. Maharana, S. K. Pradhan and S. N. Rath, *J. Biomol. Struct. Dyn.*, 2014, **32**, 1118.
93 W. Qiang, W. M. Yau and J. Schulte, *Biochim. Biophys. Acta*, 2015, **1848**, 266.
94 P. Schmidt, L. Thomas, P. Müller, H. A. Scheidt and D. Huster, *Chemistry*, 2014, **20**, 4986.
95 A. A. Petruk, S. Varriale, M. R. Coscia, L. Mazzarella, A. Merlino and U. Oreste, *Biochim. Biophys. Acta*, 2013, **1828**, 2637.
96 D. N. Langelaan, T. Reddy, A. W. Banks, G. Dellaire, D. J. Dupré and J. K. Rainey, *Biochim. Biophys. Acta*, 2013, **1828**, 1471.
97 A. H. Ahmed, C. P. Ptak, M. K. Fenwick, C. L. Hsieh, G. A. Weiland and R. E. Oswald, *J. Biol. Chem.*, 2013, **288**, 27658.
98 I. Y. Kuo, C. Keeler, R. Corbin, A. Ćelić, E. T. Petri, M. E. Hodsdon and B. E. Ehrlich, *FASEB J.*, 2014, **28**, 2332.
99 S. Ahuja, N. Jahr, S. C. Im, S. Vivekanandan, N. Popovych, S. V. Le Clair, R. Huang, R. Soong, J. Xu, K. Yamamoto, R. P. Nanga, A. Bridges, L. Waskell and A. Ramamoorthy, *J. Biol. Chem.*, 2013, **288**, 22080.
100 M. Siller, S. Goyal, F. K. Yoshimoto, Y. Xiao, S. Wei and F. P. Guengerich, *J. Biol. Chem.*, 2014, **289**, 10476.
101 M. Michalek, E. S. Salnikov and B. Bechinger, *Biophys. J.*, 2013, **105**, 699.
102 A. B. Seven, K. D. Brewer, L. Shi, Q. X. Jiang and J. Rizo, *Proc. Natl. Acad. Sci. U. S. A.*, 2013, **110**, E3243.
103 R. Huang, K. Yamamoto, M. Zhang, N. Popovych, I. Hung, S. C. Im, Z. Gan, L. Waskell and A. Ramamoorthy, *Biophys. J.*, 2014, **106**, 2126.
104 A. Sylvester, L. MacEachern, V. Booth and M. R. Morrow, *PLoS One*, 2013, **8**, e72248.
105 L. Zhang, L. Liu, S. Maltsev, G. A. Lorigan and C. Dabney-Smith, *Biochim. Biophys. Acta*, 2014, **1838**, 413.
106 T. Sinnige, M. Weingarth, M. Renault, L. Baker, J. Tommassen and M. Baldus, *J. Mol. Biol.*, 2014, **426**, 2009.

107 T. K. Gozdziewicz, C. Lugowski and J. Lukasiewicz, *J. Biol. Chem.*, 2014, **289**, 2745.
108 A. Cukkemane and M. Baldus, *ChemBioChem*, 2013, **14**, 1789.
109 G. J. Lu, Y. Tian, N. Vora, F. M. Marassi and S. J. Opella, *J. Am. Chem. Soc.*, 2013, **135**, 9299.
110 C. Ariöz, W. Ye, A. Bakali, C. Ge, J. Liebau, H. Götzke, A. Barth, A. Wieslander and L. Mäler, *Biochemistry*, 2013, **52**, 5533.
111 G. A. Cook, L. A. Dawson, Y. Tian and S. J. Opella, *Biochemistry*, 2013, **52**, 5295.
112 L. Charlier, M. Louet, L. Chaloin, P. Fuchs, J. Martinez, D. Muriaux, C. Favard and N. Floquet, *Biophys. J.*, 2014, **106**, 577.
113 W. S. Lagakos, X. Guan, S. Y. Ho, L. R. Sawicki, B. Corsico, S. Kodukula, K. Murota, R. E. Stark and J. Storch, *J. Biol. Chem.*, 2013, **288**, 19805.
114 C. Santambrogio, F. Favretto, M. D'Onofrio, M. Assfalg, R. Grandori and H. Molinari, *J. Mass Spectrom.*, 2013, **48**, 895.
115 T. Lemmin, M. Dimitrov, P. C. Fraering and M. Dal Peraro, *J. Biol. Chem.*, 2014, **289**, 6763.
116 R. L. Gill Jr., X. Wang and F. Tian, *Biochim. Biophys. Acta*, 2015, **1848**, 323.
117 A. Neuhaus, H. Kooshapur, J. Wolf, N. H. Meyer, T. Madl, J. Saidowsky, E. Hambruch, A. Lazam, M. Jung, M. Sattler, W. Schliebs and R. Erdmann, *J. Biol. Chem.*, 2014, **289**, 437.
118 J. Cai, S. Guo, J. W. Lomasney and M. F. Roberts, *J. Biol. Chem.*, 2013, **288**, 37277.
119 S. M. Lim, D. Chen, H. Teo, A. Roos, A. E. Jansson, T. Nyman, L. Trésaugues, K. Pervushin and P. Nordlund, *J. Lipid Res.*, 2013, **54**, 1630.
120 J. Prashek, T. Truong and X. Yao, *PLoS One*, 2013, **8**, e79590.
121 L. A. Sommer, M. Schaad and S. A. Dames, *J. Biol. Chem.*, 2013, **288**, 20046.
122 L. A. Sommer and S. A. Dames, *FEBS Lett.*, 2014, **588**, 1755.
123 L. Shih, Y. Chung, R. Sriram and T. Jue, *Biochim. Biophys. Acta*, 2014, **1840**, 656.
124 S. Mitra, C. A. Traughber, M. K. Brannon, S. Gomez and D. G. Capelluto, *J. Biol. Chem.*, 2013, **288**, 25780.
125 P. Lin, X. Chen, H. Moktan, E. L. Arrese, L. Duan, L. Wang, J. L. Soulages and D. H. Zhou, *Biochim. Biophys. Acta*, 2014, **1838**, 874.
126 K. F. Discola, A. Förster, F. Boulay, J. P. Simorre, I. Attree, A. Dessen and V. Job, *J. Biol. Chem.*, 2014, **289**, 3591.
127 P. Ma, M. Schwarten, L. Schneider, A. Boeske, N. Henke, D. Lisak, S. Weber, J. Mohrlüder, M. Stoldt, B. Strodel, A. Methner, S. Hoffmann, O. H. Weiergräber and D. Willbold, *J. Biol. Chem.*, 2013, **288**, 37204.
128 D. Bucher, Y. H. Hsu, V. D. Mouchlis, E. A. Dennis and J. A. McCammon, *PLoS Comput. Biol.*, 2013, **9**, e1003156.
129 A. A. Zurawel, D. J. Walsh, S. M. Fortier, T. Chidawanyika, S. Sengupta, K. Zilm and S. Supattapone, *Biochemistry*, 2014, **53**, 68.
130 G. E. Jackson, R. Gamieldien, G. Mugumbate and G. Gäde, *Peptides*, 2014, **53**, 270.
131 E. K. Lebbe, S. Peigneur, M. Maiti, P. Devi, S. Ravichandran, E. Lescrinier, C. Ulens, E. Waelkens, L. D'Souza, P. Herdewijn and J. Tytgat, *J. Biol. Chem.*, 2014, **289**, 9573.
132 P. Wadhwani, E. Strandberg, J. van den Berg, C. Mink, J. Bürck, R. A. Ciriello and A. S. Ulrich, *Biochim. Biophys. Acta*, 2014, **1838**, 940.
133 H. Yao and M. Hong, *J. Am. Chem. Soc.*, 2014, **136**, 2611.

134 S. Afonin, R. W. Glaser, C. Sachse, J. Salgado, P. Wadhwani and A. S. Ulrich, *Biochim. Biophys. Acta*, 2014, **1838**, 2260.

135 L. Caillon, J. A. Killian, O. Lequin and L. Khemtémourian, *PLoS One*, 2013, **8**, e75528.

136 D. I. Fernandez, M. A. Sani, A. J. Miles, B. A. Wallace and F. Separovic, *Biochim. Biophys. Acta*, 2013, **1828**, 1863.

137 T. Kushibiki, M. Kamiya, T. Aizawa, Y. Kumaki, T. Kikukawa, M. Mizuguchi, M. Demura, S. Kawabata and K. Kawano, *Biochim. Biophys. Acta*, 2014, **1844**, 527.

138 L. E. Marbella, H. S. Cho and M. M. Spence, *Biochim. Biophys. Acta*, 2013, **1828**, 1674.

139 S. Fanghänel, P. Wadhwani, E. Strandberg, W. P. Verdurmen, J. Bürck, S. Ehni, P. K. Mykhailiuk, S. Afonin, D. Gerthsen, I. V. Komarov, R. Brock and A. S. Ulrich, *PLoS One*, 2014, **9**, e99653.

140 M. C. Manzini, K. R. Perez, K. A. Riske, J. C. Bozelli Jr., T. L. Santos, M. A. da Silva, G. K. Saraiva, M. J. Politi, A. P. Valente, F. C. Almeida, H. Chaimovich, M. A. Rodrigues, M. P. Bemquerer, S. Schreier and I. M. Cuccovia, *Biochim. Biophys. Acta*, 2014, **1838**, 1985.

141 B. Kwon, A. J. Waring and M. Hong, *Biophys. J.*, 2013, **105**, 2333.

142 M. Fillion, M. Noël, A. Lorin, N. Voyer and M. Auger, *Biochim. Biophys. Acta*, 2014, **1838**, 2173.

143 D. Marquardt, J. A. Williams, J. J. Kinnun, N. Kučerka, J. Atkinson, S. R. Wassall, J. Katsaras and T. A. Harroun, *J. Am. Chem. Soc.*, 2014, **136**, 203.

144 K. Cieślik-Boczula, P. Swiątek, A. Jaszczyszyn, P. Zawilska, K. Gąsiorowski, W. Malinka and G. Köhler, *J. Phys. Chem. B*, 2014, **118**, 3605.

145 S. Santeusanio, O. A. Attanasi, R. Majer, M. Cangiotti, A. Fattori and M. F. Ottaviani, *Langmuir*, 2013, **29**, 11118.

146 E. M. Clop, A. K. Chattah and M. A. Perillo, *J. Phys. Chem. B*, 2014, **118**, 6150.

147 N. Jiménez-Rojo, A. R. Viguera, M. I. Collado, K. H. Sims, C. Constance, K. Hill, W. A. Shaw, F. M. Goñi and A. Alonso, *Biochim. Biophys. Acta*, 2014, **1838**, 2071.

148 M. Vermathen, M. Marzorati and P. Bigler, *J. Phys. Chem. B*, 2013, **117**, 6990.

149 M. Beaugrand, A. A. Arnold, J. Hénin, D. E. Warschawski, P. T. Williamson and I. Marcotte, *Langmuir*, 2014, **30**, 6162.

150 A. L. Furlan, A. Castets, F. Nallet, I. Pianet, A. Grélard, E. J. Dufourc and J. Géan, *Langmuir*, 2014, **30**, 5518.

151 J. C. Debouzy, D. Crouzier, F. Bourbon, M. Lahiani-Skiba and M. Skiba, *J. Drug Delivery*, 2014, **2014**, 575719.

152 D. Münch, A. Müller, T. Schneider, B. Kohl, M. Wenzel, J. E. Bandow, S. Maffioli, M. Sosio, S. Donadio, R. Wimmer and H. G. Sahl, *J. Biol. Chem.*, 2014, **289**, 12063.

153 B. Pawlikowska-Pawlęga, H. Dziubińska, E. Król, K. Trębacz, A. Jarosz-Wilkołazka, R. Paduch, A. Gawron and W. I. Gruszecki, *Biochim. Biophys. Acta*, 2014, **1838**, 254.

154 B. Pawlikowska-Pawlęga, L. E. Misiak, A. Jarosz-Wilkołazka, B. Zarzyka, R. Paduch, A. Gawron and W. I. Gruszecki, *Biochim. Biophys. Acta*, 2014, **1838**, 2127.

155 F. Gilardi, M. Giudici, N. Mitro, O. Maschi, U. Guerrini, G. Rando, A. Maggi, G. Cermenati, A. Laghezza, F. Loiodice, G. Pochetti, A. Lavecchia, D. Caruso, E. De Fabiani, K. Bamberg and M. Crestani, *J. Biol. Chem.*, 2014, **289**, 6908.

156 E. L. Dennis and P. M. Thompson, *Neuropsychol. Rev.*, 2014, **24**, 49.

157 R. J. Gropler, *J. Nucl. Cardiol.*, 2013, **20**, 1147.

158 F. M. McQueen and E. Chan, *Curr. Rheumatol. Rep.*, 2014, **16**, 388.
159 R. K. Gupta, K. J. Jobanputra and A. Yadav, *Neuroimaging Clin. N. Am.*, 2013, **23**, 475.
160 Y. Zhang, E. Guallar, Y. Qiao and B. A. Wasserman, *Arterioscler., Thromb., Vasc. Biol.*, 2014, **34**, 1341.
161 M. Mussap, R. Antonucci, A. Noto and V. Fanos, *Clin. Chim. Acta*, 2013, **426**, 127.
162 R. Fischer, P. Bowness and B. M. Kessler, *Proteomics*, 2013, **13**, 3371.
163 H. Dai, B. Hong, Z. Xu, L. Ma, Y. Chen, Y. Xiao and R. Wu, *Neural Regener Res.*, 2013, **8**, 2103.
164 A. Millon, S. D. Dickson, A. Klink, D. Izquierdo-Garcia, J. Bini, E. Lancelot, S. Ballet, P. Robert, J. Mateo de Castro, C. Corot and Z. A. Fayad, *Atherosclerosis*, 2013, **228**, 339.
165 R. Vettukattil, T. E. Hetland, V. A. Flørenes, J. Kærn, B. Davidson and T. F. Bathen, *Hum. Pathol.*, 2013, **44**, 1859.
166 X. Pan, M. Wilson, C. McConville, T. N. Arvanitis, J. L. Griffin, R. A. Kauppinen and A. C. Peet, *Metabolomics*, 2013, **9**, 722.
167 S. Akoudad, M. A. Ikram, P. J. Koudstaal, A. Hofman, W. J. Niessen, S. M. Greenberg, A. van der Lugt and M. W. Vernooij, *Cerebrovasc. Dis.*, 2014, **37**, 382.
168 C. E. Kostara, A. Papathanasiou, N. Psychogios, M. T. Cung, M. S. Elisaf, J. Goudevenos and E. T. Bairaktari, *J. Proteome Res.*, 2014, **13**, 2585.
169 S. Blüml, J. L. Wisnowski, M. D. Nelson Jr., L. Paquette and A. Panigrahy, *PLoS One*, 2014, **9**, e85829.
170 J. Z. Willey, H. Gardener, Y. P. Moon, M. Yoshita, C. DeCarli, Y. K. Cheung, R. L. Sacco, M. S. Elkind and C. B. Wright, *Cerebrovasc. Dis.*, 2014, **37**, 423.
171 S. N. Reinke, B. H. Walsh, G. B. Boylan, B. D. Sykes, L. C. Kenny, D. M. Murray and D. I. Broadhurst, *J. Proteome Res.*, 2013, **12**, 4230.
172 S. C. Forbes, R. J. Willcocks, W. T. Triplett, W. D. Rooney, D. J. Lott, D. J. Wang, J. Pollaro, C. R. Senesac, M. J. Daniels, R. S. Finkel, B. S. Russman, B. J. Byrne, E. L. Finanger, G. I. Tennekoon, G. A. Walter, H. L. Sweeney and K. Vandenborne, *PLoS One*, 2014, **9**, e106435.
173 R. Amathieu, M. N. Triba, P. Nahon, N. Bouchemal, W. Kamoun, H. Haouache, J. C. Trinchet, P. Savarin, L. Le Moyec and G. Dhonneur, *PLoS One*, 2014, **9**, e89230.
174 L. D. Whigham, D. E. Butz, H. Dashti, M. Tonelli, L. K. Johnson, M. E. Cook, W. P. Porter, H. R. Eghbalnia, J. L. Markley, S. R. Lindheim, D. A. Schoeller, D. H. Abbott and F. M. Assadi-Porter, *Curr. Metabolomics*, 2014, **2**, 269.
175 L. Wang, Y. Tang, S. Liu, S. Mao, Y. Ling, D. Liu, X. He and X. Wang, *PLoS One*, 2013, **8**, e65675.
176 G. Gouspillou, N. Sgarioto, S. Kapchinsky, F. Purves-Smith, B. Norris, C. H. Pion, S. Barbat-Artigas, F. Lemieux, T. Taivassalo, J. A. Morais, M. Aubertin-Leheudre and R. T. Hepple, *FASEB J.*, 2014, **28**, 1621.
177 T. O. Kalayci, G. Erdem, R. Kutlu, A. Kahraman and A. Alkan, *Malays. J. Med. Sci.*, 2014, **21**, 89.
178 M. I. Lassenius, V. P. Mäkinen, C. L. Fogarty, L. Peräneva, M. Jauhiainen, P. J. Pussinen, M. R. Taskinen, J. Kirveskari, O. Vaarala, J. K. Nieminen, S. Hörkkö, A. J. Kangas, P. Soininen, M. Ala-Korpela, D. Gordin, A. J. Ahola, C. Forsblom, P. H. Groop and M. Lehto, *Nutr. Metab.*, 2014, **11**, 28.
179 J. F. Gregory 3rd, Y. Park, Y. Lamers, N. Bandyopadhyay, Y. Y. Chi, K. Lee, S. Kim, V. da Silva, N. Hove, S. Ranka, T. Kahveci, K. E. Muller, R. D. Stevens, C. B. Newgard, P. W. Stacpoole and D. P. Jones, *PLoS One*, 2013, **8**, e63544.

180 S. R. El Khoudary, M. M. Brooks, R. C. Thurston and K. A. Matthews, *J. Lipid Res.*, 2014, **55**, 1498.
181 F. Qin, Z. Shen, L. Peng, R. Wu, X. Hu, G. Zhang and S. Tang, *PLoS One*, 2014, **9**, e96010.
182 P. M. Nunes, A. J. Wright, A. Veltien, J. J. van Asten, C. J. Tack, J. G. Jones and A. Heerschap, *FASEB J.*, 2014, **28**, 1988.
183 P. T. Williams, X. Q. Zhao, S. M. Marcovina, J. D. Otvos, B. G. Brown and R. M. Krauss, *Atherosclerosis*, 2014, **233**, 713.
184 L. K. Bogren, C. J. Murphy, E. L. Johnston, N. Sinha, N. J. Serkova and K. L. Drew, *PLoS One*, 2014, **9**, e107493.
185 J. J. Milner, J. Wang, P. A. Sheridan, T. Ebbels, M. A. Beck and J. Saric, *PLoS One*, 2014, **9**, e97238.
186 L. Zhao, H. Zhang, Y. Yang, Y. Zheng, M. Dong, Y. Wang, G. Bai, X. Ye, Z. Yan and H. Gao, *PLoS One*, 2014, **9**, e108678.
187 J. H. Park, H. Lee, R. Makaryus, M. Yu, S. D. Smith, K. Sayed, T. Feng, E. Holland, A. Van der Linden, T. G. Bolwig, G. Enikolopov and H. Benveniste, *PLoS One*, 2014, **9**, e94755.
188 P. Khandelwal, S. Stryker, H. Chao, N. Aranibar, R. M. Lawrence, M. Madireddi, W. Zhao, L. Chen and M. D. Reily, *J. Lipid Res.*, 2014, **55**, 1366.
189 L. Tian, H. S. Kim, H. Kim, X. Jin, H. S. Jung, K. S. Park, K. W. Cho, S. Park and W. K. Moon, *J. Proteome Res.*, 2013, **12**, 3738.
190 J. C. Ho, P. Storm, A. Rydström, B. Bowen, F. Alsin, L. Sullivan, I. Ambite, K. H. Mok, T. Northen and C. Svanborg, *J. Biol. Chem.*, 2013, **288**, 17460.
191 J. Dalli, M. Zhu, N. A. Vlasenko, B. Deng, J. Z. Haeggström, N. A. Petasis and C. N. Serhan, *FASEB J.*, 2013, **27**, 2573.
192 B. Kneidl, M. Peller, G. Winter, L. H. Lindner and M. Hossann, *Int. J. Nanomed.*, 2014, **9**, 4387.
193 K. Hadinoto, A. Sundaresan and W. S. Cheow, *Eur. J. Pharm. Biopharm.*, 2013, **85**, 427.
194 S. Langereis, T. Geelen, H. Grüll, G. J. Strijkers and K. Nicolay, *NMR Biomed.*, 2013, **26**, 728.
195 M. Han, M. R. Vakili, H. Soleymani Abyaneh, O. Molavi, R. Lai and A. Lavasanifar, *Mol. Pharm.*, 2014, **11**, 2640.
196 L. Liu, T. K. Hitchens, Q. Ye, Y. Wu, B. Barbe, D. E. Prior, W. F. Li, F. C. Yeh, L. M. Foley, D. J. Bain and C. Ho, *Biochim. Biophys. Acta*, 2013, **1830**, 3447.
197 V. Pandey, K. R. Gajbhiye and V. Soni, *Drug Delivery*, 2015, **22**, 199.
198 J. Chao, T. I. Huo, H. Y. Cheng, J. C. Tsai, J. W. Liao, M. S. Lee, X. M. Qin, M. T. Hsieh, L. H. Pao and W. H. Peng, *PLoS One*, 2014, **9**, e96969.
199 T. Schaffran, N. Jiang, M. Bergmann, E. Küstermann, R. Süss, R. Schubert, F. M. Wagner, D. Awad and D. Gabel, *Int. J. Nanomed.*, 2014, **9**, 3583.
200 R. Liang, J. Wang, X. Wu, L. Dong, R. Deng, K. Wang, M. Sullivan, S. Liu, M. Wu, J. Tao, X. Yang and J. Zhu, *Nanotechnology*, 2013, **24**, 455302.
201 E. Andreozzi, P. Wang, A. Valenzuela, C. Tu, F. Gorin, M. Dhenain and A. Louie, *Bioconjugate Chem.*, 2013, **24**, 1455.
202 A. Ketkar-Atre, T. Struys, T. Dresselaers, M. Hodenius, I. Mannaerts, Y. Ni, I. Lambrichts, L. A. Van Grunsven, M. De Cuyper and U. Himmelreich, *Biomaterials*, 2014, **35**, 1015.
203 H. Xu, W. Zhang, Y. Li, F. F. Ye, P. P. Yin, X. Yu, M. N. Hu, Y. S. Fu, C. Wang and D. J. Shang, *Pharm. Res.*, 2014, **31**, 3038.

204 S. Song, L. Z. Cheong, M. Falkeborg, L. Liu, M. Dong, H. M. Jensen, K. Bertelsen, M. Thorsen, T. Tan, X. Xu and Z. Guo, *PLoS One*, 2013, **8**, e73891.
205 A. Kasten, C. Grüttner, J. P. Kühn, R. Bader, J. Pasold and B. Frerich, *PLoS One*, 2014, **9**, e108055.
206 R. Zhang, I. D. Sahu, L. Liu, A. Osatuke, R. G. Comer, C. Dabney-Smith and G. A. Lorigan, *Biochim. Biophys. Acta*, 2015, **1848**, 329.
207 J. P. Douliez, L. Navailles, E. J. Dufourc and F. Nallet, *Langmuir*, 2014, **30**, 5075.
208 J. R. Sachleben, R. Yi, P. A. Volden and S. D. Conzen, *J. Biomol. NMR*, 2014, **59**, 161.
209 B. H. Wokke, C. Bos, M. Reijnierse, C. S. van Rijswijk, H. Eggers, A. Webb, J. J. Verschuuren and H. E. Kan, *J. Magn. Reson. Imaging*, 2013, **38**, 619.
210 J. Lu, J. Zhou, C. Cai, S. Cai and Z. Chen, *Magn. Reson. Med.*, 2015, **73**, 1615.
211 H. Zhou, J. H. Stafford, R. R. Hallac, L. Zhang, G. Huang, R. P. Mason, J. Gao, P. E. Thorpe and D. Zhao, *J. Biomed. Nanotechnol.*, 2014, **10**, 846.
212 J. M. Mitchell, T. W. Fan, A. N. Lane and H. N. Moseley, *Front. Genet.*, 2014, **5**, 237.

NMR of carbohydrates

Daniel Łowicki, Szymon Buda and Jacek Mlynarski*

DOI: 10.1039/9781782622758-00407

Recent advances in the application of nuclear magnetic resonance to the study of carbohydrates have been presented. New and modernized NMR methods both experimental and computational are collected including improvement of sugar databases. This chapter also groups literature concerning investigations of sugar structures determined on the basis of NMR spectroscopy. The applications of nuclear magnetic resonance to identifications and quantifications of carbohydrates as well as to the study of non-covalent interactions between carbohydrates and other natural and synthetic compounds are also presented in this report.

1 Introduction

Carbohydrates and their derivatives play crucial roles in living organisms such as energy storage, structural functions including maintaining of cell shape, inducing of immune responses and many others. The structures of carbohydrates involved in biochemical processes vary from monosaccharides, through oligosaccharides to polysaccharides. Nuclear Magnetic Resonance spectroscopy is an excellent tool for studying simple monosaccharide units as well as more complex structures of homo- and heteropolysaccharides in both solution and solid states. NMR methods provide also powerful approach for investigation of non-covalent interactions between carbohydrates and other bio-molecules. These interactions are of particular interest as they allow rational planning the synthesis of new drugs. This chapter is devoted to recent advances in the studies on carbohydrates which are based on the nuclear magnetic resonance spectroscopy. The papers published online in the period from July 2013 to June 2014 have been selected and included in this chapter.

2 New and modernized NMR methods

Nuclear magnetic resonance spectroscopy is one of the most commonly used methods to determine the structure of carbohydrates (most precisely and general referred as sugars) and to identify the components of mixtures including natural mono- and polysaccharides. Structural analysis and the growing demand for analysing mixtures of lower concentration poses new challenges, but also gives the motivation to improve current NMR methods and develop new ones.

Quantitative characterisation of minor components in foods is often hampered by the presence of major components such as water and sugars. Although the solvent signals can be suppressed by presaturation, WET or WATERGATE methods, simultaneous saturation of the signals of sugars and water in an extensive range remains a key issue. Tanokura

Jagiellonian University, Faculty of Chemistry, Ingardena 3, 30-060 Krakow, Poland. E-mail: jacek.mlynarski@gmail.com

et al.[1] proposed a new method that allows to solve this problem. They applied broadband WET (BB WET) technique, which is a modification of WET (Water suppression Enhanced through T_1 effects) to the simultaneous saturation of water and carbohydrate resonances in persimmon juice. The spectrum of persimmon fruit juice recorded without presaturation was compared with spectra measured using NOESYPRESAT, DPFGSE-WATERGATE, conventional WET and BB-WET. The comparison clearly shows that the best results in terms of signal-to-noise ratio are obtained for BB-WET pulse scheme. The new method can be thus successfully applied for quantitative analysis of mixtures containing a substantial amount of sugars diluted in water.

In the comprehensive review Kato *et al.*[2] summarized recent development of NMR methods for characterizing the dynamic conformations and interactions of oligosaccharides including also stable isotope labelling process. Heteronuclear $^{13}C/^{1}H$ long-range coupling constants from J-HMBC spectra can provide information on the torsion angle preferences, and hence conformation of a molecule. When the homonuclear two or three bond $^{13}C/^{13}C$ coupling constants ($^{2}J_{CC}$ and $^{3}J_{CC}$, respectively) are required, the low natural abundance of ^{13}C nuclei is a main problem. In such cases site-specifically labelled compounds are used. Unfortunately, the site-specifically labelled compounds could not simultaneously provide information on $^{13}C/^{1}H$ long-range coupling constants in the J-HMBC experiment because one-bond carbon–carbon coupling constants ($^{1}J_{CC}$) affect the carbon atoms adjacent to labelled positions and may give rise to peaks cancelation. Widmalm *et al.*[3] proposed two alternative modifications of J-HMBC sequence, that suppress the interference from one-bond $^{13}C/^{13}C$ scalar couplings. They demonstrated that for ^{13}C site-specifically labelled oligosaccharides homonuclear $^{2}J_{CC}$ and $^{3}J_{CC}$ coupling constants can be easily measured. By the use of either a selective inversion pulse or a constant-time element in the J-HMBC pulse sequence the $^{1}J_{CC}$ couplings disrupting measurement of conformationally important long-range heteronuclear coupling constants are suppressed, which significantly extends the scope of the J-HMBC experiment especially when the non-labelled analogue of a compound is not at hand.

Zujovic *et al.*[4] investigated the mobility of very mobile polysaccharides in plant cell walls by NMR method. Mobility of such polysaccharides as arabinan and galactan side chains of pectic polysaccharide rhamnogalactouronan I in two apple cultivars was evaluated by the solid-state ^{13}C NMR. The authors used two pulse phase modulation (TPPM) decoupling which improved the resolution of cross polarisation (CP) and single pulse excitation (SPE) NMR spectra. This methodology facilitated the mobility estimation of mobile polysaccharides in apples cell walls.

3 Computational NMR methods and sugar databases

Computational methods for predictions NMR parameters such as chemical shifts, spin–spin coupling constants or relaxation rates facilitate determination of carbohydrate structure and shorten the evaluation time of the results. For this reason empirical methods of NMR spectra predictions are continually being developed and databases updated with new spectroscopic data. In 2013 Toukach and Ananikov,[5] summarized the recent advances in computational methods for predictions of NMR parameters of carbohydrates.

Regarding the databases update, Egorova and Toukach[6] expanded the coverage of Carbohydrate Structure Database (CSDB) of glycans from plants and fungi. The added data was acquired from CarbBank and from the original work not deposited in CarbBank. It is worth noting that the data imported from CarbBank that contain a lot of errors were verified manually and corrected according to the original papers. The new, freely available Plant and Fungal Carbohydrate Structure Database (PFCSDB) contains 2455 assigned NMR spectra of plant and fungal carbohydrates isolated from nearly 1300 species.

Widmalm *et al.*[7] studied ^{1}H and ^{13}C NMR chemical shifts assignment of mono-, di-, tri- and tetrasaccharides (totally 43 compounds) and applied them for NMR chemical shift predictions of oligosaccharides depicted in Fig. 1 using CASPER program. The CASPER (computer assisted spectrum evaluation of regular polysaccharides) software contains chemical shift database and empirical spectra simulation routine optimized for carbohydrates. The calculated chemical shifts are in good to excellent agreement with those from ^{1}H and ^{13}C NMR experiments.

Rapid automatic structural elucidation of biological repeating unit of bacterial polysaccharides was demonstrated by Widmalm and co-authors,[8] who used predicted functions of glycosyltransferases (GTs) combined with unassigned NMR data. These two pieces of information *i.e.* about the action of GTs and ^{1}H–^{13}C HSQC spectra are submitted to the CASPER program. Predicted functions of GTs are available in the *Escherichia coli O*-antigen database (ECODAB). Comparison of the ^{1}H–^{13}C

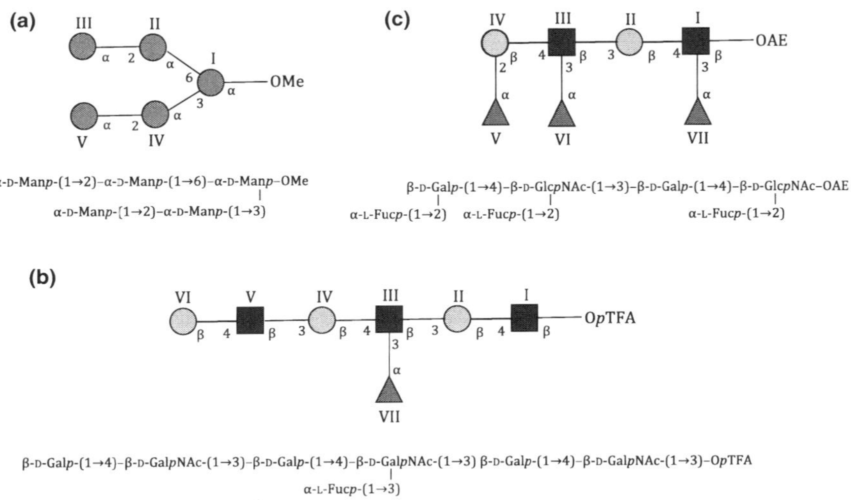

Fig. 1 Structures of oligosaccharides whose chemical shifts were predicted using CASPER program presented in CFG and 2D text notations: (a) mannose pentasaccharide (M5), (b) (LacNAc-Le^{x}-LacNAc)-heptasaccharide, (c) (Le^{y}-Le^{x})-heptasaccharide. Me – methyl, pTFA – *para*-trifluoroacetamidophenylethyl, AE – 2-azidoethyl.

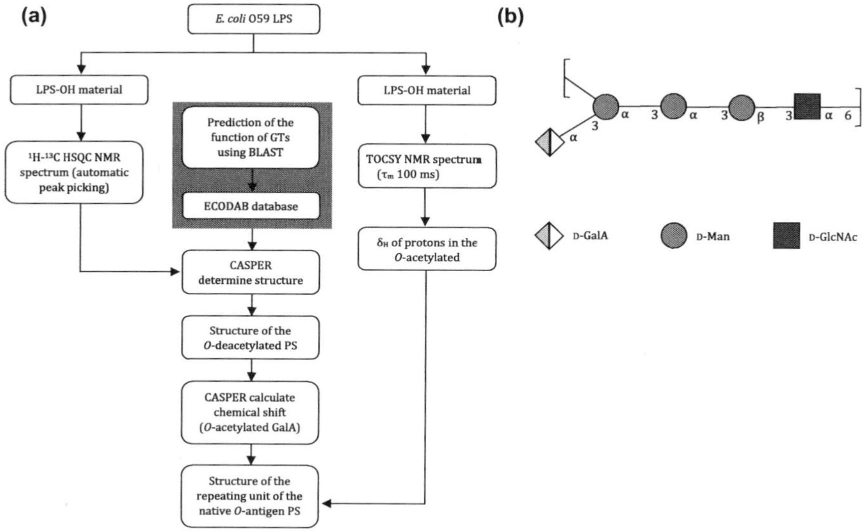

Fig. 2 (a) Flowchart showing the methodology used to determine the PS structure and (b) one of top-ranked structures of O-deacetylated repeating unit of the O-antigen PS of *E. coli* O59 obtained as CASPER output (CFG notation).

HSQC NMR spectra of the *O*-deacylated lipopolysaccharide and the lipid-free PS in conjunction with chemical shift predictions made by the CASPER program allows to identify the positions of *O*-acylation process. The new approach was successfully implemented by Widmalm *et al.* in the structural elucidation of the *O*-antigen PS of *E. coli* O59. The algorithm used in this methodology was demonstrated as flowchart in Fig. 2a and the top-ranked structure of repeating unit of *O*-deacetylated repeating unit of the *O*-antigen PS of *E. coli* O59 in Fig. 2b.

The advantages of this time-consuming methodology are:

- the results can be obtained quickly,
- information about identity and absolute configuration of the component monosacharaides is not required,
- preparation and analysis of derivatives are not necessary,
- NMR data from single 2D spectrum are sufficient to obtain valuable results.

Samsonov *et al.*[9] applied gauge invariant atomic orbital (GIAO) approach to calculate NMR parameters such as chemical shifts and proton–proton spin–spin coupling constants of glycosaminoglycan (GAG) monosaccharide blocks. The QM calculations were performed at B3LYP6311+G(2d,p) and B3LYP/aug-cc-pVDZ levels of theory for chemical shifts and spin–spin coupling constants, respectively. The calculated chemical shifts may be useful for the analysis of sulfation position of GAG monosaccharide blocks and net sulfation. In turn, $^3J_{H-H}$ could provide the information on both sulfation position and ring conformation. Although the GIAO calculations might be used to complement NMR experiment interpretation for GAGs, the authors suggest that they still require further optimisation.

4 Identification and quantification of sugars

Identification of glucans in complex mixtures of simple sugar, through oligosaccharides up to polysaccharides and their quantitative analysis is a difficult and time consuming process. It is often realised by high-performance anion exchange chromatography (HPAEC), hydrophilic interaction liquid chromatography (HILIC), capillary electrophoresis and mass spectrometry (MS). However, these methods have significant limitations and disadvantages. Reference compounds, often necessary for the calibration of instrument, are not always available. Moreover, the samples of glucan mixtures often require chemical modifications to change retention time of sugars with similar molecular weights but of different structures. Due to the lack of chromophores, fluorophores or ionisable groups in the structure of most of the carbohydrates the samples must be derivatized prior to the analysis. Although LC-MS method enables detection and analysis of compounds at very low concentration in mixture it causes destruction and loss of the sample, in contrast to NMR spectroscopy. Meier *et al.*[10] have proposed protocol for spectroscopic glycan mixture analysis method which requires neither separation of the carbohydrates mixture nor the modification of them. Fast and accurate quantification of carbohydrates in a mixture can be done by optimized heteronuclear high-resolution NMR spectroscopy. The authors used sensitivity-enhanced ^{1}H–^{13}C HSQC spectra with optimized spectral width in the ^{13}C dimension of acetal carbon atoms. This approach fully exploits the narrow line widths of glycans in the ^{13}C spectral dimension of acetal groups, which is sufficient to resolve about 20 different homopolymeric starch fragments. The reported method allows identification and quantification of at least eight linear and over twelve branched starch fragments in beer at concentration as low as 10 g mL^{-1} and in 15 minutes. Subsequently Meier *et al.*[11] published the results of their studies on the carbohydrate reference sample databases of over 80 carbohydrates. The new reference library of specific carbohydrate signals was constructed on the basis of precise measurements of ^{1}H and ^{13}C anomeric chemical shifts in two-dimensional HSQC spectra. The ppm values of both ^{1}H and ^{13}C signals are rounded to the third decimal place. Such precise chemical shift measurements in high resolved HSQC spectra with consistent referencing have a great potential for qualitative and quantitative analysis of carbohydrates in a complex mixture. Sassaki *et al.*[12] developed method, which is based on quantitative HSQC (q-HSQC) sequence for estimation and quantification of monosaccharide composition of glycans at the level of mmol l^{-1}. The glycans are natural polysaccharides and carbohydrate moieties of glycolconjugates such as glycolipids and glycoproteins, and therefore these compounds are of great biological importance. Arús *et al.*[13] applied ^{1}H high-resolution magic angle spinning NMR (HR-MAS NMR) spectroscopy for free sugar quantification in intact tissue of melon species. The content of such sugars as sucrose, glucose and fructose in mesocarp tissues from melon fruits was determined by HR-MAS qNMR and was consistent with data obtained using traditional, enzymatic procedure.

The advantage of the NMR method is relatively short time of analysis (*ca.* 1 h) in comparison to time consuming enzymatic assay (*ca.* 24 h). It should also be noted that the NMR analysis requires much less sample to provide reliable results.

Development and improvement of the quantitative determination of the components of sugar mixtures by NMR allows quick and accurate analysis of samples, so that is applicable in practice. Quan,[14] presented application of quantitative ^{1}H NMR method together with mass balance approach with high resolution liquid chromatography – evaporative light scattering detection to assign purity values to six carbohydrate certified reference materials such as glucose, fructose, galactose, lactose, xylose and sucrose, according to the ISO Guides 34 and 35. The author concludes that the qNMR method facilitate the analysis due to simple sample preparation as well as quick and easy analysis, although uncertainty is slightly greater than that of the mass balance approach.

Monakhova *et al.*[15] presented validation studies for multicomponent quantitative NMR analysis of apple fruit juice that includes, apart from other compounds, such sugars as galacturonic acid, arbutin, fructose, glucose, sucrose, quinic acid, maltotriose, maltose and xylose. The NMR analysis can be used for quality control of different fruit juices.

Proton NMR spectroscopy was used as a tool for profiling complex carbohydrate mixtures in non-fractionated beer by Meier *et al.*[16] Six samples of beer of different styles contained more than 20 small carbohydrates were tested. The authors used H_1–H_2 cross peaks in phase sensitive double quantum filtered COSY measurement as reporter signals to identify more than 50 structural motifs in beers.

Chen *et al.*[17] demonstrated quantification of wheat straw lignin structure. Combination of quantitative ^{13}C NMR and HSQC methods allowed the authors to analyse structural motifs found in milled straw lignin and cellulose-digested lignin. Möller and co-authors,[18] used ^{13}C CP/MAS NMR spectroscopy for assessment of historic *Tilia codrata* wood. Comparison of NMR results obtained from historic lime wood samples and reference wood samples demonstrated structural and chemical modifications. The changes arising in the wood structure due to natural ageing process mainly concern the decrease of the carbohydrate moiety. Naggi *et al.*[19] have shown that combined 2D NMR and LC-MS analysis permits detection of the major internal modifications introduced in low-molecular-weight heparin chains by periodate oxidation and borohydride reduction.

5 Non-covalent sugar–protein interactions studied by NMR

The non-covalent interactions in many of natural compounds commonly referred as biomolecules are responsible for very important biological processes. The molecular recognition process in biological systems is observed between such species as receptor–ligand, antigen–antibody, sugar–lectin, DNA–protein, RNA-ribosome and others. The kinetics of the majority of glycan–protein interactions is fast and with dissociation constants within the μM to mM range. For study of such transient complexes transferred NOE (tr-NOE) and saturation transfer difference (STD) NMR spectroscopic methods are the most appropriate. Both of these techniques base on the Nuclear Overhauser effect and hence allow to study in solution the interactions between large molecule called receptor and small to medium size molecule called ligand. In 1999 Mayer,[20] proposed the STD NMR technique for detection of saturation acquired by the free ligand from the transient complex. Since then it is used to determine which part of the ligand molecule is responsible for binding, because most strongly interacting groups of ligand show stronger STD effect. This technique is particularly useful when it is combined with NOESY measurements or when expected calculated values of transference are taken into account for interpretation of the results.

In 2013 Bawley and Shahzad-ul-Hussan,[21] briefly described most important approaches for characterizing carbohydrate – protein interactions by NMR methods. The review consists of two major parts: one is devoted to protein detected methods and the other to carbohydrate ligand-based approaches. Nagae and Yamaguchi,[22] focused on the three-dimensional structural aspects of the interactions between proteins and polysaccharides. In their review authors described the structural aspects of *exo*-type and *endo*-type protein-polysaccharide interactions. Examples of *exo*- and *endo*-type interactions are shown in Fig. 3.

Yamaguchi and co-authors,[23] reported studies on glycan recognition by phytohemagglutinin from *Phaselous vulgaris* (PHA-E). A novel glycan recognition mode was found. The legume lectin recognizes the biantennary *N*-glycan derivatives (Fig. 4a) using common legume lectin fold. The interactions occurring in the complex were investigated by comparison of crystal structures of the complex and PHA-E alone. Saturation transfer difference NMR measurements in solution revealed that the PHA-E protein interacts almost equally with both of the trisaccharide units of Gal-β-(1→4)GlcNAc-β-(1→2)Man sequence, recognized from X-ray study as crucial in the sugar–protein interactions (Fig. 4b).

Ravoo *et al.*[24] synthesised parallel and antiparallel isomers of receptors composed of two tripeptides Cys-His-Cys and Cys-Tyr-Cys linked by two disulfides. The noncovalent interactions of these hexapeptides with monosaccharides and disaccharides in aqueous solution were investigated by NMR titrations as well as by isothermal titration calorimetry and circular dichroism spectroscopy. According to the results obtained the parallel isomer containing histidine forms the most stable complexes

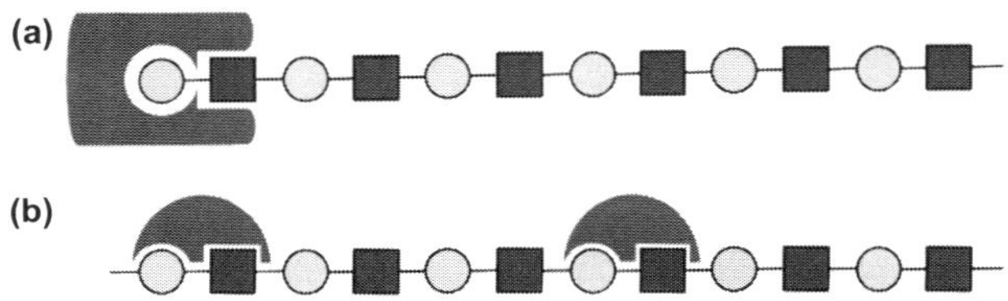

Fig. 3 Schematic representation of (a) *exo*-type and (b) *endo*-type protein-polysaccharide interactions.

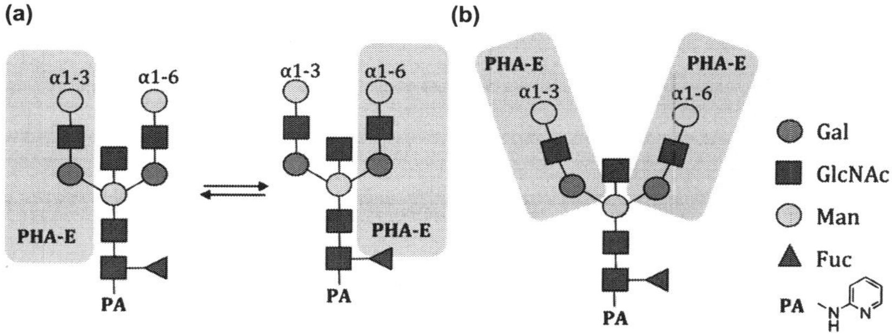

Fig. 4 Phytohemagglutin from *Phaselous vulgaris* (PHA-E) (a) interactions with biantennary glycan in two different manners determined by STD NMR analysis, (b) one biantennary glycan bridges two PHA-E molecules determined by STD NMR analysis.

with *N*-acetylneuramic acid (NANA) and with methyl-α-D-galactopyranoside. In turn both isomers containing His demonstrate high affinity to trehalose. No interaction was found between the antiparallel histidine-containing isomer and MeGal by NMR and ITC methods. Numata and Fukamizo *et al.*[25] studied the complex of a 'loopless' GH19 chitinase and chitin tetrasaccharide in crystal structure by X-ray analysis and in solution by NMR titration. The authors found that the binding mode is similar in both solid and solution states. An unusual mode of galactose recognition by a family 32 carbohydrate binding module (CBM32) was disclosed by Smith *et al.*[26] The interactions between galacto-configured sugars and CBM32-1 from *Cp*GH84A were studied by STD NMR experiments and discussed in detail. Hartmann *et al.*[27] synthesised a series of defined homo- and heteromultivalent glycooligomers and evaluated their binding behaviour toward Concanavalin A. The binding mechanisms of the multivalent glycooligomers to the model lectin Con A were studied *via* STD-NMR and 2fFCS. The authors demonstrated that galactose residues do not directly participate during the recognition of the receptors, whereas mannose interacts with the receptor upon complex formation.

Nieto *et al.*[28] investigated carbohydrate interactions with dendritic cell-specific intercellular adhesion molecule-3 grabbing non-integrin (DC-SIGN) also known as CD209. The main role of this lectin receptor occurring at the surface of macrophages is the recognition and binding of mannose-type carbohydrates being common patterns of pathogen bacteria, fungi and viruses. The authors tested both, natural mannose-based di- and trisaccharides from gp-120, and mannose-containing mimetics, in which mannose from non-terminal residue and from reducing end were replaced by cyclohexyl moiety. The gp-120 is external glycoprotein exposed on the surface of the HIV envelope and plays a crucial role in attachment to specific cell surface receptors, contributing to the entry of the virus into the cell. The NOESY experiments for Manα(1→2)[Manα(1→6)]Man and Manα(1→3)[Manα(1→6)]Man revealed that these trisaccharides are banded by DC-SIGN extracellular domain. This conclusion relies on the change of the sign of the NOE cross-peaks in comparison to the free experiments. The STD NMR measurements allowed to specify that the larger interactions with the receptor have the terminal mannose rings of both

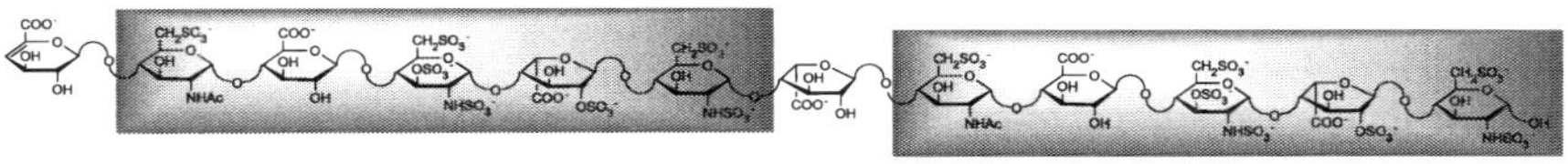

Fig. 5 Structure of dodecasaccharide studied as complex with antithrombin protein.

trisaccharides. In turn the weaker interactions with the receptor are shown by the middle mannose unit linked by $1 \rightarrow 6$ glycosidic bond.

The binding activities of heparin against variety of proteins were widely studied. Heparin is an anticoagulant used in medicine, however the physiological role of this highly sulphated glycosaminoglycan is not fully understood because anticoagulation process is realised mostly by heparan sulfate (HS). Kaundinya and co-workers,[29] studied interactions of heparin with heparin-binding proteins such as fibroblast growth factor (FGF2) and fibroblast growth factor 10 (FGF10). The STD NMR experiments revealed specific binding of octasaccharides composed of repeating iduronic acid (IdoA) and glucosamine residues and provided insight into the critical epitopes of heparin ligands involved. The authors achieved STD NMR with enhanced sensitivity for heparin–protein complex by exciting the methylene protons of protein avoiding the introduction of any bias. Heparin dodecasaccharide ability to bind with antithrombin (AT) was studied by Guerrini *et al.*[30] The AT protein is responsible for inactivation of several enzymes of the coagulation system, thus its binding properties are crucial for the full understanding of the process. It is known that AT binds specific pentasaccharide sequence shown in Fig. 5. However, the location of pentasaccharide along the heparin chain is vague. Studies of the heparin oligosaccharide with the two contiguous pentasaccharide sequences in a row are presented for the first time. It was confirmed by STD NMR spectroscopy that both pentasaccharide units interact with the antithrombin.

Rajarathnam *et al.*[31] used solution NMR spectroscopy including ^{1}H–^{15}N NOESY and HSQC experiments to characterize the binding of glycosaminoglycans heparin octasaccharide to dCXCL1 chemokine. They demonstrated that the octasaccharide binds orthogonally to the interhelical axis and spans the dimer interface.

6 Determination of sugar structure

Nuclear magnetic resonance spectroscopy is now the most powerful tool for sugars structure survey. Carbohydrates are biomolecules of great importance and play a number of key functions in living organisms. Determination of the carbohydrate structures is very important for understanding not only how they are composed, but also how the processes occur in the cells and tissues and what are the mechanisms of their control.

6.1 Conformational studies

In recent years conformational studies of sugars are widely conducted, as evidenced by the large number of publications on this topic. Recent developments in NMR techniques and their limitations were succinctly summarized by Freedberg *et al.*[32] The papers devoted to conformational studies by the use of NMR of a variety of glycosaminoglycans such as chondroitin sulfate, dermatan sulfate, heparin, heparan sulfate, and hyaluronan were reviewed by Pomin.[33]

New research on conformational properties of carbohydrates was carried out by Elli *et al.*[34] who chose fully sulfated maltotriose *alpha-beta* C–C linked dimer as the object of study. Leino *et al.*[35] focused their attention on phosphorylated β-(1→2) linked mannosides (mannotriose and tetraose) similar to those found in nature on the cell surface of *C. albicans* fungus. Conformational flexibility of xylosides with aromatic aglycon moiety was studied by Widmalm *et al.*[36] The NMR studies were carried out in different solvents in order to gain information on conformational equilibria of the saccharide units that may exist in enzyme active sites of polarity similar to appropriate solvent. Pérez and Widmalm *et al.*[37] elucidated conformational preferences of the *E. coli O*-antigen polysaccharides using NMR analysis of intra and inter-residue proton–proton distances from ^{1}H–^{1}H NOESY experiments. The linear tetrasaccharide repeating units of the two subgroups *i.e.* O5ab and O5ac of *Escherichia coli* serogroup O5, with different substitution patterns of the β-D-Qui*p*3*N*Ac (2 for O5ac and 4 for O5ab), were tested and compared. Recently, Widmalm along with MacKerell, Jr. and others,[38,39] focused their study on conformational properties of α- or β-(1→6)-linked oligosaccharides in terms of three dihedral angles, ϕ, ψ, and ω. Espinosa and Martín-Santamaría *et al.*[40] have demonstrated that human galectin-1 recognizes only *syn*-Ψ-type conformations of β-(1→3)-linked lactose and its *C*-glycosyl derivative among three existing conformers.

Auzanneau *et al.*[41] studied the dimeric Le^{X}Le^X and Le^{A}Le^X (Fig. 6) tumour-associated carbohydrate antigens and demonstrated that within the hexasaccharide structures the Lewis x (Le^X) and Lewis a (Le^A) parts adopt the rigid stacked conformations existing in the isolated trisaccharide antigens. The groundbreaking finding was made by Schubert and Ernst *et al.*[42] Based on NOESY experiments the authors demonstrated, that in addition to the previously known factors influencing the small conformational rigidity of Le^X, such as exoanomeric effects, steric effects and hydrophobic contacts, a new one that is C–H···O weak

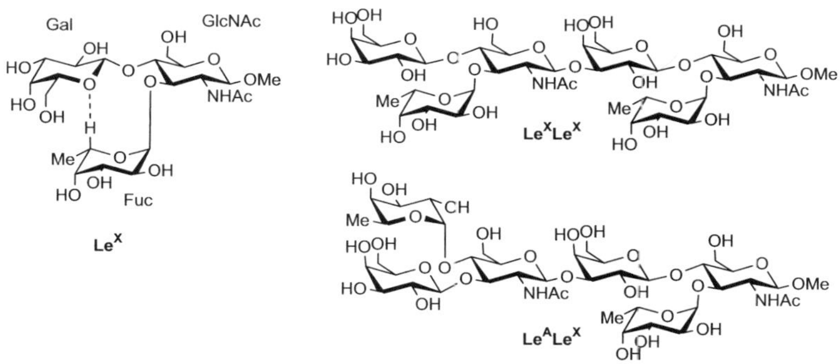

Fig. 6 Structure of methyl Lewis^X, Le^{X}Le^X and Le^{A}Le^X. The newly-identified C–H···O hydrogen bond is indicated by the dashed line.

hydrogen bond is also present in the structure. The hydrogen bond arises between C(5)-H group of L-fucose and O(5) of D-galactose (Fig. 6).

It should be emphasized that most of the studies on the conformation of sugars are usually conducted by a combination of NMR spectroscopy and MD calculations.

6.2 Sugar synthesis and modifications

6.2.1 Chemical mode. Chemical modification of carbohydrates is very demanding and time-consuming. Modification of oligo and polysaccharides is based on most common procedure which is protecting-deprotecting protocol. Winssinger and coworkers,[43] described solid phase synthesis of glucopeptides using Shoda's activation of unprotected carbohydrates. They monitored thioglycosylation reaction by ^{13}C NMR.

Unverzagt and coworkers,[44] designed a new method for the synthesis of *N*-glycosyl amides and *N*-glycopeptides by decarboxylative condensation. In this approach, unprotected sugar oximes are reacted with α-ketoacids. Synthetic polypeptides are very promising materials with potential in key biomedical application. A simple method for synthesis of that kind of molecules was presented by Schlaad's group.[45] They proposed one-pot partial glycosylation reaction and copolymerization of AGly NCA (Allylglycine *N*-carboxyanhydride). The reactions were monitored by *in situ* ^{1}H NMR and FT-IR spectroscopy. Moreover, the chemical compositions and chain length were also monitored by NMR. Ayres's group[46] presented synthesis of a glycosaminoglycan polymer mimetic using an *N*-alkyl-*N*,*N*-linked urea oligomer containing glucose pendant groups.

Bernardi and coworkers[47] presented synthesis of a multivalent inhibitor of the DC-SIGN dependent uptake of HIV-1 and Denga virus. The authors designed group of dendrimeric compounds with aliphatic core and tested inhibition properties of these substances on HIV-1 and Denga virus.

Similar synthetic approach were presented by Anandhan and Rajakumar.[48] They synthesized and subsequently characterized novel glycodendrimers and *N*-alkyldendrimers with aromatic core using NMR techniques. De Castro *et al.*[49] presented conversion of mannan-polysaccharides in mannose oligosaccharides with a thiopropargyl linker. Leino *et al.*[50] performed synthesis and conformational analysis of phosphorylated β-(1→2) linked mannosides. The stereochemistry at the anomeric centre was determined by NMR, which was a challenging task due to the fact that the H-1 and H-2 protons always displayed either axial–equatorial or equatorial–axial orientation.

Xia *et al.*[51] presented synthesis and analysis of kojic acid anchored on chitosan oligosaccharides. Bueno-Martínez *et al.*[52] showed synthesis of linear polyamide triazoles derived from glucose. Structural analysis of oligo and polysaccharide derivatives is very difficult and standard NMR techniques can be insufficient. Jan and Huang[53] presented the synthesis of carboxymethyl chitosan-*graft*-poly(γ-benzyl-L-glutamate) (PBLG) and the results of their studies on particle formation as encapsulants. Solid-phase ^{13}C NMR was chosen as the main tool in order to characterize the chain formation of PBLG. Solid-state ^{13}C NMR was used also by Wang and Liu,[54] who presented the chemical structure analysis of phosphorylated chitosan.

NMR spectra were a crucial tool for identification of synthetic derivatives containing mono and oligosaccharides. Helbert and Préchoux,[55] prepared and analyzed a series of oligo-α-carrageenans. Kaya *et al.*[56] designed the synthesis and characterized the galactochloralose based polymer. Jung *et al.*[57] analyzed supramolecular self-assembled aggregates formed by PCDA-βCD by the use of 1D and 2D NMR measurements. Porwański *et al.*[58] presented synthesis of bis-cellobiose and bis-glucose derivatives of azacrown macrocycle. They examined guest–host interaction in complex with acetylsialic acid and 4-aminophenol by NMR technique.

Jakeman and coworkers[59] carried out the synthesis of a series of phosponates and ketophosponates possessing an L-rhamnose scaffold with varying degrees of fluorination. These compounds were evaluated as potential inhibitors of L-rhamnose biosynthesis in *Streptococcus pneumoniae* cell. WaterLOGSY NMR study allowed to observe selectivity of binding between enzyme-substrate and enzyme-inhibitor.

Robina *et al.*[60] reported synthesis and biological study of *S*-galactosides containing a polyhydroxyalkylfuran as a potential inhibitors of cholera toxin (CT) and heat-labile toxin (LT). The interaction of the best candidates with CT and LT was studied by weak affinity chromatography (WAC) and saturation transfer difference NMR experiments.

Bernardi *et al.*[61] designed the synthesis of Lewis X mimics. Saturation transfer difference NMR studies allowed to obtain ligand binding epitope. Saturation transfer difference NMR technique was used also by Cui *et al.*[62] The authors designed synthesis of modified *Trichinella spiralis* disaccharide epitopes and compared their recognition by chemical mapping.

Iglesias-Guerra and coworkers,[63] prepared aziridine galactopyranoside derivatives and studied their anticancer activity. Guo *et al.*[64] presented efficient one-pot synthesis of tigogenin saponin. Tiwari *et al.*[65] used click chemistry to synthesize triazolyl ethisterone glycoconjugates. D'Andrea *et al.*[66] presented stereoselective synthesis of β-D-GlcNAc-(1→4)-D-Glc disaccharide starting from lactose. Martin and coworkers,[67] presented synthesis of L-sorbose derivatives. Authors made a critical review of the spectral data available in literature.

Bennett and Issa[68] presented the results of the low-temperature NMR experiments performed on S_N2-manifold. They have demonstrated that it is possible, using KHMDS/*p*-Ts_2O as a promoter system, to synthesize selectively β-linked 2-deoxy sugars. Varela *et al.*[69] studied synthesis of thiodisaccharides and oxidation to sulfoxides. They applied NMR techniques to assign the configuration of the sulfoxide sulfur atom. Wong *et al.*[70] presented one-pot synthesis and vaccine evaluation of the tumor-associated carbohydrate antigen RM2 from prostate cancer. Stereoselectivity of new formed glycosidic bonds was confirmed by NMR experiments.

Pirat *et al.*[71] presented synthesis of unnatural phosphonosugar analogues. The structure of phosphonosugars was confirmed by proton–proton coupling constants and NOE spectroscopy. Mong *et al.*[72] studied α-glycosylation reaction for 2-deoxy- and 2,6-dideoxyglycosyl donor based on DMF modulation concept. Using ^{1}H NMR results, they observed 2-deoxyglycosyl imidinium adduct at −40 °C and below. Shao *et al.*[73] presented the synthesis of 2,2-disubstituted perhydro[2,3-b]-pyran derivatives. Stereochemistry of these compounds was determined by X-ray crystallographic analysis and confirmed by extensive analysis of the ^{1}H NMR and NOE results.

Guo and coworkers,[74] presented synthesis of Leonisides E and F derived from *Leonurus japonicas*. NMR experiments confirmed the structure of the title trisaccharides. Schobert and Loscher,[75] presented first total synthesis of epicoccamide D. Epicoccamides A–D represent a small family of hybrid poliketide–carbohydrate–amino acid metabolites produced by various endofungal *Epicoccum* species of marine and terrestrial origin. Eighteen step synthesis lead to the desired compound with 17% of yield. Fernandes *et al.*[76] synthesized and studied cyclopentadienyl ruthenium(II) complexes containing carbohydrate-derived ligands. Scanlan and Malone,[77] applied radical cyclization for the synthesis of thiosugars. The structure and stereochemistry of these compounds were determined by 1D and 2D NMR techniques.

Glicosphingolipids are very important membrane components of eukaryotic cells and play a fundamental role in various stages of the cell cycle. Recently, the synthesis of triazol-based neoglycolipids was published by Rivera and coworkers.[78] It was based on Ugi-4CR/CuAAC combination and this type of approach was not yet explored for synthesis of triazol-linked carbohydrate–lipid conjugates. NMR analysis of the triazol-based glycolipids at room temperature showed the occurrence of two conformers and dynamic ^{1}H NMR experiment allowed to evaluate the free energy of activation of the isomerization of the *cis*/*trans* amide bond.

Hu and coworkers,[79] presented stereoselective synthesis of a sulfated tetrasaccharide corresponding to a rare sequence in the galactofucan isolated from the brown alga *Sargassum polycystum*. ^{1}H and ^{13}C signals of the synthesized molecule were fully assigned on the basis of 2D NMR techniques such as COSY, TOCSY, HSQC, HSQC-TOCSY, HMBC, and ROESY. Greatrex *et al.*[80] presented synthesis of *allo* and *epi*-inositol based on NHC-catalyzed carbocyclization of carbohydrate-derived dialdehydes. This type of aldehyde based on mannitol had characteristic symmetry in the NMR and showed the single aldehyde resonance at 9.70 ppm. After 24 h, the resonance assigned disappeared and was replaced with a complex pattern of resonances tentatively assigned to the products of hydration shown in Fig. 7.

6.2.2 Enzymatic mode. This chapter reports a selection of recent publications on the enzymatic synthesis and modification of carbohydrates. It is known that enzymes are capable of synthesizing complex carbohydrates from simple monosaccharides. Enzymatic synthesis is a very powerful tool in the modern carbohydrates chemistry. Mayer and coworkers[81] isolated two exopolysaccharides produced by *Lactobacillus johnsonii* FI9785, and the structures of the obtained EPS-1 and EPS-2 were determined using a combination of two-dimensional NMR

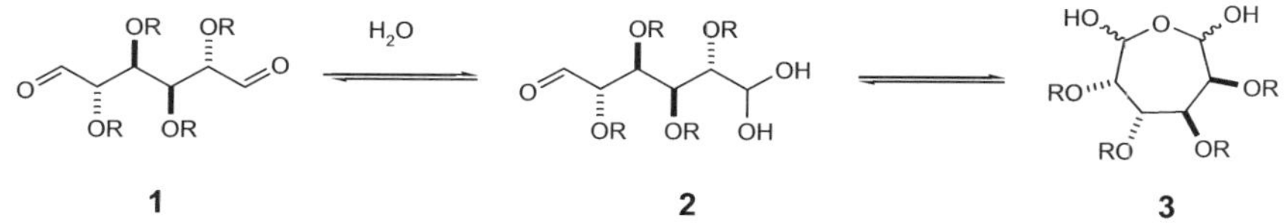

Fig. 7 Hydration of dialdehyde.

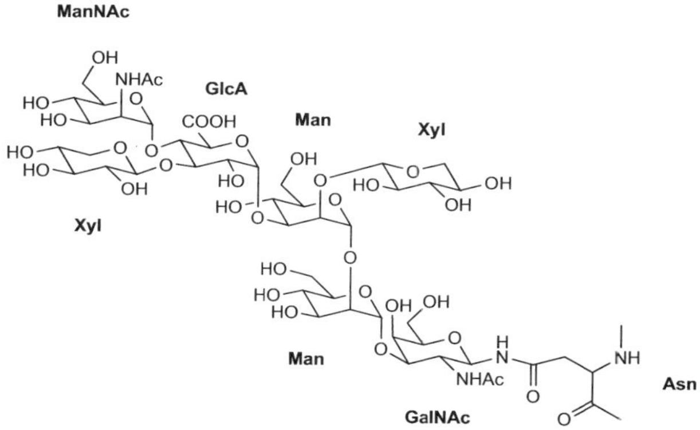

Fig. 8 Structure of *N*-linked heptasaccharide produced by oligosaccharyltransferase of the euryarchaeon, *Pyrococcus furiosus*.

methods: COSY, TOCSY, HSQC, HSQC-TOCSY to assign the ^{1}H and ^{13}C chemical shifts. Wilson *et al.*[82] performed large scale synthesis of a nematode *N*-glycan core *in vitro* using three types of fucosylotransferases. The nature of glycosidic linkages was determined by NMR. Kohda *et al.*[83] studied protein *N*-glycosylation reaction by oligosaccharyltransferase from the hyperthermophilic archaeon. The identities of the monosaccharides were deduced from the homonuclear ^{3}J coupling constants and ^{1}H chemical shifts in the 1D slices of the 2D TOCSY spectrum. The position of the glycosidic linkages were deduced using the inter-residue ^{1}H–^{1}H NOE from the anomeric protons in the 2D NOESY spectrum. The structure of the *N*-linked heptasaccharide glycan is shown in Fig. 8.

Guerardel *et al.*[84] conducted an investigation of the enzymatic properties of Bmt1 (β-mannosyltransferase 1) using NMR spectroscopy. They converted mantyl-tagged α-oligomannosides (DP3M and DP4M) to terminal β-1,2-manno product (DP5M and DP6M) using Bmt1 protein. A thorough analysis of ^{1}H–^{1}H TOCSY spectrum allowed to confirm the β-manno configuration. The NMR approach was used also by Biely *et al.*[85] for the monitoring of the rate of specific deacetylation processes for acetylxylan esterases (AcXEs). Analysis of ^{1}H NMR spectrum showed that GH10 endoxylanase efficiently deacetylates singly 2-*O*- and 3-*O*-acetylated Xyl*p* and also 2,3-di-O-acetylated Xyl*p* residues in acetyl glucuroxylan. Yoon *et al.*[86] presented biosynthesis of the 12 membered macrolide antibiotic YC-17. The authors, using 1D and 2D NMR techniques, reported fully assigned spectroscopic data for all four glycosylated derivatives. The problem of lactose intolerance and galactooligosaccharides

(GOS) production was intensively explored. The researchers were searching for commonly available and efficient β-galactosides. Polaina and coworkers,[87] reported efficient method for the hydrolysis of lactose and the biosynthesis of GOS by using recombinant β-galactosidase from *Thermotoga maritima*. The reaction mixture was very complex and the concentration of synthesized oligosaccharides was lower than 10%. Desired compounds were defined by means of 2D-HSQC spectra combined with selective COSY and TOCSY 1D NMR experiments.

6.3 Sugars isolated from natural sources

Every year a huge number of new carbohydrates are isolated from natural sources. The wealth of nature seems to be inexhaustible. Scientists around the world obtain a variety of sugar fragments from various bacteria strains and their respective organelle, as well as from the animal, plant and fungus kingdoms, and determine their chemical structures by means of NMR spectroscopy. The papers devoted to prokaryotes and eukaryotes are collected in Tables 1 and 2, respectively and as in the previous chapters the names of organisms and the general type of carbohydrate are included.

6.3.1 Carbohydrates obtained from prokaryotes. A number of carbohydrate structures from prokaryotic organisms were determined by NMR spectroscopy and often supported by other spectroscopic and spectrometric methods. In the Table 1 general types of carbohydrate are given together with the bacterial strains arranged in alphabetical order.

Table 1 Prokaryotic cells saccharides whose structures have been evaluated with the aid of NMR spectroscopy.

Polysaccharide type	Bacterial strain	Ref.
Capsular polysaccharide	*Acinetobacter baumannii* ACICU	88
O-Polysaccharide	*Acinetobacter baumannii* LAC-4	89
Exopolysaccharide	*Acinetobacter junii* BB1A	90
Lipopolysaccharide	*Aeromonas veronii Bs19* O16	91
O-Specific polysaccharide	*Azospirillum brasilense* Sp7	92
Teichoic acid	*Bacillus amyloliquefaciensis* subsp., *B. amyloliquefaciensis* VKM B-2528, *B. mojavensis* VKM B-2650, *B. sonerensis* VKM B-2652	93
Exopolysaccharide	*Bifidobacterium animalis* ssp. *lactis* AD011	94
O-Polysaccharide	*Burkholderia mallei, B. pseudomallei*	95
O-Antigen	*Burkholderia olkahomensis* E0147	96
Polysaccharide DHPD2	*Dendrobium huoshanense*	97
O-Polysaccharide with di-*N*-acetyllegionamic acid	*Enterobacter cloacae* C6285	98
O-Polysaccharide	*Enterobacter cloacae* G2277	99
O-Polysaccharide	*Escherichia coli* O30	100
O-Polysaccharide	*Escherichia coli* O39	101
O-Polysaccharide	*Escherichia coli* O91	102
Exopolysaccharide	*Gluconacetobacter diazotrophicusi* Pal5	103
Phosphonoglycans	*Glycomyces* sp. NRRL B-16210	104
Lipopolysaccharide	*Hafnia alvei* PCM 2670	105

Table 1 **(*Continued*)**

Polysaccharide type	Bacterial strain	Ref.
Polysaccharide capsule and exopolysaccharide	*Kingella kingae*	106
Exopolysaccharide	*Lactobacillus plantarum* 70810	107
Oligosaccharide	*Moraxella bovis* Epp63	108
Antagonist LPS	*Oscillatoria planktothrix* FP1	109
Lipopolysaccharide	*Pasteurella multocida*	110
O-Specific polysaccharide	*Plesiomonas shigelloides* O33:H3	111
Kdo-containing *O*-polysaccharide	*Proteus* O79	112
O-Polysaccharide	*Providencia alcalifaciens* O33	113
O-Polysaccharide (yersiniose A)	*Pseudomonas mandelii* CYar1	114
Asn-linked oligosaccharide	*Pyrococcus furiosus*	115
O-Specific polysaccharide containing D- and L-2-acetamido-2-deoxy-GalA	*Rheinheimera pacifica* KMM 1406^T	116
O-Polysaccharide	*Salmonella typhimurium*	117
O-Polysaccharide	*Salmonella* Uccle (O:3,54)	118
Heteropolysaccharide (ECAntigen)	*Shigella sonnei*	119
Phosphonoglycans	*Stackebrandtia nassauensis* NRRL B-16338	104
Capsular polysaccharide	*Streptococcus pneumonia* 47A	120
Capsular polysaccharide	*Streptococcus pneumonia* 33B, 33C, 33D	121
Exopolysaccharide	*Vibrio choleraee* O1	122

Table 2 Eukaryotic saccharides whose structures have been evaluated with the aid of NMR spectroscopy.

Polysaccharide type	Organism (organ)	Ref.
Animals		
Hyaluronic acid, chondroitin/dermatan sulfate	*Cyclopterus lumpus* (lumpsucker fish)	123
Anhydro sugar	*Plakinastrella clatharata* (marine sponge)	124
Oligosaccharides	*Phascolarctos cinereus* (koala milk)	125
Oligosaccharides	*Rangifer tarandus tarandus* (reindeer milk)	126
Fucoidan composed of tetrafucose repeating unit	*Thelenota ananas* (sea cucumber)	127
Fungi		
N-glycan ($Man_{4-18}GlcNAc_2$)	*Ashbya gossypii*	128
Galactomannan	*Aspergillus ochraceus*	129
(1→3,1→6) linked β-glucan	*Candida albicans*	130
Exopolysaccharide	*Penicillium solitum*	131
Linear (1→3) α-D-Glc*p* and branched β-D-Glc*p*	*Ramaria botrytis*	132
Mushrooms		
Heteropolysaccharide	*Boletus edulis*	133
α-(1→6) D-Glc*p* units with β-(1→3)-linked branches	*Cantharellus cibarius*	134
Heterogalactan	*Ganoderma atrum*	135
Oligosaccharides	*Ganoderma lucidum*	136
Glucan	*Lentinula edodes*	137
Glucan	*Pleurotus florida*	137
Hyper-branched cell wall polysaccharide	*Pleurotus tuber-regium*	138

Table 2 (*Continued*)

Polysaccharide type	Organism (organ)	Ref.
Algae and seaweed		
Highly branched (1→3,1→6)-β-D-glucan	*Isochrysis galbana* (microalga)	139
Laminaran	*Eisenia bicyclis* (brown alga)	140
Sulfated xylofucan	*Punctaria plantaginea* (brown alga)	141
Heteropolysaccharide	*Sargassum fusiforme* (brown alga)	142
Polysaccharide	*Sargassum* sp. (brown alga)	143
Kappa/beta-carraginan	*Ahnfeltiopsis flabelliformis* (red alga)	144
Fucoidan	*Fucus vesiculosus* (brown seaweed)	145
Fucans (sulfated polysaccharides)	*Cystoseira barbata* (brown seaweed)	146
Sulfated galactans	*Gracilaria fisheri* (red seaweed)	147
Sulfated polysaccharides	*Caulerpa cupressoides* var. *Lycopodium* (green seaweed)	148
Plants		
Pectinic polysaccharides	*Amaranthus cruentus* (aerial part)	149
Arabinan-type polysaccharides	Apple pomace (waste)	150
Type II arabinogalactan	*Artemisia absinthium, A. vulgaris* (aerial part)	151
Pectic polysaccharides	*Codonopsis pilosula* (dry powder)	152
Xylan	*Cudrania tricuspidata* (root)	153
Hemicellulose	*Dendrobium huoshanense* (protocorm-like body)	154
Water-soluble polysaccharide	*Dimocarpus longan* (pulp)	155
Pectic polysaccharides and arabinogalactan proteins	*Heracleum sosnowskyi* Manden. (aerial part)	156
Pectic polysaccharide	*Mosla chinensis* Maxim. cv. Jiangxiangru (powder of whole plant)	157
Oligosaccharides	*Prosopis juliflora* (seed)	158, 159
Pectic polysaccharide	*Prunus domestica* L. (plum without stone)	160
Rhamnogalactouronan with arabinogalactans as side chains	*Prunus domestica* L. (prunes)	161
Five natural carbohydrates	*Sauropus rostratus* (leaves)	162

6.3.2 Carbohydrates obtained from eukaryotes. General type of carbohydrates isolated from eukaryotes and evaluated with the aid of NMR spectroscopy are collected in Table 2. The organisms are divided into subsequent groups: animals, fungi, mushrooms, marine algae including red, brown and green algae, and finally plants.

Abbreviations

Monosaccharides

Ara	arabinose
Fuc	fucose
Gal	galactose
Glc	glucose
GlcA	glucuronic acid
GlcNAc	*N*-acetylglucosamine
IdoA	iduronic acid

Kdo 2-keto-3-deoxyoctonate
Man mannose
Qui quinovose
Rha rhamnose
Xyl xylose
Italic *f* or *p* sufixes indicate a ring size respectively: *f* for furanose (five membered rign), and *p* for pyranose (six-membered ring). The D and L notation refer to the relative configuration of sugars.

Amino acids
Asn asparagine
Cys cysteine
His histidine
Tyr tyrosine

References

1 K. Furihata, J. Zhang, M. Koda, T. Miyakawa and M. Tanokura, *Magn. Reson. Chem.*, 2014, **52**, 333.
2 Y. Zhang, T. Yamaguchi and K. Kato, *Chem. Lett.*, 2013, **42**, 1455.
3 R. Pendrill, O. W. Sørensen and G. Widmalm, *Magn. Reson. Chem.*, 2014, **52**, 82.
4 J. K. T. Ng, Z. D. Zujovic, B. G. Smith, J. W. Johnston and R. Schröder, *Carbohydr. Res.*, 2014, **386**, 1.
5 F. V. Toukach and V. P. Ananikov, *Chem. Soc. Rev.*, 2013, **42**, 8376.
6 K. S. Egorova and P. V. Toukach, *Carbohydr. Res.*, 2014, **389**, 112.
7 J. Rönnols, R. Pendrill, C. Fontana, C. Haark, T. Angels d'Ortoli, O. Engström, J. Ståhle, M. V. Zaccheus, E. Säwén, L. E. Hahn, S. Iqbal and G. Widmalm, *Carbohydr. Res.*, 2013, **380**, 156.
8 C. Fontana, M. Lundbord, A. Weintraub and G. Widmalm, *Glycobiology*, 2014, **24**, 450.
9 S. A. Samsonov, S. Theisgen, T. Riemer, D. Huster and M. T. Pisabarro, *BioMed Res. Int.*, 2014, **2014**, 808071.
10 M. Bøjstrup, B. O. Petersen, S. R. Beeren, O. Hindsgaul and S. Meier, *Anal. Chem.*, 2013, **85**, 8802.
11 B. O. Petersen, O. Hindsgaul and S. Meier, *Analyst*, 2014, **139**, 401.
12 G. L. Sassaki, M. Guerrini, R. V. Serrato, A. P. S. Filho, J. Carlotto, F. Simas-Tosin, T. R. Cipriani, M. Iacomini, G. Torri and P. A. J. Gorin, *Carbohydr. Polym.*, 2014, **104**, 34.
13 T. Delgado-Goñi, S. Campo, J. Martín-Sitjar, M. E. Cabañas, B. San Segundo and C. Arús, *Planta*, 2013, **238**, 397.
14 C. Quan, *Food Chem.*, 2014, **153**, 378.
15 Y. B. Monakhova, B. Schütz, H. Schäfer, M. Spraul, T. Kuballa, H. Hahn and D. W. Lachenmeier, *Accredit. Qual. Assur.*, 2014, **19**, 17.
16 B. O. Petersen, M. Nilsson, M. Bøjstrup, O. Hindsgaul and S. Meier, *Food Chem.*, 2014, **150**, 65.
17 J. Zeng, G. L. Helms, X. Gao and S. Chen, *J. Agric. Food Chem.*, 2013, **61**, 10848.
18 C.-M. Popescu, D. E. Demco and M. Möller, *Polym. Degrad. Stab.*, 2013, **98**, 2730.
19 A. Alekseeva, B. Casu, G. Cassinelli, M. Guerrini, G. Torri and A. Naggi, *Anal. Bioanal. Chem.*, 2014, **406**, 249.
20 M. Mayer and B. Meyer, *Angew. Chem., Int. Ed.*, 1999, **38**, 1784.

21 C. A. Bawley and S. Shahzad-ul-Hussan, *Biopolymers*, 2013, **99**, 796.
22 M. Nagae and Y. Yamaguchi, *Int. J. Mol. Sci.*, 2014, **15**, 3768.
23 M. Nagae, K. Soga, K. Morita-Matsumoto, S. Hanashima, A. Ikeda, K. Yamamoto and Y. Yamaguchi, *Glycobiology*, 2014, **24**, 368.
24 M. Rauschenberg, S. Bandaru, M. P. Waller and B. J. Ravoo, *Chem. – Eur. J.*, 2014, **20**, 2770.
25 T. Ohnuma, N. Umemoto, T. Nagata, S. Shinya, T. Numata, T. Taira and T. Fukamizo, *Biochim. Biophys. Acta, Proteins Proteomics*, 2014, **1844**, 793.
26 J. M. Grondin, S. Chitayat, E. Ficko-Blean, S. Houliston, C. H. Arrowsmith, A. B. Boraston and S. P. Smith, *J. Mol. Biol.*, 2014, **426**, 869.
27 D. Ponader, P. Maffre, J. Aretz, D. Pussak, N. M. Ninnemann, S. Schmidt, P. H. Seeberger, C. Rademacher, G. U. Nienhaus and L. Hartmann, *J. Am. Chem. Soc.*, 2014, **136**, 2008.
28 C. Guzzi, J. C. Muñoz-García, P. M. Enriquez-Navas, J. Rojo, J. Angulo and P. M. Nieto, *Pure Appl. Chem.*, 2013, **85**, 1771.
29 F. Yu, S. Roy, E. Arevalo, J. Schaeck, J. Wang, K. Holte, J. Duffner, N. S. Gunay, I. Capila and G. V. Kaundinya, *Anal. Bioanal. Chem.*, 2014, **406**, 3079.
30 C. Viskov, S. Elli, E. Urso, D. Gaudesi, P. Mourier, F. Herman, C. Boudier, B. Casu, G. Torri and M. Guerrini, *J. Biol. Chem.*, 2013, **288**, 25895.
31 K. M. Poluri, P. R. B. Joseph, K. V. Sawant and K. Rajarathnam, *J. Biol. Chem.*, 2013, **288**, 25143.
32 M. D. Battistel, H. F. Azurmendi, B. Yu and D. I. Freedberg, *Prog. Nucl. Magn. Reson. Spectrosc.*, 2014, **79**, 48.
33 V. H. Pomin, *Prog. Biophys. Mol. Biol.*, 2014, **114**, 61.
34 A. Coletti, S. Elli, E. Macchi, P. Galzerano, L. Zamani, M. Guerrini, G. Torri and E. Vismara, *Carbohydr. Res.*, 2014, **389**, 134.
35 J. Rahkila, F. S. Ekholm, R. Panchadhayee, A. Ardá, F. J. Cañada, J. Jiménez-Barbero and R. Leino, *Carbohydr. Res.*, 2014, **383**, 58.
36 J. Rönnols, S. Manner, A. Siegbahn, U. Ellervik and G. Widmalm, *Org. Biomol. Chem.*, 2013, **11**, 5465.
37 A. Sarkar, C. Fontana, A. Imberty, S. Pérez and G. Widmalm, *Macromolecules*, 2013, **14**, 2215.
38 R. Pendrill, E. Säwén and G. Widmalm, *J. Phys. Chem. B*, 2013, **117**, 14709.
39 D. S. Patel, R. Pendrill, S. S. Mallajsyula, G. Widmalm and A. D. MacKerell Jr., *J. Phys. Chem. B*, 2014, **118**, 2851.
40 P. Vidal, V. Roldós, M. Del Carmen Fernández-Alonso, B. Vauzeilles, Y. Bleriot, F. J. Cañada, S. André, H.-J. Gabius, J. Jiménez-Barbero, J. F. Espinosa and S. Martín-Santamaría, *Chem. – Eur. J.*, 2013, **19**, 14590.
41 T. A. Jackson, V. Robertson and F.-I. Auzanneau, *J. Med. Chem.*, 2014, **57**, 817.
42 M. Zierke, M. Smieško, S. Rabbani, T. Aeschbacher, B. Cutting, F. H.-T. Allain, M. Schubert and B. Ernst, *J. Am. Chem. Soc.*, 2013, **135**, 13464.
43 A. Novoa, S. Barluenga, C. Serba and N. Winssinger, *Chem. Commun.*, 2013, **49**, 7608.
44 C. Pöhner, V. Ullmann, R. Hilpert, E. Samain and C. Unverzagt, *Tetrahedron Lett.*, 2014, **55**, 2197.
45 K.-S. Krannig, A. Doriti and H. Schlaad, *Macromolecules*, 2014, **47**, 2536.
46 L. Taylor, X. Chen and N. Ayres, *Polym. Int.*, 2014, **63**, 127.
47 N. Varga, I. Sutkeviciute, R. Ribeiro-Viana, A. Berzi, R. Ramdasi, A. Daghetti, G. Vettoretti, A. Amara, M. Clerici, J. Rojo, F. Fieschi and A. Bernardi, *Biomaterials*, 2014, **35**, 4175.

48 P. Rajakumar and R. Anandhan, *Synth. Commun.*, 2013, **43**, 2217.
49 A. M. Marzaioli, E. Bedini, R. Lanzetta, M. Parrilli and C. De Castro, *Carbohydr. Res.*, 2014, **386**, 57.
50 J. Rahkila, F. S. Ekholm, R. Panchadhayee, A. Ardá, F. J. Cañada, J. Jiménez-Barbero and R. Leino, *Carbohydr. Res.*, 2014, **383**, 58.
51 X. Liu, W. Xia, Q. Jiang, Y. Xu and P. Yu, *J. Agric. Food Chem.*, 2014, **62**, 297.
52 I. Molina-Pinilla, M. Bueno-Martínez, K. Hakkou and J. A. Galbis, *J. Polym. Sci., Part A: Polym. Chem.*, 2014, **52**, 629.
53 Y.-C. Huang and J.-S. Jan, *Polymer*, 2014, **55**, 540.
54 K. Wang and Q. Liu, *Carbohydr. Res.*, 2014, **386**, 48.
55 A. Préchoux and W. Helbert, *Carbohydr. Polym.*, 2014, **101**, 864.
56 G. Kök, K. Ay, E. Ay, F. Doğan and İ. Kaya, *Carbohydr. Polym.*, 2014, **101**, 324.
57 E. Cho, H. Kim, J. E. Yang, B.-H. Jun, S. R. Paik and S. Jung, *Carbohydr. Res.*, 2014, **391**, 37.
58 M. Pintal, B. Kryczka, A. Marsura and S. Porwański, *Carbohydr. Res.*, 2014, **386**, 18.
59 M. W. Loranger, S. M. Forget, N. E. Mccormick, R. T. Syvitski and D. L. Jakeman, *J. Org. Chem.*, 2013, **78**, 9822.
60 J. Ramos-Soriano, U. Niss, J. Angulo, M. Angulo, A. J. Moreno-Vargas, A. T. Carmona, S. Ohlson and I. Robina, *Chem. – Eur. J.*, 2013, **19**, 17989.
61 D. Doknic, M. Abramo, I. Sutkeviciute, A. Reinhardt, C. Guzzi, M. K. Schlegel, D. Potenza, P. M. Nieto, F. Fieschi, P. H. Seeberger and A. Bernardi, *Eur. J. Org. Chem.*, 2013, 5303.
62 L. Cui, C.-C. Ling, J. Sadowska and D. R. Bundle, *Carbohydr. Res.*, 2014, **383**, 1.
63 J. M. Vega-Pérez, C. Palo-Nieto, M. Vega-Holm, P. Góngora-Vargas, J. M. Calderón-Montaño, E. Burgos-Morón, M. López-Lázaro and F. Iglesias-Guerra, *Eur. J. Med. Chem.*, 2013, **70**, 380.
64 G. Gu, L. An, M. Fang and Z. Guo, *Carbohydr. Res.*, 2014, **383**, 21.
65 D. Kumar, K. B. Mishra, B. B. Mishra, S. Mondal and V. K. Tiwari, *Steroids*, 2014, **80**, 71.
66 L. Guazzelli, G. Catelani, F. D'Andrea and T. Gragnani, *Carbohydr. Res.*, 2014, **388**, 44.
67 A. Biela-Banaś, E. Gallienne and O. R. Martin, *Carbohydr. Res.*, 2013, **380**, 23.
68 J. P. Issa and C. S. Bennett, *J. Am. Chem. Soc.*, 2014, **136**, 5740.
69 J. P. Colomer, V. E. Manzano and O. Varela, *Eur. J. Chem.*, 2013, 7343.
70 H.-Y. Chuang, C.-T. Ren, C.-A. Chao, C.-Y. Wu, S. S. Shivatare, T.-J. R. Cheng, C.-Y. Wu and C.-H. Wong, *J. Am. Chem. Soc.*, 2013, **135**, 11140.
71 B. Dayde, C. Pierra, G. Gosselin, D. Surleraux, A. T. Ilagouma, C. Laborde, J.-N. Volle, D. Virieux and J.-L. Pirat, *Eur. J. Org. Chem.*, 2014, 1333.
72 J.-H. Chen, J.-H. Ruei and K.-K. T. Mong, *Eur. J. Org. Chem.*, 2014, 1827.
73 X. Ma, Q. Tang, J. Ke, J. Zhang, C. Wang, H. Wang, Y. Li and H. Shao, *Chem. Commun.*, 2013, **49**, 7085.
74 G. Gu, Y. Zhao and Z. Guo, *Carbohydr. Res.*, 2013, **380**, 174.
75 S. Loscher and R. Schobert, *Chem. – Eur. J.*, 2013, **19**, 10619.
76 P. Florindo, I. J. Marques, C. D. Nunes and A. C. Fernandes, *J. Organomet. Chem.*, 2014, **760**, 240.
77 A. Malone and E. M. Scanlan, *J. Org. Chem.*, 2013, **78**, 10917.
78 K. Pérez-Labrada, I. Brouard, I. Méndez, C. S. Pérez, J. A. Gavín and D. G. Rivera, *Eur. J. Org. Chem.*, 2014, 3671.
79 J. Zhou, L. Yang and W. Hu, *J. Org. Chem.*, 2014, **79**, 4718.
80 K. P. Stockton, B. W. Greatrex and D. K. Taylor, *J. Org. Chem.*, 2014, **79**, 5088.
81 E. Dertli, I. J. Colquhoun, A. P. Gunning, R. J. Bongaerts, G. Le Gall, B. B. Bonev, M. J. Mayer and A. Narbad, *J. Biol. Chem.*, 2013, **288**, 31938.

82 S. Yan, S. Serna, N. C. Reichardt, K. Paschinger and I. B. H. Wilson, *J. Biol Chem.*, 2013, **288**, 21015.
83 D. Fujinami, M. Matsumoto, T. Noguchi, K. Sonomoto and D. Kohda, *Carbohydr. Res.*, 2014, **387**, 30.
84 E. Fabre, G. Sfihi-Loualia, M. Pourcelot, B. Coddeville, F. Krzewinski, J. Bouckaert, E. Maes, T. Hurtaux, R. Dubois, C. Fradin, J.-M. Mallet, D. Poulain, F. Delplace and Y. Guerardel, *Biochem. J.*, 2014, **457**, 347.
85 P. Biely, M. Cziszárová, I. Uhliariková, J. W. Agger, X.-L. Li, V. G. H. Eijsink and B. Westereng, *Biochim. Biophys. Acta*, 2013, **1830**, 5075.
86 P. B. Shinde, A. R. Han, J. Cho, S. R. Lee, Y. H. Ban, Y. J. Yoo, E. J. Kim, E. J. Kim, M. Song, J. W. Park, D. G. Lee and Y. J. Yoon, *J. Biotechnol.*, 2013, **168**, 142.
87 J. Marín-Navarro, D. Talens-Perales, A. Oude-Vrielink, F. J. Cañada and J. Polaina, *World J. Microbiol. Biotechnol.*, 2014, **30**, 989.
88 S. N. Senchenkova, A. S. Shashkov, M. M. Shneider, N. P. Arbatsky, A. V. Popova, K. A. Miroshnikov, N. V. Volozhantsev and Y. A. Knirel, *Carbohydr. Res.*, 2014, **391**, 89.
89 E. Vinogradov, L. MacLean, H. H. Xu and W. Chen, *Carbohydr. Res.*, 2014, **390**, 42.
90 I. K. Sen, A. K. Mandal, R. Chakraborty, B. Behera, K. K. Yadav, T. K. Maiti and S. S. Islam, *Carbohydr. Polym.*, 2014, **101**, 188.
91 A. Turska-Szewczuk, K. A. Duda, D. Schwudke, A. Pekala, A. Kozinska and O. Holst, *Mar. Drugs*, 2014, **12**, 1298.
92 E. N. Sigida, Y. P. Fedonenko, A. S. Shashkov, E. L. Zdorovenko, S. A. Konnova, V. V. Ignatov and Y. A. Knirel, *Carbohydr. Res.*, 2013, **380**, 76.
93 N. V. Potekhina, A. S. Shashkov, G. M. Streshinskaya, E. M. Tul'skaya, Y. I. Kozlova, S. N. Senchenkova, E. B. Kudryashova and L. I. Evtushenko, *Microbiology*, 2013, **82**, 579.
94 M. Alhudhud, P. Humphreys and A. Laws, *J. Microbiol. Methods*, 2014, **100**, 93.
95 C. Heiss, M. N. Burtnick, R. A. Roberts, I. Black, P. Azadi and P. J. Brett, *Carbohydr. Res.*, 2013, **381**, 6.
96 J. K. Stone, C. Heiss, Z. Wang, I. Black, S. A. Grasso, A. T. Koppisch, P. Azadi, P. Keim and A. Tuanyok, *Carbohydr. Res.*, 2014, **386**, 68.
97 X.-L. Li, J.-J. Xiao, X.-Q. Zha, L.-H. Pan, M.-N. Asghar and J.-P. Luo, *Carbohydr. Polym.*, 2014, **106**, 247.
98 A. V. Filatov, M. Wang, W. Wang, A. V. Perepelov, A. S. Shashkov, L. Wang and Y. A. Knirel, *Carbohydr. Res.*, 2014, **392**, 21.
99 A. V. Perepelov, M. Wang, A. V. Filatov, X. Guo, A. S. Shashkov, L. Wang and Y. A. Knirel, *Carbohydr. Res.*, 2014, **387**, 10.
100 A. V. Perepelov, X. Hao, S. N. Senchenkova, A. S. Shashkov, B. Liu, L. Wang and Y. A. Knirel, *Carbohydr. Res.*, 2014, **389**, 196.
101 A. V. Perepelov, A. V. Filatov, Q. Wang, V. L. L'vov, Y. Qian, A. S. Shashkov, L. Wang and Y. A. Knirel, *Carbohydr. Res.*, 2014, **388**, 30.
102 M. Soltesova, J. Kowalewski and G. Widmalm, *J. Biomol. NMR*, 2013, **57**, 37.
103 R. V. Serrato, C. H. S. G. Meneses, M. S. Vidal, A. P. Santana-Filho, M. Iacomini, G. L. Sassaki and J. I. Baldani, *Carbohydr. Polym.*, 2013, **98**, 1153.
104 X. Yu, N. P. J. Prince, B. S. Evans and W. W. Metcalf, *J. Bacteriol.*, 2014, **196**, 1768.
105 E. Bobko, M. Tyras and W. Jachymek, *Carbohydr. Res.*, 2013, **378**, 71.
106 K. F. Starr, E. A. Porsch, C. Heiss, I. Black, P. Azadi and J. W. St. Geme III, *PLoS One*, 2013, **8**, e75409.

107 K. Wang, W. Li, X. Rui, X. Chen, M. Jiang and M. Dong, *Int. J. Biol. Macromol.*, 2014, **67**, 71.
108 C. De Castro, I. D. Grice, T.-M. Daal, I. R. Peak, A. Molinaro and J. C. Wilson, *Carbohydr. Res.*, 2014, **388**, 81.
109 S. Carillo, G. Pieretti, E. Bedini, M. Parrilli, R. Lanzetta and M. M. Corsaro, *Carbohydr. Res.*, 2014, **388**, 73.
110 M. Harper, F. St Michael, M. John, E. Vinogradov, J. A. Steen, L. Van Dorsten, J. A. Steen, C. Turni, P. J. Blackall, B. Adler, A. D. Cox and J. D. Boyce, *J. Bacteriol.*, 2013, **195**, 4854.
111 G. Nestor, J. Lukasiewicz and C. Sandström, *Eur. J. Org. Chem.*, 2014, 1241.
112 N. P. Arbatsky, D. Drzewiecka, A. Palusiak, A. S. Shashkov, A. Zabłotni, M. Siwińska and Y. A. Knirel, *Carbohydr. Res.*, 2013, **379**, 100.
113 O. G. Ovchinnikova, A. S. Shashkov, A. O. Chizhov, M. Moryl, A. Rozalski and Y. A. Knirel, *Carbohydr. Res.*, 2014, **390**, 67.
114 A. N. Kondakova, M. S. Drutskaya, A. S. Shashkov, S. A. Nedospasov, V. N. Akimov, N. P. Arbatsky, V. F. Galchenko and Y. A. Knirel, *Carbohydr. Res.*, 2013, **381**, 138.
115 D. Fujinami, M. Matsumoto, T. Noguchi, K. Sonomoto and D. Kohda, *Carbohydr. Res.*, 2014, **387**, 30.
116 N. A. Komandrova, M. S. Kokoulin, A. I. Kalinovsky, S. V. Tomshich, L. A. Romanenko and V. E. Vaskovsky, *Carbohydr. Res.*, 2014, **394**, 1.
117 F. Micoli, N. Ravenscroft, P. Cescutti, G. Stefanetti, S. Londero, S. Rondini and C. A. MacLennan, *Carbohydr. Res.*, 2014, **385**, 1.
118 M. Paszkiwicz, E. Tokarska-Pietrzak, M. Gołębiowski, D. Kunikowska and P. Stepnowski, *J. Struct. Biol.*, 2013, **184**, 367.
119 T. K. Gozdziewicz, C. Lugowski and J. Lukasiewicz, *J. Biol. Chem.*, 2014, **289**, 2745.
120 B. O. Petersen, O. Hindsgaul, B. S. Paulsen, A. R. Redondo and I. C. Skovsted, *Carbohydr. Res.*, 2014, **386**, 62.
121 F. L. Lin, E. Vinogradov, C. Deng, S. Zeller, L. Phelan, B. A. Green, K. U. Jansen and V. Pavliak, *Carbohydr. Res.*, 2014, **383**, 97.
122 F. Yildiz, J. Fong, I. Sadovskaya, T. Grard and E. Vinogradov, *PLoS One*, 2014, **9**, e86751.
123 C. G. Panagos, D. Thomson, C. Moss, C. D. Bavington, H. G. Ólafsson and D. Uhrín, *Carbohydr. Polym.*, 2014, **106**, 25.
124 P. L. Katavic, K. W. L. Yong, J. N. Herring, M. A. Deseo, J. T. Blanchfield, V. Ferro and M. J. Garson, *Tetrahedron*, 2013, **69**, 8074.
125 T. Urashima, E. Taufik, R. Fukuda, T. Nakamura, K. Fukuda, T. Saito and M. Messer, *Glycoconjugate J.*, 2013, **30**, 801.
126 E. Taufik, K. Ganzorig, M. Nansamaa, R. Fukuda, K. Fukuda, T. Saito and T. Urashima, *Int. Dairy J.*, 2014, **34**, 104.
127 L. Yu, C. Xue, Y. Chang, X. Xu, L. Ge, G. Liu and Y. Wang, *Food Chem.*, 2014, **146**, 113.
128 T. Q. Aguiar, H. Maaheimo, A. Heiskanen, M. G. Wiebe, M. Penttilä and L. Domingues, *Carbohydr. Res.*, 2013, **381**, 19.
129 S. Guo, W. Mao, M. Yan, C. Zhao, N. Li, J. Shan, C. Lin, X. Liu, T. Guo, T. Tuo and S. Wang, *Carbohydr. Polym.*, 2014, **105**, 325.
130 D. W. Lowman, R. R. Greene, D. W. Bearden, M. D. Kruppa, M. Pottier, M. A. Monteiro, D. V. Soldatov, H. E. Ensley, S.-C. Cheng, M. G. Netea and D. L. Williams, *J. Biol. Chem.*, 2014, **289**, 3432.
131 M. Yan, W. Mao, C. Chen, X. Kong, Q. Gu, N. Li, X. Liu, B. Wang, S. Wang and B. Xiao, *Carbohydr. Polym.*, 2014, **111**, 485.

132 S. K. Bhanja, D. Rout, P. Patra, I. K. Sen, C. K. Nandan and S. S. Islam, *Bioact. Carbohydr. Diet. Fibre*, 2014, **3**, 52.
133 A.-q. Zhang, Y. Liu, N.-n. Xiao, Y. Zhang and P.-l. Sun, *Food Chem.*, 2014, **146**, 334.
134 A. Villares, A. García-Lafuente, E. Guillamón and L. Mateo-Vivaracho, *Bioact. Carbohydr. Diet. Fibre*, 2013, **2**, 15.
135 H. Zhang, S.-P. Nie, J.-Y. Yin, Y.-X. Wang and M.-Y. Xie, *Food Hydrocolloids*, 2014, **36**, 339.
136 I. Tello, V. Campos-Pena, E. Montiel, V. Rodriguez, A. Aguirre-Moreno, I. Leon-Rivera, F. Del Rio-Portilla, M. Herrera-Ruiz and J. Villeda-Hernandez, *Int. J. Med. Mushrooms*, 2013, **15**, 555.
137 P. K. Maji, I. K. Sen, K. S. P. Devi, T. K. Maiti, S. R. Sikdar and S. S. Islam, *Bioact. Carbohydr. Diet. Fibre*, 2013, **2**, 73.
138 L. Chen, B.-B. Zhang, J.-L. Chen and P. C. K. Cheung, *Food Hydrocolloids*, 2014, **38**, 48.
139 I. Sadovskaya, A. Souissi, S. Souissi, T. Grard, P. Lencel, C. M. Greene, S. Duin, P. S. Dmitrenok, A. O. Chizhov, A. S. Shashkov and A. I. Usov, *Carbohydr. Polym.*, 2014, **111**, 139.
140 R. V. Menshoa, S. P. Ermakova, S. D. Anastyuk, V. V. Isakov, Y. V. Dubrovskaya, M. I. Kusaykin, B.-H. Um and T. N. Zvyagintseva, *Carbohydr. Polym.*, 2014, **99**, 101.
141 M. I. Bilan, A. S. Shashkov and A. I. Usov, *Carbohydr. Res.*, 2014, **393**, 1.
142 P. Hu, R. Xue, Z. Li, M. Chen, Z. Sun, J. Jiang and C. Huang, *Carbohydr. Res.*, 2014, **390**, 28.
143 P. Shobharani, V. H. Nanishankar, P. M. Halami and N. M. Sachindra, *Int. J. Biol. Macromol.*, 2014, **65**, 542.
144 A. O. Kravchenko, S. D. Anastyuk, V. V. Isakov, E. V. Sokolova, V. P. Glazunov and I. M. Yermak, *Carbohydr. Polym.*, 2014, **111**, 1.
145 Z. Zhang, S. Till, C. Jiang, S. Knappe, S. Reutterer, F. Scheiflinger, C. M. Szabo and M. Dockal, *Thromb. Haemost.*, 2014, **111**, 429.
146 S. Sellimi, N. Kadri, V. Barragan-Montero, H. Laouer, M. Hajji and M. Nasri, *Int. J. Biol. Macromol.*, 2014, **66**, 281.
147 K. Wongprasert, T. Rudtanatip and J. Praiboon, *Fish Shellfish Immunol.*, 2014, **36**, 52.
148 J. A. G. Rodrigues, É. M. Neto, G. R. C. dos Santos, R. C. M. de Paula, P. A. de Souza Mourão and N. M. Barros Benevides, *Acta Sci., Technol.*, 2014, **36**, 203.
149 S. T. Minzanova, V. F. Mironov, O. V. Tsepaaeva, L. G. Mironova, A. B. Vyshtakalyuk, A. Z. Mindubaev, I. R. Nizameev, K. V. Kholin and A. V. Pashagin, *Chem. Nat. Compd.*, 2014, **50**, 54.
150 M. Walia, U. Sharma, S. Bhushan, N. Kumar and B. Singh, *Chem. Nat. Compd.*, 2013, **49**, 794.
151 M. L. Corrêa-Ferreira, G. R. Noleto and C. L. O. Petkowicz, *Carbohydr. Polym.*, 2014, **102**, 738.
152 C. Yang, Y. Gou, J. Chen, J. An, W. Chen and F. Hu, *Carbohydr. Polym.*, 2013, **98**, 886.
153 L. Shi, Q. Dong and K. Ding, *Food Chem.*, 2014, **152**, 291.
154 X.-L. Li, J.-J. Xiao, X.-Q. Zha, L.-H. Pan, M.-N. Asghar and J.-P. Luo, *Carbohydr. Polym.*, 2014, **106**, 247.
155 F.-Y. Meng, Y.-L. Ning, J. Qi, Z. He, J. Jie, J.-J. Lin, Y.-J. Huang, F.-S. Li and X.-H. Li, *Int. J. Mol. Sci.*, 2014, **15**, 5140.
156 E. G. Shakhmatov, P. V. Toukach, S. P. Kuznetsov and E. N. Makarova, *Carbohydr. Polym.*, 2014, **102**, 521.

157 J.-E. Li, S. W. Cui, S.-P. Nie and M.-Y. Xie, *Carbohydr. Polym.*, 2014, **105**, 276.
158 H. Bhatia, P. K. Gupta and P. L. Soni, *Carbohydr. Polym.*, 2014, **101**, 438.
159 H. Bhatia, *Int. J. PharmTech Res.*, 2013, **5**, 1530.
160 S. V. Popov, R. G. Ovodova, V. V. Golovchenko, D. S. Khramova, P. A. Markov, V. V. Smirnov, A. S. Shashkov and Y. S. Ovodov, *Food Chem.*, 2014, **143**, 106.
161 T. M. Cantu-Jungles, D. Maria-Ferreira, L. M. da Silva, C. H. Baggio, M. F. de Paula Werner, M. Iacomini, T. R. Cipriani and L. M. C. Cordeiro, *Food Chem.*, 2014, **146**, 492.
162 C. Wang, W. Li, H. Liu, J. Wang, G. Li, G. Wang and Y. Li, *Carbohydr. Res.*, 2014, **384**, 99.

Synthetic macromolecules

Hiromichi Kurosu*[a] and Takeshi Yamanobe*[b]
DOI: 10.1039/9781782622758-00431

In this chapter, the papers devoted to NMR application to study synthetic polymers over a period from June 2013 through May 2014 have been reviewed. It includes analysis of primary structure of polymers such as tacticity, regioregularity, end group, sequence distribution (Section 2), application of imaging, diffusion and solid-state NMR techniques to characterize the synthetic macromolecules (Sections 3 and 4). Finally in Section 5, papers devoted to dynamics of the synthetic macromolecules have been surveyed.

1 Introduction

For synthetic macromolecules, NMR has been the most powerful method to characterize and to investigate the relationship between the structure and the physical properties. In the field of synthetic macromolecules, NMR is used not only as the routine analytical method but also as the method that has the infinite possibility. In this chapter, NMR applications for synthetic polymers are reviewed.

NMR spectrum simulation techniques to estimate the NMR parameters of glycomics such as chemical shifts, coupling constants, relaxation rates and so on were reviewed.[1] Applications of NMR-MOUSE to investigate the photo-curing process of dental resins were summarized by Marchi.[2] Meier *et al.* reviewed the role of the intermolecular dipolar relaxation and the translation diffusion coefficients of liquids and polymer melts by Field cycling NMR.[3] The applications of DNP to various polymer systems were reviewed by Graf *et al.*[4] NMR analysis of the primary structure of polymers were summarized by Zhu *et al.*[5] Nyska *et al.* introduced a compact MRI as a tool for assessing biocompatibility and efficacy of implanted biodegradable materials.[6] Spiess reviewed the sideband patterns of MAS NMR and their use in elucidating non-covalent interactions.[7] Hiller *et al.* summarized the latest advanced developments of LC-NMR of polymers.[8] Reviews about MRI contrast agents,[9] poly(ethyl acrylate) filled with monodispersed spherical silica nanoparticles[10] and the characterization of crystalline cellulose[11] were published.

2 Primary structure

Characterizations of primary structures such as tacticity, regioregularity, end group, sequence distribution, and so on are investigated mainly by solution NMR. Table 1 summarizes the papers in which NMR is used to investigate the primary structure of polymers.

[a]*Nara Women's University, Kitauoya-Nishimachi, Nara, Japan. E-mail: kurosu@cc.nara-wu.ac.jp*
[b]*Department of Chemistry and Chemical Biology, Gunma University, Kiryu, Gunma, Japan. E-mail: yamanobe@gunma-u.ac.jp*

Table 1 Primary structure of polymers.

Polymer, monomer	Nucleus	Contents	Reference
Hyperbranched polyimide, 2,2-bis(3,4-dicarboxylphenyl) hexafluoropropane dianhydride, 2,4,6-triaminopyrimidine, 4,4′-oxydianiline	H	Branch	12
Poly((*R*)-3-hydroxybutyrate)		End group	13
Poly(*N*,*N*-dimethylacrylamide-*co*-3,9-divinyl-2,4,8,10-tetraoxaspiro(5.5)undecane)	H	Cross link	14
Poly(phenylacetylene), poly(butylacetylene), poly(tert-butyl acetylene), poly(4-chlorobutylacetylene), poly(cyclopropylacetylene)	H	Polymerization mechanism	15
Poly(styrene-*co*-methyl methacrylate)		Composition	16
Poly(δ-decalactone), poly(δ-decalactone-*co*-L-lactide)	H	Composition, molecular weight	17
Polyphenylene	H, C	Branch	18
Silk fibroin – rice starch	H	Cross link	19
2-hydroxypropyl-β-cyclodextrin, α,ω-ditriethylenediamino-*N*-carbamoyl-poly(ethylene oxide)-*block*-poly-(propylene oxide)-*block* -poly-(ethylene oxide)	H	Molecular weight	20
Aliphatic-aromatic copolyester, butylene adipate, butylene succinate, butylene glutarate, aromatic butylene terephthalate	H	Sequence distribution	21
Anion exchange polymer electrolyte membrane	H, C, P	Degradation	22
Carboxymethylcellulose, cyclodextrin		Composition	23
Cellulose-graft-poly(glycidyl methacrylate)		Graft	24
Chitin dipropionate, chitin diacetate	H	Composition	25
Chitin, chitosan	H	Composition	26
Copolyelectrolyte, 2,5-diethoxy-1,4-phenyleneethynylene, 2,5-diethoxy-1,4-phenyleneethynylene, -1,4-diketo-2,5-bis(4-sulfonylbutyl)-3,6-diphenylpyrrolo[3,4-*c*]pyrrole	H	Molecular weight	27
Copolyester, furanic isophthalic diester, sulfonated isophthalic diester, ethylene glycol	H	Sequence distribution	28
Cyclodextrin-calixarene co-polymer	H, C	Composition	29
Cyclopentyl trisilanol silsesquioxanes, epoxidized natural rubber	H	Composition	30
Dehydrochlorinated poly(vinyl chloride)	H	Composition	31
Dextran-graft-poly(lactic acid)		Graft	32
Epoxy-anhydride shape-memory polymer		Cross link	33
Ethylene-vinyl acetate rubber-graft-polyamide 6		Molecular weight	34
Guar gum hydrogel		Crosslink	35
Hydrolyzed polyacrylamide		Cross link	36
Hyperbranched poly(urethane-urea)	H	Branch	37
Hyperbranched polyamide	H	Branch	38
Hyperbranched polycarbosilane	H	Composition	39
Hyperbranched polyglycerol	H, C	Branch	40
Hyperbranched polyimide, 2,2-bis(3,4-dicarboxylphenyl) hexafluoropropane dianhydride, triamine monomer 2,4,6-triaminopyrimidine	H	Composition	41
Hyperbranched polymers, oxazoline monomer, succinic anhydride:	H, C	Branch	42
IPN, polyglycidylmethacrylate, acrylamide	H, C	Composition	43
Linear low density polyethylene	C	Branch	44

Table 1 (*Continued*)

Polymer, monomer	Nucleus	Contents	Reference
Methoxy-polyethylene-glycol-*block*-poly(carbonate-*co*-lactide)	H	Molecular weight	45
Multiblock copolymer, poly(arylene ether sulfone), bisphenol A, dimethylbisphenol A, tetramethylbisphenol A	H, F	Composition	46
Multiblock copolymer, styrene, methyl methacrylate, 2-vinyl pyridine	H	Composition, molecular weight	47
Nafion	F	Degradation	48
Natural rubber	H	Composition	49
Natural rubber, silicone rubber		Cross link	50
Poly(methyl methacrylate)-*N*,*N*,*N*-trimethyl chitosan	H	Composition	51
Poly(butylene succinate-*co*-cyclohexanedimethylene succinate)	H	Composition	52
Poly(D,L-lactic acid)	H	Composition	53
Poly(4,4-dimethyl 1,7-octadiyne)	H	Polymerization mechanism	54
Poly(butyl acrylate-*co*-*N*-ethylacrylate-3,4-dimethylmaleimide), poly(*N*-iso-propyl acrylamide-*co*-*N*-ethylacrylate-3,4-dimethylmaleimide)		Composition	55
Poly(1-(2-hydroxyethyl)aziridine)		End group	56
Poly(1,1,2,2-tetramethyl-3-trimethylsilyl-1,2-disilacyclobutane)	H, C, Si	Regioregurality	57
Poly(1,3-butadiene-*co*-acrylonitrile)	H	Composition	58
Poly(1,4-butylene carbonate)	H	Sequence distribution	59
Poly(1,6-heptadiene-*co*-ethylene)	H, C	Composition	60
Poly-(15*N*-(1-naphthyl)-*N*′-octadecylcarbodiimide), poly-(15*N*-(1-naphthyl)-15*N*′-octadecylcarbodiimide)	N	Conformation	61
Poly(1*H*,1*H*,2*H*,2*H*-perfluorooctyl acrylate), poly(2,2,3,3,4,4,4-heptafluorobutyl acrylate), poly(1*H*,1*H*,5*H*-octafluoropentyl methacrylate), poly(1*H*,1*H*,5*H*-octafluoropentyl acrylate)	H	End group	62
Poly(2-((2-(2-methoxyethoxy)ethoxy)methyl) oxirane-*alt*-succinic anhydride)		End group	63
Poly(2-(diisopropylamino)ethyl methacrylate)	H	End group	64
Poly(2-(methacryloyloxy)ethyl cholate-*co*-polyethylene glycol methyl ether methacrylate), poly(2-(methacryloyloxy)ethyl cholate-*co*-*N*,*N*-dimethylaminoethyl methacrylate)	H	Composition, sequence distribution	65
Poly(2-(perfluoroalkyl)ethyl methacrylate-*co*-methyl methacrylate)	H	Composition	66
Poly(2,2,2-trifluoroethyl methacrylate-*co*-butyl methacrylate), poly(1,1,1,3,3,3-hexafluoroisopropyl methacrylate-*co*-butyl methacrylate)	H	Composition	67
Poly(2,3-dicarbomethoxy-5-norbornene)	H	Polymerization mechanism	68
Poly(2,5-dibromo-3-(5′-hexylpyridine-2′-yl)thiophene), poly(5,5′-dibromo-4-(5″-hexylpyridine-2″-yl)-2,2′-bithiophene), poly(5,5′-dibromo-4-(5″-hexylpyrimidine-2″-yl)-2,2′-bithiophene)	H	Regioregularity	69

Table 1 (*Continued*)

Polymer, monomer	Nucleus	Contents	Reference
Poly(2,7-(9,9-dioctylfluorene)-*alt*-(5,5′-bithiophene/permethylated β-cyclodextrin))	H	Composition, molecular weight	70
Poly(2-ethoxy-2-oxo-1,3,2-dioxaphospholane-*co*-2-ethoxy-4-methyl-2-oxo-1,3,2-dioxaphospholane)	H, P	Sequence distribution	71
Poly(2-ethylhexyl acrylate), poly(*n*-lauryl acrylate), poly(2-(2-ethoxyethoxy)ethyl acrylate), poly(2-ethylhexyl methacrylate)	C	End group	72
Poly(2-hydroxy-2-oxo-1,3,2-dioxaphosphorane), γ-cyclodextrin	H	Composition	73
Poly(3-(4-alkoxyphenyl)thiophene)	H	Regioregularity	74
Poly(3-aminopropyltrimethoxysilane)	H, C, Si	Polymerization mechanism	75
Poly(3-hexylthiophene), poly(3-hexylselenophene)	H	Polymerization mechanism	76
Poly(3-hexylthiophene)-*block*-poly(9,9-dioctylfluorene), poly(3-dodecylthiophene)-*block*-poly(9,9-dioctylfluorene)	H	Composition	77
Poly(3-hydroxy-5-phenylvalerate), poly(3-hydroxy-5-phenylvalerate-*co*-3-hydroxydodecanoate)		Sequence distribution	78
Poly-(4-alkylthiazole)	H	Regioregularity	79
Poly(4-hydroxybutyrate)	H	Degradation	80
Poly(7-acryloyloxy-4-methylcoumarin-*co*-*N*-cyclohexyl-acrylamide)	H	Composition, sequence distribution	81
Poly(acetoxystyrene)	H	Branch	82
poly(acrylate)	H	end group	83
Poly(acrylonitrile-*co*-butadiene)	H	Sequence distribution	84
Poly(acryloxyethyltrimethyl ammonium chloride-*co*-acrylamide)	H	Sequence distribution	85
Poly(amide triazole)		Degradation	86
Poly(arylene ether sulfone)	H	End group, molecular weight	87
Poly(arylene ether sulfone), 3,8-bis(4-hydroxyphenyl)-*N*-phenyl-1,2-naphthalimide	H	Composition	88
Poly(butyl acrylate)		End group	89
Poly(butylene succinate)	H	Composition	90
Poly(cyclohexene carbonate)		Molecular weight	91
Poly(cyclopentenylene-vinylene)		Polymerization mechanism	92
Poly(dimethylsiloxane), carbon nanotube–carbon aerogel		Degradation	93
Poly(dimethylsiloxane), poly(lactide)	H	End group	94
Poly(diphenyleneoxidephthalide), copoly(arylenephthalide), terphenylenephthalide, diphenyleneoxidephthalide	H, C	Composition	95
Poly(D-lactide), poly(ε-caprolactone-*co*-lactide)		Graft	96
Poly(ether sulfone)		branch	97
Poly(ether sulfone), 4-chloro-4′-(hydroxyphenyloxy)diphenylsulfone, 4-fluoro-4′-(hydroxyphenyloxy)diphenylsulfone		Polymerization mechanism	98
Poly(ethylene glycol) monomethyl ether, poly(ethylene glycol) monomethyl etherl-*block*-oligo(glycidyl phenyl ether)		Molecular weight	99

Table 1 (*Continued*)

Polymer, monomer	Nucleus	Contents	Reference
Poly(ethylene glycol), chitosan	H	Graft	100
Poly(ethylene glycol), poly(lactic-ran-glycolic acid)	H	Molecular weight	101
Poly(ethylene glycol), poly(methyl methacrylate), ethylene glycol dimethacrylate	H	Cross link	102
Poly(ethylene glycol)-*block*-poly(2-hydroxypropyl methacrylate)	H	Composition	103
Poly(ethylene glycol-*co*-caprolactone)	H	Composition	104
Poly(ethylene oxide)	H	End group	105
Poly(ethylene terephthalate), copolyester sodium-5-sulfo-bis-(hydroxyethyl)-isophthalate, poly(ethylene glycol)		Composition	106
Poly(ethylene terephthalate)-*block*-poly(tetramethylene oxide)-*block*-poly(ethylene terephthalate)	H, C	Molecular weight	107
Poly(ethylene-*co*-4-methyl-1-pentene)	C	Branch	108
Poly(ethylene-*co*-4-penten-1-ol)	C	Sequence distribution	109
Poly(ethylene-*co*-hexene)	C	Composition	110
Poly(ethylene-*co*-propene), poly(ethylene-*co*-hexene), poly(ethylene-*co*-octene), poly(ethylene-*co*-octadecene)	H, C	Composition	111
Poly(ethylene-*co*-propylene), poly(ethylene-*co*-1-hexene), poly(ethylene-*co*-4-methyl-1-pentene, poly(propylene-*co*-1-butene)	C	Sequence distribution	112
Poly(ferrocenyl glycidyl ether-*co*-allyl glycidyl ether)	C	Sequence distribution	113
Poly(glycolide-*block*-poly(glycolide-*co*-trimethylene carbonate-*co*-ε-caprolactone)-*block*-polyglycolide	H	Degradation	114
Poly(hydroxybutyrate)	H	End group	115
Poly(isobornyl acrylate-*co*-styrene)	H, C	Stereoregularity	116
Poly(isoprene)	H	Configuration	117
Poly(lactic acid), poly(4-methyl oxazoline)	H, C	End group	118
Poly(lactic acid)-*block*-natural rubber-*block*-poly(lactic acid)	H	End group	119
Poly(limone oxide-*co*-polyethylenglycol methylacrylate)	H	Composition	120
Poly(L-lactic acid), poly(*O*-carboxyanhydride)		End group	121
Poly(L-lactide), poly(DL-lactide)		Configuration	122
Poly(l-lactide-*co*-2-(2-methoxyethoxy)ethyl methacrylate)		Sequence distribution	123
Poly(L-lysine)-graft-poly(2-methyl-2-oxazoline)		Graft, composition	124
Poly(methacrylate)	H	End group	125
Poly(methacrylate), poly(*n*-butyl acrylate), poly(methyl methacrylate)	H	End group	126
Poly(methyl acrylate)	H, C	Polymerization mechanism	127
Poly(methyl methacrylate)		Tacticity	128
Poly(methyl methacrylate), poly(acrylamide)	H	End group	129
Poly(methyl methacrylate), poly(styrene)		End group	130
Poly(methyl methacrylate), poly(ethyl methacrylate), poly(butyl methacrylate), poly(*n*-butyl methacrylate), poly(*n*-hexyl methacrylate), poly(2-ethyl hexyl methacrylate)	C	Tacticity	131

Table 1 (*Continued*)

Polymer, monomer	Nucleus	Contents	Reference
Poly(methyl methacrylate)-polysilsesquioxane	C	Composition	132
Poly(methyl methacrylate-*co*-butyl acrylate)	H, C	Sequence distribution	133
Poly(*N*(-1,1-dimethyl-3-oxobutyl)acrylamide-*co*-methyl methacrylate)	H, C	Composition	134
Poly(*N*-(2-hydroxypropyl) methacrylamide)	H, F, C	Composition	135
Poly(*N*-isopropylacrylamide), poly(*N*,*N*-dimethylacrylamide)	H	End group	136
Poly(*N*-isopropylacrylamide)-*block*-poly(L-lactic acid)-*block*-poly(*N*-isopropylacrylamide)	H	Sequence distribution	137
Poly(*N*-isopropylacrylamide-*co*-2-dimethyl-aminoethyl methacrylate-*co*-itaconic acid)	H	Composition	138
Poly(*N*-isopropylacrylamide-*co*-dopamine methacrylate)	B	Cross link	139
Poly(*N*-vinylcaprolactam), poly(*N*-isopropylacrylamide), poly(*N*-isopropylmethacrylamide)	H	Composition	140
Poly(*N*-vinylpyrrolidone-*co*-2-acrylamido-2-methylpropanesulfonate sodium)	H	Composition	141
Poly(octadecylstyrene)	H	Sequence distribution	142
Poly(organophosphazenes)	P	Degradation	143
Poly(*p*-(1-ethoxy ethoxy)styrene-*co*-4-tert-butoxystyrene), poly(styrene-*co*-*p*-(1-ethoxy ethoxy)styrene)	H	Sequence distribution	144
Poly(phenyl isocyanide)	H, C	Stereoregularity	145
Poly(phenyl vinyl sulfide, poly(4-chlorophenyl vinyl sulfide), poly(2,4-dichlorophenyl vinyl sulfide), poly(4-bromophenyl vinyl sulfide)	H	End group	146
Poly(phenylene oxide)	C, F	Branch	147
Poly-(*p*-methylstyrene)-*co*-*cis*-1,4-poly-(butadiene), poly-(*p*-methylstyrene)-*co*-*cis*-1,4-polyisoprene, poly-(*p*-methylstyrene)-*co*-*trans*-1,4-poly-(butadiene), poly-(*p*-methylstyrene)-*co*-*trans*-1,4-polyisoprene	C	Sequence distribution	148
Poly(propylene carbonate), cellulose	H	Degradation	149
Poly(propylene succinate), poly(butylene succinate), poly(ethylene terephthalate)		Composition	150
Poly(propylene-*co*-1-pentene-*co*-1-hexene)	C	Sequence distribution	151
Poly(propylene-*co*-ethylene)	C	Sequence distribution	152
Poly(styrene-*block*-isoprene)	H	Composition	153
Poly(styrene-*co*-allyl acetoacetate), poly(styrene-*co*-2-(acetoacetoxy)ethyl methacrylate)	H	Composition	154
Poly(styrene-*co*-divinylbenzene)	C	Composition	155
Poly(styrene-*co*-isoprene-*co*-butadiene)	H, C	Configuration	156
Poly(styrene-*co*-methacrylic acid), poly(styrene-*co*-dimethylamine ethyl methacrylate), poly(styrene-*co*-sodium methacrylate), poly(styrene-*co*-1-hydroxyethyl methacrylate)	C	Composition	157
Poly(styrene-*co*-*para*-butyl-3-enyl styrene)	H	Sequence distribution	158
Poly(tetrafluoroethylene)	H, C, F	End group	159
Poly(vinyl acetate-*co*-dibutyl maleate)	H	End group	160

Table 1 (*Continued*)

Polymer, monomer	Nucleus	Contents	Reference
Poly(vinyl alcohol), poly(acrylic acid)	C	Cross link	161
Poly(vinyl alcohol)-epoxypropoxy coumarin		Composition	162
Poly(vinyl chloride), poly(vinyl alcohol)	H, C	Composition	163
Poly(vinyl ether)	H	End group	164
Poly(vinylidene fluoride-*co*-hexafluoropropylene)	H, F	End group, branch	165
Poly(vinylidene fluoride-*co*-tetrafluoroethylene)	H, F	End group, branch	166
Poly(α-bromo-δ-butyrolactone-*co*-ε-caprolactone), poly(α-bromo-δ-butyrolactone-*co*-L-lactide)	C	Sequence distribution	167
Poly(δ-decalactone), polyurethane	H	Molecular weight	168
Poly(δ-valerolactone), poly(ε-caprolactone)	H	Composition	169
Poly(ε-caprolactone-*co*-lactide)-*block*-poly(ethylene glycol)-*block*-poly(ε-caprolactone-*co*-lactide)	H	Sequence distribution	170
Polyacrylonitrile	H	End group	171
Polyamide	C	Molecular weight	172
Polyamide, functional carbon nanotube		Graft	173
Polyaniline		Composition	174
Polycarbonate, poly(methyl methacrylate)	H	Polymer reaction	175
Poly(chloroprene)	H	End group	176
Polyether, phenol-amine resin, poly(propylene oxide), ethylene oxide	H	End group	177
Polyethylene	C	Branch, end group	178
Polyethylene	H	End group	179
Polyethylene	H, C	End group, molecular weight	180
Polyethylene glycol	H, C	Branch	181
Polyethylene	C	Branch	182
Polyhedral oligomeric silsesquioxanes, polystyrene	H	Composition	183
Poly(hydroxyalkanoate)	H	Degradation	184
Polyimide, polysiloxane	Si	Branch	185
Polyisocyanate		Molecular weight	186
Polyisoprene-*block*-poly(methyl methacrylate)	H	Molecular weight	187
Polymer electrolyte fuel cell	F	Degradation	188
Poly(methyl methacrylate)-*block*-poly(di-methyl siloxane)-*block*-poly(methyl methacrylate), polystyrene-*block*-poly(dimethyl siloxane)-*block*-polystyrene	H	Molecular weight	189
Polynorbornene	H, P	Configuration	190
Polynorbornene	H	End group	191
Polynorbornene		Polymerization mechanism	192
Polypropylene	C	Regioregularity, steroregularity	193
Polypropylene	C	Tacticity	194
Polypropylene		Tacticity, regioregularity	195
Polypropylene, poly(1-hexene), poly(1-octene), poly(1-decene)	C	Tacticity	196
Polyrotaxane, α-cyclodextrin, poly(ethylene glycol)	H	Composition	197
Polysaccharide	H	End group, molecular weight	198
Polysialic acid		Composition, end group	199

Table 1 (*Continued*)

Polymer, monomer	Nucleus	Contents	Reference
Polysilane		Branch	200
Polysiloxane	Si	Composition	201
Polystyrene	H	End group	202
Polystyrene	H, C	End group	203
Polystyrene	H	End group	204
Polystyrene		Stereoregularity	205
Polystyrene	C	Tacticity	206
Polystyrene, polyperoxide		Sequence distribution	207
Polystyrene-*block*-poly(1-methyl-3-(4-vinylbenzyl)-imidazolium bis(trifluoromethane sulfonyl)imide-*block*-polystyrene	H	Sequence distribution	208
Polystyrene-*block*-poly(5,6-benzo-2-methylene-1,3-dioxepane), poly(5,6-benzo-2-methylene-1,3-dioxepane)-*block*-polystyrene	P	Polymerization mechanism	209
Polytetraphenylethene	H	Configuration	210
Polyurethane, polyesteramide	H	End group	211
Redox-robust polyelectrolyte	H, C	End group	212
Segmented polyurethane, poly(lactic-*co*-glycolic acid), β-tricalcium phosphate)		Degradation	213
Silicone rubber	H	Cross link	214
Sodium polyphosphate	P	Molecular weight	215
Starch-graft-poly(butylene succinate)	H, C	Graft	216
Starch-graft-poly(sodium acrylate)	C	Graft	217
Styrene-butadiene rubber	H	Graft	218
Telechelic vinyl-functionalized poly(dimethylsiloxane), poly(dimethylsiloxane-*co*-propylmercaptomethylsiloxane)		Cross link	219
Terpolyester, isosorbide, ethylene glycol, and 1,4-cyclohexane dimethano	C	Sequence distribution	220
α,ω-di(glycerol carbonate) telechelic poly(propylene glycol), poly(ethylene glycol), poly(ester ether), poly(butadiene)	H, C	End group	221
α-chitin	C	Composition	222

3 Imaging and diffusion

NMR self-diffusion measurements were used to investigate the association of 4-aza-1-tetradecyl-1-azoniabicyclo[2.2.2]octane bromide and polyethyleneimine.[223] Water self-diffusion measuremts were carried out to investigate the correlation between the stress-induced order and the transport properties of Nafion membrane. The anisotropy of diffusion is much larger in a stretched membrane.[224] The interaction between the pryrazole-based dyes and poly(vinyl alcohol) was studied. The hydrogen bonding caused the change of the diffusion coefficients which correlates with the size of the colloidal polymer matrix.[225] New self-standing single lithium ion conductor polymer network was synthesized and the diffusion coefficients of ^{19}F and ^{7}Li were determined.[226] The morphology and performance features of polyelectrolyte-fluoropolymer blend were studied and the drastic changes of water diffusion was observed.[227] Organic–inorganic hybrid electrolyte membranes (tri-block copolymer poly(propylene glycol)-*block*-poly(ethylene glycol)-*block*-poly(propylene glycol) bis(2-aminopropyl ether), 3-isocyanatepropyltriethoxy-silane and central core 2,4,6-trichloro-1,3,5-triazine doped with $LiClO_4$) were synthesized. The correlation between diffusion coefficient and their ionic conductivities investigated.[228] A new salt-in-polymer electrolyte was prepared by the radical photopolymerization of poly(ethylene oxide) acrylate oligomers and the diffusion measurements were carried out.[229] A method to monitor the swelling and diffusion by NMR is proposed and applied to study diffusional behavior of toluene and lithium amide into polystyrene.[230] The diffusion of Na^+ decreased in aqueous solution of poly(2,2′-disulfonyl-4,4′-benzidine terephthalamide).[231] The network of acidified paper has been studied by NMR relaxometry, NMR diffusometry and NMR diffusion–diffraction. The acidifying treatment enlarges the macropore structure of paper.[232] A series of multiblock copolymers of sulfonated copolyimides were synthesized. PFG NMR results revealed that water diffusion to be faster in-plane than in the through-plane direction.[233] The self-diffusion of silicone MQ copolymers in chloroform was measured and the molecular weight effect of the diffusion behavior of the MQ resins was discussed.[234] Diffusion experiments were carried out for cyclodextrin vesicles and the mechanism of permeation was discussed.[235] Poly(*N*-[2-(2-hydroxyethoxy)ethyl]methacrylamide was prepared and the oxygen transport property was studied.[236] The dependence of proton hopping and diffusion behavior of nanostructured poly(styrenesulfonate-*block*-methylbutylene) containing imidazolium-based ionic liquid.[237] Water self-diffusion coefficients in sodium polyacrylate solution were measured. A free-volume model was found to be in good agreement with the experimental results.[238] For the system of carbon nanotubes dispersed by triblock copolymer Pluronics F127 in water, the coefficient of lateral diffusion of the polymer along the nanotube surface was determined.[239] Divinyl-11-silicotungstic acid was copolymerized with butyl acrylate and hexanediol diacrylate. Water self-diffusion was fast in this film and the water in these systems was highly associated with the hteteropoly acid clusters.[240] PFG-NMR measurements were employed to distinguish diffusion of *n*-hexane in the space between each particle, the pores in SiO_2 and the polymer shell.[241] Ion diffusion of lithium ion and fluorine anion were compared in the polymer gel electrolytes ($LiBF_4$), propylene carbonate and fluoride.[242]

The effect of MRI on physical, chemical and mechanical properties of acrylic resin denture base materials was studied.[243] The kinetics of a high-swelling ionic polymer gel was investigated by NMR microimaging. The experimental results for polyacrylamide gels were compared with the predictions of a nonlinear poroelastic theory.[244] NMR chemical shift imaging were applied to monitor the swelling and molecular transport behaviours for tablets of poly(acrylic acid).[245,246] The migration of Mn^{2+} and Cu^{2+} through a nylon-6 film was studied by MRI.[247] A lot of contrast agents were developed.[248–284]

4 Characterization of the synthetic macromolecules

Solid-state NMR measurements of ^{19}F fluorinated side-chain Azobenzene liquid crystal polymer networks (azo-LCNs) using an *in situ* visible light

(450–458 nm) LED light source are carried out. These measurements show that measurable changes in ^{19}F NMR lineshapes under light irradiation. This indicates that conformational changes can be probed by NMR.[285] ^{1}H, ^{19}F and ^{31}P solid-state NMR spectroscopy was employed to investigate crystallization in solvent cast and heat-treated samples of pure Poly[bis(trifluoroethoxy)phosphazene] (PBFP) and composites of nano-crystalline TiO_2 doped PBFP.[286] The effect of electrospinning process on morphology and conductivity of the modified poly (vinyl chloride) have been studied. Degree of dehydrochlorination and chemical structural change was confirmed by ^{1}H-NMR technique.[287] High-resolution ^{1}H NMR has been used on the thermoresponsive hyperbranched polyethylenimines (HPEIs) modified with isobutyramide (IBAm) groups (HPEI-IBAm), to study the structure and dynamics of the macromolecules in aqueous solution before and after the phase transition. By combining different NMR techniques, it is shown that the IBAm groups of two HPEI-IBAm macromolecules differ not only in the substitution degree but also in the spatial distribution in the HPEI structure. Furthermore, by using pulsed field gradient (PFG) diffusion NMR, the weight distributions of the moving particles in the solution were monitored. The β parameter used in the PFG diffusion NMR has proved to be a good way to monitor the aggregation process of the moving particles in the solution.[288] ^{29}Si solid-state NMR measurements are carried out to investigate the effects of different comb-shaped superplasticizers on the silicate structure. The analysis of ^{29}Si solid-state NMR spectra shows that the additives increase the degree of polymerization and hence the average length of the silicate chains in C–S–H.[289] Several 2D NMR techniques were applied to comprehensively elucidate cation degradation in polysulfone-based anion exchange polymer electrolyte membranes (AEMs) exposed to alkaline solutions. The 2D NMR tools utilized were: (a) homonuclear COSY, (b) ^{1}H–X (X = ^{13}C or ^{31}P) HMQC, and (c) ^{1}H–^{13}C HMBC. Analysis of the 2D NMR spectra enabled precise identification of the degradation products for several different cation chemistries. The observation of multiple degradation products highlighted the fact that the cation groups degraded through fundamentally different mechanisms. A clear understanding of the degradation mechanism is a critical requirement in the process of selecting or designing alternate cation chemistries for alkaline resilient AEMs.[290] NMR spectroscopy has been used for structural characterization of 5-fluorouracil (5FU) and daunorubicin (DAU) co-loaded poly(butylcyanoacrylate) nanoparticles(PBCN), prepared by an anionic polymerization of *n*-butylcyanoacrylate, probing two differentdrug loading approaches. Diffusion ordered spectroscopy, obtained through PFG NMR experiments, has been performed to determine the overall structure of the 5FU & DAU co-loaded PBCN and to clarify the mechanisms of drug immobilization and location in thepolymer matrix of PBCN.[291] Dynamic nuclear polarization (DNP) solid-state NMR at 14.1 T was used to obtain natural abundance ^{13}C and ^{15}N CP/MAS NMR spectra of microporous organic polymers (MOPs) with excellent signal-to-noise ratio, allowing for an enormous boost in speed and efficiency for molecular structure determination in MOPs.[292] 2D-NMR techniques were used to identify the detailed structures of chain end and branching units in poly(vinylidene fluoride-*co*-tetrafluoroethylene), poly(VDF-*co*-TFE). To study the detailed structures of low concentration chain end units, the selective ^{19}F–^{19}F COSY experiment was used together with $^{19}F\{^{1}H\}$ and $^{1}H\{^{19}F\}$ gHETCOR spectra to provide excellent quality spectral data, enabling unequivocal assignments for resonances of chain end units.[293] The order–disorder phenomenon and spatial heterogeneity of chain packing, partitions of stereodefects, and molecular dynamics of α form of isotactic polypropylene (*i*PP) samples are investigated by solid-state NMR. Solid-state NMR has been successfully applied to investigate local packing structures, defect partition, spatial correlation of different packing structures, and molecular dynamics of Zieglar–Natta *i*PP samples.[294] ^{1}H and ^{13}C NMR spectral assignments have been provided for low-molecular reference monomers, poly(diphenyleneoxidephthalide) and periodic copoly(arylenephthalide) derivatives using ^{1}H–^{1}H COSY, ^{1}H–^{13}C HSQC and HMBC NMR techniques. Assignments of chemically similar carbons of the inner and end homodiad and heterodiad have been carried out.[295] Three alternating 9,9-dialkylfluorene copolymers, having benzothiadiazole (F8BT), thiophene *S*,*S*-dioxide (PFTSO2), and phenylene (PFP) groups, have been studied by ^{1}H and ^{13}C NMR spectroscopy and density functional theory (DFT), and the behavior has been compared to that of the homopolymer poly[2,7-(9,9-bis(2′-ethylhexyl)-fluorene)] (PF2/6). In addition, time-dependent DFT has been used to obtain a better understanding of the character of selected excited states within the singlet and triplet state manifolds.[296] DNP is shown to greatly improve the solid-state NMR analysis of synthetic polymers by allowing structural assignment of intrinsically diluted NMR signals, which are typically not detected in conventional solid-state NMR. Specifically, solid-state NMR and DNP solid-state NMR were comparatively used to study functional polymers for which precise structural elucidation of chain ends is essential to control their reactivity and to eventually obtain advanced polymeric materials of complex architecture. Timeconsuming two-dimensional correlation NMR experiments can be envisioned, even for such intrinsically diluted NMR signals.[297] This solid-state ^{13}C MAS NMR spectroscopy study examines the influence of water and temperature on the hydrogen bond structures of ^{13}C-labeled poly(methacrylic acid) (PMAA) complexes with poly(ethylene oxide) (PEO) and poly(vinyl methyl ether) (PVME). The spatial variation of the inter- and intrapolymer hydrogen bonds was measured by 2D ^{13}C exchange (EXSY) NMR.[298] Four kinds of ^{13}C-labeled polyacrylonitrile (PAN) samples were prepared by solution polymerization of acrylonitrile (AN) with selective ^{13}C labeling of different molecular sites. The composition and structure of the residues were quantitatively analyzed in detail by one- and two-dimensional solid-state ^{13}C NMR experiments. Combined with the use of selectively ^{13}C-labeled PAN and a MAS spinning speed of 20 kHz, the detection sensitivity of solid-state NMR increased nearly 2 orders of magnitude and avoided the peak overlaps on the NMR spectra so that the structures of residues from the thermal treatment of PAN in argon at 250–350 °C has been identified accurately.[299] ^{13}C–^{13}C double-quantum NMR was used to determine for the detailed chain-folding structure of ^{13}C CH_3-labeled *isotactic* poly(1-butene) (*i*PB1) in solution-grown crystals blended with nonlabeled *i*PB1 across a wide range of crystallization temperatures (T_cs).[300] Experimental approaches are described which reveal the amount of rigid and soft phases in styrene–butadiene gradient copolymers with component specific resolution using fast and slow MAS NMR.[301] Conformational characteristics of poly(lactide)s have been investigated by density functional theory and ab initio MO calculations and NMR experiments on model compounds. Characteristic ratios, configurational entropies, and internal energies of poly(L-lactide) and poly(DL-lactide) were calculated under the refined rotational isomeric state scheme with conformational energies and geometrical parameters derived from the MO calculations.[302] The influence of the molar mass distribution on structural changes during solid-state deformation in the crystalline and the non-crystalline regions of UHMW-PE have been studied by various solid-state NMR methods. It reveals that differences in the deformation behavior of the mobile-amorphous, rigid-amorphous, and crystalline polymorphs have been followed in the broad and narrow molar mass UHMW-PEs.[303] Solid-state ^{13}C CP/MAS NMR measurements of order in semicrystalline, high molar mass poly(3-hexylthiophene) (P3HT) are carried out. The relative degree of crystallinity was estimated for two films with different drying conditions. Room-temperature dynamic disorder suggests that P3HT may be classified as a conformationally disordered (CONDIS) crystal.[304]

5 Dynamics of the synthetic macromolecules

^{1}H MAS, Magic Sandwich Echo NMR methods and equilibrium solvent uptake analysis are utilized to optimize the structure of a series of model PDMS networks. The influence of inter-crosslink molar mass, vinyl to silane ratio and Pt chelation effects on the dynamics of network formation have been studied using low field NMR methodologies in real time.[305] ^{1}H spin–spin relaxation time (T_2) measurement of polyampholyte hydrogel poly(methylacrylic acidacryloyloxyethyl trimethylammonium chloride)[P(MA-DAC)] in different pH, ionic strength and temperature was carried out to reveal the molecular mobility. T_2 and the proton component fraction were acquired to study the swelling behavior of the hydrogel.[306] The locally detailed structure and molecular dynamics of its structural elements in disordered and ordered phases of 1,4-bis[3,4,5-tris(alkan-1-yloxy)benzamido]benzene bisamide were investigated using various solid-state NMR techniques at the atomic level to understand its structural formation and unique phase-transition processes.[307] The nearly symetric poly(styrene-*b*-isporene) diblock copolymer PS(45 000)-*b*-PI(46 000) and its constituents PI and PS have been investigated in order to understand how the chemical crosslinks and the lamellar morphology present in the block copolymer system influence the polyisoprene molecular dynamics by NMR and dielectric spectroscopies.[308] The complex morphology of poly(urethane urea) elastomers and its effect on molecular dynamics have been investigated by solid-state ^{13}C T_1 relaxation time and ^1H wideline measurements.[309] ^{13}C high-field and ^1H low-field NMR experiments were performed at different temperatures for the semicrystalline polymer poly(ε-caprolactone) crystallized from the melt in order to elucidate the crystalline-phase dynamics over a wide range of motional rates in detail.[310] The molecular mass dependence of polymer dynamics in the melt has been investigated by field-cycling ^1H NMR relaxometry for poly-(propylene glycol) and molecular dynamics simulation for poly-(propylene oxide). These combined experimental and computational approaches revealed that different molecular dynamics at the backbone centers and at the chain ends affect the results of experimental studies, which usually probe a superposition of signals from monomers at different positions along the polymer backbone.[311] The measurement of the spin–lattic relaxation time can elucidate the molecular mobility changes in the starch-polylactic acid (PLA) matrix caused by the addition of nanoparticles and polyvinyl alcohol. Fast field cycling was used to determine the molecular motions, in addition to providing important information about the miscibility of the starch-PLA blends.[312] The segmental dynamics and orientational ordering for the free chain ends of a strongly segregated diblock copolymer, poly(styrene)-*b*-poly(dimethylsiloxane) (PS-*b*-PDMS), self-assembled into a lamellar structure, was investigated with ^2H solid-state NMR. The dynamical behavior of the free PDMS chain extremities, selectively deuterated, was probed in a temperature regime intemathrmediate between the PS and PDMS glass transition temperatures.[313] The chain packing, crystal thickness, molecular dynamics, and melting temperature of α-form*isotactic* polypropylene (*i*PP) drawn uniaxially at high temperatures of 100–150 °C were investigated using solid-state NMR and DSC. ^{13}C CP/MAS NMR detected two reverse transformations (α2 → α1 and α1 → α2) in α-form *i*PP upon hot drawing, indicating two different melting and recrystallization processes at different length scales in different engineering strain ranges.[314] The structure and dynamics of Neutral polymer bonding agent (NPBA) under different conditions of temperatures and sample processing are comprehensively investigated by solid-state NMR. The ^{13}C variable temperature CP/MAS spectra and NMR parameters of ^{13}C chemical shift anisotropy and ^{13}C–^1H derived order parameters show that main chain of the studied polymer remains relatively rigid at temperatures up to 333 K. Interestingly, the side chains are rigid at room temperature but become mobile at temperatures above Tg.[315] The segmental dynamics of rigid, intermediate, and mobile molecular components in carboxyl terminated polybutadiene (CTPB)/organo-clay (C_{18}-clay) systems was characterized by fully refocused ^1H NMR FID. In addition, ^1H DQ NMR experiments allowed semiquantitative monitoring of changes in segmental dynamics near the interface. Both methods suggest a critical concentration of 60 wt% CTPB, indicating a saturation effect for the surface-adsorbed polymer. Based on ^1H DQ NMR results, a tentative model was suggested to illustrate the cooperative effect of CTPB and organo-clay in the nanocomposites.[316] The microphase-separated structure and heterogeneous segmental dynamics of a novel thermally reversible cross-linked polyurethane were well revealed by low-field NMR spectrometer, proton T_2 relaxometry and MQ NMR experiments, which was also first utilized to *in situ* monitor the reversible cross-linking associated with Diels–Alder and retro-Diels–Alder reactions.[317] The dynamics in poly(acrylic acid) hydrogels are studied by low-field ^{1}H NMR methods. Two experimental methods allowed for quantifying the strength of residual dipolar coupling which characterizes the cross-link-induced chain motion anisotropy in these constrained systems.[318]

References

1 F. V. Toukach and V. P. Ananikov, *Chem. Soc. Rev.*, 2013, **42**, 8376–8415.
2 N. A. Marchi, J. Steinhaus, B. Hausnerova, B. Moeginger and B. Bluemich, *Appl. Magn. Reson.*, 2013, **44**, 1027–1039.
3 R. Meier, D. Kruk and E. A. Roessler, *ChemPhysChem*, 2013, **14**, 3071–3081.
4 R. Graf, M. R. Hansen, D. Hinderberger, K. Muennemann and H. W. Spiess, *Phys. Chem. Chem. Phys.*, 2014, **16**, 9700–9712.
5 Z. Zhu and A. M. Alb, *Monit. Polym. React.*, 2014, 201–227.
6 A. Nyska, Y. S. Schiffenbauer, C. T. Brami, R. R. Maronpot and Y. Ramot, *Polym. Adv. Technol.*, 2014, **25**, 461–467.
7 H. W. Spiess, *Isr. J. Chem.*, 2014, **54**, 16–24.
8 W. Hiller, P. Sinha, M. Hehn and H. Pasch, *Prog. Polym. Sci.*, 2014, **39**, 979–1016.
9 G.-L. Davies, I. Kramberger and J. J. Davis, *Chem. Commun.*, 2013, **49**, 9704–9721.
10 A. Papon, T. C. L. Guy, K. Saalwaechter, J. Oberdisse, S. Merabia, D. Long, P. Sotta, H. H. Frielinghaus, A. Radulescu, B. Deme, L. Nolrez, H. Montes and F. Lequeux, *KGK, Kautsch. Gummi Kunstst.*, 2013, **66**, 52–58.
11 S. H. Kim, C. M. Lee and K. Kafle, *Korean J. Chem. Eng.*, 2013, **30**, 2127–2141.
12 Y. Chen, Q. Zhang, W. Sun, X. Lei and P. Yao, *J. Membr. Sci.*, 2014, **450**, 138–146.
13 A. Hiroe, M. Hyakutake, N. M. Thomson, E. Sivaniah and T. Tsuge, *ACS Chem. Biol.*, 2013, **8**, 2568–2576.
14 L. E. Nita, A. P. Chiriac, M. T. Nistor and I. Neamtu, *Rev. Roum. Chim.*, 2013, **58**, 137–143.
15 O. Trhlikova, J. Zednik, H. Balcar, J. Brus and J. Sedlacek, *J. Mol. Catal. A: Chem.*, 2013, **378**, 57–66.
16 H. R. Azimi, M. Rezaei and F. Majidi, *Polym. Degrad. Stab.*, 2014, **99**, 240–248.
17 D. K. Schneiderman, C. Gilmer, M. T. Wentzel, M. T. Martello, T. Kubo and J. E. Wissinger, *J. Chem. Educ.*, 2014, **91**, 131–135.
18 I. A. Khotina, R. Consonni, N. S. Kushakova, W. Porzio, U. Giovanella, A. I. Kovalev, M. A. Babushkina, A. S. Peregudov and S. Destri, *Eur. Polym. J.*, 2013, **49**, 4224–4237.
19 A. Racksanti, S. Janhom, S. Punyanitya, R. Watanesk and S. Watanesk, *Appl. Mech. Mater.*, 2014, **446–447**, 366–372.
20 Y. A. Mondjinou, L. A. McCauliff, A. Kulkarni, L. Paul, S.-H. Hyun, Z. Zhang, Z. Wu, M. Wirth, J. Storch and D. H. Thompson, *Biomacromolecules*, 2013, **14**, 4189–4197.
21 M. Wojtczak, S. Dutkiewicz, A. Galeski and E. Piorkowska, *Eur. Polym. J.*, 2014, **55**, 86–97.
22 C. G. Arges and V. Ramani, *J. Electrochem. Soc.*, 2013, **160**, F1006–F1021.

23 H. Kono, K. Onishi and T. Nakamura, *Carbohydr. Polym.*, 2013, **98**, 784–792.
24 C. Lin, S. Qiao, D. Liu and M. Liu, *Chin. J. Chem.*, 2013, **31**, 1551–1556.
25 D. K. Jindal and S. K. Singh, *PHARMANEST*, 2013, **4**, 1177–1185.
26 E. M. Dahmane, M. Taourirte, N. Eladlani and M. Rhazi, *Int. J. Polym. Anal. Charact.*, 2014, **19**, 342–351.
27 H. Shen, C. Kou, M. He, H. Yang and K. Liu, *J. Polym. Sci., Part A: Polym. Chem.*, 2014, **52**, 739–751.
28 A. Bougarech, M. Abid, F. Gouanve, E. Espuche, S. Abid, R. El Gharbi and E. Fleury, *Polymer*, 2013, **54**, 5482–5489.
29 P. Lo Meo, G. Lazzara, L. Liotta, S. Riela and R. Noto, *Polym. Chem.*, 2014, **5**, 4499–4510.
30 R. Hamzah, M. Abu Bakar, N. N. Zulkepli, M. A. A. Mohd. Salleh and M. M. Al Bakri Abdullah, *Adv. Mater. Res.*, 2013, **795**, 251–255.
31 N. Seeponkai, P. Khunsriya, J. Wootthikanokkhan and C. Thanachayanont, *Adv. Mater. Res.*, 2013, **747**, 317–320.
32 S. Yin, M. Ji and R. Yao, *Asian J. Chem.*, 2013, **25**, 7561–7566.
33 X. L. Wu, S. F. Kang, X. J. Xu, F. Xiao and X. L. Ge, *J. Appl. Polym. Sci.*, 2014, **131**, 40559.
34 W. Wu, C. Wan, H. Zhang and Y. Zhang, *J. Appl. Polym. Sci.*, 2014, **131**, 40272.
35 H. Kono, F. Otaka and M. Ozaki, *Carbohydr. Polym.*, 2014, **111**, 830–840.
36 A. Sultan Kedir, J. Georg Seland, A. Skauge and T. Skauge, *Energy Fuels*, 2014, **28**, 2343–2351.
37 S. Gogoi, S. Barua and N. Karak, *Prog. Org. Coat.*, 2014, **77**, 1418–1427.
38 X. Tang and Q. Kong, *Adv. Mater. Res.*, 2014, **830**, 167–171.
39 S. Moreno, T. Lozano-Cruz, P. Ortega, M. P. Tarazona, F. J. de la Mata and R. Gomez, *Polym. Int.*, 2014, **63**, 1311–1323.
40 C. Schuell, T. Gieshoff and H. Frey, *Polym. Chem.*, 2013, **4**, 4730–4736.
41 Y. Chen, Q. Zhang, W. Sun, X. Lei and P. Yao, *Polym. Int.*, 2014, **63**, 788–795.
42 G. Cheng, B. W. Greenland, C. Lampard, N. Williams, M. S. Bahra and W. Hayes, *Prog. Org. Coat.*, 2014, **77**, 1516–1522.
43 A. Singh, M. Kamal and D. K. Singh, *Int. J. Plast. Technol.*, 2013, **17**, 194–208.
44 G. Xie, X. Zhang, T. Li, L. Li, G. Liu and A. Zhang, *J. Mol. Catal. A: Chem.*, 2014, **383–384**, 121–127.
45 V. Kumar, V. Mundra and R. I. Mahato, *Pharm. Res.*, 2014, **31**, 1158–1169.
46 J. R. Rowlett, Y. Chen, A. T. Shaver, G. B. Fahs, B. J. Sundell, Q. Li, Y. S. Kim, P. Zelenay, R. B. Moore, S. Mecham and J. E. McGrath, *J. Electrochem. Soc.*, 2014, **161**, F535–F543.
47 K. S. Pafiti, C. S. Patrickios, C. Abetz and V. Abetz, *J. Polym. Sci., Part A: Polym. Chem.*, 2013, **51**, 4957–4965.
48 Y. Iwai, K. Sato and T. Yamanishi, *Fusion Eng. Des.*, 2014, **89**, 1534–1538.
49 P. Saramolee, K. Sahakaro, N. Lopattananon, W. K. Dierkes and J. W. M. Noordermeer, *Rubber Chem. Technol.*, 2014, **87**, 320–339.
50 M. Li, Y. Li, J. Zhang and S. Feng, *Polym. Eng. Sci.*, 2014, **54**, 355–363.
51 T. Arpornwichanop, D. Polpanich, R. Thiramanas, T. Suteewong and P. Tangboriboonrat, *Carbohydr. Polym.*, 2014, **109**, 1–6.
52 T. Wan, T. Du and S. Liao, *J. Appl. Polym. Sci.*, 2014, **131**, 40103.
53 T. Oliveira, G. Botelho, N. M. Alves and J. F. Mano, *Colloid Polym. Sci.*, 2014, **292**, 863–871.
54 H. Park, H.-K. Lee and T.-L. Choi, *Polym. Chem.*, 2013, **4**, 4676–4681.
55 N. Foerster, A.-C. Poeppler, D. Stalke and P. Vana, *Polymers*, 2013, **5**, 706–729.

56 A. Sakalyte, J. A. Reina, M. Giamberini and A. Lederer, *Polym. Eng. Sci.*, 2014, **54**, 579–591.
57 L. E. Gusel'nikov, V. V. Volkova, E. N. Buravtseva, N. V. Ushakov, V. G. Lakhtin, L. A. Parshkova and E. A. Chernyshev, *Russ. Chem. Bull.*, 2013, **62**, 1462–1464.
58 L. Hlalele, D. R. D'hooge, C. J. Duerr, A. Kaiser, S. Brandau and C. Barner-Kowollik, *Macromolecules*, 2014, **47**, 2820–2829.
59 J. Y. Jeon, E. Y. Hwang, S. C. Eo and B. Y. Lee, *J. Polym. Sci., Part A: Polym. Chem.*, 2014, **52**, 1570–1580.
60 F. Guo, M. Nishiura, Y. Li, Z. M. Hou, *Sci. China: Chem.*, 2014, **57** , 1150-1156.
61 J. F. Reuther and B. M. Novak, *J. Am. Chem. Soc.*, 2013, **135**, 19292–19303.
62 S. R. Samanta, R. Cai and V. Percec, *Polym. Chem.*, 2014, **5**, 5479–5491.
63 L Feng, Z. Yang, Y. Liu, J. Hao, C. Xiong and X. Deng, *Iran. Polym. J.*, 2014, **23**, 217–226.
64 J. R. Gois, N. Rocha, A. V. Popov, T. Guliashvili, K. Matyjaszewski, A. C. Serra and J. F. J. Coelho, *Polym. Chem.*, 2014, **5**, 3919–3928.
65 S. Pal, S. Ghosh Roy and P. De, *Polym. Chem.*, 2014, **5**, 1275–1284.
66 S. Lin and H. Cai, *J. Coat. Technol. Res.*, 2013, **10**, 841–847.
67 D. Sakae, J. Watanabe and Y. Ikeda, *Trans. Mater. Res. Soc. Jpn.*, 2013, **38**, 643–646.
68 R. V. Ashirov, D. I. Zemlyakov, A. A. Lyapkov and S. A. Kiselev, *Kinet. Catal.*, 2013, **54**, 469–474.
69 K. Takagi, E. Kawagita and R. Kouchi, *J. Polym. Sci., Part A: Polym. Chem.*, 2014, **52**, 2166–2174.
70 A. Stefanache, M. Silion, I. Stoica, A. Fifere, V. Harabagiu and A. Farcas, *Eur. Polym. J.*, 2014, **50**, 223–234.
71 T. Steinbach, R. Schroeder, S. Ritz and F. R. Wurm, *Polym. Chem.*, 2013, **4**, 4469–4479.
72 Y. Sanai, Y. Morita, Y. Asano, K. Ishizaki and K. Kubota, *J. Polym. Sci., Part A: Polym. Chem.*, 2014, **52**, 1161–1171.
73 A. Tamura, M. Tokunaga, Y. Iwasaki and N. Yui, *Macromol. Chem. Phys.*, 2014, **215**, 648–653.
74 L. Ricciotti, F. Borbone, A. Carella, R. Centore, A. Roviello, M. Barra, G. Roviello, C. Ferone, C. Minarini and P. Morvillo, *J. Polym. Sci., Part A: Polym. Chem.*, 2013, **51**, 4351–4360.
75 P. Mazzei, L. Fusco and A. Piccolo, *Magn. Reson. Chem.*, 2014, **52**, 383–388.
76 J. H. Bannock, M. Al-Hashimi, S. H. Krishnadasan, J. J. M. Halls, M. Heeney and J. C. de Mello, *Mater. Horiz.*, 2014, **1**, 214–218.
77 K. A. Smith, D. L. Pickel, K. Yager, K. Kisslinger and R. Verduzco, *J. Polym. Sci., Part A: Polym. Chem.*, 2014, **52**, 154–163.
78 R. Shen, L. W. Cai, D. Meng, L. P. Wu, K. Guo, G. X. Dong, L. Liu, J. C. Chen, Q. Wu and G. Q. Chen, *Sci. China: Life Sci.*, 2014, **57**, 82–91.
79 F. Pammer and U. Passlack, *ACS Macro Lett.*, 2014, **3**, 170–174.
80 L.-J. Zhang, X.-X. Deng, F.-S. Du and Z.-C. Li, *Macromolecules*, 2013, **46**, 9554–9562.
81 R. Chitra, E. Kayalvizhy, P. Jeyanthi and P. Pazhanisamy, *Chem. Sci. Trans.*, 2014, **3**, 722–730.
82 Y. Shinke, H. Yamamoto, A. Kanazawa, S. Kanaoka and S. Aoshima, *J. Polym. Sci., Part A: Polym. Chem.*, 2013, **51**, 4675–4683.
83 J. A. Burns, C. Houben, A. Anastasaki, C. Waldron, A. A. Lapkin and D. M. Haddleton, *Polym. Chem.*, 2013, **4**, 4809–4813.
84 H. Ono, H. Fujiwara and S. Nishimura, *Polym. J.*, 2013, **45**, 1027–1032.

85 Q. Guan, H. Zheng, J. Zhai, C. Zhao, X. Zheng, X. Tang, W. Chen and Y. Sun, *Ind. Eng. Chem. Res.*, 2014, **53**, 5624–5635.
86 I. Molina-Pinilla, M. Bueno-Martinez, K. Hakkou and J. A. Galbis, *J. Polym. Sci., Part A: Polym. Chem.*, 2014, **52**, 629–638.
87 B. J. Sundell, K. Lee, A. Nebipasagil, A. Shaver, J. R. Cook, E.-S. Jang, B. D. Freeman and J. E. McGrath, *Ind. Eng. Chem. Res.*, 2014, **53**, 2583–2593.
88 A. K. Mohanty, S. Banerjee, H. Komber and B. Voit, *Solid State Ionics*, 2014, **254**, 82–91.
89 Y. Wu, Y. Zhou, J. Zhu, W. Zhang, X. Pan, Z. Zhang and X. Zhu, *Polym. Chem.*, 2014, **5**, 5546–5550.
90 F. Cicogna, S. Coiai, P. Rizzarelli, S. Carroccio, C. Gambarotti, I. Domenichelli, C. Yang, N. T. Dintcheva, G. Filippone, C. Pinzino and E. Passaglia, *Polym. Chem.*, 2014, **5**, 5656–5667.
91 W. Guerin, A. K. Diallo, E. Kirilov, M. Helou, M. Slawinski, J.-M. Brusson, J.-F. Carpentier and S. M. Guillaume, *Macromolecules*, 2014, **47**, 4230–4235.
92 E.-H. Kang and T.-L. Choi, *ACS Macro Lett.*, 2013, **2**, 780–784.
93 J. P. Lewicki, M. A. Worsley, R. L. F. Albo, J. A. Finnie, M. Ashmore, H. E. Mason, T. F. Baumann and R. S. Maxwell, *Polym. Degrad. Stab.*, 2014, **102**, 25–32.
94 L. M. Pitet, A. H. M. van Loon, E. J. Kramer, C. J. Hawker and E. W. Meijer, *ACS Macro Lett.*, 2013, **2**, 1006–1010.
95 V. A. Kraikin, A. A. Fatykhov, I. I. Sakhipova, E. A. Sedova, A. E. Egorov and S. N. Salazkin, *Magn. Reson. Chem.*, 2013, **51**, 621–629.
96 Y. Sun and C. He, *Macromolecules*, 2013, **46**, 9625–9633.
97 M. Jikei, D. Uchida, K. Matsumoto, R. Komuro and M. Sugimoto, *J. Polym. Sci., Part A: Polym. Chem.*, 2014, **52**, 1825–1831.
98 K. Matsumoto, H. Komuro, T. Kai and M. Jikei, *Polym. J.*, 2013, **45**, 909–914.
99 H. Morinaga, Y. Ujihara and T. Endo, *J. Polym. Sci., Part A: Polym. Chem.*, 2013, **51**, 4451–4458.
100 A. H. Najafabadi, M. Abdouss and S. Faghihi, *J. Nanopart. Res.*, 2014, **16**, 2312.
101 D. Lu, P. Duan, T. Liu, J. Li, T. Li and Z. Lei, *Polym. Int.*, 2014, **63**, 244–251.
102 L. Yu, L. Yao, J. You, Y. Guo and L. Yang, *J. Appl. Polym. Sci.*, 2014, **131**, 39623.
103 N. J. Warren, O. O. Mykhaylyk, D. Mahmood, A. J. Ryan and S. P. Armes, *J. Am. Chem. Soc.*, 2014, **136**, 1023–1033.
104 C.-S. Yang, H.-C. Wu, J.-S. Sun, H.-M. Hsiao and T.-W. Wang, *ACS Appl. Mater. Interfaces*, 2013, **5**, 10985–10994.
105 S. Park, H. Y. Cho, K. B. Wegner, J. Burdynska, A. J. D. Magenau, H. J. S. Paik and K. Matyjaszewski, *Macromolecules*, 2013, **46**, 5856–5860.
106 M.-L. Zhao, F.-X. Li, J.-Y. Yu and X.-L. Wang, *J. Appl. Polym. Sci.*, 2014, **131**, 39823.
107 W. Huang, Y. Wan, J. Chen, Q. Xu, X. Li, X. Yang, Y. Li and Y. Tu, *Polym. Chem.*, 2014, **5**, 945–954.
108 S. Losio, G. Leone, F. Bertini, G. Ricci, M. C. Sacchi and A. C. Boccia, *Polym. Chem.*, 2014, **5**, 2065–2075.
109 W. Wang, L. Hou, S. Luo, G. Zheng and H. Wang, *Macromol. Chem. Phys.*, 2013, **214**, 2245–2249.
110 S.-W. Zhang, L.-P. Lu, Y.-Y. Long and Y.-S. Li, *J. Polym. Sci., Part A: Polym. Chem.*, 2013, **51**, 5298–5306.
111 E. Cossoul, L. Baverel, E. Martigny, T. Macko, C. Boisson and O. Boyron, *Macromol. Symp.*, 2013, **330**, 42–52.
112 I. Suarez, S. Losio and B. Coto, *Eur. Polym. J.*, 2013, **49**, 3402–3409.

113 A. Alkan, A. Natalello, M. Wagner, H. Frey and F. R. Wurm, *Macromolecules*, 2014, **47**, 2242–2249.
114 Y. Marquez, L. Franco, P. Turon, A. Rodriguez-Galan and J. Puiggali, *Polym. Degrad. Stab.*, 2013, **98**, 2709–2721.
115 S.-G. Hong, H.-W. Hsu and M.-T. Ye, *J. Appl. Polym. Sci.*, 2014, **131**, 39501.
116 D. Khandelwal, S. Hooda, A. S. Brar and R. Shankar, *J. Mol. Struct.*, 2013, **1049**, 99–111.
117 O. V. Smetannikov, A. N. Tavtorkin, I. E. Nifant'ev, M. S. Chinova and I. F. Gavrilenko, *Polym. Sci., Ser. B*, 2013, **55**, 453–459.
118 D. A. L. Otte, D. E. Borchmann, C. Lin, M. Weck and K. A. Woerpel, *Org. Lett.*, 2014, **16**, 1566–1569.
119 W. Chumeka, P. Pasetto, J.-F. Pilard and V. Tanrattanakul, *Polymer*, 2014, **55**, 4478–4487.
120 H. Morinaga, Y. Kiyokawa, R. Fujikawa, D. Nagai and H. Morikawa, *Polym. Bull.*, 2014, **71**, 2421–2435.
121 M. Yuan, Z. He, H. Li, L. Jiang and M. Yuan, *Polym. Bull.*, 2014, **71**, 1331–1347.
122 Y. Sasanuma and D. Touge, *Polymer*, 2014, **55**, 1901–1911.
123 J. Xu, X. Fan, J. Yang, C. Ma, X. Ye and G. Zhang, *Colloids Surf., B*, 2014, **116**, 531–536.
124 B. Pidhatika, Y. Chen, G. Coullerez, S. Al-Bataineh and M. Textor, *Anal. Bioanal. Chem.*, 2014, **406**, 1509–1517.
125 K. E. S. Locock, L. Meagher and M. Haeussler, *Anal. Chem.*, 2014, **86**, 2131–2137.
126 C. M. R. Abreu, A. C. Serra, A. V. Popov, K. Matyjaszewski, T. Guliashvili and J. F. J. Coelho, *Polym. Chem.*, 2013, **4**, 5629–5636.
127 S. Liu, S. Srinivasan, M. C. Grady, M. Soroush and A. M. Rappe, *Int. J. Quantum Chem.*, 2014, **114**, 345–360.
128 M. Hehn, K. Maiko, H. Pasch and W. Hiller, *Macromolecules*, 2013, **46**, 7678–7686.
129 G. Soriano-Moro, J. Percino, M. Ceron, A. Banuelos, V. M. Chapela and M. E. Castro, *Macromol. Symp.*, 2014, **339**, 112–121.
130 Y. Zhang, K. Schroder, Y. Kwak, P. Krys, A. N. Morin, T. Pintauer, R. Poli and K. Matyjaszewski, *Macromolecules*, 2013, **46**, 5512–5519.
131 V. M. Ovando-Medina, M. A. Corona-Rivera, A. Marquez-Herrera, T. E. Lara-Ceniceros, R. Manriquez-Gonzalez and R. D. Peralta, *J. Appl. Polym. Sci.*, 2014, **131**, 40191.
132 H. W. P. Carvalho, A. F. Suzana, C. V. Santilli and S. H. Pulcinelli, *Polym. Degrad. Stab.*, 2014, **104**, 112–119.
133 W. Yang, D. Xie, X. Sheng and X. Zhang, *Ind. Eng. Chem. Res.*, 2013, **52**, 13466–13476.
134 D. Jagan Mohan, *Des. Monomers Polym.*, 2014, **17**, 438–444.
135 N. Mohr, M. Barz, R. Forst and R. Zentel, *Macromol. Rapid Commun.*, 2014, **35**, 1522–1527.
136 S. Wallyn, Z. Zhang, F. Driessen, J. Pietrasik, B. G. De Geest, R. Hoogenboom and F. E. Du Prez, *Macromol. Rapid Commun.*, 2014, **35**, 405–411.
137 X. Zhang, B. H. Tan and C. He, *Macromol. Rapid Commun.*, 2013, **34**, 1761–1766.
138 A. P. Chiriac, L. E. Nita, I. Neamtu, V. Balan and A. Diaconu, *J. Res. Updates Polym. Sci.*, 2014, **3**, 48–56.
139 M. Vatankhah-Varnoosfaderani, S. Hashmi, A. GhavamiNejad and F. J. Stadler, *Polym. Chem.*, 2014, **5**, 512–523.

140 A. Balaceanu, Y. Verkh, D. Kehren, W. Tillmann and A. Pich, *Z. Phys. Chem.*, 2014, **228**, 253–267.
141 H. Valle, J. Sanchez and B. L. Rivas, *J. Appl. Polym. Sci.*, 2014, **132**, 41272.
142 S. Srichan, N. Kayunkid, L. Oswald, B. Lotz and J.-F. Lutz, *Macromolecules*, 2014, **47**, 1570–1577.
143 S. Wilfert, A. Iturmendi, W. Schoefberger, K. Kryeziu, P. Heffeter, W. Berger, O. Brueggemann and I. Teasdale, *J. Polym. Sci., Part A: Polym. Chem.*, 2014, **52**, 287–294.
144 A. Natalello, M. Werre, A. Alkan and H. Frey, *Macromolecules*, 2013, **46**, 8467–8471.
145 Y.-X. Xue, Y.-Y. Zhu, L.-M. Gao, X.-Y. He, N. Liu, W.-Y. Zhang, J. Yin, Y. Ding, H. Zhou and Z. -Q Wu, *J. Am. Chem. Soc.*, 2014, **136**, 4706–4713.
146 K. Nakabayashi, Y. Abiko and H. Mori, *Macromolecules*, 2013, **46**, 5998–6012.
147 L. Luo, T. Qiu, Y. Meng, L. Guo, J. Yang, Z. Li, X. Cao and X. Li, *RSC Adv.*, 2013, **3**, 14509–14520.
148 A. Buonerba, M. Fienga, S. Milione, C. Cuomo, A. Grassi, A. Proto and C. Capacchione, *Macromolecules*, 2013, **46**, 8449–8457.
149 J.-L. Hong, X.-H. Zhang, R.-J. Wei, Q. Wang, Z.-Q. Fan and G.-R. Qi, *J. Appl. Polym. Sci.*, 2014, **131**, 39847.
150 H. Ghassemi and D. A. Schiraldi, *J. Appl. Polym. Sci.*, 2014, **131**, 39815.
151 A. Garcia-Penas, J. M. Gomez-Elvira, E. Perez and M. L. Cerrada, *J. Polym. Sci., Part A: Polym. Chem.*, 2014, **52**, 2537–2547.
152 L. Botha, P. Sinha, S. Joubert, H. Duveskog and A. J. van Reenen, *Eur. Polym. J.*, 2014, **59**, 94–104.
153 A. C. Makan, P. Sinha, N. Ngaza, W. van Aswegen and H. Pasch, *Anal. Bioanal. Chem.*, 2013, **405**, 9041–9047.
154 J. M. Giussi, I. Blaszczyk-Lezak, B. Sanz, P. E. Allegretti, C. Mijangos and M. S. Cortizo, *Eur. Polym. J.*, 2014, **59**, 84–93.
155 U. Sevsek, J. Brus, K. Jerabek and P. Krajnc, *Polymer*, 2014, **55**, 410–415.
156 Q. Xu, L. Li, F. Guo, Z. Shi, H. Ma, Y. Wang, Y. Wang and Y. Li, *Polym. Eng. Sci.*, 2014, **54**, 1858–1863.
157 M. Monroy-Barreto, M. C. Perez-Vazquez, M. Briseno-Teran and N. Esturau-Escofet, *Int. J. Polym. Anal. Charact.*, 2014, **19**, 22–30.
158 A. Natalello, A. Alkan, P. von Tiedemann, F. R. Wurm and H. Frey, *ACS Macro Lett.*, 2014, **3**, 560–564.
159 I. P. Kim, A. A. Kunitsa and A. V. Chernyak, *Russ. J. Phys. Chem. A*, 2013, **87**, 1845–1850.
160 M. Farrokhi, M. Abdollahi and M. Hemmati, *Polym. Int.*, 2014, **63**, 1494–1504.
161 S. Spoljaric, A. Salminen, N. Dang Luong, P. Lahtinen, J. Vartiainen, T. Tammelin and J. Seppaelae, *Polym. Compos.*, 2014, **35**, 1117–1131.
162 M. S. Lee and J.-C. Kim, *Polym. Eng. Sci.*, 2014, **54**, 227–233.
163 K. Miyazaki, H. Sato, S. Kikuchi and H. Nakatani, *J. Appl. Polym. Sci.*, 2014, **131**, 40760.
164 A. H. Ma'Radzi, S. Sugihara, S. Miura, N. Konegawa and Y. Maeda, *Polymer*, 2014, **55**, 1920–1930.
165 E. B. Twum, E. F. McCord, D. F. Lyons, P. A. Fox and P. L. Rinaldi, *Eur. Polym. J.*, 2014, **51**, 136–150.
166 L. Li, E. B. Twum, X. Li, E. F. McCord, P. A. Fox, D. F. Lyons and P. L. Rinaldi, *Macromolecules*, 2013, **46**, 7146–7157.
167 P. Olsen, J. Undin, K. Odelius and A.-C. Albertsson, *Polym. Chem.*, 2014, **5**, 3847–3854.
168 D. Tang, C. W. Macosko and M. A. Hillmyer, *Polym. Chem.*, 2014, **5**, 3231–3237.

169 S. Kan, Y. Jin, X. He, J. Chen, H. Wu, P. Ouyang, K. Guo and Z. Li, *Polym. Chem.*, 2013, **4**, 5432–5439.
170 A. Petit, B. Mueller, R. Meijboom, P. Bruin, F. van de Manakker, M. Versluijs-Helder, L. G. J. de Leede, A. Doornbos, M. Landin, W. E. Hennink and T. Vermonden, *Biomacromolecules*, 2013, **14**, 3172–3182.
171 F. Zhang and X. Liu, *Chin. J. Polym. Sci.*, 2013, **31**, 1613–1622.
172 A. Mirzadeh, H. Ghasemi, P. J. Bates and M. R. Kamal, *Int. Polym. Process.*, 2014, **29**, 4–12.
173 M. Shabanian, N.-J. Kang, D.-Y. Wang, U. Wagenknecht and G. Heinrich, *RSC Adv.*, 2013, **3**, 20738–20745.
174 Y. Wang, X. N. Guan, C.-Y. Wu, M.-T. Chen, H.-H. Hsieh, H. D. Tran, S.-C. Huang and R. B. Kaner, *Polym. Chem.*, 2013, **4**, 4814–4820.
175 A. K. Singh, R. Prakash and D. Pandey, *RSC Adv.*, 2013, **3**, 15411–15420.
176 J. Hui, Y. Shi and Z. Fu, *Adv. Mater. Res.*, 2013, **787**, 241–244.
177 S. Yang, Z. Zhang, F. Wang, L. Feng and T. Wang, *J. Polym. Res.*, 2013, **20**, 1–11.
178 A. A. Alsaygh, J. Al-hamidi, F. D. Alsewailem, I. M. Al-Najjar and V. L. Kuznetsov, *Appl. Petrochem. Res.*, 2014, **4**, 79–84.
179 I. E. Soshnikov, N. V. Semikolenova, V. A. Zakharov, H. M. Moeller, F. Oelscher, A. Osichow, I. Goettker-Schnettmann, S. Mecking, E. P. Talsi and K. P. Bryliakov, *Chem. – Eur. J.*, 2013, **19**, 11409–11417.
180 K. Huang, S. Zhou, D. Zhang, X. Gao, Q. Wang and Y. Lin, *J. Organomet. Chem.*, 2013, **741–742**, 83–90.
181 N. Prosa, M.-C. Scherrmann, D. Merlet and J. Farjon, *J. Magn. Reson.*, 2013, **237**, 63–72.
182 H. Jiang, F. He, J. Lu and Y. Xing, *J. Mol. Catal. A: Chem.*, 2013, **380**, 139–143.
183 I. Blanco, F. A. Bottino, G. Cicala, A. Latteri and A. Recca, *Polym. Degrad. Stab.*, 2013, **98**, 2564–2570.
184 L. Montano-Herrera, S. Pratt, M. V. Arcos-Hernandez, P. J. Halley, P. A. Lant, A. Werker and B. Laycock, *New Biotechnol.*, 2014, **31**, 357–363.
185 X. F. Lei, Y. Chen, H. P. Zhang, X. J. Li, P. Yao and Q. Y. Zhang, *ACS Appl. Mater. Interfaces*, 2013, **5**, 10207–10220.
186 N. Sakai, M. Jin, S. Sato, T. Satoh and T. Kakuchi, *Polym. Chem.*, 2014, **5**, 1057–1062.
187 M. Hehn, T. Wagner and W. Hiller, *Anal. Chem.*, 2014, **86**, 490–497.
188 C. Lim, L. Ghassemzadeh, F. Van Hove, M. Lauritzen, J. Kolodziej, G. G. Wang, S. Holdcroft and E. Kjeang, *J. Power Sources*, 2014, **257**, 102–110.
189 M. A. Semsarzadeh and S. Amiri, *Silicon*, 2014, **6**, 27–34.
190 Q. Feng, D. Chen, D. Feng, L. Jiao, Z. Peng and L. Pei, *Appl. Organomet. Chem.*, 2014, **28**, 32–37.
191 P. Kandanarachchi, C. Chang, S. Smith, P. Bradley, L. F. Rhodes, R. P. Lattimer and G. M. Benedikt, *J. Photopolym. Sci. Technol.*, 2013, **26**, 431–439.
192 J. Deng, H. Gao, F. Zhu and Q. Wu, *Organometallics*, 2013, **32**, 4507–4515.
193 R. C. Klet, C. N. Theriault, J. Klosin, J. A. Labinger and J. E. Bercaw, *Macromolecules*, 2014, **47**, 3317–3324.
194 I. Amer and A. van Reenen, *e-Polym.*, 2013, 26, **5**, 6510–6522.
195 Y.-Y. Long, Y.-X. Wang, B.-X. Li, Y.-G. Li and Y.-S. Li, *Polym. Chem.*, 2014.
196 L. A. Rishina, S. S. Lalayan, S. C. Gagieva, V. A. Tuskaev, E. O. Perepelitsyna and Y. V. Kissin, *Polymer*, 2013, **54**, 6526–6535.
197 M. A. Semsarzadeh and S. Amiri, *Bull. Mater. Sci.*, 2013, **36**, 989–996.
198 P. D. Cooper, T. G. Barclay, M. Ginic-Markovic and N. Petrovsky, *Glycobiology*, 2013, **23**, 1164–1174.

199 G. J. Ray, N. Ravenscroft, J. Siekmann, Z. Zhang, P. Sanders, U. Shaligram, C. M. Szabo and P. Kosma, *Bioconjugate Chem.*, 2014, **25**, 665–676.
200 A. Feigl, I. Chiorescu, K. Deller, S. U. H. Heidsieck, M. R. Buchner, V. Karttunen, A. Bockholt, A. Genest, N. Roesch and B. Rieger, *Chem. – Eur. J.*, 2013, **19**, 12526–12536.
201 M. Wilamowska, V. S. Pradeep, M. Graczyk-Zajac, R. Riedel and G. D. Soraru, *Solid State Ionics*, 2014, **260**, 94–100.
202 Q. Zhou, P. Gu, Y. Zhang, N. Li, Q. Xu, Y. Zhang and J. Lu, *Chin. J. Chem.*, 2014, **32**, 573–578.
203 L.-W. Zhu, W. Yang, Y. Ou, L.-S. Wan and Z.-K. Xu, *Polym. Chem.*, 2014, **5**, 3666–3672.
204 K. Khezri, V. Haddadi-Asl and H. Roghani-Mamaqani, *NANO*, 2014, **9**, 1450023.
205 J. M. Giussi, I. Blaszczyk-Lezak, M. S. Cortizo and C. Mijangos, *Polymer*, 2013, **54**, 6886–6893.
206 W. Cho, H. Cho, C. S. Lee, B. Y. Lee, B. Moon and J. Kang, *Organometallics*, 2014, **33**, 1617–1622.
207 S. Pal, A. Dhawan and P. De, *Polym. Int.*, 2014, **63**, 746–751.
208 C. Jangu, J.-H. H. Wang, D. Wang, S. Sharick, J. R. Heflin, K. I. Winey, R. H. Colby and T. E. Long, *Macromol. Chem. Phys.*, 2014, **215**, 1319–1331.
209 A. Tardy, V. Delplace, D. Siri, C. Lefay, S. Harrisson, B. F. A. Pereira, L. Charles, D. Gigmes, J. Nicolas and Y. Guillaneuf, *Polym. Chem.*, 2013, **4**, 4776–4787.
210 J.-H. Ye, *Adv. Mater. Res.*, 2013, **699**, 167–169.
211 S. D. Rajput, P. P. Mahulikar and V. V. Gite, *Prog. Org. Coat.*, 2014, **77**, 38–46.
212 Y. Wang, A. Rapakousiou and D. Astruc, *Macromolecules*, 2014, **47**, 3767–3774.
213 M. Rodriguez-Evora, A. Delgado, R. Reyes, A. Hernandez-Daranas, I. Soriano, J. San Roman and C. Evora, *Eur. J. Pharm. Sci.*, 2013, **49**, 873–884.
214 A. Formes and B. Diehl, *J. Pharm. Biomed. Anal.*, 2014, **93**, 95–101.
215 A. Momeni and M. J. Filiaggi, *J. Non-Cryst. Solids*, 2013, **382**, 11–17.
216 K. Suchao-in, P. Koombhongse and S. Chirachanchai, *Carbohydr. Polym.*, 2014, **102**, 95–102.
217 Z. Zhang, P. Chen, X. Du, Z. Xue, S. Chen and B. Yang, *Carbohydr. Polym.*, 2014, **102**, 453–459.
218 Z.-L. Gong, L. Cen, S.-T. Wang and F.-L. Chen, *J. Appl. Polym. Sci.*, 2014, **131**, 40462.
219 O. van den Berg, L.-T. T. Nguyen, R. F. A. Teixeira, F. Goethals, C. Ozdilek, S. Berghmans and F. E. Du Prez, *Macromolecules*, 2014, **47**, 1292–1300.
220 W. J. Yoon, S. Y. Hwang, J. M. Koo, Y. J. Lee, S. U. Lee and S. S. Im, *Macromolecules*, 2013, **46**, 7219–7231.
221 L. Annunziata, A. K. Diallo, S. Fouquay, G. Michaud, F. Simon, J.-M. Brusson, J.-F. Carpentier and S. M. Guillaume, *Green Chem.*, 2014, **16**, 1947–1956.
222 S. Hajji, I. Younes, O. Ghorbel-Bellaaj, R. Hajji, M. Rinaudo, M. Nasri and K. Jellouli, *Int. J. Biol. Macromol.*, 2014, **65**, 298–306.
223 E. P. Zhiltsova, V. V. Syakaev, S. S. Lukashenko, A. P. Timosheva, R. R. Kashapov, S. K. Latypov, L. Y. Zakharova and A. I. Konovalov, *Fluid Phase Equilib.*, 2013, **360**, 16–22.
224 M. Klein, J.-C. Perrin, S. Leclerc, L. Guendouz, J. Dillet and O. Lottin, *Macromolecules*, 2013, **46**, 9259–9269.
225 S. Ok, A. Altun, R. Kasimogullari and E. Sen, *J. Chem. Eng. Data*, 2013, **58**, 3521–3527.

226 G. T. M. Nguyen, A. L. Michan, A. Fannir, M. Viallon, C. Vancaeyzeele, C. A. Michal and F. Vidal, *Eur. Polym. J.*, 2013, **49**, 4108–4117.

227 J. Hou, J. Li, D. Mountz, M. Hull and L. A. Madsen, *J. Membr. Sci.*, 2013, **448**, 292–299.

228 H.-Y. Wu, Y.-H. Chen, D. Saikia, Y.-Y. Pan, J. Fang, L.-D. Tsai and H.-M. Kao, *J. Membr. Sci.*, 2013, **47**, 274–286.

229 A. Chiappone, S. Jeremias, R. Bongiovanni and M. Schoenhoff, *J. Polym. Sci., Part B: Polym. Phys.*, 2013, **51**, 1571–1580.

230 A.-C. Poeppler, S. Frischkorn, D. Stalke and M. John, *ChemPhysChem*, 2013, **14**, 3103–3107.

231 Y. Wang, J. Gao, T. J. Dingemans and L. A. Madsen, *Macromolecules*, 2014, **47**, 2984–2992.

232 A. Conti, G. Poggi, P. Baglioni and F. De Luca, *Phys. Chem. Chem. Phys.*, 2014, **16**, 8409–8417.

233 C. F. Kins, E. Sengupta, A. Kaltbeitzel, M. Wagner, I. Lieberwirth, H. W. Spiess and M. R. Hansen, *Macromolecules*, 2014, **47**, 2645–2658.

234 S. G. Vasil'ev, V. I. Volkov, E. A. Tatarinova and A. M. Muzafarov, *Appl. Magn. Reson.*, 2014, **45**, 315–328.

235 S. Himmelein, N. Sporenberg, M. Schoenhoff and B. J. Ravoo, *Langmuir*, 2014, **30**, 3988–3995.

236 V. Compan, S. Molla, S. Vallejos, F. Garcia, J. M. Garcia, J. Guzman and L. Garrido, *Macromol. Chem. Phys.*, 2014, **215**, 624–637.

237 S. Y. Kim, J. Lee and M. J. Park, *Macromolecules*, 2014, **47**, 1099–1108.

238 R. Bai, P. J. Basser, R. M. Briber and F. Horkay, *J. Appl. Polym. Sci.*, 2014, **131**, 40001.

239 R. M. F. Fernandes, M. Buzaglo, M. Shtein, I. Pri Bar, O. Regev, E. F. Marques and I. Furo, *J. Phys. Chem. C*, 2014, **118**, 582–589.

240 J. L. Horan, A. Lingutla, H. Ren, M.-C. Kuo, S. Sachdeva, Y. Yang, S. Seifert, L. F. Greenlee, M. A. Yandrasits, S. J. Hamrock, M. H. Frey and A. M. Herring, *J. Phys. Chem. C*, 2014, **118**, 135–144.

241 B. Jiang, Y. Yang, L. Du, C. Mattea, J. Wang, S. Stapf and Y. Yang, *J. Appl. Polym. Sci.*, 2014, **131**, 40161.

242 P. M. Richardson, A. M. Voice and I. M. Ward, *Int. J. Hydrogen Energy*, 2014, **39**, 2904–2908.

243 A. A. Taqa, A. K. Al-Noori and A. I. J. Al-Khyaat, *Am. J. Med. Sci. Med.*, 2013, **1**, 62–65.

244 M. Engelsberg and W. Barros Jr., *Phys. Rev. E: Stat., Nonlinear, Soft Matter Phys.*, 2013, **88**, 062602.

245 P. Knoeoes, D. Topgaard, M. Wahlgren, S. Ulvenlund and L. Piculell, *Langmuir*, 2013, **29**, 13898–13908.

246 P. Knoeoes, M. Wahlgren, D. Topgaard, S. Ulvenlund and L. Piculell, *J. Phys. Chem. B*, 2014, **118**, 9757–9767.

247 N. J. W. Reuvers, H. P. Huinink, H. R. Fischer and O. C. G. Adan, *Polymer*, 2014, **55**, 2051–2058.

248 M. Perrier, S. Kenouche, J. Long, K. Thangavel, J. Larionova, C. Goze-Bac, A. Lascialfari, M. Mariani, N. Baril, C. Guerin, B. Donnadieu, A. Trifonov and Y. Guari, *Inorg. Chem.*, 2013, **52**, 13402–13414.

249 G. Amsaveni, A. S. Farook, V. Haribabu, R. Murugesan and A. Girigoswami, *Adv. Sci., Eng. Med.*, 2013, **5**, 1340–1348.

250 S. Okada, S. Mizukami, Y. Matsumura, Y. Yoshioka and K. Kikuchi, *Dalton Trans.*, 2013, **42**, 15864–15867.

251 Z. Zheng, A. Daniel, W. Yu, B. Weber, J. Ling and A. H. E. Mueller, *Chem. Mater.*, 2013, **25**, 4585–4594.

252 J. S. Basuki, H. T. T. Duong, A. MacMillan, R. B. Erlich, L. Esser, M. C. Akerfeldt, R. M. Whan, M. Kavallaris, C. Boyer and T. P. Davis, *ACS Nano*, 2013, **7**, 10175–10189.
253 J. P. Bullivant, S. Zhao, B. J. Willenberg, B. Kozissnik, C. D. Batich and J. Dobson, *Int. J. Mol. Sci.*, 2013, **14**, 17501–17510.
254 M. H. Pablico-Lansigan, W. J. Hickling, E. A. Japp, O. C. Rodriguez, A. Ghosh, C. Albanese, M. Nishida, E. Van Keuren, S. Fricke, N. Dollahon and S. L. Stoll, *ACS Nano*, 2013, 7, 9040–9048.
255 P. Arosio, J. Thevenot, T. Orlando, F. Orsini, M. Corti, M. Mariani, L. Bordonali, C. Innocenti, C. Sangregorio, H. Oliveira, S. Lecommandoux, A. Lascialfari and O. Sandre, *J. Mater. Chem. B*, 2013, **1**, 5317–5328.
256 J. H. Bong, S.-I. Kwon, K. B. Kim, M. S. Kim, H. J. Jung, Y. H. Ji, I. O. Ko, J. A. Park and K. M. Kim, *J. Korean Phys. Soc.*, 2013, **63**, 1083–1087.
257 S. El Habnouni, B. Nottelet, V. Darcos, B. Porsio, L. Lemaire, F. Franconi, X. Garric and J. Coudane, *Biomacromolecules*, 2013, **14**, 3626–3634.
258 K. Ohno, C. Mori, T. Akashi, S. Yoshida, Y. Tago, Y. Tsujii and Y. Tabata, *Biomacromolecules*, 2013, **14**, 3453–3462.
259 J. Li, L. Zheng, H. Cai, W. Sun, M. Shen, G. Zhang and X. Shi, *Biomaterials*, 2013, **34**, 8382–8392.
260 J. S. Basuki, L. Esser, P. B. Zetterlund, M. R. Whittaker, C. Boyer and T. P. Davis, *Macromolecules*, 2013, **46**, 6038–6047.
261 D. Perego, N. Masciocchi, A. Guagliardi, J. M. Dominguez-Vera and N. Galvez, *Nanotechnology*, 2013, **24**, 075102.
262 H. Hu, J. Sun, G. Huang, X. Li, A. Dai, H. Yang and S. Yang, *J. Mater. Sci.*, 2013, **48**, 7686–7695.
263 K. Wang, H. Peng, K. J. Thurecht, S. Puttick and A. K. Whittaker, *Polym. Chem.*, 2013, **4**, 4480–4489.
264 S. Aryal, J. Key, C. Stigliano, J. S. Ananta, M. Zhong and P. Decuzzi, *Biomaterials*, 2013, **34**, 7725–7732.
265 N. M. Barkey, C. Preihs, H. H. Cornnell, G. Martinez, A. Carie, J. Vagner, L. Xu, M. C. Lloyd, V. M. Lynch, V. J. Hruby, J. L. Sessler, K. N. Sill, R. J. Gillies and D. L. Morse, *J. Med. Chem.*, 2013, **56**, 6330–6338.
266 H. Huang, Y. Li, C. Li, Y. Wang, Y. Sun and J. Wang, *Macromol. Biosci.*, 2013, **13**, 1358–1368.
267 S. Decato, E. Madsen, T. Bemis and S. Mecozzi, *Polym. Chem.*, 2014, **5**, 6461–6471.
268 X. Yang, Y. Sun, S. Kootąla, J. Hilborn, A. Heerschap and D. Ossipov, *Carbohydr. Polym.*, 2014, **110**, 95–99.
269 V. M. Suresh, S. Chatterjee, R. Modak, V. Tiwari, A. B. Patel, T. K. Kundu and T. K. Maji, *J. Phys. Chem. C*, 2014, **118**, 12241–12249.
270 Y. Zhong, F. Dai, H. Deng, M. Du, X. Zhang, Q. Liu and X. Zhang, *J. Mater. Chem. B*, 2014, **2**, 2938–2946.
271 C.-S. Tsai, W.-C. Liu, H.-Y. Chen and W.-C. Hsu, *Appl. Mech. Mater.*, 2014, **459**, 51–59.
272 C. E. Smith and H. Kong, *Langmuir*, 2014, **30**, 3697–3704.
273 S. S. Kelkar, L. Xue, S. R. Turner and T. M. Reineke, *Biomacromolecules*, 2014, **15**, 1612–1624.
274 Y. Li, S. Laurent, L. Esser, L. Vander Elst, R. N. Muller, A. B. Lowe, C. Boyer and T. P. Davis, *Polym. Chem.*, 2014, **5**, 2592–2601.
275 N. Rocha, J. Mendes, L. Duraes, H. Maleki, A. Portugal, C. F. G. C. Geraldes, A. Serra and J. Coelho, *J. Mater. Chem. B*, 2014, **2**, 1565–1575.
276 Z. Cheng, A. Al Zaki, I. W. Jones, H. K. Hall, C. A. Aspinwall and A. Tsourkas, *Chem. Commun.*, 2014, **50**, 2502–2504.

277 K. Wang, H. Peng, K. J. Thurecht, S. Puttick and A. K. Whittaker, *Polym. Chem.*, 2014, **5**, 1760–1771.
278 N. Pothayee, N. Pothayee, N. Hu, R. Zhang, D. F. Kelly, A. P. Koretsky and J. S. Riffle, *J. Mater. Chem. B*, 2014, **2**, 1087–1099.
279 S. Qazi, M. Uchida, R. Usselman, R. Shearer, E. Edwards and T. Douglas, *JBIC, J. Biol. Inorg. Chem.*, 2014, **19**, 237–246.
280 D. G. You, G. Saravanakumar, S. Son, H. S. Han, R. Heo, K. Kim, I. C. Kwon, J. Y. Lee and J. H. Park, *Carbohydr. Polym.*, 2014, **101**, 1225–1233.
281 H. Yang, C. Qin, C. Yu, Y. Lu, H. Zhang, F. Xue, D. Wu, Z. Zhou and S. Yang, *Adv. Funct. Mater.*, 2014, **24**, 1738–1747.
282 K. S. Kim, W. Park, J. Hu, Y. H. Bae and K. Na, *Biomaterials*, 2014, **35**, 337–343.
283 H.-W. Zhang, L.-Q. Wang, Q.-F. Xiang, Q. Zhong, L.-M. Chen, C.-X. Xu, X.-H. Xiang, B. Xu, F. Meng, Y.-Q. Wan and D. Y. B. Deng, *Biomaterials*, 2014, **35**, 356–367.
284 K. Yan, H. Li, P. Li, H. Zhu, J. Shen, C. Yi, S. Wu, K. W. K. Yeung, Z. Xu, H. Xu and P. K. Chu, *Biomaterials*, 2014, **35**, 344–355.
285 M. Worden, H. Wang, A. Paravastu and W. Oates, *Smart Mater. Struct.*, 2013, **22**, 094013.
286 C. Sun, P. Hazendonk, A. S. Borisov and P. G. Hayes, *Adv. Mater. Res.*, 2013, **704**, 235–240.
287 N. Seeponkai, P. Khunsriya, J. Wootthikanokkhan and C. Thanachayanont, *Adv. Mater. Res.*, 2013, **747**, 317–320.
288 S. Jiang, Y. Yao, Q. Chen and Y. Chen, *Macromolecules*, 2013, **46**, 9688–9697.
289 E. Cappelletto, S. Borsacchi, M. Geppi, F. Ridi, E. Fratini and P. Baglioni, *J. Phys. Chem. C*, 2013, **117**, 22947–22953.
290 C. G. Arges and V. Ramani, *J. Electrochem. Soc.*, 2013, **160**, F1006–F1021.
291 M. Simeonova, M. Rangel and G. Ivanova, *Phys. Chem. Chem. Phys.*, 2013, **15**, 16657–16664.
292 F. Blanc, S. Y. Chong, T. O. McDonald, D. J. Adams, S. Pawsey, M. A. Caporini and A. I. Cooper, *J. Am. Chem. Soc.*, 2013, **135**, 15290–15293.
293 L. Li, E. B. Twum, X. Li, E. F. McCord, P. A. Fox, D. F. Lyons and P. L. Rinaldi, *Macromolecules*, 2013, **46**, 7146–7157.
294 Z. Li, T. Miyoshi, M. K. Sen, T. Koga, A. Otsubo and A. Kamimura, *Macromolecules*, 2013, **46**, 6507–6519.
295 V. A. Kraikin, A. A. Fatykhov, I. I. Sakhipova, E. A. Sedova, A. E. Egorov and S. N. Salazkin, *Magn. Reson. Chem.*, 2013, **51**, 621–629.
296 L. L. G. Justino, M. L. Ramos, P. E. Abreu, A. Charas, J. Morgado, U. Scherf, B. F. Minaev, H. Ågren and H. D. Burrows, *J. Phys. Chem. C*, 2013, **117**, 17969–17982.
297 O. Ouari, T. Phan, F. Ziarelli, G. Casano, F. Aussenac, P. Thureau, D. Gigmes, P. Tordo and S. Viel, *ACS Macro Lett.*, 2013, **2**, 715–719.
298 B. Fortier-McGill, V. Toader and L. Reven, *Macromolecules*, 2014, **47**, 4298–4307.
299 Y. Wang, L. Xu, M. Wang, W. Pang and X. Ge, *Macromolecules*, 2014, **47**, 3901–3908.
300 Y. Hong and T. Miyoshi, *ACS Macro Lett.*, 2014, **3**, 556–559.
301 A. Clough, J. L. Sigle, A. Tapash, L. Gill, N. V. Patil, J. Zhou and J. L. White, *Macromolecules*, 2014, **47**, 2625–2631.
302 Y. Sasanuma and D. Touge, *Polymer*, 2014, **55**, 1901–1911.
303 Y. Yao, S. Jiang and S. Rastogi, *Macromolecules*, 2014, **47**, 1371–1382.
304 R. C. Nieuwendaal, C. R. Snyder and D. M. DeLongchamp, *ACS Macro Lett.*, 2014, **3**, 130–135.

305 J. P. Lewicki, S. J. Harley, J. A. Finnie, M. Ashmore, C. Bell and R. S. Maxwell, *ACS Symp. Ser.*, 2013, **1154**, 133–154.

306 C. Lu, K. Xu, W. Li, P. Li, Y. Tan and P. Wang, *Chem. Res. Chin. Univ.*, 2013, **29**, 1203–1207.

307 W. Chen, H.-J. Sun and T. Miyoshi, *J. Phys. Chem. B*, 2013, **117**, 13698–13709.

308 J. Jenczyk, M. Dobies, M. Makrocka-Rydzyk, A. Wypych and S. Jurga, *Eur. Polym. J.*, 2013, **49**, 3986–3997.

309 W. Hu and A. J. Hsieh, *Polymer*, 2013, **54**, 6218–6225.

310 K. Schaeler, A. Achilles, R. Baerenwald, C. Hackel and K. Saalwaechter, *Macromolecules*, 2013, **46**, 7818–7825.

311 A. Bormuth, M. Hofmann, P. Henritzi, M. Vogel and E. A. Roessler, *Macromolecules*, 2013, **46**, 7805–7811.

312 L. M. Brito, F. V. Chávez, M. I. B. Tavares and P. J. O. Sebastião, *Polym. Test.*, 2013, **32**, 1181–1185.

313 C. Lorthioir and B. Deloche, *Colloid Polym. Sci.*, 2014, **292**, 1841–1851.

314 J. Kang and T. Miyoshi, *Macromolecules*, 2014, **47**, 2993–3004.

315 W. Hu, Y. Su, L. Zhou, A. Pang, R. Cai, X. Ma and S. Li, *Molecules*, 2014, **19**, 1353–1366.

316 Y. Gao, R. Zhang, W. Lv, Q. Liu, X. Wang, P. Sun, H. H. Winter and G. Xue, *J. Phys. Chem. C*, 2014, **118**, 5606–5614.

317 R. Zhang, S. Yu, S. Chen, Q. Wu, T. Chen, P. Sun, B. Li and D. Ding, *J. Phys. Chem. B*, 2014, **118**, 1126–1137.

318 J. Hoepfner, G. Guthausen, K. Saalwaechter and M. Wilhelm, *Macromolecules*, 2014, **47**, 4251–4265.

NMR of liquid crystals and micellar solutions

Gerardino D'Errico and Luigi Paduano*
DOI: 10.1039/9781782622758-00456

This chapter presents a summary of manuscripts published in the period of June 2013–June 2014 involving NMR techniques and their applications in the field of amphiphilic systems. In Section 2 reviews and articles on general methods and models have been considered. In Section 3 the manuscripts on thermotropic and lyotropic liquid crystals, phospholipids, vesicles and bicelles have been presented. Section 4 has been devoted to micellar solutions including ionic and non ionic surfactant systems, polymer amphiphiles and mixed amphiphiles systems.

1 Introduction

This chapter presents a summary of manuscripts published in the period of June 2013–June 2014 focusing on the use of NMR techniques to elucidate the microstructure and dynamics of self-assembling systems.

It is well known that molecules having peculiar shape and properties due to unfavourable interaction with the solvent form self-assembling aggregates. Molecules that exhibit Liquid Crystal (L.C.) phases are called *mesogens* and such phase has properties between those of a conventional liquid, and those of a solid crystal. Typically, mesogens possess a rigid and anisotropic shape (*i.e.* longer in one direction than another) and orient depending on their long axis. Disk-like (discotic) mesogens are also known, and these orient in the direction of their short axis. In addition to molecules, polymers and colloidal suspensions can also form L.C. phases. Liquid crystals can be divided into thermotropic and lyotropic L.C.s. Thermotropic L.C.s exhibit a phase transition into the L.C. phase as temperature is changed, whereas lyotropic L.C.s exhibit phase transitions as a function of concentration.

Many thermotropic L.C.s exhibit a variety of phases as temperature is changed those are mainly related to the elongated shape of the molecules and to specific intermolecular interactions so that different long range order according to nematic, smectic A, B and C or cholesteric (in the presence of chiral atoms) organization can be observed.

The lyotropic L.C.s are mainly based on amphiphilic molecules, those dissolved in either a polar or apolar solvents can form different nanostructures depending on their volume balances between the hydrophilic part and hydrophobic part. At quite low concentration, amphiphilic molecules are dispersed in the solvent without any ordering. As the concentration is raised they spontaneously assemble into micelles or vesicles, and then at higher concentration, the assemblies will become ordered. A typical phase is a hexagonal columnar phase, where the

Department of Chemistry, University of Naples, Naples, Italy.
E-mail: luigi.paduano@unina.it

amphiphiles form long cylinders that arrange themselves into a roughly hexagonal lattice. At still higher concentration, a lamellar phase may form, wherein extended sheets of amphiphiles are separated by thin layers of water. For some systems, a cubic (also called viscous isotropic) phase may exist between the hexagonal and lamellar phases, wherein spheres are formed that create a dense cubic lattice. These spheres may also be connected to one another, forming a bicontinuous cubic phase. These anisotropic self-assembled nano-structures can then order themselves in much the same way as L.C. do, forming large-scale versions of all the thermotropic phases (such as a nematic phase of rod-shaped micelles).

The formation of lyotropic L.C.s, and nano-structures is driven by hydrophobic interactions along with weak intermolecular forces. The geometry of the interface that separates the hydrophobic and hydrophilic domains is related to the supramolecular architecture addressed by surfactant type and shape. It has been shown that such an interface can be exemplified by means of two different contributes: the interfacial curvature and the interfacial topology. While the first is associated to the local geometry, the second describes the global geometry through the degree of interfacial connectivity. As firstly defined by Ninham *et al.* the local constraint upon the interfacial curvature is specified by the surfactant packing parameter $P = v/al$. Here v represents the hydrophobic chain volume, a the head group area and l the chain length (taken as 80% of the fully extended chain). This parameter defines the possible nano-structures a surfactant may form. However, the curvature of the polar–apolar interface can be modulated through a suitable choice of oils and polar additives (*i.e.* electrolytes, water soluble polymers, *etc.*). The interfacial curvature is defined as positive (normal, $H > 0$, and $P < 1$) for oil-in-water (o/w) L_1 phases (normal micellar solutions), I_1 (discrete cubic L.C.) and H_1 (normal hexagonal L.C.). It is defined as negative (reverse, $H < 0$, and $P > 0$) for water-in-oil (w/o) L_2 phases (reverse micellar solutions), I_2 (reverse discrete cubic L.C.) and H_2 (reverse hexagonal L.C.). An average zero curvature $H \approx 0$, due to $P \approx 1$, is a typical feature of lamellar L_α L.C. phases and of bicontinuous cubic (V_1 and V_2) phases that are constituted by connected surfactant bilayers interwoven with a continuous water domain.

The following acronyms related to the most common NMR acquisition techniques and methods will be used in the next paragraphs: NOE for Nuclear Overhauser Effect, RDC for residual dipolar coupling, MAS for Magic Angle Spinning, SS NMR for solid-state NMR, SE for spin-echo, CPMG for Carr Purcell Meiboom Gill sequence, PFG for pulsed field gradient, PFGSE for Pulsed Field Gradient Spin-Echo sequence, and PFGSTE for Pulsed Field Gradient Stimulated Echo sequence. Throughout the chapter 1D and 2D symbols are used to indicate one- or two-dimensional NMR techniques. Moreover, dealing with microstructural characterizations, a variety of other techniques besides NMR methods are often used: typically we have Static Light Scattering (SLS), Dynamic Light Scattering (DLS), X-ray diffraction (Small and Wide Angle, SAXS and WAXRD), Small Angle Neutron Scattering (SANS), cryo-Transmission Electron Microscopy (cryo-TEM), Polarizing Optical Microscopy (POM),

Fourier Transform Infrared spectroscopy (FT-IR), Differential Scanning Calorimetry (DSC), Isothermal Titration Calorimetry (ITC), Atomic Force Microscopy (AFM) and Quasi-Elastic Light Scattering (QELS).

2 General articles: reviews, methods, models

Some papers have been published concerning methodological advancements in the NMR approach to L.C.s and micellar systems.

NMR diffusometry is nowadays a well-established and powerful technique to investigate molecular translation in fluid materials. Standard NMR diffusometry approaches are based on PFG generated by specific hardware and specially designed NMR probes. An alternative set-up that exploits the static gradient present in the fringe field of any commercial superconducting magnet has been presented. This stray field diffusometry technique can be particularly useful to study diffusional processes in fast-relaxing and slow-diffusing systems, such as thermotropic L.C., ionic liquids and polymer melts.[1] The translational self-diffusion behavior of calamitic mesogens in smectic phases measured by means of NMR diffusometry have been reviewed. In particular the results reported demonstrate that translational diffusion in smectic phases is strongly dependent on molecular packing in the layered structure.[2]

Fluorinated calamitic thermotropic L.C. represent an important class of materials for high-tech applications, especially in the field of L.C. displays. NMR spectroscopy is the most powerful technique for studying orientational order in L.C. systems at a molecular level thanks to the possibility of exploiting different anisotropic observables (chemical shift, dipolar couplings, and quadrupolar coupling) and nuclei (^{2}H, ^{13}C, and ^{19}F). In a relevant paper, the basic theory and NMR experiments useful for the investigation of orientational order on fluorinated calamitic L.C.s are reported, and a review of the literature published on this subject is given.[3]

Bent-core mesogens are an important class of thermotropic L.C.s as they exhibit unusual properties as well as morphologies distinctly different from rodlike mesogens. Two bent-core mesogens with differing center rings benzene and thiophene have been considered and studied using high-resolution oriented SS ^{13}C NMR method in their L.C. phases. The 2D separated local field (2D-SLF) NMR method was used to obtain the ^{13}C–^{1}H dipolar couplings of carbons in the center ring as well as in the side-wing Ph rings. Together with the dipolar couplings of the side-wing Ph ring carbons from which the local order parameters of the different subunits of the core could be extimated, the bent angle of the mesogenic molecule could be obtained. The ability of the ^{13}C 2D-SLF technique to provide the geometry of the bent-core mesogens in a straightforward manner through the measurement of the ^{13}C–^{1}H dipolar couplings has been highlighted.[4]

Investigations on the phase behaviors and structural properties of mono-, di- and poly-valent metal carboxylates have been reviewed with reference to developments in experimental and theoretical concepts surrounding their L.C. properties. The main methods of structural investigation such as X-ray diffraction, IR and ^{13}C-NMR spectroscopies have been examined in detail on the basis of common synthetic routes leading to the isolation of pure compounds.[5]

Biaxial thermotropic nematic L.C.s would be of great importance in L.C. display technology. Less than a decade ago, such L.C.s were suggested. The biaxiality of the phases was confirmed using ^{2}H NMR spectroscopy of deuterated probe molecules. An alternate technique that is based on the second order quadrupole shift detectable in ^{131}Xe NMR spectra of dissolved xenon has been proposed. The method has many advantages, such as the NMR spectra are taken from a static sample and the ^{131}Xe quadrupole coupling tensor is extremely sensitive to the symmetry of the phase.[6]

Other reviews have focused on NMR approaches to micellar systems. An oustanding review highlights the key role of NMR techniques in demonstrating the molecular aspects of the self-assembly of surfactant molecules that nowadays constitute the basic knowledge which modern nanoscience relies on. The story of a rigorous scientific approach to understand self-assembly in surfactant systems and biological membranes starts in the early seventies when the progresses of SAXRD and NMR facilities allowed to demonstrate the existence of ordered soft matter, and the validity of Tanford approach concerning self-assembly at a molecular level. Particularly, NMR quadrupolar splittings, NMR chemical shift anisotropy, and NMR relaxation of dipolar and quadrupolar nuclei in micellar solutions, microemulsions, and L.C.s proved the existence of an ordered polar–apolar interface, on the NMR time scale. NMR data, rationalized in terms of the two-step model of relaxation, allowed to quantify the dynamic aspects of the supramolecular aggregates in different soft matter systems. In addition, NMR techniques allowed to obtain important information on counterion binding as well as on size of the aggregate through molecular self-diffusion. Indeed NMR self-diffusion proved without any doubt the existence of bicontinuous microemulsions and bicontinuous cubic L.C.s, suggested by pioneering and brilliant interpretation of SAXRD investigations. Moreover, NMR self-diffusion played a fundamental role in the understanding of microemulsion and emulsion nanostructures, phase transitions in phase diagrams, and particularly percolation phenomena in microemulsions. Although the real applications of these systems still need major work, nevertheless new challenges are open, and perspectives based on integrated multidisciplinary approaches would enable both a deeper basic knowledge and the expected advances in biomedical field.[7]

In another review the main topics shared by the research groups leaded by Per Ekwall at Åbo Akademi University in Finland and Bjoern Lindman at Lund University in Sweden are described. NMR (shift, relaxation rate and diffusion), Laser Raman and IR spectroscopies were chosen as key instruments for molecular interaction characterization since they were used in the collaboration between the research groups in Åbo and in Lund.[8]

The effects on micellar behavior of aqueous solutions of cationic surfactants *viz.* cetyltrimethylammonium bromide (CTAB) and cetyltrimethylammonium chloride (CTAC) upon addition of ionic liquid 3-methyl-1-pentylimidazolium hexafluorophosphate [C_5mim][PF_6] have been reviewed. Particularly, ^{1}H NMR studies in micelle solutions containing [C_5mim][PF_6] have also been analysed in order to interpret the preferential solubilization site of [C_5mim][PF_6] on the studied micellar systems.[9]

The formation of reverse micelles by nonionic ethoxylated surfactants in two "dry" non polar solvents, heptane and dibutoxymethane (DBM), has been also reviewed. The self-diffusion coefficients of C_8EO_5, C_8EO_4 and $C_{10}EO_6$ in heptane, were obtained by diffusion ordered spectroscopy (DOSY ^{1}H-NMR). In all cases the presence of large and compact aggregates, with aggregation numbers going from a few dozens of monomers to a hundred of them, was shown.[10]

In a tutorial review an overview of the potential of NMR spectroscopy, and in particular SS NMR, in characterising micelle-templated mesoporous materials has been given. Particular attention is paid to recent technological and/or methodological developments.[11]

Many biological membranes consist of 50% or more (by weight) membrane proteins, which constitute approximately one-third of all proteins expressed in biological organisms. Determining the structures of membrane proteins is a daunting task because of the effects of the membrane environment; specifically, it has been difficult to combine biologically compatible environments with the requirements for the established methods of structure determination. There is strong motivation to determine the structures in their native phospholipid bilayer environment so that perturbations from nonnatural lipids and phases do not have to be taken into account. At present, the only method that can work with proteins in L.C. phospholipid bilayers is SS NMR spectroscopy.[12] As a possible alternative, a review on lipid bicelles in membranes studies using detergents has been also presented.[13]

Cell-penetrating peptides (CPPs) are a class of short, often cationic peptides that have the capability to translocate across cellular membranes, and although the translocation most likely involves several pathways, they interact directly with membranes, as well as with model bilayers. A review focuses on solution NMR as a tool for investigating CPP–lipid interactions. Structural propensities and cell-penetrating capabilities can be derived from a combination of CPP solution structures and studies of the effect that the peptides have on bilayers and the localization in a bilayer.[14]

3 Liquid crystals

The articles related to thermotropic and lyotropic L.C. phases and their use as solubilizing media are summarized in this paragraph. Self-assembly features are often investigated using the typical NMR parameters of the ^{2}H quadrupolar nucleus ($I = 1$), such as the $\Delta\nu_q$ quadrupolar splittings. Deuterium oxide (^2H_2O) is used as solvent to investigate lamellar and hexagonal lyotropic L.C. phases; alternatively, perdeuterated and/or selectively deuterated mesogens and surfactants can be used.

Topics related to the characterization of the nanostructures and their phase transitions will be mainly considered as well as the effect of solubilization of different molecules such as drugs, proteins, and peptides in the anisotropic L.C. matrices or in vesicles that align in the magnetic field. The analysis of the NMR parameters related to the degree of long-range order and also to steric and anisotropic molecular interactions generally enables a detailed characterization of the microstructural features of a solubilizate. However, other techniques are often used to fully characterise these complex fluids. For instance the different types of L.C. phases can be often distinguished on the basis of their different optical properties (such as birefringence). When viewed under a microscope using a polarized light source, a L.C. material, with the exception of cubic L.C. phases, will appear to have a distinct texture. Each 'patch' in the texture corresponds to a domain where the L.C. molecules are oriented in a different direction. Within a domain, however, the molecules are well ordered.

3.1 Thermotropic liquid crystals

NMR techniques are frequently used, combined with other experimental approaches, for the chracterization of phase behaviour of L.C.s. Microscopic diffusion processes in thermotropic 5CB L.C. with imbedded surfactant-stabilized water microemulsions have been studied using PFG NMR. The formation of the microemulsion is found to be crucial for the macroscopic separation of the mixture into L.C.- and surfactant-rich phases.[15]

Novel aromatic–aliphatic biobased polyesters showing thermotropic behavior in the melt have been presented, incorporating different biobased monomers such as 2,5-furandicarboxylic acid (2,5-FDCA), suberic acid (SuA), and vanillic acid (VA) in thermotropic L.C. polymers (TLCPs). The chemical structures, molecular weights, phase transitions, thermal behavior, and mechanical performance of the synthesized polymers are studied using polarization optical microscopy, WAXD, DSC, TGA, DMTA, SS NMR spectroscopy, rheology, and tensile tests. These materials show a low temperature transition from the crystalline to the nematic phase, and stable nematic phases up to 300 °C and higher.[16]

The stability of thermotropic ionic L.C.s is essentially due to microphase segregation between the ionic heads and the long alkyl chains. Using newly synthesized viologen dimers, it has been shown that the structure of the central core is another key parameter to play with in order to tune the mesomorphic behavior.[17]

Six novel chiral bulky styrenic monomers were synthesized and radically polymerized to yield the corresponding polymers. All of them consisted of laterally attached *p*-terphenyl pendants terminated by an identical (+)-(*S*)-2-methylbutyloxy end and an achiral end with various size. Evidenced by the results of NMR, polarimetry, CD spectroscopy, computer simulation, thermal properties, and X-ray diffractions, the chiral *p*-alkoxyphenyl group ortho to the vinyl group induced the helical conformation of polymer backbone, whereas that meta to the vinyl group failed to dictate the growth of polymer backbone. Accompanied by the existence of helical structure with a predominant screw sense, stable L.C. phases were generated for some of these polymers.[18]

A series of new low molecular organogelators (LMOGs) with thermotropic mesophase were synthesized *via* the reaction of 3,6-dimethylpyrazine-2,5-dicarboxylic acid with *p*-alkoxyl anilines. These compounds readily formed stable gels in a variety of organic solvents and their self-assembly behavior, structure–property relationship were investigated by SEM, X-ray diffraction (XRD), ^{1}H NMR, FT-IR, and UV-visible spectroscopy (UV). More importantly, all the organogelators exhibited thermotropic hexagonal column mesophase.[19]

Betaine-ester analogs of 1-*N*-alkyl-3-*N*′-methyl imidazolium salts exhibiting both thermotropic and lyotropic L.C. behaviors have been reported. These compounds are gelators, which form gels in a variety of organic solvents such as chloroform, methanol, ethanol, THF, while also exhibiting a lamellar mesophase. Results from IR and variable-temperature ^{1}H NMR spectroscopy studies show that a trace amount of water (0.3–0.5% by volume) in $CHCl_3$ plays a crucial role for developing the H-bonded network during gel formation.[20]

The first attempt to investigate the L.C. behavior of SO_3–functionalized imidazolium zwitterionic (SO_3–ImZI) salts, which display homeotropic alignment on a glass slide without the aid of any aligning approach, has been reported.[21]

A new free-standing silica-based composite film that exhibits reversible thermochromic reflection, induced by a L.C. guest in the pores of iridescent mesoporous films, has been described. The authors have demonstrated that selective reflection from the novel mesoporous organosilica material with chiral nematic organization can be reversibly switched by thermal cycling of the 8CB guest between its isotropic and L.C. states, which was proven by SS NMR experiments.[22]

3.2 Lyotropic liquid crystals

A lyotropic L.C. consists of two or more components that exhibit L.C. properties in certain concentration ranges. In the lyotropic phases, solvent molecules fill the space around the compounds to provide fluidity to the system. In contrast to thermotropic L.C.s, these lyotropics have another degree of freedom, *i.e.*, the concentration, which enables them to induce a variety of different phases. Since lyotropic L.C.s rely on a subtle balance of intermolecular interactions, it is often more difficult to analyse their structures and properties than those of thermotropic L.C.s. Lyotropic L.C. nanostructures are abundant in living systems. Accordingly, lyotropic L.C.s attract particular attention in the field of biomimetic chemistry. In particular, biological membranes and cell membranes are a form of L.C. Their constituent rod-like molecules (*e.g.*, phospholipids) are organized perpendicularly to the membrane surface, yet the membrane is fluid and elastic. The constituent molecules can flow in-plane quite easily, but tend not to leave the membrane, and can flip from one side of the membrane to the other with some difficulty. These

L.C. membrane phases can also host important proteins such as receptors freely "floating" inside, or partly outside, the membrane. ^{2}H NMR in studying lyotropic L.C. phase behaviour is of large interest in virtue of the unique performance to easily monitor the occurrence of multiphase systems (*i.e.*, the coexistence of different types of L.C. phases) in the investigations of phase diagrams. Indeed, ^{2}H NMR splitting occurs in the presence of anisotropic lamellar or hexagonal L.C. phases that align in the magnetic field, isotropic ^{2}H NMR signals are typical of isotropic L.C. phases such as cubic arrangements. ^2H_2O as solvent and fully or specifically deuterated molecules, particularly in the case of lipids and phospholipids (PLs), can be used. ^{31}P NMR is widely used for studying structural features in PL-based L.C.s. Nowadays NMR self-diffusion measurements have become a routine technique since they provide fundamental information on the dimensions of the organized nanodomains, on their spatial connectivity, *i.e.* the topology of the long-range ordered structure, and on the obstruction effects due to concentration, shape and soft boundaries. A huge number of articles have been reported on the use of PFG NMR methods to probe nanostructures and on improved methods to acquire and to model the NMR self-diffusion experiments.

Here the articles dealing with lyotropic L.C. systems are mentioned according to the following order: PL-based L.C. systems, and vesicles, bicelles, liposomes and other dispersed L.C. phases used preferentially as solubilizing matrices.

3.2.1 Phospholipids and solubilization in PL bilayers. Peptides confer interesting properties to materials, supramolecular assemblies and to lipid membranes and are used in analytical devices or within delivery vehicles. Their relative ease of production combined with a high degree of versatility make them attractive candidates to design new such products. Researchers demonstrate how CD and SS NMR spectroscopic approaches can be used to follow the reconstitution of peptides into membranes and to describe some of their fundamental characteristics. Whereas CD spectroscopy is used to monitor secondary structure in different solvent systems and thereby aggregation properties of the highly hydrophobic domain of p24, a protein involved in vesicle trafficking, SS NMR spectroscopy was used to deduce structural information and the membrane topology of a variety of peptide sequences found in nature or designed.[23]

The effect of 5th generation polyamidoamine (PAMAM G5) dendrimers on multilamellar dipalmitoylphosphocholine (DPPC) vesicles was investigated. PAMAM was added in 2 different concentrations to the lipids (10^{-3} and 10^{-2} dendrimer/lipid molar ratios). The thermal behavior of the evolved systems was characterized by DSC; while the structure and the morphology were investigated with small- and wide-angle X-ray scattering (SWAXS), freeze-fracture electron microscopy (FFTEM) and ^{31}P NMR spectroscopy, respectively. The obtained results show that the dendrimers added in 10^{-3} molar ratio to the lipids generate minor perturbations in the multilamellar structure and thermal character of liposomes, while added in 10^{-2} molar ratio dendrimers cause major disturbance in the vesicular system.[24]

The molecular dynamics at different levels of model membranes having a simulated glycoclix has been studied with the focus on the molecular crowding conditions at the lipid–water interfacial region. Binary mixtures of dipalmitoylphosphatidylcholine (dpPC) and a poly(ethylene glycol) (PEG*n*) derivative of dipalmitoylphosphatidylethanolamine (PE) (where $n = 350$, 1000, and 5000, respectively, refer to PEG molecular masses) were submitted to ^{1}H spin–lattice relaxation time (T_1) and ^{31}P NMR spectra analytical ^{1}H NMR relaxation times revealed two contributing components in each proton system (PEG, phospholipids, and water), for all the mixtures studied, exhibiting values of T_1 with very different orders of magnitude. This allowed for the identification of the confined and bulk water populations as well as PEG moieties becoming more disordered and independent from the phospholipid moiety as n increased. ^{31}P spectra showed a typical broad bilayer signal for $n = 350$ and 1000, and an isotropic signal characteristic of micelles for $n = 5000$.[25]

Bicelles are model membranes generally made of long-chain dimyristoylphosphatidylcholine (DMPC) and short-chain dihexanoyl-PC (DHPC). At low DMPC/DHPC (q) molar ratios, fast-tumbling bicelles are formed in which the DMPC bilayer is stabilized by DHPC molecules in the high-curvature rim region. Experimental constraints imposed by techniques such as CD, dynamic light scattering, or microscopy may require the use of bicelles at high dilutions. Studies have shown that such conditions induce the formation of small aggregates and alter the lipid-to-detergent ratio of the bicelle assemblies. Researchers studied the exact component of those DMPC/DHPC isotropic bicelles and study the lipid miscibility. This was done using ^{31}P NMR and exploring a wide range of lipid concentrations (2–400 mM) and q ratios (0.15–2). Our data demonstrate how dilution modifies the actual DMPC/DHPC molar ratio in the bicelles.[26]

A research dealing with tannin–lipid interactions, by mimicking both oral cavity membranes by micrometric size liposomes and lipid droplets in food by nanometric isotropic bicelles was performed. Deuterium and phosphorus SS NMR demonstrated the membrane hydrophobic core disordering promoted by catechin (C), epicatechin (EC), and epigallocatechin gallate (EGCG), the latter appearing more efficient. C and EGCG destabilize isotropic bicelles and convert them into an inverted hexagonal phase. Tannins are shown to be located at the membrane interface and stabilize the lamellar phases. These newly found properties point out the importance of lipids in the complex interactions that happen in the mouth during organoleptic feeling when ingesting tannins.[27]

Small, fast-tumbling bicelles are frequently used in solution NMR studies of protein–lipid interactions. For this purpose it is critical to have information about the organization of the lipids within the bicelle structure. reseachers have studied the morphology of small, fast-tumbling bicelles containing DMPC and DHPC as a function of temperature, lipid concentration, and the relative ratio (q value) of lipid (DMPC) to detergent (DHPC) amounts.[28]

Recombinant EF-hand domain of phospholipase C δ1 has a moderate affinity for anionic phospholipids in the absence of Ca^{2+} that is driven by interactions of cationic and hydrophobic residues in the first EF-hand sequence. This region of PLC δ1 is missing in the crystal structure. The relative orientation of recombinant EF with respect to the bilayer, established with NMR methods, shows that the N-terminal helix of EF-1 is close to the membrane interface. Specific mutations of EF-1 residues in full-length PLC δ1 reduce enzyme activity but not because of disturbing partitioning of the protein onto vesicles. The reduction in enzymic activity coupled with vesicle binding studies are consistent with a role for this domain in aiding substrate binding in the active site once the protein is transiently anchored at its target membrane.[29]

Inherent low sensitivity of NMR spectroscopy has been a major disadvantage, especially to study biomolecules like membrane proteins. Recent studies have successfully demonstrated the advantages of performing SS NMR experiments at very low and ultralow temperatures to enhance the sensitivity. However, the long spin–lattice relaxation time, T_1, at very low temperatures is a major limitation. To overcome this difficulty, it was demonstrate the use of a copper-chelated lipid for MAS SS NMR measurements on cytochrome-b5 reconstituted in multilamellar vesicles. The results suggest that multilamellar vesicles containing as small as 0.5 mol% of a copper-chelated lipid can significantly shorten T_1 of protons, which can be used to considerably reduce the data collection time or to enhance the signal-to-noise ratio.[30]

Bicellar dispersions of chain perdeuterated 1,2-dimyristoyl-*sn-glycero*-3-phosphocholine (DMPC-d54), 1,2-dimyristoyl-*sn-glycero*-3-phospho-(1′-rac-glycerol) (DMPG), and 1,2-dihexanoyl-*sn-glycero*-3-phosphocholine (DHPC), with molar ratios of 3 : 1 : 1, were studied using variable-pressure ^{2}H NMR spectroscopy at hydrostatic pressures up to 125 MPa. Upon warming of the dispersions, spectra at ambient pressure indicated a progressive coalescence from small bilayered disks undergoing isotropic reorientation to more extended micellar structures in which spectra indicated anisotropic reorientation and, under some conditions, magnetic orientation and finally to randomly oriented lamellae or multilamellar vesicles.[31]

Piscidin 1 and piscidin 3, which were discovered in the mast cells of hybrid striped sea bass, are homologous antimicrobial peptides that are active against drug-resistant bacteria. Piscidin 1, the more antimicrobial and hemolytic peptide, also has anti-HIV-1 and anti-cancer properties. The piscidins 1 and 3 genes were cloned into the TrpLE vector. The corresponding TrpLE-piscidin fusion partners were expressed in *Escherichia coli* and recovered from inclusion bodies. Following steps that included Ni-NTA chromatog., cyanogen bromide cleavage of the fusion proteins, and reverse-phase HPLC, purified piscidins 1 and 3 were recovered and characterized by NMR. High quality ^{15}N–^{1}H HSQC spectra of piscidins 1 and 3 bound to SDS micelles were collected, demonstrating the feasibility of producing and purifying the isotopically-labeled piscidin peptides required to determine their full structures by multidimensional NMR spectroscopy.[32]

Phylloseptin-1, -2, and -3 are three members of the family of linear cationic antimicrobial peptides found in tree frogs. The highly homologous peptides encompass 19 amino acids, and only differ in the amino acid component and charge at the six most carboxy-terminal residues. It was studied how such subtle changes are reflected in their membrane interactions and how these can be correlated to their biological activities. To this end, the three peptides were labeled with stable isotopes, reconstituted into oriented phospholipid bilayers, and their detailed topology detected by a combined approach using ^{2}H and ^{15}N SS NMR spectroscopy. Although phylloseptin-2 and -3 adopt perfect in-plane alignments, the tilt angle of phylloseptin-1 deviates by 8° probably to assure a more water exposed localization of the lysine-17 side chain.[33]

Sphingosine is the most common sphingoid long-chain base in sphingolipids. It is the precursor of important cell signaling molecules, such as ceramides. Researchers analyzed the mechanism by which this bioactive lipid induces vesicle content release, and the effect of negatively charged bilayers on the release process. Fluorescence lifetime measurements and confocal fluorescence microscopy were applied to observe the mechanism of sphingosine efflux from large and giant unilamellar vesicles; a graded-release efflux was detected. Because sphingosine at physiological pH has a net positive charge, its interaction with negatively charged phospholipids (*e.g.*, bilayers containing phosphatidic acid together with sphingomyelins, phosphatidylethanolamine, and cholesterol) gave rise to a release of vesicular contents, faster than with electrically neutral bilayers. Furthermore, ^{31}P-NMR and X-ray data showed the capacity of sphingosine to facilitate the formation of nonbilayer (cubic phase) intermediates in negatively charged membranes.[34]

HIV-1 Vpu and CD4(372–433), a peptide comprising the transmembrane and cytoplasmic domain of human CD4, were recombinantly expressed in *Escherichia coli*, uniformly labeled with ^{13}C and ^{15}N isotopes, and separately reconstituted into phospholipid bilayers. Highly resolved dipolar cross-polarization (CP)-based SS NMR spectra of the two transmembrane proteins were recorded under magic angle sample spinning. Partial assignment of ^{13}C resonances was achieved. Site-specific assignments were obtained for 13 amino acid residues of CD4(372–433) and two Vpu residues. Additional amino acid type-specific assignments were achieved for 10 amino acid spin systems for both CD4(372–433) and Vpu.[35]

The phenols propofol and thymol, and lately carvacrol, eugenol and chlorothymol, have been shown to act as positive allosteric modulators on GABAA receptor, which is the main inhibitory receptor of the central nervous system. GABAA receptor is an intrinsic membrane protein which activity may be affected by surface-active compounds and by physical changes in the membrane. Recently, researchers demonstrated that these phenols interacted with the lipid membrane phase. By using ^{1}H-NMR spectroscopy, the same authors have investigated the effects of the insertion and the possible preferential location of the five phenol derivatives with GABAergic activity on EPC membranes. The results indicate that all compounds are able to insert in EPC phospholipid vesicles and to locate in the region between the polar group (choline molecular), the glycerol and the first atoms of the acyl chains. The location of the phenol molecules would reduce the repulsive forces among phospholipids head groups allowing closer molecular packing.[36]

The interactions between etoposide, cytarabine and 1,2-dihexadecanoyl-*sn*-glycerol-3-phosphocholine bilayers were studied using DSC, FT-IR and NMR (NMR). In particular NMR supported the determination of the main phase transition temperatures (TC) of 1,2-dihexadecanoyl-*sn*-glycerol-3-phosphocholine (DPPC). The location of cytarabine and etoposide in liposomes was also detected by NMR.[37]

Phospholamban (PLB) is a membrane protein that regulates heart muscle relaxation rates *via* interactions with sarcoplasmic reticulum Ca^{2+} ATPase (SERCA). When PLB is phosphorylated or mutated (R9C-PLB), inhibition of SERCA is relieved. ^{13}C and ^{15}N SS NMR spectroscopy was utilized to investigate conformational changes of PLB upon phosphorylation and R9C mutation. ^{13}C:O NMR spectra of the cytoplasmic domain revealed 2 α-helical structural components with population changes upon phosphorylation and R9C mutation. The ^{15}N NMR spectra indicated an increase in backbone dynamics of the cytoplasmic domain. Wild-type PLB (WT-PLB), Ser-16-phosphorylated PLB (P-PLB), and R9C-PLB all had a very dynamic domain Ib, and the transmembrane domain had an immobile component. The ^{15}N NMR spectra indicated that the cytoplasmic domain of R9C-PLB adopted an orientation similar to P-PLB and shifted away from the membrane surface. Domain Ib (Leu-28) of P-PLB and R9C-PLB lost the alignment. R9C-PLB adopted a conformation similar to P-PLB with a population shift to a more extended and disordered state.[38]

Melittin is an amphipathic peptide which has received much attention as a model peptide for peptide–membrane interactions. It is however not suited as a transfection agent due to its cytolytic and toxicologic effects. Retro-inverso-melittin, when covalently linked to the lipid 1,2-dioleoyl-*sn*-*glycero*-3-phosphoethanolamine (riDOM), eliminates these shortcomings. RiDOM forms cationic nanoparticles with a diameter of 13 nm which are well soluble in water and bind with high affinity to DNA and lipid membranes. RiDOM-induced membrane leakiness is however much reduced compared to that of authentic melittin. The ^{31}P NMR spectrum of the nanoparticle is however transformed into a typical bilayer spectrum.[39]

Phospholamban (PLB) is a 52 amino acid integral membrane protein that interacts with the sarcoplasmic reticulum Ca^{2+} ATPase (SERCA) and helps to regulate Ca^{2+} flow. PLB inhibits SERCA impairing Ca^{2+} translocation. The Arg9 to Cys (R9C) mutation is a loss of function mutation with reduced inhibitory potency. The effect R9C PLB has on the membrane surface and the hydrophobic region dynamics was studied by ^{31}P and ^{2}H SS NMR spectroscopy in multilamellar vesicles (MLVs). The ^{31}P NMR spectra indicate that, like the phosphorylated PLB (P-PLB), the mutated R9C-PLB protein has significantly less interaction with the lipid bilayer headgroup when compared to wild-type PLB (WT-PLB). Similar to P-PLB, R9C-PLB slightly decreases ^{31}P T_1 values in the lipid headgroup region. ^{2}H SCD order parameters of ^{2}H nuclei along the lipid acyl chain decrease less dramatically for R9C-PLB and P-PLB when compared to WT-PLB. The results suggest that R9C-PLB interacts less with the membrane surface and hydrophobic region than WT-PLB.[40]

Breathing is enabled by lung surfactant, a mixture of proteins and lipids that forms a surface-active layer and reduces surface tension at the air–water interface in lungs. Surfactant protein B (SP-B) is an essential component of lung surfactant. Researchers probed the mechanism underlying the important functional contributions made by the N-terminal 7-residues of SP-B, a region sometimes called the "insertion sequence". These studies employed a construct of SP-B, SP-B (1–25,63–78), also called Super Mini-B, which is a 41-residue peptide with internal disulfide bonds comprising the N-terminal 7-residue insertion sequence and the N- and C-terminal helixes of SP-B. CD, solution NMR, and SS ^{2}H NMR were used to study the structure of SP-B (1–25,63–78) and its interactions with phospholipid bilayers. Comparison of results for SP-B (8–25,63–78) and SP-B (1–25,63–78) demonstrates that the presence of the 7-residue insertion sequence induces substantial disorder near the center of the lipid bilayer.[41]

The very amino-terminal domain of the huntingtin protein is directly located upstream of the protein's polyglutamine tract, plays a decisive role in several important properties of this large protein and in the development of Huntington's disease. This huntingtin 1–17 domain is on the one hand known to markedly increase polyglutamine aggregation rates and on the other hand has been shown to be involved in cellular membrane interactions. It was detected the high-resolution structure of huntingtin 1–17 in dodecyl phosphocholine micelles and the topology of its helical domain in oriented phosphatidylcholine bilayers. Using 2D solution NMR spectroscopy the low-energy conformations of the polypeptide were identified in the presence of dodecyl phosphocholine detergent micelles. In a next step a set of four SS NMR angular restraints was obtained from huntingtin 1–17 labeled with ^{15}N and ^{2}H at selected sites. Furthermore the SS NMR data were used to further refine the domain structure in phospholipid bilayers.[42]

The three-dimensional structure of the 81-residue mercury transporter MerF detected in L.C. phospholipid bilayers under physiological conditions by Rotationally Aligned (RA) SS NMR has two long helixes, which extend well beyond the bilayer. Truncation of the N-terminal 12 residues, which are mobile and unstructured when the protein is solubilized in micelles, results in a large structural rearrangement of the protein in bilayers. In the full-length protein, the N-terminal helix is aligned nearly parallel to the membrane normal and forms an extension of the first transmembrane helix. By contrast, this helix adopts a perpendicular orientation in the truncated protein. These results demonstrate that major changes in protein structure can result from differences in amino acid sequence.[43]

Aggregation and membrane penetration of porphyrinic photosensitizers play crucial roles for their efficacy in photodynamic therapy. Research work was aimed at comparing the aggregation behavior of selected photosensitizers and correlating it with membrane affinity.

Self-assembling properties of 15 amphiphilic free-base chlorin and porphyrin derivatives bearing carboxylate substituents were studied in phosphate buffered saline (PBS) by ^{1}H NMR spectroscopy, making use of ring current induced aggregation shifts. All compounds exhibited aggregation in PBS to a different degree with dimers or oligomers showing slow aggregate growth over time. All chlorin compounds revealed similar aggregation maps with their hydrophobic sides overlapping and their carboxylate groups protruding toward the exterior. Membrane interactions were probed using 1,2-dioleoyl-*sn-glycero*-3-phosphocholine (DOPC) bilayer vesicles and 1,2-dihexanoyl-*sn-glycero*-3-phosphocholine (DHPC) micelles as models.[44]

Multicellular organisms fight bacterial and fungal infections by producing peptide-derived broad-spectrum antibiotics; These host-defense peptides compromise the integrity of microbial cell membranes and thus evade pathways by which bacteria develop rapid antibiotic resistance. Although > 1700 host-defense peptides have been identified, the structural and mechanistic basis of their action remains speculative. This impedes the desired rational development of these agents into next-generation antibiotics. Researchers presented the X-ray crystal structure as well as SS NMR spectroscopy, electrophysiological, and molecular dynamics (MD) simulations of human dermcidin (I) in membranes that reveal the antibiotic mechanism of this major human antimicrobial, found to suppress *Staphylococcus aureus* growth on the epidermal surface.[45]

Despite arduous efforts and recent technological developments structural investigation of integral membrane proteins remains a challenge. A relatively new class of membrane mimetics, nanodiscs, is emerging as a promising alternative. Although nanodiscs have been proven successful for several biophysicals applications, they yet remain to become the system of preferred choice for structure determination Recently a noticeable improvement in the quality of NMR spectra obtained for the transmembrane and cytoplasmic domains of integrin αIIb incorporated into these smaller disks rendering them susceptible for a thorough structural investigation has been proposed.[46]

The drug:membrane interactions for the antihypertensive AT1 antagonist losartan, the prototype of the sartans class, are studied herein using an integrated approach. The pharmacophore arrangement of the drug was revealed by rotating frame nuclear Overhauser effect spectroscopy (2D ROESY) NMR spectroscopy in three different environments, namely water, dimethyl sulfoxide (DMSO), and sodium dodecyl sulfate (SDS) micellar solutions mimicking conditions of biological transport fluids and membrane lipid bilayers. Drug association with micelles was monitored by diffusion ordered spectroscopy (2D DOSY) and drug:micelle intermolecular interactions were characterized by ROESY spectroscopy. The localisation of the drug in the micellar environment was investigated by introducing 5-doxyl and 16-doxyl stearic acids. This applied integrated approach aims to explore the role of membranes in losartan's pathway towards the AT1 receptor.[47]

The cecropin–melittin hybrid antimicrobial peptide BP100 (H-KKLFKKILKYL-NH_2) is selective for Gram-negative bacteria and negatively charged membranes and is weakly hemolytic. Researchers studied BP100 conformational and functional properties upon interaction with large unilamellar vesicles (LUVs) and giant unilamellar vesicles (GUVs) containing variable proportions of phosphatidylcholine (PC) and negatively charged phosphatidylglycerol (PG). CD and NMR spectra showed that upon binding to PG-containing LUVs BP100 acquired an α-helical conformation, with the helix spanning residues 3–11. Theoretical analyses indicated that the helix was amphipathic and surface-seeking.[48]

β-Amyloid (Aβ) fibrils are the major species involved in Alzheimer's disease (AD). An atomic-resolution molecular structure of Aβ40 fibrils formed in the presence of lipid vesicles was obtained by using MAS SS NMR spectroscopy. The fibril structures formed in the presence of the lipid vesicles are remarkably different from those formed in solution. These results provide insights into the molecular mechanism of Aβ aggregation in the presence of lipid vesicles.[49]

Different types of lipid bilayers/monolayers have been used to simulate the cellular membranes in the investigation of the interactions between drugs and cells. In order to understand how do the nature and immobilization state of the biomembrane models influence the action of the local anesthetic tetracaine (TTC) upon the lipid membranes. The interaction of TTC with different biomembrane models of dimyristoylphosphatidylcholine (DMPC) with and without cholesterol (CHOL) was investigated through phosphorus NMR (^{31}P-NMR). The general conclusion was that TTC has a fluidizing effect on the lipid membranes and, above certain concentrations, induces membrane swelling or even solubilization.[50]

The conserved C-terminal FATC domain of the kinase 'target of rapamycin' is important for its regulation and was suggested to contain a peripheral membrane anchor. Researchers investigated on the interactions of the yeast TOR1 FATC domain (2438–2470 = y1fatc) and 15 mutants with membrane mimetic micelles, bicelles, and small unilamellar vesicles (SUVs) by NMR and CD spectroscopy. Replacement of up to 6–7 residues did not result in a significant abrogation of the association with micelles or bicelles. However, replacement of only one residue could result in an impairment of the interaction with SUVs that are usually used at low concentrations.[51]

It is shown that phospholipid dispersions with a few percent of diacylphosphocholine PC in water can be swollen to single-phase lyotropic L.C. L_α-phases by the addition of co-solvents like glycerol, 1,3-butyleneglycol BG or 1,2-propyleneglycol PG. The birefringent L_α-phases contain small unilamellar and multilamellar vesicles if the temperature of the samples is above the Krafft-Temperature Tm of the phospholipid. When such transparent birefringent viscous samples are cooled down below Tm the samples are transformed into birefringent gels. The bilayers of the crystal vesicles form adhesive contacts in the gel. PFG NMR measurements show that two different kinds of water or co-solvent can be distinguished in the gels. One type of solvent molecules can diffuse like normal solvent in a continuous bulk phase. A second type of water diffuses much more slowly. This type of solvent is obviously trapped in the vesicles.[52]

C99 (also known as β-CTF) is the 99-residue transmembrane C-terminal domain (residues 672–770) of the amyloid precursor protein (APP) and is the immediate precursor of amyloid-β (Aβ) polypeptides. To test the dependence of the C99 structure on the component of the host model membranes, NMR studies of C99 were conducted both in anionic lyso-myristoylphosphatidylglycerol (LMPG) micelles and in a series of 5 zwitterionic bicelle whose composition involves phosphatidylcholine and sphingomyelin with acyl chain length varying from 14 to 24 C atoms.[53]

Researchers have investigated in the present study the effect of both non-selective and selective cationic 14-*mer* peptides on the lipid orientation of DMPC bilayers by ^{31}P SS NMR spectroscopy. Depending on the position of substitution, these peptides adopt mainly either an α-helical structure able to permeabilize DMPC and DMPG vesicles (non-selective peptides) or an intermolecular β-sheet structure only able to permeabilize DMPG vesicles (selective peptides).[54]

The structure and interfacial association of the full-length vesicle SNARE, synaptobrevin (I), were compared in 4 different lipid environments using NMR and ESR spectroscopy. In micelles, segments of the SNARE motif were helical and associated with the interface. However, the fraction of helix and interfacial association decreased as I was moved from micelle to bicelle to bilayer environments, indicating that the tendency toward interfacial association was sensitive to membrane curvature. In bilayers, the SNARE motif of I transiently associated with the lipid interface, and regions that were helical in micelles were in conformational and environmental exchange in bicelles and bilayers.[55]

The general applicability of fast field-cycling NMR relaxometry in the study of dynamics in lipid bilayers is demonstrated through analytical of binary unilamellar liposomes composed of 1,2-dioleoyl-*sn-glycero*-3-posphocholine (DOPC) and cholesterol. Researchers extended an evidence-based method to simulating the NMR relaxation response, previously validated for single-component membranes, to evaluate the effect of the sterol molecular on local ordering and dynamics over multiple timescales.[56]

The effects of specific ions on the local translational diffusion of water near large hydrophilic lipid vesicle surfaces were measured by Overhauser dynamic nuclear polarization (ODNP). ODNP relies on an unpaired electron spin-containing probe located at molecular or surface sites to report on the dynamics of water protons within 10 Å from the spin probe, which give rise to spectral densities for electron–proton cross-relaxation processes in the 10 GHz regime. This pushes NMR relaxometry to more than an order of magnitude higher frequencies than conventionally feasible, permitting the measurement of water moving with picosecond to subnanosecond correlation times.[57]

Interactions among ADP-ribosylation factors (ARFs), various adaptor proteins, and membrane lipids are essential for intracellular vesicle transport of a variety of cellular materials. Researchers presented NMR (NMR)-based information on the nature of the interaction of yeast Arf1 (yArf1) and the pleckstrin homolecular (PH) domain of four-phosphate-adaptor protein 1 (Fapp1) as it occurs at a model membrane surface.[58]

Phosphatidylinositol-4,5-bisphosphate (PI(4,5)P2) is an important signaling lipid and plays a crucial role in a wide variety of cellular processes by interacting with protein targets and localizing proteins at the plasma membrane. These interactions are strongly influenced by the lateral distribution of PI(4,5)P2 as well as its ionization state. The characterization of the PI(4,5)P2 ionization state provides important information about how PI(4,5)P2 interacts with other membrane resident or associated chemical species. Researchers have studied the SS MAS ^{31}P NMR to investigate the interactions of PI(4,5)P2 with potential cluster promoting agents, divalent cations and cholesterol. Both Ca^{2+} and cholesterol were found previously to promote formation of local PI(4,5)P2 clusters *in vitro*. The NMR approach allows us to probe independently the ionization state of PI(4,5)P2 two phosphomonoester groups.[59]

The biomembrane surrounding rubber particles from the hevea latex is well known for its content of numerous allergen proteins. HbREF (Hevb1) and HbSRPP (Hevb3) are major components, linked on rubber particles, and they have been shown to be involved in rubber synthesis or quality (mass regulation), but their exact function is still to be detected. In this study we highlighted the different modes of interactions of both recombinant proteins with various membrane models (lipid monolayers, liposomes or supported bilayers, and multilamellar vesicles) to mimic the latex particle membrane.[60]

The chloroplast twin arginine translocation (cpTat) system transports highly folded precursor proteins into the thylakoid lumen using the protonmotive force as its only energy source. Hcf106, as one of the core components of the cpTat system, is part of the precursor receptor complex and functions in the initial precursor-binding step. To better understand the membrane-active capabilities of Hcf106, researchers used SS NMR spectroscopy to investigate those properties of the APH. In this study, synthesized peptides of the predicted Hcf106 APH (amino acids 28–65) were incorporated at increasing molecular% into 1-palmitoyl-2-oleoyl-*sn-glycero*-phosphocholine (POPC) and POPC/MGDG (monogalactosyldiacylglycerol; mole ratio 85 : 15) multilamellar vesicles (MLVs) to probe the peptide–lipid interaction. SS ^{31}P NMR and ^{2}H NMR spectroscopic experiments revealed that the peptide perturbs the headgroup and the acyl chain regions of phospholipids, as indicated by changes in spectral lineshape, chemical shift anisotropy (CSA) line width, and ^{2}H order SCD parameters.[61]

The chloroplast twin arginine translocation system transports highly folded precursor proteins across the thylakoid using the protonmotive force as its only energy source. Hcf106 and another thylakoid protein, cpTatC compose the precursor receptor complex. Hcf106 has been shown biochemical to insert spontaneously into thylakoid membranes; however, how this occurs is not understood. To investigate how Hcf106 inserts itself into the membrane unassisted, SS NMR spectroscopy was used to investigate the membrane activity of the TMD. A synthetic peptide of the Hcf106 TMD was incorporated into multilamellar vesicles made of 100 1-palmitoyl-2-oleoyl-*sn-glycero*-phosphocholine (POPC) or 85 : 15 ratio with monogalactosyl diacylglycerol (POPC/MGDG) to probe peptide–lipid interaction. SS ^{31}P NMR and ^{2}H NMR spectroscopic techniques were used to reveal peptide perturbations of the phospholipid membranes.[62]

In recent years, multi-lipid bicellar systems have emerged as promising membrane models. The fast orientational diffusion and magnetic alignability made these systems very attractive for NMR investigations. However, their alignment was so far achieved with a strong magnetic field, which limited their use with other methods that require macroscopic orientation. Recently, it was shown that bicelles could be aligned also by shear flow in a Couette flow cell, making it applicable to structural and biophysicals studies by polarized light spectroscopy. Researchers have characterized the flow alignment behavior of the commonly used dimyristoyl phosphatidylcholine/dicaproyl phosphatidylcholine (DMPC/DHPC) bicelle system, for various temperatures, lipid compositions, and lipid concentrations The authors concluded that at optimal flow conditions the selected bicellar systems can produce the most efficient flow alignment out of any lipid systems used so far.[63]

The protein M2 from influenza is a tetrameric membrane protein with several roles in the viral life cycle. The transmembrane helix (TMH) of M2 has proton channel activity that is required for unpackaging the viral genome. The APH interacts with membranes and is required for M2 localization to the site of viral budding. As a step toward obtaining high resolution information on the structure and lipid interactions of the M2 APH, we sought to develop a fast tumbling bicelle system, which would make studies of M2 in a membrane-like environment by solution NMR possible. Since M2 is highly sensitive to the solubilizing environment, an M2 construct containing the APH was studied under micelle and bicelle conditions while maintaining the same detergent and lipid headgroup chemistry to facilitate interpretation of the spectroscopic results.[64]

Amyloid deposits from several human diseases have been found to contain membrane lipids. Co-aggregation of lipids and amyloid proteins in amyloid aggregates, and the related extn. of lipids from cellular membranes, can influence structure and function in both the membrane and the formed amyloid deposit. Co-aggregation can therefore have important implications for the pathol. consequences of amyloid formation. In order to investigate on the mechanism behind co-aggregation and molecular structure in the formed aggregates researchers studied *in vitro* co-aggregation by incubating phospholipid model membranes with the Parkinson's disease-associated protein, α-synuclein, in monomeric form. After aggregation, spontaneous uptake of phospholipids from anionic model membranes into the amyloid fibrils was found. Phospholipid quantification, polarization transfer SS NMR and cryo-TEM together reveal co-aggregation of phospholipids and α-synuclein in a saturable manner with a strong dependence on lipid component.[65]

Dermaseptin S9 (Drs S9) is an atypical cationic antimicrobial peptide with a long hydrophobic core and with a propensity to form amyloid-like fibrils. Investigation on its membrane interaction using a variety of biophysicals Techniques was carried out. Calorimetric data together with ^{2}H- and ^{31}P-NMR experiments indicated that the peptide has a significant impact on the dynamic organization of lipid bilayers, albeit slightly less for zwitterionic than for anionic membranes. The results suggested found that Drs S9 induces efficient permeabilization in zwitterionic phosphatidylcholine (PC) vesicles, but not in anionic phosphatidylglycerol (PG) vesicles.[66]

Cytoskeleton and protein trafficking processes, including vesicle transport to synapses, are key processes in neuronal differentiation and axon outgrowth. The human protein FEZ1 (fasciculation and elongation protein zeta 1/UNC-76), SCOCO (short coiled-coil protein/UNC-69), and kinesins (*e.g.*, kinesin heavy chain/UNC116) are involved in these processes. The feature of FEZ1 protein as a bivalent adapter of transport mediated by kinesins and FEZ1 protein interaction with SCOCO (proteins involved in the same path of axonal growth) was investigated. In particular the structural aspects of intermolecular interactions involved in this complex formation by NMR, crosslinking coupled with mass spectrometry (MS), SAXS, and molecular modeling was carried out. The topology of homodimerization was accessed through NMR studies of the region involved in this process, corresponding to FEZ1.[67]

Mutations in GBA1, the gene for glucocerebrosidase (GCase), are genetic risk factors for Parkinson disease (PD). α-Synuclein (α-Syn), a protein implicated in PD, interacts with GCase and efficiently inhibits enzyme activity. GCase deficiency causes the lysosomal storage disorder Gaucher disease (GD). We show that saposin C (Sap C), a protein vital for GCase activity *in vivo*, protects GCase against α-syn inhibition. Using NMR spectroscopy, site-specific fluorescence, and Forester energy transfer probes, Sap C was observed to displace α-syn from GCase in solution and on lipid vesicles. Our results suggest that Sap C might play a crucial role in GD-related PD.[68]

Biomembranes are complex structures composed of various lipids and proteins. Different membrane compositions affect viscoelastic and hydrodynamic properties of membranes, which are critical to their functions. Cholesterol-rich vesicles are similar to cell membranes in structure and component. Therefore, cholesterol-rich vesicles can represent a typical model for studying membrane dynamics and functions. Nuclear magnetic relaxation dispersion was used to investigate the detailed molecular dynamics of membrane differences between vesicles and cholesterol vesicles in the temperature range of 278–298 K. Vesicles of two different sizes were prepared. The effect of cholesterol mainly affected the order fluctuation of membranes and the diffusional motion of lipid molecules.[69]

The hydrophobic lung surfactant SP-B is essential for respiration. SP-B promotes spreading and adsorption of surfactant at the alveolar air–water interface. SP-B63-78 is a cationic and amphipathic helical peptide containing the C-terminal helix of SP-B. ^{2}H NMR has been used to examine the effect of SP-B63-78 on the phase behavior and dynamics of bicellar lipid dispersions containing the longer chain phospholipids DMPC-d54 and DMPG and the shorter chain lipid DHPC mixed with a 3 : 1 : 1 molar ratio. Below the gel-to-liquid crystal phase transition temperature of the longer chain components, bicellar mixtures form small, rapidly reorienting disk-like particles with shorter chain lipid components predominantly found around the highly curved particle edges. With increasing temperature, the particles coalesce into larger magnetically-oriented structures and then into more extended lamellar phases.[70]

The low sensitivity inherent to both the static and MAS techniques of SS NMR spectroscopy has thus far limited the routine application of multidimensional experiments to determine the structure of membrane proteins in lipid bilayers. Researchers demonstrated the advantage of using a recently developed class of experiments, polarization optimized experiments, for both static and MAS spectroscopy to achieve higher sensitivity and substantial time-savings for 2D and 3D experiments. They sarcolipin, a single pass membrane protein, reconstituted in oriented bicelles (for oriented ssNMR) and multilamellar vesicles (for MAS ssNMR) as a benchmark. The restraints derived by these experiments are then combined into a hybrid energy function to allow simultaneous determination of structure and topology.[71]

Multidrug resistance (MDR) is a major public health problem that reduces the efficacy of antibiotics in the treatm ent of infections. One of the most common mechanisms in bacteria for conferring MDR is the coupling of drug efflux with the proton motive force. A prototype for studying ion-coupled active transport is the polytopic *E. coli* membrane protein EmrE. Here, ligand-induced conformational changes of multidrug resistance transporter EmrE has been studied by the method of oriented SS NMR spectroscopy.[72]

Microsomal cytochrome b5 (cytb5) is a membrane-bound protein that modulates the catalytic activity of its redox partner, cytochrome P 4502B4 (cytP 450). Researchers reported the structure of full-length rabbit ferric microsomal cytb5 (16 kDa), incorporated in two different membrane mimetics (detergent micelles and lipid bicelles). Differential line broadening of the cytb5 NMR resonances and site-directed mutagenesis data were used to characterize the cytb5 interaction epitope recognized by ferric microsomal cytP 450 (56 kDa).[73]

Certain membrane proteins involved in lipid synthesis can induce formation of new intracellular membranes in *Escherichia coli*, *i.e.*, intracellular vesicles. Among those, the foreign monotopic glycosyltransferase MGS from *Acholeplasma laidlawii* triggers such massive lipid synthesis when over-expressed. To examine the mechanism behind the increased lipid synthesis, the lipid binding properties of MGS were investigated *in vivo* together with the correlation between lipid synthesis and MGS over-expression levels. A good correlation between produced lipid quantities and over-expressed MGS protein was observed when standard LB medium was supplemented with four different lipid precursors that have significant roles in the lipid biosynthesis pathway. This correlation was highest concerning anionic lipid production and at the same time dependent on the selective binding of anionic lipid molecules by MGS. A selective interaction with anionic lipids was also observed *in vitro* by ^{31}P NMR binding studies using bicelles prepared with *E. coli* lipids.[74]

The interaction of DNA with salt-free tetradecyltrimethylammonium hydroxide and lauric acid lamellar vesicles with positive charges was investigated to probe potential applications of vesicles in DNA transfection. The aggregation morphology of the vesicles changes greatly with the addition of DNA due to the dissociation of anionic surfactants, as indicated by ^{1}H NMR, and the expelled surfactant molecules self-assemble into micelles at high concentrations of DNA. Salt-free cationic and anionic (catanionic) vesicles have a much higher binding saturation point with DNA at $R = 2.3$ than formerly reported salt-containing systems, implying high transfection efficiency. DNA retains its native stretched state during the interaction process. This very result shows that catanionic vesicles could help transport undisturbed and extended DNA molecules into the target cells.[75]

Neurotensin (NT) is a tridecapeptide, hormone in the periphery and neurotransmitter in the brain. High-resolution NMR (NMR) to resolve the three-dimensional structure of NT in a small unilamellar vesicle (SUV) environment was used. It was demonstrated that if the dynamic of the association–dissociation processes of peptide to SUV binding is rapid enough, structural determination can be obtained by solution NMR experiments. Thus, according to the global dynamic of the system, SUVs seem to be an effective model to mimic biological membranes, especially since the lipid component can be modified or sterols may be added to closely mimic the biological membranes studied.[76]

Cord factor (trehalose 6,6′-dimycolate, TDM) is the major lipid in the outer membrane of Corynebacteria and Mycobacteria. Although its role is well recognized in the immune response phenomena, its membrane biophysicals properties remained largely unexplored and TDM has often been described as a detergent. Researchers purified the main components of the outer membrane from *Corynebacterium glutamicum* and analyzed their membrane forming properties. In mixture with endogenous cardiolipin, but not alone, the spontaneous hydration of TDM produces liposomes. Perdeuterated cardiolipin–TDM mixtures were shown by deuterium NMR to exhibit a gel to L.C. phase transition over a 273–295 K temperature range, for cells grown at 303 K, and thus to be in a L.C. state at physiological temperature.[77]

High-pressure NMR spectroscopy has emerged as a complementary approach for investigating various structural and thermodynamic properties of macromolecules noticeably absent from the array of experimental restraints that have been employed to characterize protein structures at high hydrostatic pressure is the residual dipolar coupling, which requires the partial alignment of the macromolecular of interest. Researchers examined five alignment media that are commonly used at ambient pressure for this purpose. It was found that the spontaneous alignment of Pf1 phage, d(GpG) and a $C_{12}E_5$/*n*-hexnanol mixture in a magnetic field is preserved under high hydrostatic pressure. However, DMPC/DHPC bicelles and collagen gel are found to be unsuitable.[78]

The transport of anions across biological membranes by small molecules is a growing research field due to the potential therapeutic benefits of these compounds. However, little is known about the exact mechanism by which these drug-like molecules work and which molecular features make a good transporter. An extended series of

1-hexyl-3-phenylthioureas were synthesized, fully characterized (NMR, mass spectrometry, IR and single crystal diffraction) and their anion binding and anion transport properties were assessed using ^{1}H NMR titration techniques and a variety of vesicle-based experiments. Quantitative structure–activity relationship (QSAR) analytical revealed that the anion binding abilities of the mono-thioureas are dominated by the (hydrogen bond) acidity of the thiourea NH function.[79]

The membrane interactions of the antimicrobial peptides aurein 1.2 and caerin 1.1 were observed by ^{31}P and ^{2}H SS NMR and CD spectroscopy. Both peptides were relatively unstructured in water. In the presence of dimyristoylphosphatidylcholine (DMPC) and mixed DMPC and dimyristoylphosphatidylglycerol (DMPG) vesicles, both peptides displayed a considerable increase in helical content with the shorter aurein peptide having a higher α-helix content in both lipid systems. In fluid phase DMPC vesicles, the peptides displayed differential interactions: aurein 1.2 interacted primarily with the bilayer surface, while the longer caerin 1.1 was able to penetrate into the bilayer interior. This study suggested that the effects of the lipid phase and domains need to be considered when studying membrane interactions of antimicrobial peptides.[80]

3.2.2 Vesicles, bicelles and L.C. dispersion.

The aggregation behavior of mixtures of the amino acid L-Arginine (L-Arg) and bis(2-ethylhexyl)phosphoric acid (DEHPA) in H_2O was studied. At a fixed L-Arg concentration, a phase sequence of micellar phase (L_1 phase), vesicle phase ($L_{\alpha v}$ phase), planar lamellar phase ($L_{\alpha l}$ phase), and sponge phase (L_3 phase) was obtained with increasing DEHPA concentration due to changes in the packing parameter. The phase transition of the lamellar structures was detected by freeze-fracture TEM and ^{2}H NMR spectroscopy.[81]

Vesicles directly prepared from cyclodextrin-ftorafur supramolecular amphiphiles were reported. Ftorafur can be efficiently encapsulated in the cyclodextrin cavities embedded in the vesicle membrane. The morphologies and diameters of the vesicles were identified in detail by TEM, SEM, AFM and DLS. X-ray diffraction (XRD), FT-IR, UV-vis spectrum, ^{1}H NMR and 2D NMR ROESY were further employed to study the formation mechanism of the vesicles. Various morphologies were detected when different host molecules were employed as the hydrophilic moieties of the vesicles' building blocks.[82]

Understanding fundamental uranyl polyoxometalate chemistry in solution and the SS is the 1st step to defining its future role in the development of new actinide materials and separation processes that are vital to every step of the nuclear fuel cycle. Many SS geometries of uranyl polyoxometalates were described, but the authors are only beginning to understand their chemical behavior, which thus far includes the role of templates in their self-assembly, and the dynamics of encapsulated species in solution This study provides unprecedented detail into the exchange dynamics of the encapsulated species in the SS through MAS NMR spectroscopy.[83]

EmrE is a multidrug resistance efflux pump with specificity to a wide range of antibiotics and antiseptics. To obtain atomic-scale insight into the attributes of the native state that encodes the broad specificity, researchers used a hybrid of solution and SS NMR methods in lipid bilayers and bicelles. The results indicate that the native EmrE dimer oscillates between inward and outward facing structural conformations at an exchange rate (k_{ex}) of 300 s^{-1} at 37° (millisecond motions), which is 50-fold faster relative to the tetraphenylphosphonium (TPP+) substrate-bound form of the protein. These observables provide quantitative evidence that the rate-limiting step in the TPP+ transport cycle is not the outward–inward conformational change in the absence of drug.[84]

The formation of stable unilamellar vesicles in the aqueous mixed solution of a surface active ionic liquid (SAIL), 1-hexadecyl-3-methylimidazolium chloride ([C16mim]Cl) and cholesterol, was studied. The formation of such stable vesicles was also shown using a common cationic surfactant, benzyldimethylhexadecylammonium chloride (BHDC) which has a similar alkyl chain length but different headgroup region to that of [C16mim]Cl. DLS, transmission electron microscopy (TEM), NMR, and other optical spectroscopic techniques showed that the micelles of [C16mim]Cl and BHDC in aqueous solutions transform into stable unilamellar vesicles upon increasing concentration of cholesterol.[85]

Magnetic resonance imaging (MRI) enables high-resolution noninvasive observation of the anatomy and function of intact organisms. However, previous MRI reporters of key biologic processes tied to gene expression have been limited by the inherently low molecular sensitivity of conventional ^{1}H MRI. This limitation could be overcome through the use of hyperpolarized nuclei, such as in the noble gas xenon, but previous reporters acting on such nuclei have been synthetic. Researchers have proposed the first genetically encoded reporters for hyperpolarized ^{129}Xe MRI.[86]

The theranostic vesicle has been prepared based on an amphiphilic biocompatible and biodegradable dibock copolymer, poly(ethylene glycol)-*block*-poly(L-lactic-*co*-glycolic acid) (PEG-*b*-PLGA) and bovine serum albumin–gadolinium (BSA–Gd) complexes. The *in vitro* and *in vivo* MR imaging experiments revealed their T_1-weighted MR imaging function. Red blood cell hemolysis and cytotoxicity experiments confirmed their good blood compatibility and low cytotoxicity.[87]

Tetrameric H10/12 helix stabilization was achieved by the application of aromatic side-chains in β-peptide oligomers by intramolecular backbone-side chain CH–π interactions. Because of the enlarged hydrophobic surface of the oligomers, a further aim was the investigation of the self-assembly in a polar medium for the β-peptide H10/12 helixes. NMR, ECD, and molecular modeling results indicated that the oligomers formed by *cis*-[1*S*,2*S*]- or *cis*-[1*R*,2*R*]-1-amino-1,2,3,4-tetrahydronaphthalene-2-carboxylic acid (ATENAC) and *cis*-[1*R*,2*S*]- or *cis*-[1*S*,2*R*]-2-aminocyclohex-3-enecarboxylic acid (ACHEC) residues promote stable H10/12 helix formation with an alternating backbone configuration even at the tetrameric chain length. These results support the view that aromatic side-chains can be applied for helical structure stabilization.[88]

Cyclodextrin vesicles (CDVs) consist of a bilayer of amphiphilic cyclodextrins (CDs). CDVs exhibit CD cavities at their surface that are able to recognize and bind hydrophobic guest molecules *via* size-selective inclusion. In this study, the permeability of α- and β-CDVs was investigated by PFGSTE NMR. Diffusion experiments with water and 2 types of water-solution polymers, polyethylene glycol (PEG) and polypropylene glycol (PPG), revealed 3 main factors that influence the exchange rate and permeability of CDVs. First, the length of the hydrophobic chain of the CD amphiphile plays a crucial role. Second, the exchange rate through the bilayer membrane depended on the molecular weight of the polymer and decreased with increasing weight of the polymer. Most interestingly, a size-selective distinction of permeation due to the embedded CDs in the bilayer membrane was found.[89]

The aqueous photochemistry of 2-oxooctanoic acid (a single-tailed surfactant) results in the synthesis of a double-tailed surfactant product followed by spontaneous self-assembly into vesicles. The photochemistry mechanism is detailed here, and the reaction products are identified using mass spectrometry. Then, the self-assembled vesicles are characterized using DLS, fluorescence microscopy, and NMR.[90]

The segregation and phase sequence of semifluorinated cat-anionic surfactant membranes at different excess surface charges was investigated by freeze-fracture transmission electron microscope (FF-TEM), X-ray diffraction (XRD), and NMR. The thermal behavior of the membranes was evaluated by conductivity, rheology, and deuterium NMR (^{2}H NMR). The cat-anionic fluorinated surfactant mixtures can form faceted vesicles and punctured lamellar phase when there is excess surface charge. The cationic and anionic fluorinated surfactants are stiff in the membranes, like phospholipids in the frozen "crystalline" or "gel" phase.[91]

Sodium dicarboxylates (or Bola salts) with methylene spacers 0, 2, 4, 6, 8, and 10 were studied in aqueous solution to study their influence on the micellization of cetyltrimethylammonium bromide (CTAB). Since bolas with spacer length ≤ 12 are known not to micellize in general, the herein used sodium dicarboxylates were treated as 2 : 1 amphiphilic electrolytes which reduced surface tension of water (except sodium oxalate with zero spacer) without self-association NMR study supported the nonmixing of bolas with the CTAB micelles. They interacted in solution like amphiphilic electrolytes to influence the surface and micelle forming properties of CTAB.[92]

Two double chain cationic lipids QAS C*n*-2-S ($n = 12$, 14) derived from thio galactose and carbamate-linkage tertiary amine were synthesized and their structures were confirmed by MS, TOF-MS, ^{1}H NMR and ^{13}C NMR. The QAS C12-2-S revealed superior surface activity compared with QAS C14-2-S with lower CMC and γCMC. Though Lipo C12-2-S displayed large average particle-size with high polydispersity, positively charged Lipo C*n*-2-S can be combined with the negatively charged DNA, also negatively stained TEM images confirmed the formation of vesicles. All the above prove that the Lipo C*n*-2-S is helpful for gene transfection.[93]

The effect of architecture on polymer interactions has been investigated, and differences between branched and linear copolymers were found. Fluorescence (using pyrene and 4HP as probe molecules), IR spectra, NMR spectra and X-ray scattering, and light scattering were used to elucidate the molecular structure–property relationships. Miktoark star polymers and block copolymers were used in the study. Considerable amounts of water are entrapped within the collapsed PDMAEMA domains above 55 °C (skin effect), preventing PPO-PDMAEMA complexation within pptg. PPO-*b*-PDMAEMA. The collapsed PDMAEMA was rather polar as sensed by pyrene and 4HP.[94]

SS NMR is a promising tool for elucidating membrane-related biologic phenomena. Researchers achieved the measurement of high-resolution SS NMR spectra for a lipid-anchored protein embedded in lipid bilayers under MAS. To date, SS NMR measurements of lipid-anchored proteins have not been accomplished due to the difficulty in supplying sufficient amount of stable isotope labeled samples in the overexpression of lipid-anchored proteins requiring complex posttranslational modification. The authors designed a pseudo lipid-anchored protein in which the protein component was expressed in *E. coli* and attached to a chemically synthesized lipid-anchor mimic. Using two types of membranes, liposomes and bicelles, it was demonstrated different types of insertion procedures for lipid-anchored protein into membranes.[95]

A new kind of heterogeneous RAFT polymerization called seeded dispersion RAFT polymerization to prepare triblock copolymer corona–shell–core nanoparticles of the temperature-sensitive poly(*N*,*N*-dimethylacrylamide)-*block*-polystyrene-*block*-poly[*N*-(4-vinylbenzyl)-*N*,*N*-diethylamine] (PDMA-*b*-PS-*b*-PVEA) was proposed, and the reversible nanoparticle-to-vesicle transition of the triblock copolymer was investigated. Well-controlled seeded dispersion RAFT polymerization as indicated by the pseudo-first-order kinetics, the linear increase in the triblock copolymer molecular weight with the monomer conversion and the narrow molecular weight distribution of the synthesized triblock copolymer was performed, and triblock copolymer corona–shell–core nanoparticles, in which the temperature-sensitive poly[*N*-(4-vinylbenzyl)-*N*,*N*-diethylamine] (PVEA) block forms the core, the solvophobic polystyrene block forms the shell, and the solvophilic poly(*N*,*N*-dimethylacrylamide) block forms the corona, were prepared. The nanoparticle-to-vesicle transition is ascribed to the initial nanoparticle swelling in the alcoholic solvent and the subsequent out-migration of the core-forming PVEA block from the core layer to corona, which was confirmed by the real-time ^{1}H NMR analysis and XPS analysis. The parameters affecting the nanoparticle-to-vesicle transition were investigated, and it was found that the nanoparticle-to-vesicle transition took place in the alcoholic solvent at temperature below LCST when the triblock copolymer containing a long core-forming PVEA block.[96]

Formation and disintegration of self-assembled nanostructures in response to external stimuli are important phenomena that have been widely explored for a variety of biomedical applications. In this contribution, we report the thermally triggered assembly of block copolymer molecules in aqueous solution to form vesicles (polymersomes) and their disassembly on reduction of temperature. A new thermo-responsive diblock copolymer of poly(*N*-isopropylacrylamide) poly((3-methacrylamidopropyl)trimethylammonium chloride) (PNIPA-*b*-PMAPTAC) was synthesized by reversible addition-fragmentation chain transfer technique. The solution properties and self-assembling behavior of the block copolymer molecules were studied by turbidimetry, temperature-dependent proton NMR, fluorescence spectroscopy, DLS, and transmission electron microscopy.[97]

It was developed biodegradable polymeric vesicles as a nanocarrier system for multimodal bio-imaging and anticancer drug delivery. The poly(lactic-*co*-glycolic acid) (PLGA) vesicles were fabricated by encapsulating inorganic imaging agents of superparamagnetic iron oxide nanoparticles (SPION), manganese-doped zinc sulfide (Mn:ZnS) quantum dots (QDs) and the anticancer drug busulfan into PLGA nanoparticles *via* an emulsion-evaporation method. T_2*-weighted magnetic resonance imaging (MRI) of PLGA-SPION-Mn:ZnS phantoms exhibited enhanced negative contrast with r_2* relaxivity of approximately 523 s^{-1} mM^{-1} Fe.[98]

Recent NMR studies have successfully utilized bicelles for atomic-resolution structural and dynamic studies of antimicrobial peptides, amyloid peptides, and membrane-bound proteins. Though bicelles composed with several different types of lipids and detergents have been reported, the NMR requirement of magnetic alignment of bicelles limits the temperature range in which they can be used and subsequently their composition. Because of this restriction, low-temperature experiments desirable for heat-sensitive membrane proteins have not been conducted because bicelles could not be aligned. In this contest researchers have characterized the magnetic alignment of bicelles with various compositions for a broad range of temperatures using ^{31}P static NMR spectroscopy in search of temperature-resistant bicelles. Our systematic investigation identified a temperature range of magnetic alignment for bicelles composed of 4 : 1 DLPC : DHexPC, 4 : 1 : 0.2 DLPC : DHexPC : cholesterol, 4 : 1 : 0.13 DLPC : DHexPC : CTAB, 4 : 1 : 0.13 : 0.2 DLPC : DHexPC : CTAB : cholesterol, and 4 : 1 : 0.4 DLPC : DHexPC : cholesterol-3-sulfate. The results indicate that the presence of 75 wt% or more water is essential to achieve maximum magnetic alignment, while the presence of cholesterol and cholesterol-3-sulfate stabilizes the alignment at extreme temperatures and the positively charged CTAB avoids the mixing of bicelles.[99]

The effect of addition of a bile salt, sodium deoxycholate (NaDC), on the morphology transition in aqueous cetyl trimethylammonium bromide (CTAB) has been studied. Visual, optical density and zero-shear viscosity data of the aqueous mixtures provide some basic macroscopic information on phase behavior. NMR spectra show the peak broadening on increasing the [NaDC] in the system confirming the presence of a mixture of micelles and vesicles.[100]

Phase behavior, self-assembly, and rheology properties of 3-aminopropyltriethoxysilane (APTES) and nonionic surfactant C12EO4 in water were studied. The self-assembled structures were detected by means of small angle X-ray scattering (SAXS), cryo-transmission electron microscopy (cryo-TEM), freeze-fracture TEM (FF-TEM), and ^{2}H NMR measurements and the properties of self-assembled mixtures were obtained by rheology measurements. With a variation of the compn., different self-assembled structures were obtained.[101]

Alkyl triphenylphosphonium bromide (C16TPB), an ionic liquid-based surfactant, and β-cyclodextrin were found to be able to form into a supramolecular complex, which could further aggregate into vesicles in an aqueous solution. The vesicles transformed into precipitates as the concentration of both β-CD and the surfactant increased. Interestingly, a sheet-like hydrogel with multiple responsive properties formed upon the addition of inorganic salts. The morphology and size distribution of the supramolecular aggregates were characterized in detail by transmission electron microscopy (TEM), scanning electronic microscopy (SEM) and DLS. A possible transformation mechanism was suggested based on the results of ^{1}H NMR, FT-IR and powder X-ray diffraction (XRD). The research might be of value in applications of biochemistry and design of smart materials.[102]

During the course of their lifespan, erythrocytes actively shed phospholipid-bound microvesicles (MVs). In stored blood, the number of these erythrocyte-derived MVs has been observed to increase over time, suggesting their potential value as a quality metric for blood products. The lack of sensitive, standardized MV assays, however, poses a significant barrier to implementing MV analyses into clinical settings. Here, we report on a new nanotechnology platform capable of rapid and sensitive MV detection in packed red blood cell (pRBC) units. A filter-assisted microfluidic device was designed to enrich MVs directly from pRBC units, and label them with target-specific magnetic nanoparticles. Subsequent detection using a miniaturized NMR system enabled accurate MV quantification as well as the detection of key molecular markers (CD44, CD47, CD55).[103]

Among other perturbations, high hydrostatic pressure has proven to be a mild yet efficient way to unfold proteins. Combining pressure perturbation with NMR spectroscopy allows for a residue-per-residue description of folding reactions. Accessing the full power of NMR spectroscopy under pressure involves the investigation of conformational sampling using orientational restraints such as residual dipolar couplings (RDCs) under conditions of partial alignment. The aim of this study was to identify and characterize stable and pressure resistant alignment media for measurement of RDCs at high pressure.[104]

Carbon nanotubes (CNTs) are one of the most promising nanomaterials to be used in biomedicine for drug/gene delivery as well as biomedical imaging. A recent study develops radio-labeled, iron oxide-decorated multiwalled CNTs (MWNTs) as dual magnetic resonance (MR) and single photon emission computed tomog. (SPECT) contrast agents. Hybrids containing different amounts of iron oxide were synthesized by *in situ* generation. Physicochemical characterizations reveal the presence of superparamagnetic iron oxide nanoparticles (SPION) granted the magnetic properties of the hybrids.[105]

An asymmetric cationic ethoxylated amphiphile *N*-[10-[4-[4′-methyloxy-oligo(ethylene oxide) ethyleneoxy]oxy]decyloxybiphenyl]-*N*′,*N*″-dimethyl-*N*‴-dodecylammonium bromide ($EO_{12}BphC_{10}NC_{12}$) containing two hydrophilic groups has been prepared. Different from common amphiphiles, $EO_{12}BphC_{10}NC_{12}$ not only connects different hydrophilic units on both ends of hydrophobic spacers but also integrates the structural characters of bola- and gemini-form amphiphiles together, which brings interesting properties to the new building block. Surface activity and self-assembly of the amphiphile were studied in aqueous solution. Strong NOE between quaternary ammonium cation and the first oxyethylene group indicates that the two headgroups locate at the vesicle surface together randomly, other than selectively occupy inner or outer vesicle surface.[106]

Micromixer technology was used to prepare polymeric vesicles (Pluronic L-121) dual loaded with the anti-cancer drug camptothecin and magnetic nanoparticles. DLS measurements showed a relatively narrow size distribution of the hybrid polymersomes. Relaxometry measurements demonstrated the capacity of magnetic polymersomes to generate significant T_2-weighted MRI contrast and potentially allow for direct monitoring of the biodistribution of the polymersomes.[107]

It has long been considered that the addition of cyclodextrins (CDs) disfavors the self-assembly of surfactants in dilute solutions since the hydrophobic effect is destroyed upon the formation of the hydrophiphilic CD/surfactant inclusion complex. However, it was found that β-CD/nonionic surfactant inclusion complexes are able to self-assemble into vesicles in dilute solutions, namely in solutions with concentration lower than the CMC of surfactants. When using Tween 20 as a model surfactant, H-NMR and MS measurements indicate that the building block for the vesicles is the channel type Tween 20@2β-CD inclusion complex.[108]

Spontaneous transformation of lamellar structures, such as multilamellar vesicles from micelles or unilamellar vesicles, is an important challenge in the field of amphiphile molecules, which may serve as models to understand biologic relevant bilayer membranes. In a recent report a progressive self-assembly progress of *N*-tetradecyllactobionamide (C14G2) and tetraethylene glycol monododecyl ether (C12EO4) mixtures in aqueous solution was discussed. Increasing temperature or surfactant compositions causes spontaneous transformation from simple to high-level aggregates, *i.e.*, from unilamellar vesicles, to coexisting multilamellar vesicles, terraced planar bilayers, and finally terraced planar bilayers. D NMR (^{2}H NMR), freeze-fracture TEM (FF-TEM), and small-angle X-ray scattering (SAXS) measurements clearly demonstrate the spontaneously progressive self-assembly process.[109]

Spontaneously forming structures of a system composed of dimyristoyl phosphatidylcholine (DMPC) and 3-[(3-cholamidopropyl)-dimethylammonio]-2-hydroxy-1-propanesulfonate (CHAPSO) were studied by SANS, ^{31}P NMR, and PFGSTE ^{1}H NMR diffusion measurements. Charged lipid dimyristoyl phosphatidylglycerol (DMPG) was used to induce different surface charge densities. The structures adopted were investigated as a function of temperature and lipid concentration for samples with a constant molar ratio of long-chain to short-chain lipids (=3).[110]

Magnetic resonance imaging (MRI) is at the forefront of non-invasive medical imaging techniques. Ultrasmall superparamagnetic iron oxide (USPIO) nanoparticles are attractive MRI contrast agents due to their negative (T_2) contrast enhancement capability and biocompatibility.

Aiming to establish a relationship between the morphology of the clusters and their efficacy as MRI contrast agents (relaxometric properties), it were prepared – using three different maghemite (γ-Fe_2O_3) USPIO diameters – a series of hybrid copolymer/iron oxide CAs presenting two different geometries (micellar or vesicular). The NMR relaxometry profiles confirmed the nature of the physical mechanisms inducing the increase of nuclear relaxation rates at low (magnetic anisotropy) and high (Curie relaxation) magnetic fields. A heuristic model, first proposed by Roch, Muller, Gillis, and Brooks, allowed the fitting of the whole longitudinal relaxivity r1(ν) profile, for samples with different magnetic core sizes.[111]

A gradient copolymer of styrene (St) and Me methacrylate (MMA) is synthesized *via* reversible addition-fragmentation chain transfer living radical polymerization and its micellization behaviors in an acetone and water mixture were investigated by transmission electron microscopy, light scattering, and NMR spectroscopy. Three different kinds of transitions were found to coexist in a single system for the first time: a unimers to micelles transition, a star-like micelles to crew-cut micelles transition resulting from the shrinkage of micelles, and morphology transitions from spherical micelles to cylindrical micelles to vesicles.[112]

Rocket-like vesicles formed are composed of poly(methacrylic acid) (PMAA)/poly(ethylene glycol) (PEG) complex coated hollow silica spheres, and the structure and composition of the vesicles are characterized using TGA, ^{1}H NMR, FTIR, and TEM. Although only one-third of EG units of PEG brushes grafted to hollow silica spheres form the complex with PMAA *via* hydrogen bonding, the first "booster" layer composed of PMAA/PEG complex can provide secure encapsulation of model compound calcein blue under an acidic condition. The second "booster" layer composed of PEG brushes can be formed by changing acidic pH to 7.4 through the disassociation of the PMAA/PEG complex. A higher molecular weight PMAA exhibits a faster disassembly due to the formation of a looser PMAA/PEG complex on the surfaces of hollow silica spheres.[113]

The conformational flexibility and programmed assembly of a dumbbell-shaped polyoxometalate-organic hybrid molecule comprising two Dawson-type polyoxometalates linked by a 2,2′-bipyridine unit, which can be coordinate to metal ions, in this case of Zn^{2+}, are described. SAXS, UV/visible, and NMR spectroscopic techniques confirm that the hybrid molecules exist as the trans dumbbell in metal-ion-free solutions and can be reversibly transformed into the cis dumbbell through coordination upon the addition of $ZnCl_2$ into a DMSO solution containing the hybrid. Subsequent addition of EDTA reverses the switching process by extg. the Zn^{2+} cations from the hybrid.[114]

A study on ionically-driven self-assembly of oligomeric aminosilicones, judiciously protonated with a variety of organic acids was carried out. Depending on the length of the silicone and the strength of the associated acids, (inverse) water-in-silicone emulsions, small nanoparticles, or catanionic vesicles were prepared and characterized by conventional (TEM) or original (DIC optical microscopy, DOSY NMR) techniques. For chains longer than 40 units, a specific PEG-based sulfonic acid was synthesized and used to generate a supramolecular block-like copolymer and ensure fast and efficient emulsification. In all instances, a simple impulse such as pH increase triggered phase separation of the colloidal objects.[115]

The drug delivery system based on supramolecular vesicles that were self-assembled by a novel host–guest inclusion complex between a water-solution pillar[6]arene (WP6) and hydrophobic ferrocene derivative in water has been developed. The inclusion complexation between WP6 and ferrocene derivative in water was studied by ^{1}H NMR, UV-vis, and fluorescence spectroscopy, which showed a high binding constant of $(1.27 \pm 0.42) \times 10^{5}$ M^{-1} with 1 : 1 binding stoichiometry.[116]

Synthetic amino acid-based surfactants possess versatile aggregation properties and are typically more biocompatible and biodegradable than surfactants with conventional headgroups. This opens the possibility of a myriad of specialty applications, namely in pharmaceutics, cosmetics, biomedicine, and nanotemplating chemistry. Reseachers have investigated the interfacial and self-assembling properties in aqueous medium of novel double-chained lysine-based surfactants, with particular focus on the behavior of the dodecyl derivative, 12Lys12. Upon cooling from dilute isotropic micellar solutions, this surfactant crystallizes into micrometer-sized tubular structures that induce gelation of the system. The tubules have been characterized in terms of morphology, assembly process, thermal behavior, and stability, by using DSC, light and SEM, and deuterium NMR.[117]

Biomimetic systems based on cationic surfactants composed of an isosteviol moiety and different counterions, namely bromide (S1) and tosylate (S2), have been designed. The counterion structure was shown to play a crucial role in the surfactant association. A number of methods used provide evidence that only one type of aggregate, *i.e.*, micelles were observed in the S2 systems, while a concentration-dependent association occurred in the case of S1. The DLS and fluorescence anisotropy measurements showed that the micelle-vesicle-micelle transitions probably occur with the S1 system. The occurrence of small aggregates near the CMC with radii of 2.5 nm is supported by NMR self-diffusion data.[118]

4 Micellar solutions

Micellar systems form as a result of the surfactant self-association whenever a suitable surfactant shape occurs, that is a packing parameter $P = v/al$ not too close to unity. Surfactant molecules in micellar aggregates experience strong intermolecular interactions due to the interplay of both electrostatic and van der Waals forces, hence significant changes of NMR parameters are observed. Moreover, micellar aggregates usually form isotropic liquid systems, thus NMR experiments can be easily performed and modelled. Reliable information, at a molecular level, on critical micelle concentration (CMC), molecular conformations and interactions, counterion binding and hydration, also in mixtures of different amphiphiles, can be generally obtained.

4.1 Ionic and non ionic surfactant systems

The temperature-driven self-assembly of nonionic amphiphilic tailor-made triblock copolymers was studied by DLS, NMR, ITC, and SAXS. The composition of these triblock copolymers is more complex than that of the vast majority of poly(2-alkyl-2-oxazoline)s: a statistical thermo-responsive (iPrOx) and hydrophobic (BuOx) central block with terminal hydrophilic blocks (MeOx). Researchers made a first attempt to resolve the effects of each block on nanoparticle formation. The iPrOx/MeOx ratio dets. the value of the cloud point temperature, whereas the different BuOx–iPrOx blocks determine the character of the process. Finally, a study on the thermodynamic and structural profiles of the complexation between these triblock poly(2-alkyl-2-oxazoline)s and two ionic surfactants was presented.[119]

Effects of cationic ammonium gemini surfactant hexamethylene-1,6-bis(dodecyldimethylammonium bromide) (12-6-12) on the micellization of two triblock copolymers of poly(ethylene oxide)-poly(propylene oxide)-poly(ethylene oxide), F127 (EO97PO69EO97) and P123 (EO20PO70EO20), have been studied in aqueous solution by DSC, DLS, ITC, and NMR techniques. Compared with traditional single-chain ionic surfactants, 12-6-12 has a stronger ability of lowering the CMT of the copolymers, which should be attributed to the stronger aggregation ability and lower CMC of 12-6-12. The critical micelle temperature (CMT) of the two copolymers decreases as the 12-6-12 concentration increases and the ability of 12-6-12 in lowering the CMT of F127 is slightly stronger than that of P123.[120]

In commercial formulations, surfactants are often co-formulated with inorganic electrolytes, which are included as inexpensive thickeners. Researchers investigated the effects of alkali and alkaline-earth metal chlorides on the micellization of the nonionic surfactant hexyl penta(oxyethylene) ether, C6E5, in aqueous solution. To this aim, the C6E5 intradiffusion (also named self-diffusion) coefficients in aqueous mixtures of various alkali and alkaline-earth metal chlorides was measured by PFGSE NMR. The results show that all the considered electrolytes cause a decrease of the surfactant CMC, while the micellar size is almost unaffected. The experimental evidence can be interpreted in terms of dehydration of the apolar alkyl tails with a minor contribution arising from the dehydration of the poly(ethylene oxide) headgroups.[121]

Mixtures of ionic surfactants and the chelating surfactant 2-dodecyldiethylenetriaminepentaacetic acid (4-C12-DTPA) were examined in terms of interactions in mixed micellar systems. The amphoteric 4-C12-DTPA is zwitterionic with a negative net charge at the studied pH levels. Following the approach of Rubingh's regular solution theory, the interaction parameter β for mixed micelle formation was calculated from the cmc values detected by NMR diffusometry. As expected, the most negative β parameter was found in the mixture with DoTAC, followed by DDAO and SDS. The self-diffusion coeff. *vs.* concentration plots show two distinctly different curves, depending on the surfactant that is present in excess.[122]

The interactions between the fluorinated cationic guar gum (FCGG) and ionic surfactants including cetyl tri-Me ammonium bromide (CTAB) and sodium lauryl sulfate (SDS) were studied by light scattering, fluorescence spectroscopy, UV-spectrophotometer, ^{19}F NMR and dynamic rheometer, respectively. The results show that, with the addition of the surfactants, the stretching degree of the FCGG chains is increased in the FCGG/CTAB solutions, while the dramatical shrinking of FCGG chain, the phase separation and the re-stretched macromolecules appear successively because of the electricity neutralization reaction in the FCGG/SDS system.[123]

The focus of this report concerns nanocomposites from poly(propylene carbonate) (PPC) and multiwall carbon nanotubes (MWNTs). A solvent route using THF, ethoxylated non- ionic surfactants combined with sonication was found to be successful in deagglomerating and dispersing the nanotubes. The morphology and molecular mobility of the prepared nanocomposites (0.5, 3.0 and 5.0 wt% of nanotubes) were characterized by rheology, microscopy, low-field SS NMR, and electrical conductivity. The networking of nanotubes was highest with a stearyl alcohol ethoxylate surfactant, and was found to improve with the sonication time.[124]

For ionic surfactants in general, a change from simple to polymeric counterions leads to increasing attraction between micelles, condensing them in a concentrated phase. A two ''complex salts'' were prepared: the cationic surfactant hexadecyltrimethylammonium was neutralized by two different copolyions, both having poly(methacrylate) main chains randomly decorated with oligo(ethylene oxide) side chains. The physicochemical nature of the aggregates formed was investigated by DLS, NMR self-diffusion measurements and steady-state fluorescence spectroscopy. Much larger aggregates are formed when the temperature is increased, but the small aggregates reform at room temperature, suggesting that the soluble aggregates are equilibrium structures, much like the micelles of conventional surfactants.[125]

The assignment of NMR signals to specific components in a mixture is a challenging task. Diffusion-Ordered Spectroscopy (DOSY) has provided important progress in this area, allowing the signals originating from individual components of different molecular sizes to be distinguished. However, when the sizes of the compounds are similar and/or the spectra are overlapped, signal assignment can easily become intractable. The use of a co-solute in a matrix-assisted DOSY (MAD) experiment can be a useful solution, improving diffusional (and sometimes spectral) resolution by exploiting selective binding to the matrix. Researchers used such strategy on the study of Brij micelles in mixed solvents.[126]

Nanostructured fluids containing anionic surfactants are among the best performing systems for the cleaning of works of art. Though efficient, their application may result in the formation of a precipitate, due to the combination with divalent cations that might leach out from the artifact. Two new aqueous formulations based on nonionic surfactants, which are non-toxic, readily biodegradable and insensitive to the presence of divalent ions have been proposed. Structural information on nanofluids has been gathered by means of SANS, DLS and NMR with diffusion monitoring.[127]

NMR spectroscopy and DLS techniques were used to study the structural and dynamic properties of micellar solutions of nonionic surfactants of a homologous series of oxyethylated isononylphenols-$C_9H_{19}C_6H_4O(C_2H_4O)_nH$, where $n = 6$, 8, 9, 10, or 12-in a wide range of temperatures, including cloud points. The radii of the micelles and aggregates, as well as their compositions at different concns. of surfactant, were detected.[128]

Potential determining salts, also referred to as phase transfer agents for a future objective of electrochemistry at the oil–water interface in microemulsions are considered. Reasearchers have studied these salts, composed of a hydrophilic and a hydrophobic ion, in microemulsion stabilized by nonionic surfactants with an oligo ethylene oxide head-group. NMR measurements show that the salts preferentially dissoc. across the surfactant interface between the oil and water domains, and hence create a potential drop across the surfactant film, and back to back diffuse double layers in the oil and water phases. These observations are also supported by Poisson–Boltzmann calcns.[129]

4.2 Polymer amphiphiles

Polymer–surfactant interactions have been studied by NMR experiments. Two sensitive techniques, *i.e.*, NMR self-diffusion measurements and dielcometry are involved to test the association of cationic surfactant 4-aza-1-tetradecyl-1-azoniabicyclo[2.2.2]octane bromide (BC-14) and polyethyleneimine hydrophobized with tetradecyl moieties (PEI-14) in their chloroform solutions. The data obtained make it possible to nominate these methods as highly powerful instruments for the identification of a very uncooperative aggregation. Mixed BC-14/PEI-14 association is probably observed only at a relatively high polymer concentration, which is indicated by a decrease in the diffusivity of surfactant molecules. Aggregation is strongly promoted by the addition of water.[130] The effect of water-insoluble nonionic triblock copolymer PEO-PPO-PEO [poly-(ethylene oxide)-poly(propylene oxide)-poly(ethylene oxide)] *i.e.*, $E_6P_{39}E_6$ with molecular weight 2800, on the microstructure and self-aggregation dynamics of anionic surfactant sodium dodecylsulfate (SDS) in aqueous solution (D_2O) were investigated using high resolution NMR and SANS measurements. Variable concentration and temperature proton (^1H), carbon (^{13}C) NMR chemical shifts, ^1H self-diffusion coefficients, ^1H spin–lattice and spin–spin relaxation rates data indicate that the higher hydrophobic nature of copolymer significantly influenced aggregation characteristics of SDS. The salient features of the NMR investigations include the onset of mixed micelles at lower SDS concentrations relative to the copolymer-free case and their evolution into SDS free micelles at higher SDS concentrations, disintegration of copolymer-SDS mixed aggregate and binding of a copolymer with SDS and preferential localization of the copolymer occurred at the SDS micelle surface. The degree of counterion dissociation extimated from SANS data analysis corroborate reasonably well with those of ^1H NMR self-diffusion and sodium (^{23}Na) spin–lattice relaxation data.[131]

Block copolymers continue to be extensively investigated, eventually in combination with surfactants. The aggregation of ionic liquid-based double-tailed surfactant, 1,3-dioctylimidazolium bromide ([Doim]-Br) and its interaction with pluronic copolymer F127 were systematically investigated by NMR, surface tension, DLS, and ITC. It was found that the [Doim]Br aggregates are composed with the alkyl chains embedded in the micellar core and with the imidazolium rings parallel and staggered on the hydrophilic layer of micelles, which was generally different from the single-tailed IL [omim]-Br.[132] The effects of intermicellar interactions on the dissociation of block copolymer micelles of poly-styrene-*block*-poly-isoprene in a selective solvent, decane, have been investigated using SANS and ^{1}H NMR spectroscopy. Decane is a preferential solvent for polyisoprene (PI) and drives the association of the polystyrene (PS) blocks, resulting in spherical micelles with a PS core and a Gaussian PI corona. The dissociation of the PS–PI micelles is triggered by increasing temperature, while the intermicellar interactions are controlled by varying the polymer concentration and modulating temperature. Enhancing the intermicellar interaction delays the dissociation process of the micelles.[133] A pair of oppositely charged diblock copolymers, poly(2-(methacryloyloxy)ethyl phosphorylcholine)-*block*-poly((3-(methacryloylamino)propyl)trimethylammonium chloride) (PMPC-*b*-PMAPTAC) and poly(2-(methacryloyloxy)ethyl phosphorylcholine)-*block*-poly(sodium 2-(acrylamido)-2-methylpropanesulfonate) (PMPC-*b*-PAMPS). The mixing of aqueous solutions of PMPC-*b*-PMAPTAC and PMPC-*b*-PAMPS leads to the spontaneous formation of simple core-shell spherical polyion complex (PIC) micelles comprising of a segregated PIC core and PMPC shells. The PIC micelles were characterized using ^{1}H NMR spin–spin (T_2) and spin–lattice relaxation times (T_1), diffusion-ordered NMR spectroscopy, SLS, DLS, and TEM techniques.[134] Poly(ethylene glycol)-poly(alanine-*co*-phenylalanine) (PEG-PAF) block copolymers with similar molecular weights were synthesized to investigate the effect of the partial incorporation of D-alanine into PEG-L-PAF. ^{13}C-NMR spectra suggest the dehydration and decrease in molecular motion of PEG during the sol-to-gel transition of PEG-PAF.[135] Another paper reports a new synthetic strategy for the preparation of organometallic-polypeptide block copolymers PFS-*b*-PBLG (PFS = polyferrocenylsilane, PBLG = poly(γ-benzyl-L-glutamate). The self-assembly of obtained samples was studied in *N*,*N*-dimethylformamide, a solvent selective for PBLG. Experiments at 75 °C showed that disordered elongated structures formed initially, with fiber-like protrusions from the ends. Over time, the structures became shorter and wider, evolving into uniform truncated elliptical micelles. The process was monitored by X-ray diffraction (XRD) measurements and by ^{1}H NMR spectroscopy.[136]

Some papers have been published dealing with hydrophobically modified polymers. A combination of NMR Chemical Shift Imaging and self-diffusion experiments has been shown to give a detailed molecular picture of the events that occur when drug-loaded tablets of hydrophobically modified poly(acrylic acid) swell in water in the presence or absence of surfactant. The hydrophobic substituents on the polymer bind and trap the surfactant molecules in mixed micelles, leading to a slow effective transport of the surfactant. The surfactant effectively solubilizes the hydrophobically modified polymer, making it fully miscible with water, leading to a more pronounced swelling and a slower erosion of the polymer matrix.[137] Interaction between an anionic surfactant, sodium dodecyl sulfate (SDS), and a nonionic polymer, 2-hydroxyethyl cellulose (HEC) hydrophobically modified with benzoyl chloride (bmHEC), is studied by viscometry and 2D NOESY in a semidilute regime of bmHEC. The hydrophobicity of bmHEC was varied with different substitution of benzoyl group to HEC macromolecules. The 2D NOESY displays that the surfactants actually interact with bmHEC not only on the hydrophobes, namely benzoyl groups, but also the polymer backbone, *i.e.*, glucose units. In contrast, no interaction is revealed by 2D NOESY in the aqueous system containing SDS surfactant and HEC polymer.[138] In an interesting paper, the supramolecular assemblies of semifluorinated polymer surfactants, composed of a monomethyl poly(ethylene glycol) (mPEG) hydrophilic head group and either 1, 2, or 3 perfluoro-*tert*-Bu (PFtB) groups as the fluorophilic tail, were investigated as a potential design for theranostic nanoparticles. Polymers with three PFtB groups (PFtBTRI) solely formed stable, spherical micelles. These PFtB polymers have the additional potential to serve as F-MRI contrast agents. PFtBTRI micelles gave one narrow ^{19}F-NMR signal in D_2O, with T_1 and T_2 parameters of approximately 500 and 100 ms, respectively. Preliminary data demonstrate the potential of PFtB polymers as a basic design, which can be further modified to serve as dual drug-delivery and imaging vehicles.[139]

Self-assembly of stimuli-responsive self-associating polymers in aqueous solution has attracted considerable attention. An interesting paper deals with photoresponsive aggregates. In details, a supramolecular modular synthetic approach for the fabrication of photo-responsive block-controllable supramolecular polymers (BSPs) based on the assembly in water of two homopolymers, namely, the polystyrene modified with adamantane and azobenzene as the end groups (Ad-PS-Azo) and poly(ethylene glycol) modified with β-CD (PEG-β-CD), has been reported. The supramolecular triblock copolymer (STP) was constructed through host–guest interactions between β-CD and Ad/Azo moieties, and further self-assembled into micelles in aqueous solution. This well-defined supramolecular triblock copolymer can reversibly disassemble into a supramolecular diblock copolymer (SDP) by alternating irradiation of UV/visible light, which was revealed by TEM, UV-vis and ^{1}H NMR.[140]

Thermosensitive polymer self-assembly has been considered. A poly(*N*-isopropylacrylamide) (PNIPAm) and a bisadenine-functionalized poly(ethylene oxide) (A-PEO-A) were synthesized. When blended together in water, PNIPAm and A-PEO-A formed supramolecular aggregates stabilized through complementary multiple hydrogen bonds between the amide groups of PNIPAm and the adenine units of A-PEO-A. FT-IR spectroscopy, ^{1}H NMR spectroscopy, TEM, DLS, and rheometry revealed the thermo-sensitive behavior and viscoelastic properties of the supramolecular assemblies.[141] A series of novel stimuli-responsive AB, ABA, and BAB type block copolymers based on 6-*O*-methacryloyl-1,2:3,4-di-*O*-isopropylidene-D-galactopyranose (MAIpGP: A block) and 2-(*N*,*N*-dimethylamino)ethyl methacrylate (DMAEMA: B block) were synthesized. The PMAIpGP blocks of the AB, ABA, and BAB type linear block copolymers were converted to water soluble PMAGP blocks *via* deprotection process under mild acidic conditions. Both proton NMR and DLS studies demonstrated that block copolymers were temperature-sensitive, and that an increase on the percentage of hydrophilic PMAGP block in block copolymer caused an increase on the LCST value as expected.[142] Thermoreversible phase behaviors and rheology properties of a pentablock terpolymer solution, poly(*N*-isopropylacrylamide)-*b*-poly(ethylene oxide)-*b*-poly(propylene oxide)-*b*-poly(ethylene oxide)-*b*-poly(*N*-isopropylacrylamide) ($PNIPAM_{150}$–PEO_{136}–PPO_{45}–PEO_{136}–$PNIPAM_{150}$), have been investigated in comparison with its precursor, PEO_{136}–PPO_{45}–PEO_{136} (F108). An elastic gel property at 41 °C because of the formation of the close-packed micelle aggregates. Cryo-TEM and variable-temperature ^{1}H NMR results demonstrate that the sol–gel phase transition mechanism is mainly related to the hydrophilic/hydrophobic transition of PPO and PNIPAM groups by external temperature stimulus.[143]

Other articles have been focused on pH-responsive systems. Responsive materials that change conformation with varying pH have been prepared from a range of amphiphilic block co-polymers. The individual blocks are composed of permanently hydrophilic chains with neutral functionality and acrylate polymers with weakly basic side-chains. Variation in co-monomer content, molar mass and block ratios/compositions leads to a range of pH-responses, manifest through reversible self-assembly into micelles and/or polymersomes. These transitions can be tuned to achieve environmental responses in a pH range from 5–7, as shown by turbidimetric analysis, NMR and DLS measurements.[144] Syntheses of diblock copolymers of isobutylene (IB) with amino acid-based monomers were performed. These block copolymers can form core–shell type micellar structure in methanol, a good solvent for side-chain amino acid block but bad solvent for polyisobutylene segment. All the block copolymers exhibited "smart" pH responsiveness and amphiphilic character in aqueous media as confirmed by ^{1}H NMR, fluorescence spectroscopy, DLS, AFM, and SEM.[145] The micellization behavior was examined for a series of 3 pH-stimulable poly(2-vinylpyridine)-*b*-poly(ethylene oxide) (P2VP-*b*-PEO) copolymers. The micellar characteristics were determined by dynamic, static and electrophoretic light scattering, fluorescence spectroscopy, as well as by ^{1}H NMR in the pH range of 2–7 and in the presence or in the absence of the anionic surfactant sodium dodecylsulfate (SDS). At low pH and low copolymer concentrations, where only protonated unimers are present, the SDS induces the micellization at a given neutralization degree. A model for the micelle formation mechanism was suggested.[146]

Dual-responsive polymer-based systems have been also proposed. The design and synthesis of polymers containing cholic acid pendant groups has been reported. Random copolymers derived from 2-(methacryloyloxy)ethyl cholate (MAECA) with polyethylene glycol methyl ether methacrylate (PEGMA) and *N*,*N*-dimethylaminoethyl methacrylate (DMAEMA) were synthesized. Both thermo- and pH-responsiveness were observed for the P(MAECA-*co*-DMAEMA) copolymers. Self-assembly of random copolymers was studied by ^{1}H NMR spectroscopy, DLS, AFM and SEM (SEM), which indicated micellar structures in water from cholic acid based random copolymers.[147] A series of novel thermo- and pH-responsive block copolymers of PHis-PLGA-PEG-PLGA-PHis composed of poly(ethylene glycol) (PEG), poly(d,l-lactide-*co*-glycolide) (PLGA) and poly(l-histidine) (PHis) were synthesized and used for the construction of stimuli-responsive copolymer micelles. The copolymer micelles were constructed to have an inner core consisting of two hydrophobic blocks (PLGA and deprotonated PHis) and an outer hydrophilic PEG shell. The temperature- and pH-induced structure changes of the micelles were characterized by an alteration in particle size, a decrease in pyrene fluorescence intensity, and a variation of ^{1}H NMR spectra in D_2O.[148] Well-defined homopolymers of 2-vinyl-4,4-dimethylazlactone (VDA) and AB diblock copolymers of VDA with *N*,*N*-dimethylacrylamide (DMA) and *N*-isopropyl-acrylamide (NIPAM) were prepared. VDA homopolymers and VDA-DMA/NIPAM AB diblock copolymers reacted with *N*,*N*-dimethyl-ethylenediamine (DMEDA), *N*,*N*-diethylethylenediamine (DEEDA), and picoylamine (PA) to give novel tertiary amine functional polymers. DMEDA-DMA and DEEDA-DMA/NIPAM block copolymer derivatives undergo reversible temperature induced self-assembly in aqueous media by virtue of the inverse temperature solubility characteristics associated with these tertiary amino species. The aggregation behavior of these species was characterized using a combination of DLS, ^{1}H NMR spectroscopy and TEM (TEM). For the PA derivatives, schizophrenic behavior was demonstrated in AB block copolymers with NIPAM with normal and inverse micelles being readily accessible simply by controlling the solution pH or temperature.[149] The synthesis and self-assembly behavior are reported for a well-defined double hydrophilic brush-coil copolymer with poly(*N*-isopropylacrylamide)-*b*-poly(glycidyl methacrylate) (PNIPAM-*b*-PGMA) as backbone and poly(l-glutamic acid) (PLGA) as brush. The PNIPAM-*b*-PGMA was firstly prepared by the sequential reversible addition-fragmentation chain transfer polymerization of *N*-isopropylacrylamide and glycidyl methacrylate. The obtained diblock copolymer was reacted with ethylenediamine (EDA) yielding the aminated macroinitiator (PNIPAM-*b*-PGMA-EDA), which was then used to initiate ring-opening polymerization of γ-benzyl-l-glutamate-*N*-carboxyanhydride (BLG-NCA) to give PNIPAM-*b*-(PGMA-*g*-PBLG) copolymer. After the deprotection of benzyl groups on PBLG, double hydrophilic brush-coil copolymer, PNIPAM-*b*-(PGMA-*g*-PLGA), was obtained. The thermo- and pH-responsive micellization behaviors of PNIPAM-*b*-(PGMA-*g*-PLGA) in aqueous solution were investigated by fluorescence spectroscopy, ^{1}H NMR, DLS, SEM, and CD. It can self-assemble into PNIPAM-core micelles at pH 10 and elevated temperature and PLGA-core micelles at pH 4 and room temperature.[150] A series of well-defined thermoresponsive diblock copolymers (PEO45-*b*-PtNEA*n*, $n = 22$, 44, 63, 91, 172) were prepared by the atom transfer radical polymerization of *trans*-*N*-(2-ethoxy-1,3-dioxan-5-yl) acrylamide (tNEA) using a poly(ethylene oxide) (PEO45) macroinitiator. Acid-triggered hydrolysis behaviors of the aggregates were investigated by LLS, Nile red fluorescence, TEM, and ^{1}H NMR spectroscopy. The results revealed that the spherical nanoparticles formed from PEO45-*b*-PtNEA44 dissociated faster than the polymersomes of PEO45-*b*-PtNEA172, and both aggregates showed an enhanced hydrolysis under acidic conditions.[151]

In another work an optimized condition is identified to prepare linear poly(amido amine)s *via* polymerization of 4-(aminomethyl)piperidine (AMPD) with *N*,*N*-cystaminebis(acrylamide) (BAC). Poly(ethylene glycol) (PEG) and cholesterol (CE) are conjugated to linear poly(BAC-AMPD) to form poly(BAC-AMPD)-*g*-PEG-*g*-CE. TEM, DLS and ^{1}H NMR results show that micelles with PEG shells and hydrophobic cores composed of poly(BAC-AMPD) and CE are formed *via* self-assembly of poly(BAC-AMPD)-*g*-PEG-*g*-CE in aqueous solution, and the micelles of poly(BAC-AMPD)-*g*-PEG-*g*-CE can be degraded by the presence of L-dithiothreitol and show a limited cytotoxicity *in vitro*. The micelles of poly(BAC-AMPD)-*g*-PEG-*g*-CE show pH- and redox-responsive drug release and redox-induced formation of aggregates.[152] The fabrication of dynamic covalent shell crosslinked (SCL) micelles of amphiphilic diblock copolymers functionalized with aldehyde moieties in the hydrophilic block by utilizing difunctional crosslinkers cleavable in response to pH and thiols has been reported. Well-defined amphiphilic diblock copolymer, PCL-*b*-P(OEGMA-*co*-MAEBA), was synthesized *via* polymerization of ε-caprolactone (CL), oligo (ethylene glycol) monomethyl ether methacrylate (OEGMA) and *p*-(methacryloxyethoxy)benzaldehyde (MAEBA) comonomers. In aqueous solution, the diblock copolymer self-assembles into micelles consisting of hydrophobic PCL cores and hydrophilic P(OEGMA-*co*-MAEBA) coronas covalently anchored with aldehyde groups. The subsequent shell cross-linking reaction was conducted at pH 6.2 upon addition of difunctional dithiolbis(propanoic dihydrazide) (DTP). The obtained SCL micelles can be de-crosslinked *via* two biological relevant modes, namely, acidic pH-triggered cleavage of acylhydrazone bonds into aldehyde and hydrazide and thiol-triggered cleavage of disulfide linkages, which have been utilized for triggered release of physical encapsulated chemotherapeutic drugs. The dual-responsive dynamic covalent SCL micelles were examined by DLS, ^{1}H NMR, Ellman's assay, and enzymic degradation tests.[153]

Self-assembling polymers could be used as scaffold to prepare hydrogels, both on the macro- and on the micro- (and even nano-) scale. Novel hydrogels were prepared from CM-cellulose sodium salt by crosslinking with polyethylene glycol diglycidyl ether (PEGDE). The detailed structures of the hydrogels were detected *via* FTIR and SS NMR spectroscopic analyses.[154] Hyaluronic acid (HA) and alginate (AL) covalent nanogels cross-linked with L-lysine ethyl ester were prepared by template chemistry crosslinking of the polysaccharide in polyion complex micelle (PIC) nanoreactors. By using this method we were able to prepare HA and AL nanogels without organic solvents. The nanogels were easily purified and isolated by dialysis. The dissociation of the nanoreactors and the formation of the nanogels were confirmed by ^{1}H NMR, DLS, TEM and ζ-potential measurements.[155]

The self-aggregation of amphiphilic polymers can be combined with nanoparticles preparation to obtain interesting composite systems. The micellar self-assembly behavior of a near-monodisperse polystyrene-hydrogenated polyisoprene (PS-PEP) diblock copolymer has been examined in non-polar media (either *n*-heptane or *n*-dodecane). Direct dissolution of this diblock copolymer leads to the formation of relatively large polydisperse colloidal aggregates that are kinetically frozen artifacts of the SS copolymer morphology. Variable temperature ^1H NMR studies using deuterated *n*-alkanes showed that partial solvation (plasticization) of the polystyrene micelle cores occurs on heating. Individual PS-PEP copolymer chains adsorb onto carbon black from chloroform, whereas micellar adsorption occurs from *n*-alkanes.[156] A mechanistic study on the nucleation of polyaniline nanotubes (PANI-NT) through template-free method was explored by *in situ* solution-state ^1H NMR experiments *via* a careful analysis of the spectral evolution of the major species in the course of the reaction. Our work demonstrates that loosely packed anilinium composed micelles, highly mobile monomers within the micelle, and efficient blockage of the proton-releasing to the aqueous bulk are three key factors for the generation of tubular structures and dynamic NMR line shape analysis provides a new perspective for resolving the formation profile of nanostructured polymers.[157]

The combination of amphiphilic polymers with cyclodextrins offers the opportunity to design interesting systems. The redox switchable formation of very well-defined supramolecular graft polymers in aqueous solution driven by host–guest interactions between ferrocene (Fc) and cyclodextrin (CD) has been presented. The Fc-containing acrylic backbone copolymer (PDMA-*stat*-Fc) was prepared *via* copolymerization of *N*,*N*-dimethyl-acrylamide (DMA) and the novel monomer *N*-(ferrocenoylmethyl)acrylamide (NFMA). The β-cyclodextrin (β-CD) containing building block was synthesized employing a propargyl-functionalized chain transfer agent for the polymerization of *N*,*N*-diethylacrylamide (DEA). The polymerization product was subsequently terminated with β-CD (PDEA-βCD). Host–guest interactions between Fc and CD lead to the formation of supramolecular graft-polymers, verified *via* NOESY. The self-assembly of the statistical Fc-containing lateral polymer chain in aqueous solution leads to mono- and multi-core micelle-aggregates.[158] A series of well-defined adamantyl-terminated thermally responsive copolymers (Ad-POEGMAs) were synthesized by polymerization, in which 2-(2-methoxyethoxy) ethyl methacrylate (MEO_2MA) and oligo(ethylene glycol) methacrylate (OEGMA) served as the thermosensitive building blocks. Moreover, cyclodextrins (CDs) as bulky pendant grafted polymers (PGCD) were synthesized by homopolymerization of aminoethyl methacrylate β-cyclodextrin (GCD). The thermal-responsive behaviors were investigated by a combination of ^1H NMR, UV-vis spectroscopy, DLS, and TEM. The Ad group assembled and POEGMA chains associated to produce stable water-soluble nano-aggregates, followed by a rearrangement process at the second thermal transition. Moreover, it was found that a noncovalently connected supramolecular pseudo-graft polymer was formed *via* inclusion complexation in aqueous solution.[159] A novel kind of host–guest hydrogel possessing gel–sol phase transformation in response to temperature, oxidizing agent and glucose was prepared by ferrocene (Fc)-containing pluronic F127 and β-cyclodextrin (β-CD) linear polymer. Because of the reversible association–dissociation of Fc-β-CD inclusions and F127 micelles, the gel–sol transition was easily observed. 2D NOESY, NMR, TEM and TGA (thermogravimetric analysis) measurements were used to clarify the inclusion complexes of Fc-β-CD. The results showed that the formation of inclusion complexes affected the micelle size and stabilized the hydrogels.[160] A new supramolecular self-assembly approach to construct micelles and reverse micelles was also reported. Double-hydrophilic block copolymers poly(ethylene oxide)-*b*-poly(*N*-isopropylacrylamide) (PEO-*b*-PNIPAAm) were synthesized. The PNIPAAm-core micelles disassemble to unimers when the temperature decreases to 25 °C. But the addition of α-cyclodextrins (α-CDs) could induce the formation of PNIPAAm-shell micelles because of the "channel-type" crystallities induced by PEO/α-CDs polyrotaxanes. The assembly and disassembly procedure of micelles and reverse micelles were investigated by DLS, X-ray diffraction (XRD), ^{1}H NMR and TEM.[161] Supramolecular polymer micelles were prepared on basis of the inclusion complexation between cyclodextrin carbamates and cellulose derivatives in aqueous media. The supramolecular structure of the core–shell micelles was demonstrated by ^{1}H NMR spectra, TEM images, and fluorescence spectra.[162]

In other articles, the potentiality of polymer-based supramolecular aggregates for biomedical applications (drug delivery, imaging) has been explored. One of the main obstacles in tumor therapy is multiple drug resistance (MDR). In the synergistic treatment of siRNA and anti-cancer drug doxorubicin, it is crucial that both the siRNA and doxorubicin are simultaneously delivered to the tumor cells. To this aim a type of micelles comprising a polycationic PEI-CyD shell to condense the siRNA and hydrophobic core to package doxorubicin has been reported. The structure of the polymer is detected by ^{1}H NMR, FT-IR, DSC, and XRD and the micelles are characterized by DLS, 2D-NOESY NMR, and TEM to study the self-assembly of the micelles with siRNA and drugs.[163] Phenylboronic acid-functionalized amphiphilic block copolymer Pluronic-PMCC-BA was synthesized *via* polymerization of 5-methyl-5-benzyloxycarbonyl-1,3-dioxan-2-one (MBC) followed by the deprotection of carboxyl groups and the condensation of 3-aminophenylboronic acid with the copolymer side groups. Pluronic-PMCC-BA can form stable micelle solution by self-assembly in water. The phenylboronic acid groups are located at the shell of micelle as proved by ^{1}H NMR.[164] Click chemistry has found wide application in drug discovery, bioconjugation reactions, polymer chemistry and synthesis of amphiphilic materials with pharmaceutical and biomedical applications. Triazole substitution *via* a click reaction alters photophysical properties of coumarin. PEG supported coumarin conjugates were synthesized and the fluorescence modulation effect of the formation of triazole on coumarin was assessed. Their aggregation properties were studied by surface tension measurements, DLS, TEM, fluorescence and ^{1}H NMR spectroscopy.[165] Interactions between hydrophilic modified ibuprofen (Ibuprofen-PEG800, IP800) and copolymers

P123 or F68 (poly(ethylene oxide)–poly(propylene oxide)–poly(ethylene oxide), PEO–PPO–PEO) were systematically investigated by NMR, DLS, and freeze-fractured (FF)-TEM).[166]

Contrast agents for magnetic resonance imaging could also be delivered through polymer supramolecular aggregates. Aqueous dispersions of superparamagnetic iron oxide nanoparticles (SPIONs) have been prepared using an amphiphilic block copolymer poly(ethylene glycol)-*block*-poly(4-vinyl pyridine) (mPEG-*b*-P4VP). The hybrid SPION nanoaggregates were investigated using NMR relaxometric techniques, whereby high r_2/r_1 relaxivity ratios were achieved, making these materials potentially efficient T_2-weighted MRI contrast agents.[167] A cancer-recognizable MRI contrast agents (CR-CAs) has been developed using pH-responsive polymeric micelles. The CR-CAs with pH sensitivity were self-assembled based on well-defined amphiphilic block copolymers, consisting of methoxy poly(ethylene glycol)-*b*-poly(L-histidine) (PEG-p(L-His)) and methoxy poly(ethylene glycol)-*b*-poly(L-lactic acid)-diethylenetriaminopentaacetic acid dianhydride-gadolinium chelate (PEG-p(L-LA)-DTPA-Gd). The CR-CAs exhibit highly effective T_1 MR contrast enhancement in the tumor region, which enabled the detection of small tumors *in vivo* at 1.5 T within a few minutes.[168] Stable and cytocompatible hybrid PEGylated micelles with multimodal imaging capabilities are described. The Fe_3O_4-encapsulated polymeric micelles composed of cores containing magnetic nanoparticles and polyethylene glycol (PEG) shells are synthesized by self-assembly of amphiphilic poly(HFMA-*co*-VBK)-*g*-PEG copolymers and oleic acid stabilized Fe_3O_4 nanoparticles. *In vivo* magnetic resonance imaging (MRI) studies reveal enhanced contrast between the liver and spleen.[169] Amphiphilic copolymers prepared from methyl methacrylate and PEG methacrylate by atom transfer radical polymerization were conjugated with folic acid (for folate-receptor specificity). The folate-conjugated polymer had a low CMC, indicating stability of the micellar formulation. Superparamagnetic iron oxide nanoparticles (SPIONs)-polymeric micelle clusters were prepared by desolvation of the SPION dispersion/polymer solution in water. MRI of the formulation revealed very good contrast enhancement, with transverse (T_2) relaxivity of 260.4 mM^{-1} s^{-1}.[170]

Some self-assembling polymers or polymer-based composite have been designed to solubilise membrane protein, thus enabling spectroscopic structural investigation. Amphipathic polymers called amphipols provide a valuable alternative to detergents for keeping integral membrane proteins soluble in aqueous buffers. Spatial contacts of amphipol A8-35 with membrane proteins from two architectural classes: the 8-stranded β-barrel outer membrane protein OmpX and the α-helical protein bacteriorhodopsin have been characterized. Overall, A8-35 exhibits protein interaction properties somewhat different from detergents and lipid bilayer nanodiscs, while maintaining the structure of solubilized integral membrane proteins.[171]

4.3 Mixed amphiphiles systems

NMR is an optimally suited tool to investigate mixtures of amphiphiles. The aggregation behavior of tetradecyltrimethylammonium bromide

(TTAB), hexadecyltrimethylammonium bromide (CTAB), and hexadecyltriphe-nylphosphonium bromide (HTPB) and their mixtures in aqueous medium was investigated using conductance, fluorescence and NMR techniques. ^{1}H NMR studies suggested that when mixed with CTAB or TTAB, the bulky head group of HTPB induce steric hindrance. ^{1}H NMR results suggest that CTAB + HTPB produced compact mixed micelles.[172] The mixed system of an anionic hydrocarbon surfactant, sodium dodecyl sulfate, and a perfluorinated surfactant, perfluorononanoic acid (PFNA), was investigated by a combination of methods. The hydrodynamic radii of the aggregates were estimated through the diffusion ordered ^{19}F- and ^{1}H-NMR spectroscopy. The analysis of the data with different aggregation models suggests the formation of a non-ideal mixed micelle that is enriched in the perfluorinated surfactant when its mole fraction increases.[173] The effects of tetraalkylammonium bromides (TAABs) on the micellization of sodium dodecylsulfate (SDS) have been studied using pyrene solubilization and several NMR techniques. 2D NOESY experiments confirmed that tetraalkylammonium (TAA^+) ions associate with SDS to form mixed micelles. Using proton chemical shift analysis, the degree of hydration on the surface of the mixed micelles is detected. The self-diffusion coefficients of SDS and TAA^+ ions in the TAAB/SDS/D_2O solutions are measured by using pulse-field gradient NMR, and the fraction of TAA^+ ions associated with the SDS to form the mixed micelles is calculated. Moreover, secondary micelle formation for SDS and TAA^+ ions is observed on the basis of ^{1}H chemical shift analysis and the self-diffusion data.[174] A mixture of two amphiphiles with fluorescent head groups has been proposed as a sensing ensemble for the pattern-based analysis of aminoglycoside antibiotics.[175] Hydrogenated cardanol (HC) is known to act as an antiobesity, promising antioxidant, and eco-friendly brominating agent. In this respect, it is important to find the way to transport and protect HC into the body; a micellar structure works as the simplest membrane model and may be considered a suitable biocarrier for HC. Therefore, it is useful to analyze the impact of HC in the micellar structure and properties. A relevant study reports a computer aided Electron Paramagnetic Resonance and ^{1}H NMR investigation of structural variations of cetyltrimetylammonium bromide (CTAB) micelles upon insertion of HC at different concentration and pH variations. The results suggest a rod-like shape of the mixed micelles. The increase in pH promotes the insertion of deprotonated HC into less packed micelles, favored by the electrostatic head–head interactions between CTAB and deprotonated-HC surfactants.[176]

Effects of counterions have been also investigated. The additive-induced micellar growth and rheology modification for mixtures of anionic (sodium dodecyl sulfate, SDS) and zwitterionic (N-alkylated glycine deriatives, Empigen BB or EBB) surfactants has been studied. Two nonaromatic hydrotropic salts (hexyltrimethylammonium bromide, C_6TAB, and/or dibutylenebis(dimethylbutylammonium bromide), 4-4-4) were used as additives to induce micellar growth in these systems. NMR, photon correlation spectroscopy, TEM, the Weissenberg effect, and rheology measurements were employed to assess mixed micelle formation, micellar growth, and rheology modifications. The manner in which the surfactants and hydrotropes self-assemble into aggregates was deduced from 2D NMR NOESY measurements.[177] The correlation between interaction parameters and ion flotation efficiency in mixtures of chelating surfactant metal complexes and different foaming agents was studied. Interaction parameters for mixed micelles and mixed monolayer formation for Mg^{2+} and Ni^{2+} complexes with the chelating surfactant 4-C_{12}-DTPA and different foaming agents were calculated by Rubingh's regular solution theory. Parameters for the calculation were extimated from surface tension measurements and NMR diffusometry. The effects of metal ion coordination on the interactions between 4-C_{12}-DTPA and the foaming agents could be linked to a previously established difference in coordination chemistry between the examined metal ions.[178]

Some ionic liquids can be regarded as cosurfactants, forming mixed aggregates with other amphiphiles. Aggregates formed in an aqueous medium by three ionic liquids C_nMImIbu made up of 1-alkyl-3-methylimidazolium cation ($n = 4$, 6, 8) and ibuprofenate anion were studied. DLS, cryogenic TEM (cryo-TEM), ^{1}H NMR measurements, and atom-scale molecular dynamics simulations were used to shed light on the main interactions governing the formation of the aggregates and their composition. At high concentration, mixed micelles are formed with a composition that depends on the imidazolium alkyl chain length. Attractive interactions between these aggregates gave larger aggregates.[179] The various physicochemical properties of morpholinium based ionic liquid crystals (ILC) (*N*-methyl-*N*-tetradecylmorpholinium bromide (M14) and *N*-methyl-*N*-hexadecylmorpholinium bromide (M16)) in the absence and presence of β-cyclodextrin (β-CD) and hydroxypropyl-β-cyclodextrins (HPβ-CD) have been studied using conductance, surface tension, fluorescence and ^{1}H NMR measurements. The formation of inclusion complexes is confirmed by increase in CMC of synthesized ILC and changes in chemical shift of CDs and ILC protons in the presence of CDs.[180]

Mixtures of phospholipids, eventually including other amphiphiles, have been extensively studied in view of their possible use as systems to solubilise membrane proteins. Sticholysin I (StI), an actinoporin expressed as a water-soluble protein by the sea anemone Stichodactyla helianthus, binds to natural and model membranes, forming oligomeric pores. To date there is no high-resolution structure of the actinoporin pore or other membrane-bound form available. StI:micelle complexes of variable lipid composition have been investigated to look for a suitable model for NMR studies. Micelles of pure or mixed lysophospholipids and of dihexanoyl phosphatidylcholine (DHPC) were examined. The StI:DHPC micelle was found to be the best system, yielding a stable sample and good quality spectra.[181] A detail study on photoinduced rheology changes in nonaqueous photorheology (PR) fluids obtained with 1,2-diacyl-*sn*-*glycero*-3-phosphocholine-based reverse wormlike micelles, 1-palmitoyl-2-oleophosphatidylcholine (POPC)/cyclohexane/H_2O, POPC/isooctane/H_2O, and L-α-dioleophosphatidylcholine (DOPC)/isooctane/H_2O systems has been reported. When photosensitive molecule, *trans*-cinnamic acid (CA), is added to these mixtures, they exhibit photosensitivity. The nature of the substituent on the benzene ring of *trans*-CA influences the rheology. The different geometry obtained because of substitution and photo-induced *trans-cis* isomerization is responsible for the different rheology as confirmed by the dynamic rheology, the UV absorption, and the ^{1}H NMR spectrums.[182]

Surfactant mixtures designed for drug delivery have been studied by NMR. Mixed micelles were designed to increase oral bioavailability of Apigenin (Ap). The phospholipid (Ph) complex technology was exploited and formation of Ap–Ph complex was confirmed by FTIR and NMR spectroscopy.[183] Diethylene triamine pentaacetic acid (DTPA) bisamide derivatives functionalized with *p*-dodecylaniline (DTPA-BC_{12}PheA) and *p*-tetradecyl-aniline (DTPA-BC_{14}PheA) were coordinated to dysprosium(III) and the magnetic and optical properties of the complexes were examined in detail. The complexes assembled into mixed micelles. The efficient T_2 relaxation, especially at high magnetic field strengths, is sustained by the high magnetic moment of the dysprosium ion, the coordination of water molecules with slow water exchange kinetics and long rotational correlation times. These findings open the way to the further development of bimodal optical and MRI probes starting from single lanthanide compounds.[184] The micelle formation of dodecyltrimethylammonium chloride (DTAC) in 0.01 M methyl- and ethyl-4-hydroxybenzoate (MeP and EtP, respectively) sodium salt aqueous solutions has been investigated in the range from 278.15 to 328.15 K by ITC, viscosity, electrical conductivity measurements, and NMR spectroscopy. It was found out that hydrophobic counterion not only lowers the CMC in aqueous solution, but interacts favorably with the surfactant molecules and enables the formation of higher organized structures also. The role of counterion hydrophobicity as the main driving force for diverse rearrangements in the ionic amphiphile aqueous solutions is discussed.[185] Interactions of trianionic curcumin (Cur^{3-}) with a series of cationic surfactants, monomeric surfactant dodecyl trimethylammonium bromide (DTAB), dimeric surfactant hexamethylene-1,6-bis(dodecyldimethylammonium bromide) (12-6-12) and trimeric surfactant tri(dodecyldimethylammonioacetoxy)diethyltriamine trichloride (DTAD), have been investigated in aqueous solutions of pH 13.0. It is found that gemini surfactant 12-6-12 is the best choice to effectively suppress Cur^{3-} degradation at very low concentrations. ITC, surface tension and ^{1}H NMR results reveal that 12-6-12 and Cur^{3-} form a $(12\text{-}6\text{-}12)_2\text{-}Cur^{3-}$ complex and start to form micelles at extremely decreased concentrations, where either 12-6-12 or Cur^{3-} works as a bridge linkage and the resultant structure exhibits the characteristics of oligomeric surfactants.[186] Ultrasound is a common tool for clinical diagnosis due to its safety and economic. The addition of ultrasound contrast agents leads to a high diagnostic reliability. A nanoscaled ultrasound contrast agent (NUSCA) to improve these applications has been developed. NUSCA is based on the phospholipids 1,2-dipalmitoyl-*sn-glycero*-3-phosphocholine (DPPC) and 1,2-distearoyl-*sn-glycero*-3-phosphocholine (DSPC) and the single-chained polyethylene glycol (40) stearate (PEG40S). The effect of increasing concentrations of the single chained PEG40S on the structure of the lipid formulations was characterised using DLS, cryo-TEM, NMR spectroscopy, lipid monolayer studies and epifluorescence measurements. All PEG40S concentrations lead to a mixture of liposomes and discoid micelles. With increasing PEG40S content the amount of micelles increased.[187] Microemulsions stabilized by sucrose esters in combination with lecithin are interesting flavor delivery systems because they allow the solubilization of very high levels of oil into an aqueous phase with relatively small amounts of surfactants. PFG diffusion NMR spectroscopy revealed that at some specific ratios between both surfactants the formation of bicontinuous microemulsions takes place. In such systems, between 20% and 60% of the oil phase can be solubilized.[188]

References

1 M. Cifelli, *Magn. Reson. Chem.*, 2014, **52**, 640–648.
2 M. Cifelli and C. A. Veracini, *Mol. Cryst. Liq. Cryst.*, 2013, **576**, 127–134.
3 L. Calucci, M. Geppi and S. Urban, *Magn. Reson. Chem.*, 2014, **52**, 625–639.
4 M. K. Reddy, E. Varathan, N. P. Lobo, B. B. Das, T. Narasimhaswamy and K. V. Ramanathan, *J. Phys. Chem. C*, 2014, **118**, 15044–15053.
5 P. N. Nelson and R. A. Taylor, *Appl. Petrochem. Res.*, 2014, **4**, 253–285.
6 J. Jokisaari and J. Zhu, *Magn. Reson. Chem.*, 2014, **52**, 556–559.
7 M. Monduzzi, S. Lampis, S. Murgia and A. Salis, *Adv. Colloid Interface Sci.*, 2014, **205**, 48–67.
8 J. B. Rosenholm, *Adv. Colloid Interface Sci.*, 2014, **205**, 9–47.
9 S. Chaudhary and A. Pal, *J. Mol. Liq.*, 2014, **190**, 10–15.
10 S. V. Perez, A. F. Olea and M. P. Garate, *Curr. Top. Med. Chem.*, 2014, **14**, 774–780.
11 B. Alonso and C. Marichal, *Chem. Soc. Rev.*, 2013, **42**, 3808–3820.
12 S. J. Opella, *Annu. Rev. Anal. Chem.*, 2013, 6305–6328.
13 U. H. N. Durr, R. Soong and A. Ramamoorthy, *Prog. Nucl. Magn. Reson. Spectrosc.*, 2013, **69**, 1–22.
14 L. Maler, *Adv. Drug Delivery Rev.*, 2013, **65**, 1002–1011.
15 A. Shakhov, J. Kaerger and R. Valiullin, *Colloid Polym. Sci.*, 2014, **292**, 1961–1969.
16 C. H. R. M. Wilsens, J. M. G. A. Verhoeven, B. A. J. Noordover, M. R. Hansen, D. Auhl and S. Rastogi, *Macromolecules*, 2014, **47**, 3306–3316.
17 G. Casella, V. Causin, F. Rastrelli and G. Saielli, *Phys. Chem. Chem. Phys.*, 2014, **16**, 5048–5051.
18 R. Wang, X. Li, J. Bai, J. Zhang, A. Liu and X. Wan, *Macromolecules*, 2014, **47**, 1553–1562.
19 Y.-D. Huang, W. Tu, Y.-Q. Yuan and D.-L. Fan, *Tetrahedron*, 2014, **70**, 1274–1282.
20 J. C. W. Tseng, R. Rondla, P. Y. S. Su and I. J. B. Lin, *RSC Adv.*, 2013, **3**, 25151–25158.
21 R. Rondla, J. C. Y. Lin, C. T. Yang and I. J. B. Lin, *Langmuir*, 2013, **29**, 11779–11785.
22 M. Giese, J. C. De Witt, K. E. Shopsowitz, A. P. Manning, R. Y. Dong, C. A. Michal, W. Y. Hamad and M. J. MacLachlan, *ACS Appl. Mater. Interfaces*, 2013, **5**, 6854–6859.
23 K. P. Kemayo, C. Aisenbrey, E. Salnikov, O. Rifi and B. Bechinger, *J. Pept. Sci.*, 2014, **20**, 526–536.

24 S. Berenyi, J. Mihaly, A. Wacha, O. Toke and A. Bota, *Colloids Surf., B*, 2014, **118**, 164–171.
25 E. Clop, A. K. Chattah and M. A. Perillo, *J. Phys. Chem. B*, 2014, **118**, 6150–6158.
26 M. Beaugrand, A. A. Arnold, J. Henin, D. E. Warschawski, P. T. F. Williamson and I. Marcotte, *Langmuir*, 2014, **30**, 6162–6170.
27 A. L. Furlan, A. Castets, F. Nallet, I. Pianet, A. Grelard and E. J. Gean, *Langmuir*, 2014, **30**, 5518–5526.
28 W. Ye, J. Lind, J. Eriksson and L. Maeler, *Langmuir*, 2014, **30**, 5488–5496.
29 J. Cai, S. Guo, J. W. Lomasney and M. F. Roberts, *J. Biol. Chem.*, 2013, **288**, 37277–37288.
30 K. Yamamoto, M. A. Caporini, S. Im, L. Waskell and A. Ramamoorthy, *J. Magn. Reson.*, 2013, **237**, 175–181.
31 A. Rahmani, C. Knight and M. R. Morrow, *Langmuir*, 2013, **29**, 13481–13490.
32 W. Chen and M. L. Cotton, *Protein Expression Purif.*, 2014, **102**, 63–68.
33 J. M. Resende, R. M. Verly, C. Aisenbrey, A. Cesar, P. Bertani, D. Pilo-Veloso and B. Bechinger, *Biophys. J.*, 2014, **107**, 901–911.
34 N. Jimenez-Rojo, J. Sot, A. R. Viguera, M. I. Collado, A. Torrecillas, J. C. Gomez-Fernandez, F. M. Goni and A. Alonso, *Biophys. J.*, 2014, **106**, 2577–2584.
35 Q. Do, M. Wittlich, J. M. Glueck, L. Moeckel, D. Willbold, B. W. Koenig and H. Heise, *Biol. Chem.*, 2013, **394**, 1453–1463.
36 G. N. Reiner, L. F. Franceto, E. de Paula, M. A. Perillo and D. A. Garcia, *J. Biomater. Nanobiotechnol.*, 2013, **4**, 28–34.
37 D. Pentak, *Spectrochim. Acta, Part A*, 2014, **122**, 451–460.
38 X. Yu and G. A. Lorigan, *J. Phys. Chem. B*, 2014, **118**, 2124–2133.
39 G. Quebatte, E. Kitas and J. Seelig, *Biochim. Biophys. Acta, Biomembr.*, 2014, **1838**, 968–977.
40 X. Yu and G. A. Lorigan, *Biochim. Biophys. Acta, Biomembr.*, 2013, **1828**, 2444–2449.
41 M. Sharifahmadian, M. Sarker, D. Palleboina, A. J. Waring, F. J. Walther, M. R. Morrow and V. Booth, *PLoS One*, 2013, **8**, e72821.
42 M. Michalek, E. S. Salnikov and B. Bechinger, *Biophys. J.*, 2013, **105**, 699–710.
43 G. J. Lu, Y. Tian, N. Vora, F. M. Marassi and S. J. Opella, *J. Am. Chem. Soc.*, 2013, **135**, 9299–9302.
44 M. Vermathen, M. Marzorati and P. Bigler, *J. Phys. Chem. B*, 2013, **117**, 6990–7001.
45 C. Song, C. Weichbrodt, E. S. Salnikov, M. Dynowski, B. O. Forsberg, B. Bechinger, C. Steinem, B. L. de Groot, U. Zachariae and K. Zeth, *Proc. Natl. Acad. Sci. U. S. A.*, 2013, **110**, 4586–4591.
46 R. Puthenveetil and O. Vinogradova, *Proteins: Struct., Funct., Bioinf.*, 2013, **81**, 1222–1231.
47 M. Zervou, Z. Cournia, G. Patargias, C. Potamitis, S. Durdagi, S. G. Grdadolnik and T. Mavromoustakos, *Biochim. Biophys. Acta*, 2014, **1838**, 1031–1046.
48 M. C. Manzini, K. R. Perez, K. A. Riske, J. C. Bozelli, T. L. Santos, M. A. da Silva, G. K. V. Saraiva, M. J. Politi, A. P. Valente and F. C. L. Almeida, *Biochim. Biophys. Acta, Biomembr.*, 2014, **1838**, 1985–1999.
49 Z. Niu, W. Zhao, Z. Zhang, F. Xiao, X. Tang and J. Yang, *Angew. Chem., Int. Ed.*, 2014, **53**, 9294–9297.
50 A. P. Serro, R. Galante, A. Kozica, P. Paradiso, A. M. P. S. Goncalves da Silva, K. V. Luzyanin, A. C. Fernandes and B. Saramago, *Colloids Surf., B*, 2014, **116**, 63–71.
51 L. A. M. Sommer and S. A. Dames, *FEBS Lett.*, 2014, **588**, 1755–1766.

52 D. Graebner, H. Hoffmann, S. Foerster, S. Rosenfeldt, J. Linders, C. Mayer, Y. Talmon and J. Schmidt, *Adv. Colloid Interface Sci.*, 2014, **208**, 252–263.
53 Y. Song, K. F. Mittendorf, Z. Lu and C. R. Sanders, *J. Am. Chem. Soc.*, 2014, **136**, 4093–4096.
54 M. Fillion, M. Noel, A. Lorin, N. Voyer and M. Auger, *Biochim. Biophys. Acta, Biomembr.*, 2014, **1838**, 2173–2179.
55 B. Liang, D. Dawidowski, J. F. Ellena, L. K. Tamm and D. S. Cafiso, *Biochemistry*, 2014, **53**, 1485–1494.
56 C. C. Fraenza, C. J. Meledandri, E. Anoardo and D. F. Brougham, *ChemPhysChem*, 2014, **15**, 425–435.
57 J. Song, J. Franck, P. Pincus, M. W. Kim and S. Han, *J. Am. Chem. Soc.*, 2014, **136**, 2642–2649.
58 Y. Liu, R. A. Kahn and J. H. Prestegard, *Structure*, 2014, **22**, 421–430.
59 Z. T. Graber, A. Gericke and E. E. Kooijman, *Chem. Phys. Lipids*, 2014, **182**, 62–72.
60 K. Berthelot, S. Lecomte, Y. Estevez, V. Zhendre, S. Henry, J. Thevenot, E. J. Dufourc, I. D. Alves and F. Peruch, *Biochim. Biophys. Acta, Biomembr.*, 2014, **1838**, 287–299.
61 L. Zhang, L. Liu, S. Maltsev, G. A. Lorigan and C. Dabney-Smith, *Biochim. Biophys. Acta, Biomembr.*, 2014, **1838**, 413–418.
62 L. Zhang, L. Liu, S. Maltsev, G. A. Lorigan and C. Dabney-Smith, *Chem. Phys. Lipids*, 2013, **175–176**, 123–130.
63 M. Kogan, B. Norden and T. Beke-Somfai, *Chem. Phys. Lipids*, 2013, **175–176**, 105–115.
64 J. K. Claridge, J. Aittoniemi, D. M. Cooper and J. R. Schnell, *Biochemistry*, 2013, **52**, 8420–8429.
65 E. Hellstrand, A. Nowacka, D. Topgaard, S. Linse and E. Sparr, *PLoS One*, 2013, **8**, e77235.
66 L. Caillon, J. A. Killian, O. Lequin and L. Khemtemourian, *PLoS One*, 2013, **8**, e75528.
67 M. R. Alborghetti, F. Ariane da Silva, J. Cesar da Silva, M. L. Sforca, R. V. Honorato, D. C. Granato, D. L. dos Santos Migueleti, J. L. Neves, P. S. Lopes de Oliveira and A. F. Paes-Leme, *PLoS One*, 2013, **8**, e76602.
68 T. L. Yap, J. M. Gruschus, A. Velayati, E. Sidransky and C. Jennifer, *Biochemistry*, 2013, **52**, 7161–7163.
69 C.-J. Hsieh, Y.-W. Chen and D. W. Hwang, *Phys. Chem. Chem. Phys.*, 2013, **15**, 16634–16640.
70 A. Sylvester, L. MacEachern, V. Booth and M. R. Morrow, *PLoS One*, 2013, **8**, e72248.
71 K. R. Mote, T. Gopinath and G. Veglia, *J. Biomol. NMR*, 2013, **57**, 91–102.
72 A. Gayen, J. R. Banigan and N. J. Traaseth, *Angew. Chem., Int. Ed.*, 2013, **52**, 10321–10324.
73 S. Ahuja, N. Jahr, S.-C. Im, S. Vivekanandan, N. Popovych, S. V. Le Clair, R. Huang, R. Soong, J. Xu and K. Yamamoto, *J. Biol. Chem.*, 2013, **288**, 22080–22095.
74 C. Arioez, W. Ye, A. Bakali, C. Ge, J. Liebau, H. Goetzke, A. Barth, A. Wieslander and L. Maeler, *Biochemistry*, 2013, **52**, 5533–5544.
75 L. Xu, L. Feng, R. Dong, J. Hao and S. Dong, *Biomacromolecules*, 2013, **14**, 2781–2789.
76 G. Da Costa, A. Bondon, O. Delalande, L. Mouret and J.-P. Monti, *J. Biomol. Struct. Dyn.*, 2013, **31**, 809–817.
77 P. Rath, O. Saurel, G. Czaplicki, M. Tropis, M. Daffe, A. Ghazi, P. Demange and A. Milon, *Biochim. Biophys. Acta, Biomembr.*, 2013, **1828**, 2173–2181.

78 Y. Fu and A. J. Wand, *J. Biomol. NMR*, 2013, **56**, 353–357.
79 N. Busschaert, S. J. Bradberry, M. Wenzel, C. J. E. Haynes, J. R. Hiscock, I. L. Kirby, L. E. Karagiannidis, S. J. Moore, N. J. Wells and J. Herniman, *Chem. Sci.*, 2013, **4**, 3036–3045.
80 D. I. Fernandez, M.-A. Sani, A. J. Miles, B. A. Wallace and F. Separovic, *Biochim. Biophys. Acta, Biomembr.*, 2013, **1828**, 1863–1872.
81 S. Song, H. Wang, A. Song, S. Dong and J. Hao, *Chem. – Eur. J.*, 2014, **20**, 9063–9072.
82 M. Ma, Y. Guan, C. Zhang, J. Hao, X. Xing, J. Su, S. Li, X. Chu and A. Hao, *Colloids Surf., A*, 2014, **454**, 38–45.
83 T. M. Alam, Z. Liao, L. N. Zakharov and M. Nyman, *Chem. – Eur. J.*, 2014, **20**, 8302–8307.
84 M.-K. Cho, A. Gayen, J. R. Banigan, M. Leninger and N. J. Traaseth, *J. Am. Chem. Soc.*, 2014, **136**, 8072–8080.
85 S. Mandal, J. Kuchlyan, S. Ghosh, C. Banerjee, N. Kundu, D. Banik and N. Sarkar, *J. Phys. Chem. B*, 2014, **118**, 5913–5923.
86 M. G. Shapiro, R. M. Ramirez, L. J. Sperling, G. Sun, J. Sun, A. Pines, D. V. Schaffer and V. S. Bajaj, *Nat. Chem.*, 2014, **6**, 629–634.
87 Q. Liu, H. Zhu, J. Qin, H. Dong and J. Du, *Biomacromolecules*, 2014, **15**, 1586–1592.
88 I. M. Mandity, A. Monsignori, L. Fueloep, E. Forro and F. Fueloep, *Chem. – Eur. J.*, 2014, **20**, 4591–4597.
89 S. Himmelein, N. Sporenberg, M. Schoenhoff and B. J. Ravoo, *Langmuir*, 2014, **30**, 3988–3995.
90 E. C. Griffith, R. J. Rapf, R. K. Shoemaker, B. K. Carpenter and V. Vaida, *J. Am. Chem. Soc.*, 2014, **136**, 3784–3787.
91 Y. Shen, Z.-c. Ou-Yang, Y. Zhang, J. Hao and Z. Liu, *Langmuir*, 2014, **30**, 2632–2638.
92 A. Pan, P. Sil, S. Dutta, P. K. Das, S. C. Bhattacharya, A. K. Rakshit, V. K. Aswal and S. P. Moulik, *J. Phys. Chem. B*, 2014, **118**, 3041–3052.
93 W. Qiao, M. Zhou and L. Luo, *J. Surfactants Deterg.*, 2014, **17**, 261–268.
94 A. A. Steinschulte, B. Schulte, S. Ruetten, T. Eckert, J. Okuda, M. Moeller, S. Schneider, O. V. Borisov and F. A. Plamper, *Phys. Chem. Chem. Phys.*, 2014, **16**, 4917–4932.
95 K. Nomura, E. Harada, K. Sugase and K. Shimamoto, *J. Phys. Chem. B*, 2014, **118**, 2405–2413.
96 M. Dan, F. Huo, X. Xiao, Y. Su and W. Zhang, *Macromolecules*, 2014, **47**, 1360–1370.
97 C. Maiti, D. Dey, S. Mandal and D. Dhara, *J. Phys. Chem. B*, 2014, **118**, 2274–2283.
98 F. Ye, A. Barrefelt, H. Asem, M. Abedi-Valugerdi, I. El-Serafi, M. Saghafian, K. Abu-Salah, S. Alrokayan, M. Muhammed and M. Hassan, *Biomaterials*, 2014, **35**, 3885–3894.
99 K. Yamamoto, P. Pearcy and A. Ramamoorthy, *Langmuir*, 2014, **30**, 1622–1629.
100 S. Kumar and H. Patel, *J. Mol. Liq.*, 2014, **190**, 74–80.
101 H. Guo, D. Wang, S. Dong, Y. Dou, A. Song and J. Hao, *Huaxue Xuebao*, 2013, **71**, 1129–1135.
102 S. Li, P. Xing, Y. Hou, J. Yang, X. Yang, B. Wang and A. Hao, *J. Mol. Liq.*, 2013, **188**, 74–80.
103 J. Rho, J. Chung, H. Im, M. Liong, H. Shao, C. M. Castro, R. Weissleder and H. Lee, *ACS Nano*, 2013, **7**, 11227–11233.
104 N. Sibille, M. Dellarole, C. Royer and C. Roumestand, *J. Biomol. NMR*, 2014, **58**, 9–16.

105 J. T.-W. Wang, L. Cabana, M. Bourgognon, H. Kafa, A. Protti, K. Venner, A. M. Shah, J. K. Sosabowski, S. J. Mather, A. Roig, X. Ke, G. Van Tendeloo, R. T. M. de Rosales, G. Tobias and K. T. Al-Jamal, *Adv. Funct. Mater.*, 2014, **24**, 1880–1894.
106 Y. Wang, Z. Yu and B. Li, *Colloid Polym. Sci.*, 2014, **292**, 243–250.
107 R. Bleul, R. Thiermann, G. U. Marten, M. J. House, T. G. St. Pierre, U. O. Haefeli and M. Maskos, *Nanoscale*, 2013, **5**, 11385–11393.
108 C. Zhou, X. Cheng, Q. Zhao, Y. Yan, J. Wang and J. Huang, *Langmuir*, 2013, **29**, 13175–13182.
109 Y. Dou, P. Long, S. Dong and J. Hao, *Langmuir*, 2013, **29**, 12901–12908.
110 M. Li, H. K. Morales, J. Katsaras, N. Kucerka, Y. Yang, P. M. Macdonald and M.-P. Nieh, *Langmuir*, 2013, **29**, 15943–15957.
111 P. Arosio, J. Thevenot, T. Orlando, F. Orsini, M. Corti, M. Mariani, L. Bordonali, C. Innocenti, C. Sangregorio, H. Oliveira, S. Lecommandoux, A. Lascialfari and O. Sandre, *J. Mater. Chem. B*, 2013, **1**, 5317–5328.
112 C. Zheng, H. Huang and T. He, *Macromol. Rapid Commun.*, 2013, **34**, 1654–1661.
113 C. L. Lay, J. N. Kumar, C. K. Liu, X. Lu and Y. Liu, *Macromol. Rapid Commun.*, 2013, **34**, 1563–1568.
114 P. Yin, T. Li, R. S. Forgan, C. Lydon, X. Zuo, Z. N. Zheng, B. Lee, D. Long, L. Cronin and T. Liu, *J. Am. Chem. Soc.*, 2013, **135**, 13425–13432.
115 H. Petitjean, P. Guiglion, M. In, J. R. Aldrich-Wright, P. Castignolles, F. Ganachaud and M. Gaborieau, *J. Colloid Interface Sci.*, 2013, **408**, 87–93.
116 Q. Duan, Y. Cao, Y. Li, X. Hu, T. Xiao, C. Lin, Y. Pan and L. Wang, *J. Am. Chem. Soc.*, 2013, **135**, 10542–10549.
117 R. O. Brito, I. S. Oliveira, M. J. Araujo and E. F. Marques, *J. Phys. Chem. B*, 2013, **117**, 9400–9411.
118 M. A. Voronin, D. R. Gabdrakhmanov, R. N. Khaibullin, I. Y. Strobykina, V. E. Kataev, B. Z. Idiyatullin, D. A. Faizullin, Y. F. Zuev, L. Y. Zakharova and A. I. Konovalov, *J. Colloid Interface Sci.*, 2013, **405**, 125–133.
119 A. Bogomolova, S. K. Filippov, L. Starovoytova, B. Angelov, P. Konarev, O. Sedlacek, M. Hruby and P. Stepanek, *J. Phys. Chem. B*, 2014, **118**, 4940–4950.
120 R. Wang, Y. Tang and Y. Wang, *Langmuir*, 2014, **30**, 1957–1968.
121 R. Imperatore, G. Vitiello, D. Ciccarelli and G. D'Errico, *J. Solution Chem.*, 2014, **43**, 227–239.
122 I. Svanedal, G. Persson, M. Norgren and H. Edlund, *Langmuir*, 2014, **30**, 1250–1256.
123 C. Wang, X. Li, P. Li and Y. Niu, *Carbohydr. Polym.*, 2014, **99**, 638–645.
124 C. Barreto, A. Altskar, S. Fredriksen, E. Hansen and R. W. Rychwalski, *Eur. Polym. J.*, 2013, **49**, 2149–2161.
125 A. M. Percebom, J. Janiak, K. Schillen, L. Piculell and W. Loh, *Soft Matter*, 2013, **9**, 515–526.
126 M. G. S. Vieira, N. V. Gramosa, N. P. S. Ricardo, G. A. Morris, R. W. Adams and M. Nilsson, *RSC Adv.*, 2014, **4**, 42029–42034.
127 M. Baglioni, M. Raudino, D. Berti, U. Keiderling, R. Bordes, K. Holmberg and P. Baglioni, *Soft Matter*, 2014, **10**, 6798–6809.
128 V. P. Arkhipov, Z. S. Idiyatullin, E. F. Potapova, O. N. Antzutkin and A. V. Filippov, *J. Phys. Chem. B*, 2014, **118**, 5480–5487.
129 C. Johans, M. A. Behrens, K. E. Bergquist, U. Olsson and J. A. Manzanares, *Langmuir*, 2013, **29**, 15738–15746.
130 E. P. Zhiltsova, V. V. Syakaev, S. S. Lukashenko, A. P. Timosheva, R. R. Kashapov, S. K. Latypov, L. Y. Zakharova and A. I. Konovalov, *Fluid Phase Equilib.*, 2013, **360**, 16–22.

131 G. K. S. Prameela, B. V. N. Phani Kumar, V. K. Aswal and A. B. Mandal, *Phys. Chem. Chem. Phys.*, 2013, **15**, 17577–17586.
132 L. Ge, Q. Wang, D. Wei, X. Zhang and R. Guo, *J. Phys. Chem. B*, 2013, **117**, 15014–15022.
133 G. Cheng, B. Hammouda and D. Perahia, *Macromol. Chem. Phys.*, 2014, **215**, 341–350.
134 K. Nakai, M. Nishiuchi, M. Inoue, K. Ishihara, Y. Sanada, K. Sakurai and S. Yusa, *Langmuir*, 2013, **29**, 9651–9661.
135 M. K. Joo, D. Y. Ko, S. J. Jeong, M. H. Park, U. P. Shinde and B. Jeong, *Soft Matter*, 2013, **9**, 8014–8022.
136 G. Molev, Y. Lu, K. S. Kim, I. C. Majdalani, G. Guerin, S. Petrov, G. Walker, I. Manners and M. A. Winnik, *Macromolecules*, 2014, **47**, 2604–2615.
137 P. A. L. Knoos, M. Wahlgren, D. Topgaard, S. Ulvenlund and L. Piculell, *J. Phys. Chem. B*, 2014, **118**, 9757–9767.
138 J. Lo, H. Yen, C. Tsai, B. Chen and S. Hou, *J. Phys. Chem. B*, 2014, **118**, 6922–6930.
139 S. Decato, T. Bemis, E. Madsen and S. Mecozzi, *Polym. Chem.*, 2014, **5**, 6461–6471.
140 L. Liu, L. Rui, Y. Gao and W. Zhang, *Polym. Chem.*, 2014, **5**, 5453–5460.
141 H. Cui, W. Chu, J. Chen and S. Kuo, *Eur. Polym. J.*, 2014, **50**, 168–176.
142 H. Arslan, O. Zirtil and V. Butun, *Eur. Polym. J.*, 2013, **49**, 4118–4129.
143 Y. Wu, D. Zhu, Z. Guo and Y. Feng, *J. Polym. Sci., Part B: Polym. Phys.*, 2013, **51**, 1335–1342.
144 T. Matini, N. Francini, A. Battocchio, S. G. Spain, G. Mantovani, M. J. Vicent, J. Sanchis, E. Gallon, F. Mastrotto, S. Salmaso, P. Caliceti and C. Alexander, *Polym. Chem.*, 2014, **5**, 1626–1636.
145 K. Bauri, P. De, P. N. Shah, R. Li and R. Faust, *Macromolecules*, 2013, **46**, 5861–5870.
146 L. I. Atanase and G. Riess, *J. Colloid Interface Sci.*, 2013, **395**, 190–197.
147 S. Pal, S. Ghosh Roy and P. De, *Polym. Chem.*, 2014, **5**, 1275–1284.
148 W. Hong, D. Chen, L. Jia, J. Gu, H. Hu, X. Zhao and M. Qiao, *Acta Biomater.*, 2014, **10**, 1259–1271.
149 J. Y. Quek, Y. Zhu, P. J. Roth, T. P. Davis and A. B. Lowe, *Macromolecules*, 2013, **46**, 7290–7302.
150 Y. Tang, L. Liu, J. Wu and J. Duan, *J. Colloid Interface Sci.*, 2013, **397**, 24–31.
151 Z. Qiao, R. Ji, X. Huang, F. Du, R. Zhang, D. Liang and Z. Li, *Biomacromolecules*, 2013, **14**, 1555–1563.
152 W. Cheng, J. N. Kumar, Y. Zhang and Y. Liu, *Macromol. Biosci.*, 2014, **14**, 347–358.
153 X. Hu, H. Li, S. Luo, T. Liu, Y. Jiang and S. Liu, *Polym. Chem.*, 2013, **4**, 695–706.
154 H. Kono, *Carbohydr. Polym.*, 2014, **106**, 84–93.
155 S. De Santis, M. Diociaiuti, C. Cametti and G. Masci, *Carbohydr. Polym.*, 2014, **101**, 96–103.
156 D. J. Growney, O. O. Mykhaylyk and S. P. Armes, *Langmuir*, 2014, **30**, 6047–6056.
157 R. Wang, C. Wang, K. Liu, F. Bei, L. Lu, Q. Han and X. Wu, *J. Phys. Chem. B*, 2014, **118**, 2544–2552.
158 F. Szillat, B. V. K. J. Schmidt, A. Hubert, C. Barner-Kowollik and H. Ritter, *Macromol. Rapid Commun.*, 2014, **35**, 1293–1300.
159 Y. Li, H. Guo, Y. Zhang, J. Zheng, J. Gan, X. Guan and M. Lu, *RSC Adv.*, 2014, **4**, 17768–17779.
160 Y. Zhou, X. Fan, W. Zhang, D. Xue and J. Kong, *J. Polym. Res.*, 2014, **21**, 1–10.

161 Q. Jin, G. Liu and J. Ji, *Chin. J. Chem.*, 2014, **32**, 73–77.
162 X. Guo, X. Jia, J. Du, L. Xiao, F. Li, L. Liao and L. Liu, *Carbohydr. Polym.*, 2013, **98**, 982–987.
163 J. Shen, Q. Wang, Q. Hu, Y. Li, G. Tang and P. K. Chu, *Biomaterials*, 2014, **35**, 8621–8634.
164 X. Zhang, Z. Zhang, X. Su, M. Cai, R. Zhuo and Z. Zhong, *Biomaterials*, 2013, **34**, 10296–10304.
165 G. Behl, M. Sikka, A. Chhikara and M. Chopra, *J. Colloid Interface Sci.*, 2014, **416**, 151–160.
166 D. Wei, L. Ge, Y. Ding and R. Guo, *Colloids Surf., A*, 2013, **421**, 16–25.
167 N. Rocha, J. Mendes, L. Duraes, H. Maleki, A. Portugal, C. F. G. C. Geraldes, A. Serra and J. Coelho, *J. Mater. Chem. B*, 2014, **2**, 1565–1575.
168 K. S. Kim, W. Park, J. Hu, Y. H. Bae and K. Na, *Biomaterials*, 2014, **35**, 337–343.
169 K. Yan, H. Li, P. Li, H. Zhu, J. Shen, C. Yi, S. Wu, K. W. K. Yeung, Z. Xu, H. Xu and P. K. Chu, *Biomaterials*, 2014, **35**, 344–355.
170 S. Mahajan, V. Koul, V. Choudhary, G. Shishodia and A. Bharti, *Nanotechnology*, 2013, **24**, 015603.
171 M. Etzkorn, M. Zoonens, L. J. Catoire, J.-L. Popot and S. Hiller, *J. Membr. Biol.*, 2014, **247**, 965–970.
172 N. Dubey, *Fluid Phase Equilib.*, 2014, **368**, 51–57.
173 M. F. Torres, R. H. de Rossi and M. A. Fernandez, *J. Surfactants Deterg.*, 2013, **16**, 903–912.
174 J.-H. Lin, W.-S. Chen and S.-S. Hou, *J. Phys. Chem. B*, 2013, **117**, 12076–12085.
175 Z. Koestereli, R. Scopelliti and K. Severin, *Chem. Sci.*, 2014, **5**, 2456–2460.
176 S. Santeusanio, O. A. Attanasi, R. Majer, M. Cangiotti, A. Fattori and M. F. Ottaviani, *Langmuir*, 2013, **29**, 11118–11126.
177 K. Singh, Z. O'Toole, A. McLachlan and D. G. Marangoni, *Langmuir*, 2014, **30**, 3673–3680.
178 I. Svanedal, S. Boija, M. Norgren and H. Edlund, *Langmuir*, 2014, **30**, 6331–6338.
179 C. Tourne-Peteilh, B. Coasne, M. In, D. Brevet, J.-M. Devoisselle, A. Vioux and L. Viau, *Langmuir*, 2014, **30**, 1229–1238.
180 R. Sharma, S. Mahajan and R. K. Mahajan, *Fluid Phase Equilib.*, 2014, **361**, 104–115.
181 A. Lopez-Castilla, F. Pazos, S. Schreier and J. R. Pires, *Proteins*, 2014, **82**, 1022–1034.
182 R. G. Shrestha, N. Agari, K. Tsuchiya, K. Sakamoto, K. Sakai, M. Abe and H. Sakai, *Colloid Polym. Sci.*, 2014, **292**, 1599–1609.
183 W. L. L. Munyendo, Z. Zhang, S. Abbad, A. Y. Waddad, H. Lv, L. D. Baraza and J. Zhou, *J. Biomed. Nanotechnol.*, 2013, **9**, 2034–2047.
184 E. Debroye, S. Laurent, L. Vander Elst, R. N. Muller and T. N. Parac-Vogt, *Chem. – Eur. J.*, 2013, **19**, 16019–16028.
185 A. Kroflic, B. Sarac, J. Cerkovnik and M. Bester-Rogac, *Colloids Surf., A*, 2014, **460**, 108–117.
186 M. Wang, C. Wu, Y. Tang, Y. Fan, Y. Han and Y. Wang, *Soft Matter*, 2014, **10**, 3432–3440.
187 J. Bruessler, E. Marxer, A. Becker, R. Schubert, J. Schuemmelfeder, C. Nimsky and U. Bakowsky, *Colloids Surf., B*, 2014, **117**, 206–215.
188 V. Tchakalova, C. Bailly and W. Fieber, *Flavour Fragrance J.*, 2014, **29**, 67–74.

NMR in living systems

M. J. W. Prior

DOI: 10.1039/9781782622758-00507

This chapter reviews studies applying magnetic resonance spectroscopy to investigations of the processes occurring in living systems. In section one acquisition methods and analytical processes that are applicable to *in vivo* investigations are reviewed. In sections two and three, studies in animal models and the clinical environment are surveyed, respectively. In each of these two sections, the review is subdivided into tissue categories with each of these sub-divided further according to the category of disease or the type of metabolic investigation.

1 Experimental developments and optimizations

1.1 Pulse sequence developments

A *J*-refocused ^{1}H point resolved spectroscopy (PRESS) localized DEPT sequence has been produced for the detection of a large number of metabolites by *in vivo* ^{13}C magnetic resonance spectroscopy (MRS).[1] The sequence was applied to the detection of saturated and unsaturated fatty acids in bone marrow and skeletal muscle of healthy subjects. The method of spectroscopy with linear algebraic modelling (SLAM) has been extended to include 2D and 3D acquisition using sense methods.[2] The large improvement of acquisition times was demonstrated in twenty-four patients with brain tumours examined with ^{1}H MRS and in ^{31}P spectra acquired from the human heart. There has been further development of ^{31}P MRS methods by Mao *et al.*,[3] Parasoglou *et al.*[4] and van der Kemp *et al.*[5] Mao *et al.* used a soft-heteronuclear single quantum correlation sequence for the detection of phosphomonoesters (PME) and phosphodiesters (PDE) in mouse brain. Parasoglou *et al.* produced a 3D imaging sequence that was able to map creatine kinase rates in different muscle groups and van der Kemp *et al.* presented an adiabatic multi-echo spectroscopic imaging sequence with spherical k-space sampling and compensated phase-encoding gradients. The latter was validated in healthy calf muscle and breast tissue, and subsequently applied in a patient with breast cancer. Further refinements of two ^{1}H MRS localisation methods have been carried out by de Castro *et al.*,[6] modifying MEGA-sLASER with frequency offset corrected inversion pulses, and by de Graaf *et al.*[7] presenting a novel zero-quantum coherence method based on the stimulated echo acquisition mode (STEAM) pulse sequence. It was found by de Castro *et al.* that the detection efficiencies for γ-aminobutyric acid (GABA) and lactate were improved by 20% and 75%, respectively, whilst de Graaf demonstrated that high quality, quantitative 2D spectra can be acquired from human breast tissue at 7 T despite the

Brain and Body Centre, University of Nottingham, Nottingham NG7 2RD, UK.
E-mail: malcolm.prior@nottingham.ac.uk

presence of severe magnetic field inhomogeneity caused by water–lipid susceptibility boundaries.

1.2 Acquisition strategies

The reproducibility of MEGA-PRESS for the measurement of GABA has been addressed by Evans *et al.*[8] in the occipital, sensorimotor and dorsolateral prefrontal cortices, and by Wijtenburg *et al.*[9] in the anterior cingulate and the occipital, sensorimotor, and dorsolateral prefrontal cortices. Evans *et al.* concentrated on the effects of spectral realignment on data quality, which resulted in data handling method that reduced quantification errors. Wijtenberg *et al.* compared the reproducibility of repeated measures of GABA and ^{1}H MRS spectra acquired with STEAM. Methods for the quantitation of metabolites have been proposed by El-Sharkawy *et al.*[10] using ^{31}P MRS in the heart at 3 Tesla and by Mazzetti *et al.*[11] using ^{1}H MRS in the parietal white matter at 1.5 Tesla. El-Sharkawy *et al.* found concentrations of phosphocreatine (PCr) and ATP consistent with previous measurements made at lower fields. Mazzetti *et al.*, using spectral noise to calibrate the spectra found similar mean choline (Cho) concentrations and variance when compared to published values. Mosconi *et al.*[12] have examined the accuracy of the quantification of spectra by five methods of analysis, these were: Linear combination of Model spectra (LCModel), Advanced Method for Accurate Robust and Efficient Spectral fitting (AMARES), quantitation based on QUantum ESTimation (QUEST), Automated Quantification of Short Echo-time MRS (AQSES)-Lineshape, and peak integration. The protocols were applied to simulated spectra and to the measurement of lipid signals from obese and lean Zucker rats. Schaller *et al.*[13] have examined the effect of the macromolecular background signal on the quantification of metabolites in human brain spectra at 3 Tesla. A comparison was made between using a simulated or measured macromolecular contribution to spectral analysis using LCModel.

An automated voxel registration process has been developed and tested using measurements of tCr (creatine plus phosphocreatine), Glx (Glu plus Gln), tNAA (*N*-acetyl-aspartate plus *N*-acetyl-aspartyl-glutamate), myo-inositol (mI) and Cho in the parietal and occipital lobes.[14] Methods for the reduction of motion artefacts and field perturbations have been investigated by Wilm *et al.*[15] using field probes to provide feedback to an automated correction system and by Hock *et al.*[16] using navigator based respiratory gating. Wilm *et al.* were able to make real time corrections to field values which improved ^{1}H spectra obtained from the brain that were affected by deep breathing and arm movements. Hock *et al.* were able to improve ^{1}H MRS spectra from the liver by utilizing navigator respiratory gating during preparation as well as during acquisition; their method was validated by acquisition of ^{1}H spectra for quantitation of Cho and lipids. A comparison of methods for the acquisition of ^{31}P MRS data from the human liver has been made by Chmelik *et al.*[17] at 7 Tesla. Single voxel image selected *in vivo* spectroscopy, slab selective 1D- image selected *in vivo* spectroscopy, 2D chemical shift imaging and 3D chemical shift imaging were compared. All methods provided good quality spectra;

metabolite relaxation times were reported. The feasibility of using the PRESS sequence to obtain short echo-time ^1H MRS spectra from the mouse cervical spinal cord at 11.75 T has been demonstrated by Tachrount *et al.*[18]

Methods for combining spectroscopic data from multi-channel acquisition have been investigated by An *et al.*[19] using the generalized least square algorithm and by Hall *et al.*[20] using weighting with the ratio of signal to the square of the noise. Hall *et al.* found that this was the theoretical optimal combination provided that the noise is uncorrelated. However, the method employed by An *et al.* takes into account noise correlation between coil elements and further improves on earlier methods that also dealt with correlated noise.

1.3 Relaxation

The tranverse relaxation time for GABA has been measured in the human brain by Intrapiromkul *et al.*[21] and Andreychenko *et al.*[22] from data obtained using spectral editing techniques. Intrapiromkul *et al.* first measured the value of T_2 and the modulation of the GABA signal due to *J*-coupling at various echo times in a phantom. These data were then used to fit the *in vivo* spectra to obtain the relaxation values. Andreychenko *et al.* proposed a new method whereby they refocused the *J*-modulation of the edited signal at different echo times by using chemical shift selective refocusing. The value of T_2 for other metabolites in the human brain has been measured by Tisell *et al.*[23] in subjects aged 21 to 72 years old, by Ronen *et al.*[24] in those aged around 25 years and by Marjanska *et al.*[25] in volunteers aged around 20 and 77 years. Relaxation times were measured for tNAA, tCr and Cho by Ronen *et al.*, for tNAA, tCr, Cho and H_2O by Marjanska *et al.* and for tCr, mI, Cho, tNAA and Glx by Tisell *et al.* The latter study obtained quantitative metabolite data from white matter and the thalamus. Marjanska *et al.* made measurements in the occipital lobe and the T_2 data was acquired in 2.5 min. Ronen *et al.* used a new rapid method that involved the collection of truncated echoes for the assessment of T_2 from grey and white matter. The T_2 relaxation times for tNAA, tCr, Cho, mI, taurine and Glu have been measured at 14.1 T by Xin *et al.*[26] from a 5×5×2.5 mm voxel centred in the rat hippocampus. The relaxation of ^7Li has been measured in the human brain by Port *et al.*[27] and in the rat brain by Komoroski *et al.*[28] Port *et al.* found a bi-exponential decay curve in eleven out of thirteen patients. The average short T_2 value was 5.3 ms whilst the long T_2 component was 68.2 ms at 3 Tesla. Measurements made by Komoroski *et al.* revealed a short T_2 of 14.8 ms and a long T_2 of 295 ms in the rat brain at 7 Tesla. Komoroski *et al.* also used this bi-exponential data to estimate the concentration of ^7Li in the intracellular and extracellular compartments of the rat brain.

1.4 Measurement of pH

The measurement of intracellular pH from the chemical shift of inoranic phosphate (Pi) with reference to three independent peaks has been investigated by Rata *et al.*[29] in ten volunteers and eight patients with non-Hodgkin's lymphoma. Measurements using the PCr peak, where available, gave the best results and measurements using the α-ATP peak were better than when using the water peak as a reference. McVicar *et al.*[30] have developed a method for the assessment of intracellular pH by measuring the ratio of the chemical exchange saturation transfer effect on the amine compared to amide groups of peptides and proteins. The protocol was used to assess localised acidosis in the mouse brain following focal permanent middle cerebral artery occlusion. However, this required calibration of the procedure against intracellular pH measured by ^{31}P MRS.

1.5 Hyperpolarised metabolic studies

A method have been developed for the measurement of the arterial input function in metabolic studies using hyperpolarised [1-^{13}C]pyruvate. This consisted of a small chamber into which arterial blood was drawn for assessment within the same scans as used for recording the metabolism of the substrate in a tumour model.[31] The feasibility of metabolic studies in the rat, using multiple injections of [1-^{13}C]pyruvate, has been demonstrated by Hu *et al.*[32] Spectra were acquired from the lower abdomen following injections of [1-^{13}C]pyruvate at 5 min intervals and the effects of dichloroacetate were monitored over a similar timescale. Swisher *et al.*[33] have followed the metabolism of [1-^{13}C]pyruvate using the method known as metabolic activity decomposition-stimulated echo acquisition mode. The technique uses the STEAM pulse sequence with a specific t_1 interval that allows the separation of exchanging from non-exchanging metabolites. Spectra were recorded from mouse liver and a transgenic liver tumour model. In muscle, the apparent diffusion coefficients of muscle metabolites have been measured in the rat by Sogaard *et al.*[34] The diffusion of pyruvate, lactate and Ala were measured following the injection of hyperpolarised [1-^{13}C]pyruvate.

2 Preclinical studies

2.1 Central nervous system

2.1.1 Aging. The effects of aging on metabolite levels in the brain have been measured by Ono *et al.*[35] in dogs and by Morgan *et al.*[36] in rats. Ono *et al.* used ^{1}H MRS to measure age-related and regional differences in metabolite concentrations in the brains of healthy dogs, finding that Cho/tCr was highest in young dogs and that tNAA/Cho was lower in the youngest and oldest dogs. Morgan *et al.* used ^{1}H MRS to perform a longitudinal, within-subject investigation of the rat brain. Spectra were recorded from a voxel centred on the striatum on postnatal days 7, 35 and 60. Age related changes were found in the concentrations of tNAA, mI, Glx, Cho, tCr and taurine.

2.1.2 Perfusion and hypoxia. Models of hypoxia and ischaemia have been examined in four papers. ^{1}H MRS has been used by Hai *et al.*[37] to examine a rat model of chronic hyperperfusion. A decrease in the ratio of tNAA/tCr was observed, but no other changes in metabolites were detected. The morphological and metabolic responses to hypobaric hypoxia have been measured by Koundal *et al.*[38] using magnetic resonance imaging (MRI) and ^{1}H MRS, respectively. Hyperintensity was observed in T_2-weighted images from 2 to 14 days post-exposure, whereas ^{1}H MRS showed a decrease in mI after 1 day, an increase of Cho and a decrease of Glu after 4 days, and decreases in tNAA, tCr and Glx after seven and fourteen days. The intrastriatal microinjection of the potent vasoconstrictor, endothelin-1, has been studied by Saggu *et al.*[39] as a model of arterial ischaemic stroke or transient ischaemic attack; ^{31}P MRS, MRI and diffusion-weighted MRI (DWI) were used in the assessment of the model. Changes in diffusion-, T_2- and T_1-weighted images were observed, though no alterations in ^{31}P metabolites were seen. Ahn *et al.*[40] have examined the use of the lipid peak as a non-invasive, *in vivo* tool for quantitative detection of brain cell death following hypoxic ischaemia. ^{1}H spectra from Sprague Dawley rat pup brains were recorded with and without treatment with cycloheximide. There was a correlation between the lipid peak and indicators of cell death.

Treatment of stroke and protection against its effects have been explored in four papers. An evaluation of hypoxic and ischaemic preconditioning in rat pups has been carried out by Seo *et al.*[41] using ^{1}H MRS, and histological and morphological analysis. Here the ratios of lipids/tNAA and lipids/tCr were found to correlate with histological and morphological changes. The effects of transplanted human umbilical cord blood mesenchymal stem cells on cerebral ischaemia in rabbits has been investigated by Guan *et al.*[42] ^{1}H MRS revealed elevated lactate/tCr and decreased tNAA/tCr and Cho/tCr after 24 h and after two weeks following middle cerebral occlusion. However, treatment with mesenchymal stem cells significantly ameliorated these changes. The effectiveness of a protective ventilation procedure for new born preterm lambs has been assessed by Skiold *et al.*[43] using ^{1}H MRS, MRI and diffusion tensor imaging (DTI). Although no cerebral injury was apparent from MRI images, a higher fractional anisotropy and raised lactate were seen in lambs not ventilated with the protective procedure. The protective effects of treatment with high-dose, long-term erythropoietin following hypoxic-ischamic injury in rat pups has been assessed by van de Looij *et al.*[44] using ^{1}H MRS, DTI, histopathology and electrophysiology. However, only partial protective effects on white matter structure, cerebral metabolism and somatosensory function were observed.

2.1.3 Disease models. ^{1}H MRS has been used to characterise the neurochemical profiles of rat and mouse models of Alzheimer's disease and to assessed the effects of treatment. Differences in metabolite levels between males and female McGill-R-Thy1-APP rats have been assessed by Nilsen *et al.*[45] using ^{1}H MRS. However, no differences in metabolite levels were evident between female control and McGill-R-Thy1-APP rats, but McGill-R-Thy1-APP males had lower levels of Glu, tNAA and Cho compared with male controls. The neurochemical profile of A beta PPSswe-PS1dE9 compared to that of wild-type mice at 8 and 12 months of age has been measured by Jansen *et al.*[46] using ^{1}H MRS. At 8 months there were behavioural changes and cognitive decline in mice, but changes in ^{1}H metabolites were only detected at 12 months when further cognitive decline and behavioural changes were apparent. Zhang *et al.*[47] have evaluated the therapeutic effects of a bilateral intraventricular infusion of brain-derived neurotrophic factor (BDNF) into Alzheimer's disease APP/PS1 double transgenic mice. Mice treated with BDNF had higher tNAA and lower mI levels compared to untreated controls as well as better histological results. Marjanska *et al.*[48] have appraised the efficacy of treatment with monoclonal antibodies in APP-PS1 transgenic mice. Those mice immunized with ponezumab, or the murine version (2H6-D), had an attenuated rise of mI and decreased plaque levels.

Models of Parkinson's disease have been studied by Bagga *et al.*,[49] Hou *et al.*[50] and Fu *et al.*[51] Bagga *et al.* used ^{1}H and ^{13}C MRS to assess the mouse model of Parkinson's disease that is produced from treatment with 1-methyl-4-phenyl-1,2,3,6-tetrahydropyridin. Following the administration of ^{13}C-labelled precursors, the production of metabolites from [1,6-$^{13}C_2$]glucose was found to be decreased in the striatum, but labelling from [2-^{13}C]acetate was found to be increased. Treatment with L-DOPA did not improve neurometabolite levels detected by ^{1}H MRS, though, ^{13}C-labelling levels returned to control values and paw grip-strength recovered. Metabolite perturbations in the 6-hydroxydopamine rat model of Parkinson's disease have been measured by Hou *et al.* using ^{1}H MRS. No changes were seen in the frontal cortex until a decrease in tNAA/tCr was observed at 5 weeks post-lesion. In the substantia nigra, tNAA/tCr decreased and Cho/tCr increased significantly during 2–8 weeks post-lesion. Fu *et al.* have examined the effects of transplanted bone marrow mesenchymal stem cells on the 6-hydroxydopamine rat model of Parkinson's disease. The ratio of tNAA/tCr increased and Cho/tCr ratio decreased in response to treatment and there was an improvement in the results from the rotarod test.

Other central nervous system disease models have been studied. A knock-in mouse model of cerebellar ataxia has been characterised by Emir *et al.*[52] using ^{1}H MRS. The neurochemistry of the cerebellum in the Sca1(154Q/2Q) line was measured at 6, 12, 24, and 39 weeks. The level taurine distinguished between knock-in and wild type mice at 6 weeks, later there were differences in tCr, Gln and tCr. Totenhagen *et al.*[53] have investigated metabolite changes in a mouse model of Niemann–Pick disease type C. However, they found that the only difference between control mice and end stage disease mice were decreased taurine and increased mI; the metabolite changes found in clinical studies of the disease were not repeated in the mouse model. Models of schizophrenia have been investigated in two studies. Xuan *et al.*[54] used ^{1}H MRS to measure neurometabolites in mice treated with cuprizone; mice exhibited anxiety-like behaviour and had decreased levels of Cho, mI, tNAA and tCr. The effects of a single dose of ketamine on brain metabolite concentrations in isolated and socially reared rats has been explored by Napolitano *et al.*[55] using ^{1}H MRS. There were no differences in metabolites between the two rat groups, however, ketamine caused a decrease in GABA and a larger increase in Glu in socially-isolated rats. Chen *et al.*[56]

have studied a kainic acid-induced model of temporal lobe epilepsy in the macaque using ^{1}H MRS and MRI, as well as a electrophysiology and histology. A reduction in tNAA and an increase in Cho and tCr were observed in those treated with kaininc acid. Lodygensky *et al.*[57] have used ^{1}H MRS and DTI to study a rat model of diffuse white matter injury, which is created by a single injection of lipopolysaccharide into the corpus callosum. A decrease in the apparent diffusion coefficient was detected by DTI, which correlated with a measure of apoptosis, whilst MRS showed an increase in lactate and free lipids. Moshkin *et al.*[58] have also investigated the effects of lipopolysaccharide, but in the mouse brain, and in comparison to the effects of 2-deoxy-D-glucose. Inhibition of glycolysis by 2-deoxy-D-glucose caused a reduction in GABA, tNAA Cho in the hippocampus, whereas lipopolysaccharide increased these metabolites.

2.1.4 Intracellular metabolites. The metabolic differences between mouse strains that produce different levels of mGluR7 mRNA in the hippocampus have been investigated by Guilfoyle *et al.*[59] using ^{1}H MRS. The level of Glu was lower in the BALB/cJ strain, which produces higher levels of mGluR7 mRNA. Similarly, a congenic strain that also produce higher levels of mGluR7 mRNA had decreased Glu levels. Cudalbu *et al.*[60] have investigated the source of variations in cerebral metabolites found in C57BL/6 mice. A large increase in Gln and a decrease in mI was detected in some mice. In these mice, MRI angiography was used to demonstrate portosystemic shunting, which was not apparent in the mice with normal cerebral metabolites. Zhou *et al.*[61] have investigated neurobiological alterations during pregnancy and motherhood in primiparous rats. Spectra were recorded from the hippocampus and thalamus three days before mating, on gestational day 17, on lactation day 7 and on post-weaning day 7, respectively; age-matched nulliparous rats served as controls. Increases were observed in NAA, lactate and mI during pregnancy, decreases in Tau occurred during lactation and increased mI was observed during gestation and post-weaning. The diffusion of brain metabolites in the macaque monkey has been investigated by Najac *et al.*[62] Diffusion times of 80 ms to 1 s were used to obtain data from a region containing equal amounts white and grey matter, which indicated largely unrestricted diffusion.

2.1.5 Measurement of metabolism. The metabolism of 2-fluoro-2-deoxy-D-glucose to 2-fluoro-2-deoxyo-glucose-6-phosphate has been mapped in the rat brain by Coman *et al.*[63] Measurements made *in vitro* confirmed the splitting patterns in ^{19}F NMR spectra of the two molecules and these were then used to fit *in vivo* data. Lanz *et al.*[64] have measured the metabolism of [2-^{13}C]acetate using ^{1}H-[^{13}C] MRS in the rat brain. Data were recorded at 14.1 T, which allowed the detection of the dynamics of fractional enrichment in the positions C4 and C3 of glutamate and glutamine. Park *et al.*[65] have used hyperpolarised [2-^{13}C]pyruvate in combination with dichloroacetate to examine mitochondrial metabolism. Labelling was seen in glutamate, acetylcarnitine, citrate, pyruvate and lactate. The cerebral metabolic rate of oxygen has been measured in the mouse brain by Cui *et al.*[66] The production of $H_2{}^{17}O$ was mapped using magnetic resonance spectroscopic imaging (MRSI) at 16.4 T during a pulse-chase experiment where mice breathed ^{17}O-enriched anaesthetic gases delivered by a newly designed system. The concentration of delivered gas in the brain was calibrated against natural abundance $H_2{}^{17}O$. Just *et al.*[67] have measured changes in ^{1}H spectra recorded from the barrel cortex in rats in response to trigeminal nerve stimulation. In this brain region, which was located using data from functional MRI, there were increases in lactate and Glu along with decreases in Asp.

2.1.6 Toxicology. ^{1}H MRS has been utilized in the assessment of toxic effects. Mullins *et al.*[68] have investigated the outcome of a single exposure to the organophosphorus pesticide chlorpyrifos (300 mg kg^{-1}) in the guinea pig. One year after the exposure, MRI did not show any abnormalities, but the guinea pigs had a significant memory deficit and reduced mI detected by ^{1}H MRS in the hippocampus. Gupta *et al.*[69] have used DTI and ^{1}H MRS to assessed changes in the mouse brain in response to whole body or cranial irradiation. Both protocols produced changes in fractional anisotropy, but only whole body irradiation produced metabolite changes in the cortex. Significant reduction in the ratios of mI/tCr and taurine/tCr were observed.

2.2 Heart

Hyperpolarized pyruvate has been used by Lauritzen *et al.*[70] and Josan *et al.*[71] to examine cardiac metabolism. Lauritzen *et al.* examined the effects of glucose, insulin and potassium on the release of [^{13}C]bicarbonate from the metabolism of [1-^{13}C]pyruvate in the myocardium of fasted rats. Josan *et al.*[71] used [1-^{13}C]pyruvate and [2-^{13}C]pyruvate to examine the effects of dichloroacetate on metabolism in fed and fasted rats; the effects of dobutamine were also examined by Josan *et al.* Fasting caused a reduction in [^{13}C]bicarbonate production in both studies. Furthermore, Josan *et al.* also found increased labelling of glutamate, acetoacetate and acetylcarnitine from [2-^{13}C]pyruvate caused by dichloroacetate in fed and fasted rats. Labelling of lactate alanine and glutamate was observed following the administration of dobutamine.

Metabolism in a mouse heart disease model has been investigated by Tucci *et al.*,[72] Bakermans *et al.*[73] and Bakermans *et al.*[74] using knockout mice with long-chain acyl-CoA dehydrogenase deficiency. Tucci *et al.*[72] used ^{31}P 2D chemical shift imaging (CSI) to measure cardiac energetics. Mice showed a progressive decrease in heart function over 12 months accompanied by a reduction in PCr/ATP. Dietary supplementation with medium chain triglycerides aggravated the cardiac phenotype. Bakermans *et al.*[73] employed hyperpolarised [1-^{13}C]pyruvate detected by ^{13}C MRS, and ^{31}P MRS to measure cardiac energy metabolism. Fasting caused enhanced activity of anaplerotic pathways and a decrease in high energy phosphate status in knockout mice when compared to fasted controls. The effects of supplementation with L-carnitine were also studied by Bakermans *et al.*[74] using ^{1}H MRS and MRI to demonstrate an increase in cardiac triglycerides and hypertrophy, respectively. Supplementation with L-carnitine in knockout mice reduced the accumulation of cardiac triglycerides.

In other models of heart disease, the effects of coronary infarct, and cardiac hypertrophy and failure have been investigated. The effects of acute and chronic coronary infarct were examined using ^{31}P MRS, hyperpolarised ^{13}C MRS and MRI in the perfused rat heart.[75] In the acute infarct region, lactate production increased and bicarbonate decreased, and in the chronically infarcted region, there was a significant reduction in both signals as well as the ratio of PCr/ATP. Hyperpolarised ^{13}C MRS and MRI showed that perfusion was reduced in the both types of infarct region. ^{31}P MRS has been used to assess the energy status of the rat heart in an investigation that assessed the initial stages of development of cardiac hypertrophy and failure following chronic administration of isoproterenol.[76] The study also included the use of multi-modal cardiac MRI, biochemical analysis and histology. The mechanism of cardiotoxicity from doxorubicin has been studied with ^{31}P MRS in mice overexpressing the myofibrillar isoform of creatine kinase.[77] Animals exposed to doxorubicin had decreased PCr at 7 weeks and a reduced left ventricular ejection fraction at 8 weeks. Overexpression of creatine kinase resulted in better creatine kinase levels and higher creatine kinase flux. The effects of dichloroacetate and dobutamine on the metabolism of hyperpolarised [2-^{13}C]pyruvate in fed and starved rat hearts has been studied by Josan *et al.*[71] Dichloroacetate led to a modest increase in ^{13}C-labelling of Glu and higher labelling in acetylcarnitine and acetoacetate, whereas dobutamine resulted in increased labelling of lactate, Ala and Glu.

2.3 Liver

^{1}H MRS has been used for the assessment of hepatic fat in a mouse model of glycogen storage disease,[78] in the ob/ob mouse model of obesity and diabetes,[79] in a study of the effects of a high fat diet,[80] in a model of non-alcoholic fatty liver disease[81] in study of the effects dietary fructose[82] and in lean and obese cats.[83] Ramamonjisoa *et al.*[78] examining a mouse model of glycogen storage disease found that control mice fed on a high-fat, high-sucrose diet had raised hepatic lipids, whereas mice with glycogen storage disease had raised lipids that were not further affected by diet. The latter also had higher olefinic signals and no noticeable diallylic signals from the liver. Ye *et al.*[79] investigating the ob/ob mouse found correlations between fraction of polyunsaturated lipids and fasting plasma insulin and glucose levels, and between fractional lipid mass and fasting plasma insulin levels in adolescent mice. However, these correlations were lost in adult ob/ob mice. Yaligar *et al.*[80] found that rats fed on a high fat diet had increases in liver fat content with lower unsaturated fat indices. Lee *et al.*,[81] determining hepatic lipid composition in rat model of non-alcoholic fatty liver disease, found that steatosis and fibrosis had a positive correlation with monounsaturated fatty acids and a negative correlation with diunsaturated fatty acids. Nunes *et al.*[82] found that mice fed with fructose had 1.5-fold higher hepatic triglyceride content than glucose-fed mice. However, there was no effect on ATP levels measured by ^{31}P MRS. In the assessment of hepatic lipids in lean and obese cats, Clark *et al.*[83] found that the liver fat content was 1.3% and 6.8%, respectively.

In other studies of in the liver, Laurent *et al.*[84] examined the action of a glucokinase activator (LCZ960)using ^{13}C MRS in obese mouse and rat models. Measurements indicated an increased glycogen turnover, but no effect on glucose stimulated glucose-6-phosphate accumulation. Josan *et al.*[71] have mapped the production of [1-^{13}C]lactate from hyperpolarised [1-^{13}C]pyruvate in the rat liver in the presence of ethanol and disulfiram. Hanpanich *et al.*[85] used ^{1}MRS and MRI to investigate hamsters infected with liver fluke (*Opisthorchis viverrini*). In infected hamsters, the ratio of Cho/lipid steadily increased during the 21 weeks of the study period.

2.4 Muscle

The impact of a spinal cord contusion model on muscle bioenergetics in rat hind limb muscle have been studied by Shah *et al.*[86] using ^{31}P MRS. A reduction in PCr content was apparent one week after injury. Furthermore, the rate of PCr synthesis, mitochondrial capacity and the synthesis of ATP, measured following electrical stimulation of the sciatic nerve, were all reduced. ^{31}P MRS has been used to measure muscle energy metabolism in the Tg(ACTA1)(Asp286Gly) mouse model of nemaline myopathy.[87] An increased anaerobic energy cost of contraction was detected, along with altered muscle T_2 and diffusion characteristics. The effects of transplantation of adipose tissue-derived mesenchymal stromal cells into the ischemic muscular tissue of diabetic animals has been investigated by Madonna *et al.*[88] using MRI protocols and ^{1}H MRS. There was a decreased in the ratio of tCr/water in ischaemic limbs, which was improved by the presence of transplanted cells or conditioned media. The detection of the presence of phosphoarginine in murine hind limbs following gene delivery of arginine kinase from an adeno-associated virus has been demonstrated by Forbes *et al.*[89] using ^{31}P MRS at 11.7 and 17.6 T.

2.5 Tumour

The metabolism of tumour models and their response to treatment have been evaluated using ^{1}H MRS, and ^{13}C MRS to detected the metabolism of hyperpolarised substrates. Xu *et al.*[90] have measured the metabolism of hyperpolarised [1-^{13}C]-pyruvate in mouse xenografts of human breast cancer cells lines. Three methods of analysis were used on the data, which all showed a reduced production of lactate from pyruvate in the highly metastatic cell line. The use of hyperpolarised [1-^{13}C]α-ketoglutarate has been assessed by Chaumeil *et al.*[91] as a tool to measure the status of the gene for isocitrate dehydrogenase 1 in low-grade gliomas and secondary glioblastomas. The production of hyperpolarised [1-^{13}C]2-hydroxyglutarate was observed in tumours with a mutant gene.

^{1}H MRS has been used by Weidensteiner *et al.*,[92] Pisanu *et al.*[93] and Hamans *et al.*[94] in the assessment of the response of tumours to treatment. Weidensteiner *et al.* investigated the reduction of T_1 in mouse tumours in response to chemotherapy. In those tumours responding to treatment, there was a reduction in T_1 measured by MRI, an increase in apoptosis detected by histology and reduction in tumour Cho assessed by ^{1}H MRS. Pisanu *et al.* have investigated the effects of cisplatin on HER2(+) ovary cancer cells and xenon grafts. Measurements were made using multiple methods including diffusion weight-MRI, high resolution ^{1}H MRS of extracts and ^{1}H MRS *in vivo*. However, no effects of treatment were detected by the latter method. The use of MRSI and contrast enhance MRI to determine the response of diffuse-infiltrative glioblastoma to anti-angiogenic therapy have been compared by Hamans *et al.* Metabolites maps of Cho/tNAA were found to be more sensitive than contrast enhanced MRI for the measurement of response to therapy.

Hyperpolarised substrates have been used by Mignion *et al.*,[95] Lin *et al.*[96] and Sandulache *et al.*[97] to examine the response of tumours to treatment. Measurements of the conversion of hyperpolarised ^{13}C-fumarate to malate was investigated by Mignion *et al.* as a means of determining tumour response to therapy with sorafenib; changes in the apparent diffusion coefficient of tumour water were also measured. Both parameters indicated a response, which was confirmed by histology, however the conversion of fumarate was the more sensitive indicator. Hyperpolarised [1-^{13}C]pyruvic acid has been used by Lin *et al.* in an assessment of autophagy induced in colorectal HT29 and HCT1-16 Bax knock-out cells by serum and amino acid deprivation or treatment with a class I PI3K/mTOR inhibitor. Both regimes reduced the production of lactate, which then increased above control levels after 48 h of recovery. Sandulache *et al.* have evaluated the measurement of the conversion of hyperpolarised pyruvate to lactate to determine the effects of radiotherapy. The conversion of pyruvate was used to infer intracellular NADH levels as an assay of the level of reducing equivalents in anaplastic thyroid carcinoma orthotopic xenografts treated with ionizing radiation. Irradiation of the xenografts resulted in an acute drop in reducing potential.

3 Clinical studies

3.1 Bone marrow

^{1}H MRS has been used to measure the fat content of bone marrow as a means of assessing bone health. Karampinos *et al.*[98] have compared chemical shift imaging and ^{1}H MRS for the assessment of the fat fraction of bone marrow in the presence of trabecular bone. Ojanen *et al.*[99] have compared phase-separated water-fat imaging and ^{1}H MRS for the assessment of vertebral bone marrow fat. Bolacchi *et al.*[100] have used ^{1}H MRS to measure the level of polyunsaturated lipid in bone marrow of patients with acute Charcot neuro-osteoarthropathy. A decrease in polyunsaturated lipid was negatively correlated with the disease recovery time. In a study of the utility of DTI imaging in the assessment of osteoporosis, Manenti *et al.*[101] have also used ^{1}H MRS to measure the fraction of fat in bone marrow. ^{1}H MRS could distinguish between healthy and osteoporotic subjects, whereas DTI could discriminate healthy from osteopenic and healthy from osteoporotic subjects.

3.2 Central nervous system

3.2.1 Effects of drugs. The consequences of exposure to alcohol, methamphetamine, cannabis and ketamine have been investigated using ^{1}H MRS. Metabolite levels in the frontal white matter of non-treatment-seeking heavy drinkers has been measured by Ende *et al.*[102] An increase in Cho was observed and a negative association of Glu levels with severity of alcohol dependence and loss of control of drinking was observed. The outcome of prenatal alcohol exposure on metabolite concentrations in the deep cerebellar nuclei of children who were 8 to 10 years old have been characterised by du Plessis *et al.*[103] High maternal alcohol consumption around time of conception was associated with lower tNAA, when during pregnancy it was associated with lower Cho, and when at conception and during pregnancy it was associated with higher Glx. Subjects with methamphetamine dependence, with and without psychosis, have been assessed in comparison to control subjects by Howells *et al.*[104] Methamphetamine users had lower tNAA/tCr in right anterior cortex and right dorsolateral prefrontal cortex, and a correlation was seen between onset of use and tNAA levels in users with psychosis. Prescot *et al.*[105] have determined metabolite levels in the anterior cingulate cortex of adolescent cannabis users. Significantly lower GABA levels were found, when compared to control subjects, that paralleled lower Glu levels. Stone *et al.*[106] have compared metabolite ratios in the anterior cingulate cortex, left thalamus and left medial temporal cortex of chronic ketamine users to those found in poly-drug users witn no exposure to ketamine. Chronic ketamine users had higher levels of subthreshold psychotic symptoms and lower thalamic tNAA/Cr.

3.2.2 Measurement of metabolites. The feasibility of measuring GABA using short echo time MRS has been explored by Moheet *et al.*[107] using STEAM in the hypothalamus and occipital cortex at 7 Tesla and by Near *et al.*[108] using simulations and SPECIAL, compared to MEGA-SPECIAL, for measurements in the occipital cortex at 4 Tesla. Moheet *et al.* made measurements under conditions of euglycaemic and hypoglycaemic clamp and found GABA levels of 3.5 μmol g^{-1} and 3.0 μmol g^{-1}, respectively, in the hypothalamus, whereas in the occipital lobe the level of GABA was 1.4 μmol g^{-1} in both conditions. Near *et al.* found that GABA measurements made with SPECIAL were significantly correlated with measurements made by MEGA-SPECIAL, but the estimation of the concentration of GABA is strongly dependent on the signal to noise ratio and linewidth. The level of GABA, Glu and Gln have been measured by Dou *et al.*[109] in four sub-regions of the human cingulate cortex. The pregenual region had the highest level of GABA, Glu and Gln and variations were seen for each metabolite across other regions. The technique of chemical exchange saturation transfer has been used to map the distribution of Glu in the spinal cord[110] and in a coronal slice of the human brain encompassing the amygdala and hippocampus.[111] The level of HCO_3^- in the human brain has been measured by Sailasuta *et al.*[112] using ^{13}C MRS in healthy subjects following four hours or 12 hours of fasting. Fasting levels of HCO_3^- were reduced compared to those found in the fed state. No difference in brain pH was found in measurements made using ^{31}P MRS.

3.2.3 Function, metabolites and blood flow. The relationship between neurometabolite levels and brain function have been explored using measurements of IQ,[113] reading ability,[114] and genetic variation.[115] Patel *et al.*[113] did not find significant correlations between IQ measurements and the ratios of tNAA/tCr detected in the occipito-parietal and the frontal cortices of healthy volunteers. Pugh *et al.*[114] measuring the ratios of Cho/tCr, tNAA/tCr, Glu/tCr and GABA/tCr in the midline occipital cortex of children found that Cho/tCr and Glu/tCr were negatively correlated with reading and related linguistic measures in phonology and vocabulary. Levels of GABA in the prefrontal cortex have been measured by Preuss *et al.*[115] in relation to single nucleotide polymorphisms in the TPH2 gene, which is a panic disorder risk allele in women. A significant association between increased GABA concentration with the T-allele frequencies of two polymorphisms, and with the C-allele frequency of another polymorphism was found in female subjects.

Measurements of GABA and Glu in the anterior and posterior cingulate have been assessed in relation to cerebral perfusion.[116] The level of GABA in the anterior cingulate were inversely correlated with global cerebral blood flow, whilst GABA levels in the posterior cingulate, and glutamate levels from both regions, were not significantly correlated with blood flow. The ratio of GABA/tNAA measured in the visual cortex of cognitively matched, healthy volunteers has been examined in relation to measurements of perfusion, fractional grey matter volume, and arterial arrival time measured by MRI.[117] Analysis revealed an inverse correlation between perfusion and GABA/tNAA.

3.2.4 The effects of age and cognitive impairment. ^1H MRS spectra have been recorded from the basal ganglia, centrum semiovale and the cerebellum of neonates.[118] Increases in tCr, tNAA and Glx were seen with greater postconception age in all regions, whereas mI increased in the cerebellum and decreased in the basal ganglia. In adolescents and emerging adults, the level of GABA has been recorded from the anterior cingulate and the parieto-occipital cortices.[119] GABA/tCr levels in the anterior cingulate cortex were lower in adolescents relative to emerging adults, whereas no age differences were observed in the parieto-occipital cortex. The level of GABA has also been measured in the frontal and parietal regions in subjects aged from 20 to 76 years.[120] Negative correlations were observed between age and GABA in both regions and the decrease in GABA with age in the frontal region was more rapid in women than men. Changes in Glu in the medial frontal cortex have been recorded in adults aged 18 to 31 years;[121] the levels of Glu were lower in older adults. Metabolite levels in the anterior cingulate cortex, posterior cingulate cortex, and left and right hippocampi have been assessed in subjects aged 22 to 82 years.[122] A positive correlation with age was found for Cho, tCr and tNAA in the anterior and posterior cingulate cortices , and for tNAA in the left hippocampus. Metabolites have also been measured in the left frontal white matter of subjects aged 21 to 82 years.[123] A negative association of tNAA, and positive associations of mI, Cr and Cho were seen with age. Furthermore, geneotyping revealed subjects with alterations in the sodium channel voltage gated type I, alpha subunit SCN1A variant rs10930201 had higher mI levels. Genotyping has also been used to measure APOE polymorphism in subjects aged 50 to 86 years who were assessed for ^{1}H metabolite levels in the posterior cingulate and precuneus region.[124] Older APOE E4 carriers had higher Cho/tCr and mI/tCr ratios compared to those who were APOE E3 homozygotes.

Ge *et al.*[125] have founds a positive correlation between C-reactive protein levels and mI in the left hippocampus in patients with cognitive impairment. In a search for predictors of cognitive impairment, Kantarci *et al.*[126] have measured metabolite levels in the posterior cingulate gyri and assessed hippocampal volumes and white matter hyperintensity volumes in 1156 adults. Hippocampal volume, white matter hyperintensity volume and tNAA/mI were found to be significant predictors of mild cognitive impairment in older adults. The effects of tesamorelin (an analogue of growth hormone releasing hormone) has been tested on patients with mild cognitive impairment.[127] After 20 weeks, GABA levels were increased in the dorsolateral frontal, posterior cingulate and posterior parietal regions, *N*-acetyl-aspartyl-glutamate (NAAG) levels were increased in the dorsolateral frontal cortex, and mI levels were decreased in the posterior cingulate. Patients with suspected vascular cognitive impairment have been assessed with ^{1}H MRS, and MRI to quantify white matter hyperintensities.[128] A correlation between tCr and tNAA with standardised neuropsychological test scores was found.

In a study of risk factors for Alzheimer's disease, subjects with mild cognitive impairment have been assessed with ^{1}H MRS in the posterior cingulate gyrus, left hippocampus, and the cortical area of the right and left parietal lobe.[129] The ratio of mI/tCr in the right parietal lobe predicted the conversion to Alzheimer's disease with sensitivity 70% and specificity 85%. Neurometabolite correlations of unawareness of deficits in Alzheimer's disease have been sought in data obtained from the anterior cingulate and right orbitofrontal areas.[130] The condition was negatively correlated with tNAA/tCr in the anterior cingulate area and positively correlated with tNAA/tCr in the right orbitofrontal area.

3.2.5 Mental health disorders. The level of GABA has been measured with ^{1}H MRS in subjects with autism spectrum disorder. Gaetz *et al.*[131] made measurements in the motor, visual and auditory areas and found reduced GABA/tCr in the motor and auditory regions in comparison to subjects with normal development. Rojas *et al.*[132] found reduced GABA/tCr in voxels centred on the auditory cortex of the left hemisphere of subjects with autism spectrum disorder and their siblings. The level of ^{1}H metabolites has been measured by Horder *et al.*[133] in the basal ganglia, dorosolateral prefrontal cortex and parietal cortex of subjects with autism spectrum disorder. A decrease Glx was seen in the basal ganglia along with a reduction in Cho, tCr and tNAA. In the dorsolateral prefrontal cortex tCr and tNAA were lower compared to control subjects. Golomb *et al.*[134] used ^{31}P MRS to investigate muscle and brain metabolites in subjects with autism spectrum disorder.

In schizophrenia patients, ^{31}P polarisation transfer 3D MRSI has been used to measure phosphocholine (PC), Phosphoethanolamine (PE), glycerophosphocholine (GPC) and glycerophosphoethanolamine (GPE) levels.[135] Correlations were found for psychotic symptoms with PC levels in the cerebral cortex, depression with PC levels in the cerebellum and executive functioning with GPC levels in the insular and temporal cortices. ^{1}H MRS has also been used to measure neurometabolite levels in subjects with schizpophrenia. Granata *et al.*[136] have evaluated tNAA, Cho and tCr in the prefrontal cortex, thalami and vermis and found no differences between patients and controls. Reid *et al.*[137] found that there was no difference in the ratio of tNAA/tCr, Glx/tCr and Cho/tCr in the left substatia nigra between patients and controls. Bustillo *et al.*[138] assessed Glu, Gln and Gln/Glu in the dorsal anterior cingulate in relation to symptoms and cognition in schizophrenia. There was an increase in Gln and Gln/Glu and Glu levels were positively correlated with severity of psychotic symptoms. Rowland *et al.*[139] compared GABA, Glx, and NAAG levels in the anterior cingulate and centrum semiovale regions of older and younger schizophrenics. Glx was reduced in all schizophrenia patients compared to healthy controls. In the centrum semiovale, NAAG was higher in younger schizophrenics, and lower in older schizophrenics, compared to control subjects. Szulc *et al.*[140] have examined patients who responded to treatment in comparison to non-responders and healthy control subjects. Those that responded to treatment had lower frontal Glx/Cr compared to non-responders. Patients had lower tNAA/Cr in the frontal lobe than the controls, and non-responders had lower tNAA/Cr in the thalamus. Tandon *et al.*[141] determined metabolite levels in the thalamus, caudate, and anterior cingulate cortex of offspring of schizophrenia patients. Compared to control subjects, there was a reduction in tNAA in all regions, increases in Glx in the thalamus and caudate, and increases in Cho in the anterior cingulate.

^{1}H MRS has been used to investigated patients with bipolar disorder. Xu *et al.*[142] measured neurometabolite levels in the thalamus and in the anterior and posterior cingulate cortices of bipolar manic and bipolar depressed subjects. Brain lactate/tCr was significantly increased in all patients compared to controls. Depressed patients had increased Glx/tCr in the left thalamus, but maniac patients had decreased Glu/tCr and Glx/tCr in the posterior cingulate cortex. Gigante *et al.*,[143] examining metabolite levels in the hippocampus of first episode manic patients, found no differences compared to control subjects. Godlewska *et al.*[144] measuring neurometabolites, including GABA and glutathione, in the prefrontal and occipital cortices of young bipolar patients also found no differences in levels when compared to control subjects.

Chitty *et al.*[145] have measured glutathione levels in the anterior cingulate cortex of bipolar patients who were considered to be high risk or low-risk drinkers. Patients who were high risk drinkers had reduced glutathione levels compared to low-risk drinkers, but there were no differences in glutathione levels between patients and controls. Singh *et al.*[146] have examined metabolite levels in the dorsolateral prefrontal cortex of the offspring of patients with bipolar disorder. Subjects were scanned at baseline and at either 8, 10, 12, 52, 104, 156, 208, or 260 weeks. However, no group differences were found for tNAA/tCr or mI/tCr compared to healthy controls.

^{31}P MRS has been used to measure PC, PE, GPC and GPE levels in subjects with late life depression.[147] The level of GPE was elevated in the white matter of depressed subjects, but no other differences were detected. Ezzati *et al.*[148] have also examined depression in older subjects, but have used ^{1}H MRS to measure neurometabolite levels in the hippocampus. Although depression was associated with a smaller hippocampal volume, there was no such association with tNAA/tCr in the hippocampus. ^{1}H MRS has been used by Lapidus *et al.*[149] to measure glutathione levels in the occipital cortex of patients with major depressive disorder. It was found that severity of anhedonia and occipital glutathione levels were negatively correlated. Taylor *et al.*[150] have examined cortical metabolites in depression associated with treatment with cytokine interferon-alpha. Treatment led to an increase in Glu and a significant elevation in the ratio of Gln/Glu. Furthermore, changes in Gln level correlated significantly with ratings of depression and anxiety.

^{1}H MRS has been used to examine patients with post-traumatic stress disorder. Rosso *et al.*[151] measured GABA levels in the right anterior insula and dorsal anterior cingulate cortex. The level of GABA was lower in the right anterior insula in patients compared to controls and was associated with greater anxiety. GABA levels have also been measured by Meyerhoff *et al.*[152] Subjects with post-traumatic stress disorder had lower levels in the parieto-occipital and temporal cortices. The level of GABA in the anterior cingulate cortex and the occipital cortex of subjects with panic disorder have been measured by Long *et al.*[153] A decrease in the ratio of GABA/tCr was detected in the anterior cingulate cortex in panic disorder patients, which tended to be pronounced in patients with a panic disorder family history. Shin *et al.*[154] have also used ^{1}H MRS to measure GABA levels in the anterior cingulate cortex, which they had found to have a higher connectivity with the precuneous in patients with panic disorder. The functional connectivity, measured by functional MRI, was negatively correlated with the GABA concentration of the anterior cingulate cortex. Metabolite levels in the left and right prefrontal white matter have been measured by Weber *et al.*[155] examining adolescent subjects with obsessive compulsive disorder. Compared to controls, there was a higher level of Cho and tNAA in the right prefrontal white matter. Furthermore, the level of tNAA, tCr and mI in that region were positively correlated with severity of symptoms. The effects of treatment with methylphenidate or atomoxetine in patients with attention deficit/hyperactivity disorder have been investigated by Husarova *et al.*[156] Metabolite levels were recorded in dorsolateral prefrontal grey and white matter of subjects who responded to two months of treatment.

The ratio of NAA/tCr decreased in the left dorsolateral prefrontal cortex and Cho/tCr increased in the right after atomoxetine medication. The ratio of Glx/tCr increased in the left prefrontal white matter after methylphenidate medication.

3.2.6 The effects of infections. The use of MRSI compared to single voxel MRS for the assessment of pyogenic brain abscesses has been evaluated by Hsu *et al.*[157] Data from MRSI was very similar to that obtained from the single voxel method, but had the advantage of providing a metabolic profile from surrounding tissue. Peluso *et al.*[158] have assessed neurochemistry with ^{1}H MRS and measured biomarkers of neuronal injury in the cerebrospinal fluid of subjects with primary HIV infection. It was found that neurofilament light chain levels correlated with decreased ratios of tNAA/tCr and Glu/tCr in the anterior cingulate, frontal white matter and parietal grey matter.

3.2.7 Genetic conditions. Two studies have examined neurometabolite levels in patients with fragile X syndrome. Utine *et al.*[159] used CSI and single voxel MRS to assess brain white matter the cerebellar vermis, respectively. Lower levels of Cho/NAA were found in the right parietal white matter. Bruno *et al.*,[160] using single voxel ^{1}H MRS, found decreased levels of Cho and Glx in the caudate nucleus in patients with fragile X syndrome compared to healthy controls. The effects of hyperarginineamia, a rare autosomal recessive metabolic disorder, has been investigated by Carvalho *et al.*[161] using ^{1}H MRS and MRI. Although atrophy was present in all but one patient, no effects on neurometabolite levels were observed in tested patients. In an examination of patients with Wilson's disease, Pulai *et al.*[162] used ^{1}H MRS and MRI methods to characterise the condition. A reduction in tNAA/tCr and Cho/tCr was observed in patients compared to control subjects. Roos *et al.*[163] have used ^{1}H MRS to examine neurometabolites in four patients with SPG5A subtype of Hereditary Spastic Paraplegia. Blood biochemical analysis, electrophysiology and brain MRI were also used in the study. The value of ^{1}H MRS in the assessment of neuronopathic Gaucher's disease has been evaluated by Razek *et al.*[164] An increase in the ratio of Cho/tCr was found in the frontal white matter, which was higher in acute *versus* chronic patients, and was higher in homozygote *versus* heterozygote patients. The ratio of Cho/tCr was negatively correlated with disease severity and in 70% of patients, a lipid peak was detected. ^{1}H MRS has been used in the characterisation of three other genetic disorders. Kimura *et al.*[165] examined a patient with static encephalopathy of childhood with neurodegeneration in adulthood, Tada *et al.*[166] reported results from a 5 year-old boy with 18q(−) syndrome and Moghadam *et al.*[167] have investigated two asymptomatic daughters of a patient with autosomal dominant cerebellar ataxia. Furthermore, ^{1}H MRS has been used by Nicita *et al.*[168] to characterise brain metabolite levels in patients with neurofibromatosis type 1. Spectra recorded from the right and left caudate nucleus, globus pallidum, putamen and thalamus showed a decrease in tNAA, and changes in the levels of Cho and tCr compared to control subjects. ^{1}H and ^{31}P MRS have been used to measure brain temperature and phosphorous metabolite, respectively, in patients with mitochondrial diseases.[169] There was a small increase in brain temperature in control subjects during visual stimulation and a small rise in high energy phosphate following this. However, there was no such changes in patients with mitochondrial disease.

3.2.8 Autoimmune diseases. Patients with Sjogren syndrome[170] have been examined to evaluate the prevalence of signs and symptoms in the central nervous system and to identify possible biomarkers of damage. ^{1}H MRS found a reduction of NAA levels in subcortical frontal and basal ganglia white matter whilst other clinical manifestation of the disease were recorded. The metabolic consequences of systemic lupus erythematosus have been investigated in two papers. Cagnoli *et al.*[171] examined the frontal white matter, right insula and occipital grey matter and found reduced tNAA/tCr in the right insula of patients with neurological symptoms compared to patients without neurological symptoms and control subjects. Zimny *et al.*,[172] using ^{1}H MRS to examine the posterior cingulate cortex and left parietal white matter, found lower tNAA/tCr in both regions in patients with or without neurological symptoms compared to control subjects. Furthermore, DTI revealed reduce fractional anisotropy within the cingulum of patients.

3.2.9 Neurodegenerative diseases. Metabolite levels in the substantia nigra of patients with Parkinson's disease have been measured using ^{1}H MRS by Zhou *et al.*[173] who compared patients with asymmetric Parkinson's disease to healthy controls, and by Groeger *et al.*[174] studying patients with idiopathic Parkinson's disease and those with atypical Parkinsonian syndrome. Zhou *et al.* found reduced tNAA in patients, which was also lower on the most affected side, and differences were also seen between patients with mild or severe Parkinson's disease. Groeger *et al.* reported differences in tNAA/tCr between controls and idiopathic patients, as well as between idiopathic and atypical patients. Neurometabolite levels in grey and white matter of the temporal lobe, occipital lobe and parietal lobe of patients with Parkinson's disease have been measured by Levin *et al.*[175] Significant differences were found in metabolite levels between patients and control, though no correlations were found between neurometabolites and neuropsychological measures. Differences in the energy metabolism in the striatum and temporo-parietal cortical grey matter between male and female patients with Parkinson's disease have been reported by Weiduschat *et al.*[176] using ^{31}P MRS. Lower ATP values were recorded in male patients compared to female patients. Patients with Huntington's disease has been investigated by Casseb *et al.*[177] and Padowski *et al.*[178] using ^{1}H MRS to measure metabolite levels in the thalamus and in the caudate and frontal lobe, respectively. Casseb *et al.* found reduced tNAA/tCr and increased Cho/tCr, but there were no correlations between these and clinical features. Padowski *et al.* found reduced tNAA, tCr, Cho and Glx in the caudate that correlated with caudate volume. However, the ratio of Glx to other neurometabolites was the only significant difference seen between patients and controls. Changes in metabolite concentrations in the geniculocalcarine and the striate area of occipital lobe have been investigated in patients with glaucoma.[179] Compared to control subjects, there were decreases in NAA/tCr and Cho/tCr in the both regions. However, no effects were seen using T_1-weighted or T_2-weighted magnetic resonance imaging.

3.2.10 Neurological disorders. The use of ^{1}H MRS in the clinic investigation of epilepsy has been the subject of four research papers. Pan *et al.*[180] have used MRSI to analyse metabolism prior to resection and found that the best post-surgical outcomes were where the region with metabolic anomalies were completely removed. Peng *et al.*[181] analysed neurometabolites in the hippocampus in relation to scores of depression rating in patients with epilepsy. The ratio of Glx/tCr in the right hippocampus was increased in patients with moderate depression and correlated positively with depression rating. Zhang *et al.*[182] investigated the use of MRI and MRS, in combination with EEG measurements, in the assessment of patients with bilateral or unilateral temporal lobe epilepsy with and without lesions. However, MRS did not offer benefit to the prediction of outcomes in this study. Munakata *et al.*[183] have examined the effects of treatment with a synthetic analogue of adrenocorticotropic hormone in West syndrome, which is an intractable epileptic encephalopathy of infancy. Following treatment Glx/tCr and tNAA/tCr were reduced, and the latter was correlated with outcome.

Neurometabolite changes in leukoencephalopathy have been recorded using ^{1}H MRS. Kassem *et al.*[184] also used a battery of MRI techniques to characterise the condition of leukoencephalopathy with brain stem and spinal cord involvement and lactate elevation. Mori *et al.*[185] examined a patient with Pelizaeus–Merzbacher disease at ages 2, 6, 14, and 25 months. Finnsson *et al.*[186] measured metabolite levels in the supraventricular white matter, which included an area under the sensory motor cortex. Absolute concentration of tNAA, Cho, and tCr, using internal water as a reference, were all reduced in patients compared to control subjects. However, there were no differences in the ratios of tNAA/tCr or Cho/tCr between patients and controls.

Patients with multiple sclerosis have been investigated by Bellenberg *et al.*,[187] Tur *et al.*,[188] Pokryszko-Dragen *et al.*[189] and Vingara *et al.*[190] Bellenberg *et al.* examined patients with focal and diffuse cervical lesions without acute relapses. Compared with focal lesions, diffuse lesions resulted in more cord atrophy, longer T_2 times and reduced tNAA/tCr. The latter was associated with cord atrophy and with clinical progression during the follow-up exam. Tur *et al.* determined changes in metabolite concentrations along the cortico-spinal tract of patients with relapsing-remitting and primary-progressive multiple sclerosis. Relapsing-remitting patients had higher Cho than control subjects and mI than primary-progressive patients. Pokryszko-Dragen *et al.* measured metabolite ratios in the parietal white matter and the posterior cingulate gyrus with regard to measurements of fatigue and cognition. Patients had decreased tNAA/tCr and increased mI/tCr in these regions, respectively. However, no correlations were found between these ratios and measures of fatigue or cognition. Vingara *et al.* used multivariate statistical methods to analyse ^{1}H MRS data from the frontal lobe of patients with relapsing-remitting multiple sclerosis. Significant differences were found in comparison to data from control subjects and correlations were seen with measures of stress, verbal memory and neurological impairment.

In other neurological disorders, the metabolic and physical characteristics of white matter hyperintensities in patients with migraine have been investigated by Aradi *et al.*[191] and altered neurometabolites in patients with restless leg syndrome have been assessed by Allen *et al.*[192] Aradi *et al.* found that white matter hyperintensities had higher apparent diffusion coefficient values, prolonged T_2 relaxation times and decreased tNAA and tCr compared to contralateral white matter and white matter in control subjects. Allen *et al.* used ^{1}H MRS to compare the ratio of Glx/tCr in the right thalamus of patients with restless leg syndrome to that in control subjects. The Glx/Cr was higher in patients and correlated significantly with the wake time during sleep and all other polysomnographic sleep variables related to restless leg syndrome.

3.2.11 Injury. Changes in brain metabolites caused by traumatic brain injury has been assessed by Sivak *et al.*[193] within three days of injury, by Kierans *et al.*[194] around 22 days after, by Dhandapani *et al.*[195] at 3 months after, by Chamard *et al.*[196] at least 7 months following trauma and by Tremblay *et al.*[197] assessing subjects three decades post-injury. Sivak *et al.* found a decrease in tNAA in the frontal lobes of patients with mild traumatic brain injury, and in the upper brain stem for those subjects having had unconsciousness. Correlations were found between metabolite levels and neurological testing. Kierans *et al.* made measurements in caudate, globus pallidus, putamen, and thalamus of subject with mild traumatic brain injury and found increase mI in the putamen. Dhandapani *et al.* assessed patients with mild to moderate brain injury using single photon emission computed tomography (SPECT) and MRS. A general hypoperfusion was seen with SPECT, and MRS detected decreased tNAA/tCr in the frontal white matter. Chamard *et al.* used MRS and DTI in patients who had suffered a concussion to show reduced mI in the primary motor cortex and hippocampus and increased mean diffusivity in the large white matter tracts. Tremblay *et al.* found effects on verbal fluency and episodic memory, which correlated with increased mI in left medial temporal lobe.

3.2.12 Stroke, hypoxia and Apnoea. Stroke patients have been assessed by Lin *et al.*[198] using ^{1}H MRS in the infarct centre, the infarction border region and the contralateral brain area within 12 h after the onset of stroke and during 1–2 months post-stroke. Changes in metabolites were recorded and compared with patient outcomes. A comparison has been made of the impact of cerebral infarct in patients with type 2 diabetes and non-diabetic controls.[199] Following injury, the ratio of lactate/tCr was higher in the infarct area of both groups. However, the diabetic group had lower tNAA/tCr in contralateral brain and the infarct area when compared to the non-diabetic brain. The prognostic value of MRI and ^{1}H MRS in perinatal hypoxic-ischaemic encephalopathy has been assessed. Goergen *et al.*[200] found a significant association between outcome and lactate/NAA and diffusion-weighted imaging (DWI) scores for the lentiform nucleus, thalamus, cortex, paracentral white matter and posterior limb of the internal capsule, with the latter giving 100% specificity and sensitivity for poor outcome. The effects of hypothermia treatment on infants with hypoxic-ischaemic encephalopathy has been investigated two papers. Ancora *et al.*[201] examined patients treated with selective brain cooling and followed up the study two years later. It was found that ^{1}H MRS and DTI showed higher prognostic accuracy than conventional MRI. Thakur *et al.*[202] studied the effects of whole body hypothermia treatment and found abnormal results from MRI in 32% of neonates and increased lactate in 18%. The effects of mild and severe obstructive sleep apnoea have been assessed by ^{1}H MRS.[203] The ratios of tNAA/tCr was increased in the hippocampus, and Cho/tCr was increased in the hippocampus and putamen, in patients with severe apnoea compared to those with mild apnoea.

3.2.13 Foetal and neonatal brain. The effects of intrauterine growth restriction on foetal brain metabolism has been investigated by Story *et al.*[204] using ^{1}H MRS. However, the ratios of mI/Cho, mI/tCr and Cho/tCr, measured from a $20\times20\times20$ mm voxel, showed no difference when compared to results from foetuses that had grown appropriately. The value of ^{1}H MRS in differentiating bilirubin encephalopathy from severe hyperbilirubinaemia in neonates has been investigated by Wu *et al.*[205] Spectra obtained from the thalamus showed reduced ratios of tNAA/Cho and tNAA/tCr, however no changes in these metabolite ratios were seen in the basal ganglia.

3.2.14 Effects of diabetes and body mass index. The impact of type 1 diabetes mellitus on the neurochemical profile of the human brain has been measured by Mangia *et al.*[206] Spectra recorded under hyperglycaemic clamp (16.7 mM) showed lower levels of tNAA and Glu in the grey matter of diabetic patients compared to controls. The effects of euglycaemic-hyperinsulinaemic clamp on metabolite levels in the left frontal lobe, left temporal lobe and left thalamus have been investigated in non-diabetic subjects assigned to a group of either high or low high insulin sensitivity.[207] Increases in tNAA and Glx, and decreases in Cho and mI were seen in patients with high insulin sensitivity. The role of hypertension, hyperglycaemia and dyslipidaemia in the disruption of brain chemistry in overweight or obese subjects has been investigated by Haley *et al.*[208] A higher body mass index was found to be related to a higher ratio of mI/tCr in the occipitoparietal cortex, independent of age and sex. Furthermore, dyslipidaemia was also independently related to increased mI/tCr.

3.2.15 Cushion's disease. Resmini *et al.*[209] have examined neurometabolite levels in the hippocampus of patients that have been cured of Cushions syndrome. Compared to control subjects, patients had reduced levels of tNAA and higher levels of Glx.

3.2.16 Effects of spinal conditions. The effects of unilateral pain on metabolite levels in the anterior cingulate cortex, prefrontal cortex and thalamus of patients with lumbar spine diseases have been measured by Yabuki *et al.*[210] Effects were seen in the contralateral thalamus, where tNAA was reduced relative to tCr and Cho, and there was a correlation between tNAA levels and perceived pain. Salamon *et al.*[211] have used ^{1}H MRS to characterise the metabolic alterations in the spinal cord in patients with spondylosis with or without hyperintensity seen in the spinal cord with T_2-weighted imaging. Compared to controls, patients had higher Cho/tNAA, and higher Cho/tCr in those with T_2-hyperintensity.

3.3 Heart

^{31}P MRS has been used to investigate energy metabolism in the heart. The associations between normal cardiac energy metabolism, skeletal muscle energy metabolism, moderate-to-vigorous physical activity and age have been explored in adolescents and young adults.[212] The ratio of PCr/ATP for the heart at rest was positively associated with PCr recovery half-time in skeletal muscle following 60 s of exercise. The effects of beta-blockers on the ratio of PCr/ATP in the left ventricle of patients with heart failure has been investigated by Spoladore *et al.*[213] After beta-blockade, measure of cardiac function improved and the ratio of PCr/ATP increased. The influence of Chagas disease on cardiac energy metabolism has been investigated by Betim Paes Leme *et al.*[214] Patients had reduced high-energy phosphates at rest and had an exercise-induced decline in cardiac high-energy phosphates consistent with myocardial ischemia. The progress of heart failure in patients with cardiomyopathy has been followed by Bottomley *et al.*[215] Patients were investigated for a median of 4.7 years. It was found that the flux through creatine kinase was a significant predictor of heart failure.

3.4 Lipids

^{1}H MRS has been used as a validation standard in studies designed to investigate the utility of other methods for the assessment of intrahepatocellular lipids (IHCL). Oresic *et al.*[216] investigated the use of ultra-performance liquid chromatography in the diagnosis of non-alcoholic fatty liver disease from blood samples; the chromatography method had a sensitivity of 69.1% and specificity of 73.8%. Cusi *et al.*[217] have tested the use of the measurement of plasma caspase-generated cytokeratin-18 fragments for diagnosing and staging non-alcoholic steatohepatitis. The method was compared to results obtained using histological as well as the ^{1}H MRS assessments. Martin-Rodriguez *et al.*[218] appraised the value ultra-sound measurements in the diagnosis of steatosis and found some validity in this method. Karlas *et al.*[219] comparing ^{1}H MRS and an ultrasound method for the assessment of steatosis, found a comparable accuracy for both methods against histological findings. Wu *et al.*[220] compared double-echo MRI, triple echo-MRI and MRS for the assessment of steatosis in patients scheduled for liver resection. Triple-echo MRI and ^{1}H MRS were in greatest agreement with histological measures. However, Noureddin *et al.*[221] also compared MRI methods, histology and ^{1}H MRS in the assessment of IHCL found that MRI and MRS methods were able to measure changes in IHCL as small as 1%, which could not be quantified by histology. Abrigo *et al.*[222] have used ^{1}H and ^{31}P MRS to assess patients with non-alcoholic fatty liver disease. Using measures of phosphodiesters, phosphomonoesters, nucleoside triphosphate and inorganic phosphate it was possible to distinguish between patients and controls, and between patients with, or without, non-alcoholic steatohepatitis.

Kato *et al.*[223] have used ^{1}H MRS to assess intramyocellular lipids (IMCL) and IHCL whilst looking at the associations between ectopic fat and organ-specific insulin resistance in patients with non-alcoholic fatty liver disease. ^{1}H MRS has been used to measure IHCL levels by Lebensztejn *et al.*[224] in an assessment of serum fetuin A concentration as a potential marker of subclinical atherosclerosis in obese children with, and without, non-alcoholic fatty liver disease. Forbes *et al.*[225] used ^{1}H MRS to assess IHCL in study of non-alcoholic fatty liver disease in women with a history of gestational diabetes mellitus. Zhang *et al.*[226] have measured IHCL in relation to serum irisin concentrations in obese Chinese adults. Bian *et al.*[227] measured IHCL in a comparison of the effects of weight loss induced either by a short-term low-carbohydrate or by a long-term hypocaloric diet. Ryan *et al.*[228] found a reduction in hepatic steatosis in a study of the effects of a Mediterranean diet, whilst Ryberg *et al.*[229] examined the effects of a Palaeolithic-type diet on obese, post-menopausal women; ^{1}H MRS detected a reduction in IHCL, but there was no effect on IMCL.

^{1}H MRS has been used by Sprung *et al.*[230] to measure level of IHLC and IMCL in women with polycystic ovary syndrome. Phielix *et al.*[231] also used ^{1}H MRS to measure IMCL and IHCL, and used ^{31}P MRS to measure glucose uptake and phosphorylation, in an investigation of the action of pioglitazone plus nateglinide in non-obese patients with type 2 diabetes mellitus. The levels of IHCL, IMCL and extra-myocellular lipids have been measured using ^{1}H in patients with type 2 diabetes.[232] Glycogen levels were also assessed by ^{13}C MRS in this study of metabolite variability and the reproducibility of measurement. The level of triglyceride accumulation in the heart of patients with generalized lipodystrophy has been measured by Nelson *et al.*[233] A comparison of Dixon-based MRI and MRS for the quantification of intramuscular fat has been carried out by Noble *et al.*[234] A good correlation was found between the two techniques. Kim *et al.*[235] have compared the use of T_2-mapping with ^{1}H MRS for the measurement of fatty infiltration into muscle.

3.5 Liver

In the liver, a method for the quantification of phosphorous metabolites has been developed by Laufs *et al.*[236] The concentrations of PME, PDE, Pi and ATP were obtained from eighty-five subjects. The effects of hyponatraemia on cognition, quality of life and brain metabolites has been investigated by Ahluwalia *et al.*[237] in cirrhotic patients with or without hepatic encephalopathy. The level of Glx was found to be lowest in patients with hyponatraemia and highest in patients hepatic encephalopathy.

3.6 Muscle

3.6.1 Exercise and energy metabolism. ^{31}P MRS has been used in the assessment of muscle energy metabolism and the effects of exercise. Layec *et al.*[238] found that training for 5 days had no effects on oxidative capacity and PCr recovery, but there was an increase in the Hill coefficient. Chidnok *et al.*[239] have examined the role of the length of the recovery period between bouts of high intensity training on the recovery of PCr and the level of work achieved. Yoshida *et al.*[240] measured a faster time constant for PCr off-kinetics following a bout of 60% maximal exercise in the plantar flexion, compared to the knee flexion, muscle. Kappenstein *et al.*[241] detected a lower depletion of PCr following short periods of exercise in children compared to adults, which was attributed to reduced oxidative formation of ATP in adults. However, there was a later criticism of the measurement of oxidative capacity of muscle using this method,[242] though this criticism was not accepted by Kappenstein *et al.*[243] in reply. Valkovic *et al.*[244] observed a linear correlation between measurements of PCr recovery following exercise and the synthesis rate of PCr from magnetization transfer experiments in obese subjects. Layec *et al.*[245] have used ^{31}P MRS to compare oxidative formation of ATP in young and older sedentary adults. End-exercise pH and the depletion of PCr was the same in both groups and was unaffected by hyperoxia. However, the recovery of PCr was slower in older subjects, though this was also unaffected by hyperoxia. Layec *et al.*[246] have also examined the effects of reactive hyperaemia after a period of exercise on oxygen delivery and mitochondrial function in the calf muscle of young adults. Reactive hyperaemia increased oxygen delivery and peak ATP synthesis, but the PCr recovery time constant was unaffected. A new method for the assessment of muscle energetics has been proposed by van Oorschot *et al.*[247] The assessment of muscle oxidative capacity by the recovery kinetics of PCr following exercise was compared to the proposed method of the assessment of oxidative capacity using a measure of mitochondrial density estimated from the ^{31}P MRS signal indicating Pi in an alkali environment.

The level of lactate in the human forearm following exercise has been detected in ^{1}H spectra obtained without water suppression. Data was obtained by Ren *et al.*[248] using the STEAM sequence with an echo-time of 140 ms and demonstrated the rapid decline in lactate levels during recovery from exercise to fatigue. ^{1}H chemical exchange saturation transfer has been used by Kogan *et al.*[249] to measure tCr levels in human calf muscle following exercise. The recovery of creatine levels post-exercise were at a similar rate to the recovery of PCr measured by ^{31}P MRS. Schmid *et al.*[250] have investigated the use of changes in intensity of images from

MRI of muscle as an means of assessing energy metabolism during exercise. Correlations were found between image intensity and intracellular pH, the time constant for PCr recovery and maximum oxidative phosphorylation.

3.6.2 The measurement of metabolites and metabolism. The effects of foot flexion on the appearance of ^{1}H spectra obtained from tibialis anterior and soleus muscles have been studied in humans and rats by Agarwal *et al.*[251] Variations in the tCr peak were seen with foot flexion, including a four-way split of the methylene peak of creatine and phosphocreatine. Magnetization transfer between the water resonance and tCr has been explored by MacMillan *et al.*[252] Magnetization transfer rates from water to the creatine methylene resonances were 0.56 and 0.28 s^{-1}, and for the methyl resonances were 0.39 and 0.30 s^{-1} for tibialis and soleus anterior muscles, respectively. The effects of training on carnosine loading in leg and arm muscle has been studied by Bex *et al.*[253] using ^{1}H MRS. Greater muscle loading of carnosine from beta-alanine supplementation was observed in trained muscle groups. Ryan *et al.*[254] have compared the measurement of skeletal muscle oxidative capacity made with near-infrared spectroscopy to those made with ^{31}P MRS. A good agreement was seen between the two methods. ^{31}P MRS has been used by Slattery *et al.*[255] to measure mitochondrial function in obese adolescent girls with and without with insulin resistance. ^{1}H MRS was also used to quantify IMCL levels. Girls with insulin-resistance had a lower rate of post-exercise PCr recovery, although no difference was seen in IMCL levels. The oxidative capacity of different muscle groups within the same, untrained subjects has also been investigated by Layec *et al.*[256] The influence of end-exercise acidosis was seen to be greater in finger muscle compared to quadriceps and plantar flexion muscles.

3.6.3 Disease and conditions. The effects of disease and other conditions on muscle metabolites and metabolism have been investigated. Gerdle *et al.*[257] measured phosphorous metabolism in patients with fibromyalgia. There were lower levels of PCr and ATP in patients compared to controls, though no correlations were found between muscle metabolites and pain intensity. The post-exercise kinetics of quadriceps muscle in subjects with Down's syndrome has been investigated in comparison to that in subjects with intellectual disability.[258] The recovery rate constant for PCr was found to be decreased in subjects with Down's syndrome. Muscle metabolism in patients with Friedreich ataxia has been investigated by Nachbauer *et al.*[259] Patients showed a more rapid depletion of PCr and an increased accumulation of Pi during incremental exercise compared to control subjects. Furthermore, they also had a prolonged regeneration of PCr and a slowed decline in Pi post-exercise. Patients were also tested for the effects of 8 weeks treatment with recombinant human erythropoietin, though this showed no effect. The influence of anti-viral therapy on muscle energetics has been investigated by Payne *et al.*[260] using ^{31}P MRS in HIV-infected patients. Post-exercise ATP synthesis was similar in patients and controls, though patients had a higher resting ADP/ATP ratio. PCr recovery following exercise has been measured by Koslik *et al.*[261] in subjects with Gulf War syndrome. Seven veterans were compared to matched controls and showed a prolonged recovery of PCr post-exercise.

3.7 Placenta

^{31}P MRS has been used to study the metabolism of the placenta during late and early preeclampsia.[262] In early onset preeclampsia, the level of PDE compared to total phosphorous signal, and the ratio of PDE/PME, were higher compared to levels in control placenta. In normal pregnancy, PDE and PDE/PME increased from 22 to 41 weeks gestational age. No differences were found in metabolites levels between late preeclampsia and controls.

3.8 Tumour

3.8.1 Brain. Metabolite levels in children's brain tumours have been examined by Gill *et al.*[263] before treatment compared to following relapse, by Orphanidou-Vlachou *et al.*[264] assessing differences between low grade gliomas and related tumours, and by Novaka *et al.*[265] looking at the value of ^{31}P MRS for tumour grading compared to ^{1}H MRS and MRI. Gill *et al.* found that a relapsed tumour, whether local or distant, had the same metabolic profile in ^{1}H spectra as the tumour prior to treatment. Orphanidou-Vlachou *et al.* found differences in ^{1}H metabolite concentrations between glioneuronal and glial tumours, and between most of the individual histological subtypes of low grade gliomas. Novaka *et al.* found that, in three optic gliomas examined, the ratio of PC/GPC indicated low grade tumours in keeping with their clinical behaviour and histology.

In adult tumours, the role of ^{1}H MRS and DWI in the characterisation of germinomas has been evaluated by Saito *et al.*[266] Data were retrospectively analysed from thirty-four patients. ^{1}H MRS and DWI provided useful data to distinguish between germinomas and other tumour types. Jaskolski *et al.*[267] have compared ^{1}H MRS spectra from four patients with meningiomas, six with schwannomas, two with lymphomas, two with haemangiopericytomas; control spectra were taken from healthy hemispheres. Bisdas *et al.*[268] have examined the use of PET-detected methionine uptake and ^{1}H MRS assessment of metabolite ratios for the grading of gliomas in twenty-eight patients. The method provided useful information on the location of tumour masses. The use of ^{1}H MRS in conjunction with MRI measurements of perfusion and contrast agent-enhancement has been assessed by Seeger *et al.*[269] in patients with glioma. The method was useful in the distinction between recurrent gliomas and stable disease. A comparison has been made by Choi *et al.*[270] of short and long TE ^{1}H MRS for the detection of 2-hydroxyglutarate. Measurements made in phantoms, healthy subjects and patients with IDH-mutated gliomas indicated that a TE of 97 ms gave the best results at 3 Tesla. Single and multivoxel MRS with dynamic susceptibility contrast

MRI for the differentiation of glioblastoma multiforme from metastases has been appraised by Tsolaki *et al.*[271] in forty-nine patients with solitary brain tumours. Balos *et al.*[272] have retrospectively analysed the role of ^{1}H MRS and DWI in distinguishing between intracranial glial tumours and tumour-like non-neoplastic lesions. Data taken from twenty-five non-neoplastic lesions, fourteen low-grade and eight anaplastic glial tumours could be used to distinguish between tumours and lesions. The use of ^{1}H MRS in the pre-operative assessment of twenty-three patients with brain tumours has been evaluated by Rao *et al.*[273] It was found that ^{1}H distinguished between abscess and high-grade tumour, and between high-grade glioma and low-grade glioma, but was not able to reliably distinguish between recurrent glioma and radiation necrosis. The detection of the uptake of 2-fluoro-deoxyglucose by positron emission tomography in combination with ^{1}H MRS for the distinction between tumour progression and radiation damage has been studied by Imani *et al.*;[274] an automated method for tumour detection provided some useful diagnostic information. The utility of MRS in differentiating lymphoma from tumefactive demyelination has been assessed in sixty-five patients by Lu *et al.*[275] The results indicated that lymphoma could be implied when the Cho/tCr ratio was >2.58, the Cho/tNAA ratio was >1.73, and a high lipid and/or lactate peak was seen.

3.8.2 Breast. Stehouwer *et al.*[276] have investigated phosphorous metabolite ratios in glandular breast tissue and the potential effect of the menstrual cycle on these. However, no effect of was observed. The value of the measurement of the level of Cho in breast lesion detected by dynamic contrast-enhanced MRI has been assessed by Vassiou *et al.*,[277] Basara *et al.*[278] and Suppiah *et al.*[279] Vassiou *et al.* found the combination of MRI and ^{1}H MRS provided 96.4% accuracy in diagnosis, whilst Basara *et al.* found ^{1}H MRS gave some false negative results. Suppiah *et al.* reported a good correlation between Cho peaks and malignancy. However, the accuracy of the results were questioned by Battal *et al.*,[280] though this was dismissed by Suppiah *et al.* in reply.[281]

3.8.3 Reproductive. Takeuchi *et al.*[282] have retrospectively evaluated the significance of lipid and Cho detected by ^{1}H MRS in thirty-eight pathologically diagnosed uterine corpus tumours. They concluded that ^{1}H MRS may be useful in distinguishing uterine sarcomas from benign leiomyomas. The use of MRS in combination with various MRI methods for the diagnosis of prostate cancer has been assessed by Chang *et al.*[283] in eighty-eight patients with biopsy-proven tumours, by Kuru *et al.*[284] in fifty patients assessed with biopsy and by Petrillo *et al.*[285] in two hundred patients with either prostate-specific antigen levels between 4 to 10 ng cm^{-3} or between 2.5 to 4 ng cm^{-3} and an abnormal digital-rectal examination. Chang *et al.* were able to recognise patients suitable for radiation therapy, whilst Petrillo *et al.* identified patients with the highest prostate-specific antigen score. Kuru *et al.* found that using multiparametric magnetic resonance methods may be helpful in the initial diagnosis of prostate cancer. Weis *et al.*[286] have examine the relaxation properties and concentrations of metabolites in the normal prostate. The T_1 and T_2 relaxation times for water, Cho, tCr and citrate were reported along with the concentrations of the metabolites, using tissue water as an internal reference.

3.8.4 Bone. The value of ^{1}H MRS in the discrimination of benign and malignant bone tumours has been evaluated by Zhang *et al.*[287] Cho and lipids were detected in malignant tumours, whereas no Cho was seen in normal bone marrow and most benign tumours.

References

1 X. Chen, P. Boesiger and A. Henning, *NMR Biomed.*, 2013, **26**, 1113–1124.
2 Y. Zhang, R. E. Gabr, J. Zhou, R. G. Weiss and P. A. Bottomley, *J. Magn. Reson.*, 2013, **237**, 125–138.
3 X.-a. Mao, B. Jiang, L. Jiang and M. Liu, *Mol. Imaging Biol.*, 2013, **15**, 245–249.
4 P. Parasoglou, D. Xia, G. Chang, A. Convit and R. R. Regatte, *NMR Biomed.*, 2013, **26**, 1142–1151.
5 W. J. M. van der Kemp, V. O. Boer, P. R. Luijten, B. L. Stehouwer, W. B. Veldhuis and D. W. J. Klomp, *NMR Biomed.*, 2013, **26**, 1299–1307.
6 C. S. A. de Castro, V. O. Boer, A. Andreychenko, J. P. Wijnen, U. A. van der Heide, P. R. Luijten and D. W. J. Klomp, *NMR Biomed.*, 2013, **26**, 1213–1219.
7 R. A. de Graaf, D. W. J. Klomp, P. R. Luijten and V. O. Boer, *Magn. Reson. Med.*, 2014, **71**, 451–457.
8 C. J. Evans, N. A. J. Puts, S. E. Robson, F. Boy, D. J. McGonigle, P. Sumner, K. D. Singh and R. A. E. Edden, *J. Magn. Reson. Imaging*, 2013, **38**, 970–975.
9 S. A. Wijtenburg, L. M. Rowland, R. A. E. Edden and P. B. Barker, *J. Magn. Reson. Imaging*, 2013, **38**, 460–467.
10 A.-M. M. El-Sharkawy, R. E. Gabr, M. Schaer, R. G. Weiss and P. A. Bottomley, *NMR Biomed.*, 2013, **26**, 1363–1371.
11 S. Mazzetti, C. Bracco, D. Regge, R. Caivano, F. Russo and M. Stasi, *Phys. Med.: Eur. J. Med. Phys.*, 2013, **29**, 677–683.
12 E. Mosconi, D. M. Sima, M. I. O. Garcia, M. Fontanella, S. Fiorini, S. Van Huffel and P. Marzola, *NMR Biomed.*, 2014, **27**, 431–443.
13 B. Schaller, L. Xin, C. Cudalbu and R. Gruetter, *NMR Biomed.*, 2013, **26**, 593–599.
14 H. Lee, E. Caparelli, H. Li, A. Mandal, S. D. Smith, S. Zhang, T. V. Bilfinger and H. Benveniste, *Magn. Reson. Imaging*, 2013, **31**, 1197–1205.
15 B. J. Wilm, Y. Duerst, B. E. Dietrich, M. Wyss, S. J. Vannesjo, T. Schmid, D. O. Brunner, C. Barmet and K. P. Pruessmann, *Magn. Reson. Med.*, 2014, **71**, 1657–1662.
16 A. Hock, L. Valkovic, A. Geier, T. Kuntzen, P. Boesiger and A. Henning, *NMR Biomed.*, 2014, **27**, 348–355.
17 M. Chmelik, M. Povazan, M. Krssak, S. Gruber, M. Tkacov, S. Trattnig and W. Bogner, *NMR Biomed.*, 2014, **27**, 478–485.
18 M. Tachrount, G. Duhamel, J. Laurin, T. Marqueste, A. M. de Paula, P. Decherchi, P. J. Cozzone and V. Callot, *Magn. Reson. Med.*, 2013, **69**, 1226–1232.
19 L. An, J. W. van der Veen, S. Li, D. M. Thomasson and J. Shen, *J. Magn. Reson. Imaging*, 2013, **37**, 1445–1450.
20 E. L. Hall, M. C. Stephenson, D. Price and P. G. Morris, *NeuroImage*, 2014, **86**, 35–42.

21 J. Intrapiromkul, H. Zhu, Y. Cheng, P. B. Barker and R. A. E. Edden, *J. Magn. Reson. Imaging*, 2013, **38**, 1224–1229.
22 A. Andreychenko, D. W. J. Klomp, R. A. de Graaf, P. R. Luijten and V. O. Boer, *NMR Biomed.*, 2013, **26**, 1596–1601.
23 A. Tisell, O. D. Leinhard, J. B. M. Warntjes and P. Lundberg, *Magn. Reson. Med.*, 2013, **70**, 905–915.
24 I. Ronen, E. Ercan and A. Webb, *NMR Biomed.*, 2013, **26**, 1291–1298.
25 M. Marjanska, U. E. Emir, D. K. Deelchand and M. Terpstra, *PLoS One*, 2013, **8**(10), e77572.
26 L. Xin, G. Gambarota, C. Cudalbu, V. Mlynarik and R. Gruetter, *Magn. Reson. Mater. Phys., Biol. Med.*, 2013, **26**, 549–554.
27 J. D. Port, K. E. Rampton, Y. Shu, A. Manduca and M. A. Frye, *J. Magn. Resonance Imaging*, 2013, **37**, 1451–1459.
28 R. A. Komoroski, D. M. Lindquist and J. M. Pearce, *NMR Biomed.*, 2013, **26**, 1152–1157.
29 M. Rata, S. L. Giles, N. M. deSouza, M. O. Leach and G. S. Payne, *NMR Biomed.*, 2014, **27**, 158–162.
30 N. McVicar, A. X. Li, D. F. Goncalves, M. Bellyou, S. O. Meakin, M. A. M. Prado and R. Bartha, *J. Cereb. Blood Flow Metab.*, 2014, **34**, 690–698.
31 S. M. Kazan, S. Reynolds, A. Kennerley, E. Wholey, J. E. Bluff, J. Berwick, V. J. Cunningham, M. N. Paley and G. M. Tozer, *Magn. Reson. Med.*, 2013, **70**, 943–953.
32 S. Hu, P. E. Z. Larson, M. VanCriekinge, A. M. Leach, I. Park, C. Leon, J. Zhou, P. J. Shin, G. Reed, P. Keselman, C. von Morze, H. Yoshihara, R. A. Bok, S. J. Nelson, J. Kurhanewicz and D. B. Vigneron, *Magn. Reson. Imaging*, 2013, **31**, 490–496.
33 C. L. Swisher, P. E. Z. Larson, K. Kruttwig, A. B. Kerr, S. Hu, R. A. Bok, A. Goga, J. M. Pauly, S. J. Nelson, J. Kurhanewicz and D. B. Vigneron, *Magn. Reson. Med.*, 2014, **71**, 1–11.
34 L. V. Sogaard, F. Schilling, M. A. Janich, M. I. Menzel and J. H. Ardenkjaer-Larsen, *NMR Biomed.*, 2014, **27**, 561–569.
35 K. Ono, M. Kitagawa, D. Ito, N. Tanaka and T. Watari, *Am. J. Vet. Res.*, 2014, **75**, 179–186.
36 J. J. Morgan, G. A. Kleven, C. D. Tulbert, J. Olson, D. A. Horita and A. E. Ronca, *NMR Biomed.*, 2013, **26**, 683–691.
37 J. Hai, Y.-F. Wu, Q. Lin, X.-S. Huang and G.-Y. Zhang, *Acta Neurol. Belg.*, 2013, **113**, 313–317.
38 S. Koundal, S. Gandhi, T. Kaur and S. Khushu, *NMR Biomed.*, 2014, **27**, 341–347.
39 R. Saggu, *Transl. Stroke Res.*, 2013, **4**, 351–367.
40 S. Y. Ahn, H. S. Yoo, J. H. Lee, D. K. Sung, Y. J. Jung, S. I. Sung, K. H. Lim, Y. S. Chang, J. H. Lee, K. S. Kim and W. S. Park, *J. Korean Med. Sci.*, 2013, **28**, 1071–1076.
41 H. Seo, K. H. Limb, J.-H. Choi and S.-M. Jeong, *Int. J. Dev. Neurosci.*, 2013, **31**, 616–623.
42 Y.-M. Guan, Y. Zhu, X.-C. Liu, H.-L. Huang, Z.-W. Wang, B. Liu, Y.-Z. Zhu and Q.-S. Wang, *BMC Neurosci.*, 2014, **15**, 616–623.
43 B. Skiold, Q. Wu, S. B. Hooper, P. G. Davis, R. McIntyre, M. Tolcos, J. Pearson, R. Vreys, G. F. Egan, S. K. Barton, J. L. Y. Cheong and G. R. Polglase, *PLoS One*, 2014, **9**, e95804.
44 Y. van de Looij, A. Chatagner, C. Quairiaux, R. Gruetter, P. S. Hueppi and S. V. Sizonenko, *PLoS One*, 2014, **9**, e95643.
45 L. H. Nilsen, T. M. Melo, M. P. Witter and U. Sonnewald, *Neurochem. Res.*, 2014, **39**, 305–312.

46 D. Jansen, V. Zerbi, C. I. F. Janssen, P. J. W. C. Dederen, M. P. C. Mutsaers, A. Hafkemeijer, A.-L. Janssen, C. L. M. Nobelen, A. Veltien, J. J. Asten, A. Heerschap and A. J. Kiliaan, *PLoS One*, 2013, **8**, e63643.
47 W. Zhang, P.-j. Wang, M.-h. Li, G.-l. Wang, P. Li and X.-l. Gao, *J. Mol. Neurosci.*, 2013, **50**, 434–42.
48 M. Marjanska, S. D. Weigand, G. Preboske, T. M. Wengenack, R. Chamberlain, G. L. Curran, J. F. Poduslo, M. Garwood, D. Kobayashi, J. C. Lin and C. R. Jack Jr., *Neuroscience*, 2014, **259**, 94–100.
49 P. Bagga, A. N. Chugani, K. S. Varadarajan and A. B. Patel, *J. Neurochem.*, 2013, **127**, 365–377.
50 Z. Hou, Z. Zhang, H. Meng, X. Lin, B. Sun, H. Lei, K. Fang, F. Fang, M. Liu and S. Liu, *Int. J. Neurosci.*, 2014, **124**, 125–132.
51 W. Fu, Z. Zheng, W. Zhuang, D. Chen, X. Wang, X. Sun and X. Wang, *Int. J. Neurosci.*, 2013, **123**, 883–891.
52 U. E. Emir, H. B. Clark, M. L. Vollmers, L. E. Eberly and G. Oez, *J. Neurochem.*, 2013, **127**, 660–668.
53 J. W. Totenhagen, E. S. Yoshimaru, R. P. Erickson and T. P. Trouard, *J. Magn. Reson. Imaging*, 2013, **37**, 1195–1201.
54 Y. Xuan, G. Yan, H. Peng, R. Wu and H. Xu, *Neurochem. Int.*, 2014, **69**, 20–27.
55 A. Napolitano, K. Shah, M. I. Schubert, V. Porkess, K. C. F. Fone and D. P. Auer, *Schizophrenia Bull.*, 2014, **40**, 566–574.
56 N. Chen, C. Liu, N. Yan, W. Hu, J.-g. Zhang, Y. Ge and F.-g. Meng, *PLoS One*, 2013, **8**, e72336.
57 G. A. Lodygensky, N. Kunz, E. Perroud, E. Somm, V. Mlynarik, P. S. Hueppi, R. Gruetter and S. V. Sizonenko, *Pediatr. Res.*, 2014, **75**, 415–423.
58 M. P. Moshkin, A. E. Akulov, D. V. Petrovski, O. V. Saik, E. D. Petrovskiy, A. A. Savelov and I. V. Koptyug, *NMR Biomed.*, 2014, **27**, 399–405.
59 D. N. Guilfoyle, S. Gerum and C. Vadasz, *J. Mol. Neurosci.*, 2014, **53**, 135–141.
60 C. Cudalbu, V. A. McLin, H. Lei, J. M. N. Duarte, A.-L. Rougemont, G. Oldani, S. Terraz, C. Toso and R. Gruetter, *PLoS One*, 2013, **8**, e69782.
61 I. Y. Zhou, R. W. Chan, L. C. Ho and E. X. Wu, *Neurosci. Lett.*, 2013, **553**, 170–175.
62 C. Najac, C. Marchadour, M. Guillermier, D. Houitte, V. Slavov, E. Brouillet, P. Hantraye, V. Lebon and J. Valette, *NeuroImage*, 2014, **90**, 374–380.
63 D. Coman, B. G. Sanganahalli, D. Cheng, T. McCarthy, D. L. Rothman and F. Hyder, *Magn. Reson. Imaging*, 2014, **32**, 305–313.
64 B. Lanz, L. Xin, P. Millet and R. Gruetter, *J. Neurochem.*, 2014, **128**, 125–139.
65 J. M. Park, S. Josan, T. Grafendorfer, Y.-F. Yen, R. E. Hurd, D. M. Spielman and D. Mayer, *NMR Biomed.*, 2013, **26**, 1197–1203.
66 W. Cui, X.-H. Zhu, M. L. Vollmers, E. T. Colonna, G. Adriany, B. Tramm, J. M. Dubinsky and G. Oez, *J. Cereb. Blood Flow Metab.*, 2013, **33**, 1846–1849.
67 N. Just, L. Xin, H. Frenkel and R. Gruetter, *NeuroImage*, 2013, **74**, 343–351.
68 R. J. Mullins, S. Xu, E. F. R. Pereira, J. Mamczarz, E. X. Albuquerque and R. P. Gullapalli, *Neurotoxicology*, 2013, **36**, 42–48.
69 M. Gupta, P. Rana, R. Trivedi, B. S. H. Kumar, A. R. Khan, R. Soni, R. K. S. Rathore and S. Khushu, *NMR Biomed.*, 2013, **26**, 1733–1741.
70 M. H. Lauritzen, C. Laustsen, S. A. Butt, P. Magnusson, L. V. Sogaard, J. H. Ardenkjaer-Larsen and P. Akeson, *NMR Biomed.*, 2013, **26**, 1496–1500.
71 S. Josan, J. M. Park, R. Hurd, Y.-F. Yen, A. Pfefferbaum, D. Spielman and D. Mayer, *NMR Biomed.*, 2013, **26**, 1680–1687.
72 S. Tucci, U. Floegel, S. Hermann, M. Sturm, M. Schaefers and U. Spiekerkoetter, *Biochim. Biophys. Acta, Mol. Basis Dis.*, 2014, **1842**, 677–685.

73 A. J. Bakermans, M. S. Dodd, K. Nicolay, J. J. Prompers, D. J. Tyler and S. M. Houten, *Cardiovasc. Res.*, 2013, **100**, 441–449.
74 A. J. Bakermans, M. van Weeghel, S. Denis, K. Nicolay, J. J. Prompers and S. M. Houten, *J. Inherited Metab. Dis.*, 2013, **36**, 973–981.
75 D. R. Ball, R. Cruickshank, C. A. Carr, D. J. Stuckey, P. Lee, K. Clarke and D. J. Tyler, *NMR Biomed.*, 2013, **26**, 1441–1450.
76 M. Desrois, F. Kober, C. Lan, C. Dalmasso, M. Cole, K. Clarke, P. J. Cozzone and M. Bernard, *NMR Biomed.*, 2014, **27**, 529–538.
77 A. Gupta, C. Rohlfsen, M. K. Leppo, V. P. Chacko, Y. Wang, C. Steenbergen and R. G. Weiss, *PLoS One*, 2013, **8**, e74675.
78 N. Ramamonjisoa, H. Ratiney, E. Mutel, H. Guillou, G. Mithieux, F. Pilleul, F. Rajas, O. Beuf and S. Cavassila, *J. Lipid Res.*, 2013, **54**, 2010–2022.
79 Q. Ye, C. F. Danzer, A. Fuchs, D. Vats, C. Wolfrum and M. Rudin, *NMR Biomed.*, 2013, **26**, 1079–1088.
80 J. Yaligar, V. Gopalan, O. W. Kiat, S. Sugii, G. Shui, B. D. Lam, C. J. Henry, M. R. Wenk, E. S. Tai and S. S. Velan, *PLoS One*, 2014, **9**, e91436.
81 Y. Lee, H.-J. Jee, H. Noh, G.-H. Kang, J. Park, J. Cho, J.-H. Cho, S. Ahn, C. Lee, O.-H. Kim, B.-C. Oh and H. Kim, *Magn. Reson. Med.*, 2013, **70**, 620–629.
82 P. M. Nunes, A. J. Wright, A. Veltien, J. J. A. van Asten, C. J. Tack, J. G. Jones and A. Heerschap, *FASEB J.*, 2014, **28**, 1988–1997.
83 M. H. Clark, R. Larsen, W. Lu and M. Hoenig, *Res. Vet. Sci.*, 2013, **95**, 678–680.
84 D. Laurent, B. Yerby, B. Zhang, X.-H. Chen, J. Gounarides, J. Zhang, J. Gao, G. Bebernitz and A. Duttaroy, *Eur. J. Pharmacol.*, 2013, **715**, 89–95.
85 P. Hanpanich, S. Pinlaor, L. Charoensuk, P. Yongvanit, C. Thomas, S. Kothan and E. Mairiang, *Magn. Reson. Imaging*, 2013, **31**, 1418–1425.
86 P. K. Shah, F. Ye, M. Liu, A. Jayaraman, C. Baligand, G. Walter and K. Vandenborne, *Eur. J. Appl. Physiol.*, 2014, **114**, 847–858.
87 C. Gineste, G. Duhamel, Y. Le Fur, C. Vilmen, P. J. Cozzone, K. J. Nowak, D. Bendahan and J. Gondin, *PLoS One*, 2013, **8**, e72294.
88 R. Madonna, S. D. Pizzi, A. Tartaro and R. De Caterina, *Mol. Biotechnol.*, 2014, **56**, 438–448.
89 S. C. Forbes, L. T. Bish, F. Ye, J. Spinazzola, C. Baligand, D. Plant, K. Vandenborne, E. R. Barton, H. L. Sweeney and G. A. Walter, *Gene Ther.*, 2014, **21**, 387–392.
90 H. N. Xu, S. Kadlececk, H. Profka, J. D. Glickson, R. Rizi and L. Z. Li, *Acad. Radiol.*, 2014, **21**, 223–231.
91 M. M. Chaumeil, P. E. Z. Larson, H. A. I. Yoshihara, O. M. Danforth, D. B. Vigneron, S. J. Nelson, R. O. Pieper, J. J. Phillips and S. M. Ronen, *Nat. Commun.*, 2013, **4**, 2429.
92 C. Weidensteiner, P. R. Allegrini, M. Sticker-Jantscheff, V. Romanet, S. Ferretti and P. M. J. McSheehy, *BMC Cancer*, 2014, **14**, 88.
93 M. E. Pisanu, A. Ricci, L. Paris, E. Surrentino, L. Liliac, M. Bagnoli, S. Canevari, D. Mezzanzanica, F. Podo, E. Iorio and R. Canese, *Br. J. Cancer*, 2014, **110**, 625–635.
94 B. Hamans, A. C. Navis, A. Wright, P. Wesseling, A. Heerschap and W. Leenders, *Neuro-Oncology*, 2013, **15**, 1615–1624.
95 L. Mignion, P. Dutta, G. V. Martinez, P. Foroutan, R. J. Gillies and B. F. Jordan, *Cancer Res.*, 2014, **74**, 686–694.
96 G. Lin, G. Andrejeva, A.-C. W. Te Fong, D. K. Hill, M. R. Orton, H. G. Parkes, D.-M. Koh, S. P. Robinson, M. O. Leach, T. R. Eykyn and Y.-L. Chung, *PLoS One*, 2014, **9**, e92645.

97 V. C. Sandulache, Y. Chen, J. Lee, A. Rubinstein, M. S. Ramirez, H. D. Skinner, C. M. Walker, M. D. Williams, R. Tailor, L. E. Court, J. A. Bankson and S. Y. Lai, *PLoS One*, 2014, **9**, e87031.
98 D. C. Karampinos, G. Melkus, T. Baum, J. S. Bauer, E. J. Rummeny and R. Krug, *Magn. Reson. Med.*, 2014, **71**, 1158–1165.
99 X. Ojanen, R. J. H. Borra, M. Havu, S. M. Cheng, R. Parkkola, P. Nuutila, M. Alen and S. Cheng, *Osteoporosis Int.*, 2014, **25**, 653–662.
100 F. Bolacchi, L. Uccioli, S. Masala, L. Giurato, V. Ruotolo, M. Meloni, E. Baffari, E. Cinelli, M. Cadioli, E. Squillaci, G. Simonetti and A. Bergamini, *Eur. Radiol.*, 2013, **23**, 2807–2813.
101 G. Manenti, S. Capuani, E. Fanucci, E. P. Assako, S. Masala, R. Sorge, R. Iundusi, U. Tarantino and G. Simonetti, *Bone*, 2013, **55**, 7–15.
102 G. Ende, D. Hermann, T. Demirakca, M. Hoerst, N. Tunc-Skarka, W. Weber-Fahr, S. Wichert, J. Rabinstein, U. Frischknecht, K. Mann and S. Vollstaedt-Klein, *Alcohol.: Clin. Exp. Res.*, 2013, 37, 1643–1649.
103 L. du Plessis, J. L. Jacobson, S. W. Jacobson, A. T. Hess, A. van der Kouwe, M. J. Avison, C. D. Molteno, M. E. Stanton, J. A. Stanley, B. S. Peterson and E. M. Meintjes, *Alcohol.: Clin. Exp. Res.*, 2014, **38**, 1330–1338.
104 F. M. Howells, A. Uhlmann, H. Temmingh, H. Sinclair, E. Meintjes, D. Wilson and D. J. Stein, *Schizophrenia Res.*, 2014, **153**, 122–128.
105 A. P. Prescot, P. F. Renshaw and D. A. Yurgelun-Todd, *Drug Alcohol Depend.*, 2013, **129**, 232–239.
106 J. M. Stone, F. Pepper, J. Fam, H. Furby, E. Hughes, C. Morgan and O. D. Howes, *Psychopharmacology*, 2014, **231**, 2107–2116.
107 A. Moheet, U. E. Emir, M. Terpstra, A. Kumar, L. E. Eberly, E. R. Seaquist and G. Oez, *Magn. Reson. Med.*, 2014, **71**, 12–18.
108 J. Near, J. Andersson, E. Maron, R. Mekle, R. Gruetter, P. Cowen and P. Jezzard, *NMR Biomed.*, 2013, **26**, 1353–1362.
109 W. Dou, N. Palomero-Gallagher, M.-J. van Tol, J. Kaufmann, K. Zhong, H.-G. Bernstein, H.-J. Heinze, O. Speck and M. Walter, *J. Neurosci.*, 2013, **33**(31), 12698–12704.
110 F. Kogan, A. Singh, C. Debrosse, M. Haris, K. Cai, R. P. Nanga, M. Elliott, H. Hariharan and R. Reddy, *NeuroImage*, 2013, **77**, 262–267.
111 K. Cai, A. Singh, D. R. Roalf, R. P. R. Nanga, M. Haris, H. Hariharan, R. Gur and R. Reddy, *NMR Biomed.*, 2013, **26**, 1278–1284.
112 N. Sailasuta, K. C. Harris, T. T. Tran, O. Abulseoud and B. D. Ross, *J. Magn. Reson. Imaging*, 2014, **39**, 398–401.
113 T. Patel, J. C. Blyth, G. Griffiths, D. Kelly and J. B. Talcott, *Front. Hum. Neurosci.*, 2014, **8**, 00039.
114 K. R. Pugh, S. J. Frost, D. L. Rothman, F. Hoeft, S. N. Del Tufo, G. F. Mason, P. J. Molfese, W. E. Mencl, E. L. Grigorenko, N. Landi, J. L. Preston, L. Jacobsen, M. S. Seidenberg and R. K. Fulbright, *J. Neurosci.*, 2014, **34**, 4082–4089.
115 N. Preuss, B. Salehi, J. W. van der Veen, J. Shen, W. C. Drevets, C. Hodgkinson, D. Goldman and G. Hasler, *Int. J. Neuropsychopharmacol.*, 2013, **16**, 1707–1717.
116 B. W. Krause, S. A. Wijtenburg, H. H. Holcomb, P. Kochunov, D. J. J. Wang, L. E. Hong and L. M. Rowland, *Neurosci. Lett.*, 2014, **561**, 188–191.
117 M. J. Donahue, S. Rane, E. Hussey, E. Mason, S. Pradhan, K. W. Waddell and B. A. Ally, *J. Cereb. Blood Flow Metab.*, 2014, **34**, 532–541.
118 M. Tomiyasu, N. Aida, M. Endo, J. Shibasaki, K. Nozawa, E. Shimizu, H. Tsuji and T. Obata, *PLoS One*, 2013, **8**, e82746.
119 M. M. Silveri, J. T. Sneider, D. J. Crowley, M. J. Covell, D. Acharya, I. M. Rosso and J. E. Jensen, *Biol. Psychiatry*, 2013, **74**, 296–304.

120 F. Gao, R. A. E. Edden, M. Li, N. A. J. Puts, G. Wang, C. Liu, B. Zhao, H. Wang, X. Bai, C. Zhao, X. Wang and P. B. Barker, *NeuroImage*, 2013, **78**, 75–82.
121 A. Marsman, R. C. W. Mandl, M. P. van den Heuvel, V. O. Boer, J. P. Wijnen, D. W. J. Klomp, P. R. Luijten and H. E. H. Pol, *Eur. Neuropsychopharmacol.*, 2013, **23**, 1484–1490.
122 P.-W. Chiu, H. K.-F. Mak, K. K.-W. Yau, Q. Chan, R. C.-C. Chang and L.-W. Chu, *Age*, 2014, **36**, 251–264.
123 N. Tunc-Skarka, S. Meier, T. Demirakca, M. Sack, W. Weber-Fahr, W. Brusniak, I. Wolf, F. Matthaeus, T. G. Schulze, C. Diener and G. Ende, *NMR Biomed.*, 2014, **27**, 228–234.
124 J. J. Gomar, M. L. Gordon, D. Dickinson, P. B. Kingsley, A. M. Ulug, L. Keehlisen, S. Huet, J. J. Buthorn, J. Koppel, E. Christen, C. Conejero-Goldberg, P. Davies and T. E. Goldberg, *Biol. Psychiatry*, 2014, **75**, 686–692.
125 X. Ge, X.-y. Xu, C.-h. Feng, Y. Wang, Y.-l. Li and B. Feng, *BMC Neurol.*, 2013, **13**, 110.
126 K. Kantarci, S. D. Weigand, S. A. Przybelski, G. M. Preboske, V. S. Pankratz, P. Vemuri, M. L. Senjem, M. C. Murphy, J. L. Gunter, M. M. Machulda, R. J. Ivnik, R. O. Roberts, B. F. Boeve, W. A. Rocca, D. S. Knopman, R. C. Petersen and C. R. Jack Jr., *Neurology*, 2013, **81**, 126–133.
127 S. D. Friedman, L. D. Baker, S. Borson, J. E. Jensen, S. M. Barsness, S. Craft, G. R. Merriam, R. K. Otto, E. J. Novotny and M. V. Vitiello, *JAMA Neurol.*, 2013, **70**, 883–890.
128 C. Gasparovic, J. Prestopnik, J. Thompson, S. Taheri, B. Huisa, R. Schrader, J. C. Adair and G. A. Rosenberg, *J. Neurol., Neurosurg. Psychiatry*, 2013, **84**, 715–721.
129 M. G. Targosz-Gajniak, J. S. Siuda, M. M. Wicher, T. J. Banasik, M. A. Bujak, A. M. Augusciak-Duma and G. Opala, *J. Neurol. Sci.*, 2013, **335**, 58–63.
130 Y.-C. Yeh, C.-F. Yen, C.-W. Li, Y.-T. Kuo, C.-H. Chen, C.-C. Lee, G.-C. Liu, M.-F. Huang, T.-L. Liu and C.-S. Chen, *Int. Psychogeriatr.*, 2014, **26**, 393–402.
131 W. Gaetz, L. Bloy, D. J. Wang, R. G. Port, L. Blaskey, S. E. Levy and T. P. L. Roberts, *NeuroImage*, 2014, **86**, 1–9.
132 D. C. Rojas, D. Singel, S. Steinmetz, S. Hepburn and M. S. Brown, *NeuroImage*, 2014, **86**, 28–34.
133 J. Horder, T. Lavender, M. A. Mendez, R. O'Gorman, E. Daly, M. C. Craig, D. J. Lythgoe, G. J. Barker and D. G. Murphy, *Transl. Psychiatry*, 2013, **3**, e239.
134 B. A. Golomb, L. C. Erickson, A. A. Scott-Van Zeeland, S. Koperski, R. H. Haas, D. C. Wallace, R. K. Naviaux, A. J. Lincoln, G. E. Reiner and G. Hamilton, *J. Child Neurol.*, 2014, **29**, 187–193.
135 W. Weber-Fahr, S. Englisch, A. Esser, N. Tunc-Skarka, A. Meyer-Lindenberg, G. Ende and M. Zink, *Psychiatry Res., Neuroimaging*, 2013, **214**, 365–373.
136 F. Granata, G. Pandolfo, S. Vinci, C. Alafaci, N. Settineri, R. Morabito, A. Pitrone and M. Longo, *Neuroradiol. J.*, 2013, **26**, 277–283.
137 M. A. Reid, N. V. Kraguljac, K. B. Avsar, D. M. White, J. A. den Hollander and A. C. Lahti, *Schizophrenia Res.*, 2013, **147**, 348–354.
138 J. R. Bustillo, H. Chen, T. Jones, N. Lemke, C. Abbott, C. Qualls, J. Canive and C. Gasparovic, *JAMA Psychiatry*, 2014, **71**, 265–272.
139 L. M. Rowland, K. Kontson, J. West, R. A. Edden, H. Zhu, S. A. Wijtenburg, H. H. Holcomb and P. B. Barker, *Schizophrenia Bull.*, 2013, **39**, 1096–1104.
140 A. Szulc, B. Konarzewska, B. Galinska-Skok, J. Lazarczyk, N. Waszkiewicz, E. Tarasow, R. Milewski and J. Walecki, *Neurosci. Lett.*, 2013, **547**, 37–41.
141 N. Tandon, N. R. Bolo, K. Sanghavi, I. T. Mathew, A. N. Francis, J. A. Stanley and M. S. Keshavan, *Schizophrenia Res.*, 2013, **148**, 59–66.

142 J. Xu, U. Dydak, J. Harezlak, J. Nixon, M. Dzemidzic, A. D. Gunn, H. S. Karne and A. Anand, *Psychiatry Res., Neuroimaging*, 2013, **213**, 235–241.

143 A. D. Gigante, B. Lafer and L. N. Yatham, *World J. Biol. Psychiatry*, 2014, **15**, 145–154.

144 B. R. Godlewska, S. W. Yip, J. Near, G. M. Goodwin and P. J. Cowen, *Psychopharmacology*, 2014, **231**, 327–332.

145 K. M. Chitty, J. Lagopoulos, I. B. Hickie and D. F. Hermens, *J. Affective Disord.*, 2013, **150**, 1238–1241.

146 M. K. Singh, B. Jo, N. E. Adleman, M. Howe, L. Bararpour, R. G. Kelley, D. Spielman and K. D. Chang, *Psychiatry Res., Neuroimaging*, 2013, **214**, 153–160.

147 D. G. Harper, J. E. Jensen, C. Ravichandran, Y. Sivrioglu, M. Silveri, D. V. Iosifescu, P. F. Renshaw and B. P. Forester, *Am. J. Geriatr. Psychiatry*, 2014, **22**, 499–509.

148 A. Ezzati, M. E. Zimmerman, M. J. Katz and R. B. Lipton, *Hippocampus*, 2013, **23**, 1137–1142.

149 K. A. B. Lapidus, V. Gabbay, X. Mao, A. Johnson, J. W. Murrough, S. J. Mathew and D. C. Shungu, *Neurosci. Lett.*, 2014, **569**, 74–79.

150 M. J. Taylor, B. Godlewska, J. Near, D. Christmas, J. Potokar, J. Collier, P. Klenerman, E. Barnes and P. J. Cowen, *Psychol. Med.*, 2014, **44**, 789–795.

151 I. M. Rosso, M. R. Weiner, D. J. Crowley, M. M. Silveri, S. L. Rauch and J. E. Jensen, *Depression Anxiety*, 2014, **31**, 115–123.

152 D. J. Meyerhoff, A. Mon, T. Metzler and T. C. Neylan, *Sleep*, 2014, **37**, 893–900.

153 Z. Long, C. Medlock, M. Dzemidzic, Y.-W. Shin, A. W. Goddard and U. Dydak, *Prog. Neuro-Psychopharmacol. Biol. Psychiatry*, 2013, **44**, 131–135.

154 Y.-W. Shin, M. Demidzic, H. J. Jo, Z. Long, C. Medlock, U. Dydak and A. W. Goddard, *J. Affective Disord.*, 2013, **150**, 1091–1095.

155 A. M. Weber, N. Soreni, J. A. Stanley, A. Greco, S. Mendlowitz, P. Szatmari, R. Schachar, K. Mannasis, P. Pires, R. Swinson and M. D. Noseworthy, *Psychiatry Res., Neuroimaging*, 2014, **222**, 67–74.

156 V. Husarova, M. Bittsansky, I. Ondrejka and D. Dobrota, *Psychiatry Res., Neuroimaging*, 2014, **222**, 75–83.

157 S.-H. Hsu, M.-C. Chou, C.-W. Ko, S.-S. Hsu, H.-S. Lin, J.-H. Fu, P.-C. Wang, H.-B. Pan and P.-H. Lai, *Eur. J. Radiol.*, 2013, **82**, 1299–1307.

158 M. J. Peluso, D. J. Meyerhoff, R. W. Price, J. Peterson, E. Lee, A. C. Young, R. Walter, D. Fuchs, B. J. Brew, P. Cinque, K. Robertson, L. Hagberg, H. Zetterberg, M. Gisslen and S. Spudich, *J. Infect. Dis.*, 2013, **207**, 1703–1712.

159 G. E. Utine, B. Akpinar, U. Arslan, P. O. S. Kiper, B. Volkan-Salanci, Y. Alanay, D. Aktas, G. Haliloglu, K. K. Oguz, K. Boduroglu and M. Alikasifoglu, *Am. J. Med. Genet., Part A*, 2014, **164**, 99–105.

160 J. L. Bruno, E. W. Shelly, E.-M. Quintin, M. Rostami, S. Patnaik, D. Spielman, D. Mayer, M. Gu, A. A. Lightbody and A. L. Reiss, *J. Neurodev. Disorders*, 2013, **5**, 20.

161 D. R. Carvalho, L. Farage, B. J. A. F. Martins, J. M. Brum, C. E. Speck-Martins and R. Pratesi, *J. Neuroimaging*, 2014, **24**, 155–160.

162 S. Pulai, A. Biswas, A. Roy, D. S. Guin, A. Pandit, G. Gangopadhyay, P. K. Ghorai, S. Sarkhel and A. K. Senapati, *Neurol. India*, 2014, **62**, 153–158.

163 P. Roos, K. Svenstrup, E. R. Danielsen, C. Thomsen and J. E. Nielsen, *Acta Neurol. Scand.*, 2014, **129**, 330–334.

164 A. A. K. A. Razek, A. Abdalla, N. A. Gaber, A. Fathy, A. Megahed, T. Barakat and M. A. L. Alsayed, *Eur. Radiol.*, 2013, **23**, 3005–3011.

165 Y. Kimura, N. Sato, K. Sugai, S. Maruyama, M. Ota, K. Kamiya, K. Ito, Y. Nakata, M. Sasaki and H. Sugimoto, *Brain Dev.*, 2013, **35**, 458–461.
166 H. Tada and J.-i. Takanashi, *Brain Dev.*, 2014, **36**, 57–60.
167 K. K. Moghadam, F. Pizza, C. Tonon, R. Lodi, V. Carelli, F. Poli, C. Franceschini, P. Barboni, M. Seri, S. Ferrari, C. La Morgia, C. Testa, F. Cornelio, R. Liguori, J. Winkelmann, L. Lin, E. Mignot and G. Plazzi, *Sleep Med.*, 2014, **15**, 582–585.
168 F. Nicita, C. Di Biasi, S. Sollaku, S. Cecchini, V. Salpietro, A. Pittalis, L. Papetti, F. Ursitti, F. Ulgiati, A. M. Zicari, G. F. Gualdi, E. Properzi, M. Duse, M. Ruggieri and A. Spalice, *Childs Nerv. Syst.*, 2014, **30**, 319–325.
169 M. Rango, A. Arighi, C. Bonifati, R. Del Bo, G. Comi and N. Bresolin, *J. Cereb. Blood Flow Metab.*, 2014, **34**, 915–920.
170 M. Morreale, P. Marchione, P. Giacomini, S. Pontecorvo, M. Marianetti, C. Vento, E. Tinelli and A. Francia, *PLoS One*, 2014, **9**, e84605.
171 P. Cagnoli, R. E. Harris, D. Frechtling, G. Berkis, R. H. Gracley, C. C. Graft, S. E. Lowe, T. L. Chenevert, W. J. McCune, S. Gebarski and P. C. Sundgren, *Acad. Radiol.*, 2013, **20**, 1286–1296.
172 A. Zimny, M. Szmyrka-Kaczmarek, P. Szewczyk, J. Bladowska, A. Pokryszko-Dragan, E. Gruszka, P. Wiland and M. Sasiadek, *Lupus*, 2014, **23**, 10–19.
173 B. Zhou, F. Yuan, Z. He and C. Tan, *Brain Imaging Behav.*, 2014, **8**, 97–101.
174 A. Groeger, B. Bender, I. Wurster, G. L. Chadzynski, U. Klose and D. Berg, *J. Neurol., Neurosurg. Psychiatry*, 2013, **84**, 644–649.
175 B. E. Levin, H. L. Katzen, A. Maudsley, J. Post, C. Myerson, V. Govind, F. Nahab, B. Scanlon and A. Mittel, *J. Neuroimaging*, 2014, **24**, 39–44.
176 N. Weiduschat, X. Mao, M. F. Beal, M. J. Nirenberg, D. C. Shungu and C. Henchcliffe, *Parkinsonism Related Disord.*, 2014, **20**, 545–548.
177 R. F. Casseb, A. D'Abreu, H. H. Ruocco, I. Lopes-Cendes, F. Cendes and G. Castellano, *Braz. J. Med. Biol. Res.*, 2013, **46**, 722–727.
178 J. M. Padowski, K. E. Weaver, T. L. Richards, M. Y. Laurino, A. Samii, E. H. Aylward and K. E. Conley, *Mov. Disord.*, 2014, **29**, 327–335.
179 Y. Zhang, X. Chen, G. Wen, G. Wu and X. Zhang, *PLoS One*, 2013, **8**, e73197.
180 J. W. Pan, R. B. Duckrow, J. Gerrard, C. Ong, L. J. Hirsch, S. R. Resor, Jr., Y. Zhang, O. Petroff, S. Spencer, H. P. Hetherington and D. D. Spencer, *Epilepsia*, 2013, **54**, 1668–1678.
181 W.-F. Peng, J. Ding, L.-Y. Mao, X. Li, L. Liang, C.-Z. Chen, W.-Z. Cheng, W. Fan and X. Wang, *Epilepsy Behav.*, 2013, **29**, 144–149.
182 J. Zhang, Q. Liu, S. Mei, X. Zhang, W. Liu, H. Chen, H. Xia, Z. Zhou, X. Wang and Y. Li, *Neuropsychiatr. Dis. Treat.*, 2014, **10**, 71–81.
183 M. Munakata, N. Togashi, O. Sakamoto, K. Haginoya, Y. Kobayashi, A. Onuma, K. Iinuma and S. Kure, *Tohoku J. Exp. Med.*, 2014, **232**, 277–283.
184 H. Kassem, A. Wafaie, S. Abdelfattah and T. Farid, *Eur. J. Radiol.*, 2014, **83**, 191–196.
185 T. Mori, K. Mori, H. Ito, A. Goji, M. Miyazaki, M. Harada, K. Kurosawa and S. Kagami, *J. Child Neurol.*, 2014, **29**, 283–288.
186 J. Finnsson, A. Melberg and R. Raininko, *Neuroradiology*, 2013, **55**, 933–939.
187 B. Bellenberg, M. Busch, N. Trampe, R. Gold, A. Chan and C. Lukas, *Eur. Radiol.*, 2013, **23**, 3379–3392.
188 C. Tur, C. A. M. Wheeler-Kingshott, D. R. Altmann, D. H. Miller, A. J. Thompson and O. Ciccarelli, *Human Brain Mapping*, 2014, **35**, 993–1003.
189 A. Pokryszko-Dragan, J. Bladowska, A. Zimny, K. Slotwinski, M. Zagrajek, E. Gruszka, M. Bilinska, M. Sasiadek and R. Podemski, *J. Neurol. Sci.*, 2014, **339**, 35–40.

190 L. K. Vingara, H. J. Yu, M. E. Wagshul, D. Serafin, C. Christodoulou, I. Pelczer, L. B. Krupp and M. Maletic-Savatic, *NeuroImage*, 2013, **82**, 586–594.
191 M. Aradi, A. Schwarcz, G. Perlaki, G. Orsi, N. Kovacs, A. Trauninger, D. O. Kamson, S. Erdelyi-Botor, F. Nagy, S. A. Nagy, T. Doczi, S. Komoly and Z. Pfund, *Headache*, 2013, **53**, 752–763.
192 R. P. Allen, P. B. Barker, A. Horska and C. J. Earley, *Neurology*, 2013, **80**, 2028–2034.
193 S. Sivak, M. Bittsansky, J. Grossmann, V. Nosal, E. Kantorova, J. Sivakova, A. Demkova, P. Hnilicova, D. Dobrota and E. Kurca, *Brain Injury*, 2014, **28**, 341–346.
194 A. S. Kierans, I. I. Kirov, O. Gonen, G. Haemer, E. Nisenbaum, J. S. Babb, R. I. Grossman and Y. W. Lui, *Neurology*, 2014, **82**, 521–528.
195 S. Dhandapani, A. Sharma, K. Sharma and L. Das, *J. Clin. Neurosci.*, 2014, **21**, 745–750.
196 E. Chamard, M. Lassonde, L. Henry, J. Tremblay, Y. Boulanger, L. De Beaumont and H. Theoret, *Brain Injury*, 2013, **27**, 1038–1046.
197 S. Tremblay, L. De Beaumont, L. C. Henry, Y. Boulanger, A. C. Evans, P. Bourgouin, J. Poirier, H. Theoret and M. Lassonde, *Cereb. Cortex*, 2013, **23**, 1159–1166.
198 A.-Q. Lin, J.-X. Shou, X.-Y. Li, L. Ma and X.-H. Zhu, *Exp. Ther. Med.*, 2014, **7**, 451–455.
199 M. Zhang, X. Sun, Z. Zhang, Q. Meng, Y. Wang, J. Chen, X. Ma, H. Geng and L. Sun, *Int. J. Neurosci.*, 2014, **124**, 37–41.
200 S. K. Goergen, H. Ang, F. Wong, E. A. Carse, M. Charlton, R. Evans, G. Whiteley, J. Clark, D. Shipp, D. Jolley, E. Paul and J. L. Y. Cheong, *Clin. Radiol.*, 2014, **69**, 72–81.
201 G. Ancora, C. Testa, S. Grandi, C. Tonon, F. Sbravati, S. Savini, D. N. Manners, L. L. Gramegna, G. Tani, E. Malucelli, L. T. Corvaglia, G. Faldella and R. Lodi, *Neuroradiology*, 2013, **55**, 1017–1025.
202 N. H. Thakur, A. J. Spencer, H. W. Kilbride and L. H. Lowe, *Southern Med. J.*, 2013, **106**, 350–355.
203 A. Alkan, R. Sharifov, M. E. Akkoyunlu, R. Kilicarslan, H. Toprak, A. Aralasmak and L. Kart, *Clin. Imaging*, 2013, **37**, 989–992.
204 L. Story, M. S. Damodaram, V. Supramaniam, J. M. Allsop, A. McGuinness, A. Patel, M. Wylezinska, S. Kumar and M. A. Rutherford, *Eur. J. Obstet. Gynecol. Reprod. Biol.*, 2013, **170**, 77–81.
205 W. Wu, P. Zhang, X. Wang, A. Chineah and M. Lou, *J. Magn. Reson. Imaging*, 2013, **38**, 634–640.
206 S. Mangia, A. F. Kumar, A. A. Moheet, R. J. Roberts, L. E. Eberly, E. R. Seaquist and I. Tkac, *J. Cereb. Blood Flow Metab.*, 2013, **33**, 754–759.
207 M. Karczewska-Kupczewska, E. Tarasow, A. Nikolajuk, M. Stefanowicz, N. Matulewicz, E. Otziomek, M. Gorska, M. Straczkowski and I. Kowalska, *Diabetes Care*, 2013, **36**, 2787–2793.
208 A. P. Haley, M. M. Gonzales, T. Tarumi and H. Tanaka, *Obesity*, 2013, **21**, 2007–2013.
209 E. Resmini, A. Santos, B. Gomez-Anson, O. Lopez-Mourelo, P. Pires, Y. Vives-Gilabert, I. Crespo, M. J. Portella, M. de Juan-Delago and S. M. Webb, *Clin. Endocrinol.*, 2013, **79**, 700–707.
210 S. Yabuki, S.-i. Konno and S.-i. Kikuchi, *J. Orthop. Sci.*, 2013, **18**, 363–368.
211 N. Salamon, B. M. Ellingson, R. Nagarajan, N. Gebara, A. Thomas and L. T. Holly, *Spinal Cord*, 2013, **51**, 558–563.
212 L. Banks, G. D. Wells and B. W. McCrindle, *Appl. Physiol., Nutr., Metab.*, 2014, **39**, 363–368.

213 R. Spoladore, G. Fragasso, G. Perseghin, F. De Cobelli, A. Esposito, F. Maranta, G. Calori, M. Locatelli, G. Lattuada, P. Scifo, A. Del Maschio and A. Margonato, *Fundam. Clin. Pharmacol.*, 2013, **27**, 455–464.

214 A. M. Betim Paes Leme, V. M. Cury Salemi, R. G. Weiss, J. R. Parga, B. M. Ianni, C. Mady and R. Kalil-Filho, *J. Card. Failure*, 2013, **19**, 454–460.

215 P. A. Bottomley, G. S. Panjrath, S. Lai, G. A. Hirsch, K. Wu, S. S. Najjar, A. Steinberg, G. Gerstenblith and R. G. Weiss, *Sci. Transl. Med.*, 2013, **5**, 215re3.

216 M. Oresic, T. Hyotylainen, A. Kotronen, P. Gopalacharyulu, H. Nygren, J. Arola, S. Castillo, I. Mattila, A. Hakkarainen, R. J. H. Borra, M.-J. Honka, A. Verrijken, S. Francque, P. Iozzo, M. Leivonen, N. Jaser, A. Juuti, T. I. A. Sorensen, P. Nuutila, L. Van Gaal and H. Yki-Jarvinen, *Diabetologia*, 2013, **56**, 2266–2274.

217 K. Cusi, Z. Chang, S. Harrison, R. Lomonaco, F. Bril, B. Orsak, C. Ortiz-Lopez, J. Hecht, A. E. Feldstein, A. Webb, C. Louden, M. Goros and F. Tio, *J. Hepatol.*, 2014, **60**, 167–174.

218 J. L. Martin-Rodriguez, J. P. Arrebola, J. J. Jimenez-Moleon, N. Olea and J. L. Gonzalez-Calvin, *Eur. J. Gastroenterol. Hepatol.*, 2014, **26**, 88–94.

219 T. Karlas, D. Petroff, N. Garnov, S. Boehm, H. Tenckhoff, C. Wittekind, M. Wiese, I. Schiefke, N. Linder, A. Schaudinn, H. Busse, T. Kahn, J. Moessner, T. Berg, M. Troeltzsch, V. Keim and J. Wiegand, *PLoS One*, 2014, **9**, e91987.

220 C.-H. Wu, M.-C. Ho, Y.-M. Jeng, C.-Y. Hsu, P.-C. Liang, R.-H. Hu, H.-S. Lai and T. T.-F. Shih, *J. Gastroenterol. Hepatol.*, 2014, **29**, 807–813.

221 M. Noureddin, J. Lam, M. R. Peterson, M. Middleton, G. Hamilton, L. Thuy-Anh, R. Bettencourt, C. Changchien, D. A. Brenner, C. Sirlin and R. Loomba, *Hepatology*, 2013, **58**, 1930–1940.

222 J. M. Abrigo, J. Shen, V. W. S. Wong, D. K. W. Yeung, G. L. H. Wong, A. M. L. Chim, A. W. H. Chan, P. C. L. Choi, F. K. L. Chan, H. L. Y. Chan and W. C. W. Chu, *J. Hepatol.*, 2014, **60**, 809–815.

223 K.-i. Kato, T. Takamura, Y. Takeshita, Y. Ryu, H. Misu, T. Ota, K. Tokuyama, S. Nagasaka, M. Matsuhisa, O. Matsui and S. Kaneko, *PLoS One*, 2014, **9**, e92170.

224 D. M. Lebensztejn, I. Bialokoz-Kalinowska, M. Klusek-Oksiuta, E. Tarasow, M. Wojtkowska and M. Kaczmarski, *Adv. Med. Sci.*, 2014, **59**, 81–84.

225 S. Forbes, I. F. Godsland, S. D. Taylor-Robinson, J. D. Bell, E. L. Thomas, N. Patel, G. Hamilton, K. H. Parker, I. Marshall, C. D. Gray, D. Bedford, M. Caslake, B. R. Walker and D. G. Johnston, *Diabetologia*, 2013, **56**, 2021–2033.

226 H.-J. Zhang, X.-F. Zhang, Z.-M. Ma, L.-L. Pan, Z. Chen, H.-W. Han, C.-K. Han, X.-J. Zhuang, Y. Lu, X.-J. Li, S.-Y. Yang and X.-Y. Li, *J. Hepatol.*, 2013, **59**, 557–562.

227 H. Bian, A. Hakkarainen, N. Lundbom and H. Yki-Jarvinen, *Obesity*, 2014, **22**, 989–995.

228 M. C. Ryan, C. Itsiopoulos, T. Thodis, G. Ward, N. Trost, S. Hofferberth, K. O'Dea, P. V. Desmond, N. A. Johnson and A. M. Wilson, *J. Hepatol.*, 2013, **59**, 138–143.

229 M. Ryberg, S. Sandberg, C. Mellberg, O. Stegle, B. Lindahl, C. Larsson, J. Hauksson and T. Olsson, *J. Intern. Med.*, 2013, **274**, 67–76.

230 V. S. Sprung, H. Jones, C. J. A. Pugh, N. F. Aziz, C. Daousi, G. J. Kemp, D. J. Green, N. T. Cable and D. J. Cuthbertson, *Clin. Sci.*, 2014, **126**, 67–74.

231 E. Phielix, A. Brehm, E. Bernroider, M. Krssak, C. H. Anderwald, M. Krebs, A. I. Schmid, P. Nowotny and M. Roden, *Diabetes, Obes. Metab.*, 2013, **15**, 915–922.

232 M. C. Stephenson, E. Leverton, E. Y. H. Khoo, S. M. Poucher, L. Johansson, J. A. Lockton, J. W. Eriksson, P. Mansell, P. G. Morris and I. A. MacDonald, *NMR Biomed.*, 2013, **26**, 1518–1526.
233 M. D. Nelson, R. G. Victor, E. W. Szczepaniak, V. Simha, A. Garg and L. S. Szczepaniak, *Am. J. Cardiol.*, 2013, **112**, 1019–1024.
234 J. J. Noble, S. F. Keevil, J. Totman and G. D. Charles-Edwards, *Br. J. Radiol.*, 2014, **87**, 20130761.
235 H. K. Kim, S. Serai, A. C. Merrow, L. Wang, P. S. Horn and T. Laor, *Pediatr. Radiol.*, 2014, **44**, 149–157.
236 A. Laufs, R. Livingstone, B. Nowotny, P. Nowotny, F. Wickrath, G. Giani, J. Bunke, M. Roden and J.-H. Hwang, *Magn. Reson. Med.*, 2014, **71**, 1670–1675.
237 V. Ahluwalia, J. B. Wade, L. Thacker, K. A. Kraft, R. K. Sterling, R. T. Stravitz, M. Fuchs, I. Bouneva, P. Puri, V. Luketic, A. J. Sanyal, H. Gilles, D. M. Heuman and J. S. Bajaj, *J. Hepatol.*, 2013, **59**, 467–473.
238 G. Layec, L. J. Haseler, J. Hoff, C. R. Hart, X. Liu, Y. Le Fur, E. K. Jeong and R. S. Richardson, *Acta Physiol.*, 2013, **208**, 376–386.
239 W. Chidnok, F. J. DiMenna, J. Fulford, S. J. Bailey, P. F. Skiba, A. Vanhatalo and A. M. Jones, *Am. J. Physiol.: Regul., Integr. Comp. Physiol.*, 2013, **305**, R1085–R1092.
240 T. Yoshida, D. Abe and Y. Fukuoka, *Scand. J. Med. Sci. Sports*, 2013, **23**, E313–E319.
241 J. Kappenstein, A. Ferrauti, B. Runkel, J. Fernandez-Fernandez, K. Mueller and J. Zange, *Eur. J. Appl. Physiol.*, 2013, **113**, 2769–2779.
242 S. Ratel, A. Tonson, P. J. Cozzone and D. Bendahan, *Eur. J. Appl. Physiol.*, 2014, **114**, 217–218.
243 J. Kappenstein, A. Ferrauti, B. Runkel, J. Fernandez-Fernandez, K. Mueller and J. Zange, *Eur. J. Appl. Physiol.*, 2014, **114**, 219–220.
244 L. Valkovic, B. Ukropcova, M. Chmelik, M. Balaz, W. Bogner, A. I. Schmid, I. Frollo, E. Zemkova, I. Klimes, J. Ukropec, S. Trattnig and M. Krssak, *NMR Biomed.*, 2013, **26**, 1714–1722.
245 G. Layec, L. J. Haseler and R. S. Richardson, *Age*, 2013, **35**, 1183–1192.
246 G. Layec, L. J. Haseler, J. D. Trinity, C. R. Hart, X. Liu, Y. Le Fur, E.-K. Jeong and R. S. Richardson, *J. Appl. Physiol.*, 2013, **115**, 803–811.
247 J. W. M. van Oorschot, J. P. J. Schmitz, A. Webb, K. Nicolay, J. A. L. Jeneson and H. E. Kan, *PLoS One*, 2013, **8**, e76628.
248 J. Ren, A. D. Sherry and C. R. Malloy, *Magn. Reson. Med.*, 2013, **70**, 610–619.
249 F. Kogan, M. Haris, A. Singh, K. Cai, C. Debrosse, R. P. R. Nanga, H. Hariharan and R. Reddy, *Magn. Reson. Med.*, 2014, **71**, 164–172.
250 A. I. Schmid, K. Schewzow, G. B. Fiedler, S. Goluch, E. Laistler, M. Wolzt, E. Moser and M. Meyerspeer, *NMR Biomed.*, 2014, **27**, 553–560.
251 N. Agarwal, G. Pages, L. D' Silva, N. Said, P. W. Kuchel and S. S. Velan, *Magn. Reson. Imaging*, 2014, **32**, 379–384.
252 E. L. MacMillan, C. Boesch and R. Kreis, *Magn. Reson. Med.*, 2013, **70**, 916–924.
253 T. Bex, W. Chung, A. Baguet, S. Stegen, J. Stautemas, E. Achten and W. Derave, *J. Appl. Physiol.*, 2014, **116**, 204–209.
254 T. E. Ryan, W. M. Southern, M. A. Reynolds and K. K. McCully, *J. Appl. Physiol.*, 2013, **115**, 1757–1766.
255 M. J. Slattery, M. A. Bredella, H. Thakur, M. Torriani and M. Misra, *Metab. Syndr. Relat. Disord.*, 2014, **12**, 56–61.
256 G. Layec, E. Malucelli, Y. Le Fur, D. Manners, K. Yashiro, C. Testa, P. J. Cozzone, S. Iotti and D. Bendahan, *NMR Biomed.*, 2013, **26**, 1403–1411.

257 B. Gerdle, M. F. Forsgren, A. Bengtsson, O. D. Leinhard, B. Soren, A. Karlsson, V. Brandejsky, E. Lund and P. Lundberg, *Eur. J. Pain*, 2013, **17**, 1205–1215.
258 A. C. Phillips, A. Sleigh, C. J. McAllister, S. Brage, T. A. Carpenter, G. J. Kemp and A. J. Holland, *PLoS One*, 2013, **8**, e84031.
259 W. Nachbauer, S. Boesch, R. Schneider, A. Eigentler, J. Wanschitz, W. Poewe and M. Schocke, *PLoS One*, 2013, **8**, e69229.
260 B. A. I. Payne, K. G. Hollingsworth, J. Baxter, E. Wilkins, V. Lee, D. A. Price, M. Trenell and P. F. Chinnery, *PLoS One*, 2014, **9**, e84678.
261 H. J. Koslik, G. Hamilton and B. A. Golomb, *PLoS One*, 2014, **9**, e92887.
262 S. Sohlberg, A. K. Wikstrom, M. Olovsson, P. Lindgren, O. Axelsson, A. Mulic-Lutvica, J. Weis and J. Wikstrom, *Placenta*, 2014, **35**, 318–323.
263 S. K. Gill, M. Wilson, N. P. Davies, L. MacPherson, M. English, T. N. Arvanitis and A. C. Peet, *Neuro-Oncology*, 2014, **16**, 156–164.
264 E. Orphanidou-Vlachou, D. Auer, M. A. Brundler, N. P. Davies, T. Jaspan, L. MacPherson, K. Natarajan, Y. Sun, T. N. Arvanitis, R. G. Grundy and A. C. Peet, *Eur. J. Radiol.*, 2013, **82**, E295–E301.
265 J. Novaka, M. Wilson, L. MacPherson, T. N. Arvanitis, N. P. Davies and A. C. Peet, *Eur. J. Radiol.*, 2014, **83**, E106–E112.
266 R. Saito, T. Kumabe, M. Kanamori, Y. Sonoda, M. Watanabe, S. Mugikura, S. Takahashi and T. Tominaga, *J. Clin. Neurosci.*, 2014, **21**, 124–130.
267 D. J. Jaskolski, J. Fortuniak, A. Majos, W. Gajewicz, W. Papierz, P. P. Liberski, B. Sikorska and L. Stefanczyk, *Neurol. Neurochir. Pol.*, 2013, **47**, 438–449.
268 S. Bisdas, R. Ritz, B. Bender, C. Braun, C. Pfannenberg, M. Reimold, T. Naegele and U. Ernemann, *Invest. Radiol.*, 2013, **48**, 295–301.
269 A. Seeger, C. Braun, M. Skardelly, F. Paulsen, J. Schittenhelm, U. Ernemann and S. Bisdas, *Acad. Radiol.*, 2013, **20**, 1557–1565.
270 C. Choi, S. Ganji, K. Hulsey, A. Madan, Z. Kovacs, I. Dimitrov, S. Zhang, K. Pichumani, D. Mendelsohn, B. Mickey, C. Malloy, R. Bachoo, R. DeBerardinis and E. Maher, *NMR Biomed.*, 2013, **26**, 1242–1250.
271 E. Tsolaki, P. Svolos, E. Kousi, E. Kapsalaki, K. Fountas, K. Theodorou and I. Tsougos, *Int. J. Comput. Assist. Radiol. Surg.*, 2013, **8**, 751–761.
272 D. R. Balos, S. Gavrilovic, S. Lavrnic, B. Vasic, M. Macvanski, D. Damjanovic and T. S. Opincal, *Vojnosanit. Pregl.*, 2013, **70**, 637–644.
273 P. J. Rao, R. Jyoti, P. J. Mews, P. Desmond and V. G. Khurana, *Br. J. Neurosurg.*, 2013, **27**, 646–653.
274 F. Imani, F. E. Boada, F. S. Lieberman, D. K. Davis and J. M. Mountz, *Eur. J. Radiol.*, 2014, **83**, E100–E105.
275 S. S. Lu, S. J. Kim, H. S. Kim, C. G. Choi, Y. M. Lim, E. J. Kim, D. Y. Kim and S. H. Cho, *Am. J. Neuroradiol.*, 2014, **35**, 270–277.
276 B. L. Stehouwer, W. J. M. van der Kemp, P. R. Luijten, M. A. A. J. van den Bosch, W. B. Veldhuis, J. P. Wijnen and D. W. J. Klomp, *Breast Cancer Res. Treat.*, 2014, **144**, 583–589.
277 K. Vassiou, I. Tsougos, E. Kousi, M. Vlychou, E. Athanasiou, K. Theodorou, D. L. Arvanitis and I. V. Fezoulidis, *Acta Radiol.*, 2013, **54**, 380–388.
278 I. Basara, S. Orguc and T. Coskun, *Diagn. Interventional Radiol.*, 2013, **19**, 221–226.
279 S. Suppiah, K. Rahmat, M. N. Mohd-Shah, C. A. Azlan, L. K. Tan, Y. F. A. Aziz, A. Vijayananthan, A. L. Wui and C. H. Yip, *Clin. Radiol.*, 2013, **68**, e502–e510.
280 B. Battal, Y. Bozkurt, S. Hamcan and V. Akgun, *Clin. Radiol.*, 2014, **69**, e108–e109.

281 S. Suppiah, K. Rahmat, F. I. Rozalli and C. A. Azlan, *Clin. Radiol.*, 2014, **69**, e110–e111.

282 M. Takeuchi, K. Matsuzaki and M. Harada, *Eur. Radiol.*, 2013, **23**, 2358–2363.

283 S. T. Chang, A. C. Westphalen, P. Jha, A. J. Jung, P. R. Carroll, J. Kurhanewicz and F. V. Coakley, *J. Magn. Reson. Imaging*, 2014, **39**, 519–525.

284 T. H. Kuru, M. C. Roethke, P. Rieker, W. Roth, M. Fenchel, M. Hohenfellner, H.-P. Schlemmer and B. A. Hadaschik, *BJU Int.*, 2013, **112**, 1080–1087.

285 A. Petrillo, R. Fusco, S. V. Setola, F. M. Ronza, V. Granata, M. Petrillo, G. Carone, M. Sansone, R. Franco, F. Fulciniti and S. Perdona, *J. Magn. Reson. Imaging*, 2014, **39**, 1206–1212.

286 J. Weis, F. Ortiz-Nieto and H. Ahlstrom, *Magn. Reson. Med. Sci.*, 2013, **12**, 289–296.

287 J. Zhang, K. Cheng, Y. Ding, W. Liang, D. Vanel and X. Cheng, *Eur. J. Radiol.*, 2013, **82**, 2124–2128.

Oriented molecules

Sandeep Kumar Mishra and N. Suryaprakash*
DOI: 10.1039/9781782622758-00547

Over the years number of books, comprehensive reviews and biennial reports have been published on the study of oriented molecules using NMR spectroscopy. The present specialists periodical report provides a short account of the published works, on the NMR spectroscopic investigations of oriented molecules during the biennial period July 2012 to June 2014. The report deviates significantly from the earlier volumes and the topics covered include studies of small molecules oriented in liquid crystals, new aligning media, new NMR techniques, weak ordering in membrane mimicking environment, measurement of residual dipolar couplings in lipids and membranes, quantum computing, synthesis and characterization of new liquid crystal materials, general studies on liquid crystals and the related works. The report gives the salient points of the published works in one or two sentences, most often borrowed directly from the cited papers, which obviously will not be exhaustive and clearly communicative. Readers are thus requested to visit the original cited published papers for getting deeper insights into the very important contributions made by the authors. The literature on the NMR spectroscopy of oriented molecules is vast and it is very likely that many papers might have missed our attention.

1 Introduction

The comprehensive report of works published during two years period ending June 2014 on the study of oriented molecules using NMR spectroscopy is being regularly updated in number of earlier volumes of this biennial series.[1,2] As is well known the study of the oriented molecules using NMR spectroscopy took a phenomenal growth soon after the initial and the seminal work of Alfred Saupe.[3–5] Over the years number of books, comprehensive reviews and biennial reports provided details on the account of the literature available, growth of methodology, structure and orientational parameters of small molecules aligned in liquid crystalline media, the development of NMR experimental methodologies to analyze the complex spectra dominated by dipolar couplings. It is impossible and also not the aim of the present abridged report to cite all the published work in the vast literature available. A sequential walk through the earlier specialist biennial periodical reports and some of the recent reviews, books and book chapters will provide a complete access to the readers about published work in the area of oriented molecules.[6–9] In this report, an attempt is made to give the excerpts of the published works on the study of oriented molecules using NMR spectroscopy during the period July 2012 to June 2014. From the vast literature available, it is very likely that many papers might have missed our attention. Furthermore, wherever possible the report gives the important points directly borrowed from the cited papers only in one or two sentences, which obviously will not be exhaustive and clearly communicative. We therefore appeal to the

NMR Research Centre and Solid State and Structural Chemistry Unit, Indian Institute of Science, Bangalore 560 012, India. E-mail: nsp@nrc.iisc.ernet.in

readers to visit the original published papers for getting more insights into the important contributions made by the authors.

During the initial days of its discovery numerous studies were carried out by different groups, focusing mainly on the determination of geometrical and orientational parameters of small molecules aligned in liquid crystals. As a consequence a wealth of information is available on the structural and orientational parameters derived by the partially averaged dipolar couplings obtained by the analyses of NMR spectra of small molecules embedded in strongly ordered thermotropic or lyotropic liquid crystals. The NMR spectrum of molecules embedded in liquid crystals, being dominated by dipolar couplings is enormously complex with number of detected trasitions arising due to large number of pairwise interaction of nuclear spins. As an example for N interacting spin 1/2 nuclei, the number of theoretically allowed transitions is given by ${}^{2N}C_{N-1}$. It implies that the ^{1}H NMR spectrum of molecules aligned in liquid crystals become highly complex with the increase in the number of interacting spins. The first order analyses of the spectra similar to that of liquids are generally not possible due to large strengths of homonuclear proton–proton couplings, and one has to diagonalize the Hamiltonian numerically, adapting the least-square-fit techniques using computers. In this direction number of iterative computer programs was developed for the analyses of the spectra either by interactively or automatically. There was also number of one and two dimensional NMR experimental strategies developed to simplify the spectral complexity or to derive the parameters in an orchestrated manner. Nevertheless the studies were generally restricted to small molecules containing very few interacting spins. In the recent times the interest in the area of determination of structure and orientational parameters of small molecules aligned in liquid crystalline media has substantially reduced.

There is enormous shift in the research interests towards the determination of structure and conformational analyses of biomolecules in aqueous media. The discovery of bicelles revolutionized the area of research in low ordered media. The anisotropic NMR parameters, such as, reduced dipolar couplings (RDCs), chemical shift anisotropies, though reflected in the NMR spectrum, retained the first order characteristic of the spectrum, enabling the straightforward analysis. This paved the way for the analysis of bigger molecules aligned in such weakly orienting media. The combined utility of RDCs measured from such weak aligning medium together with the nOe restraints aided the determination of structure and conformation of biomolecules. Several water and organic solvent compatible orienting media with very low order parameters have been reported. Some of these media are also chiral and are able to impart differential orientational order on the enantiomeric pairs. This brings into effect the anisotropic NMR interactions, such as, the dipolar and the quadrupolar couplings as well as the chemical shift anisotropy, making enantiomeric discrimination and quantification of excess of one form over the other relatively easy. Another area which has been of significant interest over the years is the study of oriented lipids and membranes and molecules embedded in them. The dipolar coupled spins have also been used as quantum bits for NMR quantum computational studies. With large number of experimental, computational and theoretical works, review articles, new liquid crystals being reported, the area of NMR spectroscopy of oriented molecules continues to be an active area of research. Thus the present report is broadly classified into different subsections, depending on the various published works.

The classification of the topics reported differs from the earlier volumes and the available literature is presented under the following headings: Studies of Small Molecules Oriented in liquid crystals, New Aligning Media, New NMR Techniques, Weak Ordering in membrane mimicking environment, Measurement of Residual Dipolar Couplings in lipids and membranes, Quantum Computing, Synthesis and characterization of new liquid crystal materials, General studies on Liquid Crystals and the related works. Efforts have been made to put the relevant publications under these categories. However, more often, there could be large overlap of research results of many areas and hence such publications have been mentioned under most relevant area.

2 Studies on small molecules aligned in liquid crystals

As discussed previously, the interests in this area of research is substantially reduced and only few groups are working. The r_α structure of C_6F_5Br is determined by dissolving it in liquid crystal ZLI-1695. The extraction of ^{13}C–^{19}F dipolar couplings from the ^{13}C spectrum is more advantageous compared to those determined from ^{13}C satellites in ^{19}F spectrum.[10] The ^{1}H NMR spectra of ethane, propane, and *n*-butane codissolved in two liquid-crystal solvents that exhibit both nematic (N) and smectic A (SmA) phases (one of which also has a reentrant nematic (RN) phase) are analyzed using covariance-matrix adaptation evolution strategy.[11] The NMR spectra of *n*-pentane are obtained in the liquid crystal 5CB at several temperatures. Predicted dipolar couplings by the atomistic molecular dynamics simulations are compared with the NMR experimentally derived ones. The simulation predictions provide an excellent starting point for analysis of the experimental NMR spectra using the covariance matrix adaptation evolutionary strategy.[12] The orientational order properties of two nematogens containing a fluoro- and isothiocyanate-substituted biphenyl moiety have been investigated by means of ^{13}C NMR spectroscopy.[13] The location, orientation and dynamics of two molecules, 4,4-dimethyl-5,8-dihydroanthracene-1,9,10(4*H*)-trione, and 9,10-dihydroxy-4,4-dimethyl-5,8-dihydro-1(4*H*)-anthracenone, have been studied by ^{2}H NMR by dissolving them in discotic nematic lyotropic liquid crystal.[14] Solute molecules dissolved in the liquid crystal 4-cyano-4′-*n*-octyloxybiphenyl (8OCB), known to form a partial bilayer smectic-A phase. Using orientational order parameters and statistical thermodynamic density functional theory, values of the solvent's positional order parameters and solutes' positional-orientational distribution functions are obtained.[15] Solid state ^{19}F-NMR studies has been carried out on three different hexafluorinated druglike compounds containing CF_3 groups, dissolved in a lipid membrane. From the observed dipolar couplings the molecular orientational order is determined.[16] The alignment of biphenyl in the liquid-crystal phases of 4-*n*-pentyl-4″-cyanobiphenyl (5CB) and 4-*n*-octyl-4′-cyanobiphenyl (8CB) is investigated by using a combination of predictive atomistic molecular dynamics simulations and ^{1}H RDC measurements.[17] A ^{2}H NMR and thermodynamics study on the insertion, distribution, interactions, and thermodynamics of two biologically active molecules, 9,10-dihydroxy-4,4-dimethyl-1,4,5,8-tetrahydroanthracen-1-one, with anticancer activity, and 4,4-dimethyl-1,4,5,8,9,10-hexahydroanthracen-1,9,10-trione a fungicide, dissolved in a nematic discotic lyotropic liquid crystal is reported.[18] NMR study of a bimesogenic liquid crystal with two nematic phases is carried out by dissolving the rigid solutes *para*-, *meta*- and *ortho*-dichlorobenzene. The nematic potential that is modelled by two independent Maier–Saupe terms is successful in fitting the observed dipolar couplings in both nematic phases of 39 wt.% 4-*n*-pentyl-40-cyanobiphenyl (5CB) in α,ω-bis (4-40-cyanobiphenyl) nonane (CBC9CB) to better than the 5% level.[19]

3 New aligning media

Several aligning media are available for the measurement of anisotropic NMR parameters, such as, poly-γ-benyl-L-glutamate, collagen gels, gelatins, cross linked poly(acrylonitrile), polysaccharide gels, fragmented DNA, *etc.* The desirable property of the aligning media is that the order parameter should be sufficiently weak to reflect the anisotropic parameters in the NMR spectrum, while retaining the high resolution analogous to that in the liquid state. The different research groups have introduced new and weak alignment media for specific applications. A new class of alignment medium suitable for organic molecules based on helically chiral polyisocyanides has been introduced. The medium is useful for the study of molecules that are soluble in chloroform and THF or CD_2Cl_2. This liquid crystal has a very low order compared to PBLG (poly-γ-benzyl-L-glutamate) giving a deuterium quadrupole splitting of 272 Hz for chloroform compared to 423 Hz in PBLG. The number of heteronuclear RDCs is determined for the molecule strychnine in $CDCl_3$/Poly-L-1.[20] The same group also has reported two other liquid crystals. Alignment medium is derived from polyacetylenes to measure residual dipolar couplings (RDCs). The medium can also be used to study the molecules that are soluble in benzene, toluene, THF, pyridine, and CD_2Cl_2. This medium has an advantage that the quadrupolar splitting of the deuterium signal of $CDCl_3$ in this medium is 107 Hz. The medium also is used to discriminate two enantiomers of menthol.[21] The ionic liquid *N*-dodecyl-*N*-methyl pyrrolidinium bromide [C_{12}MPB] has also been reported to exhibit liquid crystal alignment in D_2O at higher concentrations (70%) with a large quadrupolar splitting of 690 Hz. At lower concentrations the medium becomes isotropic. The three component 1 : 1 : 1 mixture of [C_{12}MPB]/D_2O/decanol shows liquid crystal alignment with a tiny quadrupolar splitting of 84 Hz. The 1 : 1 : 1 : 1 mixture of [C_{12}MPB]/D_2O/decanol/DMSO-d_6 shows liquid crystal alignment with a quadrupolar splitting of 52 Hz. Five independent RDCs have been measured for glucose in this medium.[22] The 4′-mercapto-2,2′,6′,2″-terpyridine-6,6″-dicarboxylic acid (4MTDA) has been introduced as a small and rigid lanthanide binding tag for site-specific binding to a protein by the formation of disulphide bridge. Anisotropic paramagnetism partially aligns the molecules in an external magnetic field. RDCs were determined on a 600 MHz NMR spectrometer, for the lanthanide complexes of 4MTDA-tagged ubiquitin.[23] A novel weakly ordered chiral lyotropic alignment medium, derived by the self-assembly of guanosine 5′-monophosphate (5′-GMP): guanosine, which is inexpensive, easy to prepare, requires less time, possesses weak alignment strength, sustainable, and is reversible over a broad range of temperature (280–330 K) and concentration of the mixture has been reported. This medium has been demonstrated to possess enormous potential for the measurement of RDCs in small and medium sized water-soluble molecules and for the differentiation of enantiomers and enantiotopic directions. The order parameters can be systematically tuned by iterating the physical parameters to scale RDCs to the desired value within its permissible range of mesophase for many exciting applications of NMR structure calculations and/or chiral discrimination.[24] A new type of weak chiral liquid crystalline medium of disodium folate, a member of vitamin 'B' family, for enantiomeric and enantiotopic discrimination of organic molecules in an aqueous environment. The first indication of the enantiomeric discrimination ability of this chiral mesophase obtained from a self-assembly of molecules is reported.[25] It is shown that a butadiyne-linked zinc porphyrin tetramer forms a trimolecular aggregates, in a parallel stacked arrangement in chloroform solvent. The aggregates gave sharp NMR resonances and are in slow exchange on the NMR time scale. Solutions of this aggregate experience magnetic field-induced alignment and the NMR spectra exhibit RDC. The magnitude of the observed RDCs also confirms that the aggregate is trimolecular.[26] Polyurethane (PU) has been introduced as a novel alignment medium for organic solvents. The medium is robust, cheap, easy-to-handle, can be applied with or without covalent cross-linking. The alignment has been successfully demonstrated on strychnine, norcamphor. The material is perfectly suited for large scale production.[27] Lyotropic liquid crystalline (LLC) phases of a phenylalanine-based polyacetylene in $CDCl_3$ are introduced as new enantiodifferentiating alignment media. It was possible to align small molecules like isopinocampheol and determine one bond C–H RDCs. The phases display unusual temperature dependence of the quadrupolar splitting of the solvent by which, depending on the polymer concentration, up to three states of the LLC phase can be identified. Within each state (temperature domain), the orientation of isopinocampheol is different.[28] A protocol has been described for robust, large-scale synthesis of detergent-resistant DNA nanotubes that can be assembled into dilute liquid crystals for application as weak-alignment media in solution. This detergent-resistant liquid-crystal medium offers a number of properties conducive for membrane protein alignment.[29] Crosslinked Poly(ethylene oxide) has been reported as a versatile alignment medium for the Measurement of residual anisotropic NMR parameters.[30] Fast, homogeneous and uniform orientational alignment of mesogenic ionic-liquid electrolytes made up of long-chain imidazolium cations mixed with lithium salts can be obtained by cooling the mixture from the isotropic melt under magnetic constraint. The local orientational behaviour of the different species inside the electrolyte and the ionic diffusional properties in oriented phases of these new materials are investigated by NMR.[31] Disodium cromoglycate nematic LC phase has been demonstrated to be a good weakly aligning medium for small water soluble organic molecules. The medium is compatible with a variety of samples regarding their polarity, and the degree of alignment can be tuned by varying the mesogen concentration, the temperature and by addition of brine.[32] Crosslinked PBLG gels, can be used as a strain induced alignment media in organic solvents for the reliable measurement of RDCs. Alignment strength could be scaled by varying the diameter of the gel sticks. These gels permitted enantiodiscrimination.[33] L-and D-valine-derived polyacetylenes have shown to be excellent alignment media for enantiodifferentiating capability and also the measurement of residual dipolar couplings in $CDCl_3$.[34] A chiral alignment medium for dimethyl sulfoxide, methanol, and water as solvents is developed using chiral polyacrylamide-based gels. The degree of alignment can be adjusted, and the director changes from horizontal to vertical.[35]

4 Development and utility of NMR techniques

The report includes NMR instrumentation, the efficient utility of the existing one and multidimensional NMR techniques, the design of new experiments and processing strategies for the measurement of anisotropic NMR parameters. The only method that can work for the study of proteins in liquid crystalline phospholipid bilayers is solid-state NMR spectroscopy. An account of the solid-state NMR methods available for structure determination of membrane proteins in phospholipid bilayers is given with examples from two bacterial outer membrane proteins (OmpX and attachment invasion locus (Ail) that form integral membrane β-barrels.[36] A review discusses the structure determination of membrane proteins by NMR Spectroscopy.[37] The combined utility of Non-Uniform Sampling (NUS) and compressed sensing algorithm is shown to improve the measurement of $^1J_{CH}$ and $^1T_{CH}$ ($T_{CH} = J_{CH} + D_{CH}$) on small organic molecules from different types of ω_1-coupled HSQC experiments without any loss of spectral quality.[38] The strong coupling effect is known to have detrimental effect on the accurate measurement of $^1J_{CH}$. In circumventing this problem a new J modulated constant-time INEPT CT-HSQC (CTi-CT-HSQC) experiment is reported to accurately measure one-bond scalar coupling constant and RDCs without interference from strong coupling. As a consequence the RDCs can be measured with greater accuracy using this method.[39] A methodology, specially applicable to carbohydrates, has been developed to extract $^1J_{CC}$ from a series of constant time ^{13}C–^{13}C COSY experiments with variable constant time. The accurate values of $^1J_{CC}$ thus measured find utility in the accurate measurement of $^1D_{CC}$.[40] Novel NMR experimental methodologies are developed for the measurement of heteronuclear dipolar couplings in molecules with severe overlapping resonances. As a result 31 RDCs for α-D-cellobiose and 24 RDCs for β-D-cellobiose could be determined in the alignment medium which is a ternary mixture of pentaethylene glycol monododecyl ether, $C_{12}E_5$, *n*-hexanol and D_2O. It is also observed that the order parameters calculated using RDCs of rings I and II differed slightly indicating that the glycosidic bond of cellobiose is not completely rigid. The conformations of the molecules obtained using measured RDCs and interglycosidic ^3J coupling constants, have close resemblance to the X-ray determined structures.[41] An ultrafast HSQC NMR pulse sequence is reported for the measurement of RDCs from single-scan two-dimensional NMR. Ultrafast spectra are obtained in 60 s on an organic compound (+)-isopinocampheol at natural abundance.[42] ADRF-CP experiment carried out on an oriented liquid crystalline sample was found to provide ^{13}C spectra which have signal enhanced by as much as two times compared to the signal obtained with HHCP. The experiment combines the features that are advantageous in both the polarization transfer techniques to provide an improved pulse scheme.[43] The improvement in the efficiency of polarization transfer has been demonstrated by adiabatic demagnetization and remagnetization techniques on stationary aligned samples, and proton assisted insensitive nuclei cross-polarization in magic angle sample spinning samples.[44] Motion-adapted pulse sequences for the study of oriented sample of biopolymers have been reported. The design of separated local field pulse sequence and the resulting motion-adapted SAMPI4 is tested experimentally.[45] Selective double quantum resolved correlation experiment is used for the complete separation of entire proton NMR spectra of enantiomers aligned in the chiral liquid crystal. The experiment is a blend of selective double quantum refocusing, for resolving couplings and chemical shift interactions along two dimensions followed by correlation of the selectively excited protons to the entire coupled spin network.[46] Natural abundance deuterium 2D-NMR (NAD 2D-NMR) is utilized to analyse the spectra of saturated fatty acids in polypeptide aligning media. The method is utilized for analyzing the saturated C14 to C18 fatty acid methyl esters.[47] The band selective homonuclear decoupling (BASH) for enhancing resolution and sensitivity in NOE and RDC measurements of peptides and proteins is reported. The method is particularly useful for enhancing spectral resolution in the H^N–H^α region of NOESY spectra of peptides and proteins. The method also prevents the generation of zero quantum and $H_z{}^N$–$H_z{}^\alpha$ terms.[48] The assignment of solid-state NMR spectra of Pf1 coat protein reconstituted in magnetically aligned bicelles is significantly improved by the combination of the spin exchange under mismatched Hartmann–Hahn conditions and a sensitivity-enhancement repetitive cross-polarization (REP-CP) sequence.[49] The 2D $^1H/^1H$ radio frequency-driven recoupling (RFDR) and nuclear Overhauser effect spectroscopy (NOESY) techniques are utilized under magic angle spinning to study an isotopically unlabelled membrane bound antimicrobial peptide MSI-78 (pexiganan). The proton resonances are assigned and structural constraints, NOE and ^1H–^1H dipolar couplings, are measured without the need for any isotopic enrichment.[50] The high-resolution solid-state NMR spectra for a lipid-anchored protein embedded in lipid bilayers is obtained under magic angle spinning. The different types of insertion procedures for lipid-anchored protein into liposomes and bicelles membranes are demonstrated. In liposome sample, ^{13}C–^{13}C chemical shift correlation spectra under MAS could be obtained. Scalar based experiments gave sufficient quality spectra in bicelle samples.[51] The *in situ* and real-time monitoring of the interconversion of L- and D-alanine-d_3 by alanine racemase from *Bacillus stearothermophilus* is directly observed by ^2H NMR spectroscopy in anisotropic phase.[52] The solid-state ^{19}F-NMR is developed for structural studies of membrane active peptides in lipid bilayers, and how this can be translated to measurements in native biomembranes. The chapter also reviews essentials of the methodology and discusses the key objectives in the practice of ^{19}F-labelling of peptides.[53] Broadband ^1H decoupled CLIP/CLAP-HSQC experiment is reported for the accurate determination of one-bond heteronuclear couplings and small RDCs. The unwanted additional splittings caused by co-evolution of proton–proton couplings are eliminated by the incorporation of isotope-selective BIRD-based broadband proton decoupling scheme during evolution.[54] An ultrafast HSQC NMR pulse sequence is reported for the measurement of RDCs from single-scan two-dimensional NMR. The spectrum is obtained in 60 sec on an organic compound(+)-isopinocampheol, at natural abundance.[55] By combining covariance spectroscopy with non-uniform sampling (NUS), many different three-dimensional homonuclear spin-exchange/separated-local-field (SLF) experiments have been performed with substantial increase in overall sensitivity. The method is also applicable to membrane proteins studied by oriented sample (OS) solid-state NMR in stationary, aligned samples or in unoriented samples undergoing magic angle spinning.[56] The substantial improvement in the ^1H amide resonance line widths of <300 Hz (over typically observable line widths of ≈ 1 kHz) in $^1H/^{15}N$ heteronuclear correlation spectra of membrane proteins in aligned phospholipid bilayers is achieved. The decrease in line widths occurs only for membrane proteins that undergo fast rotational diffusion around the bilayer normal, but not in immobile molecules.[57] The non-uniform sampling compressed sensing is incorporated into oriented sample solid-state NMR for stationary aligned samples and magic angle spinning solid-state NMR for unoriented 'powder' samples.[58] By using a combination of solid-state NMR techniques that improve the spectral resolution and provide correlations between residues and resonances, the complete resonance assignment of the full-length mercury transporter, MerF, an 81-residue protein is reported.[59] An NMR method for determining the three-dimensional structures of membrane proteins in proteoliposomes is demonstrated by determining the structure of MerFt, the 60-residue helix–loop–helix integral membrane core of the 81-residue mercury transporter MerF.[60] Deuterium NMR imaging has been used to evaluate the spatial distribution of the degree of alignment in different types of alignment media by monitoring the deuterium quadrupolar splitting using spatially resolved NMR techniques.[61] The ^2H NMR in chiral liquid crystalline solvents is used to measure enantiomeric excess using exchangeable deuterons in alcohols.[62] Both pulsed field gradient (PFG) and exchange spectroscopy (EXSY) methods are used for NMR measurements of lateral diffusion in membranes. An overview is given, highlighting advantages and limitations of each, with particular emphasis on results obtained from using ^{1}H PFG NMR measurements in magnetically aligned bicelles and ^{31}P CODEX (Centreband-Only-Detection-of-Exchange) measurements in spherical phospholipid vesicles.[63] A theoretical model is developed for calculating the signal intensities in ^{1}H–^{13}C CP and INEPT MAS NMR of liquid crystals.[64] The methods for measurement of base ^{13}C–^{1}H and ^{15}N–^{1}H RDCs in protonated nucleic acids, based on quantitative analysis of intensities in ^{1}H–^{15}N and ^{13}C–^{1}H TROSY-HSQC spectra, are introduced and is illustrated for a 71-nucleotide adenine riboswitch.[65] A simple and effective selective *J*-Scaled HSQC NMR experiment that allows the measurement of long-range CH RDCs at very high precision is applied to sesquiterpene lactone with 5 stereocenters.[66] The one bond RDCs are obtained from the spectra of sucrose in stretched gelatin/D_2O using broadband homonuclear decoupled CLIP/CLAP-RESET experiments based on an isotope-selective BIRD filter element using an improved version of Zangger–Sterk data chunking.[67] Low-ε static probe is developed for an 800 MHz narrow bore magnet, for ^{15}N–^{1}H solid-state NMR study of membrane proteins oriented in hydrated and dielectrically lossy lipid environments.[68] A 500 MHz narrow bore ^{19}F–^{13}C double resonance solid-state NMR probe with a flat-square coil has been built and the first application of this probe for the *in situ* analysis of liquid crystal display panel samples has been reported.[69] The use and limitations of the natural abundance deuterium two-dimensional NMR spectroscopy in polypeptide chiral and achiral aligning media in the studies of homogenous triglycerides has been reported.[70] Orientation of water when confined in the interlayer space of smectites has been detected and quantified using NMR nutation experiments.[71] Investigations are carried out using biradicals, such as TOTAPOL and bTbK, for solid-state NMR signal enhancement using DNP in the context of lipid membranes.[72]

5 Membranes, membrane proteins and lipids

Structures of membrane proteins have been determined by solution-state, oriented solid-state or magic-angle spinning solid-state NMR techniques. Large number of interesting papers has been published in this area. A comprehensive review of the NMR experiments and strategies for the determination of NMR parameters that enable the generation of detailed structural models of intrinsically disordered proteins is reported.[73] The multiple strategies available for expression of isotopically labeled membrane proteins are summarized in a review. Different environments for mimicking lipid bilayers are presented, along with the most important NMR methods and labeling schemes used to generate high-quality NMR spectra.[74] The theory, the technical details, as well as the recent applications of new techniques developed for the measurement of dipolar couplings, chemical shift anisotropy parameters that are used to determine structure and orientation of membrane proteins in lipid bilayers is reviewed.[75] The ^{2}H NMR is used to investigate the way in which SP-B63–78, the C-terminal helix of the lung surfactant protein SP-B, perturbs the phase behavior and dynamics of the bicellar mixture DPMC-d_{54}/DMPG/DHPC (3 : 1 : 1).[76] A mixture of 1,2-dimyristoyl-*sn-glycero*-3-phosphocholine (DMPC) with the short-chain detergent *n*-dodecylphosphocholine (DPC) is introduced as a new membrane-mimetic bicelle system for solid state NMR structure analysis of membrane proteins in oriented samples.[77] A general method for determining the structures of unmodified membrane proteins in phospholipid bilayers by rotationally aligned (RA) solid-state NMR spectroscopy has been reported.[78] A study on bilitranslocase (BTL), a membrane protein primarily involved in the transport of bilirubin from blood to liver cells is reported. By the use of multidimensional NMR spectroscopy and molecular dynamics it is shown that the second transmembrane segment, TM2, also appears in a form of α-helix.[79] The detailed structural and dynamic properties of cationic peptide [KIGAKI] in lipid membranes is characterized using solid-state ^{19}F NMR.[80] The structure of the PhoD signal peptide from *Bacillus subtilis* is characterized in suitable membrane-mimicking environments. High-resolution $^{13}C/^{15}N$ NMR analysis in detergent micelles revealed a helical stretch in the signal peptide between positions 5 and 15.[81] The solution NMR, solid-state NMR, EPR and other methods available to obtain the orientation and immersion depth of bioactive peptides in the membrane or a membrane-mimetic system is reviewed.[82] Comparative analysis of ^{2}H-NMR and separated local field $^{15}N–^{1}H$ polarization inversion spin exchange at magic angle (PISEMA) spectra, in order to identify their strengths and drawbacks for correctly determining the orientation and mobility of α-helical transmembrane peptides, is reported. Both approaches can yield reliable helix tilt angles and can distinguish different degrees of mobility in transmembrane peptides.[83] The nearly complete backbone ^{1}H, ^{13}C and ^{15}N resonance assignments and C_{β} resonance assignments of the two transmembrane helical protein TM0026 Thermotoga maritime is reported.[84] The interactions between two paddle domains, KvAPp from the Kv channel from *Aeropyrum pernix* and HsapBKp from the BK channel from Homo sapiens, and membrane models have been studied. NMR relaxation measurements for HsapBKp show that overall the sequence has anisotropic motions.[85] The membrane interaction properties of two single-residue variants, R6W and L5S, of the 17-amino acid neuropeptide dynorphin A (DynA) are studied by circular dichroism (CD) and nuclear magnetic resonance (NMR) spectroscopy. The results show that R6W-DynA has a small α-helical fraction in buffer, which increases in the presence of bicelles, while L5S-DynA is mainly unstructured.[86] The FKBP–Rapamycin Binding domain of human target of rapamycin (TOR) undergoes strong conformational changes in the presence of membrane mimetics with and without the Regulator Phosphatidic Acid.[87] The methodology for the efficient preparative scale construction of covalent transmembrane complexes and the concomitant high-throughput selection of membrane mimics is introduced. The method facilitates structure determination by NMR spectroscopy by the measurement of RDCs and ^{15}N relaxation parameters.[88] It is shown that the conserved GFFKR motif in α integrins forms a structural motif, a backbone reversal, that can be observed when transmembrane domain, cytoplasmic tail and a lipid environment consisting of membranes or bicelles are present.[89] The set of ganglioside-embedding small bicelles serve as nanoscale standardized membrane mimics for probing membrane-landing processes of intrinsically disordered proteins that are supposed to undergo subsequent conformational transition coupled with their membrane anchoring on a sufficiently large ganglioside cluster.[90] A unified theory for the NMR line shapes of aligned membrane proteins arising from mosaic spread and global rotational diffusion about the director axis is reported. The experimental ssNMR spectra of Pf1 coat protein support the conclusions of the theory for two limiting cases. The static disorder dominates the ^{15}N NMR spectra of Pf1 aligned on a phage, while fast uniaxial diffusion provides a line narrowing mechanism for the Pf1 protein reconstituted in magnetically aligned bicelles.[91] The isotopic labelling strategies to obtain high quality spectra of biological samples, with a special emphasis to integral membrane proteins are summarized. The discussion on how isotopic labeling schemes that can be exploited for studying protein–protein interactions in micelles and lipid vesicles, techniques to engineer spin-labels and isotopically labelled chemical groups to large mammalian membrane proteins is given.[92] The asymmetric antiparallel EmrE exchanges between inward and outward-facing states are identical except that they have opposite orientation in the membrane. The global conformational exchange between these two states for substrate-bound EmrE in bicelles is quantified using solution NMR dynamics experiments.[93] The spatial structure and backbone dynamics of Antiamoebin I has been studied in the DMPC/DHPC bicelles by heteronuclear ^{1}H, ^{13}C and ^{15}N-NMR spectroscopy. The experimental data are in agreement with the barrel-stave model of transmembrane pore formation, similarly to the mechanism proposed for Zrv-IIB and other peptaibols.[94] The curvature properties of a membrane are a key factor in the interactions of amphiphilic helical peptides in general, whose re-alignment and immersion preferences may be inferred in a straightforward manner from the lipid-shape concept. The orientation of α-helical model peptide MSI-103 (sequence $[KIAGKIA]_3$-NH_2) is examined by solid state ^{2}H-NMR in different lipid systems and correlates very well with the shape of the lipid molecules constituting the membrane.[95] A strategy is demonstrated for assigning the resonances of transmembrane sites in SLF spectra of membrane protein Rv1861 in a synthetic lipid bilayer. The strategy is suitable for interpreting spectra from membrane proteins with three and potentially more transmembrane helices. Sequence specific assignments for all labeled transmembrane amino acid sites are obtained, which provide a set of orientational restraints and helix orientations in the bilayer.[96] Drug–membrane interactions of the candesartan cilexetil (TCV-116) are studied by various experimental techniques, including solution ^{1}H and ^{13}C NMR and solid state ^{13}C and ^{31}P NMR spectroscopies. The effects of this prodrug are compared with other AT_1 receptor antagonists.[97] By incorporating specific ^{2}H and ^{15}N labels in the core sequence of Y^{5}GWALP23, it was possible to use solid-state NMR spectroscopy to examine the peptide orientation in hydrated lipid bilayer membranes. The peptide orients well in membranes and gives well-defined ^{2}H quadrupolar splittings and ^{15}N/^{1}H dipolar couplings throughout the core helical sequence between the aromatic residues.[98] By using selected ^{2}H and ^{15}N labels, the influence of a central proline residue on the properties of a defined peptide GWALP23 (acetyl-GGALW5LALALALALA-LALW19LAGA-ethanolamide) that spans lipid bilayer membranes is examined by solid state NMR.[99] The high-resolution structures of the homologous 22-residue long antimicrobial peptides piscidin 1 and piscidin 3 are determined by solid state NMR in fluid-phase 3 : 1 phosphatidylcholine/phosphatidylglycerol (PC/PG) and 1 : 1 phosphatidylethanolamine/phosphatidylglycerol (PE/PG) bilayers.[100] The molecular mobility of the human neuropeptide Y receptor type 2 reconstituted into dimyristoylphosphatidylcholine (DMPC) membranes was investigated by solid-state NMR. The results suggest that the Y2 receptor is highly mobile in the biologically active liquid-crystalline phase state of the membrane.[101] Sensitivity enhancement and contrasting information is provided by free radicals in oriented-sample NMR of DMPC/DHPC bicelle-reconstituted membrane proteins. Up to 74% signal enhancement between the control and radical samples is obtained during constant experimental time at "optimal" relaxation delays.[102] The combination of oriented ssNMR ^{1}H–^{15}N dipolar couplings and ^{15}N chemical shift anisotropy data, and the first-principles MD simulations is exploited to describe the conformational dynamics and topology of the 31-residue membrane protein sarcolipin and the 53-residue membrane protein phospholamban in 1,2-dioleoyl-*sn-glycero*-3-phosphocholine (DOPC) lipid bilayers.[103] A method to calculate ^{31}P solid-state NMR spectra based on the dynamic input from extended molecular dynamics (MD) simulations is reported and the method is applied to study the dynamics of lipid bilayers containing the antimicrobial peptide alamethicin.[104] The three-dimensional structure of human chemokine receptor CXCR1 is determined by NMR spectroscopy. The receptor is in liquid crystalline phospholipid bilayers, without modification of its amino acid sequence and under physiological conditions. Three-dimensional ^{13}C-detected separated local field (SLF) experiments were used to measure the ^{1}H–^{15}N and ^{1}H–^{13}Cα dipolar coupling frequencies that provide orientation restraints for structure determination.[105] The properties of lipid bicelle phases with an emphasis on NMR experimental measurements has been reviewed. In addition, some of the most exciting recent applications of bicelles in the structural and dynamic studies of membrane proteins has been discussed.[106] The ability of the GROMOS 54A8 force field to accurately model the structural properties of lipid bilayers, proteins, and electrolyte solutions has been investigated.[107] The immersion properties of the oxidized and reduced FATC domain of yeast target of rapamycin in membrane mimetics has been characterized in detail.[108] Lipid–protein nanodiscs of different lipid compositions (DMPC, DMPG, POPC, POPC/DOPG) have been compared with classical membrane mimicking media such as detergent micelles, lipid/detergent bicelles and liposomes in their ability to support cell free synthesis of integral membrane proteins in a folded and soluble state.[109] The quantitative ^{1}H MAS NMR measurements are carried out to determine the location and orientation of the local anesthetics lidocaine, dibucaine, tetracaine, and procaine hydrochloride in phospholipid bilayers. The results revealed that the aromatic ring structures of all the investigated local anesthetics are localized in the lipid–water interface of the membrane defined by the upper chain/glycerol/headgroup region.[110] N-terminal (SP-B_{8-25}) and C-terminal (SP-B_{63-78}) helices of lung surfactant protein SP-B, either linked as Mini-B or unlinked but present in equal amounts, were incorporated into either model phospholipid mixtures or into bovine lipid extract surfactant in the form of vesicle dispersions or mechanically oriented bilayer samples. The effects of these peptides on phospholipid chain orientational order, headgroup orientation, and the response of lipid–peptide mixtures to mechanical orientation by mica plates has been characterized by ^{2}H and ^{31}P NMR.[111] The preferential binding of the overexpressed monotopic glycosyltransferase MGS protein to anionic lipids triggers a response promoting the total lipid production in the cell, which in turn results in an interactive cycle between phospholipid synthesis and protein production.[112] It is suggested that solubilization of swM2 (construct of protein M2) into bicelles results in an increased ordering of the transmembrane helix (TMH) – amphipathic helix (APH) linker resulting in a more fixed orientation between TMH and APH. Furthermore, it is shown that the swM2 construct, is particularly amenable to high resolution solution NMR studies in the presence of lipids.[113] The methods and challenges in the characterization of the atomic-level structure and dynamics of glycolipids at membrane surfaces is reported.[114] The lipid bicelles as employed in the study of protein structure is reviewed and covers both the isotropic and the aligned bicelles as used in NMR studies.[115] The atomic resolution structure of a myristoylated cytosolic tail of αX in perdeuterated dodecylphosphocholine (DPC) is determined by NMR spectroscopy. The studies reveal that 35-residue long CT of aX adopts a helical conformation for residues F4-N17 at the N-terminal region.[116] Studies using different detergent/lipid mixtures lead to the identification of two different families of bicelles for which it is possible to lower the total amphiphile concentration to <1% (w/v) while retaining isotropic assemblies.[117] The effect of varying concentrations of divalent cations, Ca^{2+}, Mg^{2+}, Zn^{2+}, and Cd^{2+}, on cholesterol sulfate-stabilized DMPC/DHPC bicelles is investigated. The result indicates that divalent cations are not only required to maintain the biological activity of proteins and nucleic acids, they may also be used to manipulate the behavior of the magnetically aligned phase.[118] Spontaneously forming structures of a system composed of DMPC, and 3-[(3-cholamidopropyl)dimethylammonio]-2-hydroxy-1-propanesulfonate (CHAPSO) were studied by small-angle neutron scattering (SANS), ^{31}P NMR, and ^{1}H NMR diffusion measurements. Charged lipid dimyristoyl phosphatidylglycerol (DMPG) was used to induce different surface charge densities. The structures adopted were investigated as a function of temperature and lipid concentration while keeping the constant molar ratio of long-chain to short-chain lipids.[119] The magnetic alignment of bicelles with various compositions for a broad range of temperatures is characterized using ^{31}P static NMR. The results indicate that the presence of 75 wt% or more water is essential to achieve maximum magnetic alignment, while the presence of cholesterol and cholesterol-3-sulfate stabilizes the alignment at extreme temperatures and the positively charged CTAB avoids the mixing of bicelles.[120] The model lipid membranes made of phosphatidylcholine, phosphatidylserine, and ceramide and rat alveolar macrophages are used to show how lipid bilayer properties probed by SAXS and solid-state ^{2}H NMR correlate with engulfment rates measured by flow cytometry.[121] An improved reconstitution of integral membrane proteins into small isotropic bicelles that allows reconstitution into bicelles with a much wider range of lipid compositions is reported.[122] A mixture of 1,2-dimyristoyl-*sn-glycero*-3-phosphocholine (DMPC) with the short-chain detergent n-dodecylphosphocholine (DPC) is introduced as a new membrane-mimetic bicelle system for solidstate NMR structure analysis of membrane proteins in oriented samples. The applicability of this medium is demonstrated on ^{15}N-labeled transmembrane protein TatA.[123] The dimerization of the ErbB4 transmembrane domain in membrane-mimicking lipid bicelles DMPC/DHPC is investigated by solution NMR.[124] The ^{31}P solid-state NMR is used to study the effect of both non-selective and selective cationic 14-mer peptides on the lipid orientation of DMPC bilayers. The results indicate that the secondary structure of the peptides, and hence their selectivity towards bacterial membranes, greatly influence the types of deformation that these peptides induce in oriented lipid systems.[125] A review discusses the number of useful NMR-based techniques for measuring lateral diffusion in membranes.[126] The changes that occur in morphology, fluidity as well as in the phase behavior of bicelles with increase in number of sialic acids is reported.[127] The change of neurotensin solution structure induced by isotropic 3-[(3-Cholamidopropyl) dimethylammonio]-1-propanesulfonate CHAPS-PC bicelles with and without Ganglioside Monosialo 1 (GM1) is investigated using solution state NMR spectroscopy.[128] By changing the ratio of Sphingomyelin (SM)/DHPC it is demonstrated that SM forms both oriented and isotropic bicelles. Utilizing the bicelles as a substitute for a lipid bilayer membrane, the conformation of SM has been elucidated on the basis of vicinal ^{1}H–^{1}H coupling constants and NOEs.[129] The measurement of signal intensities in solution NMR spectra of membrane proteins that are solubilized in DHPC micelles and DMPC:DHPC isotropic bicelles with q (where q is the molar ratio of the long-chain lipids to short-chain lipids) values <1.0 serves as an effective tool for the qualitative characterization of the structure, dynamics, and topology of helical membrane proteins in micelle and isotropic bicelle environments.[130] NMR spectra of cell-free synthesis of membrane subunits of ATP synthase in phospholipid bicelles and from the bacterial cell membranes are similar, indicating that cell-free synthesis produces protein structurally similar to that prepared from the cell membranes.[131] The direct utility of GB1 fusion proteins for NMR interaction studies using lipids or membrane mimetics is studied. NMR binding studies using only the GB1 part reveal that this fusion tag does not significantly interact with different membrane-mimetics such as micelles, bicelles, or liposomes. Thus spectral changes observed using GB1-fusion proteins indicate lipid- and membrane interactions of the target protein.[132] The pH-dependent interaction between C-peptide and phospholipid bicelles has been studied by CD and NMR.[133] The site-specifically deuterated 1-palmitoyl-2-stearoyl-*sn-glycero*-3-phosphocholines that match the acyl chain length of stearoyl-SM (SSM) is synthesised, and compared their deuterium quadrupole coupling profiles.[134] Model lipid membranes (made of phosphatidylcholine, phosphatidylserine, and ceramide) and rat alveolar macrophages is used to show how lipid bilayer properties probed by small-angle X-ray scattering and solid-state ^{2}H NMR correlate with engulfment rates measured by flow cytometry.[135] When dissolved in bilayer membranes, retroinverso-melittin, when covalently linked to the lipid 1,2-dioleoyl-*sn-glycero*-3-phosphoethanolamine (riDOM) nanoparticles dissociate and form transient pores. riDOM-induced membrane leakiness is however much reduced compared to that of authentic melittin. The secondary structure of the ri-melittin is not changed when riDOMis transferred from water to the membrane and displays a large fraction of β-structure. The ^{31}P NMR spectrum of the nanoparticle is however transformed into a typical bilayer spectrum.[136] A thorough conformational analysis of AT_1 antagonist losartan is performed by ROESY NOESY NMR studies in three different environments, mimicking conditions of biological transport fluids and membrane lipid bilayers.[137] The structure and dynamics of caveolin-1 (D82–S136; Cav182–136) is characterised in a DMPC bilayer. Results show that Cav182–136 prefers to adopt a slightly dynamic U-shaped conformation in a lipid bilayer.[138] A solution NMR structure of membrane protein diacylglycerol kinase is obtained in detergent micelles and three crystal structures are obtained in a monoolein cubic phase. All the structures are trimeric with each monomer having three transmembrane and one amphipathic helices.[139] The transmembrane structure, topology, and dynamics of the FMN binding domain of CYPOR in a native membrane-like environment is probed. The solid-state NMR results reveal that the N-terminal transmembrane domain of CYPOR adopts a α-helical conformation in the lipid membrane environment.[140] The reconstitution of full-length BamA in proteoliposomes at low lipid-to-protein ratio, leads to high sensitivity and resolution in solid-state NMR (ssNMR) experiments. The POTRA domains are detected in ssNMR experiments probing rigid protein segments in our preparations. The results suggest that the periplasmic region of BamA is firmly attached to the β-barrel and does not experience fast global motion around the angle between POTRA 2 and 3.[141] The bilayer phase transitions from the ripple gel phase ($P_{\beta'}$) to the liquid-crystal phase (L_α) of 1,2-dipalmitoyl-*sn*-glycerol-3-phosphocholine (DPPC) and 1,2-dimyristoyl-*sn-glycero*-3-phosphocholine (DMPC) were studied by different techniques. The proposed methods allow for determining the T_C temperature with a high degree of accuracy.[142] Membrane binding by amphibian peptide agonists of CCK2R is investigated using ^{31}P and ^{2}H solid-state NMR. Sttudies in eukaryotic phospholipid models at 25 °C shows that rothein 1 does not interact with the membrane at all. In contrast, other cyclic disulfides signiferin 1 and riparin 1.1 interact with phospholipid head groups and partially penetrate into the upper leaflet of the model bilayer, but to different extents.[143] Localization and interaction of hydroxyflavones with lipid bilayer model membranes is studied using DSC and multinuclear NMR.[144] NMR studies suggested that flurbiprofen molecules were entrapped in the surfactant based vesicles. Flurbiprofen permeability was improved by encapsulation in vesicles. The progression of flurbiprofen permeation was reduced by the transition to a lamellar liquid crystalline state, which was associated with water evaporation.[145] The effect of the local anaesthetic tetracaine (TTC) on the lipid bilayer of DMPC and DMPC + CHOL unilamellar vesicles is investigated using supported liposomes by QCM-D, and liposomes in solution by ^{31}P-NMR and DSC.[146] The effect of elasticity on the skin permeation of ketoprofen from surfactant-based vesicular formulations and elastic liposomes is determined. The vesicles enhance ketoprofen transport across human skin, while no enhancement of ketoprofen was observed when loaded in elastic liposomes.[147] The background information on membrane proteins, and why this environment is both difficult to model and has such a significant impact on protein structure is discussed. A justification is given for ssNMR characterizations of membrane protein structures through a discussion of two structures determined by ssNMR compared with characterizations by both solution NMR and X-ray crystallography.[148] Solution-NMR Characterization of Outer-Membrane Protein A (OmpA) from *E. coli* in Lipid Bilayer Nanodiscs and Detergent Micelles is reported. It is shown that the NMR-invisible ring around the β-barrel of OmpA is also present in lipid bilayer nanodiscs and in mixed micelles with a third detergent.[149] Intermolecular distances that define the location, orientation, and conformation of *sn*-2-Arachidonoylglycerol in palmitoyl-oleoyl-phosphatidylcholine (POPC) lipid bilayers is determined using rotational-echo double-resonance (REDOR) NMR.[150] The structure and configuration of synaptobrevin 2 has been examined using both NMR and EPR spectroscopy in bilayer and bicelle environments and the result is compared with those found previously by NMR in DPC micelles.[151] By focusing on the lipid and protein binding properties of the detergent, proteoliposome preparations can be readily produced, which provide double the signal-to-noise ratios for both the oriented sample and magic angle spinning solid state NMR.[152] Comparisons of the influence of interfacial Trp, Tyr, or Phe residues on the properties of tilted helical transmembrane peptides are made. The results suggest that a dynamic interplay between lipid membrane thickness and protein helix rotation may regulate aspects of biological function.[153] A new method for the direct detection of lipid bilayer sulphate transport is developed based on ^{33}S NMR that can be used to monitor sulfate transport, using ^{33}S-labelled sulfate and paramagnetic agents such as Mn^{2+} and Fe^{3+} to discriminate between intra- and extra vesicular sulphate.[154] Evidence is presented which explains some of the conflicting and inexplicable data found in the literature regarding α-tocopherol's (aToc's) behavior in dimyristoyl phosphatidylcholine (di-14:0PC) bilayers.[155] NMR studies of C99 were conducted both in anionic lysomyristoylphosphatidylglycerol (LMPG) micelles and in a series of five zwitterionic bicelle compositions involving phosphatidylcholine and sphingomyelin in which the acyl chain lengths of these lipid components varied from 14 to 24 carbons.[156] Effect of Cholesterol on the Phase Change of Lipid Membranes by Antimicrobial Peptides is investigated using solid state ^{2}H NMR.[157] A novel method for membrane protein topology and immersion depth based on site-directed spin labeling and paramagnetic relaxation enhancement ^{31}P NMR is reported. The method can be utilized as a molecular ruler to measure positions of labeled amino acids within the lipid membrane.[158] Molecular conformation and bilayer pores in a nonionic surfactant lamellar phase are studied using ^{1}H–^{13}C solid-state NMR and molecular dynamics simulations. The simulation model is used to interpret the magnitude and temperature dependence of experimental C–H order parameter profiles in terms of the molecular conformation and orientation.[159] The conformation, membrane alignment and dynamical behavior of multifunctional membrane active 11 residue peptide BP100 is studied in lipid bilayers (DMPC/DMPG), using oriented circular dichroism (OCD) and solid-state ^{19}F and ^{15}N NMR.[160] Low q DMPC/DHPC bicelles were studied by ^{31}P NMR, FTIR spectroscopy, and molecular dynamics. The critical bicellar concentration (CBC) was introduced as an important parameter which allows determining the exact long-to-short chain lipid ratio in the system and the diameter of the bicelle.[161] Two sterols desmosterol and lanosterol, are used to demonstrate the increased molecular mobility inside the bilayer, that could allow the detection of their proton resonances.[162] The membrane protein bond orientations in nanodiscs is obtained by measuring RDCs with the outer membrane protein OmpX embedded in nanodiscs using Pf1 phage as an alignment medium.[163] Actinoporin Sticholysin I (StI):DHPC micelle complexes of variable lipid composition is evaluated to search for a suitable model for NMR studies. The StI:DHPC micelle was found to be the best system for characterization of their bound form, yielding a stable sample and good quality spectra.[164] Solid-state NMR spectroscopy provides a valuable tool to analyse the topology of membrane-associated peptides, the packing and morphology of the lipid membranes and thereby control the design of the peptide sequences.[165] Static oriented and magic angle spinning solid state NMR, along with molecular dynamics simulation was used to investigate its structure, orientation, and depth in lipid bilayers of several compositions, namely POPC, DPPC, DPPC/POPC/POPG, and bovine lung surfactant extract (BLES).[166] Various solid-state NMR approaches and the basic underlying concept as to how angular information can be obtained from oriented samples is reviewed. It is also demonstrated how this information is used to calculate structures and topologies of peptides in their native liquid-disordered phospholipid bilayer environment.[167] First insights into the functional and structural properties of the human cytotoxic mitogaligin [31–53] fragment are described. Because the insertion of the [31–53] fragment into SDS micelles appeared to be comparable to that in DMPC/DHPC/Cardiolipin bicelles, it is considered that

SDS micelles to be a convenient membrane model for determination of a 3D structure of the fragment.[168] Using sarcolipin, a single pass membrane protein, reconstituted in oriented bicelles the advantage of polarization optimized experiments is demonstrated for both static and MAS spectroscopy to achieve higher sensitivity and substantial time-savings for 2D and 3D experiments.[169] A new fluorescent L-amino acid (AFaa) was synthesized and introduced into the melittin peptide at two alternative positions by solid-phase synthesis and the insertion and orientation of the labeled melittin peptide in cell membranes is determined.[170] Aliphatic oligoureas, labeled at selected positions with ^{15}N are prepared, and reconstituted into oriented phospholipid bilayers. The ^{15}N chemical shift and ^{1}H–^{15}N dipolar couplings of two labeled sites were determined by solid-state NMR spectroscopy.[171] Solution NMR studies on the orientation of membrane-bound peptides and proteins by paramagnetic Probes is reviewed. The examples discussed ranges from the large pool of antimicrobial peptides, bacterial toxins, cell penetrating peptides to domains of larger proteins or the calcium regulating protein phospholamban.[172] Mechanisms of peptide-induced pore formation in lipid bilayers is investigated by oriented ^{31}P solid-state NMR spectroscopy. For two antimicrobial peptides, alamethicin and novicidin, it is observed that the majority of the lipids remain in a planar bilayer conformation but a number of lipids are involved in the peptide anchoring.[173] A detailed description of a robust protocol that results in high-quality membrane protein samples for both magic-angle spinning and oriented-sample solid-state NMR is demonstrated using two proteins: CrgA(two transmembrane helices) and Rv1861 (three transmembrane helices), both from *Mycobacterium tuberculosis*.[174] The current state of membrane protein structure determination using solid-state NMR is reviewed.[175] Solid-state NMR spectroscopy is used to investigate the conformation, topology, and lipid and water interactions of the fusion peptide of the PIV5 virus F protein in three lipid membranes, POPC/POPG, DOPC/DOPG, and DOPE.[176] The ^{31}P and ^{2}H solid-state NMR static spectra lineshape, CSA widths, and ^{2}H order parameters are utilized to investigate lipid–peptide interaction and the results are complimentary to each other.[177] Solid-state NMR and EPR spectroscopy in liquid crystal DMPC lipid bilayer preparations were used to demonstrate the formation of a heterodimer between *S. Typhi* MgtR and the transmembrane helix 4 of Mtb MgtC.[178] The combination of oriented solid-state NMR and cross-linking experiments demonstrated that the ligand-free form of *E. Coli* membrane protein EmrE is an anti-parallel dimer with asymmetric tilt angles relative to the lipid bilayer normal.[179] Magic Angle Spinning and Oriented Sample Solid-State NMR structural restraints combine for functional insights of influenza A M2 protein are reported.[180] It is demonstrated that alignment of a structured peptide or small protein solubilized in mixed phospholipid: detergent micelles or bicelles, when embedded in a compressed gel or liquid crystalline medium, can be altered by either changing the phospholipid aggregate shape, charge, or both together.[181] The ^{31}P NMR spectroscopic study of planar glass-plate-oriented multi-bilayers of dimyristoylphosphatidylcholine (DMPC) with addition of polyethylene oxide (PEO) has been reported. The changes in the phospholipid membrane induced by the presence of small amounts of PEO, was revealed in ^{31}P NMR spectra.[182] Bicellar dispersions of chain perdeuterated 1,2-dimyristoyl-*sn-glycero*-3-phosphocholine (DMPC-d_{54}), 1,2-dimyristoyl-*sn-glycero*-3-phospho-(1′-rac-glycerol) (DMPG), and 1,2-dihexanoyl-*sn-glycero*-3-phosphocholine (DHPC), with molar ratios of 3 : 1 : 1, is studied using variable-pressure ^{2}H NMR spectroscopy at hydrostatic pressures up to 125 MPa.[183]

6 Measurement and utility of residual dipolar couplings

In arriving at the molecular structure and conformation of biological macromolecules in a solution like environment, the utility of residual dipolar couplings, an anisotropic NMR parameter, is a routine practice. The RDCs in conjunction with nOe restraints are invaluable in the determination of biomolecular structure and function. Due to inverse cube dependence on the distance between two interacting nuclear spins RDCs provide long range orientational information. For the experimental determination of RDCs, there are several types of aligning media are available, such as, lipid bicelles, liquid crystalline media, bacteriophage, anisotropic gels, *etc.* During this review period, there are large number of reviews and research articles published on the measurement and utility of RDCs obtained using different weak alignment media. The recent advances with the combined use of NMR and the state-of the art molecular simulation that provides new insight into the molecular motions on timescales from nanoseconds to milliseconds is reviewed. It focuses mainly on the mapping of the protein energy landscape using RDCs.[184] A detailed analysis of how to extract protein dynamics encoded in RDCs and how to relate this information to protein functionality within the previously inaccessible supra-τ_c window is reviewed (where τ_c is the overall tumbling time of a molecule). It also discusses the current state of the art of relaxation dispersion measurements.[185] The recent progress in the development of alignment media to derive RDCs to obtain accurate structure of proteins is reviewed. The mathematical expressions for the RDCs and the common descriptions of the alignment tensor are discussed. The particular focus is given to the lipid-compatible media that allow the measurement of RDCs for membrane proteins.[186] A method has been described to determine the conformational fluctuations of RNA using RDCs as replica-averaged structural restraints in molecular dynamics simulations. For the purpose of calculation of RDCs, the alignment tensor is estimated from the shape.[187] Using the strain induced alignment of PDMS/$CDCl_3$ gels and variable angle assisted scaling technique, the one bond C–H RDCs are determined in Rifamycin-S. Using these refined RDCs the exact relative configuration as well as relative configuration of Rifamycin-S from the 256 possible configurations has been singled out.[188] By developing an iterative sampling protocol that enriches native-like features by resolution-adapted structural recombination (RASREC), the size limitation Rosetta *de novo* structure generation has been overcome. The improved sampling has made greater contribution in obtaining accurate structures over a benchmark set of 11 proteins in the 15–25 kDa size range using chemical shifts, backbone RDCs and H^N–H^N NOE data.[189] Experimental RDCs are also sensitive to the specific geometry of each amino acid as defined by energy minimized internal co-ordinates. The specific geometries incorporated in the modified the FM (flexible-Meccano) alogorithm resulted in a better reproduction of experimental RDCs, and are applicable to characterize the conformational properties of intrinsically disordered proteins.[190] Generating protein backbone models from backbone only NMR data based on molecular fragment replacement (MFR) requires only backbone chemical shifts and RDCs and no or only minimal requirement of NOE restraints. As far as the required completeness of RDC data sets is concerned, this new method is reported to be more robust than the previous algorithms.[191] It is possible to extract information provided by RDCs about the conformational fluctuations in the native states of proteins. The accuracy of the method is demonstrated by the analysis of the Q factors for RDCs not used as restraints in the calculations.[192] In Pf1 phage, under strong alignment conditions, the very large Main-chain $^1H^N$–^{15}N residual dipolar couplings varying from −200 to 200 Hz are measured in ubiquitin. Compared to weakly aligned molecules, this indicates an order of magnitude larger in the degree of alignment, achieved by high level of deuteration of the protein with partial back-substitution of labile protons from 25% H_2O/75% D_2O buffer. The signs of the couplings are determined based on identification of the trosy/anti-trosy components in the $^1H^N$–^{15}N quartet.[193] The model calculations have been carried out to quantify the extent to which mobility of the metal-binding tag can compromise the quality of the magnetic susceptibility tensor ($\Delta\chi$) that can be determined from the pseudo contact shifts observed in the protein. Any tag that generates large pseudo contact shifts also produces significant RDCs.[194] The structural information derived by the measurement of RDCs is extended for the measurement of protein folding under pressure. Four different stable and pressure resistant alignment media have been identified and characterized for measurement of RDCs at high hydrostatic pressure. It is concluded that alignment medium A C12E5/*n*-hexanol alcohol mixture is stable from 1 to 2500 bar, the media Pf1 phage and DNA nanotubes undergo a reversible transition between 300 and 900 bar. Phospholipid bicellear media are stable only up to 300 bar at ambient temperature.[195] The mutation of Arg76 to Ser (R76S) that is correlated with childhood cataract in an Indian family has been investigated by expressing R76S γD-crystallin protein in *E. coli.* No significant biochemical or biophysical differences were observed between the wild-type HGD protein and the R76S variant. R76S γD-crystallin solution structure, using RDCs, and of its motional properties by relaxation measurements, also revealed a close resemblance to wild-type crystalline.[196] The combination of contrast-matched SAXS, aqueous SAXS/WAXS, and NMR-derived RDCs is sufficient to define the positions of the two domains of calmodulin (CaM) within a precision of 1 Å in a calmodulin – myosin lightchain kinase (MLCK) peptide complex in which the four coordinated Ca^{2+} ions have been substituted by Pb^{2+}.[197] High precision RDCs measured for the backbone ^{1}H–^{15}N, ^{15}N–$^{13}C'$, $^{1}H\alpha$–$^{13}C\alpha$, and $^{13}C'$–$^{13}C\alpha$ one-bond interactions in the squalamine medium fit well to the static structural model previously derived from NMR data.[198] Replica Exchange Molecular Dynamics (REMD) simulations are used with temperature as a proxy for solvent quality to examine how well the polymer theories developed for homopolymeric chains describe hetero-polymeric α-synuclein. The results indicate that α-synuclein behaves like a homopolymer at the extremes of solvent quality, while in the intermediate solvent regime, the uneven distribution of charged residues along the sequence strongly influences the conformations adopted by the chain. The ensemble extracted from the REMD simulations of α-synuclein was refined by fitting to the experimental RDCs and Paramagnetic Relaxation Enhancements (PREs).[199] An approach is developed by combining highly efficient conformational sampling with ensemble selection to map the backbone conformational sampling of intrinsically disordered proteins (IDPs) on a residue specific level. By the systematic analyses it is identified that the combination of RDCs and chemical shifts can be used to raise conformational degeneracies inherent to different data types. This approach has been applied to characterize the conformational behaviour of two intrinsically disordered proteins, the K18 domain from Tau protein and N_{TAIL} from measles virus nucleoprotein.[200] The reliability of molecular simulation methods to study protein structure–dynamics–function relationships at an atomic level is largely dependent on the accuracy of the empirical force field (FF) and its ability to reproduce and predict experimental observables. It is shown that the recent updated version CHARMM36 (C36) FF leads to better correlation with experimental data compared to the CHARMM22/CMAP FF. It provides better correlation between experimental and computed hydrogen bond scalar couplings in five proteins. It also led to better reproduction of backbone RDCs in ubiquitin when a set of extensive alignment medium is considered.[201] A new approach is reported that broadens the scope of using RDCs in characterizing the dynamics of nucleic acids. This strategy enables the use of RDCs measured in variably elongated RNAs for constructing ensembles. The approach predicts RDCs for a given RNA conformer based on its overall shape using the program PALES. Authors are able to successfully construct and validate an atomic-resolution ensemble of HIV-1 TAR, based on four sets of RDCs.[202] RDC restraints have been used to study the organization of a six domain fragment of filamin A, consisting of domains 16–21 (FLNa16–21). Four hundred and thirty $^{1}H^{N}$–^{15}N RDCs have been extracted from a 3D MQ-HNCO TROSY spectrum. Using the previously published structures of FLNa domain pairs 16–17, 18–19, and 20–21, a model of the 60 kDa FLNa16–21 domain sextet has been built through rigid-body modelling. Additional information about the interdomain dynamics has been derived from ^{15}N relaxation data.[203] A scheme for analyzing experimental RDCs based on the standard tensorial perspective, which allows for arbitrary orientation of the local director in the protein and of the Principal Axis System of local ordering tensor in the probe, is formulated. The new insights into RDC-detected local ordering in proteins are obtained by virtue of the clear physical meaning and internal consistency of the tensorial representation.[204] The influence of solvent and salt concentration on the alignment properties of acrylamide copolymer gels for the measurement of RDCs is reported. It is shown that the solution composition influences the gel swelling and hence its alignment strength. The alignment of *N*-methylcodeinium iodide was higher with Na^+ than with H^+ counterions in the gel and was reduced by high concentration of Na^+ ions in water or by the decreased solvent polarity of DMSO-d6.[205] A new strategy to assign diastereotopic protons is developed on the basis of RDCs. The approach uses the combination of the *J*-scaled BIRD HSQC (JSB-HSQC) and *J*-scaled HMQC/HSQC (JSB-HMQC/HSQC) experiments. The strategy is demonstrated on strychnine molecule that contains five methylene groups with diastereotopic protons.[206] The tetra-substituted pyrrolidine with four chiral carbon atoms is analyzed using classical methods based on ^3J and NOE data in solution and compared and verified with alternative approaches *via* RDCs in two weak anisotropic alignment media. Twenty one RDCs have been measured in the two alignment media, PBLG and stretched PDMS, including dipolar couplings to ^{31}P of a phosphine moiety.[207] The residue specific ^{15}N relaxation and ^1H–^{15}N RDC data has been collected for 70 kDa protein, NADPH cytochrome P450 reductase (CPR). The analysis of these NMR data revealed that the CPR exists in a unique and predominant conformation that highly resembles the closed conformation observed in the crystalline state.[208] The structure and dynamics of λ lysozyme Bacteriophage Lambda in solution is characterized in solution by ^{15}N relaxation studies, ^{15}N–$^1H^N$ RDCs and analyses of chemical shift, NOE, coupling constants and hydrogen exchange data.[209] The structure of the yeast U2/U6 snRNA complex is analysed by SAXS and NMR. Partial alignment for RDC measurements was achieved by the addition of Pf1 filamentous bacteriophage to a ^{13}C, ^{15}N G- and U-labeled sample.[210] Five different alignment media that are commonly used for the measurement of RDCs of macromolecules at ambient pressure have been studied. The spontaneous alignment of Pf1 phage, d(GpG) and a C12E5/*n*-hexnanol mixture in a magnetic field is preserved under high hydrostatic pressure. However, DMPC/DHPC bicelles and collagen gel are found to be unsuitable.[211] By aligning 4,6-diacetylhygrophorone A^{12} in PBLG it was possible to measure RDCs, and unambiguously determine the relative configuration of all three stereogenic centers.[212] The conformation of efrapeptin C is studied in detail by NMR and molecular modeling. RDCs observed upon partial alignment with stretched PDMS gels are used for the conformational analysis and confirmed the 3_{10}-helical conformation.[213] The sets of ^{13}C–^1H RDCs obtained from the proton-encoded ^{13}C 2D NMR experiment are used to investigate the conformational changes which occur when the achiral symmetric liquid crystal dimer CB7CB changes from the achiral nematic to the chiral twist bend nematic phase.[214] A new methodology combining RDC in microtubes, with a motif-inspired biological assessment strategy is reported. Using only one milligram (*ca.* 1.5 mmol) of sample, the new protocol established the bioactivity as well as the relative and absolute configuration of vatiparol obtained from *Vatica parvifolia*.[215] The experimental

^{13}C–^{1}H coupling values of paramagnetic organometallic complexes were dependent on both the temperature and the magnetic field, and deviated significantly from the $^{1}J_{CH}$ couplings. This deviation is attributed to RDCs that arise from the partial alignment of the complexes in the magnetic field, aided the ^{13}C assignments and permitted the refinement of the conformation of the complexes in the solution state based on dipolar couplings.[216] Additional features have been included in the REDCAT software package, previously developed in the Object Oriented C++ environment, for the analysis of molecular structures using RDC data.[217] A methodology for the combined analysis of pseudocontact shifts and RDCs of lanthanide single molecular magnets (SMMs) is developed. In a SMM tris(octabutoxyphthalocyaninato) diterbium strong pseudocontact shifts as well as large residual dipolar couplings have been observed in both ^{13}C and ^{1}H NMR. The RDCs arise due to partial alignment of tris(octabutoxyphthalocyaninato) diterbium in the strong magnetic field.[218] A Critical Assessment as whether it is possible for Enantiomer Assignment Possible by NMR Spectroscopy Using Residual Dipolar Couplings has been made.[219] A new algorithm, INVEDCIN is reported, which models the protein backbone structure as a robot kinematics problem constrained geometrically mainly from RDC data. The benefit of this method is that complete and accurate solutions are obtained overcoming the local minima problems of heuristic procedures.[220] A new iterative procedure entitled Optimized RDC-based Iterative and Unified Model-free analysis (ORIUM) is reported to determine these structural and dynamic parameters.[221] Methods of calculating the RDCs for the study of protein dynamics that do not require the definition of an alignment tensor is reported.[222]

7 Quantum computation

The experimental digital quantum simulation for time-evolution of a quantum particle under various potentials using NMR techniques on a 5-qubit system is reported. The simulator consisted of a mutually interacting 5-spin system 1-bromo-2,4,5-trifluorobenzene partially oriented in liquid crystal *N*-(4-methoxybenzaldehyde)-4-butylaniline (MBBA).[223] NMR implementation of a quantum algorithm to find the parity of a permutation is reported. A qutrit is qualitatively more quantum than a qubit because of the existence of quantum contextuality and a single qutrit can be used for computing. The deuterium nucleus oriented in a liquid crystal as the experimental qutrit.[224] An effective and fast quantum control method for both the state preparation and the realization of quantum gates in quadrupolar systems is reported. All the four pseudo pure states in a quadrupolar spin-3/2 system is experimentally prepared by the application of GRAPE pulses. The sample of sodium decyl sulfate in lyotropic liquid crystalline phase is used for such a purpose.[225] An improved adiabatic factoring algorithm and its experimental realization to factor the number 143 on a liquid-crystal NMR quantum processor with dipole–dipole couplings is reported. It is believed to be the largest number factored in quantum-computation realizations, which shows the practical importance of adiabatic quantum algorithms.[226] The advantages and experimental applications of strong modulating pulses (SMP) for quantum information processing in a three q-bit quadrupolar systems composed of 7/2 spin nuclei derived from the liquid crystalline phase of cesium pentadecauorooctanoate (CsPFO) in D_2O is reported. The experimental data showed that SMPs permit the logic operations in a time scale ten times smaller than with conventional pulses.[227]

8 Liquid crystals, their properties, structures, phase behaviour, related NMR studies

Large number of interesting studies have been carried out on the study of liquid crystals, their properties, phase transitions, *etc.* using NMR. The ^{13}C chemical shift anisotropy tensors and molecular order of a nematic liquid crystal Hexyloxybenzoic Acid is determined using ^{13}C NMR studies in liquid, solid and liquid crystalline phases.[228] The ^{31}P NMR spectroscopic study of planar glass-plate-oriented multi-bilayers of dimyristoylphosphatidylcholine (DMPC) with addition of polyethylene oxide (PEO) has been reported. The changes in the phospholipid membrane induced by the presence of small amounts of PEO, was revealed in ^{31}P NMR spectra.[229] A simple and versatile pathway is reported to incorporate redox-active 4,4′-bipyridinium ("viologen") units with two, three, or six charges into dynamically ordered liquid crystalline structures by ionic interactions. The formed complexes, $MV^{2+}(DOBS)_2$, $V^{3+}(DOBS)_3$ and $MV^{6+}(DOBS)_6$, were characterized by NMR spectroscopy.[230] One GHz ^1H frequency is used to study mesogenic molecules *N*-(4-methoxybenzylidene)-4-butylaniline (MBBA) and 4-pentyloxy-4′-cyanobiphenyl (5OCB) and their approximately equimolar mixture at temperatures above the isotropic to nematic phase transition. The magnetic field induced dipolar couplings between proton (H-1), and between H-1 and carbon nuclei are used to investigate the structure, conformational distribution and orientational order in both the pure compounds and in their mixtures.[231] The permanent electric dipole moment of a chiral molecule is partially oriented through the antisymmetric part of the nuclear magnetic shielding tensor, causing the electric dipole to precess with the nuclear magnetic moment and producing a much larger temperature-dependent electric polarization with better prospects of detection.[232] The use and limitations of the natural abundance deuterium two-dimensional NMR spectroscopy in polypeptide chiral and achiral aligning media in the studies of homogenous triglycerides is reported.[233] Understanding the mechanism and membrane selectivity of antimicrobial, peptides (AMPs) and the recent advances in the rational design of synthetic mimics of antimicrobial peptides (SMAMPs) is reviewed.[234] A short review describes the continuous and tightly coupled analysis chain, starting with structural determination of glycans, often present as glycoproteins or glycolipids, from analysis of the primary structure, followed by conformational and dynamics analysis to the study of interaction with proteins.[235] Phase behavior and structure of a board-shaped tetrathiafulvalene (TTF)-based liquid crystal molecule (symTTF11) are investigated by the combined experimental techniques, which includes solid state ^{13}C NMR.[236] The (Cholesteryl)benzoateethynylene oligomers having 3, 5 and 7 repeat units and the homologue polymers are synthesized and their chemical structures are analysed.[237] Alignment and phase behavior in intact wood cell walls is probed using ^{2}H NMR by small deuterium-labeled probe molecules absorbed into bulk wood.[238] The first experimental detection of naturally abundant ^{2}H–^{13}C isotopomers in liquids and oriented solvents using a 21.1 T B_0-field, inverse cryoprobe, and 2D heteronuclear correlation experiments is demonstrated.[239] The application of NAD 2D NMR in aligned media enables both detailed description of the fractionation occurring in the pro-R/pro-S enantiotopic hydrogen positions in linoleic acid pertinent to possible reaction mechanisms, and of the pro-*R*,12*S*,13*R*/pro-*S*,12*S*,13*R* diastereotopic hydrogen positions in vernoleic acid.[240] The experimental evidence for several dipolar quasi-invariants in liquid crystals is reported. The time-domain signals yielded by Jeener–Broekaert pulse sequence as a function of the preparation times, in a variety of dipolar networks, solids, and liquid crystals is analysed.[241] The effect of ionic liquid pretreatment on enzymatic hydrolysis of cellulose was investigated in terms of the changes in the chemical and physical structure of the preparation. The chemical and physical changes between the original and reconstituted cellulose were thoroughly characterized by various analytical techniques.[242] Conformational properties and orientational order of a de Vries liquid crystal (9HL) are investigated. The whole rigid aromatic core constituted of three rings and the carbonyl groups bonded to them is studied and it was possible to characterize the conformational and orientational properties in the two smectic phases.[243] The formation of rich lamellar structures by a single-tailed amphiphilic ionic liquid, 1-dodecyl-3-methylimidazoliumb-naphthalene sulfonate ([C_{12}mim][Nsa]), in aqueous solutions without any additives, is observed. It is reported that with increasing ionic liquid concentration, the spontaneous transition from micelles to unilamellar and multilamellar vesicles, planar bilayers, and then to lamellar liquid crystals occurs.[244] The structural behavior of cholesteric liquid crystals of 4-(hexyloxy)-4′-cyanobiphenyl (6OCB) is investigated by solid-state ^{13}C NMR. Sharp signals observed in the ^{13}C CP stationary NMR spectra of 6OCB in the nematic liquid crystal phase, indicated that the director was parallel with the magnetic field.[245] Mesophase evolution in heat-treated solid petroleum pitches is studied by various analytical techniques including, ^{13}C, ^{1}H NMR and T_1 measurements.[246] Morphology, phase behavior, and relaxation processes in a series of aromatic–aliphatic thermotropic polymer biomaterials are studied. Solid-state wide-line ^{1}H NMR T_1 and $T_{1\rho}$ measurements are used to evaluate both the homogeneity of the polymer morphologies and molecular motions.[247] The initially predicted transmembrane regions of BTL is reexamined and substantiated that 3D structure of one of the transmembrane peptides (TM 3) is alpha helical, by MD simulations in the standard DPPC (dipalmitoyl phosphatidyl choline) membrane. This result has been further confirmed by the means of NMR spectroscopic study carried out in SDS (sodium dodecyl sulfate) micelle environment as well.[248] The behavior of A6/6 (The T-shaped amphiphilic molecule that forms a columnar hexagonal liquid-crystalline phase between the crystalline and the isotropic liquid when studied in bulk) has been studied in water and also in mixtures with bilayer-forming phospholipid, DPPC, using various techniques including solid state NMR.[249] A method for the preparation of bicelles that contain galactolipids, monogalactosyl-diacylglycerol (MGDG) and digalactosyl-diacylglycerol (DGDG), together with a characterization of the bicelles, and the lipids within the bicelles is reported.[250] The segmental motions of flexible chains in the lamellar structure of a strongly segregated poly(styrene)-poly(dimethylsiloxane) (PS-PDMS) diblock are investigated over a time scale of a few tens of microseconds. ^{2}H NMR, T_2 relaxation and ^{1}H double quantum experiments have been performed on selectively perdeuterated PDMS block. T_2 experiments reveal that main part of PDMS show anisotropic orienational motions.[251] A detailed quantitative analysis of the dynamic properties of thrombin across a broad range of time scales using a combination of NMR relaxation parameters, heteronuclear nOes and RDCs from Pf1 phage aligned samples and state-of-the-art molecular dynamics (MD) simulation is reported.[252] The enantiomeric excess of the chiral fatty acids is measured by natural abundance deuterium (NAD) 2D-NMR in a polypeptide based chiral liquid crystal.[253] Chiral solutes can seed the formation of enantiomorphic domains in a twist-bend nematic liquid crystal. This seeding of the domains in an enantiomorphic nematic conglomerate is revealed by deuterium NMR spectroscopy.[254] The chirality of a twist–bend nematic phase formed by the methylene-linked liquid crystal dimer 1″,7″-bis(4-cyanobiphenyl-4′-yl) heptane (CB7CB), is identified by NMR.[255] ^{2}H NMR spectroscopy is used to characterize changes in the liquid crystal phase behavior and orientational order resulting from the dispersion of graphene nanoplatelets within the nematic 4-cyano-4′-*n*-pentyl-biphenyl (5CB) liquid crystal.[256] NMR diffusometry, ^{2}H NMR spectroscopy and other techniques are employed to determine relations between counterion transport, dynamic coupling of water, and molecular alignment in aqueous solutions of a rigid rod sulfonated-aramid polyelectrolyte: poly(2,2′-disulfonyl-4,4′-benzidine terephthalamide) (PBDT).[257] A simplified theoretical model is developed for predicting the signal intensities obtained with CP and INEPT under MAS for anisotropic liquid crystalline systems, where the dynamics and the anisotropy of C–H bond reorientation are quantified with the rotational correlation time τ_c and the order parameter S_{CH}, respectively.[258] A variety of molecular dynamics simulations are carried out on disaccharide sucrose using the GLYCAM06 force field and various water models. Comparison with RDCs and spin–spin couplings show that sucrose is indeed a dynamic molecule.[259] A review on understanding and exploiting the phase behavior of mixtures of oppositely charged polymers and surfactants in water that come in many varieties, including liquid-crystalline materials, soluble complexes, structured nanoparticles, and water-insoluble surface layers is reported.[260] The complex morphology of the prototypic discotic liquid crystal charge-transfer compound 2,3,6,7,10,11-hexakis(hexyloxy)triphenylene

(HAT6)-2,4,7-trinitro-9-fluorenone(TNF) is elucidated by combining density measurements, detailed neutron diffraction studies, and sophisticated NMR techniques.[261] The field and temperature dependence of the ^{1}H NMR T_1 is measured in the isotropic and columnar phases of both chain-end fluorinated triphenylene disk like and fully hydrogenated molecules.[262] Using a hybrid of NMR approaches, the dynamics of native EmrE is probed and found that the transporter undergoes rapid conformational switching at a rate of $\sim$300 s^{-1} at 37 °C and a pH of 6.9 ($\sim$50-fold faster than TPP^{+}-bound EmrE).[263] Observations on the phase structure transition of C14G2 (*N* tetradecyllactobionamide)/C12EO4 (tetraethylene glycol monododecyl ether) mixtures in water induced by magnetic field are reported. The phase was quickly transferred from lamellar to isotropic phases.[264] The restricted isometry constants allow prediction of the optimal sampling time constant for exponentially decaying Non Uniform Sampling. The criteria for the choice of the maximal indirect evolution period is introduced and analysed the influence of the maximal indirect evolution period and the sampling time constant on the NUS sensitivity enhancement. Its application is shown for natural abundance deuterium two dimensional NMR experiments using polypeptide lyotropic chiral liquid crystals (CLCs) as aligning media.[265] A very simple system made of fatty acids (dodecanoic and tetradecanoic) and ethanolamine in water, depending on the molar ratio and the concentration, is shown to self-assemble at room temperature into a direct hexagonal phase that is oriented by moderate magnetic fields of a few tesla.[266] The effect of the shape of the mesogenic group on the structure and phase behaviour of 2,3,4-tris(dodecyloxy)benzenesulfonates with alkaline cations is reported. With increasing ion size, the mesogen shape becomes more tapered, and columnar mesophases appear to be more stable.[267] A computer program developed within the XPLOR-NIH package for the assignment of dipolar coupling (DC) and chemical shift anisotropy (CSA) restraints derived from the solid-state NMR spectra of protein samples with uniaxial order.[268] Solid state dipolar ^{13}C–^{1}H NMR spectroscopy is applied to quantitatively characterize the conformational dynamics of organic surfactants in the mesostructured composite CTAB–MCM41. The degree of order and the order parameter profile is similar to that observed in hexagonal liquid crystalline phases.[269] The exploitation of NMR spectroscopy to investigate the order, structure and dynamic of some selected examples of "Complex Systems" in the context of the liquid crystalline materials is reviewed.[270] An azobenzene derivative, namely diheptylazobenzene, showing the nematic and smectic A liquid crystalline phases, was investigated by means of a combined approach based on NMR and DFT calculations. The main outcome of this work is the determination of a relatively high and temperature-dependent molecular biaxiality of the trans state of this azomesogen.[271] The translational self-diffusion behavior of calamitic mesogens in smectic phases measured by means of NMR diffusometry is reviewed.[272] Solid State H-2 NMR was applied to the structural and orientational studies of a selectively labelled 'de Vries' liquid crystal.[273] The results obtained in the field of translational self-diffusion studies by means of H-1 NMR diffusometry in smectic phases formed by calamitic chiral liquid crystals is reported.[274] Proton NMR relaxation study of molecular dynamics of chromonic liquid crystal Edicol Sunset Yellow is reported. In the nematic phase, the collective motion are attributed to columnar undulations within the stacked molecules, while, in the isotropic phase, the results are explained by local order fluctuations due to the formation of the stacks.[275]

9 Synthesis and characterization

Large number of papers is published on the synthesis of new liquid crystals that are characterized by various analytical techniques including NMR. Mesogen-jacketed liquid crystalline polymers (MJLCPs) based on the octyl substituted biphenyl mesogenic core through different linkage groups synthesized and the molecular characterization of polymers is carried out.[276] Two novel series of first and second generation Frechet type dendritic compounds bearing amino-nitro-substituted azobenzene units and tetra(ethylene glycol) spacers are synthesized. Some of these dendrons showed a liquid crystalline behaviour.[277] A series of star mesogens based on 1,3,5-benzenetricarboxylic acid central core, side arms containing two phenyl rings, linked through azomethine groups and hexamethylene spacers, which connect the central core and side arms[278] and also mesogens based on isophthalic acid and side arms containing two and three phenyl rings linked through azomethine or ester groups[279] are synthesised *via* divergent approach. A series of 1,3-dimethyl-5-(3,4,5-tris(alkoxy)benzoyl) barbituric acid derivatives and their difluoroboron complexes are synthesized and characterized by various techniques including multinuclear NMR.[280] Organogelators based on pyrazine-2,5-dicarboxylic acid derivatives and their mesomorphic behaviors are investigated.[281] The novel calixarene-linked discotic triphenylene are synthesized *via* click chemistry. The structural and conformational characterization of new compounds have been achieved by NMR, MS, and elemental analysis.[282] NewSchiff base dimers based on phenothiazine heterocycle have been obtained. The obtained compounds have high thermal stability and some of them self-assemble into cubic mesophase and form stable molecular glass.[283] Novel liquid crystal trimers with a flexible tris(2-aminoethyl)amine (TREN) core and cyanobiphenyl arms are designed. These non-traditional mesogens form nematic mesophase over a broad temperature range depending on the alkyl chain length.[284] A new series of liquid-crystalline tapered hydrazide derivatives with an amino head group, *e.g.* *N*-(3,4,5-trialkoxylbenzoyl)-*N*_-(4_-amino-benzoyl) hydrazine (D*n*, where *n* is the number of carbon atoms in the alkyl chains, $n=6$, 8, 16), were designed and synthesized. D6 is non-mesomorphic, D8 exhibits a monotropic hexagonal columnar mesophase, while D16 shows enantiotropic oblique columnar phases.[285] A new series of Ni(II)–salphen complexes bearing differently substituted aromatic spacer are synthesized and characterized.[286] A linear hydrogen bonded liquid crystalline homologous series has been synthesized and characterized. All the seven synthesized complexes exhibit rich liquid crystalline mesomorphism.[287] New room-temperature thermotropic perylene-based bisimides are synthesized which possess liquid crystalline properties at room temperature and a single mesophase structure exists in wide range of temperatures.[288] A mesogenic Schiff-base, *N*,*N*′-di-(4′-octyloxybenzoatesalicylidene)-l″,8″-diamino-3″,6″-dioxaoctane; H-$_2$dobsdd (H2L3), that shows nematogenic mesophase is synthesized.[289] The same group also synthesized another mesogenic Schiff-base, *N*,*N*′-di-(4-decyloxysalicylidene)-2′,6′-diaminopyridine, H2ddsdp (abbreviated as H2L3) that exhibits nematic mesophase.[290] A novel kind of metallophthalocyanine derivative, zinc 2,9,16,23-tetraaminophthalocyanine perfluorooctanoate (Zn-APc-pFOA), was synthesized from zinc tetraaminophthalocyanine and perfluorooctanoic acid. The chemical structure, liquid crystalline behavior, and electrorheological properties are characterized by the use of various experimental techniques.[291] New lanthanide complexes, Tb(III), Dy(III) and Gd(III) of *N*-aryl based Schiff bases exhibiting mesomorphism are synthesized. The crossover phenomena of calamitic lamellar to discotic columnar phase behaviour in these lanthanidomesogens is found to be controlled by the variation in chain length of the single alkyl substituent on the terminal *N*-aryl ring.[292] A novel set of discotic molecules comprising a rigid OPE core, three amide groups, and peripheral solubilising wedge groups featuring C3 and C2 core symmetry is synthesised. In dilute methylcyclohexane solution, one-dimensional supramolecular polymers are formed stabilised by threefold intermolecular hydrogen bonding and π–π interaction.[293] Two triazine-based unconventional dendrimers were prepared. These dendrimers display columnar liquid-crystalline phases during thermal treatment.[294] New liquid crystalline materials based on two generations of dendronised cyclophosphazenes are synthesized.[295] The synthesis, structure, and properties of four new chiral LC monomers and homopolymers based on menthol are reported.[296] New class of photo and electrically switchable azobenzene containing pendant bent-core liquid crystalline monomers and their polymers are synthesized and characterized.[297] New cholesteric liquid-crystalline copolymers containing chiral mesogenic groups derived from menthol derivatives are synthesed and their phase behavior is investigated.[298] Cholesteric monomer and elastomers containing menthyl groups are synthesised and phase behavior is investigated.[299] A series of new chalcone derivatives, 1-(4′-undecylcarbonyloxyphenyl)-3-(X-substituted phenyl)-2-propen-1-one, is synthesized. The physical properties as well as their mesomorphic behaviors are studied by using spectroscopic techniques.[300] Cyanoethyl konjac glucomannan (CKGM) products with various degree of substitution (DS) are synthesized.[301] Unsymmetrical *N*-[4-(alkoxy)-2-hydroxybenzylidene]-*N*′-(4-alkoxybenzylidene) azines and their copper (ii) complexes are synthesized and their mesomorphic properties have been studied.[302] A new series of acid dimers, *N*-(4′-carboxybenzylidene)-4-alkoxybenzoyloxyaniline [(4″-HOOCC6H4CH=N)-4-C6H4OCOC6H4OR, where $R = C_mH_{2m+1}$ and $m = 6$, 8, 10, 12, 14, and 16] are synthesized. The compounds with $m = 6$ and 8 show a characteristic enantiotropic nematic mesophase, whereas those with $m = 12$ and 14 exhibit monotropic behavior. The compounds with $m = 10$ and 16 exhibit non-mesomorphic behaviour.[303] A series of side-chain liquid crystalline oligomers (P1–P7) with cyclo(methylhydrogeno)siloxane and two cholesteric liquid crystalline monomers cholesteryl 4-(10-undecylen-1-yloxy)benzoate (M1) and cholesterol 4-{6-[(4-(allyloxyl)-benzoyl]-hexanoxocarbonyl}-benzoate (M2) are synthesized.[304] The chemical structures and phase behaviour of a series of side-chain liquid crystalline polysiloxanes bearing the cholesteric mesogens and the fluorinated nematic mesogens have been investigated. All chiral polymers show wide mesophase temperature ranges and a high thermal stability with decomposition temperatures (Td) at 5% weight loss greater than 300 °C.[305] The photo and electrical switching properties of new series of synthesized bent-core liquid crystals are investigated.[306] A new polymerizable macromer containing naphthalene ring in the mesogen, 4-butoxyphenyl 6-(6-(methacryloyloxy)hexyl)-2-naphthoate, has been synthesized.[307] Side-chain liquid-crystal polymers (SCLCPs) with different molecular weights were successfully prepared by radical polymerization.[308] Liquid crystalline cholesterol-based side-chain (Chol-C6) and a series of poly(monomethyl itaconate) (PMMI) with various contents of mesogenic side-chains, having even methylene spacers and ester linkages, were successfully prepared and characterized in detail.[309] Fifteen new branched triethanolamine esters of the benzoic acid derivatives were designed and the mesomorphism typical of discotic mesogens was forecast for them. In eight of the obtained compounds a manifestation of liquid crystal properties was predicted, for four derivatives equiprobable prognosis was given, and only three compounds were predicted not to form a mesophase.[310] A macroscopically oriented lamello-columnar mesophase was obtained from a symmetric tetrathiafulvalene-based molecule containing four flexible alkyl chains and its phase behaviors and physical properties are investigated by the combined experimental techniques.[311] The synthesis and self-assembly properties of a fluorinated C_3-symmetrical 3,3 bis(acylamino)-2,2-bipyridine discotic in the mesophase and in solution are described. The columnar liquid crystallinity is displayed over a temperature range of more than 350 K.[312] A series of main chain thermotropic liquid crystalline copolyesters containing a non-coplanar phosphorus monomer were synthesized and characterized.[313] Unexpected liquid crystalline behaviour of three-ring bent-core mesogens: bis(4-subst.-phenyl) 2-methyl-iso-phthalates, exhibiting nematic, SmA and SmC phases is reported. These types of molecules represent the borderline between bent-shaped and calamitic liquid crystals.[314] A pyridine based nematogenic acceptor has been synthesized and mixed with non-mesogenic 4-methoxy benzoic acid to get a hydrogen bonded mesogen. Studies using 2D separated local field NMR experiments confirm the molecular structure in the mesophase and yield the local orientational order parameters.[315]

The phase behavior of two photoresponsive side-chain liquid crystalline polymers bearing branched azobenzene-containing side-chains are explored. Both polymers formed the smectic-A mesophase in the LC states, while the polymer containing three azobenzene-containing branches in each repeating unit showed a higher order phase in the lower temperature range.[316] Betaine-ester analogues of 1-*N*-alkyl-3-*N*′-methyl imidazolium salts exhibiting both thermotropic and lyotropic liquid crystal behaviors are reported. These compounds are gelators, which form gels in a variety of organic solvents such as chloroform, methanol, ethanol, tetrahydrofuran, while also exhibiting a lamellar mesophase.[317] Two new sets of tetrazole liquid crystals were synthesised and fully characterized.[318] Synthesis, the structure and mesophase characterization of star mesogens have been carried out using 1D and 2D solution NMR techniques. The nematic phase was observed for the mesogens with short terminal chain length, whereas smectic polymorphism was observed on increasing the terminal chain length.[319] An efficient route to stabilize alumina mesophases derived from evaporation-induced self-assembly is reported.[320] A series of *ortho*-metallated palladium(II) complexes with two dimeric liquid crystals Schiff base (methoxy and decyloxy as terminal groups) as cyclometallated ligands and *N*-aryl-*N*′-benzoyl thiourea derivatives as co-ligands are investigated for their liquid crystalline properties.[321] The mesogenic amine-capped nematic gold nanoparticles with surface-enhanced plasmonic resonances are thermally and chemically stable and show nematic phase behavior close to room temperature.[322] Investigations on supramolecularly engineered perylene bisimide assemblies exhibiting thermal transition from columnar to multilamellar structures suggested that they form well-defined columnar assemblies in organic solvents through the formation of hydrogen-bonded rosettes.[323] The complex morphology of the prototypic discotic liquid crystal charge-transfer compound HAT6-TNF is elucidated and found that the liquid crystalline HAT6 columns are still present in the charge transferred compound, although the neighboring molecules are slid laterally leading to a tilted column.[324] Structural characterizations of two rod like mesogens consisting of three phenyl ring core with a polar cyano terminal and four phenyl ring core with flexible dodecyl terminal chain are reported. The three-ring-core mesogen with cyano terminal exhibits enantiotropic smectic A phase while the four-ring mesogen reveals polymesomorphism and shows enantiotropic nematic, smectic C, and tilted hexatic phases.[325] Proton/Deuteron Exchange in Aqueous Solutions of Ionic Liquids Forming Hydrogen bond has been monitored and the role of anions, self-aggregation, and mesophase formation is discussed.[326] Novel thiophene mesogens built with two phenyl ring core and alkoxy terminal chains are synthesized by multistep route and characterized.[327] Oxymethylene-bridged bent core liquid crystals are synthesized and their structures were confirmed.[328] Symmetrical bent-core liquid crystalline dimers with diacetylene spacer are designed, synthesised, and characterised.[329] synthesis, characterisation and liquid crystalline properties of some Schiff base and cinnamate central linkages involving 1,3,5-trisubstituted pyrazolone ring system and their Cu(II) complexes are reported.[330] New chiral liquid crystal monomers and cholesteric polyacrylates[331] and new side chain liquid crystalline oligomers containing salicylaldimine mesogenic groups are reported.[332] Evaluation of liquid crystalline properties of a novel homologous series:-3-[4-*n*-alkoxy benzoyloxy] phenyl-4-nitro benzoyl ethylenes[333] And a seriese of symmetrical three-armed star-shaped mesogens based on phloroglucinol (1,3,5-trihydroxybenzene) as a core unit containing 94-benzylidene-substituted-aniline-4-oxy)-6-bromohexane as peripheral arm where the 4-position at one end consisting different terminal substituent, X (X = fluoro, chloro, bromo, iodo and ethyl) are reported.[334] Supramolecular hydrogen bonded liquid crystals comprising of *p-n*-alkyloxy benzoic acids with suberic acid and pimelic acid are synthesized and characterized.[335] Two series of hydrogen-bonded liquid crystal complexes are obtained from binary mixtures of 4,4′-bipyridine and 2-fluoro-6-X-4-alkoxybenzoic acids (where X = H or F) with a molar ratio of 2 : 1.[336] The side-chain liquid crystalline polymers containing terminally substituted biphenyl cinnamoyl esters were synthesized and characterized.[337] A novel photoluminescent room-temperature liquid-crystalline salicylaldimine Schiff base with a short alkoxy substituent and a series of lanthanide (III) complexes of the type [$Ln(LH)_3(NO_3)_3$] (Ln = La, Pr, Sm, Gd, Tb, Dy; LH = (*E*)-5-(hexyloxy)-2-[{2-(2-hydroxyethylamino)ethylimino}-methyl}-phenol) are synthesized and characterized.[338] The chiral amino acid derived ionic liquid crystals with amine and amide moieties as spacers between the imidazolium head group and the alkyl chain are synthesised.[339] The synthesis of star-shaped discotic liquid crystal trimers using $Co_2(CO)_8$-catalyzed terminal alkyne [2 + 2 + 2] cycloaddition reaction is reported.[340] Self-assembly of Silver(I) and Gold(I) *n*-heterocyclic carbene complexes in solid state, mesophase, and solution is reported. The technique of NMR DOSY is found for the first time to be a good tool to study the phase transition of gels.[341] Liquid crystalline azobenzene containing triped and tetraped monomers were designed and synthesized.[342] Synthesis of three stereoisomers of disodium chromonyl carboxylate, 5′DSCG-diviol, and the correlation between the molecular structure, bulk assembly, and liquid crystal formation is reported. It was observed that the chiral isomers (enantiomers 5′DSCG-(*R*,*R*)-diviol and 5′DSCG-(*S*,*S*)-diviol) formed liquid crystals while the achiral isomer 5′DSCG-meso-diviol did not.[343] The 4-(cyanomethoxy)phenyl 4-alkoxybenzoates, 4-(cyanomethoxy)-4′-alkoxyazo- and –azoxybenzenes were synthesised and their composition and structure were proved by elemental analysis and ^{1}H NMR spectra. Only four among the compounds obtained exhibited the thermotropic nematic mesomorphism.[344] Synthesis, physico-chemical characterization and mesomorphic properties of some novel azoderivatives is reported.[345] The synthesis and self-assembly behavior of coil-rod-coil molecules, consisting of three biphenyls linked through a vinylene unit as a conjugated rod segment and poly(ethylene oxide) (PEO) with a degree of polymerization of 7, 12 and 17, incorporating lateral methyl groups between the rod and coil segments as the coil segment is reported.[346] Syntheses and mesomorphic properties of novel triphenylene liquid crystals with 15-crown-5 unit as side-chain.[347] and of chiral liquid crystalline elastomers containing four-arm crosslinking agent are reported.[348] The synthesis and mesomorphic behavior of new mesogenic compounds possessing a biphenyl ester moiety with a 6-amino-1,3-dimethyluracil[349] new family of photoactive liquid crystalline polyesters based on alpha-methylstilbene[350] the influence of multiple H-bonding on mesomorphic properties of novel triphenylene derivatives with acylthiosemicarbazide group[351] and the mesophase characterization of three ring based thiophene liquid crystals are reported.[352] New Columnar Liquid Crystal Containing 1,3,5-triiminebenzene with Pendant 1,3,4-thiadiazole Group[353] liquid crystalline alkylthia-substituted novel phthalocyanines[354] a series of several new families of ferrocene-containing imidazolium salts[355] also a series of aromatic liquid crystalline copolyesters P-BPAx with lower melting temperature and higher glass transition temperature derived from hydroxybenzoic acid (HBA), 6-hydroxy-2-naphthoic acid (HNA), bisphenol A (BPA) and terephthalic acid (TA) were synthesized and characterized.[356] A series of new symmetrical trimeric star-shaped mesogens in which the molecular architecture composed of three 4-[(4-substituted-phenyl)diazenyl]phenoxy-6-bromohexane connected as the pheripheral units to 1,3,5-positions of the benzene core group[357] and of reactive liquid crystals containing an azo group are characterized.[358]

References

1 K. V. Ramanathan, N. P. Lobo and C. L. Khetrapal, in *Nuclear Magnetic Resonance*, ed. G. A. Webb and K. Kamienska-Trela, 2010, vol. 39, p. 456.
2 Y. J. Reddy and K. V. Ramanathan, *Nuclear Magnetic Resonance*, ed. K. Kamienska-Trela and J. Wojcik, 2013, vol. 42, p. 535.
3 A. Saupe and G. Englert, *Phys. Rev. Lett.*, 1963, **11**, 462.
4 A. Saupe, *Z. Naturforsch.*, 1964, **19a**, 161.
5 H. Pleiner, *Liq. Cryst.*, 2010, **37**, 619.
6 Y. Ronald, Dong in Nuclear Magnetic Resonance and Electron Spin Resonance Spectroscopy, *Encyclopaedia of Analytical Chemistry*, John Wiley & Sons Ltd, 2006.
7 Anil Kumar and N. Suryaprakash, in *Multidimensional NMR Methods for the Solution State*, ed. G. A. Morris and J. W. Emsley, John Wiley & Sons Ltd, 2010.
8 *Thermotropic Liquid Crystals: Recent Advances*, ed. A. Ramamoorthy, Springer, Dordrecht, The Netherlands, Chichester, UK 2007, p. 401.
9 B. Baishya, U. R. Prabhu and N. Suryaprakash, *Annu. Rep. NMR Spectrosc.*, 2009, **67**, 331.
10 A. Haloui and E. Haloui, *J. Mol. Struct.*, 2014, **1068**, 140.
11 A. C. J. Weber, R. Y. Dong, W. L. Meerts, X. Yang and E. E. Burnell, *J. Phys. Chem. A*, 2013, **117**, 9224.
12 A. C. J. Weber, A. Pizzirusso, L. Muccioli, C. Zannoni, W. L. Meerts, C. A. de Lange and E. E. Burnell, *J. Chem. Phys.*, 2012, **136**, 174506.
13 L. Calucci, E. Carignani, M. Geppi, S. Macchi, B. Mennucci and S. Urban, *J. Phys. Chem. B*, 2014, **118**, 3469.
14 V. E. Bahamonde-Padilla, M. Martínez-Cifuentes, D. muñoz-masson, Á. Ruiz, H. Ahumada, R. Araya-Maturana, J. Soto-Delgado and B. E. Weiss-López, *J. Chil. Chem. Soc.*, 2012, **57**, 1295.
15 M. E. Di Pietro, G. Celebre, G. De Luca, H. Zimmermann and G. Cinacchi, *Eur. Phys. J. E: Soft Matter Biol. Phys.*, 2012, **35**, 112.
16 U. H. N. Dürr, S. Afonin, B. Hoff, G. de Luca, J. W. Emsley and A. S. Ulrich, *J. Phys. Chem. B*, 2012, **116**, 4769.
17 A. Pizzirusso, M. E. Di Pietro, G. De Luca, G. Celebre, M. Longeri, L. Muccioli and C. Zannoni, *ChemPhysChem*, 2014, **15**, 1356.

18 V. E. Bahamonde-Padilla, J. J. López-Cascales, R. Araya-Maturana, M. Martínez-Cifuentes and B. E. W. López, *ChemPhysChem*, 2014, **15**, 1422.
19 R. Y. Dong, A. Kohlmeier, M. G. Tamba, G. H. Mehl and E. E. Burnell, *Chem. Phys. Lett.*, 2012, **552**, 44.
20 M. Dama and S. Berger, *Org. Letts.*, 2012, **14**, 241.
21 M. Dama and S. Berger, *Tetrahedron Lett.*, 2012, **53**, 6439.
22 M. Dama and S. Berger, *Carbohydr. Res.*, 2013, **377**, 44.
23 F. Huang, Y. Pei, H. Zuo, J. Chen, Y. Yang and X. Su, *Chem. – Eur. J.*, 2013, **19**, 17141.
24 Lokesh and N. Suryaprakash, *Chem. Commun.*, 2013, **49**, 2049.
25 Lokesh and N. Suryaprakash, *Chem. – Eur. J.*, 2012, **18**, 11560.
26 Marie Hutin, J. K. Sprafke, B. Odell, H. L. Anderson and T. D. W. Claridge, *J. Am. Chem. Soc.*, 2013, **135**, 12798.
27 P. Kaden, J. C. Freudenberger and B. Luy, *Magn. Reson. Chem.*, 2012, **50**, S22.
28 A. Krupp and M. Reggelin, *Magn. Reson. Chem.*, 2012, **50**, S45.
29 G. Bellot, M. A. McClintock, J. J. Chou and W. M. Shih, *Nat. Protoc.*, 2013, **8**, 755.
30 C. Merle, G. Kummerlöwe, J. C. Freudenberger, F. Halbach, W. Stöwer, C. L. v. Gostomski, J. Hçpfner, T. Beskers, M. Wilhelm and B. Luy, *Angew. Chem., Int. Ed.*, 2013, **52**, 10309.
31 P. Judeinstein, S. Huet and P. Lesot, *RSC Adv.*, 2013, **3**, 16604.
32 E. Troche-Pesqueira, M. Cid and A. Navarro-Vázquez, *Org. Biomol. Chem.*, 2014, **12**, 1957.
33 T. Montag and C. M. Thiele, *Chem. – Eur. J.*, 2013, **19**, 2271.
34 N. Meyer, A. Krupp, V. Schmidts, C. M. Thiele and M. Reggelin, *Angew. Chem., Int. Ed.*, 2012, **51**, 8334.
35 M. Schmidt, H. Sun, A. Leonov, C. Griesinger and U. M. Reinscheid, *Magn. Reson. Chem.*, 2012, **50**, S38.
36 Y. Ding, Y. Yao and F. M. Marassi, *Acc. Chem. Res.*, 2013, **46**, 2182.
37 S. J. Opella, *Annu. Rev. Anal. Chem.*, 2013, **6**, 305.
38 C. M. Thiele and W. Bermel, *J. Magn. Reson.*, 2012, **216**, 134.
39 B. Yu, H. van Ingen and D. I. Freedberg, *J. Magn. Reson.*, 2013, **228**, 159.
40 H. F. Azurmendi and D. I. Freedberg, *J. Magn. Reson.*, 2013, **228**, 130.
41 N. G. A. Bell, G. Rigg, S. Masters, J. Bella and D. Uhrín, *Phys. Chem. Chem. Phys.*, 2013, **15**, 18223.
42 P. Giraudeau, T. Montag, B. Charrier and C. M. Thiele, *Magn. Reson. Chem.*, 2012, **50**, S53.
43 N. P. Lobo and K. V. Ramanathan, *Chem. Phys. Lett.*, 2012, **536**, 155.
44 B. B. Das, E. C. Lin and S. J. Opella, *J. Phys. Chem. B*, 2013, **117**, 12422.
45 G. J. Lu and S. J. Opella, *J. Chem. Phys.*, 2013, **139**, 084203.
46 S. Hebbar, U. R. Prabhu and N. Suryaprakash, *J. Magn. Reson.*, 2012, **215**, 23.
47 Z. Serhan, A. Borgogno, I. Billault, A. Ferrarini and P. Lesot, *Chem. – Eur. J.*, 2012, **18**, 117.
48 J. Ying, J. Roche and A. Bax, *J. Magn. Reson.*, 2014, **241**, 97.
49 W. Tang, R. W. Knox and A. A. Nevzorov, *J Biomol. NMR*, 2012, **54**, 307.
50 A. Ramamoorthy and J. Xu, *J. Phys. Chem. B*, 2013, **117**, 6693.
51 K. Nomura, E. Harada, K. Sugase and K. Shimamoto, *J. Phys. Chem. B*, 2014, **118**, 2405.
52 M. Chan-Huot, P. Lesot, P. Pelupessy, L. Duma, G. Bodenhausen, P. Duchambon, M. D. Toney, U. V. Reddy and N. Suryaprakash, *Anal. Chem.*, 2013, **85**, 4694.
53 K. Koch, S. Afonin, M. Ieronimo, M. Berditsch and A. S. Ulrich, *Top. Curr. Chem.*, 2012, **306**, 89.

54 I. Timári, L. Kaltschnee, A. Kolmer, R. W. Adams, M. Nilsson, C. M. Thiele, G. A. Morris and K. E. Kövér, *J. Magn. Reson.*, 2014, **239**, 130.
55 P. Giraudeau, T. Montag, B. Charriera and C. M. Thiele, *Magn. Reson. Chem.*, 2013, **51**, 192.
56 E. C. Lin and S. J. Opella, *J. Magn. Reson.*, 2014, **239**, 57.
57 G. J. Lu, S. H. Park and S. J. Opella, *J. Magn. Reson.*, 2012, **220**, 54.
58 E. C. Lin and S. J. Opella, *J. Magn. Reson.*, 2013, **237**, 40.
59 G. J. Lu and S. J. Opella, *J. Biomol. NMR*, 2014, **58**, 69.
60 B. B. Das, H. J. Nothnagel, G. J. Lu, W. S. Son, Y. Tian, F. M. Marassi and S. J. Opella, *J. Am. Chem. Soc.*, 2012, **134**, 2047.
61 P. Trigo-Mouriño, C. Merle, M. R. M. Koos, B. Luy and R. R. Gil, *Chem. - Eur. J.*, 2013, **19**, 7013.
62 M. Palomino, H. Khudr, J. Courtieu, D. Merlet and A. Meddoura, *Magn. Reson. Chem.*, 2012, **50**, S12.
63 P. M. Macdonalda, Q. Saleema, A. Lai and H. H. Moralesa, *Chem. Phys. Lipids*, 2013, **166**, 31.
64 A. Nowacka, N. A. Bongartz, O. H. S. Ollila, T. Nylander and D. Topgaard, *J. Magn. Reson.*, 2013, **230**, 165.
65 J. Ying, J. Wang, A. Grishaev, P. Yu, Y. Wang and A. Bax, *J. Biomol. NMR.*, 2011, **51**, 89.
66 P. Trigo-Mouriño, A. Navarro-Vázquez, J. Ying, R. R. Gil and A. Bax, *Angew Chem., Int. Ed.*, 2011, **50**, 7576.
67 T. Reinsperger and B. Luy, *J. Magn. Reson.*, 2014, **239**, 110.
68 T. Park, S. Choi, J. Jung, Y. Park and Y. Kim, *Bull. Korean Chem. Soc.*, 2013, **34**, 823.
69 S. Choi, J. Jung, Y. Park, T. Park, G. H. J. Park and Y. Kim, *Bull. Korean Chem. Soc.*, 2012, **33**, 1577.
70 P. Lesot, Z. Serhan, C. Aroulanda and I. Billault, *Magn. Reson. Chem.*, 2012, **50**, S2; P. Lesot, Z. Serhan, C. Aroulanda and I. Billault, *Magn. Reson. Chem.*, 2013, **51**, 444.
71 M. Fleury and D. Canet, *J. Phys. Chem. C*, 2014, **118**, 4733.
72 E. S. Salnikov, O. Ouari, E. Koers, H. Sarrouj, T. Franks, M. Rosay, S. Pawsey, C. Reiter, P. Bandara, H. Oschkinat, P. Tordo, F. Engelke and B. Bechinger, *Appl. Magn. Reson.*, 2012, **43**, 91.
73 S. Kosol, S. Contreras-Martos, C. Cedeño and P. Tompa, *Molecules*, 2013, **18**, 10802.
74 A. Gautier, *Biochim. Biophys. Acta*, 2014, **1837**, 578.
75 T. Gopinath, K. R. Mote and G. Veglia, *Prog. Nucl. Magn. Reson. Spectrosc.*, 2013, **75**, 50.
76 A. Sylvester, L. MacEachern, V. Booth and M. R. Morrow, *PLoS One*, 2013, **8**, e72248.
77 O. V. Nolandt, T. H. Walther, S. L. Grage and A. S. Ulrich, *Biochim. Biophys. Acta*, 2012, **1818**, 1142.
78 S. J. Opella, *Acc. Chem. Res.*, 2013, **46**, 2145.
79 A. R. Choudhury, A. Perdih, Š. Župerl, E. Sikorska, T. Solmajer, S. J. I. Zhukov and M. Novič, *Biochim. Biophys. Acta*, 2013, **1828**, 2609.
80 P. Wadhwani, E. Strandberg, N. Heidenreich, J. Bürck, S. Fanghänel and A. S. Ulrich, *J. Am. Chem. Soc.*, 2012, **134**, 6512.
81 M. J. Klein, S. L. Grage, C. Muhle-Goll, J. Bürck, S. Afonin and A. S. Ulrich, *Biochim. Biophys. Acta*, 2012, **1818**, 3025.
82 W. Hohlweg, S. Kosol and K. Zangger, *Curr. Protein Pept. Sci.*, 2012, **13**, 267.
83 S. L. Grage, E. Strandberg, P. Wadhwani, S. Esteban-Martín, J. Salgado and A. S. Ulrich, *Eur. Biophys. J.*, 2012, **41**, 475.

84 B. M. Kroncke and L. Columbus, *Biomol. NMR Assignments*, 2013, 7, 203.
85 S. Unnerståle, F. Madani, A. Gräslund and L. Mäler, *Biochemistry*, 2012, **51**, 3982.
86 J. Björnerås, A. Gräslund and L. Mäler, *Biochemistry*, 2013, **52**, 4157.
87 D. C. R. Camargo, N. M. Link and S. A. Dames, *Biochemistry*, 2012, **51**, 4909.
88 J. Suk, A. J. Situ and T. S. Ulmer, *J. Am. Chem. Soc.*, 2012, **134**, 9030.
89 W. Surya, Y. Li, O. Millet, T. Diercks and J. Torres, *PLoS One*, 2013, **8**, e74281.
90 T. Yamaguchi, T. Uno, Y. Uekusa, M. Yagi-Utsumi and K. Kato, *Chem. Commun.*, 2013, **49**, 1235.
91 A. A. Nevzorov, *J. Phys. Chem. B*, 2011, **115**, 15406.
92 R. Verardi, N. J. Traaseth, L. R. Masterson, V. V. Vostrikov and G. Veglia, *Adv. Exp. Med. Biol.*, 2012, **992**, 35.
93 E. A. Morrison, G. T. DeKoster, S. Dutta, R. Vafabakhsh, M. W. Clarkson, A. Bahl, D. Kern, T. Ha and K. A. Henzler-Wildman, *Nature*, 2012, **481**, 45.
94 Z. O. Shenkarev, A. S. Paramonova, E. N. Lyukmanova, A. K. Gizatullina, A. V. Zhuravleva, A. A. Tagaev, Z. A. Yakimenko, I. N. Telezhinskaya, M. P. Kirpichnikov, T. V. Ovchinnikova and A. S. Arseniev, *Chem. Biodiversity*, 2013, **10**, 838.
95 E. Strandberg, D. Tiltak, S. Ehni, P. Wadhwani and A. S. Ulrich, *Biochim. Biophys. Acta*, 2012, **1818**, 1764.
96 D. T. Murray, I. Hung and T. A. Cross, *J. Magn. Reson.*, 2014, **240**, 34.
97 C. Fotakis, G. Megariotis, D. Christodouleas, E. Kritsi, P. Zoumpoulakis, D. Ntountaniotis, M. Zervou, C. Potamitis, A. Hodzic, G. Pabst, M. Rappolt, G. Mali, J. Baldus, C. Glaubitz, M. G. Papadopoulos, A. Afantitis, G. Melagraki and T. Mavromoustakos, *Biochim. Biophys. Acta*, 2012, **1818**, 3107.
98 N. J. Gleason, V. V. Vostrikov, D. V. Greathouse, C. V. Grant, S. J. Opella and R. E. Koeppe II, *Biochemistry*, 2012, **51**, 2044.
99 J. M. Rankenberg, V. V. Vostrikov, C. D. DuVall, D. V. Greathouse, R. E. Koeppe II, C. V. Grant and S. J. Opella, *Biochemistry*, 2012, **51**, 3554.
100 B. S. J. Perrin, Y. Tian, R. Fu, C. V. Grant, E. Y. Chekmenev, W. E. Wieczorek, A. E. Dao, R. M. Hayden, C. M. Burzynski, R. M. Venable, M. Sharma, S. J. Opella, R. W. Pastor and M. L. Cotton, *J. Am. Chem. Soc.*, 2014, **136**, 3491.
101 P. Schmidt, L. Thomas, P. Müller, H. A. Scheidt and D. Huster, *Chem. – Eur. J.*, 2014, **20**, 4986.
102 D. M. Tesch and A. A. Nevzorov, *J. Magn. Reson.*, 2014, **239**, 9.
103 A. D. Simone, K. R. Mote and G. Veglia, *Biophys. J.*, 2014, **106**, 2566.
104 S. K. Hansen, M. Vestergaard, L. Thøgersen, B. Schiøtt, N. C. Nielsen and T. Vosegaard, *J. Phys. Chem. B*, 2014, **118**, 5119.
105 S. H. Park, B. B. Das, F. Casagrande, Y. Tian, H. J. Nothnagel, M. Chu, H. Kiefer, K. Maier, A. A. D. Angelis, F. M. Marassi and S. J. Opella, *Nature*, 2012, **491**, 779.
106 U. H. N. Dürr, R. Soong and A. Ramamoorthy, *Prog. Nucl. Magn. Reson. Spectrosc.*, 2013, **69**, 1.
107 M. M. Reif, M. Winger and C. Oostenbrink, *J. Chem. Theory Comput.*, 2013, **9**, 1247.
108 L. A. M. Sommer, J. J. Janke, W. F. D. Bennett, J. Bürck, A. S. Ulrich, D. P. Tieleman and S. A. Dames, *J. Phys. Chem. B*, 2014, **118**, 4817.
109 E. N. Lyukmanova, Z. O. Shenkarev, N. F. Khabibullina, G. S. Kopeina, M. A. Shulepko, A. S. Paramonov, K. S. Mineev, R. V. Tikhonov,

L. N. Shingarova, L. E. Petrovskaya, D. A. Dolgikh, A. S. Arseniev and M. P. Kirpichnikov, *Biochim. Biophys. Acta*, 2012, **1818**, 349.
110 N. Weizenmann, D. Huster and H. A. Scheidt, *Biochim. Biophys. Acta*, 2012, **1818**, 3010.
111 D. Palleboina, A. J. Waring, R. H. Notter, V. Booth and M. Morrow, *Eur. Biophys. J.*, 2012, **41**, 755.
112 C. Ariöz, W. Ye, A. Bakali, C. Ge, J. Liebau, H. Götzke, A. Barth, Å. Wieslander and L. Mäler, *Biochemistry*, 2013, **52**, 5533.
113 J. K. Claridge, J. Aittoniemi, D. M. Cooper and J. R. Schnell, *Biochemistry*, 2013, **52**, 8420.
114 L. Mari and DeMarco, *Biochemistry*, 2012, **51**, 5725.
115 U. H. N. Dürr, M. Gildenberg and A. Ramamoorthy, *Chem. Rev.*, 2012, **112**, 6054.
116 G. Chua, X. Tang, A. T. Patra, S. Tan and S. Bhattacharjya, *PLoS One*, 2012, **7**, e41924.
117 Z. Lu, W. D. V. Horn, J. Chen, S. Mathew, R. Zent and C. R. Sanders, *Mol. Pharmaceutics*, 2012, **9**, 752.
118 A. J. Brindley and R. W. Martin, *Langmuir*, 2012, **28**, 7788.
119 M. Li, H. H. Morales, J. Katsaras, N. Kučerka, Y. Yang, P. M. Macdonald and M. Nieh, *Langmuir*, 2013, **29**, 15943.
120 K. Yamamoto, P. Pearcy and A. Ramamoorthy, *Langmuir*, 2014, **30**, 1622.
121 M. J. Justice, D. N. Petrusca, A. L. Rogozea, J. A. Williams, K. S. Schweitzer, I. Petrache, S. R. Wassall and H. I. Petrache, *Biophys. J.*, 2014, **106**, 598.
122 E. A. Morrison and K. A. Henzler-Wildman, *Biochim. Biophys. Acta*, 2012, **1818**, 814.
123 O. V. Nolandt, T. H. Walther, S. L. Grage and A. S. Ulrich, *Biochim. Biophys. Acta*, 2012, **1818**, 1142.
124 E. V. Bocharov, K. S. Mineev, M. V. Goncharuk and A. S. Arseniev, *Biochim. Biophys. Acta*, 2012, **1818**, 2158.
125 M. Fillion, M. Noël, A. Lorin, N. Voyer and M. Auger, *Biochim. Biophys. Acta*, 2014, **1838**, 2173.
126 P. M. Macdonald, Q. Saleem, A. Lai and H. H. Morales, *Chem. Phys. Lipids*, 2013, **166**, 31.
127 U. L. Khatun, A. Gayen and C. Mukhopadhyay, *Chem. Phys. Lipids*, 2013, **170–171**, 8.
128 U. L. Khatun, S. K. Goswami and C. Mukhopadhyay, *Biophys. Chem.*, 2012, **168–169**, 48.
129 T. Yamaguchi, T. Suzuki, T. Yasuda, T. Oishi, N. Matsumori and M. Murata, *Bioorg. Med. Chem.*, 2012, **20**, 270.
130 W. S. Son, S. H. Park, H. J. Nothnagel, G. J. Lu, Y. Wang, H. Zhang, G. A. Cook, S. C. Howell and S. J. Opella, *J. Magn. Reson.*, 2012, **214**, 111.
131 E. E. Uhlemann, H. E. Pierson, R. H. Fillingame and O. Y. Dmitriev, *Protein Sci.*, 2012, **21**, 279.
132 L. A. M. Sommer, M. A. Meier and S. A. Dames, *Protein Sci.*, 2012, **21**, 1566.
133 S. Unnerståle and LenaMäler, *J. Biophys.*, 2012, **2012**, 1.
134 T. Yasuda, M. Kinoshita, M. Murata and N. Matsumori, *Biophys. J.*, 2014, **106**, 631.
135 M. J. Justice, D. N. Petrusca, A. L. Rogozea, J. A. Williams, K. S. Schweitzer, I. Petrache, S. R. Wassall and H. I. Petrache, *Biophys. J.*, 2014, **106**, 598.
136 G. Québatte, E. Kitas and J. Seelig, *Biochim. Biophys. Acta*, 2014, **1838**, 968.
137 M. Zervou, Z. Cournia, C. Potamitis, G. Patargias, S. Durdagi, S. G. Grdadolnik and T. Mavromoustakos, *Biochim. Biophys. Acta*, 2014, **1838**, 1031.

138 H. Rui, K. T. Root, J. Lee, K. J. Glover and W. Im, *Biophys. J.*, 2014, **106**, 1371.
139 D. T. Murray, C. Li, F. P. Gao, H. Qin and T. A. Cross, *Biophys. J.*, 2014, **106**, 1559.
140 R. Huang, K. Yamamoto, M. Zhang, N. Popovych, I. Hung, S. Im, Z. Gan, L. Waskell and A. Ramamoorthy, *Biophys. J.*, 2014, **106**, 2126.
141 T. Sinnige, M. Weingarth, M. Renault, L. Baker, J. Tommassen and M. Baldus, *J. Mol. Biol.*, 2014, **426**, 2009.
142 D. Pentak, *Thermochim. Acta*, 2014, **584**, 36.
143 P. J. Shermana, F. Separovicb and J. H. Bowiea, *Peptides*, 2014, **55**, 98.
144 R. Sinha, A. Joshi, U. J. Joshi, S. Srivastava and G. Govil, *Eur. J. Med. Chem.*, 2014, **80**, 285.
145 T. Uchino, Y. Matsumoto, A. Murata, T. Oka, Y. Miyazaki and Y. Kagawa, *Int. J. Appl. Pharm.*, 2014, **464**, 75.
146 A. P. Serroa, R. Galante, A. Kozica, P. Paradiso, A. M. P. S. Gonçalves da Silva, K. V. Luzyanin, A. C. Fernandes and B. Saramago, *Colloids Surf. B*, 2014, **116**, 63.
147 T. Uchino, F. Lefeber, G. Gooris and J. Bouwstra, *Eur. J. Pharm. Biopharm.*, 2014, **86**, 156.
148 T. A. Cross, V. Ekanayake, J. Paulino and A. Wright, *J. Magn. Reson.*, 2014, **239**, 100.
149 L. Sušac, R. Horst and K. Wüthrich, *ChemBioChem*, 2014, **15**, 995.
150 E. K. Tiburu and L. Shen, *J. Membr. Biol.*, 2014, **247**, 231.
151 B. Liang, D. Dawidowski, J. F. Ellena, L. K. Tamm and D. S. Cafiso, *Biochemistry*, 2014, **53**, 1485.
152 D. T. Murray, J. Griffin and T. A. Cross, *Biochemistry*, 2014, **53**, 2454.
153 K. A. Sparks, N. J. Gleason, R. Gist, R. Langston, D. V. Greathouse and R. E. Koeppe II, *Biochemistry*, 2014, **53**, 3637.
154 N. Busschaert, L. E. Karagiannidis, M. Wenzel, C. J. E. Haynes, N. J. Wells, P. G. Young, D. Makuc, J. Plavec, K. A. Jolliffe and P. A. Gale, *Chem. Sci.*, 2014, **5**, 1118.
155 D. Marquardt, J. A. Williams, J. J. Kinnun, N. Kučerka, J. Atkinson, S. R. Wassall, J. Katsaras and T. A. Harroun, *J. Am. Chem. Soc.*, 2014, **136**, 203.
156 Y. Song, K. F. Mittendorf, Z. Lu and C. R. Sanders, *J. Am. Chem. Soc.*, 2014, **136**, 4093.
157 H. Choi and C. Kim, *Bull. Korean Chem. Soc.*, 2014, **35**, 1317.
158 S. Maltsev, S. M. Hudson, I. D. Sahu, L. Liu and G. A. Lorigan, *J. Phys. Chem. B*, 2014, **118**, 4370.
159 T. M. Ferreira, D. Topgaard and O. H. S. Ollila, *Langmuir*, 2014, **30**, 461.
160 P. Wadhwani, E. Strandberg, J. van den Berg, C. Mink, J. Bürck, R. A. M. Ciriello and A. S. Ulrich, *Biochim. Biophys. Acta*, 2014, **1838**, 940.
161 M. Beaugrand, A. A. Arnold, J. Hénin, D. E Warschawski, P. T. F. Williamson and I. Marcotte, *Langmuir*, 2014, **30**, 6162.
162 L. Mouret, G. D. Costaa and A. Bondon, *Magn. Reson. Chem.*, 2014, **52**, 339.
163 S. Bibow, M. G. Carneiro, T. M. Sabo, C. Schwiegk, S. Becker, R. Riek and D. Lee, *Protein Sci.*, 2014, **23**, 851.
164 A. López-Castilla, F. Pazos, S. Schreier and J. R. Pires, *Proteins*, 2014, **82**, 1022.
165 P. K. Koumkoua, C. Aisenbrey, E. Salnikov, O. Rifi and B. Bechinger, *J. Pept. Sci.*, 2014, **20**, 526.
166 P. Bertani, V. Vidovic, T. Yang, J. Rendell, L. M. Gordon, A. J. Waring, B. Bechinger and V. Booth, *Biochim. Biophys. Acta*, 2012, **1818**, 1165.
167 B. Bechinger and E. S. Salnikov, *Chem. Phys. Lipids*, 2012, **165**, 282.

168 V. Senille, D. Lelievre, F. Paquet, N. Garnier, N. Lamb, A. Legrand, A. F. Delmas and C. Landon, *ChemBioChem*, 2013, **14**, 711.
169 K. R. Mote, T. Gopinath and G. Veglia, *J. Biomol. NMR*, 2013, **57**, 91.
170 V. Y. Postupalenko, O. M. Zamotaiev, V. V. Shvadchak, A. V. Strizhak, V. G. Pivovarenko, A. S. Klymchenko and Y. Mely, *Bioconjugate Chem.*, 2013, **24**, 1998.
171 C. Aisenbrey, N. Pendem, G. Guichard and B. Bechinger, *Org. Biomol. Chem.*, 2012, **10**, 1440.
172 E. Schrank, G. E. Wagner and K. Zangger, *Molecules*, 2013, **18**, 7407.
173 K. Bertelsen, J. Dorosz, S. K. Hansen, N. C. Nielsen and T. Vosegaard, *PLoS One*, 2012, **7**, e47745.
174 N. Das, D. T. Murray and T. A. Cross, *Nat. Protoc.*, 2013, **8**, 2256.
175 M. Hong, Y. Zhang and F. H. Hu, *Ann. Rev. Phys. Chem.*, 2012, **63**, 1.
176 H. Yao and M. Hong, *J. Am. Chem. Soc.*, 2014, **136**, 2611.
177 L. Zhang, L. Liu, S. Maltsev, G. A. Lorigan and C. Dabney-Smith, *Biochim. Biophys. Acta*, 2014, **1838**, 413.
178 F. L. Jean-Francois, J. Dai, L. Yu, A. Myrick, E. Rubin, P. G. Fajer, L. Song, H. Zhou and T. A. Cross, *J. Mol. Biol.*, 2014, **426**, 436.
179 A. Gayen, J. R. Banigan and N. J. Traaseth, *Angew. Chem., Int. Ed.*, 2013, **52**, 10321.
180 T. V. Can, M. Sharma, I. Hung, P. L. Gor'kov, W. W. Brey and T. A. Cross, *J. Am. Chem. Soc.*, 2012, **134**, 9022.
181 J. L. Lorieau, A. S. Maltsev, J. M. Louis and A. Bax, *J. Biomol. NMR*, 2013, **55**, 369.
182 A. Filippov, B. Munavirov, B. Gizatullina and O. N. Antzutkinb, *Magn. Reson. Chem.*, 2013, **51**, 1.
183 A. Rahmani, C. Knight and M. R. Morrow, *Langmuir*, 2013, **29**, 13481.
184 P. Guerry, L. Mollica and M. Blackledge, *ChemPhysChem*, 2013, **14**, 3046.
185 D. Ban, T. M. Sabo, C. Griesinger and D. Lee, *Molecules*, 2013, **18**, 11904.
186 K. Chen and N. Tjandra, *Top. Curr. Chem.*, 2012, **326**, 47.
187 A. N. Borkar, A. D. Simone, R. W. Montalvao and M. Vendruscolo, *J. Chem. Phys.*, 2013, **138**, 215103.
188 V. M. R. Kakita, K. Rachineni and J. Bharatam, *J. Mol. Struct.*, 2013, **1053**, 122.
189 O. F. Lange and D. Baker, *Proteins*, 2012, **80**, 884.
190 J. Huang, M. Gentner, N. Vajpai, S. Grzesiek and M. Blackledge, *Biochem. Soc. Trans.*, 2012, **40**, 989.
191 G. Kontaxis, *J. Biomol. NMR*, 2012, **53**, 149.
192 R. W. Montalvao, A. D. Simone and M. Vendruscolo, *J. Biomol. NMR*, 2012, **53**, 281.
193 J. M. Ward and N. R. Skrynnikov, *J. Biomol. NMR*, 2012, **54**, 53.
194 D. Shishmarev and G. Otting, *J. Biomol. NMR*, 2013, **56**, 203.
195 N. Sibille, M. Dellarole, C. Royer and C. Roumestand, *J. Biomol. NMR*, 2014, **58**, 9.
196 F. Ji, J. Jung and A. M. Gronenborn, *Biochemistry*, 2012, **51**, 2588.
197 A. Grishaev, N. J. Anthis and G. M. Clore, *J. Am. Chem. Soc.*, 2012, **134**, 14686.
198 A. S. Maltsev, A. Grishaev, J. Roche, M. Zasloff and A. Bax, *J. Am. Chem. Soc.*, 2014, **136**, 3752.
199 C. Narayanan, D. S. Weinstock, K. Wu, J. Baum and R. M. Levy, *J. Chem. Theory Comput.*, 2012, **8**, 3929.
200 V. Ozenne, R. Schneider, M. Yao, J. Huang, L. Salmon, M. Zweckstetter, M. R. Jensen and M. Blackledge, *J. Am. Chem. Soc.*, 2012, **134**, 15138.
201 J. Huang and A. D. MacKerell Jr, *J. Comput. Chem.*, 2013, **34**, 2135.

202 L. Salmon, G. Bascom, I. Andricioaei and H. M. Al-Hashimi, *J. Am. Chem. Soc.*, 2013, **135**, 5457.
203 H. Tossavainen, O. Koskela, P. Jiang, J. Ylänne, I. D. Campbell, I. Kilpeläinen and P. Permi, *J. Am. Chem. Soc.*, 2012, **134**, 6660.
204 E. Meirovitch, D. Lee, K. F. A. Walter and C. Griesinger, *J. Phys. Chem. B*, 2012, **116**, 6106.
205 P. Trigo-Mouriño, A. Navarro-Vázquez and V. M. Sánchez-Pedregal, *Magn. Reson. Chem.*, 2012, **50**, S29.
206 J. D. Snider, E. Troche-Pesqueira, S. R. Woodruff, C. Gayathri, N. V. Tsarevsky and R. R. Gil, *Magn. Reson. Chem.*, 2012, **50**, S86.
207 P. Tzvetkova, B. Luy and S. Simova, *Magn. Reson. Chem.*, 2012, **50**, S92.
208 B. Vincent, N. Morellet, F. Fatemi, L. Aigrain, G. Truan, E. Guittet and E. Lescop, *J. Mol. Biol.*, 2012, **420**, 296.
209 L. J. Smith, A. M. Bowen, A. D. Paolo, A. Matagne and C. Redfield, *ChemBioChem*, 2013, **14**, 1780.
210 J. E. Burke, D. G. Sashital, X. Zuo, Y. Wang and S. E. Butcher, *RNA*, 2012, **18**, 673.
211 Y. Fu and A. J. Wand, *J. Biomol. NMR*, 2013, **56**, 353.
212 V. Schmidts, M. Fredersdorf, T. Lübken, A. Porzel, N. Arnold, L. Wessjohann and C. M. Thiele, *J. Nat. Prod.*, 2013, **76**, 839.
213 S. Weigelt, T. Huber, F. Hofmann, M. Jost, M. Ritzefeld, B. Luy, C. Freudenberger, Z. Majer, E. Vass, J. Greie, L. Panella, B. Kaptein, Q. B. Broxterman, H. Kessler, K. Altendorf, M. Hollósi and N. Sewald, *Chem. – Eur. J.*, 2012, **18**, 478; S. Weigelt, T. Huber, F. Hofmann, M. Jost, M. Ritzefeld, B. Luy, C. Freudenberger, Z. Majer, E. Vass, J. Greie, L. Panella, B. Kaptein, Q. B. Broxterman, H. Kessler, K. Altendorf, M. Hollósi and N. Sewald, *Chem. – Eur. J.*, 2013, **19**, 3256.
214 J. W. Emsley, M. Lelli, A. Lesage and G. R. Luckhurst, *J. Phys. Chem. B*, 2013, **117**, 6547.
215 H. M. Ge, H. Sun, N. Jiang, Y. H. Qin, H. Dou, T. Yan, Y. Y. Hou, C. Griesinger and R. X. Tan, *Chem. – Eur. J.*, 2012, **18**, 5213.
216 M. Kruck, H. Wadepohl, M. Enders and L. H. Gade, *Chem. – Eur. J.*, 2013, **19**, 1599.
217 C. Schmidt, S. J. Irausquin and H. Valafar, *BMC Bioinf.*, 2013, **14**, 302.
218 M. Damjanovic, K. Katoh, M. Yamashita and M. Enders, *J. Am. Chem. Soc.*, 2013, **135**, 14349.
219 R. Berger, J. Courtieu, R. R. Gil, C. Griesinger, M. Köck, P. Lesot, B. Luy, D. Merlet, A. Navarro-Vázquez, M. Reggelin, U. M. Reinscheid, C. M. Thiele and M. Zweckstetter, *Angew. Chem., Int. Ed.*, 2012, **51**, 8388.
220 S. I. Pantos and E. Tiligada, *Int. J. Quantum Chem.*, 2013, **113**, 1095.
221 T. M. Sabo, C. A. Smith, D. Ban, A. Mazur, D. Lee and C. Griesinger, *J. Biomol. NMR*, 2014, **58**, 287.
222 R. Montalvao, C. Camilloni, A. De Simone and M. Vendruscolo, *J. Biomol. NMR*, 2014, **58**, 233.
223 R. Shankar, S. S. Hegde and T. S. Mahesh, *Phys. Lett. A*, 2014, **378**, 10.
224 S. Dogra, Arvind and K. Dorai, *Quantum. Phys.*, 2014, **2014**, 1.
225 T. Yi-Peng, N. Xin-Fang, L. Jun, C. Hong-Wei, Z. Xian-Yi, P. Xin-Hua and D. Jiang-Feng, *Chin. Phys. Lett.*, 2012, **29**, 127601.
226 N. Xu, J. Zhu, D. Lu, X. Zhou, X. Peng and J. Du, *Phys. Rev. Lett.*, 2012, **108**, 130501.
227 A. G. Araujo-Ferreira, C. A. Brasil, D. O. Soares-Pinto, E. R. deAzevedo, T. J. Bonagamba and J. Teles, *Int. J. Quantum Inf.*, 2012, **10**, 1250016.

228 N. P. Lobo, M. Prakash, T. Narasimhaswamy and K. V. Ramanathan, *J. Phys. Chem. A*, 2012, **116**, 7508.
229 A. Filippov, B. Munavirov, B. Gizatullina and O. N. Antzutkinb, *Magn. Reson. Chem.*, 2013, **51**, 1.
230 S. Asaftei, M. Ciobanu, A. M. Lepadatu, E. Song and U. Beginn, *J. Mater. Chem.*, 2012, **22**, 14426.
231 G. D. Luca, J. W. Emsley, A. Lesage and D. Merlet, *Liq. Cryst.*, 2012, **39**, 211.
232 A. D. Buckingham, *J. Chem. Phys.*, 2014, **140**, 011103.
233 P. Lesot, Z. Serhan, C. Aroulanda and I. Billault, *Magn. Reson. Chem.*, 2012, **50**, S2; P. Lesot, Z. Serhan, C. Aroulanda and I. Billault, *Magn. Reson. Chem.*, 2013, **51**, 444.
234 J. M. Henderson and K. Y. C. Lee, *Curr. Opin. Solid State Mater. Sci.*, 2013, **17**, 175.
235 G. Widmalm, *Carbohydr. Res.*, 2013, **378**, 123.
236 D. Kim, L. Wang, Y. Cao, X. Yu, S. Z. D. Cheng, S. Kuo, D. Song, S. H. Lee, M. Lee and K. Jeong, *J. Mater. Chem.*, 2012, **22**, 16382.
237 G. Castruita, V. García, E. Arias, I. Moggio, R. Ziolo, A. Ponce, V. Gonzílez, J. E. Haley, J. L. Flikkema and T. Cooper, *J. Mater. Chem.*, 2012, **22**, 3770.
238 S. Chowdhury, L. A. Madsen and C. E. Frazier, *Biomacromolecules*, 2012, **13**, 1043.
239 P. Lesot and O. Lafon, *Anal. Chem.*, 2012, **84**, 4569.
240 I. Billault, A. L. Du, M. Ouethrani, Z. Serhan, P. Lesot and R. J. Robins, *Anal. Bioanal. Chem.*, 2012, **402**, 2985.
241 C. J. Bonin, C. E. González, H. H. Segnorile and R. C. Zamar, *J. Chem. Phys.*, 2013, **139**, 144907.
242 J. Bian, F. Peng, X. Peng, X. Xiao, P. Peng, F. Xu and R. Sun, *Carbohydr. Polym.*, 2014, **100**, 211.
243 V. Domenici, M. Lelli, M. Cifelli, V. Hamplova, A. Marchetti and C. A. Veracini, *ChemPhysChem.*, 2014, 1485.
244 L. Shi, Y. Wei, N. Suna and L. Zheng, *Chem. Commun.*, 2013, **49**, 11388.
245 K. Yamada, K. Marumo, S. Kang, K. Deguchi, T. Nakai, T. Shimizu and J. Watanabe, *J. Phys. Chem. B*, 2013, **117**, 16325.
246 A. L. dos, S. Lima, K. dos, S. C. Lima, T. C. C. França, M. I. B. Tavares, R. A. da, S. San-Gil, M. N. Eberlind and A. C. Pinto, *J. Braz. Chem. Soc.*, 2012, **23**, 1355.
247 C. S. Lovell, M. E. Ries, I. M. Ward, H. M. de Oca and D. Farrar, *Macromolecules*, 2013, **46**, 1201.
248 A. Perdih, A. R. Choudhury, Š. Župerl, E. Sikorska, I. Zhukov, T. Solmajer and M. Novič, *PLoS One*, 2012, **7**, 1.
249 P. Scholtysek, A. Achilles, C. Hoffmann, B. Lechner, A. Meister, C. Tschierske, K. Saalwächter, K. Edwards and A. Blume, *J. Phys. Chem. B*, 2012, **116**, 4871.
250 W. Ye, J. Liebau and L. Mäler, *J. Phys. Chem. B*, 2013, **117**, 1044.
251 C. Lorthioir, S. Randriamahefa and B. Deloche, *J. Chem. Phys.*, 2013, **139**, 224903.
252 B. Fuglestad, P. M. Gasper, M. Tonelli, J. A. McCammon, P. R. L. Markwick and E. A. Komives, *Biophys. J.*, 2012, **103**, 79.
253 A. Lemétais, Y. Bourdreux, P. Lesot, J. Farjon and J. Beau, *J. Org. Chem.*, 2013, **78**, 7648.
254 W. E. James, L. Philippe, R. L. Geoffrey, M. Abdelkrim and M. Denis, *Phys. Rev. E*, 2013, **87**, 040501.
255 L. Beguin, J. W. Emsley, M. Lelli, A. Lesage, G. R. Luckhurst, B. A. Timimi and H. Zimmermann, *Phys. Rev. B: Condens. Matter Mater. Phys.*, 2012, **116**, 7940.

256 T. M. Alam and C. J. Pearce, *Chem. Phys. Lett.*, 2014, **592**, 7.
257 Y. Wang, J. Gao, T. J. Dingemans and L. A. Madsen, *Macromolecules*, 2014, **47**, 2984.
258 A. Nowacka, N. A. Bongartz, O. H. S. Ollila, T. Nylander and D. Topgaard, *J. Magn. Reson.*, 2013, **230**, 165.
259 J. Xia and D. A. Case, *Biopolymers*, 2012, **97**, 276.
260 L. Piculell, *Langmuir*, 2013, **29**, 10313.
261 L. A. Haverkate, M. Zbiri, M. R. Johnson, B. Deme, H. J. M. de Groot, F. Lefeber, A. Kotlewski, S. J. Picken, F. M. Mulder and G. J. Kearley, *J. Phys. Chem. B*, 2012, **116**, 13098.
262 F. V. Chávez, P. J. Sebastião, Y. Miyake, H. Monobe and Y. Shimizu, *J. Phys. Chem. B*, 2012, **116**, 2339.
263 M. Cho, A. Gayen, J. R. Banigan, M. Leninger and N. J. Traaseth, *J. Am. Chem. Soc.*, 2014, **136**, 8072.
264 Y. Dou, S. Dong and J. Hao, *Langmuir*, 2014, **30**, 1266.
265 K. Kazimierczuk, O. Lafon and P. Lesot, *Analyst*, 2014, **139**, 2702.
266 J. Douliez, L. Navailles, E. J. Dufourc and F. Nallet, *Langmuir*, 2014, **30**, 5075.
267 M. A. Shcherbina, A. V. Bakirov, A. N. Yakunin, U. Beginn, L. Yan, M. Möller and S. N. Chvalun, *Soft Matter*, 2014, **10**, 1746.
268 Y. Tian, C. D. Schwieters, S. J. Opella and F. M. Marassi, *J. Magn. Reson.*, 2012, **214**, 42.
269 B. B. Kharkov and S. V. Dvinskikh, *Phys. Chem. Chem. Phys.*, 2013, **15**, 18620.
270 M. Cifelli, V. Domenici and C. A. Veracini, *Curr. Opin. Colloid Interface Sci.*, 2013, **18**, 190.
271 A. Marini, B. Zupančič, V. Domenici, B. Mennucci, B. Zalar and C. A. Veracini, *ChemPhysChem*, 2012, **13**, 3958.
272 M. Cifelli and C. A. Veracini, *Mol. Cryst. Liq. Cryst.*, 2013, **576**, 127.
273 D. Valentina, C. Mario, M. Alessandro, L. Moreno, H. Vera, K. Miroslav and V. C. Alberto, *Mol. Cryst. Liq. Cryst.*, 2012, **553**, 103.
274 C. Mario, D. Valentina, D. V. Sergey, V. C. Alberto and Z. Herbert, *Phase Transitions*, 2012, **85**, 861.
275 H. Cachitas, P. J. Sebastiao, G. Feio and F. V. Chavez, *Liq. Cryst.*, 2014, **41**, 1080.
276 L. Zhang and Y. Zhang, *J. Polym. Sci., Part A: Polym. Chem.*, 2013, **51**, 2545.
277 J. Ortíz-Palacios, E. Rodríguez-Alba, M. Avelar, A. Martínez, M. del, P. Carreón-Castro and E. Rivera, *Molecules*, 2013, **18**, 1502.
278 A. Shanavas, T. Narasimhaswamy and E. R. Sadiku, *J. Mol. Struct.*, 2013, **18**, 1054.
279 A. Shanavas, S. Sathiyaraj, A. Chandramohan, T. Narasimhaswamy and A. S. Nasar, *J. Mol. Struct.*, 2013, **1038**, 126.
280 E. Giziroglu, A. Nesrullajev and N. Orhan, *J. Mol. Struct.*, 2014, **246**, 1056.
281 Y. Huang, W. Tu, Y. Yuan and D. Fan, *Tetrahedron*, 2014, **70**, 1274.
282 B. Hong, F. Yang, H. Guo and Z. Jiao, *Tetrahedron Lett.*, 2014, **55**, 252.
283 A. Zabulica, M. Balan, D. Belei, M. Sava, B. C. Simionescu and L. Marin, *Dyes Pigm.*, 2013, **96**, 686.
284 S. Malkondu and A. Kocak, *J. Mol. Liq.*, 2013, **188**, 167.
285 B. Baia, C. Zhao, H. Wang, X. Ran, D. Wang and M. Li, *Mater. Chem. Phys.*, 2012, **133**, 232.
286 C. R. Bhattacharjee, C. Datta, G. Das, R. Chakrabarty and P. Mondal, *Polyhedron*, 2012, **33**, 417.
287 C. Kavitha, N. P. S. Prabu and M. L. N. M. Mohan, *Phys. B*, 2012, **407**, 859.
288 K. Bijak, H. Janeczek, M. Grucela-Zajac and E. Schab-Balcerzak, *Opt. Mater.*, 2013, **35**, 1042.

289 S. Kumari, A. K. Singh and T. R. Rao, *Mater. Science and Engineering C*, 2012, **32**, 1146.
290 P. R. Shakya, A. K. Singh and T. R. Rao, *Materials Sci. Eng., C*, 2012, **32**, 1906.
291 F. Meng, N. Zhou, N. Diao and C. Du, *Smart Mater. Struct.*, 2013, **22**, 127002.
292 N. V. S. Rao, T. D. Choudhury, R. Deb, M. K. Paul, T. R. Rao, T. Francis and I. I. Smalyukh, *Liq. Cryst.*, 2010, **37**, 1393.
293 F. Wang, M. A. J. Gillissen, P. J. M. Stals, A. R. A. Palmans and E. W. Meijer, *Chem. - Eur. J.*, 2012, **18**, 11761.
294 L. Lai, S. Wang, K. Cheng, J. Lee, T. Wang and H. Hsu, *Chem. - Eur. J.*, 2012, **18**, 15361.
295 J. Jiménez, A. Laguna, E. Gascón, J. A. Sanz, J. L. Serrano, J. Barberá and L. Oriol, *Chem. - Eur. J.*, 2012, **18**, 16801.
296 J. Hu, D. Li, W. Zhang and Q. Meng, *J. Polym. Sci., Part A: Polym. Chem.*, 2012, **50**, 5049.
297 M. V. Srinivasan, P. Kannan and A. Roy, *J. Polym. Sci. Part A: Polym. Chem.*, 2013, **51**, 936.
298 J. Hu, S. Ren, B. Zhang and C. Chao, *J. Appl. Polym. Sci.*, 2008, **109**, 2187.
299 J. Hu, D. Li, Q. Meng and D. Yao, *J. Appl. Polym. Sci.*, 2012, **125**, 3849.
300 S. T. Ha and Y. W. Low, *J. Chem.*, 2013, **2013**, 1.
301 C. Shen, W. Li, L. Zhang, C. Wan and S. Gao, *J. Polym. Res.*, 2012, **19**, 9758.
302 A. Pandey and B. Singh, *Mol. Cryst. Liq. Cryst.*, 2012, **552**, 123.
303 P. Ashwini and S. Bachcha, *Mol. Cryst. Liq. Cryst.*, 2012, **552**, 43.
304 J. Wang and B. Zhang, *J. Appl. Polym. Sci.*, 2013, **130**, 1321.
305 J. Wang and B. Zhang, *Colloid Polym. Sci*, 2013, **291**, 2917.
306 M. V. Srinivasan, P. Kannan and A. Roy, *J. Mater Sci.*, 2013, **48**, 2433.
307 N. M. Salleh, M. R. K. Sheikh, R. Yahya, M. R. Karim and A. Hassan, *J. Polym Res.*, 2013, **20**, 131.
308 N. M. Salleh, M. Rezaul, K. Sheikh, R. Yahya, M. R. Karim, A. D. Azzahari and A. Hassan, *J. Polym. Res.*, 2013, **20**, 296.
309 M. Bagheri, Z. Pourmoazzen and A. A. Entezami, *Iran. Polym. J.*, 2013, **22**, 303.
310 M. S. Gruzdev, O. B. Akopova and T. V. Frolova, *Russ. J. Gen. Chem.*, 2013, **83**, 652.
311 D. Kim, L. Wang, Y. Cao, X. Yu, S. Z. D. Cheng, S. Kuo, D. Song, S. H. Lee, M. Lee and K. Jeong, *J. Mater. Chem.*, 2012, **22**, 16382.
312 M. H. C. J. van Houtem, F. Benaskar, C. F. C. Fitié, R. Martín-Rapún, J. A. J. M. Vekemans and E. W. Meijer, *Org. Biomol. Chem.*, 2012, **10**, 5898.
313 R. Yang, L. Chen, R. Jin and Y. Wang, *Polym. Chem.*, 2013, **4**, 329.
314 W. Weissflog, U. Baumeister, M. Tamba, G. Pelzl, H. Kresse, R. Friedemann, G. Hempel, R. Kurz, M. Roos, K. Merzweiler, A. Jákli, C. Zhang, N. Diorio, R. Stannarius, A. Eremine and U. Kornek, *Soft Matter*, 2012, **8**, 2671.
315 M. K. Reddy, K. S. Reddy, T. Narasimhaswamy, B. B. Das, N. P. Lobo and K. V. Ramanathan, *New J. Chem.*, 2013, **37**, 3195.
316 Y. Zhu and X. Wang, *Polym. Chem*, 2013, **4**, 5108.
317 J. C. W. Tseng, R. Rondla, P. Y. S. Su and I. J. B. Lin, *RSC Adv.*, 2013, **3**, 2515.
318 M. Tariq, S. Hameed, I. H. Bechtold, A. J. Bortoluzzi and A. A. Merlo, *J. Mater. Chem. C*, 2013, **1**, 5583.
319 A. Shanavas, T. Narasimhaswamy, B. V. N. P. Kumar and A. S. Nasar, *Can. J. Chem.*, 2013, **91**, 196.
320 L. L. Pérez, S. Perdriau, G. ten Brink, B. J. Kooi, H. J. Heeres and I. Melián-Cabrera, *Chem. Mater.*, 2013, **25**, 848.
321 V. Cîrcu, A. S. Mocanu, C. Roşu, D. Manaila-Maximean and F. Dumitraşcu, *J. Therm. Anal. Calorim.*, 2012, **107**, 877.

322 C. H. Yu, C. P. J. Schubert, C. Welch, B. J. Tang, M. Tamba and G. H. Mehl, *J. Am. Chem. Soc.*, 2012, **134**, 5076.
323 S. Yagai, M. Usui, T. Seki, H. Murayama, Y. Kikkawa, S. Uemura, T. Karatsu, A. Kitamura, A. Asano and S. Seki, *J. Am. Chem. Soc.*, 2012, **134**, 7983.
324 L. A. Haverkate, M. Zbiri, M. R. Johnson, B. Deme, H. J. M. de Groot, F. Lefeber, A. Kotlewski, S. J. Picken, F. M. Mulder and G. J. Kearley, *J. Phys. Chem. B*, 2012, **116**, 13098.
325 M. K. Reddy, K. S. Reddy, K. Yoga, M. Prakash, T. Narasimhaswamy, A. B. Mandal, N. P. Lobo, K. V. Ramanathan, D. S. S. Rao and S. K. Prasad, *J. Phys. Chem. B*, 2013, **117**, 5718.
326 V. Klimavicius, Z. Gdaniec, J. Kausteklis, V. Aleksa, K. Aidas and V. Balevicius, *J. Phys. Chem. B*, 2013, **117**, 10211.
327 M. K. Reddy, K. S. Reddy, M. Prakash and T. Narasimhaswamy, *Mol. Cryst. Liq. Cryst.*, 2013, **582**, 1.
328 W. Yongli, D. Qiong, W. Guohua, Z. Zhiyong, D. Zhiqun, G. Jintao and X. Ying, *Chin. J. Org. Chem.*, 2013, **9**, 2010.
329 Z. Lan-Ying, Z. Qi-Kai and Z. Yu-Dong, *Liq. Cryst.*, 2013, **40**, 1263.
330 B. T. Thaker, D. B. Solanki, B. S. Patel and N. B. Patel, *Liq. Cryst.*, 2013, **40**, 1296.
331 J. Hu, Q. Li, Y. Lin, Y. Song and W. Li, *Liq. Cryst.*, 2013, **40**, 1095.
332 H. E. Gulbas, H. Ocak, Y. H. Gursel and B. Bilgin-Eran, *Mol. Cryst. Liq. Cryst.*, 2013, **574**, 40.
333 D. M. Suthar, A. A. Doshi and A. V. Doshi, *Mol. Cryst. Liq. Cryst.*, 2013, **574**, 75.
334 G. Yeap, Y. Ooi, K. Kenji and M. M. Ito, *Phase Transitions*, 2012, **86**, 365.
335 P. Subhapriya, P. S. Vijayanand and M. L. N. M. Mohan, *Mol. Cryst. Liq. Cryst.*, 2013, **571**, 40.
336 W. Qiang, G. Xiaonan and Y. Huai, *Mol. Cryst. Liq. Cryst.*, 2012, **557**, 1.
337 J. Sivakumar and S. Kathavarayan, *Mol. Cryst. Liq. Cryst.*, 2012, **552**, 53.
338 H. A. R. Pramanik, G. Das, C. R. Bhattacharjee, P. C. Paul, P. Mondal, S. K. Prasad and D. S. S. Rao, *Chem. – Eur. J.*, 2013, **19**, 13151.
339 M. Mansueto, W. Frey and S. Laschat, *Chem. – Eur. J.*, 2013, **19**, 16058.
340 B. Han, P. Hu, B. Wang, C. Redshaw and K. Zhao, *Beilstein J. Org. Chem.*, 2013, **9**, 2852.
341 T. H. T. Hsu, J. J. Naidu, B. Yang, M. Jang and I. J. B. Lin, *Inorg. Chem.*, 2012, **51**, 98.
342 C. G. Nardele and S. K. Asha, *J. Phys. Chem. B*, 2014, **118**, 1670.
343 S. Yang, B. Wang, D. Cui, D. Kerwood, S. Wilkens, J. Han and Y. Luk, *J. Phys. Chem. B*, 2013, **117**, 7133.
344 S. A. Kuvshinova, V. A. Burmistrov, I. V. Novikov, K. M. Litov, V. V. Aleksandriiskii and O. I. Koifman, *Russ. J. Org. Chem.*, 2014, **50**, 615.
345 G. Rău, G. D. Mogoşanu, C. G. Pisoschi and C. E. Stănciulescu, *Farmacia*, 2014, **62**, 494.
346 Z. Wang, Y. Lan, K. Zhong, Y. Liang, T. Chen and L. Y. Jin, *Int. J. Mol. Sci.*, 2014, **15**, 5634.
347 H. Y. Guo, J. Q. Ye and F. F. Yang, *Liq. Cryst.*, 2014, **41**, 946.
348 X. Z. He, L. J. Zhang, F. B. Meng, M. Tian and B. Y. Zhang, *Liq. Cryst.*, 2014, **41**, 530.
349 A. T. Mohammad, G. Y. Yeap and H. Osman, *Mol. Cryst. Liq. Cryst.*, 2014, **590**, 130.
350 B. Tylkowski, K. A. Bogdanowicz, V. Ambrogi, A. Lederer, V. Patroniak and M. Giamberini, *Polym. Int.*, 2014, **63**, 315.

351 F. F. Yang, J. Yuan, C. C. Li, H. Y. Guo and X. Y. Yan, *Liq. Cryst.*, 2014, **41**, 137.
352 M. K. Reddy, K. S. Reddy, B. V. N. P. Kumar and T. Narasimhaswamy, *Mol. Cryst. Liq. Cryst.*, 2014, **593**, 1.
353 M. K. S. Al-Malki, A. S. Hameed and A. H. Al-Dujaili, *Mol. Cryst. Liq. Cryst.*, 2014, **593**, 34.
354 B. S. Erdogan, D. Atilla, A. G. Gurek and V. J. Ahsen, *Porphyrins and Phthalocyanines*, 2014, **18**, 139.
355 B. Gharib and A. Hirsch, *Eur. J. Org. Chem.*, 2014, **2014**, 4123.
356 P. Wei, M. Cakmak, Y. Chen, X. Wang, Y. Wang and Y. Wang, *J. Appl. Polym. Sci.*, 2014, **131**, 40487.
357 Y. H. Ooi, G. Y. Yeap, C. C. Han, H. C. Lin, T. Shinomiya and M. M. Ito, *Liq. Cryst.*, 2014, **41**, 1017.
358 J. Q. Xie, G. M. Zhu, Y. S. Tang and Y. K. Wang, *Liq.Cryst.*, 2014, **41**, 36.

Combining MD simulations and NMR spectroscopy for molecular insight and methodological synergy: the integrated MD-NMR method

Francesca Mocci*[a,b,c] and Aatto Laaksonen*[b,c,d]

DOI: 10.1039/9781782622758-00592

NMR spectroscopy and MD simulations are highly complementary techniques to study molecular structures, interactions and dynamics. MD simulations are currently reaching the millisecond timescales covering a great variety of dynamical processes. Faster computers and new efficient sampling techniques allow calculations of NMR averages for a more reliable comparison with experiment, while new generations of force fields give better and better agreement between simulated and measured quantities. We review in this Chapter studies where close combination of these two techniques is the method itself to obtain the results and draw conclusions on the dynamics of bio, organic and inorganic systems.

1 Introduction

No modern Science laboratory exists without a park of computers where data is collected, stored, shared, analysed and visualized. However, in many laboratories, hosting also theoretical groups, computers have obtained a status of pure research equipment and used as virtual laboratories where real molecular structures, dynamical processes and reactions can be modelled and simulated side by side with experiments. Historically one of the most powerful combinations of experimental and computational techniques is that of Nuclear Magnetic Resonance (NMR) spectroscopy and Molecular Dynamics (MD) simulations which has been successfully used to obtain molecular information of nearly all possible systems from glasses to biological complexes. NMR and MD can be combined in many different ways simply because both methods provide information on molecular level, giving access to structure, interactions and dynamics better than any other techniques.[1] Whereas values of static NMR parameters (*i.e.* chemical shielding and *J* couplings) can be routinely computed from energy-minimized structures using most Quantum Chemistry packages, we focus in this chapter on the importance in connecting molecular motional degrees of freedom and structural and conformational dynamics with observed spectral quantities. Force-field based (and to some extent *ab initio*) MD simulations and NMR

[a]*Department of Chemical and Geological Sciences, Cittadella Universitaria di Monserrato, University of Cagliari, Cagliari, Italy. E-mail: fmocci@unica.it*
[b]*Department of Materials and Environmental Chemistry, Arrhenius Laboratory, Stockholm University, Stockholm, Sweden. E-mail: aatto@mmk.su.se*
[c]*Science for Life Laboratory, Solna, Sweden*
[d]*Stellenbosch Institute of Advanced Study (STIAS), Wallenberg Research Centre at Stellenbosch University, Stellenbosch, South Africa*

spectroscopy can be combined to study dynamical behaviour of molecular spin systems at mutually common (ps–ns–μs–ms) time scales to both techniques. The symbiosis NMR-MD is always mutual as the both methods benefit from it. NMR provides data to calibrate simulation methods, molecular models and force fields, while MD opens the window to the molecular world where the experimental observations can be connected to various microscopic phenomena. In both methods the physical conditions and environment can be varied allowing further evaluation of the used models, both theoretical and experimental. This Chapter covers the literature from the period of SPR on NMR volume 44 (June 1st 2013 until May 31st 2014) but since it is a new Chapter it also contains some older references needed in defining the field according to the title.

The chapter is organized as follows: first are reviewed the force fields developed or improved during the last year making use of NMR data in the parameterization and/or in validation, then we summarize the latest works in combining MD and NMR to study the structure and dynamics of Biomolecules, including proteins, sugars and polysaccharides, lipids. Then we review the papers where the combined approach has been used to study solvent and mixtures, surfaces, and porous materials.

2 Force field development and validation through NMR

When carrying out molecular simulations the quality of the results is at the very end directly dependent on the quality of the applied force field (FF). FF is a conceptually simple mimic of the quantum mechanical interactions within a molecular system made of nuclei and electrons (atoms and bonds). In its basic form it contains a set of balls and springs (like in a molecule kit) and simple mathematical functions with a few parameters, reasonably flexible to describe separately the short-range Van der Waals and long-range Coulombic intermolecular interactions between the non-bonded atoms, as well as, bond vibrations, angle bending and rotations around covalent bonds. In other words it should allow the molecular internal degrees of freedom and intermolecular interactions same way as in real systems. It is very robust and made transferable from molecule to molecule, at least in case of common molecules. The user should be prepared for situations where some parameters are missing and have to be created and inserted in the force field. Also, it is common that atomic charges need to be computed using special quantum mechanical schemes, most often based on electrostatic potentials.

NMR and MD simulations are both the most powerful techniques to study the structure and dynamics of macromolecules at atomistic level. When combined together they are highly complementary. Both techniques benefit greatly from a close collaboration, because there are so many inter-related quantities that can be obtained using both methods at mutual time scales. The molecular models and force fields can be refined and validated based on them. As a result of this the MD simulations have become more accurate and can better provide detailed insight into highly complex heterogeneous structures and dynamical processes in proteins, including the roles of specific solute–solvent interactions, as well as, ligand binding. The number of available general atomistic force fields in the scientific literature is too large to be listed here. For a review of different common force fields the reader is referred for example to the paper of Karttunen and coworkers[2] and those for protein simulations in particular to the chapter of Guvench and MacKerell.[3] Here we only mention the most recent developments of force fields parameterized and/or validated against NMR data.

MacKerell and coworkers[4] presented a polarizable force field, called Drude-2013, based on the classical Drude oscillator framework, currently implemented in simulation programs CHARMM and NAMD, for modeling of peptides and proteins. It can be used for several hundreds of ns simulations without requiring too much computing power. In its first generation version it gives improved order parameters S^2 for certain residues, for which they were underestimated using additive force fields, but does not perform so well yet for NMR chemical shifts.

Huang and MacKerell[5] have validated the most recent all-atom additive protein force field CHARMM36 (C36) on a variety of NMR data, including backbone scalar couplings across hydrogen bonds, residual dipolar couplings (RDCs) relaxation order parameters S^2, as well as vicinal scalar couplings. C36 FF generated results agree better with experimental data than CHARMM22/CMAP FF. The studied NMR properties have not been used in the parameterization of CHARMM36 FF.

Widmalm and coworkers combine NMR transglycosidic J coupling constants and explicit solvent Hamiltonian replica exchange simulations to improve and validate the CHARMM36 force field as well as their simulation method on a set of α- or β-(1→6)-linked oligosaccharides.[6] The experimental trends in rotamer distributions are reproduced in simulations. Walker and coworkers report a new AMBER14 FF for lipids which is validated against NMR data. The philosophy is to be compatible with AMBER FFs for proteins and nucleic acids.[7]

Zhang and coworkers[8] have developed an effective polarizable bond (EPB) force field for computer simulation of proteins where only all polar groups of amino acids are treated as polarizable, by fitting to electrostatic properties of the polar groups from DFT calculations. It is applied on 9 different proteins and found to be robust in describing the structure and dynamics of these proteins correctly. The agreement with NMR experiments is found better compared to results obtained from simulations using non-polarizable force fields. The model avoids problems of overpolarization (polarization catastrophe)[9] and is numerically stable adding only 5% to the computing compared to when standard non-polarizable FFs are used.

Ren and coworkers[10] report a new polarizable development and parameterization of their atomic multipole-based optimized energetics for biomolecular simulation, AMOEBA2013 force field for proteins. It uses atomic multipole-based electrostatics and has explicit treatment of dipole polarization. Based on a mutual induction model with Thole damping it describes both intra- and intermolecular polarization.

Chemical shifts, J-couplings and S^2 order parameters are well described for several peptides and proteins.

Jurecka and coworkers[11] have optimized the Amber ff99bsc0 force field for the backbone torsional parameters epsilon and zeta of DNA, observing an overall improvement in the structure, using the NMR and X-ray data as reference.

Langella and coworkers[12] validate against NMR and X-ray data their parameterizations of Peptide Nucleic Acids (PNAs) by performing MD simulations of two model systems constituted by a PNA–PNA duplex and a PNA–DNA hetero-duplex. New conformational features, related to the dynamics of the system, are found, in good agreement with the NMR data and with some difference to the crystal structure, due to packing effects.

To optimize force fields for long time scale motions Aliev *et al.*[13] propose a new robust approach to use ^{13}C NMR spin-lattice relaxation times T_1 of both backbone and sidechain carbons. This allows a selective determination of both overall molecular and intramolecular motional time scales. In addition they use motionally averaged experimental J coupling constants for torsional FF parameters. The force constants in the FFs and the correlation times are fitted in an Arrhenius-type of equation.

Jiang, Zhou and Wu[14] report a residue-specific force field (RSFF1) for proteins based on statistical analysis of protein coil library and parameterize all rotable torsional angles on free energy surface rather than potential energy surfaces. They modify the OPLS-AA/L force field and scale 1–4 Van der Waals interactions but leave the electrostatic 1–4 interactions untouched. They did also add 1–5 interactions. Their force field performs better than several very recent force fields parameterized for proteins in reproducing the ^3J of dipeptides. In a subsequent paper[15] they fold fourteen small proteins using replica-exchange MD simulations demonstrating the quality of their protein force field.

3 MD in NMR prediction of 3D structures of biomolecules

NMR has in last decades developed to a powerful method to study and determine structures of biomacromolecules. The reader is advised to familiarize themselves with a recent book edited by Bertini, McGreevy and Parigi,[16] a comprehensive guide to the community of protein NMR. Force-field-based MD simulations clearly play an important role in the NMR determination of biomolecular 3D structures. When used together with highly efficient MD techniques to enhance the sampling, such as the replica exchange and parallel tempering combined NMR-MD is an ultimate tool in structure determination Nevertheless, structure determination of biomacromolecules with NMR is a tedious procedure containing several steps, many flavors and a variety of protocols. It starts from NMR experiment and resonance assignment and continuing by imposing of structural restraints to calculate the structure – or rather an ensemble of structures satisfying the experimental data. From liquid state NMR we can obtain parameters such as chemical shifts (can be related to local electronic structure and environment, distances, torsional angles, hydrogen bonds), scalar coupling constants (torsional angles, hydrogen bonds), NOE & ROE (proton–proton distances), paramagnetic nuclei (distances and orientation) residual dipolar couplings (bond orientation). In solid state NMR chemical shifts, spin–spin coupling and quadrupolar coupling data are the primary sources of structural information. The data is often incomplete and empirical relationships and QM calculations can be used to complement it. 1D and 2D NMR data need to be converted to a 3D model using force field and other techniques. The structure needs to be energy-minimized. Simulated annealing and MD simulations with enhanced sampling are important. Finally the structure(s) need(s) to be validated with simulations in water solution or against known structures within the same protein family or homolog modeling or statistically (Bayesian). These last steps have to be repeated in iterative manner to improve the structure. NMR can be used to determine structures for only small and medium sized proteins (regularly below 25 kDa) whereas X-ray and electron microscopy can be used to study biomolecular structures over 100 kDa.[17] To integrate NMR, X-ray crystallography, cryo-TEM and modeling would give an even better integrative structural biology approach as visioned by Lengyel *et al.*[18]

We wish to refer to a number of excellent recent reviews we found important in the context of this chapter as they contain discussions of how the techniques and methodologies have evolved together with MD simulations and computational chemistry. Although this chapter collects works from the last year we have taken the freedom cite several reviews from the year before because of their relevance here. Allison discusses assessing and refining of MD simulations of proteins with NMR data stressing the importance of ever longer simulations exposing deficiencies in existing force fields.[19] Salvatella and coworkers review the synergetic use of MD simulations and NMR spectroscopy to study the structural heterogeneity of proteins.[20] Widmalm presents a perspective on the primary and 3D structures of carbohydrates.[21] Montelione and coworkers discuss how obtain high quality NMR structures for proteins.[22] Sanders and coworkers survey the technical advances behind the quiet renaissance of protein NMR.[23] Vögeli reviews NOE from quantitative perspective.[24] Vögeli and coworkers reviews the perspective on the ensemble-based structure determination using exact NOEs.[25] Showalter[26] reviews methods for structure and dynamics studies of intrinsically disordered proteins by combining MD simulations and NMR data. Vuister *et al.* gives an overview of tools to validate protein NMR structures.[27] Marsh and Teichmann discuss the dynamic aspects of protein structures from evolutionary point of view.[28] Im *et al.* presents an ensemble dynamics approach to decipher ssNMR observables of membrane proteins.[29] Comellas and Rienstra review protein structure determination by MAS ssNMR in studies amyloid fibrils.[30] Watt and Rienstra review the recent advances in ssNMR to quantify biomolecular dynamics.[31] Orozco has written an excellent review giving a theoretical view to protein dynamics.[32] Gryff-Keller turns to the experimentalists giving advices about how to obtain best possible results from theoretical calculations to compare with experimental results.[33] Finally we wish to acknowledge the work on nucleic acids done the group of Olson. Olson and coworkers[34] demonstrate their software package 3DNA to analyze, construct and visualize 3D DNA structures. The latest version of 3DNA has new features to analyze and manipulate ensembles of structures as those from NMR structure prediction and MD simulations.

4 MD and prediction of NMR parameters from 3D structures

The flow discussed above can also be reversed and protein (and other biomolecules) structures can be used to predict NMR parameters. This is an important area of its own but has also many applications. Here we only choose prediction of chemical shifts and *J* couplings although any other NMR parameter can be in principle predicted from the molecular structure. NMR chemical shifts are sensitive to several structural changes and conformational dynamics in the close vicinity. Currently there is no direct relationship between a molecular structure and NMR parameters we can use other than more or less empirical models, with limited validity, and Quantum Chemistry calculations for static parameters of systems of limited size and MD simulations for dynamic parameters. During the last decade several groups have developed methods, algorithms and software to predict NMR chemical shifts and *J*-couplings from protein structures. These packages are based on a variety of mathematical and statistical approaches, including Principal Component Analysis, k Nearest Neighbors and Random forest. The regularly contain a training data base from which the actual NMR shift is calculated assuming all contributions to be additive to values obtained from intrinsically disordered structures used normally as the basis on which all corrections are added. Although there are numerous packages for chemical shift prediction only a few of them rely on MD simulations in the process. Chemical shifts obtained in NMR spectroscopy are long time averages of conformations populated in a protein during the experiment. By averaging the calculated values over an ensemble of structure the prediction should be improved.

An excellent overview to the field is given Lehtivarjo where many of the software packages are reviewed in Table 2.[35] Lehtivarjo *et al.* have developed a software package 4DSPOT to predict chemical shifts from ensembles of protein structures obtained in from MD simulations[36] to add "fourth" dimension to include the dynamical effects.

Ochsenfeld and coworkers[37] have compared the sensitivity of *ab initio* methods against empirical methods in calculations of NMR chemical shifts from molecular structures of several prototypical peptides. They find that although simple empirical models can give rather good agreement with experimental results they are highly insensitive for structural (and electronic) changes. On the other hand the cheap empirical methods are continuously improved while *ab initio* are not feasible at all to use. *Ab initio* methods can also be used to improve the empirical methods same way as it is done for the empirical force fields.

It should be also mentioned that the effect from neighboring residues to both chemical shifts and *J*-coupling constants is not negligible at all as shown by Jung *et al.*[38] for intrinsically unfolded/disordered proteins.

Again, we also wish to mention a selection of somewhat older reviews about prediction of chemical shifts from protein structures taking into account the dynamical nature of proteins as they give a good overview of the field. Kalbitzer and coworkers discuss mapping of protein structural ensembles on chemical shifts.[39] Vendruscolo and coworkers investigate how to include time averaging of chemical shifts in DFT calculations to better compare with experiments.[40] Nielsen *et al.* presents a robust method for chemical shift prediction from protein structures reviewing the field.[41] Palmer and coworkers present their semi-empirical dynamically averaged chemical shift prediction tool discussing the use of MD simulations to improve the results.[42] Tantillo and coworkers review computational predictions of ^{1}H and ^{13}C chemical shifts discussing various sources of errors.[43] Case reviews calculations of chemical shifts in biomolecules.[44]

5 Multiple time scales in MD-NMR studies of protein systems

Although time scales are a somewhat artificial concept it is useful to keep in mind that experimental techniques applied on macroscopic amounts of substances give results averaged over typical time scales the techniques used. This can be a problem for MD simulations not able to follow long time dynamics to sample corresponding time windows. Encouraging, however, is that ensemble sampling over shorter time than is characteristic for NMR measurements improves the agreement as can be seen in many papers included in this Chapter.

An increasing number of studies is pointing the attention to the possible functional role of biomolecules dynamics.[45–47] NMR techniques constitute a very powerful mean to obtain information either on relatively fast timescale, in the range of ps to ns – where the fast backbone and side chain motion occurs – using the relaxation times R_1, R_2, and the steady state heteronuclear NOE or on larger time scale, in the μs to ms realm where conformational exchange typically occurs – through Carr–Purcell–Meiboom–Gill (CPMG) R_2 relaxation dispersion experiment R_2, or by $R_{1\rho}$ spin lock experiments. The reader is referred to the excellent reviews of Shapiro,[45] Jarymowycz and Stone[47] and Boher, Dyson and Wright[48] for a detailed discussion of the NMR relaxation methods to study protein dynamics. As discussed below, also the chemical shifts can be used to study several aspects in the system dynamics.

The importance of treating the NMR observables and from them derived parameters as ensembles over relevant time scales rather than individual conformers is highlighted by Ángyán and Gáspári[49] who discuss several MD simulation techniques to interpret the structural and dynamical data from NMR. Internal dynamics in proteins take place at several different time scales. It is therefore important to sample the ensembles covering the specific time scales.

Brüschweiler and coworkers[50] use MD simulations to calculate order parameters and to compare them with model-free parameters obtained in NMR relaxation measurements. They present a more accurate method based on ^{15}N R_1 and R_2 as well as heteronuclear $\{^1H\}$–^{15}N NOE NMR relaxation parameters they compute from long (0.5 μs) MD trajectories of ten different proteins. The experimental model-free order parameters are compared to those obtained in isotropic reorientational eigen mode dynamics (iRED) analysis method. Very good agreement is obtained between the experimental and simulated order parameters. The proposed comparative scheme should be valuable in the refinement of force fields in order to make them more accurate in calculations of a variety of properties in MD simulations. In a related work Brüschweiler and Li[51] present a new robust protocol to obtain native like protein structures with a proper choice of temperature and NOE restraint strength testing it for several proteins.

Calligari and Abergel[52] have studied multiple time scale dynamics in proteins to investigate the internal motion in proteins based on fluctuations in ^{15}N chemical shifts. To improve the simple two-site jump process model they present an alternative model to interpret the stochastic processes behind chemical shift fluctuations tested here on the small BPTI protein. From MD simulations of time length 200 ns and 1 ms, they calculate chemical shift time correlation functions to identify the different regimes along the distribution of the time scales and to access the fractional diffusion processes. Differently from the two site exchange model, commonly employed to interpret the fluctuations of the chemical shift, such an approach is capable of accounting for a continuous distribution of time scales, while keeping the number of model parameters low. Their method is expected to improve the current analysis used in NMR dispersion measurements.

Esposito and coworkers[53] report a comparative ^{15}N NMR relaxation study of wild-type and W60G β2m microglobulin to gain detailed insight into the dynamics of both systems where various types of motions span a broad range of amplitudes and time scales. MD simulations and thermodynamic integration to obtain the free energy together with hydrodynamic calculations were performed and support the data. The analysis of the mutant protein is consistent showing a reduced aggregation compared to the wild-type due to an increased conformational rigidity probed by both NMR relaxation studies and MD simulations.

García and coworkers[54] study amyloid Aβ peptides, implicated in the etiology of Alzheimer disease, using a combined MD/NMR approach. *J*-coupling data are calculated from the REMD trajectories and compared to corresponding NMR-derived values. All simulations are found to converge on the order of hundreds of nanoseconds-per-replica toward ensembles yielding a good agreement with experiment.

Sekhar *et al.*[55] present a study using NMR methods to define a length scale for intermediate-native transition of four-helix bundle FF module (a 60 amino acid residue module found in a variety of eukaryotic proteins) that provides an average measure of the size of the structural units participating in the transition.

Zachariae, Zeth and coworkers[56] combine ssNMR and MD simulations of human dermcidin in membranes that reveal the antibiotic mechanism of this major human antimicrobial, found to suppress Staphylococcus aureus growth on the epidermal surface, forming a foundation for the structure-based design of peptide antibiotics.

Rienstra and coworkers[57] have developed a high-resolution structural model of a bacterial disulfide oxidoreductase (DsbB) by combining experimental solid-state NMR, embedding the high-resolution structure into the lipid bilayer and performed MD simulations to provide a mechanistic view of DsbB function in the membrane. They revealed the membrane topology of DsbB by selective proton spin diffusion experiments, directly probing the correlations of DsbB with water and lipid acyl chains.

Westlund *et al.*[58] have studied the hydration of Peroxiredoxin 5 protein combining MD simulations and data from the water proton R_1 Nuclear Magnetic Relaxtion Dispersion (NMRD) experiments using simulated orientational order parameters and residence times of buried water molecules as parameters in the NMRD model. They find the dynamics of the water molecules associated with the protein, both with short or long residence time, to be complex and beyond the reach of the NMRD analysis alone. This again supports the use of MD and NMR in combination to obtain a more detailed analysis.

Niccolai and coworkers[59] use paramagnetic fragment-based NMR and MD simulations to find binding sites transient pocket openings on protein surfaces and in protein–protein interfaces for which MD simulations provide solvent accessible surface areas (SASA) and intramolecular hydrogen bonds as descriptors. 4-Hydroxy-TEMPO (TEMPOL) is used to induce paramagnetic perturbations on the ^{1}H–^{15}N HSQC signals and the backbone amides are analyzed for surface hotspots and to validate MD predicted pockets. They apply this technique on human chemokine protein CXCL12 connected to tumor progression. By combining their MD data with their paramagnetic profiles, two CXCL12 sites suitable for the binding of small molecules were identified. This MD-assisted paramagnetic-based NMR technique as dynamic drug discovery procedure is found very promising for detecting binding sites.

Palmer III and coworkers[60] have studied the partially disordered yeast transcription factor GCN4 by combining spin-relaxation and back-calculated chemical shifts, which provide information about residual secondary structure and helix capping interactions. MD simulations provide the order parameters (in excellent agreement with the experimental generalized order parameters) and spectral density functions calculated from autocorrelations functions over the second order Legendre polynomials *via* Fourier transform where the leucine zipper is used as the reference frame. Analysis of salt bridges and hydrogen bonds is added to the study. Detailed insight into secondary structure elements was obtained, again demonstrating the power of combining MD and NMR in structure prediction and studies of structural and conformational dynamics.

Rosetti and coworkers[61] present a methodology where MD simulations are used to interpret 2D NMR data and apply it on the naturally unfolded human α-synuclein protein while it interacts with dopamine, an inhibitor of fibril formation, and with its oxidation products in water solutions.

They find preferential binding sites to the ligands. The ligands may also cause conformational rearrangements of distal residues. Results from simulations are found to be consistent with all available experimental results and helpful in interpreting the NMR data.

Orozco and coworkers[62] employed NMR, small angle X-ray scattering (SAXS) data and MD simulations to characterize the urea-denatured state of a the small regulatory protein ubiquitin. Comparison between experimental parameters, including ^{3}J and RDC and SAXS data, with those calculated from MD is used to validate the simulation. The validation is particularly relevant when using force field parameters developed to well reproduce folded states in simulation of unfolded states with denaturant. The combined approach allowed to characterize the unfolded state ensemble, to energetic of the unfolded state, and to describe the interactions of the unfolded protein with the solvents urea and water.

Chary *et al.*[63] study and mutate to functionalize the EhCaBP1 calcium-binding protein (two-domain EF hand protein) using paramagnetic NMR and MD simulation. The two domains are different from each other but important for structural stability. To understand the nature of structural and functional changes in the protein NMR spectroscopy and MD simulations are used. ^{15}N relaxation is used to probe backbone motions on the ps–ms time-scale giving information about the overall and internal motions in proteins. The 3D structure is characterized both by multidimensional NMR and MD simulation to study the overall structural stability, dynamics, Ca^{2+} binding properties. The ^{15}N-longitudinal relaxation and [^{15}N,^{1}H]-NOE measurements are used to obtain information about fast motion on ps–ns time scale. ^{15}N-transverse relaxation probe slower motions μs–ms time scale. Conformational changes in the ms time scale or fast exchange with solvent enhance the R_2 values.

Griesinger and coworkers[64] present a simple but efficient tool to probe and describe interdomain dynamics in multidomain proteins. They apply it on a complex of calmodulin (CaM) with the IQ-recognition motif (which adopts three different interdomain orientations in the crystal) by paramagnetic NMR. To collect the pseudo-contact shifts and paramagnetically induced residual dipolar couplings for six different lanthanide ions, N60D mutant of calmodulin was used. An ensemble-based approach is used for the conformational space showing moderate interdomain translational and rotational dynamics. Pools of conformations were generated by MD in the crystal structure. Principal component analysis (PCA) is used in the analysis and the four largest vectors describe more 90% of the interdomain motion.

Ryde and coworkers[65] use NMR relaxation data to estimate conformational entropy changes and analyze the results against trajectories from long MD simulations in attempt to generate an "order-parameter-to-entropy dictionary", which they find to depend more on the studied protein and sampling frequency than the force field they use making the transferability of such dictionary rather poor but still useful within the same protein.

Fuglestad *et al.*[66] presented a combined approach of MD, accelerated MD calibrated on experimental RDCs, and residual local frustration analysis to study the ns to ms dynamics of apo-thrombin and active-site-bound (PPACK-thrombin) in solution. A community analysis of MD simulations allowed verifying how the binding at the active site affects correlated motions, also at distal allosteric sites.

Sharp and coworkers[67] have studied protein motion using the Lipari-Szabo model-free squared generalized methyl axis order parameter technique connected with Bayesian statistics and by combining NMR and MD simulations with the focus on the motion of the methyl groups in seven different proteins and conformational entropy. With new better optimized force fields the accuracy of the computed order parameters has increased markedly and as a result from the study a three-class model is proposed. This study promotes the dynamical view of proteins obtained in a multidimensional-NMR/MD approach.

Brooks and coworkers[68] use atomistic simulations and NMR spectroscopy to study the mechanism of action of a small molecule inhibitor that covalently binds to cysteine residues on Regulator of G-Protein Signaling Protein RGS4. They apply temperature-accelerated molecular dynamics (TAMD) to carry out enhanced conformational sampling. Both NMR experiments and MD simulations reveal chemical shift perturbations in residues in the vicinity of inhibitor binding site as well as in the RGS4-Gα binding interface.

Smith *et al.*[69] have combined measured chemical shifts, residual dipolar couplings, NOEs and H^N–H^α J coupling constants and ^{15}N NMR relaxation studies with MD simulations to characterize the dynamical behavior of the enzyme lysozyme from bacteriophage lambda (λ lysozyme) in solution. The focus is on the different conformations for the upper and lower lip regions (observed also in the X-ray structures) and their dynamic nature allowing substrates to enter the active site in connection to the enzyme activity. In another study Smith *et al.*[70] have studied the white hen egg lysozyme in both water and TFE/water solution using MD simulations and NMR NOE measurements to obtain reliable structures as NMR studies alone could not determine the structure in the TFE state.

Song and coworkers[71] did study the structure and dynamics of ephrin type-A receptor 5 (EPHA5) ligand binding domain (LBD) using NMR relaxation data within the model-free approach at a variety of time scales and combined the studies with MD simulations to obtain insight into the intrinsic dynamics of the LBD. They find the ligand binding dynamics different from other EPHA4 receptors in spite of their high sequence identity homology and same overall architecture.

In a beautiful methodological study and application investigation Levy and coworkers[72] studied the HIV protease, which undergoes motions on multiple time scales, coupling a Markov state kinetic network model (MSM) with a model-free approach[73] suitably extended to include fast motions occurring within two exchanging macrostates. The MSM was built from several short (20 ns) MD, and was used to generate a 10 μs stochastic trajectory, a time length which covers both the fast and slow HIV protease equilibria. The trajectory was used to calculate the order parameters and the internal rotational correlation functions for the N–H vector, and on the pertinent time scale a good agreement was found with those obtained from the original short MD simulations. The MSM obtained correlation functions are in good agreement with the author proposed extension of the model free approach, and the authors outline how the model could be used to interpret the experimental data to gain insight on the effect of drug binding to the studied protein, structural and dynamical effects due to drug resistant mutations.

Im and coworkers[29] use orientation-dependent NMR observables (ssNMR CSA & DC, solution RDC) to restrain MD simulations aimed at refining the structure of the Pf1 Coat Protein in Explicit Membranes. The used approach gives access to information not obtainable with only NMR data, such as a detailed description of the interaction of the side chains with other side chains or with the lipids, and the depth of the protein insertion in the membrane. In order to avoid conflicting structural results and to reach a converged structure in the simulations, it is found that restraints measured in very different environment should be avoided, *i.e.* NMR data obtained in micelles were found not to be proper to restrain the simulation in bilayers.

Vogel and coworkers[74] study cell-penetrating peptides using ssNMR and MD simulations. The simulation was too short for a penetration but the peptides are observed to induce an enhanced flexibility in membranes assumed to increase the affinity of peptides to the membrane.

The group of Vendruscolo has been developing in recent years many novel techniques where NMR parameters are used as restraints in the molecular dynamics simulations to characterise the dynamics of highly flexible proteins and nucleic acids. In ref. 75 Vendruscolo and coworkers describe a method to use methyl chemical shifts as restraints in replica-averaged MD simulations to follow the conformational fluctuations of proteins and applied it on four proteins demonstrating their approach to be useful for generating ensembles consistent with those obtained from NMR measurements. They found for the HU protein dimer a clamping mechanism to bind to DNA.[75] The proposed method is an extension to side chains of the method presented earlier by the same group, that used chemical shifts of backbone atoms as restraints in replica-averaged MD simulations,[76] and showed the capability of characterizing conformational fluctuations in proteins native states. Chemical shift are predicted using the CamShift[77] method, based on a differentiable function of the atomic coordinates, which makes possible the determination of protein structures from chemical shift information same way as other standard NMR observables, such as NOEs, scalar couplings, and residual dipolar couplings. In another work they generate an ensemble of conformations representing the free energy surface of RNase using chemical shifts as replica averaged restraints.[76] The same approach has been shown to work also to study interdomain motions of multidomain proteins in solution, using calcium-bound state of calmodulin (Ca^{2+}-CaM) as benchmark system.[78] The method developed by Vendruscolo and co-workers for structure determination from chemical shift (CS) has been included by Cavalli and co-workers in the code ALMOST, which contains several tools to determine, using experimentally derived restraints from CS, RDC,

SAXS and other techniques, and analyze the structure and dynamics of complex molecular systems.[79] In ref. 80 Vendruscolo, Laio and coworkers show how the chemical shifts can be used as collective variable in metadynamics to obtain the free-energy landscapes of proteins with relatively limited computational resources.

Vendruscolo and coworkers[81] did also characterize the breathing motion in two inter-domains inside Hen lysozyme enzyme involved in substrate binding and release using NMR RDCs as structural restraints in MD simulations. They nicely illustrate this technique to explore large-amplitude fluctuations in proteins. The restraint MD simulations are capable of sampling highly heterogeneous conformational ensembles in the spirit of maximum entropy principle providing experimental measurements as Boltzmann distributions for equilibrium properties. The use of RDCs as replica-averaged restraints in the MD simulations has also been used to study the conformational fluctuations of RNA.[82] Further recent developments of the work of Vendruscolo and coworkers include a simple method to use residual dipolar couplings as restraints for the structural and dynamical refinement of proteins in such a way that they don't need to define an overall alignment tensor but rather use the direct information from the RDCs about the angles between the internuclear vectors and the external magnetic field.[83,84] They then apply the method on ubiquitin and obtain the same accuracy as in other more standard NMR methods.

Using the above described methods, Christodoulou and coworkers[85] report in-cell NMR characterization of the secondary structure populations of a disordered conformation of α-Synuclein within *E. coli* Cells. The small protein α-Synuclein populates a highly dynamic state that has the same characteristics as the disordered monomeric form observed in aqueous solution studied by them previously using MD simulations.

6 Saccharides and lipids

Carbohydrates are flexible molecular systems and ideal to be studied with both NMR and MD simulations. Hydration structure and dynamics around sugars is an integral part and often responsible to populating conformations around glycosidic bonds. Phospholipids as amphiphilic systems, both flexible and ordered are frequently using both solution and solid-state NMR. Lipid bilayers are complicated and highly interesting structures themselves with intriguing phase behavior. However, they host other molecules not at least cholesterol and membrane proteins. Order parameters can be obtained from both NMR and MD simulations. The behavior of the head groups can be simulated and compared to NMR relaxations studies (phosphorus, nitrogen, carbon and proton).

Widmalm and coworkers have developed a software called CASPER for ^{1}H and ^{13}C NMR chemical shift assignments polysaccharides.[86,87] Polysaccharides have a high intrinsic flexibility and in particular the glycosidic linkage is difficult to describe in different models. It has been devised several times in the force models and MD simulations have been important in refining the distribution of the conformational populations.

Combined NMR and MD has been successfully used by Widmalm cow-orkers to improve the models for carbohydrates and to gain detailed insight into structure and dynamics of these systems in aqueous solution.[21,87]

D'Amelio *et al.*[88] did study chitlac (a chitosan derivative) in solution to obtain the average structure and the dynamics of the side chains using NMR relaxation and MD simulations. Both NMR and MD agree well with each other showing a high mobility in the side chains exhibiting rich variety of distinct conformations while being fully hydrated.

Losartan, an AT1 antagonist drug, were studied in several biological fluids and DPPC membrane bilayer by Zervou *et al.*[89] using NMR (DOSY,NOESY/ROESY) and MD simulations with an implicit solvation model to investigate its diffusion, conformational dynamics and self-assembly behavior at higher concentrations in DPPC with a general focus on the transportation mechanisms and permeability of the drug. *In silico*-NMR approach is found powerful with an excellent agreement between experiment and modeling.

Ferreira *et al.*[90] have reported cholesterol and POPC segmental order parameters in lipid membranes using solid state R-type proton-detected local field (R-PDLF) ^1H–^{13}C NMR with CP and INEPT as polarization methods and MD simulations. Good agreement between experiments and simulations was found for the cholesterol alignment in the bilayer and for the $|S_{CH}|$ profiles of acyl chains below 15 mol% cholesterol, whereas the agreement was not so good above this concentration calling for a better description of the cholesterol alkyl chain in the force field for interactions between cholesterol and other molecules at the center of the bilayers. In fact the NMR data in this study could be used to improve the FF models.

Topgaard and coworkers[91] suggest and experimentally verify an improved theoretical model for calculating the CP and INEPT ^{13}C signal intensities for alkyl maltosides (C8G2) anisotropic liquids and glasses under MAS conditions. They find MD simulations useful to interpret the experimental trends in CP/INEPT intensities at different temperatures and hydration levels. The model is applied on C8G2/water lamellar liquid crystals but is generally applicable on other ordered systems.

Cotten and coworkers[92] have studied two antimicrobial peptides, piscidin 1 & 3 in two different bilayers: (3 : 1 PC/PG and 1 : 1 PE/PG) to investigate their structure and dynamics as well as their role in membrane destabilization in bacterial cell membrane mimics and their anti-microbial capacity. They combine oriented sample solid-state NMR by measuring the ^1H–^{15}N dipolar couplings and ^{15}N chemical shifts and MD to obtain structural and orientational information of about these two interfacial α-helical peptides. High resolution structures are obtained for both peptides in biologically relevant conditions as well as very detailed information about the binding modes by combing MD and NMR.

Vosegaard and coworkers[93] study lipid bilayer dynamics and calculate ^{31}P solid-state NMR spectra from ensembles generated in extended MD simulations providing more comprehensive information than traditional NMR dynamical models. They have included the antimicrobial peptide alamethicin. The calculated ssNMR spectra is in good agreement with the experimental one in spite of the fact that dynamical averaging from MD simulation is on ns–ms time scale. Although the averaging using MD did not cover the ms time scale it was enough to capture the essential dynamics Combination MD simulations and ssNMR is very promising for studies of individual lipids as well as vesicles and other ordered structures.

7 Small organic/bio-organic molecules

In contrast to large biomolecules where the dynamical behavior can be characterized by large-amplitude conformational changes for smaller molecules the dynamical aspects deal more about translational and rotational motion at much shorter time scales. Translational and rotational diffusion can be studied with both MD and NMR relaxation studies. Also, quantum mechanical studies and *ab initio* MD simulations become feasible for small molecules. Also hydration can be studied in more details compared to that in large biomolecular systems.

Boutis and coworkers[94] report measurements of the relaxation times of water hydrated *N. clavipes* and *A. aurantia* spider silks as a function of temperature by deuterium 2D T_1–T_2 inverse Laplace transform (ILT) NMR to study the distribution, population and dynamics of hydration water at different temperatures finding correlation times much longer than those for free water and in some cases increasing with increased T. MD simulations reveal that peptides prepared from a number of repeating motifs show inverse temperature transition behavior found for example in protein elastin.

Vogiatzis and Theodorou[95] study local segmental dynamics and stresses in polystyrene (PS)–C_{60} mixtures using MD simulations. Results indicate that when C_{60} is added to PS it leads to a slower segmental dynamics. This can be estimated by the characteristic decay times from the orientational time-autocorrelation functions for selected vectors. The dynamics for C–H bonds is found in good agreement with spin–lattice relaxation and solid echo NMR measurements on molten polystyrene.

Bowers *et al.*[96] present *in situ* ^{13}C and ^{23}Na MAS NMR investigation of supercritical CO_2 incorporation in smectite–natural organic composites under CO_2 storage conditions (90 bar and 50 °C) showing that supercritical CO_2 interacts more strongly with the composite than with the clay. The supercritical CO_2 increases the ^{23}Na signal intensity, reducing the line width at half height and increasing the basal width giving more rapid ^{23}Na T_1 relaxation rates, and a redshift. Based on recent MD simulations of similar polymer–Na–smectite composites the observed changes may be due to an elevated rate of Na^+ site hopping in the presence of supercritical CO_2.

Exner and coworkers[97] have used *N*-methyl acetamide (NMA) as a test system for amide groups in protein backbones to calculate NMR chemical shifts with the Car-Parrinello MD method with explicit solvent molecules and quantum-chemical calculations of NMR parameters and compare with classical MD simulations. For example the C–P calculations give in general shorter solute–solvent H-bonds which in turn give a better agreement with experimental shielding constants. They obtain an almost perfect agreement with their simulated results and experiment. There are significant differences for the solvation shell structure modeled by *ab initio* and classical MD and the distribution of bond distances differs considerably due to the fact that *ab initio* MD allows for anharmonic molecular vibrations, whereas in current empirical force fields bond vibrations are based on the harmonic approximation.

Gerig[98] has evaluated force field models for trifluoroethanol–water mixture in order to perform a reliable MD simulation of the system. In this study it was very important to obtain good description of solvent fluorine/solute hydrogen NMR cross-relaxation obtained from solute–solvent nuclear Overhauser effects (NOE). An additional requirement was good agreement between experimental and simulated diffusion coefficients as NMR cross relaxation depends on the diffusion of the components in the mixture. It was found that OPLS-AA (TFE) and TIP5P-E (H_2O) did do a reasonably good work. The mixture will be used as a solvent for small peptides in the future.

Kathmann *et al.*[99] did determine reorientational correlation times for several amine bases in organic solvents based on both NMR relaxation and MD simulations as a test ability MD simulations and DFT calculations to explain the mechanism in complex reactions catalyzed by frustrated Lewis pairs (FLP) in metal free scission of H_2. As the Debye–Stokes–Einstein (DSE) model gives only qualitative predictions, MD simulations are found valuable to validate the spectroscopic studies.

Prisner and coworkers[100] present a combined approach of dynamic nuclear polarization (DNP) and nuclear magnetic relaxation dispersion (NMRD) measurements to detect fast dynamics between radicals and solvent molecules at high fields. TEMPOL radical in several solvents (water, toluene, acetone and DMSO) all with different viscosities and diffusion constants are used in evaluation. MD simulations can be used to compute the coupling factors between DNP and NMRD and give insight into the molecular origins of fast dynamics as well as help in predicting and optimizing suitable DNP agent-target systems for different field strengths.

Sezer[101] has developed an approach to quantitatively predict the contribution of molecular motions to the dynamic nuclear polarization (DNP) enhancement to increase the NMR signal of nuclear spins at high magnetic fields based on the use of MD simulations and applies it on a small model system nitroxide 4-hydroxy-2,2,6,6-tetramethylpiperidin-1-oxyl (TEMPOL) in liquid toluene. The DNP coupling between the protons on toluene and the unpaired electron of a small nitroxide radical was targeted using the so called force-free hard-sphere (FFHS) model. The FFHS model did agree well with the atomistic simulation results. Sezer discuss the possibilities to predict the DNP coupling factors from MD simulations of biological macromolecules such as nitroxide-doped lipids and proteins.

8 Room temperature ionic liquids

First known room temperature ionic liquid (RTIL), ethylammonium nitrate, was reported 100 years ago. Interest in RTILs started in 1970's with imidazolium & pyridinium cations but they finally started to find applications in 1990's when thermally and chemically stable ILs were synthetized. Today they appear virtually in every field as RTILs offer a large number of attractive properties and nearly limitless possibilities to explore their structure–property relationships. The highly complex landscape of interactions of RTILs makes it highly challenging to study these organic–inorganic salt-structures either experimentally or using computer modeling. To obtain insight into RTILs as pure liquids or in solution, combination of experimental and computational techniques is of great importance. Nuclear Magnetic Resonance (NMR) spectroscopy and Molecular Dynamics (MD) simulations as well as Quantum Chemistry provide a powerful tool to obtain a detailed picture of the structure and dynamics of RTILs predict and design RTILs for various applications. For the development up to the beginning of 2013 the reader is referred to a recent publication[102] where combined modeling and NMR work on RTIL is reviewed.

Borodin *et al.*[103] study the specific influence of the solvent on ions aggregation and transport in the *N*-methyl-*N*-pentylpyrrolidinium bis(trifluoromethanesulfonyl)imide (PY15TFSI) ionic liquid in three aprotic solvent mixtures using MD simulations and many-body polarizable APPLE&P force field. MD simulations predict the density, viscosity, and ionic conductivity in good agreement with the experiments. Diffusion was compared with that obtained in PFG-NMR experiments. Using ^{1}H T_1 and T_2 relaxation measurements and ^{13}C NMR and validated with MD simulations Hagaman and coworkers[104] did study the rotational dynamics of the ionic liquid 1-butyl-3-methyl-imidazolium bistrifluoromethyl-sulfonimide, [C4mim][Tf2N] both as a neat liquid and confined in mesoporous silica. Also the self-diffusion was followed using ^{1}H pulsed field gradient (PFG) NMR. Both the rotational and translational motions were reduced on the silica surface. The same IL was also studied by Feng and coworkers[105] in both silica and carbon mesopores with NMR and MD to obtain temperature-dependent diffusion coefficients for confined ILs and related to surface heterogeneity and interfacial microstructures. Venkanathan *et al.* have combined MD simulations with NMR and other spectroscopic techniques to study the structure and dynamics of Ammonium-based benzyl-NX3 (X = methyl, ethyl) trifluoromethanesulfonate (TFA) ionic liquids as potential electrolytes in batteries and fuel cells finding these ILs thermally stable with good conductivity.[106] Solano *et al.*[107] report a joint experimental study on structure and dynamics of *N*-butyl-*N*-methyl pyrrolidinium + bis([tri]fluoro[methane] sulfonyl)-imide − (PYR_{14}^{+} − [T]FSI −) ionic liquids doped with Li(T)FSI to be applied as electrolytes for lithium batteries. NMR is mainly used to measure the diffusion. MD simulations are used for the structure and various transport properties including shear viscosity and conductivity in addition to diffusion. The authors use a polarizable force field and obtain good agreement with experiments. Herring and coworkers[108] have studied ion transport and solvation by combining NMR and MD simulations by characterizing the transport processes for several quaternary ammonium cations paired with several anions experimentally based on the ionic conductivity (with and without CO_2) and self-diffusion coefficients from NMR and interpreted using results from MD simulations for cations and anions. Self-diffusion coefficients and residence times give a link between experiment and simulation both at short and long time scales. Combination of NMR and MD is valuable as both Nernst–Einstein and Debye–Huckel–Onsager conductivity theories are clearly insufficient.

Canongia Lopes, Vaca Chávez and coworkers[109] study 1-ethyl-3-methylimidazolium bis(trifluoromethylsulfonyl)imide ionic liquid in several aromatic solvents (benzene and its fluorinated derivatives). The structural changes in the IL organization around the solvent molecules were studied combining the analysis of MD simulations with that of ^{1}H and ^{13}C chemical shifts, and ^{1}H–^{1}H NOESY. CS variations and 2D NOESY spectra confirm the molecular organization observed in the simulations, indicating that with increasing number of fluorine atoms in the ring the IL cation is shifted from above and below the aromatic plane to its equatorial region.

The rate of reaction of a Menschutkin process in a range of ionic liquids with different cations was investigated by Harper and coworkers[110] by ^{1}H variable temperature NMR, using MD simulation to rationalize the results.

Combination of NMR spectroscopy and MD simulations were used when Figueiredo *et al.*[111] study protein destabilisation in ionic liquids and the role of preferential interactions in the denaturation process when anions and cations of the ionic liquids (ILs) 1-butyl-3-methylimidazolium ($[C4mim]^{+}$) and 1-ethyl-3 methylimidazolium ($[C2mim]^{+}$) chloride and dicyanamide (dca^{-}) in aqueous solutions were binding to the small alpha-helical protein Im7. Direct ion interactions were found crucial to understand the effects of ILs on the stability of proteins and that an anion effect is dominant. Direct cation–protein interactions also mediate stability; cation size and hydrophobicity are relevant to account for destabilization. The specificity in the interaction of IL ions with protein residues, established by weak favorable interactions, is confirmed by NMR chemical shift perturbation, amide hydrogen exchange data and MD simulations.

To obtain atomistic insight into orthoborate-based ionic liquids Wang *et al.*[112] have first developed a force field for this new class of halogen-free chelated orthoborate-phosphonium ionic liquids. NMR data is used to partially calibrate the FF. The diffusion coefficients from simulations show dynamically three different regimes telling about the high complexity of these systems.

9 Interfaces: inorganic, organic and porous

Solid–liquid and solid–gas interfaces are important in heterogeneous catalysis. A special type of interface is provided by porous materials, such as zeolites, metal organic frameworks (MOF) and porous silica. Porous materials are used for capture and separation of small molecules. Not at least in carbon capture of CO_2 from flue gas or purification of natural gas components from containing nitrogen and water. NMR and MD simulations are ideal to study diffusion of guest molecules inside the pores.

Ramalho and coworkers[113] have studied dynamics and calculated NMR parameters of the (100) face of magnetite (Fe_3O_4) at water interface and its implications as MRI probes. MD simulations were performed using the ReaxFF model and structures were selected for calculations of NMR parameters. They found MD simulations and inclusion of solute–solvent interactions important to obtain good spectroscopic properties as water molecules and hydrogen bonds play critical role as well as thermal effects. The MD averaged ^2J(H,H) were in good agreement with the experimental ones. Vega and coworkers[114] did study the mobility of (O)3Si–CH_2–CH_2–CO–NH–CD_2–CD_2–NH_2, a *N*-(2-aminoethyld4) propanamide moiety with two deuterated methylene groups toward the end terminus, grafted at the inner surface of mesoporous silica SBA-15. Results from the accompanied MD simulations suggest a binding with silanol groups on surface of SBA-15 through C(O)···HO, N(H)···OH, or NH_2···OH hydrogen bonding. The focus was to obtain dynamic information at different hydration levels. No dynamical models were assumed. MD simulations did support the experiment giving insight to structural dynamics and hydration mechanisms.

Maurin and coworkers[115] have studied the dynamics of xylene isomers inside the metal–organic framework MIL-47(v) by combining MD simulations and 2H NMR to follow the diffusion of the isomers at different temperatures finding a non-monotonous seen both in MD and in NMR. They find two regimes: one at low temperature where xylenes are in the vicinity of the pore wall with a high activation energy barrier for the diffusion and the other at high temperature where the xylenes are found in the center with lower activation energy for the diffusion. NMR gives details about packing effects and guest–guest interactions. Although the simulations could describe the overall trends very well they failed to give the quantitative differences between *o*- and *p*-isomers of xylene. This problem can be solved with a better force field.

Stallmach and coworkers[116] have studied ferrocene adsorbed in MOF-5 using solid-state relayed ^{13}C–^1H heteronuclear correlation (HETCOR) 2D NMR and MD simulations to obtain more details of the specific interactions. They find short guest–host contacts (less than 5Å) to the linkers close to the aromatic rings mediated through proton diffusion. Results from MD simulations agree well with the conclusions from the 2D ^1H and ^{13}C NMR studies. There is a residual mobility of the linker and ferrocene rather than well-defined stationary positions.

To evaluate a numerical simulation methodology and previously suggested analytical models Faux *et al.*[117] used MD simulations to determine the spin-pair correlation function G*(t) needed to determine the T_1 and T_2 relaxation times for water diffusion in liquid and inside pore. Both the rotational and translation motion of water is obtained from MD simulations and compared with corresponding NMR relaxation data. MD simulations were found as a powerful tool to explain the diffusive behavior on atomistic scale and did provide parameterization for subsequent simulations to cover longer time and length scales.

10 Water, ions and salts

Water is the most important solvent but also among the most complicated. Therefore it is of great value to calibrate methods, both experimental and theoretical, on water and aqueous systems containing ions or other solutes. For water there exist hundreds of interaction potentials to use in MD simulations. None of them is able to reproduce all the anomalous properties of water. This make very important to validate the simulation results against experimental data, and the coupling with NMR data is of great use.

Vaara and coworkers[118] have studied contributions to paramagnetic nuclear spin relaxation enhancement for Ni^{2+} in aqueous solution as a prototype system from Curie-type relaxation mechanism (analogous to chemical shielding anisotropy (CSA) in diamagnetic systems). They use a broad spectrum of computational techniques starting with MD simulations and selecting snap shot configurations to compute EPR parameters such as hyperfine coupling constants and calculation paramagnetic NMR shielding tensor based on non-doublet theory. After that follows calculation of time correlation functions of the shielding tensors and also spectral density functions. At the end the Redfield theory is applied to obtain the Curie contribution.

Sebastiani and coworkers[119] used *ab initio* molecular dynamics and experimental and simulated CS to study to the hydrogen bond network dynamics in low and high salt water solutions of LiI. The combined approach allows reaching a considerable insight on the competitive chaotropic effects of the anion and the kosmotropic effect of the cation, indicating that the former has longer range effects.

Aidas *et al.*[120] have been using a combined QM/MD scheme to study the electric field gradient fluctuations for monoatomic species in condensed phases, using as benchmark case Na^+ in aqueous solution. The results reveal the importance of properly treating by QM method both the ion and the first solvation layer, and question the applicability of the Sternheimer approximation to predict correctly the quadrupolar coupling constant. In a following investigation Badu *et al.*[121] employ Car–Parrinello MD to predict quadrupolar monoatomic ions relaxation, obtaining a reasonable agreement with the experimental, in all cases remaining in the same order of magnitude with experimental data. However the agreement can be improved for some of the studied anions.

Ohkubo *et al.*[122] study electronic properties structure and transport coefficients (conductivity and self-diffusion) of a molten xLi$_2$O-$(1-x)$B$_2$O$_3$ system using first-principles MD simulations performed with their own finite element density functional theory code, FEMTECK and PFG NMR measurements. For diffusion the first-principles simulation results were in better agreement with experiment than that obtained from classical simulations.

References

1 M. Odelius and A. Laaksonen, in *Molecular Dynamics: From Classical to Quantum Methods*, ed. J. Balbuena, Theoretical and Computational Chemistry, Elsevier BV, Perla, Seminario, 1999, vol. 7, pp. 281–324.

2 E. A. Cino, W. Choy and M. Karttunen, *J. Chem. Theory Comput.*, 2012, **8**, 2725–2740.
3 O. Guvench and A. D. J. MacKerell, in *Molecular Modeling of Proteins*, ed. A. Kukol, Methods in Molecular Biology, Humana Press, 2008, vol. 443, pp. 63–88.
4 P. E. M. Lopes, J. Huang, J. Shim, Y. Luo, H. Li, B. Roux and A. D. Mackerell, *J. Chem. Theory Comput.*, 2013, **9**, 5430–5449.
5 J. Huang and A. D. MacKerell, *J. Comput. Chem.*, 2013, **34**, 2135–2145.
6 D. S. Patel, R. Pendrill, S. S. Mallajosyula and G. Widmalm, *J. Phys. Chem. B*, 2014, **118**, 2851–2871.
7 C. J. Dickson, B. D. Madej, A. A. Skjevik, R. M. Betz, K. Teigen, I. R. Gould and R. C. Walker, *J. Chem. Theory Comput.*, 2014, **10**, 865–879.
8 X. Xiao, T. Zhu, C. G. Ji and J. Z. H. Zhang, *J. Phys. Chem. B*, 2013, **117**, 14885–14893.
9 P. Cieplak, F.-Y. Dupradeau, Y. Duan and J. Wang, *J. Phys. Condens. Matter*, 2009, **21**, 333102.
10 Y. Shi, Z. Xia, J. Zhang, R. Best, C. Wu, J. W. Ponder and P. Ren, *J. Chem. Theory Comput.*, 2013, **9**, 4046–4063.
11 M. Zgarbova, F. J. Luque, S. Jir, T. E. Cheatham, M. Otyepka and P. Jurecka, *J. Chem. Theory Comput.*, 2013, **9**, 2339–2354.
12 I. Autiero, M. Saviano and E. Langella, *Phys. Chem. Chem. Phys.*, 2014, **16**, 1868–1874.
13 A. E. Aliev, M. Kulke, H. S. Khaneja, V. Chudasama, T. D. Sheppard and R. M. Lanigan, *Proteins*, 2014, **82**, 195–215.
14 F. Jiang, C. Zhou and Y. Wu, *J. Phys. Chem. B*, 2014, **118**, 6983–6998.
15 F. Jiang and Y. Wu, *J. Am. Chem. Soc.*, 2014, **136**, 9536–9539.
16 *NMR of Biomolecules: Towards Mechanistic Systems Biology – Wiley Online Library*, ed. I. Bertini, K. S. McGreevy, and G. Parigi, Wiley-VCH Verlag GmbH & Co. KGaA, 2012.
17 L. Ferella, A. Rosato and P. Turano, in *NMR of Biomolecules: Towards Mechanistic Systems Biology*, ed. I. Bertini, K. S. McGreevy and G. Parigi, Wiley-VCH Verlag GmbH & Co. KGaA, 2012, pp. 33–45.
18 J. Lengyel, E. Hnath, M. Storms and T. Wohlfarth, *J. Struct. Funct. Genomics*, 2014, **15**, 117–124.
19 J. R. Allison, *Biophys. Rev.*, 2012, **4**, 189–203.
20 S. Esteban-Martín, R. Bryn Fenwick and X. Salvatella, *Wiley Interdiscip. Rev.: Comput. Mol. Sci.*, 2012, **2**, 466–478.
21 G. Widmalm, *Carbohydr. Res.*, 2013, **378**, 123–132.
22 A. Rosato, R. Tejero and G. T. Montelione, *Curr. Opin. Struct. Biol.*, 2013, **23**, 715–724.
23 P. J. Barrett, J. Chen, M.-K. Cho, J.-H. Kim, Z. Lu, S. Mathew, D. Peng, Y. Song, W. D. Van Horn, T. Zhuang, F. D. Sönnichsen and C. R. Sanders, *Biochemistry*, 2013, **52**, 1303–1320.
24 B. Vögeli, *Prog. Nucl. Magn. Reson. Spectrosc.*, 2014, **78**, 1–46.
25 B. Vögeli, J. Orts, D. Strotz, C. Chi, M. Minges, M. A. Wälti, P. Güntert and R. Riek, *J. Magn. Reson.*, 2014, **241**, 53–59.
26 S. A. Showalter, *eMagRes*, 2014, **3**, 181–190.
27 G. W. Vuister, R. H. Fogh, P. M. S. Hendrickx, J. F. Doreleijers and A. Gutmanas, *J. Biomol. NMR*, 2014, **58**, 259–285.
28 J. a. Marsh and S. a. Teichmann, *BioEssays*, 2014, **36**, 209–218.
29 W. Im, S. Jo and T. Kim, *Biochim. Biophys. Acta*, 2012, **1818**, 252–262.
30 G. Comellas and C. M. Rienstra, *Annu. Rev. Biophys.*, 2013, **42**, 515–536.
31 E. D. Watt and C. M. Rienstra, *Anal. Chem.*, 2014, **86**, 58–64.

32 M. Orozco, *Chem. Soc. Rev.*, 2014, **43**, 5051–5066.
33 A. Gryff-Keller, *Concepts Magn. Reson., Part A*, 2011, **38A**, 289–307.
34 A. V. V. Colasanti, X.-J. Lu and W. K. Olson, *J. Visualized Exp.*, 2013, e4401.
35 J. Lehtivarjo, PhD Thesis, Predicting NMR Parameters from the Molecular Structure, School of Pharmacy, Faculty of Health Sciences, University of Eastern Finland, Kuopio, 2014.
36 J. Lehtivarjo, K. Tuppurainen, T. Hassinen, R. Laatikainen and M. Peräkylä, *J. Biomol. NMR*, 2012, **52**, 257–267.
37 C. V. Sumowski, M. Hanni, S. Schweizer and C. Ochsenfeld, *J. Chem. Theory Comput.*, 2014, **10**, 122–133.
38 Y. Jung, K. Oh, G. Hwang and M. Cho, *Chirality*, 2014, **26**, 443–452.
39 K. Baskaran, K. Brunner, C. E. Munte and H. R. Kalbitzer, *J. Biomol. NMR*, 2010, **48**, 71–83.
40 I. De Gortari, G. Portella, X. Salvatella, V. S. Bajaj, P. C. A. Van der Wel, J. R. Yates, M. D. Segall, C. J. Pickard, M. C. Payne and M. Vendruscolo, *J. Am. Chem. Soc.*, 2010, **139**, 5993–6000.
41 J. T. Nielsen, H. R. Eghbalnia and N. C. Nielsen, *Prog. Nucl. Magn. Reson. Spectrosc.*, 2012, **60**, 1–28.
42 P. Robustelli, K. a. Stafford and A. G. Palmer, *J. Am. Chem. Soc.*, 2012, **134**, 6365–6374.
43 M. W. Lodewyk, M. R. Siebert and D. J. Tantillo, *Chem. Rev.*, 2012, **112**, 1839–1862.
44 D. A. Case, *Curr. Opin. Struct. Biol.*, 2013, **23**, 172–176.
45 Y. E. Shapiro, *Prog. Biophys. Mol. Biol.*, 2013, **112**, 58–117.
46 K. A. Henzler-Wildman, M. Lei, V. Thai, S. J. Kerns, M. Karplus and D. Kern, *Nature*, 2007, **450**, 913–916.
47 V. A. Jarymowycz and M. J. Stone, *Chem. Rev.*, 2006, **106**, 1624–1661.
48 D. D. Boehr, H. J. Dyson and P. E. Wright, *Chem. Rev.*, 2006, **106**, 3055–3079.
49 A. F. Ángyán and Z. Gáspári, *Molecules*, 2013, **18**, 10548–10567.
50 Y. Gu, D. Li and R. Brüschweiler, *J. Chem. Theory Comput.*, 2014, **10**, 2559–2607.
51 Da-wei Li and R. Brüschweiler, *J. Chem. Theory Comput.*, 2014, **10**, 1781–1787.
52 P. Calligari and D. Abergel, *J. Phys. Chem. B*, 2014, **118**, 3823–3831.
53 D. Gümral, F. Fogolari, A. Corazza, P. Viglino, S. Giorgetti, M. Stoppini, V. Bellotti and G. Esposito, *Magn. Reson. Chem.*, 2013, **51**, 795–807.
54 D. J. Rosenman, C. R. Connors, W. Chen, C. Wang and A. E. García, *J. Mol. Biol.*, 2013, **425**, 3338–59.
55 A. Sekhar, P. Vallurupalli and L. E. Kay, *Proc. Natl. Acad. Sci. U. S. A.*, 2013, **110**, 11391–11396.
56 C. Song, C. Weichbrodt, E. S. Salnikov, M. Dynowski, B. O. Forsberg, B. Bechinger, C. Steinem, B. L. de Groot, U. Zachariae and K. Zeth, *Proc. Natl. Acad. Sci. U. S. A.*, 2013, **110**, 4586–4591.
57 M. Tang, A. E. Nesbitt, L. J. Sperling, D. A. Berthold, C. D. Schwieters, R. B. Gennis and C. M. Rienstra, *J. Mol. Biol.*, 2013, **425**, 1670–1682.
58 Y. Huang, K. Nam and P.-O. Westlund, *Phys. Chem. Chem. Phys.*, 2013, **15**, 14089–14097.
59 A. Bernini, L. Henrici De Angelis, E. Morandi, O. Spiga, A. Santucci, M. Assfalg, H. Molinari, S. Pillozzi, A. Arcangeli and N. Niccolai, *Biochim. Biophys. Acta*, 2014, **1844**, 561–566.
60 P. Robustelli, N. Trbovic, R. A. Friesner and A. G. Palmer III, *J. Chem. Theory Comput.*, 2013, **9**, 5190–5200.

61 D. Dibenedetto, G. Rossetti, R. Caliandro and P. Carloni, *Biochemistry*, 2013, **52**, 6672–6683.
62 M. Candotti, S. Esteban-Martín, X. Salvatella and M. Orozco, *Proc. Natl. Acad. Sci. U. S. A.*, 2013, **110**, 5933–5938.
63 A. K. Rout, S. Patel, Somlata, M. Shukla, D. Saraswathi, A. Bhattacharya and K. V. R. Chary, *J. Biol. Chem.*, 2013, **288**, 23473–23487.
64 L. Russo, M. Maestre-martinez, S. Becker, C. Griesinger and S. Wolff, *J. Am. Chem. Soc.*, 2013, **135**, 17111–17120.
65 S. Genheden, M. Akke and U. Ryde, *J. Chem. Theory Comput.*, 2014, **10**, 432–438.
66 B. Fuglestad, P. M. Gasper, J. A. McCammon, P. R. L. Markwick and E. a. Komives, *J. Phys. Chem. B*, 2013, **117**, 12857–12863.
67 K. A. Sharp, V. Kasinath and A. J. Wand, *Proteins*, 2014, **82**, 2106–2117.
68 H. Vashisth, A. J. Storaska, R. R. Neubig and C. L. Brooks, *ACS Chem. Biol.*, 2013, **8**, 2778–2784.
69 L. J. Smith, A. M. Bowen, A. Di Paolo, A. Matagne and C. Redfield, *ChemBioChem*, 2013, **14**, 1780–1788.
70 A. P. Eichenberger, W. F. van Gunsteren and L. J. Smith, *J. Biomol. NMR*, 2013, **55**, 339–353.
71 X. Huan, J. Shi, L. Lim, S. Mitra, W. Zhu, H. Qin, E. B. Pasquale and J. Song, *PLoS One*, 2013, **8**, e74040.
72 J. Xia, N. Deng and R. M. Levy, *J. Phys. Chem. B*, 2013, **117**, 6625–6634.
73 G. Lipari and A. Szabo, *J. Am. Chem. Soc.*, 1982, **104**, 4546–4559.
74 K. Witte, B. E. S. Olausson, A. Walrant, I. D. Alves and A. Vogel, *Biochim. Biophys. Acta*, 2013, **1828**, 824–833.
75 A. Kannan, C. Camilloni, A. B. Sahakyan, A. Cavalli and M. Vendruscolo, *J. Am. Chem. Soc.*, 2014, **136**, 2204–2207.
76 C. Camilloni, P. Robustelli, A. De Simone, A. Cavalli and M. Vendruscolo, *J. Am. Chem. Soc.*, 2012, **134**, 3968–3971.
77 K. J. Kohlhoff, P. Robustelli, A. Cavalli, X. Salvatella and M. Vendruscolo, *J. Am. Chem. Soc.*, 2009, **131**, 13894–13895.
78 P. Kukic, C. Camilloni, A. Cavalli and M. Vendruscolo, *J. Mol. Biol.*, 2014, **426**, 1826–1838.
79 B. Fu, A. B. Sahakyan, C. Camilloni, G. G. Tartaglia, E. Paci, A. Caflisch, M. Vendruscolo and A. Cavalli, *J. Comput. Chem.*, 2014, **35**, 1101–1105.
80 D. Granata, C. Camilloni, M. Vendruscolo and A. Laio, *Proc. Natl. Acad. Sci. U. S. A.*, 2013, **110**, 6817–6822.
81 A. De Simone, R. W. Montalvao, C. M. Dobson and M. Vendruscolo, *Biochemistry*, 2013, **52**, 6480–6486.
82 A. N. Borkar, A. De Simone, R. W. Montalvao and M. Vendruscolo, *J. Chem. Phys.*, 2013, **138**, 215103.
83 C. Camilloni and M. Vendruscolo, *J. Am. Chem. Soc.*, 2014, **134**, 3968–3971.
84 R. Montalvao, C. Camilloni, A. De Simone and M. Vendruscolo, *J. Biomol. NMR*, 2014, **58**, 233–238.
85 C. A. Waudby, C. Camilloni, A. W. P. Fitzpatrick, L. D. Cabrita, C. M. Dobson, M. Vendruscolo and J. Christodoulou, *PLoS One*, 2013, **8**, e72286.
86 M. Lundborg and G. Widmalm, *Anal. Chem.*, 2011, **83**, 1514–1517.
87 J. Rönnols, R. Pendrill, C. Fontana, C. Hamark, T. A. A. D'Ortoli, O. Engström, J. Ståhle, M. V. V. Zaccheus, E. Säwén, L. E. E. Hahn, S. Iqbal and G. Widmalm, *Carbohydr. Res.*, 2013, **380**, 156–166.
88 N. D'Amelio, C. Esteban, A. Coslovi, L. Feruglio, F. Uggeri, M. Villegas, J. Benegas, S. Paoletti and I. Donati, *J. Phys. Chem. B*, 2013, **117**, 13578–13587.

89 M. Zervou, Z. Cournia, C. Potamitis, G. Patargias, S. Durdagi, S. Golic Grdadolnik and T. Mavromoustakos, *Biochim. Biophys. Acta*, 2014, **1838**, 1031–1046.
90 T. M. Ferreira, F. Coreta-Gomes, O. H. S. Ollila, M. J. Moreno, W. L. C. Vaz and D. Topgaard, *Phys. Chem. Chem. Phys.*, 2013, **15**, 1976–1989.
91 A. Nowacka, N. A. Bongartz, O. H. S. Ollila, T. Nylander and D. Topgaard, *J. Magn. Reson.*, 2013, **230**, 165–175.
92 B. S. Perrin, Y. Tian, R. Fu, C. V Grant, E. Y. Chekmenev, W. E. Wieczorek, A. E. Dao, R. M. Hayden, C. M. Burzynski, R. M. Venable, M. Sharma, S. J. Opella, R. W. Pastor and M. L. Cotten, *J. Am. Chem. Soc.*, 2014, **136**, 3491–3504.
93 S. K. Hansen, M. Vestergaard, L. Thøgersen, B. Schiøtt, N. C. Nielsen and T. Vosegaard, *J. Phys. Chem. B*, 2014, **118**, 5119–5129.
94 O. T. Ukpebor, A. Shah, E. Bazov and G. S. Boutis, *Soft Matter*, 2014, **10**, 773–785.
95 G. G. Vogiatzis and D. N. Theodorou, *Macromolecules*, 2014, **47**, 387–404.
96 M. G. Bowers, D. W. Hoyt, S. D. Burton, B. O. Ferguson, T. Varga and R. J. Kirkpatrick, *J. Phys. Chem. C*, 2014, **118**, 3564–3573.
97 M. Dračínský, H. M. Möller and T. E. Exner, *J. Chem. Theory Comput.*, 2013, **9**, 3806–3815.
98 J. T. Gerig, *J. Phys. Chem. B*, 2014, **118**, 1471–1480.
99 S. M. Kathmann, H. Cho, T. Chang, G. K. Schenter, K. Parab and T. Autrey, *J. Phys. Chem. B*, 2014, **118**, 4883–4888.
100 P. Neugebauer, J. G. Krummenacker, V. P. Denysenkov, C. Helmling, C. Luchinat, G. Parigi and T. F. Prisner, *Phys. Chem. Chem. Phys.*, 2014, **16**, 18781–18787.
101 D. Sezer, *Phys. Chem. Chem. Phys.*, 2013, **15**, 526–540.
102 F. Mocci, A. Laaksonen, Y. Wang, G. Saba, A. Lai, and F. Cesare Marincola, in *The Structure of Ionic Liquids*, ed. R. Caminiti and L. Gontrani, Springer International Publishing, Switzerland 2014, pp. 97–126.
103 O. Borodin, W. A. Henderson, E. T. Fox, M. Berman, M. Gobet and S. Greenbaum, *J. Phys. Chem. B*, 2013, **117**, 10581–10588.
104 K. S. Han, X. Wang, S. Dai and E. W. Hagaman, *J. Phys. Chem. C*, 2013, **117**, 15754–15762.
105 S. Li, K. S. Han, G. Feng, E. W. Hagaman, L. Vlcek and P. T. Cummings, *Langmuir*, 2013, **29**, 9744–9749.
106 A. P. Sunda, V. M. Dhavale, S. Kurungot and A. Venkatnathan, *J. Phys. Chem. B*, 2014, **118**, 1831–1838.
107 C. J. F. Solano, S. Jeremias, E. Paillard, D. Beljonne and R. Lazzaroni, *J. Chem. Phys.*, 2013, **139**, 034502.
108 H. N. Sarode, G. E. Lindberg, Y. Yang, L. E. Felberg, G. A. Voth and A. M. Herring, *J. Phys. Chem. B*, 2014, **118**, 1363–1372.
109 N. Dias, K. Shimizu, P. Morgado, E. J. M. Filipe, J. N. Canongia Lopes and F. Vaca Chávez, *J. Phys. Chem. B*, 2014, **118**, 5772–5780.
110 E. E. L. Tanner, H. M. Yau, R. R. Hawker, A. K. Croft, and J. B. Harper, *Org. Biomol. Chem.*, 2013, **11**, 6170–6175.
111 A. M. Figueiredo, J. Sardinha, G. R. Moore and E. J. Cabrita, *Phys. Chem. Chem. Phys.*, 2013, **15**, 19632–19643.
112 Y.-L. Wang, F. U. Shah, S. Glavatskih, O. N. Antzutkin and A. Laaksonen, *J. Phys. Chem. B*, 2014, **118**, 8711–8723.
113 M. a. Gonçalves, F. C. Peixoto, E. F. F. da Cunha and T. C. Ramalho, *Chem. Phys. Lett.*, 2014, **609**, 88–92.

114 S. Jayanthi, M. Werner, Y. Xu, G. Buntkowsky and S. Vega, *J. Phys. Chem. C*, 2013, **117**, 13114–13121.
115 G. Rives, S. Jobic, H. Kolokolov, D. I. Gabrienko, A. A. Stepanov, A. G. Ke, Y. Frick, B. Devic, T. Ferey and G. Maurin, *J. Phys. Chem. C*, 2013, 6293–6302.
116 M. Wehring, P. C. M. M. Magusin, S. Amirjalayer, R. Schmid and F. Stallmach, *Microporous Mesoporous Mater.*, 2014, **186**, 130–136.
117 D. A. Faux, P. J. McDonald, N. C. Howlett, J. S. Bhatt and S. V. Churakov, *Phys. Rev. E: Stat., Nonlinear, Soft Matter Phys.*, 2013, **87**, 062309.
118 J. Mareš, M. Hanni, P. Lantto, J. Lounila, J. Vaara and V. A. Online, *Phys. Chem. Chem. Phys.*, 2014, **16**, 6916–6924.
119 C. Allolio, N. Salas-Illanes, Y. S. Desmukh, M. R. Hansen and D. Sebastiani, *J. Phys. Chem. B*, 2013, **117**, 9939–9946.
120 K. Aidas, H. Ågren, J. Kongsted, A. Laaksonen and F. Mocci, *Phys. Chem. Chem. Phys.*, 2013, **15**, 1621–1631.
121 S. Badu, L. Truflandier and J. Autschbach, *J. Chem. Theory Comput.*, 2013, **9**, 4074–4086.
122 T. Ohkubo, E. Tsuchida, M. Gobet, V. Sarou-Kanian, C. Bessada and Y. Iwadate, *J. Phys. Chem. B*, 2013, **117**, 5668–5674.